Molecular Biology of B Cells

Molecular Biology of B Cells

Third Edition

Edited by

Tasuku Honjo
Department of Immunology and Genomic Medicine, Center for Cancer Immunology and Immunobiology, Kyoto University Graduate School Medicine, Kyoto, Japan

Michael Reth
Faculty of Biology, Institute of Biology III, University of Freiburg, Freiburg, Germany; Signaling Research Centers CIBSS and BIOSS, University of Freiburg, Freiburg, Germany

Andreas Radbruch
German Rheumatism Research Centre (DRFZ), A Leibniz Institute, Berlin, Germany; Charité – Universitätsmedizin Berlin, Berlin, Germany

Frederick Alt
Howard Hughes Medical Institute, Program in Cellular and Molecular Medicine, Boston Children's Hospital, Boston, MA, United States; Department of Genetics, Harvard Medical School, Boston, MA, United States

Alberto Martin
Department of Immunology, University of Toronto, Toronto, ON, Canada

ELSEVIER

ACADEMIC PRESS
An imprint of Elsevier

Academic Press is an imprint of Elsevier
125 London Wall, London EC2Y 5AS, United Kingdom
525 B Street, Suite 1650, San Diego, CA 92101, United States
50 Hampshire Street, 5th Floor, Cambridge, MA 02139, United States

Notices

Knowledge and best practice in this field are constantly changing. As new research and experience broaden our understanding, changes in research methods, professional practices, or medical treatment may become necessary.

Practitioners and researchers must always rely on their own experience and knowledge in evaluating and using any information, methods, compounds, or experiments described herein. In using such information or methods they should be mindful of their own safety and the safety of others, including parties for whom they have a professional responsibility.

To the fullest extent of the law, neither the Publisher nor the authors, contributors, or editors, assume any liability for any injury and/or damage to persons or property as a matter of products liability, negligence or otherwise, or from any use or operation of any methods, products, instructions, or ideas contained in the material herein.

ISBN: 978-0-323-95895-0

For information on all Academic Press publications visit our website at
https://www.elsevier.com/books-and-journals

Publisher: Stacy Masucci
Acquisitions Editor: Kattie Washington
Editorial Project Manager: Pat Gonzalez
Production Project Manager: Swapna Srinivasan
Cover Designer: Mark Rogers

Typeset by TNQ Technologies

Contents

18. Anti-Tumour Necrosis Factor and New Paradigms for Therapies Using Antibodies

Sergei A. Nedospasov, Marina S. Drutskaya and Andrey A. Kruglov

19. Therapeutic Targeting of B Cells and Plasma Cells with a Focus on Multiple Sclerosis and Other Autoimmune Conditions

Andreia C. Lino and Thomas Dörner

20. IMGT Immunoglobulin Repertoire Analysis and Antibody Humanization

Marie-Paule Lefranc

Contributors

Frederick W. Alt, Howard Hughes Medical Institute, Program in Cellular and Molecular Medicine, Boston Children's Hospital, Boston, MA, United States; Department of Genetics, Harvard Medical School, Boston, MA, United States

Yemil Atisha-Fregoso, Center for Autoimmune Musculoskeletal and Hematopoietic Diseases, The Feinstein Institutes for Medical Research, Northwell Health, Manhasset, NY, United States

Philip Barbulescu, Department of Immunology, University of Toronto, Toronto, ON, Canada

Uttiya Basu, Department of Microbiology and Immunology, Vagelos College of Physicians and Surgeons, Columbia University, New York, NY, United States

Nasim A. Begum, Department of Immunology and Genomic Medicine, Center for Cancer Immunology and Immunobiology, Kyoto University Graduate School Medicine, Kyoto, Japan

Thomas Boehm, Department of Developmental Immunology, Max-Planck Institute of Immunobiology and Epigenetics, Freiburg, Germany; Faculty of Medicine, University of Freiburg, Freiburg, Germany

Stylianos Bournazos, Laboratory of Molecular Genetics and Immunology, The Rockefeller University, New York, NY, United States

Rita Carsetti, B Cell Unit, Immunology Research Area, Bambino Gesù Children's Hospital, Rome, Italy

Jean-Laurent Casanova, Laboratory of Human Genetics of Infectious Diseases, Paris, France; University of Paris Cité, Imagine Institute, Paris, France; St. Giles Laboratory of Human Genetics of Infectious Diseases, The Rockefeller University, New York, NY, United States; Howard Hughes Medical Institute, New York, NY, United States

Jayanta Chaudhuri, Immunology and Microbial Pathogenesis Program, Weill Cornell Graduate School of Medical Sciences, New York, NY, United States; Immunology Program, Sloan Kettering Institute, Memorial Sloan Kettering Cancer Center, New York, NY, United States

Max D. Cooper, Department of Pathology and Laboratory Medicine, Emory University, Atlanta, GA, United States; Emory Vaccine Center, Emory University, Atlanta, GA, United States

Riccardo Dalla-Favera, Institute for Cancer Genetics, Columbia University, New York, NY, United States; Department of Pathology & Cell Biology, Columbia University, New York, NY, United States; Herbert Irving Comprehensive Cancer Center, Vagelos College of Physicians and Surgeons, Columbia University, New York, NY, United States; Department of Genetics & Development, Columbia University, New York, NY, United States; Department of Microbiology & Immunology, Columbia University, New York, NY, United States

Sabyasachi Das, Department of Pathology and Laboratory Medicine, Emory University, Atlanta, GA, United States; Emory Vaccine Center, Emory University, Atlanta, GA, United States

Pantaleo De Simone, Institute for Cancer Genetics, Columbia University, New York, NY, United States

Betty Diamond, Center for Autoimmune Musculoskeletal and Hematopoietic Diseases, The Feinstein Institutes for Medical Research, Northwell Health, Manhasset, NY, United States

Javier M. Di Noia, Institut de Recherches Cliniques de Montréal, Montréal, QC, Canada; Division of Experimental Medicine, Department of Medicine, McGill University, Montréal, QC, Canada; Department of Medicine, University of Montreal, Montréal, QC, Canada

Thomas Dörner, Department of Rheumatology and Clinical Immunology, Charité Universitätsmedizin Berlin, Berlin, Germany; German Rheumatism Research Center Berlin-Leibniz Institute, Berlin, Germany

Marina S. Drutskaya, Engelhardt Institute of Molecular Biology, Russian Academy of Sciences, Moscow, Russia; Sirius University of Science and Technology, Federal Territory Sirius, Russia

Sidonia Fagarasan, Laboratory for Mucosal Immunity, Center for Integrative Medical Sciences, RIKEN Yokohama Institute, Yokohama, Japan

Ann J. Feeney, Department of Immunology and Microbial Science, The Scripps Research Institute, La Jolla, CA, United States

Keith C. Fernandez, Immunology and Microbial Pathogenesis Program, Weill Cornell Graduate School of Medical Sciences, New York, NY, United States

Michael R. Gold, Department of Microbiology and Immunology, University of British Columbia, Vancouver, BC, Canada; Life Sciences Institute, University of British Columbia, Vancouver, BC, Canada

Jennifer L. Gommerman, Department of Immunology, University of Toronto, Toronto, ON, Canada

Shabirul Haque, Center for Autoimmune Musculoskeletal and Hematopoietic Diseases, The Feinstein Institutes for Medical Research, Northwell Health, Manhasset, NY, United States

Masayuki Hirano, Department of Pathology and Laboratory Medicine, Emory University, Atlanta, GA, United States; Emory Vaccine Center, Emory University, Atlanta, GA, United States

Tasuku Honjo, Department of Immunology and Genomic Medicine, Center for Cancer Immunology and Immunobiology, Kyoto University Graduate School Medicine, Kyoto, Japan

Ellen Hsu, Department of Physiology and Pharmacology, The State University of New York Health Science Center at Brooklyn, Brooklyn, NY, United States

Peter K. Jani, German Rheumatism Research Centre Berlin (DRFZ), A Leibniz Institute, Berlin, Germany

Julia Jellusova, Institute of Clinical Chemistry and Pathobiochemistry, School of Medicine and Health, Technical University of Munich, Munich, Germany; TranslaTUM, Center for Translational Cancer Research, Technical University Munich, Munich, Germany

Lucia M. Kato, Laboratory for Mucosal Immunity, Center for Integrative Medical Sciences, RIKEN Yokohama Institute, Yokohama, Japan

Kathrin Kläsener, Faculty of Biology, Institute of Biology III, University of Freiburg, Freiburg, Germany; Signalling Research Centers CIBSS and BIOSS, University of Freiburg, Freiburg, Germany

Maki Kobayashi, Department of Immunology and Genomic Medicine, Center for Cancer Immunology and Immunobiology, Kyoto University Graduate School Medicine, Kyoto, Japan

Andrey A. Kruglov, Belozersky Institute of Physico-Chemical Biology and Biological Faculty, Lomonosov Moscow State University, Moscow, Russia; German Rheumatism Research Center (DRFZ), A Leibniz Institute, Berlin, Germany

Marie-Paule Lefranc, IMGT, The International ImMunoGeneTics Information System, Université Montpellier 2, CNRS, Institut de Génétique Humaine, Montpellier, France

Susanna M. Lewis, Department of Pathology and Laboratory Medicine, Perelman School of Medicine, University of Pennsylvania, Philadelphia, PA, United States

Andreia C. Lino, Department of Rheumatology and Clinical Immunology, Charité Universitätsmedizin Berlin, Berlin, Germany; German Rheumatism Research Center Berlin-Leibniz Institute, Berlin, Germany

Joseph S. Lucas, Division of Biological Sciences, Department of Molecular Biology, University of California, San Diego, CA, United States

Cindy S. Ma, Garvan Institute of Medical Research, Darlinghurst, NSW, Australia; School of Clinical Medicine, Faculty of Medicine and Health, UNSW Sydney, Sydney, NSW, Australia

Fabienne Mackay, Department of Immunology, Monash University, Clayton, VIC, Australia

Andrew J. Macpherson, Department of Visceral Surgery and Medicine, Bern University Hospital, University of Bern, Bern, Switzerland

Alberto Martin, Department of Immunology, University of Toronto, Toronto, ON, Canada

Mir-Farzin Mashreghi, German Rheumatism Research Centre Berlin (DRFZ), A Leibniz Institute, Berlin, Germany

Mairi Anne McGrath, German Rheumatism Research Centre Berlin (DRFZ), A Leibniz Institute, Berlin, Germany

Fritz Melchers, German Rheumatism Research Centre Berlin (DRFZ), A Leibniz Institute, Berlin, Germany

Fei-Long Meng, State Key Laboratory of Molecular Biology, Shanghai Institute of Biochemistry and Cell Biology, Center for Excellence in Molecular Cell Science, Chinese Academy of Sciences, University of Chinese Academy of Sciences, Shanghai, China; Shanghai Huashen Institute of Microbes and Infections, Shanghai, China

Ryo Morimoto, Department of Developmental Immunology, Max-Planck Institute of Immunobiology and Epigenetics, Freiburg, Germany

Rhiannon Morris, Blood Cell and Blood Cancer Division, The Walter and Eliza Hall Institute of Medical Research, Melbourne, VIC, Australia; The Department of Medical Biology, The University of Melbourne, Melbourne, VIC, Australia

Cornelis Murre, Division of Biological Sciences, Department of Molecular Biology, University of California, San Diego, CA, United States

Hitoshi Nagaoka, Department of Molecular Pathobiochemistry, Graduate School of Medicine, Gifu University, Gifu, Japan

Lekha Nair, Department of Microbiology and Immunology, Vagelos College of Physicians and Surgeons, Columbia University, New York, NY, United States

Sergei A. Nedospasov, Engelhardt Institute of Molecular Biology, Russian Academy of Sciences, Moscow, Russia; Sirius University of Science and Technology, Federal Territory Sirius, Russia; Belozersky Institute of Physico-Chemical Biology and Biological Faculty, Lomonosov Moscow State University, Moscow, Russia

Ashley P. Ng, Immunology Division, The Walter and Eliza Hall Institute of Medical Research, Melbourne, VIC, Australia; The Department of Medical Biology, The University of Melbourne, Melbourne, VIC, Australia; Clinical Haematology Department, The Royal Melbourne Hospital and Peter MacCallum Cancer Centre, Melbourne, VIC, Australia; Sir Peter MacCallum Department of Oncology, The University of Melbourne, Melbourne, VIC, Australia

Lars Nitschke, Division of Genetics, Department of Biology, University of Erlangen-Nürnberg, Erlangen, Germany

Stephen L. Nutt, Immunology Division, The Walter and Eliza Hall Institute of Medical Research, Melbourne, VIC, Australia; The Department of Medical Biology, The University of Melbourne, Melbourne, VIC, Australia

Laura Pasqualucci, Institute for Cancer Genetics, Columbia University, New York, NY, United States; Department of Pathology & Cell Biology, Columbia University, New York, NY, United States; Herbert Irving Comprehensive Cancer Center, Vagelos College of Physicians and Surgeons, Columbia University, New York, NY, United States

Eva Piano Mortari, B Cell Unit, Immunology Research Area, Bambino Gesù Children's Hospital, Rome, Italy

Rita Pozovskiy, Center for Autoimmune Musculoskeletal and Hematopoietic Diseases, The Feinstein Institutes for Medical Research, Northwell Health, Manhasset, NY, United States; Zucker School of Medicine at Hofstra/Northwell, Hempstead, NY, United States

Andreas Radbruch, German Rheumatism Research Centre Berlin (DRFZ), A Leibniz Institute, Berlin, Germany; Charité — Universitätsmedizin Berlin, Berlin, Germany

Jonathan P. Rast, Department of Pathology and Laboratory Medicine, Emory University, Atlanta, GA, United States; Emory Vaccine Center, Emory University, Atlanta, GA, United States

Jeffrey V. Ravetch, Laboratory of Molecular Genetics and Immunology, The Rockefeller University, New York, NY, United States

Michael Reth, Faculty of Biology, Institute of Biology III, University of Freiburg, Freiburg, Germany; Signalling Research Centers CIBSS and BIOSS, University of Freiburg, Freiburg, Germany

Roy Riblet, Torrey Pines Institute for Molecular Studies, San Diego, CA, United States

Jana Ridani, Institut de Recherches Cliniques de Montréal, Montréal, QC, Canada; Division of Experimental Medicine, Department of Medicine, McGill University, Montréal, QC, Canada

Tim Rollenkse, Department of Visceral Surgery and Medicine, Bern University Hospital, University of Bern, Bern, Switzerland

David G. Schatz, Department of Immunobiology, Yale School of Medicine, New Haven, CT, United States; Department of Molecular Biophysics and Biochemistry, Yale University, New Haven, CT, United States

Mark Shlomchik, Department of Immunology, School of Medicine, University of Pittsburgh, Pittsburgh, PA, United States

Stuart G. Tangye, Garvan Institute of Medical Research, Darlinghurst, NSW, Australia; School of Clinical Medicine, Faculty of Medicine and Health, UNSW Sydney, Sydney, NSW, Australia

Manami Watanabe, Center for Autoimmune Musculoskeletal and Hematopoietic Diseases, The Feinstein Institutes for Medical Research, Northwell Health, Manhasset, NY, United States

Florian Weisel, Department of Immunology, School of Medicine, University of Pittsburgh, Pittsburgh, PA, United States

Jianxiong Xiao, Department of Immunobiology, Yale School of Medicine, New Haven, CT, United States

Shan Zha, Institute for Cancer Genetics, College of Physicians and Surgeons, Columbia University, New York City, NY, United States; Division of Pediatric Oncology, Hematology and Stem Cell Transplantation, Department of Pediatrics, College of Physicians and Surgeons, Columbia University, New York City, NY, United States

Yiwen Zhang, Howard Hughes Medical Institute, Program in Cellular and Molecular Medicine, Boston Children's Hospital, Boston, MA, United States; Department of Genetics, Harvard Medical School, Boston, MA, United States

Yu Zhang, Department of Microbiology and Molecular Genetics, Michigan State University, East Lansing, MI, United States

Yong-Rui Zou, Center for Autoimmune Musculoskeletal and Hematopoietic Diseases, The Feinstein Institutes for Medical Research, Northwell Health, Manhasset, NY, United States

Preface

B lymphocytes are an essential part of the humoral immune system, and once activated, they develop into plasma cells that produce the specific antibodies generated during infection or vaccination. Soon after their discovery as antitoxins more than a hundred years ago by Emil Behring and Shibasaburo Kitasato, it became clear that antibodies are enormously diverse and can be generated against many different structures. The elucidation of the mechanism underlying the generation of diversity of the antibody system kept immunologists busy over the last hundred years. With the discovery of the V(D)J recombination mechanism of the immunoglobulin (Ig) genes and the somatic hypermutation (SHM) processes, many aspects of this problem were solved. The discovery of the V(D)J recombinase machinery consisting of the lymphocyte lineage-specific protein recombination activating gene (RAG) 1 and RAG2 which provide the specific endonuclease function and the generally expressed non-homologous DNA end joining factors that join RAG-generated DNA breaks was a fascinating development. The discovery of activation-induced deaminase (AID) and its associated protein machinery mediating SHM, class switch recombination (CSR), as well as gene conversion, processes that are restricted to the B cell lineage, furthered our understanding and appreciation of the complexities required to generate an effective humoral response. The progress in the B-cell field in this area is well documented by the different editions of this book series which started in 1989 under the title *Immunoglobulin Genes* and in 2004 became *Molecular Biology of B cells.*

This third edition of the *Molecular Biology of B Cells* brings the reader up to date with these Ig diversifying processes. However, this book has now a much broader scope as it covers many different aspects of B cell biology. B cells are a well-characterized and easily assessable cellular system that can be used to answer many issues of general cell biology. Furthermore, due to the availability of several different B cell-specific Cre deleter mice, the B cell system is also useful for general gene function studies. These are the reasons why many scientists worldwide start to work on B cells although they are not regarded as classical 'B cell immunologists' and for these this book is particularly useful. In fact, B cells have served as excellent model systems to study many common biological questions including transcriptional regulation of differentiation, signal transduction and tumorigenesis. Furthermore, the mechanism behind the generation of diversity has been considered to be unique to lymphocytes especially B cells as the expression of AID is specific to B cells. Subsequently, however, the DNA repair phases that occur during the generation of diversity were shown to be common between B cells and other types of cells and that are entrusted with preserving genome stability. In addition, IgA synthesis has been shown to influence the whole body metabolic regulation through symbiotic interaction with the gut microbiota. These striking developments convince us that molecular biology of B cells does not only mean molecular biology 'for' B cells but also molecular biology 'through' B cells to understand the whole biology.

The chapters of this new third edition of *Molecular Biology of the B Cells* have been written by an international faculty of experts in their fields and the readers can expect to get a comprehensive overview about the current status and the future development of the B-cell field. In particular, the reader will learn how a B-cell-specific transcriptional network drives differentiation of haematopoietic stem cells (HSCs) through B lymphopoiesis and how during their development B cells repeatedly switch between molecular programs promoting proliferation and those involved in differentiation. B-cell research in the age of genomics also means that we now know many more details about the gene expression program of different B-cell developmental stages. We also learned much more about the signalling mechanisms controlling the development and activation of B cells.

New topics that were not at all covered in the last edition of this book are, for example, on how metabolism affects B cell fate and function, and how B cells contribute to various pathological diseases including multiple sclerosis (MS) and how new discoveries in B cell targeting therapies are being used to treat MS.

Apart from being a useful tool for basic cell biological and signalling studies, B cells are also playing an increasingly important role in the clinic. Genetic defects of the B-cell system are the cause of important human immunodeficiency diseases. A dysregulation of this system causes autoimmunity or tumour diseases, topics which are both well covered in this new edition. Furthermore, antibodies are highly versatile tools that are not only used in clinical diagnostics but also play an important role as therapeutic agents. Thus, most drug companies have antibody departments and factories that produce antibodies and bring them into the clinic for the treatment of various diseases.

In spite of the importance of B cells in basic research and the clinic, there are still many questions that needs to address in the future. We still have to learn more about how the B cell system distinguishes between self and foreign antigens, in particular after the finding that many newly generated B cells have a certain level of autoreactivity. What is the role of B cells in specialized compartments such as mucosal tissues? The maintenance of B-cell memory and antibody production over a long period of time is also a topic of active research where major breakthroughs are to be expected.

Frederick W. Alt
Tasuku Honjo
Radbruch Andreas
Michael Reth
Alberto Martin

October 2023.

Chapter 1

The Structure and Regulation of the Immunoglobulin Loci

Joseph S. Lucas[1], **Cornelis Murre**[1], **Ann J. Feeney**[2] **and Roy Riblet**[3]

[1] *Division of Biological Sciences, Department of Molecular Biology, University of California, San Diego, CA, United States;* [2] *Department of Immunology and Microbial Science, The Scripps Research Institute, La Jolla, CA, United States;* [3] *Torrey Pines Institute for Molecular Studies, San Diego, CA, United States*

1. INTRODUCTION

Our adaptive immune system relies heavily on the use of antibodies, produced by B cells, to eliminate foreign pathogens and toxins. It is estimated that mammalian organisms have the ability to generate on the order of 10^{11} different antibodies [1]. This staggering number is made possible by an elaborate mechanism that assembles complete antigen binding site sequences from arrays of gene segments encoding portions of the complete antibody proteins. This assembly involves combinatorial selection of each type of gene segment and creation of short novel sequences at the junctions of these DNA segments, thus generating a vast variety of assembled gene sequences and antigen binding specificities. Sequence diversity is further increased by the somatic hypermutation process that is active in these sequences.

Antibodies are proteins made up of two identical heavy chains and two identical light chains. All heavy chains are expressed from the Igh locus, whereas light chains are expressed from one of two loci, Igκ or Igλ [2]. The DNA segments that rearrange to create antibody heavy and light chain genes in developing B cells include variable (V) and joining (J) elements, and heavy chain genes include a third, diversity (Dh), gene segment [3]. Single light chain V and J and heavy chain V, D, and J segments are selected and joined in a seemingly random process out of the many such genomic segments that span as much as 3 Mbp of DNA.

Over the past 30 years, a large number of studies have been performed to describe the process of V(D)J recombination in molecular terms, including how it is regulated during development; how it is controlled by cell signalling; how it is modulated by transcription; and how chromatin modifications, nuclear positioning and three-dimensional (3D) topology affect rearrangement. Collectively, these experiments aimed to address a number of critical questions. Some of the prominent questions follow: How do DNA sequences separated by millions of base pairs interact to allow recombination? How are gene segments selected randomly to allow for the generation of antibody diversity?

2. GENOMIC ORGANIZATION OF THE MOUSE IMMUNOGLOBULIN HEAVY CHAIN LOCUS

The Igh locus is made up of multiple variable (Vh), Dh, joining (Jh), and constant (Ch) segments, arrayed in adjacent regions of the Igh locus. In mice the locus spans 2.75 million base pairs, and there are about 100 Vh segments that have seemingly functional coding sequences and nearly as many nonfunctional Vh segments, 10−15 Dh segments depending on mouse strain, and four Jh segments. Eight constant regions encode the Igh isotypes [4,5]. Vh genes encode most of the heavy chain variable region including the first two hypervariable loops that form the binding site for antigen [6,7]. Dh segments are very short, 10−15 nucleotides of coding sequence, but they are critically important in the generation of antibody diversity. Dh segments determine most of the heavy chain's third hypervariable region or complementarity determining region 3 (CDR3) [8−10]. In the folded protein's antigen binding site, this region makes major, sometimes dominant, contacts with antigen.

Each set of gene segments has evolved, and continues to evolve, via gene addition by duplication, divergence through mutation, and loss by nonsense or frameshift mutations or by partial or total deletion. A prominent example

Molecular Biology of B Cells. https://doi.org/10.1016/B978-0-323-95895-0.00016-7

is the ancestral duplication of the constant region IgG2a gene and divergence and alternate gene deletion leaving the present IgG2a gene in BALB/c and many other mouse strains and the IgG2c gene in C57BL/6 and related strains [11,12]. A larger scale case in the Igh-V (Vh) array is the duplication of the entire proximal region of the Vh array comprising the interspersed Vh7183 and VhQ52 genes (Ighv5 and Ighv2 families in Figure 1.1) from 400 kb in C57BL/6 to 800 kb in 129 and BALB/c [5].

Alignment of Vh gene sequences and grouping them by similarity yields an evolutionary structure of the gene population, revealing three major branches or groups, splitting further into 15 families of genes that share 80% sequence identity. The three groups of Vh genes are an ancient evolutionary division; Vh genes from any vertebrate species fall into several or all three of these groups. Vh gene family structure is apparently not generally shared beyond closely related species [13], although the Vh7183 or Ighv5 family is found widely [14]. In mice, at least, a family of Vh genes occupies a restricted region of the Vh array and are interspersed there with other families; the proximal, 3′ 300 kb region of the Vh locus contains only the interspersed Vh7183 and VhQ52 (Ighv5 and Ighv2) families, the distal, 5′ 1.5 Mb portion contains only VhJ558 and Vh3609 (Ighv1 and Ighv8) genes, and the central 680 kb contains the remaining families.

In developing B cells, Vh, Dh, and Jh segments recombine to encode the antigen-binding domain of the antibody heavy chain [3]. Each Vh, Dh, and Jh segment is flanked by a short DNA sequence called the recombination signal sequence (RSS). This sequence is recognized by recombination activating gene enzymes RAG1 and RAG2. RAG1 and RAG2 form a complex with additional proteins and act to create a loop between the two RSSs, bringing them into proximity. The RAG complex then induces double-strand DNA breaks and promotes ligation of

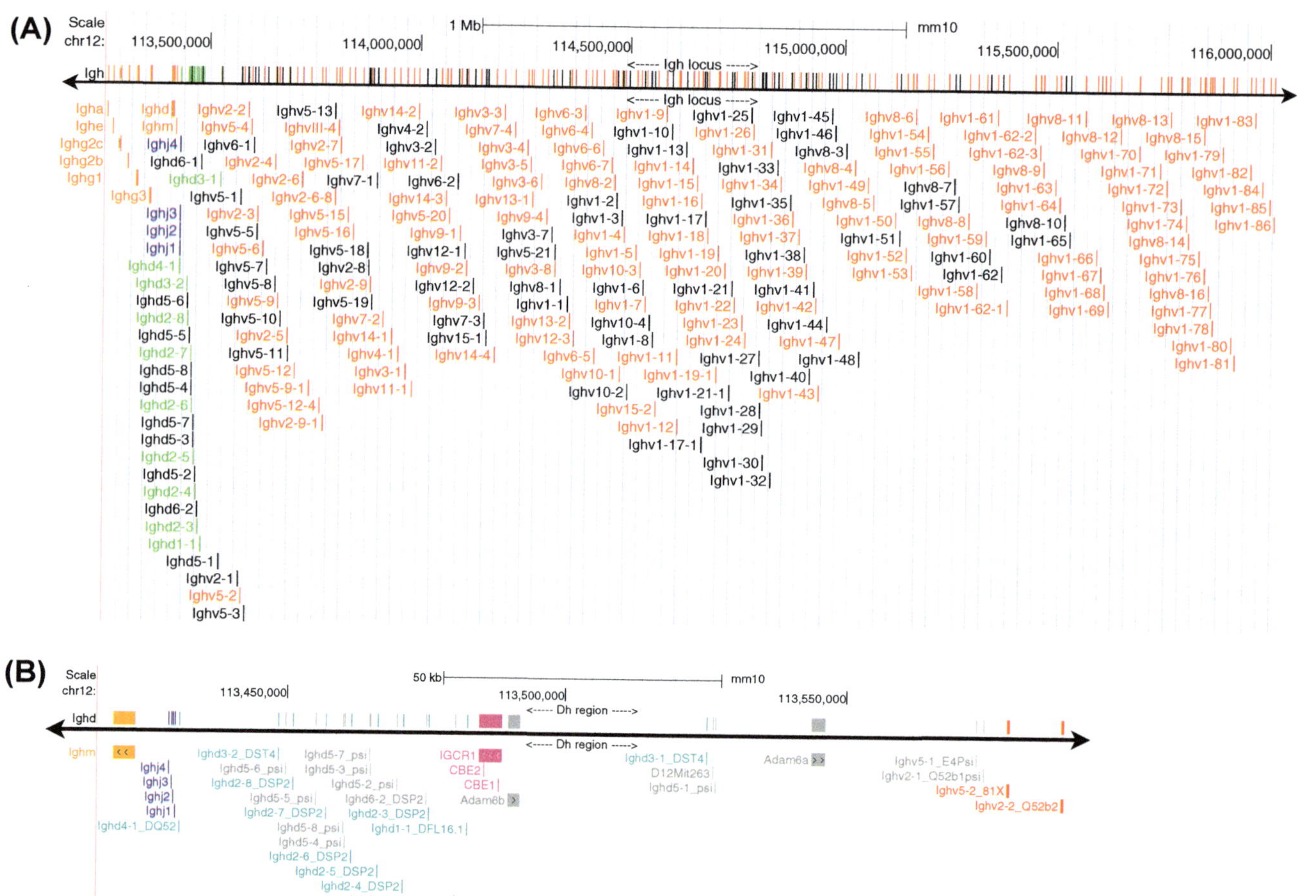

FIGURE 1.1 **The genomic structure of the immunoglobulin heavy chain locus.** The 2.75 Mb Igh locus is located on the distal region of mouse chromosome 12. (A) The various types of gene segments are represented in different colours: Functional Vh segments are *red*, functional Dh segments are *green*, the four Jh segments are *blue*, and the eight Ch loci are *orange*. Nonfunctional Vh and Dh pseudogenes are *black*. The sequence is from mm10, the December 2011 *Mus musculus* assembly (Genome Reference Consortium Mouse Build 38), documented in NCBI Gene ID: 111507 (http://ncbi.nlm.nih.gov/gene/?term=111507). (B) Expanded view of the Dh region; the IgM constant region at the left is *orange*; the Jh regions are *blue*; functional Dh segments are *cyan*; nonfunctional Dh, Vh, and Adam6 pseudogenes are *grey*; the intergenic control region 1 [88] with its two CCCTC-binding factor binding elements, CBE1 and CBE2, are *magenta*; and functional Vh genes at the right are *red*.

appropriate coding segments via the nonhomologous end-joining machinery [15–17].

In the Igh locus, V, D, J, and C gene segments are in separate clusters, and all in the same 5′ → 3′ polarity. VDJ rearrangements proceed via serial excisions of intervening sequences, D to J, then V to DJ, and loss of the excised sequences as free closed DNA circles. VDJ recombination is irreversible, although a secondary V–V replacement mechanism occurs at low frequency [18].

3. GENOMIC ORGANIZATION OF THE MOUSE IMMUNOGLOBULIN KAPPA LIGHT CHAIN LOCUS

The kappa light chain locus (Figure 1.2) is even larger than the heavy chain locus, extending 3.17 Mb; it too contains approximately 100 functional and 60 nonfunctional V segments and four functional (plus one nonfunctional) J segments [19]. Light chain genes do not use D segments. Interestingly, half of all Vk genes are in the opposite polarity to the Jk and Ck loci and must rearrange through a nondestructive inversion of the portion of the Vk gene array between the joining Vk and Jk segments. As a result, many Vk segments are retained rather than excised, for potential use in a secondary rearrangement with any remaining downstream Jk segments. This conserves a larger set of Vk genes for "B cell editing," i.e., replacement of an initial Vk-Jk choice that yields an antiself antibody or is otherwise a poor partner for the rearranged heavy chain. There seems to be no bias between the excision and inversion mechanisms because both sets of Vk genes contain similar frequencies of both functional genes and highly expressed genes [19].

4. GENOMIC ORGANIZATION OF THE MOUSE IMMUNOGLOBULIN LAMBDA LIGHT CHAIN LOCUS

In mice the lambda light chain locus is quite small, 200 kb, and of limited diversity (Figure 1.3). There are two adjacent VJC clusters, 5′-V2-V3 (originally termed VλX)-J2C2 followed by V1-J3C3-J1C1. The major rearranged combinations expressed in serum antibodies are λ1, V1J1C1, λ2, V2J2C2, and λ3, V1J3C3. Rarely, λx, V3J2C2, and V2J3C3 and V2J1C1 products have been noted [20]. The large difference in germline-encoded diversity between kappa and lambda light chains is reflected in the population of serum antibodies that is 95% kappa [20].

5. B CELL DEVELOPMENT AND REGULATION OF V(D)J RECOMBINATION

Recombination steps define the stages of development in B cells. Before V(D)J recombination, the cells are called pre-pro-B cells and are not fully committed to the B cell lineage [21–23]. However, at the Igh locus Dh to Jh recombination can occur at this stage [24]. Vh to DJh rearrangement occurs in the pro-B cell stage, and the successful production of a heavy chain protein is necessary to permit the transition from the pro-B stage to the pre-B cell stage [25,26]. Upon productive Igh locus V(D)J gene rearrangement, cycling large pre-B cells express the pre-B cell receptor that, in turn, functions to suppress RAG1/2 activity, preventing continued Igh locus rearrangement [27]. Large pre-B cells divide several times to increase the population of cells with functional heavy chains. Large pre-B cells differentiate into resting small pre-B cells [26]. In small pre-B cells, the RAG genes are expressed again to allow for Ig light chain gene rearrangement [27]. At this point, assuming successful recombination, the cells will express a B cell receptor. Before the cells are released from the bone marrow, they undergo negative selection. Self-antigens are presented by antigen-presenting cells such as dendritic cells. B-cells expressing receptors with high affinity towards self-antigens will undergo apoptosis or additional recombination of the light chains [28,29]. Surviving B cells migrate to the periphery as naïve cells, waiting to be activated by their specific antigens [30].

V(D)J recombination is a highly regulated and ordered event. Dh to Jh recombination always precedes Vh to DhJh recombination, and recombination of the heavy chain locus always precedes recombination of the light chain loci [31,32]. Furthermore, Vh to DhJh recombination occurs on only one allele at a time, although both alleles have previously undergone Dh to Jh rearrangement [33]. Such a strict order of events is crucial to ensure that each B cell produces antibodies with affinity towards a single antigen. Although the expression of RAG genes is tightly regulated, the fact that they catalyze all recombination steps means that additional mechanisms need to be in place to ensure the ordered progression of rearrangement [27]. Hence only at the correct developmental stage does an antigen receptor locus become accessible to recombination.

The Igh locus is, by default, nonpermissive to V(D)J recombination. This state is achieved by several mechanisms, many of which resemble general mechanisms used in transcriptional repression. First, the locus is tethered at the

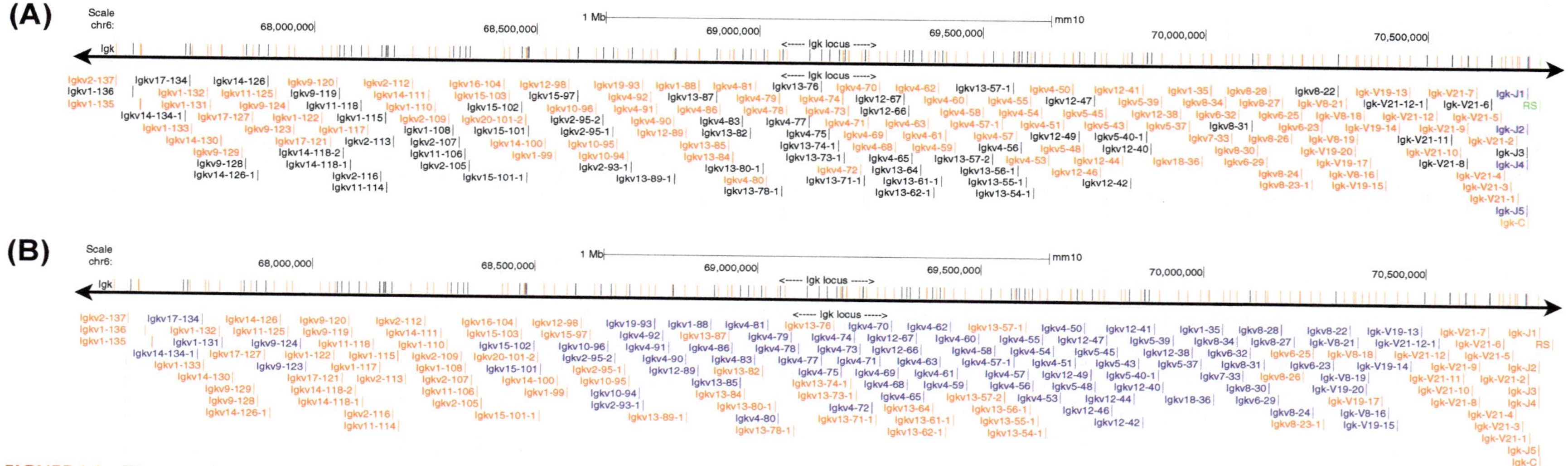

FIGURE 1.2 **The genomic structure of the immunoglobulin kappa light chain locus.** The 3.2 Mb Igk locus is located centrally on mouse chromosome 6. (A) The various types of gene segments are represented in different colours: Functional Vk segments are *red*, the four functional Jk segments are *blue*, and the single Ck locus is *orange*. Nonfunctional Vk and Jk pseudogenes are *black*. The downstream "recombining sequence" (RS) is *green*. (B) On this map gene segments on the plus strand are coloured *red*, and those Vk segments on the minus strand that rearrange by inversion are *blue*. The sequence is from mm10, the December 2011 *Mus musculus* assembly (Genome Reference Consortium Mouse Build 38), documented in NCBI Gene ID: 243469 (http://ncbi.nlm.nih.gov/gene/?term=243469).

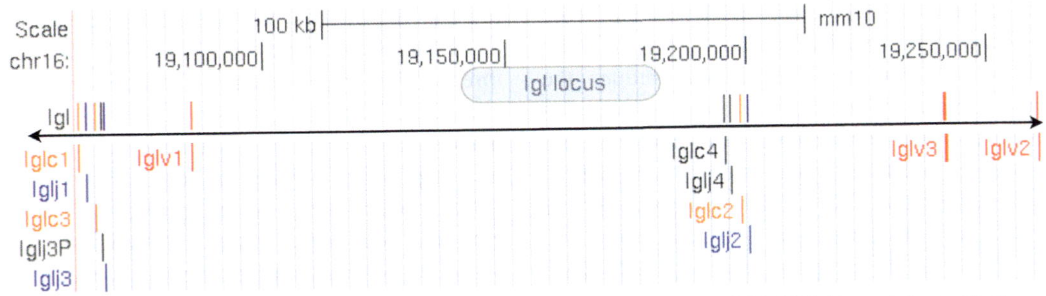

FIGURE 1.3 **The genomic structure of the immunoglobulin lambda light chain locus.** The small 200 kb Igl locus is located in the proximal region of mouse chromosome 16. The various types of gene segments are represented in different colours: The three Vl segments are *red*, the three functional Jl segments are *blue*, and the three functional Cl loci are *orange*. Nonfunctional Jl and Cl pseudogenes are *black*. The sequence is from mm10, the December 2011 *Mus musculus* assembly (Genome Reference Consortium Mouse Build 38), documented in NCBI Gene ID: 111519 (http://ncbi.nlm.nih.gov/gene/?term=111519).

transcriptionally repressive nuclear periphery [34]. The Igh locus is highly enriched for nucleosomes that are dimethylated at lysine nine of histone 3, correlating with transcriptionally inactive genomic regions [35]. Furthermore, the RSS sequences may act as weak nucleosome-positioning sequences, causing the recombination sequences to be buried within the histones and inaccessible to the RAG complex [36]. In addition, the Igh locus is decontracted, positioning many of the Vh gene segments away from the DhJh elements [34,37]. Recombination of the Igh locus requires a reversal of the above-mentioned trends. The mechanisms by which this occurs vary for different regions of the locus.

6. JUNCTIONAL DIVERSITY

The primary antibody repertoire derives its diversity from two sources: combinatorial diversity (mix-and-match usage of V, D, and J gene segments) and junctional diversity [38]. Junctional diversity occurs because the RAG1/2 recombinase makes a covalent hairpin at the coding end. Because the hairpin can be opened at any location near its end, this can result in a random loss of a small number of nucleotides at the coding end, or the gain of a few base pairs from the opposite strand that is now covalently linked to the coding end. These latter nucleotides are called "P" nucleotides because they are palindromic to the coding end. In addition, tremendous diversity is added to the junctions by the nontemplated addition of nucleotides to the coding ends by terminal deoxynucleotidyl transferase [38–41]. This "N" region diversity provides extensive diversity to all junctions of Ig and of T cell receptors, with the exception of light chains. In mice, there are no N nucleotides at the light chain V–J junctions, and in humans there is less N addition than in the heavy chain. N region diversity is also absent early in ontogeny, greatly restricting the foetal repertoire of antibodies [42,43].

7. COMBINATORIAL DIVERSITY

Initial estimates of the diversity of the antibody repertoire assumed that all V, D, and J genes recombined with equal frequencies, thus making combinatorial association of the large number of V, D, and J gene segments a major aspect of creating a diverse repertoire. However, the advent of polymerase chain reaction (PCR) technology provided the means to determine that this was not the case, but rather Vh and Vk genes were shown to rearrange at very different intrinsic frequencies [44–48]. D and J genes also were used at different frequencies, but the differences were much more modest. Early repertoire studies consisted of PCR, cloning, and sequencing, but after 30 years of this approach, next generation sequencing now is producing vastly more detailed studies [49]. Nonetheless, early studies demonstrating nonrandom V gene usage on a small scale are now substantiated and extended by the deep sequencing approach [19,50]. The reasons for this unequal V gene usage are myriad.

The quality of the flanking RSS is one factor in the nonrandom usage of V genes in the primary repertoire [51]. The RSS consists of a conserved heptamer and nonamer separated by a spacer of approximately 12 or 23 bp. Although there is a consensus, very few RSSs actually have the full consensus heptamer and nonamer sequence. In vitro plasmid-based recombination substrate assays have shown the effect of variation from the consensus at each position [52,53]. These studies demonstrated that the first 3 bp of the heptamer are essential. The remaining 4 bp of the heptamer can tolerate variation to a greater extent, with reduction in recombination frequency of 2- to 50-fold. The nonamer has more flexibility and should have a core of 3–5 A's, flanked by non-A's. Again, recombination substrates have demonstrated the importance of each site. The 12 and

23 bp spacers are one or two turns of the DNA helix, respectively. Hence, they can tolerate only a slight variation in length, and such variation in length will reduce the recombination efficiency. The sequence of the spacer is much less conserved than the heptamer and nonamer, but some spacer sequences support more recombination than others [54,55]. Because the RSS is the binding site for the recombinase, one might expect that variation from the consensus sequence might support less efficient RAG binding, and thus less efficient recombination of that gene. Indeed, this has been shown for allelic variants of a V gene [56]. However, it has also been demonstrated that V genes with identical RSS can rearrange at very different frequencies, thus demonstrating that other factors play important roles in V, D, and J gene use [47].

8. NONCODING TRANSCRIPTION AND IMMUNOGLOBULIN LOCUS REARRANGEMENT

It was demonstrated many years ago that J gene and constant region genes undergo robust "sterile" or "germline" transcription preceding V(D)J rearrangement at that locus [39,57,58]. For many years it was not clear whether this made those regions accessible for rearrangement, or if the transcription was a byproduct of the fact that the regions had been made accessible. Key studies using transcriptional terminators inserted into the genome demonstrated that the J genes downstream of the terminator rearranged less well, providing evidence that the germline transcription directly facilitated the accessibility of those downstream gene segments [59]. We know now that noncoding RNA plays many regulatory roles, and these "germline transcripts" through the J/C regions at the Igh and Igk loci were among the first noncoding RNAs to be identified [60].

In addition to the high abundance of transcripts through the J/C region, it was also shown many years ago that the V genes underwent germline transcription [61]. These transcripts started at the V gene promoter, proceeded through the coding region of the V gene segment, and ended at variable sites a little further downstream of the gene. These "sense" V gene transcripts were easily observed at the VhJ558 genes, but they were much more difficult to observe in other V genes. Much more recently, it was demonstrated that there is antisense transcription also in the Igh locus [62]. The D genes undergo antisense transcription precisely at the stage of D to J rearrangement [63], whereas the V gene part of the locus is transcribed only after DJ rearrangement has completed [62]. The precise timing of the antisense transcripts to the stage at which those segments are rearranging led to the proposal that the transcription was making the region accessible for rearrangement, although other hypotheses were also proposed [62–64]. Recently, RNA-sequencing (RNA-seq) analysis of the Igh locus revealed that there were three regions of extensive antisense transcription in the Vh locus, all at the distal end of the locus [65]. These originated at three of the "PAIR" elements, sites that bound Pax5, E2A, and CCCTC-binding factor (CTCF) [66]. RNA-seq also demonstrated that there is extensive, but low level, sense transcription of V genes in the distal half of the locus, with lower transcription in the proximal half of the locus. Nonfunctional pseudogenes had significantly lower level of sense transcripts than functional V genes.

9. THE PROCESS OF DH–JH RECOMBINATION

Dh to Jh recombination of the Igh locus is perhaps the least regulated step of V(D)J recombination as it is found on both alleles in a majority of B cells [31]. DhJh joints can also be detected in the common lymphoid progenitor compartment and in a fraction of developing T-lineage cells [67–69]. At the time of Dh to Jh recombination, the locus is still tethered at the nuclear periphery on its 5′ side, allowing the 3′ end, containing the Dh and Jh elements, to extend towards the cell's interior, perhaps allowing it to interact with transcription factories [34,70]. The intronic enhancer (Eμ) located just downstream of the Jh segments is involved in switching the Dh and Jh regions of the locus from a repressive to an active state, allowing for germline transcription across the region [71]. Several transcription factors are involved, including E2A, EBF, and Pax5 [72–74]. Transcription, along with recruitment of the Swi/Snf chromatin-remodelling complex is thought to expose RSS sequences on the 5′ end of Jh segments and the 3′ end of Dh segments, allowing for their recombination [75]. A successful recombination exposes the 5′ RSS of the used Dh segment, targeting it for recombination with an upstream Vh segment [35].

10. EPIGENETICS AND IMMUNOGLOBULIN LOCUS REARRANGEMENT

The epigenetic status of V, D, and J genes can also impact rearrangement. Histone proteins can be posttranslationally modified by acetylation and methylation, and indeed there is significant variation in the local epigenetic environment of each V, D, and J gene. Actively transcribed regions acquire trimethylation of lysine four on histone 3 (H3K4me3) because the H3K4 methyltransferase travels with the RNA polymerase II complex [76]. Not surprisingly, therefore, the highly transcribed J and C genes have the highest amount of H3K4me3 [77,78]. This is of great importance for repertoire development, and for ordered rearrangement of DJ genes before V to DJ gene segments, because it was demonstrated that RAG2 specifically binds to H3K4me3 [79,80]. Thus, RAG2 will initially be bound at the J genes. However, there are also regions of H3K4me3 throughout the V regions at sites of higher transcription [65]. In addition, acetylation is found more generally throughout actively rearranging loci [81]. Indeed, V genes have been demonstrated to have more acetylation when that locus is undergoing rearrangement than when the locus is inactive [82]. Histone posttranslational modifications such as acetylation and methylation can act as docking sites for other chromatin-modifying enzymes, e.g., ATP-dependent chromatin-remodelling complexes. These could be important because the RSS needs to be exposed to bind the RAG complex, and thus nucleosomes may need to be moved to achieve this goal [83,84]. Bioinformatic analysis of the epigenetic status of each V gene demonstrated that there was a correlation of the extent of histone acetylation or methylation and rearrangement frequency for the distal half of the Igh locus [50].

11. INSULATORS AND IMMUNOGLOBULIN LOCUS REARRANGEMENT

The transition from a repressive to active chromatin state that occurs during Dh to Jh recombination is highly localised and does not extend past the 5′ most Dh segment. Recent studies have identified an insulator element located 5′ of the DhJh cluster that interferes with rearrangement of Vh gene elements before DhJh joining [85,86]. The insulator element is characterised by the presence of two CTCF binding sites, also named the CTCF-binding element (CBE). In pre-pro-B cells, these CTCF sites interact with CTCF sites downstream of the 3′ regulatory regions, creating a looped domain that includes all of the Dh and Jh genes, but excludes all Vh genes, thus facilitating DhJh rearrangement before Vh to DhJh [87]. Deletion of the CBE results in a loss of ordered and lineage-specific Igh locus recombination [88]. This leads to the cardinal conclusion that the process of ordered rearrangement is controlled, at least in part, by an insulator element, acting to suppress VhDhJh gene rearrangement before the formation of a DhJh joint. We suggest that after Dh to Jh recombination, this insulator is inactivated, plausibly through the removal of cohesin, a binding partner of CTCF [85,87,88].

12. 3D STRUCTURE AND COMPACTION OF THE IMMUNOGLOBULIN HEAVY CHAIN LOCUS

It is well established that the folding of the chromatin fibre plays a critical role in genome function. For more than a century, the organization of the chromatin fibre has been extensively analysed using light microscopy. Each chromosome is folded into its own territory [89–91]. The DNA from each chromosome can intermingle with other chromosomes at the boundaries, but in general it is kept within its own territory [92]. In eukaryotes, DNA is restricted to the nucleus throughout most of the cell cycle where it is highly compacted. The first level of compaction is achieved by wrapping of the DNA around core histones (H2A, H2B, H3, and H4) to form nucleosomes [93,94]. Addition of histone H1 brings the nucleosomes together in a zigzag pattern to form a fibre approximately 30 nm in diameter [95]. The exact architecture of the chromatin past the 30 nm fibre is still a topic of intense study. It has been well established that the level of DNA compaction varies throughout the nucleus. Gene poor regions and repressed genes tend to be located in highly compacted heterochromatin, whereas highly expressed genes are present in less compacted euchromatin [96,97]. Furthermore, heterochromatin tends to aggregate at the nuclear periphery. Highly expressed genes are often found near the centre of the nucleus where they are thought to interact with complexes of polymerases and other factors, termed transcription factories [98,99]. Other highly expressed genes, though, are known to associate with the nuclear pores during transcription to expedite nuclear export of transcripts [100].

Electron microscopic studies have demonstrated the presence of ~90 kb loops in mitotic cells that assemble into rosettes that are comprised of ~18 loops [101–103]. Similar bundles of loops have been seen in interphase nuclei [104]. Despite the progress in studying the genomic organization of the Igh locus, the topology of the Igh locus in developing B cells remains unresolved. As a first approach to address this critical issue, recent studies have used 3D fluorescence in situ hybridisation to measure end-to-end physical distances separating multiple markers across the Igh locus. These studies showed that in developing B cells, the distance distributions are compatible with a multiloop-subcompartment (MLS) model [37]. The MLS model predicts that the chromatin fibre is structured as rosettes of small loops, folding into 1 Mbp domains, connected by flexible linkers [105,106]. Furthermore, using computational geometry, this study revealed the striking conformational compaction of the Igh topology during the transition from the pre-pro-B to the pro-B cell stage [37].

Recent observations have identified CTCF binding located directly adjacent to many of the proximal Vh segments. This has raised the interesting possibility that CTCF helps to position Vh segments at the centre of a rosette. Such a configuration would permit DhJh segments to have equal access to any Vh segment [85,107,108]. The comparison of experimental and simulated measurements suggests that the distal Vh segments also adopt a rosette-like structure and are repositioned during B-lineage progression to merge with the proximal Vh region cluster [37,73,109]. In theory, this should allow the DhJh regions equal access to nearly any Vh segment regardless of genomic separation. Although the exact mechanism remains to be determined, it has been shown that interleukin-7 signalling as well as Pax5, YY1, and Ezh2 are required to promote the merging of proximal and distal Vh segments [73,110,111].

13. CONCLUSION

During the past 2 decades, much has been learnt as to how the immunoglobulin loci are organised in terms of their genomic structure. Recent studies also have revealed as to how the immunoglobulin heavy chain locus is spatially structured. However, the majority of these studies have used populations of studies to generate statistical topologies. What is needed now is to reveal the distribution of structures in single cells and to determine the mechanics that permit V, D, and J elements to encounter each other with the appropriate frequencies.

REFERENCES

[1] Nussenzweig MC, Alt FW. Antibody diversity: one enzyme to rule them all. Nat Med 2004;10(12):1304–5.

[2] Jung D, Giallourakis C, Mostoslavsky R, Alt FW. Mechanism and control of V(D)J recombination at the immunoglobulin heavy chain locus. Annu Rev Immunol 2006;24:541–70.

[3] Early P, Huang H, Davis M, Calame K, Hood L. An immunoglobulin heavy chain variable region gene is generated from three segments of DNA: VH, D and JH. Cell 1980;19(4):981–92.

[4] Brodeur PH, Riblet R. The immunoglobulin heavy chain variable region (Igh-V) locus in the mouse. I. One hundred Igh-V genes comprise seven families of homologous genes. Eur J Immunol 1984;14(10):922–30.

[5] Retter I, Chevillard C, Scharfe M, Conrad A, Hafner M, Im TH, et al. Sequence and characterization of the Ig heavy chain constant and partial variable region of the mouse strain 129S1. J Immunol 2007;179(4):2419–27.

[6] Chothia C, Lesk AM. Canonical structures for the hypervariable regions of immunoglobulins. J Mol Biol 1987;196:901–17.

[7] Al-Lazikani B, Lesk AM, Chothia C. Standard conformations for the canonical structures of immunoglobulins. J Mol Biol 1997;273:927–48.

[8] Wilson IA, Stanfield RL. Antibody-antigen interactions: new structures and new conformational changes. Curr Opin Struct Biol 1994;4:857–67.

[9] Morea V, Tramontano A, Rustici M, Chothia C, Lesk AM. Conformations of the third hypervariable region in the VH domain of immunoglobulins. J Mol Biol 1998;275:269–94.

[10] Larimore K, McCormick MW, Robins HS, Greenberg PD. Shaping of human germline IgH repertoires revealed by deep sequencing. J Immunol 2012;189(6):3221–30.

[11] Fukui K, Hamaguchi Y, Shimizu A, Nakai S, Moriwaki K, Wang CH, et al. Duplicated immunoglobulin gamma 2a genes in wild mice. J Mol Cell Immunol 1984;1(5):321–30.

[12] Morgado MG, Cam P, Gris-Liebe C, Cazenave PA, Jouvin-Marche E. Further evidence that BALB/c and C57BL/6 gamma 2a genes originate from two distinct isotypes. EMBO J 1989;8(11):3245–51.

[13] Tutter A, Riblet R. Evolution of the immunoglobulin heavy chain variable region (Igh-V) locus in the genus Mus. Immunogenetics 1989;30(5):315–29.

[14] Tutter A, Riblet R. Conservation of an immunoglobulin variable-region gene family indicates a specific, noncoding function. Proc Natl Acad Sci USA 1989;86(19):7460–4.

[15] Oettinger MA, Schatz DG, Gorka C, Baltimore D. RAG-1 and RAG-2, adjacent genes that synergistically activate V(D)J recombination. Science 1990;248(4962):1517–23.

[16] van Gent DC, McBlane JF, Ramsden DA, Sadofsky MJ, Hesse JE, Gellert M. Initiation of V(D)J recombination in a cell-free system. Cell 1995;81(6):925–34.

[17] Hendrickson EA, Qin XQ, Bump EA, Schatz DG, Oettinger M, Weaver DT. A link between double-strand break-related repair and V(D)J recombination: the scid mutation. Proc Natl Acad Sci USA 1991;88(10):4061–5.

[18] Kleinfield R, Hardy RR, Tarlinton D, Dangl J, Herzenberg LA, Weigert M. Recombination between an expressed immunoglobulin heavy-chain gene and a germline variable gene segment in a Ly 1+ B-cell lymphoma. Nature 1986;322(6082):843–6.

[19] Aoki-Ota M, Torkamani A, Ota T, Schork N, Nemazee D. Skewed primary Igk repertoire and V-J joining in C57BL/6 Mice: implications for recombination accessibility and receptor editing. J Immunol 2012;188(5):2305–15.

[20] Eisen HN, Reilly EB. Lambda chains and genes in inbred mice. Annu Rev Immunol 1985;3:337–65.

[21] Ikawa T, Kawamoto H, Wright LY, Murre C. Long-term cultured E2A-deficient hematopoietic progenitor cells are pluripotent. Immunity 2004;20(3):349–60.

[22] Osmond DG, Rolink A, Melchers F. Murine B lymphopoiesis: towards a unified model. Immunol Today 1998;19(2):65–8.

[23] Hardy RR, Carmack CE, Shinton SA, Kemp JD, Hayakawa K. Resolution and characterization of pro-B and pre-pro-B cell stages in normal mouse bone marrow. J Exp Med 1991;173(5):1213–25.

[24] Li YS, Hayakawa K, Hardy RR. The regulated expression of B lineage associated genes during B cell differentiation in bone marrow and fetal liver. J Exp Med 1993;178(3):951–60.

[25] Opstelten D, Osmond DG. Pre-B cells in mouse bone marrow: immunofluorescence stathmokinetic studies of the proliferation of cytoplasmic mu-chain-bearing cells in normal mice. J Immunol 1983;131(6):2635–40.

[26] Osmond DG, Owen JJ. Pre-B cells in bone marrow: size distribution profile, proliferative capacity and peanut agglutinin binding of cytoplasmic mu chain-bearing cell populations in normal and regenerating bone marrow. Immunology 1984;51(2):333–42.

[27] Grawunder U, Leu TM, Schatz DG, Werner A, Rolink AG, Melchers F, et al. Down-regulation of RAG1 and RAG2 gene expression in preB cells after functional immunoglobulin heavy chain rearrangement. Immunity 1995;3(5):601–8.

[28] Goodnow CC, Crosbie J, Adelstein S, Lavoie TB, Smith-Gill SJ, Brink RA, et al. Altered immunoglobulin expression and functional silencing of self-reactive B lymphocytes in transgenic mice. Nature 1988;334(6184):676–82.

[29] Hartley SB, Crosbie J, Brink R, Kantor AB, Basten A, Goodnow CC. Elimination from peripheral lymphoid tissues of

self-reactive B lymphocytes recognizing membrane-bound antigens. Nature 1991;353(6346):765–9.
[30] Allman DM, Ferguson SE, Lentz VM, Cancro MP. Peripheral B cell maturation. II. Heat-stable antigen(hi) splenic B cells are an immature developmental intermediate in the production of long-lived marrow-derived B cells. J Immunol 1993;151(9):4431–44.
[31] Alt FW, Yancopoulos GD, Blackwell TK, Wood C, Thomas E, Boss M, et al. Ordered rearrangement of immunoglobulin heavy chain variable region segments. EMBO J 1984;3(6):1209–19.
[32] Maki R, Kearney J, Paige C, Tonegawa S. Immunoglobulin gene rearrangement in immature B cells. Science 1980;209:1366–9.
[33] Chowdhury D, Sen R. Mechanisms for feedback inhibition of the immunoglobulin heavy chain locus. Curr Opin Immunol 2004;16(2):235–40.
[34] Kosak ST, Skok JA, Medina KL, Riblet R, Le Beau MM, Fisher AG, et al. Subnuclear compartmentalization of immunoglobulin loci during lymphocyte development. Science 2002;296(5565):158–62.
[35] Subrahmanyam R, Du H, Ivanova I, Chakraborty T, Ji Y, Zhang Y, et al. Localized epigenetic changes induced by DH recombination restricts recombinase to DJH junctions. Nat Immunol 2012;13(12):1205–12.
[36] Baumann M, Mamais A, McBlane F, Xiao H, Boyes J. Regulation of V(D)J recombination by nucleosome positioning at recombination signal sequences. EMBO J 2003;22(19):5197–207.
[37] Jhunjhunwala S, van Zelm MC, Peak MM, Cutchin S, Riblet R, van Dongen JJ, et al. The 3D structure of the immunoglobulin heavy-chain locus: implications for long-range genomic interactions. Cell 2008;133(2):265–79.
[38] Desiderio SV, Yancopoulos GD, Paskind M, Thomas E, Boss MA, Landau N, et al. Insertion of N regions into heavy-chain genes is correlated with expression of terminal deoxytransferase in B cells. Nature 1984;311(5988):752–5.
[39] Yancopoulos GD, Alt FW. Regulation of the assembly and expression of variable-region genes. Annu Rev Immunol 1986;4:339–68.
[40] Komori T, Okada A, Stewart V, Alt FW. Lack of N regions in antigen receptor variable region genes of TdT-deficient lymphocytes. Science 1993;261(5125):1171–5.
[41] Gilfillan S, Dierich A, Lemeur M, Benoist C, Mathis D. Mice lacking TdT: mature animals with an immature lymphocyte repertoire. Science 1993;261(5125):1175–8.
[42] Feeney AJ. Lack of N regions in fetal and neonatal mouse immunoglobulin V-D-J junctional sequences. J Exp Med 1990;172(5):1377–90.
[43] Gu H, Förster I, Rajewsky K. Sequence homologies, N sequence insertion and JH gene utilization in VHDJH joining: implications for the joining mechanism and the ontogenetic timing of Ly1 B cell and B-CLL progenitor generation. EMBO J 1990;9(7):2133–40.
[44] Yancopoulos GD, Desiderio SV, Paskind M, Kearney JF, Baltimore D, Alt FW. Preferential utilization of the most JH-proximal VH gene segments in pre-B-cell lines. Nature 1984;311(5988):727–33.
[45] Perlmutter RM, Kearney JF, Chang SP, Hood LE. Developmentally controlled expression of immunoglobulin VH genes. Science 1985;227(4694):1597–601.
[46] Feeney AJ, Lugo G, Escuro G. Human cord blood kappa repertoire. J Immunol 1997;158(8):3761–8.
[47] Williams GS, Martinez A, Montalbano A, Tang A, Mauhar A, Ogwaro KM, et al. Unequal VH gene rearrangement frequency within the large VH7183 gene family is not due to recombination signal sequence variation, and mapping of the genes shows a bias of rearrangement based on chromosomal location. J Immunol 2001;167(1):257–63.
[48] Feeney AJ, Goebel P, Espinoza CR. Many levels of control of V gene rearrangement frequency. Immunol Rev 2004;200:44–56.
[49] Boyd SD, Marshall EL, Merker JD, Maniar JM, Zhang LN, Sahaf B, et al. Measurement and clinical monitoring of human lymphocyte clonality by massively parallel VDJ pyrosequencing. Sci Transl Med 2009;1(12):12–23.
[50] Choi NM, Loguercio S, Verma-Gaur J, Degner SC, Torkamani A, Su AI, et al. Deep sequencing of the murine IgH repertoire reveals complex regulation of nonrandom v gene rearrangement frequencies. J Immunol 2013;191(5):2393–402.
[51] Feeney AJ. Genetic and epigenetic control of V gene rearrangement frequency. Adv Exp Med Biol 2009;650:73–81.
[52] Hesse JE, Lieber MR, Mizuuchi K, Gellert MV(D. J recombination: a functional definition of the joining signals. Genes Dev 1989;3(7):1053–61.
[53] Akamatsu Y, Tsurushita N, Nagawa F, Matsuoka M, Okazaki K, Imai M, et al. Essential residues in V(D)J recombination signals. J Immunol 1994;153(10):4520–9.
[54] Montalbano A, Ogwaro KM, Tang A, Matthews AG, Larijani M, Oettinger MA, et al. V(D)J recombination frequencies can be profoundly affected by changes in the spacer sequence. J Immunol 2003;171(10):5296–304.
[55] Lee AI, Fugmann SD, Cowell LG, Ptaszek LM, Kelsoe G, Schatz DG. A functional analysis of the spacer of V(D)J recombination signal sequences. PLoS Biol 2003;1(1):E1.
[56] Feeney AJ, Atkinson MJ, Cowan MJ, Escuro G, Lugo G. A defective Vkappa A2 allele in Navajos which may play a role in increased susceptibility to haemophilus influenzae type b disease. J Clin Invest 1996;97(10):2277–82.
[57] Van Ness BG, Weigert M, Coleclough C, Mather EL, Kelley DE, Perry RP. Transcription of the unrearranged mouse C kappa locus: sequence of the initiation region and comparison of activity with a rearranged V kappa-C kappa gene. Cell 1981;27(3 Pt 2):593–602.
[58] Lennon GG, Perry RP. C mu-containing transcripts initiate heterogeneously within the IgH enhancer region and contain a novel 5'-nontranslatable exon. Nature 1985;318(6045):475–8.
[59] Abarrategui I, Krangel MS. Regulation of T cell receptor-alpha gene recombination by transcription. Nat Immunol 2006;7(10):1109–15.
[60] Perry RP, Kelley DE, Coleclough C, Seidman JG, Leder P, Tonegawa S, et al. Transcription of mouse kappa chain genes: implications for allelic exclusion. Proc Natl Acad Sci USA 1980;77(4):1937–41.
[61] Yancopoulos G, Alt FW. Developmentally controlled and tissue-specific expression of unrearranged VH gene segments. Cell 1985;40:271–81.
[62] Bolland DJ, Wood AL, Johnston CM, Bunting SF, Morgan G, Chakalova L, et al. Antisense intergenic transcription in V(D)J recombination. Nat Immunol 2004;5(6):630–7.
[63] Bolland DJ, Wood AL, Afshar R, Featherstone K, Oltz EM, Corcoran AE. Antisense intergenic transcription precedes Igh D-to-J recombination and is controlled by the intronic enhancer Emu. Mol Cell Biol 2007;27(15):5523–33.
[64] Chakraborty T, Chowdhury D, Keyes A, Jani A, Subrahmanyam R, Ivanova I, et al. Repeat organization and epigenetic regulation of

the DH-Cmu domain of the immunoglobulin heavy-chain gene locus. Mol Cell 2007;27(5):842–50.
[65] Verma-Gaur J, Torkamani A, Schaffer L, Head SR, Schork NJ, Feeney AJ. Noncoding transcription within the Igh distal V(H) region at PAIR elements affects the 3D structure of the Igh locus in pro-B cells. Proc Natl Acad Sci USA 2012;109(42):17004–9.
[66] Ebert A, McManus S, Tagoh H, Medvedovic J, Salvagiotto G, Novatchkova M, et al. The distal V(H) gene cluster of the Igh locus contains distinct regulatory elements with Pax5 transcription factor-dependent activity in pro-B cells. Immunity 2011;34(2):175–87.
[67] Kurosawa Y, von Boehmer H, Haas W, Sakano H, Trauneker A, Tonegawa S. Identification of D segments of immunoglobulin heavy-chain genes and their rearrangement in T lymphocytes. Nature 1981;290(5807):565–70.
[68] Allman D, Sambandam A, Kim S, Miller JP, Pagan A, Well D, et al. Thymopoiesis independent of common lymphoid progenitors. Nat Immunol 2003;4(2):168–74.
[69] Borghesi L, Hsu LY, Miller JP, Anderson M, Herzenberg L, Herzenberg L, et al. B lineage-specific regulation of V(D)J recombinase activity is established in common lymphoid progenitors. J Exp Med 2004;199(4):491–502.
[70] Yang Q, Riblet R, Schildkraut CL. Sites that direct nuclear compartmentalization are near the 5' end of the mouse immunoglobulin heavy-chain locus. Mol Cell Biol 2005;25(14):6021–30.
[71] Perlot T, Alt FW, Bassing CH, Suh H, Pinaud E. Elucidation of IgH intronic enhancer functions via germ-line deletion. Proc Natl Acad Sci USA 2005;102(40):14362–7.
[72] Romanow WJ, Langerak AW, Goebel P, Wolvers-Tettero IL, van Dongen JJ, Feeney AJ, et al. E2A and EBF act in synergy with the V(D)J recombinase to generate a diverse immunoglobulin repertoire in nonlymphoid cells. Mol Cell 2000;5:343–53.
[73] Fuxa M, Skok J, Souabni A, Salvagiotto G, Roldan E, Busslinger M. Pax5 induces V-to-DJ rearrangements and locus contraction of the immunoglobulin heavy-chain gene. Genes Dev 2004;18(4):411–22.
[74] Seet CS, Brumbaugh RL, Kee BL. Early B cell factor promotes B lymphopoiesis with reduced Interleukin 7 responsiveness in the absence of E2A. J Exp Med 2004;199:1689–700.
[75] Osipovich OA, Subrahmanyam R, Pierce S, Sen R, Oltz EM. Cutting edge: SWI/SNF mediates antisense Igh transcription and locus-wide accessibility in B cell precursors. J Immunol 2009;183(3):1509–13.
[76] Milne TA, Dou Y, Martin ME, Brock HW, Roeder RG, Hess JL. MLL associates specifically with a subset of transcriptionally active target genes. Proc Natl Acad Sci USA 2005;102(41):14765–70.
[77] Xu CR, Feeney AJ. The epigenetic profile of Ig genes is dynamically regulated during B cell differentiation and is modulated by pre-B cell receptor signaling. J Immunol 2009;182(3):1362–9.
[78] Ji Y, Resch W, Corbett E, Yamane A, Casellas R, Schatz DG. The in vivo pattern of binding of RAG1 and RAG2 to antigen receptor loci. Cell 2010;141(3):419–31.
[79] Matthews AG, Kuo AJ, Ramón-Maiques S, Han S, Champagne KS, Ivanov D, et al. RAG2 PHD finger couples histone H3 lysine 4 trimethylation with V(D)J recombination. Nature 2007;450(7172):1106–10.
[80] Liu Y, Subrahmanyam R, Chakraborty T, Sen R, Desiderio S. A plant homeodomain in RAG-2 that binds hypermethylated lysine 4 of histone H3 is necessary for efficient antigen-receptor-gene rearrangement. Immunity 2007;27(4):561–71.
[81] McMurry MT, Krangel MS. A role for histone acetylation in the developmental regulation of VDJ recombination. Science 2000;287(5452):495–8.
[82] Chowdhury D, Sen R. Stepwise activation of the immunoglobulin mu heavy chain gene locus. EMBO J 2001;20(22):6394–403.
[83] Kwon J, Imbalzano AN, Matthews A, Oettinger MA. Accessibility of nucleosomal DNA to V(D)J cleavage is modulated by RSS positioning and HMG1. Mol Cell 1998;2(6):829–39.
[84] Golding A, Chandler S, Ballestar E, Wolffe AP, Schlissel MS. Nucleosome structure completely inhibits in vitro cleavage by the V(D)J recombinase. EMBO J 1999;18(13):3712–23.
[85] Degner SC, Wong TP, Jankevicius G, Feeney AJ. Cutting edge: developmental stage-specific recruitment of cohesin to CTCF sites throughout immunoglobulin loci during B lymphocyte development. J Immunol 2009;182(1):44–8.
[86] Featherstone K, Wood AL, Bowen AJ, Corcoran AE. The mouse immunoglobulin heavy chain V-D intergenic sequence contains insulators that may regulate ordered V(D)J recombination. J Biol Chem 2010;285(13):9327–38.
[87] Degner SC, Verma-Gaur J, Wong TP, Bossen C, Iverson GM, Torkamani A, et al. CCCTC-binding factor (CTCF) and cohesin influence the genomic architecture of the Igh locus and antisense transcription in pro-B cells. Proc Natl Acad Sci USA 2011;108(23):9566–71.
[88] Guo C, Yoon HS, Franklin A, Jain S, Ebert A, Cheng HL, et al. CTCF-binding elements mediate control of V(D)J recombination. Nature 2011;477(7365):424–30.
[89] Cremer T, Lichter P, Borden J, Ward DC, Manuelidis L. Detection of chromosome aberrations in metaphase and interphase tumor cells by in situ hybridization using chromosome-specific library probes. Hum Genet 1988;80(3):235–46.
[90] Lichter P, Cremer T, Borden J, Manuelidis L, Ward DC. Delineation of individual human chromosomes in metaphase and interphase cells by in situ suppression hybridization using recombinant DNA libraries. Hum Genet 1988;80(3):224–34.
[91] Pinkel D, Landegent J, Collins C, Fuscoe J, Segraves R, Lucas J, et al. Fluorescence in situ hybridization with human chromosome-specific libraries: detection of trisomy 21 and translocations of chromosome 4. Proc Natl Acad Sci USA 1988;85(23):9138–42.
[92] Branco MR, Pombo A. Intermingling of chromosome territories in interphase suggests role in translocations and transcription-dependent associations. PLoS Biol 2006;4(5):e138.
[93] Kornberg RD, Thomas JO. Chromatin structure; oligomers of the histones. Science 1974;184(4139):865–8.
[94] Oudet P, Gross-Bellard M, Chambon P. Electron microscopic and biochemical evidence that chromatin structure is a repeating unit. Cell 1975;4(4):281–300.
[95] Bednar J, Horowitz RA, Grigoryev SA, Carruthers LM, Hansen JC, Koster AJ, et al. Nucleosomes, linker DNA, and linker histone form a unique structural motif that directs the higher-order folding and compaction of chromatin. Proc Natl Acad Sci USA 1998;95(24):14173–8.
[96] Hsu TC. Differential rate in RNA synthesis between euchromatin and heterochromatin. Exp Cell Res 1962;27:332–4.
[97] Frenster JH, Allfrey VG, Mirsky AE. Repressed and active chromatin isolated from interphase lymphocytes. Proc Natl Acad Sci USA 1963;50:1026–32.

[98] Boyle S, Gilchrist S, Bridger JM, Mahy NL, Ellis JA, Bickmore WA. The spatial organization of human chromosomes within the nuclei of normal and emerin-mutant cells. Hum Mol Genet 2001;10(3):211−9.

[99] Sadoni N, Langer S, Fauth C, Bernardi G, Cremer T, Turner BM, et al. Nuclear organization of mammalian genomes. Polar chromosome territories build up functionally distinct higher order compartments. J Cell Biol 1999;146(6):1211−26.

[100] Chan JK, Park PC, De Boni U. Association of DNAse sensitive chromatin domains with the nuclear periphery in 3T3 cells in vitro. Biochem Cell Biol 2000;78(2):67−78.

[101] Paulson JR, Laemmli UK. The structure of histone-depleted metaphase chromosomes. Cell 1977;12(3):817−28.

[102] Paulson JR. Scaffolding and radial loops: the structural organization of metaphase chromosomes. In: Adolph KW, editor. Chromosomes and chromatin, vol III. Boca Raton, FL: CRC Press; 1988. p. 3−36.

[103] Pienta KJ, Coffey DS. A structural analysis of the role of the nuclear matrix and DNA loops in the organization of the nucleus and chromosome. J Cell Sci 1984;1(Suppl. 1):123−35.

[104] Okada TA, Comings DE. Higher order structure of chromosomes. Chromosoma 1979;72(1):1−14.

[105] Münkel C, Langowski J. Chromosome structure described by a polymer model. Phys Rev E Stat Phys Plasmas Fluids Relat Interdiscip Top 1998;57(5B):5888−96.

[106] Knoch TA. Approaching the three-dimensional organization of the human genome: structural-, scaling- and dynamic-properties in the simulation of interphase chromosomes and cell nuclei, long-range correlations in complete genomes, in vivo quantification of the chromatin distribution, construct conversions in simultaneous co-transfections. Mannheim: TAK Press; 2002.

[107] Lin YC, Jhunjhunwala S, Benner C, Heinz S, Welinder E, Mansson R, et al. A global network of transcription factors, involving E2A, EBF1 and Foxo1, that orchestrates B cell fate. Nat Immunol 2010;11:635−43.

[108] Lucas JS, Bossen C, Murre C. Transcription and recombination factories: common features? Curr Opin Cell Biol 2011;23(3):318−24.

[109] Sayegh CE, Jhunjhunwala S, Riblet R, Murre C. Visualization of looping involving the immunoglobulin heavy-chain locus in developing B cells. Genes Dev 2005;19(3):322−7.

[110] Su IH, Basavaraj A, Krutchinsky AN, Hobert O, Ullrich A, Chait BT, et al. Ezh2 controls B cell development through histone H3 methylation and Igh rearrangement. Nat Immunol 2003;4:124−31.

[111] Liu H, Schmidt-Supprian M, Shi Y, Hobeika E, Barteneva N, Jumaa H, et al. Yin Yang 1 is a critical regulator of B-cell development. Genes Dev 2007;21(10):1179−89.

Chapter 2

The Mechanism, Regulation and Evolution of V(D)J Recombination

David G. Schatz[1,2], Yiwen Zhang[3,4], Jianxiong Xiao[1], Shan Zha[5,6], Yu Zhang[7] and Frederick W. Alt[3,4]

[1]*Department of Immunobiology, Yale School of Medicine, New Haven, CT, United States;* [2]*Department of Molecular Biophysics and Biochemistry, Yale University, New Haven, CT, United States;* [3]*Howard Hughes Medical Institute, Program in Cellular and Molecular Medicine, Boston Children's Hospital, Boston, MA, United States;* [4]*Department of Genetics, Harvard Medical School, Boston, MA, United States;* [5]*Institute for Cancer Genetics, College of Physicians and Surgeons, Columbia University, New York City, NY, United States;* [6]*Division of Pediatric Oncology, Hematology and Stem Cell Transplantation, Department of Pediatrics, College of Physicians and Surgeons, Columbia University, New York City, NY, United States;* [7]*Department of Microbiology and Molecular Genetics, Michigan State University, East Lansing, MI, United States*

1. INTRODUCTION

During lymphocyte development, diverse primary repertoires of antigen receptors are created by V(D)J recombination. In the first phase of the V(D)J recombination reaction, double-stranded (ds) DNA breaks are generated by the recombinase machinery made up of the lymphocyte lineage-specific proteins recombination-activating gene (RAG) 1 and RAG2 (collectively, RAG), which are aided by the ubiquitously expressed DNA-binding/bending protein high mobility group box protein 1 (HMGB1). Extensive mechanistic insights into the V(D)J recombination cleavage phase have been obtained from structures of RAG bound to substrate DNA. The second phase of V(D)J recombination, end-processing and end-joining, is carried out by DNA repair proteins of the non-homologous end joining (NHEJ) DNA repair pathway. Various mechanisms associated with NHEJ diversify V(D)J junctions, while the lymphocyte-specific TdT factor further diversifies junctions via N-region addition. V(D)J recombination is tightly regulated during lymphocyte development. RAG proteins initiate the reaction from their location in recombination centres (RCs) in various antigen receptor loci. Chromatin loop extrusion then presents RAG with chromatin containing substrate segments embedded within megabases of DNA segments. Chromatin loop extrusion facilitates various mechanisms, including transcription, to mediate chromatin accessibility of particular variable region gene segments by increasing their interaction with RC-bound RAG during scanning. The loop extrusion mechanism confines RAG activity to loop domains containing antigen receptor loci, thereby dampening aberrant V(D)J recombinase activity that can promote oncogenic chromosomal translocations. In this chapter, we consider in detail our knowledge of the V(D)J recombination reaction, including the biochemistry of the cleavage reactions, activities and evolution of the RAG proteins and regulation of recombination in the context of chromatin loop extrusion.

2. RAG 1/2 ACTIVITY IN V(D)J RECOMBINATION

2.1 Overview

In 1976, the revolutionary discovery was made that DNA in lymphoid cells encoding the antigen receptors is altered from that of other somatic tissues and germline cells [1]. This DNA rearrangement is at the heart of the ability of B and T cells to generate a diverse collection of antigen receptor molecules, allowing a virtually unlimited set of antigen molecules to be recognised with a high degree of specificity. A series of site-specific recombination events, collectively termed V(D)J recombination, assembles antigen receptor genes from arrays of gene segments (Figures 2.1 and 2.2A) (for additional reviews on this topic, see Refs. [2–7]). Defects in the V(D)J recombination mechanism can result in immunodeficiencies, such as severe combined immunodeficiency (SCID) and Omenn syndrome [8].

The process of V(D)J recombination can be divided into two highly coordinated stages: (i) DNA binding and cleavage, and (ii) DNA repair. In the first stage, chromatin

Molecular Biology of B Cells. https://doi.org/10.1016/B978-0-323-95895-0.00004-0

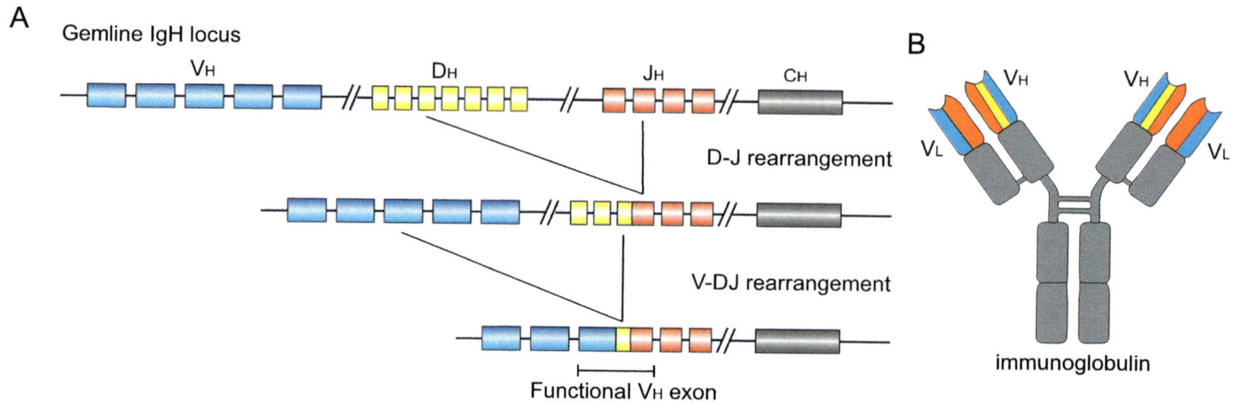

FIGURE 2.1 Schematic of V(D)J recombination in the immunoglobulin heavy chain locus. (A) The germ-line immunoglobulin heavy chain (IgH) locus is composed of many V (variable), D (diversity) and J (joining) gene segments upstream of the constant (C) regions. For simplicity, only one C region is depicted. During the stochastic gene rearrangement process of V(D)J recombination, one D gene segment is rearranged to join one J gene segment (D–J rearrangement). A second recombination event rearranges one V gene segment to join the previously rearranged DJ gene segment (V–DJ rearrangement), creating the functional V_H exon that encodes the variable domain of the immunoglobulin heavy chain. (B) Schematic of an immunoglobulin showing the contribution of the V, D and J gene segments to the variable domains of the IgH and IgL chains. Disulfide bonds are shown as grey lines.

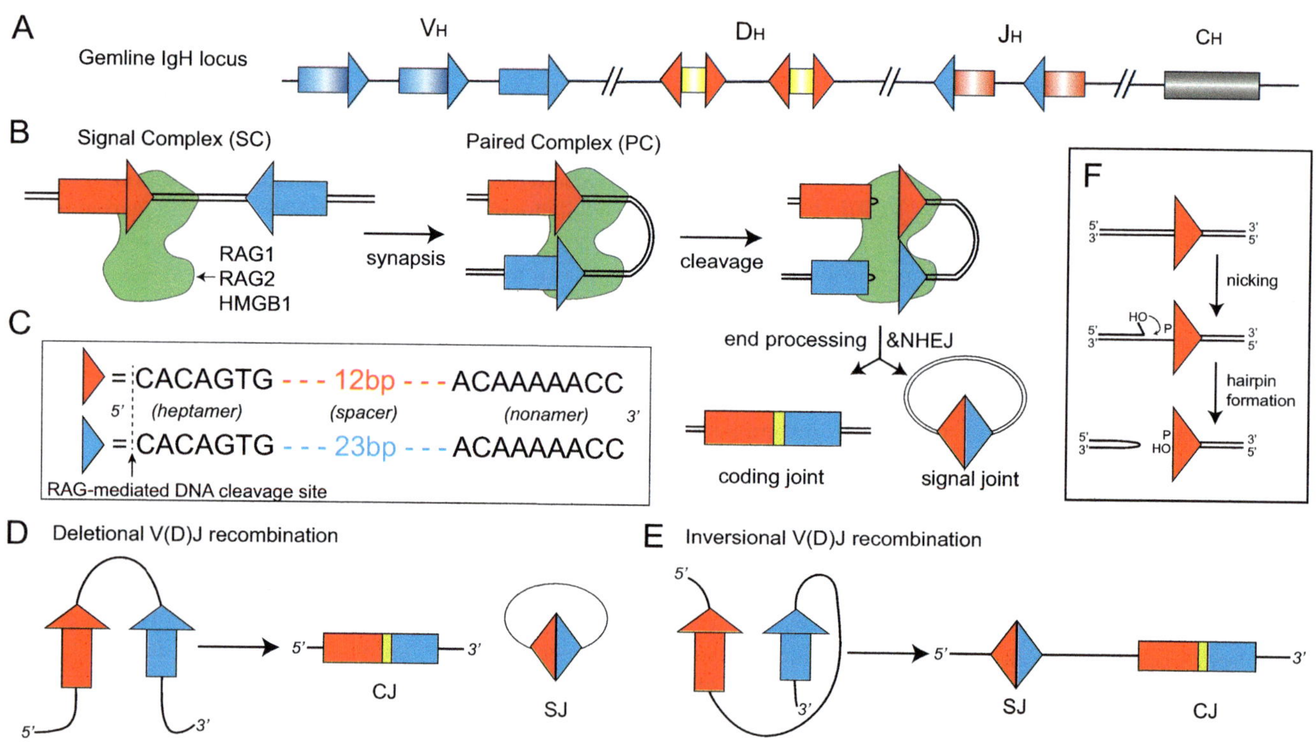

FIGURE 2.2 Steps in the V(D)J recombination reaction. (A) *Igh* locus showing 23RSSs and 12RSSs (blue and red triangles, respectively) flanking the V, D and J gene segments. (B) A RAG1/RAG2/HMGB1 complex can bind a single RSS, forming a signal complex (SC), which then captures another RSS to form the paired complex (PC). Under physiological conditions, nicking can occur in the SC or PC, but hairpin formation can occur only in the context of the PC. After cleavage, hairpin coding ends and blunt signal ends are processed by the non-homologous end-joining pathway, creating an imperfectly joined coding joint (CJ) (imperfect joining indicated by yellow section between coding gene segments) and a perfectly joined signal joint (SJ). (C) 12RSS (*red*) and 23RSS (*blue*) showing consensus heptamer and nonamer sequences. For simplicity, only top-strand sequences are shown. The heptamer and nonamer are separated by a poorly conserved 12 or 23-bp spacer. (D) The intervening sequences between the recombining coding segments are deleted when the 12 and 23 RSSs are oriented as depicted in convergent orientation, forming a CJ on the chromosome and an SJ on an extra-chromosomal circle. (E) RSSs oriented in the same direction along the chromosome, as depicted, lead to inversional V(D)J recombination, in which both CJ and SJ remain in the chromosome. SJs are typically joined precisely end to end, and CJs are imperfectly joined, as indicated by the green section between coding gene segments. (F) RAG-mediated DNA cleavage occurs in two steps, nicking and hairpin formation. Triangle depicts RSS; flanking coding gene segment is omitted for clarity. HO, hydroxyl group; P, phosphate group.

architectural/looping factors or random diffusion bring two DNA substrates into close proximity to facilitate their engagement by the RAG1 and 2 proteins [9,10]. Juxtaposition of the two substrates allows their simultaneous binding by the RAG1-RAG2 protein complex (hereafter, RAG) to form a synaptic or 'paired' complex (Figure 2.2B). Synaptic complex formation is aided by the DNA-binding/bending proteins high mobility group box protein (HMGB) 1 or 2 [11]. The DNA sequences bound by RAG, known as recombination signal sequences (RSSs) (Figure 2.2C), immediately flank the gene segments. Once the paired complex (PC) is formed, RAG introduces a pair of double-strand breaks (DSBs) between the RSSs and their flanking coding DNA segments (Figure 2.2B). The DNA ends generated by cleavage are asymmetrical, with the coding end covalently sealed into a hairpin and the signal end present as a 5′ phosphorylated blunt DNA end [12–14]. This first stage of V(D)J rearrangement is the point at which most of the regulation of the recombination reaction is imposed.

The second stage of V(D)J recombination, end-processing and end-joining, involves RAG and DNA repair proteins of the non-homologous (NHEJ) repair pathway (reviewed in Refs. [2,15]). The broken DNA generated by RAG cleavage can be resolved through several different pathways. The first is standard V(D)J joining, in which the hairpins are opened and joined imprecisely to each other to form a coding joint (CJ) and the two signal ends are ligated together to generate a signal joint (SJ) (Figure 2.2B). Alternatively, two non-standard products of V(D)J recombination can be generated by the joining of signal ends to coding ends [16]. Rejoining of a signal to its original coding flank yields an open and shut joint, while joining of a signal to its reaction partner's coding flank generates a hybrid joint (HJ). Finally, RAG bound to the signal ends can catalyse the transpositional attack of the signal ends on unrelated target DNA [17,18], though this process is strongly suppressed in cells.

This chapter will consider what is known about V(D)J recombination including the processes that bring two RSSs together, RAG structure, activities and evolution, the biochemistry of the cleavage reaction and the regulation of recombination.

2.2 Antigen Receptor Gene Assembly

Immunoglobulin (Ig) genes and T cell receptor (TCR) genes exist in the germline as linear arrays of clustered gene segments (Figures 2.1 and 2.2A). During V(D)J recombination, the many germline V (variable), D (diversity) in some cases and J (joining) segments undergo stochastic rearrangement, with one V, one D in some cases and one J segment being brought together by coordinated DNA cleavage and repair steps, resulting in extensive combinatorial diversity (Figure 2.1A). Lymphocyte antigen receptor diversity is greatly enhanced by the imperfect joining of the V, D and J segments, resulting in junctional diversity. Each Ig molecule is made up of two identical heavy chains (IgH) and two identical light chains (IgL) (Figure 2.1B), while each TCR is made up of two polypeptides, either TCRα and TCRβ or TCRγ and TCRδ. The many different possible combinations of individual IgH and IgL, TCRα and TCRβ, or TCRγ and TCRδ polypeptide chains that pair up to form a functional lymphocyte antigen receptor add an additional layer of 'chain pairing' diversity [19–21].

Seven antigen receptor loci exist: *TCR α, β, γ* and *δ* and *Ig h, κ* and *λ*. *Ig κ* and *λ* are the two IgL loci. All loci contain V and J segments, while three (*TCR β* and *δ* and *Igh*) also contain D segments between the V and J clusters. Antigen receptors are always composed of one polypeptide derived from a locus containing V, D and J elements and one from a locus with just V and J elements. During gene expression at each locus, the variable region exon consisting of VJ or VDJ elements created during V(D)J recombination is fused to a C (constant) region through RNA splicing. There are multiple *Igh* constant regions, each of which corresponds to a different Ig isotype. The incorporation of some *Igh* constant regions is determined by a separate somatic DNA rearrangement process, class switch recombination (CSR), which is discussed in later chapters.

In the germline configuration, gene segments to be recombined are flanked by RSSs, each of which contains two highly conserved sequence elements, a heptamer (consensus sequence 5′-CACAGTG-3′) and a nonamer (consensus sequence 5′-ACAAAAACC-3′), separated by a spacer of conserved length (12 ± 1 or 23 ± 1 bp) but poorly conserved sequence (Figure 2.2C). These are referred to as the 12RSS and 23RSS, respectively. Efficient recombination occurs only between a 12RSS and a 23RSS (the '12/23 rule'), helping to ensure that the correct types of gene segments are joined. All segments of a particular type (e.g. V_κ segments) are flanked by one type of signal and all the segments to which they could be joined (J_κ) are flanked by the opposite type. In the *Igh* locus, the 12/23 rule ensures that V segments recombine with D segments, but not directly to J segments (Figure 2.2A) [5,20].

However, the 12/23 rule is not sufficient to account for the rearrangement patterns observed at the *TCRβ* locus, where V_β segments are flanked by 23RSSs, J_β segments by 12RSSs and D_β segments have a 5′ 12RSS and a 3′ 23RSS. Direct V_β to J_β joining is rarely observed in vivo, although it would be in accordance with the 12/23 rule. In vivo, V_β gene segments preferentially recombine with the 12RSS 5′ of a D_β gene segment over any of the 12RSS flanking the J_β gene segments, a restriction termed the 'beyond 12/23' rule [22,23]. The beyond 12/23 restriction is largely enforced by the intrinsic properties of interactions within the functional cleavage complexes (discussed later) containing the *TCRβ* DNA substrates, the RAG proteins and HMGB1 or 2 [24–28].

The RSS sequences are all that is required to render a piece of DNA a substrate for V(D)J recombination in *in vitro* reactions and on extrachromosomal recombination substrates. The orientation of the RSSs with respect to each other determines the outcome of the reaction. Deletional V(D)J recombination occurs when the flanking RSSs are oriented convergently (nonamers pointing towards one another), resulting in retention of the CJ in the chromosome and deletion of a circular molecule containing the SJ. Alternatively, recombination can lead to inversion of the DNA between the RSSs with retention of both the SJ and CJ in the chromosome (Figure 2.2D and E). Both of these arrangements are found in vivo [29–31].

2.3 Recombination Signal Sequences

RSS substrates containing consensus heptamer and nonamer sequences (Figure 2.2C) are cleaved and recombined efficiently, and most in vitro experiments have used such substrates. However, virtually all endogenous RSS sequences vary from the consensus in some positions [32]. Only the first three nucleotides of the heptamer (5′-CAC) are 99%–100% conserved and are absolutely required for cleavage. The remaining positions exhibit 77%–91% conservation [32,33]. Similarly, not all positions of the nonamer are equally conserved. In the nonamer, the cytosines at the second and eighth positions and several adenines in the A tract are over 85% conserved. The remaining nonamer positions show 56%–77% conservation [32]. This distribution of sequence conservation has functional correlates, with the most highly conserved positions of the nonamer being most important for initial protein binding and the conserved heptamer positions being required for efficient RAG-mediated cleavage [32,33]. Low levels of DNA cutting can be detected with substrates containing only a heptamer or only a nonamer [34,35], an important observation given that cryptic RSSs, which sometimes consist of no more than the 5′-CAC of the heptamer, can support RAG cutting and genome instability [6,36,37].

Spacer sequences are much less conserved than heptamer or nonamer sequences, though some positions exhibit conservation at levels exceeding 50%. The sequence of the spacer can have a substantial effect on RSS recombination efficiency [38], and an 'anticonsensus' spacer sequence can strongly inhibit RSS function [39]. Spacer length (deviating no more than one bp from 12 or 23) is critical for RSS function [33], though some function can be restored if one full turn of the DNA helix is added to a 23 bp spacer to create a 33 or 34 bp spacer [34,35].

In addition to the RSS, the first two or three nucleotides of the coding flank immediately abutting the RSS heptamer can have considerable effects on the efficiency of the recombination reaction. Although most sequences are essentially neutral, certain nucleotide combinations can be favourable or unfavourable to the reaction. The effect of the coding flank sequence appears to be primarily at the cleavage phase of V(D)J recombination [34,35,40]. This might be related to structural distortion of the coding flank and heptamer regions that occurs during cleavage, as discussed later.

Variation in endogenous RSS sequences and coding flanks has an effect on cleavage efficiency, and this has an impact on gene segment usage in the antibody or TCR repertoire [38,41,42]. Numerous other factors contribute to the control of V(D)J recombination targeting and the tightly regulated order of recombination events in vivo [4,43], discussed below.

2.4 Biochemistry of RAG-Mediated DNA Cleavage

DNA cleavage requires only RAG1 and RAG2, a divalent metal ion, and a DNA substrate containing the RSS, and no external source of energy is needed [44,45]. DNA cleavage in V(D)J recombination occurs in two separable steps. First, the top strand is nicked 5′ of the RSS precisely at the heptamer-coding flank junction, generating a 3′ hydroxyl group. In the second step, this 3′ hydroxyl attacks the phosphate on the bottom strand of the DNA, performing direct transesterification and resulting in a double-stranded DNA break (DSB) with a hairpin coding end and a blunt signal end [44,46] (Figure 2.2F). During hairpin formation, the energy required for the formation of the new bond is derived from the breakage of the old one. Stereochemical studies have shown that a covalent bond between the RAG proteins and DNA is not formed [46]. This distinguishes the RAG cleavage reaction from that of a number of other site-specific recombinases, such as Cre, Flp, and Lambda-Int, which rely on a covalent intermediate. Instead, the mechanism used by RAG1/RAG2 is similar to that of bacterial transposases and HIV integrase [46–48].

The V(D)J recombinase machinery can first assemble on a single 12 or 23RSS, creating a signal complex (SC) in which nicking can occur. Biochemical [49–52] and structural [53,54] evidence indicates that the PC, which contains both a 12 and 23RSS, is then formed by the 'capture model', in which one SC captures the appropriate partner RSS to form the PC, maintaining the same protein content in both the SC and the PC. Under physiological conditions, dsDNA breaks are generated by 'coupled cleavage' in the PC [55] (Figure 2.2B), which requires Mg^{2+} divalent metal ions and HMGB1 or HMGB2.

After coupled cleavage, the recombinase proteins and both the coding and signal ends are contained in a post-cleavage complex, from which the coding ends are shepherded to the NHEJ proteins for end-processing and joining [56,57]. The hairpin-sealed coding ends are first opened by a complex of the proteins Artemis and DNA-dependent protein kinase catalytic subunit (DNA-PK_{CS}), generally resulting in P (palindromic) nucleotides due to hairpin opening two nucleotides 3′ of the tip of the hairpin [58]. As discussed later in this chapter, the ends are then processed by other ubiquitous proteins of the NHEJ pathway and DNA polymerases, resulting in the addition of N (non-templated) nucleotides and deletion of nucleotides by nucleolytic trimming, before being joined to form a CJ. This coding end processing is responsible for the imprecise joining of the V, D and J segments, which creates junctional diversity. After the coding ends are released, the signal ends are retained by RAG/HMGB1/2 in a stable signal end complex (SEC) [56], which shields the ends from the action of DNA ligase IV together with other purified NHEJ factors [59]. In vivo, signal ends are generally precisely joined to each other by the NHEJ pathway, forming an SJ that is in many cases lost from the genome [60]. The mechanism by which the signal ends are liberated from the SEC to allow end joining is not known.

2.5 RAG1 and RAG2

The *Rag1* and *Rag2* genes are not homologous to each other, and highly conserved homologues of both genes have been identified in all jawed vertebrates studied, with the intriguing exception of certain species of anglerfish that have dispensed with adaptive immunity and deleted or inactivated their *Rag* genes [61]. In all species examined thus far, *Rag1* and *Rag2* are tightly linked, and in most species, the entire open reading frame of each gene is contained in a single exon, features that are more commonly found in prokaryotic than eukaryotic organisms. As discussed below, there is now compelling evidence that the *Rag* genes evolved from a transposable element. The amino acid sequences of murine and human RAG1 and RAG2 are 90% and 88% identical, respectively [9,62], and most studies of RAG protein biochemistry have used the mouse proteins. In addition, because the minimal functional truncation mutants, core RAG1 and core RAG2, are more soluble in vitro than the full-length proteins, most in vitro biochemical experiments have used the core proteins. However, mice expressing core RAG1 or core RAG2 exhibit reduced V(D)J recombination efficiency and fidelity as well as defective ordering of gene segment rearrangement [63–67].

Full-length murine RAG1 is a 1040 amino acid polypeptide, with the minimal core protein capable of recombination encoded by residues 384–1008 [68,69] (Figure 2.3A). The RAG1 core is composed of the NBD, which forms a tight dimer and binds to the RSS nonamer [70,71] and a large region composed of multiple structurally defined subdomains that mediate nuclear localisation, binding to RAG2 and DNA and DNA cleavage [3,72]. DNA cleavage is achieved using an active site containing

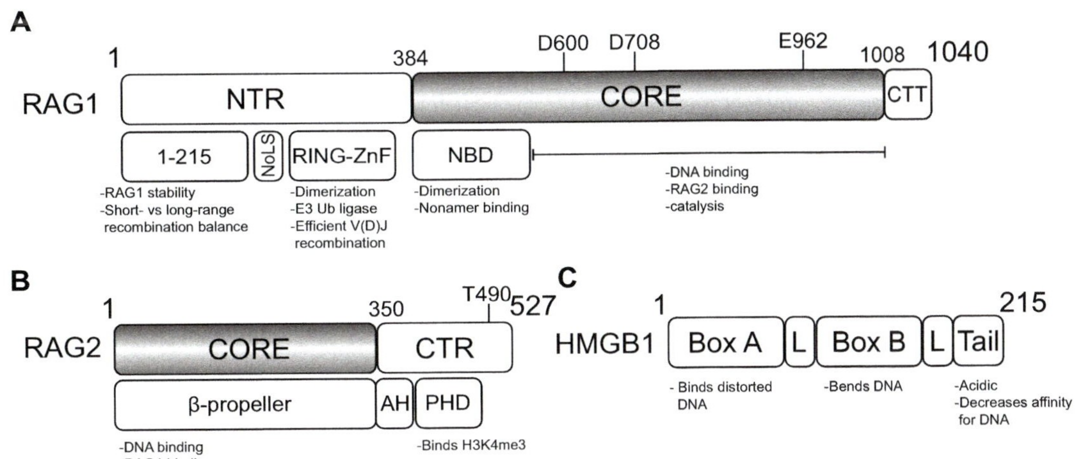

FIGURE 2.3 **RAG1, RAG2 and HMGB1 proteins.** (A) Schematic of RAG1. The essential core region contains the nonamer-binding domain (NBD) and the DDE catalytic triad. The N-terminal region (NTR) contains a RING-ZnF (zinc finger) domain, a nucleolar localisation signal (NoLS) and a poorly understood N-terminal 215 amino acids. The RAG1 C-terminal tail (CTT) is not required for activity, and its contribution to V(D)J recombination in vivo is not well understood. (B) Schematic of RAG2. The essential core region folds into a 6-bladed β-propeller, while the C-terminal region (CTR) contains an acidic hinge (AH) implicated in numerous functions, a plant homeodomain (PHD) and threonine 490 (T490), which is important in regulating RAG2 levels in the cell cycle. Amino acid and domain boundary numbers derive from mouse RAG1 and RAG2. (C) HMGB1 protein. HMGB1 contains two HMG boxes (Box A and Box B) and an acidic tail region, with basic linkers (L) separating the domains. Functions of many of the protein domains and regions are indicated below. See text for additional information.

an Asp/Asp/Glu (DDE) catalytic triad (D600, D708 and E962 in mouse RAG1) that coordinates two essential divalent metal ions [73–75] (Figure 2.3A). Mg^{2+} is thought to be the physiological ion because it supports properly regulated coupled cleavage. Ca^{2+} supports DNA binding but not nicking or hairpin formation, while Mn^{2+} allows for efficient uncoupled cleavage even when RAG is bound to only a single RSS [3]. Core RAG1 forms a tight dimer in solution [76–79].

The large non-core N-terminal region (NTR) of RAG1 (aa 1-383) plays numerous roles (Figure 2.3A). The best characterised portion of the NTR is the zinc dimerisation domain (ZDD) (aa 287-351), located just N-terminal to the NBD of core RAG1, which dimerises and contains a RING (really interesting new gene) finger and a C_2H_2 zinc finger [80,81]. The ZDD is an E3 ubiquitin ligase that can ubiquitylate RAG1 itself, KPNA1, histone H3 and histone variant H3.3 [82–85]. Recent studies of RAG1 ZDD point-mutant mice (Pro 326 mutated to Gly) reveal that E3 ligase activity downregulates RAG1 protein levels (likely through autoubiquitylation of Lys 233) and is important for efficient V(D)J recombination through an as yet undetermined mechanism [86,87]. RAG1 NTR residues 1-215 downregulate RAG1 protein levels through interaction with an E3 ubiquitin ligase complex containing VprBP, DDB1, Cul4A and Roc1 [88,89] and also control the balance between short-range (diffusional) and long-range (chromatin scanning) modes of recombination at the Ig heavy chain (*Igh*) locus [86]. Nucleolar sequestration of RAG1, which requires NTR aa 243-249, appears to repress RAG function [90]. It is likely that additional functions will be discovered for the RAG1 NTR.

Murine RAG2 is a 527 amino acid polypeptide (Figure 2.2B). Like RAG1, RAG2 has a minimal core region necessary for catalysis (amino acids 1–350) [72,91–93]. The RAG2 core is attached via a flexible acidic hinge to a structurally characterised plant homeodomain (PHD) that specifically binds trimethylated lysine 4 on histone H3 (H3K4me3) [94,95]. Binding of H3K4me3 by the RAG2 PHD not only helps guide RAG to transcriptionally active regions of chromatin, including V(D)J RCs [96–98] but also relieves repression of RAG catalytic activity through an autoregulatory, allosteric mechanism [99–102]. The most C-terminal portion of RAG2 contains Thr490, which is phosphorylated in a cell-cycle–dependent manner (in the S, G2 and M phases of the cell cycle), resulting in ubiquitylation and proteosomal degradation [103–106]. This C-terminal region also contains a nuclear localisation signal (NLS), which interacts with importin 5 and can mediate RAG2 nuclear transport [107]. Studies using RAG2 core knock-in mice have shown that the RAG2 C-terminus plays an important role in maintaining genome integrity during V(D)J recombination in vivo [108,109], a function that can be attributed at least in part to the cell cycle regulation imposed by Thr490 [110]. RAG1 and RAG2 have been found to associate by co-immunoprecipitation, co-purification, and pulldown assays [111–114], and the molecular nature of their interaction is now understood from structural studies (see below).

2.6 HMGB1 and HMGB2 in V(D)J Recombination

HMGB1 and HMGB2 are sequence-nonspecific DNA-binding proteins composed of two HMG-box domains, which bind in the DNA minor groove, connected by basic linkers to each other and to an acidic C-terminal tail domain (Figure 2.3C) [115]. HMGB1/2 were identified as a component of the V(D)J recombinase through an effort to isolate factors in whole-cell extract that stimulate RSS cleavage [11,116]. In vitro, RAG1 and RAG2 can efficiently assemble on a 12RSS where nicking and, in the presence of a 23RSS, low levels of hairpin formation can occur. However, the RAG proteins alone bind and nick the 23RSS at very low levels, and coupled hairpin cleavage at the 23RSS is nearly undetectable [3,44,117]. HMGB1 and 2 enhance 23SC formation, incorporating stably into the complex and stimulating nicking while suppressing aberrant nicking at inappropriate sites within the 23RSS. With the addition of partner 12RSS, HMGB1 stimulates coupled cleavage of both the 12 and 23RSS by more than 100-fold in vitro, achieving a level of coupled cleavage that would be physiologically relevant for V(D)J recombination in vivo [11]. HMGB1 and HMGB2 are very similar in sequence, structure and function and are functionally redundant in vitro as a cofactor in V(D)J cleavage [11,118,119]. For simplicity, we refer only to HMGB1 hereafter.

HMGB1 binds B-form DNA weakly, preferring non-canonical DNA structures such as ssDNA and four-way junctions [120]. HMGB1 binding can also introduce DNA bends via partial intercalation of hydrophobic residues [121,122], and these residues play a role in stimulating RAG-mediated RSS binding and cleavage [123,124]. The acidic tail domain makes extensive contacts with the DNA-binding surface of both HMG boxes which regulates HMGB1 DNA binding by acting as a DNA mimetic, resulting in a 'closed' HMGB1 conformation incapable of binding DNA [125]. The HMGB1 acidic tail is not required for enhancing 23RSS cleavage. HMGB1 associates with the RAG1 nonamer-binding domain (NBD) in the absence of an RSS [119] and with the RSS in the absence of RAG1 [118], but HMGB1 is recruited into a ternary HMGB1-RAG1-RSS complex much more efficiently than to either RAG1 or RSS alone [126].

Genetic evidence for the role of HMGB1 in V(D)J recombination in vivo has been difficult to obtain due to functional redundancy and the diverse but essential roles of the HMGB1 and HMGB2 proteins in a variety of nuclear processes. Single knockout mice do not display obvious defects in V(D)J recombination or lymphocyte development [127,128], and *Hmgb1*$^{-/-}$:*Hmgb2*$^{-/-}$ double knockouts are lethal [129]. Overexpression, knockdown studies and analyses using mutants of RAG1 defective in binding HMGB1 argue that HMGB1 is required for efficient V(D)J recombination in vivo [119,130].

2.7 The Structure of RAG-RSS Complexes and the Mechanism of the 12/23 Rule

For 25 years after their discovery, RAG1 and RAG2 remained poorly understood structurally, with only isolated subdomains of the proteins resolved at near atomic resolution [71,81,94]. During this period, biochemical analyses, particularly DNA footprinting, interference and ultraviolet (UV) crosslinking experiments, provided significant insights into the stoichiometry and organisation of RAG-DNA complexes (reviewed in Refs. [3,5,131]). Particularly important were findings showing that: (i) the SC contains a tight dimer of RAG1 that cannot reassort in solution and either one or two molecules of RAG2 that can reassort in solution [52,76,77,132]; (ii) the PC likely contains the same proteins as the SC, in support of the capture model [49–52,118]; (iii) RAG1 makes contacts with almost all portions of the DNA substrate including the coding flank and can recognise the nonamer with some sequence specificity in the absence of RAG2 [70,133–136]; (iv) RAG2 is important for enabling RAG1 to interact with the heptamer and might itself closely approach the heptamer and coding flank [133–135,137]; (v) HMGB1 interacts with the RSS spacer, particularly of the 23RSS [3,118]; (vi) in interactions with a given RSS, the RAG1 DDE active site motif is contributed by one RAG1 protomer (of a RAG1 dimer), while the NBD is contributed by the other RAG1 protomer, leading to a *trans* cleavage model in which each RSS is contacted by both RAG1 protomers in a dimer [132]; (vii) DNA unpairing and structural distortion—including base flipping of the bottom strand nucleotide of the coding flank immediately adjacent to the heptamer—likely play an important role in V(D)J cleavage [34,35,138,139] and (viii) RSSs in the PC are unlikely to be planar but rather are bent and cross over each other [140–142].

Beginning in 2015, a succession of studies led to a cornucopia of atomic or near-atomic resolution structural information for RAG, the SC, the PC and even RAG in the act of performing transposition [53,54,72,143–146]. The foundational study used X-ray crystallography to determine the structure of the core regions of RAG1 and RAG2 in the apo RAG complex [72]. The Y-shaped core RAG structure (Figure 2.4A) is composed of a dimer of RAG1 with a RAG2 monomer at the tip of each arm of the Y. The pedestal of the Y is formed by a dimer of the NBD, which is connected to the rest of the RAG1 dimer by a flexible linker. RAG1 dimerisation is mediated by the NBD and the adjacent dimerisation and DNA-binding domain (DDBD) which lies at the branch point of the Y (Figure 2.4B and C). The active site is contained in an RNase H-like fold domain (RNH) that brings the DDE motif residues (D600, D708 and E962) of one RAG1 monomer into close proximity. The RAG1 RNH domain is structurally similar to the RNase H-like domain of DDE-family transposase and retroviral integrase enzymes [147]. The large intervening region between D708 and E962 is composed of two subdomains, ZnC2 and ZnH2 (hereafter, referred to together as ZnB), named for two Cys and two His residues they contribute that coordinate a Zn^{2+} atom (Figure 2.4C). ZnB plays numerous roles including stabilising the domain architecture of RAG1, interacting with RAG2 and undergoing large rotational and pivoting movements critical for DNA binding and cleavage.

RAG2 core is a doughnut-shaped structure that adopts a six-bladed β-propeller fold, as predicted by modelling studies [148,149]. One surface of the RAG2 doughnut extends multiple loops that protrude into and around RAG1 (Figure 2.4A), forming an extensive interaction surface of complementary 'ridges and canyons' involving several RAG1 subdomains and interacting with RAG1 residues near the active site [72].

Subsequent structural studies revealed how RAG and HMGB1 bind RSS substrates, the mechanism of the 12/23 rule and numerous insights into the mechanism and regulation of DNA cleavage [53,54,143,144]. The PC retains the overall Y shape seen for apo RAG (Figure 2.5A). The 12RSS and 23RSS pass through the RAG complex with the coding flanks emerging at the top and the spacers emerging and undergoing large bends to position the nonamers for binding by the NBD dimer at the bottom. Each monomer of RAG1 contacts both DNA substrates, crossing back and forth between them multiple times. The DNA surrounding the site of cleavage (the first 5 bp of the coding flank and the first 3 bp of the heptamer) is engaged by one monomer of RAG1, referred to as the cis monomer because this monomer's active site will execute cleavage of this RSS. The remaining residues of the heptamer and the first ~7 bp of the spacer are engaged by the other (trans) RAG1 monomer. The cis RAG1 monomer returns to bind the final 5 bp of the spacer and the beginning of the nonamer, leaving the bulk of nonamer contacts to be made by the trans RAG1 monomer. For each RSS, cleavage is mediated by the cis RAG1 monomer, while most of the DNA contacts are made by the trans RAG1 monomer, consistent with the trans model of cleavage [53,54,132].

The coding flank is cradled in a positively charged channel created by the cis RAG1 subunit and more distally,

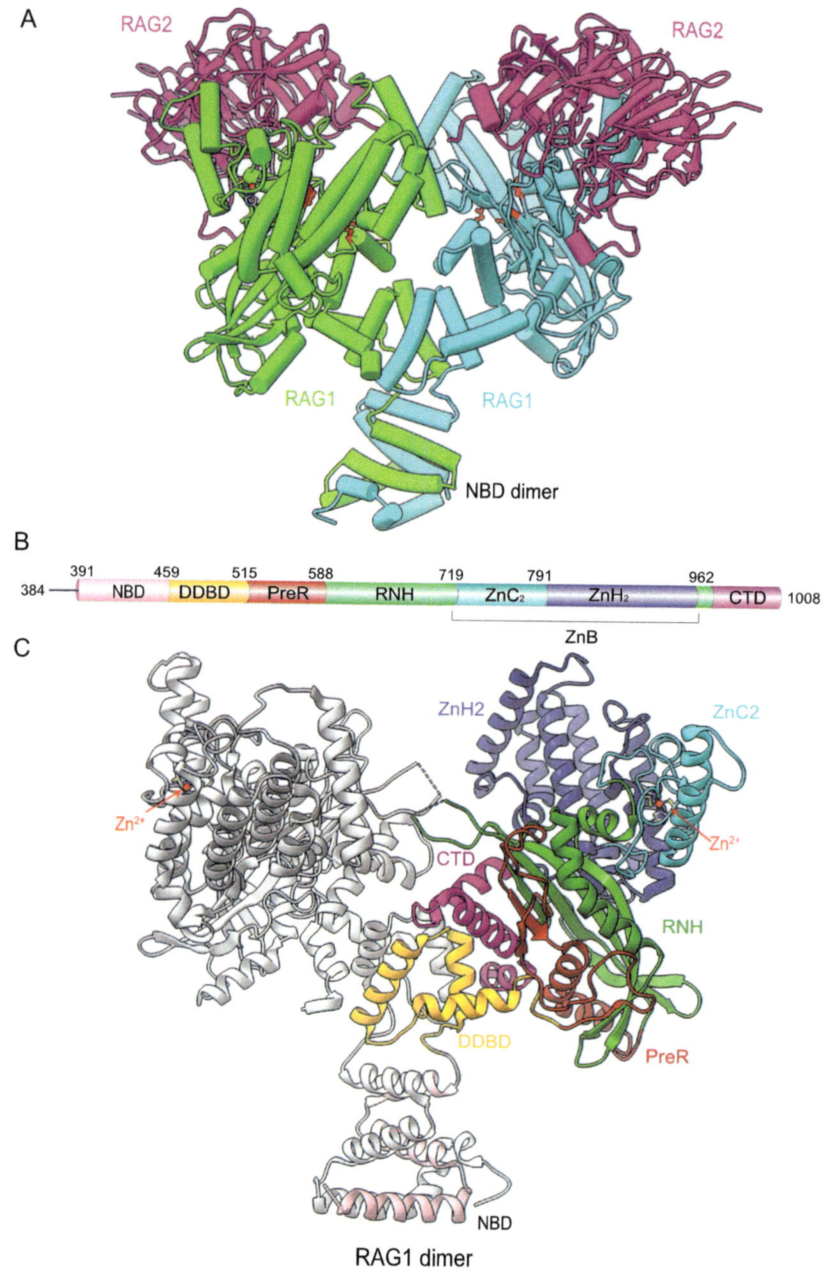

FIGURE 2.4 Structure of the RAG1-RAG2 heterotetramer. (A) RAG1-RAG2 heterotetramer (crystal structure [72]; PDB:4WWX), with RAG1 subunits in green and blue and RAG2 subunits in purple. The structure encompasses the RAG1 and RAG2 core regions. The nonamer-binding domain (NBD) dimer is indicated at the bottom. Active site DDE residues are indicated as red stick figures. Red sphere, zinc atom. (B) Linear depiction of subdomains of the RAG1 core region. DDBD, dimerisation and DNA-binding domain; preR, pre-RNase H domain; RNH, RNase H domain; ZnC_2 and ZnH_2, domains that together constitute a zinc-binding domain (ZnB) that undergoes large movements during DNA binding and catalysis; CTD, C-terminal domain. (C) RAG1 dimer with one subunit in white and the other with subdomains coloured as in panel B, and zinc atoms as red spheres. Figure panels redrawn based on figures from Ref. [72].

its associated RAG2 monomer. Overall, the RAG proteins bind 12RSS and 23RSS substrates in a similar manner, engaging about 12 bp of coding flank, the heptamer, the entire 12RSS spacer (and, using equivalent contacts, the first 7 bp and last 5 bp of the 23RSS spacer) and the nonamer.

Taking advantage of the flexible hinge connecting it to the rest of the RAG1 core, the NBD dimer tilts towards the 12RSS and away from the 23RSS in the PC relative to apo RAG (Figure 2.5A and B) [53,54]. This tilt and the overall architecture of the complex provide a ready explanation of the 12/23 rule: the tilt towards a bound 12RSS positions the NBD such that a 23RSS but not a second 12RSS can be accommodated, and, reciprocally, the tilt away from a bound 23RSS positions the NBD such that a 12RSS but not a second 23RSS can be bound (Figure 2.5A and B). This model for the 12/23 rule is supported by both biochemical [150] and structural [53,54] considerations.

In the NBD dimer, the two nonamers align in an almost antiparallel configuration [71]. This, combined with the

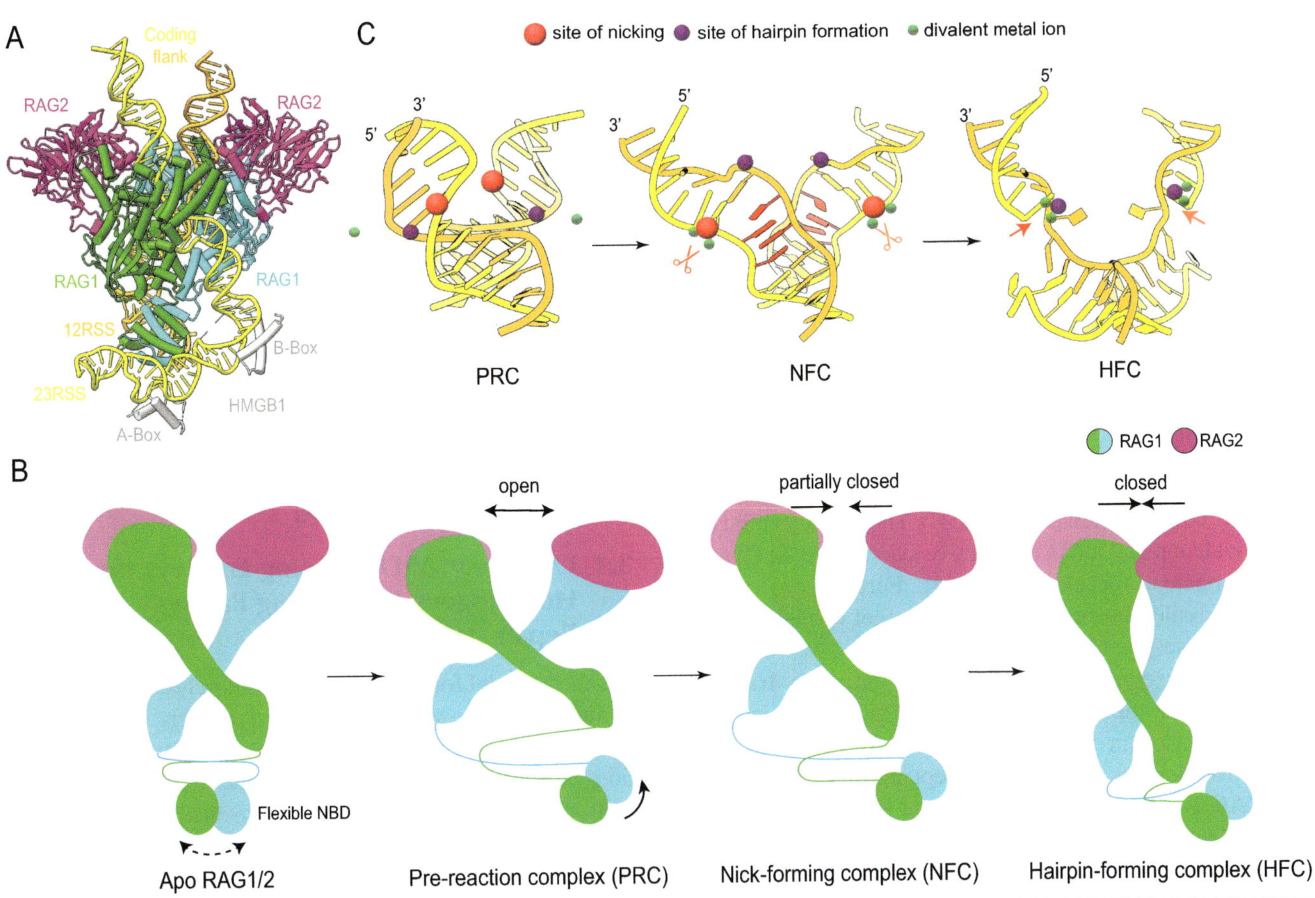

FIGURE 2.5 **Structure of the RAG-RSS synaptic complex and structural alterations during cleavage.** (A) RAG1-RAG2-HMGB1-12RSS-23RSS hairpin-forming complex (cryo-EM structure [54]; PDB:6CG0), with RAG1 subunits in green and blue, RAG2 subunits in purple, HMGB1 A and B boxes in white, and 12 and 23RSSs in orange and yellow, respectively. Panel redrawn based on a figure from Ref. [54]. (B) Cartoon showing domain movements during DNA binding and cleavage. The complex opens during initial DNA binding to form the pre-reaction complex (PRC), then partially closes during formation of the nick-forming complex (NFC) and completely closes during formation of the hairpin-forming complex (HFC). (C) DNA surrounding the cleavage site showing structural changes that occur during the transition from the PRC (PDB:6CIK) to NFC (PDB:6OEO) to HFC (PDB:6CG0). Divalent metal ions (green spheres) indicate the location of the active site, while red and purple spheres indicate the locations of the bonds that will be broken during nicking and hairpin formation, respectively. The site of nicking is far from the active site in the PRC, and it moves into the active site in the NFC as a result of a 180° DNA rotation accompanied by unpairing of heptamer positions 2 and 3 (red stick figures), which form an interstrand 'zipper' structure. Subsequent DNA and protein conformational changes position the scissile phosphate for hairpin formation in the active site in the HFC. Panel redrawn based on a figure from Ref. [144].

short (6 aa) tether connecting the NBD to the remainder of core RAG1, forces the two RSSs to bend significantly (Figure 2.5A). Both RSSs undergo a ~60-degree bend at the nonamer/spacer border, while the 23RSS undergoes an additional ~90-degree bend in the centre of its spacer (the only portion of the RSS not contacted by RAG) [53,54]. The two HMG boxes of HMGB1 each bind one of the bends in the 23RSS, thereby stabilising the extremely (~150-degree) bent 23RSS and explaining why HMGB1 is critical for PC formation and coupled cleavage (Figure 2.5A) [53,54]. A second molecule of HMGB1 can use one HMG box to bind the 60-degree bend in the 12RSS, but it is unlikely that this is required for PC formation or cleavage. The elaborate series of bends that occur in the spacers, a kink between positions 6 and 7 of the heptamer, and the dearth of sequence-specific protein-DNA interactions in these regions, argues that preferred sequence context in the distal heptamer and spacer (particularly incorporating CA and TG dinucleotides), rather than conservation of individual DNA residues, helps dictate RSS functionality based on sequence propensity for deformation and bending [54]. This idea is supported by a high-throughput analysis of 12RSSs randomised in the last four positions of the heptamer and the first two positions of the spacer [151]. Because changing the magnitude or location of deformation in one RSS imposes constraints on the allowed structure of the partner RSS, this model likely has important implications for understanding RSS incompatibilities that underlie the beyond 12/23 rule [54].

2.8 Structural Insights into the Mechanism of DNA Cleavage by RAG

Comparisons of structures of apo RAG and RAG bound to intact and nicked RSSs reveal major protein and DNA conformational changes required for DNA nicking and hairpin formation, explain how a single active site can execute both steps of DNA cleavage, and highlight possible mechanisms ensuring coupled cleavage. In apo RAG, the DDE catalytic residues are not properly oriented for divalent metal binding and catalysis. Initial binding of intact 12/23RSS substrates to create the pre-reaction complex (PRC) triggers tilting of the NBD dimer (as described above), an outward 'opening' movement of the arms of the Y, outward rotation of the RAG1 ZnB domains and movement of the two D residues (but not E962) of the catalytic triad into their catalytic configuration (Figure 2.5B). The PRC can bind only one divalent metal ion in each active site (Figure 2.5C) and is not competent for cleavage [144]. Furthermore, the top strand scissile phosphate that is the target of nicking is far from the catalytic triad (Figure 2.5C).

Dramatic conformational changes occur to transform the PRC into the nick-forming complex (NFC) in which nicking can occur [143,144]. The DNA undergoes a dramatic 180 degrees untwisting at heptamer positions 2 and 3 (5′-CACAGTG), thereby rotating the site of cleavage and the coding flank by 180 degrees and moving the scissile phosphate into the active site (Figure 2.5C). Heptamer positions 2 and 3 become unpaired and the bases from the two strands become interleaved and stacked upon one another to form a novel interstrand 'zipper' structure [144] (Figure 2.5C). The DNA rotation and movement of the scissile phosphate is accompanied by an inward rotation of ZnB, proper positioning of E962, coordination of a second divalent metal ion and creation of a functional active site ready to perform nicking (Figure 2.5B and C) [144].

Subsequent nicking of both RSSs to give the hairpin-forming complex (HFC) is accompanied by further 'closure' of the arms of the Y and inward rotation of ZnB, additional tilting and rotation of the NBD dimer, restoration of base pairing in the heptamer and a scissors-like movement of the heptamer and coding flank that positions the bottom strand scissile phosphate in the active site for hairpin formation (Figure 2.5B and C) [54,143,144]. The structures of the first 3 bp of the heptamer are strongly distorted in the HFC with the first C base flipped out of the helix, and the distortion of the coding flank needed to position the hairpin scissile phosphate is accompanied by coding flank melting and base flipping at the first position of bottom strand (Figure 2.5C). The inward movement of the arms of the Y in the HFC results in the formation of new arm bridging interactions involving both RAG1 and RAG2. In the NFC and HFC (but not PRC), E962 lies at the end of a long alpha helix, α23. The other end of α23 contributes residues that bind the heptamer of the partner RSS, and hence this alpha helix runs from one site of cleavage to the other. Formation of α23 as well as structural alterations near D600 are thought to be important mechanisms for the crosstalk between the two active sites needed to ensure coupled cleavage [53,54].

Together, RAG structural analyses have revealed an elegant, tightly regulated molecular machine that opens widely for initial binding of its DNA substrates followed by inward domain movements and massive DNA restructuring that together create a functional active site and position the DNA substrate appropriately for nicking and hairpin formation in the same active site.

3. THE JOINING PHASE OF V(D)J RECOMBINATION

3.1 Overview: V(D)J Joining Employs the Non-Homologous End Joining Pathway of DSB Repair Pathway

An early model for V(D)J recombination [20,152] predicted a number of now well-recognised aspects of the V(D)J recombination reaction, including the initiation of the reaction via site-specific DSBs between the two coding and RSS ends of the participating gene segments. The model also predicted that the two coding ends would be joined to each other and the two RSS ends joined to each other to complete the reaction. While the model also predicted that RSS ends would be simply joined to each other without modification, it predicted that coding ends could be further processed before being joined. One such coding end processing event predicted was the addition of non-templated 'N-nucleotides' to some coding ends by early lymphocyte-specific terminal deoxynucleotidyl transferase (TdT) enzyme before they were joined [152]. The proposed TdT N-region addition activity in V(D)J recombination was subsequently proven and revealed to be a major mechanism by which Ig and TCR variable regions are diversified [153,154]. In this context, while TdT is not required for efficient V(D)J recombination, it was, in retrospect, the first identified component of the overall V(D)J recombinase. The subsequent discovery and characterisation of the RAG1/2 endonuclease revealed that this key V(D)J recombination component initiates the reaction by generating the site-specific DSBs between participating coding and RSS ends [9,10,155]; beyond that, additional studies revealed that RAG1/2 is also involved in orchestrating the differential processing and joining of the two distinct types of DNA ends generated by RAG activity [12,44,45,109,156,157].

Early work on the *SCID* mouse revealed a major impairment in V(D)J recombination [158], with subsequent work demonstrating that the *SCID* mutation affected a downstream step in the reaction [159–161]. Moreover, even non-lymphoid cells from *SCID* mice were significantly sensitive to ionizing radiation, indicating that the gene mutated in the *SCID* mice might encode a product involved in DNA end-joining more generally [162]. To identify such a pathway, an extensive screen was carried out on a series of DNA repair mutant Chinese hamster ovary (CHO) cell lines in which V(D)J recombination had been activated on extra-chromosomal substrates via the introduction of RAG1/2 expression constructs [163–165]. These latter studies revealed that V(D)J recombination was severely impaired in three X-ray-sensitive, DSB response-deficient mutant CHO cell lines denoted XR-1, xrs-6 and V-3 [163–165]. Complementation studies had already indicated that a different gene was mutated in each of these three lines, and the putative genes were named X-ray cross-complementation (XRCC) genes 4, 5 and 7, respectively [166–169]. Together, these studies strongly indicated that V(D)J recombination harnesses a, at the time uncharacterised, multi-component cellular DNA DSB repair pathway that operates independently of sequence homology. Subsequent studies rapidly identified the genes mutated in these three mutant CHO cell lines, which are now known to encode XRCC4 [170], KU80 (XRCC5) [171,172] and the 'DNA-PKcs' (XRCC7) [167]. Notably, the DNA-PKcs was also found to be the gene mutated in the *SCID* mice [165,173]. Together, these studies opened the door to the elucidation of a major mammalian DNA DSB repair pathway termed NHEJ (Figure 2.6). NHEJ

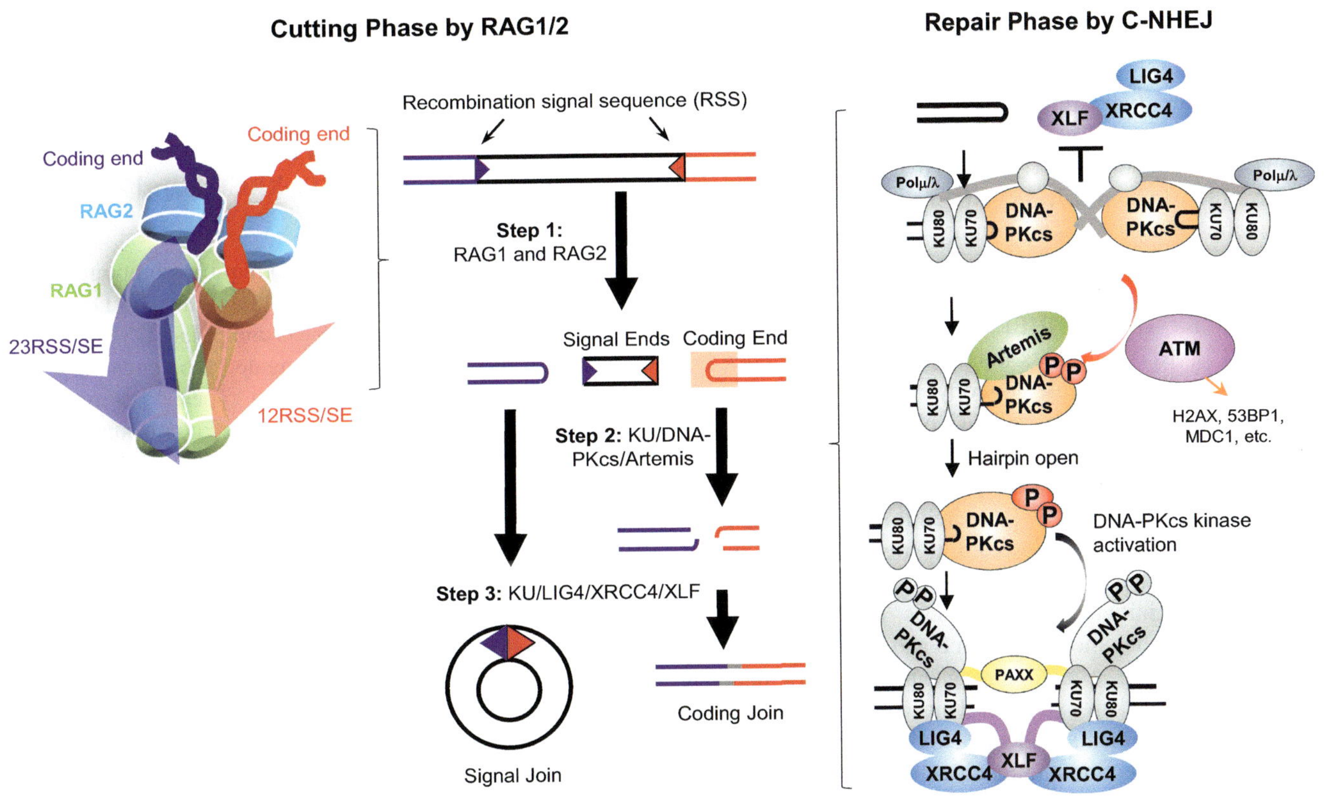

FIGURE 2.6 V(D)J recombination at the cutting phase by RAG and the repair phase by C-NHEJ. RAG1 (*green*) and RAG2 (*blue*) protein form a heterodimer that recognises the 12 and 23 RSSs (red and blue triangles). Two RAG1-RAG2 heterodimers then bind to each other to form the active cleavage complex. RAG1 binds to the conserved nonamer (further away from the breaks) and conserved heptamer (close to the break). RAG1/2 tilts towards one side to accommodate the 12 and 23 RSSs followed by the generation of DNA DSBs between RSSs (triangle) and coding gene segments to generate signal ends and coding ends (step 2). The four DNA DSBs are initially held in a post-cleavage synaptic complex by RAG (Left, cutting phase). Through unclear mechanisms, the reaction is handed off to the C-NHEJ machinery for repair (right, repair phase), where the KU70/KU80 heterodimer forms a ring to initiate the reaction by binding both coding and signal ends. At this point, coding ends (hairpin) and signal ends (blunt) are handled in a different manner by the C-NHEJ machinery. The blunt signal ends can be directly ligated by KU/Lig4/XRCC4/XLF complex. The hairpinned coding ends need to be opened by DNA-PKcs and Artemis prior to final ligation. Repair phase by C-NHEJ: KU70/KU80 binds to the signal/coding ends and thereby recruits DNA-PKcs. The C-terminal domain and tails of KU80 (grey balls and line) reach out to DNA-PKcs to ensure recruitment and could also reach out to DNA-PKcs at the other end to bring the two ends together. In this configuration, DNA-PKcs caps the ends and prevents access by Lig4. Phosphorylation of DNA-PKcs by itself or ATM triggers an allosteric change in DNA-PKcs, which now allows the recruitment of Artemis endonuclease and presents the DNA ends (hairpin or not) to Artemis. After Artemis cleavage, DNA-PKcs must be activated to move away from the ends and allow the ligation complex to access the ends. In the short-range synaptic complex, DNA-PKcs either leaves or moves out of the way. XLF and PAXX use their flexible KU-binding motif to drag KU from the two ends together. Meanwhile, XLF interacts with XRCC4 on each end and brings Lig4 to the DNA breaks. XLF, PAXX and DNA-PKcs all have synapsis function, likely explaining their redundancy during C-NHEJ. The breaks also activate the ATM-dependent DNA double-strand break response (DSBR), including ATM itself, H2AX, 53BP1, MDC1 and many others to promote end-joining during chromosomal V(D)J recombination.

operates broadly in mammalian cells, is particularly important in post-mitotic mammalian cells and uniquely provides the end-joining component of the V(D)J recombination reaction [174—177].

The V(D)J recombination reaction evolved to occur strictly in the G1 phase of the mammalian cell cycle, including both the RAG cleavage and subsequent processing and joining of RAG cleaved RSS sequences and coding-end sequences [178,179]. This regulation is ensured by the specific degradation of RAG2 at the G1/S transition stage of the cell cycle [103,180]. In the context of this mechanism, RAG2 is phosphorylated on threonine 490 by cyclin A/CDK2 upon S phase entry, which targets RAG2 for cytoplasmic export and phosphorylation-dependent ubiquitination and degradation [104—106,180,181]. Together these pre- and post-translational mechanisms ensure that the initiation of V(D)J recombination occurs in G1 cells before DNA replication.

There are two well-characterised, major DNA DSB repair pathways that operate in mammalian cells, homologous recombination (HR) and NHEJ. The latter is often referred to as the classical C-NHEJ to prevent confusing it with the alternative end-joining (A-EJ) pathways that operate at a lower level in mammalian cells [182]. A-EJ is most accurately defined as end-joining occurring in cells deficient for core C-NHEJ components (e.g. Ligase 4, XRCC4 or KU). HR uses long stretches of sequence homology as the template to copy and repair DSBs and operates largely in the S and G2 phases of the cell cycle when homologous template sequences are available in the form of sister chromatids or homologous chromosomes in the case of meiosis [183]. To search for the homology, the DSB ends need to be resected to reveal the 3′ single-stranded overhang, which invades the homologous sequence and initiates DNA synthesis. DNA end-resection activities are much more active in S and G2 cells than in G1 phase cells. Also, several components of the HR pathway are not expressed in G1 cells. Thus, both with respect to template requirements and also due to the lack of both requisite end-resection and repair activities in G1 cells, HR does not contribute to end-joining during V(D)J recombination. In contrast, C-NHEJ operates throughout the cell cycle, including in the G1 phase, in which HR is not available [184,185]. In the latter context, C-NHEJ is broadly critical for the repair of DSBs in post-replicative mammalian cells (e.g. terminally differentiated neurons) and is exclusively required for the repair of RAG-initiated DSBs that occur in the G1 cell cycle phase [174,182].

C-NHEJ can join DSB ends that lack homologous nucleotides (referred to as blunt or direct end-joining), such as the blunt 5′ phosphorylated RSS ends generated by RAG1/2, as well as ends that become blunt after end-trimming or filled-in on single-strand extensions by the broadly expressed DNA polymerases μ and λ [186—188]. C-NHEJ can also readily join ends with short (1-4 nucleotide) complementary overhangs resulting in overlapping nucleotide homologies, referred to as ‘micro-homologies’. Such junctions encompass a subset of microhomology (MH)-mediated end-joining events [189,190]. Both blunt and MH junctions are regularly generated in the context of processing RAG-generated hairpin coding ends [15,191,192]. In this regard, there are also mammalian A-EJ pathways that may, in some cases, employ some core C-NHEJ components or may represent totally distinct pathways that also could, in theory, operate in the G1 phase and participate in the joining of at least a subset of ends joined by C-NHEJ [15,193—198]. However, while A-EJ pathways can, particularly in the absence of core C-NHEJ factors, contribute substantially to *Igh* CSR in mammalian cells [199], they are excluded from contributing measurably to V(D)J recombination in G1 phase cells [157,200] as will be discussed further below.

3.2 Evolutionarily Conserved Core C-NHEJ Factors (KU70, KU80, XRCC4 and Ligase 4)

The repair of DSB ends by C-NHEJ requires components that recognise the DSBs, components that tether the DSB ends together, components that process and polish DSB ends and eventually components that ligate the ends together [194,201]. We will consider these types of components separately, but it is clear that some may contribute to more than one of these functions. Two of the first three identified C-NHEJ factors (i.e., KU80 and XRCC4) and their respective interaction partners are widely considered to be the core C-NHEJ factors because they are both required to complete all known C-NHEJ joining events (e.g. for both RSS and coding join formation; Figure 2.6) and also because they are evolutionarily conserved in eukaryotes (for example, in yeast). The first C-NHEJ factor identified was KU80, encoded by the *XRCC5* gene [171,172,202]. KU80, together with its obligate partner KU70 protein, forms a heterodimer (KU) that recognises and broadly binds to DNA DSB ends, including both blunt V(D)J RSS ends and hair-pinned V(D)J coding ends [60] (Figure 2.6). KU is generally considered to be the DSB recognition component of the C-NHEJ pathway and is responsible for guiding DSB repair in the G1 phase, specifically into C-NHEJ [200]. The XRCC4 protein, which is the product of the *XRCC4* gene, was discovered by complementation cloning [170]. It was subsequently found to interact specifically with mammalian DNA Ligase 4 (Lig4) to form the specific, evolutionarily conserved C-NHEJ ligation complex [203] (Figure 2.6). XRCC4 is essential for the stability of the Lig4 protein [204—206]. Each of these four factors is required for C-NHEJ repair of both RAG-generated coding and RSS ends during V(D)J recombination [207—212].

3.3 Non-Evolutionarily Conserved C-NHEJ Factors with Specialised Functions: DNA-PKcs and Artemis

As described earlier in this chapter, RAG generates 'blunt and 5′-phosphorylated' RSS ends and 'hairpin' coding ends. During V(D)J recombination, the RSS ends are directly ligated together by the core C-NHEJ components to generate the RSS joins (sometimes referred to as 'signal join') of V(D)J recombination. In contrast, the hairpin coding ends require opening and processing before they can be joined. In this case, C-NHEJ incorporates activities of several specialised proteins that are not evolutionarily conserved (i.e., not present in yeast and drosophila) and that are not requisite for directly ligating RSS ends [58,174,213–215] (Figure 2.6). However, these proteins are critically required for processing hairpin coding ends before subsequent joining. In this context, the effect of truncation mutations of DNA-PKcs, as initially found for *SCID* mice and also for V3 CHO cells, and later confirmed by gene-targeted mutations in mice [216–218] on coding versus RSS end joining was quite revealing. Thus, the loss of DNA-PKcs protein expression caused by such mutations essentially abrogated coding end joining but had little or no effect on RSS joining [216–218]. Indeed, in the absence of DNA-PKcs, coding ends accumulate as un-opened hairpins [58,219]. This finding was particularly intriguing since DNA-PKcs binds to KU70/80 to form the DNA-PK holo-enzyme (Figure 2.6), suggesting that DNA-PKcs after recruitment to DSBs via binding to KU provides an additional C-NHEJ function more specialised for generating coding joins.

In this regard, another specialised C-NHEJ factor, termed Artemis, was identified based on its mutation in certain human SCID patients [214]. Intriguingly, Artemis deficiency in humans [214,220] and later, as shown in knockout mice [215] had a very similar impact as DNA-PKcs deficiency on coding versus RSS end joining during V(D)J recombination. Artemis was found to be an endonuclease that is required to open coding end hairpins for further processing [58] (Figure 2.6). Initial biochemical analyses with purified DNA-PKcs and Artemis showed that the activation of Artemis endonuclease activity requires both DNA-PKcs and its kinase activity since non-hydrolysable ATP cannot support endonuclease cleavage [58]. Subsequent genetics studies show that in vivo, hairpin opening in cells requires DNA-PKcs protein and the kinase activity from either DNA-PKcs or the related ATM kinase [219]. Mouse model expressing the kinase-dead (KD) DNA-PKcs accumulates unrepaired opened hairpins during V(D)J recombination [219]. Thus, ATM kinase inhibitors block hairpin opening in cells expressing the KD form of DNA-PKcs [219,221]. Furthermore, DNA-PKcs auto-phosphorylation site mutants also support hairpin opening [222,223]. More recently, cryo-EM analyses provided visual support for a clear role of DNA-PKcs protein as a platform to recruit Artemis [224,225]. Indeed, at the hairpin, Artemis only interacts with 4-6 bp at the apex and with DNA-PKcs holding the rest of the hairpin stem [224]. Artemis endonuclease activity also can be involved in other types of end-processing events in the context of C-NHEJ DSB repair [226].

3.4 Diversification of Antigen Receptor Repertoires by C-NHEJ Activities

The primary antigen contact regions of Ig heavy and light chain variable regions are encoded by the three complementarity-determining region (CDR) sequences. Each V_H or V_L gene segment contains distinct CDR1 or CDR2 encoding sequences [227]. CDR3, however, is encoded by the HC V_HDJ_H and LC V_LJ_L junctional sequences [228,229]. The primary B cell receptor (BCR) repertoire (and, by extension, antibody repertoire) is generated by combinatorial assortment of HC and LC Vs along with each of their distinct sets of CDR1 and 2 sequences along with the sets of Ds and/or Js to which they join [229,230]. However, by providing an immense set of de novo-generated CDR3s, V(D)J junctional diversification mechanisms provide by far the greatest impact on the diversification of the primary BCR (antibody) repertoires [21]. A similar mechanism contributes to the diverse repertoire of the TCR genes [21,231]. In addition, all three CDRs can be further diversified in the context of the secondary SHM/affinity mature process that occurs in the context of antigen-specific responses in germinal centres, as is described later in this volume.

To produce a functional variable region via the V(D)J recombination reaction, the triplet DNA reading frame prescribed by the ATG start codon at the beginning of the V region-encoding segment must be aligned in frame with the open reading frame of the J segment to which they are fused. Thus, RNA splicing from the downstream J splice donor to an upstream C region splice acceptor puts the J and the downstream constant region exon sequences in the same productive reading frame. In this regard, due to frequent, extensive junctional diversification during HC V(D)J or LC VJ joining, only about one in three junctions place the ATG-determined reading frame of the V segment in the same reading as the J (and C). The other two out of three assembled variable region exons are out of frame (non-productive) and cannot encode IgH or IgL chains that can contribute to further development. However, developing lymphocytes have at least two chances to form a productive rearrangement on their two HC alleles and even more chances to form productive *Igκ* alleles (due to secondary Vκ-to-Jκ joins if the first one is non-productive) or to subsequent Vλ-to-Vλ joins if *Igκ* rearrangements do not succeed. While this process might, at first glance, appear inefficient, the loss of a portion of developing B cells due to non-productive rearrangements is a small price to pay for the enormous level of antigen receptor diversification that it enables. Indeed, it is likely that a much larger fraction of developing B cells that form productive rearrangements are lost to the central tolerance mechanism associated with the formation of auto-reactive receptors or to the formation of H or L chains that do not properly pair [232].

The DNA-PKcs/Artemis complex forms an activated endonuclease that can cleave the hairpin coding ends either at their apex or at positions up to approximately 4 nucleotides 5′ or 3′ of the apex [58] (Figure 2.6). Joining of hairpin coding ends cleaved at the apex, in the absence of further processing, could yield 'blunt' coding end joins. However, cleavage of the hairpin beyond on either side of the apex would generate a single-stranded overhang, which could be filled in by polymerases and, if subsequently ligated to the other coding-end, could provide a 1 to approximately 4 nt palindromic sequence, which if incorporated into the V(D)J junction is referred to as a palindromic ('P') element (Figure 2.7). Thus, addition of P elements to D-to-J_H, V_H-to-D or V_L-to-J_L junctions provides a significant source of non-germline-encoded junctional diversification of primary BCR repertoires. P elements and N regions also contribute to TCR gene diversification [21].

The addition of non-templated TdT-generated N regions to D-to-J_H, V_H-to-DJ_H and V_L-to-J_L junctions provides an even greater source of BCR diversity (Figure 2.7). Indeed, it has been estimated that N-region addition can increase the diversity of BCR repertoires by multiple orders of magnitude and that it may have an even greater enhancement of TCR repertoires [21]. Notably, TdT not only increases the sequence diversity of CDR3 sequences, it also can increase the CDR3 length, providing another important functional attribute of certain antibodies [233,234]. Beyond these mechanisms, TdT also contributes to repertoire diversification by an additional mechanism. As mentioned above, C-NHEJ uses short MHs at two coding ends to align them for joining. Indeed, due to the absence of TdT expression during foetal B and T cell development in mice and humans, foetal BCR and TCR V(D)J repertories are much less complex than those of postnatal adult repertoires

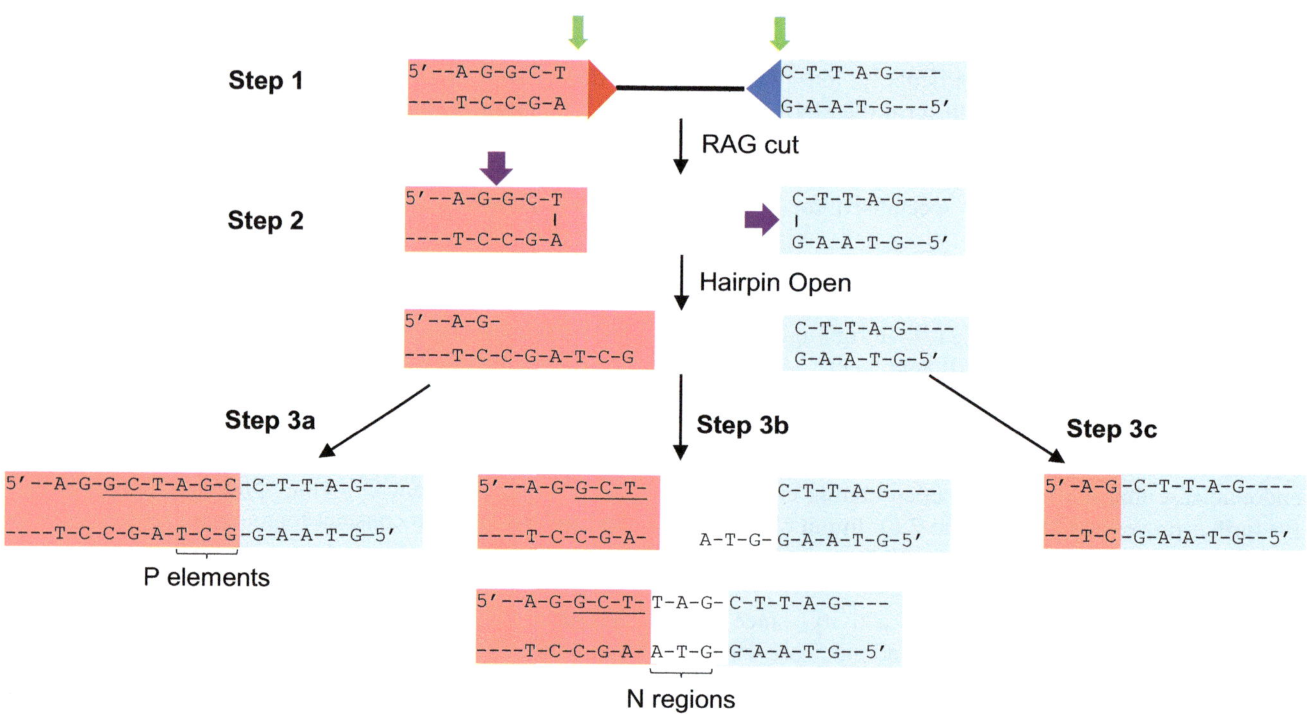

FIGURE 2.7 **V(D)J recombination coding join diversification.** RAG generates DNA DSBs (*green arrow*) between an RSS and a coding gene segment, which generates a hairpin structure at the coding end (step 2). At this point, the Artemis nuclease (denoted by *purple arrows*) can cleave the hairpin structure at, for example, the apex (right) or several nucleotides from the apex (left), leaving either an overhang or blunt end. After hairpin cleavage, there are a series of further end processing steps (step 3) which may lead to either addition of non-germline-derived nucleotides or loss of nucleotides. Step 3a: after polymerase fill-in of the overhang and ligation, a palindromic sequence ("P element") which was not originally present in the germline sequence, appears at the final junction. Step 3b: TdT may add random nucleotides to the 3′ terminal of a coding end, which, after polymerase fill-in and repair, leads to non-germline-encoded nucleotides at the coding join (N elements). Step 3c: Nucleotides may be deleted prior to joining by various exonuclease activities of polymerase μ or λ or endonucleolytic activity from MRE11 and others, resulting in a deletion at the coding junction. *Figure adapted from Sekiguchi, Alt, and Oettinger, Molecular Biology of B Cells, 2004.*

that are generated in the presence of TdT expression [231,235–241]. In this regard, in TCR and BCR repertoires generated in TdT-deficient mice, N regions are absent, and MH-mediated joins involving terminal MHs found at V, D or J coding ends are abundant [153,154]. Thus, TdT also diversifies BCR and TCR repertoires by impeding the utilisation of micro-homologies within the coding ends of two gene segments being joined. Notably, certain recurrent MH-mediated V(D)J joins in foetal T or B cell repertoires generate TCRs or BCRs that are critical for certain physiological responses and, in this sense, some TCRs so generated may be considered innate-like in origin [235,242–244]. It is also notable that while TdT is expressed during LC V(D)J recombination in postnatal human pre-B cells [245], it is not expressed in those of mice [233,246,247], leading to decreased junctional diversity and much more abundant MH-mediated joins in primary mouse LC repertoires compared to those of humans [248,249]. In this regard, the generation of certain HIV1-broadly neutralising antibodies requires an Igκ LC CDR3 in which a critical encoded binding residue is generated via N-region addition [233].

3.5 Impact of Defects in C-NHEJ Factors on Lymphocyte Development and Cancer

Mice have been generated that are deficient for each of the currently known C-NHEJ factors. In mice lacking any one of the four evolutionarily conserved factors, B and T lymphocyte development is blocked at the progenitor lymphocyte stage due to the inability to join RAG-generated DSB ends to form functional Ig and TCR variable region exons required to drive B and T cell development. Mice harbouring KU70- or KU80-inactivating mutations survive into adulthood, and besides their SCID phenotype, they have growth defects, general DSB repair deficiency and increased apoptosis of newly generated post-mitotic neurons [208,211,250]. In contrast, mice deficient for XRCC4 or Lig4 die during late embryonic development in association with severe p53 tumour suppressor-induced apoptotic death of newly generated neurons in response to unrepaired DSBs [207–212]. Thus, the neuronal apoptosis phenotype and embryonic lethality of the XRCC4- or Lig4-deficiency can be rescued by a p53 tumour suppressor gene inactivation, which rescues neuronal apoptosis, but the resulting mice still lack lymphocytes due to the inability to repair V(D)J recombination-associated DSBs [251,252]. Remarkably the neuronal apoptosis and embryonic lethality of Lig4-deficient mice can also be rescued by KU deficiency [253,254], likely because KU binding to DSBs in newly generated neurons prevents their resection and subsequent repair by alternative DNA end-joining pathways that allow sufficient repair to rescue much of the apoptosis phenotype to levels in KU-deficient mice [200]. Like the classical *SCID* mice, DNA-PKcs and Artemis-deficient mice also have severely impaired B cell development due to their impaired ability to open hairpin coding ends and form V(D)J rearrangements [215–217]. Otherwise, they are relatively normal beyond a mild sensitivity to ionizing radiation, a variable defect in *Igh* CSR, but no evidence of neuronal defects likely due to their ability to repair many types of DSB lesions by C-NHEJ [215–217]. Notably, however, expression of a KD DNA-PKcs blocks end-ligation but not hairpin opening, creating a phenotype that closely assembles the Lig4 null [219]. In this context, human patients with almost complete Artemis deficiency have relatively mild microcephaly and one DNA-PKcs mutation disrupting Artemis activation without affecting DNA-PKcs kinase activity also leads to somewhat restricted immunodeficiency [214,255,256]. On the other hand, DNA-PKcs mutations that do impact kinase activity lead to severe microcephaly and mortality [175,257].

When C-NHEJ-deficient mice are bred into p53-deficient backgrounds, they routinely succumb to progenitor B cell lymphomas that harbour translocations between the J_H-containing region of the *Igh* locus and sequences between 10 and 100s of kb downstream of the *c-Myc* locus [258–260]. These dicentric translocations then undergo breakage–fusion–bridge cycles resulting in the amplification of the *c-Myc* oncogene in resulting progenitor B cell lymphomas [258]. The development of these B cell lymphomas is dependent on RAG1/2 and results from RAG-generated DSBs in the *Igh* J_H locus that translocate to DSBs spread over a long distance downstream of the *c-Myc* gene [258]. Subsequent detailed analyses indicated that the translocation target DSBs generated downstream of *c-Myc* are likely also generated by RAG [36,259]. These studies provided an initial basis for the discovery of cohesin-loop extrusion-mediated RAG chromatin scanning [36], as will be described in a subsequent section of this chapter.

3.6 Double-Strand Break Response and Factors That Have Functional Overlap with C-NHEJ Factors

In recent years, several additional factors that contribute to C-NHEJ during V(D)J recombination have been identified (Figure 2.6). Foremost is the XRCC4-like factor (XLF; also known as Cernunnos) that was identified in the context of immunodeficient patients [261–263]. While XLF deficiency has a significant impact on V(D)J recombination in extra-chromosomal substrates introduced into non-lymphoid cells, XLF deficiency in mice has little, if any, impact on chromosomal V(D)J recombination in developing lymphocytes [264]. This apparent discrepancy led to a search for potential compensating factors [265], and many were identified, including ATM itself and multiple ATM-dependent DSB response factors and several additional potential C-NHEJ factors [266–268]. In this regard, deficiency for the ATM kinase has a very modest impact on V(D)J recombination, but it clearly can allow V(D)J DSBs to escape the normal RAG post-synaptic complex and translocate to other DSBs in the genome, potentially leading to oncogenic translocations [269–271]. However, combined ATM and XLF deficiency in mice resulted in an SCID-like phenotype due to a severe block in V(D)J recombination [265]. Indeed, deficiency for several other members of the ATM-dependent DSB response, including 53BP1, MDC1 and H2AX, also showed a severe V(D)J recombination defect when combined with XLF deficiency [266–268]. While the precise shared function implicated by the combined defects is still under investigation, one possibility might be a role in end-tethering [272,273]. Two other potential C-NHEJ factors, termed PAXX and MRI, interact with KU but, when deficient in progenitor B cells, have little impact on V(D)J recombination [201,274–277]. Yet, when deficiency of either is combined with XLF deficiency, they result in severe inactivation of V(D)J recombination [201,274–277]. Similarly, loss of XLF in DNA-PKcs null cells also abrogates end-ligation [278]. The mechanism(s) by which XLF complements this diverse set of factors in promoting V(D)J recombination is of great interest. In this regard, single-molecule and cryo-EM analyses elegantly show that one XLF interacts with two XRCC4-Lig4 at each DSB end to facilitate end-ligation [273,279–281]. Thus, XLF might share functional redundancy with other end-tethering proteins, including DNA-PKcs and ATM-mediated DNA damage responses (Figure 2.6).

3.7 RAG2 and KU Both Function in Restricting V(D)J Recombination to the G1 Phase

In the context of V(D)J recombination, the post-synaptic RAG complex itself has been implicated in shepherding V(D)J RSS and coding end DSBs into the C-NHEJ versus A-EJ pathways in progenitor B cells [109,145,157]. Restriction of V(D)J joining to C-NHEJ by RAG could, in theory, prevent end-resection and exclude MH-biased A-EJ pathways that could restrict diversity generation. Further support for a role for RAG in steering V(D)J recombination into C-NHEJ comes from the remarkable discovery that a C-terminal truncation mutation of RAG2 expressed during B cell development in XLF-deficient mice results in severely impaired V(D)J recombination, and in a p53-deficient background results in pro-B cell lymphomas similar with *Igh-c-Myc* translocations described above [282]. Finally, recent studies have demonstrated that binding of KU to V(D)J recombination-generated or other types of DSBs in G1 phase pro-B cells prevents their joining by A-EJ. Thus, DSBs generated in Lig4-deficient G1-arrested pro-B cells accumulate as free DSB ends in G1-arrested pro-B cell lines [200]. Notably, while Lig4-deficiency abrogates Vκ-to-Jκ recombination in G1-arrested pro-B cell lines, Vκ-to-Jκ joins are readily recovered when these Lig4-deficient pro-B cell lines are also made KU-deficient [200]. However, the VκJκ joins, while generating a subset that appears as normal V(D)J recombination joins, are actually generated by a chromosomal translocation-like process that is mediated by A-EJ. Thus, there are several mechanisms by which limiting V(D)J recombination to the G1 phase restricts the joining phase of this process to C-NHEJ.

4. REGULATION OF V(D)J RECOMBINATION

4.1 Overview of V(D)J Recombination Regulation During Lymphocyte Development

V(D)J recombination is tightly regulated to only occur in the early stages of B and T lymphocyte development. The developing lymphocyte specificity of the reaction is enforced by the strict expression of RAG1 plus RAG2 in the appropriate stages of developing B and T lymphocytes and not in other cell types [283]. Indeed, ectopic expression of RAG1 and RAG2 can promote V(D)J recombination on various substrates introduced into non-lymphoid cells [9,10] since all of the other requisite components of the reaction are broadly expressed as outlined in the preceding section. During lymphocyte development, V(D)J recombination is regulated at multiple different levels [182,284]. First, V(D)J recombination is regulated in the context of lineage specificity as the complete assembly of *Igh* and *Igκ/λ* variable region exons only occurs in developing B cells, while complete assembly of the various TCR variable region exons only occurs in developing T cells. Within B and T cell lineages, V(D)J recombination is also regulated at the level of developmental stage-specificity. Thus, *Igh* variable region exons are assembled in progenitor B (pro-B) cells, while *Igκ* and *Igλ* variable region exons are assembled in the subsequent precursor B (pre-B) cell stage (Figure 2.8A). Within progenitor and precursor B cells, there is further temporal ordering of V(D)J recombination events. Thus, during *Igh* locus V(D)J recombination, D-to-J_H joining usually occurs first, and on both alleles, before a V_H is appended to a pre-existing DJ_H complex (Figure 2.8B) [284]. Also, for Ig light chain loci, Vκ-to-Jκ recombination usually precedes Vλ-to-Jλ joining, as evidenced by findings in both mice and humans that Igλ-producing B cells usually have either inactivated both *Igκ* loci through non-productive VκJκ rearrangements or have deleted one or both Cκ exons through V(D)J recombination events to isolated RSSs that serve as *Igκ* 'deletion elements'

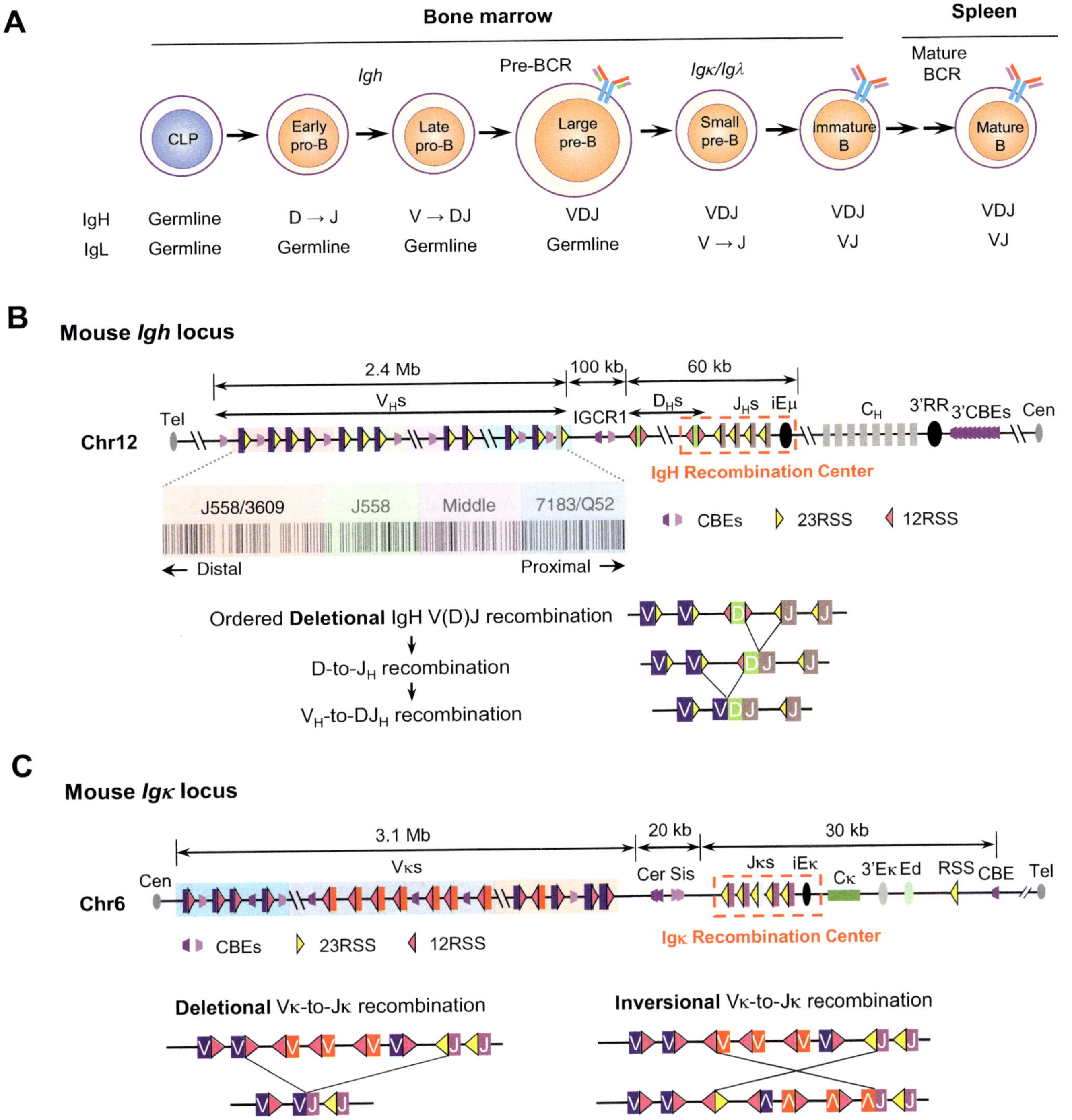

FIGURE 2.8 V(D)J recombination in *Igh* and *Igκ* loci during B cell development. (A) Schematic showing the ordered sequence of B cell developmental stages with V(D)J recombination occurring in immunoglobulin heavy chain and light chain loci. (B) Diagram of the mouse *Igh* locus on chromosome 12 shown in the direction from telomere (Tel) to centromere (Cen). Top: approximately 100 V_H segments in the 2.4 Mb region near the telomere of chromosome 12 lie upstream of 9–14 D segments (depending on mouse strain) and 4 J_H segments (not to scale). The V_H segments are clustered based on V_H family type and position into four colour-coded V_H groups (J558/3609, J558, middle, 7183/Q52), which are also shown with the same colour code in several subsequent figures to maintain context. The CTCF-binding element (CBE)-based intergenic control region 1 (IGCR1) lies within the approximately 100 kb intervening sequence between the most downstream V_H (V_H5-1) and the most upstream D (DFL16.1). The most downstream D (DQ52) lies within 1 kb of the 4 J_Hs and the iEμ intronic enhancer which together form the nascent *Igh* recombination centre (RC) to which RAG binds and from which RAG orchestrates V(D)J recombination (highlighted in the dashed red box). The region further downstream of the RC contains the various C_H exons, the 3′ *Igh* regulatory region and the 3′ *Igh* CBEs. Position and orientation of CBEs (pink and purple trapezoids), as well 12RSSs upstream and downstream of Ds (*red triangles*), 23RSSs upstream of J_Hs (*yellow triangles*) and 23RSSs downstream of V_Hs (*yellow triangles*) are indicated. Bottom: Illustration of ordered *Igh* recombination with D-to-J_H joining generally preceding V_H-to-DJ_H joining. As indicated, in the *Igh* locus, D-to-J_H joining and V_H-to-DJ_H joining occurs via a deletional mechanism mediated by convergent 23 and 12 RSSs. (C) The mouse *Igκ* locus on chromosome 6 is illustrated in the orientation from centromere to telomere. Top: 194 functional Vκ segments in the upstream (centromeric) 3.1 Mb region lie approximately 20 kb upstream of the four functional Jκ segments, which lie just upstream of the intronic iEκ enhancer followed by a single Cκ exon. Two downstream *Igκ* enhancers are also illustrated. The Jκs and the iEκ together constitute the *Igκ* RC that recruits RAG (highlighted in the dashed red box). It is possible that the downstream *Igκ* enhancers become associated with the RC via loop extrusion and, thereby, contribute to RC activity. Bottom: Unlike the *Igh* locus in which all V_Hs are oriented for deletional joining (panel B), Vκs mostly occur in proximal and distal blocks (*blue rectangles*) oriented for deletional joining and a large middle block (*red rectangles*) oriented for inversional joining to Jκs. See text for more details.

[43,285]. Similar temporal ordering and stage-specific regulation of V(D)J recombination occurs in developing T cells [286]. Another notable form of V(D)J recombination regulation comes from the finding that both normal on-target RAG1/2 endonuclease activity and very low-level RAG1/2 off-target activity is largely confined under normal circumstances to occur within the chromatin domains of the particular antigen receptor locus developmentally activated for rearrangement [36].

Early studies led to the model that in pro-B cells with two DJ_H rearranged alleles, feedback regulation from a μ heavy chain produced from a productive V_HDJ_H rearrangement suppresses V_H-to-DJ_H rearrangement on the other DJ_H-rearranged allele of the cell [287]. Such a process has been proposed to enforce 'allelic exclusion', a phenomenon in which a given B cell expresses a productive heavy chain from only one of its two alleles [287]. In this regard, Ig μ heavy chains encoding the membrane-bound (μ_m) form of the protein associate with a surrogate light chain and are expressed on the surface of pro-B cells (Figure 2.8A) [288,289]. In this context, the μ_m chains can signal the development of pro-B cells to the pre-B cell stage in which IgL V(D)J recombination is activated (Figure 2.8A) [290,291]. The proposed feed-back functions of μ_m chains in the context of allelic exclusion might be related to mechanisms by which they signal developmental progression [290]. In this regard, while the expression of a transgenic μ_m chain drives progression of pro-B cell to precursor stages, to some extent, this occurs in the absence of pro-B cells forming DJ_H rearrangements on both alleles as observed in normal pre-B cells [292]. Thus, the apparent feedback regulation of *Igh* V(D)J recombination by transgenic μ_m chains could, in theory, be a by-product of rapid progression through the progenitor stage [292]. However, the precise molecular mechanism by which a μ_m chain may signal to mediate feedback regulation of V_H-to-DJ_H rearrangement and, correspondingly *Igh* allelic exclusion, separate from its role in signalling developmental progression requires further investigation. Finally, generation of a light chain that can pair with the μ_m heavy chain leads to the expression of a BCR and differentiation to the B cell stage (Figure 2.8A). B cells down-regulate RAG expression and migrate to the peripheral lymphoid tissues where they may be activated by a cognate antigen.

The lineage-specific, stage-specific and temporal aspects of V(D)J recombination regulation outlined above have all been implicated as being mediated in large part by epigenetically modulating 'accessibility' of various antigen receptor loci or particular portions of these loci to the RAG1/2 endonuclease [284,293]. This so-called 'accessibility hypothesis' was first proposed based on the finding of non-coding transcripts from a number of the most distal V_Hs before they had undergone *Igh* locus V(D)J recombination [293]. Indeed, it is now known that these distal V_H transcripts are even generated in RAG-deficient pro-B cells [294–298]. Direct evidence for the accessibility hypothesis was provided by the finding that pro-B cell lines that undergo robust endogenous *Igh* D-to-J_H rearrangements, but do not undergo endogenous TCR gene rearrangements, could undergo robust rearrangements of TCR gene segments ectopically introduced into their genome [284,299]. Thus, this study provided the first evidence that Ig and TCR variable region exons were assembled by a common 'V(D)J recombinase' and also correlated accessibility of the substrate gene segments with their location in open chromatin as measured by DNase1 sensitivity [299]. The common recombinase was, of course, identified by the discovery of RAG1 and 2 [9,10] and its common activity verified by studies that showed that elimination of either of these genes in mice abrogates all V(D)J recombination [300,301]. Additional correlates of V(D)J recombination substrate accessibility included DNA hypomethylation, various histone modifications, chromosome positioning in the nucleus, locus contraction and anti-sense transcription [295,302–310].

Despite the attractiveness of the accessibility mechanism to explain how RAG activity is guided to different targets, the actual mechanism and clear-cut role of any of the correlates had, until recently, remained enigmatic. For example, while germline V_H transcription was the first accessibility correlate described, it later became evident that many of the frequently rearranged proximal V_Hs were barely, if at all, transcribed [294–298]. Likewise, a major question for the field was how the various substrate accessibility correlates could attract RAG to sequences up to several Mbs apart in the genome [182,284]. In this regard, beyond substrate accessibility per se, some accessibility correlates, including active chromatin marks and transcription, have also been implicated in attracting RAG binding to focal V(D)J RCs from which it orchestrates all of the various V(D)J recombination events [96,97] (discussed further below). This finding, coupled with the finding that *Igh* RC-bound RAG can scan very long segments of adjacent chromatin presented to it by linear cohesin-mediated chromatin loop extrusion, has provided answers to many prior enigmas [6]. This latter work showed how RAG can actively locate substrate V_H, D and J_H sequences spread over a 2.7 Mb *Igh* locus. New insights related to RAG scanning also have finally elucidated a mechanism by which transcription and other 'accessibility' factors influence V(D)J recombinational accessibility [6,311], as well as how RAG activity is largely confined within antigen receptor loci [6,36]. The fundamental role of cohesin-mediated loop extrusion in mediating V(D)J recombination and how accessibility is regulated in that context has been studied in most detail for the mouse *Igh* locus as will be discussed in depth below. Beyond that we will also discuss more briefly the regulation of V(D)J recombination in the *Igκ* locus and TCR loci including the potential involvement of loop extrusion, which is a focus of many ongoing investigations.

4.2 Cohesin-Mediated Loop Extrusion

Mammalian chromosomes are folded into a hierarchy of structures to direct genetic functions. Genomic regions are segregated into large transcriptionally active or repressive compartments [312] which are further organised into topologically associated domains (TADs), characterised by their enriched contact frequency within the domain [313,314]. Many TADs contain chromatin loop domains up to megabases long with well-defined boundaries harbouring convergent CTCF-binding elements (CBEs) co-bound by the structural protein CTCF and cohesin ring protein complex (Figure 2.9A) [315]. In this regard, CTCF binds to the non-palindromic CBEs in an orientation-dependent manner [316,317]. The numerous CBEs dispersed across the genome can be in the same, divergent orientation, or convergent orientation relative to neighbouring CBEs [315]. Upon loading onto chromatin, the cohesin complex mediates loop extrusion, forming contact loops when extrusion in each direction reaches CTCF-bound CBEs that impede extrusion (Figure 2.9A). Notably, such CBE-anchored chromatin loops were found to be most abundant when they involved convergently oriented CBEs [315,318,319]. In that regard, various lines of evidence have indicated that CBEs in convergent orientation form the most stable loop anchors [320−323], although single CTCF-bound CBEs in any direction can, to varying degrees, impede loop extrusion [324]. Co-enrichment of CTCF and cohesin at loop end points and the strong preference for the convergent orientation of boundary CBEs led to a model that proposed the cohesin complex linearly extrudes chromatin into larger loops until arrested in each direction by CTCF bound to convergent CBEs [325−331] (Figure 2.9A).

The loop extrusion model is now widely accepted and supported by multiple lines of evidence. Deleting or mutating the boundary convergent CBEs leads to fusion of adjacent loop domains [331−334]. Depletion of CTCF [335,336] or cohesin [336−339] or inhibition of cohesin ATPase activity [340] all eliminated chromatin loops genome-wide. Moreover, several studies demonstrated in vitro that cohesin can extrude naked and nucleosome-bound DNA into loops [341−343]. Cohesin is unloaded from chromatin by the WAPL-PDS5 complex [344−346]. Notably, inhibition of cohesin removal from chromatin upon loss of WAPL greatly promotes extension of chromatin loops [336,337,347], indicating that chromatin retention time of cohesin determines the processivity of loop extrusion and size of chromatin loops [348,349]. As will be discussed below, these findings in non-lymphoid cells were important in understanding the mechanism by which WAPL down-regulation in pro-B cells promotes long-range RAG scanning of the *Igh* locus [6,297,298,350]. More details on the cohesin-mediated loop extrusion process can be found in several recent reviews [351,352].

4.3 *Igh* Locus Structure

The 2.7 Mb mouse *Igh* locus lies near the telomere of chromosome 12 and is organised from the telomere to centromere of: V_H segments, D segments, J_H segments and constant region exons (C_Hs) [353] (Figure 2.8B). After assembly of a V_HDJ_H exon by V(D)J recombination, transcription starting from a V_H promoter runs downstream through the first C_H exon (Cμ) after which the V_HDJ_H exon and $C_μ$ exon are fused via RNA splicing to make a mature μ heavy chain transcript. The V_H-containing domain of the locus spans approximately 2.4 Mb and has approximately 100 V_Hs located within it. The individual V_Hs occur in clusters of related V_Hs in four sub-domains termed in order from 'distal' to 'proximal' relative to the D domain as J558/3609, J558, middle and 7183/Q52 (Figure 2.8B) [296]. As discussed in depth below, different accessibility mechanisms appear to target distal versus proximal V_Hs as they are extruded past the RC [6]. Downstream of the V_H domain, there is an approximately 100 kb intervening region followed by a 50 kb domain containing 9 Ds (or more depending on the mouse strain) (Figure 2.8B). Just downstream of the last D (DQ52), there are 4 J_Hs followed by an intronic region containing the iEμ enhancer and then a series of eight different sets of C_H exons contained within an approximately 250 kb C_H domain (Figure 2.8B). In mature B cells, the different C_Hs can be expressed by a CSR process described in a subsequent chapter of this book. CSR also occurs in a CSR centre that is assembled by chromatin loop extrusion [6,354]. At the downstream end of the locus there is a 3′ *Igh* regulatory region super-enhancer required for CSR and somatic hypermutation [355], and beyond that a cluster of 10 CBEs termed the 3′ *Igh* CBEs and also referred to as the 3′ *Igh* super-anchor [340,356] (Figure 2.8B). The 3′ *Igh* CBEs insulate CSR activities within the C_H domain [357] and have been speculated to have roles in *Igh* V(D)J recombination [350,358−362] (discussed below).

The J_Hs are flanked directly upstream by 23RSSs, Ds are directly flanked on both sides by 12RSSs and V_Hs are directly flanked downstream by 23RSSs (Figure 2.8B). Thus, due to 12/23 restricted joining, V_Hs cannot be joined directly to J_Hs and Ds are required to mediate the 12/23 restricted D-to-J_H and V_H-to-DJ_H joins that form a complete $V_H(D)J_H$ (Figure 2.8B, *bottom*). The *Igh* intronic enhancer iEμ, J_Hs and the J_H-proximal DQ52 all lie within a region of active chromatin that accumulates high levels of RAG1/2 binding to form an RC (Figure 2.8B) [96,97].

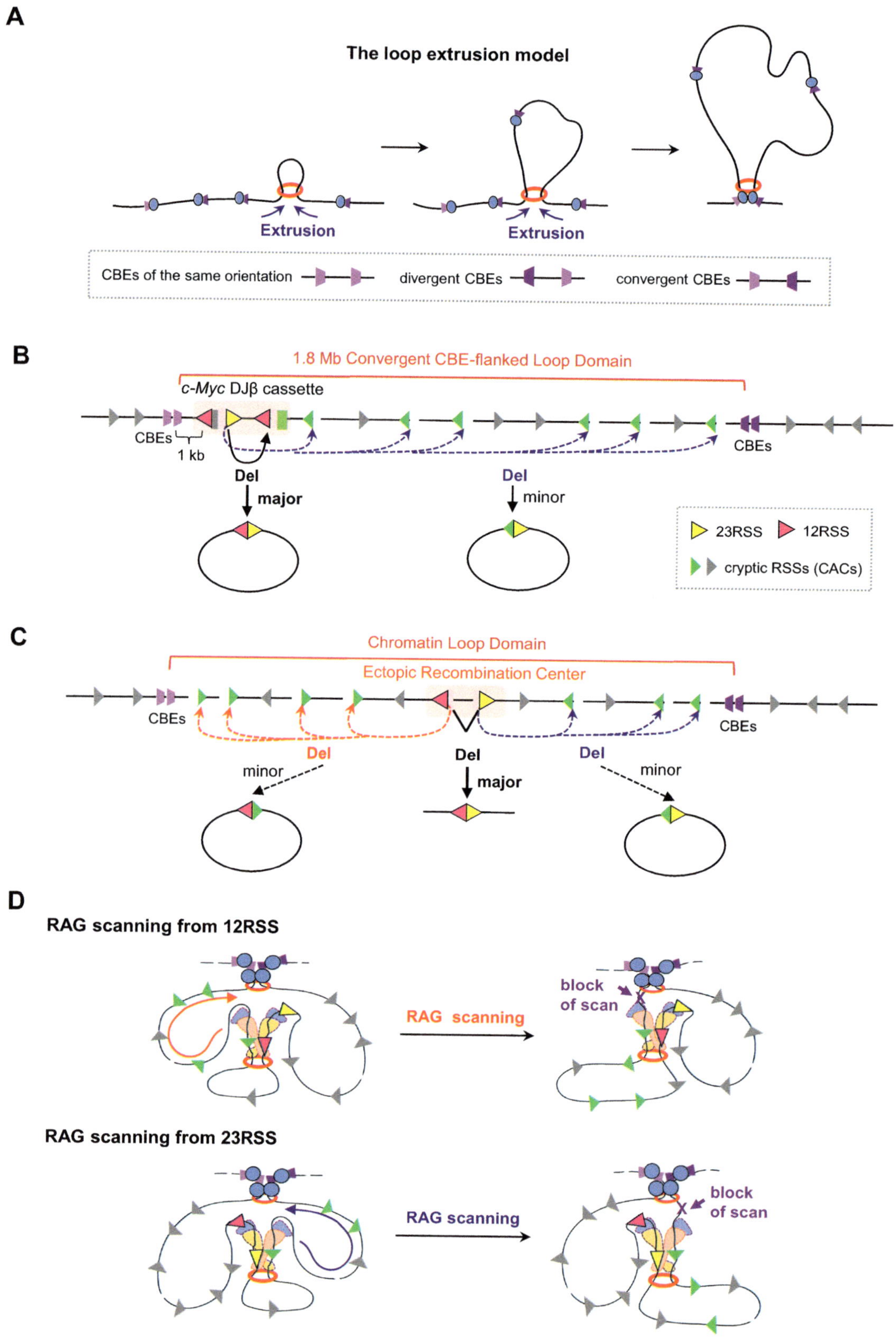

FIGURE 2.9 **Chromatin loop extrusion drives RAG chromatin scanning.** (A) The cohesin-mediated loop extrusion model. Cohesin (red ring) is a highly conserved ring-shaped protein complex. The CTCF chromatin-looping factor (blue ovals) binds to CBEs (pink and purple trapezoids) in an orientation-dependent fashion. Thus, genomic CTCF-binding elements, relative to each other, can occur in the same, divergent or convergent orientation. During cohesin-mediated chromatin loop extrusion, cohesin is loaded onto chromatin and then progressively extrudes a chromatin loop until extrusion is

Deletion of iEμ nearly abrogates V_H-to-DJ_H recombination and substantially impairs D-to-J_H recombination [363], likely by impacting RC activity (unpublished data). This active chromatin region occurs independent of RAG-binding and in this form has been referred to as a nascent RC [361]. Temporally, the active RAG1/2-bound RC acquires a bound J_H-23RSS and subsequently a D-12RSS, which leads to D-to-J_H recombination to form the DJ_H complex. The DJ_H complex, together with the adjacent iEμ, then serves as a new DJ_H-RC for subsequent V_H-to-DJ_H recombination (Figure 2.10B) [6].

The *Igh* locus contains a remarkable number of CBEs that also have a notable organisation [358–360,364,365]. The intergenic control region 1 (IGCR1) is a major V(D)J recombination control region that lies within the 100 kb intergenic region between the V_H and D domains at a position just beyond the most upstream D (DFL16.1) [360]. IGCR1 contains two divergently oriented CBEs, referred to as CBE1 (upstream) and CBE2 (downstream) (Figure 2.8B). There are 100s of CBEs located across the V_H domain, all of which point downstream and are in convergent orientation with IGCR1 CBE1 [365] (Figure 2.8B). The several dozen most D-proximal V_Hs contain a single CBE located just downstream of their RSSs [295,296] (Figures 2.8B and 2.10B). While distal V_Hs rarely have immediately adjacent CBEs, nearly all have a CBE located within 10 kb and often much closer [295,296] (Figure 2.8B). Strikingly, mutational inactivation of both IGCR1 CBEs in mice results in dramatically increased rearrangement of the most proximal V_Hs, in particular V_H5-2 (also known as 'V_H81X') which is the very most proximal functional V_H [360]. The tremendously increased V_H81X rearrangement is associated with greatly reduced rearrangement of V_Hs upstream of the four most proximal V_Hs, allowing the proximal V_Hs, again mostly V_H81X, to dominate the newly generated IgH variable region repertoire [360,362,366]. Further studies revealed that this remarkable phenotype could largely be observed by the inactivation of CBE1 alone, with a far lesser effect of the inactivation of CBE2 alone [362,366]. This striking *Igh* CBE organisation led to several different models for how IGCR1 might orchestrate ordered D-to-J_H versus V_H-to-DJ_H rearrangement [358–360,362]. Ultimately, the mechanism of IGCR1 action was clarified after the discovery of RAG linear chromatin scanning, as will be described below.

4.4 Discovery of RAG Chromatin Scanning

RAG scanning was discovered on the basis of mechanistic insights raised from studies of RAG off-target activity [36] as it related to recurrent RAG-dependent oncogenic translocations in a mouse B cell lymphoma model [36,259]. In these studies, a TCR Dβ1-Jβ1-1 RC cassette was inserted into intron 1 of the *c-Myc* gene (generating the *c-Myc*-DJβ allele) (Figure 2.9B) and found to promote, in developing B and T lineage cells, robust V(D)J recombination between the convergently oriented bona fide 3′-Dβ1 23RSS (pointing downstream in the *c-Myc* gene) and Jβ1-1 12RSS (pointing upstream in the *c-Myc* gene) [367]. Strikingly, when bred into an ATM-deficient background, these mice routinely developed B cell lymphomas that harboured RAG-dependent oncogenic translocations between their J_H locus and sequences spread over a large distance downstream of the *c-Myc* gene [259]. Moreover, in mice heterozygous for the cassette allele, all of the oncogenic translocations involved the cassette-containing alleles [259]. This finding indicated that robust RAG activity initiated from a V(D)J recombination cassette within the *c-Myc* gene could lead to lower level DSBs far downstream on the allele that could provide oncogenic translocation targets.

The mechanistic basis for this startling finding was elucidated by the application of a highly sensitive, next-generation sequencing-based V(D)J repertoire sequencing ('HTGTS-V(D)J-seq') assay [36]. This assay was developed based on the earlier linear amplification-mediated high-throughput genome-wide translocation sequencing (LAM-HTGTS) method [368,369]. HTGTS-V(D)J-seq employs the target sequences of endogenous RAG-initiated breaks (e.g. an RSS or coding segment) as 'bait' to capture junctions of the RAG-generated break at the bait site and target DSBs during V(D)J recombination. This assay has much greater sensitivity than prior assays. Thus,

strongly impeded in each direction by convergently oriented CTCF-bound CBEs which form stronger loop anchors than CBEs in other orientations. (B) Long-range RAG exploration of the *c-Myc* domain. RAG activity initiated upon binding bona fide D and Jβ RSSs in the *Myc*-DJβ cassette, which, beyond the major bona fide 12/23-restricted deletional D-to-Jβ joining within the cassette, also results in low-level Dβ-23RSS-mediated joining to 100s of convergent cryptic RSSs located throughout the 1.8 Mb convergent CBE-based loop domain of *Myc*. Blue dashed arrows indicate that, beyond the immediate cassette locale, such joining events dominantly utilise cryptic RSSs (*green triangles*) that are convergently oriented with the Dβ-23RSS (*yellow triangle*); cryptic RSSs (*grey triangles*) in the same orientation as the Dβ-23RSS generally are not used for such cryptic V(D)J recombination events. (C) Random insertion of pairs of bona fide RSS-based RCs into 12 different convergent CBE-based loop domains in different genomic locations activates RAG scanning across these domains in which they are located. Red and blue dashed arrows indicate upstream and downstream utilisation of cryptic RSSs in convergent orientation with the bona fide 12RSS and 23RSS, respectively. (D) Orientation of the initial RAG-binding RSS determines the direction by which chromatin will be presented to RAG during scanning, which continues until terminated by convergent CBE-based loop anchors. See text for more details. *Parts (A-C) are adapted with permission from Ref. [6]*

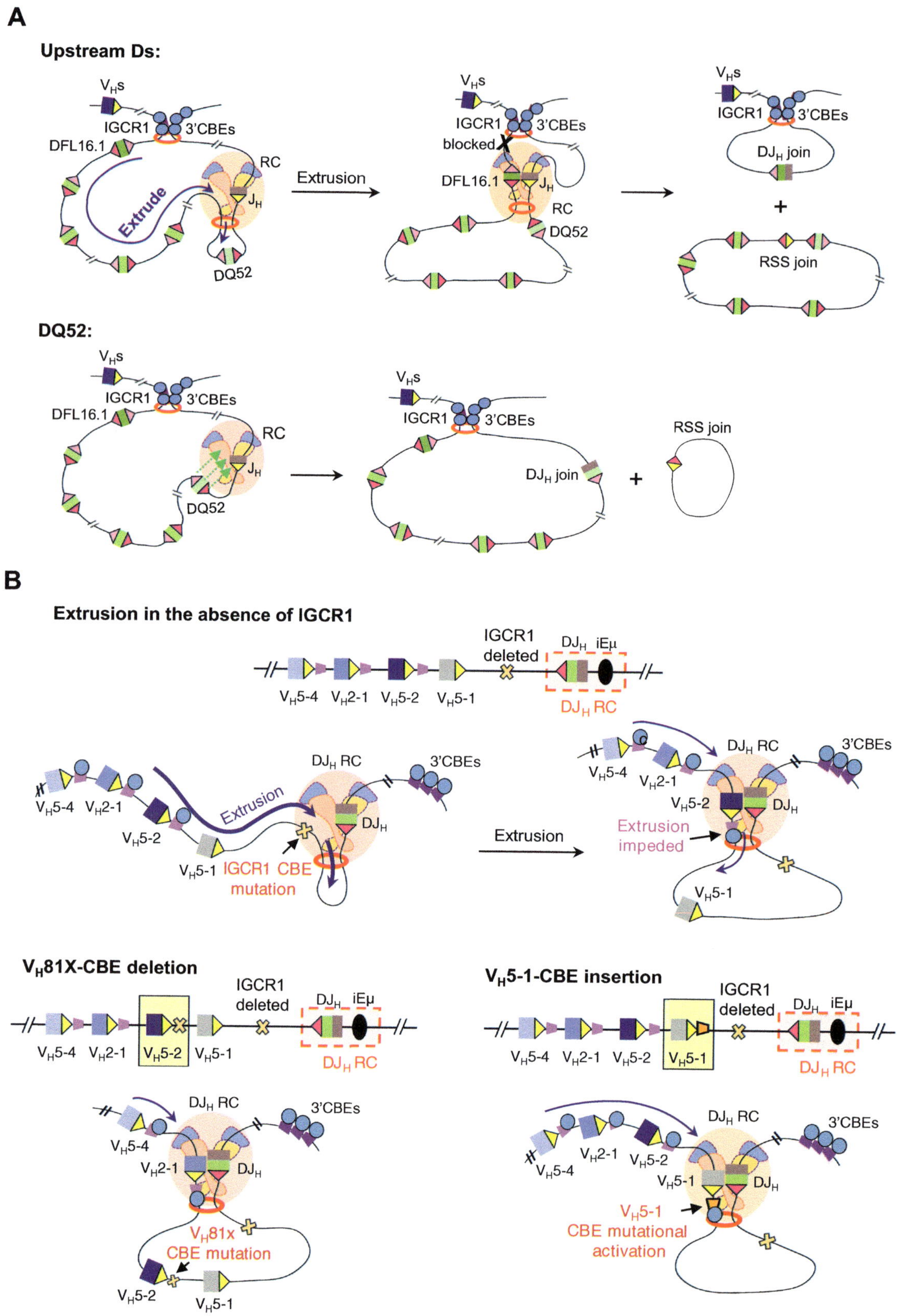

FIGURE 2.10 **Loop extrusion-mediated RAG scanning drives D-to-J_H and proximal V_H-to-DJ_H recombination.** (A) Mechanisms of deletional D-to-J_H recombination within the 3′ *Igh* loop domain anchored by IGCR1 and 3′CBEs. Top: For Ds upstream of DQ52, cohesin (*red ring*)-mediated chromatin loop extrusion, anchored at the J_H-based RC, sequentially presents upstream D-RSSs to the RC-bound RAG. During this linear RAG scanning process, only the downstream D-RSSs (*red triangles*) can be aligned with the convergent J_H-RSS (*yellow triangle*), thereby enforcing predominant

beyond identifying junctions formed by normal, robust 12/23-restricted RAG on-target activity mediated by bona fide 12/23 matched RSSs, HTGTS-V(D)J-seq also reveals far lower activity at cryptic RSSs. When the cassette Dβ-23RSS was used as bait, the vast majority of recovered junctions were from bona fide Dβ-Jβ joins. However, the sensitive HTGTS-V(D)J-seq assay also revealed at much lower level (1%–3% of the total) RAG-dependent DJβ-23RSS joins, mediated through a reaction with all of the hallmarks of normal V(D)J recombination, to many different cryptic RSSs spread widely over the 1.8 Mb downstream end of the convergent CBE-anchored chromatin loop domain containing the *c-Myc* gene [36] (Figure 2.9B). While most of these junctions likely occurred in different cells, when viewed together they provided a map of the chromatin region that RAG can explore from its RC location and what it functionally 'saw' during the exploration process [6]. The first notable observation from the assay was that these downstream joins between the DJβ-23RSS and cryptic RSSs abruptly terminated at the downstream CBE loop anchor (Figure 2.9B). Moreover, baiting from the Jβ-12RSS revealed similar low-level joining to cryptic RSSs in the short 1.2 kb sequence upstream of the cassette to the other CBE anchor of the *c-Myc* loop domain, where it again stopped [36].

The identified RAG 'off-targets' contained cryptic RSSs that most frequently represented just the 3 bp 'CAC' found within the RSS heptamer [36] (Figure 2.9B). In this regard, CACs can equally occur in an orientation convergent to the bait RSS (resulting in deletional joining) or in the same orientation as the bait RSS (resulting in inversional joining). Remarkably, the vast majority of these cryptic junctions involved CACs in convergent orientation to the bait RSS [36] (Figure 2.9B). This strong orientation bias is not readily explained by diffusion-based mechanisms that are expected to be orientation-independent, but rather supports a linear scanning mechanism [6,370]. In the context of such a model, RAG would linearly scan chromatin for targets and thus would only 'see' substrates in the convergent orientation with the initiating RAG-bound RSS [36]. Additional studies revealed that such linear RAG chromatin exploration activity occurred in 12 other convergent CBE-based chromatin loop domains into which paired bona fide RSSs were randomly introduced via retroviral integrations [36] (Figure 2.9C). RAG chromatin exploration is clearly directional in that the orientation of the initial RAG-bound RSS prescribes the direction of RAG exploration [36,311,371] (Figure 2.9D). For example, when the DJ_H-based *Igh* RC is inverted, RAG now scans through the downstream C_H locus for targets [311]. Finally, deletion of a given CBE anchor in such domains allows RAG to continue scanning to the next convergent CBE anchor, clearly demonstrating the ability of such CBE anchors to terminate the RAG exploration process [36,311].

The initial RAG linear chromatin exploration studies referred to the process as 'tracking' and proposed cohesin or transcription as potential 'motors' [36]. Two studies published contemporaneously with the RAG tracking study provided evidence for the cohesin-mediated chromatin loop extrusion model discussed above [328,331]. Notably, RAG tracking and cohesin-mediated loop extrusion share striking similarities: both processes are linear, both employ the 'convergent rule' and both are restricted by the same loop anchors, strongly indicating a shared mechanistic basis [327,361,372] (Figure 2.9). RAG tracking was subsequently referred to as 'RAG chromatin scanning' to reflect chromatin loop extrusion as the underlying mechanism [361,372]. The fundamental role of cohesin in mediating RAG scanning was experimentally demonstrated [294]. The studies employed G1-arrested *v-Abl*-transformed pro-B cells that undergo robust *Igh* locus V(D)J recombination upon G1 arrest [179]. In these G1-arrested cells, cohesin depletion disrupted all *Igh* locus chromatin loops and essentially all V(D)J recombination, except for DQ52-to-J_H joining, which was used as a control for RC activity, because this D lies in the RC and accesses RAG via short-range diffusion (see below) [294].

The RAG scanning discovery also had important implications for mechanisms that restrict the dangerous RAG DSB-inducing activity to the *Igh* locus. In this regard, the convergent CBE-based chromatin loop domains that contain *Igh* and other antigen receptor loci restrict RAG

deletional D-to-J_H recombination. The upstream D-RSSs (*light red triangles*) are not utilised because they are in the same orientation with the J_H-RSS. Loop extrusion-mediated RAG scanning of upstream D locus terminates at IGCR1, which lies just upstream of distal DFL16.1 and strongly impedes extrusion further upstream. Bottom: Due to its location in the RC where RAG binds J_H-RSSs, the J_H-proximal D segment DQ52-RSSs access the open RAG1 active site by diffusion (indicated by green dashed arrows), with deletional joining dominating owing to the strong downstream DQ52-RSS. (B) CBEs mediate accessibility of proximal V_Hs during RAG chromatin scanning. Top: Upon inactivation of IGCR1, loop extrusion-mediated RAG scanning, anchored at the DJ_H RC, can proceed into the proximal V_H domain before being impeded by the V_H5-2-associated CBE, which is the first CBE in line of scanning. The strong loop extrusion impediment activity of V_H5-2-associated CBE stabilises interactions between V_H5-2-RSS and the RAG-bound DJ_H RC, thereby promoting the nearly dominant utilisation of V_H5-2. Bottom left: Deletion of V_H5-2-associated CBE abrogates rearrangement of V_H5-2 and allows RAG scanning to proceed to the next V_H2-1. V_H2-1-associated CBE again promotes dominant utilisation of V_H2-1 accompanied by greatly increased interactions with the DJ_H RC. Bottom right: Restoration of a vestigial CBE immediately downstream of the first V_H, V_H5-1, generates a strong loop extrusion impediment and makes V_H5-1 the most frequently used V_H. See text for more details. *Parts (A) and (B) are adapted with permission from Ref. [6]*

off-target activity by impeding scanning beyond their boundaries [36]. On the other hand, NHEJ and DSB response factors such as ATM restrict RAG activity to a loop by enforcing robust V(D)J recombination joining and even lower-level cryptic RSS joining to sequences within the loop. In the absence of NHEJ or ATM-associated DSB response factors, RAG-generated DSBs can escape the RAG-bound RC and resident loop domains and translocate to other dominant DSBs genome-wide [36]. Moreover, RAG scanning analyses defined 107 cryptic RSSs genome-wide that translocate out of loop domains in ATM-deficient cells. These cryptic RSSs nearly all had a conserved heptamer, were found within enhancers or super-enhancers and, notably, were convergently paired with another cryptic RSS in the same convergent CBE-based loop domain [36]. Overall, the findings offer a mechanistic explanation for a number of the recurrent deletions and translocations that have been implicated in cancers of developing lymphocytes [36].

4.5 RAG Scanning Mediates Physiological D-to-J_H Recombination

Over 40 years ago, a linear recombinase tracking model was proposed to explain the observation that, while Ds have 12RSSs on both sides, they appeared to preferentially use the downstream D-12RSS for D-to-J_H joining [370]. However, subsequent studies of extra-chromosomal V(D)J recombination substrates led to the proposal that deletional joining is mediated by the downstream D-12RSS being stronger than the upstream D-12RSS [373]. In this context, recent HTGTS-V(D)J-seq studies clearly documented the nearly complete dominance of the downstream D-12RSSs in mediating joining of all Ds across the 50 kb D locus to J_Hs [311]. Ultimately, comprehensive mutational analysis of the entire chromosomal D locus revealed the critical role for convergent D-RSS orientation in enforcing deletional D-to-J_H rearrangement in the context of linear RAG chromatin scanning, demonstrating that physiological bona fide RSSs are subject to the same dominant convergent RSS utilisation during RAG chromatin scanning [311], as was first shown by the cryptic RSS studies outlined above.

Various lines of evidence support the following general model for how cohesin-mediated loop extrusion initiates *Igh* locus V(D)J recombination [6] (Figure 2.10A, *top*). In pro-B cells, cohesin is loaded at an unknown point or points upstream of the *Igh* V(D)J RC and then extrudes a chromatin loop. On the downstream side, the extrusion of this loop is impeded by the highly transcribed, active chromatin of V(D)J RC and immediately downstream sequences, which like other such sequences across the genome [374–377], serves as a dynamic, non-CBE-based loop anchor [6,361]. Upon binding to the RC, RAG recruits one of the J_H-23RSSs into one of its two active sites. Continued extrusion of upstream chromatin past the RC sequentially presents Ds to RC-bound RAG. During this scanning process, RAG can capture D-12RSSs in its other active site and then mediate 12/23-restricted cleavage to generate the DJ_H intermediate that will serve as the new RC for the subsequent V_H-to-DJ_H joining process [6,311] (Figure 2.10B). Consistent with the properties of RAG scanning from ectopic RCs, during RAG linear scanning of the D locus, only the downstream D-12RSSs, which are convergently oriented to the J_H-23RSSs, are used. This orientation-specific joining process results in 'deletional' D-to-J_H joining that deletes all sequences between the participating D and J_H [311] (Figure 2.10A, *top*). This mechanism applies to recombination of all Ds except the RC-located DQ52, which, as proposed [373], relies on a stronger downstream RSS for deletion-biased joining [311]. DQ52 likely evolved the strong downstream RSS due its location in the RC, which allows it to access RAG predominantly by short-range diffusion (Figure 2.10A, *bottom*). In support of this interpretation, inverting DQ52 in its normal location allows it to join predominantly by inversion. However, when the distal DFL16.1 is replaced with DQ52 in inverted orientation, its weaker downstream RSS dominates rearrangement, clearly demonstrating the dominance of convergently oriented RSSs during RAG scanning of the D locus [311].

The IGCR1 element serves as a CBE-based upstream anchor during the D-to-J_H rearrangement process, and, thereby, essentially terminates extrusion of the D locus loop and terminates RAG scanning at the end of the D locus (Figure 2.10A, *top*). In this way, IGCR1 can largely prevent this initial RAG scanning process from entering the V_H domain [36,361]. Thus, while V_Hs have 23RSSs that are compatible with joining to D-12RSSs, the IGCR1 CBE-based barrier largely prevents access of V_Hs to Ds during the D-to-J_H rearrangement stage of the reaction and thereby contributes to 'ordered' rearrangement [311,361]. As noted above and discussed in the next section, mutating the two IGCR1 CBEs allows RAG scanning to proceed into the proximal V_H domain, resulting in dominant rearrangement of proximal V_Hs along with enhanced interactions between proximal V_Hs and the RC [360–362,366]. In this regard, IGCR1 inactivation also leads to an, albeit low level, increase in direct V_H-to-D rearrangements [360,362].

Finally, as mentioned above, there is a cluster of 10 3′ *Igh* CBEs at the very downstream end of the locus that all point upstream in convergent orientation to IGCR1 CBE2 (Figure 2.8B). One attractive early model for IGCR1 function was that it sequestered the V_H domain from the downstream D and J_H-containing domain by forming a convergent-CBE-based contact loop with the 3′ *Igh* CBEs [358–362]. However, more recent studies showed that deletion of the 3′CBEs does not lead to an IGCR1-inactivation phenotype [297,366]. Thus, it seems possible that, like other CBEs [324], CTCF-bound IGCR1 CBEs may be able to function independently as a loop extrusion (and RAG scanning) impediment. However, some recent studies suggest that, upon juxtaposition by loop extrusion, the 3′CBEs could reinforce the RC as an extrusion anchor [366]. Also, in the absence of 3′CBEs, further downstream CBEs might, at least partially, compensate for the potential 3′CBEs activity [366], as observed for *Igh* CSR [357].

4.6 CBEs Mediate Accessibility of Proximal V_Hs During RAG Scanning

In normal developing pro-B cells, the IGCR1 impediment was thought to somehow be neutralised to allow V_Hs to access the DJ_H RC for V_H-to-DJ_H recombination [360,362]. Groundbreaking 3D-fluorescence in situ hybridisation (3D-FISH) and chromatin conformation capture (3C)-based studies demonstrated that the V_H locus undergoes a contraction process in pro-B cells that is associated with providing distal V_Hs access to DJ_H RC [304–306,309,358,378–381]. One attractive model posited that locus contraction would bring all of the V_Hs into proximity with the DJ_H RC where they would gain access [364,365]. In this regard, proximal V_Hs were found to interact with the DJ_H domain and undergo active rearrangement in PAX5- or YY1-deficient pro-B cells that did not undergo V_H locus contraction [304,378,382,383]. More recent studies revealed that the mechanism of V_H locus contraction is based on cohesin-mediated loop extrusion [294,297,298,350] and that V_Hs become associated with the DJ_H RC via linear RAG scanning [297,298].

V_H5-2 (V_H81X) is the first functional and most frequently rearranging proximal V_H in pro-B cells. This observation led to the early proposal of 'recombinase scanning' mechanism for V_H-to-DJ_H recombination [384]. A unique feature of the several dozen proximal V_Hs is that a single CBE lies just downstream of each functional V_H-RSS [295,296] (Figures 2.8B and 2.10B). Notably, a loss of function mutation of V_H5-2-associated CBE almost abrogated the utilisation of V_H5-2 and led to robust utilisation of the next upstream V_H2-2 (Figure 2.10B, *bottom left*). Indeed, subsequent mutation of the V_H2-2-associated CBE again drastically reduced its usage and led to over-utilisation of the next V_H upstream [361]. Moreover, V_H5-1 that lies just downstream of V_H5-2 normally recombines at very low frequency despite its consensus RSS. However, V_H5-1 lacks a functional CBE. Remarkably, restoration of a vestigial CBE adjacent to V_H5-1 made it the most-frequently utilised V_H [361] (Figure 2.10B, *bottom right*). These data, together with complementary results from the high-resolution chromatin interaction assay LAM-HTGTS-3C-Seq, showed that proximal V_H-RSS-associated CBEs greatly augmented their interaction with the RC, suggesting that V_H-associated CBEs increase proximal V_H recombinational accessibility by promoting prolonged V_H interactions with RC during linear RAG chromatin scanning [361]. These studies provided some of the earliest evidence in support of the RC-based RAG linear scanning model [361,372]. Notably, inversion of the V_H81X-associated CBE only led to a modest twofold reduction in V_H81X recombination, consistent with the notion that impediment function of CBEs is not absolutely orientation-dependent [361]. Moreover, these studies further showed that the nearly exclusive utilisation of V_H5-2 and its robust association with the RC during direct scanning, upon IGCR1 inactivation, was almost totally dependent on its RSS-associated CBE [361]. More generally, these studies also indicated that the high density of proximal V_H locus CBEs presents a strong barrier to the utilisation of more distal V_Hs. In this regard, pro-B cells indeed employ specific mechanisms that modulate CBE impediment activity in the *Igh* locus to promote a diverse V_H repertoire [297,298,350] (see below).

4.7 Cohesin-Mediated Loop Extrusion Promotes V_H Locus Contraction

Unlike the *Igh*-contracted primary pro-B cells from the bone marrow, *v-Abl* pro-B cells lack *Igh* contraction and undergo little distal V_H-to-DJ_H recombination [294,297,298,361]. Notably, degradation of CTCF via an auxin-inducible degron system in G1-arrested *v-Abl* cells activated significant RC interactions across the distal V_H locus with a pattern reminiscent of that observed in primary pro-B cells [294]. Correspondingly, CTCF depletion increased distal V_H utilisation and long-range RAG scanning as evidenced both by bona fide RSS-mediated V_H rearrangements and targeting of convergently oriented cryptic RSSs across the V_H locus. Thus, dampening the IGCR1 and V_H locus CBE-based loop extrusion impediments activated V_H locus contraction and distal V_H usage [294]. Although most CTCF binding was removed from chromatin upon CTCF depletion, there remained an uneven residual CTCF binding at various V_H locus CBEs. This residual CTCF activity correlated with V_H utilisation frequency, indicating again that residual CBE impediment activity upon CTCF depletion can target proximal V_H accessibility for V(D)J recombination [294]. Finally, the ability to establish V_H locus contraction in *v-Abl* cells via modulating loop extrusion impediments provided strong evidence that physiological locus contraction could, in principle, be mediated by loop extrusion [294].

4.8 Cryptic RSS Usage Supports Long-Range V_H Locus Scanning

In the wild-type mouse *Igh* locus, all V_H-23RSSs are oriented convergently with the upstream 12RSS of the DJ_H RC. Remarkably, inversion of the entire 2.4 Mb V_H locus, except for the most proximal V_H5-2 (to provide a control), in primary pro-B cells abolished utilisation of all inverted V_Hs, as well as utilisation of all of the many normally convergent cryptic RSSs contained within the inverted domain [297,298] (Figure 2.11A and B). A smaller inversion of an 809 kb portion of the upstream V_H locus similarly affected rearrangement of V_Hs in the inverted region [350]. The abrogation of V_H rearrangements in the 2.4 Mb inversion did not appear to reflect changes in V_H accessibility per se, as the inverted V_H locus still underwent locus contraction and had normal (albeit inverted) germline transcription patterns [297,298]. While it was considered that reversed orientation of the V_H-associated CBEs might contribute majorly to the phenotype [350], the reversal of the orientation of proximal V_H-RSS-associated CBEs only modestly affected rearrangement of associated V_Hs [361]. Also, more recent studies revealed that inversion of the DJ_H RC led to a similar abrogation of V_H rearrangements, without any modification of CBE organisation within the *Igh* locus or beyond (unpublished data).

The major question regarding the interpretation of the V_H locus inversion experiments was whether or not DJ_H RC-bound RAG still functionally scanned across the V_H locus. This question was answered by the finding that RAG indeed robustly utilises the newly convergent cryptic RSSs within the inverted V_H locus for joining to the bona fide 12RSS of the DJ_H RC, confirming that RAG linearly scans the entire inverted V_H locus but hardly recognises inverted V_H-RSSs (Figure 2.11B). Together, these finding revealed that linear RAG scanning of upstream chromatin from the DJ_H RC is the dominant mechanism for V_H utilisation across the V_H locus [297,298]. Finally, it was notable that, in the absence of dominant V_H-RSS-mediated rearrangements, RAG scanning continued past the inverted V_H locus linearly through multiple convergent CBE-based domains to the telomere (Figure 2.11B). This latter finding indicated that long-range RAG scanning of the V_H locus may involve a mechanism that more generally down-regulates CBE impediments in pro-B cells [297,298]. A contemporaneous study directly documented such a phenomenon in pro-B cells via Hi-C genomic interaction experiments [350].

4.9 WAPL Down-Regulation Promotes Long-Range Scanning of the V_H Locus

The finding that CBE impediments may be more generally down-regulated in pro-B cells led to studies aimed at identifying potential cohesin complex factors that could be involved based on their down-regulation in pro-B cells. Two independent studies, based on different criteria, identified the WAPL-cohesin unloading protein as being down-regulated in pro-B cells [297,298,350]. As mentioned above, it had already been shown that WAPL depletion in non-lymphoid cells can extend the size of loop domains genome-wide [336,337,347], with the mechanistic interpretation that increased retention of cohesin on the chromatin enhances its ability to extend through dynamic CBE loop anchors [352,385] and potentially other scanning impediments. Indeed, one of the above-mentioned studies clearly revealed that WAPL down-regulation is employed physiologically in pro-B cells to activate locus contraction and distal V_H-to-DJ_H recombination [350]. This study demonstrated that the *Wapl* transcription is repressed in pro-B cells by the B cell-specific PAX5 transcription factor. PAX5 suppresses *Wapl* transcription by binding to the *Wapl* promoter and recruiting the Polycomb Repressive Complex 2 (PRC2) to induce the repressive histone mark H3K27me3. Indeed, deletion of the PAX5-binding site in the *Wapl* promoter resulted in loss of *Igh* contraction and distal V_H rearrangement [350]. Notably, it was noted long ago that PAX5 expression was required for locus contraction and distal V_H rearrangement [304,382], but it was generally thought that this function may be mediated through PAX5 regulation of the so-called PAIR elements spread across the distal V_H locus [386,387]. Whether or not PAX5 also has direct functions in the context of transcriptional regulation of distal V_H locus elements remains to be determined. However, the main function of Pax5 in promoting locus contraction and distal V_H rearrangement appears to be down-regulating *Wapl* expression [350].

Contemporaneously with the study outlined just above, a different group also identified WAPL down-regulation in pro-B cells and tested the notion that such down-regulation could mediate locus contraction and distal V_H utilisation by degrading WAPL via the auxin-inducible degron system in *v-Abl* cells as outlined above for CTCF degradation [297,298]. Indeed, essentially complete WAPL depletion in *v-Abl* pro-B cells resulted in V_H locus contraction involving the same set of robust RC interactions with sites across the V_H locus as observed in normal pro-B cells. Correspondingly, RAG scanning and V_H rearrangements were extended across the locus [297,298]. In *v-Abl* cells in which the V_H locus was inverted, WAPL depletion again revealed the same phenotype as observed in pro-B cell with an inverted V_H locus. These phenotypes included loss of bona fide V_H-to-DJ_H joining across the locus even though RAG scanning, based on cryptic RSS utilisation, extended through the locus to the telomere [297,298].

Thus, this body of work identified WAPL down-regulation in pro-B cells as the mechanism of V_H locus contraction and the ability of pro-B cells to activate rearrangement of distal V_H-to-DJ_H rearrangements. Moreover, one of the studies clearly documented that WAPL down-regulation allows RAG to linearly scan through the entire V_H locus [297,298]. In this regard, WAPL down-regulation substantially neutralises the various CBE scanning impediments including IGCR1 CBEs and proximal V_H-RSS-associated CBEs to facilitate long range scanning. Recent

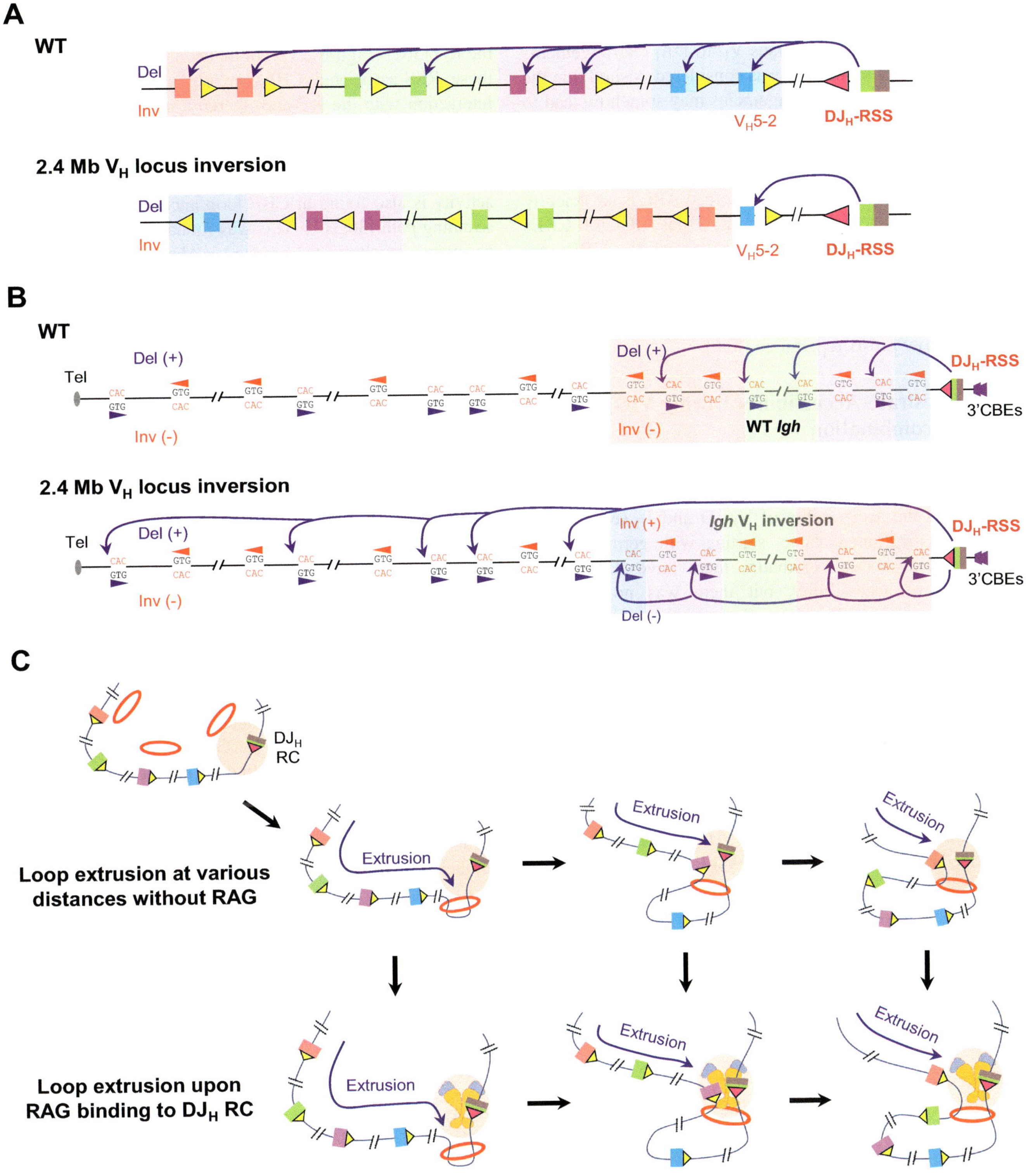

FIGURE 2.11 **Long-range RAG scanning mediates V_H utilisation across the *Igh* locus.** (A) Schematic showing joining events between bona fide DJ_H-RSS and V_H-RSSs in WT (top) and *Igh* V_H inversion (bottom) primary pro-B cells and WAPL-downregulated *v-Abl* lines. The inversion abrogated utilisation of all inverted V_Hs but permitted utilisation of the most proximal V_H5-2 (also known as V_H81X) which was left in its normal orientation as a control. (B) Possible joining outcomes between bona fide DJ_H-RSS and cryptic RSSs mostly represented by CAC motifs in the upstream D, V_Hs and domains upstream of V_Hs to the telomere in WT (top) and *Igh* V_H inversion (bottom) primary pro-B cells and WAPL-downregulated *v-Abl* lines. The inversion did not affect RAG chromatin scanning, as evidenced by the utilisation of convergent cryptic RSSs relative to the DJ_H-RSS throughout the inverted V_H locus up to the telomere. (C) Model for loop-extrusion-mediated locus contraction for distal V_H utilisation in the absence (top) or presence (bottom) of RAG. In the absence of RAG binding, cohesin (red ring)-mediated chromatin loop extrusion, anchored at the downstream DJ_H RC, can proceed to varying positions in the V_H locus from cell to cell. Upon RAG binding to the DJ_H RC, loop extrusion-mediated RAG scanning can thus initiate from different extrusion points across the V_H locus to avoid over-representation of proximal V_H rearrangements and provide a more balanced V_H repertoire. See text for more details. *Parts (A-C) are adapted with permission from Ref. [6]*

studies have also given some insight into how WAPL down-regulation in pro-B cells mediates the apparent ordered process of D-to-J_H versus V_H-to-DJ_H rearrangements [366]. One early model that was considered posited that the formation of D-to-J_H rearrangements may somehow lead to down-regulation of IGCR1 impediments [388]. However, current studies better support a model in which gradual reduction of WAPL levels in pro-B cells initially allows robust D-to-J_H rearrangements before WAPL is sufficiently down-regulated to allow scanning to pass IGCR1 CBEs into the V_H locus. Also, before proximal V_H CBEs are neutralised via WAPL down-regulation, they are still relatively dominantly utilised until down-regulation is complete, which explains why they are over-utilised in normal bone marrow pro-B repertoires [366].

4.10 Substrate Accessibility During V(D)J Recombination

As outlined in the beginning of this section on regulation, *Igh* locus V(D)J recombination regulation is achieved by controlling the accessibility of V_H, D and J_H segments to RAG. As mentioned above, starting with germline transcription and active chromatin, there have been many correlates of accessibility, but there was no clear-cut mechanism of how these correlates could actually function to mediate accessibility [182,284]. However, at least for the *Igh* locus, a number of these correlates can now be mechanistically explained when put into the context of loop extrusion-mediated RAG scanning process that establishes physical juxtaposition of the compatible RSSs for RAG cleavage [6].

The role of CBE impediments in targeting RAG scanning activity was clearly demonstrated in studies of the mechanisms that target proximal V_Hs for V(D)J recombination. As described in detail above, proximal V_H-RSS-associated CBEs can increase the V(D)J recombination activity of the V_H-RSSs by 50- to 100-fold, if they are first in line to be scanned through the RC [361] (Figure 2.12). In that context, during scanning, these RSS-associated CBEs greatly increase the level of interaction with the RC and, thereby, make their associated RSSs much stronger targets for capture by RAG and V(D)J recombination [6,361]. This phenomenon is not just limited to V_H-RSS-associated CBEs. Indeed, increased RAG cryptic activity is also found at CBE loop anchors that impede the scanning path [36,311]. This was found most notably in experiments involving G1-arrested *v-Abl* cells. In such cells, inversion of the DJ_H-RSS programs RC-bound RAG to scan downstream chromatin extruded past it from the downstream C_H locus, which lacks any bona fide RSS, until scanning is finally impeded by the 3′ *Igh* CBEs [311]. Indeed, cryptic RSSs in the 3′ *Igh* CBEs, which have robust interactions with the RC, become RAG activity hotspots [311].

This same system also clearly demonstrated the role of transcription in targeting substrate accessibility via impeding loop extrusion-mediated RAG scanning. In this case, the study took advantage of the fact that some of these lines highly transcribe the Cγ2b switch (Sγ2b) regions and Cγ2b exons in the middle of the several hundred kb C_H domain. During such downstream scanning from the RAG-bound inverted DJ_H RC, transcription across Sγ2b targets substantial, dynamic interactions with the inverted DJ_H RC and robust RAG activity on cryptic RSSs across the Sγ2b sequence [311]. Indeed, the role of transcription in this process was proven by inactivation of the Iγ2b transcriptional promoter, which eliminated robust RC interactions with this region along with RAG scanning activity within it. The interpretation of this finding is that transcription of the long repetitive S region can increase accessibility of cryptic RSSs within it by increasing the duration that these sequences interact with the RAG-bound RC [311]. In another experiment with the same system, nuclease-dead

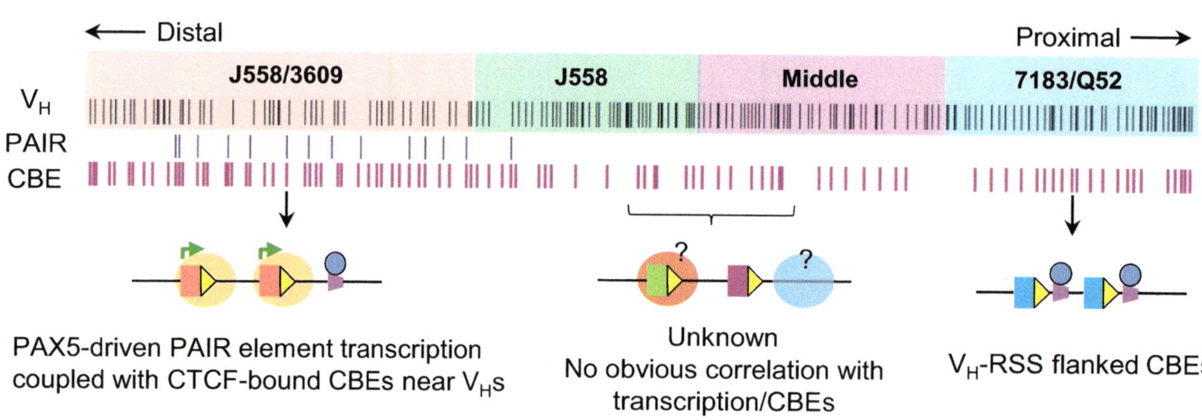

FIGURE 2.12 **Mechanisms promoting V_H accessibility during RAG scanning.** Proximal V_Hs rely dominantly on V_H-associated CBEs for accessibility. PAX5-driven sense/antisense transcription from PAIR elements and/or V_H-associated promoters, coupled with CTCF-bound CBEs contribute to promoting distal V_H usage. Mechanisms for access of many J558 V_Hs involve transcription, while many middle V_Hs have RSS-associated CBEs. However, some highly utilised V_Hs in these two domains have no clear-cut correlation with transcription or CBEs. See text for more details.

Cas9 (dCas9) binding, ectopically targeted to the downstream RAG scanning path by focussing dCas9 binding to multiple sites at the Sγ1 switch region, led to the accumulation of cohesin at Sγ1, enhanced interactions between RC and Sγ1, and substantial RAG targeting of cryptic RSSs within the dCas9-targeted Sγ1 sequence. These effects were accompanied by decreased RAG-targeting activity in and RC interactions with the 3′CBEs further downstream of the dCas9-bound Sγ1 [311].

Together, the above findings clearly showed that a variety of different impediments to cohesin-mediated loop extrusion can increase the level or length of time of RC-based interactions with sequences near the impediments and thereby target the accessibility of bonafide RSSs or cryptic RSSs associated with the impeded region to RAG (Figure 2.12). Additional studies based on these new insights help clarify various long-standing mysteries. For one, the most distal DFL16.1 is the most highly used of the D segments [311]. This observation can be explained by the finding that there are relatively limited impediments across the D locus and, correspondingly most of the Ds in the middle of the locus have low utilisation. However, DFL16.1 is located just downstream of IGCR1, which greatly increases its interaction time with the RC [311,389]. Likewise, the finding that proximal V_Hs are highly utilised but barely, if at all, transcribed, provided some difficulty over the years in explaining the correlation of transcription with accessibility [43,364]. However, we now know that these proximal V_Hs use RSS-associated CBEs to increase interactions with the RC and RAG accessibility and, thus, do not require transcription for this purpose [361] (Figure 2.12). On the other hand, the most distal V_Hs do not have RSS-associated CBEs, but they are nearly all substantially transcribed. Indeed, the transcription of these V_Hs correlates quite positively with their rearrangement during RAG scanning [294,297,298]. However, for some far upstream V_Hs, it seems likely that a combination of transcription and relatively proximal CBEs may also be needed to sufficiently impede scanning to provide requisite accessibility [297,298] (Figure 2.12). There may still be other, yet uncharacterised, factors that promote accessibility during scanning, as illustrated by the dCas9 experiments [311], and also by the fact that some V_Hs with robust utilisation are not obviously associated with known extrusion impediments [294,297,298] (Figure 2.12). Finally, locus contraction was also considered an accessibility correlate [182,284]. We now know that this process occurs by loop extrusion, which in WAPL down-regulated pro-B cells allow RAG to scan the length of the V_H locus [297,298,350]. Without loop-extrusion-mediated RAG scanning of the V_H locus, none of the accessibility mechanisms described above could function. Indeed, several novel enhancers have recently been identified within the V_H locus that impede loop extrusion and promote utilisation of associated V_Hs [390].

4.11 Summary of Loop Extrusion-Based Models for V(D)J Recombination in the *Igh* Locus

Recent studies strongly support the fundamental role of the loop extrusion-mediated RAG scanning model in both D-to-J_H and V_H-to-DJ_H recombination [6]. These studies also provided one of the first examples of a transcription-based loop extrusion anchor, the *Igh* RC [297,298,361]. The importance of transcription-based loop extrusion impediments had largely remained hypothetical with respect to genome-wide studies until recently [374,391]. Targeted mutagenesis of the endogenous D locus demonstrated a physiological role for RAG scanning in promoting deletional D-to-J_H recombination with an important exception of DQ52 which lies within the RC and accesses RAG by short-range diffusion [311]. Primary pro-B cells undergo physical contraction of the 2.4 Mb V_H region which is mechanistically driven by cohesin-mediated loop extrusion [294,297,298,350]. Such extrusion activity across the long, CBE-rich V_H locus is promoted by down-regulation of the WAPL cohesin unloading protein [297,298,350]. The proposed mechanism is that increased cohesin density allows frequent passing of dynamic CBE [352,385], and potentially other, impediments to extend loops. WAPL down-regulation also clears a path for linear RAG scanning of the entire V_H locus [297,298]. During scanning, loop extrusion impediments, including CBE anchors, nascent transcription and potentially other DNA-binding protein complexes, focus RAG activity on targets within impeded regions to promote their utilisation [6,294,297,298,311,361] (Figure 2.12). How does a linear extrusion-mediated RAG scanning process promote a diverse V_H repertoire without being dominated by the RC-proximal V_Hs? One possible mechanism is that RC-anchored loop extrusion occurs independently of RAG binding and extends a variety of loop sizes in individual pro-B cells in which RAG is not bound [297,298]. Subsequent RAG binding to the RC (or to a J_H-23RSS) would then initiate extrusion-mediated scanning and recombination from various extrusion points, promoting a more balanced V_H repertoire that is not so dominated by the most proximal V_Hs [297,298] (Figure 2.11C).

4.12 Programmed Inversional V(D)J Recombination in the *Igκ* Locus?

The 3.2 Mb *Igκ* locus undergoes direct Vκ-to-Jκ rearrangement between the 92 functional Vκs and 4 functional Jκs [365,392] (Figure 2.8C). The Jκs are located about 20 kb downstream of the most proximal Vκ and function with several enhancers to form a Jκ-RC [96,97]. Notably, the most proximal and distal Vκs are mainly convergently oriented for deletional joining to Jκs, whereas the middle Vκs are mainly oriented for inversional joining [365]

(Figure 2.8C). Yet, most of the Vκs are robustly used in the primary VκJκ repertoire [392]. After primary deletional or inversional Vκ-to-Jκ joining, complex secondary deletional or inversional Vκ-to-Jκ joins can occur both in the chromosome and within extrachromosomal deletion circles [392]. Such joining complexity has to some degree hindered in-depth analyses of primary Vκ-to-Jκ joining. In theory, more insights into the primary rearrangement mechanisms might be obtained by modifying the *Igκ* locus in a fashion that would dampen or prevent secondary rearrangements, for example, by deletion of all but one Jκ.

Assembly of *Igκ* variable region exons occurs mainly in pre-B cells that develop from pro-B cells upon expression of a μ IgH chain [392] (Figure 2.8A). Although the Vκ locus is already contracted in pro-B cells [309,393], which is likely driven by cohesin-mediated loop extrusion, they have only low-level Vκ-to-Jκ joining [394], suggesting that other events in addition to locus contraction activate robust Vκ-to-Jκ joining in pre-B cells. The short Vκ–Jκ intergenic region contains a regulatory segment with two major components, separated by 4 kb, referred to as Cer [395] and Sis [396] (Figure 2.8C). Both Cer and Sis contain two CBEs, with the Cer CBEs pointing towards the Vκs and the Sis CBEs pointing towards Jκs. Deletion of either Cer or Sis leads to increased proximal Vκ utilisation and decreased distal Vκ utilisation, with a much more pronounced phenotype observed for Cer deletion [395,397,398]. The dramatically increased use of proximal, deletional-oriented Vκs upon inactivation of Cer and Sis suggests, in analogy to studies of V_H usage in the absence of IGCR1 [36], the activation of direct RAG scanning from the Jκ-RC to proximal deletional-oriented Vκs [397]. However, unlike IGCR1, which has no known role in *Igh* locus contraction [360,378], Cer appears to have a more important role in *Igκ* contraction [395], potentially by acting as a dynamic loop extrusion anchor. Thus, decreased usage of distal Vκs upon Cer deletion could reflect a role for Cer in increasing interactions between distal Vκs and the Jκ-RC-proximal portion of the locus, potentially in a locus contraction process driven by cohesin-mediated loop extrusion [6].

It is notable that the deletion of Cer and Sis mimics the chromosomal context in which secondary Vκ-to-Jκ joins occur, because the Vκ–Jκ intergenic region will be deleted or displaced by primary Vκ-to-Jκ joins [285]. In this case, in theory, secondary rearrangements following a primary deletional Vκ-to-Jκ join could access upstream deletional-oriented Vκs via direct linear scanning. Moreover, primary rearrangements to an inverted Vκ could place a series of downstream inverted Vκs in deletional orientation upstream of the Jκ RC, allowing them to also be accessed by linear RAG scanning. Long-range primary inversional joining of Vκs to Jκs is not consistent with direct linear RAG scanning from the Jκ to an inverted Vκ. Conceivably, loop extrusion might mediate locus contraction to bring the inverted Vκ-RSSs close enough to the Jκ-RC for a short-range diffusion-mediated joining [311,350,364,399], as has been shown to occur at low level for inverted D-to-J_H joining [6,311]. In such a context, local interactions have been suggested to potentially be mediated by phase separation [400]. Potential roles of WAPL down-regulation in *Igκ* locus contraction are not evident. Thus, although contraction of the *Igκ* locus occurs in pro-B cells [309,393], pre-B cells in which *Igκ* is predominantly rearranged [392] have similarly reduced *Wapl* levels [350]. Perhaps continued reduction of *Wapl* levels maintains *Igκ* contraction at the pre-B cell stage. However, WAPL levels sufficient to impede V_H locus contraction and distal V_H usage allow for normal patterns of V(D)J recombination across the 3 Mb Vκ locus in *v-Abl* cells [297]. In this regard, G1-phase arrest of *v-Abl* cells decreases WAPL levels by two to three-fold [297], which, in theory, might be sufficient to promote Vκ locus contraction owing to the substantially lower CBE density in the Vκ locus versus the V_H locus [297,365]. Several enhancers within the Vκ locus recently have been implicated in promoting Vκ-to-Jκ recombination [401,402]. It will be of interest to determine if they might also function in the context of a loop extrusion mechanism.

4.13 T Cell Receptor Loci V(D)J Recombination

The *TCRα/δ* and *TCRβ* loci also share organisational features with the *Igh* locus, including long length, locus contraction, having their RC near the D and/or J segments, and having Vs mostly sequestered from the Ds and/or Js within domains containing numerous CBEs [286,365,403]. Early studies implicated a fundamental role for cohesin in *TCRβ* locus V(D)J recombination [404], and more recent studies have implicated roles of CBE-based elements in insulating Vβ domains from D and J domains [405]. Additional studies implicated potential roles for RAG scanning within CBE-based domains in the *TCRα/δ* locus [371,406,407]. The large body of relevant work on this topic has been covered in a recent review [286]. If loop extrusion is used by TCR loci to mediate locus contraction and V(D)J recombination, mechanisms other than WAPL downregulation may regulate their potential CBE-mediated impediments to loop extrusion, given that WAPL expression levels remain unchanged during T cell development [350]. In this regard, given the similar effects of CTCF depletion and WAPL depletion on *Igh* locus contraction in *v-Abl* lines [294,297,298], it is conceivable that down-regulation or modification of cohesin-complex factors other than WAPL could contribute to contraction of other loci.

5. EVOLUTION OF V(D)J RECOMBINATION

5.1 Evolutionary Origins of RAG and V(D)J Recombination

The genomic organisation of *Rag1* and *Rag2* (see above) and biochemical parallels between RAG and DNA transposases led to the proposal that RAG and V(D)J recombination evolved from a DNA transposon [408]. There is now extensive evidence supporting this hypothesis (reviewed in Refs. [7,409,410]). Of particular importance was the finding that the RAG proteins are capable of performing transposition in biochemical assays [17,18]. In addition, numerous *RAG-like* (*RAGL*) transposons, containing *RAG1-like* (*RAG1L*) and *RAG2-like* (*RAG2L*) genes flanked by terminal inverted repeats (TIRs) with similarity to the RSS heptamer, have been identified in the genomes of metazoans including single-celled algae, cnidarians (e.g. jellyfish and corals), protostomes (e.g. mussels, oysters and ribbon worms) and deuterostomes (e.g. sea urchins, acorn worms and amphioxus) [411–414]. One such element, the *ProtoRAG* transposon from the cephalochordate amphioxus, encodes RAG1L and RAG2L proteins that closely resemble RAG structurally and in biochemical activities despite having only modest sequence similarity with RAG [415,416]. In addition, *Transib* transposons, composed of a single open reading frame with sequence similarity to *Rag1* flanked by TIRs that in some cases resemble a 12/23 RSS pair, are present in the genomes of bacteria and many metazoans [414,417,418]. The one known active *Transib* element, from the corn earworm moth *Helicoverpa zea*, encodes a transposase that has strong structural similarities to core RAG1, cleaves DNA by a nick-hairpin mechanism similar to that of RAG and ProtoRAG and can interact physically and functionally with mouse RAG2 [419–421].

These and other findings support a transposon 'molecular domestication' model for the evolution of RAG and V(D)J recombination (see Ref. [7] for additional details) that begins with an ancient bacterial [414] *Transib* transposon, which was transmitted to both prokaryotic and eukaryotic descendants. In an early eukaryote, a *Transib* transposon is proposed to have gained a *RAG2L* gene to yield the first *RAGL* transposon. The *RAGL* transposon subsequently diversified into several families [412,413] and was transmitted (primarily by vertical transmission) into the genomes of many invertebrates and also into the genomes of jawed vertebrates (whether this occurred by vertical transmission or horizontal gene transfer remains uncertain). The model proposes that in an early jawed vertebrate, a pivotal evolutionary event occurred in which the RAGL transposon inserted into an exon of an Ig domain receptor gene to create the first 'split' antigen receptor gene. This split gene subsequently gave rise to the Ig and TCR genes of all extant jawed vertebrates, while the *RAG1L* and *RAG2L* genes of the transposon gave rise to *Rag1* and *Rag2* encoding the RAG recombinase.

Early in jawed vertebrate evolution, as RAG1 and RAG2 began to function as a recombinase, it would have been important to suppress their transposase activity to protect genome integrity. Consistent with potent suppression of transposase activity, no examples have yet been described of RAG-mediated transposition in mice or humans and RAG transposase activity is low or undetectable in cell line assays [416,422–424]. Four evolutionary adaptations, found in all known jawed vertebrate RAG proteins, have been identified that suppress RAG transposase activity and likely contributed to RAG domestication (reviewed in Ref. [7]). Two of these suppress transposition potently: a single amino acid change from Met to Arg at position 848 of RAG1, and the presence of an acidic hinge in the C terminal region of RAG2. The Met → Arg-848 change suppresses transposition in vitro and in vivo and does so through a mechanism that involves loss of key protein–DNA interactions needed to form a stable complex with target DNA capable of undergoing the integration step of transposition [146]. The RAG2 acidic hinge appears to suppress transposition only in vivo. Its mechanism of action is not known. Two other adaptations suppress RAG-mediated transposition to a small (two- to threefold) extent: the presence of Glu at RAG1 position 649 (mechanism unknown) [146] and an inhibitory loop of RAG2 (aa 333-342) that might interfere with target DNA binding. Overall, existing data indicate powerful selection for the suppression of RAG-mediated transposition during jawed vertebrate evolution, with multiple mechanisms having arisen to provide multilayered protection of genome integrity.

REFERENCES

[1] Hozumi N, Tonegawa S. Evidence for somatic rearrangement of immunoglobulin genes coding for variable and constant regions. Proc Natl Acad Sci U S A 1976;73:3628–32.

[2] Bassing CH, Swat W, Alt FW. The mechanism and regulation of chromosomal V(D)J recombination. Cell 2002;109(Suppl. 1):S45–55. https://doi.org/10.1016/s0092-8674(02)00675-x.

[3] Swanson PC. The bounty of RAGs: recombination signal complexes and reaction outcomes. Immunol Rev 2004;200:90–114. https://doi.org/10.1111/j.0105-2896.2004.00159.x.

[4] Schatz DG, Ji Y. Recombination centres and the orchestration of V(D)J recombination. Nat Rev Immunol 2011;11:251–63. https://doi.org/10.1038/nri2941.

[5] Schatz DG, Swanson PC. V(D)J recombination: mechanisms of initiation. Annu Rev Genet 2011;45:167–202. https://doi.org/10.1146/annurev-genet-110410-132552.
[6] Zhang Y, Zhang X, Dai H-Q, Hu H, Alt FW. The role of chromatin loop extrusion in antibody diversification. Nat Rev Immunol 2022;22:550–66. https://doi.org/10.1038/s41577-022-00679-3.
[7] Liu C, Zhang Y, Liu CC, Schatz DG. Structural insights into the evolution of the RAG recombinase. Nat Rev Immunol 2022;22:353–70. https://doi.org/10.1038/s41577-021-00628-6.
[8] Bosticardo M, Pala F, Notarangelo LD. RAG deficiencies: recent advances in disease pathogenesis and novel therapeutic approaches. Eur J Immunol 2021;51:1028–38. https://doi.org/10.1002/eji.202048880.
[9] Schatz DG, Oettinger MA, Baltimore D. The V(D)J recombination activating gene, RAG-1. Cell 1989;59:1035–48. https://doi.org/10.1016/0092-8674(89)90760-5.
[10] Oettinger MA, Schatz DG, Gorka C, Baltimore D. RAG-1 and RAG-2, adjacent genes that synergistically activate V(D)J recombination. Science 1990;248:1517–23.
[11] van Gent DC, Hiom K, Paull TT, Gellert M. Stimulation of V(D)J cleavage by high mobility group proteins. EMBO J 1997;16:2665–70. https://doi.org/10.1093/emboj/16.10.2665.
[12] Roth DB, Menetski JP, Nakajima PB, Bosma MJ, Gellert M. V(D)J recombination: broken DNA molecules with covalently sealed (hairpin) coding ends in scid mouse thymocytes. Cell 1992;70:983–91. https://doi.org/10.1016/0092-8674(92)90248-b.
[13] Roth DB, Zhu C, Gellert M. Characterization of broken DNA molecules associated with V(D)J recombination. Proc Natl Acad Sci U S A 1993;90:10788–92. https://doi.org/10.1073/pnas.90.22.10788.
[14] Schlissel M, Constantinescu A, Morrow T, Baxter M, Peng A. Double-strand signal sequence breaks in V(D)J recombination are blunt, 5'-phosphorylated, RAG-dependent, and cell cycle regulated. Genes Dev 1993;7:2520–32. https://doi.org/10.1101/gad.7.12b.2520.
[15] Chang HHY, Pannunzio NR, Adachi N, Lieber MR. Non-homologous DNA end joining and alternative pathways to double-strand break repair. Nat Rev Mol Cell Biol 2017;18:495–506. https://doi.org/10.1038/nrm.2017.48.
[16] Lewis SM, Hesse JE, Mizuuchi K, Gellert M. Novel strand exchanges in V(D)J recombination. Cell 1988;55:1099–107. https://doi.org/10.1016/0092-8674(88)90254-1.
[17] Agrawal A, Eastman QM, Schatz DG. Transposition mediated by RAG1 and RAG2 and its implications for the evolution of the immune system. Nature 1998;394:744–51. https://doi.org/10.1038/29457.
[18] Hiom K, Melek M, Gellert M. DNA transposition by the RAG1 and RAG2 proteins: a possible source of oncogenic translocations. Cell 1998;94:463–70. https://doi.org/10.1016/s0092-8674(00)81587-1.
[19] Early P, Huang H, Davis M, Calame K, Hood L. An immunoglobulin heavy chain variable region gene is generated from three segments of DNA: VH, D and JH. Cell 1980;19:981–92. https://doi.org/10.1016/0092-8674(80)90089-6.
[20] Tonegawa S. Somatic generation of antibody diversity. Nature 1983;302:575–81. https://doi.org/10.1038/302575a0.
[21] Davis MM, Bjorkman PJ. T-cell antigen receptor genes and T-cell recognition. Nature 1988;334:395–402. https://doi.org/10.1038/334395a0.
[22] Bassing CH, Alt FW, Hughes MM, D'Auteuil M, Wehrly TD, Woodman BB, et al. Recombination signal sequences restrict chromosomal V(D)J recombination beyond the 12/23 rule. Nature 2000;405:583–6. https://doi.org/10.1038/35014635.
[23] Sleckman BP, Bassing CH, Hughes MM, Okada A, D'Auteuil M, Wehrly TD, et al. Mechanisms that direct ordered assembly of T cell receptor beta locus V, D, and J gene segments. Proc Natl Acad Sci U S A 2000;97:7975–80. https://doi.org/10.1073/pnas.130190597.
[24] Jung D, Bassing CH, Fugmann SD, Cheng H-L, Schatz DG, Alt FW. Extrachromosomal recombination substrates recapitulate beyond 12/23 restricted VDJ recombination in nonlymphoid cells. Immunity 2003;18:65–74. https://doi.org/10.1016/s1074-7613(02)00507-1.
[25] Tillman RE, Wooley AL, Khor B, Wehrly TD, Little CA, Sleckman BP. Cutting edge: targeting of V beta to D beta rearrangement by RSSs can be mediated by the V(D)J recombinase in the absence of additional lymphoid-specific factors. J Immunol 2003;170:5–9. https://doi.org/10.4049/jimmunol.170.1.5.
[26] Olaru A, Patterson DN, Villey I, Livák F. DNA-Rag protein interactions in the control of selective D gene utilization in the TCR beta locus. J Immunol 2003;171:3605–11. https://doi.org/10.4049/jimmunol.171.7.3605.
[27] Drejer-Teel AH, Fugmann SD, Schatz DG. The beyond 12/23 restriction is imposed at the nicking and pairing steps of DNA cleavage during V(D)J recombination. Mol Cell Biol 2007;27:6288–99. https://doi.org/10.1128/MCB.00835-07.
[28] Banerjee JK, Schatz DG. Synapsis alters RAG-mediated nicking at Tcrb recombination signal sequences: implications for the "beyond 12/23" rule. Mol Cell Biol 2014;34:2566–80. https://doi.org/10.1128/MCB.00411-14.
[29] Fujimoto S, Yamagishi H. Isolation of an excision product of T-cell receptor alpha-chain gene rearrangements. Nature 1987;327:242–3. https://doi.org/10.1038/327242a0.
[30] Okazaki K, Davis DD, Sakano H. T cell receptor beta gene sequences in the circular DNA of thymocyte nuclei: direct evidence for intramolecular DNA deletion in V-D-J joining. Cell 1987;49:477–85. https://doi.org/10.1016/0092-8674(87)90450-8.
[31] Zachau HG. The immunoglobulin kappa locus-or-what has been learned from looking closely at one-tenth of a percent of the human genome. Gene 1993;135:167–73. https://doi.org/10.1016/0378-1119(93)90062-8.
[32] Ramsden DA, Baetz K, Wu GE. Conservation of sequence in recombination signal sequence spacers. Nucleic Acids Res 1994;22:1785–96.
[33] Hesse JE, Lieber MR, Mizuuchi K, Gellert M. V(D)J recombination: a functional definition of the joining signals. Genes Dev 1989;3:1053–61. https://doi.org/10.1101/gad.3.7.1053.
[34] Ramsden DA, McBlane JF, van Gent DC, Gellert M. Distinct DNA sequence and structure requirements for the two steps of V(D)J recombination signal cleavage. EMBO J 1996;15:3197–206.
[35] Cuomo CA, Mundy CL, Oettinger MA. DNA sequence and structure requirements for cleavage of V(D)J recombination signal sequences. Mol Cell Biol 1996;16:5683–90. https://doi.org/10.1128/MCB.16.10.5683.
[36] Hu J, Zhang Y, Zhao L, Frock RL, Du Z, Meyers RM, et al. Chromosomal loop domains direct the recombination of antigen receptor genes. Cell 2015;163:947–59. https://doi.org/10.1016/j.cell.2015.10.016.

[37] Papaemmanuil E, Rapado I, Li Y, Potter NE, Wedge DC, Tubio J, et al. RAG-mediated recombination is the predominant driver of oncogenic rearrangement in ETV6-RUNX1 acute lymphoblastic leukemia. Nat Genet 2014;46:116–25. https://doi.org/10.1038/ng.2874.

[38] Feeney AJ, Goebel P, Espinoza CR. Many levels of control of V gene rearrangement frequency. Immunol Rev 2004;200:44–56. https://doi.org/10.1111/j.0105-2896.2004.00163.x.

[39] Lee AI, Fugmann SD, Cowell LG, Ptaszek LM, Kelsoe G, Schatz DG. A functional analysis of the spacer of V(D)J recombination signal sequences. PLoS Biol 2003;1:E1. https://doi.org/10.1371/journal.pbio.0000001.

[40] Yu K, Lieber MR. Mechanistic basis for coding end sequence effects in the initiation of V(D)J recombination. Mol Cell Biol 1999;19:8094–102. https://doi.org/10.1128/MCB.19.12.8094.

[41] Gopalakrishnan S, Majumder K, Predeus A, Huang Y, Koues OI, Verma-Gaur J, et al. Unifying model for molecular determinants of the preselection Vβ repertoire. Proc Natl Acad Sci USA 2013;110:E3206–15. https://doi.org/10.1073/pnas.1304048110.

[42] Wu GS, Yang-Iott KS, Klink MA, Hayer KE, Lee KD, Bassing CH. Poor quality Vβ recombination signal sequences stochastically enforce TCRβ allelic exclusion. J Exp Med 2020;217. https://doi.org/10.1084/jem.20200412. e20200412.

[43] Cobb RM, Oestreich KJ, Osipovich OA, Oltz EM. Accessibility control of V(D)J recombination. Adv Immunol 2006;91:45–109. https://doi.org/10.1016/S0065-2776(06)91002-5.

[44] McBlane JF, van Gent DC, Ramsden DA, Romeo C, Cuomo CA, Gellert M, et al. Cleavage at a V(D)J recombination signal requires only RAG1 and RAG2 proteins and occurs in two steps. Cell 1995;83:387–95. https://doi.org/10.1016/0092-8674(95)90116-7.

[45] van Gent DC, McBlane JF, Ramsden DA, Sadofsky MJ, Hesse JE, Gellert M. Initiation of V(D)J recombination in a cell-free system. Cell 1995;81:925–34. https://doi.org/10.1016/0092-8674(95)90012-8.

[46] van Gent DC, Mizuuchi K, Gellert M. Similarities between initiation of V(D)J recombination and retroviral integration. Science 1996;271:1592–4. https://doi.org/10.1126/science.271.5255.1592.

[47] Mizuuchi K, Adzuma K. Inversion of the phosphate chirality at the target site of Mu DNA strand transfer: evidence for a one-step transesterification mechanism. Cell 1991;66:129–40. https://doi.org/10.1016/0092-8674(91)90145-O.

[48] Engelman A, Mizuuchi K, Craigie R. HIV-1 DNA integration: mechanism of viral DNA cleavage and DNA strand transfer. Cell 1991;67:1211–21. https://doi.org/10.1016/0092-8674(91)90297-c.

[49] Curry JD, Geier JK, Schlissel MS. Single-strand recombination signal sequence nicks in vivo: evidence for a capture model of synapsis. Nat Immunol 2005;6:1272–9. https://doi.org/10.1038/ni1270.

[50] Jones JM, Gellert M. Ordered assembly of the V(D)J synaptic complex ensures accurate recombination. EMBO J 2002;21:4162–71. https://doi.org/10.1093/emboj/cdf394.

[51] Mundy CL, Patenge N, Matthews AGW, Oettinger MA. Assembly of the RAG1/RAG2 synaptic complex. Mol Cell Biol 2002;22:69–77. https://doi.org/10.1128/MCB.22.1.69-77.2002.

[52] Swanson PC. A RAG-1/RAG-2 tetramer supports 12/23-regulated synapsis, cleavage, and transposition of V(D)J recombination signals. Mol Cell Biol 2002;22:7790–801. https://doi.org/10.1128/MCB.22.22.7790-7801.2002.

[53] Ru H, Chambers MG, Fu T-M, Tong AB, Liao M, Wu H. Molecular mechanism of V(D)J recombination from synaptic RAG1-RAG2 complex structures. Cell 2015;163:1138–52. https://doi.org/10.1016/j.cell.2015.10.055.

[54] Kim M-S, Chuenchor W, Chen X, Cui Y, Zhang X, Zhou ZH, et al. Cracking the DNA code for V(D)J recombination. Mol Cell 2018;70:358–70. https://doi.org/10.1016/j.molcel.2018.03.008.

[55] Eastman QM, Leu TM, Schatz DG. Initiation of V(D)J recombination in vitro obeying the 12/23 rule. Nature 1996;380:85–8. https://doi.org/10.1038/380085a0.

[56] Agrawal A, Schatz DG. RAG1 and RAG2 form a stable post-cleavage synaptic complex with DNA containing signal ends in V(D)J recombination. Cell 1997;89:43–53. https://doi.org/10.1016/s0092-8674(00)80181-6.

[57] Hiom K, Gellert M. Assembly of a 12/23 paired signal complex: a critical control point in V(D)J recombination. Mol Cell 1998;1:1011–9. https://doi.org/10.1016/s1097-2765(00)80101-x.

[58] Ma Y, Pannicke U, Schwarz K, Lieber MR. Hairpin opening and overhang processing by an Artemis/DNA-dependent protein kinase complex in nonhomologous end joining and V(D)J recombination. Cell 2002;108:781–94. https://doi.org/10.1016/s0092-8674(02)00671-2.

[59] Jones JM, Gellert M. Intermediates in V(D)J recombination: a stable RAG1/2 complex sequesters cleaved RSS ends. Proc Natl Acad Sci USA 2001;98:12926–31. https://doi.org/10.1073/pnas.221471198.

[60] Lieber MR. The mechanism of double-strand DNA break repair by the nonhomologous DNA end-joining pathway. Annu Rev Biochem 2010;79:181–211. https://doi.org/10.1146/annurev.biochem.052308.093131.

[61] Swann JB, Holland SJ, Petersen M, Pietsch TW, Boehm T. The immunogenetics of sexual parasitism. Science 2020;369:1608–15. https://doi.org/10.1126/science.aaz9445.

[62] Ichihara Y, Hirai M, Kurosawa Y. Sequence and chromosome assignment to 11p13-p12 of human RAG genes. Immunol Lett 1992;33:277–84. https://doi.org/10.1016/0165-2478(92)90073-w.

[63] Liang H-E, Hsu L-Y, Cado D, Cowell LG, Kelsoe G, Schlissel MS. The "dispensable" portion of RAG2 is necessary for efficient V-to-DJ rearrangement during B and T cell development. Immunity 2002;17:639–51. https://doi.org/10.1016/s1074-7613(02)00448-x.

[64] Akamatsu Y, Monroe R, Dudley DD, Elkin SK, Gärtner F, Talukder SR, et al. Deletion of the RAG2 C terminus leads to impaired lymphoid development in mice. Proc Natl Acad Sci USA 2003;100:1209–14. https://doi.org/10.1073/pnas.0237043100.

[65] Dudley DD, Sekiguchi J, Zhu C, Sadofsky MJ, Whitlow S, DeVido J, et al. Impaired V(D)J recombination and lymphocyte development in core RAG1-expressing mice. J Exp Med 2003;198:1439–50. https://doi.org/10.1084/jem.20030627.

[66] Talukder SR, Dudley DD, Alt FW, Takahama Y, Akamatsu Y. Increased frequency of aberrant V(D)J recombination products in core RAG-expressing mice. Nucleic Acids Res 2004;32:4539–49. https://doi.org/10.1093/nar/gkh778.

[67] Curry JD, Schlissel MS. RAG2's non-core domain contributes to the ordered regulation of V(D)J recombination. Nucleic Acids Res 2008;36:5750–62. https://doi.org/10.1093/nar/gkn553.

[68] Sadofsky MJ, Hesse JE, McBlane JF, Gellert M. Expression and V(D)J recombination activity of mutated RAG-1 proteins. Nucleic Acids Res 1993;21:5644–50.

[69] Silver DP, Spanopoulou E, Mulligan RC, Baltimore D. Dispensable sequence motifs in the RAG-1 and RAG-2 genes for plasmid V(D)J recombination. Proc Natl Acad Sci U S A 1993;90:6100–4. https://doi.org/10.1073/pnas.90.13.6100.

[70] Spanopoulou E, Zaitseva F, Wang FH, Santagata S, Baltimore D, Panayotou G. The homeodomain region of Rag-1 reveals the parallel mechanisms of bacterial and V(D)J recombination. Cell 1996;87:263–76. https://doi.org/10.1016/s0092-8674(00)81344-6.

[71] Yin FF, Bailey S, Innis CA, Ciubotaru M, Kamtekar S, Steitz TA, et al. Structure of the RAG1 nonamer-binding domain with DNA reveals a dimer that mediates DNA synapsis. Nat Struct Mol Biol 2009;16:499–508. https://doi.org/10.1038/nsmb.1593.

[72] Kim M-S, Lapkouski M, Yang W, Gellert M. Crystal structure of the V(D)J recombinase RAG1–RAG2. Nature 2015;518:507–11. https://doi.org/10.1038/nature14174.

[73] Kim DR, Dai Y, Mundy CL, Yang W, Oettinger MA. Mutations of acidic residues in RAG1 define the active site of the V(D)J recombinase. Genes Dev 1999;13:3070–80.

[74] Landree MA, Wibbenmeyer JA, Roth DB. Mutational analysis of RAG1 and RAG2 identifies three catalytic amino acids in RAG1 critical for both cleavage steps of V(D)J recombination. Genes Dev 1999;13:3059–69.

[75] Fugmann SD, Villey IJ, Ptaszek LM, Schatz DG. Identification of two catalytic residues in RAG1 that define a single active site within the RAG1/RAG2 protein complex. Mol Cell 2000;5:97–107. https://doi.org/10.1016/S1097-2765(00)80406-2.

[76] Bailin T, Mo X, Sadofsky MJ. A RAG1 and RAG2 tetramer complex is active in cleavage in V(D)J recombination. Mol Cell Biol 1999;19:4664–71.

[77] Swanson PC, Desiderio S. RAG-2 promotes heptamer occupancy by RAG-1 in the assembly of a V(D)J initiation complex. Mol Cell Biol 1999;19:3674–83. https://doi.org/10.1128/MCB.19.5.3674.

[78] Ciubotaru M, Ptaszek LM, Baker GA, Baker SN, Bright FV, Schatz DG. RAG1-DNA binding in V(D)J recombination. Specificity and DNA-induced conformational changes revealed by fluorescence and CD spectroscopy. J Biol Chem 2003;278:5584–96. https://doi.org/10.1074/jbc.M209758200.

[79] Rodgers KK, Villey IJ, Ptaszek L, Corbett E, Schatz DG, Coleman JE. A dimer of the lymphoid protein RAG1 recognizes the recombination signal sequence and the complex stably incorporates the high mobility group protein HMG2. Nucleic Acids Res 1999;27:2938–46. https://doi.org/10.1093/nar/27.14.2938.

[80] Rodgers KK, Bu Z, Fleming KG, Schatz DG, Engelman DM, Coleman JE. A zinc-binding domain involved in the dimerization of RAG1. J Mol Biol 1996;260:70–84. https://doi.org/10.1006/jmbi.1996.0382.

[81] Bellon SF, Rodgers KK, Schatz DG, Coleman JE, Steitz TA. Crystal structure of the RAG1 dimerization domain reveals multiple zinc-binding motifs including a novel zinc binuclear cluster. Nat Struct Mol Biol 1997;4:586–91. https://doi.org/10.1038/nsb0797-586.

[82] Jones JM, Gellert M. Autoubiquitylation of the V(D)J recombinase protein RAG1. Proc Natl Acad Sci USA 2003;100:15446–51. https://doi.org/10.1073/pnas.2637012100.

[83] Yurchenko V, Xue Z, Sadofsky M. The RAG1 N-terminal domain is an E3 ubiquitin ligase. Genes Dev 2003;17:581–5. https://doi.org/10.1101/gad.1058103.

[84] Grazini U, Zanardi F, Citterio E, Casola S, Goding CR, McBlane F. The RING domain of RAG1 ubiquitylates histone H3: a novel activity in chromatin-mediated regulation of V(D)J joining. Mol Cell 2010;37:282–93. https://doi.org/10.1016/j.molcel.2009.12.035.

[85] Jones JM, Bhattacharyya A, Simkus C, Vallieres B, Veenstra TD, Zhou M. The RAG1 V(D)J recombinase/ubiquitin ligase promotes ubiquitylation of acetylated, phosphorylated histone 3.3. Immunol Lett 2011;136:156–62. https://doi.org/10.1016/j.imlet.2011.01.005.

[86] Beilinson HA, Glynn RA, Yadavalli AD, Xiao J, Corbett E, Saribasak H, et al. The RAG1 N-terminal region regulates the efficiency and pathways of synapsis for V(D)J recombination. J Exp Med 2021;218. https://doi.org/10.1084/jem.20210250.

[87] Burn TN, Miot C, Gordon SM, Culberson EJ, Diamond T, Kreiger PA, et al. The RAG1 ubiquitin ligase domain stimulates recombination of TCRβ and TCRα genes and influences development of αβ T cell lineages. J Immunol 2022;209:938–49. https://doi.org/10.4049/jimmunol.2001441.

[88] Kassmeier MD, Mondal K, Palmer VL, Raval P, Kumar S, Perry GA, et al. VprBP binds full-length RAG1 and is required for B-cell development and V(D)J recombination fidelity. EMBO J 2012;31:945–58. https://doi.org/10.1038/emboj.2011.455.

[89] Schabla NM, Perry GA, Palmer VL, Swanson PC. VprBP (DCAF1) regulates RAG1 expression independently of dicer by mediating RAG1 degradation. J Immunol 2018;201:930–9. https://doi.org/10.4049/jimmunol.1800054.

[90] Brecht RM, Liu CC, Beilinson HA, Khitun A, Slavoff SA, Schatz DG. Nucleolar localization of RAG1 modulates V(D)J recombination activity. Proc Natl Acad Sci USA 2020;117:4300–9. https://doi.org/10.1073/pnas.1920021117.

[91] Cuomo CA, Oettinger MA. Analysis of regions of RAG-2 important for V(D)J recombination. Nucleic Acids Res 1994;22:1810–4. https://doi.org/10.1093/nar/22.10.1810.

[92] Sadofsky MJ, Hesse JE, Gellert M. Definition of a core region of RAG-2 that is functional in V(D)J recombination. Nucleic Acids Res 1994;22:1805–9.

[93] Gigi V, Lewis S, Shestova O, Mijušković M, Deriano L, Meng W, et al. RAG2 mutants alter DSB repair pathway choice in vivo and illuminate the nature of "alternative NHEJ.". Nucleic Acids Res 2014;42:6352–64. https://doi.org/10.1093/nar/gku295.

[94] Matthews AGW, Kuo AJ, Ramón-Maiques S, Han S, Champagne KS, Ivanov D, et al. RAG2 PHD finger couples histone H3 lysine 4 trimethylation with V(D)J recombination. Nature 2007;450:1106–10. https://doi.org/10.1038/nature06431.

[95] Liu Y, Subrahmanyam R, Chakraborty T, Sen R, Desiderio S. A plant homeodomain in rag-2 that binds hypermethylated lysine 4 of histone H3 is necessary for efficient antigen-receptor-gene rearrangement. Immunity 2007;27:561–71. https://doi.org/10.1016/j.immuni.2007.09.005.

[96] Ji Y, Resch W, Corbett E, Yamane A, Casellas R, Schatz DG. The in vivo pattern of binding of RAG1 and RAG2 to antigen receptor loci. Cell 2010;141:419–31. https://doi.org/10.1016/j.cell.2010.03.010.

[97] Teng G, Maman Y, Resch W, Kim M, Yamane A, Qian J, et al. RAG represents a widespread threat to the lymphocyte genome. Cell 2015;162:751–65. https://doi.org/10.1016/j.cell.2015.07.009.

[98] Maman Y, Teng G, Seth R, Kleinstein SH, Schatz DG. RAG1 targeting in the genome is dominated by chromatin interactions

mediated by the non-core regions of RAG1 and RAG2. Nucleic Acids Res 2016;44:9624–37. https://doi.org/10.1093/nar/gkw633.
[99] Shimazaki N, Tsai AG, Lieber MR. H3K4me3 stimulates the V(D)J RAG complex for both nicking and hairpinning in trans in addition to tethering in cis: implications for translocations. Mol Cell 2009;34:535–44. https://doi.org/10.1016/j.molcel.2009.05.011.
[100] Grundy GJ, Yang W, Gellert M. Autoinhibition of DNA cleavage mediated by RAG1 and RAG2 is overcome by an epigenetic signal in V(D)J recombination. Proc Natl Acad Sci U S A 2010;107:22487–92. https://doi.org/10.1073/pnas.1014958107.
[101] Lu C, Ward A, Bettridge J, Liu Y, Desiderio S. An autoregulatory mechanism imposes allosteric control on the V(D)J recombinase by histone H3 methylation. Cell Rep 2015;10:29–38. https://doi.org/10.1016/j.celrep.2014.12.001.
[102] Bettridge J, Na CH, Pandey A, Desiderio S. H3K4me3 induces allosteric conformational changes in the DNA-binding and catalytic regions of the V(D)J recombinase. Proc Natl Acad Sci USA 2017;114:1904–9. https://doi.org/10.1073/pnas.1615727114.
[103] Lin WC, Desiderio S. Cell cycle regulation of V(D)J recombination-activating protein RAG-2. Proc Natl Acad Sci U S A 1994;91:2733–7. https://doi.org/10.1073/pnas.91.7.2733.
[104] Lee J, Desiderio S. Cyclin A/CDK2 regulates V(D)J recombination by coordinating RAG-2 accumulation and DNA repair. Immunity 1999;11:771–81. https://doi.org/10.1016/s1074-7613(00)80151-x.
[105] Jiang H, Chang F-C, Ross AE, Lee J, Nakayama K, Nakayama K, et al. Ubiquitylation of RAG-2 by Skp2-SCF links destruction of the V(D)J recombinase to the cell cycle. Mol Cell 2005;18:699–709. https://doi.org/10.1016/j.molcel.2005.05.011.
[106] Mizuta R, Mizuta M, Araki S, Kitamura D. RAG2 is down-regulated by cytoplasmic sequestration and ubiquitin-dependent degradation. J Biol Chem 2002;277:41423–7. https://doi.org/10.1074/jbc.M206605200.
[107] Ross AE, Vuica M, Desiderio S. Overlapping signals for protein degradation and nuclear localization define a role for intrinsic RAG-2 nuclear uptake in dividing cells. Mol Cell Biol 2003;23:5308–19. https://doi.org/10.1128/MCB.23.15.5308-5319.2003.
[108] Chaumeil J, Micsinai M, Ntziachristos P, Roth DB, Aifantis I, Kluger Y, et al. The RAG2 C-terminus and ATM protect genome integrity by controlling antigen receptor gene cleavage. Nat Commun 2013;4:2231. https://doi.org/10.1038/ncomms3231.
[109] Deriano L, Chaumeil J, Coussens M, Multani A, Chou Y, Alekseyenko AV, et al. The RAG2 C terminus suppresses genomic instability and lymphomagenesis. Nature 2011;471:119–23. https://doi.org/10.1038/nature09755.
[110] Zhang L, Reynolds TL, Shan X, Desiderio S. Coupling of V(D)J recombination to the cell cycle suppresses genomic instability and lymphoid tumorigenesis. Immunity 2011;34:163–74. https://doi.org/10.1016/j.immuni.2011.02.003.
[111] Leu TM, Schatz DG. rag-1 and rag-2 are components of a high-molecular-weight complex, and association of rag-2 with this complex is rag-1 dependent. Mol Cell Biol 1995;15:5657–70.
[112] McMahan CJ, Sadofsky MJ, Schatz DG. Definition of a large region of RAG1 that is important for coimmunoprecipitation of RAG2. J Immunol 1997;158:2202–10.
[113] Arbuckle JL, Fauss LA, Simpson R, Ptaszek LM, Rodgers KK. Identification of two topologically independent domains in RAG1 and their role in macromolecular interactions relevant to V(D)J recombination. J Biol Chem 2001;276:37093–101. https://doi.org/10.1074/jbc.M105988200.
[114] Aidinis V, Dias DC, Gomez CA, Bhattacharyya D, Spanopoulou E, Santagata S. Definition of minimal domains of interaction within the recombination-activating genes 1 and 2 recombinase complex. J Immunol 2000;164:5826–32. https://doi.org/10.4049/jimmunol.164.11.5826.
[115] Thomas JO, Travers AA. HMG1 and 2, and related "architectural" DNA-binding proteins. Trends Biochem Sci 2001;26:167–74. https://doi.org/10.1016/s0968-0004(01)01801-1.
[116] Sawchuk DJ, Weis-Garcia F, Malik S, Besmer E, Bustin M, Nussenzweig MC, et al. V(D)J recombination: modulation of RAG1 and RAG2 cleavage activity on 12/23 substrates by whole cell extract and DNA-bending proteins. J Exp Med 1997;185:2025–32. https://doi.org/10.1084/jem.185.11.2025.
[117] Hiom K, Gellert M. A stable RAG1–RAG2–DNA complex that is active in V(D)J cleavage. Cell 1997;88:65–72. https://doi.org/10.1016/S0092-8674(00)81859-0.
[118] Swanson PC. Fine structure and activity of discrete RAG-HMG complexes on V(D)J recombination signals. Mol Cell Biol 2002;22:1340–51. https://doi.org/10.1128/MCB.22.5.1340-1351.2002.
[119] Aidinis V, Bonaldi T, Beltrame M, Santagata S, Bianchi ME, Spanopoulou E. The RAG1 homeodomain recruits HMG1 and HMG2 to facilitate recombination signal sequence binding and to enhance the intrinsic DNA-bending activity of RAG1-RAG2. Mol Cell Biol 1999;19:6532–42. https://doi.org/10.1128/MCB.19.10.6532.
[120] Stros M. HMGB proteins: interactions with DNA and chromatin. Biochim Biophys Acta 2010;1799:101–13. https://doi.org/10.1016/j.bbagrm.2009.09.008.
[121] Klass J, Murphy FV, Fouts S, Serenil M, Changela A, Siple J, et al. The role of intercalating residues in chromosomal high-mobility-group protein DNA binding, bending and specificity. Nucleic Acids Res 2003;31:2852–64.
[122] Saito K, Kikuchi T, Yoshida M. The mechanism of sequence non-specific DNA binding of HMG1/2-box B in HMG1 with DNA. Protein Eng 1999;12:235–42. https://doi.org/10.1093/protein/12.3.235.
[123] Bergeron S, Madathiparambil T, Swanson PC. Both high mobility group (HMG)-boxes and the acidic tail of HMGB1 regulate recombination-activating gene (RAG)-mediated recombination signal synapsis and cleavage in vitro*[boxs]. J Biol Chem 2005;280:31314–24. https://doi.org/10.1074/jbc.M503063200.
[124] Dai Y, Wong B, Yen Y-M, Oettinger MA, Kwon J, Johnson RC. Determinants of HMGB proteins required to promote RAG1/2-recombination signal sequence complex assembly and catalysis during V(D)J recombination. Mol Cell Biol 2005;25:4413–25. https://doi.org/10.1128/MCB.25.11.4413-4425.2005.
[125] Stott K, Watson M, Howe FS, Grossmann JG, Thomas JO. Tail-mediated collapse of HMGB1 is dynamic and occurs via differential binding of the acidic tail to the A and B domains. J Mol Biol 2010;403:706–22. https://doi.org/10.1016/j.jmb.2010.07.045.
[126] Little AJ, Corbett E, Ortega F, Schatz DG. Cooperative recruitment of HMGB1 during V(D)J recombination through interactions with RAG1 and DNA. Nucleic Acids Res 2013;41:3289–301. https://doi.org/10.1093/nar/gks1461.
[127] Calogero S, Grassi F, Aguzzi A, Voigtländer T, Ferrier P, Ferrari S, et al. The lack of chromosomal protein Hmg1 does not disrupt cell growth but causes lethal hypoglycaemia in newborn mice. Nat Genet 1999;22:276–80. https://doi.org/10.1038/10338.

[128] Ronfani L, Ferraguti M, Croci L, Ovitt CE, Schöler HR, Consalez GG, et al. Reduced fertility and spermatogenesis defects in mice lacking chromosomal protein Hmgb2. Development 2001;128:1265–73. https://doi.org/10.1242/dev.128.8.1265.
[129] Bianchi ME, Agresti A. HMG proteins: dynamic players in gene regulation and differentiation. Curr Opin Genet Dev 2005;15:496–506. https://doi.org/10.1016/j.gde.2005.08.007.
[130] Thwaites DT, Carter C, Lawless D, Savic S, Boyes JM. A novel RAG1 mutation reveals a critical in vivo role for HMGB1/2 during V(D)J recombination. Blood 2019;133:820–9. https://doi.org/10.1182/blood-2018-07-866939.
[131] Little AJ, Matthews A, Oettinger M, Roth DB, Schatz DG. Chapter 2 - the mechanism of V(D)J recombination. In: Alt FW, Honjo T, Radbruch A, Reth M, editors. Molecular Biology of B cells. 2nd ed. London: Academic Press; 2015. p. 13–34. https://doi.org/10.1016/B978-0-12-397933-9.00002-3.
[132] Swanson PC. The DDE motif in RAG-1 is contributed in trans to a single active site that catalyzes the nicking and transesterification steps of V(D)J recombination. Mol Cell Biol 2001;21:449–58. https://doi.org/10.1128/MCB.21.2.449-458.2001.
[133] Swanson PC, Desiderio S. V(D)J recombination signal recognition: distinct, overlapping DNA-protein contacts in complexes containing RAG1 with and without RAG2. Immunity 1998;9:115–25. https://doi.org/10.1016/s1074-7613(00)80593-2.
[134] Akamatsu Y, Oettinger MA. Distinct roles of RAG1 and RAG2 in binding the V(D)J recombination signal sequences. Mol Cell Biol 1998;18:4670–8.
[135] Nagawa F, Ishiguro K, Tsuboi A, Yoshida T, Ishikawa A, Takemori T, et al. Footprint analysis of the RAG protein recombination signal sequence complex for V(D)J type recombination. Mol Cell Biol 1998;18:655–63.
[136] Difilippantonio MJ, McMahan CJ, Eastman QM, Spanopoulou E, Schatz DG. RAG1 mediates signal sequence recognition and recruitment of RAG2 in V(D)J recombination. Cell 1996;87:253–62. https://doi.org/10.1016/S0092-8674(00)81343-4.
[137] Eastman QM, Villey IJ, Schatz DG. Detection of RAG protein-V(D)J recombination signal interactions near the site of DNA cleavage by UV cross-linking. Mol Cell Biol 1999;19:3788–97. https://doi.org/10.1128/MCB.19.5.3788.
[138] Grundy GJ, Hesse JE, Gellert M. Requirements for DNA hairpin formation by RAG1/2. Proc Natl Acad Sci USA 2007;104:3078–83. https://doi.org/10.1073/pnas.0611293104.
[139] Bischerour J, Lu C, Roth DB, Chalmers R. Base flipping in V(D)J recombination: insights into the mechanism of hairpin formation, the 12/23 rule, and the coordination of double-strand breaks. Mol Cell Biol 2009;29:5889–99. https://doi.org/10.1128/MCB.00187-09.
[140] Ciubotaru M, Kriatchko AN, Swanson PC, Bright FV, Schatz DG. Fluorescence resonance energy transfer analysis of recombination signal sequence configuration in the RAG1/2 synaptic complex. Mol Cell Biol 2007;27:4745–58. https://doi.org/10.1128/MCB.00177-07.
[141] Ciubotaru M, Trexler AJ, Spiridon LN, Surleac MD, Rhoades E, Petrescu AJ, et al. RAG and HMGB1 create a large bend in the 23RSS in the V(D)J recombination synaptic complexes. Nucleic Acids Res 2013;41:2437–54. https://doi.org/10.1093/nar/gks1294.
[142] Ciubotaru M, Surleac MD, Metskas LA, Koo P, Rhoades E, Petrescu AJ, et al. The architecture of the 12RSS in V(D)J recombination signal and synaptic complexes. Nucleic Acids Res 2015;43:917–31. https://doi.org/10.1093/nar/gku1348.
[143] Ru H, Mi W, Zhang P, Alt FW, Schatz DG, Liao M, et al. DNA melting initiates the RAG catalytic pathway. Nat Struct Mol Biol 2018;25:732–42. https://doi.org/10.1038/s41594-018-0098-5.
[144] Chen X, Cui Y, Best RB, Wang H, Zhou ZH, Yang W, et al. Cutting antiparallel DNA strands in a single active site. Nat Struct Mol Biol 2020;27:119–26. https://doi.org/10.1038/s41594-019-0363-2.
[145] Chen X, Cui Y, Wang H, Zhou ZH, Gellert M, Yang W. How mouse RAG recombinase avoids DNA transposition. Nat Struct Mol Biol 2020;27:127–33. https://doi.org/10.1038/s41594-019-0366-z.
[146] Zhang Y, Corbett E, Wu S, Schatz DG. Structural basis for the activation and suppression of transposition during evolution of the RAG recombinase. EMBO J 2020;39:e105857. https://doi.org/10.15252/embj.2020105857.
[147] Montaño SP, Rice PA. Moving DNA around: DNA transposition and retroviral integration. Curr Opin Struct Biol 2011;21:370–8. https://doi.org/10.1016/j.sbi.2011.03.004.
[148] Callebaut I, Mornon JP. The V(D)J recombination activating protein RAG2 consists of a six-bladed propeller and a PHD fingerlike domain, as revealed by sequence analysis. Cell Mol Life Sci 1998;54:880–91. https://doi.org/10.1007/s000180050216.
[149] Aravind L, Koonin EV. Gleaning non-trivial structural, functional and evolutionary information about proteins by iterative database searches. J Mol Biol 1999;287:1023–40. https://doi.org/10.1006/jmbi.1999.2653.
[150] Lapkouski M, Chuenchor W, Kim M-S, Gellert M, Yang W. Assembly pathway and characterization of the RAG1/2-DNA paired and signal-end complexes. J Biol Chem 2015;290:14618–25. https://doi.org/10.1074/jbc.M115.641787.
[151] Hoolehan W, Harris JC, Byrum JN, Simpson DA, Rodgers KK. An updated definition of V(D)J recombination signal sequences revealed by high-throughput recombination assays. Nucleic Acids Res 2022;50:11696–711. https://doi.org/10.1093/nar/gkac1038.
[152] Alt FW, Baltimore D. Joining of immunoglobulin heavy chain gene segments: implications from a chromosome with evidence of three D-JH fusions. Proc Natl Acad Sci U S A 1982;79:4118–22.
[153] Gilfillan S, Dierich A, Lemeur M, Benoist C, Mathis D. Mice lacking TdT: mature animals with an immature lymphocyte repertoire. Science 1993;261:1175–8. https://doi.org/10.1126/science.8356452.
[154] Komori T, Okada A, Stewart V, Alt FW. Lack of N regions in antigen receptor variable region genes of TdT-deficient lymphocytes. Science 1993;261:1171–5. https://doi.org/10.1126/science.8356451.
[155] Teng G, Schatz DG. Regulation and evolution of the RAG recombinase. Adv Immunol 2015;128:1–39. https://doi.org/10.1016/bs.ai.2015.07.002.
[156] Melek M, Gellert M, van Gent DC. Rejoining of DNA by the RAG1 and RAG2 proteins. Science 1998;280:301–3. https://doi.org/10.1126/science.280.5361.301.
[157] Corneo B, Wendland RL, Deriano L, Cui X, Klein IA, Wong S-Y, et al. Rag mutations reveal robust alternative end joining. Nature 2007;449:483–6. https://doi.org/10.1038/nature06168.
[158] Bosma M, Schuler W, Bosma G. The SCID mouse mutant. Curr Top Microbiol Immunol 1988;137:197–202. https://doi.org/10.1007/978-3-642-50059-6_29.
[159] Blackwell TK, Ferrier P, Malynn BA, Pollock RR, Covey LR, Suh HY, et al. The effect of the SCID mutation on mechanism and control of immunoglobulin heavy and light chain gene rearrangement. Curr Top Microbiol Immunol 1989;152:85–94. https://doi.org/10.1007/978-3-642-74974-2_12.

[160] Lieber MR, Hesse JE, Lewis S, Bosma GC, Rosenberg N, Mizuuchi K, et al. The defect in murine severe combined immune deficiency: joining of signal sequences but not coding segments in V(D)J recombination. Cell 1988;55:7–16. https://doi.org/10.1016/0092-8674(88)90004-9.

[161] Malynn BA, Blackwell TK, Fulop GM, Rathbun GA, Furley AJ, Ferrier P, et al. The SCID defect affects the final step of the immunoglobulin VDJ recombinase mechanism. Cell 1988;54:453–60. https://doi.org/10.1016/0092-8674(88)90066-9.

[162] Fulop GM, Phillips RA. The SCID mutation in mice causes a general defect in DNA repair. Nature 1990;347:479–82. https://doi.org/10.1038/347479a0.

[163] Taccioli GE, Rathbun G, Shinkai Y, Oltz EM, Cheng H, Whitmore G, et al. Activities involved in V(D)J recombination. Curr Top Microbiol Immunol 1992;182:107–14. https://doi.org/10.1007/978-3-642-77633-5_13.

[164] Taccioli GE, Rathbun G, Oltz E, Stamato T, Jeggo PA, Alt FW. Impairment of V(D)J recombination in double-strand break repair mutants. Science 1993;260:207–10. https://doi.org/10.1126/science.8469973.

[165] Taccioli GE, Cheng HL, Varghese AJ, Whitmore G, Alt FW. A DNA repair defect in Chinese hamster ovary cells affects V(D)J recombination similarly to the murine scid mutation. J Biol Chem 1994;269:7439–42.

[166] Giaccia AJ, Denko N, MacLaren R, Mirman D, Waldren C, Hart I, et al. Human chromosome 5 complements the DNA double-strand break-repair deficiency and gamma-ray sensitivity of the XR-1 hamster variant. Am J Hum Genet 1990;47:459–69.

[167] Blunt T, Finnie NJ, Taccioli GE, Smith GC, Demengeot J, Gottlieb TM, et al. Defective DNA-dependent protein kinase activity is linked to V(D)J recombination and DNA repair defects associated with the murine scid mutation. Cell 1995;80:813–23. https://doi.org/10.1016/0092-8674(95)90360-7.

[168] Jeggo PA, Hafezparast M, Thompson AF, Broughton BC, Kaur GP, Zdzienicka MZ, et al. Localization of a DNA repair gene (XRCC5) involved in double-strand-break rejoining to human chromosome 2. Proc Natl Acad Sci U S A 1992;89:6423–7.

[169] Hafezparast M, Kaur GP, Zdzienicka M, Athwal RS, Lehmann AR, Jeggo PA. Subchromosomal localization of a gene (XRCC5) involved in double strand break repair to the region 2q34-36. Somat Cell Mol Genet 1993;19:413–21. https://doi.org/10.1007/BF01233246.

[170] Li Z, Otevrel T, Gao Y, Cheng HL, Seed B, Stamato TD, et al. The XRCC4 gene encodes a novel protein involved in DNA double-strand break repair and V(D)J recombination. Cell 1995;83:1079–89. https://doi.org/10.1016/0092-8674(95)90135-3.

[171] Smider V, Rathmell WK, Lieber MR, Chu G. Restoration of X-ray resistance and V(D)J recombination in mutant cells by Ku cDNA. Science 1994;266:288–91. https://doi.org/10.1126/science.7939667.

[172] Taccioli GE, Gottlieb TM, Blunt T, Priestley A, Demengeot J, Mizuta R, et al. Ku80: product of the XRCC5 gene and its role in DNA repair and V(D)J recombination. Science 1994;265:1442–5. https://doi.org/10.1126/science.8073286.

[173] Blunt T, Gell D, Fox M, Taccioli GE, Lehmann AR, Jackson SP, et al. Identification of a nonsense mutation in the carboxyl-terminal region of DNA-dependent protein kinase catalytic subunit in the scid mouse. Proc Natl Acad Sci U S A 1996;93:10285–90. https://doi.org/10.1073/pnas.93.19.10285.

[174] Wang XS, Lee BJ, Zha S. The recent advances in non-homologous end-joining through the lens of lymphocyte development. DNA Repair 2020;94:102874. https://doi.org/10.1016/j.dnarep.2020.102874.

[175] Woodbine L, Gennery AR, Jeggo PA. The clinical impact of deficiency in DNA non-homologous end-joining. DNA Repair 2014;16:84–96. https://doi.org/10.1016/j.dnarep.2014.02.011.

[176] Ochi T, Wu Q, Blundell TL. The spatial organization of non-homologous end joining: from bridging to end joining. DNA Repair 2014;17:98–109. https://doi.org/10.1016/j.dnarep.2014.02.010.

[177] Rooney S, Chaudhuri J, Alt FW. The role of the non-homologous end-joining pathway in lymphocyte development. Immunol Rev 2004;200:115–31. https://doi.org/10.1111/j.0105-2896.2004.00165.x.

[178] Desiderio S, Lin W-C, Li Z. The cell cycle and V(D)J recombination. In: Jessberger R, Lieber MR, editors. Molecular analysis of DNA rearrangements in the immune system. Berlin, Heidelberg: Springer; 1996. p. 45–59. https://doi.org/10.1007/978-3-642-50140-1_4.

[179] Bredemeyer AL, Sharma GG, Huang C-Y, Helmink BA, Walker LM, Khor KC, et al. ATM stabilizes DNA double-strand-break complexes during V(D)J recombination. Nature 2006;442:466–70. https://doi.org/10.1038/nature04866.

[180] Lin WC, Desiderio S. Regulation of V(D)J recombination activator protein RAG-2 by phosphorylation. Science 1993;260:953–9. https://doi.org/10.1126/science.8493533.

[181] Li Z, Dordai DI, Lee J, Desiderio S. A conserved degradation signal regulates RAG-2 accumulation during cell division and links V(D)J recombination to the cell cycle. Immunity 1996;5:575–89. https://doi.org/10.1016/s1074-7613(00)80272-1.

[182] Alt FW, Zhang Y, Meng F-L, Guo C, Schwer B. Mechanisms of programmed DNA lesions and genomic instability in the immune system. Cell 2013;152:417–29. https://doi.org/10.1016/j.cell.2013.01.007.

[183] Chen C-C, Feng W, Lim PX, Kass EM, Jasin M. Homology-directed repair and the role of BRCA1, BRCA2, and related proteins in genome integrity and cancer. Annu Rev Cell Biol 2018;2:313–36. https://doi.org/10.1146/annurev-cancerbio-030617-050502.

[184] Mills KD, Ferguson DO, Essers J, Eckersdorff M, Kanaar R, Alt FW. Rad54 and DNA Ligase IV cooperate to maintain mammalian chromatid stability. Genes Dev 2004;18:1283–92. https://doi.org/10.1101/gad.1204304.

[185] Mills KD, Ferguson DO, Alt FW. The role of DNA breaks in genomic instability and tumorigenesis. Immunol Rev 2003;194:77–95. https://doi.org/10.1034/j.1600-065x.2003.00060.x.

[186] Zhao B, Watanabe G, Lieber MR. Polymerase μ in non-homologous DNA end joining: importance of the order of arrival at a double-strand break in a purified system. Nucleic Acids Res 2020;48:3605–18. https://doi.org/10.1093/nar/gkaa094.

[187] Kaminski AM, Chiruvella KK, Ramsden DA, Bebenek K, Kunkel TA, Pedersen LC. Analysis of diverse double-strand break synapsis with Polλ reveals basis for unique substrate specificity in nonhomologous end-joining. Nat Commun 2022;13:3806. https://doi.org/10.1038/s41467-022-31278-4.

[188] Chandramouly G, Jamsen J, Borisonnik N, Tyagi M, Calbert ML, Tredinnick T, et al. Polλ promotes microhomology-mediated end-joining. Nat Struct Mol Biol 2023;30:107–14. https://doi.org/10.1038/s41594-022-00895-4.

[189] Sfeir A, Symington LS. Microhomology-mediated end joining: a back-up survival mechanism or dedicated pathway? Trends Biochem Sci 2015;40:701–14. https://doi.org/10.1016/j.tibs.2015.08.006.

[190] McVey M, Lee SE. MMEJ repair of double-strand breaks (director's cut): deleted sequences and alternative endings. Trends Genet 2008;24:529–38. https://doi.org/10.1016/j.tig.2008.08.007.
[191] Kim DR, Park SJ, Oettinger MA. V(D)J recombination: site-specific cleavage and repair. Mol Cell 2000;10:367–74.
[192] Brandt VL, Roth DB. V(D)J recombination: how to tame a transposase. Immunol Rev 2004;200:249–60. https://doi.org/10.1111/j.0105-2896.2004.00161.x.
[193] Lee GS, Neiditch MB, Salus SS, Roth DB. RAG proteins shepherd double-strand breaks to a specific pathway, suppressing error-prone repair, but RAG nicking initiates homologous recombination. Cell 2004;117:171–84. https://doi.org/10.1016/s0092-8674(04)00301-0.
[194] Boboila C, Oksenych V, Gostissa M, Wang JH, Zha S, Zhang Y, et al. Robust chromosomal DNA repair via alternative end-joining in the absence of X-ray repair cross-complementing protein 1 (XRCC1). Proc Natl Acad Sci U S A 2012;109:2473–8. https://doi.org/10.1073/pnas.1121470109.
[195] Boboila C, Jankovic M, Yan CT, Wang JH, Wesemann DR, Zhang T, et al. Alternative end-joining catalyzes robust IgH locus deletions and translocations in the combined absence of ligase 4 and Ku70. Proc Natl Acad Sci U S A 2010;107:3034–9. https://doi.org/10.1073/pnas.0915067107.
[196] Zha S, Boboila C, Alt FW. Mre11: roles in DNA repair beyond homologous recombination. Nat Struct Mol Biol 2009;16:798–800. https://doi.org/10.1038/nsmb0809-798.
[197] Lieber MR, Raghavan SC, Yu K. Mechanistic aspects of lymphoid chromosomal translocations. J Natl Cancer Inst Monogr 2008:8–11. https://doi.org/10.1093/jncimonographs/lgn012.
[198] Lieber MR. The mechanism of human nonhomologous DNA end joining. J Biol Chem 2008;283:1–5. https://doi.org/10.1074/jbc.R700039200.
[199] Yan CT, Boboila C, Souza EK, Franco S, Hickernell TR, Murphy M, et al. IgH class switching and translocations use a robust non-classical end-joining pathway. Nature 2007;449:478–82. https://doi.org/10.1038/nature06020.
[200] Liang Z, Kumar V, Le Bouteiller M, Zurita J, Kenrick J, Lin SG, et al. Ku70 suppresses alternative end joining in G1-arrested progenitor B cells. Proc Natl Acad Sci USA 2021;118:e2103630118. https://doi.org/10.1073/pnas.2103630118.
[201] Kumar V, Alt FW, Frock RL. PAXX and XLF DNA repair factors are functionally redundant in joining DNA breaks in a G1-arrested progenitor B-cell line. Proc Natl Acad Sci U S A 2016;113:10619–24. https://doi.org/10.1073/pnas.1611882113.
[202] Weaver DT, Alt FW. From RAGs to stitches. Nature 1997;388:428–9. https://doi.org/10.1038/41225.
[203] Grawunder U, Zimmer D, Kulesza P, Lieber MR. Requirement for an interaction of XRCC4 with DNA ligase IV for wild-type V(D)J recombination and DNA double-strand break repair in vivo. J Biol Chem 1998;273:24708–14. https://doi.org/10.1074/jbc.273.38.24708.
[204] Bryans M, Valenzano MC, Stamato TD. Absence of DNA ligase IV protein in XR-1 cells: evidence for stabilization by XRCC4. Mutat Res 1999;433:53–8. https://doi.org/10.1016/s0921-8777(98)00063-9.
[205] Teo SH, Jackson SP. Lif1p targets the DNA ligase Lig4p to sites of DNA double-strand breaks. Curr Biol 2000;10:165–8. https://doi.org/10.1016/s0960-9822(00)00317-1.
[206] Herrmann G, Lindahl T, Schär P. *Saccharomyces cerevisiae* LIF1: a function involved in DNA double-strand break repair related to mammalian XRCC4. EMBO J 1998;17:4188–98. https://doi.org/10.1093/emboj/17.14.4188.
[207] Li Z, Alt FW. Identification of the XRCC4 gene: complementation of the DSBR and V(D)J recombination defects of XR-1 cells. Curr Top Microbiol Immunol 1996;217:143–50. https://doi.org/10.1007/978-3-642-50140-1_10.
[208] Gu Y, Seidl KJ, Rathbun GA, Zhu C, Manis JP, van der Stoep N, et al. Growth retardation and leaky SCID phenotype of Ku70-deficient mice. Immunity 1997;7:653–65. https://doi.org/10.1016/s1074-7613(00)80386-6.
[209] Frank KM, Sekiguchi JM, Seidl KJ, Swat W, Rathbun GA, Cheng HL, et al. Late embryonic lethality and impaired V(D)J recombination in mice lacking DNA ligase IV. Nature 1998;396:173–7. https://doi.org/10.1038/24172.
[210] Gao Y, Sun Y, Frank KM, Dikkes P, Fujiwara Y, Seidl KJ, et al. A critical role for DNA end-joining proteins in both lymphogenesis and neurogenesis. Cell 1998;95:891–902. https://doi.org/10.1016/s0092-8674(00)81714-6.
[211] Nussenzweig A, Chen C, da Costa Soares V, Sanchez M, Sokol K, Nussenzweig MC, et al. Requirement for Ku80 in growth and immunoglobulin V(D)J recombination. Nature 1996;382:551–5. https://doi.org/10.1038/382551a0.
[212] Zhu C, Bogue MA, Lim D-S, Hasty P, Roth DB. Ku86-Deficient mice exhibit severe combined immunodeficiency and defective processing of V(D)J recombination intermediates. Cell 1996;86:379–89. https://doi.org/10.1016/S0092-8674(00)80111-7.
[213] Smith GC, Jackson SP. The DNA-dependent protein kinase. Genes Dev 1999;13:916–34. https://doi.org/10.1101/gad.13.8.916.
[214] Moshous D, Callebaut I, de Chasseval R, Corneo B, Cavazzana-Calvo M, Le Deist F, et al. Artemis, a novel DNA double-strand break repair/V(D)J recombination protein, is mutated in human severe combined immune deficiency. Cell 2001;105:177–86. https://doi.org/10.1016/s0092-8674(01)00309-9.
[215] Rooney S, Sekiguchi J, Zhu C, Cheng HL, Manis J, Whitlow S, et al. Leaky Scid phenotype associated with defective V(D)J coding end processing in Artemis-deficient mice. Mol Cell 2002;10:1379–90. https://doi.org/10.1016/s1097-2765(02)00755-4.
[216] Gao Y, Chaudhuri J, Zhu C, Davidson L, Weaver DT, Alt FW. A targeted DNA-PKcs-null mutation reveals DNA-PK-independent functions for KU in V(D)J recombination. Immunity 1998;9:367–76. https://doi.org/10.1016/s1074-7613(00)80619-6.
[217] Taccioli GE, Amatucci AG, Beamish HJ, Gell D, Xiang XH, Torres Arzayus MI, et al. Targeted disruption of the catalytic subunit of the DNA-PK gene in mice confers severe combined immunodeficiency and radiosensitivity. Immunity 1998;9:355–66. https://doi.org/10.1016/s1074-7613(00)80618-4.
[218] Kurimasa A, Ouyang H, Dong L, Wang S, Li X, Cordon-Cardo C, et al. Catalytic subunit of DNA-dependent protein kinase: impact on lymphocyte development and tumorigenesis. Proc Natl Acad Sci USA 1999;96:1403–8. https://doi.org/10.1073/pnas.96.4.1403.
[219] Jiang W, Crowe JL, Liu X, Nakajima S, Wang Y, Li C, et al. Differential phosphorylation of DNA-PKcs regulates the interplay between end-processing and end-ligation during nonhomologous end-joining. Mol Cell 2015;58:172–85. https://doi.org/10.1016/j.molcel.2015.02.024.
[220] Moshous D, Pannetier C, de Chasseval R, le Deist F, Cavazzana-Calvo M, Romana S, et al. Partial T and B lymphocyte

immunodeficiency and predisposition to lymphoma in patients with hypomorphic mutations in Artemis. J Clin Invest 2003;111:381–7. https://doi.org/10.1172/JCI16774.

[221] Lee B-S, Gapud EJ, Zhang S, Dorsett Y, Bredemeyer A, George R, et al. Functional intersection of ATM and DNA-dependent protein kinase catalytic subunit in coding end joining during V(D)J recombination. Mol Cell Biol 2013;33:3568–79. https://doi.org/10.1128/MCB.00308-13.

[222] Crowe JL, Wang XS, Shao Z, Lee BJ, Estes VM, Zha S. DNA-PKcs phosphorylation at the T2609 cluster alters the repair pathway choice during immunoglobulin class switch recombination. Proc Natl Acad Sci USA 2020;117:22953–61. https://doi.org/10.1073/pnas.2007455117.

[223] Jiang W, Estes VM, Wang XS, Shao Z, Lee BJ, Lin X, et al. Phosphorylation at S2053 in murine (S2056 in human) DNA-PKcs is dispensable for lymphocyte development and class switch recombination. J Immunol 2019;203:178–87. https://doi.org/10.4049/jimmunol.1801657.

[224] Liu L, Chen X, Li J, Wang H, Buehl CJ, Goff NJ, et al. Autophosphorylation transforms DNA-PK from protecting to processing DNA ends. Mol Cell 2022;82:177–89. https://doi.org/10.1016/j.molcel.2021.11.025.

[225] Chen X, Xu X, Chen Y, Cheung JC, Wang H, Jiang J, et al. Structure of an activated DNA-PK and its implications for NHEJ. Mol Cell 2021;81:801–10. https://doi.org/10.1016/j.molcel.2020.12.015.

[226] Riballo E, Kühne M, Rief N, Doherty A, Smith GCM, Recio M-J, et al. A pathway of double-strand break rejoining dependent upon ATM, Artemis, and proteins locating to gamma-H2AX foci. Mol Cell 2004;16:715–24. https://doi.org/10.1016/j.molcel.2004.10.029.

[227] Jackson K, Kidd M, Wang Y, Collins A. The shape of the lymphocyte receptor repertoire: lessons from the B cell receptor. Front Immunol 2013;4.

[228] Imkeller K, Wardemann H. Assessing human B cell repertoire diversity and convergence. Immunol Rev 2018;284:51–66. https://doi.org/10.1111/imr.12670.

[229] Oltz EM. Regulation of antigen receptor gene assembly in lymphocytes. Immunol Res 2001;23:121–33. https://doi.org/10.1385/IR:23:2–3:121.

[230] Jung D, Giallourakis C, Mostoslavsky R, Alt FW. Mechanism and control of V(D)J recombination at the immunoglobulin heavy chain locus. Annu Rev Immunol 2006;24:541–70. https://doi.org/10.1146/annurev.immunol.23.021704.115830.

[231] Lafaille JJ, DeCloux A, Bonneville M, Takagaki Y, Tonegawa S. Junctional sequences of T cell receptor gamma delta genes: implications for gamma delta T cell lineages and for a novel intermediate of V-(D)-J joining. Cell 1989;59:859–70. https://doi.org/10.1016/0092-8674(89)90609-0.

[232] Pelanda R, Greaves SA, Alves da Costa T, Cedrone LM, Campbell ML, Torres RM. B-cell intrinsic and extrinsic signals that regulate central tolerance of mouse and human B cells. Immunol Rev 2022;307:12–26. https://doi.org/10.1111/imr.13062.

[233] Luo S, Jing C, Ye AY, Kratochvil S, Cottrell CA, Koo J-H, et al. Humanized V(D)J-rearranging and TdT-expressing mouse vaccine models with physiological HIV-1 broadly neutralizing antibody precursors. Proc Natl Acad Sci USA 2023;120. https://doi.org/10.1073/pnas.2217883120.

[234] Komori T, Pricop L, Hatakeyama A, Bona CA, Alt FW. Repertoires of antigen receptors in Tdt congenitally deficient mice. Int Rev Immunol 1996;13:317–25. https://doi.org/10.3109/08830189609061755.

[235] Feeney AJ. Predominance of the prototypic T15 anti-phosphorylcholine junctional sequence in neonatal pre-B cells. J Immunol 1991;147:4343–50.

[236] Feeney AJ. Predominance of VH-D-JH junctions occurring at sites of short sequence homology results in limited junctional diversity in neonatal antibodies. J Immunol 1992;149:222–9.

[237] Gu H, Förster I, Rajewsky K. Sequence homologies, N sequence insertion and JH gene utilization in VHDJH joining: implications for the joining mechanism and the ontogenetic timing of Ly1 B cell and B-CLL progenitor generation. EMBO J 1990;9:2133–40. https://doi.org/10.1002/j.1460-2075.1990.tb07382.x.

[238] Aguilar LK, Belmont JW. V gamma 3 T cell receptor rearrangement and expression in the adult thymus. J Immunol 1991;146:1348–52.

[239] Elliott JF, Rock EP, Patten PA, Davis MM, Chien YH. The adult T-cell receptor delta-chain is diverse and distinct from that of fetal thymocytes. Nature 1988;331:627–31. https://doi.org/10.1038/331627a0.

[240] Bogue M, Candéias S, Benoist C, Mathis D. A special repertoire of alpha:beta T cells in neonatal mice. EMBO J 1991;10:3647–54.

[241] Feeney AJ. Lack of N regions in fetal and neonatal mouse immunoglobulin V-D-J junctional sequences. J Exp Med 1990;172:1377–90.

[242] Zhang Y, Cado D, Asarnow DM, Komori T, Alt FW, Raulet DH, et al. The role of short homology repeats and TdT in generation of the invariant gamma delta antigen receptor repertoire in the fetal thymus. Immunity 1995;3:439–47. https://doi.org/10.1016/1074-7613(95)90173-6.

[243] Benedict CL, Kearney JF. Increased junctional diversity in fetal B cells results in a loss of protective anti-phosphorylcholine antibodies in adult mice. Immunity 1999;10:607–17. https://doi.org/10.1016/s1074-7613(00)80060-6.

[244] Aono A, Enomoto H, Yoshida N, Yoshizaki K, Kishimoto T, Komori T. Forced expression of terminal deoxynucleotidyl transferase in fetal thymus resulted in a decrease in gammadelta T cells and random dissemination of Vgamma3Vdelta1 T cells in skin of newborn but not adult mice. Immunology 2000;99:489–97. https://doi.org/10.1046/j.1365-2567.2000.00987.x.

[245] Thai T-H, Kearney JF. Distinct and opposite activities of human terminal deoxynucleotidyltransferase splice variants. J Immunol 2004;173:4009–19. https://doi.org/10.4049/jimmunol.173.6.4009.

[246] Thai T-H, Purugganan MM, Roth DB, Kearney JF. Distinct and opposite diversifying activities of terminal transferase splice variants. Nat Immunol 2002;3:457–62. https://doi.org/10.1038/ni788.

[247] Li YS, Hayakawa K, Hardy RR. The regulated expression of B lineage associated genes during B cell differentiation in bone marrow and fetal liver. J Exp Med 1993;178:951–60. https://doi.org/10.1084/jem.178.3.951.

[248] Heller M, Owens JD, Mushinski JF, Rudikoff S. Amino acids at the site of V kappa-J kappa recombination not encoded by germline sequences. J Exp Med 1987;166:637–46. https://doi.org/10.1084/jem.166.3.637.

[249] Victor KD, Vu K, Feeney AJ. Limited junctional diversity in kappa light chains. Junctional sequences from CD43+B220+ early B cell progenitors resemble those from peripheral B cells. J Immunol 1994;152:3467–75.

[250] Gu Y, Sekiguchi J, Gao Y, Dikkes P, Frank K, Ferguson D, et al. Defective embryonic neurogenesis in Ku-deficient but not DNA-dependent protein kinase catalytic subunit-deficient mice. Proc Natl Acad Sci USA 2000;97:2668–73. https://doi.org/10.1073/pnas.97.6.2668.

[251] Frank KM, Sharpless NE, Gao Y, Sekiguchi JM, Ferguson DO, Zhu C, et al. DNA ligase IV deficiency in mice leads to defective neurogenesis and embryonic lethality via the p53 pathway. Mol Cell 2000;5:993–1002. https://doi.org/10.1016/s1097-2765(00)80264-6.

[252] Gao Y, Ferguson DO, Xie W, Manis JP, Sekiguchi J, Frank KM, et al. Interplay of p53 and DNA-repair protein XRCC4 in tumorigenesis, genomic stability and development. Nature 2000;404:897–900. https://doi.org/10.1038/35009138.

[253] Karanjawala ZE, Adachi N, Irvine RA, Oh EK, Shibata D, Schwarz K, et al. The embryonic lethality in DNA ligase IV-deficient mice is rescued by deletion of Ku: implications for unifying the heterogeneous phenotypes of NHEJ mutants. DNA Repair 2002;1:1017–26. https://doi.org/10.1016/s1568-7864(02)00151-9.

[254] Adachi N, Ishino T, Ishii Y, Takeda S, Koyama H. DNA ligase IV-deficient cells are more resistant to ionizing radiation in the absence of Ku70: implications for DNA double-strand break repair. Proc Natl Acad Sci U S A 2001;98:12109–13. https://doi.org/10.1073/pnas.201271098.

[255] de Villartay J-P. V(D)J recombination and DNA repair: lessons from human immune deficiencies and other animal models. Curr Opin Allergy Clin Immunol 2002;2:473–9. https://doi.org/10.1097/00130832-200212000-00001.

[256] van der Burg M, IJspeert H, Verkaik NS, Turul T, Wiegant WW, Morotomi-Yano K, et al. A DNA-PKcs mutation in a radiosensitive T−B− SCID patient inhibits Artemis activation and nonhomologous end-joining. J Clin Invest 2009;119:91–8. https://doi.org/10.1172/JCI37141.

[257] Woodbine L, Neal JA, Sasi N-K, Shimada M, Deem K, Coleman H, et al. PRKDC mutations in a SCID patient with profound neurological abnormalities. J Clin Invest 2013;123:2969–80. https://doi.org/10.1172/JCI67349.

[258] Zhu C, Mills KD, Ferguson DO, Lee C, Manis J, Fleming J, et al. Unrepaired DNA breaks in p53-deficient cells lead to oncogenic gene amplification subsequent to translocations. Cell 2002;109:811–21. https://doi.org/10.1016/s0092-8674(02)00770-5.

[259] Tepsuporn S, Hu J, Gostissa M, Alt FW. Mechanisms that can promote peripheral B-cell lymphoma in ATM-deficient mice. Cancer Immunol Res 2014;2:857–66. https://doi.org/10.1158/2326-6066.CIR-14-0090.

[260] Hu J, Tepsuporn S, Meyers RM, Gostissa M, Alt FW. Developmental propagation of V(D)J recombination-associated DNA breaks and translocations in mature B cells via dicentric chromosomes. Proc Natl Acad Sci USA 2014;111:10269–74. https://doi.org/10.1073/pnas.1410112111.

[261] Ahnesorg P, Smith P, Jackson SP. XLF interacts with the XRCC4-DNA ligase IV complex to promote DNA nonhomologous end-joining. Cell 2006;124:301–13. https://doi.org/10.1016/j.cell.2005.12.031.

[262] Buck D, Malivert L, de Chasseval R, Barraud A, Fondanèche M-C, Sanal O, et al. Cernunnos, a novel nonhomologous end-joining factor, is mutated in human immunodeficiency with microcephaly. Cell 2006;124:287–99. https://doi.org/10.1016/j.cell.2005.12.030.

[263] Dai Y, Kysela B, Hanakahi LA, Manolis K, Riballo E, Stumm M, et al. Nonhomologous end joining and V(D)J recombination require an additional factor. Proc Natl Acad Sci USA 2003;100:2462–7. https://doi.org/10.1073/pnas.0437964100.

[264] Li G, Alt FW, Cheng H-L, Brush JW, Goff P, Murphy MM, et al. Lymphocyte-specific compensation for XLF/cernunnos end-joining functions in V(D)J recombination. Mol Cell 2008;31:631–40. https://doi.org/10.1016/j.molcel.2008.07.017.

[265] Zha S, Guo C, Boboila C, Oksenych V, Cheng H-L, Zhang Y, et al. ATM damage response and XLF repair factor are functionally redundant in joining DNA breaks. Nature 2011;469:250–4. https://doi.org/10.1038/nature09604.

[266] Liu X, Jiang W, Dubois RL, Yamamoto K, Wolner Z, Zha S. Overlapping functions between XLF repair protein and 53BP1 DNA damage response factor in end joining and lymphocyte development. Proc Natl Acad Sci USA 2012;109:3903–8. https://doi.org/10.1073/pnas.1120160109.

[267] Oksenych V, Alt FW, Kumar V, Schwer B, Wesemann DR, Hansen E, et al. Functional redundancy between repair factor XLF and damage response mediator 53BP1 in V(D)J recombination and DNA repair. Proc Natl Acad Sci U S A 2012;109:2455–60. https://doi.org/10.1073/pnas.1121458109.

[268] Beck C, Castañeda-Zegarra S, Huse C, Xing M, Oksenych V. Mediator of DNA damage checkpoint protein 1 facilitates V(D)J recombination in cells lacking DNA repair factor XLF. Biomolecules 2019;10:60. https://doi.org/10.3390/biom10010060.

[269] Zha S, Bassing CH, Sanda T, Brush JW, Patel H, Goff PH, et al. ATM-deficient thymic lymphoma is associated with aberrant tcrd rearrangement and gene amplification. J Exp Med 2010;207:1369–80. https://doi.org/10.1084/jem.20100285.

[270] Jiang W, Lee BJ, Li C, Dubois RL, Gostissa M, Alt FW, et al. Aberrant TCRδ rearrangement underlies the T-cell lymphocytopenia and t(12;14) translocation associated with ATM deficiency. Blood 2015;125:2665–8. https://doi.org/10.1182/blood-2015-01-622621.

[271] Liyanage M, Weaver Z, Barlow C, Coleman A, Pankratz DG, Anderson S, et al. Abnormal rearrangement within the alpha/delta T-cell receptor locus in lymphomas from Atm-deficient mice. Blood 2000;96:1940–6.

[272] Zhao B, Watanabe G, Morten MJ, Reid DA, Rothenberg E, Lieber MR. The essential elements for the noncovalent association of two DNA ends during NHEJ synapsis. Nat Commun 2019;10:3588. https://doi.org/10.1038/s41467-019-11507-z.

[273] Graham TGW, Walter JC, Loparo JJ. Two-stage synapsis of DNA ends during non-homologous end joining. Mol Cell 2016;61:850–8. https://doi.org/10.1016/j.molcel.2016.02.010.

[274] Lescale C, Lenden Hasse H, Blackford AN, Balmus G, Bianchi JJ, Yu W, et al. Specific roles of XRCC4 paralogs PAXX and XLF during V(D)J recombination. Cell Rep 2016;16:2967–79. https://doi.org/10.1016/j.celrep.2016.08.069.

[275] Liu X, Shao Z, Jiang W, Lee BJ, Zha S. PAXX promotes KU accumulation at DNA breaks and is essential for end-joining in XLF-deficient mice. Nat Commun 2017;8:13816. https://doi.org/10.1038/ncomms13816.

[276] Hung PJ, Johnson B, Chen B-R, Byrum AK, Bredemeyer AL, Yewdell WT, et al. MRI is a DNA damage response adaptor during classical non-homologous end joining. Mol Cell 2018;71:332–42. https://doi.org/10.1016/j.molcel.2018.06.018.

[277] Xing M, Oksenych V. Genetic interaction between DNA repair factors PAXX, XLF, XRCC4 and DNA-PKcs in human cells. FEBS Open Bio 2019;9:1315–26. https://doi.org/10.1002/2211-5463.12681.

[278] Oksenych V, Kumar V, Liu X, Guo C, Schwer B, Zha S, et al. Functional redundancy between the XLF and DNA-PKcs DNA repair factors in V(D)J recombination and nonhomologous DNA end joining. Proc Natl Acad Sci USA 2013;110:2234–9. https://doi.org/10.1073/pnas.1222573110.

[279] Chen S, Lee L, Naila T, Fishbain S, Wang A, Tomkinson AE, et al. Structural basis of long-range to short-range synaptic transition in NHEJ. Nature 2021;593:294–8. https://doi.org/10.1038/s41586-021-03458-7.

[280] Carney SM, Moreno AT, Piatt SC, Cisneros-Aguirre M, Lopezcolorado FW, Stark JM, et al. XLF acts as a flexible connector during non-homologous end joining. Elife 2020;9:e61920. https://doi.org/10.7554/eLife.61920.

[281] Graham TGW, Carney SM, Walter JC, Loparo JJ. A single XLF dimer bridges DNA ends during nonhomologous end joining. Nat Struct Mol Biol 2018;25:877–84. https://doi.org/10.1038/s41594-018-0120-y.

[282] Lescale C, Abramowski V, Bedora-Faure M, Murigneux V, Vera G, Roth DB, et al. RAG2 and XLF/Cernunnos interplay reveals a novel role for the RAG complex in DNA repair. Nat Commun 2016;7:10529. https://doi.org/10.1038/ncomms10529.

[283] Kuo TC, Schlissel MS. Mechanisms controlling expression of the RAG locus during lymphocyte development. Curr Opin Immunol 2009;21:173–8. https://doi.org/10.1016/j.coi.2009.03.008.

[284] Sleckman BP, Oltz EM. Preparing targets for V(D)J recombinase: transcription paves the way. J Immunol 2012;188:7–9. https://doi.org/10.4049/jimmunol.1103195.

[285] Gorman JR, Alt FW. Regulation of immunoglobulin light chain isotype expression. In: Dixon FJ, editor. Advances in immunology, vol. 69. Academic Press; 1998. p. 113–81. https://doi.org/10.1016/S0065-2776(08)60607-0.

[286] Allyn BM, Lee KD, Bassing CH. Genome topology control of antigen receptor gene assembly. J Immunol 2020;204:2617–26. https://doi.org/10.4049/jimmunol.1901356.

[287] Mostoslavsky R, Alt FW, Rajewsky K. The lingering enigma of the allelic exclusion mechanism. Cell 2004;118:539–44. https://doi.org/10.1016/j.cell.2004.08.023.

[288] Melchers F, Kong XC, Yamagami T, Onishi K, Shimizu T, Rolink AG, et al. Repertoire selection by pre-B-cell receptors and B-cell receptors, and genetic control of B-cell development from immature to mature B cells. Immunol Rev 2000;175:33–46.

[289] Rajewsky K. Clonal selection and learning in the antibody system. Nature 1996;381:751–8. https://doi.org/10.1038/381751a0.

[290] Nussenzweig MC, Shaw AC, Sinn E, Danner DB, Holmes KL, Morse HC, et al. Allelic exclusion in transgenic mice that express the membrane form of immunoglobulin μ. Science 1987;236:816–9.

[291] Reth MG, Ammirati P, Jackson S, Alt FW. Regulated progression of a cultured pre-B-cell line to the B-cell stage. Nature 1985;317:353–5. https://doi.org/10.1038/317353a0.

[292] Chang Y, Bosma MJ, Bosma GC. Extended duration of DH–JH rearrangement in immunoglobulin heavy chain transgenic mice: implications for regulation of allelic exclusion. J Exp Med 1999;189:1295–305. https://doi.org/10.1084/jem.189.8.1295.

[293] Yancopoulos G, Alt FW. Developmentally controlled and tissue-specific expression of unrearranged VH gene segments. Cell 1985;40:271–81. https://doi.org/10.1016/0092-8674(85)90141-2.

[294] Ba Z, Lou J, Ye AY, Dai H-Q, Dring EW, Lin SG, et al. CTCF orchestrates long-range cohesin-driven V(D)J recombinational scanning. Nature 2020;586:305–10. https://doi.org/10.1038/s41586-020-2578-0.

[295] Bolland DJ, Koohy H, Wood AL, Matheson LS, Krueger F, Stubbington MJT, et al. Two mutually exclusive local chromatin states drive efficient V(D)J recombination. Cell Rep 2016;15:2475–87. https://doi.org/10.1016/j.celrep.2016.05.020.

[296] Choi NM, Loguercio S, Verma-Gaur J, Degner SC, Torkamani A, Su AI, et al. Deep sequencing of the murine Igh repertoire reveals complex regulation of nonrandom V gene rearrangement frequencies. J Immunol 2013;191:2393–402. https://doi.org/10.4049/jimmunol.1301279.

[297] Dai H-Q, Hu H, Lou J, Ye AY, Ba Z, Zhang X, et al. Loop extrusion mediates physiological Igh locus contraction for RAG scanning. Nature 2021;590:338–43. https://doi.org/10.1038/s41586-020-03121-7.

[298] Dai H-Q, Hu H, Lou J, Ye AY, Chapdelaine-Williams AM, Kyritsis N, et al. Loop extrusion mediates physiological locus contraction for V(D)J recombination. bioRxiv 2020:181222. https://doi.org/10.1101/2020.06.30.181222.

[299] Yancopoulos GD, Blackwell TK, Suh H, Hood L, Alt FW. Introduced T cell receptor variable region gene segments recombine in pre-B cells: evidence that B and T cells use a common recombinase. Cell 1986;44:251–9. https://doi.org/10.1016/0092-8674(86)90759-2.

[300] Mombaerts P, Iacomini J, Johnson RS, Herrup K, Tonegawa S, Papaioannou VE. RAG-1-deficient mice have no mature B and T lymphocytes. Cell 1992;68:869–77. https://doi.org/10.1016/0092-8674(92)90030-G.

[301] Shinkai Y, Rathbun G, Lam K-P, Oltz EM, Stewart V, Mendelsohn M, et al. RAG-2-deficient mice lack mature lymphocytes owing to inability to initiate V(D)J rearrangement. Cell 1992;68:855–67. https://doi.org/10.1016/0092-8674(92)90029-C.

[302] Bolland DJ, Wood AL, Johnston CM, Bunting SF, Morgan G, Chakalova L, et al. Antisense intergenic transcription in V(D)J recombination. Nat Immunol 2004;5:630–7. https://doi.org/10.1038/ni1068.

[303] Featherstone K, Wood AL, Bowen AJ, Corcoran AE. The mouse immunoglobulin heavy chain V-D intergenic sequence contains insulators that may regulate ordered V(D)J recombination. J Biol Chem 2010;285:9327–38. https://doi.org/10.1074/jbc.M109.098251.

[304] Fuxa M, Skok J, Souabni A, Salvagiotto G, Roldan E, Busslinger M. Pax5 induces V-to-DJ rearrangements and locus contraction of the immunoglobulin heavy-chain gene. Genes Dev 2004;18:411–22. https://doi.org/10.1101/gad.291504.

[305] Jhunjhunwala S, van Zelm MC, Peak MM, Cutchin S, Riblet R, van Dongen JJM, et al. The 3D structure of the immunoglobulin heavy-chain locus: implications for long-range genomic interactions. Cell 2008;133:265–79. https://doi.org/10.1016/j.cell.2008.03.024.

[306] Kosak ST, Skok JA, Medina KL, Riblet R, Beau MML, Fisher AG, et al. Subnuclear compartmentalization of immunoglobulin loci during lymphocyte development. Science 2002;296:158–62.

[307] McMurry MT, Krangel MS. A role for histone acetylation in the developmental regulation of V(D)J recombination. Science 2000;287:495–8. https://doi.org/10.1126/science.287.5452.495.

[308] Mostoslavsky R, Singh N, Kirillov A, Pelanda R, Cedar H, Chess A, et al. κ chain monoallelic demethylation and the establishment of allelic exclusion. Genes Dev 1998;12:1801–11.

[309] Rother MB, Palstra R-J, Jhunjhunwala S, van Kester KAM, van IJcken WFJ, Hendriks RW, et al. Nuclear positioning rather than contraction controls ordered rearrangements of immunoglobulin loci. Nucleic Acids Res 2016;44:175–86. https://doi.org/10.1093/nar/gkv928.

[310] Stanhope-Baker P, Hudson KM, Shaffer AL, Constantinescu A, Schlissel MS. Cell type–specific chromatin structure determines the targeting of V(D)J recombinase activity in vitro. Cell 1996;85:887–97. https://doi.org/10.1016/S0092-8674(00)81272-6.

[311] Zhang Y, Zhang X, Ba Z, Liang Z, Dring EW, Hu H, et al. The fundamental role of chromatin loop extrusion in physiological V(D)J recombination. Nature 2019;573:600–4. https://doi.org/10.1038/s41586-019-1547-y.

[312] Lieberman-Aiden E, Berkum NL van, Williams L, Imakaev M, Ragoczy T, Telling A, et al. Comprehensive mapping of long-range interactions reveals folding principles of the human genome. Science 2009;326:289–93.

[313] Dixon JR, Selvaraj S, Yue F, Kim A, Li Y, Shen Y, et al. Topological domains in mammalian genomes identified by analysis of chromatin interactions. Nature 2012;485:376–80. https://doi.org/10.1038/nature11082.

[314] Nora EP, Lajoie BR, Schulz EG, Giorgetti L, Okamoto I, Servant N, et al. Spatial partitioning of the regulatory landscape of the X-inactivation centre. Nature 2012;485:381–5. https://doi.org/10.1038/nature11049.

[315] Rao SSP, Huntley MH, Durand NC, Stamenova EK, Bochkov ID, Robinson JT, et al. A 3D map of the human genome at kilobase resolution reveals principles of chromatin looping. Cell 2014;159:1665–80. https://doi.org/10.1016/j.cell.2014.11.021.

[316] MacPherson MJ, Sadowski PD. The CTCF insulator protein forms an unusual DNA structure. BMC Mol Biol 2010;11:101. https://doi.org/10.1186/1471-2199-11-101.

[317] Nakahashi H, Kwon K-RK, Resch W, Vian L, Dose M, Stavreva D, et al. A genome-wide map of CTCF multivalency redefines the CTCF code. Cell Rep 2013;3:1678–89. https://doi.org/10.1016/j.celrep.2013.04.024.

[318] de Wit E, Vos ESM, Holwerda SJB, Valdes-Quezada C, Verstegen MJAM, Teunissen H, et al. CTCF binding polarity determines chromatin looping. Mol Cell 2015;60:676–84. https://doi.org/10.1016/j.molcel.2015.09.023.

[319] Vietri Rudan M, Barrington C, Henderson S, Ernst C, Odom DT, Tanay A, et al. Comparative Hi-C reveals that CTCF underlies evolution of chromosomal domain architecture. Cell Rep 2015;10:1297–309. https://doi.org/10.1016/j.celrep.2015.02.004.

[320] Li Y, Haarhuis JHI, Sedeño Cacciatore Á, Oldenkamp R, van Ruiten MS, Willems L, et al. The structural basis for cohesin–CTCF-anchored loops. Nature 2020;578:472–6. https://doi.org/10.1038/s41586-019-1910-z.

[321] Nishana M, Ha C, Rodriguez-Hernaez J, Ranjbaran A, Chio E, Nora EP, et al. Defining the relative and combined contribution of CTCF and CTCFL to genomic regulation. Genome Biol 2020;21:108. https://doi.org/10.1186/s13059-020-02024-0.

[322] Nora EP, Caccianini L, Fudenberg G, So K, Kameswaran V, Nagle A, et al. Molecular basis of CTCF binding polarity in genome folding. Nat Commun 2020;11:5612. https://doi.org/10.1038/s41467-020-19283-x.

[323] Pugacheva EM, Kubo N, Loukinov D, Tajmul M, Kang S, Kovalchuk AL, et al. CTCF mediates chromatin looping via N-terminal domain-dependent cohesin retention. Proc Natl Acad Sci USA 2020;117:2020–31. https://doi.org/10.1073/pnas.1911708117.

[324] Rowley MJ, Corces VG. Organizational principles of 3D genome architecture. Nat Rev Genet 2018;19:789–800. https://doi.org/10.1038/s41576-018-0060-8.

[325] Alipour E, Marko JF. Self-organization of domain structures by DNA-loop-extruding enzymes. Nucleic Acids Res 2012;40:11202–12. https://doi.org/10.1093/nar/gks925.

[326] Bouwman BA, de Laat W. Getting the genome in shape: the formation of loops, domains and compartments. Genome Biol 2015;16:154. https://doi.org/10.1186/s13059-015-0730-1.

[327] Dekker J, Mirny L. The 3D genome as moderator of chromosomal communication. Cell 2016;164:1110–21. https://doi.org/10.1016/j.cell.2016.02.007.

[328] Fudenberg G, Imakaev M, Lu C, Goloborodko A, Abdennur N, Mirny LA. Formation of chromosomal domains by loop extrusion. Cell Rep 2016;15:2038–49. https://doi.org/10.1016/j.celrep.2016.04.085.

[329] Nasmyth K. Disseminating the genome: joining, resolving, and separating sister chromatids during mitosis and meiosis. Annu Rev Genet 2001;35:673–745. https://doi.org/10.1146/annurev.genet.35.102401.091334.

[330] Nichols MH, Corces VG. A CTCF code for 3D genome architecture. Cell 2015;162:703–5. https://doi.org/10.1016/j.cell.2015.07.053.

[331] Sanborn AL, Rao SSP, Huang S-C, Durand NC, Huntley MH, Jewett AI, et al. Chromatin extrusion explains key features of loop and domain formation in wild-type and engineered genomes. Proc Natl Acad Sci USA 2015;112:E6456–65. https://doi.org/10.1073/pnas.1518552112.

[332] Lupiáñez DG, Kraft K, Heinrich V, Krawitz P, Brancati F, Klopocki E, et al. Disruptions of topological chromatin domains cause pathogenic rewiring of gene-enhancer interactions. Cell 2015;161:1012–25. https://doi.org/10.1016/j.cell.2015.04.004.

[333] Merkenschlager M, Nora EP. CTCF and cohesin in genome folding and transcriptional gene regulation. Annu Rev Genom Hum Genet 2016;17:17–43. https://doi.org/10.1146/annurev-genom-083115-022339.

[334] Narendra V, Rocha PP, An D, Raviram R, Skok JA, Mazzoni EO, et al. CTCF establishes discrete functional chromatin domains at the Hox clusters during differentiation. Science 2015;347:1017–21. https://doi.org/10.1126/science.1262088.

[335] Nora EP, Goloborodko A, Valton A-L, Gibcus JH, Uebersohn A, Abdennur N, et al. Targeted degradation of CTCF decouples local insulation of chromosome domains from genomic compartmentalization. Cell 2017;169:930–44. https://doi.org/10.1016/j.cell.2017.05.004.

[336] Wutz G, Várnai C, Nagasaka K, Cisneros DA, Stocsits RR, Tang W, et al. Topologically associating domains and chromatin loops depend on cohesin and are regulated by CTCF, WAPL, and PDS5 proteins. EMBO J 2017;36:3573–99. https://doi.org/10.15252/embj.201798004.

[337] Gassler J, Brandão HB, Imakaev M, Flyamer IM, Ladstätter S, Bickmore WA, et al. A mechanism of cohesin-dependent loop extrusion organizes zygotic genome architecture. EMBO J 2017;36:3600–18. https://doi.org/10.15252/embj.201798083.

[338] Rao SSP, Huang S-C, Glenn St Hilaire B, Engreitz JM, Perez EM, Kieffer-Kwon K-R, et al. Cohesin loss eliminates all loop domains. Cell 2017;171:305–20. https://doi.org/10.1016/j.cell.2017.09.026.
[339] Schwarzer W, Abdennur N, Goloborodko A, Pekowska A, Fudenberg G, Loe-Mie Y, et al. Two independent modes of chromatin organization revealed by cohesin removal. Nature 2017;551:51–6. https://doi.org/10.1038/nature24281.
[340] Vian L, Pekowska A, Rao SSP, Kieffer-Kwon K-R, Jung S, Baranello L, et al. The energetics and physiological impact of cohesin extrusion. Cell 2018;173:1165–78. https://doi.org/10.1016/j.cell.2018.03.072.
[341] Davidson IF, Bauer B, Goetz D, Tang W, Wutz G, Peters J-M. DNA loop extrusion by human cohesin. Science 2019;366:1338–45. https://doi.org/10.1126/science.aaz3418.
[342] Kim Y, Shi Z, Zhang H, Finkelstein IJ, Yu H. Human cohesin compacts DNA by loop extrusion. Science 2019;366:1345–9. https://doi.org/10.1126/science.aaz4475.
[343] Pradhan B, Barth R, Kim E, Davidson IF, Bauer B, Laar T van, et al. SMC complexes can traverse physical roadblocks bigger than their ring size. Cell Rep 2022;41. https://doi.org/10.1016/j.celrep.2022.111491.
[344] Buheitel J, Stemmann O. Prophase pathway-dependent removal of cohesin from human chromosomes requires opening of the Smc3–Scc1 gate. EMBO J 2013;32:666–76. https://doi.org/10.1038/emboj.2013.7.
[345] Chan K-L, Roig MB, Hu B, Beckouët F, Metson J, Nasmyth K. Cohesin's DNA exit gate is distinct from its entrance gate and is regulated by acetylation. Cell 2012;150:961–74. https://doi.org/10.1016/j.cell.2012.07.028.
[346] Eichinger CS, Kurze A, Oliveira RA, Nasmyth K. Disengaging the Smc3/kleisin interface releases cohesin from Drosophila chromosomes during interphase and mitosis. EMBO J 2013;32:656–65. https://doi.org/10.1038/emboj.2012.346.
[347] Haarhuis JHI, van der Weide RH, Blomen VA, Yáñez-Cuna JO, Amendola M, van Ruiten MS, et al. The cohesin release factor WAPL restricts chromatin loop extension. Cell 2017;169:693–707. https://doi.org/10.1016/j.cell.2017.04.013.
[348] Gandhi R, Gillespie PJ, Hirano T. Human Wapl is a cohesin-binding protein that promotes sister-chromatid resolution in mitotic prophase. Curr Biol 2006;16:2406–17. https://doi.org/10.1016/j.cub.2006.10.061.
[349] Kueng S, Hegemann B, Peters BH, Lipp JJ, Schleiffer A, Mechtler K, et al. Wapl controls the dynamic association of cohesin with chromatin. Cell 2006;127:955–67. https://doi.org/10.1016/j.cell.2006.09.040.
[350] Hill L, Ebert A, Jaritz M, Wutz G, Nagasaka K, Tagoh H, et al. Wapl repression by Pax5 promotes V gene recombination by Igh loop extrusion. Nature 2020;584:142–7. https://doi.org/10.1038/s41586-020-2454-y.
[351] Davidson IF, Peters J-M. Genome folding through loop extrusion by SMC complexes. Nat Rev Mol Cell Biol 2021:1–20. https://doi.org/10.1038/s41580-021-00349-7.
[352] Mirny L, Dekker J. Mechanisms of chromosome folding and nuclear organization: their interplay and open questions. Cold Spring Harbor Perspect Biol 2022;14:a040147. https://doi.org/10.1101/cshperspect.a040147.
[353] Kumari G, Sen R. Chapter two - chromatin interactions in the control of immunoglobulin heavy chain gene assembly. In: Murre C, editor. Advances in immunology, vol. 128. Academic Press; 2015. p. 41–92. https://doi.org/10.1016/bs.ai.2015.08.001.
[354] Zhang X, Zhang Y, Ba Z, Kyritsis N, Casellas R, Alt FW. Fundamental roles of chromatin loop extrusion in antibody class switching. Nature 2019;575:385–9. https://doi.org/10.1038/s41586-019-1723-0.
[355] Pinaud E, Marquet M, Fiancette R, Péron S, Vincent-Fabert C, Denizot Y, et al. Chapter 2 - the IgH locus 3′ regulatory region: pulling the strings from behind. In: Alt FW, Austen KF, Honj T, Melchers F, Uhr JW, Unanue ER, editors. Advances in immunology, vol. 110. Academic Press; 2011. p. 27–70. https://doi.org/10.1016/B978-0-12-387663-8.00002-8.
[356] Birshtein B. The role of CTCF binding sites in the 3′ immunoglobulin heavy chain regulatory region. Front Genet 2012;3.
[357] Zhang X, Yoon HS, Chapdelaine-Williams AM, Kyritsis N, Alt FW. Physiological role of the 3′IgH CBEs super-anchor in antibody class switching. Proc Natl Acad Sci USA 2021;118. https://doi.org/10.1073/pnas.2024392118.
[358] Benner C, Isoda T, Murre C. New roles for DNA cytosine modification, eRNA, anchors, and superanchors in developing B cell progenitors. Proc Natl Acad Sci USA 2015;112:12776–81. https://doi.org/10.1073/pnas.1512995112.
[359] Degner SC, Verma-Gaur J, Wong TP, Bossen C, Iverson GM, Torkamani A, et al. CCCTC-binding factor (CTCF) and cohesin influence the genomic architecture of the Igh locus and antisense transcription in pro-B cells. Proc Natl Acad Sci U S A 2011;108:9566–71. https://doi.org/10.1073/pnas.1019391108.
[360] Guo C, Yoon HS, Franklin A, Jain S, Ebert A, Cheng H-L, et al. CTCF-binding elements mediate control of V(D)J recombination. Nature 2011;477:424–30. https://doi.org/10.1038/nature10495.
[361] Jain S, Ba Z, Zhang Y, Dai H-Q, Alt FW. CTCF-binding elements mediate accessibility of RAG substrates during chromatin scanning. Cell 2018;174:102–16. https://doi.org/10.1016/j.cell.2018.04.035.
[362] Lin SG, Guo C, Su A, Zhang Y, Alt FW. CTCF-binding elements 1 and 2 in the Igh intergenic control region cooperatively regulate V(D)J recombination. Proc Natl Acad Sci USA 2015;112:1815–20. https://doi.org/10.1073/pnas.1424936112.
[363] Perlot T, Alt FW, Bassing CH, Suh H, Pinaud E. Elucidation of IgH intronic enhancer functions via germ-line deletion. Proc Natl Acad Sci USA 2005;102:14362–7. https://doi.org/10.1073/pnas.0507090102.
[364] Bossen C, Mansson R, Murre C. Chromatin topology and the regulation of antigen receptor assembly. Annu Rev Immunol 2012;30:337–56. https://doi.org/10.1146/annurev-immunol-020711-075003.
[365] Proudhon C, Hao B, Raviram R, Chaumeil J, Skok JA. Long-range regulation of V(D)J recombination. Adv Immunol 2015;128:123–82. https://doi.org/10.1016/bs.ai.2015.07.003.
[366] Liang Z, Zhao L, Ye AY, Lin SG, Zhang Y, Guo C, et al. Contribution of the IGCR1 regulatory element and the 3′*Igh* CTCF-binding elements to regulation of *Igh* V(D)J recombination. Proc Natl Acad Sci USA 2023;120(26). https://doi.org/10.1073/pnas.2306564120. e2306564120.
[367] Ranganath S, Carpenter AC, Gleason M, Shaw AC, Bassing CH, Alt FW. Productive coupling of accessible Vβ14 segments and DJβ complexes determines the frequency of Vβ14 Rearrangement1. J Immunol 2008;180:2339–46. https://doi.org/10.4049/jimmunol.180.4.2339.
[368] Chiarle R, Zhang Y, Frock RL, Lewis SM, Molinie B, Ho Y-J, et al. Genome-wide translocation sequencing reveals mechanisms of

chromosome breaks and rearrangements in B cells. Cell 2011;147:107–19. https://doi.org/10.1016/j.cell.2011.07.049.

[369] Frock RL, Hu J, Meyers RM, Ho Y-J, Kii E, Alt FW. Genome-wide detection of DNA double-stranded breaks induced by engineered nucleases. Nat Biotechnol 2015;33:179–86. https://doi.org/10.1038/nbt.3101.

[370] Wood C, Tonegawa S. Diversity and joining segments of mouse immunoglobulin heavy chain genes are closely linked and in the same orientation: implications for the joining mechanism. Proc Natl Acad Sci USA 1983;80:3030–4. https://doi.org/10.1073/pnas.80.10.3030.

[371] Zhao L, Frock RL, Du Z, Hu J, Chen L, Krangel MS, et al. Orientation-specific RAG activity in chromosomal loop domains contributes to Tcrd V(D)J recombination during T cell development. J Exp Med 2016;213:1921–36. https://doi.org/10.1084/jem.20160670.

[372] Lin SG, Ba Z, Alt FW, Zhang Y. RAG chromatin scanning during V(D)J recombination and chromatin loop extrusion are related processes. Adv Immunol 2018;139:93–135. https://doi.org/10.1016/bs.ai.2018.07.001.

[373] Gauss GH, Lieber MR. The basis for the mechanistic bias for deletional over inversional V(D)J recombination. Genes Dev 1992;6:1553–61. https://doi.org/10.1101/gad.6.8.1553.

[374] Banigan EJ, Tang W, van den Berg AA, Stocsits RR, Wutz G, Brandão HB, et al. Transcription shapes 3D chromatin organization by interacting with loop-extruding cohesin complexes. Genomics 2022. https://doi.org/10.1101/2022.01.07.475367.

[375] Hsieh T-HS, Cattoglio C, Slobodyanyuk E, Hansen AS, Rando OJ, Tjian R, et al. Resolving the 3D landscape of transcription-linked mammalian chromatin folding. Mol Cell 2020;78:539–53. https://doi.org/10.1016/j.molcel.2020.03.002.

[376] Kraft K, Magg A, Heinrich V, Riemenschneider C, Schöpflin R, Markowski J, et al. Serial genomic inversions induce tissue-specific architectural stripes, gene misexpression and congenital malformations. Nat Cell Biol 2019;21:305–10. https://doi.org/10.1038/s41556-019-0273-x.

[377] Thiecke MJ, Wutz G, Muhar M, Tang W, Bevan S, Malysheva V, et al. Cohesin-dependent and -independent mechanisms mediate chromosomal contacts between promoters and enhancers. Cell Rep 2020;32:107929. https://doi.org/10.1016/j.celrep.2020.107929.

[378] Medvedovic J, Ebert A, Tagoh H, Tamir IM, Schwickert TA, Novatchkova M, et al. Flexible long-range loops in the VH gene region of the Igh locus facilitate the generation of a diverse antibody repertoire. Immunity 2013;39:229–44. https://doi.org/10.1016/j.immuni.2013.08.011.

[379] Montefiori L, Wuerffel R, Roqueiro D, Lajoie B, Guo C, Gerasimova T, et al. Extremely long-range chromatin loops link topological domains to facilitate a diverse antibody repertoire. Cell Rep 2016;14:896–906. https://doi.org/10.1016/j.celrep.2015.12.083.

[380] Roldán E, Fuxa M, Chong W, Martinez D, Novatchkova M, Busslinger M, et al. Locus "decontraction" and centromeric recruitment contribute to allelic exclusion of the immunoglobulin heavy-chain gene. Nat Immunol 2005;6:31–41. https://doi.org/10.1038/ni1150.

[381] Sayegh C, Jhunjhunwala S, Riblet R, Murre C. Visualization of looping involving the immunoglobulin heavy-chain locus in developing B cells. Genes Dev 2005;19:322–7. https://doi.org/10.1101/gad.1254305.

[382] Hesslein DGT, Pflugh DL, Chowdhury D, Bothwell ALM, Sen R, Schatz DG. Pax5 is required for recombination of transcribed, acetylated, 5′ IgH V gene segments. Genes Dev 2003;17:37–42. https://doi.org/10.1101/gad.1031403.

[383] Liu H, Schmidt-Supprian M, Shi Y, Hobeika E, Barteneva N, Jumaa H, et al. Yin Yang 1 is a critical regulator of B-cell development. Genes Dev 2007;21:1179–89. https://doi.org/10.1101/gad.1529307.

[384] Yancopoulos GD, Desiderio SV, Paskind M, Kearney JF, Baltimore D, Alt FW. Preferential utilization of the most JH-proximal VH gene segments in pre-B-cell lines. Nature 1984;311:727–33. https://doi.org/10.1038/311727a0.

[385] Hansen AS, Cattoglio C, Darzacq X, Tjian R. Recent evidence that TADs and chromatin loops are dynamic structures. Nucleus 2018;9:20–32. https://doi.org/10.1080/19491034.2017.1389365.

[386] Ebert A, McManus S, Tagoh H, Medvedovic J, Salvagiotto G, Novatchkova M, et al. The distal VH gene cluster of the Igh locus contains distinct regulatory elements with Pax5 transcription factor-dependent activity in pro-B cells. Immunity 2011;34:175–87. https://doi.org/10.1016/j.immuni.2011.02.005.

[387] Verma-Gaur J, Torkamani A, Schaffer L, Head SR, Schork NJ, Feeney AJ. Noncoding transcription within the Igh distal VH region at PAIR elements affects the 3D structure of the Igh locus in pro-B cells. Proc Natl Acad Sci USA 2012;109:17004–9. https://doi.org/10.1073/pnas.1208398109.

[388] Subrahmanyam R, Du H, Ivanova I, Chakraborty T, Ji Y, Zhang Y, et al. Localized epigenetic changes induced by D H recombination restricts recombinase to DJ H junctions. Nat Immunol 2012;13:1205–12. https://doi.org/10.1038/ni.2447.

[389] Guo C, Gerasimova T, Hao H, Ivanova I, Chakraborty T, Selimyan R, et al. Two forms of loops generate the chromatin conformation of the immunoglobulin heavy-chain gene locus. Cell 2011;147:332–43. https://doi.org/10.1016/j.cell.2011.08.049.

[390] Bhat KH, Priyardarshi S, Naiyer S, Qu X, Farooq H, Kleiman E, et al. An Igh novel enhancer modulates antigen receptor diversity by determining locus conformation. Nat Commun 2022:492988. https://doi.org/10.1101/2022.05.23.492988.

[391] Banigan EJ, Mirny LA. Loop extrusion: theory meets single-molecule experiments. Curr Opin Cell Biol 2020;64:124–38. https://doi.org/10.1016/j.ceb.2020.04.011.

[392] de Almeida CR, Hendriks RW, Stadhouders R. Dynamic control of long-range genomic interactions at the immunoglobulin κ light-chain locus. Adv Immunol 2015;128:183–271. https://doi.org/10.1016/bs.ai.2015.07.004.

[393] Stadhouders R, de Bruijn MJW, Rother MB, Yuvaraj S, de Almeida CR, Kolovos P, et al. Pre-B cell receptor signaling induces immunoglobulin κ locus accessibility by functional redistribution of enhancer-mediated chromatin interactions. PLoS Biol 2014;12:e1001791. https://doi.org/10.1371/journal.pbio.1001791.

[394] Novobrantseva TI, Martin VM, Pelanda R, Müller W, Rajewsky K, Ehlich A. Rearrangement and expression of immunoglobulin light chain genes can precede heavy chain expression during normal B cell development in mice. J Exp Med 1999;189:75–88. https://doi.org/10.1084/jem.189.1.75.

[395] Xiang Y, Park S-K, Garrard WT. Vκ gene repertoire and locus contraction are specified by critical DNase I hypersensitive sites within the Vκ-Jκ intervening region. Jikken Igaku 2013;190:1819–26. https://doi.org/10.4049/jimmunol.1203127.

[396] Liu Z, Widlak P, Zou Y, Xiao F, Oh M, Li S, et al. A recombination silencer that specifies heterochromatin positioning and ikaros association in the immunoglobulin κ locus. Immunity 2006;24:405–15. https://doi.org/10.1016/j.immuni.2006.02.001.

[397] Luo S, Zhang J, Kreutzberger AJB, Eaton A, Edwards RJ, Jing C, et al. An antibody from single human VH-rearranging mouse neutralizes all SARS-CoV-2 variants through BA.5 by inhibiting membrane fusion. Science Immunology 2022;7:eadd5446. https://doi.org/10.1126/sciimmunol.add5446.

[398] Xiang Y, Zhou X, Hewitt SL, Skok JA, Garrard WT. A multifunctional element in the mouse Igκ locus that specifies repertoire and Ig loci subnuclear location. J Immunol 2011;186:5356–66. https://doi.org/10.4049/jimmunol.1003794.

[399] Lucas JS, Zhang Y, Dudko OK, Murre C. 3D trajectories adopted by coding and regulatory DNA elements: first-passage times for genomic interactions. Cell 2014;158:339–52. https://doi.org/10.1016/j.cell.2014.05.036.

[400] Khanna N, Zhang Y, Lucas JS, Dudko OK, Murre C. Chromosome dynamics near the sol-gel phase transition dictate the timing of remote genomic interactions. Nat Commun 2019;10:2771. https://doi.org/10.1038/s41467-019-10628-9.

[401] Barajas-Mora EM, Lee L, Lu H, Valderrama JA, Bjanes E, Nizet V, et al. Enhancer-instructed epigenetic landscape and chromatin compartmentalization dictate a primary antibody repertoire protective against specific bacterial pathogens. Nat Immunol 2023;24:320–36. https://doi.org/10.1038/s41590-022-01402-z.

[402] Barajas-Mora EM, Kleiman E, Xu J, Carrico NC, Lu H, Oltz EM, et al. A B-Cell-Specific enhancer orchestrates nuclear architecture to generate a diverse antigen receptor repertoire. Mol Cell 2019;73:48–60. https://doi.org/10.1016/j.molcel.2018.10.013.

[403] Ebert A, Hill L, Busslinger M. Chapter three - spatial regulation of V–(D)J recombination at antigen receptor loci. In: Murre C, editor. Advances in immunology, vol. 128. Academic Press; 2015. p. 93–121. https://doi.org/10.1016/bs.ai.2015.07.006.

[404] Seitan VC, Hao B, Tachibana-Konwalski K, Lavagnolli T, Mira-Bontenbal H, Brown KE, et al. A role for cohesin in T-cell-receptor rearrangement and thymocyte differentiation. Nature 2011;476:467–71. https://doi.org/10.1038/nature10312.

[405] Majumder K, Koues OI, Chan EAW, Kyle KE, Horowitz JE, Yang-Iott K, et al. Lineage-specific compaction of Tcrb requires a chromatin barrier to protect the function of a long-range tethering element. J Exp Med 2014;212:107–20. https://doi.org/10.1084/jem.20141479.

[406] Chen L, Zhao L, Alt FW, Krangel MS. An ectopic CTCF binding element inhibits tcrd rearrangement by limiting contact between Vδ and Dδ gene segments. J Immunol 2016;197:3188–97. https://doi.org/10.4049/jimmunol.1601124.

[407] Chen L, Carico Z, Shih H-Y, Krangel MS. A discrete chromatin loop in the mouse Tcra-Tcrd locus shapes the TCRδ and TCRα repertoires. Nat Immunol 2015;16:1085–93. https://doi.org/10.1038/ni.3232.

[408] Thompson CB. New insights into V(D)J recombination and its role in the evolution of the immune system. Immunity 1995;3:531–9. https://doi.org/10.1016/1074-7613(95)90124-8.

[409] Carmona LM, Schatz DG. New insights into the evolutionary origins of the recombination-activating gene proteins and V(D)J recombination. FEBS J 2017;284:1590–605. https://doi.org/10.1111/febs.13990.

[410] Fugmann SD. The origins of the RAG genes – from transposition to V(D)J recombination. Semin Immunol 2010;22:10–6. https://doi.org/10.1016/j.smim.2009.11.004.

[411] Fugmann SD, Messier C, Novack LA, Cameron RA, Rast JP. An ancient evolutionary origin of the Rag1/2 gene locus. Proc Natl Acad Sci USA 2006;103:3728–33. https://doi.org/10.1073/pnas.0509720103.

[412] Morales Poole JR, Huang SF, Xu A, Bayet J, Pontarotti P. The RAG transposon is active through the deuterostome evolution and domesticated in jawed vertebrates. Immunogenetics 2017;69:391–400. https://doi.org/10.1007/s00251-017-0979-5.

[413] Martin EC, Vicari C, Tsakou-Ngouafo L, Pontarotti P, Petrescu AJ, Schatz DG. Identification of RAG-like transposons in protostomes suggests their ancient bilaterian origin. Mobile DNA 2020;11:17. https://doi.org/10.1186/s13100-020-00214-y.

[414] Tao X, Huang Z, Chen F, Wang X, Zheng T, Yuan S, et al. The RAG key to vertebrate adaptive immunity descended directly from a bacterial ancestor. Natl Sci Rev 2022;9:nwac073. https://doi.org/10.1093/nsr/nwac073.

[415] Huang S, Tao X, Yuan S, Zhang Y, Li P, Beilinson HA, et al. Discovery of an active RAG transposon illuminates the origins of V(D)J recombination. Cell 2016;166:102–14. https://doi.org/10.1016/j.cell.2016.05.032.

[416] Zhang Y, Cheng TC, Huang G, Lu Q, Surleac MD, Mandell JD, et al. Transposon molecular domestication and the evolution of the RAG recombinase. Nature 2019;569:79–84. https://doi.org/10.1038/s41586-019-1093-7.

[417] Kapitonov VV, Jurka J. Molecular paleontology of transposable elements in the *Drosophila melanogaster* genome. Proc Natl Acad Sci USA 2003;100:6569–74. https://doi.org/10.1073/pnas.0732024100.

[418] Kapitonov VV, Jurka J. RAG1 core and V(D)J recombination signal sequences were derived from Transib transposons. PLoS Biol 2005;3:e181. https://doi.org/10.1371/journal.pbio.0030181.

[419] Hencken CG, Li X, Craig NL. Functional characterization of an active Rag-like transposase. Nat Struct Mol Biol 2012;19:834–6. https://doi.org/10.1038/nsmb.2338.

[420] Liu C, Yang Y, Schatz DG. Structures of a RAG-like transposase during cut-and-paste transposition. Nature 2019;575:540–4. https://doi.org/10.1038/s41586-019-1753-7.

[421] Carmona LM, Fugmann SD, Schatz DG. Collaboration of RAG2 with RAG1-like proteins during the evolution of V(D)J recombination. Genes Dev 2016;30:909–17. https://doi.org/10.1101/gad.278432.116.

[422] Chatterji M, Tsai C-L, Schatz DG. Mobilization of RAG-generated signal ends by transposition and insertion in vivo. Mol Cell Biol 2006;26:1558–68. https://doi.org/10.1128/MCB.26.4.1558-1568.2006.

[423] Reddy YVR, Perkins EJ, Ramsden DA. Genomic instability due to V(D)J recombination-associated transposition. Genes Dev 2006;20:1575–82. https://doi.org/10.1101/gad.1432706.

[424] Curry JD, Schulz D, Guidos CJ, Danska JS, Nutter L, Nussenzweig A, et al. Chromosomal reinsertion of broken RSS ends during T cell development. J Exp Med 2007;204:2293–303. https://doi.org/10.1084/jem.20070583.

Chapter 3

The Origin of V(D)J Diversification

Ellen Hsu[1] and Susanna M. Lewis[2]

[1]*Department of Physiology and Pharmacology, The State University of New York Health Science Center at Brooklyn, Brooklyn, NY, United States;*
[2]*Department of Pathology and Laboratory Medicine, Perelman School of Medicine, University of Pennsylvania, Philadelphia, PA, United States*

1. INTRODUCTION

The vertebrate immune system can respond to a broad and ever-changing spectrum of pathogens, and V(D)J recombination plays a central role in achieving this stunning versatility. V(D)J recombination has ancient origins; its first appearance in vertebrates is estimated at about 450 million years ago. The selective forces that set the stage and directed the evolutionary process have intrigued investigators ever since the discovery that the DNA of immunoglobulin (Ig) and T cell receptor (TCR) genes rearranged. V(D)J recombination appears to have emerged apparently without antecedent in evolution, an observation suggesting that perhaps the essentials of the recombination machinery were acquired through horizontal transfer. The recent availability of fully sequenced genomes from a variety of organisms has enabled closer examination of this idea. In the first part of this chapter we examine the hypothesis that the V(D)J recombination machinery is derived from a DNA transposon that was "exapted" [1]. The second section discusses the possible effect of germline V(D)J recombination events on the creation of different Ig and TCR loci. The last part of the chapter considers the nature of the ur-V gene and reviews current thinking on how V(D)J recombination fits into the evolution of adaptive immunity as a whole.

2. THE ALIEN SEED

The discovery of V(D)J recombination was a breakthrough. In rapid succession, papers outlined the basic features of the process and defined sequence motifs marking recombination sites [2,3]. Noting similarities between V(D)J recombination targets and those of prokaryotic insertion sequences, Susumu Tonegawa first articulated the evolutionary implications:

> *We propose that such a mechanism was initiated when an IS-like DNA element was accidently inserted into one of the multiple V DNA copies of an ancestral polymeric gene, splitting it into two portions, one corresponding to the present day embryonic V DNA segment and the other to the J DNA segment. Thus the V-J joining may well be a reversal of an ancient accidental insertion of an IS-like DNA element which was subsequently exploited by vertebrates to increase the V-region diversity within the general framework of the intron-exon structure of higher cell genes [2].*

The idea was gradually fleshed out as more was learnt about V(D)J joining, but the essential features have held up (reviewed in Refs. [4–6]). After entry into the genome and perhaps no longer under selection for transposon function, it is thought that the DNA element decayed. Possibly multiple transposition events were involved, but ultimately terminal recognition sites of the transposon became the recombination signal sequences (RSS) seen as present-day V, D, and J gene segments. The transposase enzyme responsible for element mobilisation was recruited as the V(D)J recombinase, encoded by two recombination-activating genes, RAG1 and RAG2. The proposed correlation is shown in Figure 3.1. In a sense, the transposon theory is especially fitting in that an integral aspect of the adaptive immune system—chance and necessity—provided its own diversification machinery. Three mechanisms underlying the model: transposition, horizontal transfer and exaptation are presented in the sections to follow.

2.1 Transposons and Transposition

Transposable elements are mobile genetic elements that include (1) retrotransposons, where an RNA transcript of the element is copied by reverse transcriptase and the DNA copy is inserted at random locations into new sites of the genome, and (2) DNA transposons, which are segments of DNA able to move to new sites in a genome, generally in a "cut-and-paste" mode. Some transposons have a transient existence as infectious phage or virus-like particles when they cycle in and out

Molecular Biology of B Cells. https://doi.org/10.1016/B978-0-323-95895-0.00017-9

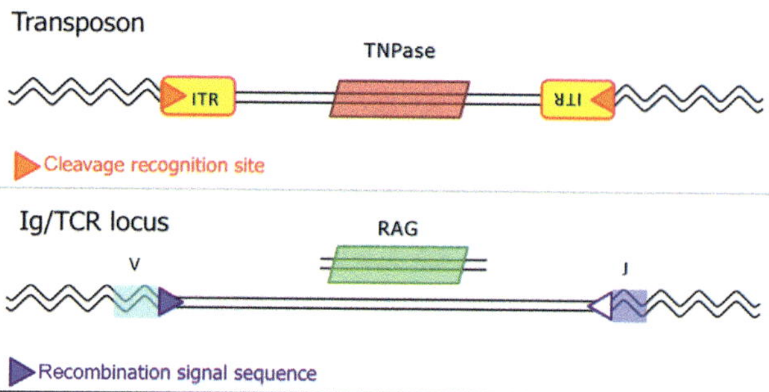

FIGURE 3.1 Similarities between transposition and VDJ recombination. *Top* panel shows a minimal autonomous transposon. The termini contain inverted terminal repeats (ITR, also called TIR) that contain transposase-binding sites. The outer edges of the ITRs also contain sequence motifs that define the two cleavage sites at the elements borders. In some transposons, the ITRs and the cleavage recognition motifs are the same. *Bottom* panel illustrates the basic features of a rearranging Ig or TCR locus. It has the features of a non-autonomous transposon because the transposase (RAG) is not part of the structure. The RSSs correspond to cleavage recognition sites that can still be site-specifically cleaved by RAG, encoded "in trans" as shown.

of genomes, examples being bacteriophage Mu or the HERV-K human endogenous retrotransposon. In DNA transposons that contain a minimal amount of coding sequence, their sole gene encodes the catalytic transposase, the protein(s) essential for mobility. Certain transposons have ancient origins, as implied by a ubiquitous distribution among living organisms. In a recent bioinformatic survey involving over 10 million protein-encoding genes from numerous species, transposase genes topped the list with respect to both abundance and representation in all sampled biomes [7]. Most were of the cut-and-paste class and have roots that extend far back in the history of life. The representation of certain structural folds in their transposases and the ability to bind Mn^{2+} suggest they originated in an Archaean world over 2.5 billion years ago [8].

In addition to a transposase gene, autonomous or fully functional DNA transposons must also carry recognition sequences for transposase binding and cleavage, and these are situated within inverted terminal repeats (ITR) at either end of the element (Figure 3.1, upper panel). Although a transposon may contain other genes, their main function requires simply a transposase gene flanked by ITRs. Typically the number of transposon copies will wax and wane after invading a genome, while decayed "non-autonomous" versions of the elements steadily accumulate. The corrupted non-autonomous elements often have internal defects that destroy the transposase gene, and they can move only if transposase is available "in trans," that is, from other autonomous copies. A copy that lost one or both ITR is no longer able to move at all and subject to further mutational degeneration. Eventually, production of transposase by the remaining few autonomous copies becomes insufficient for self-mobilisation in the presence of far larger numbers of non-autonomous elements. When the invasion subsides, the genome is left littered with non-autonomous element "fossils" [9].

The current understanding of the mechanism of transposition for the major class of DNA cut-and-paste transposons, including all known DNA transposons in eukaryotes [10], is based on close analysis of a few key examples [11–13]. The relevant transposases cleave and join DNA strands via a sequence of metal-dependent phosphoryl transfers with water or the 3′ OH group of a broken DNA strand serving as the nucleophile. The catalytic site of the transposases contains a "DDE" or "DDD" amino acid motif comprising three negatively charged residues (Asp, Asp, Glu/Asp) incorporated into a characteristic RNase H-like fold. Though details vary for different DDE/D transposons, a successful round of transposition invariably initiates with a pair of hydrolytic nicks at the outside edge of each cleavage motif, that is, at the left and right boundary of the ITRs. According to which set of strands is broken, 3′ hydroxyls are associated either with the flanking DNA or with the element ends. This process is illustrated in Figure 3.2A and explained in the legend. Depending on the position of the first nick, excision by transesterification will create hairpins either at the flanking DNA ends or at transposon ends. Once a transposon with exposed 3′ hydroxyls is created (Figure 3.2A, step 2), DNA cut-and-paste transposases carry out the integration step the same way. A random site in the host genome is captured, and the transposon is inserted into the recipient target DNA (Figure 3.2A, step 3).

Intermediates in transposition occur in the context of a protein-DNA complex that undergoes structural rearrangements during the course of the integration process. The initial paired end complex comprises a multimer of transposase molecules in contact with the transposon DNA ends. Protein–protein contacts may shift, and different DNAs may move in and out of the catalytic site(s). At the integration step, a "strand transfer complex" has formed with the genomic DNA that in some cases is so stable the transposase must be proteolytically removed [12].

Host repair factors ligate broken DNA strands left by the transposase. One leftover break remains at the exit site, and two single strand gaps are found at each 5′ end of the transposon. Target site duplications (TSDs) arise from repair of the two single strand gaps that are left at either end of the transposon (Figure 3.2A, step 4). The length of the TSDs (~9 bp or less) reflects the staggered nature of the strand interruptions in the new target site and is a characteristic feature of a given transposon family. In general, the presence of TSDs helps identify novel transposons where

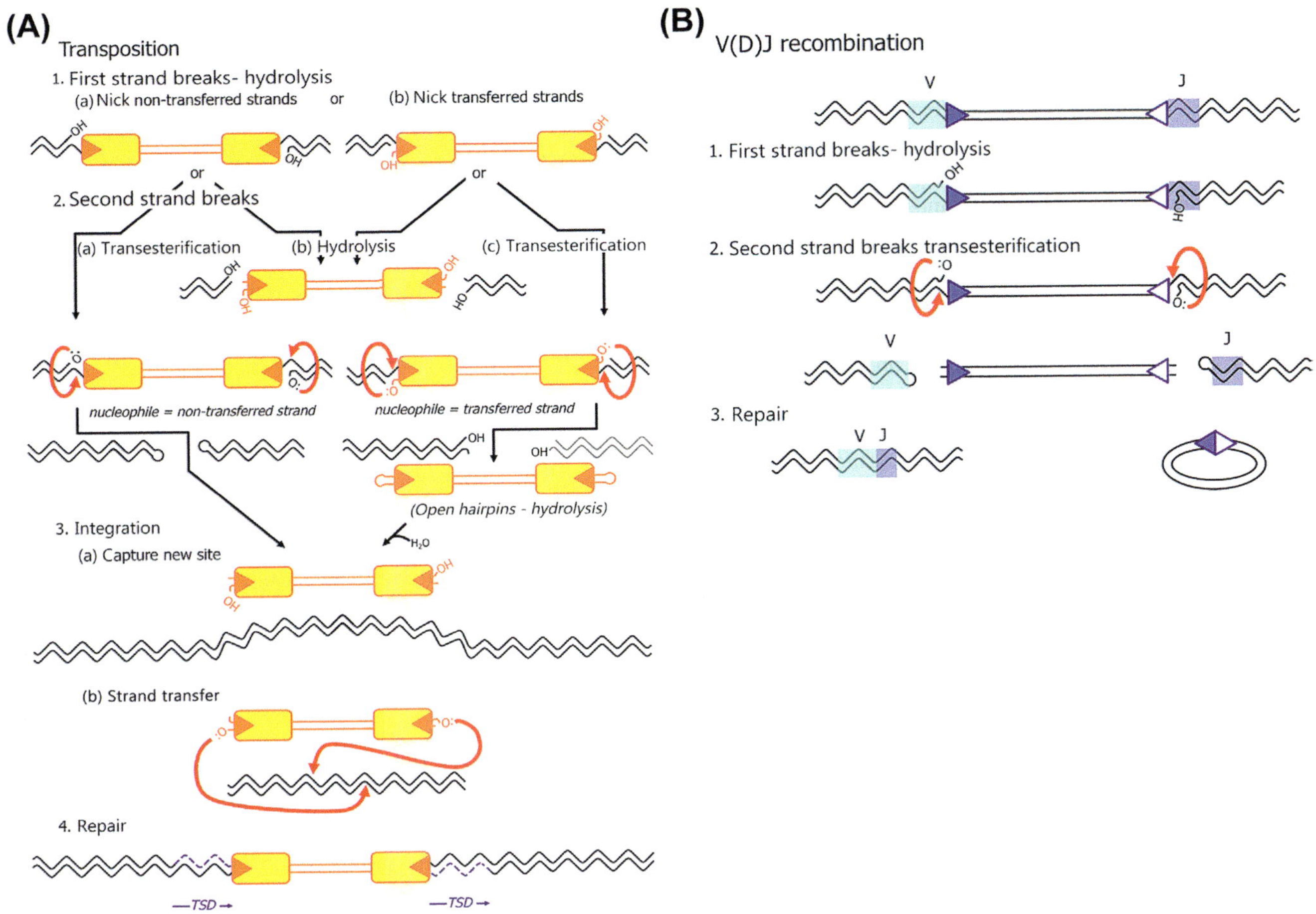

FIGURE 3.2 **Transposition compared to V(D)J recombination.** (A) **Transposition**. Transposons mobilised by DDE family transposases exit and enter DNA through phosphoryl transfers. Excision and integration take place in the context of a protein-DNA complex made up of the transposase (not depicted) and the DNA ends of the element. Step 1. Hydrolytic strand breaks. The two strands at either end of a transposon are differentiated by the designation "non-transferred strand" or "transferred strand," referring to which strand becomes chemically linked to the new integration site in the final strand transfer step (see below). Step 2. The second strands are cut by hydrolysis or transesterification [2(a) or (b)]. (a) If the first nick occurred on the non-transferred strand, transesterification will release a linear transposon DNA with conventional termini but leave hairpin structures at the donor break site. (b) Alternatively, the second strand break is achieved by hydrolysis. Both liberated transposon and donor break site will have 3′ hydroxyls and 5′ phosphates. (c) Where the first nick was made on the transferred strand, the transposon's 3′ hydroxyls can serve as the nucleophiles in transesterification. If so, hairpins are formed at the transposon's ends, while conventional DNA ends occur at the donor break site. An additional step (hydrolysis) is required to nick open the hairpin termini prior to integration. Step 3. Integration by transesterification. (a) Once free 3′ hydroxyls are created at each transposon end, the complex can capture the new integration target site. (b) Transesterification uses the available transposon hydroxyls to interrupt the target site. The breaks in the integration target are usually staggered, so that the integrated transposon has single strand gaps at each side, typically of 2–9 bp. Step 4. Repair is carried out by host polymerase and ligase functions, with the result that 2–9 bp TSDs, characteristic of a given transposon, are created and reflect the placement of the strand breaks at the integration step. (B) **V(D)J recombination**. Step 1. The catalytic RAG protein (not depicted) makes a first strand break by hydrolysis. The first cleavage is on the non-transferred strand; the 3′ hydroxyls are associated with the flanking V and J gene segments rather that the RSS cleavage targets. Step 2. The second break is introduced by transesterification, with the coding ends bearing hairpin termini. Step 3. Processing and joining of ends are carried out by components of the non-homologous end-joining pathway of DNA double strand break repair.

they appear as non-annotated repeats in genome assemblies of plants, animals, and microbes.

2.2 Horizontal Transfer

Horizontal or "lateral" transfer is a term used when genetic material is passed between different species or even phyla instead of being vertically transmitted from parent to offspring [14]. Foreign DNA such as transposons can be introduced when brought in by a virus, parasite, or symbiont, or even perhaps in the diet [15]. Horizontal transfer is generally suspected when phylogenetic relationships are in conflict, such as when sequences in two different species seem more similar than anticipated or when a class of repeats has a patchy distribution, being present or wholly absent from related species in a way that is incongruent with established kinships [16]. The possibility of horizontal transfer can be tested in various ways such as by comparing mutation rates in the suspect sequences with nuclear genes of the organism.

Initially, there were two reasons to suspect that horizontal transfer played a part in creating the V(D)J recombination apparatus, though neither fits the above criteria. One reason, as mentioned, was that site-directed recombination is unusual in vertebrates, and features of the V(D)J recombination system seemed reminiscent of prokaryotic transposons. Another was that V(D)J recombination and adaptive immunity appeared to have arisen somewhat abruptly in vertebrate evolution, constituting what has been termed a "big bang" [17]. There is an estimated 50–70 million year interval between the divergence of jawless fishes and the jawed fishes, today represented respectively by lampreys and hagfish and by cartilaginous fishes; intervening lineages are extinct. In this period an immune system arose with all of the basic elements, including V(D)J recombination, and is found from sharks to human beings. Wholesale import of a pre-formed recombination capacity seemed to greatly simplify the possible schemes for how this could have taken form.

Unfortunately the theory that V(D)J recombination was acquired from elsewhere does not lend itself to standard tests for horizontal transfer. Tens of millions of years can obscure evidence of transmission so early in vertebrate evolution. There are no detectable differences between nucleotide substitution rates in RAG1 or RAG2 and the genomes in which they reside, and in fact evolutionists have used the RAG1 and RAG2 genes as representative markers in phylogeny studies [18]. However, in the next sections other observations are presented, supporting the idea that V(D)J recombination originated outside the vertebrate lineage in which it became a fixture.

2.3 Exaptation

In contrast to adaptation, where incremental improvements are selected to build a structure over time, exaptation describes the process of using pre-formed components to create a new function [1]. As such, transposons and/or their proteins have been exapted for new purposes. Transposons have been transformed into structural components of centromeres or telomeres, have provided transcription factors, and became recombinases in mating type switching and macronuclear differentiation as well as contributing regulatory elements [13,19]. How the putative ancestral RAG transposon was domesticated is perhaps the biggest mystery surrounding the origin of V(D)J recombination. Although we lack a detailed idea of what sort of organism was the involuntary host of the ancient RAG transposon, new information has begun to suggest profitable directions to pursue.

2.4 Modus Operandi

Parallels to transposition were strengthened once RAG proteins were identified and site-specific cleavage activity was reconstructed in vitro [20–24]. The first strand breaks in V(D)J recombination are hydrolytic nicks that occur at the 5′ side of the RSS edge. These nicks are positionally equivalent to an initial hydrolysis of a transposon's nontransferred strand (step 1 in Figure 3.2A,B). The second strand break is accomplished by transesterification [22,25], with the 3′hydroxyl generated at the first break acting as the acceptor. This is again consistent with a transposase-catalysed cleavage pattern. The second break establishes hairpin connections between the top and bottom strands at the V, D, or J ends. The coding end hairpins correspond to hairpins on flanking broken ends in transposition, and as such, VDJ recombination is overall analogous to transposon excision as it occurs in the eukaryotic DNA transposons, Hermes and Hztransib [26,27].

Testing of experimentally generated RAG mutants has established that three amino acids, two aspartic and one glutamic acid, are critical for RAG1 nuclease function, and they occur with spacing consistent with a legitimate DDE motif [28–30]. DDE motifs are characteristic of enzymes that carry out metal-dependent phosphoryl transferase reactions, which accords with the Mg^{2+} dependence of RAG1 and RAG2 cleavage activity in vitro.

Regardless of the parallels, a fundamental difference between V(D)J recombination and transposition is that the former involves manipulation of two DNA segments, but the latter involves three. V(D)J recombination corresponds to transposition events that stop before any integration-related activities takes place. In this view of V(D)J recombination as aborted transposition, the RSS-containing "signal ends" correspond to the liberated transposon that would normally be held in a post-cleavage complex. However, the cleaved signal ends as well as that of the V or J coding ends are processed by the cellular DNA repair operations.

Would the hypothetical RAG transposase have preserved transposition-specific functions after 450 million years of disuse? The answer was provided by two groups that independently demonstrated RAG1 and RAG2 could indeed carry out complete transposition in vitro [31,32]. Given appropriate DNA substrates and conditions, RAG proteins cleaved a pair of RSSs, excise the intervening DNA segment, captured a DNA target site, and transferred the RSS 3′ ends. The transposition products exhibited short but somewhat variable TSDs of about 5 bp. Such duplications are not associated with any aspect of normal V(D)J recombination and provided an unambiguous transposition signature. These observations were subsequently extended to the in vivo context. The occurrence of reproducible low-frequency transposition events was observed when expressing RAG1 and RAG2 in yeast cells [33], and in vivo transposition in mouse cells was confirmed with a specifically designed detection system [34]. In the latter study, the ratio of transposition to V(D)J recombination was measured at about 1: 50,000. By extrapolation to humans, an adult will have hosted 10,000 transposition events during each day of lymphocyte production; this estimate raises interesting questions about how transposition by RAG1/2 is normally curtailed [35]. In retrospect, the retention of vestigial transposase functions by the RAG proteins is probably not accidental and indicates instead an inseparability of cleavage and integration functions [12,36].

2.5 Structure and Linkage of RAG1 and RAG2 Genes

The RAG1 and RAG2 genes are closely linked and in convergent transcriptional orientation in animals representing a variety of taxa: chickens, rabbits, humans, mice, *Xenopus*, pufferfish, zebrafish, and sharks [37]. The coding sequences of the RAG1 and RAG2 genes are each fully contained in a single exon in most of these animals. The exception is that an intron is present in the middle of the RAG1 coding sequences of ray-finned fish [38].

Both of the above features, the tight and conserved linkage between the structurally unrelated RAG1 and RAG2 genes and a paucity of introns, are consistent with a prokaryotic origin [21,39]. In bacteria, genes for proteins belonging to the same metabolic pathway are often grouped together in coregulated operons. Their introns, especially those that are not self-splicing, are rare whereas the coding regions in human genes on average are interrupted by 7.2 introns each [40]. In short, the structural features of the RAG1/RAG2 locus are exceptional and contribute to the evidence that favours their horizontal transfer from a prokaryotic source.

2.6 Fossil Transposons

Evidence presented thus has led to a general acceptance of the hypothesis that key elements of the V(D)J recombination system were indeed derived from a transposon. What was the structure of the "RAG transposon" and at what point in animal evolution did the presumed horizontal transmission event(s) occur?

When genome analyses began on species representing key radiations, there arose an opportunity to pursue the origins of V(D)J recombination from a paleogenomic perspective. At present somewhere between one to two dozen eukaryotic transposon superfamilies have been discovered, one of which is called Transib [9]. Fossil Transib elements are present in insects, hydra, echinoderms, and fungi, though detected in neither plants nor vertebrates. A transposase sequence reconstructed from multiple fossil copies revealed an identifiable DDE motif and exhibited some similarity to RAG1 [41]. Notably, some Transib fossils are flanked by 5 bp duplications, as seen in the majority of in vitro and in vivo RAG1/2 transposition products. The sequences of some Transib ITRs were similar to the RSSs in that abutting the cleavage site is a heptamer-like sequence followed by nonamer-like sequence after an interval of 23 bp. Some Transib transposons carry a 23 bp RSS-like sequence in one ITR and a 12 bp spacer RSS in the other, suggestive of the 12/23 rule. RAG1 similarity has also been reported in transposons belonging to other eukaryotic transposon superfamilies, ([42] and citations therein), but additional RAG1 signature features distinguish Transib [10,41].

An intact Transib element was identified in the cotton bollworm, *Helicoverpa zea*, with terminal motifs (CACGGTGG) including underlined heptamer-like sequence and an open reading frame encoding 507 amino acids [43]. The Hztransib transposase was expressed, purified from *Escherichia coli*, and investigated for biochemical activities in vitro [27]. Just as for RAG1/2-mediated transposition, the Hztransib transposase was able to nick an appropriate substrate DNA between the ITR end and the flanking sequences and to transfer those ends to a new DNA site. The formation of hairpin ends at the flanking host break was also demonstrated, as were the expected 5 bp TSDs, visualised upon bacterial replication of the in vitro generated transposition products. Mutations at the DDE triad affected nicking and hairpinning activities to varying extents. The features of the Hztransib transposon suggest that some type of RAG1 transposon could have existed at one time. How RAG2 might fit into this picture is not clear.

The Transib sequences had little sequence identity (14%−17%) to the RAG1 core, which is the 625 amino acid portion of RAG1 containing its catalytic activity [44]. A separate set of sequences with greater similarity to RAG1 core (26%−35%) was detected in lancelet, hydra, and sea anemone; on the other hand, they lacked any evidence of ITRs or other indicators of transposition [41]. The relationship between these two sets, which are non-overlapping in their RAG1 homology, is not clear.

More than two dozen RAG1 core-like sequences were detected in sea urchin genomic sequence. They were nearly all fragmented or disrupted; however, one encoded an apparent RAG1 homologue with a large open reading frame. The gene, named SpRAG1L, is transcribed and encodes a 983 amino acid sequence that is 31% identical to the mouse RAG1 core [45]. SpRAG1L moreover possessed sequence similarities necessary for the function of the

RAG1/2 complex. Not only did these include the DDE motif and a domain implicated in binding the RSS, but also conserved was a RAG1 zinc finger that is thought to be important for interaction with RAG2. These similarities and others involve a considerably larger portion of RAG1 than the Transib transposases.

Previously there had been fruitless searches for RAG2 genes in diverse organisms. Taking a different tack, Fugmann and coworkers [45] focused attention specifically on the region downstream from SpRAG1L and discovered a sequence with low but discernible identity to RAG2. This open reading frame was 3181 bp away from SpRAG1L, and as in the vertebrate RAG1/2 locus, was oriented in a convergent transcriptional orientation.

2.7 RAG2-like Gene in Sea Urchin

The mammalian RAG2 protein has two distinct features. In silico analysis revealed a series of "kelch" repeats in RAG2 [46], and structural studies in other systems have shown that kelch repeats can fold into a propeller-like structure that supports intermolecular protein–protein and protein–DNA interactions. Consistent with this model, spontaneous and directed mutations of RAG2 in the β propeller can reduce or fully abrogate both cleavage and RAG1/RAG2 association in vitro whereas mutations outside the β propellers tend to have a much milder effect [47,48]. A plant homeodomain (PHD) motif occurs near the C terminus of RAG2 [46]. The PHD motif occurs in a region of the protein that is not essential for in vitro V(D)J recombination [44], but mutations in this motif are pathogenic in humans. Several groups have determined that the PHD domain in RAG2 binds hypermethylated histone H3 modified on lysine 4 (H3K4), which is a general mark for active chromatin (reviewed in Ref. [49]).

The sea urchin SpRAG2L gene encodes a predicted protein that, as with vertebrate RAG2, has a six-bladed propeller structure comprising kelch repeats. C-terminal to the kelch repeats is an identifiable PHD domain. There are many proteins in the sea urchin genome that contain either a predicted propeller *or* a PHD motif, but SpRAG2L is unique in possessing both. It has been demonstrated that the PHD domain of SpRAG2L, like RAG2, binds hypermethylated lysine 4 in H3 tails, although the former recognises the dimethyl modification, whereas the latter favours the trimethylated form [50]. The identities of the sea urchin proteins as homologues of RAG1 and RAG2 are supported by additional studies showing regulated co-expression of their genes, as well as evidence of protein–protein associations in vitro [45]. At this time the function of SpRAG1L and SpRAG2L is an unsolved question.

SpRAG1L/2L shows no evidence of having been recently brought into the sea urchin genome by a transposition event. Unlike most vertebrate RAG genes, the protein-coding region of both SpRAG1L and SpRAG2L apparently have been in place long enough to have acquired multiple introns [45]. Recent work suggests that such genes are likely to be present throughout eleutherozoan echinoderms (sea urchins to starfish) (J. Rast, pers. commun.). If the SpRAG1L/2L genes are a sea urchin equivalent of the RAG1/2 locus in jawed vertebrates, are these loci related by descent?

At the moment, no animal evolving from a lineage that diverged between echinoderms and cartilaginous fishes harbours both RAG1 and RAG2-like genes, and certainly none has evidence of a locus where they are linked in a tail-to-tail arrangement (Figure 3.3). The genome of amphioxus (cephalochordate) contains RAG1 core-like genes [61,62] and those of two divergent tunicate (urochordate) species, *Ciona intestinalis* and *Oikopleura dioica*, do not have any at all (Ref. [63] and L. Du Pasquier, pers. commun.). About 85% of the sea lamprey genome has been analysed to date without uncovering RAG homologues [53], but since this is a crucial position—jawless fish are the only surviving vertebrate precedents before the emergence of cartilaginous fishes—no conclusion can be drawn until the lamprey genome assembly has been completely examined.

One recent hypothesis is that during the time RAG1 core-like sequences were inserted into various lineages by a RAG1 transposon, one integrated next to an existing RAG2 gene in an ancestral deuterostome [5,6]. The primordial RAG1 and RAG2 coupling took place with this insertion event, and SpRAG1L/2L is a descendent; in this scenario no RAG1/2 transposon ever existed as such. However, if this unique RAG1/2 confection had been established at that time and was thus incapable of subsequent movement, the expectation would be to detect a RAG1/2 locus in some animals whose divergence preceded jawed vertebrates. If it turns out that there is no RAG1/2 locus in the lamprey—just as there are no Ig, TCR, or major histocompatibility (MHC) genes—then it is necessary to explain how RAG1/2 apparently was discarded or mutationally degraded in so many organisms and yet survived to evolve into a critical role in the stem vertebrate.

The alternate idea is that the RAG1/2 locus was contained in a RAG1/2 transposon that independently inserted into echinoderms and into an ancestral vertebrate [5]. Then the absence of a RAG1/2 locus in the lamprey genome, if established, would constitute the type of "patchy" distribution that is suggestive of horizontal transfer [16]. This scheme places no constraint on whether or not the RAG1/2 locus should be present in jawless fish, but asks for some

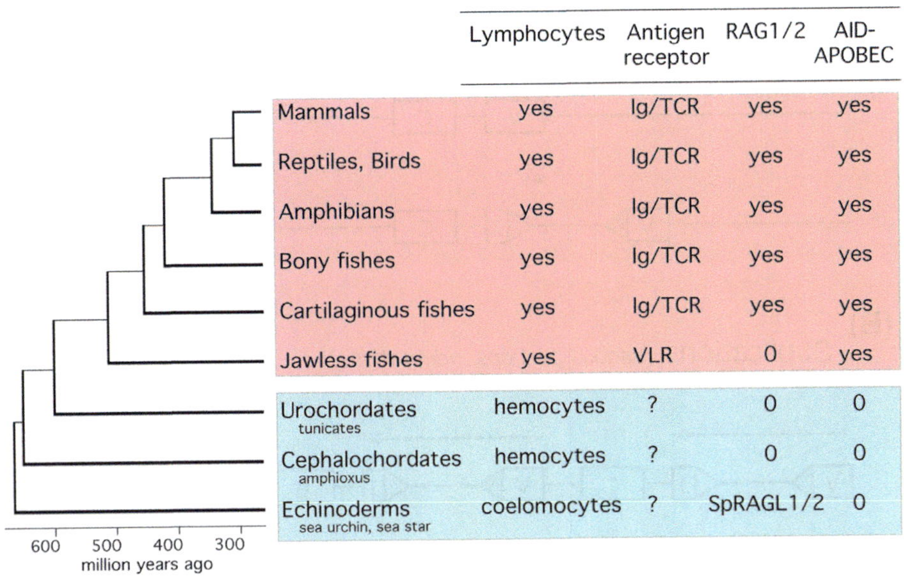

FIGURE 3.3 **Adaptive immune system components in deuterostome animals.** Vertebrates (top box) include gnathostomes and jawless fishes; invertebrates (bottom box) protochordates and echinoderms. The scale shows when taxa emerged according to molecular clock estimations [51,52]. Jawless fishes include both hagfish and lamprey, which express variable lymphocyte receptors (VLRs). Information on jawless fishes, protochordates and sea urchin is from Refs. [45,53–59]. *After Ref. [60], p. 2, copyright (2011), with permission from Elsevier.*

explanation for its relationship to the RAG1 core-like genes in a wide variety of non-vertebrate animals [41].

2.8 Current Questions: Transposon Origins, Horizontal Transfer, and Exaptation

There are many aspects of RAG structure–function to solve before we can begin to understand exaptation. Biochemical experiments have shown that RAG1 and RAG2 comprise a functional transposase. How a transposase can be converted to a V(D)J recombinase without also shedding the (unused) ability to capture and integrate into DNA is a question worthy of attention, as it may provide information on protein evolution, defence mechanisms, and RAG1/2 catalysis. In tracing the steps of how these proteins came to have their current role in antigen receptor diversification, discovering the function of the sea urchin SpRAG1L and SpRAG2L gene products would be especially informative. The presence or absence of RAG homologues in the lamprey and other species may close a gap (or establish one) in the evolutionary history of V(D)J recombination and should help clarify if and when horizontal transfer took place. All of these provide a way to test notions of horizontal transfer and exaptation. Where once the field had to wait on chance observations and what might be considered circumstantial evidence, it has been transformed, and definite experimental lines of inquiry can be pursued. Ultimately, it may be possible to define the transposon contribution to V(D)J recombination at a level of detail that includes a description of the transposons involved, the recipient organisms, and the status of the immune system at the time.

3. THE EVOLUTION OF BCR AND TCR LOCI

At the outset, as commonly depicted, a V-like gene belonging to the Ig superfamily was interrupted by a transposon insertion (Figure 3.4A). The transposing element possessed RSS-containing termini, but may or may not have contained RAG genes as well. It is not necessary that the actual RAG1/2 transposon itself inserted into the V gene; as discussed previously, a fossil transposon with intact termini can be mobilised “in trans” by the RAG transposase. Once the ability of RAG1/2 to carry out transposition was suppressed and redirected to V region assembly, the recombinase appears to have gone to work on the germ line and likely was instrumental in forming the different Ig and TCR loci. Figure 3.4 sketches out the RAG-mediated manipulations that are either known or postulated to have modified the antigen receptor loci during vertebrate evolution.

A preliminary step (Figure 3.4B) would have been duplications of the split gene. This must be assumed in order to create substrates for generating different configurations. There is plenty of support for ancient and ongoing segmental duplications throughout evolution [65], and duplications are unrestricted in regard to gene segment or locus boundaries. With respect to antigen receptor loci, there exist examples that involved only a sub-region of an

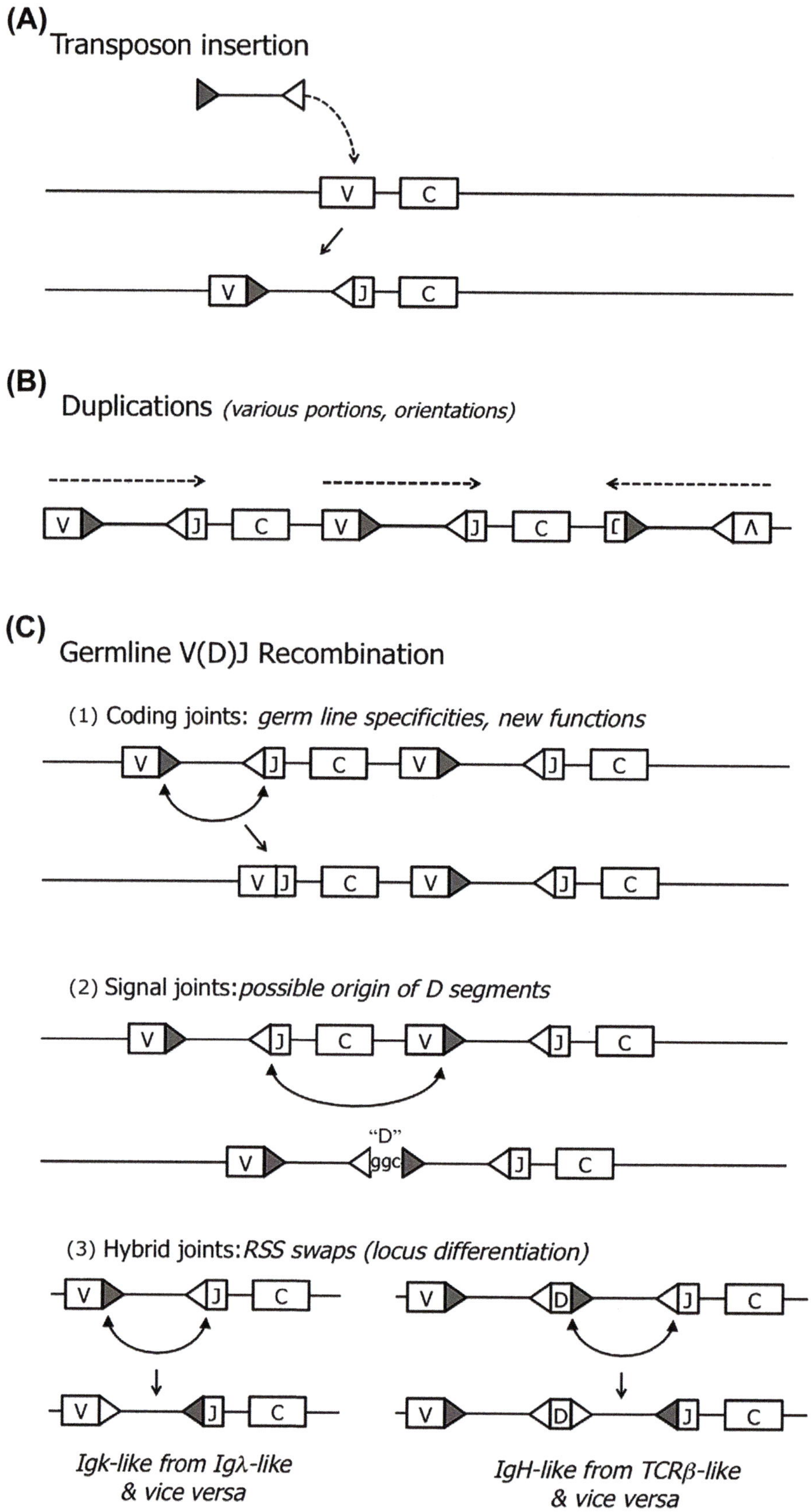

FIGURE 3.4 **Germ line rearrangement gives rise to different locus configurations.** (A) A transposon splits the ur-V gene, placing RSSs adjacent to the interrupted coding sequences. (B) Duplications of the interrupted unit take place. (C) (1) standard germ line V-to-J recombination creates fused germ line genes. (2) some types of recombination between the repeated clusters, such as J-to-V recombination as shown, can create a signal joint; signal joints are also created by V-to-J recombination between clusters that are inverted with respect to one another. Such events, with the insertion of non-templated "N-region" nucleotides into the signal joint, generate a proto-D segment. (3) A non-standard V(D)J recombination outcome, termed a "hybrid joint" is known to occur in vivo, and evidence exists that they arise in the germ line as well [64]. As depicted, hybrid joints effectively swap the 12 and 23 RSSs between two gene segments. The change in RSS affiliation brought about by aberrant hybrid joint formation may have helped drive the differentiation of Ig/TCR loci in evolution.

Ig locus [66] or alternatively an entire functional V(D)J/C unit, giving rise to Ig clusters in cartilaginous fish [67]. Once the split V-like gene was established in multiple copies, the basic elements for the creation of different loci and alternative configurations were in place.

3.1 Germline Coding Joint Formation: V(D)J Gene Segment Fusions

The strongest evidence that V(D)J recombination takes place in the germ line comes from examining Ig locus structures in sharks. Shark IgH genes occur as duplications of a simple [V-D-J/CH] recombination modules; likewise IgL genes are multiply reiterated [V-J/CL] clusters as depicted in Figure 3.4B [67,68]. In the genome of every elasmobranch species examined so far a number of Ig genes seem to have recombined elements [69]; for instance, some IgL genes have fused VJ instead of the "split" V-J or non-joined configuration (Figure 3.4C-(1)). It appeared as if these had recombined in the germ line. Since the leader intron was retained in the germline-rearranged IgH or IgL, they were not the result of retrotransposed mRNA. One suggestion was that these phenomena may have arisen from recombinase activity in germ cells, although V(D)J rearrangement was established as an activity crucial to lymphocyte development [64].

The germ line joining hypothesis is supported by studies that examined the nature of the VJ junctions, demonstrating that they were generated from non-joined gene segments [70]. By determining the phylogenetic relationship among joined and split IgL gene clusters among two shark species, it was possible to estimate roughly how often and when germ line joining events occurred and with respect to species divergence. One particular V to J recombination event was proposed to have occurred within the last 7 million years. Evidence also suggests that such germ line joining occurred not once but several times in the nurse shark genome. Thus in some species germ line V(D)J recombination, creating fused VJ and VDJ genes, is perhaps still ongoing.

What are consequences of germ line joining events? Joined genes in shark, catfish, chicken, and opossum are available for consideration. Although fully joined V(D)J segments were described in cartilaginous fishes a while ago [64], the role of their products as antibodies remains unclear. However, in the nurse shark there is one IgH cluster carrying a joined, in-frame VDJ that is unique in that its 3 C exons are derived but highly divergent from the 4 Cμ exons. The IgM1gj polypeptide associates with shark L chains, is prominent in neonatal sera, and is probably expressed exclusively as a secreted molecule [68,71]. Because it is not a surface receptor, IgM1gj is not expected to participate directly in activating the cell that expresses it. Without diversity and without the ability to induce clonal expansion, the IgM1gj has lost the main features of adaptive immunity and appears to have actively evolved to another function.

In the channel catfish there are three duplicated IgH with large numbers of V, D, and J segments at two of them [72]; at one there is an assembled VDJ [73]. This VDJ exon prevents further rearrangement at its IgH because all other D gene segments were eliminated at its formation. The VDJ has apparently fallen into disuse, as it does not seem to be expressed [74]. Instead, a leader-like sequence arose downstream of the VDJ unit, enabling a V-less C region transcript to be processed for that locus. The secreted C-alone protein binds catfish granulocyte blood cells, similar to the binding of IgD on mammalian basophils [75], raising the possibility of a specialised immune function. Like IgM1gj, the C-alone polypeptide may have passed into the realm of innate immunity. In both instances germ line rearrangement, by pre-empting somatic recombination, is the first step in initiating changes in the fate of an Ig locus. The joined gene may become defunct or alternatively the recombination event can initiate an evolutionary exploration of new functional possibilities.

Germ line joining could well have occurred in species like humans and mice that do not have redundant or duplicated recombination modules as in fish, but since tetrapods have Ig loci configured in an extended array with linked groupings of Vs, Ds and Js, germline V(D)J recombination would be expected to delete much of the locus and negatively impact the diversity of T or B cell receptors. Yet, there are two animals in which evidence of germ line joining can be seen—opossum and chickens [76,77]. In the former, a VD fusion was observed amidst a VH gene segment array in *Monodelphis domestica*. The VD can undergo rearrangement, generating unusually long CDR3s due to its extended flank, but as there is little compositional variation it is not clear what role such H chains play in the opossum antibody repertoire. In chickens this kind of VD fusion was incorporated into its unique Ig diversification process. There is only one functional, rearranging VH gene at the chicken IgH, and all others are present as VD fusions. The series of pseudo V genes are contiguous with D-like sequences and sometimes also with recognisable J segments at their 3′ ends. In chickens, the germ line-joined copies augment antibody diversity by providing lengthier donor sequence for somatic gene conversion at the IgH locus [78].

In conclusion, where germ line-joining events have not resulted in inactivation, the outcomes are remarkable for the transformative changes that are induced. Some examples described above suggest that sometimes germ line rearrangement of V gene segments does not affect the antibody repertoire so much as *catalyze other events* shaping the fate of the Ig locus or cluster. In shark and catfish, novel immune molecules were generated. In chicken, the locus configuration was altered for improved templates used in gene conversion diversification.

3.2 Germ Line Signal Joint Formation: the Creation of D Segments

If RAG can act on RSSs in germ cells, it also has the potential to change the organization of antigen receptor loci. It is possible that this may indeed have happened at multiple points in the ancestral vertebrate lineage. In Figure 3.4C-(2) we consider how RAG action may have driven significant alterations like creating three-segment loci (V, D, and J) from two segments (V and J).

V(D)J recombination creates a reciprocal product to coding joints, termed signal joints. These comprise RSS-to-RSS junctions, and although normally excised from the genome when formed, signal joints are chromosomally retained when rearrangement involves inverted gene segments. Inversional joining is common and occurs for the many V genes in the mouse and human Ig kappa locus that are in the opposite transcriptional orientation to the J genes. In Figure 3.4C-(2) it is suggested that V(D)J recombination in germ cells could likewise produce a chromosomally retained signal joint between inverted gene segments, and if such a signal joint includes nucleotides as in N region, it becomes structurally analogous to a D segment [79]. Thereafter, positive selection for diversity (as discussed in the next section) would work to ensure such changes became fixed in a population. There is no reason to suppose that D segments were invented only once or only rarely by germ line V(D)J recombination; N region insertion into signal joints is not uncommon [80] and inversional joining would provide a ready source of signal joint formation. Thus a three-segment locus could be derived by germ line joining from simpler kappa or lambda-like loci.

3.3 Germ Line Hybrid Joint Formation: RSS Swaps and Locus "Speciation"

Conflicting family relationships between Ig and TCR genes [4] are even harder to understand when the RSS affiliations of the gene segments are taken into consideration. For example, the V genes of kappa and lambda are more similar to each other (and to IgH) than they are to TCR V genes, yet Ig VH and Vλ and TCR Vβ all have 23 spacer RSSs, whereas Ig Vκ has a 12 spacer RSS. Likewise, JH and Jκ have 23 RSSs whereas 12 RSSs occur at both Jλ and Jβ. Some of this complexity can be unravelled by schemes that allow for germ line joining; then it is possible to interchange one type of RSS for another through V(D)J recombination.

Figure 3.4C-(3) illustrates how any RSS configuration can derive from another by germ line hybrid joint formation [64,79]. RSS swapping is seen in vivo in lymphoid cells where hybrid joint formation takes place instead of the usual (and productive) coding joint/signal joint outcome of V(D)J recombination [79–83]. Germ line hybrid joint formation, similarly resulting in an RSS swap, would set aside the affected gene segments for separate evolution ("speciation" at the molecular level) because swapped units constrained by the 12/23 rule, can profitably rearrange only with one another. Recombination events with any of the original gene segments will result in aberrant V–V and J–J joining.

The above schemes provide a simple basis for much of the observed variation in locus configuration; in some animals, evidence of these rearrangements has been preserved in the genome. Distinctive RSS configurations may have been selected for at some point, differentiating the various loci as regulatory mechanisms became more complex. Although the odd germ line joining event allows exploration of new functional possibilities, inducing large rearrangements in the germ line is probably hazardous, and its likelihood of being retained must be on the same order as any spontaneous mutation in the population [70].

4. CONSIDERATIONS ON THE UR-V GENE

The previous sections have described the possible transposon origin for V(D)J recombination and the subsequent sculpting of the Ig and TCR gene organization by germ line rearrangements during evolution. If we postulate the first rearrangeable antigen receptor gene indeed arose from an incident of splitting through transposon insertion, several questions come to mind—how and why was this one particular insertion selected for, and what might have been the nature of the archaic "ur-V gene"?

The ur-V gene would have encoded a domain—an autonomous folding unit—of the Ig superfamily, whose members include cell surface molecules (Ig, TCR, MHC, β2-microglobulin, Thy-1, CD4, CD8) as well as cell adhesion and muscle components (for reviews, see Refs. [84,85]). Four sets of domain structures have been distinguished: V (variable), I (intermediary), C1, and C2 (constant). These structures are differentiated from each other by sequence patterns and lengths but they all share the immunoglobulin fold, which is formed by two β-pleated sheets that are packed face-to-face and consist of antiparallel β-strands. The disulfide bridge characteristic of Ig-related molecules is not necessary to the formation of this sandwich structure. Figure 3.5 shows ribbon diagrams of the V, I, and C structures. V domains are distinguished by the presence of two strands, C′ and C″. The antigen-combining portions are mainly situated in loops, those connecting B to C strand (CDR1), F to G strand (CDR3). In the prototype Ig gene, the hypothetical transposition event would have disrupted the region encoding the loop between F and G strands in the V domain.

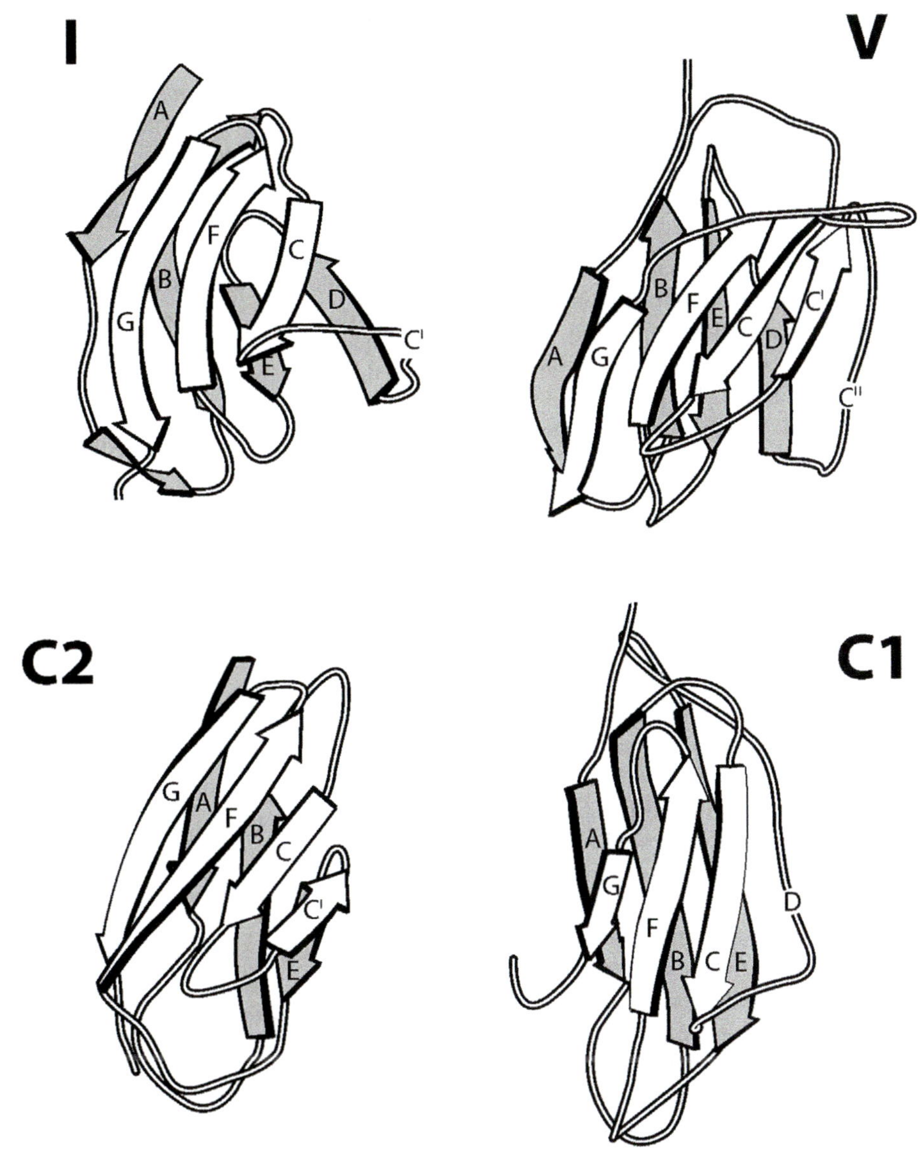

FIGURE 3.5 **Immunoglobulin superfamily domain structures.** The region between C and E strands varies among the sets. C1 set structures appear only late in evolution, in gnathostomes, and differ from I set by the absence of C ′. V set differs the addition of C″ strand and C2 set by the deletion of D strand. Ribbon diagram of I set is based on telokin structure [84]. *Adapted from Ref. [84] with permission from Elsevier.* The V set is based on Fab NEW VH, C2 set on human CD4 domain 2, and C1 set on β-2 microglobulin [86]. *Adapted from Ref. [86], p. 54, copyright (1997) with permission from Elsevier.*

The relationships between the four Ig superfamily structural sets are not entirely clear as yet, but the I set is probably the earliest and the C1 the most recent in evolution. Du Pasquier [87] analysed a receptor tyrosine kinase in the sponge *Geodia cydonium*, concluding that it possessed the V frame that is common to V and I set structures [84]. With completion of the *Caenorhabditis elegans* genome, an analysis was made to identify the repertoire of the Ig superfamily domains in the nematode. Out of 64 proteins, all 488 domains were of the I set, establishing that this was the ancestral domain structure [88]. Among molecules reported in later invertebrates, such as mollusks and arthropods, domains with V or C2 features, together or alone, could be distinguished, such as fibrinogen-related proteins (FREPs) from the snail or amalgam in Drosophila [87,89].

TCR and Ig domains consist of V and C1 type domains, and C1 domains are also part of other immune response-associated molecules like class I and class II MHC, β2-microglobulin, and tapasin, all of which also apparently have no homologue earlier than in cartilaginous fishes (for a review see Ref. [90]). Current analyses of the available sea lamprey genome revealed a few C1-like sequences (J. Rast and K. Buckley, pers. commun.) and suggest that such structures primarily evolved after divergence of jawless fishes.

4.1 Rearrangement Generates Sequence Length Diversity

We reason that the ur-V gene probably was one member of a multigene family. Since the immediate effect of the transposition would have been disruption of the target gene, redundancy of the gene function would have prevented immediate elimination of the to-be founder gene. Amongst multiple duplicated members that compensated for its loss of function, the split gene could have hitchhiked along, eventually to be eliminated by recombination or possibly altered by gene conversion, as has been observed in the "birth-and-death" evolution of multigene families [91]. Fixation of this novel but null (because silenced) mutant could have been by chance, but the recruitment for function must have arisen soon after the integration event. Had there not been selection for the split gene, it would have acquired unfavourable changes in the coding region, or the RSS would have mutated out of recognition for the RAG recombinase. It has been observed that TCR V pseudogenes with frameshift mutations tend to accumulate additional changes, possibly because reversion to a functional gene is less apt to occur than for V pseudogenes with point mutations that generate premature termination; and the transposition in this case is effectively a frameshift mutation.

The selecting feature would have included RAG activity, and this activity would have to be expressed in somatic cells. In the somatic cell, the RAG proteins would excise the RSSs, along with the intervening DNA, that had been incorporated into the ur-V gene in the germ line. With this removal the cleaved DNA ends at the exit site would be rejoined, most likely by RAG-independent endogenous repair mechanisms. Even discounting genetic and enzymatic elaborations, like the D gene segments and N region addition by terminal deoxynucleotidyl transferase (TdT), heterogeneity can be generated at the new junction. The opening of hairpin DNA as formed by RAG recombinase action, followed by limited exonuclease trimming and ligation, all are carried out by functions that are part of the endogenous non-homologous end-joining pathway of repair (reviewed in Ref. [92]).

The net effect of the removal of the intervening DNA from the split gene is that the re-formed gene sequences, if ligated in-frame, encode proteins with both sequence and length heterogeneity. Although there are somatic mechanisms that change DNA sequence—mutation and gene conversion among them—there are none that generate sequence length diversity at high frequency. Gene conversion will incorporate insertions or deletions, but according to the set of templates available in the germ line. Somatic hypermutation can create insertions and deletions (in/del) in V regions but does so with low efficiency in functioning genes [93,94].

Whereas in/dels occur at different locations throughout non-productive V(D)J, they have survived mainly in CDR in functional rearrangements [95]. Sequence length variability alters not only the chemical nature but, more drastically, the physical shape of the region involved. The loops best support this type variability, without interfering with the main-chain folding that creates the Ig fold (see Figure 3.5), and illustrate it by being the least conserved feature in sequence alignments of evolutionarily related proteins [85]. The RAG recombinase activity is site-specific, and this ensured localization of sequence length variability at both a tolerated place and on a regular basis. The subsequent creation of D gene segments (possibly via germ line joining, as above) to further increase heterogeneity at the CDR3 loop supports the idea that junctional diversity is a very strong selecting feature for rearranging genes [96].

TdT is a unique DNA polymerase whose function in developing lymphocytes is to elaborate antigen receptor junctional diversity by addition of non-templated bases [97]. TdT and polymerase μ are members of the pol X family of DNA polymerases; their similarity (61%) and identical exon-intron organization suggest a common ancestry [98] that is supported by the presence of only a single copy gene in the sea lamprey, tunicate, amphioxus, and sea urchin genomes [53,54,62,63]. Sharks represent the earliest vertebrates with both pol μ and TdT and whose TdT expression is restricted to lymphoid tissues ([99], E. Hsu unpublished data). The appearance of TdT with V(D)J recombination supports the notion that the generation of the unusual molecular heterogeneity provided by RAG action may have been the selecting factor for the split gene.

4.2 Considering Jawless Fishes

If the RAG transposon arrived after the divergence of jawless fishes, what sort of immune system did the ancestral jawed vertebrate possess? The existing representatives of jawless fishes (cyclostomes), hagfish and lamprey, have adaptive immune systems based on lymphocytes that express diverse molecules called variable lymphocyte receptors (VLR) that contain leucine-rich-repeat (LLR) elements (see Chapter 8). Could the jawed (gnathostome) ancestor have once used the VLR system?

Some part of the answer rests on the phylogenetic position of lampreys, that is, if the lamprey diverged from the hagfish or the stem vertebrate lineage (Figure 3.6A,B, resp.) ([100] and references therein). The short interval between these events has made their relationships a difficult and highly debated issue. Since hagfish and lamprey have similar immune systems, the VLR gene system evolved prior to their diversification, and the arrows in Figure 3.6 indicate the latest points in the phylogeny trees it could have arisen. In Figure 3.6B the inference is that the ancestor of jawed vertebrates possessed a VLR system because it and lampreys shared lineage after divergence of hagfishes.

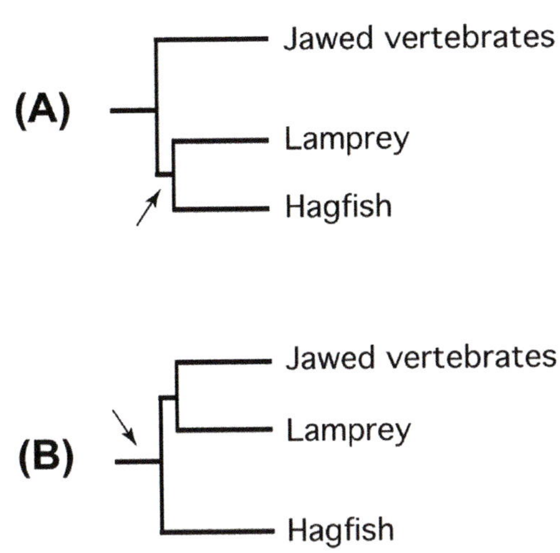

FIGURE 3.6 **Hypothetical trees for relationships among hagfishes, lampreys, and jawed vertebrates.** Tree (A) shows hagfishes and lampreys as sister groups. Tree (B) shows lampreys and jawed vertebrates as sister groups. Arrows indicate the latest point at which the VLR system could have arisen in order for it to exist in both hagfishes and lampreys.

The story is less clear in Figure 3.6A; the ancestor would not have carried VLR only if VLR did not arise at an earlier point than indicated.

The question is even more interesting for what answers may reveal about the circumstances under which lymphocyte receptors evolved. There are at least two kinds of cyclostome lymphocyte subsets with reciprocally expressed VLR receptors, one subset that secretes VLR molecules like B cells whereas the other resembles T cells and derives from a thymus-like structure [55,101]. This discovery was most unexpected, because it presents two interesting possibilities; either lymphocyte subsets arose in jawed vertebrates independently from cyclostomes or else subsets were already present much earlier, in the stem vertebrate [60].

If the relationships are as depicted in Figure 3.6B, the VLR-based immune system was present in gnathostomes but superseded by Ig/TCR. As inferred from Figure 3.6A, the ancestors of jawed and jawless vertebrates already possessed lymphocyte subsets; the respective Ig/TCR and VLR evolved independently in the different lineages. In such case, perhaps what now appears to be convergent evolution was the result of selection for molecules most effectively transacting the existing lymphocytes' immune functions [60].

The temporal relationship of Ig/TCR to VLR, if any, may be resolved when the hagfish genome is available and can be assessed relative to those of lamprey and jawed vertebrates. Janvier [102] commented that whereas physiological and morphological studies supported monophyly of lampreys and gnathostomes (Figure 3.6B), molecular analyses tended to favour the hagfish and lamprey being sister groups (Figure 3.6A). If examination of the hagfish genome eventually supports the latter hypothesis, then VLR and Ig/TCR could have evolved in parallel. The ancestral vertebrates may have had an immune system with lymphocytes subsets, perhaps using another gene system (discussed below). They may have possessed both LRR and Ig superfamily elements, the former predominating after hagfish divergence, and the other after acquisition of RAG1/2 genes.

4.3 Hypermutation in Evolution

If VLR was not the kind of lymphocyte antigen receptor utilised by the ancestral gnathostome, we speculate that there may have already existed a mechanism whereby the archaic non-rearranging V genes possessed the ability to somatically diversify. If the novel advantage introduced by RAG was extensive sequence diversification, the original gene function of the unsplit ur-genes would have also required this feature.

Rearrangement and somatic mutation both exist in cartilaginous fishes (sharks, skates, rays), representatives of the earliest jawed vertebrates [103]. Somatic hypermutation (SHM) at Ig genes is instigated by activation-induced cytidine deaminase (AID), which converts cytidine in DNA to uracil [104,105]. Repair of the site leads to gene conversion in some species and in others insertion of mismatches, resulting in mutated sequence. AID is a recently evolved member of the APOBEC family of cytidine deaminases and is highly conserved among jawed vertebrates [106]. Two members of the AID-APOBEC family have been identified and detected in lamprey lymphocytes. Their function is not known, but PmCDA1 has been shown to induce transition mutations when expressed in *E. coli*, which suggests a similar ability to cause deamination at C:G sites [56]. It is hypothesised that assembly of the VLR antigen receptor cassette through a homology-pairing recombination mechanism is initiated by the AID-APOBEC-induced DNA break.

No AID-APOBEC members were found in tunicate or amphioxus [56]. The human AID protein shares 66% identity and 79% similarity with shark AID and 25% identity and 50% similarity with lamprey pmCDA1. If the jawless fish and shark genes had a common AID precursor that affected DNA, it evolved in the stem vertebrate.

What could have been the role of AID in this ancestral vertebrate? SHM produces primarily point mutations, in mouse and human Ig as well as Ig from the amphibian *Xenopus* [105,107]. However, in cartilaginous fish Ig half of the substitutions are in tandem, as stretches of 2–5 bp; the difference is most likely due to species-specific processing of the AID-induced lesion [96,108]. Recent experiments have revealed that the translesion synthesis DNA polymerase zeta is capable of extending a mismatch with a mismatch during SHM [109,110]; possibly pol zeta is more prominent or of considerably higher activity during shark

SHM. It is not clear why sharks have a mutational pathway that operates more efficiently per AID hit, but that it is so in cartilaginous fishes suggests there could exist a system where, in the absence of a VDJ recombination mechanism, very extensive antigen receptor sequence diversity was possible through somatic mutation [96].

We suggest that the function of the ur-V genes in the ancestral jawed vertebrate was presumably immune in nature and effected by circulating cells. In addition to expression of AID-APOBEC, both jawless and jawed vertebrate lymphocytes share the ability to undergo cell division upon activation. In the latter, antigen receptor engagement is coupled to pathways taking the lymphocyte out of G_0 into the G_1 state and upregulating genes that enable mitosis; usually a second receptor sensing pathogen presence contributes the crucial signal. In B cells there are Toll receptors or receptors for factors produced by accessory cells such as helper T cells and dendritic cells that have already been activated by pathogen. Proliferation is specific to the activated cells. The greater the antigen receptor diversity, the lower the frequency for any specificity (estimated at one in 10^5–10^6) and the fewer the initial cell numbers to combat a particular pathogen. Hence, clonal proliferation is not only fundamental to adaptive immunity, but also necessary from the very beginning for a diversification mechanism to have selective value. During the evolution of both the VLR and the transposon-split V gene, certain hemocytes must already have possessed some ability to proliferate.

4.4 An Innate Defence V Gene

It has been proposed that a transposition event involving a RAG transposon and an archaic V gene initiated the creation of rearranging antigen receptors. We suggest that the prototype V genes was part of a multigene family that may have diversified by hypermutation, and introduction of a recombining mechanism that generated loop length variability greatly augmented diversification in a way more revolutionary than base substitutions. This was the selecting factor for the split gene created by RAG.

There have been attempts to identify prototype V lineage genes in jawless fish or other animals; some examples are referenced here ([111,112,113,114], discussed in Ref. [115]) but will not be examined. One difficulty in identifying the "actual" lineage that became disrupted in jawed vertebrates is that, with the acquisition of junctional diversification, the rest of the sequence will have evolved to accommodate or enhance the useful new structures generated by rearrangement. Hence, the modern vertebrate V gene is sufficiently diverged so that its sequence no longer serves much more than as a general guideline to searches in jawless fish or protochordates. For example, large sets of non-rearranging V-like genes were isolated from amphioxus [116], but so far their function is not clear.

We have speculated that the ancient V gene mutated and may have had some sort of defence function. Although in itself not proof of any immune function, somatic mutation should be a fairly restricted phenomenon. One candidate for a cousin of the Ig prototype may be the FREPs isolated from the haemolymph of freshwater snails, *Biomphalaria glabrata*. FREPs are a family of molecules consisting of one or two Ig superfamily V-set-like domains connected with a carbohydrate-binding lectin domain [87,117]. There is extensive sequence heterogeneity that may be produced by somatic diversification. The FREPs do not rearrange in the manner of Ig/TCR. They are a multigene family consisting of sets of tandemly arranged V domain-like exons with fibrinogen exons, and the fibrinogen domains can be construed as serving a C region-type function. Some genes are upregulated in response to infection by flatworm parasites as *Schistosoma mansoni* and *Echinostoma paraensi* [118]. These characteristics suggest that some similar kind of ancestral innate defence gene in a primitive vertebrate could have been the target of the postulated RAG transposition.

5. CONCLUDING REMARKS

Although we have made large claims for RAG1/2 manipulating the genome, there is little actual experimental evidence for such activity. There are few animal species where both RAG1 and RAG2 mRNA are present in ovary or testes tissue, and it is not clear that there are active proteins in the germ cells themselves. Those species possessing the possibility for ongoing germ line rearrangement, such as cartilaginous fishes, are not typical laboratory models, and their genomic content is not (yet) characterised. Defined reporter sequences are needed to monitor for changes, and a sensitive or high throughput assay system must be devised in expectation that germ line rearrangement events are of very low frequency.

If the genomic DNA in germ cells of any animal expressing RAG proteins in gonads can be closely tracked, one can try to ascertain the frequency of RAG-induced recombination per germ cell. Does RAG act at all germ cell stages? If so, could RAG-induced changes accumulate? Is all chromatin equally susceptible to RAG? In studying RAG action on germ line DNA, its role in antigen receptor evolution can be investigated: as provider of parts and mechanism of a recombining gene system, sculptor of gene organization, and mutator of genetic material to inactivation or transformation.

ACKNOWLEDGEMENTS

We would like to thank the following individuals for answering many and varied questions: Jonathan Rast, Kate Buckley, Rob Miller, Maja Nowakowski, Zahra Zakeri, Chris Amemiya. The authors also thank Jürgen Brosius for reviewing the manuscript and Louis Du Pasquier for his comments and advice throughout. This work was supported in part by the National Institutes of Health (R01-GM068095).

REFERENCES

[1] Gould SJ, Vrba ES. Exaptation – a missing term in the science of form. Paleobiology 1982;8:4–15.
[2] Sakano H, Hüppi K, Heinrich G, Tonegawa S. Sequences at the somatic recombination sites of immunoglobulin light-chain genes. Nature 1979;280:288–94.
[3] Siu G, Clark SP, Yoshikai Y, Malissen M, Yanagi Y, Strauss E, et al. The human T cell antigen receptor is encoded by variable, diversity, and joining gene segments that rearrange to generate a complete V gene. Cell 1984;37:393–401.
[4] Flajnik MF, Du Pasquier L. Evolution of the immune system. In: Paul WE, editor. Fundamental immunology. 7th ed. Philadelphia, PA: Wollers Kluwer Health/Lippincott Williams and Wilkins; 2013. p. 57–124.
[5] Litman GW, Rast JP, Fugmann SD. The origins of vertebrate adaptive immunity. Nat Rev Immunol 2010;10:543–53.
[6] Fugmann SD. The origins of the RAG genes-from transposition to V(D)J recombination. Semin Immunol 2010;22:10–6.
[7] Aziz RK, Breitbart M, Edwards RA. Transposases are the most abundant, most ubiquitous genes in nature. Nucleic Acids Res 2010;38:4207–17.
[8] Abrusan G, Zhang Y, Szilagyi A. Structure prediction and analysis of DNA transposon and LINE retrotransposon proteins. J Biol Chem 2013;288:16127–38.
[9] Ladeveze V, Chaminade N, Lemeunier F, Periquet G, Aulard S. General survey of hAT transposon superfamily with highlight on hobo element in Drosophila. Genetica 2012;140:375–92.
[10] Yuan YW, Wessler SR. The catalytic domain of all eukaryotic cut-and-paste transposase superfamilies. Proc Natl Acad Sci USA 2011;108:7884–9.
[11] Hickman AB, Chandler M, Dyda F. Integrating prokaryotes and eukaryotes: DNA transposases in light of structure. Crit Rev Biochem Mol Biol 2010;45:50–69.
[12] Montano SP, Rice PA. Moving DNA around: DNA transposition and retroviral integration. Curr Opin Struct Biol 2011;21:370–8.
[13] Dyda F, Chandler M, Hickman AB. The emerging diversity of transpososome architectures. Q Rev Biophys 2012;45:493–521.
[14] Bushman F. Lateral DNA transfer. Cold Spring Harbor: Cold Spring Harbor Laboratory Press; 2002.
[15] Wijayawardena BK, Minchella DJ, DeWoody JA. Hosts, parasites, and horizontal gene transfer. Trends Parasitol 2013;29:329–38.
[16] Wallau GL, Ortiz MF, Loreto ELS. Horizontal transposon transfer in Eukarya: detection, bias, and perspectives. Genome Biol Evol 2012;4:801–11.
[17] Bernstein RM, Schluter SF, Bernstein H, Marchalonis JJ. Primordial emergence of the recombination activating gene 1 (RAG1): sequence of the complete shark gene indicates homology to microbial integrases. Proc Natl Acad Sci USA 1996;93:9454–9.
[18] Shen XX, Liang D, Feng YJ, Chen MY, Zhang P. A versatile and highly efficient toolkit including 102 nuclear markers for vertebrate phylogenomics, tested by resolving the higher level relationships of the caudata. Mol Biol Evol 2013;30:2235–48.
[19] Oliver KR, Greene WK. Transposable elements: powerful facilitators of evolution. Bioessays 2009;31:703–14.
[20] Schatz DG, Oettinger MA, Baltimore D, The V(D)J. Recombination activating gene, RAG-1. Cell 1989;59:1035–48.
[21] Oettinger MA, Schatz DG, Gorka C, Baltimore D. Rag-1 and Rag-2, adjacent genes that synergistically activate V(D)J recombination. Science 1990;248:1517–23.
[22] McBlane JF, van Gent DC, Ramsden DA, Romeo C, Cuomo CA, Gellert M, et al. Cleavage at a V(D)J recombination signal requires only RAG1 and RAG2 proteins and occurs in two steps. Cell 1995;83:387–95.
[23] Eastman QM, Leu TMJ, Schatz DG. Initiation of V(D)J recombination in vitro obeying the 12/23 rule. Nature 1996;380:85–8.
[24] Cortes P, Weis-Garcia R, Misulovin Z, Nussenzweig A, Lai J-S, Li G, et al. In vitro V(D)J recombination: signal joint formation. Proc Natl Acad Sci USA 1996;93:14008–13.
[25] van Gent DC, Mizuuchi D, Gellert M. Similarities between initiation of V(D)J recombination and retroviral integration. Science 1996;271:1592–4.
[26] Zhou L, Mitra R, Atkinson PW, Hickman AB, Dyda F, Craig NL. Transposition of hAT elements links transposable elements and V(D)J recombination. Nature 2004;432:995–1001.
[27] Hencken CG, Li XH, Craig NL. Functional characterization of an active Rag-like transposase. Nat Struct Mol Biol 2012;19:834–6.
[28] Landree MA, Wibbenmeyer JA, Roth DB. Mutational analysis of RAG1 and RAG2 identifies three catalytic amino acids in RAG1 critical for both cleavage steps of V(D)J recombination. Genes Dev 1999;13:3059–69.
[29] Kim DR, Dai Y, Mundy CL, Yang W, Oettinger MA. Mutations of acidic residues in RAG1 define the active site of the V(D)J recombinase. Genes Dev 1999;13:3070–80.
[30] Fugmann SD, Villey IJ, Ptaszek LM, Schatz DG. Identification of two catalytic residues in RAG1 that define a single active site within the RAG1/RAG2 protein complex. Mol Cell 2000;5:97–107.
[31] Agrawal A, Eastman QM, Schatz DG. Transposition mediated by RAG1 and RAG2 and its implications for the evolution of the immune system. Nature 1998;394:744–51.
[32] Hiom K, Melek M, Gellert M. DNA transposition by the RAG1 and RAG2 proteins: a possible source of oncogenic translocations. Cell 1998;94:463–70.
[33] Clatworthy AE, Valencia MA, Haber JE, Oettinger MA. V(D)J recombination and RAG-mediated transposition in yeast. Mol Cell 2003;12:489–99.
[34] Ramsden DA, Weed BD, Reddy YV. V(D)J recombination: born to be wild. Semin Cancer Biol 2010;20:254–60.
[35] Matthews AGW, Oettinger MA. Regulation of RAG transposition. In: V(D)J recombination. Ferrier P, editorvol 650; 2009. p. 16–31.
[36] Montano SP, Pigli YZ, Rice PA. The Mu transpososome structure sheds light on DDE recombinase evolution. Nature 2012;491:413–7.

[37] Schluter SF, Marchalonis JJ. Cloning of shark RAG2 and characterization of the RAG1/RAG2 gene locus. Faseb J 2003;17:470–2.

[38] Venkatesh B, Ning Y, Brenner S. Late changes in spliceosomal introns define clades in vertebrate evolution. Proc Natl Acad Sci USA 1999;96:10267–71.

[39] Oettinger MA. Cutting apart V(D)J recombination. Curr Opin Genet Dev 1996;6:141–5.

[40] Hong X, Scofield DG, Lynch M. Intron size, abundance, and distribution within untranslated regions of genes. Mol Biol Evol 2006;23:2392–404.

[41] Kapitonov VV, Jurka J. RAG1 core and V(D)J recombination signal sequences were derived from Transib transposons. PLoS Biol 2005;3:998–1011.

[42] Panchin Y, Moroz LL. Molluscan mobile elements similar to the vertebrate – recombination-activating genes. Biochem Biophys Res Commun 2008;369:818–23.

[43] Chen S, Li X. Molecular characterization of the first intact Transib transposon from *Helicoverpa zea*. Gene 2008;408:51–63.

[44] Sadofsky MJ, Hesse JE, McBlane JF, Gellert M. Expression and V(D)J recombination activity of mutated RAG-1 proteins. Nucleic Acids Res 1993;21:5644–50.

[45] Fugmann SD, Messier C, Novack LA, Cameron RA, Rast JP. An ancient evolutionary origin of the Rag1/2 gene locus. Proc Natl Acad Sci USA 2006;103:3728–33.

[46] Callebaut I, Mornon JP. The V(D)J recombination activating protein RAG2 consists of a six-bladed propeller and a PHD fingerlike domain, as revealed by sequence analysis. Cell Mol Life Sci 1998;54:880–91.

[47] Corneo B, Moshous D, Callebaut I, de Chasseval R, Fischer A, de Villartay JP. Three-dimensional clustering of human RAG2 gene mutations in severe combined immune deficiency. J Biol Chem 2000;275:12672–5.

[48] Gomez CA, Ptaszek LM, Villa A, Bozzi F, Sobacchi C, Brooks EG, et al. Mutations in conserved regions of the predicted RAG2 kelch repeats block initiation of V(D)J recombination and result in primary immunodeficiencies. Mol Cell Biol 2000;20:5653–64.

[49] Jones JM, Simkus C. The roles of the RAG1 and RAG2 "non-core" regions in V(D)J recombination and lymphocyte development. Arch Immunol Ther Exp 2009;57:105–16.

[50] Wilson DR, Norton DD, Fugmann SD. The PHD domain of the sea urchin RAG2 homolog, SpRAG2L, recognizes dimethylated lysine 4 in histone H3 tails. Dev Comp Immunol 2008;32:1221–30.

[51] Benton MJ, Donoghue PC. Paleontological evidence to date the tree of life. Mol Biol Evol 2007;24:26–53.

[52] Erwin DH, Laflamme M, Tweedt SM, Sperling EA, Pisani D, Peterson KJ. The Cambrian conundrum: early divergence and later ecological success in the early history of animals. Science 2011;334:1091–7.

[53] Smith JJ, Kuraku S, Holt C, Sauka-Spengler T, Jiang N, Campbell MS, et al. Sequencing of the sea lamprey (*Petromyzon marinus*) genome provides insights into vertebrate evolution. Nat Genet 2013;45:415–21.

[54] Hibino T, Loza-Coll M, Messier C, Majeske AJ, Cohen AH, Terwilliger DP, et al. The immune gene repertoire encoded in the purple sea urchin genome. Dev Biol 2006;300:349–65.

[55] Guo P, Hirano M, Herrin BR, Li J, Yu C, Sadlonova A, et al. Dual nature of the adaptive immune system in lampreys. Nature 2009;459:796–801.

[56] Rogozin IB, Iyer LM, Liang L, Glazko GV, Liston VG, Pavlov YI, et al. Evolution and diversification of lamprey antigen receptors: evidence for involvement of an AID-APOBEC family cytosine deaminase. Nat Immunol 2007;8:647–56.

[57] Huang S, Yuan S, Guo L, Yu Y, Li J, Wu T, et al. Genomic analysis of the immune gene repertoire of amphioxus reveals extraordinary innate complexity and diversity. Genome Res 2008;18:1112–26.

[58] Azumi K, Sabau SV, Fujie M, Usami T, Koyanagi R, Kawashima T, et al. Gene expression profile during the life cycle of the urochordate Ciona intestinalis. Dev Biol 2006;308:572–82.

[59] Denoeud F, Henriet S, Mungpakdee S, Aury JM, Da Silva C, Brinkmann H, et al. Plasticity of animal genome architecture unmasked by rapid evolution of a pelagic tunicate. Science 2010;330:1381–5.

[60] Hsu E. The invention of lymphocytes. Curr Opin Immunol 2011;23:1–7.

[61] Dong ML, Fu YG, Yu CL, Su J, Huang SF, Wu XK, et al. Identification and characterisation of a homolog of an activation gene for the recombination activating gene 1 (RAG 1) in amphioxus. Fish Shellfish Immun 2005;19:165–74.

[62] Holland LZ, Albalat R, Azumi K, Benito-Gutierrez E, Blow MJ, Bronner-Fraser M, et al. The amphioxus genome illuminates vertebrate origins and cephalochordate biology. Genome Res 2008;18:1100–11.

[63] Azumi K, De Santis R, De Tomaso A, Rigoutsos I, Yoshizaki F, Pinto MR, et al. Genomic analysis of immunity in a Urochordate and the emergence of the vertebrate immune system: "waiting for Godot". Immunogenetics 2003;55:570–81.

[64] Kokubu F, Litman R, Shamblott MJ, Hinds K, Litman GW. Diverse organization of immunoglobulin VH gene loci in a primitive vertebrate. EMBO J 1988;11:3413–22.

[65] Bailey JA, Gu Z, Clark RA, Reinert K, Samonte RV, Schwartz S, et al. Recent segmental duplications in the human genome. Science 2002;297:1003–7.

[66] Matsuda F, Shin EK, Hirabayashi Y, Nagaoka H, Yochida MC, Zong SQ, et al. Organization of variable region segments of the human immunoglobulin heavy chain: duplication of the D5 cluster within the locus and interchromosomal translocation of variable region segments. EMBO J 1990;9:2501–6.

[67] Litman GW, Anderson MK, Rast JP. Evolution of antigen binding receptors. Annu Rev Immunol 1999;17:109–47.

[68] Hsu E, Pulham N, Rumfelt LL, Flajnik MF. The plasticity of immunoglobulin gene systems in evolution. Immunol Rev 2006;210:8–26.

[69] Yoder JA, Litman GW. Immune-type diversity in the absence of somatic rearrangement. Curr Top Microbiol 2000;248:271–82.

[70] Lee SS, Fitch D, Flajnik MF, Hsu E. Rearrangement of immunoglobulin genes in shark germ cells. J Exp Med 2000;191:1637–48.

[71] Rumfelt LL, Avila D, Diaz M, Bartl S, McKinney EC, Flajnik MF. A shark antibody heavy chain encoded by a nonsomatically rearranged VDJ is preferentially expressed in early development and is

convergent with mammalian IgG. Proc Natl Acad Sci USA 2001;98:1775–80.

[72] Edholm ES, Bengtén E, Wilson M. Insights into the function of IgD. Dev Comp Immunol 2011;35:1309–16.

[73] Ghaffari SH, Lobb CJ. Structure and genomic organization of a second cluster of immunoglobulin heavy chain gene segments in the channel catfish. J Immunol 1999;162:1519–29.

[74] Edholm E-S, Bengtén E, Stafford JL, Sahoo M, Taylor EB, Miller NW, et al. Identification of two IgD(+) B cell populations in channel catfish, *Ictalurus punctatus*. J Immunol 2010;185:4082–94.

[75] Chen K, Xu W, Wilson M, He B, Miller NW, Bengtén E, et al. Immunoglobulin D enhances immune surveillance by activating antimicrobial, proinflammatory and B cell-stimulating programs in basophils. Nat Immunol 2009;10:889–98.

[76] Wang X, Miller RD. Recombination, transcription, and diversity of a partially germline-joined VH in a mammal. Immunogenetics 2012;64:713–7.

[77] Reynaud CA, Dahan A, Anquez V, Weill JC. Somatic hyperconversion diversifies the single Vh gene of the chicken with a high incidence in the D region. Cell 1989;59:171–83.

[78] Reynaud C-A, Bertocci B, Dahan A, Weill J-C. Formation of the chicken B-cell repertoire: ontogenesis, regulation of Ig gene rearrangement, and diversification by gene conversion. Adv Immunol 1994;57:353–78.

[79] Lewis S, Hesse JE, Mizuuchi K, Gellert M. Novel strand exchanges in V(D)J recombination. Cell 1988;55:1099–107.

[80] Lieber MR, Hesse JE, Mizuuchi K, Gellert M. Lymphoid V(D)J recombination: nucleotide insertion at signal joints as well as coding joints. Proc Natl Acad Sci USA 1988;85:8588–92.

[81] Morzycka-Wroblewska E, Lee FEH, Desiderio SV. Unusual immunoglobulin gene rearrangement leads to replacement of recombinational signal sequences. Science 1988;242:261–3.

[82] Sollbach AE, Wu GE. Inversions produced during V(D)J rearrangement at IgH, the immunoglobulin heavy-chain locus. Mol Cell Biol 1995;15:671–81.

[83] VanDyk L, Meek K. Assembly of IgH CDR3: mechanism, regulation, and influence on antibody diversity. Int Rev Immunol 1992;8:123–33.

[84] Harpaz Y, Chothia C. Many of the immunoglobulin superfamily domains in cell adhesion molecules and surface receptors belong to a new structural set which is close to that containing variable domains. J Mol Biol 1994;238:528–39.

[85] Williams AF, Barclay AN. The immunoglobulin superfamily—domains for cell surface recognition. Annu Rev Immunol 1988;6:381–405.

[86] Barclay AN, Birkeland ML, Brown MH, Beyers AD, Davis SJ, Somoza C, et al. The leucocyte antigen facts book. London: Academic Press, Harcourt Brace Jovanovich; 1997.

[87] Du Pasquier L, Zucchetti I, De Santis R. Immunoglobulin superfamily receptors in protochordates: before RAG time. Immunol Rev 2004;198:233–48.

[88] Teichmann SA, Chothia C. Immunoglobulin superfamily proteins in *Caenorhabditis elegans*. J Mol Biol 2000;296:1367–83.

[89] Seeger MA, Haffley L, Kaufman TC. Characterization of amalgam: a member of the immunoglobulin superfamily from Drosophila. Cell 1988;55:589–600.

[90] Flajnik MF, Kasahara M. Comparative genomics of the MHC: glimpses into the evolution of the adaptive immune system. Immunity 2001;15:351–62.

[91] Ota T, Nei M. Divergent evolution and evolution by the birth-and-death process in the immunoglobulin VH gene family. Mol Biol Evol 1994;11:469–82.

[92] Lieber MR. The mechanism of double-strand DNA break repair by the nonhomologous DNA end-joining pathway. Annu Rev Biochem 2010;79:181–211.

[93] Goossens T, Klein U, Kuppers R. Frequent occurrence of deletions and duplications during somatic hypermutation: implications for oncogene translocations and heavy chain disease. Proc Natl Acad Sci USA 1998;95:2463–8.

[94] Wilson PC, de Bouteiller O, Liu YJ, Potter K, Banchereau J, Capra JD, et al. Somatic hypermutation introduces insertions and deletions into immunoglobulin V genes. J Exp Med 1998;187:59–70.

[95] de Wildt RM, van Venrooij WJ, Winter G, Hoet RM, Tomlinson IM. Somatic insertions and deletions shape the human antibody repertoire. J Mol Biol 1999;294:701–10.

[96] Lee SS, Tranchina D, Ohta Y, Flajnik MF, Hsu E. Hypermutation in shark immunoglobulin light chain genes results in contiguous substitutions. Immunity 2002;16:571–82.

[97] Gilfillan S, Benoist C, Mathis D. Mice lacking terminal deoxynucleotidyl transferase: adult mice with a fetal antigen receptor repertoire. Immunol Rev 1995;148:201–19.

[98] Aoufouchi S, Flatter E, Dahan A, Faili A, Bertocci B, Storck S, et al. Two novel human and mouse DNA polymerases of the polX family. Nucleic Acids Res 2000;28:3684–93.

[99] Bartl S, Miracle AL, Rumfelt LL, Kepler TB, Mochon E, Litman GW, et al. Terminal deoxynucleotidyl transferases from elasmobranchs reveal structural conservation within vertebrates. Immunogenetics 2003;55:594–604.

[100] Near TJ. Conflict and resolution between phylogenies inferred from molecular and phenotypic data sets for hagfish, lampreys, and gnathostomes. J Exp Zool B Mol Dev Evol 2009;312B:749–61.

[101] Hirano M, Guo P, McCurley N, Schorpp M, Das S, Boehm T, et al. Evolutionary implications of a third lymphocyte lineage in lampreys. Nature 2013;501:435–9.

[102] Janvier P. Modern look for ancient lamprey. Nature 2006;443:921–4.

[103] Flajnik MF. Comparative analyses of immunoglobulin genes: surprises and portents. Nat Rev Immunol 2002;2:688–98.

[104] Muramatsu M, Kinoshita K, Fagarasan S, Yamada S, Shinkai Y, Honjo T. Class switch recombination and hypermutation require activation-induced cytidine deaminase (AID), a potential RNA editing enzyme. Cell 2000;102:553–63.

[105] Di Noia JM, Neuberger MS. Molecular mechanisms of antibody somatic hypermutation. Annu Rev Biochem 2007;76:1–22.

[106] Conticello SG, Thomas CJ, Petersen-Mahrt SK, Neuberger MS. Evolution of the AID/APOBEC family of polynucleotide (deoxy) cytidine deaminases. Mol Biol Evol 2005;22:367–77.

[107] Wilson M, Hsu E, Marcuz A, Courtet M, Du Pasquier L, Steinberg C. What limits affinity maturation of antibodies in *Xenopus*—the rate of somatic mutation or the ability to select mutants? EMBO J 1992;11:4337–47.

[108] Zhu C, Hsu E. Error-prone DNA repair activity during somatic hypermutation in shark B lymphocytes. J Immunol 2010;185:5336–59.

[109] Saribasak H, Maul RW, Cao Z, Yang WW, Schenten D, Kracker S, et al. DNA polymerase ζ generates tandem mutations in immunoglobulin variable regions. J Exp Med 2012;209:1075–81.

[110] Daly J, Bebenek K, Watt DL, Richter K, Jiang C, Zhao ML, et al. Altered Ig hypermutation pattern and frequency in complementary mouse models of DNA polymerase ζ activity. J Immunol 2012;18:5528–37.

[111] Pancer Z, Mayer WE, Klein J, Cooper MD. Prototypic T cell receptor and CD4-like coreceptor are expressed by lymphocytes in the agnathan sea lamprey. Proc Natl Acad Sci USA 2004;101:13273–8.

[112] Cannon JP, Haire RN, Pancer Z, Mueller MG, Skapura D, Cooper MD, et al. Variable domains and a VpreB-like molecule are present in a jawless vertebrate. Immunogenetics 2005;56:924–9.

[113] Suzuki T, Shin-I T, Fujiyama A, Kohara Y, Kasahara M. Hagfish leukocytes express a paired receptor family with a variable domain resembling those of antigen receptors. J Immunol 2005;174:2885–91.

[114] Haruta C, Suzuki T, Kasahara M. Variable domains in hagfish: NICIR is a polymorphic multigene family expressed preferentially in leukocytes and is related to lamprey TCR-like. Immunogenetics 2006;58:216–25.

[115] Zucchetti I, De Santis R, Grusea S, Pontarotti P, Du Pasquier L. Origin and evolution of the vertebrate leukocyte receptors: the lesson from tunicates. Immunogenetics 2009;61:463–81.

[116] Cannon JP, Haire RN, Litman GW. Identification of diversified genes that contain immunoglobulin-like variable regions in a protochordate. Nat Immunol 2002;3:1200–7.

[117] Zhang SM, Adema CM, Kepler TB, Loker ES. Diversification of Ig superfamily genes in an invertebrate. Science 2004;305:251–4.

[118] Hanington PC, Forys MA, Dragoo JW, Zhang SM, Adema CM, Loker ES. Role for a somatically diversified lectin in resistance of an invertebrate to parasite infection. Proc Natl Acad Sci USA 2010;107:21087–92.

Chapter 4

The Variable Lymphocyte Receptor B System of the Jawless Vertebrates

Thomas Boehm[1,2], Max D. Cooper[3,4], Masayuki Hirano[3,4], Sabyasachi Das[3,4], Ryo Morimoto[1] and Jonathan P. Rast[3,4]

[1]*Department of Developmental Immunology, Max-Planck Institute of Immunobiology and Epigenetics, Freiburg, Germany;* [2]*Faculty of Medicine, University of Freiburg, Freiburg, Germany;* [3]*Department of Pathology and Laboratory Medicine, Emory University, Atlanta, GA, United States;* [4]*Emory Vaccine Center, Emory University, Atlanta, GA, United States*

1. INTRODUCTION

Recent work suggests that two separate systems of adaptive immune receptors emerged with the dawn of vertebrate evolution: one in an ancestor of the living jawed vertebrates and the other in an ancestor of the living jawless vertebrates. These sister groups emerged from a common ancestor about 500 million years ago [1–3]. The Rag-based adaptive immune system of jawed vertebrates is well-studied [4,5], whereas the analysis of the immune system of the jawless vertebrates is in its infancy [6]. Nonetheless, several features have already been discovered that indicate substantial similarities in general design, but this work has also highlighted critical differences, attesting to evolutionary flexibility in solving the needs of adaptive immunity.

Immunologists as well as evolutionary biologists and developmental biologists have long been fascinated by the unusual life cycle and anatomical characteristics of jawless fishes, lampreys and hagfishes [6–10]. Importantly, the phylogenetic position of the agnathans represents a key window into a critical period of the evolution of the vertebrate immune system and in the structuring of the vertebrate genome [4,11–13]. It was shown in the last century that both hagfishes and lampreys can respond to immunisation with bacteria and heterologous erythrocytes by producing specific agglutinins [14–19], much like jawed vertebrates respond to the same challenges by producing specific immunoglobulins. However, since immunoglobulin-based antibodies could not be found in jawless vertebrates, the molecular underpinnings of these responses remained a mystery.

Intriguingly, lampreys [14,20] and hagfishes [21] were also found to reject allogeneic skin graft, with second-set transplants being rejected at an accelerated rate. These initial findings were re-addressed following the identification of lymphocyte-like cells by flow cytometry in hematopoietic tissues of sea lampreys [22]. However, an analysis of their transcriptomes indicated the absence of classical components of adaptive immunity of jawed vertebrates, such as MHC, T cell receptor or immunoglobulin genes [22,23].

The discovery of a somatically generated family of diverse leucine-rich repeat (LRR)-encoding genes [24,25], now referred to as variable lymphocyte receptor B (*VLRB*), finally resolved the puzzle of the nature of agglutinins in jawless fishes. The upregulated expression of these genes in lamprey lymphocyte-like cells after immunisation and the remarkable diversity of the encoded variable lymphocyte receptor (VLR) molecules indicated that they account for the antigen-specific agglutinins demonstrated decades earlier. Two related *VLR* genes, *VLRA* and *VLRC*, were subsequently identified and shown to be expressed on lymphocytes resembling the two principal T cell lineages, αβ and γδ T lymphocytes [26,27].

Several recent reviews cover the wider aspects of the agnathan VLR system [6,28,29]. This chapter will focus on the lamprey VLRB genes and proteins, representing the best characterised of these receptors in terms of structure and function, as well as on the B cell-like lymphocytes and plasma-like cells that express them. The study of this system offers fresh insight into the origins and evolution of

Molecular Biology of B Cells. https://doi.org/10.1016/B978-0-323-95895-0.00015-5

vertebrate adaptive immunity, and, importantly, opens a path to the development of an entirely novel form of antibodies for applications in bioscience and medicine.

2. THE STRUCTURE OF VLRB PROTEIN

2.1 The Structure of the VLRB Monomer

Attention was first drawn to VLRB following the isolation of a highly diverse set of cDNA sequences encoding LRR proteins from activated lymphocytes [25]. The deduced VLRB protein sequences exhibit a conserved general structure composed of highly variable LRR motifs sandwiched between invariant N-terminal and C-terminal constant regions (Figure 4.1A). Mature *VLRB* transcripts encode a signal peptide, an LRRNT motif, two to eight LRR loops and an LRRCT motif followed by an unstructured, flexible stalk region and a cysteine-rich C-terminus.

Because the VLRB structure is built from several types of genomic donor cassettes, there are discernible characteristic sequences within the variable portion of the mature VLRB LRR structures. These include (1) a diverse region corresponding to the C-terminal portion of LRRNT, (2) an 18 amino acid LRR1 segment just downstream of the LRRNT domain, (3) zero to seven 24 (or more rarely 23) amino acid LRRV loops, (4) a specialised LRRVe (end) sequence with a 12 amino acid-long diverse stretch followed by a more constrained C-terminal sequence, (5) a 13 amino acid-long connecting peptide (CP) that completes the last LRR loop and leads into (6) a variable 5′LRRCT portion that makes up about two-thirds of the LRRCT motif, including a variable length loop that often contributes to antigen binding. Excluding the LRRe elements, most (~80%) of VLRB sequences have 0, 1, or 2 LRRV sequences [24]. Notably, the signal peptide and first six amino acids of LRRNT and the last third of the LRRCT structure are invariant and contributed by the *VLRB* germline gene.

Structural studies have shown that VLRs conform to the right-handed solenoid structure of many other LRR-containing proteins [30–39], such as TLRs; in these structures, LRRNT and LRRCT modules cap the ends of the solenoid to protect the leucine-rich hydrophobic core. The solenoid structure is bent with a concave face made up of β-strands of the LRRs: two antiparallel strands from the LRRNT and one parallel strand each contributed by LRR1, the LRRVs if present, LRRVe and CP (Figure 4.1A–C). The solvent-exposed side chains of the concave surface contribute key elements of the VLRB antigen-binding surface. The hydrophobic core of the solenoid is composed of the conserved leucine residues (sometimes substituted by other hydrophobic substitutions [isoleucine, valine]) that give the LRR its name, and by a largely conserved phenylalanine residue (Figure 4.1D). The structure is further stabilised by hydrogen bonds between the β-strands and between the LRR loops. The LRRNT and LRRCT motifs each are stabilised by two disulfide bonds formed by eight conserved cysteine residues [24,25,27,30–33,35,36,38–49].

Of particular interest is the fact that although VLRB proteins usually possess only a few LRRV modules [25,33,40,41,43,45–48,50], a loop of variable length located between helix one and two of the three LRRCT helices is often directly involved in antigen binding [32,36–39] (Figure 4.1E). This loop projects forward towards the concave binding surface formed by the LRR β-strands. The considerable variability in the length and sequence composition of this loop region is evident in alignments of 5′LRRCT donor cassettes (Figure 4.1F).

2.2 The Structure of Secreted VLRB Antibody

VLRBs are secreted in a different form after lymphocyte activation and plasma cell differentiation [24–27,30,42,51,52]. Unlike VLRA and VLRC, the C-terminal region of VLRB does not encode a typical transmembrane (TM) α-helix. Instead, the cell surface form of VLRB is anchored to the cell membrane by a glycosyl–phosphatidyl–inositol (GPI) linkage and can be cleaved from the cell surface with bacterial phospholipase C [25]. A GPI cleavage site (aspartate–glycine–glycine) is located within the VLRB C-terminal region. Note that the corresponding VLRA and VLRC receptors resemble the TCRs of jawed vertebrates, in that they are transmembrane proteins and are not secreted [26,27].

Soluble VLRB antibodies isolated either from lamprey serum or from the culture medium of *VLRB*-transfected HEK-293T cells migrate in a native gel at a size of about ~400 kDa, with a minor band often migrating slightly faster [30]. With partial reduction, bands at ~80 and ~40 kDa are also evident. Under conditions of complete reduction of disulfide bonds, only the ~40 kDa fragment remains. The C-terminal domain of VLRB contains seven cysteine residues, four of which are closely spaced near the C-terminus. Tandem mass spectrometry indicates that this particular region, which is located C-terminal to the GPI cleavage site, is retained in secreted VLRB antibodies. Truncation of this region by introduction of a stop codon just downstream of the GPI cleavage site results in exclusive expression of monomers; this finding suggests that the formation of a polymeric antibody depends on the presence of this region in facilitating interchain disulfide bonding [30]. Proteomic analyses are also consistent with the presence of VLRB dimers as intermediates *en route* to the pentameric (or sometimes tetrameric) multimeric structures. Images from negative stain transmission electron microscopy of purified secreted VLRB preparations reveal pentamers with a radial form akin to the multimeric IgMs secreted by plasma cells of mammals [30,51]; (Figure 4.1G).

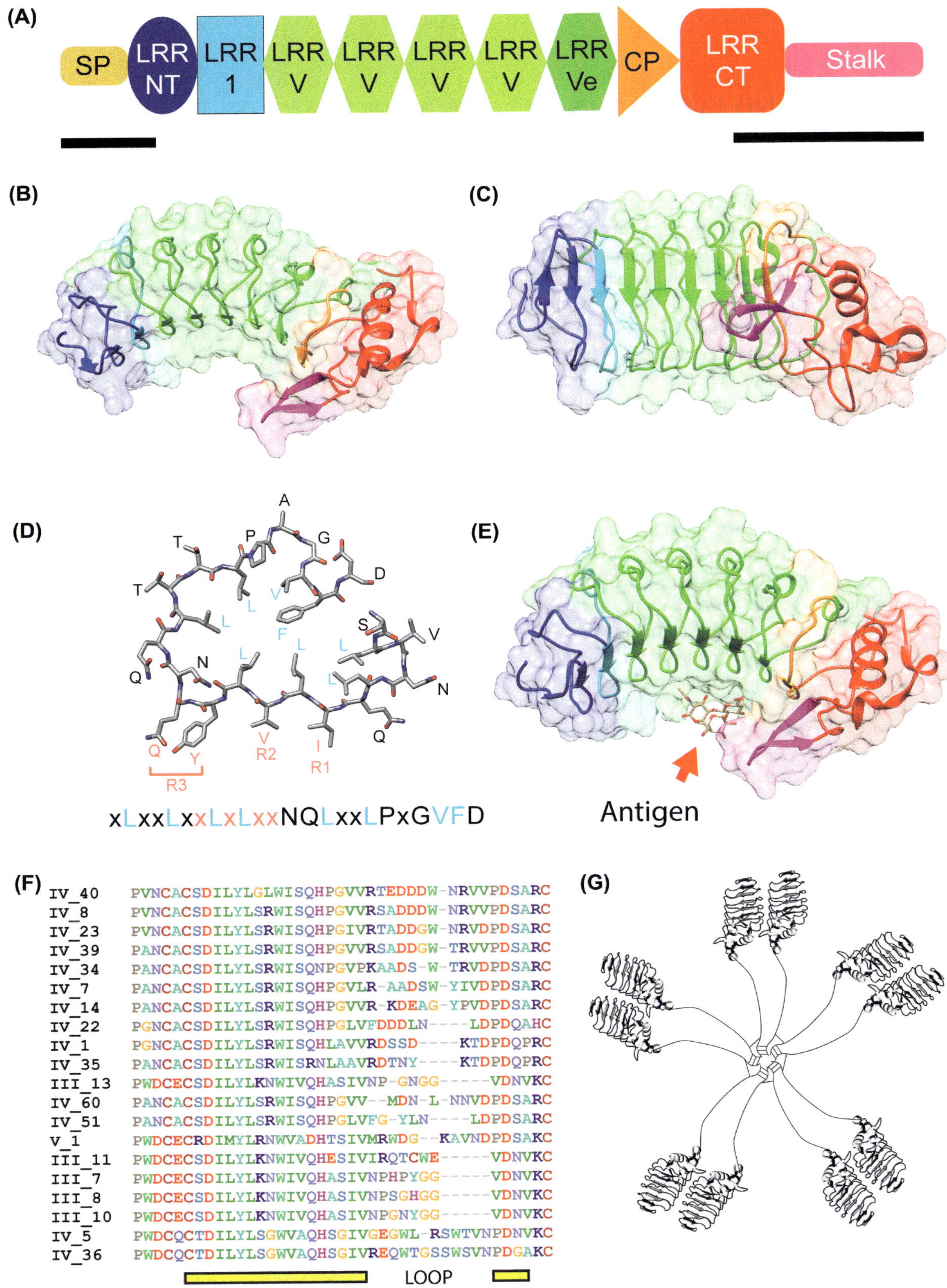

FIGURE 4.1 The structure of the VLRB protein. (A) A cartoon of the linear makeup of a typical VLRB chain. *SP*, Signal peptide; *LRRNT*, N-terminal LRR cap motif; *LRR1*, first LRR; *LRRV*, variable LRR motif; *LRRVe*, last LRRV motif; *CP*, connecting peptide; *LRRCT*, C-terminal LRR cap motif. Black bars under the diagram indicate the position of invariant VLRB sequence. (B, C) Folded ribbon structure of the a VLRB showing solvent-exposed surface representation (based on crystal structure of PDB:5UFC, VLR Tn4-22 with H-trisaccharide bound). (B) Bottom view of concave surface showing β strands contributed by LRRNT (*blue*), LRR1 (*light blue*), LRRV motifs (*green*) and CP (*orange*). (C) Side view showing protruding loop (magenta) from the LRRCT motif (*red*). (D) A single LRRV coil showing the hydrophobic core made up of leucines, phenylalanine and valine residues (*blue*) and highly variable β strand positions (*red*). The LRR consensus sequence is shown at the bottom. (E) A structural representation derived from the crystal structure of a VLRB monomer bound to the H-trisaccharide blood group antigen (PDB Code: 5UFC). Contacts to the antigen are contributed by amino acids in the LRR β-strands and LRRCT loop region. (F) An alignment of the amino acids encoded by a selection of 20 genomic 5′LRRCT donor cassettes showing length and sequence diversity in the region of the LRRCT loop known to be important for antigen binding. The yellow bars under the alignment indicate the position of conserved LRRCT alpha helices. (G) An illustration of the multivalent sea lamprey VLRB antibody after [30].

3. STRUCTURE AND EVOLUTION OF THE GERMLINE *VLRB* GENE

Like all *VLR* genes, the *VLRB* gene is incomplete in its germline configuration; it consists of an N-terminal coding region followed by an intervening sequence and a C-terminal coding region. Before the assembly into a mature gene, it is incapable of encoding any functional protein. The *VLRB* gene of the sea lamprey (*Petromyzon marinus*) was the first *VLRB* locus to be studied in detail; the hagfish germline *VLRB* was identified and characterised later [47]. A recent comparative analysis of germline *VLRB* genes in two hagfish and six lamprey species [53] revealed that the germline configurations of *VLRB* genes are quite different across agnathans, most notably with respect to the length and composition of the non-coding intervening regions. As is true for the *VLRA* and *VLRC* genes in all jawless vertebrate species studied to date, the germline *VLRB* genes in hagfish have a short intervening sequence (104 bp in the pacific hagfish and 118 bp in the inshore hagfish), whereas intervening sequences of the germline genes for *VLRB* in lampreys are much larger (ranging from 6.9 to 19.3 kb). The large intervening sequences in lamprey *VLRB* genes are characterised by the insertion of numerous transposable elements (TEs) [53] (Figure 4.2A). Among these diverse TEs, non-LTR retrotransposons are the most abundant, although LTR retrotransposons, DNA transposons and endogenous retroviruses are also found. The diversity and richness of TEs in the lamprey *VLRB* intervening sequences suggest that multiple species-specific modifications have occurred in this region during lamprey evolution (Figure 4.2A).

The species-specific diversity of lamprey germline *VLRB* genes is also evident in the donor cassettes of the intervening region. The pouched lamprey (Geotriinae subfamily) and the lampreys that belong to Petromyzontinae subfamily (sea lamprey, Far Eastern brook lamprey and European brook lamprey) have only one copy of a *5′LRRCT* cassette located between the long stretches of non-coding intervening sequence (Figure 4.2A), whereas the Japanese lamprey (also belonging to the Petromyzontinae subfamily) has two such cassettes [53]. In contrast, the germline *VLRB*-intervening sequence of the short-headed lamprey (Mordaciinae subfamily) contains no intervening *5′LRRCT* cassette; instead, it has a multiplex cassette encoding a composite LRRNT-LRR1-LRRV-N-terminal LRRV region. Another remarkable feature of the lineage-specific changes in lamprey germline *VLRB* genes is the partial or complete duplication of germline *VLRB* genes in some species [25,53]. For example, a partial duplication of germline *VLRB* gene (located on chromosome 21) is found in the sea lamprey genome about 214 kb downstream of the intact germline *VLRB*; the duplicate contains the 5′UTR and N-terminal VLRB coding region and a *5′LRRCT* cassette (Figure 4.2A). The Far Eastern brook lamprey genome contains a complete duplication of the germline *VLRB* gene [53]. This duplicate copy also has a *5′LRRCT* cassette in the intervening region (Figure 4.2A). Analysis of assembled genes indicates that the *5′LRRCT* cassette in the sea lamprey partial *VLRB* duplicate actively participates in the formation of mature *VLRB* assemblies [53].

Only *VLRB* germline genes in lampreys appear to have these intervening sequence modifications. A plausible explanation is that a TE was inserted into the intervening sequence of the germline *VLRB* before the divergence of modern lamprey lineages and was fixed in the ancestral lamprey population, followed by subsequent duplications and insertions of additional TEs (including the TE carrying genomic donor cassette sequences), leading to the lineage-specific modifications of extant lamprey *VLRB* genes. A short intervening sequence is the likely ancestral state of *VLRB* genes, given the small size of hagfish germline *VLRB* genes and those of *VLRA* and *VLRC* germline genes in all of the agnathan representatives examined (Figures 4.2B and C).

3.1 Genomic Organisation of the *VLRB* Locus

Recent analyses of chromosome scale genome sequence assemblies reveal 428 *VLRB* donor cassettes in five different genomic clusters [53] (Figure 4.2D). Three of these clusters (Clusters I, II and III) are located on chromosome 21. Cluster II spans 539 kb and contains the *VLRB* germline genes; a partial duplicate of the 5′ UTR, signal peptide, N-terminal region and intervening 5′LRRCT cassette; and 178 donor cassettes. Cluster I is located 2 Mb from Cluster II and contains 167 donor cassettes distributed over a 785 kb region. Cluster III is located 7.8 Mb to the other side of Cluster II and contains 13 5′LRRCT cassettes spanning a region of 37 kb. Cluster IV, located on an unplaced genomic scaffold, contains 63 donor cassettes (55 5′LRRCT and 8 LRRV-CP cassettes) over a 345 kb region (interrupted by a 128 kb region lacking donor cassettes). Finally, Cluster V contains seven nearly identical 5′LRRCT donor cassettes in an 11 kb region of chromosome 54. Donor cassettes encoding LRRNT, LRR1 and LRRV motifs are found only in Clusters I and II. The majority of 5′LRRCT donor cassettes (75 of the total of 89) are located in tandem arrays in Clusters III, IV and V, although 14 5′LRRCT cassettes are located in Clusters I and II. The widespread distribution of *VLRB* cassettes is observed also in the genomes of Japanese lamprey and Far Eastern brook lamprey, suggesting that the *VLRB* assembly process must involve the co-localisation of distant regions on chromosome 21, and sometimes even the apposition of clusters situated on separate chromosomes; such trans-chromosomal assembly of antigen receptor genes is exceedingly rare in jawed vertebrates [55], and often associated with genetic disorders, such as ataxia telangiectasia [56]. Other genes are notably excluded from the regions spanned by these donor cassette clusters.

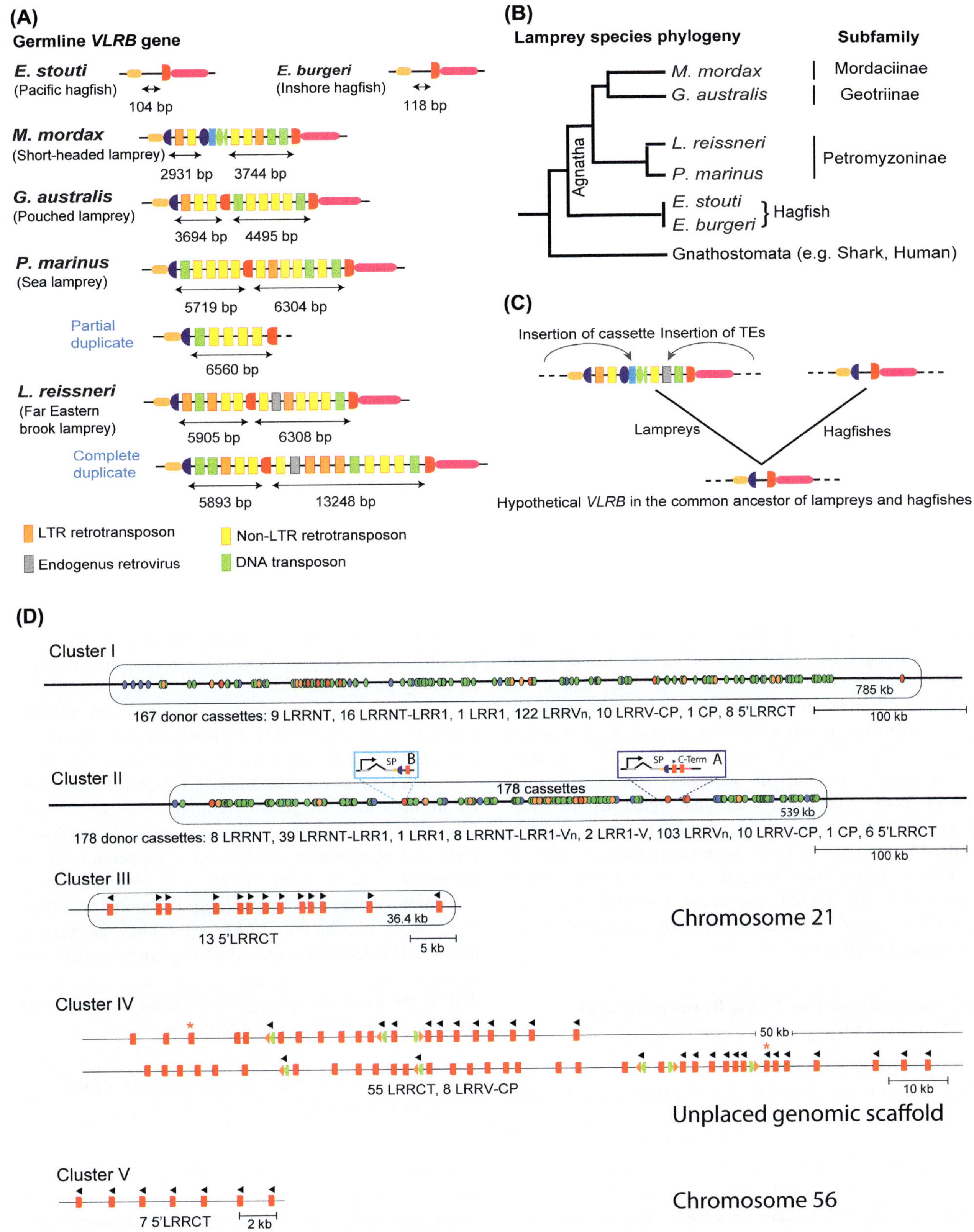

FIGURE 4.2 Genomic organisation of *VLRB* locus. (A) Genomic configurations of germline *VLRB* genes in lampreys and hagfishes. The intervening region of lamprey *VLRB* is considerably larger than its counterpart in hagfishes and contains insertions of multiple transposable elements. Except for the germline *VLRB* in *Mordacia mordax* (short-headed lamprey), which possesses a multiplex cassette encoding *LRRNT-LRR1-LRRV-3′LRRV*, the *VLRB* genes in other lamprey species all contain *5′LRRCT*-LRRCTm cassette. (B) The phylogenetic relationships of the six representative jawless vertebrates (agnathan) species are adapted from Ref. [54]. (C) A plausible evolutionary scenario of germline VLRB genes. Considering short intervening sequences of all *VLRs* except lamprey *VLRB*, a short intervening sequence seems likely the ancestral *VLRB* state. Longer intervening sequences found in all lamprey *VLRB* genes likely evolved after the divergence of the lamprey and hagfish lineages. (D) The sea lamprey VLRB locus. Five clusters of donor cassettes contribute the assembly of mature VLRB. Clusters I, II and III are distributed across chromosome 21 (NC_046089.1 in the kPetMar1.pri genome assembly). The VLRB germline gene and a partial duplicate are present in Cluster II. Several donor cassette clusters are entirely of largely made up of 5′LRRCT cassette. These include Cluster III (chromosome 21), Cluster IV (present on an unplaced scaffold NW_022639073.1) and Cluster V (present on chromosome 56 (NC_046122.1).

Comparisons of the chromosomes containing *VLRB* donor sequences from sea lamprey with Far Eastern brook lamprey (*Lethenteron reissneri*) and the Arctic lamprey (*Lethenteron camtschaticum*) reveal generally conserved organisations of *VLRB* donor cassettes. Over evolutionary time, regions within cassettes appear to be the subject of inversions and regional duplications. As observed in the sea lamprey genome, the Cluster IV cassettes are located on an unplaced scaffold in both species. A single Cluster V-like 5′LRRCT donor cassette is located in a region on a chromosomal scaffold (Hic_chr_52) in the Far Eastern brook lamprey that is syntenic with sea lamprey chromosome 54. Although these Cluster V donor cassettes are on a different chromosome in both species than that containing the *VLRB* germline gene and major Clusters I, II and III, they are used in *VLRB* assemblies in the sea lamprey.

The structure of individual *VLRB* donor cassettes can be variable in terms of the number of encoded LRR motif units and the combinations of LRR motifs that are present in compound cassettes. The 89 LRRCT cassettes are relatively homogeneous in structure, with most elements extending from a small overlap with the CP region that is presumably used in the assembly process. Some shorter LRRCT cassettes are also present. The CP-encoding cassettes also tend to be homogeneous, typically encoding about half of an LRRV region upstream of the complete 13 amino acid CP region and a few amino acids overlapping the N-terminal extension of the LRRCT region. The LRRNT cassettes can be highly variable in sequence, encoding partial LRRNT motifs with some N-terminal signal peptide overlap to cassettes encoding LRRNT, LRR1 and two LRRV motifs. More typically, these cassettes contain an LRRNT and LRR1 motif (55/80). Several cassettes contain isolated LRR1 motifs. The LRRV cassettes vary in size, from partial LRRV motifs to cassettes encoding eight and a half contiguous LRRV units.

3.2 Assembly of the *VLRB* Repertoire and Cassette Usage

In the process of assembly, sequences for the 3′ LRRNT, LRR1, LRRVs, LRRVe, CP and 5′ LRRCT are brought into the germline gene intervening region [24] (Figure 4.3A). While the number of LRRVs incorporated can range up to eight, a majority of assembled VLRBs contain zero to two LRRV sequences (excluding the last LRRVe motif). The number of LRRV elements incorporated into the assembled VLRB genes can be fitted to a binomial distribution model consistent with an equal probability of about 15% of successful incorporation at each stage, explaining the scarcity of VLRB assemblies with more than 3LRRVs ($\sim$0.3% probability of successful assembly of three consecutive LRRVs) [48]. The incorporation of cassette sequences does not conform to the borders of protein units, especially for LRRV and LRRVe motifs, where 40% of these protein segments encoded in assembled genes are the result of intra-module recombination events [50]. *VLRB* gene assembly has been investigated by analysing partially assembled genes isolated from transcripts and genomic DNA of the germline *VLRB* gene [24]. Deletions in the intervening sequence and, in some cases, the insertion of noncoding sequences are characteristic of partial assemblies. Assembly takes place by adding elements at both ends of the incomplete germline gene in a stepwise fashion. Analysis of the incorporation of specific germline cassette elements suggests that a gene conversion-like process underlies *VLRB* assembly [46]. The reciprocal exchange between the intervening region of the germline gene and the donor element, which would be predicted if assembly occurred by homologous recombination, was not evident in PCR-based analyses. The short regions of sequence identity (10–30 nucleotides) among cassettes are also inconsistent with usual forms of eukaryotic homologous recombination, although bacterial recombination can be driven by short sequences [57]. Partial assembly is associated with deletions in the germline *VLRB* intervening DNA, thus suggesting that strand breakage in this region, possibly initiated by cytidine deaminases, may initiate the assembly process [24,46].

The prevalence of cassette usage in *VLRB* assemblies can vary significantly [53]. LRRNT-containing donor cassettes in Cluster II tend to be used more often than those in Cluster I. Collectively the 5′LRRCT cassettes from Cluster III are used most frequently, although the use of cassettes from Cluster IV is also substantial. Several germline cassettes are interrupted by stop codons or deletions that cause frameshifts, but fragments of these cassettes appear to be used in the assembly of potentially functional *VLRB* genes [53]. Most translatable *VLRB* assemblies isolated from lamprey B cells are composed of sequences contributed by donor cassettes from several different clusters, indicating that in the assembly process, individual genomic regions are brought together over large distances.

3.3 Role of Cytidine Deaminases in *VLRB* Gene Assembly

In certain gnathostomes, such as chicken and rabbit, a gene conversion mechanism underlies the generation of the preimmune repertoire of immunoglobulin heavy and light chain genes [58–61]. This process depends on the activity of activation-induced cytidine deaminase (AID) [62,63], a member of the AID/APOBEC family of cytidine deaminases (AADs) [64,65]. Cytidine deaminases represent a large and functionally diverse family of enzymes that all convert cytidine to uridine, in either mononucleotides or single-stranded nucleic acid polymers. Detailed phylogenetic analyses suggest that biological conflicts of AADs

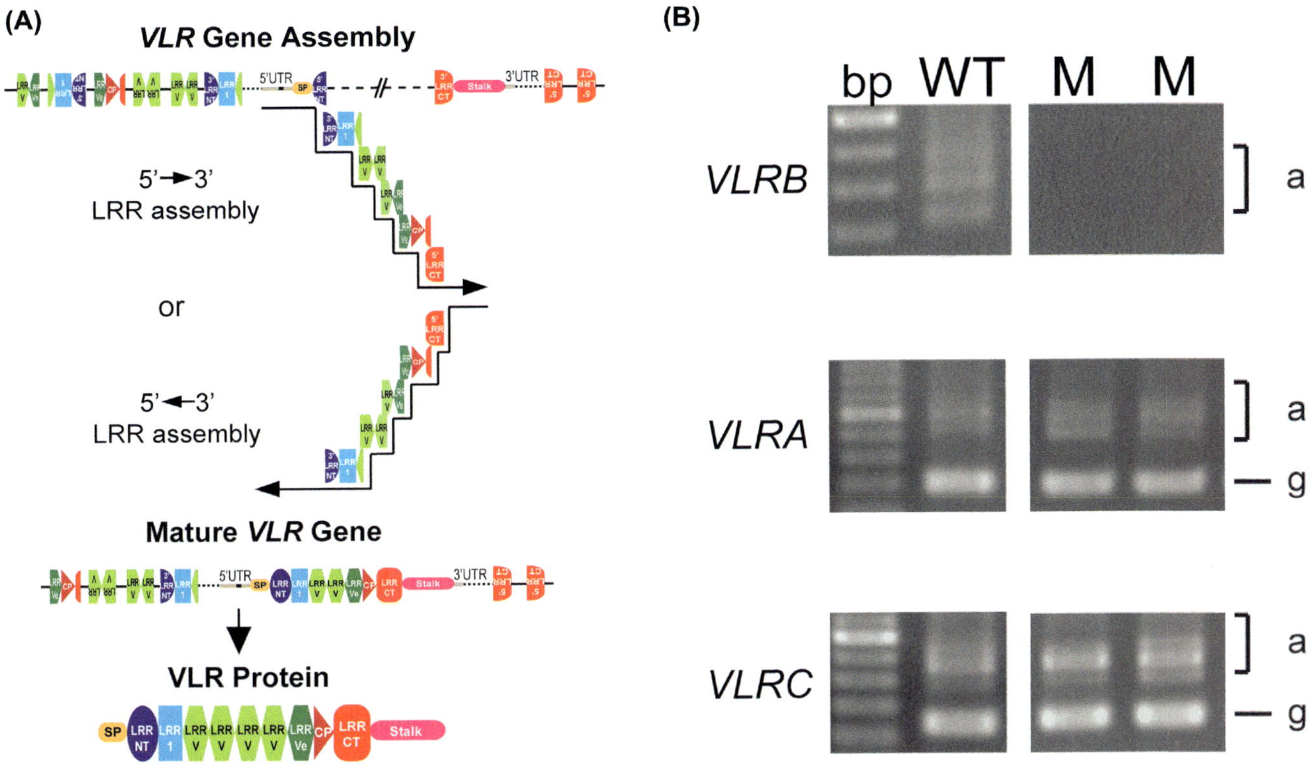

FIGURE 4.3 B cell-like lymphocyte development and *VLRB* gene assembly in lampreys. (A) *VLRB* gene assembly is catalysed by an AID-APOBEC family cytidine deaminase, *CDA2*. The incomplete germline *VLR* genes are flanked by hundreds of LRR cassettes that serve as templates for the stepwise assembly of a mature *VLR* gene, which can begin at either the LRRNT or LRRCT end. (B) Lineage-specific requirement for *CDA2* in the assembly of lamprey *VLR* genes. Lack of *VLRB* antibody gene assembly, but not *VLRA* and *VLRC* assembly in *CDA2*-deficient larvae; *a*, assembled *VLR* gene; *gl*, germline configuration of VLR gene; *M1, M2*, mutant larvae; *WT*, wildtype larva. *Modified from Ref. [77].*

with viruses and genomic retroelements are drivers of their rapid evolution [65]. It has long been recognised that exogenous viral genomes are targets of AADs [66], although the activity of APOBEC enzymes can also be directed against self-DNA in certain tumours [67]. Recent work has identified a second type of antiviral defence executed by AADs, namely the depletion of mononucleotides, effectively blocking the synthesis of viral DNA after infection [68,69]. Cytidine deaminases whose purpose is to target self-DNA, such as AID or CDA2, can be thought of as 'institutionalised' or 'domesticated' versions of these primordial genome mutators [70].

An initial search for AAD homologues in the genome of *P. marinus* identified two genes encoding presumptive AID/APOBEC cytidine deaminase-related genes, designated *CDA1* and *CDA2* [50]; these genes were also found in the genomes of other lamprey species [71], with evidence of rapid diversification of the *CDA1* clade [72]. Unlike the situation in jawed vertebrates, wherein expression of *RAG1* and *RAG2* genes occurs in both T and B lymphocytes [73], the expression patterns of *CDA1* and *CDA2* differ among lymphocyte populations of the sea lamprey [26]. *CDA2* expression is associated with the VLRB-expressing lymphocyte lineage [26,27], whereas *CDA1* expression is preferentially observed in lamprey lymphocytes of VLRA and VLRC lineages [74] in the thymoid, the presumptive thymus equivalent in lamprey larvae. By contrast, *CDA2* expression is found primarily in cells of the typhlosole and the kidneys [74], the major hematopoietic sites of lamprey larvae [75]. Of note, whereas CDA1-like proteins exhibit mutagenic activity in heterologous systems [50,72,76], no such activity has yet been reported for CDA2; however, the loss of *VLRB* assemblies in *CDA2* mutants suggests that CDA2 has activity in vivo (see below; Figure 4.3B).

Gathering information about the ontogeny of the adaptive immune system in lampreys has proven difficult, because of the many challenges confronting the maintenance in the laboratory of in vitro-fertilised lampreys for extended periods of time. However, recent studies have partially overcome this problem and provided the first clues about the onset of *CDA2* expression relative to that of *VLR* genes in *Lampetra planeri*, the European brook lamprey [77]. Low levels of *VLRB* expression are detectable in the second month after fertilisation, preceding the detection of assembled *VLRB* genes by about four–six weeks. This represents an interesting parallel to the situation in jawed vertebrates, wherein the transcriptional activity of antigen receptor genes is considered to be a prerequisite of recombination and subsequent modification (in immunoglobulin loci). Whereas transcripts of fully assembled *VLRA* and *VLRC* appear even later, *CDA2* transcripts are markedly upregulated in synchrony with the detection of *VLRB* assemblies. Inactivation of the *CDA2* gene by CRISPR/Cas9 is associated with the loss of *VLRB* assembly, whereas the maturation of *VLRA* and *VLRC* genes proceeds in a normal fashion [77] (Figure 4.3B). These studies established a B lineage-specific function of *CDA2*, unlike the pan-lymphocytic activity of *RAG* genes in jawed vertebrates. Given that of the two principal lamprey *CDA* genes, *CDA2* most closely resembles *AID* of jawed vertebrates, these results reveal a remarkable convergence of AAD activity in establishing antibody repertoires in all vertebrates.

4. CHARACTERISTICS OF LAMPREY B CELLS

Lamprey VLRB cells and the B cells of jawed vertebrates exhibit striking functional and molecular similarities [24,30,51]. Although members of all three lamprey lymphocyte lineages, VLRA$^+$, VLRB$^+$ and VLRC$^+$ cells, respectively, undergo proliferation and lymphoblastic transformation in response to immunisation with antigens (e.g. bacteria, viruses, fungi and mammalian cells), only VLRB$^+$ lymphocytes differentiate into plasma cells that produce their antigen receptors as disulfide-linked, multivalent antibodies similar to the IgM antibodies of jawed vertebrates [30,51]. Antigen-specific VLRB antibodies are detectable in the serum around one week after immunisation [51]. VLRB antibody-secreting plasma cells with expanded cytoplasm and secretory organelles characteristic of jawed vertebrate plasma cells are observed in the blood, kidney and typhlosole.

VLRB cells express homologues of many of the genes that are expressed by B cells of jawed vertebrates, including the hematopoietic progenitor homing receptor CXCR4; the herpesvirus entry mediator/tumour necrosis factor receptor superfamily, member 14 (TNFRSF14) that binds to LIGHT on T cells; two components of the BCR-mediated signalling cascade, Syk and the B cell adaptor protein (BCAP); the chemotactic inflammatory cytokine IL-8; the IL-17 receptor and the Toll-like receptor orthologues TLR2 and TLR10, the ligation of which is important for jawed vertebrate B cell activation [26,27] (Figure 4.4A). Recent characterisation of interleukin receptor genes and their corresponding signalling ligands in the sea lamprey genome expands the description of the VLRB cells and will aid understanding of their interactions with other cell types [78].

The VLRB antibodies may engage effector mechanisms analogous to those employed by the Ig-based antibodies of jawed vertebrates for protective immunity. Specific VLRB antibodies from immunised lampreys can trigger complement-dependent lysis of bacteria, heterologous erythrocytes and tumour cell lines via interaction with a C1q-like complex and the C3 complement component [79]. These findings indicate that orthologs of the classical complement components, which allow Ig-based antibodies to link adaptive and innate immune defence mechanisms in jawed vertebrates, have analogous functions in jawless vertebrates. Whether or not the VLRB antibodies can bind to functional equivalents of the Ig Fc receptors expressed on effector cells is still unknown.

5. SITES OF LAMPREY B CELL DEVELOPMENT

Functional information about the development of the hematopoietic system of lampreys is still limited due to the shortage of suitable discriminating reagents. The hematopoietic and lymphopoietic tissues in the immune system of jawed vertebrates have been well defined by classical histological studies, complemented by the use of numerous cell-type−specific monoclonal antibodies for flow cytometry and immunohistology, and gene-specific reagents for RNA in situ hybridisation. Using a combination of related methodological approaches, the typhlosole and kidneys have emerged as major sites of blood formation in lamprey larvae; haemopoiesis in these sites is terminated during metamorphosis, and the supraneural body becomes the major hematopoietic site during the short life-span of adult animals [75].

Until recently, the presence of a thymus equivalent in lamprey larvae was unclear [74]. However, small cellular aggregates, characterised by the presence of tight associations of lymphocytes and epithelial cells, and situated at the tips of the larval gill filaments are now considered the site of development of the VLRA$^+$ and VLRC$^+$ lineages [27,74]. These 'thymoid' structures exhibit the following characteristics: (1) a preponderance of partial and nonfunctional *VLRA* and *VLRC* gene assemblies, indicative of the presence of immature lymphocytes; (2) the presence of large numbers of apoptotic cells, suggesting − by way of analogy to the thymus in jawed vertebrates − ongoing selection processes and (3) the expression of a cytidine deaminase gene, *CDA1* which is believed to be functionally linked to the assembly of *VLRA* and *VLRC* genes. However, conclusive evidence for the lymphopoietic function of the thymoid awaits the demonstration of T-lymphopenia in lampreys mutant for T-lineage genes, such as *CDA1*.

Based on the observed co-expression of *VLRB* and *CDA2* genes, the hematopoietic aspects of typhlosole and kidney are considered to be sites of B lymphocyte development in lamprey larvae [74]. The most compelling

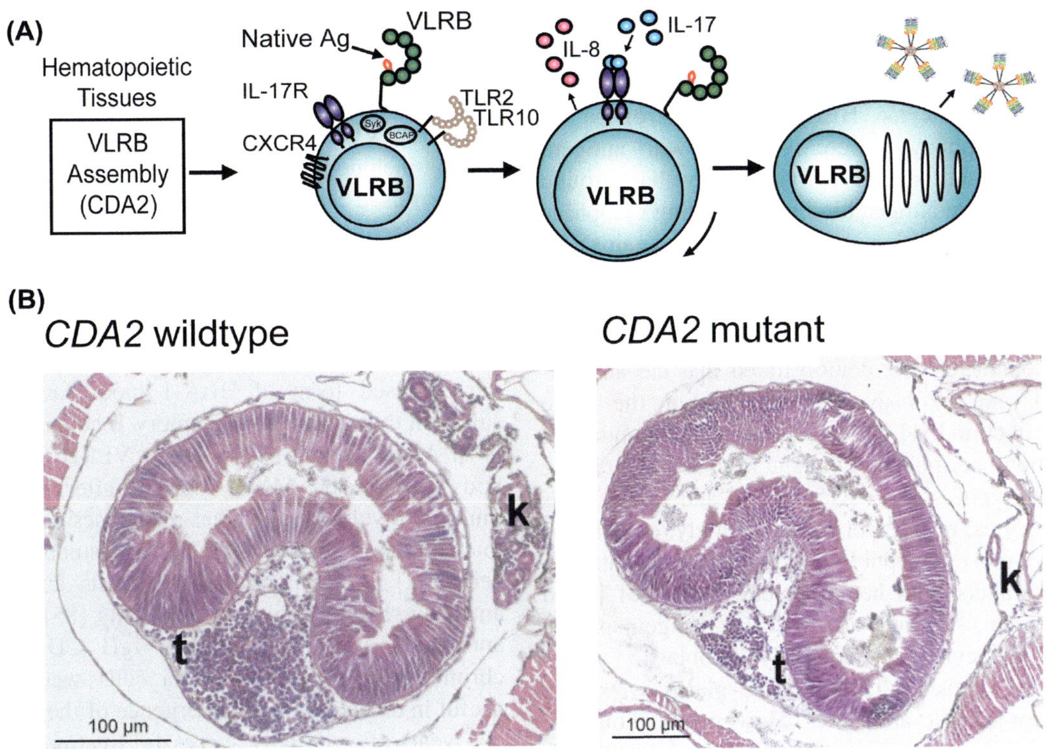

FIGURE 4.4 Development of VLRB-expressing cells. (A) *CDA2* is expressed by B-like VLRB cells in the hematopoietic tissues. VLRB cells recognise native antigen, proliferate and secrete multimeric VLRB antibodies. VLRB cells express IL-17 receptor (IL-17R), TLR2, TLR10 and CXCR4 on the cell surface, cytokine IL-8 and signalling molecules Syk and BCAP. (B) Lymphopenia in *CDA2*-deficient typhlosole is evident in histological sections (haematoxylin and eosin staining) depicting typhlosole (t) and kidney (k) tissues taken from *CDA2* wild-type and *CDA2* mutant animals at 26 weeks of age.

evidence comes from the analysis of *CDA2* mutants, which were generated by the use of CRISPR/Cas9 mutagenesis after in vitro fertilisation [77]. In *L. planeri CDA2* crispants, maturation of B cells is entirely blocked, as evidenced by the absence of cells expressing assembled *VLRB* genes. Remarkably, this genotype features profound hypocellularity in typhlosole and kidneys; cells remaining in these tissues exhibit the cytological features of myeloid and erythroid-lineage cells. Thus, B lymphocytes represent a major fraction of the hematopoietic cells in typhlosole and kidneys [77] (Figure 4.4B). Moreover, the absence of *VLRB*-expressing cells in these sites of the *CDA2*-mutant animals suggests that only immature B cells (that is, cells in which only low levels of non-functional germline *VLRB* transcripts are present) survive under these conditions, analogous to the requirement of the BCR for survival of B cells in jawed vertebrates [80].

6. THE SEA LAMPREY VLRB ANTIBODY RESPONSE

Several investigations have directly evaluated the kinetics of the lamprey VLRB response. After two biweekly inoculations with the exosporium (spore coat) of *Bacillus anthracis*, the proportion of large VLRB$^+$ lymphocytes increased almost 25-fold in blood and approximately10-fold in kidney and typhlosole [24]. After inoculation with both *B. anthracis* exosporium and influenza virus particles, total VLRB concentrations in the plasma increased by 7—10-fold [24,52]. Serum VLRB specific for *B. anthracis* spores, but not those reacting with spores of *Bacillus subtilis* increased with each boost during the eight-week period [24]. In another study, it was demonstrated that VLRB-binding specificity across the BclA proteins of *B. anthracis, Bacillus cereus* and *Bacillus thuringiensis* depends either on BclA packing in the spore structure or on differential glycosylation, rather than on the primary amino acid sequence [36].

When heterologous erythrocytes were used as inoculum, agglutination titres peaked at ~19 days after exposure and increased 20-fold with a two-week after a second inoculation [51]. In this investigation, agglutination responses were dose-dependent, distinguished between mouse and human RBCs and depended on serum VLRB; remarkably, the discriminatory specificity of anti-sera focused on the H-trisaccharide.

A specific VLRB response is most readily induced with the use of particulate antigen. In most cases, a response cannot be elicited with soluble protein antigen; this failure is reflected in the lack of lymphoblastoid transformation even upon the hyperimmunisation regimes that readily elicit a response with *B. anthracis* inoculation [51]. Nonetheless, when combined with complete Freund's adjuvant, responses could be obtained against a soluble protein, hen egg lysozyme, HEL [39]. In control experiments, naïve lymphocytes were incubated with antigen-specific plasma and shown to yield only background levels of lymphocyte antigen binding, excluding cytophilic binding of VLRB (e.g. though an Fc receptor analog) for the detection of antigen-specific VLRB lymphocytes in cell-based experiments [51].

Elispot analyses after sorting lamprey leucocytes on light scatter characteristics demonstrated that the antigen-specific cells are exclusively found within the large $VLRB^+$ cell fraction [51]. These cells are most abundant in the kidney and in peripheral blood and least abundant in the hematopoietic typhlosole (~215, ~270 and ~100 cells per 10^6 leucocytes, respectively). Interestingly, cells in the typhlosole are most susceptible to corticosteroid treatment, suggesting parallels with hematopoietic tissues of jawed vertebrates, such as bone marrow. Thus, antigen-specific $VLRB^+$ lymphocytes undergo lymphoblastic transformation into surface VLRB-bearing plasmacytes that secrete antigen-specific antibodies. However, it should be noted that to become efficient secretors of soluble VLRB antibodies, mature plasma cells have lost cell surface VLRB expression.

7. VLRB MONOCLONAL ANTIBODIES

As described below, monoclonal VLRB antibodies have been invaluable in defining the structural mechanisms of VLRB-antigen specificity. A promising application of VLRB antibodies is their use as an alternative form of reagents with unique structural features and binding specificities. Since lamprey VLRB antibodies are structurally unrelated to immunoglobulins, different binding specificities may be expected relative to traditional antibody reagents. Furthermore, the phylogenetic distance separating mammals and lampreys may overcome constraints imposed by tolerance induction when generating antibodies to mammalian targets. Monoclonal VLRB antibodies have been isolated after inoculation of ammocoete larvae by intraperitoneal injection and boosting with antigens of different molecular nature (described in more detail below). Leucocytes are then isolated, and expression libraries are prepared by the insertion of assembled *VLRB* message cDNA sequences into expression vectors. Monoclonal VLRB proteins are then produced in yeast and mammalian cells that can then be screened for specific binding. Affinity purification of cells from these heterogeneous mixtures via surface display of VLRB antibodies directly links antigen specificity to the molecular sequence. A summary of monoclonal isolation strategies is shown in Figure 4.5. This approach and several variants have been described in detail [30,49,81–93].

7.1 Monoclonal VLRB to Cell Surface Antigens

Lamprey VLRB antibodies have been raised against sorted cell subsets to identify known and novel cell targets. For example, VLRB antibodies generated against human T cells included those specific for CD5 [93]. VLRB antibodies raised against previously uncharacterised cell surface antigens and/or their posttranslational modifications indicated their potential to reveal new aspects of lymphocyte biology. For instance, when sea lamprey ammocoete larvae were immunised with a mixture of plasmacytoma cell lines, one VLRB antibody identified a formerly unrecognised form of HLA-1 modification by tyrosine sulfation that is found on memory B cells and plasma cells [94]. An additional example is a VLRB monoclonal antibody, designated MM3, isolated after inoculation with multiple myeloma cells, that recognises a tetrameric enzymatically active form of CD38 on plasmablasts and plasma cells [92]. A tetrameric form of this anti-CD38 VLRB antibody showed enhanced sensitivity [85]. Finally, VLRB antibodies directed against an IgH CDR3 idiotope on chronic lymphocytic leukemia cells were shown to be useful in detecting the re-emergence of the malignant clone ~5 years after initially successful treatment [95].

7.2 VLRB Antigen Interaction With Defined Protein and Glycans

In addition to cell surface markers, VLRB monoclonals have been isolated for a variety of antigens, including extracellular matrix components [96], glycans [32,38,83], virus particles [52] and purified proteins [36,39,97,98].

Twenty-two structures of sea lamprey and inshore hagfish (*Eptatretus burgeri*) VLRB proteins have been solved, 12 of these in complex with antigens. As predicted from positional amino acid variability and evidence for positive selection [24], interactions with antigen are primarily mediated through solvent exposed residues on the concave β-strand surface of VLRB and by residues in the LRRCT loop. Three rows of residues within the concave VLRB surface (termed R1, R2 and R3) contribute to this hypervariable binding surface (Figure 4.1D) [39]. While the concave β-strand surface provides a fairly flat binding surface, the LRRCT loop can extend into crevasses of the antigenic structure, as exemplified by its interaction with the catalytic site of HEL [39]. Whereas most Ag-VLRB interactions involve the hypervariable residues contributed by LRR1, LRRV, CP regions and the LRRCT loop, the interactions of two VLRB monoclonal antibodies with zebrafish TLR5 include residues in one or both LRRNT β-strands [97]. Compared to immunoglobulins, VLRB antibodies have a very rigid structure, a feature which contributes to the exquisite specificity particularly for glycan antigens, a class of antigens that is notoriously

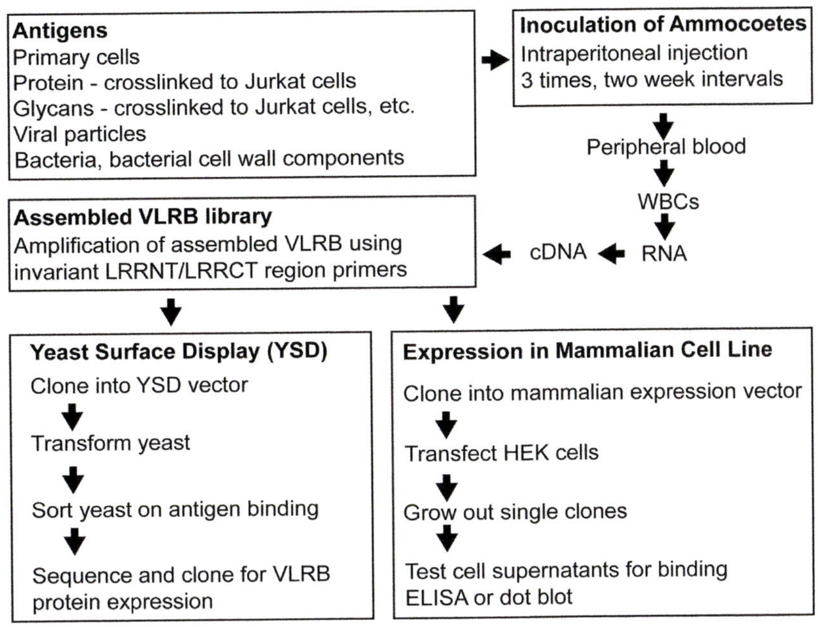

FIGURE 4.5 Flow diagram showing strategies for the production of VLRB monoclonal antibodies. Lamprey ammocoete larvae (2–4 years old) are inoculated intraperitoneally with particulate antigen (generally heterologous cells or proteins/glycans crosslinked to heterologous carrier cells). After boosting, peripheral leucocytes are isolated, assembled *VLRB* cDNA is amplified and introduced into mammalian cell or yeast display expression systems for specificity screening.

difficult to discriminate using immunoglobulins [83]. For VLRB monoclonal antibodies that bind carbohydrates, some recurrent molecular features have been noted, including the presence of tryptophan residues which exhibit stacking interactions with glycans, and an aspartic acid-glutamine motif that supports hydrogen bonding with sugar hydroxyl groups [83]. In a VLRB antibody specific for the Thomsen–Friedenreich glycan tumour antigen, the glycan moiety is engaged by a hydrophobic cage formed by four tryptophans, of which two are supplied by LRRV motifs, one by CP and one by the LRRCT loop [38].

The dissociation constants for VLRB monomer–antigen interactions vary by about three orders of magnitude, from 13 nM to 45 μM [97]. The buried VLRB surface in antigen binding ranges from 200 to 210 $Å^2$ for smaller glycan antigens (e.g. H-trisaccharide) to 700–850 $Å^2$ for protein antigens [97]; as judged by the nature of four Ag-VLRB structures, 10–19 VLRB contact residues are defined, with their number scaling to the size of the antigen [52].

Although LRR-based VLRs have an entirely different structure than immunoglobulins, comparative analyses of some antigen interactions suggest strikingly common features of antigenicity. Similar to the mammalian antibody response to influenza virus particles, the VLRB response is dominated by the recognition of epitopes on the globular region of HA [52]. Strikingly, similar contact areas are seen for VLRB and immunoglobulin interactions with protein antigens [36,39,52,83].

8. COMMONALITIES AND DIFFERENCES OF VERTEBRATE ADAPTIVE IMMUNE SYSTEMS

Despite many differences in molecular detail, the overall design principles of adaptive immune systems of jawless and jawed vertebrates are surprisingly similar [4,28,70,99,100]. The dichotomy of lymphocyte lineages, that is, the presence of antibody-secreting B cells and T-like cells exhibiting membrane-bound receptors is clearly an ancestral feature of vertebrates, as is the reliance on somatic assembly to generate functional antigen receptor genes. Moreover, the secreted antibodies of jawless and jawed vertebrates are present in multimeric form in the serum. With respect to the mechanism by which functional antigen receptor genes are generated, it is of note that both branches of vertebrates evolved domesticated forms of ancient genome mutators; even more remarkably, the repertoire formation of antibody genes relies on the activities of dedicated cytidine deaminases that are structurally closely related [4,28,77,99,100]. These findings suggest that the activities of cytidine deaminases were a defining feature of the fledgling adaptive immune system of early vertebrates and may perhaps represent an ancient component of metazoan genome modification systems.

The similarities between jawless and jawed vertebrates extend to lymphopoietic organs; most notable in this respect are the presence of a thymus equivalent associated with T cell development and the general hematopoietic tissue in typhlosole and kidney associated with B cell development [74,101].

The most conspicuous difference between the two alternative adaptive immune systems of vertebrates can be found in the chemical nature of antigen receptors. Whereas jawless vertebrates rely on LRRs as their main building block [25], the receptors of jawed vertebrates are built from immunoglobulin-fold domains [102]. While it is theoretically possible that a common ancestor of vertebrates possessed both LRR- and immunoglobulin-based antigen receptor systems, independent but convergent evolution of the two types diversified receptors seems more likely than reciprocal loss of the cardinal elements of each receptor system.

9. OUTSTANDING QUESTIONS ABOUT THE JAWLESS VERTEBRATE VLRB SYSTEM

One of the most pressing issues in advancing our understanding of the agnathan VLRB antibody system concerns the structure of the composite B cell receptor and its signalling mechanisms. Findings in this area would lend a more complete understanding of VLRB function and more precisely clarify the relationship between the jawless and jawed vertebrate B cell systems. Solutions to the current inability to maintain primary VLRB cells in culture and to generate lamprey B cell lines are clearly needed.

A more detailed understanding of the development of lamprey B cells will come from further manipulation of candidate regulatory genes in the lamprey embryo and larva using CRISPR/Cas9 mutation strategies and related sequence insertion approaches. Single-cell RNA sequencing (scRNA-seq) carried out on cells from key lymphopoietic and secondary immune tissues, from embryos and larvae at critical developmental stages will be useful in identifying candidate regulators for further mutational study. Importantly, scRNAseq will help to define VLRB cell developmental cell states and potential lamprey B cell subsets. These approaches will provide insight into the evolutionary relationship of lineage-determining factors for lampreys B cells and allow more detailed comparison to those of jawed vertebrates.

Finally, on a genomic scale, an understanding of the forces that shape the organisation of the complex *VLRB* locus is needed. Why has the size of the *VLRB* germline gene intervening region expanded in lampreys, and why have there been partial and total duplications of the *VLRB* germline genes in some lamprey species? Why are most *VLRB 5′LRRCT* donor cassettes clustered in tandem arrays separately from germline gene? Insight into these questions will come from more detailed comparative genomic analyses among lamprey and hagfish species using new and soon-to-be available genome assemblies. Questions about how CDA2 is targeted to *VLRB* germline gene and donor cassettes, and what cofactors define locus accessibility can be approached using a more comprehensive understanding of the *VLRB* locus in combination with information derived from cultured lamprey B cells, assembled *VLRB* repertoire comparisons and scRNAseq investigations.

COMPETING INTEREST STATEMENT

M.D.C. is a cofounder and shareholder of NovAb, Inc., which produces lamprey antibodies for biomedical purposes, and J.P.R. is a consultant for NovAb. The other authors have no financial conflicts of interest.

REFERENCES

[1] Brazeau MD, Friedman M. The origin and early phylogenetic history of jawed vertebrates. Nature 2015;520:490–7.

[2] Janvier P. Facts and fancies about early fossil chordates and vertebrates. Nature 2015;520:483–9.

[3] Kuratani S. Evo-devo studies of cyclostomes and the origin and evolution of jawed vertebrates. Curr Top Dev Biol 2021;141:207–39.

[4] Flajnik MF, Kasahara M. Origin and evolution of the adaptive immune system: genetic events and selective pressures. Nat Rev Genet 2010;11:47–59.

[5] Liu C, Zhang Y, Liu CC, Schatz DG. Structural insights into the evolution of the RAG recombinase. Nat Rev Immunol 2022;22:353–70.

[6] Boehm T, Hirano M, Holland SJ, Das S, Schorpp M, Cooper MD. Evolution of alternative adaptive immune systems in vertebrates. Annu Rev Immunol 2018;36:19–42.

[7] Bone Q, Marshall NB, Blaxter JHS. Biology of Fishes. Boston, MA: Springer US; 1995.

[8] Hall BK. Bones and cartilage. New York, NY: Elsevier; 2015.

[9] Hirano M, Das S, Guo P, Cooper MD. The evolution of adaptive immunity in vertebrates. Adv Immunol 2011;109:125–57.

[10] Xu Y, Zhu S-W, Li Q-W. Lamprey: a model for vertebrate evolutionary research. Zool Res 2016;37:263–9.

[11] Litman GW, Cannon JP, Dishaw LJ. Reconstructing immune phylogeny: new perspectives. Nat Rev Immunol 2005;5:866–79.

[12] Nakatani Y, Shingate P, Ravi V, Pillai NE, Prasad A, McLysaght A, et al. Reconstruction of proto-vertebrate, proto-cyclostome and proto-gnathostome genomes provides new insights into early vertebrate evolution. Nat Commun 2021;12:4489.

[13] Simakov O, Marlétaz F, Yue J-X, O'Connell B, Jenkins J, et al. Deeply conserved synteny resolves early events in vertebrate evolution. Nat. Ecol. Evol. 2020;4:820–30.

[14] Finstad J, Good RA. The evolution of the immune response: 3. Immunologic responses in lamprey. J Exp Med 1964;120:1151–68.

[15] Fujii T, Nakagawa H, Murakawa S. Immunity in lamprey. I. Production of haemolytic and haemagglutinating antibody to sheep red blood cells in Japanese lampreys. Dev Comp Immunol 1979;3:441–51.

[16] Hagen M, Filosa MF, Youson JH. The immune response in adult sea lamprey (*Petromyzon marinus* L.): the effect of temperature. Comp Biochem Physiol A 1985;82:207–10.

[17] Linthicum DS, Hildemann WH. Immunologic responses of Pacific hagfish. 3. Serum antibodies to cellular antigens. J Immunol 1970;105:912–8.

[18] Litman GW, Finstad FJ, Howell J, Pollara BW, Good RA. The evolution of the immune response. VIII. Structural studies of the lamprey immuoglobulin. J Immunol 1970;105:1278–85.

[19] Pollara B, Litman GW, Finstad J, Howell J, Good RA. The evolution of the immune response. VII. Antibody to human "O" cells and properties of the immunoglobulin in lamprey. J Immunol 1970;105:738–45.

[20] Fujii T, Hayakawa I. A histological and electron-microscopic study of the cell types involved in rejection of skin allografts in ammocoetes. Cell Tissue Res 1983;231:301–12.

[21] Hildemann WH, Thoenes GH. Immunological responses of Pacific hagfish. I. Skin transplantation immunity. Transplantation 1969;7:506–21.

[22] Mayer WE, Uinuk-ool T, Tichy H, Gartland LA, Klein J, Cooper MD. Isolation and characterization of lymphocyte-like cells from a lamprey. Proc Natl Acad Sci USA 2002;99:14350–5.

[23] Uinuk-ool T, Mayer WE, Sato A, Dongak R, Cooper MD, Klein J. Lamprey lymphocyte-like cells express homologs of genes involved in immunologically relevant activities of mammalian lymphocytes. Proc Natl Acad Sci USA 2002;99:14356–61.

[24] Alder MN, Rogozin IB, Iyer LM, Glazko GV, Cooper MD, Pancer Z. Immunology: diversity and function of adaptive immune receptors in a jawless vertebrate. Science 2005;310:1970–3.

[25] Pancer Z, Amemiya CT, Ehrhardt GRA, Ceitlin J, Larry Gartland G, Cooper MD. Somatic diversification of variable lymphocyte receptors in the agnathan sea lamprey. Nature 2004;430:174–80.

[26] Guo P, Hirano M, Herrin BR, Li J, Yu C, Sadlonova A, et al. Dual nature of the adaptive immune system in lampreys. Nature 2009;459:796–801.

[27] Hirano M, Guo P, McCurley N, Schorpp M, Das S, Boehm T, et al. Evolutionary implications of a third lymphocyte lineage in lampreys. Nature 2013;501:435–8.

[28] Cooper MD, Alder MN. The evolution of adaptive immune systems. Cell 2006;124:815–22.

[29] Pancer Z, Cooper MD. The evolution of adaptive immunity. Annu Rev Immunol 2006;24:497–518.

[30] Herrin BR, Alder MN, Roux KH, Sina C, Boydston JA, Turnbough CL, et al. Structure and specificity of lamprey monoclonal antibodies. Proc Natl Acad Sci USA 2008;105:2040–5.

[31] Deng L, Velikovsky CA, Xu G, Iyer LM, Tasumi S, et al. A structural basis for antigen recognition by the T cell-like lymphocytes of sea lamprey. Proc Natl Acad Sci USA 2010;107:13408–13.

[32] Han BW, Herrin BR, Cooper MD, Wilson IA. Antigen recognition by variable lymphocyte receptors. Science 2008;321:1834–7.

[33] Holland SJ, Gao M, Hirano M, Iyer LM, Luo M, et al. Selection of the lamprey VLRC antigen receptor repertoire. Proc Natl Acad Sci USA 2014;111:14834–9.

[34] Kanda R, Sutoh Y, Kasamatsu J, Maenaka K, Kasahara M, Ose T. Crystal structure of the lamprey variable lymphocyte receptor C reveals an unusual feature in its N-terminal capping module. PLoS One 2014;9:e85875.

[35] Ho MK, Se CO, Ki JL, Kasamatsu J, Jin YH, et al. Structural diversity of the hagfish variable lymphocyte receptors. J Biol Chem 2007;282:6726–32.

[36] Kirchdoerfer RN, Herrin BR, Han BW, Turnbough CL, Cooper MD, Wilson IA. Variable lymphocyte receptor recognition of the immunodominant glycoprotein of bacillus anthracis spores. Structure 2012;20:479–86.

[37] Lee S-C, Park K, Han J, Lee J, Kim HJ, et al. Design of a binding scaffold based on variable lymphocyte receptors of jawless vertebrates by module engineering. Proc Natl Acad Sci USA 2012;109:3299–304.

[38] Luo M, Velikovsky CA, Yang X, Siddiqui MA, Hong X, et al. Recognition of the thomsen-friedenreich pancarcinoma carbohydrate antigen by a lamprey variable lymphocyte receptor. J Biol Chem 2013;288:23597–606.

[39] Velikovsky CA, Deng L, Tasumi S, Iyer LM, Kerzic MC, et al. Structure of a lamprey variable lymphocyte receptor in complex with a protein antigen. Nat Struct Mol Biol 2006;16:725–30.

[40] Das S, Li J, Holland SJ, Iyer LM, Hirano M, et al. Genomic donor cassette sharing during VLRA and VLRC assembly in jawless vertebrates. Proc Natl Acad Sci USA 2014;111:14828–33.

[41] Das S, Hirano M, Aghaallaei N, Bajoghli B, Boehm T, Cooper MD. Organization of lamprey variable lymphocyte receptor C locus and repertoire development. Proc Natl Acad Sci USA 2013;110:6043–8.

[42] Im SP, Lee JS, Kim SW, Yu JE, Kim YR, et al. Investigation of variable lymphocyte receptors in the alternative adaptive immune response of hagfish. Dev Comp Immunol 2016;55:203–10.

[43] Kasamatsu J, Sutoh Y, Fugo K, Otsuka N, Iwabuchi K, Kasahara M. Identification of a third variable lymphocyte receptor in the lamprey. Proc Natl Acad Sci USA 2010;107:14304–8.

[44] Kishishita N, Matsuno T, Takahashi Y, Takaba H, Nishizumi H, Nagawa F. Regulation of antigen-receptor gene assembly in hagfish. EMBO Rep 2010;11:126–32.

[45] Li J, Das S, Herrin BR, Hirano M, Cooper MD. Definition of a third VLR gene in hagfish. Proc Natl Acad Sci USA 2013;110:15013–8.

[46] Nagawa F, Kishishita N, Shimizu K, Hirose S, Miyoshi M, et al. Antigen-receptor genes of the agnathan lamprey are assembled by a process involving copy choice. Nat Immunol 2007;8:206–13.

[47] Pancer Z, Saha NR, Kasamatsu J, Suzuki T, Amemiya CT, Kasahara M, et al. Variable lymphocyte receptors in hagfish. Proc Natl Acad Sci USA 2005;102:9224–9.

[48] Sutoh Y, Kasahara M. Copy number and sequence variation of leucine-rich repeat modules suggests distinct functional constraints operating on variable lymphocyte receptors expressed by agnathan T cell-like and B cell-like lymphocytes. Immunogenetics 2014;66:403–9.

[49] Tasumi S, Velikovsky CA, Xu G, Gai SA, Wittrup KD, et al. High-affinity lamprey VLRA and VLRB monoclonal antibodies. Proc Natl Acad Sci USA 2009;106:12891–6.

[50] Rogozin IB, Iyer LM, Liang L, Glazko GV, Liston VG, et al. Evolution and diversification of lamprey antigen receptors: evidence for involvement of an AID-APOBEC family cytosine deaminase. Nat Immunol 2007;8:647–56.

[51] Alder MN, Herrin BR, Sadlonova A, Stockard CR, Grizzle WE, et al. Antibody responses of variable lymphocyte receptors in the lamprey. Nat Immunol 2008;9:319–27.

[52] Altman MO, Bennink JR, Yewdell JW, Herrin BR. Lamprey VLRB response to influenza virus supports universal rules of immunogenicity and antigenicity. Elife 2015;4:e07467.

[53] Das S, Rast JP, Li J, Kadota M, Donald JA, et al. Evolution of variable lymphocyte receptor B antibody loci in jawless vertebrates. Proc Natl Acad Sci USA 2021;118. e2116522118.

[54] Kuraku S, Kuratani S. Time scale for cyclostome evolution inferred with a phylogenetic diagnosis of hagfish and lamprey cDNA sequences. Zool Sci (Tokyo) 2006;23:1053–64.

[55] Tycko B, Palmer JD, Sklar J. T cell receptor gene trans-rearrangements: chimeric gamma-delta genes in normal lymphoid tissues. Science 1989;245:1242–6.

[56] Lipkowitz S, Stern MH, Kirsch IR. Hybrid T cell receptor genes formed by interlocus recombination in normal and ataxia-telangiectasis lymphocytes. J Exp Med 1990;172:409–18.

[57] Fujimoto R, Osakabe T, Saito M, Kurosawa N, Isobe M. Minimum length of homology arms required for effective Red/ET recombination. Biosci Biotechnol Biochem 2009;73:2783–6.

[58] Becker RS, Knight KL. Somatic diversification of immunoglobulin heavy chain VDJ genes: evidence for somatic gene conversion in rabbits. Cell 1990;63:987–97.

[59] Parng CL, Hansal S, Goldsby RA, Osborne BA. Gene conversion contributes to Ig light chain diversity in cattle. J Immunol 1996;157:5478–86.

[60] Reynaud CA, Anquez V, Grimal H, Weill JC. A hyperconversion mechanism generates the chicken light chain preimmune repertoire. Cell 1987;48:379–88.

[61] Thompson CB, Neiman PE. Somatic diversification of the chicken immunoglobulin light chain gene is limited to the rearranged variable gene segment. Cell 1897;48:369–78.

[62] Muramatsu M, Kinoshita K, Fagarasan S, Yamada S, Shinkai Y, Honjo T. Class switch recombination and hypermutation require activation-induced cytidine deaminase (AID), a potential RNA editing enzyme. Cell 2000;102:553–63.

[63] Revy P, Muto T, Levy Y, Geissmann F, Plebani A, et al. Activation-induced cytidine deaminase (AID) deficiency causes the autosomal recessive form of the Hyper-IgM syndrome (HIGM2). Cell 2000;102:565–75.

[64] Conticello SG, Langlois M-A, Yang Z, Neuberger MS. DNA deamination in immunity: AID in the context of its APOBEC relatives. Adv Immunol 2007;94:37–73.

[65] Krishnan A, Iyer LM, Holland SJ, Boehm T, Aravind L. Diversification of AID/APOBEC-like deaminases in metazoa: multiplicity of clades and widespread roles in immunity. Proc Natl Acad Sci USA 2018;115:E3201–10.

[66] Harris RS, Dudley JP. APOBECs and virus restriction. Virology 2015;479–480:131–45.

[67] Petljak M, Green AM, Maciejowski J, Weitzman MD. Addressing the benefits of inhibiting APOBEC3-dependent mutagenesis in cancer. Nat Genet 2022;54:1599–608.

[68] Hsueh BY, Severin GB, Elg CA, Waldron EJ, Kant A, et al. Phage defence by deaminase-mediated depletion of deoxynucleotides in bacteria. Nat Microbiol 2022;7:1210–20.

[69] Tal N, Millman A, Stokar-Avihail A, Fedorenko T, Leavitt A, et al. Bacteria deplete deoxynucleotides to defend against bacteriophage infection. Nat Microbiol 2022;7:1200–9.

[70] Trancoso I, Morimoto R, Boehm T. Co-evolution of mutagenic genome editors and vertebrate adaptive immunity. Curr Opin Immunol 2020;65:32–41.

[71] Hirano M. Evolution of vertebrate adaptive immunity: immune cells and tissues, and AID/APOBEC cytidine deaminases. Bioessays 2015;37:877–87.

[72] Holland SJ, Berghuis LM, King JJ, Iyer LM, Sikora K, et al. Expansions, diversification, and interindividual copy number variations of AID/APOBEC family cytidine deaminase genes in lampreys. Proc Natl Acad Sci USA 2018;115:E3211–20.

[73] Schatz DG, Oettinger MA, Schlissel MS. V(D)J recombination: molecular biology and regulation. Annu Rev Immunol 1992;10:359–83.

[74] Bajoghli B, Guo P, Aghaallaei N, Hirano M, Strohmeier C, et al. A thymus candidate in lampreys. Nature 2011;470:90–5.

[75] Ardavin CF, Gomariz RP, Barrutia MG, Fonfria J, Zapata A. The lympho-hemopoietic organs of the anadromous sea lamprey, *Petromyzon marinus*. A comparative study throughout its life span. Acta Zool 1984;65:1–15.

[76] Lada AG, Dhar A, Boissy RJ, Hirano M, Rubel AA, Rogozin IB, et al. AID/APOBEC cytosine deaminase induces genome-wide kataegis. Biol Direct 2012;7:47.

[77] Morimoto R, O'Meara CP, Holland SJ, Trancoso I, Souissi A, et al. Cytidine deaminase 2 is required for *VLRB* antibody gene assembly in lampreys. Sci Immunol 2020;5:eaba0925.

[78] Boulay J-L, Du Pasquier L, Cooper MD. Cytokine receptor diversity in the lamprey predicts the minimal essential cytokine networks of vertebrates. J Immunol 2022;209:1013–20.

[79] Wu F, Chen L, Liu X, Wang H, Su P, et al. Lamprey variable lymphocyte receptors mediate complement-dependent cytotoxicity. J Immunol 2013;190:922–30.

[80] Mackay F, Figgett WA, Saulep D, Lepage M, Hibbs ML. B-cell stage and context-dependent requirements for survival signals from BAFF and the B-cell receptor. Immunol Rev 2010;237:205–25.

[81] Bommakanti G. Lamprey variable lymphocyte receptor monoclonal antibodies for whole-cell surface antigens. Methods Mol Biol 2022;2421:115–25.

[82] Collins BC, Nakahara H, Acharya S, Cooper MD, Herrin BR, Wilson IA. Crystal structure of an anti-idiotype variable lymphocyte receptor. Acta Crystallogr Sect F Structural Biol Commun 2017;73:682–7.

[83] Collins BC, Gunn RJ, McKitrick TR, Cummings RD, Cooper MD, Herrin BR, et al. Structural insights into VLR fine specificity for blood group carbohydrates. Structure 2017;25:1667–78.

[84] Hassan KMA, Hansen JD, Herrin BR, Amemiya CT. Generation of lamprey monoclonal antibodies (lampribodies) using the phage display system. Biomolecules 2019;9:868.

[85] Khan S, Liu Y, Ernst LM, Leung LYT, Budylowski P, et al. Detection of human CD38 using variable lymphocyte receptor (VLR) tetramers. Cells 2020;9:950.

[86] Lajoie JM, Katt ME, Waters EA, Herrin BR, Shusta EV. Identification of lamprey variable lymphocyte receptors that target the brain vasculature. Sci Rep 2022;12:6044.

[87] McKitrick TR, Hanes MS, Rosenberg CS, Heimburg-Molinaro J, Cooper MD, Herrin BR, et al. Identification of glycan-specific

variable lymphocyte receptors using yeast surface display and glycan microarrays. Methods Mol Biol 2022;2421:73–89.

[88] McKitrick TR, Bernard SM, Noll AJ, Collins BC, Goth CK, et al. Novel lamprey antibody recognizes terminal sulfated galactose epitopes on mammalian glycoproteins. Commun Biol 2021;4:674.

[89] McKitrick TR, Eris D, Mondal N, Aryal RP, McCurley N, Heimburg-Molinaro J, et al. Antibodies from lampreys as smart anti-glycan reagents (SAGRs): perspectives on their specificity, structure, and glyco-genomics. Biochemistry 2020;59:3111–22.

[90] McKitrick TR, Goth CK, Rosenberg CS, Nakahara H, Heimburg-Molinaro J, et al. Development of smart anti-glycan reagents using immunized lampreys. Commun Biol 2020;3:91.

[91] Waters EA, Shusta EV. The variable lymphocyte receptor as an antibody alternative. Curr Opin Biotechnol 2018;52:74–9.

[92] Yu C, Liu Y, Chan JTH, Tong J, Li Z, et al. Identification of human plasma cells with a lamprey monoclonal antibody. JCI Insight 2016;1:e84738.

[93] Yu C, Ali S, St-Germain J, Liu Y, Yu X, et al. Purification and identification of cell surface antigens using lamprey monoclonal antibodies. J Immunol Methods 2012;386:43–9.

[94] Chan JTH, Liu Y, Khan S, St-Germain JR, Zou C, et al. A tyrosine sulfation-dependent HLA-I modification identifies memory B cells and plasma cells. Sci Adv 2018;4:eaar7653.

[95] Nakahara H, Herrin BR, Alder MN, Catera R, Yan XJ, Chiorazzi N, et al. Chronic lymphocytic leukemia monitoring with a lamprey idiotope-specific antibody. Cancer Immunol Res 2013;1:223–8.

[96] Umlauf BJ, Clark PA, Lajoie JM, Georgieva JV, Bremner S, et al. Identification of variable lymphocyte receptors that can target therapeutics to pathologically exposed brain extracellular matrix. Sci Adv 2019;5:eaau4245.

[97] Gunn RJ, Herrin BR, Acharya S, Cooper MD, Wilson IA. VLR recognition of TLR5 expands the molecular characterization of protein antigen binding by non-ig-based antibodies. J Mol Biol 2018;430:1350–67.

[98] Leung LYT, Khan S, Budylowski P, Li Z, Goroshko S, et al. Detection and neutralization of SARS-CoV-2 using non-conventional variable lymphocyte receptor antibodies of the evolutionarily distant sea lamprey. Front Immunol 2021;12:659071.

[99] Boehm T. Design principles of adaptive immune systems. Nat Rev Immunol 2011;11:307–17.

[100] Litman GW, Rast JP, Fugmann SD. The origins of vertebrate adaptive immunity. Nat Rev Immunol 2010;10:543–53.

[101] Boehm T, Hess I, Swann JB. Evolution of lymphoid tissues. Trends Immunol 2012;33:315–21.

[102] Williams AF, Barclay AN. The immunoglobulin superfamily—domains for cell surface recognition. Annu Rev Immunol 1998;6:381–405.

Chapter 5

Structure and Signalling Function of the B-Cell Antigen Receptor and Its Coreceptors

Michael Reth[1,2], Kathrin Kläsener[1,2], Lars Nitschke[3] and Michael R. Gold[4,5]

[1]*Faculty of Biology, Institute of Biology III, University of Freiburg, Freiburg, Germany;* [2]*Signalling Research Centers CIBSS and BIOSS, University of Freiburg, Freiburg, Germany;* [3]*Division of Genetics, Department of Biology, University of Erlangen-Nürnberg, Erlangen, Germany;* [4]*Department of Microbiology and Immunology, University of British Columbia, Vancouver, BC, Canada;* [5]*Life Sciences Institute, University of British Columbia, Vancouver, BC, Canada*

1. INTRODUCTION

The B-cell antigen receptor (BCR) plays a central role in the clonal selection of B cells and the generation of humoral immunity. In addition to eliciting a specific immune response, BCR expression is also required for proper B-cell development and maintenance of the peripheral B-cell pool. Each mature B cell carries roughly 120,000 BCR complexes with identical antigen-binding sites on its cell surface. Based on the highly variable immunoglobulin (Ig) gene rearrangement programme, most of the 10^{11} B cells of the human immune system differ from each other in their exact antigen-binding specificity of their BCR. The enormous binding variability of their antigen receptors is characteristic of both B and T cells. However, unlike T cells, whose antigen receptors are restricted in binding to a defined antigen structure, namely the major histocompatibility complex (MHC)/peptide complex, B cells can be activated by a large group of structurally highly diverse molecules. In this respect, the BCR is quite different from most other receptors on the cell surface that have only one or a few specific ligands. A successful model of B-cell regulation and activation should take into account these unique properties of the BCR and answer two major questions: How do the many BCR-complexes on the B-cell surface remain silent in the absence of a cognate antigen, and how can the BCR be activated by so many different antigenic structures? These features of the BCR are required for the proper function of B cells in the process of clonal selection. In this chapter, we describe the basic structure of the BCR in light of recent cryo-electron microscopy (cryo-EM) studies and discuss the regulation, antigen-sensing and intracellular signalling mechanisms of this receptor. We also discuss the interaction of the BCR with signal-transducing kinases, focussing on the Src-family kinase (SFK) Lyn and the spleen tyrosine kinase (Syk). We discuss the structure and function of the B-cell coreceptors that either regulate (CD45), amplify (CD19) or inhibit (CD22 and CD32) BCR signalling. Once B cells are activated, signals from the BCR are processed through multiple signalling pathways. In our discussion of these processes, we will focus on the phospholipase C2 (PLCγ2) calcium release machinery and the phosphoinositide 3-kinase (PI3K) signalling pathway, controlling B-cell proliferation and survival. We also discuss how the cytoskeleton is associated with the BCR signal processes. Together, these pathways drive the clonal expansion of B cells and their differentiation into antibody-producing plasma cells.

2. BASIC STRUCTURE OF THE BCR COMPLEX

In his famous Croonian Lecture in 1900, Paul Ehrlich postulated the existence of an antigen receptor as part of his side-chain theory. It was not until 1960 that membrane-bound antibodies and, in 1989, the subunits of the BCR signalling pathway were discovered (Figure 5.1A). The BCRs of all major Ig classes (IgM, IgD, IgG, IgE and IgA) have a similar structural design [1,2]. They consist of the membrane-bound Ig (mIg) molecule and the Igα/Igβ (CD79a/CD79b) heterodimer, which mediate antigen

Molecular Biology of B Cells. https://doi.org/10.1016/B978-0-323-95895-0.00003-9

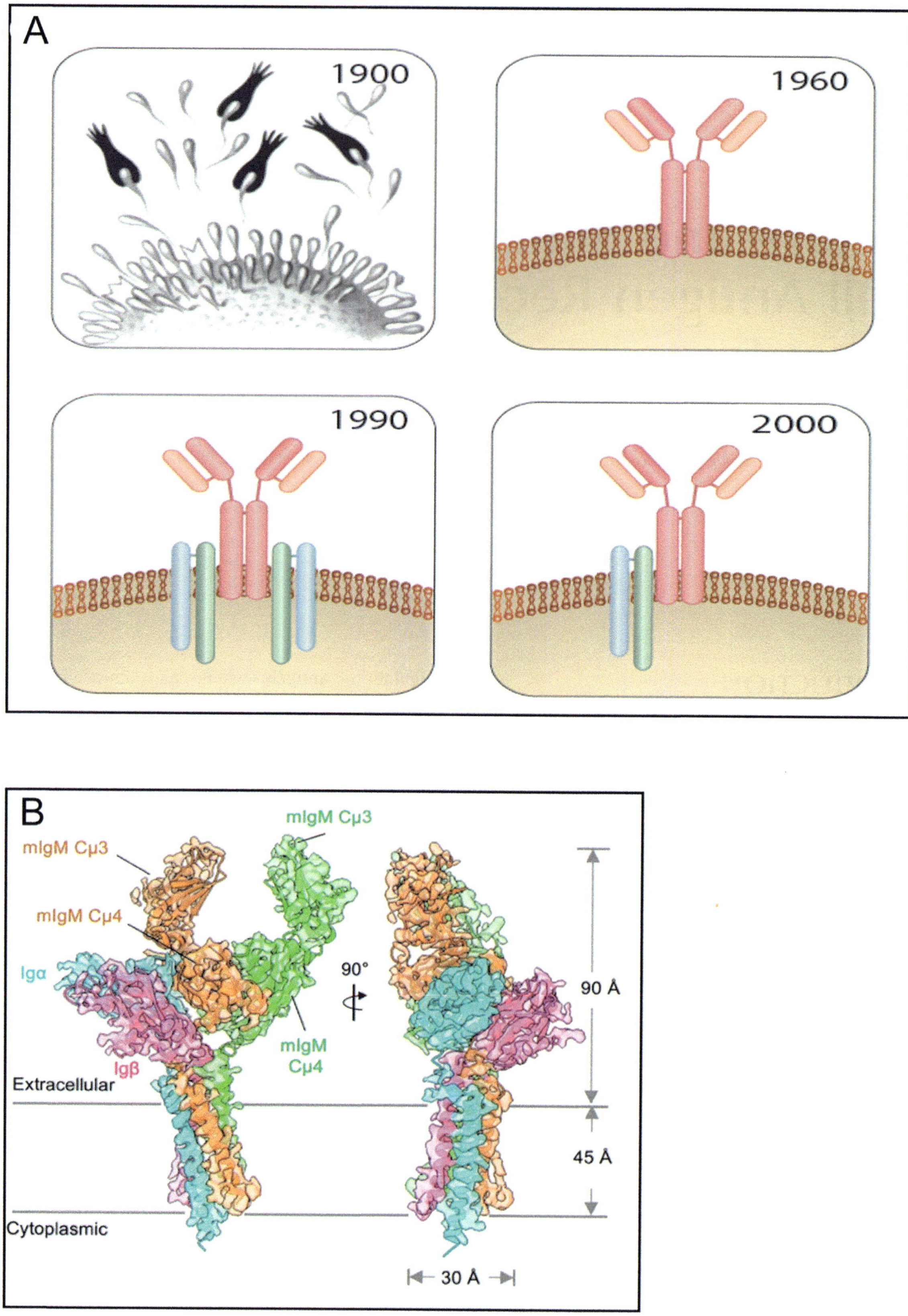

FIGURE 5.1 The basic structure of the monomeric B-cell antigen receptor (BCR). (A) History of the BCR models. In 1900, Paul Ehrlich proposed the first receptor-ligand model in his side-chain theory to explain the specific activation of cells for antibody production. The discovery that B cells carry membrane-bound antibody molecules on their cell surface was made in 1960, and immunoglobulin (Ig) α and Igβ, the signalling subunits of the BCR, were identified in 1989. The original model of the BCR suggested a symmetric 1:2 interaction between the two heavy chains of the membrane-bound Ig molecule and the Igα/Igβ heterodimer. In 2000, a biochemical study demonstrated that the mIg molecule and the Igα/Igβ heterodimer form a 1:1 complex. (B) The membrane-proximal part of the 3.3 Å resolution cryo-EM structure of the monomeric IgM-BCR complex in two orientations. The two truncated HCs (Cμ3-Cμ4-CP-TMD) of the mIgM molecule are shown in *orange* and *green*, and the Igα and Igβ proteins are presented in *cyan* and *pink*, respectively. The cryo-EM structure confirms the 1:1 model and shows an asymmetric IgM-BCR complex. The Igα/Igβ heterodimer dominantly interacts with only one (*orange*) of the 2 HCs of the mIgM molecule with Igα (*cyan*) being the dominant mIgM interactor. The symmetry of the mIgM molecule is broken by the CP of the *orange* HC, which is located between the CPs of Igα and Igβ. The four TMDs form a tight 4 α-helical bundle. The structure of the ITAM-containing cytoplasmic tails of the Igα/Igβ heterodimer could not be resolved by this technique.

binding and signal transduction, respectively [3,4]. The mIg molecule is a disulfide-linked tetramer consisting of two identical heavy chains (HCs) and two identical light chains (LCs). The two components of the Igα/β heterodimer share many structural features. Both proteins carry a glycosylated extracellular Ig domain, a linker region containing a heterodimer-forming cysteine, an evolutionarily highly conserved α-helical transmembrane domain (TMD) and a cytoplasmic tail sequence of either 61 (Igα) or 48 (Igβ) amino acids. While Igα is covalently bound to Igβ via a disulfide bridge, the Igα/Igβ heterodimer is non-covalently associated with the mIg molecule [5]. Thus, the purification of the 440 kDa IgM-BCR complex requires special detergents such as digitonin [6].

Because the membrane-bound IgM (mIgM) molecule is a homodimer, it was initially thought that an Igα/Igβ heterodimer is bound on each side of the mIgM molecule, forming a symmetric 1:2 complex. However, a biochemical study in the year 2000 showed that the mIgM molecule and the Igα/Igβ heterodimer form a 1:1 complex [7]. The asymmetric IgM-BCR architecture has been confirmed by the recent cryo-EM studies of the human and mouse IgM-BCR [8–10], which reveal the molecular details of the interaction between mIgM and the Igα/Igβ heterodimer (Figure 5.1B). The complex formation between these receptor components involved three interaction sites namely the extracellular Ig-domains, the connecting peptides (CPs) and the TMDs with the latter two playing a dominant role. Interestingly, the CP of only one of the two mIgM heavy chains (μHC and μHC') intervenes in between the CPs of Igα and Igβ, thus breaking the symmetry of the mIgM molecule. The binding of the Igα/Igβ heterodimer to the mIgM molecule is mainly mediated by Igα. The extracellular Ig-domain of Igα is associated on one side with the Ig-domain of Igβ and at the other side with the Cμ4 domains of μHC while leaving the μHC' unengaged. The TMDs of the IgM-BCR complex form a tight asymmetric 4-helix bundle that is stabilised by specific interactions of their amino acids. It has been previously described that the μHC TMD has two distinct sides with one of them highly conserved in all five different mIg classes [11]. The cryo-EM structure shows that several amino acids at the conserved side are involved in μHC/μHC' dimerisation as well as the binding to the Igα TMD. The TMDs of the Igα/Igβ heterodimer also display a conserved side with an array of amino acids that are identical or similar for both signalling components. The cryo-EM structure shows that it is this conserved side that forms the Igα/Igβ TMD interphase. While the extracellular and TMD part of Igα and the μHC are prominently involved in IgM-BCR complex formation, the μHC' and Igβ have free sides that are accessible for higher complex formation and binding to lateral BCR interactors. Interestingly, the free TMD side of δHC', which is equivalent to that of μHC' is implicated in the formation of IgD-BCR oligomers [12]. The cytoplasmic sequences of Igα and Igβ contain a highly conserved consensus sequence D/Ex$_7$D/ExxYxxLx$_7$YxxL/I, called the immunoreceptor tyrosine-based activation motif (ITAM) [13]. This motif is found in most signalling subunits of the multicomponent immune receptor family [14,15] and in proteins of viruses that cause B-cell tumours [13]. The ITAM motif is required for the interaction of the BCR with downstream signalling elements [16] (see below).

3. THE RESTING BCR AND THE THREE-DIMENSIONAL TOPOGRAPHY OF THE B-CELL SURFACE

At present, we know more about the function and interactions of the activated BCR than the resting BCR. Without information about the organisation of the BCR on the surface of resting B cells, our knowledge of the B-cell activation process will remain incomplete. This problem applies not only to the BCR but also to most receptors on the B-cell surface. The fluid mosaic model described by Singer and Nicolson [17] suggested free diffusion of membrane proteins in the plasma membrane lipid bilayer. Light microscopy studies with fluorescent anti-receptor antibodies also suggested a random distribution of many receptors on the B-cell surface [18,19]. However, it is now clear that the lymphocyte membrane is dynamically structured and compartmentalised into membrane domains with different protein/lipid compositions and that the equal distribution of receptors on resting B cells is an illusion of a diffraction limited light microscopy [20,21]. At nanoscale distances, many receptors are clustered and enriched at specific location [22]. This is demonstrated by EM studies of the inner leaflet of the plasma membrane, which show that membrane proteins of T lymphocytes are confined to so-called protein islands with a size of 80–150 nm [23,24]. A similar EM study of the B-cell membrane showed that the two BCR classes that are co-expressed on mature, naive B cells namely the IgD-BCR and IgM-BCR both form clusters that are localised at different membrane areas. The nanocluster formation of the IgD-BCR and IgM-BCR on the B-cell surface was confirmed by the super-resolution two-colour direct stochastic optical reconstruction microscopy (dSTORM), which showed differential localisation for both classes of antigen receptors [25,26]. Another important technique to study the nanoscale organisation of receptors on the B-cell surface is the proximity ligation assay (PLA) and its extension in form of the Fab-based PLA with a resolution of 10–20 nm [27]. Indeed, a Fab-PLA study confirmed not only the separate clustering of the IgM and IgD-BCRs but also show the distinct composition of the class-specific protein/lipid nanoclusters.

On resting B cells, the transmembrane phosphatase CD45 (PTP1C) is localised in close proximity of the IgM-BCR, whereas the IgD-type nanocluster contains CD19, CD20, CD40, BAFFR and the chemokine receptor CXCR4. These different receptors are not only colocalised but also functionally connected. For example, CXCR4 signalling is defective on the surface of IgD-BCR-deficient murine B cells [28]. Loss of the B-cell–specific tetraspanin protein CD20 on a human B-cell line results not only in a reorganisation of the IgD-type nanocluster but also in B-cell activation. Thus, CD20 seems to be an organiser of the nanoscale IgD-BCR organisation and a gatekeeper of the resting state of B cells [29].

The crystal structure of the CD20:CD20 dimer shows that the CD20 TMD are associated with several cholesterol molecules [30,31]. This is in line with the finding that both, the IgD-BCR and CD20, are associated with a raft-type lipid domain [32]. Another raft-associated protein is the 'phosphoprotein associated with glycosphingolipid-enriched microdomains' (PAG), also known as Csk-binding protein (Cbp). This transmembrane adaptor is constitutively phosphorylated in resting lymphocytes and is with the C-terminal SRC-kinase (CSK), which inhibits SFKs such as Lyn. Within the IgD-type nanocluster, the PAG-CSK complex can inhibit the kinase activity of Lyn by phosphorylating a C-terminal tyrosine that locks Lyn in an autoinhibited state (Figure 5.3A).

Other tetraspanins also seem to support the nanocluster formation and the location of coreceptors within such clusters. It thus has been found that CD19 and CD45 are associated with the tetraspanin proteins CD81 and CD53, respectively [33,34]. Furthermore, the nanoclustering of glycoproteins on the surface of resting or activated B cells is also supported by galectins. These β-galactoside-binding lectins contain one or two carbohydrate-recognition domains and, interestingly, partly share together with the diverse tetraspanins binding partners such as CD19 and CD45 [35–37].

The resting state of the BCR is also controlled by the protein-arginine-methyltransferase1 (PRMT1), an enzyme that methylates a highly conserved arginine (R198) within the cytosolic tail of Igα [38]. R198 methylation is the first protein modification found to be associated with the resting BCR, and B-cell activation is accompanied by the demethylation of R198. PRMT1 also has substrates at other cellular locations like the nucleus and cytosol and may be a general regulator of the resting state of B cells [39].

Most immunological textbooks depict lymphocytes as round, spherical cells with a smooth surface bearing randomly distributed receptors. The notion of a smooth unstructured B-cell surface has already been challenged by EM images of fixed and frozen lymphocytes, which reveal thin membrane protrusions called microvilli (MV) on the cell surface [40]. The dynamic movement of the tentacle-like MV structures may facilitate antigen sensing by lymphocytes. Using lattice light sheet microscopy (LLSM), it is now possible to study the 3-D topography of living lymphocytes embedded inside a matrigel (Figure 5.2A). An LLSM study combined with machine learning image analysis revealed that the B-cell surface harbours a dynamic

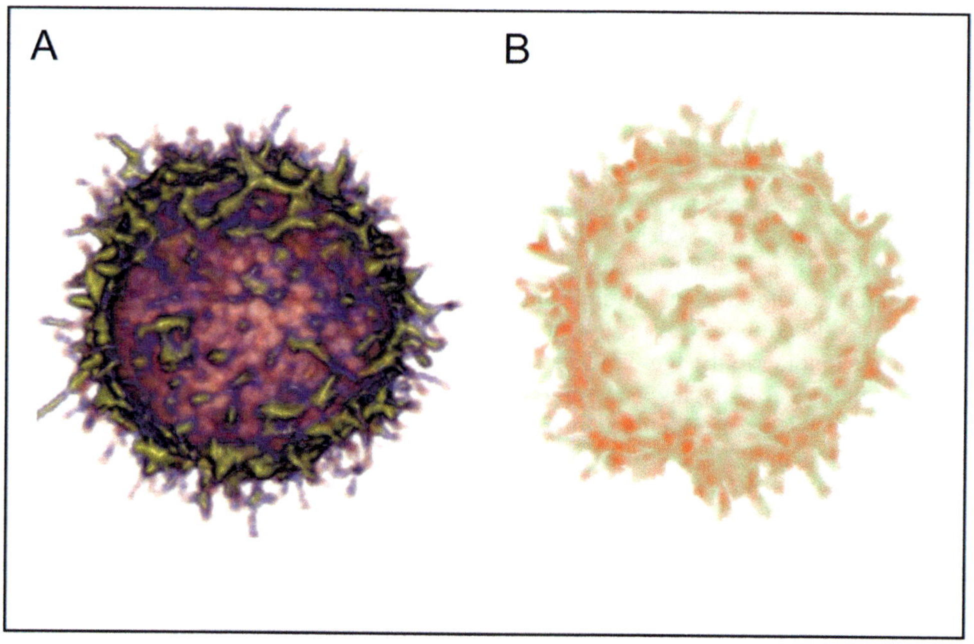

FIGURE 5.2 **Three-dimensional (3-D) topography of the plasma membrane of a human B cell (Ramos)** and the location of IgM-BCR clusters studied by high-end lattice light sheet microscopy (LLSM) combined with machine-learning image analysis (for details see Ref. [41]). (A) A computer-generated 3-D image of the LLSM membrane intensity data shows protruding structures and microvilli on the segmented B-cell surface with a colour code (the highest elevations are shown in *yellow*). (B) The location of IgM-BCR clusters (*orange*) correlated with the 3-D topography of the B-cell membrane (*green*). The SNAP-tagged IgM-BCR complexes were visualised by fluorescent SNAP ligand and the membrane by the expression of a membrane-bound GFP-CaaX fusion protein. IgM-BCR clusters are most frequently associated with the base and the tip of the microvilli.

network of elevated ridges, with MV growing from the intersections of the network [41]. This 3-D image analysis also detects IgM-BCR clusters, which are interestingly enriched at the nodes of an actin-based ridge network and at the tip of MV (Figure 5.2B). A previous 2-D study of live B cells using total internal reflection fluorescence microscopy (TIRFM) suggested that most BCRs on resting B cells are mobile monomers whose diffusion is limited by the underlying actin bundles [42,43]. The 3-D imaging analysis of B cells suggests that the actin bundles also can act as roads that guide antigen receptor clusters to the antigen-sensing sites, namely the tip of the MV protrusions.

4. BCR ACTIVATION MODELS

There are currently two opposing models for BCR activation. These models differ only in the organisation of the BCR on resting B cells and not in the aggregated state of the receptor on activated B cells. The cross-linking model (CLM) suggests that the numerous BCR complexes on the resting B-cell surface are randomly distributed monomers and that it is the cross-linking of two monomers that initiates the BCR activation process [44]. In contrast, the dissociation activation model (DAM) proposes that the BCR forms auto-inhibited dimers, oligomers and clusters that are opened and activated upon contact with cognate antigens [25]. The cross-linking hypothesis has gained wide acceptance as it once was thought that most receptors on the cell surface are monomers that are activated by dimerisation. However, with more and more structural information revealing a multimeric organisation of the BCR, this support is falling apart. In addition, the CLM fails to explain how the many BCR-complexes remain silent on the surface of resting B cells and how the BCR is activated by monovalent antigens that per se cannot cross-link the BCR. In contrast, the DAM postulates a closed autoinhibited BCR structure that cannot signal on resting B cells. Furthermore, the DAM is supported by Fab-PLA studies showing that monovalent antigens can dissociate the oligomeric BCR [32].

According to the new information about the nanoscale organisation of BCR clusters and the 3-D topography of the B cell, the following scenario for the antigen-dependent BCR activation can be drawn. On mature, naive B cells, the IgD-BCR is the dominant receptor and organiser of a nanocluster containing CD20, CD40, CxCR4 and BAFF receptors (Figure 5.3A). In addition, CD19 bound to the CD81 tetraspanin is associated with this nanocluster that seems to be enriched at shallow invaginations of the B-cell plasma cell membrane. In contrast, the IgM-BCR clusters are negatively regulated by CD45 and associated with the ridge network that directs their movement to the base of the MV. Exposure of the B cells to innate immune signals such as LPS or chemokines results in the MV elongation and the movement of the IgM-BCR clusters to the tip of the MV. On T and B cells, CD45 has been found to be excluded from the tip of the MV, suggesting a lower threshold for antigen-dependent activation of IgM-BCR on this site [45]. Binding of a cognate antigen results in dissociation of the antigen receptor oligomers in to monomers, followed by reorganisation of the IgD- and IgM-BCR type nanoclusters. Opening the closed BCR oligomer could increase the accessibility of the ITAM sequence, thus allowing interaction with BCR signal-transducing kinases. The opened IgM-BCR clusters move down the MV and are found in closer proximity to the IgD-BCR cluster. Interestingly, the CD19/CD81 complex has been found to dissociate upon B-cell activation, which may allow the CD19 coreceptor to bind to the opened IgM-BCR complex [46]. Indeed, on antigen-stimulated B cells, CD19 is found in close proximity to the IgM-BCR where the pITAM/Syk and PI-3 kinase modules can cooperate and amplify BCR signalling resulting in the proliferation and further differentiation of the activated B cell into either memory B or plasma cells [32]. The antigen-bound BCR complexes are then internalised and transported to the MHC peptide-loading compartment.

5. THE BCR CORECEPTORS CD19 AND CD22

The signalling from the BCR is modulated by accessory transmembrane molecules or coreceptors on the B-cell surface. This modulation of signalling is achieved by the recruitment of additional intracellular signalling molecules. Coreceptors can not only increase or decrease the BCR signalling strength but can also change the quality of the induced signalling pathways. Very often, the impact of coreceptors on the BCR signalling pathways is regulated by ligand binding. Important examples of B-cell coreceptors are CD19, which enhances BCR signalling and CD22 which inhibits it. CD19 comes on the B-cell surface only in association with the tetraspanin CD81 [47] (Figure 5.3A). CD22 carries an immunoreceptor tyrosine-based inhibition motif (ITIM) that recruits the tyrosine phosphatase SHP-1 which inhibits BCR signalling. The inhibition by CD22 is regulated via binding to CD22 ligands, which are sialic acids attached to cellular glycoproteins. The presence or absence of these ligands regulates the CD22 association to the BCR and thus influences the inhibition strength [48] (Figure 5.3B).

5.1 CD19 Functions in Complex with CD21 and CD21-Independent Functions

CD19 has been identified as part of a membrane protein complex on B cells with the complement receptor 2 (CR2) or CD21 (or its alternatively spliced form CD35) and the tetraspanin CD81. The association of CD19 with CD21 is

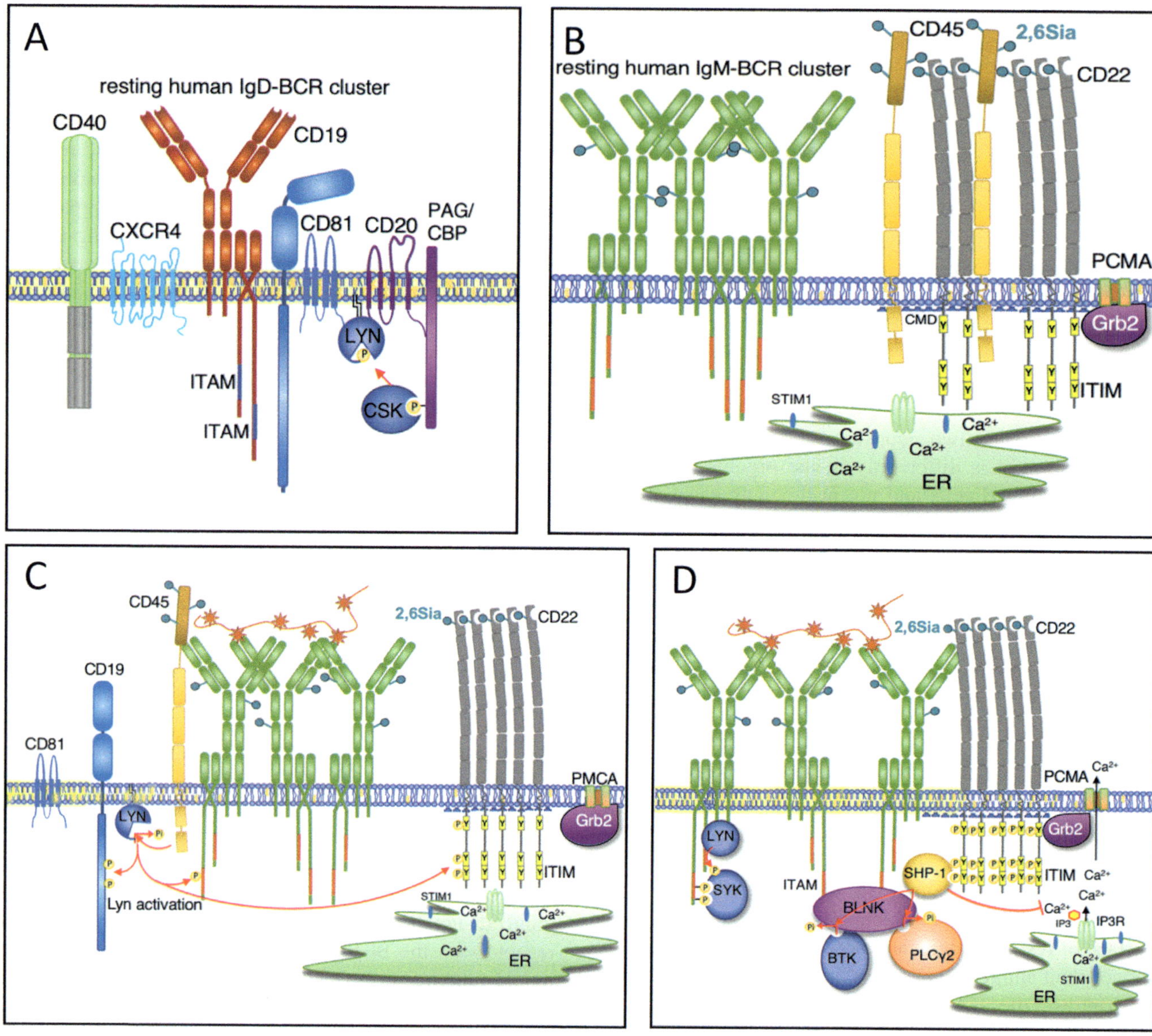

FIGURE 5.3 **Localisation of CD19 and CD22 in resting and activated B cells.** (A) In resting B cells, the IgD-BCR complex is located in lipid raft domains. Lyn is autoinhibited by CSK-mediated C-terminal phosphorylation. (B) The unstimulated IgM-BCR is located in a different membrane domain than the resting IgD BCR cluster and is separated from CD22. CD22 binds to other CD22 molecules via 2,6-linked sialic acids (2,6Sia) forming CD22 oligomers that are found in clathrin-rich membrane domains (CMD, shown in *blue*). Phosphatase CD45 molecules are also part of this complex. (C) During B-cell activation, the BCR opens up its structure upon antigen binding and is recruited into raft domains (cholesterol in *yellow*). CD45 delocalises from the CD22 complex and activates Lyn which in turn can phosphorylate the ITAM motif of the IgM-BCR, CD19 and CD22. (D) CD22 clusters are recruited to the IgM-BCR domains, leading to dephosphorylation of BLNK and inhibition of Ca^{2+} signalling by recruitment of the phosphatase SHP-1 to phosphorylated inhibitory ITIM motifs on the cytoplasmic tail of CD22. CD22 also regulates the Ca^{2+} pump PMCA, which terminates Ca^{2+} signalling in a Grb-2-dependent mechanism.

crucial for transducing signals triggered by combined binding of complement C3d-coupled antigens to the BCR and CD21 [49], but there are also CD21-independent functions of CD19, as these two proteins are not always co-expressed on all B-cell subtypes. The association of CD19 with CD81 is important for CD19 surface expression and links the membrane protein complex to the cytoskeleton, as described below. The CD19/CD21/CD81 complex has been discovered when it was observed that co-ligation of CD21 and the BCR can greatly enhance BCR signalling [50]. While CD21 has a short intracellular tail, the 230 amino acid long-intracellular tail of associated CD19 was soon discovered to be the responsible signal transduction element. This was strongly supported by the finding that crosslinking of CD19 with the BCR enhanced both B-cell Ca^{2+} signalling, as well as proliferation. Thus, C3d-coupled antigens can enhance signalling by co-ligating CD21 with the BCR and transmitting activation signals via CD19 [49].

CD19-deficient mice had a phenotype affecting mainly peripheral B-cell populations. While conventional B2 cells developed normally, there was a defect in B1-cell and MZ B-cell populations in these mice. In contrast, transgenic mice over-expressing human CD19 had an enlarged B1 cell population [51,52]. The subpopulation of B1 cells is dependent on BCR signalling for its maintenance, and therefore these results of CD19-deficient or CD19 over-expressing mice can be explained by changed BCR signalling. Upon immunisation with thymus-dependent (TD) antigens, there was a defect of $CD19^{-/-}$ mice in the generation of switched Ig responses, associated with a defective germinal centre (GC) formation [51,52]. Also, B-cell memory was impaired. VSV virus-infected CD19-deficient animals could form GCs, however, but had a defect in memory responses and in high titre antibody responses [53]. CD21-deficient mice show a similar, but weaker defect in GC responses to TD antigens [54]. This may indicate that co-recognition of complement Cd3-opsonised antigen by CD21 and BCR plays a role in GC formation. However, not all soluble protein antigens or immune complexes contain attached Cd3 and therefore the GC defect of $CD19^{-/-}$ mice may also be explainable by impaired BCR signalling during B-cell activation. CD19 is expressed from the pro-B-cell developmental stage onwards in the bone marrow, while CD21 is expressed later, at the stage when transitional B cells leave the bone marrow and migrate to the spleen [47]. Therefore, CD21-independent functions of CD19 exist, particularly in early B-cell development and selection. An impaired proliferation of cells at the transition from the pro-B to small resting pre-B-cell stage has been found in CD19-deficient mice [55]. Also, these mice show impaired positive selection processes during transition from immature to mature B cells.

Further insight into the function of CD19 came by high-resolution and super-resolution microscopy. Membrane clusters of IgM-BCR bound by antigen were substantially, but not fully co-localising with CD19 clusters. With lipid bilayer systems as surrogate for cellular membranes, it was shown that CD19 is crucial for responses to membrane-bound antigen, but not to soluble antigen [56]. This role of CD19 in enhancing B-cell activation by membrane-bound antigen was independent of CD21. It was also shown that CD19 is found on the plasma membrane in nanoclusters that are regulated by the tetraspanin CD81 and the actin cytoskeleton [26]. On activated B cells, CD19 is also associated with the interferon-induced transmembrane protein 1 (IFITM1) also known as Leu-13 or CD225. A related protein of the CD225 superfamily namely IFITM3 has a cytosolic PIP_3 binding site and can enlarge the PIP_3-enriched nanodomain of CD19 and thus support its signalling function [57].

B-cell activation caused by the disruption of the actin cytoskeleton required the expression of the BCR as well as CD19 and CD81. This study also suggests that on resting B cells, IgM, IgD and CD19 are organised in different nanoclusters that are not altered during B-cell activation [26]. However, by PLAs, it was found that on resting B cell, CD19 is specifically colocalised together with the IgD-BCR, whereas after B-cell activation, CD19 is found in association with the IgM-BCR [58]. This change in localisation of CD19 from IgD-BCR to IgM-BCR may be a consequence of the IgD and IgM protein islands that concatenate after B-cell activation [59].

5.2 Signalling by the CD19 Tail

The cytoplasmic tail of CD19 carries nine tyrosine residues that are well conserved between human and mouse and some of which are phosphorylated upon BCR crosslinking [47]. The phosphorylation of the CD19 tail is augmented when CD19 is co-ligated to the IgM-BCR. The importance of these tyrosines for signal transduction processes has been studied individually by mutants in cell lines, and its physiological role has been studied by transgenic mice expressing CD19 with mutated tyrosines. Of these nine tyrosines, Y482 and Y513 seem to be particularly important as Y→F mutations at these positions prevent phosphorylation of the other seven tyrosines of the CD19 cytoplasmic tail [60–62]. One study showed that Lyn is activated by binding to Y513 and subsequently phosphorylates Y482 and Y513 [62]. In contrast, another study reported normal CD19 tyrosine phosphorylation in $Lyn^{-/-}$ mice [63]. This suggests that another SFK or Syk can compensate when Lyn is absent.

Transgenic mice with tandem Y→F mutations in the CD19 tail were created in order to examine the in vivo contribution of all the CD19 tyrosines. These transgenic mice were then crossed onto the CD19-deficient background. Mice with Y482F/Y513F mutations developed the same phenotype as $CD19^{-/-}$ mice, whereas mutations of the other tyrosines were dispensable for the in vivo function of CD19 [64,65]. It is not very likely that this phenotype can be explained by grossly impaired Lyn activation, as $Lyn^{-/-}$ and $CD19^{-/-}$ mice have quite opposite phenotypes [66]. Y482F and Y513 are both part of YxxM PI3 kinase (PI3K)-binding domains, and $CD19^{-/-}$ mice show impaired PI3K signalling [67,68]. Since the deletion of the PI3K components p85α or p110δ leads to similar B-cell phenotypes as in $CD19^{-/-}$ mice, the main in vivo function of CD19 is likely augmentation of PI3K signalling [69–72]. This is supported by data from $PTEN^{-/-}$ mice, with a genetic deletion in the inositol phosphatase that directly reverses the PI3K reaction that can rescue all phenotypes of CD19-deficient mice [73]. However, CD19-independent IgM-BCR-induced activation of PI3K also exists, as discussed below. An additional signalling protein which binds to the CD19 tail is the guanine nucleotide exchange factor (GEF) for the Rho GTPases Vav. Vav proteins are both important for initiating Ca^{2+} responses in B cells, as well as for regulating the actin cytoskeleton [74]. Thus, the functional regulation of actin polymerisation by CD19 may be mediated by this recruited protein. Since crucial functions for CD19 in regulating PI3K signalling and in controlling antibody responses were found in mouse models, it is not surprising that patients with homozygous CD19 mutations develop an antibody-deficiency syndrome [75]. The same is true for patients with CD81 gene defects, which have impaired CD19 surface expression and an antibody deficiency [76]. These findings show a conservation of CD19/CD81 functions between mouse and man.

6. CD22: AN INHIBITORY RECEPTOR

CD22 (Siglec-2) is an inhibitory receptor of the Siglec family that is predominantly expressed on B cells. It binds to its ligands alpha2,6-linked sialic acids (2,6Sia), which are abundant terminal sugars on cell surface-expressed glycoproteins [48]. CD22 is to some extent associated with the IgM-BCR; however, this association is regulated by 2,6Sia-carrying *cis*-ligands on glycoproteins on the B-cell surface, as will be discussed below. CD22 contains six tyrosines in its cytoplasmic tail. Three tyrosines (Y783, Y843 and Y863 of murine CD22) are parts of classical inhibitory ITIM motifs, one (Y817) is in an ITIM-like motif and one is needed for the recruitment of Grb-2 (Y828) (Figure 5.3B). CD22 is rapidly phosphorylated after IgM-BCR crosslinking [48]. The kinase Lyn is crucial for this phosphorylation, as Lyn-deficient mice show strongly decreased CD22 tyrosine phosphorylation [77,78]. Upon tyrosine phosphorylation of CD22, the tyrosine phosphate SHP-1 is recruited to phosphorylated CD22 ITIM motifs [79] which is the crucial negative regulator of inhibitory signalling by CD22 [80,81] (Figure 5.3B–D). CD22-deficient mice show increased BCR-induced Ca^{2+} signalling due to the loss of this negative regulation [82–85]. Also, CD22 knock-in mice with Y→F mutations of all three ITIMs show a similar increased Ca^{2+} signalling [86]. As a consequence of the lacking inhibition, CD22-deficient B cells show an increased tyrosine phosphorylation status of important signalling molecules such as SLP65/BLNK or CD19 [87,88]. This can explain the negative regulation of the initiation of Ca^{2+} signalling. CD22 also associates with a calcium pump called PMCA-4, which is involved in pumping out Ca^{2+} from the cytosol after termination of signalling. This pump is activated by CD22 via Grb-2 binding in an SHP-1/ITIM-independent pathway [80,89]. Consequently, B-cell-specific Grb-2 deficient mice show elevated Ca^{2+} signalling in their B cells [90].Thus, loss of CD22 in CD22-deficient mice affects both the initiation, as well as the termination of Ca^{2+} signalling in B cells.

6.1 Regulation of CD22 Signalling by Ligand Interactions of Glycoproteins *in cis*

CD22 is a transmembrane protein with seven extracellular Ig domains. The first extracellular Ig domain of CD22 binds to 2,6Sia, which are sialic acids that are present on the surface of many cells, including B cells and T cells. The structure determination of soluble CD22 and modelling predicts a rod-like structure of this protein on the B-cell surface [91]. CD22 can bind ligands *in trans* on adjacent cell surfaces or *in cis* on the same cell surface. The *cis* ligands need to be on larger glycosylated membrane proteins, as predicted from the CD22 structure. These interactions to ligands are low-affinity interactions and therefore CD22 needs a structural vicinity to ligands on larger glycoproteins on the same plasma membrane *in cis* or multimeric ligands on other cells for binding to several CD22 molecules *in trans*. Of note, CD22 on the B-cell surface cannot bind to monomeric soluble sialylated proteins. The strongly *cis*-interacting ligands have to be removed first, e.g. by sialidase treatment, before CD22 can bind to external sialylated proteins [92,93].

CD22 is organised in clathrin-rich microclusters consisting of CD22 homooligomers that are distinct from IgM-BCR clusters [94]. These homooligomers form by binding of CD22 to 2,6Sia attached to neighbouring CD22 molecules. This has first been shown by photo-crosslinking glycan ligands to CD22 [95]. As N-linked glycosylation sites are found in the first Ig domains of CD22 and as CD22 has a rod-like structure, neighbouring CD22 molecules are optimally accessible in these CD22 clusters [91]. High-resolution TIRFM revealed that CD22 microclusters are close, but apart from microclusters formed by the IgM-BCR or IgD-BCR (Figure 5.3B). By single particle tracking, it was also found that CD22 has a higher lateral mobility in the membrane, when compared to that of the IgM-BCR or IgD-BCR clusters [96]. Upon BCR activation, CD22 is recruited to the IgM-BCR and is phosphorylated by Lyn which initiates the inhibitory signalling cascade (Figure 5.3C,D). The role of 2,6Sia containing *cis*-ligands for CD22 was also examined by genetic mouse models. ST6GalI-deficient mice lack the sialyltransferase that attaches sialic acid in the α2,6-linkage to galactose as terminal glycan. In these ST6GalI-deficient mice, CD22 is not bound to *cis*-ligands anymore and the CD22 homooligomers are disrupted [94,97]. Interestingly, this mouse line showed reduced IgM-BCR signalling with lower Ca^{2+} mobilisation, due to stronger association of CD22 with the IgM-BCR. Similarly, in a mouse line in which the ligand-binding domain of CD22 was mutated (CD22-R130E mice), there was a lower IgM-BCR-induced Ca^{2+} mobilisation detected. A higher CD22 phosphorylation, stronger SHP-1 recruitment and a higher association of CD22 with the IgM-BCR were found [86]. Super-resolution microscopy determined smaller CD22 microclusters and a higher lateral mobility of CD22 in CD22-R130E mice [96]. Thus, it seems that the formation of CD22 clusters limits the accessibility of this inhibitory receptor to the IgM-BCR, and without ligand binding, there is higher CD22 lateral mobility, resulting in a stronger association of CD22 with the IgM-BCR and therefore, stronger signal inhibition.

CD22 microclusters also contain CD45 molecules, which are abundant, highly glycosylated proteins on leucocytes. CD45 is an important transmembrane protein with an intracellular tyrosine phosphatase domain, but the extracellular domain seems to be more important in regulating CD22 activity. It was postulated that CD22 *cis*-binding to 2,6Sia on the extracellular domain of CD45 also prevents the CD22-BCR association [98]. This affected particularly the association of CD22 and the IgM-BCR on resting cells [98,99]. Super-resolution microscopy revealed that CD22 microclusters are larger in CD45-deficient mice, indicating that more CD22−CD22 homotypic interactions may take place [96]. Thus, there could be a competition for CD22 in binding to sialylated CD45 and to form oligomers with itself, but further research is necessary to reveal the details. Recent studies also show a role for galectin-9, a soluble lectin with two glycan-binding domains that is produced by B cells. Galectin-9 binds to N-acetyllactosamine (LacNAc) containing glycans. These glycans are highly abundant as poly-LacNAc on B cells. CD45 is a binding partner for galectin-9, and galectin-9 influences the CD45 and the IgM-BCR membrane organisation. Treatment of B cells with galectin-9 induced Lyn and CD22 phosphorylation and inhibited anti-IgM-induced Ca^{2+} mobilisation [100,101]. Therefore, galectin-9 seems to be an additional glycan-binding regulator of CD22/CD45 functions. In addition to *cis*-binding, CD22 was shown to be able to bind to ligands on other cells, *in trans* [102,103]. However, hardly any biological functions of CD22 that are dependent on binding to 2,6Sia-carrying ligands *in trans* have been demonstrated so far.

6.2 The Role of CD22 in Germinal Centres and in Immune Responses

Murine GC B cells can be stained with the GL7 antibody. This antibody detects N-acetyl-neuraminic acid (Neu5Ac). Naïve B cells, however, express N-glycolyl-neuraminic acid (Neu5Gc), another form of sialic acid, that is, the high-affinity ligand of CD22. Thus, GC B cells replace the high-affinity ligand of CD22 with a 20-fold lower affinity form of sialic acid, whereas memory B cells re-express the Neu5Gc form [104]. This differential expression is caused by a downregulation of the responsible sialic acid-modifying enzyme cytidine monophospho-N-acetylneuraminic acid hydroxylase (CMAH) in GC B cells [105]. Human GC B cells also downregulate high-affinity CD22 ligands, but by a different mechanism [104]. Recently, two groups showed that during an immune response, CD22-deficient GC B cells decrease over time when competing with wild-type GC B cells. There was a lower number of GC B cells derived from $CD22^{-/-}$ mice in bone marrow chimaeras, resulting in lower output of memory B cells and plasma cells. This was due to higher BCR-induced Ca^{2+} signalling in $CD22^{-/-}$ GC B cells, leading to increased apoptosis levels [106,107]. Mice with overexpressed CMAH in GC B cells (thus maintaining Neu5Gc expression) showed the same phenotype [106]. As normal GC B cells show lower BCR signalling than naïve B cells, this resulted in the model, that the lower BCR signalling of GC B cells is due to less *cis*-binding of CD22 and less CD22 homooligomer formation in GC B cells and therefore a stronger BCR inhibition by CD22. This seems to be necessary to prevent high apoptosis induction in GC B cells.

Recently, it was found that CD22 not only inhibits IgM-BCR signals, but it also controls B cell homing to the gut. CD22 interacts in a sialic acid-dependent manner with α4β7 integrin and targets SHP-1 to the β7 chain of this integrin [108]. CD22-deficient mice, or mice with mutated ITIM or ligand-binding motifs of CD22, showed lower α4β7 integrin expression and were impaired in homing to gut-associated lymphoid tissue. This led to impaired intestinal antibody and pathogen responses [108]. A further surprising function of CD22 was detected in *S. pneumoniae* infections of the lung. CD22-deficient mice showed a high susceptibility to this bacterial infection, resulting in uncontrolled bacterial replication and early death. A deficiency in GM-CSF-producing and IgM-secreting plasmablasts was found to be responsible for this phenotype [109].

6.3 The Role of CD22 in Preventing Autoimmunity

Hyper-responsive B cells resulting from a lack of inhibition of BCR signalling can lead to the development of autoimmunity. This has been shown in mice lacking the inhibitory FcγRIIb receptor [110] and in Lyn [66] or SHP-1 [111]-deficient mice whose B cells have hyperactive signalling pathways. CD22-deficient mice do not spontaneously develop autoimmunity on the C57BL/6 background. However, ageing CD22-deficient mice on a mixed C57BL/6 × 129 or on a Y-linked autoimmune accelerator (*Yaa*) background do develop autoimmunity [112,113]. This is also the case in several natural mouse strains prone to autoimmunity such as NZW, MRL and BXSB, which carry the *Cd22a* allele with a truncated first CD22 Ig domain [114]. CD22 protein expressed from the *Cd22a* allele shows impaired ligand binding. These data indicate that the mouse *Cd22* gene could be a susceptibility gene for autoimmunity. In addition to CD22, mouse B cells express the inhibitory receptor Siglec-G, which is from the same Siglec family as CD22. In contrast to CD22-deficient mice, ageing CD22 × Siglec-G double-deficient mice developed a lupus-like autoimmune disease with high-affinity autoantibody formation and glomerulonephritis [115]. This indicates a redundancy of CD22 and Siglec-G, with one being able to replace the function of the other and only loss of both Siglecs leading to autoimmunity. Interestingly, there was no autoimmune disease detected in double-mutant mice with defective CD22 and Siglec-G ligand-binding domains [116]. This finding excluded the possibility that *trans*-ligand binding of Siglecs to sialic acids on other cells is the dominant mechanism involved to prevent autoimmunity.

6.4 Targeting CD19 and CD22

As CD19 and CD22 are widely expressed in B-cell leukemias and in B-cell lymphomas, they are prominent targets for cancer therapy. Additionally, they are targets in B-cell-mediated autoimmune diseases. For CD19, therapeutic monoclonal antibodies have been developed that mediate antibody-dependent cellular cytotoxicity [117]. Furthermore, antibody-drug conjugates, as well as bispecific antibodies against CD19 and CD3 (so called BITEs) have been developed, in order to recruit cytotoxic T cells to CD19-expressing leukaemic cells [118,119]. Furthermore, CD19 was the first target and prototype for a chimaeric antigen receptor (CAR) T cell therapy against B-cell leukemias and lymphomas. CD19-targeted CAR T-cell therapy was very successful, as it led to remission in about 80% of B-ALL and about 50% in CLL leukemias, with high fractions of complete remission [120]. Recently, CD19-targeted CAR T-cell therapy was also successfully used in six patients with severe systemic lupus erythematosus (SLE) with a long-lasting effect on disease activity [121]. CD22-specific monoclonal antibodies like epratuzumab have been developed that in contrast to CD19 antibodies usually do not deplete B cells, but modify signalling, by inhibiting BCR-induced Ca^{2+} signalling. However, a phase 3 clinical trial to treat SLE with epratuzumab was not successful [122]. CD22-specific antibodies are also tested as drug candidates for various B-cell leukemias and lymphomas [123]. For CD22, synthetic sialic acid-related ligands have been developed. Whereas natural sialic acids bind CD22 in the mM range with low affinity, several groups could increase the affinity by generating synthetic, modified sialic acid-binding CD22 in the nanometre range [93,124,125]. These artificial high-affinity CD22 ligands can be used to modify BCR signalling, immune responses or as drug conjugates [126,127]. Finally, clinical trials with CD22-targeted CAR T-cell therapies for B-cell leukemias have been started as well, in order to have an additional target available for the group of patients that are resistant to CD19 T-cell CAR therapy [128].

7. INITIAL ANTIGEN-INDUCED EVENTS IN BCR SIGNALLING

Phosphorylation of the two tyrosine residues in the Igα/Igβ ITAMs, and the recruitment and activation of the Syk tyrosine kinase, are the critical initial events in BCR signalling. BCR signalling is completely ablated in B cells in which the ITAM tyrosine residues are mutated [129]. Only dually phosphorylated ITAMs that can engage both SH2 domains of Syk can efficiently bind Syk, recruit it to the BCR and act as an allosteric activator that increases the kinase activity of Syk [130–134]. Syk-mediated tyrosine phosphorylation then nucleates the assembly of the BLNK-Btk-phospholipase Cγ2 (PLCγ2) signalosome that activates major BCR signalling pathways (see below).

The Igα/Igβ ITAMs can be phosphorylated by both the SFK Lyn and Syk. However, Lyn predominantly phosphorylates only the first ITAM tyrosine (Y182) of Igα, whereas Syk can efficiently phosphorylate both ITAM tyrosine residues (Y182 and Y193) [135–137]. Moreover, ITAM-bound Syk can then amplify BCR signalling by phosphorylating the ITAMs of other nearby BCRs [32,133,138]. This suggests a critical role for Syk in initiating BCR signalling. The roles of Lyn and the other SFKs expressed by B cells (e.g. Blk) in the initiation of BCR signalling are not fully understood, but Lyn-mediated phosphorylation of Syk promotes its full activation [134].

In vitro, Syk can readily phosphorylate ITAM-containing peptides. Hence, a key question is how Syk-mediated phosphorylation of the BCR ITAMs is limited in resting B cells, and conversely, how antigen binding allows Syk to phosphorylate the BCR ITAMs and initiate BCR signalling. Although recent cryo-EM analyses of the resting BCR revealed interactions between the extracellular and TMDs of the Ig HCs, Igα and Igβ [8–10], the structure of the Igα/Igβ cytoplasmic domains was not determined. Whether antigen binding alters the conformation or interactions of the Igα/Igβ cytoplasmic domains is not known. In resting cells, the Igα/Igβ ITAMs may be shielded from Syk, perhaps via interactions between the Igα/Igβ cytoplasmic domains or via interactions with other proteins or with the plasma membrane, as has been shown for the CD3ε chain [139]. Antigen binding could cause conformational changes in the mIg subunit of the BCR that are translated via interactions with the Igα/Igβ extracellular domain and transmembrane interactions into conformational changes in the Igα/Igβ cytoplasmic domains that expose the ITAMs to Syk. Indeed, a Förster resonance energy transfer (FRET) study suggests that the conformation of the tails is altered during BCR activation [140].

The submembrane cortical actin cytoskeleton clearly plays a role in restricting BCR signalling in resting B cells. Treating B cells with the actin-depolymerising drug latrunculin increases BCR mobility within the plasma membrane and is sufficient to recruit Syk to the BCR and induce robust BCR signalling [32,42,141]. In contrast, when the enzymatic activity of Syk is inhibited, latrunculin-induced recruitment of Syk to the BCR is not observed [32]. This is consistent with a role for Syk in phosphorylating the Igα/Igβ ITAMs and supports the existence of a positive feedback loop in which activated Syk promotes further recruitment of Syk to BCRs.

The cortical cytoskeleton consists mainly of branched actin networks that are nucleated by the actin-related protein 2/3 (Arp2/3) complex. Inhibition of Arp2/3 complex activity, as well as human loss-of-function mutations in the Arpc1B components of the complex, results in increased BCR mobility, increased antigen-independent BCR signalling and B-cell hyperactivation [142,143].

Whereas the cortical actin network limits BCR signalling in resting B cells, antigen-induced BCR signalling stimulates actin network disassembly by the actin-severing proteins cofilin [144] and gelsolin [145] and also transiently uncouples the submembrane actin cytoskeleton from the plasma membrane via dephosphorylation of the linker protein ezrin [42,146,147]. Phosphorylated ezrin links actin filaments to the cytoplasmic domains of membrane proteins such as CD44. Localised actin network disassembly and uncoupling from the plasma membrane removes actin-based diffusion barriers, resulting in increased lateral mobility of BCRs, an increased frequency of BCR-BCR collisions and the formation of BCR microclusters [141,144,147,148]. BCR—BCR collisions and microcluster formation amplify BCR signalling by allowing activated Syk that is associated with antigen-bound BCRs to phosphorylate the ITAMs in nearby BCRs such that those BCRs can now recruit Syk and initiate signalling [58,149]. This ‘signal spreading’ may include BCRs that have not bound antigen but which can act as signalling platforms once their ITAMs are dually phosphorylated. By amplifying BCR signalling (i.e. the amount of signalling generated per unit of bound antigen), microcluster formation reduces the amount of antigen required for B-cell activation [56,150,151].

As discussed earlier, the BCRs on resting B cells exist as a spectrum of monomers, small oligomers and larger nanoclusters, with IgM- and IgD-containing BCRs forming separate nanoscale protein islands with distinct protein and lipid compositions [59]. In this context, actin-based diffusion barriers may limit interactions between protein islands in resting B cells. The removal of such barriers in response to antigen binding enables nanocluster coalescence or concatenation, resulting in the formation of microclusters that contain both IgM-BCRs and IgD-BCRs [26]. At the same time, the mobility and spatial organisation of positive and negative regulators of BCR signalling (e.g. CD19, CD22 and CD45) may be altered in a manner that favours BCR signalling. Gasparrini et al. showed that highly mobile CD22 nanoclusters limit BCR signalling in resting B cells and that CD45 can interact with CD22 and shield the BCR from the inhibitory functions of CD22 [96].

Once Syk has been recruited to the BCR ITAMs and BCR microclusters form, protein—protein interactions mediated by SH2 domains binding to phosphotyrosine-containing sequence motifs initiate the activation of multiple BCR signalling pathways. This leads to the assembly of the BCR signalosome, the synthesis and cleavage of phosphoinositide lipids involved in signal transduction and the activation of membrane-associated GTPases that recruit and activate their downstream effectors (Figure 5.4A). These membrane-proximal BCR signalling events are described in the next sections.

8. THE BCR SIGNALOSOME AND PHOSPHOLIPASE Cγ2 ACTIVATION

The core components of the BCR signalosome are the cytoplasmic adaptor protein BLNK (SLP-65), the Bruton tyrosine kinase (Btk) and PLCγ2 (Figure 5.4B). Assembly of the BCR signalosome at the inner face of the plasma membrane promotes the activation of Btk and PLCγ2 and brings PLCγ2 close to its substrate, the membrane lipid phosphatidylinositol 4,5-bisphosphate (PIP_2). Cleavage of PIP_2 by PLCγ2 yields two critical second messengers, inositol 1,4,5-trisphosphate (IP_3) and diacylglycerol (DAG) (Figure 5.4D).

IP_3, once liberated, binds to receptors on the endoplasmic reticulum (ER) (and perhaps other organelles such as mitochondria), causing the release of stored Ca^{2+} ions into the cytoplasm. Ca^{2+} signalling in B cells has been reviewed in detail [152,153]. Depletion of intracellular Ca^{2+} stores promotes their capacitative refiling with Ca^{2+} ions from the extracellular medium. This store-operated Ca^{2+} entry (SOCE) involves close juxtaposition of intracellular Ca^{2+} storage compartments with the plasma membrane such that Ca^{2+} ions can sequentially pass through the Orai1 plasma membrane Ca^{2+} channel and the ER-localised STIM1/STIM2 Ca^{2+} channels. Depletion of ER Ca^{2+} stores cause STIM1 to aggregate into puncta that interact directly with Orai1 at ER-plasma membrane contact sites. The resulting BCR-induced increases in intracellular-free Ca^{2+} concentrations promote the activation of conventional protein kinase C (PKC) isoforms and the nuclear factor of activated T cells (NFAT) transcription factors.

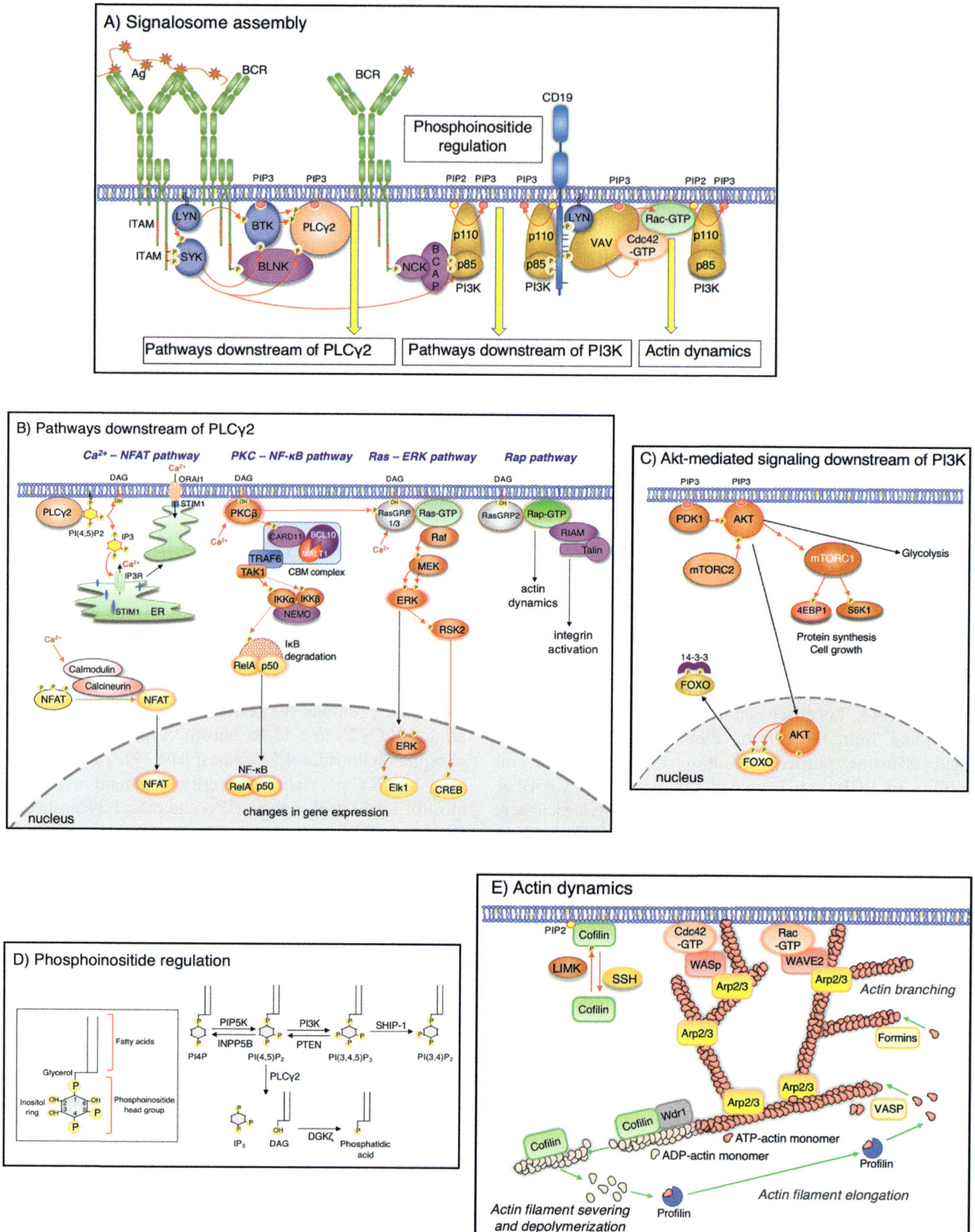

FIGURE 5.4 **BCR signalling pathways.** (A) BCR-induced activation of Lyn and Syk, assembly of the BLNK-Btk-PLCγ2 signalosome, recruitment of PI3K and Vav to the plasma membrane, PI3K-mediated phosphorylation of PIP_2 (*yellow hexagons*) to yield PIP_3 (*red hexagons*) and activation of the Rac

DAG remains attached to the inner leaflet of the plasma membrane and is a binding target and activator of proteins carrying a C1 domain [154]. These are most prominently the conventional PKC isoforms (PKCα, PKCβII) and novel PKC isoforms (PKCδ, PKCε) that are activated by BCR signalling. PKCβII has a critical role in BCR-induced activation of the NF-κB pathway [155]. Other C1 domain-containing proteins that are the RasGRP family GEFs that activate the Ras and Rap GTPases [156] (see Section 14).

The formation of BLNK-Btk-PLCγ2 signalosomes in the proximity to the BCR ITAMs and to PIP_2 in the plasma membrane is a critical step in BCR signalling [157]. Signalosome formation is nucleated by BLNK/SLP-65, which is recruited to the BCR and then phosphorylated by Syk [158]. The C-terminal SH2 domain of BLNK binds to the non-ITAM tyrosine (Y204) in the cytoplasmic tail of Igα that is phosphorylated in response to BCR activation [159]. This positions BLNK close to Syk, resulting in the phosphorylation of BLNK on nine tyrosine resides [160] and the generation of phosphotyrosine-containing binding motifs for the SH2 domains of Btk and PLCγ2. BCR-induced translocation of PLCγ2 from the cytosol to the inner face of the plasma membrane is dependent on BLNK [161] and PLCγ2 activation is dependent on its co-localisation with, and phosphorylation by Btk [162]. After binding via their SH2 domains to BLNK and being brought close to the plasma membrane, both PLCγ2 and Btk bind the PI3K-generated membrane lipid phosphatidylinositol 3,4,5-trisphosphate (PIP_3) via their pleckstrin homology (PH) domains. In this way, these enzymes are activated and colocalised within a PIP_3-domain of the plasma membrane [163]. Btk activation also requires its phosphorylation on Y551 by either Lyn or Syk [164]. Subsequent autophosphorylation of Btk on Y223 results in full activation [165]. Activated Btk can then phosphorylate PLCγ2 on Y753 and Y759, modifications that increase the enzymatic activity of PLCγ2 [166,167].

Additional mechanisms for recruiting BLNK to the BCR may be important for amplifying BCR signalling. BLNK associates with VAMP7-containing cytoplasmic vesicles where it forms large complexes with the CIN85 (Cbl-interacting protein of 85 kDa) adaptor protein and undergoes phase separation into droplets that may be analogous to stress granules [168–170]. An N-terminal leucine zipper motif in BLNK mediates its association with the vesicles, the curvature of which is essential for droplet formation. Proline-rich motifs in BLNK mediate its constitutive association with the SH3 domains of the CIN85. CIN85 forms stable homotrimers via its coiled-coil domains, resulting in the formation of large phase-separated BLNK-CIN85 complexes. Initial BCR signalling may stimulate the translocation of vesicle-associated BLNK-CIN85 complexes to the plasma membrane, thereby amplifying signalosome assembly and PLCγ2 activation. If so, then components required for translocating these complexes to the membrane (e.g. microtubule motor proteins) may regulate the magnitude of PLCγ2-dependent BCR signalling. Indeed, loss of CIN85, as well as disrupting the ability of BLNK to bind CIN85, impairs BCR-induced Ca^{2+} signalling and NF-κB activation [171–173].

Because the signalosome is a central nexus of initial BCR signalling, as well as the pre-B-cell receptor signalling that drives B-cell development, loss-of-function mutations in signalosome components have deleterious effects on B-cell development, survival and activation. In mice, Syk deficiency leads to complete absence of mature B cells [174,175]. Human loss-of-function mutations in Syk have not been reported, but mutations resulting in hyper-activation of Syk lead to reduced B-cell numbers and circulating antibody levels in both humans and mice [176], presumably due to increased apoptosis. B-cell development is impaired in BLNK knockout mice, and BLNK-deficient mature B cells exhibit reduced BCR signalling and B-cell activation [177]. Loss of BLNK appears to have a more profound defect on B-cell development in humans. Pre-B cells and mature B cells were completely absent in a patient lacking the BLNK protein [178]. B-cell-specific disruption of the gene encoding CIN85 (*SH3KBP1*) in mice impairs T-dependent antibody responses [172], whereas a germline deletion of *SH3KBP1* in humans results in an X-linked antibody deficiency that is associated with an

and Cdc42 GTPases. The pathways indicated in the boxes are expanded on in panels (B–E). (B) Pathways downstream of PLCγ2-generated second messengers. IP_3-mediated Ca^{2+} flux activates the NFAT transcription factors. DAG and Ca^{2+} activate PKCβ, which initiates the NK-κB pathway. DAG-mediated membrane recruitment of RasGRP GEFs leads to activation of the Ras-ERK and Rap pathways. NFAT, NF-κB and ERK promote changes in gene expression. Activated Rap promotes integrin activation and stimulates actin remodelling. (C) The Akt kinase is a multifunctional mediator of PI3K signalling. Following phosphorylation by PDK-1 and mTORC2, Akt increases protein synthesis by activating mTORC1, promotes cell survival and cell cycle progression by inhibiting the FOXO transcription factors and promotes cell growth by increasing glucose uptake and glycolysis. (D) Phosphoinositide structure with the carbons of the inositol ring numbered (*left panel*). PLCγ2, PI3K and a network of lipid kinases and phosphatases regulate the levels of phosphoinositides and their derivatives that act as second messengers (*right panel*). Phosphate groups on the inositol rings are shown, hydroxyl groups accordingly to the *left panel*. (E) Branched actin networks promote the formation of lamellipodia and generate the actin retrograde flow that drives immune synapse formation. The Cdc42-WASp and Rac-WAVE2 modules activate the Arp2/3 complex, which nucleates the formation of actin branches. Uncapped actin filaments grow via the addition of actin monomers by VASP and formins. Activated cofilin severs and depolymerises older actin filaments to promote actin network treadmilling and remodelling and to liberate actin monomers that can be loaded with ATP by profilin and then used for actin polymerisation. See text for details.

impaired NF-κB activation in B cells [173]. Human loss-of-function mutations in Btk cause a severe block in B-cell development, resulting in X-linked gammaglobulinemia (XLA). Patients with these Btk mutations have <1% of the normal number of circulating B cells and very low levels of circulating antibodies [164]. In contrast, Btk knockout mice, as well as *xid* CBA/N mice that have a mutation that disables the Btk PH domain, exhibit a mild immunodeficiency with impaired antigen-induced proliferation and activation of mature B cells [164]. In PLCγ2 knockout mice, the number of mature B cells in the spleen is dramatically reduced, B1a cells in the peritoneal cavity are absent and the levels of circulating antibodies are reduced [179–181]. Selectively deleting the PLCγ2 gene in GC B cells showed that PLCγ2 is important for the development and maintenance of memory B cells [182]. Human loss-of-function mutations in PLCγ2 have not been reported. However, both patients and mice with gain-of-function mutations resulting in PLCγ2 hyperactivity exhibit increased BCR-induced Ca^{2+} signalling and autoimmunity [183,184].

In the next sections, we discuss two major signalling pathways that are activated downstream of PLCγ2 and link the BCR to changes in gene expression. IP_3-induced Ca^{2+} mobilisation promotes activates the NFAT transcription factor, whereas DAG- and Ca^{2+}-dependent activation of PKCβ initiates the activation of NF-κB.

9. NFAT ACTIVATION AND FUNCTION

B cells express three NFAT family members, NFATc1 (also called NFAT2 or NFATc), NFATc2 (NFAT1 or NFATp) and NFATc3 (NFAT4 or NFATx) [185]. NFAT regulation is well-defined [186]. NFAT is cytosolic in resting B cells, but sustained increases in intracellular Ca^{2+} levels induce translocation of NFAT to the nucleus (Figure 5.4B). NFAT proteins contain both a nuclear localisation signal (NLS) and a nuclear export signal (NES). Receptor-induced Ca^{2+} increases allow calmodulin to activate the serine/threonine phosphatase calcineurin. Calcineurin dephosphorylates multiple residues on NFAT, which exposes the NLS and promotes translocation of NFAT into the nucleus. In the nucleus, NFAT interacts with AP-1 and other transcription factors to drive target gene expression. NFAT phosphorylation within the nucleus exposes the NES and occludes the NLS, resulting in nuclear export. Hence, sustained BCR-induced Ca^{2+} increases are required to maintain NFAT in the nucleus [187,188]. NFAT family members act in complex ways to decode the magnitude and duration of BCR-induced Ca^{2+} increases. In particular, NFAT is activated by the low, sustained Ca^{2+} signalling in anergic B cells that continually encounter self-antigens [187]. Antigen-specific anergy is impaired in mice lacking NFATc2 [189], and NFATc2-deficient B cells exhibit reduced production of the immunosuppressive cytokine IL-10 [190]. Consistent with NFATc2 being a negative regulator in B cells, a patient with a loss-of-function mutation in NFATc2 exhibited increased numbers of B cells and recurrent B-cell lymphomas [191]. In contrast, NFATc1 is a positive regulator of B cell survival and activation. NFATc1 ablation in B cells results in impaired BCR-stimulated proliferation and increased activation-induced cell death [192]. The number of peritoneal B1a cells is also substantially decreased in the absence of NFATc1 [193]. The differential regulation and functions of the various NFAT family members are not fully understood.

10. NF-κB ACTIVATION

NF-κB activation, as well as human mutations in the CARMA1/BCL10/MALT1 (CBM) complex that activates NF-kB, has been reviewed elsewhere [194–196]. The BCR activates the canonical NF-κB pathway that involves NF-kB dimers consisting of a p50 subunit (encoded by the *NFκB1* gene) and either p65/RelA or c-Rel, both of which have transactivation domains and function as transcriptional activators. In resting cells, these NF-κB dimers are restrained in the cytosol by inhibitory IκB proteins that mask the NLS in p65 or c-Rel. Activation of the canonical NF-κB pathway requires degradation of IκB. Activation of the IκB kinase (IKK) complex, which consists of the IKKα and IKKβ catalytic subunits and the NEMO (IKKγ) regulatory subunit, allows it to phosphorylate IκBα. Phosphorylated IκBα is then ubiquitinated, which targets it for degradation by the proteasome. Once freed from IκB, the NF-κB dimer can translocate into the nucleus and bind to the promoters of its target genes.

The pathway by which the BCR activates the IKK complex (Figure 5.4B) is initiated by PLCγ2-dependent activation of PKCβ, which requires both DAG and Ca^{2+}. Ablation of PKCβ prevents activation of the canonical NF-κB pathway and impairs B-cell activation [197]. PKCβ promotes the assembly of the CBM complex by phosphorylating serine residues in the CARD11 adaptor protein, which induces a conformational change in CARD11 that allows it to bind pre-existing BCL10-MALT1 complexes. This enables the formation of BCL10 filaments that promote MALT1 oligomerisation and binding of the TRAF6 ubiquitin ligase to MALT1 [198,199]. Ubiquitination of BCL10, MALT1 and NEMO by TRAF6 (and other ubiquitinases) allows the CBM complex to bind the IKK complex and recruit TAK1, the kinase that phosphorylates and activates IKKα and IKKβ [200].

MALT1 also promotes NF-κB activation via a second mechanism. MALT1 is a paracaspase with proteolytic activity, and its major substrates are negative regulators of

NF-κB [201]. MALT1-dependent cleavage of the deubiquitinases A20 and cylindromatosis (CYLD) prevents them from opposing ubiquitin-dependent activation of the canonical NF-κB pathway. Thus, MALT1 acts both as a scaffold for the activation of the CBM complex and as a protease that sustains NF-κB activation.

The NF-κB pathway is essential for the activation of mature B cells by the BCR. Disruption of the genes encoding p50, c-Rel or both in mice impairs BCR-stimulated B-cell survival, proliferation and differentiation into antibody-producing cells [202–207]. This reflects a role for NF-κB in BCR-induced expression of the pro-survival proteins Bcl2A1 and Bcl-xL, as well as the E2F3 and c-Myc transcription factors that drive cell cycle progression [206,208,209]. Similarly, BCR-induced NF-κB activation, B-cell proliferation and antibody responses are impaired in cells lacking MALT1 [210–212], CARMA1 [213,214] or BCL10 [215,216]. Reflecting the essential role of NF-κB in T- and B-cell activation, patients with loss-of-function mutations in CARD11, BCL10 or MALT1 exhibit combined immunodeficiency and decreased levels of circulating antibodies [196]. Conversely, patients with CARD11 gain-of-function mutations develop polyclonal B-cell lymphocytosis [196].

11. ACTIVATION OF PHOSPHOINOSITIDE 3-KINASE (PI3K)

PI3K is a lipid kinase that phosphorylates PIP_2, on the 3-position of the inositol ring to yield PIP_3. PIP_3 generation promotes the accumulation of PH domain-containing proteins at the plasma membrane and helps colocalise proteins to form signalling complexes. The class I PI3K enzymes consist of a 110-kDa catalytic subunit (p110α, p110β, p110γ or p110δ) and a regulatory subunit (p85 or p101) that links the PI3K to receptor activation. Mammals express three class IA PI3K complexes, PI3Kα (p85/p110α), PI3Kβ (p85/p110β) and PI3Kδ (p85/p110δ) as well as the class IB PI3Kγ (p101/p110γ). There are two p85 isoforms, p85α and p85β. p85α is essential for B-cell development as a patient with a loss-of-function mutation in p85α exhibited a very early block in B-cell development and virtually no mature B cells [217]. Of the p110 subunits, p110δ is expressed primarily in hematopoietic cells. B cells have high levels of PI3Kδ and low levels of PI3Kα, PI3Kγ and PI3Kβ [218]. PI3Kα and PI3Kδ both support early B-cell development, whereas PI3Kδ is responsible for most PI3K signalling in mature B cells [218].

The roles of PI3Kδ in B cells are well characterised. In mice in which the gene encoding p110δ has been disrupted or replaced with the D910A catalytically inactive mutant, B-cell development in the bone marrow is relatively normal, but further B-cell maturation in the periphery is defective [69,71,72]. The number of follicular B2 cells in the spleen is dramatically reduced, and the development of peritoneal B1 cells and marginal zone B cells is almost completely blocked. In the follicular B2 cells that develop in p110δ-deficient mice, BCR-induced Ca^{2+} flux is greatly reduced, reflecting the role of BCR-induced PIP_3 production in Btk activation, stabilisation of PLCγ2 at the plasma membrane and phosphorylation of PLCγ2 by Btk. BCR-induced phosphorylation of the Akt kinase and FoxO transcription factors (see Section 12.3) is also reduced in p110δ-deficient B cells. These signalling defects result in impaired BCR-induced cell cycle progression, proliferation and differentiation into antibody-producing cells [69,71,72,219]. Similarly, selective inhibitors of p110δ catalytic activity inhibit BCR signalling and B-cell activation [219–222], resulting in impaired B-cell survival and homeostasis [223]. Conversely, expression of constitutively active p110δ rescues B cells from apoptosis upon inducible deletion of the BCR [224]. This indicates that PI3K is a primary component of the tonic BCR signalling required for the maintenance of peripheral B cells.

Given the critical role of PI3K in B-cell survival and activation, it is important to understand how antigen binding stimulates PIP_3 production. In resting cells, PI3Ks are present mainly in the cytoplasm and the regulatory subunit inhibits the activity of the catalytic subunit. In B cells, the activation and membrane recruitment of PI3K is mediated by CD19 and by the B-cell adaptor for PI3K (BCAP). After BCR engagement, the cytoplasmic domain of CD19 is phosphorylated on two YxxM motifs that can recruit the p85α subunit [225], and BCAP is phosphorylated on four YxxM motifs [226]. Once the tandem SH2 domains in the p85 regulatory subunit bind to tyrosine-phosphorylated YxxM motifs, inhibition of the associated catalytic subunit is relieved.

Deletion of the *CD19* gene in mice impairs B-cell development but not to the same extent as deletion of p85α [51,52,70,227]. However, when the genes encoding both CD19 and BCAP are disrupted, B-cell maturation is almost completely blocked [228]. This block can be partly overcome by expressing a constitutively active form of PI3K or PDK1 [228], a PI3K-activated kinase that phosphorylates and activates Akt. This shows that pro-survival pathways downstream of PI3K are essential for B-cell development and that PI3K signalling in B-lineage cells depends on both CD19 and BCAP.

How the cytosolic adaptor protein BCAP recruits PI3K to the plasma membrane was elucidated by Castello et al. [229]. A proline-rich region in BCAP binds the SH3 domain of the adaptor protein Nck. Upon BCR stimulation, Nck binds via its SH2 domain to the phosphorylated non-ITAM tyrosine (Y204) in Igα, thereby recruiting BCAP to the BCR signalosome. BCAP is then phosphorylated by

Syk on YxxM motifs and can recruit PI3K. Nck is essential for BCR-induced tyrosine phosphorylation of BCAP, and Nck-deficient B cells exhibit reduced BCR-induced Ca^{2+} mobilisation, Akt phosphorylation and proliferation in vitro [229]. In vivo, Nck-deficient mice have reduced antibody responses [229].

12. AKT, A MULTIFUNCTIONAL TRANSDUCER OF PI3K SIGNALLING

The serine/threonine kinase Akt (also called protein kinase B) depends on PI3K for its activation and controls multiple pathways that regulate cell survival, growth and metabolism, as well as cell cycle progression. There are three Akt isoforms: Akt1, Akt2 and Akt3. From these, Akt1 and Akt2 have critical and partly overlapping functions in B-cell development, survival and activation [230,231]. A role for Akt3 in B cells has not been reported.

Akt is in the cytosol of resting B cells, but in response to BCR stimulation, Akt is recruited to the plasma membrane via binding of its PH domain to PIP_3 [232]. Full activation of Akt requires sequential phosphorylation of Akt on T308 by the phosphoinositide-dependent kinase-1 (PDK1) and on S473 by the mammalian target of rapamycin (mTOR) kinase that is part of the mammalian target of rapamycin complex 2 (mTORC2) [233]. PDK1 also has a PH domain and is recruited to plasma membrane regions where PI3K has generated PIP_3. It is not clear how mTORC2-mediated phosphorylation of Akt is regulated. Nevertheless, mTORC2 activity in B cells is required for Akt to stimulate B-cell growth by activating its downstream effector, the mammalian target of rapamycin complex 1 (mTORC1) [234] (see below).

Ablating the PDK1 gene in the B-lineage substantially impairs B-cell development [235], whereas combined deletion of Akt1 and Akt2 does not prevent the maturation of follicular B2 cells [230]. This suggests that PDK1 has additional functions that are not dependent on Akt. However, Akt1/Akt2 depletion blocks the development of marginal zone and B1 B cells [230], reflecting the critical role of PI3Kδ activity in the development of these innate-like B cells [69,71,72]. For mature B2 cells, both PDK1 and Akt1/Akt2 are important for their homeostatic survival and activation [230,235]. Recent work showed that mature B cells from Akt2 knockout mice exhibit decreased BCR signalling, GC responses and antigen-specific antibody responses [236]. These defects were associated with reduced phosphorylation of CD19 and Btk in response to BCR stimulation, suggesting that Akt2 acts both upstream and downstream of the phospho-CD19→PI3K→Btk signalling module.

12.1 Regulation of Cell Growth via Akt-Dependent Activation of mTORC1

Cell growth and proliferation require increased protein synthesis, membrane synthesis and energy production, all of which Akt can trigger by activating mTORC1. The multifaceted role of mTORC1 in promoting cell growth and metabolism has been reviewed [237]. The BCR activates mTORC1 via the PI3K-Akt pathway [238]. Activated Akt phosphorylates the tuberin proteins, Tsc1 and Tsc2, and prevents them from inactivating the Rheb GTPase [239]. This allows active GTP-bound Rheb to bind to and activate the mTOR kinase within mTORC1. Activated mTORC1 increases protein synthesis by phosphorylating eukaryotic initiation factor 4E-(eIF4E)-binding protein 1 (4E-BP1) and p70 ribosomal S6 kinase (S6K1). Phosphorylated 4E-BP1 dissociates from eIF4E, allowing eIF4E to bind the 5′ cap of mRNAs and promote cap-dependent mRNA translation. Phosphorylation-dependent activation of S6K1 increases cap-dependent mRNA translation and elongation. Activated S6K1 also increases the translation of ribosomal proteins, which promotes ribosome biogenesis. The BCR activates S6K1 in a PI3K-dependent manner [238]. In addition to increasing protein synthesis, mTORC1 activation promotes lipid biosynthesis for new membrane formation, increases nucleotide synthesis and increases both mitochondrial activity and biogenesis [237,240].

12.2 Akt Enhances Glucose Uptake and Metabolism

When naive B cells are stimulated via the BCR, they enter the cell cycle. Prior to entering S phase, they undergo a substantial increase in size that reflects increased protein and nucleic acid synthesis. This depends on glycolysis, which generates ATP to meet the energy demands for cell growth while generating precursors for macromolecule synthesis. BCR stimulation causes a PI3K-dependent increase in expression of the Glut1 glucose transporter, increased glucose uptake and increased glycolytic activity [241]. Moreover, inducible activation of Akt is sufficient to increase glucose uptake and utilisation [241]. Hence, BCR-induced activation of the PI3K-Akt pathway is essential for meeting the metabolic demands of B-cell activation.

12.3 The Akt-FOXO Pathway

The evolutionarily conserved Forkhead box O (FOXO) transcription factors, which drive the expression of genes that promote cell cycle arrest (e.g. cyclin G2, p130Rbl2, p27kip1 and GADD45) and apoptosis (e.g. Bim and FasL), are inhibited by Akt-mediated phosphorylation [231,242]. FOXO1 and FOXO3a are expressed in B cells. Their role in B-cell development, activation and differentiation has been reviewed [231]. In resting cells, FOXO proteins are primarily in the nucleus where they bind the promoters of their target genes to enforce quiescence. Ectopic expression of FOXO1 or FOXO3a in activated B cells results in cell cycle arrest and increased cell death [243]. In response to sustained BCR signalling, membrane-bound Akt is released into the cytosol and translocates into the nucleus [232] where it phosphorylates FOXO proteins [243]. This creates binding sites for the adaptor protein 14-3-3 which mask both the FOXO DNA-binding domain and NLS [244], causing Akt-phosphorylated FOXO proteins to be exported to the cytosol (Figure 5.4C). Consistent with this, BCR signalling downregulates the expression of FOXO target genes in a PI3K-dependent manner [243,245]. In activated B cells, the extent of BCR-induced inhibition of FOXO transcriptional activity impacts the formation of short-lived plasmablasts versus entry into GCs as well as positive selection and class switch recombination within the GC [246,247].

13. PHOSPHOINOSITIDE REGULATION

As the substrate for both PLCγ2 and PI3K, the membrane lipid $PI(4,5)P_2$ plays a central role in BCR signalling. Hydrolysis of $PI(4,5)P_2$ into IP_3 and DAG by PLCγ2 leads to Ca^{2+}-dependent activation of the NFAT transcription factors, PKCβ-dependent activation of NF-κB and activation of the Ras and Rap GTPases by RasGRP family members that are recruited to the plasma membrane by DAG. At the same time, PI3K-mediated phosphorylation of $PI(4,5)P_2$ generates $PI(3,4,5)P_3$, which recruits and activates PH domain-containing proteins including the Akt kinase.

BCR-stimulated $PI(4,5)P_2$ cleavage also directly influences actin dynamics. $PI(4,5)P_2$ binds a number of actin-binding proteins and regulates their membrane association and function [248]. $PI(4,5)P_2$ binding supports the ability of ezrin to couple the submembrane actin cytoskeleton to the plasma membrane and maintains the actin-severing protein cofilin in an inactive phosphorylated form. PLCγ2-mediated hydrolysis of $PI(4,5)P_2$ would release ezrin and cofilin from the membrane. Once cofilin is released from the membrane, it can be dephosphorylated, which is required for it to bind to and sever actin filaments. In this way, BCR signalosome assembly and PLCγ2-mediated $PI(4,5)P_2$ hydrolysis reduce the local density of the submembrane actin cytoskeleton [249,250].

The levels of $PI(4,5)P_2$ and $PI(3,4,5)P_3$, as well as their spatial organisation within the membrane, are controlled by a network of enzymes [251] (Figure 5.4D). PI3K signalling is limited by the SH2 domain-containing inositol polyphosphate 5-phosphatase 1 (SHIP-1) and phosphatase and tensin homologue (PTEN) lipid phosphatases. SHIP-1 dephosphorylates the 5 position of the $PI(3,4,5)P_3$ inositol ring, yielding $PI(3,4)P_2$. This limits the BCR-induced membrane recruitment and activation of multiple PH domain-containing signalling enzymes including Akt and Btk [252,253]. SHIP-1 is recruited to the site of BCR signalling via its interaction with the BLNK/Grb2 adaptor complex [254]. B cells lacking SHIP1 are hyperresponsive to BCR stimulation, exhibiting increased release of intracellular Ca^{2+}, increased activation of Akt and ERK and increased proliferation [252,255,256]. The SHIP-1 SH2 domain can also bind phosphotyrosine-containing motifs in the cytoplasmic domains of CD22 and FcγRIIb and inhibit BCR signalling when these receptors are co-clustered with the BCR [253,257].

PTEN opposes PI3K signalling by dephosphorylating PIP_3 on the 3 position of the inositol ring and converting PIP_3 back to $PI(4,5)P_2$. Consistent with a role for PTEN in limiting PI3K signalling, PTEN-deficient B cells exhibit increased BCR-stimulated PIP_3 levels, Akt activation and proliferation compared to wild-type cells [73]. B-cell-specific deletion of the *PTEN* gene in mice results in increased numbers of innate-like B cells [73], the opposite of what is observed in p110δ-deficient mice [69,71]. The regulation of PTEN during BCR signalling is not fully understood. In other cell types, the transient phosphorylation of the C-terminal tail of PTEN prevents PIP_3 dephosphorylation at the membrane [258]. This may allow initial PIP_3 accumulation after receptor stimulation. Subsequent recruitment of PTEN to the plasma membrane may limit the magnitude and duration of signalling.

Because BCR signalling depletes $PI(4,5)P_2$ via hydrolysis by PLCγ2 and phosphorylation by PI3K, sustained BCR signalling requires synthesis of new $PI(4,5)P_2$. This occurs largely through the phosphorylation of phosphatidylinositol-4-phosphate (PI4P) by PI4P 5-kinases (PIP5Ks). PIP5Ks constitutively associate with Btk, and upon BCR stimulation, Btk shuttles PIP5Ks to the plasma membrane [259]. When PIP5Kβ is overexpressed in B-cell lines, BCR-induced generation of both IP_3 and PIP_3 was increased [259], suggesting that the limited availability of $PI(4,5)P_2$ restricts BCR signalling.

The effects of PIP5Ks are opposed by inositol polyphosphate 5-phosphatase B (INPP5B), which converts $PI(4,5)P_2$ back to PI4P. Surprisingly, Droubi et al. found that deleting INPP5B in B cells did not increase $PI(4,5)P_2$ levels or BCR signalling but instead resulted in impaired BCR signalling and BCR-induced actin remodeling [260]. This conundrum was resolved by live-cell imaging of the distribution of PIP_2 and BCRs in the plasma membrane, which highlighted the importance of precise spatiotemporal regulation of PIP_2 levels in BCR signalling. In normal B cells, PIP_2 was selectively depleted at sites where nascent BCR microclusters were forming. In B cells lacking INPP5B or expressing a catalytically inactive version of

INPP5B, the rate at which PIP_2 was removed from the areas surrounding BCR microclusters was substantially reduced, as was the dephosphorylation/activation of cofilin and the dephosphorylation/inactivation of ezrin that normally occurs after BCR stimulation. Thus, INPP5B-mediated depletion of PIP_2, specifically at nascent BCR microclusters, is an important early event in BCR signalling. However, it is not known whether INPP5B is rapidly recruited to activated BCRs.

The levels of DAG, one of the products of PIP_2 hydrolysis by PLCγ2, are also regulated. By recruiting Ras activators and PKCβ to the plasma membrane, DAG controls the activation of both the Ras-ERK pathway and the NF-kB pathway. These pathways drive B-cell survival and proliferation and must therefore be tightly regulated. After BCR stimulation, DAG is rapidly phosphorylated by diacylglycerol kinases (DGKs) to yield phosphatidic acid [261]. B cells deficient for the DGKζ isoform exhibit increased ERK activation and proliferation in response to BCR stimulation in vitro and increased antibody responses in vivo. Thus, DGK activity may normally limit DAG-mediated BCR signalling and B-cell activation. However, phosphatidic acid can also act as a second messenger. Phosphatidic acid species with unsaturated fatty acids activate mTORC1 [262,263], a key regulator of protein synthesis and cell growth. Whether this occurs in B cells is not known. DGKζ and phosphatidic acid also collaborate to activate the Rac GTPases, and this is important for BCR-induced actin polymerisation and LFA-1 activation at the immune synapse [264].

14. GTPASE ACTIVATION AND GTPASE-REGULATED SIGNALLING PATHWAYS

Small GTPases act as molecular switches that can be interconverted between an inactive GDP-bound form and an active GTP-bound form. GTPases are activated by GEFs that promote the release of bound GDP, which allows binding of the more abundant GTP. GTP binding causes a conformational change that exposes effector-binding sites on the GTPase, enabling them to recruit and activate signalling proteins. The small GTPases are generally membrane associated due to lipid modifications, and their activation requires recruitment of cytosolic GEFs to the relevant cellular membrane. BCR engagement leads to activation of the Ras [265,266], Rap1 and Rap2 [267,268], Rac1 and Rac2 [269,270] and Cdc42 GTPases [271], all of which have important roles in B-cell development and function, which may depend on their precise location relative to the 3-D topography of the B-cell surface.

14.1 The Ras/ERK Pathway

Ras controls the Raf-1 → MEK (mitogen-activated protein kinase (MAPK) kinase) → extracellular signal-regulated kinase (ERK) cascade (Figure 5.4B). The sequential activation of the components of this kinase cascade has been reviewed [272,273]. The Ras-ERK pathway is essential for early B-cell development, acting at checkpoints both before and after expression of the pre-BCR [274–276]. In peripheral B cells, Ras-dependent activation of ERK1 and ERK2 (collectively referred to as ERK) [266,276,277] is important for B-cell survival, proliferation and transcriptional regulation [278,279]. ERK supports the survival of activated B cells by phosphorylating Bim, a pro-apoptotic member of the Bcl2 family, and targeting it for degradation [278,280]. In resting cells, ERK is maintained in an inactive state in the cytosol via its binding to MEK. MEK activation downstream of Ras and Raf-1 allows MEK to phosphorylate ERK on the threonine and tyrosine residues within a conserved TEY motif. Activated ERK then dissociates from MEK. In addition to phosphorylating cytoplasmic substrates, activated ERK translocates into the nucleus and induces transcriptional changes by phosphorylating multiple transcription factors [272,273]. ERK phosphorylates ETS-family transcription factors such as Elk1, which interacts with serum response factor to drive expression of immediate early genes (e.g. *c-fos*), many of which encode transcription factors that promote cell cycle entry and progression through G1 phase. c-Fos is a component of the AP-1 transcription factor complex, which both induces pro-proliferative genes and represses anti-proliferative genes [281].

ERK also regulates transcription, translation and cell survival by phosphorylating and activating the p90 ribosomal S6 kinase (p90Rsk, now referred to as RSK2) [282]. RSK2-mediated phosphorylation on S133 activates the cAMP-response element-binding (CREB) protein, which regulates immediate early gene expression [283]. A S133A mutant of CREB is associated with reduced B-cell survival and proliferation after BCR stimulation [284]. In addition to RSK2, ERK also phosphorylates and activates the MAPK-interacting kinases 1 and 2 (Mnk1/2), which promote mRNA translation and protein synthesis by phosphorylating eIF-4E [285].

The BCR activates Ras via two different GEFs namely RasGRP1 and RasGRP3, with RasGRP3 having a dominant role in BCR-induced ERK activation and B-cell proliferation [286,287]. PLCγ2-generated DAG can bind the C1 domain in RasGRP proteins and recruit them to the plasma membrane where Ras is located [286]. In order to active Ras, RasGRP3 needs to be phosphorylated on T133 by PKC [288], which is also recruited to the plasma membrane by DAG. Consistent with this, BCR-induced

Ras activation is greatly reduced in PLCγ2-deficient B cells [286]. BCR-induced Ras activation is amplified by another Ras-GEF, namely the mammalian Son-of-Sevenless protein (mSOS), which undergoes allosteric activation when it binds Ras-GTP [289].

14.2 Pathways Downstream of the Rac and Cdc42 GTPases

In addition to ERK, the BCR activates the c-Jun N-terminal kinase (JNK) and the p38 MAPKs via analogous kinase cascades that are controlled by the Rac GTPases [276,290] (Figure 5.4B). The BCR can also selectively activate the JNK2 isoform, but not the JNK1 isoform, via CBM complex-mediated activation of the TAK1 kinase, the kinase that phosphorylates the IKK complex to activate NF-κB [291]. JNK phosphorylates and activates a wide variety of transcription factors including c-Jun [292]. c-Jun is a component of multiple AP-1 transcription factor complexes that regulate cell survival, cell cycle progression and proliferation [293]. Activated p38 phosphorylates the MEF2c transcription factor in its transactivation domain. B-cell-specific deletion of MEF2C in mice significantly impairs anti-Ig-induced B-cell proliferation, antibody responses to T-dependent antigens and the expansion of GC B cells [294].

The active GTP-bound form of Rac can also bind to PI3K and promote its activation. This was confirmed in the DT40 chicken B-cell line [295]. Murine B cells deficient in Rac1 and Rac2 exhibit reduced activation of Akt and reduced Ca^{2+} fluxes in response to BCR stimulation [296]. The Rac and Cdc42 GTPases are major regulators of actin polymerisation that act upstream of the Arp2/3 complex to control cell morphology, cell polarity and cell motility. Activation of Rac and Cdc42 GTPases allows the BCR to stimulate branched actin polymerisation that is nucleated by Arp2/3 complex. At the plasma membrane, the Arp2/3 complex is activated by the Cdc42→WASp/N-WASp module and the Rac→WAVE2 module [297] (Figure 5.4E). WASp, N-WASp and WAVE are nucleation-promoting factors (NPFs) that exist in autoinhibited conformations until they bind active GTP-bound Cdc42 or Rac. Once activated by an NPF, the Arp2/3 complex nucleates the formation of a new filament [298]. Branched actin networks can exert outward force on the plasma membrane to drive the formation of broad membrane protrusions such as lamellipodia [299]. As described below, this allows B cells to scan the surface of antigen-presenting cells (APCs) for antigens and form an immune synapse. Arp2/3 complex activity is also important for directing IgM-BCR clusters to membrane ridges and MV [41], which enhances antigen detection by B cells.

The Rac1 and Rac2 GTPases have key roles in B-cell development and activation. B-cell-specific ablation of both Rac1 and Rac2 results in a near absence of mature B cells [296]. Loss of Rac1 alone has little effect on B-cell numbers, while loss of Rac2 alone results in decreased numbers of marginal zone and B1 cells. Thus, Rac1 and Rac2 have both overlapping and unique roles in B-cell development. BCR signalling and anti-IgM-induced proliferation are reduced in Rac2-deficient mature B cells and to a greater extent when these cells have only one functional allele of Rac1 [296]. Rac2, but not Rac1, is important for the BCR-induced actin remodelling that drives immune synapse formation and for BCR-induced integrin activation [270].

Burbage et al. conducted a thorough investigation of the role of the Cdc42 GTPase in B cells [271]. Selectively ablating the Cdc42 gene in B-lineage cells results in the development of significantly fewer mature follicular B2 B cells, the loss of marginal zone B cells and greatly reduced numbers of peritoneal B1 B cells. Mature B cells lacking Cdc42 have less filamentous actin (F-actin) content than wild-type cells, consistent with a role for Cdc42 in stimulating Arp2/3 complex-dependent actin polymerisation. Cdc42-deficient B cells display reduced BCR signal transduction, perhaps reflecting a role for actin in stabilising BCR microclusters as well as BCR interactions with CD19 that amplify BCR signalling. The WASp NPF, which is activated by Cdc42, plays a critical role in the actin remodelling that enables B cells to spread on antigen-bearing membranes and form BCR microclusters [300,301]. Consistent with Cdc42 being an upstream activator of WASp, Cdc42-deficient B cells exhibit significantly reduced spreading and microcluster formation [271]. As a consequence of these defects, mice with B-cell-specific loss of Cdc42 have impaired GC expansion and reduced antibody responses [271,302].

The BCR can activate Rac and Cdc42 via multiple GEFs, most notably the Vav proteins. Analysis of B-cell development and BCR signalling in mice lacking one, two or all three of the Vav proteins revealed that single knockouts had minimal effect, Vav1/2 and Vav1/3 double knockouts exhibited significant reductions in mature B-cell numbers and BCR signalling responses, and more severe defects were observed in Vav1/2/3 triple knockouts [303]. BCR-induced activation of Vav involves binding of the Vav SH2 domain to phosphotyrosine-containing sequences on CD19 or BLNK, which recruits Vav to the plasma membrane. Subsequent binding of PIP_3 to the PH domain of Vav, combined with phosphorylation of Vav on key tyrosine residues, increases the ability of Vav to activate Rac and Cdc42 [304].

DOCK2 is a Rac-specific GEF that contains a DHR-1 domain that binds PIP_3 [305]. When B cells interact with anti-Ig antibodies tethered to lipid bilayers, DOCK2 is recruited to the contact site in a PIP_3-dependent manner and contributes to the BCR-induced actin remodelling that drives immune synapse formation and the sustained growth of BCR microclusters [250].

Another GEF, DOCK8, may also contribute to BCR-induced activation of Cdc42. DOCK8 activates Cdc42 in T cells and myeloid cells and links multiple receptors to actin-dependent processes [306]. DOCK8 also acts as a scaffold protein that colocalises activated Cdc42 with WASp [307]. B cells from DOCK8 knockout mice or from patients with mutations that ablate DOCK8 expression exhibit impaired immune synapse formation and defective GC responses [308,309]. Although consistent with a role for DOCK8-induced Cdc42 activation, DOCK8 also promotes CD19 expression [309] and CD19 has an important role in BCR-induced activation of PI3K and Vav.

14.3 The Rap GTPases

The BCR, as well as multiple chemokine receptors expressed by B cells, activates the Rap1 and Rap2 GTPases [267,268,310,311]. The Rap GTPases are evolutionarily conserved regulators of actin organisation, cell polarity, cell motility and integrin-mediated cell adhesion [312–318]. As such, the Rap GTPases and their effectors play important roles in B-cell development, trafficking and activation. There are five Rap isoforms in mammals, Rap1a, Rap1b, Rap2a, Rap2b and Rap2c, each encoded by a separate gene. In human and mice, the two Rap1 proteins are >95% identical at the amino acid level and the three Rap2 isoform are 90%–95% identical [319]. The in vivo functions of Rap1a and Rap1b have been studied via the generation of single isoform knockout mice and tissue-specific ablation of both genes. Overexpression of a Rap-specific GTPase-activating protein (GAP), which converts both Rap1 and Rap2 to their inactive GDP-bound state, has been used to identify processes that may depend on either Rap1 or Rap2.

Rap1 regulates many biological processes via its ability to activate integrins via 'inside out' signalling. For B cells, integrin activation is important for trafficking into lymph nodes, interacting with APCs and T cells and immune synapse formation. As for other GTPases, Rap activation involves a conformational change that reveals an effector-binding domain. Active GTP-bound Rap1 binds the Rap1-GTP-interacting-adaptor molecule (RIAM; encoded by the *APBB1IP* gene), which in turn recruits talin to the plasma membrane. The talin head domain binds to integrin β chains and induces a conformational change that converts integrins such as LFA-1 from a low-affinity to a high-affinity 'activated' state [314]. The talin rod domain links integrins to actin filaments, promotes integrin clustering and recruits proteins such as vinculin and focal adhesion kinase (FAK) that organise adhesion complexes and initiate 'outside-in' integrin signalling. The RapL adaptor protein is another Rap1 effector that binds Rap1-GTP [320]. When T cells interact with APCs, RapL is recruited to the T cell immune synapse and contributes to integrin activation [320]. The role of RapL at the B-cell immune synapse has not been established, but RapL is essential for integrin-dependent B-cell trafficking in vivo [321].

Activated Rap is also a central regulator of actin remodelling and cell polarity. In migrating adherent cells, a polarised distribution of activated Rap1 establishes the leading edge where Rac-GTP and PI3K signalling are concentrated and branched actin polymerisation generates lamellipodia [322]. Similarly, Rap activation is essential for leucocyte polarisation and chemotaxis [323,324]. In B cells, suppressing Rap activation, as well as ablating the genes encoding Rap1a and Rap1b, greatly impairs chemotaxis [310,311,325]. Rap activation is also essential for the actin remodelling that underlies the ability of B cells to spread on antigen-bearing surfaces such as anti-Ig-coated coverslips, antigen-functionalised lipid bilayers and APCs [144,326]. This spreading response, which facilitates further antigen encounter, involves the formation of a peripheral ring of branched actin that drives membrane protrusion and the concomitant depletion of F-actin from the centre of the B-cell-substrate contact site. Consistent with the impaired spreading and actin clearance observed when Rap activation is blocked, the rates of both actin polymerisation and actin disassembly at the cell periphery are reduced and actin structures are more stable [144]. Hence, Rap activation is critical for the actin turnover dynamics and actin network remodelling that support B-cell spreading and chemokine-induced B-cell motility.

The actin remodelling and integrin activation involved in immune synapse formation (see Sections 15 and 16) is strongly dependent on Rap activation. Compared to control cells, when B cells in which Rap activation is blocked are added to antigen-bearing lipid bilayers or APCs, they exhibit decreased actin polymerisation, an impaired ability to form BCR-antigen microclusters and reduced microcluster-associated BCR signalling [144,326]. The BCR microclusters that do form when Rap activation is blocked fail to undergo centripetal movement and coalesce into a central supramolecular activation cluster (cSMAC), a characteristic feature of an immune synapse [327]. Moreover, the ability of B cells to form a ring of LFA-1 clusters surrounding the cSMAC (i.e. at the peripheral SMAC (pSMAC)) was completely abolished when Rap activation was blocked [326]. This reflects the requirement for Rap-GTP to activate LFA-1 so that it can bind ICAM-1 [328]. Thus, BCR-induced Rap activation plays a critical role in the actin remodelling and integrin activation that drives immune synapse formation. During immune synapse formation, the microtubule-organising centre (MTOC) and the microtubule network emanating from it move towards the antigen contact site. This polarisation of the microtubule network, which contributes to antigen acquisition and delivery to MHC II-loading compartments, is also dependent on Rap1 activation [327].

How activated Rap regulates actin dynamics is not fully understood. Cofilin-mediated actin filament severing and depolymerisation disassembles existing actin networks and liberates actin monomers to fuel new actin polymerisation and enable actin network remodelling. Freeman et al. showed that blocking Rap activation prevents TCR- and BCR-induced activation of cofilin [144]. The RapL adaptor protein also couples Rap1-GTP to actin dynamics by binding Mst1/STK4 [329], a serine/threonine kinase that regulates multiple actin-regulatory proteins. In B cells, loss of Mst1 decreases the amount of phosphorylated (i.e. activated) WASp at the plasma membrane and impairs B-cell spreading and microcluster-based BCR signalling [301]. This may reflect the ability of Mst1 to phosphorylate L-plastin and Mob1, which has been shown in other cell types [330,331]. L-plastin is an actin-bundling protein that stabilises lamellipodial F-actin [330], and Mob1 activates the Cdc42 GEF DOCK8 [331].

The role of Rap1 in promoting cell motility, integrin activation and immune synapse formation makes it an important regulator of B-cell function in vivo. In Rap1a/Rap1b double knockout mice, the trafficking of B cells into lymph nodes is dramatically impaired and very few B1a cells develop [325]. Similarly, RapL-deficient B cells exhibit impaired trafficking into and within lymph nodes [321,332]. Moreover, marginal zone B cells, which require integrin-mediated adhesion for their proper localisation, are absent in RapL knockout mice. The GC response is strongly dependent on B-cell motility, integrin-mediated interactions with follicular dendritic cells (FDCs) and T follicular helper cells and immune synapse formation. Consistent with a role for Rap1 in all of these processes, GC formation and antibody responses to T cell-dependent antigens are greatly impaired when B cells lack Rap1a and Rap1b [325]. As Rap1b is expressed at higher levels than Rap1a in B cells, similar defects in B-cell trafficking and function are observed in Rap1b knockout mice [333,334].

It is not clear how the BCR activates Rap1 and Rap2. There are multiple Rap-GEFs, two or which are members of the RasGRP family. RasGRP2 is a Rap-specific GEF, whereas RasGRP3 activates Rap as well as the Ras [335]. The RasGRP proteins are potent activators of Rap in B cells [267]. This suggests that one of the RasGRP GEFs links the BCR to Rap activation. RasGRP2 has been linked to Rap activation in a human B-cell chronic lymphocytic leukemia cell line [336]. C3G (also called RapGEF1) is a Rap-GEF that binds the adaptor protein CrkL, and in activated B cells, the CrkL/C3G complex interacts with a variety of scaffolding proteins such as Cbl and Cas [337]. However, a role for C3G in BCR-induced Rap activation or B-cell function has yet to be defined. Finally, in primary B cells lacking the Rac2 GTPase, BCR-induced activation of Rap1 is reduced [270], indicating that Rac2 may amplify Rap1 activation via an unknown mechanism.

15. BCR SIGNALLING TO THE CYTOSKELETON

Multiple BCR signalling pathways converge on the actin cytoskeleton. There are two main modes of actin polymerisation [297,338–340]. Formin proteins, which are activated by the RhoA GTPase, mediate linear actin polymerisation that generates thin membrane protrusions such as filopodia [341]. In contrast, branched actin polymerisation is initiated by the 7-subunit Arp2/3 complex and generates broad sheet-like lamellipodial protrusions. As shown in Figure 5.4E, the Arp2/3 complex is activated the Cdc42→WASp/N-WASp module and the Rac→WAVE module. WAVE2 is the most abundant WAVE isoform in immune cells and is part a 5-subunit complex. When activated by Cdc42-GTP or Rac-GTP, WASp, N-WASp and WAVE2 can induce a conformational change in the Arp2/3 complex, which nucleates a new actin filaments that grows at a 70 degree angle relative to the existing filament. Formins, as well as other elongation proteins such as VASP, then transfer ATP-bound actin monomers to the ends of uncapped growing filaments. Capping proteins limit the growth of actin filaments and also couple these 'barbed' ends of the filament to the plasma membrane so that they can exert outward force. Because the availability of ATP-actin monomers is limited, the growth of actin filaments at the cell membrane needs to be sustained by the depolymerisation of actin filaments and the recycling of actin monomers, which is referred to as 'actin treadmilling' [342]. Actin filaments have a built-in timer as the ATP bound to constituent actin monomers is hydrolysed to ADP. The actin-severing protein cofilin, together with its co-factor Wdr1 [343–345], binds to and disassembles older ADP-actin-rich filament segments that are further from the membrane. Cofilin activity is dynamically regulated by phosphorylation and dephosphorylation of S3. Dephosphorylation by serine/threonine phosphatases such as Slingshot (SSH1-3), PP1 or PP2A enables cofilin to bind and sever F-actin, whereas phosphorylation on S3 by LIM domain kinases (LIMK) inactivates cofilin [346,347]. Precise spatiotemporal regulation of cofilin activity determines where actin networks are expanded, remodelled or disassembled. Actin monomers that are released from filaments by cofilin are bound by profilin, which facilitates the exchange of ATP for ADP [348] so that ATP-actin monomers can be added to the ends of growing filaments.

16. IMMUNE SYNAPSE FORMATION AND FUNCTION

In vivo, B cells are often activated by APCs such as FDCs that reside within the B-cell follicles of lymphoid organs and subcapsular sinus macrophages that surround the follicles [349–352]. These APCs capture antigens via Fc receptors, complement receptors and cell surface lectins and retain them on their surface in an intact form. Recent work has shown that the B-cell follicles are a region of low proteolytic activity, which allows FDCs to present intact antigens to B cells [353]. By concentrating antigens on their surface, and presenting them as a multivalent array, APCs greatly increase the ability of B cells to be activated by small amounts of antigen [150]. Moreover, the actin-dependent spatial reorganisation of BCRs, integrins, coreceptors (e.g. CD19) and negative regulators (e.g. CD45, CD148) at the immune synapse (reviewed in Refs. [354–358]) amplifies BCR signalling and reduces the amount of antigen required for B-cell activation [142,150,358]. It should be noted that the actin-dependent amplification of BCR signalling during the first 5–15 min of the B cell-APC interaction have a profound effect on B cell activation responses that occur many hours later [142]. The efficacy of APC-induced B-cell activation is now being applied to vaccine development by coupling antigens to ligands for 'capture receptors' expressed on subcapsular macrophages or FDCs [359,360].

As reviewed previously, the immune synapse formation that occurs within minutes after B cell contact with an APC is driven by BCR-induced actin remodelling [354–358] (Figure 5.5). When B cells bind their cognate antigen on the surface of an APC, localised breakdown of the cortical actin cytoskeleton by cofilin, as well as uncoupling of the cytoskeleton from the membrane via ezrin dephosphorylation, increases BCR mobility and BCR–BCR collisions, leading to the formation of BCR microclusters that assemble signalosomes [26,42,144,147,157]. As actin filaments are depleted from the centre of the B cell-APC contact site, a peripheral ring of branched actin that is nucleated by the Arp2/3 complex forms at the cell periphery. Actin polymerisation at the cell periphery allows the B cell to extend actin-rich membrane protrusions that scan the APC surface for additional antigen molecules [142,361]. The extent to which the B cell spreads across an antigen-bearing membrane is dependent on the affinity of the BCR for the antigen, the density of the membrane-bound antigen and the strength of BCR signalling [361]. BCR microclusters form on these protrusions and are moved towards the cell body as the protrusions are retracted, a process driven by myosin-mediated contractile forces and by actin retrograde flow. Actin retrograde flow occurs when membrane-proximal actin polymerisation within branched actin networks exerts outward force on the membrane and the elastic resistance of the membrane exerts opposing forces that cause centripetal movement of peripheral actin structures. In addition to promoting the retraction of membrane protrusions, actin retrograde flow mediates the initial movement of BCR microclusters towards the centre of the B cell-APC contact site.

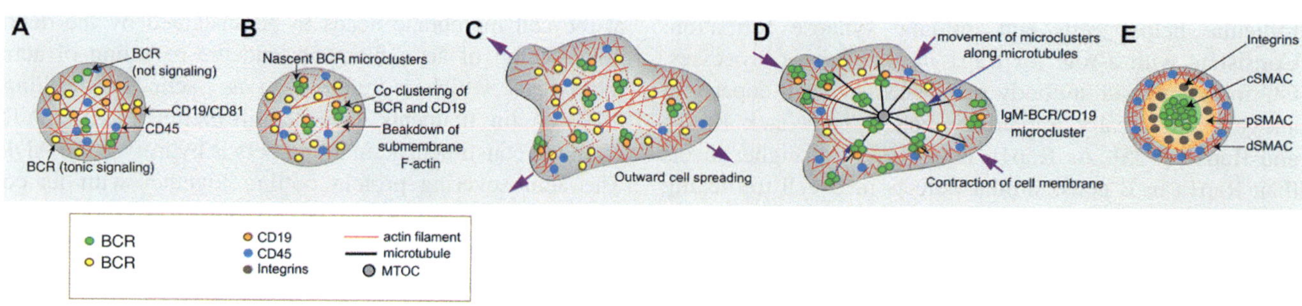

FIGURE 5.5 BCR-induced cytoskeletal reorganisation drives B-cell spreading and immune synapse formation. The diagrams show the organisation of the indicated membrane proteins, as well as cytoskeletal elements that are in close proximity to the plasma membrane, at the site where the B cell contacts the APC. (A) In resting B cells, the mobility of the BCR is restricted by the actin cytoskeleton such that only a small fraction of BCRs are signalling (*yellow circles*). (B) When B cells contact APCs displaying captured antigens on their surface, BCR signalling promotes localised disassembly of the submembrane actin cytoskeleton. This increases BCR mobility and leads to the formation of BCR microclusters that assemble signalosomes and interact with CD19. (C) F-actin clearance at the centre of the B cell:APC contact site is accompanied by actin polymerisation at the cell periphery. This drives B-cell spreading, allowing the BCR to encounter more antigen molecules on the surface of the APC and form additional microclusters. (D) As membrane protrusions are retracted, BCR microclusters move towards the centre of the B cell:APC contact site. This centripetal movement is driven first by actin retrograde flow, then by the sweeping motion of actomyosin arcs and finally by microtubule-based transport. During immune synapse formation, the microtubule network is reoriented towards the B cell:APC contact site and the microtubule-organising centre (MTOC) moves close to the plasma membrane. (E) In the final phase of immune synapse formation, the coalescence of BCR microclusters results in the formation of a cSMAC that is surrounded by clusters of activated integrins in the pSMAC. Integrin-mediated adhesion stabilises the B cell:APC interaction and increases B-cell spreading and microcluster formation. Membrane proteins with large extracellular domains (e.g. the CD45 and CD148 tyrosine phosphatases) are excluded from the central region of the B cell:APC contact site where the two cell membranes are closely apposed and accumulate at the distal SMAC (dSMAC), where they are segregated away from the BCR. *Adapted with permission from Abraham et al. [357].*

At the same time, BCR signalling activates the LFA-1 integrin in a Rap-dependent manner, converting it to its active confirmation [326]. The binding of LFA-1 to ICAM-1 on the APC strengthens the B cell–APC interaction and integrin signalling enhances B-cell spreading, which is important when the antigen density on the APC is low [362,363]. Activated LFA-1 undergoes actin-dependent clustering, and actin retrograde flow drives the centripetal movement of LFA-1 microclusters. At the inner face of the peripheral ring of branched actin, LFA-1 signalling stimulates formin-mediated linear actin polymerisation that leads to the formation of actin arcs that recruit myosin [364]. Contractile forces generated by these actoymyosin arcs sweep BCR microclusters further towards the centre of the B cell-APC contact site. In contrast, the centripetal movement of LFA-1 clusters, as well as CD19 clusters, is arrested at these actomyosin arcs, establishing the peripheral supramolecular activation cluster (pSMAC) that is characteristic of an immune synapse [56,363]. At the pSMAC, inwardly moving BCR microclusters interact transiently with CD19 microclusters, facilitating CD19 phosphorylation and the recruitment of PI3K and Vav to CD19, which amplifies BCR signalling and further enhances B-cell spreading [56] and immune synapse formation [270].

In contrast to the centripetal movement of BCR, LFA-1 and CD19 microclusters, proteins with large extracellular domains remain at the peripheral actin ring (also called the distal SMAC or dSMAC) because they extend too far from the cell surface to enter the central region of the B cell–APC contact site where the cells are in close apposition. This 'kinetic segregation' separates the inwardly moving BCR microclusters from the CD45 and CD148 transmembrane tyrosine phosphatases that are negative regulators of BCR signalling [150,365].

BCR microclusters are not arrested at the pSMAC and continue to undergo centripetal motion across the central actin-depleted region of the contact site until they form a large central SMAC (or cSMAC). This third phase of BCR microcluster centralisation is mediated by microtubules [366]. When a B cell contacts an antigen-bearing surface, the MTOC, from which the majority of microtubules originate, moves close to the contact site such that microtubules extend along the inner face of the plasma membrane [327,366,367]. Binding of the Grb2, Dok3 and Cbl adaptor proteins to BCR microclusters recruits dynein, which couples BCR microclusters to these juxtamembrane microtubules [366]. Processive movement of the dynein/dynactin motor complex towards the MTOC then drives the centralisation of BCR microclusters and their coalescence into a cSMAC [366].

At the immune synapse, microcluster-based BCR signalling is amplified by the combined effects of Arp2/3 complex-dependent actin retrograde flow [142,364] and the progressive coalescence of the microclusters [249,250]. Within microclusters, activated Syk that is associated with a BCR can phosphorylate and activate nearby BCRs. This may enable signal spreading by activating BCRs within the microcluster that are not bound to antigen or by reactivating BCRs in which the ITAMs have been dephosphorylated. As well, in larger microclusters a greater number of BCRs would be in the interior of the cluster and shielded from negative regulators. The BCR is also a mechanosensitive receptor, and actin retrograde flow could increase the mechanical tension on Ag-bound BCRs, which increases their ability to recruit Syk and other signalling effectors [368–370].

The actin and microtubule dynamics at the immune synapse also enable B cell to extract antigens from the surface of an APC and deliver them to MHC II-loading compartments. This allows B cells to present antigens to T cells and elicit T cell help. The antigen-presentation capability of B cells is especially important during the GC reaction where antigen-activated B cells compete to receive survival and differentiation signals from T follicular helper cells.

B cells are very effective at extracting antigens from surfaces (e.g. polystyrene beads), lipid bilayers and APCs [150,367,371,372]. When BCRs engage surface-bound antigens, the activated form of myosin II in which the myosin LC is phosphorylated accumulates near the BCR [373]. Forces exerted on the BCR by actomyosin contractility then facilitate antigen extraction [373–378]. If the antigen is non-covalently bound to a capture receptor on the APC, these forces can overcome the affinity of the antigen for the capture receptor, resulting in the antigen being transferred to the BCR [375,379]. When these forces are insufficient to extract the antigen, B cells secrete lysosomal contents into the synapse between the B cell and the APC [367]. Maeda et al. showed that myosin II activity at the site where the BCR has bound antigen results in transient membrane permeabilisation and an influx of Ca^{2+} that stimulates lysosomal exocytosis [373], a well-characterised membrane repair response [380]. Within the synaptic space between the B cell and the APC, the actions of lysosomal proteases can release the antigen from the APC surface so that it can be bound by the BCR [381]. BCR-bound antigens are then internalised via receptor-mediated endocytosis and delivered to MHC II loading compartments in which antigen-derived peptides are loaded onto MHC II molecules [372,381–383]. Lysosomes and other vesicles are often associated with microtubules. Hence, the BCR-induced polarisation and movement of the microtubule network towards the immune synapse positions lysosomes close to the immune synapse where they are poised for exocytosis [327,367]. The microtubule network also brings MHC II-containing vesicles close to the immune synapse in order to facilitate formation of the multivesicular MHC II loading compartment and increase the efficiency of antigen presentation [367,381,384]. The movement of the MTOC and the microtubule network towards the immune synapse depends on coupling of microtubule plus ends to the peripheral actin network by the IQGAP1 scaffolding protein as well as dynein activity and proteasome-mediated degradation of the actin filaments that link the MTOC to the nucleus and prevent its polarisation [327,385].

Immune synapse formation enhances the two major functions of the BCR, signal transduction and antigen presentation to T helper cells. The dynamic cytoskeletal reorganisation that drives immune synapse formation involves the coordinated function of a large network of proteins [386]. In addition to BCR signalosome components, integrins and the CD19 coreceptor, loss-of-function approaches and human mutations have shown that B-cell immune synapse formation and antigen extraction require (a) GTPases (Rap1, Rac2 and Cdc42) and their activators (Vav and DOCK8) [270,271,296,302,308,309], (b) NPFs and their interactors (WASp, WIP, the Hem-1 component of the WAVE-regulatory complex) [300,301,387,388], (c) actin-nucleation and polymerisation proteins (the Arp2/3 complex and formins) [142,364], (d) actin disassembly proteins and their regulators that allow continual remodelling of actin networks (cofilin, Wdr1, LIMK and the actin-debranching protein GMFγ) [389,390], (e) non-muscle myosins that generate actomyosin contractile forces [364,391] and (f) proteins that link BCR microclusters to microtubules and mediate the polarisation and movement of the microtubule network towards the immune synapse [327,366]. Hence, it is not surprising that human loss-of-function mutations in DOCK8, WASp, WIP, Hem-1, the Arpc1B component of the Arp2/3 complex and the cofilin co-factor Wdr1 result in immune dysfunction syndromes that are associated with defective or aberrant B-cell development or activation [392–394]. Highlighting the key role of actin in B-cell function, these immune dysfunction syndromes have been termed 'actinopathies' [393].

ACKNOWLEDGEMENTS

The authors thank Lise Leclercq for critical reading of this manuscript and Libin Abraham for Figure 5.5. This study was supported by the German Science Foundation through (CIBSS-EXC-2189) (MR), by TRR130 (P02 and P04) (MR and LN), and by grants from the Canadian Institutes of Health Research and the Natural Sciences and Engineering Research Council (Canada) (MRG).

REFERENCES

[1] Neuberger MS, Patel KJ, Dariavach P, Nelms K, Peaker CJ, Williams GT. The mouse B-cell antigen receptor: definition and assembly of the core receptor of the five immunoglobulin isotypes. Immunol Rev April 1993;132:147–61.

[2] Reth M. Antigen receptors on B lymphocytes. cited 2015 July 19 Annu Rev Immunol January 1992;10:97–121. Available from: http://www.ncbi.nlm.nih.gov/pubmed/1591006.

[3] Kurosaki T. Molecular dissection of B cell antigen receptor signaling (review). Int J Mol Med March 1998;1(3):515–27.

[4] Wienands J, Engels N. Multitasking of Ig-alpha and Ig-beta to regulate B cell antigen receptor function. Int Rev Immunol 2001;20(6):679–96.

[5] Hombach J, Tsubata T, Leclercq L, Stappert H, Reth M. Molecular components of the B-cell antigen receptor complex of the IgM class. Nature February 1990;343(6260):760–2.

[6] Schamel WW, Reth M. Stability of the B cell antigen receptor complex. Mol Immunol April 2000;37(5):253–9.

[7] Schamel WW, Reth M. Monomeric and oligomeric complexes of the B cell antigen receptor. Immunity July 2000;13(1):5–14.

[8] Dong Y, Pi X, Bartels-Burgahn F, Saltukoglu D, Liang Z, Yang J, et al. Structural principles of B cell antigen receptor assembly. Nature December 2022;612(7938):156–61.

[9] Su Q, Chen M, Shi Y, Zhang X, Huang G, Huang B, et al. Cryo-EM structure of the human IgM B cell receptor. Science August 2022;377(6608):875–80.

[10] Ma X, Zhu Y, Dong D, Chen Y, Wang S, Yang D, et al. Cryo-EM structures of two human B cell receptor isotypes. Science August 2022;377(6608):880–5.

[11] Gottwick C, He X, Hofmann A, Vesper N, Reth M, Yang J. A symmetric geometry of transmembrane domains inside the B cell antigen receptor complex. Proc Natl Acad Sci U S A July 2019;116(27):13468–73.

[12] Yang J, Reth M. Oligomeric organization of the B-cell antigen receptor on resting cells. cited 2014 September 25 Nature September 23, 2010;467(7314):465–9. Available from: http://www.ncbi.nlm.nih.gov/pubmed/20818374.

[13] Reth M. Antigen receptor tail clue. cited 2015 July 7 Nature March 30, 1989;338(6214):383–4. Available from: http://www.ncbi.nlm.nih.gov/pubmed/2927501.

[14] Cambier JC. Antigen and Fc receptor signaling. The awesome power of the immunoreceptor tyrosine-based activation motif (ITAM). J Immunol October 1995;155(7):3281–5.

[15] Cambier JC. New nomenclature for the Reth motif (or ARH1/TAM/ARAM/YXXL). Immunol Today 1995;16:110.

[16] Johnson SA, Pleiman CM, Pao L, Schneringer J, Hippen K, Cambier JC. Phosphorylated immunoreceptor signaling motifs (ITAMs) exhibit unique abilities to bind and activate Lyn and Syk tyrosine kinases. J Immunol November 1995;155(10):4596–603.

[17] Singer SJ, Nicolson GL. The fluid mosaic model of the structure of cell membranes. cited 2015 April 8 Science February 18, 1972;175(4023):720–31. Available from: http://www.ncbi.nlm.nih.gov/pubmed/4333397.

[18] Kusumi A, Nakada C, Ritchie K, Murase K, Suzuki K, Murakoshi H, et al. Paradigm shift of the plasma membrane concept from the two-dimensional continuum fluid to the partitioned fluid: high-speed single-molecule tracking of membrane molecules. Annu Rev Biophys Biomol Struct 2005;34:351–78.

[19] Sako Y, Kusumi A. Compartmentalized structure of the plasma membrane for receptor movements as revealed by a nanometer-level motion analysis. J Cell Biol June 1994;125(6):1251–64.

[20] Simons K, Ikonen E. Functional rafts in cell membranes. cited 2014 July 12 Nature June 5, 1997;387(6633):569–72. Available from: http://www.ncbi.nlm.nih.gov/pubmed/9177342.

[21] Simons K, Sampaio JL. Membrane organization and lipid rafts. cited 2014 December 31 Cold Spring Harb Perspect Biol October 1, 2011;3(10):a004697. Available from: http://cshperspectives.cshlp.org/content/3/10/a004697.full#ref-76.

[22] Wilson BS, Pfeiffer JR, Oliver JM. Observing FcepsilonRI signaling from the inside of the mast cell membrane. J Cell Biol May 2000;149(5):1131–42.

[23] Lillemeier BF, Mörtelmaier MA, Forstner MB, Huppa JB, Groves JT, Davis MM. TCR and Lat are expressed on separate protein islands on T cell membranes and concatenate during activation. Nat Immunol January 2010;11(1):90–6.

[24] Lillemeier BF, Pfeiffer JR, Surviladze Z, Wilson BS, Davis MM. Plasma membrane-associated proteins are clustered into islands attached to the cytoskeleton. cited 2015 October 16 Proc Natl Acad Sci U S A December 4, 2006;103(50):18992–7. Available from: http://www.pnas.org/content/103/50/18992.long.

[25] Yang J, Reth M. The dissociation activation model of B cell antigen receptor triggering. cited 2015 July 18 FEBS Lett December 15, 2010;584(24):4872–7. Available from: http://www.ncbi.nlm.nih.gov/pubmed/20920502.

[26] Mattila PK, Feest C, Depoil D, Treanor B, Montaner B, Otipoby KL, et al. The actin and tetraspanin networks organize receptor nanoclusters to regulate B cell receptor-mediated signaling. Immunity 2013;38(3):461–74. Available from: http://doi.org/10.1016/j.immuni.>2012.11.019.

[27] Kläsener K, Yang J, Reth M. Study B cell antigen receptor nanoscale organization by in situ fab proximity ligation assay. Methods Mol Biol 2018;1707:171–81.

[28] Becker M, Hobeika E, Jumaa H, Reth M, Maity PC. CXCR4 signaling and function require the expression of the IgD-class B-cell antigen receptor. cited 2020 May 21 Proc Natl Acad Sci U S A May 1, 2017. Available from: https://www.pnas.org/content/early/2017/04/27/1621512114.

[29] Kläsener K, Jellusova J, Andrieux G, Salzer U, Böhler C, Steiner SN, et al. CD20 as a gatekeeper of the resting state of human B cells. cited 2021 February 11 Proc Natl Acad Sci U S A February 16, 2021;118(7). Available from: http://www.ncbi.nlm.nih.gov/pubmed/33563755.

[30] Semac I, Palomba C, Kulangara K, Klages N, van Echten-Deckert G, Borisch B, et al. Anti-CD20 therapeutic antibody rituximab modifies the functional organization of rafts/microdomains of B lymphoma cells. Cancer Res January 2003;63(2):534–40.

[31] Rougé L, Chiang N, Steffek M, Kugel C, Croll TI, Tam C, et al. Structure of CD20 in complex with the therapeutic monoclonal antibody rituximab. Science 2020;367(6483):1224–30.

[32] Klasener K, Maity PC, Hobeika E, Yang J, Reth M. B cell activation involves nanoscale receptor reorganizations and inside-out signaling by Syk. Elife June 2014;3:e02069.

[33] Dunlock V-ME, Arp AB, Singh SP, Charrin S, Nguyen V, Jansen E, et al. Tetraspanin CD53 controls T cell immunity through regulation of CD45RO stability, mobility, and function. Cell Rep June 2022;39(13):111006.

[34] Susa KJ, Rawson S, Kruse AC, Blacklow SC. Cryo-EM structure of the B cell co-receptor CD19 bound to the tetraspanin CD81. Science January 2021;371(6526):300–5.

[35] Garcia-Parajo MF, Cambi A, Torreno-Pina JA, Thompson N, Jacobson K. Nanoclustering as a dominant feature of plasma membrane organization. J Cell Sci December 2014;127(Pt 23):4995–5005.

[36] Termini CM, Gillette JM. Tetraspanins function as regulators of cellular signaling. Front Cell Dev Biol 2017;5(Apr):1–14.

[37] Yáñez-Mó M, Barreiro O, Gordon-Alonso M, Sala-Valdés M, Sánchez-Madrid F. Tetraspanin-enriched microdomains: a functional unit in cell plasma membranes. Trends Cell Biol September 2009;19(9):434–46.

[38] Infantino S, Benz B, Waldmann T, Jung M, Schneider R, Reth M. Arginine methylation of the B cell antigen receptor promotes differentiation. J Exp Med April 2010;207(4):711–9.

[39] Nicholson TB, Chen T, Richard S. The physiological and pathophysiological role of PRMT1-mediated protein arginine methylation. Pharmacol Res December 2009;60(6):466–74.

[40] Polliack A, Gamliel H, Ben Bassat H, Gurfel D, Leizerowitz R, Minowada J. Surface morphology and membrane phenotype of cultured human leukemia-lymphoma cells. A scanning electron microscopic study of 36 cell lines. Cancer January 1983;51(1):72–9.

[41] Saltukoglu D, Özdemir B, Holtmannspötter M, Reski R, Piehler J, Kurre R, et al. Plasma membrane topography governs the 3D dynamic localization of IgM B cell antigen receptor clusters. EMBO J February 2023;42(4):e112030.

[42] Treanor B, Depoil D, Gonzalez-Granja A, Barral P, Weber M, Dushek O, et al. The membrane skeleton controls diffusion dynamics and signaling through the B cell receptor. Immunity February 2010;32(2):187–99.

[43] Li J, Yin W, Jing Y, Kang D, Yang L, Cheng J, et al. The coordination between B cell receptor signaling and the actin cytoskeleton during B cell activation. Front Immunol 2019;9:3096. Available from: https://www.frontiersin.org/article/10.3389/fimmu.2018.03096.

[44] Metzger H. Transmembrane signaling: the joy of aggregation. J Immunol September 1992;149(5):1477–87.

[45] Jung Y, Wen L, Altman A, Ley K. CD45 pre-exclusion from the tips of T cell microvilli prior to antigen recognition. Nat Commun 2021;12(1):3872. Available from: https://doi.org/10.1038/s41467-021-23792-8.

[46] Susa KJ, Seegar TCM, Blacklow SC, Kruse AC. A dynamic interaction between cd19 and the tetraspanin CD81 controls B cell co-receptor trafficking. Elife 2020;9:1–20.

[47] Del Nagro CJ, Otero DC, Anzelon AN, Omori SA, Kolla RV, Rickert RC. CD19 function in central and peripheral B-cell development. cited 2015 July 9 Immunol Res January 2005;31(2):119–31. Available from: http://www.ncbi.nlm.nih.gov/pubmed/15778510.

[48] Meyer SJ, Linder AT, Brandl C, Nitschke L. B cell Siglecs—News on signaling and its interplay with ligand binding. Front Immunol 2018;9. Available from: https://www.frontiersin.org/articles/10.3389/fimmu.2018.02820.

[49] Fearon DT, Carroll MC. Regulation of B lymphocyte responses to foreign and self-antigens by the CD19/CD21 complex. Annu Rev Immunol 2000;18:393–422.

[50] Carter RH, Fearon DT. CD19: lowering the threshold for antigen receptor stimulation of B lymphocytes. Science April 1992;256(5053):105–7.

[51] Engel P, Zhou LJ, Ord DC, Sato S, Koller B, Tedder TF. Abnormal B lymphocyte development, activation, and differentiation in mice

that lack or overexpress the CD19 signal transduction molecule. Immunity July 1995;3(1):39–50.

[52] Rickert RC, Rajewsky K, Roes J. Impairment of T-cell-dependent B-cell responses and B-1 cell development in CD19-deficient mice. Nature July 1995;376(6538):352–5.

[53] Fehr T, Rickert RC, Odermatt B, Roes J, Rajewsky K, Hengartner H, et al. Antiviral protection and germinal center formation, but impaired B cell memory in the absence of CD19. J Exp Med July 1998;188(1):145–55.

[54] Ahearn JM, Fischer MB, Croix D, Goerg S, Ma M, Xia J, et al. Disruption of the Cr2 locus results in a reduction in B-1a cells and in an impaired B cell response to T-dependent antigen. Immunity March 1996;4(3):251–62.

[55] Otero DC, Rickert RC. CD19 function in early and late B cell development. II. CD19 facilitates the pro-B/pre-B transition. J Immunol December 2003;171(11):5921–30.

[56] Depoil D, Fleire S, Treanor BL, Weber M, Harwood NE, Marchbank KL, et al. CD19 is essential for B cell activation by promoting B cell receptor-antigen microcluster formation in response to membrane-bound ligand. cited 2015 May 5 Nat Immunol January 2008;9(1):63–72. Available from: http://www.ncbi.nlm.nih.gov/pubmed/18059271.

[57] Lee J, Robinson ME, Ma N, Artadji D, Ahmed MA, Xiao G, et al. IFITM3 functions as a PIP3 scaffold to amplify PI3K signalling in B cells. Nature 2020;588(7838):491–7.

[58] Kläsener K, Maity PC, Hobeika E, Yang J, Reth M, Aiba Y, et al. B cell activation involves nanoscale receptor reorganizations and inside-out signaling by Syk. cited 2017 January 31 Elife June 24, 2014;3:e02069. Available from: http://www.ncbi.nlm.nih.gov/pubmed/24963139.

[59] Gold MR, Reth MG. Antigen receptor function in the context of the nanoscale organization of the B cell membrane. Annu Rev Immunol April 2019;37(1):97–123.

[60] Brooks SR, Li X, Volanakis EJ, Carter RH. Systematic analysis of the role of CD19 cytoplasmic tyrosines in enhancement of activation in Daudi human B cells: clustering of phospholipase C and Vav and of Grb2 and Sos with different CD19 tyrosines. J Immunol March 2000;164(6):3123–31.

[61] Buhl AM, Cambier JC. Phosphorylation of CD19 Y484 and Y515, and linked activation of phosphatidylinositol 3-kinase, are required for B cell antigen receptor-mediated activation of Bruton's tyrosine kinase. J Immunol April 1999;162(8):4438–46.

[62] Fujimoto M, Poe JC, Hasegawa M, Tedder TF. CD19 amplification of B lymphocyte Ca2+ responses: a role for Lyn sequestration in extinguishing negative regulation. J Biol Chem November 2001;276(48):44820–7.

[63] Xu Y, Beavitt S-JE, Harder KW, Hibbs ML, Tarlinton DM. The activation and subsequent regulatory roles of Lyn and CD19 after B cell receptor ligation are independent. cited 2015 June 9 J Immunol December 15, 2002;169(12):6910–8. Available from: http://www.ncbi.nlm.nih.gov/pubmed/12471124.

[64] Wang Y, Brooks SR, Li X, Anzelon AN, Rickert RC, Carter RH. The physiologic role of CD19 cytoplasmic tyrosines. cited 2015 July 13 Immunity October 2002;17(4):501–14. Available from: http://www.ncbi.nlm.nih.gov/pubmed/12387743.

[65] Wang Y, Carter RH. CD19 regulates B cell maturation, proliferation, and positive selection in the FDC zone of murine splenic germinal centers. Immunity June 2005;22(6):749–61.

[66] Hibbs ML, Tarlinton DM, Armes J, Grail D, Hodgson G, Maglitto R, et al. Multiple defects in the immune system of Lyn-deficient mice, culminating in autoimmune disease. Cell October 1995;83(2):301–11.

[67] Buhl AM, Pleiman CM, Rickert RC, Cambier JC. Qualitative regulation of B cell antigen receptor signaling by CD19: selective requirement for PI3-kinase activation, inositol-1,4,5-trisphosphate production and Ca2+ mobilization. J Exp Med 1997;186(11):1897–910.

[68] Otero DC, Omori SA, Rickert RC. Cd19-dependent activation of Akt kinase in B-lymphocytes. cited 2015 June 18 J Biol Chem January 12, 2001;276(2):1474–8. Available from: http://www.ncbi.nlm.nih.gov/pubmed/11042164.

[69] Clayton E, Bardi G, Bell SE, Chantry D, Downes CP, Gray A, et al. A crucial role for the p110delta subunit of phosphatidylinositol 3-kinase in B cell development and activation. J Exp Med September 2002;196(6):753–63.

[70] Fruman DA, Snapper SB, Yballe CM, Davidson L, Yu JY, Alt FW, et al. Impaired B cell development and proliferation in absence of phosphoinositide 3-kinase p85alpha. Science January 1999;283(5400):393–7.

[71] Jou ST, Carpino N, Takahashi Y, Piekorz R, Chao JR, Carpino N, et al. Essential, nonredundant role for the phosphoinositide 3-kinase p110delta in signaling by the B-cell receptor complex. Mol Cell Biol December 2002;22(24):8580–91.

[72] Okkenhaug K, Bilancio A, Farjot G, Priddle H, Sancho S, Peskett E, et al. Impaired B and T cell antigen receptor signaling in p110delta PI 3-kinase mutant mice. Science August 2002;297(5583):1031–4.

[73] Anzelon AN, Wu H, Rickert RC. Pten inactivation alters peripheral B lymphocyte fate and reconstitutes CD19 function. Nat Immunol March 2003;4(3):287–94.

[74] Tedford K, Nitschke L, Girkontaite I, Charlesworth A, Chan G, Sakk V, et al. Compensation between Vav-1 and Vav-2 in B cell development and antigen receptor signaling. Nat Immunol June 2001;2(6):548–55.

[75] van Zelm MC, Reisli I, van der Burg M, Castaño D, van Noesel CJM, van Tol MJD, et al. An antibody-deficiency syndrome due to mutations in the CD19 gene. N Engl J Med May 4, 2006;354(18):1901–12. Available from: http://www.ncbi.nlm.nih.gov/pubmed/16672701.

[76] Van Zelm MC, Smet J, Adams B, Mascart F, Schandené L, Janssen F, et al. CD81 gene defect in humans disrupts CD19 complex formation and leads to antibody deficiency. J Clin Invest 2010;120(4):1265–74.

[77] Smith KGC, Tarlinton DM, Doody GM, Hibbs ML, Fearon DT. Inhibition of the B cell by CD22: a requirement for Lyn. J Exp Med 1998;187(5):807–11.

[78] Chan VW, Lowell CA, DeFranco AL. Defective negative regulation of antigen receptor signaling in Lyn-deficient B lymphocytes. Curr Biol May 1998;8(10):545–53.

[79] Doody GM, Justement LB, Delibrias CC, Matthews RJ, Lin J, Thomas ML, et al. A role in B cell activation for CD22 and the protein tyrosine phosphatase SHP. Science July 1995;269(5221):242–4.

[80] Chen J, McLean PA, Neel BG, Okunade G, Shull GE, Wortis HH. CD22 attenuates calcium signaling by potentiating plasma membrane calcium-ATPase activity. Nat Immunol June 2004;5(6):651–7.

[81] Nadler MJ, McLean PA, Neel BG, Wortis HH. B cell antigen receptor-evoked calcium influx is enhanced in CD22-deficient B cell lines. J Immunol November 1997;159(9):4233–43.

[82] Nitschke L, Carsetti R, Ocker B, Köhler G, Lamers MC. CD22 is a negative regulator of B-cell receptor signalling. Curr Biol February 1997;7(2):133–43.

[83] O'Keefe TL, Williams GT, Davies SL, Neuberger MS. Hyperresponsive B cells in CD22-deficient mice. Science November 1996;274(5288):798–801.

[84] Otipoby KL, Andersson KB, Draves KE, Klaus SJ, Farr AG, Kerner JD, et al. CD22 regulates thymus-independent responses and the lifespan of B cells. Nature December 1996;384(6610):634–7.

[85] Sato S, Miller AS, Inaoki M, Bock CB, Jansen PJ, Tang ML, et al. CD22 is both a positive and negative regulator of B lymphocyte antigen receptor signal transduction: altered signaling in CD22-deficient mice. Immunity December 1996;5(6):551–62.

[86] Müller J, Obermeier I, Wöhner M, Brandl C, Mrotzek S, Angermüller S, et al. CD22 ligand-binding and signaling domains reciprocally regulate B-cell Ca2+ signaling. cited 2014 October 17 Proc Natl Acad Sci U S A July 23, 2013;110(30):12402–7. Available from: http://www.pubmedcentral.nih.gov/articlerender.fcgi?artid=3725054&tool=pmcentrez&rendertype=abstract.

[87] Gerlach J, Ghosh S, Jumaa H, Reth M, Wienands J, Chan AC, et al. B cell defects in SLP65/BLNK-deficient mice can be partially corrected by the absence of CD22, an inhibitory coreceptor for BCR signaling. Eur J Immunol December 2003;33(12):3418–26.

[88] Sato S, Jansen PJ, Tedder TF. CD19 and CD22 expression reciprocally regulates tyrosine phosphorylation of Vav protein during B lymphocyte signaling. Proc Natl Acad Sci U S A November 1997;94(24):13158–62.

[89] Chen J, Wang H, Xu W-P, Wei S-S, Li HJ, Mei Y-Q, et al. Besides an ITIM/SHP-1-dependent pathway, CD22 collaborates with Grb2 and plasma membrane calcium-ATPase in an ITIM/SHP-1-independent pathway of attenuation of Ca2+i signal in B cells. Oncotarget August 2016;7(35):56129–46.

[90] Ackermann JA, Radtke D, Maurberger A, Winkler TH, Nitschke L. Grb2 regulates B-cell maturation, B-cell memory responses and inhibits B-cell Ca2+ signalling. EMBO J April 2011;30(8):1621–33.

[91] Ereño-Orbea J, Sicard T, Cui H, Mazhab-Jafari MT, Benlekbir S, Guarné A, et al. Molecular basis of human CD22 function and therapeutic targeting. Nat Commun October 2017;8(1):764.

[92] Danzer C-P, Collins BE, Blixt O, Paulson JC, Nitschke L. Transitional and marginal zone B cells have a high proportion of unmasked CD22: implications for BCR signaling. Int Immunol October 2003;15(10):1137–47.

[93] Kelm S, Gerlach J, Brossmer R, Danzer C-P, Nitschke L. The ligand-binding domain of CD22 is needed for inhibition of the B cell receptor signal, as demonstrated by a novel human CD22-specific inhibitor compound. J Exp Med May 2002;195(9):1207–13.

[94] Collins BE, Smith BA, Bengtson P, Paulson JC. Ablation of CD22 in ligand-deficient mice restores B cell receptor signaling. cited 2015 September 11 Nat Immunol February 2006;7(2):199–206. Available from: http://www.ncbi.nlm.nih.gov/pubmed/16369536.

[95] Han S, Collins BE, Bengtson P, Paulson JC. Homomultimeric complexes of CD22 in B cells revealed by protein-glycan cross-linking. Nat Chem Biol July 2005;1(2):93–7.

[96] Gasparrini F, Feest C, Bruckbauer A, Mattila PK, Muller J, Nitschke L, et al. Nanoscale organization and dynamics of the siglec CD22 cooperate with the cytoskeleton in restraining BCR signalling. EMBO J February 2016;35(3):258–80.

[97] Ghosh S, Bandulet C, Nitschke L. Regulation of B cell development and B cell signalling by CD22 and its ligands alpha2,6-linked sialic acids. cited 2015 July 19 Int Immunol April 2006;18(4):603–11. Available from: http://www.ncbi.nlm.nih.gov/pubmed/16497829.

[98] Coughlin S, Noviski M, Mueller JL, Chuwonpad A, Raschke WC, Weiss A, et al. An extracatalytic function of CD45 in B cells is mediated by CD22. Proc Natl Acad Sci U S A November 2015;112(47):E6515–24.

[99] Akatsu C, Alborzian Deh Sheikh A, Matsubara N, Takematsu H, Schweizer A, Abdu-Allah HHM, et al. The inhibitory coreceptor CD22 restores B cell signaling by developmentally regulating Cd45(-/-) immunodeficient B cells. Sci Signal March 2022;15(723):eabf9570.

[100] Cao A, Alluqmani N, Buhari FHM, Wasim L, Smith LK, Quaile AT, et al. Galectin-9 binds IgM-BCR to regulate B cell signaling. Nat Commun August 2018;9(1):3288.

[101] Giovannone N, Antonopoulos A, Liang J, Geddes Sweeney J, Kudelka MR, King SL, et al. Human B cell differentiation is characterized by progressive remodeling of O-linked glycans. Front Immunol 2018;9(December):2857.

[102] Collins BE, Blixt O, DeSieno AR, Bovin N, Marth JD, Paulson JC. Masking of CD22 by cis ligands does not prevent redistribution of CD22 to sites of cell contact. Proc Natl Acad Sci U S A 2004;101(16):6104–9.

[103] Lanoue A, Batista FD, Stewart M, Neuberger MS. Interaction of CD22 with alpha2,6-linked sialoglycoconjugates: innate recognition of self to dampen B cell autoreactivity? Eur J Immunol February 2002;32(2):348–55.

[104] Macauley MS, Kawasaki N, Peng W, Wang SH, He Y, Arlian BM, et al. Unmasking of CD22 co-receptor on germinal center B-cells occurs by alternative mechanisms in mouse and man. J Biol Chem 2015;290(50):30066–77.

[105] Naito Y, Takematsu H, Koyama S, Miyake S, Yamamoto H, Fujinawa R, et al. Germinal center marker GL7 probes activation-dependent repression of N-glycolylneuraminic acid, a sialic acid species involved in the negative modulation of B-cell activation. Mol Cell Biol April 2007;27(8):3008–22.

[106] Enterina JR, Sarkar S, Streith L, Jung J, Arlian BM, Meyer SJ, et al. Coordinated changes in glycosylation regulate the germinal center through CD22. Cell Rep March 2022;38(11):110512.

[107] Meyer SJ, Steffensen M, Acs A, Weisenburger T, Wadewitz C, Winkler TH, et al. CD22 controls germinal center B cell receptor signaling, which influences plasma cell and memory B cell output. J Immunol August 2021;207(4):1018–32.

[108] Ballet R, Brennan M, Brandl C, Feng N, Berri J, Cheng J, et al. A CD22-Shp1 phosphatase axis controls integrin β(7) display and B cell function in mucosal immunity. Nat Immunol March 2021;22(3):381–90.

[109] Fernandes VE, Ercoli G, Bénard A, Brandl C, Fahnenstiel H, Müller-Winkler J, et al. The B-cell inhibitory receptor CD22 is a major factor in host resistance to Streptococcus pneumoniae infection. PLoS Pathog April 2020;16(4):e1008464.

[110] Bolland S, Ravetch JV. Spontaneous autoimmune disease in Fc(gamma)RIIB-deficient mice results from strain-specific epistasis. Immunity August 2000;13(2):277–85.

[111] Pao LI, Lam KP, Henderson JM, Kutok JL, Alimzhanov M, Nitschke L, et al. B cell-specific deletion of protein-tyrosine phosphatase Shp1 promotes B-1a cell development and causes systemic autoimmunity. Immunity 2007;27:35–48.

[112] Mary C, Laporte C, Parzy D, Santiago ML, Stefani F, Lajaunias F, et al. Dysregulated expression of the Cd22 gene as a result of a short interspersed nucleotide element insertion in Cd22a lupus-prone mice. J Immunol September 2000;165(6):2987–96.

[113] O'Keefe TL, Williams GT, Batista FD, Neuberger MS. Deficiency in CD22, a B cell-specific inhibitory receptor, is sufficient to predispose to development of high affinity autoantibodies. J Exp Med April 1999;189(8):1307–13.

[114] Lajaunias F, Nitschke L, Moll T, Martinez-Soria E, Semac I, Chicheportiche Y, et al. Differentially regulated expression and function of CD22 in activated B-1 and B-2 lymphocytes. J Immunol 2002;168(12):6078–83.

[115] Jellusova J, Wellmann U, Amann K, Winkler TH, Nitschke L. CD22 x Siglec-G double-deficient mice have massively increased B1 cell numbers and develop systemic autoimmunity. J Immunol April 2010;184(7):3618–27.

[116] Özgör L, Meyer SJ, Korn M, Terörde K, Nitschke L. Sialic acid ligand binding of CD22 and siglec-G determines distinct B cell functions but is dispensable for B cell tolerance induction. J Immunol October 2018;201(7):2107–16.

[117] Horton HM, Bernett MJ, Pong E, Peipp M, Karki S, Chu SY, et al. Potent in vitro and in vivo activity of an Fc-engineered anti-CD19 monoclonal antibody against lymphoma and leukemia. Cancer Res October 2008;68(19):8049–57.

[118] Baeuerle PA, Reinhardt C. Bispecific T-cell engaging antibodies for cancer therapy. Cancer Res June 2009;69(12):4941–4.

[119] Zammarchi F, Corbett S, Adams L, Tyrer PC, Kiakos K, Janghra N, et al. ADCT-402, a PBD dimer-containing antibody drug conjugate targeting CD19-expressing malignancies. Blood March 2018;131(10):1094–105.

[120] Porter DL, Hwang W-T, Frey NV, Lacey SF, Shaw PA, Loren AW, et al. Chimeric antigen receptor T cells persist and induce sustained remissions in relapsed refractory chronic lymphocytic leukemia. Sci Transl Med September 2015;7(303):303ra139.

[121] Mackensen A, Müller F, Mougiakakos D, Böltz S, Wilhelm A, Aigner M, et al. Anti-CD19 CAR T cell therapy for refractory systemic lupus erythematosus. Nat Med October 2022;28(10):2124–32.

[122] Clowse MEB, Wallace DJ, Furie RA, Petri MA, Pike MC, Leszczyński P, et al. Efficacy and safety of epratuzumab in moderately to severely active systemic lupus erythematosus: results from two phase III randomized, double-blind, placebo-controlled trials. Arthritis Rheumatol February 2017;69(2):362–75.

[123] Kantarjian H, Thomas D, Jorgensen J, Jabbour E, Kebriaei P, Rytting M, et al. Inotuzumab ozogamicin, an anti-CD22-calecheamicin conjugate, for refractory and relapsed acute lymphocytic leukaemia: a phase 2 study. Lancet Oncol April 2012;13(4):403–11.

[124] Abdu-Allah HHM, Watanabe K, Completo GC, Sadagopan M, Hayashizaki K, Takaku C, et al. CD22-antagonists with nanomolar potency: the synergistic effect of hydrophobic groups at C-2 and C-9 of sialic acid scaffold. Bioorg Med Chem March 2011;19(6):1966–71.

[125] Prescher H, Schweizer A, Kuhfeldt E, Nitschke L, Brossmer R. Discovery of multifold modified sialosides as human CD22/Siglec-2 ligands with nanomolar activity on B-cells. ACS Chem Biol July 2014;9(7):1444–50.

[126] Macauley MS, Pfrengle F, Rademacher C, Nycholat CM, Gale AJ, von Drygalski A, et al. Antigenic liposomes displaying CD22 ligands induce antigen-specific B cell apoptosis. J Clin Invest July 2013;123(7):3074–83.

[127] Schweizer A, Wöhner M, Prescher H, Brossmer R, Nitschke L. Targeting of CD22-positive B-cell lymphoma cells by synthetic divalent sialic acid analogues. Eur J Immunol October 2012;42(10):2792–802.

[128] Fry TJ, Shah NN, Orentas RJ, Stetler-Stevenson M, Yuan CM, Ramakrishna S, et al. CD22-targeted CAR T cells induce remission in B-ALL that is naive or resistant to CD19-targeted CAR immunotherapy. Nat Med January 2018;24(1):20–8.

[129] Pao LI, Famiglietti SJ, Cambier JC. Asymmetrical phosphorylation and function of immunoreceptor tyrosine-based activation motif tyrosines in B cell antigen receptor signal transduction. J Immunol April 1998;160(7):3305–14.

[130] Rowley RB, Burkhardt AL, Chao HG, Matsueda GR, Bolen JB. Syk protein-tyrosine kinase is regulated by tyrosine-phosphorylated Ig alpha/Ig beta immunoreceptor tyrosine activation motif binding and autophosphorylation. J Biol Chem May 1995;270(19):11590–4.

[131] Kurosaki T, Johnson SA, Pao L, Sada K, Yamamura H, Cambier JC. Role of the Syk autophosphorylation site and SH2 domains in B cell antigen receptor signaling. J Exp Med December 1995;182(6):1815–23.

[132] Futterer K, Wong J, Grucza RA, Chan AC, Waksman G. Structural basis for Syk tyrosine kinase ubiquity in signal transduction pathways revealed by the crystal structure of its regulatory SH2 domains bound to a dually phosphorylated ITAM peptide. J Mol Biol August 1998;281(3):523–37.

[133] Rolli V, Gallwitz M, Wossning T, Flemming A, Schamel WW, Zurn C, et al. Amplification of B cell antigen receptor signaling by a Syk/ITAM positive feedback loop. Mol Cell November 2002;10(5):1057–69.

[134] Mansueto MS, Reens A, Rakhilina L, Chi A, Pan BS, Miller JR. A reevaluation of the spleen tyrosine kinase (SYK) activation mechanism. J Biol Chem May 2019;294(19):7658–68.

[135] Flaswinkel H, Reth M. Dual role of the tyrosine activation motif of the Ig-alpha protein during signal transduction via the B cell antigen receptor. EMBO J January 1994;13(1):83–9.

[136] Schmitz R, Baumann G, Gram H. Catalytic specificity of phosphotyrosine kinases Blk, Lyn, c-Src and Syk as assessed by phage display. J Mol Biol August 1996;260(5):664–77.

[137] Pao LI, Cambier JC. Syk, but not Lyn, recruitment to B cell antigen receptor and activation following stimulation of CD45- B cells. cited 2015 October 27 J Immunol March 15, 1997;158(6):2663–9. Available from: http://www.ncbi.nlm.nih.gov/pubmed/9058799.

[138] Mukherjee S, Zhu J, Zikherman J, Parameswaran R, Kadlecek TA, Wang Q, et al. Monovalent and multivalent ligation of the B cell receptor exhibit differential dependence upon Syk and Src family kinases. Sci Signal January 2013;6(256):ra1.

[139] Xu C, Gagnon E, Call ME, Schnell JR, Schwieters CD, Carman CV, et al. Regulation of T cell receptor activation by dynamic membrane binding of the CD3epsilon cytoplasmic tyrosine-based motif. Cell November 2008;135(4):702–13.

[140] Lee W-Y, Tolar P. Activation of the B cell receptor leads to increased membrane proximity of the Igα cytoplasmic domain. PLoS One 2013;8(11):e79148.
[141] Freeman SA, Jaumouillé V, Choi K, Hsu BE, Wong HS, Abraham L, et al. Toll-like receptor ligands sensitize B-cell receptor signalling by reducing actin-dependent spatial confinement of the receptor. cited 2017 October 11 Nat Commun February 3, 2015;6:6168. Available from: http://www.nature.com/doifinder/10.1038/ncomms7168.
[142] Bolger-Munro M, Choi K, Scurll J, Abraham L, Chappell R, Sheen D, et al. Arp2/3 complex-driven spatial patterning of the B cell receptor (BCR) enhances immune synapse formation, BCR signaling, and B cell activation. Elife 2019;8:e44574.
[143] Leung G, Zhou Y, Ostrowski P, Mylvaganam S, Boroumand P, Mulder DJ, et al. ARPC1B binds WASP to control actin polymerization and curtail tonic signaling in B cells. JCI Insight December 2021;6(23):e149376.
[144] Freeman SA, Lei V, Dang-Lawson M, Mizuno K, Roskelley CD, Gold MR. Cofilin-mediated F-actin severing is regulated by the Rap GTPase and controls the cytoskeletal dynamics that drive lymphocyte spreading and BCR microcluster formation. J Immunol December 2011;187(11):5887−900.
[145] Liu C, Miller H, Orlowski G, Hang H, Upadhyaya A, Song W. Actin reorganization is required for the formation of polarized B cell receptor signalosomes in response to both soluble and membrane-associated antigens. J Immunol April 2012;188(7):3237−46.
[146] Gupta N, Wollscheid B, Watts JD, Scheer B, Aebersold R, DeFranco AL. Quantitative proteomic analysis of B cell lipid rafts reveals that ezrin regulates antigen receptor-mediated lipid raft dynamics. Nat Immunol June 2006;7(6):625−33.
[147] Treanor B, Depoil D, Bruckbauer A, Batista FD. Dynamic cortical actin remodeling by ERM proteins controls BCR microcluster organization and integrity. J Exp Med May 2011;208(5):1055−68.
[148] Wan Z, Liu W. The growth of B cell receptor microcluster is a universal response of B cells encountering antigens with different motion features. Protein Cell July 2012;3(7):545−58.
[149] Ketchum C, Miller H, Song W, Upadhyaya A. Ligand mobility regulates B cell receptor clustering and signaling activation. Biophys J January 2014;106(1):26−36.
[150] Batista FD, Iber D, Neuberger MS. B cells acquire antigen from target cells after synapse formation. Nature May 2001;411(6836):489−94.
[151] Weber M, Treanor B, Depoil D, Shinohara H, Harwood NE, Hikida M, et al. Phospholipase C-gamma2 and Vav cooperate within signaling microclusters to propagate B cell spreading in response to membrane-bound antigen. J Exp Med April 2008;205(4):853−68.
[152] Baba Y, Matsumoto M, Kurosaki T. Calcium signaling in B cells: regulation of cytosolic Ca2+ increase and its sensor molecules, STIM1 and STIM2. cited 2015 July 1 Mol Immunol December 2014;62(2):339−43. Available from: http://www.sciencedirect.com/science/article/pii/S0161589013005324.
[153] Baba Y, Kurosaki T. Role of calcium signaling in B cell activation and biology. Curr Top Microbiol Immunol 2016;393:143−74.
[154] Colon-Gonzalez F, Kazanietz MG. C1 domains exposed: from diacylglycerol binding to protein-protein interactions. Biochim Biophys Acta August 2006;1761(8):827−37.
[155] Turvey SE, Durandy A, Fischer A, Fung SY, Geha RS, Gewies A, et al. The CARD11-BCL10-MALT1 (CBM) signalosome complex: stepping into the limelight of human primary immunodeficiency. J Allergy Clin Immunol August 2014;134(2):276−84.
[156] Stone JC. Regulation of Ras in lymphocytes: get a GRP. Biochem Soc Trans November 2006;34(Pt 5):858−61.
[157] Treanor B, Harwood NE, Batista FD. Microsignalosomes: spatially resolved receptor signalling. Biochem Soc Trans October 2009;37(Pt 5):1014−8.
[158] Dal Porto JM, Gauld SB, Merrell KT, Mills D, Pugh-Bernard AE, Cambier J. B cell antigen receptor signaling 101. Mol Immunol July 2004;41(6−7):599−613.
[159] Kabak S, Skaggs BJ, Gold MR, Affolter M, West KL, Foster MS, et al. The direct recruitment of BLNK to immunoglobulin alpha couples the B-cell antigen receptor to distal signaling pathways. Mol Cell Biol April 2002;22(8):2524−35.
[160] Oellerich T, Gronborg M, Neumann K, Hsiao HH, Urlaub H, Wienands J. SLP-65 phosphorylation dynamics reveals a functional basis for signal integration by receptor-proximal adaptor proteins. Mol Cell Proteomics July 2009;8(7):1738−50.
[161] Ishiai M, Kurosaki M, Pappu R, Okawa K, Ronko I, Fu C, et al. BLNK required for coupling Syk to PLC gamma 2 and Rac1-JNK in B cells. Immunity January 1999;10(1):117−25.
[162] Hashimoto S, Iwamatsu A, Ishiai M, Okawa K, Yamadori T, Matsushita M, et al. Identification of the SH2 domain binding protein of Bruton's tyrosine kinase as BLNK–functional significance of Btk-SH2 domain in B-cell antigen receptor-coupled calcium signaling. Blood October 1999;94(7):2357−64.
[163] Saito K, Scharenberg AM, Kinet JP. Interaction between the Btk PH domain and phosphatidylinositol-3,4,5-trisphosphate directly regulates Btk. J Biol Chem May 2001;276(19):16201−6.
[164] Corneth OBJ, Klein Wolterink RGJ, Hendriks RW. BTK signaling in B cell differentiation and autoimmunity. Curr Top Microbiol Immunol 2016;393:67−105.
[165] Park H, Wahl MI, Afar DE, Turck CW, Rawlings DJ, Tam C, et al. Regulation of Btk function by a major autophosphorylation site within the SH3 domain. Immunity May 1996;4(5):515−25.
[166] Rodriguez R, Matsuda M, Perisic O, Bravo J, Paul A, Jones NP, et al. Tyrosine residues in phospholipase Cgamma 2 essential for the enzyme function in B-cell signaling. J Biol Chem December 2001;276(51):47982−92.
[167] Kim YJ, Sekiya F, Poulin B, Bae YS, Rhee SG. Mechanism of B-cell receptor-induced phosphorylation and activation of phospholipase C-gamma2. Mol Cell Biol November 2004;24(22):9986−99.
[168] Engelke M, Pirkuliyeva S, Kuhn J, Wong L, Boyken J, Herrmann N, et al. Macromolecular assembly of the adaptor SLP-65 at intracellular vesicles in resting B cells. Sci Signal August 2014;7(339):ra79.
[169] Kuhn J, Wong LE, Pirkuliyeva S, Schulz K, Schwiegk C, Funfgeld KG, et al. The adaptor protein CIN85 assembles intracellular signaling clusters for B cell activation. Sci Signal June 2016;9(434):ra66.
[170] Wong LE, Bhatt A, Erdmann PS, Hou Z, Maier J, Pirkuliyeva S, et al. Tripartite phase separation of two signal effectors with vesicles priming B cell responsiveness. Nat Commun February 2020;11(1):848.

[171] Oellerich T, Bremes V, Neumann K, Bohnenberger H, Dittmann K, Hsiao HH, et al. The B-cell antigen receptor signals through a preformed transducer module of SLP65 and CIN85. EMBO J August 2011;30(17):3620–34.

[172] Kometani K, Yamada T, Sasaki Y, Yokosuka T, Saito T, Rajewsky K, et al. CIN85 drives B cell responses by linking BCR signals to the canonical NF-kappaB pathway. J Exp Med July 2011;208(7):1447–57.

[173] Keller B, Shoukier M, Schulz K, Bhatt A, Heine I, Strohmeier V, et al. Germline deletion of CIN85 in humans with X chromosome-linked antibody deficiency. J Exp Med May 2018;215(5):1327–36.

[174] Cheng AM, Rowley B, Pao W, Hayday A, Bolen JB, Pawson T. Syk tyrosine kinase required for mouse viability and B-cell development. Nature November 1995;378(6554):303–6.

[175] Turner M, Mee PJ, Costello PS, Williams O, Price AA, Duddy LP, et al. Perinatal lethality and blocked B-cell development in mice lacking the tyrosine kinase Syk. Nature November 1995;378(6554):298–302.

[176] Wang L, Aschenbrenner D, Zeng Z, Cao X, Mayr D, Mehta M, et al. Gain-of-function variants in SYK cause immune dysregulation and systemic inflammation in humans and mice. Nat Genet April 2021;53(4):500–10.

[177] Jumaa H, Wollscheid B, Mitterer M, Wienands J, Reth M, Nielsen PJ. Abnormal development and function of B lymphocytes in mice deficient for the signaling adaptor protein SLP-65. cited 2015 November 27 Immunity November 1999;11(5):547–54. Available from: http://www.ncbi.nlm.nih.gov/pubmed/10591180.

[178] Minegishi Y, Rohrer J, Coustan-Smith E, Lederman HM, Pappu R, Campana D, et al. An essential role for BLNK in human B cell development. Science December 1999;286(5446):1954–7.

[179] Hashimoto A, Takeda K, Inaba M, Sekimata M, Kaisho T, Ikehara S, et al. Cutting edge: essential role of phospholipase C-gamma 2 in B cell development and function. J Immunol August 2000;165(4):1738–42.

[180] Wang D, Feng J, Wen R, Marine JC, Sangster MY, Parganas E, et al. Phospholipase Cgamma2 is essential in the functions of B cell and several Fc receptors. Immunity July 2000;13(1):25–35.

[181] Bell SE, Vigorito E, McAdam S, Reynolds HM, Caraux A, Colucci F, et al. PLCgamma2 regulates Bcl-2 levels and is required for survival rather than differentiation of marginal zone and follicular B cells. Eur J Immunol August 2004;34(8):2237–47.

[182] Hikida M, Casola S, Takahashi N, Kaji T, Takemori T, Rajewsky K, et al. PLC-gamma2 is essential for formation and maintenance of memory B cells. J Exp Med March 2009;206(3):681–9.

[183] Novice T, Kariminia A, Del Bel KL, Lu H, Sharma M, Lim CJ, et al. A germline mutation in the C2 domain of PLCgamma2 associated with gain-of-function expands the phenotype for PLCG2-related diseases. J Clin Immunol February 2020;40(2):267–76.

[184] Yu P, Constien R, Dear N, Katan M, Hanke P, Bunney TD, et al. Autoimmunity and inflammation due to a gain-of-function mutation in phospholipase C gamma 2 that specifically increases external Ca2+ entry. Immunity April 2005;22(4):451–65.

[185] Timmerman LA, Healy JI, Ho SN, Chen L, Goodnow CC, Crabtree GR. Redundant expression but selective utilization of nuclear factor of activated T cells family members. J Immunol September 1997;159(6):2735–40.

[186] Macian F. NFAT proteins: key regulators of T-cell development and function. Nat Rev Immunol June 2005;5(6):472–84.

[187] Dolmetsch RE, Lewis RS, Goodnow CC, Healy JI. Differential activation of transcription factors induced by Ca2+ response amplitude and duration. Nature April 1997;386(6627):855–8.

[188] Muller MR, Rao A. NFAT, immunity and cancer: a transcription factor comes of age. Nat Rev Immunol September 2010;10(9):645–56.

[189] Barrington RA, Borde M, Rao A, Carroll MC. Involvement of NFAT1 in B cell self-tolerance. J Immunol August 2006;177(3):1510–5.

[190] Matsumoto M, Fujii Y, Baba A, Hikida M, Kurosaki T, Baba Y. The calcium sensors STIM1 and STIM2 control B cell regulatory function through interleukin-10 production. Immunity May 2011;34(5):703–14.

[191] Sharma M, Fu MP, Lu HY, Sharma AA, Modi BP, Michalski C, et al. Human complete NFAT1 deficiency causes a triad of joint contractures, osteochondromas, and B-cell malignancy. Blood October 2022;140(17):1858–74.

[192] Bhattacharyya S, Deb J, Patra AK, Thuy Pham DA, Chen W, Vaeth M, et al. NFATc1 affects mouse splenic B cell function by controlling the calcineurin–NFAT signaling network. J Exp Med April 2011;208(4):823–39.

[193] Berland R, Wortis HH. Normal B-1a cell development requires B cell-intrinsic NFATc1 activity. Proc Natl Acad Sci U S A November 2003;100(23):13459–64.

[194] Zhang Q, Lenardo MJ, Baltimore D. 30 years of NF-kappaB: a blossoming of relevance to human pathobiology. Cell January 2017;168(1–2):37–57.

[195] Sasaki Y, Iwai K. Roles of the NF-kappaB pathway in B-lymphocyte biology. Curr Top Microbiol Immunol 2016;393:177–209.

[196] Lu HY, Biggs CM, Blanchard-Rohner G, Fung SY, Sharma M, Turvey SE. Germline CBM-opathies: from immunodeficiency to atopy. J Allergy Clin Immunol May 2019;143(5):1661–73.

[197] Saijo K, Mecklenbrauker I, Santana A, Leitger M, Schmedt C, Tarakhovsky A. Protein kinase C beta controls nuclear factor kappaB activation in B cells through selective regulation of the IkappaB kinase alpha. J Exp Med June 2002;195(12):1647–52.

[198] Qiao Q, Yang C, Zheng C, Fontan L, David L, Yu X, et al. Structural architecture of the CARMA1/Bcl10/MALT1 signalosome: nucleation-induced filamentous assembly. Mol Cell September 2013;51(6):766–79.

[199] David L, Li Y, Ma J, Garner E, Zhang X, Wu H. Assembly mechanism of the CARMA1-BCL10-MALT1-TRAF6 signalosome. Proc Natl Acad Sci U S A February 2018;115(7):1499–504.

[200] Lu HY, Bauman BM, Arjunaraja S, Dorjbal B, Milner JD, Snow AL, et al. The CBM-opathies-A rapidly expanding spectrum of human inborn errors of immunity caused by mutations in the CARD11-BCL10-MALT1 complex. Front Immunol 2018;9:2078.

[201] Afonina IS, Elton L, Carpentier I, Beyaert R. MALT1–a universal soldier: multiple strategies to ensure NF-kappaB activation and target gene expression. FEBS J September 2015;282(17):3286–97.

[202] Pohl T, Gugasyan R, Grumont RJ, Strasser A, Metcalf D, Tarlinton D, et al. The combined absence of NF-kappa B1 and c-Rel reveals that overlapping roles for these transcription factors in the B cell lineage are restricted to the activation and function of mature cells. Proc Natl Acad Sci U S A April 2002;99(7):4514–9.

[203] Grumont RJ, Rourke IJ, O'Reilly LA, Strasser A, Miyake K, Sha W, et al. B lymphocytes differentially use the Rel and nuclear factor kappaB1 (NF-kappaB1) transcription factors to regulate cell cycle progression and apoptosis in quiescent and mitogen-activated cells. J Exp Med March 1998;187(5):663–74.

[204] Doi TS, Takahashi T, Taguchi O, Azuma T, Obata Y. NF-kappa B RelA-deficient lymphocytes: normal development of T cells and B cells, impaired production of IgA and IgG1 and reduced proliferative responses. J Exp Med March 1997;185(5):953–61.
[205] Heise N, De Silva NS, Silva K, Carette A, Simonetti G, Pasparakis M, et al. Germinal center B cell maintenance and differentiation are controlled by distinct NF-κB transcription factor subunits. J Exp Med 2014;211(10):2103–18.
[206] Grumont RJ, Strasser A, Gerondakis S. B cell growth is controlled by phosphatidylinosotol 3-kinase-dependent induction of Rel/NF-kappaB regulated c-myc transcription. Mol Cell December 2002;10(6):1283–94.
[207] Gugasyan R, Grumont R, Grossmann M, Nakamura Y, Pohl T, Nesic D, et al. Rel/NF-kappaB transcription factors: key mediators of B-cell activation. Immunol Rev August 2000;176:134–40.
[208] Grumont RJ, Rourke IJ, Gerondakis S. Rel-dependent induction of A1 transcription is required to protect B cells from antigen receptor ligation-induced apoptosis. Genes Dev February 1999;13(4):400–11.
[209] Cheng S, Hsia CY, Leone G, Liou HC. Cyclin E and Bcl-xL cooperatively induce cell cycle progression in c-Rel-/- B cells. Oncogene November 2003;22(52):8472–86.
[210] Ruefli-Brasse AA, French DM, Dixit VM. Regulation of NF-kappaB-dependent lymphocyte activation and development by paracaspase. Science November 2003;302(5650):1581–4.
[211] Bornancin F, Renner F, Touil R, Sic H, Kolb Y, Touil-Allaoui I, et al. Deficiency of MALT1 paracaspase activity results in unbalanced regulatory and effector T and B cell responses leading to multiorgan inflammation. J Immunol April 2015;194(8):3723–34.
[212] Hara H, Wada T, Bakal C, Kozieradzki I, Suzuki S, Suzuki N, et al. The MAGUK family protein CARD11 is essential for lymphocyte activation. Immunity June 2003;18(6):763–75.
[213] Egawa T, Albrecht B, Favier B, Sunshine MJ, Mirchandani K, O'Brien W, et al. Requirement for CARMA1 in antigen receptor-induced NF-kappa B activation and lymphocyte proliferation. Curr Biol July 2003;13(14):1252–8.
[214] Jun JE, Wilson LE, Vinuesa CG, Lesage S, Blery M, Miosge LA, et al. Identifying the MAGUK protein Carma-1 as a central regulator of humoral immune responses and atopy by genome-wide mouse mutagenesis. Immunity June 2003;18(6):751–62.
[215] Ruland J, Duncan GS, Elia A, del Barco Barrantes I, Nguyen L, Plyte S, et al. Bcl10 is a positive regulator of antigen receptor-induced activation of NF-kappaB and neural tube closure. Cell January 2001;104(1):33–42.
[216] Xue L, Morris SW, Orihuela C, Tuomanen E, Cui X, Wen R, et al. Defective development and function of Bcl10-deficient follicular, marginal zone and B1 B cells. Nat Immunol September 2003;4(9):857–65.
[217] Conley ME, Dobbs AK, Quintana AM, Bosompem A, Wang YD, Coustan-Smith E, et al. Agammaglobulinemia and absent B lineage cells in a patient lacking the p85alpha subunit of PI3K. J Exp Med March 2012;209(3):463–70.
[218] Ramadani F, Bolland DJ, Garcon F, Emery JL, Vanhaesebroeck B, Corcoran AE, et al. The PI3K isoforms p110alpha and p110delta are essential for pre-B cell receptor signaling and B cell development. Sci Signal August 2010;3(134):ra60.
[219] Bilancio A, Okkenhaug K, Camps M, Emery JL, Ruckle T, Rommel C, et al. Key role of the p110delta isoform of PI3K in B-cell antigen and IL-4 receptor signaling: comparative analysis of genetic and pharmacologic interference with p110delta function in B cells. Blood January 2006;107(2):642–50.
[220] Durand CA, Hartvigsen K, Fogelstrand L, Kim S, Iritani S, Vanhaesebroeck B, et al. Phosphoinositide 3-kinase p110 delta regulates natural antibody production, marginal zone and B-1 B cell function, and autoantibody responses. J Immunol November 2009;183(9):5673–84.
[221] Puri KD, Gold MR. Selective inhibitors of phosphoinositide 3-kinase delta: modulators of B-cell function with potential for treating autoimmune inflammatory diseases and B-cell malignancies. Front Immunol 2012;3:256.
[222] Chiu H, Mallya S, Nguyen P, Mai A, Jackson LV, Winkler DG, et al. The selective phosphoinoside-3-kinase p110delta inhibitor IPI-3063 potently suppresses B cell survival, proliferation, and differentiation. Front Immunol 2017;8:747.
[223] Delgado P, Cubelos B, Calleja E, Martínez-Martín N,Ciprés A, et al. Essential function for the GTPase TC21 in homeostatic antigen receptor signaling. Nat Immunol 2009;10(8):880–8.
[224] Srinivasan L, Sasaki Y, Calado DP, Zhang B, Paik JH, DePinho RA, et al. PI3 kinase signals BCR-dependent mature B cell survival. Cell October 2009;139(3):573–86.
[225] Tuveson DA, Carter RH, Soltoff SP, Fearon DT. CD19 of B cells as a surrogate kinase insert region to bind phosphatidylinositol 3-kinase. Science May 1993;260(5110):986–9.
[226] Okada T, Maeda A, Iwamatsu A, Gotoh K, Kurosaki T. BCAP: the tyrosine kinase substrate that connects B cell receptor to phosphoinositide 3-kinase activation. cited 2015 November 10 Immunity December 2000;13(6):817–27. Available from: http://www.ncbi.nlm.nih.gov/pubmed/11163197.
[227] Suzuki H, Terauchi Y, Fujiwara M, Aizawa S, Yazaki Y, Kadowaki T, et al. Xid-like immunodeficiency in mice with disruption of the p85alpha subunit of phosphoinositide 3-kinase. Science January 1999;283(5400):390–2.
[228] Aiba Y, Kameyama M, Yamazaki T, Tedder TF, Kurosaki T. Regulation of B-cell development by BCAP and CD19 through their binding to phosphoinositide 3-kinase. cited 2015 October 30 Blood March 1, 2008;111(3):1497–503. Available from: http://www.bloodjournal.org/content/111/3/1497.abstract.
[229] Castello A, Gaya M, Tucholski J, Oellerich T, Lu KH, Tafuri A, et al. Nck-mediated recruitment of BCAP to the BCR regulates the PI(3)K-Akt pathway in B cells. Nat Immunol September 2013;14(9):966–75.
[230] Calamito M, Juntilla MM, Thomas M, Northrup DL, Rathmell J, Birnbaum MJ, et al. Akt1 and Akt2 promote peripheral B-cell maturation and survival. Blood May 2010;115(20):4043–50.
[231] Limon JJ, Fruman DA. Akt and mTOR in B cell activation and differentiation. Front Immunol 2012;3:228.
[232] Astoul E, Watton S, Cantrell D. The dynamics of protein kinase B regulation during B cell antigen receptor engagement. J Cell Biol June 1999;145(7):1511–20.
[233] Manning BD, Toker A. AKT/PKB signaling: navigating the network. Cell 2017;169(3):381–405. Available from: http://doi.org/10.1016/j.cell.2017.04.001.

[234] Li M, Lazorchak AS, Ouyang X, Zhang H, Liu H, Arojo OA, et al. Sin1/mTORC2 regulate B cell growth and metabolism by activating mTORC1 and Myc. Cell Mol Immunol September 2019;16(9):757–69.
[235] Baracho GV, Cato MH, Zhu Z, Jaren OR, Hobeika E, Reth M, et al. PDK1 regulates B cell differentiation and homeostasis. Proc Natl Acad Sci U S A July 2014;111(26):9573–8.
[236] Du Z, Yang D, Zhang Y, Xuan X, Li H, Hu L, et al. AKT2 deficiency impairs formation of the BCR signalosome. Cell Commun Signal April 2020;18(1):56.
[237] Saxton RA, Sabatini DM. mTOR signaling in growth, metabolism, and disease. Cell March 2017;168(6):960–76.
[238] Donahue AC, Fruman DA. Distinct signaling mechanisms activate the target of rapamycin in response to different B-cell stimuli. Eur J Immunol October 2007;37(10):2923–36.
[239] Dibble CC, Cantley LC. Regulation of mTORC1 by PI3K signaling. Trends Cell Biol September 2015;25(9):545–55.
[240] Morita M, Gravel SP, Chenard V, Sikstrom K, Zheng L, Alain T, et al. mTORC1 controls mitochondrial activity and biogenesis through 4E-BP-dependent translational regulation. Cell Metab November 2013;18(5):698–711.
[241] Doughty CA, Bleiman BF, Wagner DJ, Dufort FJ, Mataraza JM, Roberts MF, et al. Antigen receptor-mediated changes in glucose metabolism in B lymphocytes: role of phosphatidylinositol 3-kinase signaling in the glycolytic control of growth. Blood June 2006;107(11):4458–65.
[242] Haftmann C, Stittrich AB, Sgouroudis E, Matz M, Chang HD, Radbruch A, et al. Lymphocyte signaling: regulation of FoxO transcription factors by microRNAs. Ann N Y Acad Sci January 2012;1247:46–55.
[243] Yusuf I, Zhu X, Kharas MG, Chen J, Fruman DA. Optimal B-cell proliferation requires phosphoinositide 3-kinase-dependent inactivation of FOXO transcription factors. Blood August 2004;104(3):784–7.
[244] Tzivion G, Dobson M, Ramakrishnan G. FoxO transcription factors; regulation by AKT and 14-3-3 proteins. Biochim Biophys Acta November 2011;1813(11):1938–45.
[245] Fruman DA, Ferl GZ, An SS, Donahue AC, Satterthwaite AB, Witte ON. Phosphoinositide 3-kinase and Bruton's tyrosine kinase regulate overlapping sets of genes in B lymphocytes. Proc Natl Acad Sci U S A January 2002;99(1):359–64.
[246] Omori SA, Cato MH, Anzelon-Mills A, Puri KD, Shapiro-Shelef M, Calame K, et al. Regulation of class-switch recombination and plasma cell differentiation by phosphatidylinositol 3-kinase signaling. Immunity October 2006;25(4):545–57.
[247] Luo W, Weisel F, Shlomchik MJ. B cell receptor and CD40 signaling are rewired for synergistic induction of the c-Myc transcription factor in germinal center B cells. Immunity February 2018;48(2):313–326 e5.
[248] Janmey PA, Bucki R, Radhakrishnan R. Regulation of actin assembly by PI(4,5)P2 and other inositol phospholipids: an update on possible mechanisms. Biochem Biophys Res Commun November 2018;506(2):307–14.
[249] Xu C, Xie H, Guo X, Gong H, Liu L, Qi H, et al. A PIP(2)-derived amplification loop fuels the sustained initiation of B cell activation. Sci Immunol November 2017;2(17).
[250] Wang J, Xu L, Shaheen S, Liu S, Zheng W, Sun X, et al. Growth of B cell receptor microclusters is regulated by PIP(2) and PIP(3) equilibrium and Dock2 recruitment and activation. Cell Rep November 2017;21(9):2541–57.
[251] Posor Y, Jang W, Haucke V. Phosphoinositides as membrane organizers. Nat Rev Mol Cell Biol December 2022;23(12):797–816.
[252] Helgason CD, Kalberer CP, Damen JE, Chappel SM, Pineault N, Krystal G, et al. A dual role for Src homology 2 domain-containing inositol-5-phosphatase (SHIP) in immunity: aberrant development and enhanced function of b lymphocytes in ship -/- mice. J Exp Med March 2000;191(5):781–94.
[253] Bolland S, Pearse RN, Kurosaki T, Ravetch JV. SHIP modulates immune receptor responses by regulating membrane association of Btk. Immunity April 1998;8(4):509–16.
[254] Wienands J, Schweikert J, Wollscheid B, Jumaa H, Nielsen PJ, Reth M. SLP-65: a new signaling component in B lymphocytes which requires expression of the antigen receptor for phosphorylation. J Exp Med August 1998;188(4):791–5.
[255] Liu Q, Oliveira-Dos-Santos AJ, Mariathasan S, Bouchard D, Jones J, Sarao R, et al. The inositol polyphosphate 5-phosphatase ship is a crucial negative regulator of B cell antigen receptor signaling. J Exp Med October 1998;188(7):1333–42.
[256] Okada H, Bolland S, Hashimoto A, Kurosaki M, Kabuyama Y, Iino M, et al. Role of the inositol phosphatase SHIP in B cell receptor-induced Ca2+ oscillatory response. J Immunol November 1998;161(10):5129–32.
[257] Ono M, Okada H, Bolland S, Yanagi S, Kurosaki T, Ravetch JV. Deletion of SHIP or SHP-1 reveals two distinct pathways for inhibitory signaling. Cell July 1997;90(2):293–301.
[258] Shi Y, Paluch BE, Wang X, Jiang X. PTEN at a glance. J Cell Sci October 2012;125(Pt 20):4687–92.
[259] Saito K, Tolias KF, Saci A, Koon HB, Humphries LA, Scharenberg A, et al. BTK regulates PtdIns-4,5-P2 synthesis: importance for calcium signaling and PI3K activity. Immunity November 2003;19(5):669–78.
[260] Droubi A, Wallis C, Anderson KE, Rahman S, de Sa A, Rahman T, et al. The inositol 5-phosphatase INPP5B regulates B cell receptor clustering and signaling. J Cell Biol September 2022;221(9).
[261] Gold MR, DeFranco AL. Phorbol esters and dioctanoylglycerol block anti-IgM-stimulated phosphoinositide hydrolysis in the murine B cell lymphoma WEHI-231. J Immunol February 1987;138(3):868–76.
[262] Avila-Flores A, Santos T, Rincon E, Merida I. Modulation of the mammalian target of rapamycin pathway by diacylglycerol kinase-produced phosphatidic acid. J Biol Chem March 2005;280(11):10091–9.
[263] Yoon MS, Rosenberger CL, Wu C, Truong N, Sweedler JV, Chen J. Rapid mitogenic regulation of the mTORC1 inhibitor, DEPTOR, by phosphatidic acid. Mol Cell May 2015;58(3):549–56.
[264] Merino-Cortes SV, Gardeta SR, Roman-Garcia S, Martinez-Riano A, Pineau J, Liebana R, et al. Diacylglycerol kinase zeta promotes actin cytoskeleton remodeling and mechanical forces at the B cell immune synapse. Sci Signal April 2020;13(627).
[265] Lazarus AH, Kawauchi K, Rapoport MJ, Delovitch TL. Antigen-induced B lymphocyte activation involves the p21ras and ras.GAP signaling pathway. J Exp Med November 1993;178(5):1765–9.

[266] Tordai A, Franklin RA, Patel H, Gardner AM, Johnson GL, Gelfand EW. Cross-linking of surface IgM stimulates the Ras/Raf-1/MEK/MAPK cascade in human B lymphocytes. J Biol Chem March 1994;269(10):7538–43.
[267] McLeod SJ, Ingham RJ, Bos JL, Kurosaki T, Gold MR. Activation of the Rap1 GTPase by the B cell antigen receptor. J Biol Chem October 1998;273(44):29218–23.
[268] Christian SL, Lee RL, McLeod SJ, Burgess AE, Li AH, Dang-Lawson M, et al. Activation of the Rap GTPases in B lymphocytes modulates B cell antigen receptor-induced activation of Akt but has no effect on MAPK activation. J Biol Chem October 2003;278(43):41756–67.
[269] Grill B, Schrader JW. Activation of Rac-1, Rac-2, and Cdc42 by hemopoietic growth factors or cross-linking of the B-lymphocyte receptor for antigen. Blood November 2002;100(9):3183–92.
[270] Arana E, Vehlow A, Harwood NE, Vigorito E, Henderson R, Turner M, et al. Activation of the small GTPase Rac2 via the B cell receptor regulates B cell adhesion and immunological-synapse formation. Immunity January 2008;28(1):88–99.
[271] Burbage M, Keppler SJ, Gasparrini F, Martinez-Martin N, Gaya M, Feest C, et al. Cdc42 is a key regulator of B cell differentiation and is required for antiviral humoral immunity. J Exp Med January 2015;212(1):53–72.
[272] Eblen ST. Extracellular-regulated kinases: signaling from ras to ERK substrates to control biological outcomes. Adv Cancer Res 2018;138:99–142.
[273] Guo YJ, Pan WW, Liu SB, Shen ZF, Xu Y, Hu LL. ERK/MAPK signalling pathway and tumorigenesis. Exp Ther Med March 2020;19(3):1997–2007.
[274] Iritani BM, Forbush KA, Farrar MA, Perlmutter RM. Control of B cell development by Ras-mediated activation of Raf. EMBO J December 1997;16(23):7019–31.
[275] Yasuda T, Sanjo H, Pages G, Kawano Y, Karasuyama H, Pouyssegur J, et al. Erk kinases link pre-B cell receptor signaling to transcriptional events required for early B cell expansion. Immunity April 2008;28(4):499–508.
[276] Sutherland CL, Heath AW, Pelech SL, Young PR, Gold MR. Differential activation of the ERK, JNK, and p38 mitogen-activated protein kinases by CD40 and the B cell antigen receptor. J Immunol October 1996;157(8):3381–90.
[277] Gold MR, Sanghera JS, Stewart J, Pelech SL. Selective activation of p42 mitogen-activated protein (MAP) kinase in murine B lymphoma cell lines by membrane immunoglobulin cross-linking. Evidence for protein kinase C-independent and -dependent mechanisms of activation. Biochem J October 1992;287(Pt 1):269–76.
[278] O'Reilly LA, Kruse EA, Puthalakath H, Kelly PN, Kaufmann T, Huang DC, et al. MEK/ERK-mediated phosphorylation of Bim is required to ensure survival of T and B lymphocytes during mitogenic stimulation. J Immunol July 2009;183(1):261–9.
[279] Richards JD, Dave SH, Chou CH, Mamchak AA, DeFranco AL. Inhibition of the MEK/ERK signaling pathway blocks a subset of B cell responses to antigen. J Immunol March 2001;166(6):3855–64.
[280] Adem J, Hamalainen A, Ropponen A, Eeva J, Eray M, Nuutinen U, et al. ERK1/2 has an essential role in B cell receptor- and CD40-induced signaling in an in vitro model of germinal center B cell selection. Mol Immunol October 2015;67(2 Pt B):240–7.
[281] Eferl R, Wagner EF. AP-1: a double-edged sword in tumorigenesis. Nat Rev Cancer November 2003;3(11):859–68.
[282] Cho YY. RSK2 and its binding partners in cell proliferation, transformation and cancer development. Arch Pharm Res (Seoul) March 2017;40(3):291–303.
[283] Blois JT, Mataraza JM, Mecklenbrauker I, Tarakhovsky A, Chiles TC. B cell receptor-induced cAMP-response element-binding protein activation in B lymphocytes requires novel protein kinase Cdelta. J Biol Chem July 2004;279(29):30123–32.
[284] Zhang CY, Wu YL, Boxer LM. Impaired proliferation and survival of activated B cells in transgenic mice that express a dominant-negative cAMP-response element-binding protein transcription factor in B cells. J Biol Chem December 2002;277(50):48359–65.
[285] Proud CG. Mnks, eIF4E phosphorylation and cancer. Biochim Biophys Acta July 2015;1849(7):766–73.
[286] Oh-hora M, Johmura S, Hashimoto A, Hikida M, Kurosaki T. Requirement for Ras guanine nucleotide releasing protein 3 in coupling phospholipase C-gamma2 to Ras in B cell receptor signaling. J Exp Med December 2003;198(12):1841–51.
[287] Coughlin JJ, Stang SL, Dower NA, Stone JC. RasGRP1 and RasGRP3 regulate B cell proliferation by facilitating B cell receptor-Ras signaling. J Immunol December 2005;175(11):7179–84.
[288] Aiba Y, Oh-hora M, Kiyonaka S, Kimura Y, Hijikata A, Mori Y, et al. Activation of RasGRP3 by phosphorylation of Thr-133 is required for B cell receptor-mediated Ras activation. Proc Natl Acad Sci U S A November 2004;101(47):16612–7.
[289] Roose JP, Mollenauer M, Ho M, Kurosaki T, Weiss A. Unusual interplay of two types of Ras activators, RasGRP and SOS, establishes sensitive and robust Ras activation in lymphocytes. Mol Cell Biol April 2007;27(7):2732–45.
[290] Hashimoto A, Okada H, Jiang A, Kurosaki M, Greenberg S, Clark EA, et al. Involvement of guanosine triphosphatases and phospholipase C-gamma2 in extracellular signal-regulated kinase, c-Jun NH2-terminal kinase, and p38 mitogen-activated protein kinase activation by the B cell antigen receptor. J Exp Med October 1998;188(7):1287–95.
[291] Blonska M, Lin X. CARMA1-mediated NF-kappaB and JNK activation in lymphocytes. Immunol Rev March 2009;228(1):199–211.
[292] Zeke A, Misheva M, Remenyi A, Bogoyevitch MA. JNK signaling: regulation and functions based on complex protein-protein partnerships. Microbiol Mol Biol Rev September 2016;80(3):793–835.
[293] Shaulian E, Karin M. AP-1 as a regulator of cell life and death. Nat Cell Biol May 2002;4(5):E131–6.
[294] Khiem D, Cyster JG, Schwarz JJ, Black BL. A p38 MAPK-MEF2C pathway regulates B-cell proliferation. Proc Natl Acad Sci U S A November 2008;105(44):17067–72.
[295] Inabe K, Ishiai M, Scharenberg AM, Freshney N, Downward J, Kurosaki T. Vav3 modulates B cell receptor responses by regulating phosphoinositide 3-kinase activation. J Exp Med January 2002;195(2):189–200.
[296] Walmsley MJ, Ooi SK, Reynolds LF, Smith SH, Ruf S, Mathiot A, et al. Critical roles for Rac1 and Rac2 GTPases in B cell development and signaling. Science October 2003;302(5644):459–62.
[297] Rottner K, Faix J, Bogdan S, Linder S, Kerkhoff E. Actin assembly mechanisms at a glance. J Cell Sci October 2017;130(20):3427–35.
[298] Goley ED, Welch MD. The ARP2/3 complex: an actin nucleator comes of age. Nat Rev Mol Cell Biol October 2006;7(10):713–26.

[299] Pollard TD, Borisy GG. Cellular motility driven by assembly and disassembly of actin filaments. Cell February 2003;112(4):453–65.
[300] Liu C, Bai X, Wu J, Sharma S, Upadhyaya A, Dahlberg CI, et al. N-wasp is essential for the negative regulation of B cell receptor signaling. PLoS Biol November 2013;11(11):e1001704.
[301] Huang L, Sun X, Yang D, Dai X, Jiang P, Bai X, et al. WASP and Mst1 coregulate B-cell development and B-cell receptor signaling. Blood Adv February 2020;4(3):573–85.
[302] Gerasimcik N, Dahlberg CI, Baptista MA, Massaad MJ, Geha RS, Westerberg LS, et al. The rho GTPase Cdc42 is essential for the activation and function of mature B cells. J Immunol May 2015;194(10):4750–8.
[303] Fujikawa K, Miletic AV, Alt FW, Faccio R, Brown T, Hoog J, et al. Vav1/2/3-null mice define an essential role for Vav family proteins in lymphocyte development and activation but a differential requirement in MAPK signaling in T and B cells. J Exp Med November 2003;198(10):1595–608.
[304] Han J, Luby-Phelps K, Das B, Shu X, Xia Y, Mosteller RD, et al. Role of substrates and products of PI 3-kinase in regulating activation of Rac-related guanosine triphosphatases by Vav. Science January 1998;279(5350):558–60.
[305] Cote JF, Motoyama AB, Bush JA, Vuori K. A novel and evolutionarily conserved PtdIns(3,4,5)P3-binding domain is necessary for DOCK180 signalling. Nat Cell Biol August 2005;7(8):797–807.
[306] Kearney CJ, Randall KL, Oliaro J. DOCK8 regulates signal transduction events to control immunity. Cell Mol Immunol May 2017;14(5):406–11.
[307] Janssen E, Tohme M, Hedayat M, Leick M, Kumari S, Ramesh N, et al. A DOCK8-WIP-WASp complex links T cell receptors to the actin cytoskeleton. J Clin Invest October 2016;126(10):3837–51.
[308] Randall KL, Lambe T, Johnson AL, Treanor B, Kucharska E, Domaschenz H, et al. Dock8 mutations cripple B cell immunological synapses, germinal centers and long-lived antibody production. Nat Immunol December 2009;10(12):1283–91.
[309] Sun X, Wang J, Qin T, Zhang Y, Huang L, Niu L, et al. Dock8 regulates BCR signaling and activation of memory B cells via WASP and CD19. Blood Adv February 2018;2(4):401–13.
[310] McLeod SJ, Li AH, Lee RL, Burgess AE, Gold MR. The Rap GTPases regulate B cell migration toward the chemokine stromal cell-derived factor-1 (CXCL12): potential role for Rap2 in promoting B cell migration. J Immunol August 2002;169(3):1365–71.
[311] Durand CA, Westendorf J, Tse KW, Gold MR. The Rap GTPases mediate CXCL13- and sphingosine1-phosphate-induced chemotaxis, adhesion, and Pyk2 tyrosine phosphorylation in B lymphocytes. Eur J Immunol August 2006;36(8):2235–49.
[312] Jeon TJ, Lee DJ, Merlot S, Weeks G, Firtel RA. Rap1 controls cell adhesion and cell motility through the regulation of myosin II. J Cell Biol March 2007;176(7):1021–33.
[313] Frische EW, Zwartkruis FJ. Rap1, a mercenary among the Ras-like GTPases. Dev Biol April 2010;340(1):1–9.
[314] Sun H, Lagarrigue F, Ginsberg MH. The connection between Rap1 and Talin1 in the activation of integrins in blood cells. Front Cell Dev Biol 2022;10:908622.
[315] Raman R, Pinto CS, Sonawane M. Polarized organization of the cytoskeleton: regulation by cell polarity proteins. J Mol Biol September 2018;430(19):3565–84.
[316] Bakal C, Aach J, Church G, Perrimon N. Quantitative morphological signatures define local signaling networks regulating cell morphology. Science June 2007;316(5832):1753–6.
[317] Chant J. Cell polarity in yeast. Annu Rev Cell Dev Biol 1999;15:365–91.
[318] Schwamborn JC, Puschel AW. The sequential activity of the GTPases Rap1B and Cdc42 determines neuronal polarity. Nat Neurosci September 2004;7(9):923–9.
[319] Paganini S, Guidetti GF, Catricala S, Trionfini P, Panelli S, Balduini C, et al. Identification and biochemical characterization of Rap2C, a new member of the Rap family of small GTP-binding proteins. Biochimie April 2006;88(3–4):285–95.
[320] Katagiri K, Maeda A, Shimonaka M, Kinashi T. RAPL, a Rap1-binding molecule that mediates Rap1-induced adhesion through spatial regulation of LFA-1. Nat Immunol August 2003;4(8):741–8.
[321] Katagiri K, Ohnishi N, Kabashima K, Iyoda T, Takeda N, Shinkai Y, et al. Crucial functions of the Rap1 effector molecule RAPL in lymphocyte and dendritic cell trafficking. Nat Immunol October 2004;5(10):1045–51.
[322] Freeman SA, McLeod SJ, Dukowski J, Austin P, Lee CC, Millen-Martin B, et al. Preventing the activation or cycling of the Rap1 GTPase alters adhesion and cytoskeletal dynamics and blocks metastatic melanoma cell extravasation into the lungs. Cancer Res June 2010;70(11):4590–601.
[323] Shimonaka M, Katagiri K, Nakayama T, Fujita N, Tsuruo T, Yoshie O, et al. Rap1 translates chemokine signals to integrin activation, cell polarization, and motility across vascular endothelium under flow. J Cell Biol April 2003;161(2):417–27.
[324] Gerard A, Mertens AE, van der Kammen RA, Collard JG. The Par polarity complex regulates Rap1- and chemokine-induced T cell polarization. J Cell Biol March 2007;176(6):863–75.
[325] Ishihara S, Sato T, Sugioka R, Miwa R, Saito H, Sato R, et al. Rap1 is essential for B-cell locomotion, germinal center formation and normal B-1a cell population. Front Immunol 2021;12:624419.
[326] Lin KB, Freeman SA, Zabetian S, Brugger H, Weber M, Lei V, et al. The rap GTPases regulate B cell morphology, immune-synapse formation, and signaling by particulate B cell receptor ligands. Immunity January 2008;28(1):75–87.
[327] Wang JC, Lee JY, Christian S, Dang-Lawson M, Pritchard C, Freeman SA, et al. The Rap1-cofilin-1 pathway coordinates actin reorganization and MTOC polarization at the B cell immune synapse. J Cell Sci March 2017;130(6):1094–109.
[328] McLeod SJ, Shum AJ, Lee RL, Takei F, Gold MR. The Rap GTPases regulate integrin-mediated adhesion, cell spreading, actin polymerization, and Pyk2 tyrosine phosphorylation in B lymphocytes. J Biol Chem March 2004;279(13):12009–19.
[329] Katagiri K, Imamura M, Kinashi T. Spatiotemporal regulation of the kinase Mst1 by binding protein RAPL is critical for lymphocyte polarity and adhesion. Nat Immunol September 2006;7(9):919–28.
[330] Xu X, Wang X, Todd EM, Jaeger ER, Vella JL, Mooren OL, et al. Mst1 kinase regulates the actin-bundling protein L-plastin to promote T cell migration. J Immunol September 2016;197(5):1683–91.
[331] Mou F, Praskova M, Xia F, Van Buren D, Hock H, Avruch J, et al. The Mst1 and Mst2 kinases control activation of rho family

GTPases and thymic egress of mature thymocytes. J Exp Med April 2012;209(4):741–59.
[332] Ebisuno Y, Katagiri K, Katakai T, Ueda Y, Nemoto T, Inada H, et al. Rap1 controls lymphocyte adhesion cascade and interstitial migration within lymph nodes in RAPL-dependent and -independent manners. Blood January 2010;115(4):804–14.
[333] Chu H, Awasthi A, White GC, Chrzanowska-Wodnicka M, Malarkannan S. Rap1b regulates B cell development, homing, and T cell-dependent humoral immunity. J Immunol September 2008;181(5):3373–83.
[334] Chen Y, Yu M, Podd A, Wen R, Chrzanowska-Wodnicka M, White GC, et al. A critical role of Rap1b in B-cell trafficking and marginal zone B-cell development. Blood May 2008;111(9):4627–36.
[335] Ksionda O, Limnander A, Roose JP. RasGRP Ras guanine nucleotide exchange factors in cancer. Front Biol October 2013;8(5):508–32.
[336] Mele S, Devereux S, Pepper AG, Infante E, Ridley AJ. Calcium-RasGRP2-Rap1 signaling mediates CD38-induced migration of chronic lymphocytic leukemia cells. Blood Adv July 2018;2(13):1551–61.
[337] Ingham RJ, Krebs DL, Barbazuk SM, Turck CW, Hirai H, Matsuda M, et al. B cell antigen receptor signaling induces the formation of complexes containing the Crk adapter proteins. J Biol Chem December 1996;271(50):32306–14.
[338] Bezanilla M, Gladfelter AS, Kovar DR, Lee WL. Cytoskeletal dynamics: a view from the membrane. J Cell Biol May 2015;209(3):329–37.
[339] Chhabra ES, Higgs HN. The many faces of actin: matching assembly factors with cellular structures. Nat Cell Biol October 2007;9(10):1110–21.
[340] Michelot A, Drubin DG. Building distinct actin filament networks in a common cytoplasm. Curr Biol July 2011;21(14):R560–9.
[341] Mattila PK, Lappalainen P. Filopodia: molecular architecture and cellular functions. Nat Rev Mol Cell Biol June 2008;9(6):446–54.
[342] Dimchev G, Steffen A, Kage F, Dimchev V, Pernier J, Carlier MF, et al. Efficiency of lamellipodia protrusion is determined by the extent of cytosolic actin assembly. Mol Biol Cell May 2017;28(10):1311–25.
[343] Rodal AA, Tetreault JW, Lappalainen P, Drubin DG, Amberg DC. Aip1p interacts with cofilin to disassemble actin filaments. J Cell Biol June 1999;145(6):1251–64.
[344] Li J, Brieher WM, Scimone ML, Kang SJ, Zhu H, Yin H, et al. Caspase-11 regulates cell migration by promoting Aip1-cofilin-mediated actin depolymerization. Nat Cell Biol March 2007;9(3):276–86.
[345] Chen Q, Courtemanche N, Pollard TD. Aip1 promotes actin filament severing by cofilin and regulates constriction of the cytokinetic contractile ring. J Biol Chem January 2015;290(4):2289–300.
[346] Ohashi K. Roles of cofilin in development and its mechanisms of regulation. Dev Growth Differ May 2015;57(4):275–90.
[347] Mizuno K. Signaling mechanisms and functional roles of cofilin phosphorylation and dephosphorylation. Cell Signal February 2013;25(2):457–69.
[348] Kotila T, Kogan K, Enkavi G, Guo S, Vattulainen I, Goode BL, et al. Structural basis of actin monomer re-charging by cyclase-associated protein. Nat Commun May 2018;9(1):1892.
[349] Batista FD, Harwood NE. The who, how and where of antigen presentation to B cells. Nat Rev Immunol January 2009;9(1):15–27.
[350] Cyster JG. B cell follicles and antigen encounters of the third kind. Nat Immunol November 2010;11(11):989–96.
[351] Heesters BA, Chatterjee P, Kim YA, Gonzalez SF, Kuligowski MP, Kirchhausen T, et al. Endocytosis and recycling of immune complexes by follicular dendritic cells enhances B cell antigen binding and activation. Immunity June 2013;38(6):1164–75.
[352] Moran I, Grootveld AK, Nguyen A, Phan TG. Subcapsular sinus macrophages: the seat of innate and adaptive memory in murine lymph nodes. Trends Immunol January 2019;40(1):35–48.
[353] Aung A, Cui A, Maiorino L, Amini AP, Gregory JR, Bukenya M, et al. Low protease activity in B cell follicles promotes retention of intact antigens after immunization. Science January 2023;379(6630):eabn8934.
[354] Harwood NE, Batista FD. The cytoskeleton coordinates the early events of B-cell activation. cited 2014 October 15 Cold Spring Harb Perspect Biol February 2011;3(2). Available from: http://www.pubmedcentral.nih.gov/articlerender.fcgi?artid=3039531&tool=pmcentrez&rendertype=abstract.
[355] Song W, Liu C, Seeley-Fallen MK, Miller H, Ketchum C, Upadhyaya A. Actin-mediated feedback loops in B-cell receptor signaling. Immunol Rev November 2013;256(1):177–89.
[356] Song W, Liu C, Upadhyaya A. The pivotal position of the actin cytoskeleton in the initiation and regulation of B cell receptor activation. Biochim Biophys Acta February 2014;1838(2):569–78.
[357] Abraham L, Wang JC, Bolger-Munro M, Gold MR. Structure, function, and spatial organization of the B cell receptor (BCR). In: Ratcliffe MJ, editor. Encycl immunobiol, vol. 2; 2016. p. 40–54.
[358] Bhanja A, Rey-Suarez I, Song W, Upadhyaya A. Bidirectional feedback between BCR signaling and actin cytoskeletal dynamics. FEBS J August 2022;289(15):4430–46.
[359] Tokatlian T, Read BJ, Jones CA, Kulp DW, Menis S, Chang JYH, et al. Innate immune recognition of glycans targets HIV nanoparticle immunogens to germinal centers. Science February 2019;363(6427):649–54.
[360] Veninga H, Borg EG, Vreeman K, Taylor PR, Kalay H, van Kooyk Y, et al. Antigen targeting reveals splenic CD169+ macrophages as promoters of germinal center B-cell responses. Eur J Immunol March 2015;45(3):747–57.
[361] Fleire SJ, Goldman JP, Carrasco YR, Weber M, Bray D, Batista FD. B cell ligand discrimination through a spreading and contraction response. Science May 2006;312(5774):738–41.
[362] Carrasco YR, Batista FD. B-cell activation by membrane-bound antigens is facilitated by the interaction of VLA-4 with VCAM-1. EMBO J February 2006;25(4):889–99.
[363] Carrasco YR, Fleire SJ, Cameron T, Dustin ML, Batista FD. LFA-1/ICAM-1 interaction lowers the threshold of B cell activation by facilitating B cell adhesion and synapse formation. Immunity May 2004;20(5):589–99.
[364] Wang JC, Yim YI, Wu X, Jaumouille V, Cameron A, Waterman CM, et al. A B-cell actomyosin arc network couples integrin co-stimulation to mechanical force-dependent immune synapse formation. Elife April 2022;11.
[365] Cordoba SP, Choudhuri K, Zhang H, Bridge M, Basat AB, Dustin ML, et al. The large ectodomains of CD45 and CD148 regulate their segregation from and inhibition of ligated T-cell receptor. Blood May 2013;121(21):4295–302.
[366] Schnyder T, Castello A, Feest C, Harwood NE, Oellerich T, Urlaub H, et al. B cell receptor-mediated antigen gathering requires

ubiquitin ligase Cbl and adaptors Grb2 and Dok-3 to recruit dynein to the signaling microcluster. Immunity June 2011;34(6):905–18.
[367] Yuseff MI, Reversat A, Lankar D, Diaz J, Fanget I, Pierobon P, et al. Polarized secretion of lysosomes at the B cell synapse couples antigen extraction to processing and presentation. Immunity September 2011;35(3):361–74.
[368] Wan Z, Chen X, Chen H, Ji Q, Chen Y, Wang J, et al. The activation of IgM- or isotype-switched IgG- and IgE-BCR exhibits distinct mechanical force sensitivity and threshold. Elife 2015;4:e06925.
[369] Wan Z, Xu C, Chen X, Xie H, Li Z, Wang J, et al. PI(4,5)P2 determines the threshold of mechanical force-induced B cell activation. J Cell Biol July 2018;217(7):2565–82.
[370] Wan Z, Zhang S, Fan Y, Liu K, Du F, Davey AM, et al. B cell activation is regulated by the stiffness properties of the substrate presenting the antigens. J Immunol 2013;190(9).
[371] Batista FD, Neuberger MS. B cells extract and present immobilized antigen: implications for affinity discrimination. EMBO J February 2000;19(4):513–20.
[372] Hoogeboom R, Tolar P. Molecular mechanisms of B cell antigen gathering and endocytosis. Curr Top Microbiol Immunol 2016;393:45–63.
[373] Maeda FY, van Haaren JJ, Langley DB, Christ D, Andrews NW, Song W. Surface-associated antigen induces permeabilization of primary mouse B-cells and lysosome exocytosis facilitating antigen uptake and presentation to T-cells. Elife October 2021;10:e66984.
[374] Spillane KM, Tolar P. Mechanics of antigen extraction in the B cell synapse. Mol Immunol September 2018;101:319–28.
[375] Natkanski E, Lee WY, Mistry B, Casal A, Molloy JE, Tolar P. B cells use mechanical energy to discriminate antigen affinities. Science June 2013;340(6140):1587–90.
[376] Kumari A, Pineau J, Saez PJ, Maurin M, Lankar D, San Roman M, et al. Actomyosin-driven force patterning controls endocytosis at the immune synapse. Nat Commun June 2019;10(1):2870.
[377] Wang JC, Hammer JA. The role of actin and myosin in antigen extraction by B lymphocytes. Semin Cell Dev Biol June 2020;102:90–104.
[378] Roper SI, Wasim L, Malinova D, Way M, Cox S, Tolar P. B cells extract antigens at Arp2/3-generated actin foci interspersed with linear filaments. Elife December 2019;8:e48093.
[379] Tolar P, Spillane KM. Force generation in B-cell synapses: mechanisms coupling B-cell receptor binding to antigen internalization and affinity discrimination. Adv Immunol 2014;123:69–100.
[380] Andrews NW, Almeida PE, Corrotte M. Damage control: cellular mechanisms of plasma membrane repair. Trends Cell Biol December 2014;24(12):734–42.
[381] Yuseff MI, Pierobon P, Reversat A, Lennon-Dumenil AM. How B cells capture, process and present antigens: a crucial role for cell polarity. Nat Rev Immunol July 2013;13(7):475–86.
[382] Vascotto F, Le Roux D, Lankar D, Faure-Andre G, Vargas P, Guermonprez P, et al. Antigen presentation by B lymphocytes: how receptor signaling directs membrane trafficking. Curr Opin Immunol February 2007;19(1):93–8.
[383] McShane AN, Malinova D. The ins and outs of antigen uptake in B cells. Front Immunol 2022;13:892169.
[384] Lankar D, Vincent-Schneider H, Briken V, Yokozeki T, Raposo G, Bonnerot C. Dynamics of major histocompatibility complex class II compartments during B cell receptor-mediated cell activation. J Exp Med February 2002;195(4):461–72.
[385] Ibanez-Vega J, Del Valle Batalla F, Saez JJ, Soza A, Yuseff MI. Proteasome dependent actin remodeling facilitates antigen extraction at the immune synapse of B cells. Front Immunol 2019;10:225.
[386] Tolar P. Cytoskeletal control of B cell responses to antigens. Nat Rev Immunol October 2017;17(10):621–34.
[387] Keppler SJ, Burbage M, Gasparrini F, Hartjes L, Aggarwal S, Massaad MJ, et al. The lack of WIP binding to actin results in impaired B cell migration and altered humoral immune responses. Cell Rep July 2018;24(3):619–29.
[388] Salzer E, Zoghi S, Kiss MG, Kage F, Rashkova C, Stahnke S, et al. The cytoskeletal regulator HEM1 governs B cell development and prevents autoimmunity. Sci Immunol July 2020;5(49):eabc3979.
[389] Bolger-Munro M, Choi K, Cheung F, Liu YT, Dang-Lawson M, Deretic N, et al. The Wdr1-LIMK-cofilin axis controls B cell antigen receptor-induced actin remodeling and signaling at the immune synapse. Front Cell Dev Biol 2021;9:649433.
[390] Deretic N, Bolger-Munro M, Choi K, Abraham L, Gold MR. The actin-disassembly protein glia maturation factor gamma enhances actin remodeling and B cell antigen receptor signaling at the immune synapse. Front Cell Dev Biol 2021;9:647063.
[391] Hoogeboom R, Natkanski EM, Nowosad CR, Malinova D, Menon RP, Casal A, et al. Myosin IIa promotes antibody responses by regulating B cell activation, acquisition of antigen, and proliferation. Cell Rep May 2018;23(8):2342–53.
[392] Tangye SG, Bucciol G, de Las Casas JM, Pillay B, Ma CS, Moens L, et al. Human inborn errors of the actin cytoskeleton affecting immunity: way beyond WAS and WIP. Immunol Cell Biol February 2019;97:389–402.
[393] Sprenkeler EGG, Webbers SDS, Kuijpers TW. When actin is not actin' like it should: a new category of distinct primary immunodeficiency disorders. J Innate Immun 2021;13:3–25.
[394] Tur-Gracia S, Martinez-Quiles N. Emerging functions of cytoskeletal proteins in immune diseases. J Cell Sci February 2021;134(3):jcs253534.

Chapter 6

Fc Receptors

Jeffrey V. Ravetch and Stylianos Bournazos
Laboratory of Molecular Genetics and Immunology, The Rockefeller University, New York, NY, United States

1. INTRODUCTION

Throughout their development, B cells express specialised receptors with the capacity to interact with the Fc domain of antibodies. Engagement of these Fc receptors (FcRs) mediates pleiotropic signalling events that readily impact numerous cellular processes. As a result, Fc–FcR interactions on B cells have significant effects on the development of humoral immune responses through modulation of B-cell selection and survival, regulation of affinity maturation, as well as control of antibody production and plasma cell apoptosis. Given the critical importance of Fc–FcR interactions during the humoral immune response, such interactions are regulated through the controlled expression of the various FcRs expressed on B cells at different stages of B-cell development and activation, as well as through modulation of the Fc domain structural heterogeneity, which in turn influences the Fc domain affinity for FcRs. In this chapter, we discuss the role and biological significance of specific FcR pathways in the regulation of B-cell function, focussing on the determinants that regulate Fc–FcR interactions, the downstream signalling events that result from FcR crosslinking and the biological consequences of FcR engagement in the various B-cell subsets and populations. Lastly, we discuss studies that have characterised the phenotype of mouse strains deficient in specific FcR expression, as well as clinical studies examining the association of aberrant FcR expression and activity with autoimmune disease pathologies.

2. STRUCTURAL DETERMINANTS OF IgG–FcR INTERACTIONS

IgG antibodies represent the most abundant immunoglobulin isotype in circulation with the capacity to mediate pleiotropic effector functions upon engagement to FcγRs expressed on the surface of various innate and adaptive leucocyte populations during an immune response. Given the central role of IgG antibodies in immunity against foreign pathogens, the determinants that regulate their affinity for the various classes of FcRs have been extensively studied. Antibody responses are characterised by inherent polyclonality, resulting in effective targeting of multiple antigens through diversification of their antigen-binding Fab domains, as well as through modulation of their Fc domain structure and function. For the Fab domains, somatic recombination and hypermutation of their variable regions give rise to antibodies with almost unlimited specificity against diverse foreign antigens, whereas the Fc domain also exhibits significant heterogeneity, which accounts for differential capacity to engage and activate specific FcRs on effector leucocytes [1]. Indeed, similar to any aspect of the immune system, there are multiple layers of immunoregulation that fine tune the affinity of the Fc domain for FcRs, thereby significantly reducing the possibility for aberrant activation of pro-inflammatory pathways, or the emergence of autoreactive antibody clones.

The intrinsic structure and heterogeneity of the Fc domain of the IgG molecule represent the key determinants of Fc–FcR interactions. Such heterogeneity stems from the differences in the primary amino acid sequence among the four subclasses of human IgG (IgG1-4), as well as from the structure and composition of the complex, biantennary N-linked glycan attached at Asn297, a site that is conserved for all IgG subclasses across multiple mammalian species [1,2] (Figure 6.1). Such heterogeneity translates into hundreds of different Fc structures with differential conformational flexibility and capacity to interact with the various FcR types (Figure 6.2). Based on the conformational states that the Fc domain can adopt, FcRs are broadly categorised into two main types, each with distinct structural properties, IgG specificity and capacity to interact with the IgG Fc domain at specific regions [3]. Type I FcRs comprise six structurally related FcRs that belong to the immunoglobulin receptor superfamily, as their extracellular, ligand-binding domain comprises two or three Ig domains. Type I FcRs

Molecular Biology of B Cells. https://doi.org/10.1016/B978-0-323-95895-0.00005-2

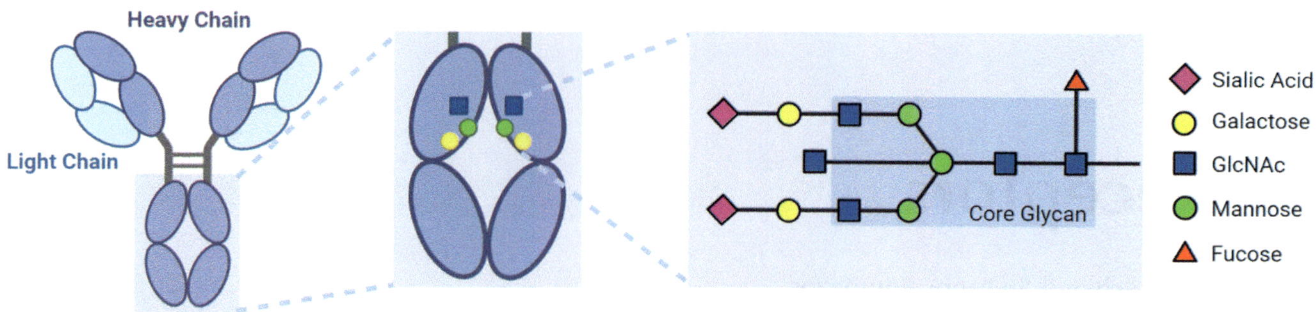

FIGURE 6.1 **Overview of the structure and composition of the Fc-associated glycan.** An N-linked glycan is conjugated at a conserved site at the CH2 domain of the Fc region of IgG antibodies. This complex heptasaccharide structure comprises mannose and *N*-acetylglucosamine residues. Selective addition of other saccharide units, such as fucose, galactose, *N*-acetylglucosamine and sialic acid to the core glycan structure (noted in darker shade), results in the generation of diverse Fc glycoforms with differential Fc domain structural flexibility. As a result, Fc glycoforms exhibit unique binding profiles to type I and type II FcRs. *Created with BioRender.com.*

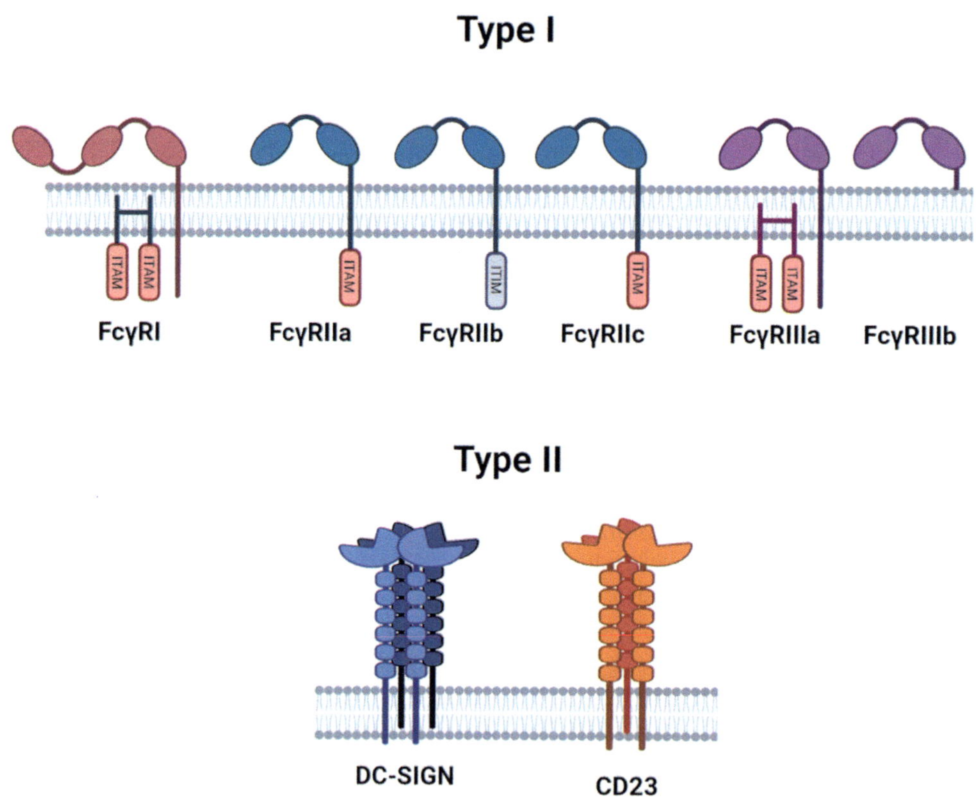

	Type I	Type II
Structural Receptor Family	IgSF	C-type lectins
Receptor:Ligand Stoichiometry	1:1	2:1
Ligand Specificity	High	Low
Fc Binding Site	Hinge-CH2	CH2-CH3

FIGURE 6.2 **Characteristics of type I and type II FcRs.** FcRs are broadly divided into two main types – type I and type II, based on their capacity to interact with different conformational states of the Fc domain. Type I FcRs belong to the Ig superfamily (IgSF), having two or three Ig domains that facilitate binding of the Fc region at the hinge–CH2 interface. In contrast, type II FcRs are heavily glycosylated C-type lectin receptors, which can engage a diverse range of ligands, including sialylated IgG Fc glycoforms. Cross-linking of type I and type II FcRs results in pleiotropic biological consequences that modulate several aspects of the innate and adaptive immune response. *Created with BioRender.com.*

include FcγRI, FcγRIIa, FcγRIIb, FcγRIIc, FcγRIIIa and FcγRIIIb and interact with the IgG Fc domain at the hinge-proximal region of the CH2 domain at a 1:1 complex [4–8]. By contrast, Type II FcγRs engage the Fc domain at the CH2–CH3 interface in a complex with a stoichiometry of 2:1 (receptor:antibody) [9–11]. Type II FcRs comprise the C-type lectin receptors DC-SIGN and CD23, which in contrast to Type I FcRs, have the capacity to interact with multiple, diverse ligands, including IgE for CD23, as well as glycoproteins and carbohydrate structures for DC-SIGN.

Although crystallographic studies have demonstrated that the Fc-associated glycan structure does not make any direct contacts with Type I or Type II FcRs, engagement of these receptors requires at a minimum the presence of the core glycan structure, suggesting that the glycan likely determines the affinity of the Fc–FcR interaction by modulating the conformational state of the Fc domain [6,7,12,13]. Indeed, the presence of the Fc-associated glycan is necessary to stabilise the Fc domain at a conformation that is permissive for FcR binding. Crystal structures of aglycosylated Fc fragments suggested that the lack of the glycan structure results in a closed conformation, where the CH2 domains collapse, thereby occluding the Type I FcR binding site [14]. Given the biantennary nature of the Fc-associated glycan, the α1,6 arm folds along the CH2 protein backbone, making several non-covalent contacts with amino acid side chains, whereas the α1,3 arm extends into the cavity between the CH2 domains, interacting with the other α1,3 arm conjugated to the opposite heavy chain. This unique orientation of the Fc-associated glycan limits the interdomain flexibility of the CH2 domains, stabilising the Fc domain at a conformation that allows for binding to Type I FcRs [3,9,15].

The presence of the core glycan structure is therefore a critical determinant for the control Fc domain structure and Fc–FcγR interactions; however, the affinity of the Fc domain for Type I and Type II FcRs is largely determined by the precise composition of the Fc-associated glycan structure [1,15]. The core glycan structure is a complex heptasaccharide comprising mannose and N-acetylglucosamine residues (Figure 6.1). Selective addition of other saccharide units, such as fucose, galactose, N-acetylglucosamine and sialic acid to the core glycan structure has a significant impact on the flexibility of the Fc domain, thereby modulating its affinity for Type I and Type II FcRs. For example, the presence of terminal sialic acid residues at the Fc-associated glycan enhances the flexibility of the Fc domain through destabilisation of its quaternary structure, resulting in the exposure of the Type II FcR binding site at the CH2–CH3 interface of the Fc domain [3,9,16–19]. As a result, sialylated Fc glycoforms acquire the capacity to interact with Type II FcRs, such as DC-SIGN and CD23, to modulate innate and adaptive immune responses.

The importance of the Fc-glycan composition during an immune response has been highlighted in multiple studies on the heterogeneity of the Fc glycan structure in health and disease, as well as in studies focussing on the mechanisms that drive these glycan modifications and their biological impact on immune responses. Analyses of the IgG Fc glycan structure revealed that pregnancy, chronic inflammatory diseases, systemic metabolic disorders and infection often result in changes in the serum levels of specific Fc glycoforms, suggesting the presence of active immunoregulatory pathways that control the composition of the Fc glycan structure during an immune response [20–37]. This assumption is further strengthened by a number of mechanistic studies in human vaccination cohorts and in transgenic mouse strains that provided novel insights into the dynamic changes in the Fc glycan composition during an immune response. Influenza vaccination of humans results in the induction of specific Fc glycoforms at specific time points following immunisation, with specific glycan changes being associated with enhanced vaccine efficacy [23]. Likewise, studies in mouse strains encompassing conditional B-cell genetic deletion of the enzyme that catalyses the addition of terminal sialic acid residues to the Fc glycan structure (*St6gal1*) (AID-Cre, *St6gal1*$^{fl/fl}$) revealed that in models of collagen-induced arthritis (CIA), loss of sialylation of anti-collagen II IgGs elicited upon immunisation further exacerbated joint inflammation, whereas transfer of ex vivo sialylated IgGs abrogated their pathogenic potency [25]. In summary, the findings of these studies highlight the critical importance of the Fc glycan structure in modulating the Fc domain affinity for Type I and Type II FcRs and consequently the capacity of the IgG molecule to mediate immunomodulatory activities.

3. B-CELL IgG Fc RECEPTORS

3.1 FcγRIIb

The Type I FcR, FcγRIIb, shares the same structural properties and has a high degree of sequence similarity with the other members of the FcγRII family (FcγRIIa and FcγRIIc), featuring a characteristic structure that includes functional signalling motifs in the intracellular domains. As a result, in contrast to other FcγRs, FcγRII have the capacity to directly transduce intracellular signals following receptor crosslinking and engagement without the need for co-expression of accessory signalling subunits (e.g. FcR γ-chain as in the case of FcγRI and FcγRIIIa). At its extracellular, ligand-binding region, FcγRIIb comprises two Ig domains, whereas its intracellular domain features immunoreceptor tyrosine-based inhibition motifs (ITIMs) that account for FcγRIIb signalling activity [5,38,39].

Alternative splicing of the *FCGR2B* transcript gives rise to two distinct isoforms of FcγRIIb: FcγRIIb1, which represents the full-length transcript, and FcγRIIb2, which is generated by skipping an exon that encodes a membrane proximal intracellular region [40–43]. Since this region includes a signal sequence that putatively inhibits receptor internalisation, FcγRIIb2 is readily internalised following receptor crosslinking. In contrast, the full-length FcγRIIb1 isoform is incapable of internalisation [42–44]. Apart from their differential internalisation properties, FcγRIIb isoforms exhibit lineage-specific expression; FcγRIIb1 expression is restricted to B cells, while cells of the myeloid lineage including macrophages, dendritic cells (DCs) and monocytes express the FcγRIIb2 isoform [42,44].

At all stages of B-cell development, including pre-, pro- and mature B-cell populations, FcγRIIb is constitutively expressed. Of note, FcγRIIb expression is maintained on memory B cells and plasma cells, on which most of the other B lineage molecules are either absent or expressed at very low levels [45,46]. Expression levels of FcγRIIb are modulated on different B-cell populations, displaying higher levels on peripheral B cells and reduced expression on germinal centre (GC) and activated B cells [47,48]. Interleukin-4 (IL-4) further downregulates FcγRIIb expression on GC B cells [48]. FcγRIIb deficiency does not appear to perturb the development of any of these lineages, suggesting that FcγRIIb functions specifically in response to immune complex engagement during the active immune response. Consistent with this scenario, several studies indicate that FcγRIIb limits the expansion of autoreactive plasmablasts and plasma cells that are present at increased frequencies in autoimmune-prone mouse strains expressing reduced levels of the inhibitory FcγRIIb [45,49]. FcγRIIb is not expressed on natural killer (NK) cells, T cells or stromal cells.

Similar to other members of the Type I FcγRs, FcγRIIb interacts with the IgG Fc domain at the hinge-proximal region of the CH2 domain at a 1:1 stoichiometry. This interaction is low affinity (KD for human IgG1 is approximately 10^{-6} M), and therefore, FcγRIIb is incapable of binding monomeric IgG molecules [50]. Instead, IgG immune complexes can sufficiently engage and crosslink FcγRIIb through multimeric low-affinity yet high-avidity interactions.

3.2 CD23

The Type II FcR, CD23, is encoded by the *FCER2* gene and processed as two distinct splicing variants – CD23a and CD23b. These variants are differentially expressed and regulated among the various leucocyte types and exhibit distinct signalling activity [51]. CD23a is constitutively expressed throughout B-cell development with a very low expression on marginal zone B cells and T1 transitional B cells in the spleen. By contrast, CD23b is expressed by B cells as well as by cells of the non-B-cell lineage, including neutrophils, eosinophils, macrophages, monocytes and T cells and its expression is induced by IL-4 [51]. The cytoplasmic domain of CD23a and CD23b differs by six amino acids, resulting in differential receptor signalling activity and receptor endocytosis. Crosslinking of either isoform results in an intracellular increase in the levels of cAMP; however, CD23a can uniquely activate the PLC pathway and modulate the activity of multiple signalling pathways (NF-κB and MAPK) [52–55]. Likewise, the presence of a tyrosine-based endocytosis signal at the cytoplasmic tail of CD23a results in receptor endocytosis into clathrin-coated pits, whereas crosslinking of CD23b by IgE immune complexes leads to phagocytosis [10,11,56].

Unlike FcγRIIb, CD23 does not belong to the immunoglobulin superfamily of proteins but rather to the family of C-type lectin receptors, which includes DC-SIGN (CD209). Both receptors share a characteristic oligomeric structure (trimeric or tetrameric) that is stabilised by an α-helical coiled-coil stalk domain at the extracellular, ligand-binding region. *FCER2*, the gene that encodes CD23, is mapped at the same locus with *CD209* (19*p*13), reflecting the similarities in the function and IgG-binding activity between these two receptors. As member of the Type II FcR family, CD23 interacts with IgG at the CH2:CH3 interface at a 2:1 (receptor:IgG) stoichiometry [9,10]. This binding is dependent on the presence of terminal sialic acid residues at the Fc-associated glycan structure, which increases the flexibility of the IgG Fc domain, thereby allowing it to interact with sufficient affinity with CD23 in a fashion analogous to IgE [9].

4. FcγR FUNCTION AND PROPERTIES

4.1 Signalling Pathways

Type I FcRs are functionally categorised into activating or inhibitory based on the presence of ITAM or ITIM motifs at their intracellular domains [57]. The singular inhibitory receptor, FcγRIIb, is expressed by several leucocyte cell types of the myeloid and lymphoid lineage, including macrophages, DCs, granulocytes, monocytes and B cells. Myeloid cell populations typically co-express activating and inhibitory FcRs, and the outcome of IgG-mediated inflammation represents the outcome of the opposing signalling activity of activating and inhibitory FcRs [58]. In contrast, B cells represent a unique cell population that expresses the inhibitory FcγRIIb, while they are negative for activating FcR expression. As a result, on B cells, the main signalling function of FcγRIIb is to modulate the activity of the B-cell receptor (BCR), thereby impacting many aspects of B-cell activation, selection, differentiation and survival [58].

The cytoplasmic domain of FcγRIIb features the inhibitory motif (ITIM), composed of a 13 amino acid sequence (AENTITY**Y**SLLKHP), which is necessary and sufficient to mediate the inhibition of BCR-generated calcium mobilisation and cellular proliferation [39,59]. The inhibitory activity of FcγRIIb is achieved through co-ligation with the BCR, which induces phosphorylation of the ITIM tyrosine (highlighted above in bold). ITIM phosphorylation generates a Src homology 2 (SH2) recognition domain that serves as the binding site for the inhibitory signalling molecule SH2 domain-containing inositol polyphosphate 5-phosphatase (SHIP) [60,61].

FcγRIIb displays three distinct inhibitory activities, two of which are dependent on the ITIM motif and one that is independent of this motif (Figure 6.3). Co-engagement of FcγR to an ITAM-containing receptor (e.g. BCR for B cells) leads to tyrosine phosphorylation of the ITIM by the lyn kinase, recruitment of SHIP and the inhibition of ITAM-triggered calcium mobilisation and cellular proliferation [38,61]. These two activities result from different signalling pathways, with calcium inhibition requiring the phosphatase activity of SHIP to hydrolyse phosphatidylinositol (3,4,5)-triphosphate (PIP3) and the ensuing dissociation of pleckstrin homology (PH) domain–containing proteins such as Btk and phospholipase Cγ [62]. The net effect is to block calcium influx and prevent sustained calcium signalling. Calcium-dependent processes such as degranulation, phagocytosis, antibody-dependent cell-mediated cytotoxicity, cytokine release and proinflammatory activation are all blocked. The arrest of proliferation in B cells is also dependent upon the ITIM pathway, through the activation of the adaptor protein, Dok, and subsequent inactivation of mitogen-activated protein kinases [63,64]. The role of SHIP in this process has not been fully defined, because it can affect proliferation in several ways. SHIP, through its catalytic phosphatase domain, can prevent activation of the PH domain survival factor Akt by hydrolysis of PIP3 [65,66]. SHIP also contains phosphotyrosine-binding domains that could act to recruit Dok to the membrane and provide access to the Lyn kinase that is involved in its activation. Dok-deficient B cells are unable to mediate the FcγRIIb-triggered arrest of BCR-induced proliferation, while retaining their ability to inhibit a calcium influx, thus demonstrating the dissociation of these two ITIM-dependent pathways (Figure 6.3).

The third inhibitory activity displayed by FcγRIIb is independent of the ITIM sequence and is displayed upon homoaggregation of the receptor. Under these conditions of FcγRIIb clustering by IgG immune complexes, a proapoptotic signal is generated through the transmembrane sequence. This proapoptotic signal is blocked by the recruitment of SHIP, which occurs upon co-ligation of FcγRIIb to the BCR, due to the Btk requirement for this apoptotic pathway. In addition to Btk, c-Jun NH2-terminal kinase and the c-Abl kinase were also involved in this pathway [67] (Figure 6.3B). This novel activity has only been reported in B cells and has been proposed to act as a means of maintaining peripheral tolerance for B cells that have undergone somatic hypermutation. Support for this model comes from the in vivo studies of FcγRIIB-deficient mice in induced and spontaneous models of autoimmunity. Studies have provided evidence that FcγRIIb is involved in affinity maturation, Ig enhancement and suppression, and

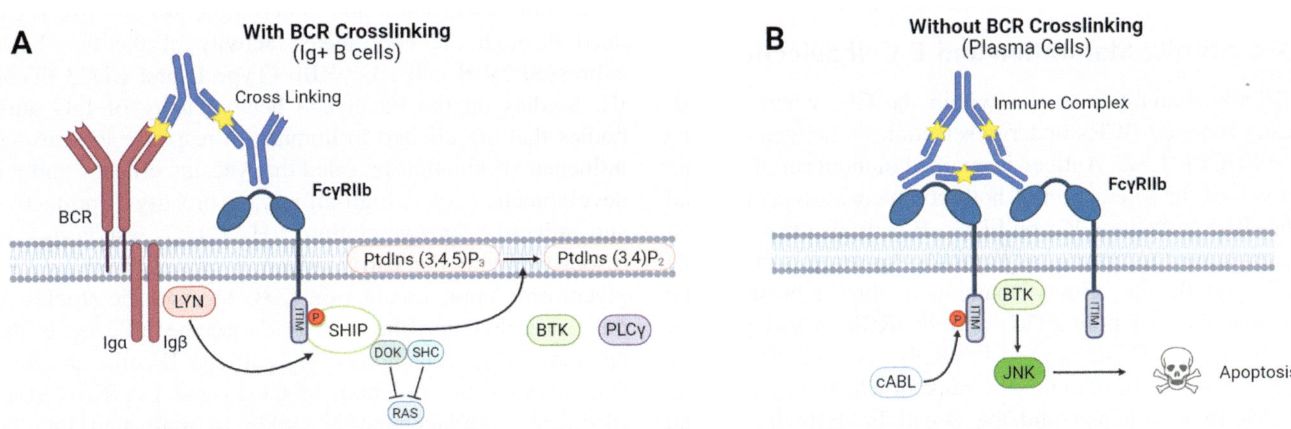

FIGURE 6.3 **Differential signalling events following FcγRIIb engagement on B cells by IgG immune complexes.** Depending on the expression of the B-cell receptor (BCR), engagement of FcγRIIb triggers distinct signalling events following crosslinking. (A) In the presence of BCR, co-crosslinking of the BCR with FcγRIIb results in the phosphorylation of the ITIM motif that is present at the cytoplasmic tail. This phosphorylation is mediated by Lyn and results in the recruitment of SHIP phosphatases. SHIP contains phosphotyrosine-binding domains that could act to recruit Dok to the membrane, resulting in the inactivation of mitogen-activated protein kinases. Additionally, SHIP, through its catalytic phosphatase domain, induces the hydrolysis of $PtdIns(3,4,5)P_3$ into $PtdIns(4,5)P_2$, thereby inhibiting the recruitment and activation of PH-containing proteins, such as BTK and PLCγ. (B) In the absence of BCR co-ligation, crosslinking of FcγRIIb by IgG immune complexes induces cell apoptosis through SHIP-independent mechanisms that involve the activation of cABL, BTK and Jun N-terminal kinase (JNK). *Created with BioRender.com.*

generation of the memory and plasma cell response [46,49,68]. These various roles contribute to the overall role of FcγRIIb in the maintenance of peripheral tolerance because perturbation in one or more of these pathways can contribute to the emergence and amplification of the autoimmune response.

Although numerous studies have defined the exact downstream events that are initiated upon FcγRIIb crosslinking, the signalling activities of Type II FcRs, such as DC-SIGN and CD23 are poorly defined. Since CD23 has been originally described as the low-affinity receptor for IgE, studies on CD23 signalling have mostly focused on characterising the activity of signalling pathways upon engagement and crosslinking of CD23 by IgE immune complexes. IgE–CD23 interactions trigger an increase in the intracellular cAMP levels, as well as activation of several kinases, including those of the MAPK kinase family [53,55,69]. Although these effects have been observed to occur following engagement and crosslinking of CD23 by IgE immune complexes, it remains to be characterised whether binding of sialylated IgG antibodies also results in comparable effects.

Additionally, CD23 lacks any functional intracellular motifs at its cytoplasmic tail, suggesting that any signalling events that are initiated following receptor crosslinking might be mediated through putative accessory signalling proteins that are associated with CD23. The exact nature of these proteins remains unknown; however, it has been previously demonstrated that CD23 can interact with CD21 on B cells and with CD11b and CD11c on myeloid cells [11,69,70]. Whether or not the physical association of CD23 with these receptors is necessary for the downstream effects following CD23 crosslinking is unclear.

4.2 Affinity Maturation and B-Cell Selection

Affinity maturation occurs within the GC, where somatically mutated BCRs undergo selection on antigen retained on FDCs [71,72]. Antigen is retained in the form of ICs and involves the interaction of both complement receptors and FcγRIIb with these ICs on FDCs. B cells also express both complement receptors and FcγRIIb. An analysis of the role of FcγRIIb in affinity maturation thus represents the contribution of both FDCs and FcγRIIb to this process. The role of FDC-expressed FcγRIIb has been clarified by the development of chimaeric mice in which only the FDC FcγRIIb is deficient and the B-cell FcγRIIb is retained [71,73]. In these mice, potentiation of affinity maturation is seen. This potentiation likely results from the increased stringency of selection that occurs in the GC that, in turn, results from unopposed access of FDC ICs to B-cell FcγRIIb, as has been proposed previously [65]. Coligation of the BCR and FcγRIIB attenuates the BCR signal by the negative signalling role of SHIP. Such negative signal would impose a requirement for a higher threshold for effective BCR stimulation and thus favour higher affinity BCRs. In addition, under those conditions where co-ligation to BCR is ineffective, ligation of FcγRIIb alone results in an apoptotic signal, resulting in the elimination of those somatically mutated B cells with low affinity for antigen (Figure 6.3).

The situation is reversed when B cells that lack FcγRIIb expression are transferred. In that case, decreased apoptosis is observed, and the resulting B cells display a reduced affinity for antigen. This is perhaps due to the persistence of low-affinity B cells that arise as a consequence of somatic mutation and are normally eliminated by a negative selection mechanism involving FcγRIIb crosslinking and induction of apoptosis. Because FcγRIIb is regulated on both FDCs and B cells, the potential is provided for fine-tuning the survival and selection of B cells in the GC by the interaction with ICs on FDCs. This role of FcγRIIb on affinity maturation could provide part of the explanation for the loss of tolerance in FcγRIIb-deficient animals. Because FcγRIIb would provide one mechanism for the elimination of low-affinity and potentially autoreactive BCRs that may arise through somatic mutation during the GC reaction, decreasing its expression on GC B cells could favour the persistence and eventual positive selection of these cells. This property could dominate over loss of FcγRIIb on FDCs, because its contribution to the retention of antigen is minimal compared with that of complement receptors. Enhancement of this autoimmune phenotype would be mediated through DC maturation and the resulting T-cell activation [72]. The analysis of conditional deficiency of FcγRIIb on B cells, FDCs and DCs will clarify these activities.

Affinity maturation and B-cell selection are also regulated through the coordinated activity of the two FcRs expressed by B cells: FcγRIIb (Type I) and CD23 (Type II). Studies on the Fc glycan heterogeneity of IgG antibodies that are elicited in humans in response to seasonal influenza vaccination revealed that vaccine efficacy and the development of high-affinity, broadly protective anti-influenza haemagglutinin (HA) IgG responses are associated with the early induction of sialylated IgG Fc glycoforms upon vaccination [23]. Mechanistic studies in mice revealed that the improved vaccine efficacy is the outcome of the selection of high-affinity B cells, an effect that involves the interplay of CD23 and FcγRIIb. More specifically, engagement of CD23 by sialylated IgG Fc glycoforms induced the upregulation of FcγRIIb on B cells in an autocrine manner [23]. As a result of the increased FcγRIIb expression, the threshold for B-cell selection is elevated, resulting in the elimination of low-affinity B cells, while B cells with high-affinity BCRs are selected (Figure 6.4). Novel immunisation approaches have therefore been developed based on the novel

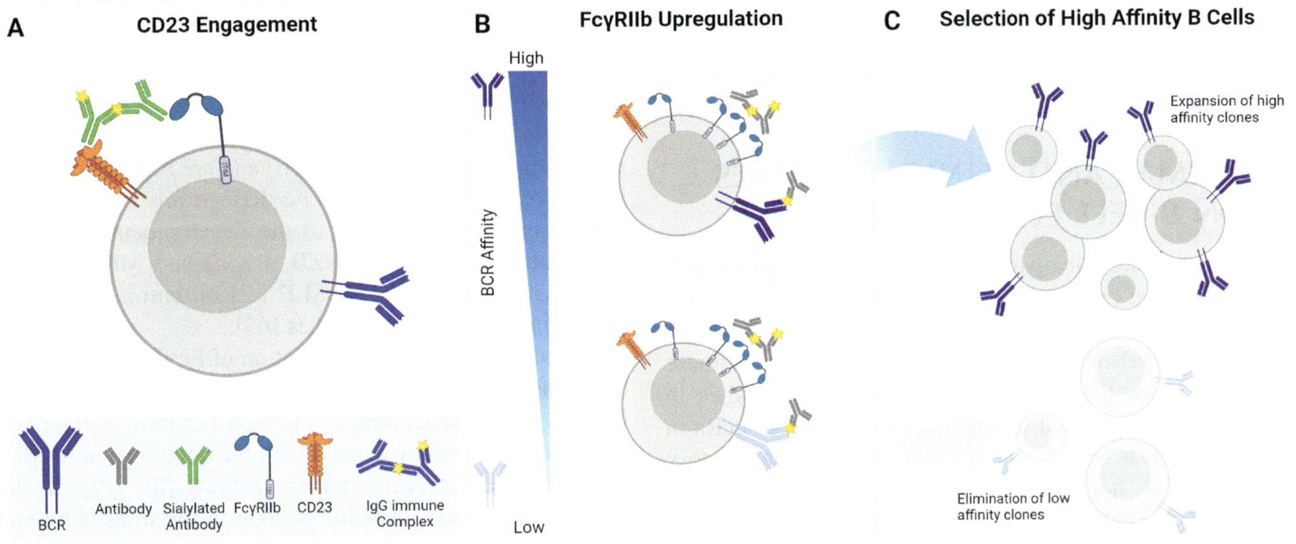

FIGURE 6.4 **Interplay of type I and type II FcR signalling in the regulation of B-cell selection and the development of high-affinity IgG responses.** B cells co-express the type I FcR, FcγRIIb, as well as the type II FcR, CD23. (A) Engagement of CD23 by sialylated IgG results in the subsequent upregulation of FcγRIIb on B cells (B). Increased FcγRIIb expression raises the threshold for BCR-mediated signalling and activation. As a result, B cells with higher affinity and signalling activity through BCR can overcome the inhibitory signals from FcγRIIb, leading consequently to the selection of high-affinity B-cell clones and the elimination of B cells with low BCR affinity (C). *Created with BioRender.com.*

immunomodulatory activity of the sialylated Fc–CD23–FcγRIIb pathway [1,23,74–76]. Indeed, using influenza HA as a model antigen, administration of sialylated Fc anti-HA immune complexes resulted in the induction of higher affinity anti-HA IgG responses, with potent and broad neutralising activity, as well as capacity to protect in vivo against homologous and heterologous influenza challenge [23,75]. By contrast, administration of asialylated Fc IgG immune complexes or vaccination of CD23$^{-/-}$ mice failed to result in the induction of high-affinity anti-HA antibody responses, thereby highlighting the key importance of the CD23- FcγRIIb pathway in the regulation of B-cell selection [23,75].

4.3 Ig Suppression

The ability of IgG to mediate the suppression of antibody responses is perhaps the best-known clinical application of feedback suppression. The administration of IgG anti-erythrocyte-specific antibody completely prevents the emergence of anti-erythrocyte antibody response [72,77] and prevents the emergence of haemolytic disease of the newborn [78]. The responsible mechanism is still not defined, although experimental support for both epitope masking and FcγRIIb-mediated inhibition has been presented. In experimental mouse models, monoclonal IgG-anti-TNP suppressed the primary response to trinitrophenyl-conjugated sheep erythrocytes in mice that lacked FcγRIIb, suggesting an FcγRIIb-independent mechanism [79]. Other studies demonstrated decreased suppression for IgG antibodies with reduced FcR-binding ability or through blocking FcR engagement by antibody blockade [80]. These differences may reflect the selective effects of decreased FcγRIIb binding on either B cells, FDCs or antigen-presenting cells (APCs) where different effects can be elicited depending upon the pathway affected. Discrimination between these pathways is necessary to define the principal mechanism underlying IgG suppression.

4.4 Plasma Cell Homoeostasis

Upon B-cell activation and affinity maturation in the spleen and lymph nodes (LNs), short-lived plasmablasts and long-lived plasma cells develop of which the latter populations home to niches in the bone marrow that provide factors for the long-term survival of these cells. The limited number of niches available in the bone marrow creates the conceptual problem that once all the niches are occupied with plasma cells, no further plasma cells generated during a subsequent antigen encounter or vaccination would have the capacity to become long lived and generate protective antibody titres [81]. As this is not the case, there have to be mechanisms in place that generate free niches for newly generated plasma cells. There is convincing evidence that signalling via FcγRIIb might be involved in this pathway via its proapoptotic function. Thus, plasma cells downregulate BCR expression while upregulating FcγRIIb expression levels. ICs generated during an ongoing immune response would therefore selectively trigger FcγRIIb-dependent proapoptotic signalling pathway on plasma cells that would induce cell death on at least a portion of the current plasma cell pool, allowing newly formed plasma cells to occupy these niches [46,82].

5. ANIMAL AND CLINICAL STUDIES ON B-CELL FcR FUNCTION

The role and biological significance of Type I and Type II FcRs expressed by B cells in the development of antibody responses is best exemplified by the phenotype and functional characteristics of animal strains deficient for these receptors, as well as from genetic association studies in humans that established that contribution of these FcRs in the pathogenesis of autoimmune and chronic inflammatory disorders. The targeted disruption of FcγRIIb results in animals with amplified antibody responses to both T-independent and T-dependent antigens, with Ig titres increased by five- to ten-fold [83]. Baseline serum immunoglobulin levels are unaffected, and the mice display a normal Ig half-life. This amplification is apparent on all genetic backgrounds investigated. Backcrossing onto the C57BL/6 background further resulted in animals with severe autoimmune disease [84,85]. These animals displayed a spontaneous loss of tolerance to nuclear antigens, with high titres of IgG antibodies to double-stranded DNA (dsDNA) and histone H2A/2B by six months of age. Autoimmune disease was apparent in these animals, with prominent vasculitis and glomerulonephritis resulting from IC deposition. The defect in these animals appears to reside, in part, in the B-cell compartment, as determined by reconstitution of the autoimmune phenotype through the transfer of FcγRIIb-deficient bone marrow into B-cell— or T- and B-cell—deficient recipients who have otherwise normal FcγRIIb expression on their myeloid cells. Moreover, restoring FcγRIIb expression either ubiquitously, or selectively on B cells, was able to restore humoral tolerance in autoimmune-prone mouse strains with impaired FcγRIIb expression, including FcγRIIb knockout mice [86,87]. Of note, treatment of mice and human patients with chronic inflammatory demyelinating polyneuropathy (CIDP) with pooled polyclonal IgG preparations (intravenous immunoglobulin therapy (IVIG)) was shown to upregulate the inhibitory FcR on myeloid cells and B cells, which may heighten the threshold for autoantibody production [88—91]. Further evidence that FcγRIIb-deficient animals have a defect in the maintenance of peripheral tolerance was demonstrated by the ability of these animals to develop autoantibodies to murine type II or type IV collagen after immunisation with bovine type II or IV collagen, respectively, on a non-susceptible H-2b background [60,92]. The consequence of this induced breakage of tolerance to these autoantigens was the development of CIA in non-susceptible backgrounds in response to type II collagen immunisation, or Goodpasture's syndrome in the case of type IV collagen immunisation. These data suggest that decreased levels of FcγRIIb correlate with a susceptibility to the development of autoimmune diseases such as systemic lupus erythematosus (SLE) and are further supported by the reports of decreased FcγRIIb levels on B cells in mouse strains predisposed to the development of antinuclear antibodies, such as NZB, BXSB and MRL [93,94], and by its tight linkage to SLE loci on chromosome 1 in both human and mouse studies [63].

In humans, a failed upregulation of FcγRIIb from naïve to memory B cells was noted in SLE and CIDP patients [88,95], and a functionally impaired FcγRIIb variant was found to be highly associated with the development of SLE in Asian and Caucasian patient cohorts [96—99]. In this allelic variant, an isoleucine residue in the transmembrane domain of FcγRIIb is replaced by a charged threonine residue (FcγRIIb$_{I232T}$), resulting in an impaired recruitment to lipid rafts and a strongly reduced function of this receptor variant. More direct evidence that this functionally impaired FcγRIIb allele indeed might be responsible for a loss of humoral tolerance in humans was afforded by studies in which immunodeficient mice were reconstituted with human hematopoietic stem cells from donors carrying either the functionally intact FcγRIIb-232I or the altered FcγRIIb-232T variant [45]. Indeed, a variety of autoantibody species including dsDNA antibodies associated with the development of autoimmune disease in humans could be detected in humanised mice, in which a human immune system developed on the nonfunctional FcγRIIb-I232T background. In addition, polymorphisms at the promoter of the *FCGR2B* gene that decrease FcγRIIb expression levels have been shown to be associated with SLE susceptibility and likely contribute to disease pathogenesis [100,101]. These data indicate that B-cell FcγRIIb expression and activity contributes to the maintenance of peripheral tolerance to nuclear antigens and to the suppression of autoantibody production.

5.1 Non-IgG Fc Receptors

5.1.1 IgE Engagement of CD23

CD23 engagement on B cells by IgE can result in either positive or negative responses. For example, injection of small soluble antigens complexed with IgE was demonstrated to enhance the antibody response in a CD23-dependent manner [77]. Additionally, CD23 engagement is required for the IgE-mediated enhancement of T-cell responses [69,102]. These enhanced T cell responses result from more efficient capture of IgE immune complexes by B cells and transfer to CD11c$^+$ APCs [78]. In contrast, IgE binding to CD23 on B cells results in reduced

IgE production. As a consequence of this activity, CD23-deficient mice have elevated serum IgE levels. Apart from B cells, CD23 expression has been described on FDCs; however, the aforementioned enhancement of antibody responses was suggested to be dependent on CD23 expressed on B cells, indicating that B cells may directly act as APCs to enhance T helper (Th) cell responses [78].

5.1.2 FcR-like (FCRL) Proteins

In addition to the inhibitory FcγRIIb, several members of the family of FCRL proteins are expressed on mouse and human B cells [103,104]. As most of our current knowledge about the function of these proteins comes from human cells, we focus on the human system here. The family of human FCRLs consists of eight members of which FcRL1−FcRL6 are transmembrane proteins, whereas FcRLA and FcRLB may represent intracellular or secreted proteins. Similar to Type I FcγRs, the FcRL receptors can be divided into activating and inhibitory family members depending on the presence of cytoplasmatic ITAM (FcRL1) or ITIM (FcRL4 and FcRL6) motifs; FcRL2, FcRL3 and FcRL5 have both, ITAM-like and ITIM motifs. On B cells, expression of FcRL1−FcRL5 can be detected, although differences with respect to the various B-cell differentiation stages were noted. FcRL1, FcRL3 and FcRL5 are already expressed at the pre-B-cell stage, whereas expression of FcRL2 and FcRL4 is limited to memory and plasma cells. In general, FcRL expression is most abundant on memory B cells, whereas on plasma cells, FcRL5 expression is dominant.

Consistent with the presence of ITAM motifs in FcRL1, crosslinking of this receptor on B cells was shown to enhance BCR-induced calcium flux and B-cell proliferation [105]. In contrast, triggering of FcRL4 and FcRL5 delivered potent inhibitory signals, leading to reduced BCR-induced calcium flux in an SHP-1-dependent manner, suggesting that the noncanonical ITAM motif of FcRL5 is dominated by the ITIM motif also present in this receptor [106,107]. Despite the identification of this receptor family due to their sequence similarity to classical FcγRs, so far only FcRL4 and FcRL5 have been shown to be bona fide FcRs. Thus, FcRL4 was able to bind to immunoglobulin A, whereas FcRL5 could bind IgG. Interestingly, the molecular mechanism of binding seems to differ from IgG binding to classical FcγRs [108,109]. In both cases, the binding affinity was in the micromolar range, suggesting that immune complexes may be the physiological ligands of these receptors. At present, one can only speculate about the role of FcRL triggering via IgA- or IgG-containing immune complexes, but considering the rather restricted expression of FcRL4 on a subset of memory B cells, a negative feedback regulation via IgA complexes or dimers may be anticipated. With respect to FcRL5, one may expect a broader negative feedback on B-cell developmental stages due to the ubiquitous expression pattern throughout B-cell development, but more work will be necessary to fully understand the function of these proteins.

5.1.3 FcμR

Although the existence of an FcR for IgM was anticipated for a long time, the cloning and molecular characterisation of this molecule only started in 2009 [110]. Originally identified as a molecule modulating FAS-mediated apoptosis and hence called Fas apoptotic inhibitory molecule 3 (or TOSO), it was demonstrated to have the capacity to bind to polymeric IgM recently. The FcμR is located in proximity to the poly-Ig receptor and the Fca/μR on chromosome 1, but it only shows a distant relationship to these receptors based on amino acid sequence [110]. Its intracellular sequence does not contain any ITIM or ITAM motifs; however, it contains conserved Ser and Tyr residues that become phosphorylated upon IgM binding to the receptor. The expression of the FcμR is largely restricted to T cells and B cells in humans and shows a cell-type−specific modulation upon activation. Whereas it is upregulated on B cells upon BCR activation, it becomes downmodulated on T cells after activation. In mice and humans, FcμR is expressed from the pre-B/immature B cell to the plasma cell stage, with a notable reduction of expression on GC B cells. Besides B2-B cells, B1 cells also show a prominent expression of FcμR. In contrast to humans, FcμR expression in mice is restricted to B cells, and the first analysis of a mouse strain deficient for FcμR expression revealed no major phenotype for B-cell development [111]. However, these mice had increased levels of serum IgM and IgG3 and an increase in splenic B1 cells. In contrast, the amount of marginal zone B cells was reduced. Moreover, although responding well to nonprotein antigens such as phosphorylcholine, the antibody response to protein antigens was reduced in the absence of the FcμR. Of note, the heightened level of natural IgM antibodies was paralleled by enhanced levels of anti-chromatin and anti-DNA antibodies of IgM and IgG isotypes. Although much more work needs to be done to fully elucidate the function of FcμR on B cells, the available data suggest that it may be involved in modulating natural serum IgM and at least in part the B-cell response towards protein antigens.

References

[1] Wang TT, Ravetch JV. Functional diversification of IgGs through Fc glycosylation. J Clin Invest 2019;129:3492−8.

[2] Anthony RM, Wermeling F, Ravetch JV. Novel roles for the IgG Fc glycan. Ann N Y Acad Sci 2012;1253:170−80.

[3] Pincetic A, Bournazos S, DiLillo DJ, Maamary J, Wang TT, et al. Type I and type II Fc receptors regulate innate and adaptive immunity. Nat Immunol 2014;15:707−16.

[4] Radaev S, Motyka S, Fridman WH, Sautes-Fridman C, Sun PD. The structure of a human type III Fcgamma receptor in complex with Fc. J Biol Chem 2001;276:16469–77.

[5] Shields RL, Namenuk AK, Hong K, Meng YG, Rae J, et al. High resolution mapping of the binding site on human IgG1 for Fc gamma RI, Fc gamma RII, Fc gamma RIII, and FcRn and design of IgG1 variants with improved binding to the Fc gamma R. J Biol Chem 2001;276:6591–604.

[6] Sondermann P, Kaiser J, Jacob U. Molecular basis for immune complex recognition: a comparison of Fc-receptor structures. J Mol Biol 2001;309:737–49.

[7] Sondermann P, Huber R, Oosthuizen V, Jacob U. The 3.2-A crystal structure of the human IgG1 Fc fragment-Fc gammaRIII complex. Nature 2000;406:267–73.

[8] Maxwell KF, Powell MS, Hulett MD, Barton PA, McKenzie IF, Garrett TP, et al. Crystal structure of the human leukocyte Fc receptor, Fc gammaRIIa. Nat Struct Biol 1999;6:437–42.

[9] Sondermann P, Pincetic A, Maamary J, Lammens K, Ravetch JV. General mechanism for modulating immunoglobulin effector function. Proc Natl Acad Sci U S A 2013;110:9868–72.

[10] Dhaliwal B, Yuan D, Pang MO, Henry AJ, Cain K, et al. Crystal structure of IgE bound to its B-cell receptor CD23 reveals a mechanism of reciprocal allosteric inhibition with high affinity receptor FcεRI. Proc Natl Acad Sci U S A 2012;109:12686–91.

[11] Hibbert RG, Teriete P, Grundy GJ, Beavil RL, Reljic R, et al. The structure of human CD23 and its interactions with IgE and CD21. J Exp Med 2005;202:751–60.

[12] Ferrara C, Grau S, Jäger C, Sondermann P, Brünker P, et al. Unique carbohydrate-carbohydrate interactions are required for high affinity binding between FcgammaRIII and antibodies lacking core fucose. Proc Natl Acad Sci U S A 2011;108:12669–74.

[13] Krapp S, Mimura Y, Jefferis R, Huber R, Sondermann P. Structural analysis of human IgG-Fc glycoforms reveals a correlation between glycosylation and structural integrity. J Mol Biol 2003;325:979–89.

[14] Borrok MJ, Jung ST, Kang TH, Monzingo AF, Georgiou G. Revisiting the role of glycosylation in the structure of human IgG Fc. ACS Chem Biol 2012;7:1596–602.

[15] Wang TT. IgG Fc glycosylation in human immunity. Curr Top Microbiol Immunol 2019;423:63–75.

[16] Fiebiger BM, Maamary J, Pincetic A, Ravetch JV. Protection in antibody- and T cell-mediated autoimmune diseases by antiinflammatory IgG Fcs requires type II FcRs. Proc Natl Acad Sci U S A 2015;112:E2385–2394.

[17] Anthony RM, Kobayashi T, Wermeling F, Ravetch JV. Intravenous gammaglobulin suppresses inflammation through a novel T(H)2 pathway. Nature 2011;475:110–3.

[18] Anthony RM, Wermeling F, Karlsson MC, Ravetch JV. Identification of a receptor required for the anti-inflammatory activity of IVIG. Proc Natl Acad Sci U S A 2008;105:19571–8.

[19] Kaneko Y, Nimmerjahn F, Ravetch JV. Anti-inflammatory activity of immunoglobulin G resulting from Fc sialylation. Science 2006;313:670–3.

[20] Bournazos S, Vo HTM, Duong V, Auerswald H, Ly S, et al. Antibody fucosylation predicts disease severity in secondary dengue infection. Science 2021;372:1102–5.

[21] Wang TT, Sewatanon J, Memoli MJ, Wrammert J, Bournazos S, et al. IgG antibodies to dengue enhanced for FcγRIIIA binding determine disease severity. Science 2017;355:395–8.

[22] Yamin R, Kao K, MacDonald M, C T, R CM, R JV, et al. Human FcγRIIIa activation on splenic macrophages drives the in vivo pathogenesis of dengue disease. biorxv 2022. https://doi.org/10.1101/2022.11.02.514909.

[23] Wang TT, Maamary J, Tan GS, Bournazos S, Davis CW, et al. Anti-HA glycoforms drive B cell affinity selection and determine influenza vaccine efficacy. Cell 2015;162:160–9.

[24] Theodoratou E, Thaçi K, Agakov F, Timofeeva MN, Štambuk J, et al. Glycosylation of plasma IgG in colorectal cancer prognosis. Sci Rep 2016;6:28098.

[25] Ohmi Y, Ise W, Harazono A, Takakura D, Fukuyama H, et al. Sialylation converts arthritogenic IgG into inhibitors of collagen-induced arthritis. Nat Commun 2016;7:11205.

[26] Selman MH, de Jong SE, Soonawala D, Kroon FP, Adegnika AA, et al. Changes in antigen-specific IgG1 Fc N-glycosylation upon influenza and tetanus vaccination. Mol Cell Proteomics 2012;11. M111.014563.

[27] Espy C, Morelle W, Kavian N, Grange P, Goulvestre C, et al. Sialylation levels of anti-proteinase 3 antibodies are associated with the activity of granulomatosis with polyangiitis (Wegener's). Arthritis Rheum 2011;63:2105–15.

[28] Larsen MD, de Graaf EL, Sonneveld ME, Plomp HR, Nouta J, et al. Afucosylated IgG characterizes enveloped viral responses and correlates with COVID-19 severity. Science 2021;371.

[29] Kao KS, Gupta A, Zong G, Li C, Kerschbaumer I, et al. Synthetic nanobodies as tools to distinguish IgG Fc glycoforms. Proc Natl Acad Sci U S A 2022;119. e2212658119.

[30] Gonzalez JC, Chakraborty S, Thulin NK, Wang TT. Heterogeneity in IgG-CD16 signaling in infectious disease outcomes. Immunol Rev 2022;309:64–74.

[31] Chakraborty S, Gonzalez J, Edwards K, Mallajosyula V, Buzzanco AS, et al. Proinflammatory IgG Fc structures in patients with severe COVID-19. Nat Immunol 2021;22:67–73.

[32] Thulin NK, Brewer RC, Sherwood R, Bournazos S, Edwards KG, et al. Maternal anti-dengue IgG fucosylation predicts susceptibility to dengue disease in infants. Cell Rep 2020;31:107642.

[33] Offersen R, Yu WH, Scully EP, Julg B, Euler Z, et al. HIV antibody Fc N-linked glycosylation is associated with viral rebound. Cell Rep 2020;33:108502.

[34] Štambuk J, Nakić N, Vučković F, Pučić-Baković M, Razdorov G, et al. Global variability of the human IgG glycome. Aging (Albany NY) 2020;12:15222–59.

[35] Gudelj I, Lauc G, Pezer M. Immunoglobulin G glycosylation in aging and diseases. Cell Immunol 2018;333:65–79.

[36] de Man YA, Dolhain RJ, Hazes JM. Disease activity or remission of rheumatoid arthritis before, during and following pregnancy. Curr Opin Rheumatol 2014;26:329–33.

[37] van de Geijn FE, Wuhrer M, Selman MH, Willemsen SP, de Man YA, et al. Immunoglobulin G galactosylation and sialylation are associated with pregnancy-induced improvement of rheumatoid arthritis and the postpartum flare: results from a large prospective cohort study. Arthritis Res Ther 2009;11:R193.

[38] Daëron M, Latour S, Malbec O, Espinosa E, Pina P, Pasmans S, et al. The same tyrosine-based inhibition motif, in the intracytoplasmic domain of Fc gamma RIIB, regulates negatively BCR-, TCR-, and FcR-dependent cell activation. Immunity 1995;3.

[39] Muta T, Kurosaki T, Misulovin Z, Sanchez M, Nussenzweig M, Ravetch J. A 13-amino-acid motif in the cytoplasmic domain of Fc

gamma RIIB modulates B-cell receptor signalling. Nature 1994;368.

[40] Brooks DG, Qiu WQ, Luster AD, Ravetch JV. Structure and expression of human IgG FcRII(CD32). Functional heterogeneity is encoded by the alternatively spliced products of multiple genes. J Exp Med 1989;170:1369–85.

[41] Lewis VA, Koch T, Plutner H, Mellman I. A complementary DNA clone for a macrophage-lymphocyte Fc receptor. Nature 1986;324:372–5.

[42] Latour S, Fridman WH, Daëron M. Identification, molecular cloning, biologic properties, and tissue distribution of a novel isoform of murine low-affinity IgG receptor homologous to human Fc gamma RIIB1. J Immunol 1996;157:189–97.

[43] Hibbs ML, Bonadonna L, Scott BM, McKenzie IF, Hogarth PM. Molecular cloning of a human immunoglobulin G Fc receptor. Proc Natl Acad Sci U S A 1988;85:2240–4.

[44] Miettinen HM, Rose JK, Mellman I. Fc receptor isoforms exhibit distinct abilities for coated pit localization as a result of cytoplasmic domain heterogeneity. Cell 1989;58:317–27.

[45] Baerenwaldt A, Lux A, Danzer H, Spriewald BM, Ullrich E, et al. Fcγ receptor IIB (FcγRIIB) maintains humoral tolerance in the human immune system in vivo. Proc Natl Acad Sci USA 2011;108.

[46] Xiang Z, Cutler AJ, Brownlie RJ, Fairfax K, Lawlor KE, et al. FcgammaRIIb controls bone marrow plasma cell persistence and apoptosis. Nat Immunol 2007;8:419–29.

[47] Rao SP, Vora KA, Manser T. Differential expression of the inhibitory IgG Fc receptor FcgammaRIIB on germinal center cells: implications for selection of high-affinity B cells. J Immunol 2002;169:1859–68.

[48] Rudge UE, Cutler JA, Pritchard RN, Smith GK. Interleukin 4 reduces expression of inhibitory receptors on B cells and abolishes CD22 and Fc gamma RII-mediated B cell suppression. J Exp Med 2002;195.

[49] Fukuyama H, Nimmerjahn F, Ravetch JV. The inhibitory Fcgamma receptor modulates autoimmunity by limiting the accumulation of immunoglobulin G+ anti-DNA plasma cells. Nat Immunol 2005;6.

[50] Bournazos S, Corti D, Virgin HW, Ravetch JV. Fc-optimized antibodies elicit CD8 immunity to viral respiratory infection. Nature 2020;588:485–90.

[51] Yokota A, Kikutani H, Tanaka T, Sato R, Barsumian EL, Suemura M, et al. Two species of human Fc epsilon receptor II (Fc epsilon RII/CD23): tissue-specific and IL-4-specific regulation of gene expression. Cell 1988;55:611–8.

[52] Sancho D, Santis AG, Alonso-Lebrero JL, Viedma F, Tejedor R, Sánchez-Madrid F. Functional analysis of ligand-binding and signal transduction domains of CD69 and CD23 C-type lectin leukocyte receptors. J Immunol 2000;165:3868–75.

[53] Bonnefoy JY, Lecoanet-Henchoz S, Aubry JP, Gauchat JF, Graber P. CD23 and B-cell activation. Curr Opin Immunol 1995;7:355–9.

[54] Gordon J. B-cell signalling via the C-type lectins CD23 and CD72. Immunol Today 1994;15:411–7.

[55] Kolb JP, Abadie A, Paul-Eugene N, Capron M, Sarfati M, Dugas B, et al. Ligation of CD23 triggers cyclic AMP generation in human B lymphocytes. J Immunol 1993;150:4798–809.

[56] Ten RM, McKinstry MJ, Trushin SA, Asin S, Paya CV. The signal transduction pathway of CD23 (Fc epsilon RIIb) targets I kappa B kinase. J Immunol 1999;163:3851–7.

[57] Bournazos S, Gupta A, Ravetch JV. The role of IgG Fc receptors in antibody-dependent enhancement. Nat Rev Immunol 2020;20:633–43.

[58] Bournazos S, Wang TT, Dahan R, Maamary J, Ravetch JV. Signaling by antibodies: recent progress. Annu Rev Immunol 2017;35:285–311.

[59] Amigorena S, Bonnerot C, Drake J, Choquet D, Hunziker W, et al. Cytoplasmic domain heterogeneity and functions of IgG Fc receptors in B lymphocytes. Science (New York, NY) 1992;256.

[60] Tridandapani S, Pradhan M, LaDine J, Garber S, Anderson C, Coggeshall K. Protein interactions of Src homology 2 (SH2) domain-containing inositol phosphatase (SHIP): association with Shc displaces SHIP from FcgammaRIIb in B cells. J Immunol 1999;162.

[61] Ono M, Bolland S, Tempst P, Ravetch J. Role of the inositol phosphatase SHIP in negative regulation of the immune system by the receptor Fc(gamma)RIIB. Nature 1996;383.

[62] Bolland S, Pearse R, Kurosaki T, Ravetch J. SHIP modulates immune receptor responses by regulating membrane association of Btk. Immunity 1998;8.

[63] Tamir I, Stolpa J, Helgason C, Nakamura K, Bruhns P, Daeron M, et al. The RasGAP-binding protein p62dok is a mediator of inhibitory FcgammaRIIB signals in B cells. Immunity 2000;12.

[64] Yamanashi Y, Tamura T, Kanamori T, Yamane H, Nariuchi H, Yamamoto T, et al. Role of the rasGAP-associated docking protein p62(dok) in negative regulation of B cell receptor-mediated signaling. Gene Dev 2000;14.

[65] Liu Q, Sasaki T, Kozieradzki I, Wakeham A, Itie A, Dumont D, et al. SHIP is a negative regulator of growth factor receptor-mediated PKB/Akt activation and myeloid cell survival. Gene Dev 1999;13.

[66] Aman J, Lamkin T, Okada H, Kurosaki T, Ravichandran K. The inositol phosphatase SHIP inhibits Akt/PKB activation in B cells. J Biol Chem 1998;273.

[67] Tzeng S-J, Bolland S, Inabe K, Kurosaki T, Pierce SK. The B cell inhibitory Fc receptor triggers apoptosis by a novel c-Abl family kinase-dependent pathway. J Biol Chem 2005;280.

[68] Tiller T, Kofer J, Kreschel C, Busse CE, Riebel S, et al. Development of self-reactive germinal center B cells and plasma cells in autoimmune Fc gammaRIIB-deficient mice. J Exp Med 2010;207.

[69] Chan MA, Gigliotti NM, Matangkasombut P, Gauld SB, Cambier JC, Rosenwasser LJ. CD23-mediated cell signaling in human B cells differs from signaling in cells of the monocytic lineage. Clin Immunol 2010;137:330–6.

[70] Weskamp G, Ford JW, Sturgill J, Martin S, Docherty AJ, et al. ADAM10 is a principal 'sheddase' of the low-affinity immunoglobulin E receptor CD23. Nat Immunol 2006;7:1293–8.

[71] MacLennan I. Germinal centers. Annu Rev Immunol 1994;12.

[72] Tew J, Wu J, Qin D, Helm S, Burton G, Szakal A. Follicular dendritic cells and presentation of antigen and costimulatory signals to B cells. Immunol Rev 1997;156.

[73] Burton G, Masuda A, Heath S, Smith B, Tew J, Szakal A. Follicular dendritic cells (FDC) in retroviral infection: host/pathogen perspectives. Immunol Rev 1997;156.

[74] Wang TT, Bournazos S, Ravetch JV. Immunological responses to influenza vaccination: lessons for improving vaccine efficacy. Curr Opin Immunol 2018;53:124–9.

[75] Maamary J, Wang TT, Tan GS, Palese P, Ravetch JV. Increasing the breadth and potency of response to the seasonal influenza virus vaccine by immune complex immunization. Proc Natl Acad Sci U S A 2017;114:10172–7.

[76] Bournazos S, Ravetch JV. Fcγ receptor function and the design of vaccination strategies. Immunity 2017;47:224–33.

[77] Hjelm F, Carlsson F, Getahun A, Heyman B. Antibody-mediated regulation of the immune response. Scand J Immunol 2006;64.
[78] Gustavsson S, Wernersson S, Heyman B. Restoration of the antibody response to IgE/antigen complexes in CD23− deficient mice by CD23+ spleen or bone marrow cells. J Immunol 2000;164.
[79] Karlsson M, Getahun A, Heyman B. FcgammaRIIB in IgG-mediated suppression of antibody responses: different impact in vivo and in vitro. J Immunol 2001;167.
[80] Kumpel MB. In vivo studies of monoclonal anti-D and the mechanism of immune suppression. Transfus Clin Biol 2002;9.
[81] Radbruch A, Muehlinghaus G, Luger EO, Inamine A, Smith KGC, Dörner T, et al. Competence and competition: the challenge of becoming a long-lived plasma cell. Nat Rev Immunol 2006;6.
[82] Ravetch JV, Nussenzweig M. Killing some to make way for others. Nat Immunol 2007;8:337−9.
[83] Takai T, Ono M, Hikida M, Ohmori H, Ravetch JV. Augmented humoral and anaphylactic responses in Fc gamma RII-deficient mice. Nature 1996;379:346−9.
[84] Bolland S, Yim Y, Tus K, Wakeland E, Ravetch J. Genetic modifiers of systemic lupus erythematosus in FcgammaRIIB(−/−) mice. J Exp Med 2002;195.
[85] Bolland S, Ravetch J. Spontaneous autoimmune disease in Fc(gamma)RIIB-deficient mice results from strain-specific epistasis. Immunity 2000;13.
[86] Brownlie R, Lawlor K, Niederer H, Cutler A, Xiang Z, et al. Distinct cell-specific control of autoimmunity and infection by FcgammaRIIb. J Exp Med 2008;205.
[87] McGaha T, Sorrentino B, Ravetch J. Restoration of tolerance in lupus by targeted inhibitory receptor expression. Science (New York, NY) 2005;307.
[88] Tackenberg B, Jelcic I, Baerenwaldt A, Oertel W, Sommer N, Nimmerjahn F, et al. Impaired inhibitory Fcgamma receptor IIB expression on B cells in chronic inflammatory demyelinating polyneuropathy. Proc Natl Acad Sci USA 2009;106.
[89] Bruhns P, Samuelsson A, Pollard J, Ravetch J. Colony-stimulating factor-1-dependent macrophages are responsible for IVIG protection in antibody-induced autoimmune disease. Immunity 2003;18.
[90] Kaneko Y, Nimmerjahn F, Madaio M, Ravetch J. Pathology and protection in nephrotoxic nephritis is determined by selective engagement of specific Fc receptors. J Exp Med 2006;203.
[91] Samuelsson A, Towers T, Ravetch J. Anti-inflammatory activity of IVIG mediated through the inhibitory Fc receptor. Science (New York, NY) 2001;291.
[92] Nakamura A, Yuasa T, Ujike A, Ono M, Nukiwa T, Ravetch J, et al. Fcgamma receptor IIB-deficient mice develop Goodpasture's syndrome upon immunization with type IV collagen: a novel murine model for autoimmune glomerular basement membrane disease. J Exp Med 2000;191.
[93] Jiang Y, Hirose S, Abe M, Sanokawa-Akakura R, Ohtsuji M, et al. Polymorphisms in IgG Fc receptor IIB regulatory regions associated with autoimmune susceptibility. Immunogenetics 2000;51.
[94] Pritchard N, Cutler A, Uribe S, Chadban S, Morley B, Smith K. Autoimmune-prone mice share a promoter haplotype associated with reduced expression and function of the Fc receptor FcgammaRII. Curr Biol: CB 2000;10.
[95] Mackay M, Stanevsky A, Wang T, Aranow C, Li M, et al. Selective dysregulation of the FcgammaIIB receptor on memory B cells in SLE. J Exp Med 2006;203.
[96] Chu Z, Tsuchiya N, Kyogoku C, Ohashi J, Qian Y, et al. Association of Fcgamma receptor IIb polymorphism with susceptibility to systemic lupus erythematosus in Chinese: a common susceptibility gene in the Asian populations. Tissue Antigens 2004;63.
[97] Siriboonrit U, Tsuchiya N, Sirikong M, Kyogoku C, Bejrachandra S, et al. Association of Fcgamma receptor IIb and IIIb polymorphisms with susceptibility to systemic lupus erythematosus in Thais. Tissue Antigens 2003;61.
[98] Kyogoku C, Dijstelbloem HM, Tsuchiya N, Hatta Y, Kato H, et al. Fcgamma receptor gene polymorphisms in Japanese patients with systemic lupus erythematosus: contribution of FCGR2B to genetic susceptibility. Arthritis Rheum 2002;46.
[99] Willcocks LC, Carr EJ, Niederer HA, Rayner TF, Williams TN, et al. A defunctioning polymorphism in FCGR2B is associated with protection against malaria but susceptibility to systemic lupus erythematosus. Proc Natl Acad Sci USA 2010;107.
[100] Blank MC, Stefanescu RN, Masuda E, Marti F, King PD, et al. Decreased transcription of the human FCGR2B gene mediated by the -343 G/C promoter polymorphism and association with systemic lupus erythematosus. Hum Genet 2005;117:220−7.
[101] Su K, Wu J, Edberg JC, Li X, Ferguson P, et al. A promoter haplotype of the immunoreceptor tyrosine-based inhibitory motif-bearing FcgammaRIIb alters receptor expression and associates with autoimmunity. I. Regulatory FCGR2B polymorphisms and their association with systemic lupus erythematosus. J Immunol 2004;172:7186−91.
[102] Henningsson F, Ding Z, Dahlin JS, Linkevicius M, Carlsson F, et al. IgE-mediated enhancement of CD4+ T cell responses in mice requires antigen presentation by CD11c+ cells and not by B cells. PLoS One 2011;6:e21760.
[103] Davis SR. Fc receptor-like molecules. Annu Rev Immunol 2007;25.
[104] Ehrhardt G, Leu C, Zhang S, Aksu G, Jackson T, et al. Fc receptor-like proteins (FCRL): immunomodulators of B cell function. Adv Exp Med Biol 2007;596.
[105] Leu C-M, Davis SR, Gartland AL, Fine WD, Cooper DM. FcRH1: an activation coreceptor on human B cells. Blood 2005;105.
[106] Haga LC, Ehrhardt RG, Boohaker JR, Davis SR, Cooper DM. Fc receptor-like 5 inhibits B cell activation via SHP-1 tyrosine phosphatase recruitment. Proc Natl Acad Sci USA 2007;104.
[107] Ehrhardt RG, Davis SR, Hsu TJ, Leu C-M, Ehrhardt A, Cooper DM. The inhibitory potential of Fc receptor homolog 4 on memory B cells. Proc Natl Acad Sci USA 2003;100.
[108] Wilson JT, Fuchs A, Colonna M. Cutting edge: human FcRL4 and FcRL5 are receptors for IgA and IgG. J Immunol 2012;188.
[109] Franco A, Damdinsuren B, Ise T, Dement-Brown J, Li H, Nagata S, et al. Human Fc receptor-like 5 binds intact IgG via mechanisms distinct from those of Fc receptors. J Immunol 2013;190.
[110] Kubagawa H, Oka S, Kubagawa Y, Torii I, Takayama E, et al. Identity of the elusive IgM Fc receptor (FcmuR) in humans. J Exp Med 2009;206.
[111] Li F, Kubagawa Y, McCollum M, Wilson L, Motohashi T, et al. Enhanced levels of both the membrane-bound and soluble forms of IgM Fc receptor (FcμR) in patients with chronic lymphocytic leukemia. Blood 2011;118.

Chapter 7

Transcriptional Regulation of Early B-Cell Development

Ashley P. Ng[1,3,4,5], Rhiannon Morris[2,3] and Stephen L. Nutt[1,3]

[1]*Immunology Division, The Walter and Eliza Hall Institute of Medical Research, Melbourne, VIC, Australia;* [2]*Blood Cell and Blood Cancer Division, The Walter and Eliza Hall Institute of Medical Research, Melbourne, VIC, Australia;* [3]*The Department of Medical Biology, The University of Melbourne, Melbourne, VIC, Australia;* [4]*Clinical Haematology Department, The Royal Melbourne Hospital and Peter MacCallum Cancer Centre, Melbourne, VIC, Australia;* [5]*Sir Peter MacCallum Department of Oncology, The University of Melbourne, Melbourne, VIC, Australia*

1. INTRODUCTION

Early B-cell development is dependent on complex transcriptional changes that enable lineage committed, functional and immunologically diverse B-cell clones to develop from multipotent progenitor cells. Rapid genomic changes occur during the proB and preB stages of development that are linked to the essential and ordered immunoglobulin gene rearrangements required to establish pre-B cell receptor (pre-BCR) and B-cell receptor (BCR) signalling. Successful completion of this programme permits functional B-cell clones to transit these critical clonal selection checkpoints and contribute to the peripheral humoral immune repertoire. Early B-cell development requires the contribution of multiple sequence-specific transcription factors for precise control of chromatin accessibility, chromatin reorganisation and ultimately command of gene transcription that can be organised into gene regulatory networks which are often directed by more than one factor.

Transcription factors are critical regulators of cell fate and function. Gene transcription of mRNA from protein coding genes is dynamically regulated through the pre-initiation complex (PIC) assemblage of the 'general' multi-subunit transcription factors (e.g. TFIIB-, D, E, F and H) that together with RNA polymerase II (Pol II) [1] bind and open promoter DNA [2] (Figure 7.1). The specificity and timing of gene expression by the PIC and Pol II requires interaction with sequence-specific transcription factors. Following 'closed chromatin' nucleosome binding by the small subset of transcription factors with pioneering function [3] and recruitment of chromatin remodelers [4] such as P300 and histone acetyl transferases [5], gene regulatory elements in accessible chromatin regions can be permissively bound by other transcription factors in a DNA sequence-specific manner with transient high affinity [6–8]. While the PIC at promoter regions initiates gene transcription, the transcriptional efficiency is strongly influenced by enhancer–promoter interactions [9]. Such looping interactions are facilitated by the CTCF/cohesin complex modulation of chromatin architecture [10]. In addition, the mediator complex also bridges enhancer-bound transcription factors to Pol II [11], and YY1 dimers have been shown to mediate interactions of enhancer and proximal promoter elements [12].

The transcription factors essential for early B-cell development derive from diverse structural families that can recognise and bind to DNA sequence motifs across the genome. Remarkably, convergence of several motifs within single gene regulatory elements proximal and distal to a gene body is observed for multiple genes required for early B-cell development [13], permitting diverse, cooperative and stage-specific roles for different transcriptional regulators along the B-lineage trajectory.

Here, we discuss the identity and functions of the transcription factors that are known to play important roles during early B-cell development. We will emphasise the technical approaches that have provided increasingly detailed insights into their specific mechanisms of action and highlight how these factors act in complex gene regulatory networks.

Molecular Biology of B Cells. https://doi.org/10.1016/B978-0-323-95895-0.00014-3

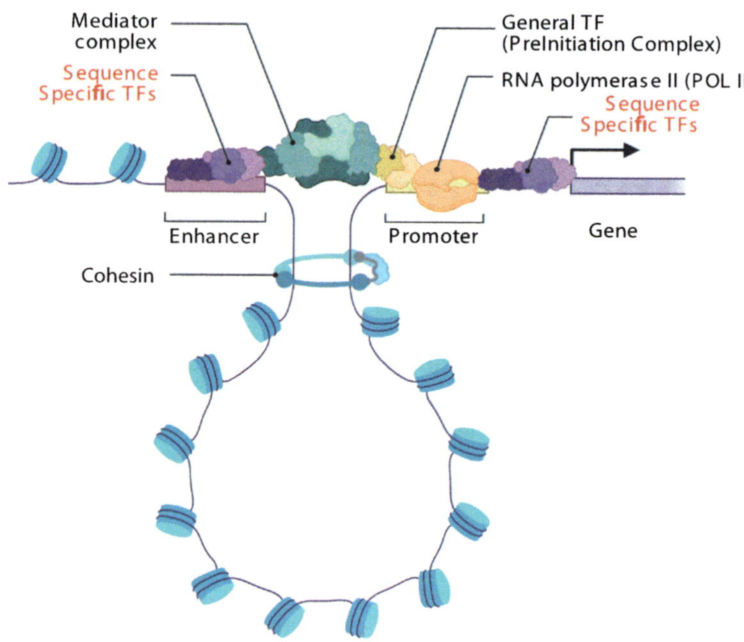

FIGURE 7.1 Schematic of roles for sequence specific transcription factors (TFs) in transcriptional regulation of gene expression. *Created with BioRender.com Agreement Number BV252MQZA5.*

2. THE TRAJECTORY OF B-CELL DEVELOPMENT

B cells derive from multipotent haematopoietic progenitors that progressively commit to the lymphoid and ultimately, the B-cell lineage. In the adult, the earliest 'lymphoid primed progenitors' can be identified within the heterogeneous lineage-negative cKit$^+$Sca1$^+$ stem cell and progenitor cell population [14–16], before acquisition of Flt3 [17] and IL7R [18], two important cytokine receptors that immunophenotypically define and drive development from common lymphoid progenitors (CLPs [19]). CLPs can be further divided into all-lymphoid (ALP) and B-lymphoid committed (BLP) progenitors by Ly6D expression [20]. Early pre-proB-cell progenitors that retain some myeloid potential further commit to the B-cell lineage [21] and activate a distinct transcriptional network. The unique B-cell-specific processes of immunoglobulin gene rearrangements in pro-B (Ig heavy chain, *Igh*) and preB cells (Ig light chain, *Igl*) [22,23] are the two important checkpoints where clonally unique pre-BCR and mature BCR are formed via V(D)J rearrangements. Appropriate signalling via pre-BCR (with a functionally rearranged *Igh* and a germline encoded surrogate light chain) and BCR (with functionally rearranged *Igh* and *Igl*) allows selection, initial clonal expansion and emergence of quiescent B-cell clones that circulate to undergo maturation in the spleen and are able to contribute to adaptive immunity (Figure 7.2).

3. EARLY B-LINEAGE SPECIFICATION REGULATED BY TRANSCRIPTION FACTORS

Mouse studies using constitutional or conditional knockout models of single genes or compound gene deficiencies [24] have identified several transcription factors that are required to progress the B-cell developmental trajectory. For example, analysis of *Ikzf1* [25] and *Spi1* (PU.1) [26,27], *Foxo3* [28] and *Mef2C* [29] knockout models identified a requirement for these transcription factors for the development of lymphoid progenitors from multipotent progenitors. Knockout of *Tcf3* (previously named *Tcf2a*) that encodes E2A proteins E12 and E47 [30], *Myb* [31] and dual *Foxo3* [28] and *Foxo1* [32] deletions resulted in the failure of B-cell development from the CLP developmental stage. *Ebf1* [33] and *Erg* [34] knockout models demonstrated a developmental arrest at the pre-proB-cell stage of B-cell development, while other gene knockout models, such as *Pax5* [35], demonstrated later proB-cell developmental arrest. From these functional studies, a hierarchy can be constructed that broadly assigns requirements for specific transcription factors across the B-lineage trajectory (Figure 7.2, Table 7.1).

Analysis of cells arrested at these developmental stages provides broad insights into biological processes that may be regulated by a given transcriptional regulator, such as lineage commitment and maintenance of B-cell identity

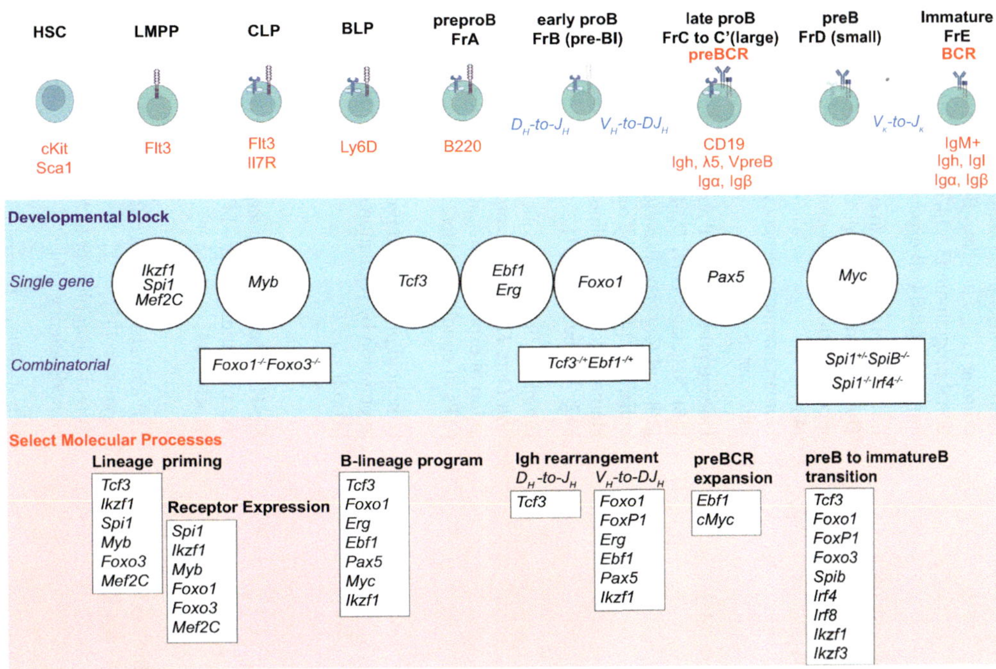

FIGURE 7.2 **Transcriptional regulation of stages of B-lineage development.** Stages of B-lineage development are shown: haematopoietic stem cells (HSCs), lymphoid myeloid primed progenitor (LMPP), common lymphoid progenitor (CLP), B-lymphoid progenitor (BLP), with select surface receptors shown in *red* and stages of V(D)J recombination. *Blue box*: stages of B-cell developmental block are indicated from examples of single gene or combination gene knockout models as indicated. *Pink box*: transcription factors implicated in select B-cell developmental molecular processes (see main text and Table 7.1 for details).

[35,80], or immunoglobulin gene rearrangements [33,34,81], in addition to allowing changes in gene expression at developmentally perturbed stages to be characterised. However, defining whether the effect of a transcription factor occurs via primary or indirect modulation of other genes requires identification of gene regulatory element binding, genomic colocalisation with other transcription factors and knowledge of potential interactions with chromatin-modifying proteins [82]. All of these attributes can help define the position of a particular regulatory protein within the transcriptional network orchestrating B-cell development.

4. DEFINING TRANSCRIPTION FACTOR FUNCTION

Structural studies, particularly the analysis and classification of DNA-binding domains (Figure 7.3), provide a classification framework for transcription factors that strongly correlates with their DNA-binding profiles (Figure 7.4, Table 7.1) [83,84]. While early analysis of specific B-cell gene loci was instrumental to identifying and defining critical B-cell regulatory factors [54,85–87], the advent of chromatin immunoprecipitation with deep sequencing (ChIPseq) [88] permitted genome-wide identification of transcription factor binding to gene regulatory elements [89]. This technical advance has been instrumental in refining the consensus DNA-binding motif sequences [84] and defining gene regulatory networks acting across the entire B-cell genome [34,38]. Importantly, analysis of chromatin accessibility and modification [90], and more recently, three-dimensional chromatin conformational analysis [43], has also permitted new insights into transcription factor regulation and function.

Several classes of transcription factors will now be discussed with a focus on those with an instructive role in early B-cell development. We will highlight recent insights into the key players that collectively provide the molecular mechanisms that power the unfolding transcriptional programme of developing B cells (Figure 7.4, Table 7.1).

5. EBF1: A COE FAMILY TRANSCRIPTION FACTOR

EBF1 is a critical zinc finger helix–loop–helix (HLH) transcription factor that plays a central role in B-cell lineage specification [80], regulating multiple genomic and biological processes required for B-cell development [93]. A member of the Collier-Olf1-Ebf (COE) family of transcription factors, *Ebf1* is highly expressed in early B-cell progenitors. The N-terminal DNA-binding zinc finger HLH domain structure is unique to the COE family

TABLE 7.1 Summary of Transcription Factors Involved With Early B-Cell Development

Gene	Family	Motif (JASPAR)	Interaction	Select functions in early B cell development
Ebf1	COE Rel homology region (RHR) factors			Pioneer transcription factor [36], lineage priming [37], *Ebf1-Pax5* B-lineage transcriptional programme [32,34,38] for CLP to proB transition, proB-cell expansion [39]
Pax5	Paired box factors	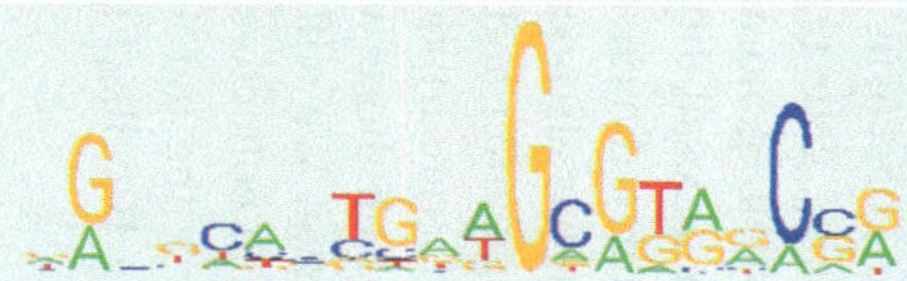	Ets family proteins at composite *Ets:Pax5* element [40] BAF complex, TFIID, HDAC containing NCoR1 complex [41]	B-lineage programme from proB cells [35], *Ebf1-Pax5* network [36]. Distal V_H-DJ_H [42] recombination, global chromatin organisation [43]. Human immunodeficiency and due to biallelic mutation [44]
E2A (*Tcf3*)	Basic helix–loop–helix factors (class I)		p300 acetyl transferase [45]	Lineage priming, LMPP and CLP formation [46]. Regulates *Ebf1*, D_H-J_H recombination [30] and *Rag* gene expression [47] via novel enhancers [48], and *Igl* rearrangement [49]. Human agammaglobulinemia [50] due to $E47^{E555K}$ mutation [51]
Mef2c	MADS box factors	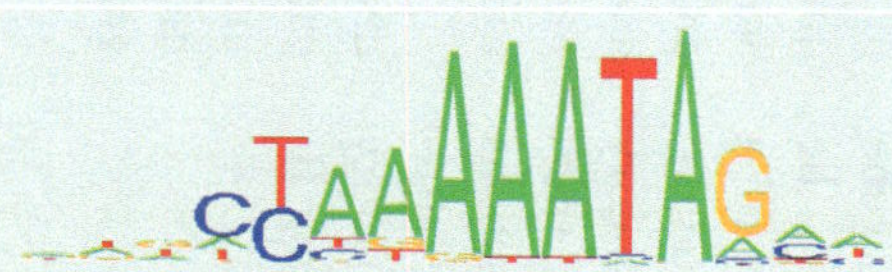		Lymphoid lineage specification [29]: regulation *Dntt, Flt3, Ikzf1* expression, direct regulation of expression *Il7r* and *Rag1*. Reduced CLPs in conditional *Mef2c* knockout
Myb	Tryptophan cluster factors Myb/SANT domain factors			Lymphoid priming and promotion of proB differentiation by regulation of *Il7r* expression [31,52]
Erg	ETS-related			*Erg-Ebf1-Pax5* network required for V_H-DJ_H recombination and pre-BCR signalling components [34]
PU.1 (*Spi1*)	ETS-related		SWI/SNF complex, Brg1 [53] Irf4:PU.1 ternary complex [54]	CLP formation, required for *Flt3* and *IL7r* expression [55] Non-classical pioneer factor [53]. Human immunodeficiency [56]
Spi-B (*Spib*)	ETS-related			BCR signalling, transcription of Blnk/SLP-65 [57]

Foxo1	Fork head/winged helix factors		Akt (phosphorylated) for nuclear export and inactivation [58]	CLP development [32], *Ebf1* expression [38]. ProB development, regulation of *Il7r* [59]. *Rag1/2* expression for *Igh* [60] and *Igl* [60] recombination. Negative regulation of pre-BCR signalling and Akt.
FoxP1	Fork head/winged helix factors			V_H-DJ_H recombination, regulation of *Rag1/Rag2* expression, *Erag* enhancer binding [61]
Foxo3	Fork head/winged helix factors		Phospho-Akt for inactivation. Pre-BCR signalling promoted by BLNK [62]	LMPP and CLP development cooperating with *Foxo1* [28]. *Igl* recombination promoted by pre-BCR signalling, regulated by PI3K and Akt [62]
Ikaros (*Ikzf1*)	Krüppel-type zinc finger C2H2 zinc finger factors		Heterodimerisation with other C2H2 zinc finger factors [63] NuRD, SWI/SNF complex (T lymphocytes) [64] CHD4 [65]	LMPP, CLP, proB development: regulation *Flt3* and *IL7r* expression and preBCR signalling. *Igh* and [60] *Igl* recombination with binding to *Rag1,2* promoter and *Erag* enhancer. Gene silencing including *Myc* by Ikzf3 [66,67]. Mutation associated with immunodeficiency [68].
Aiolos (*Ikzf3*)	Krüppel-type zinc finger C2H2 zinc finger factors		Heterodimerisation with other C2H2 zinc finger factors [63]	PreB maturation to a quiescent state [69], *Ikzf1/Ikzf3* termination pre-BCR signalling, silencing of *Igll1* [70] *Myc* repression [67]. Human immunodeficiency [71]
Irf4 (*and Irf8*)	Tryptophan cluster factors Interferon regulatory factors		Irf4:PU.1 ternary complex [54] Irf4:E2A ternary complex [49]	Promotes preB-cell maturation to immature B cells [72]. Downregulation of preBCR signalling, associated with upregulation of *Ikzf1,3* expression [73] *Igl* recomb [49,72]
Bcl6	C2H2 zinc finger factors	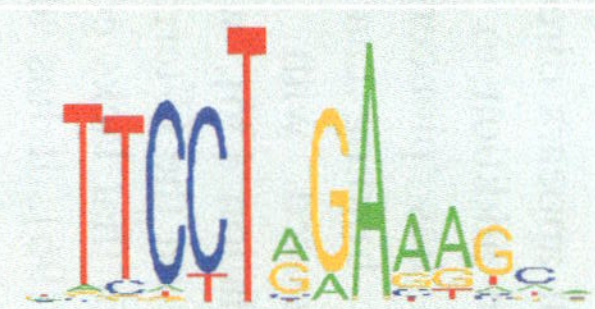	Bcor, Ncor, NuRD and CtBP [74]	Positive selection of preB cells after pre-BCR checkpoint, promotion of *Igl* recombination antagonised by *Myc* [75]
Bach2	Basic leucine zipper factors (bZIPs)			Negative selection of early B cells with failed V_H-DJ_H rearrangements, increasing expression and activity of *Rag1,2* [76]
Myc	Basic helix–loop–helix factors bHLH-ZIP		Heterodimerisation with *Max* [77] *Wdr5* [78]	Target of *Ebf1* [39] for the regulation of B-cell identity [79], biogenesis, proliferation and expansion of large preB cells

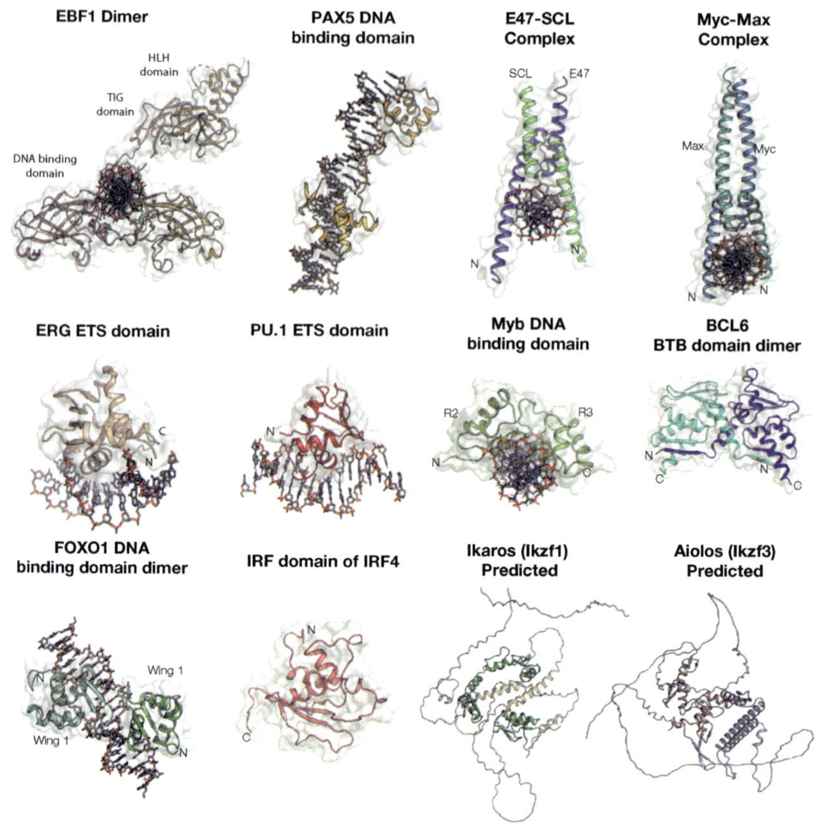

FIGURE 7.3 Structures of transcription factors with important roles in B-cell development. Select examples of transcription factors with resolved structures and AlphaFold predictions. EBF1 dimer bound to DNA including DNA-binding domain, immunoglobulin-like fold domain (TIG) and helix−loop−helix (HLH) domain (PDB ID: 3MLP); PAX5 DNA-binding domain (PDB ID: 1K78); E47-SCL complex bound to DNA (PDB ID: 6LBI); Myc-Max complex bound to DNA (PDB ID: 5I50); ERG ETS domain bound to DNA (PDB ID: 4IRI); PU.1 ETS domain bound to DNA (PDB ID: 1PUE); MYB DNA-binding domain (PDB ID: 1MSF); BCL6 BTB domain dimer (PDB ID: 1RB); FOXO1 DNA-binding domain dimer (PDB ID); IRF4 IRF domain (PDB ID: 3DLL); structures of IKAROS and AIOLOS as predicted by AlphaFold. Both proteins comprise zinc fingers (*green* in IKAROS, *pink* in AIOLOS) linked by unstructured regions interspersed with alpha helices and beta strands.

(zinc knuckle) that recognises the palindromic 5-TCCCNNGGGA motif [94] via a symmetric dimeric clamp. ChIP-on-ChIP and ChIPseq analysis identified EBF1 binding to gene regulatory elements enriched for lymphoid genes and genes involved in signalling pathways, a quarter whose expression appears directly regulated by EBF1 [94]. Importantly, in addition to a role as a transcriptional activator, EBF1 has 'pioneering' transcription factor activity [95]. EBF1 binding can induce dynamic epigenetic changes associated with locus activation including H3K4 di-methylation [36], which, for certain genomic loci, is dependent on the C-terminal domain that can bind to naïve chromatin that lacks co-occupancy with other transcription factors [96]. It has also been hypothesised that EBF1 binding at poised and partially activated B-cell enhancers through recruitment of the p300 histone acetylase may subsequently recruit, via addition of H2K27Ac marks, the SWI/SNF complex member BRG1 to facilitate the activation of B-cell enhancers, *Igh* locus contraction and *Myc* expression [97].

In keeping with its pioneering activity, overexpression of *Ebf1* restricts haematopoietic progenitor developmental potential to the B-cell lineage [98,99]. EBF1 binding to cis-regulatory elements has been shown to regulate the expression of multiple genes required for B-cell development, including repressing expression of genes associated with alternative lineage fates. This regulation by EBF1 contributes to lymphoid priming of multipotent progenitor cells by suppressing *Cebpa* and myeloid lineage development [37], as well as suppressing T-lineage and innate lymphoid development from committed proB cells by the suppression of genes such as *Gata3*, *Id2* and *Tcf7* [80]. EBF1 is also required for initiating and maintaining the expression of B-cell transcription factors and coactivators such as *Foxo1*, *Pax5*, *Erg*, *Irf4* and *Pou2af1* [34], as well as other essential genes involved in immunoglobulin gene rearrangement and pre-BCR and BCR signalling that are required for proB- and preB-cell development [34,94]. With this broad repertoire of essential B-cell target genes, it is therefore perhaps unsurprising that EBF1 has been shown to be able to partially rescue B-cell development in the absence of TCF3 [100], PU.1 [27], IKAROS [25] and IL7R [18].

EBF1 has been implicated as a key player in several transcription factor networks throughout B-cell development, where coregulation through binding of complementary cis-gene regulatory elements reinforces transcription factor expression, as well as expression of other target genes of the gene regulatory network. These include E2A, EBF1 and FOXO1 [38] in early B-lineage development, the

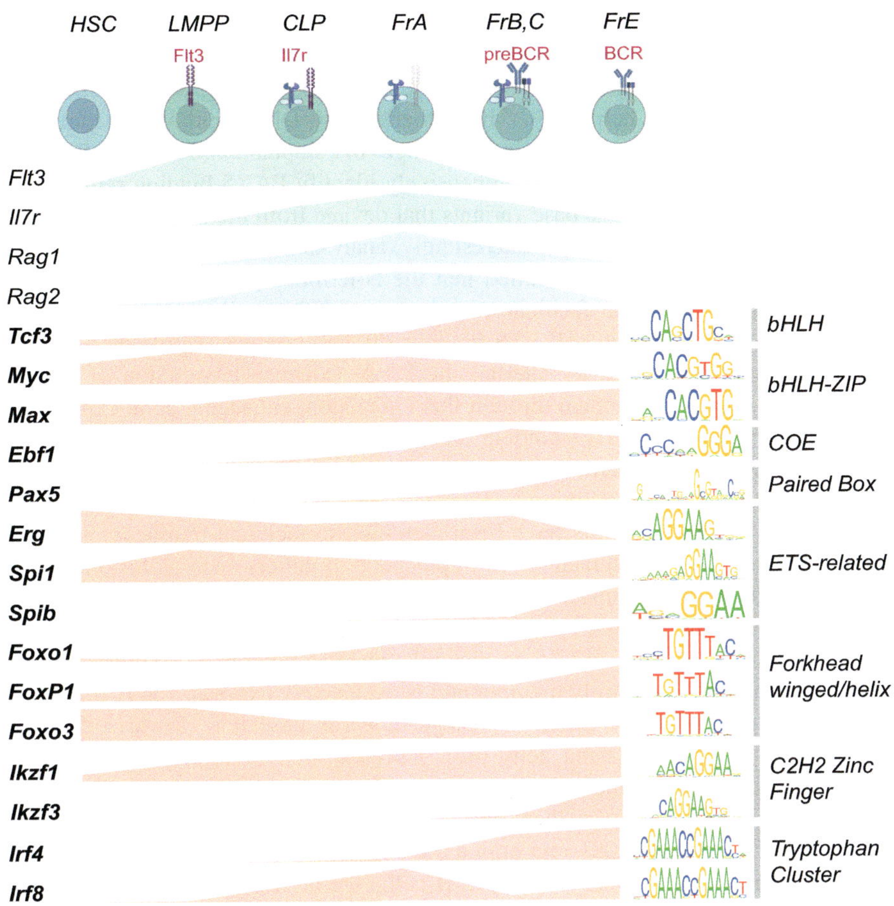

FIGURE 7.4 Transcription factor expression during early B-cell development. B-lineage development from haematopoietic stem cells (HSCs) with schematic of stages showing important cellular receptors (Flt3, IL7R, preBCR and BCR) and examples of B-lineage genes (*Flt3*, *Il7r*, *Rag1* and *Rag2*) with instructive B-lineage transcription factors (Entrez gene symbol) that are discussed in the text shown below. The relative transcription factor expression levels across the stages of differentiation were obtained from the Immgen project [91] and normalised to unity against the highest expression level. DNA motifs were obtained from the JASPAR database [92].

recently described ERG, EBF1 and PAX5 network required for *Igh* gene rearrangement and pre-BCR formation and signalling in proB cells [34] and EBF1, PAX5 and MYC for clonal expansion from proB cells [39]. The intimate relationship between EBF1 and PAX5 has been notably demonstrated through several orthogonal studies [27,34,36,100] that highlight the importance of EBF1, which is not only required for *Pax5* expression but also through its pioneering transcription factor role, permits PAX5 binding to colocalised gene regulatory elements [36].

6. E PROTEINS

E2A (*Tcf3*) is a transcription factor within the class I basic helix–loop–helix (bHLH) family of E proteins that are widely expressed throughout haematopoiesis [101]. Structurally, the C-terminal 'basic' arginine residue-rich DNA-binding domain and the adjacent HLH domain that forms closely apposed alpha-helices that mediate dimerisation, facilitate binding to the canonical palindromic E-Box 'CANNTG' DNA motif [87] as well as interaction with the p300 histone acetyl transferase [45]. Alternative splicing of the E2A gene, *Tcf3* gives rise to the 'immunoglobulin enhancer-binding factors' E12 and E47 that differ in their DNA-binding domain [87], with E47 binding the E box enhancer as a dimer [87].

Tcf3 expression is upregulated at the CLP stage of lymphoid development and remains elevated in proB, preB and immature B cells in the bone marrow [102]. E2A is absolutely required for B-cell development with knockout mice lacking both E12 and E47 transcripts failing to develop B-lymphocytes with early developmental arrest at an early CLP to proB cell stage with a manifest lack of immunoglobulin gene rearrangements [30,81]. This phenotype is phenocopied by E47 knockout models (which also result in low-level E12 expression), where $E47^{-/-}$ foetal livers fail to develop early B-cell progenitors and notably, do not express *Ebf1* [100]. Consistent with the observation that E2A is required to promote *Ebf1* expression at this stage of development, B-cell formation from E47-deficient foetal livers can be rescued by *Ebf1* expression, including D_H-J_H and V_H-DJ_H rearrangements and expression of B-lineage genes [100]. Supporting this relationship between E2A and EBF1 was the severe proB-cell phenotype and reduction of lymphoid specific transcripts in mice haploinsufficient for *Tcf3* and *Ebf1* [24]. Furthermore, analysis of cis-regulatory sequences identified binding of E2A to *Ebf1* locus regulatory elements, as well as combinatorial E2A and EBF1 binding to *Foxo1* regulatory elements in a transcriptional network [38]. In a complimentary model of E2A and EBF1 interaction, EBF1 was shown to downregulate expression of the antagonistic HLH DNA-binding protein inhibitors *Id2* and *Id3*, thereby increasing E47 activity at a post-transcriptional level that could overcome the pre-proB-cell differentiation block observed in IL7R-deficient bone marrow [103].

Interestingly, lymphoid defects are apparent in mice deficient in E2A at the $Flt3^{hi}$ lymphoid myeloid primed progenitor (LMPP) and CLP stages of development, prior to significant *Ebf1* expression. Here, E2A was shown to promote the formation of LMPPs through facilitating the expression of *PU.1* (*Spi1*), *Ikaros* (*Ikzf1*) and *Gfi1*, thereby contributing to lymphoid gene priming and antagonising myeloid differentiation [46], a role also proposed for E2A at the CLP stage of development [18,46,47]. Given the interaction of E2A with the P300 acetyl transferase [45], it is interesting to speculate about the potential role that E2A may play in facilitating BRG1 binding and lymphoid priming via epigenetic mechanisms.

The requirement for E2A to initiate immunoglobulin gene rearrangements, including the initial *Igh* D_H-J_H rearrangements [30], is of particular interest. While E47 has been shown to bind to cis-acting *Erag* enhancer in CLPs that is required for V(D)J recombinase activity [47], E2A has also been reported to activate novel lineage-specific enhancers that drive *Rag1/2* gene expression in developing T and B cells [48], including recently described *RB1/RB2* enhancer elements in proB cells. A role for E2A for *Igκ* locus activation through binding to the intronic enhancer (iEκ), and together with IRF4, binding to the κE3′ enhancer to facilitate *Igl* recombination for BCR formation has also been proposed [49].

The early developmental block seen in E2A-deficient mice appears to be closely mirrored in humans as a rare syndrome of agammaglobulinemia with a B-cell differentiation block at the CLP and proB stage with profound B-cell lymphopenia is caused by *TCF3* gene mutations [50]. These patients carry a recurrent mutation within the DNA-binding domain (TCF^{E555K}) that acts as a dominant negative, inhibiting the function of the remaining functional *TCF3* allele [51].

7. PAX TRANSCRIPTION FACTORS

PAX5 is a member of the PAX family of transcription factors characterised by the conserved DNA-binding 'paired' domain: a bipartite DNA-binding region consisting of an N- and a C-terminal subdomains [104]. Each subregion contains a homeodomain-like motif that binds to a distinct half-site of the canonical 'GNCCANTGAAGCGTGAC' PAX5 recognition sequence [105] in adjacent major grooves of the DNA helix [40]. This rather unique DNA-binding structure provides ongoing challenges to computational methods that may fail to comprehensively identify PAX5-binding regions that have base variants that deviate from the consensus sequence [84,106]. Interestingly, analysis using the *Cd79a* gene locus [40] identified that the N-terminal-paired domain of PAX5 can facilitate recruitment of ETS family proteins to a 'sub-optimal' ETS-binding site in a composite promoter element and by altering the DNA sequence recognition of the Ets domain through this interaction, enhancing gene expression. PAX5 binding has also been shown to induce active chromatin at binding sites through recruitment of histone-modifying enzymes [41,107]. Central and C-terminal motifs have been identified to facilitate transactivation, and interaction with epigenetic modifiers such as histone acetyl transferases and histone deacetylases that may also modulate PAX5 transcriptional function [105].

PAX5 is the only haematopoietically expressed PAX family member and is exclusively expressed in B cells from the pro-B cell stage onwards until it is downregulated in plasma cells [42]. ChIPseq analysis demonstrates PAX5 binds ~20,000 sites in the mouse B cell genome with fewer than half of these located in genes or promoters [108], and only a fraction appearing to be directly regulated by PAX5 in pro-B cells. Unlike E2A and EBF1, PAX5 is not required for the initial stages of B-cell specification [35,109]. PAX5, however, is essential to maintaining B-cell identity. $Pax5^{-/-}$ pro-B cells can differentiate into other haematopoietic cell lineages including T cells [110], natural killer cells and myeloid cells [35] but are unable to differentiate beyond the proB-cell stage. This phenotype was also observed with conditional inactivation of *Pax5* in pro-B cells highlighting the stage-specific role for PAX5 mediating B-lineage-commitment [35].

PAX5 has been shown to be important for repressing critical non-B-lineage genes such as *Flt3* receptor, *Notch1* and *Mcsfr* (*Csf1r*), the lack of suppression contributing to the multipotential $Pax5^{-/-}$ proB-cell phenotype [35,111–116]. The subsequent discovery of multilineage differentiation in *E2A* [117] and *Ebf1* [80] deficient cell lines is likely explained by the secondary loss of *Pax5* expression. Importantly, positively regulated B-cell-specific genes regulated by PAX5 [116] significantly overlap with E2A and EBF1 targets [38], with the close and dynamic relationship between *Ebf1* and *Pax5* gene expression and gene regulation previously discussed [36]. Taken together, these data demonstrate that PAX5 is critical to establishing and maintaining the B-cell identity from proB cells and critically requires EBF1 to facilitate its transactivating function for its regulation of gene expression.

PAX5 also has a unique role in the regulation of chromatin structure, including promoting *Igh* locus contraction and distal V_H-DJ_H recombination. Conserved Pax5-activated intergenic repeat domains are marked by active chromatin when bound by PAX5 in proB cells and are clustered predominantly across the distal *Igh* V_H gene cluster, with PAX5 colocalising with E2A, CTCF and RAD21 [118]. These domains provide an underlying genomic basis for the role of PAX5 in *Igh* locus contraction and efficient distal V_H gene recombination [42] that contributes to immunological diversity. Recent evidence also demonstrates *Wapl*, a cohesin release factor, is directly downregulated in proB cells by PAX5 binding to the *Wapl* promoter, that is followed by recruitment and locus inactivation by the polycomb repressive complex 2. This reduction of *Wapl* by PAX5 increases the residence time of cohesin, facilitating cohesion-mediated loop extrusion at the chromatin loops anchored by the CTCF zinc-finger protein at CTCF-binding elements across the V_H gene cluster, and V_H-DJ_H recombination across the *Igh* locus in proB cells [119]. A recent study of three-dimensional chromatin architecture using a HiC approach [120] has also established that PAX5 through its multiple binding sites across the genome was also essential for the establishment and maintenance of a general B-cell chromatin structure until the plasmablast stage that was independent of its transactivating activity [43].

Recently, a primary immunodeficiency associated with autism spectrum disorder due to biallelic mutation in *PAX5* has been described [44]. The phenotype due to a missense mutation R31Q in the N-terminal DNA-binding paired domain in one allele and nonsense mutation (E242Stop) in the partial homeodomain of the second *PAX5* allele was recapitulated in a mouse model that led to neurological abnormalities and a profound reduction in preB and immature and mature B cells, with the R31Q mutation shown to interfere with PAX5 DNA binding. This recapitulation of the Pax5-deficient mouse phenotype in an immunodeficiency patient demonstrates the conservation of the gene regulatory network of B-cell development in both species.

8. ETS FAMILY TRANSCRIPTION FACTORS

ERG is a member of the ETS family transcription factor that possess the winged helix–turn–helix ETS domain that transactivates transcription following binding to the canonical ‘GGA(A/T)’ DNA sequence [121,122]. The ETS domain of **ERG** has also been shown to mediate interaction with the AP1 family member Jun in a ternary complex, as well as homodimerisation along with the amino terminal pointed domain [123].

Erg is expressed throughout the haematopoietic system and plays a role in embryonic [124] and adult [125] haematopoietic stem cells (HSCs) maintenance and self-renewal [126]. ERG also regulates megakaryocyte/erythroid lineage specification from early multipotent progenitors [127]. While other ETS factors such as ETS1, PU.1 and SPI-B are known to have roles in immune development and function [128,129], the importance of ERG in B-cell development was only recently uncovered. One outstanding question was how stage-specific induction of *Pax5* expression [42] and the significant upregulation of *Ebf1* expression [32,130] at the proB-cell stage occurred. Conditional knockout throughout lymphoid development identified ERG as the critical ETS factor that regulated *Ebf1* and *Pax5* expression at the proB-cell stage of development [34]. Loss of ERG demonstrated a developmental block at pre-proB cells, with lack of V_H-DJ_H gene rearrangements [34,131]. We established ERG was an essential stage-specific factor required to initiate a coregulated ERG-EBF1-PAX5 gene regulatory network to drive *Igh* gene rearrangement and pre-BCR signalling that permits B-cell clones to proceed through the proB developmental checkpoint [34]. These studies also confirmed the dependence of *Pax5* expression and transcriptional activation on EBF1, with transduction of *Ebf1* able to partially overcome the loss of *Erg* expression in pre-proB cells, with the induction of *Pax5* expression, initiation of V_H-DJ_H recombination and expression of pre-BCR components, attributes that *Pax5* transduction alone could not rescue. The colocalisation of PAX5 and ERG at B-cell-specific genes was of interest given the prior data that demonstrated PAX5 can facilitate recruitment of ETS family proteins to a composite promoter element in *Cd79a* [40], a locus at which PAX5 and ERG colocalised in the transcriptional network [34]. An interesting consequence of the ERG-EBF1-PAX5 gene regulatory network was the activation of expression of several transcriptional regulators, including *Myc*, *Bach2* and *Irf4* that have important functions at the pre-B-cell stage of development. Notably, ERG did not appear to be required for further development from preB cells and beyond as *Erg*-deficient preB cells could be rescued by the expression of the V_H10^{tar} *Igh* knock-in allele, with *Igl* recombination and BCR signalling proceeding without *Erg* expression. These findings suggested other factors are required to sustain *Ebf1* and *Pax5* expression beyond the proB-cell stage.

Two other members of the ETS domain transcription factor family, PU.1 and its closely related family member SPI-B, have also been implicated in early B-cell development.

PU.1 is encoded by the gene *Spi1* and has been shown to have multiple roles in haematopoiesis [132] with particular importance for the formation of myeloid and lymphoid lineages in germline deletion models [133,134].

PU.1 expression can be identified in early haematopoietic stem and progenitor cell populations, as well as more restricted early myeloid and lymphoid progenitors [135]. Graded expression of PU.1 in haematopoietic progenitors also appears to have an important role in lymphoid versus macrophage lineage specification, with high PU.1 driving macrophage differentiation and low PU.1 expression favouring B-lymphocyte formation [136]. While conditional *Spi1* inactivation in adult HSCs demonstrated expanded granulopoiesis, there was a lack of B-cell formation with the absence of CLPs [55]. Importantly, PU.1 has been shown to regulate *Flt3* [137] and *Il7r* [138] expression, two key cytokine receptors that are expressed by early lymphoid progenitors and are essential for the LMPP [139] as well as CLPs, with the combined knockout model failing to generate B-cell progenitors [140]. PU.1 appears dispensable in the mouse for later lymphopoiesis, as CLP cells in which *Spi1* is conditionally inactivated can form B-cell progenitors in vitro [141]. CD19-cre-mediated inactivation in committed B-cell progenitors also demonstrated PU.1 did not appear essential for B-cell development from proB cells in vivo with PU.1-deficient B cells being capable of BCR signalling and proliferation upon mitogenic stimulation in murine models [142]. Beyond CLPs, the requirement for other transcription factors such as MYB [31,52] for *Il7r* expression and proB-cell development has been demonstrated, in addition to roles in priming in earlier lymphoid progenitors. Although how MYB activity integrates with other transcriptional regulators in the B-cell gene regulatory network remains unclear.

Intriguingly, a human monogenic immunodeficiency disease caused by *SPI1* haploinsufficiency (PU.MA) has been recently described, with the phenotype suggesting, in contrast to mouse, a pre-eminent role for *PU.1* at the pro to preB transition phase in human that was sensitive to the loss of *PU.1* function, leading to proB-cell arrest, failure of B-cell development and agammaglobulinemia [56].

In its role as a haematopoietic lineage transcriptional regulator, PU.1 had been noted to function as a pioneer transcription factor maintaining H3K4me1 marks at macrophage and splenic B-cell-specific enhancers [143,144]. More recent studies have proposed PU.1 as representing a new class of 'non-classical' pioneer transcription factors that while not binding to nucleosome-associated DNA, binding and colocalisation with other transcription factor partners induces chromatin modification and redistribution of partner proteins. Interestingly, the N-terminal acidic activation domain of PU.1 was also shown to mediate interaction with members of the SWI/SNF complex, including BRG1 that was required for PU.1 to access and remodel chromatin de novo, leading to a possible model where PU.1 and SWI/SNF establish and maintain lineage-specific enhancers across several haematopoietic cell types [53].

SPI-B (*Spib*), in contrast to PU.1, does not appear to have an essential role in early haematopoietic or lymphoid development [145]. Knockin of *Spib* into the *PU.1* locus also definitively demonstrated a non-redundant role for PU.1 in early lymphopoiesis [146]. *Spib* expression in B cells is under the regulation of PAX5 [116,147], and loss of *Spib* was demonstrated to be important for both T-cell-independent and T-cell-dependent BCR-mediated proliferative responses. Interestingly, combinatorial loss of *PU.1* and *Spib* leads to a preB-cell developmental arrest through defective BCR signalling [57]. This latter finding suggests that while PU.1 and SPI-B are clearly non-redundant in early lymphoid development, they have redundant roles at later B-cell developmental stages in mediating BCR signalling responses that may in part be related to transcription of the B-cell linker protein BLNK/SLP-65 [148].

9. FORKHEAD O (FOXO) TRANSCRIPTION FACTORS

FOXO proteins are members of a large group of ubiquitously expressed transcription factors characterised by the forkhead DNA-binding domain that recognises a canonical 'TTGTTTAC' DNA motif. FOXO proteins regulate a wide number of cellular processes such as cell cycle progression, metabolism, apoptosis and oxidative defence in several different cell types [149]. Intriguingly, they are post-translationally and negatively regulated by cellular signalling pathways as direct targets of the serine-threonine Akt kinase, whereby phosphorylation of the FOXO proteins with consensus Akt phosphorylation sites induces nuclear export and inactivation of transcriptional activity [58,150,151]. FOXO1, FOXO3 and FOXP1 have been shown to be important in early B-cell development, with roles in early lymphoid lineage specification, as well as FOXO1 and FOXO3 being downstream of pre-BCR signalling for regulation of the immunoglobulin recombination machinery [152].

Foxo1 is expressed from the CLP stage of development [28]. B-lineage-restricted inactivation of FOXO1 resulted in a developmental block at the proB-cell stage and accumulation of pre-proB cells, with reduced expression of *Rag1* and *Rag2*, failure to upregulate *Il7r* expression and increased apoptosis, with increased expression of the pro-apoptotic BH3 protein *Bcl2l11* (encoding BIM) and reduced expression of the anti-apoptotic *Bcl2l1* (encoding BCL-XL) [59]. While apoptosis was rescued by the overexpression of *Bcl2*, the proliferation defect persisted. In keeping with the reduced expression of *Rag1* and *Rag2*, *Foxo1*-deficient proB cells exhibited defects in V_H-DJ_H recombination and reduced *Igh* expression, as well as defects in $V\kappa$-to-$J\kappa$ recombination in preB cells. ChIP identified binding of FOXO1 to the B-cell-specific *Erag* enhancer that is also bound by FOXP1 [60]. In addition, genome-wide analysis identified E2A and EBF1 binding to *Foxo1* [38] with co-occupancy of E2A and FOXO1 to B-cell gene targets such as *Blk*, *Dntt*, *Erg* and *Bcl2l11*. Mice that are compound heterozygous for *Tcf3* (E2A) and conditionally induced *Foxo1* haploinsufficiency resulted in reduction of the preB-cell compartment, supporting a coordinate role for E2A and FOXO1 in the proB to preB transition.

Foxp1 is an essential regulator of B-cell development, with inactivation of FOXP1 resulting in a proB developmental block associated with significantly diminished expression of *Rag1* and *Rag2*, while expression of many components of early B-cell development including *Tcf3*, *Ebf1*, *Pax5*, *Cd79a*, *Cd79b*, *Igll1* and *Vpreb1* were maintained. Notably, impaired *Igh* rearrangement with markedly impaired V_H-DJ_H and to a lesser extent D_H-J_H recombination was noted in FOXP1-deficient proB cells. Explaining this phenotype, FOXP1 was shown to bind and activate the B-lineage-specific *Erag* enhancer [61]. That both FOXO1 and FOXP1 are important regulators of immunoglobulin recombination in early B-cell development that can be influenced by Akt phosphorylation, provides a molecular pathway by which pre-BCR signalling can be linked to clonal selection and *Igl* recombination [152].

Foxo3 is expressed in HSCs and LMPPs before its expression declines in the CLP stage of development as *Foxo1* expression rises. Recently, a Vav-iCre deletion model of FOXO1 or FOXO3 inactivation throughout haematopoiesis demonstrated stage-specific requirements for both factors in keeping with their pattern of expression. This analysis showed loss of FOXO3 impaired the development of LMPPs, CLPs and B-cell precursors, while mature B-cell numbers were not significantly affected. In contrast, loss of FOXO1 resulted in a proB developmental block associated with dramatic reduction in *Rag* gene expression and phenocopying *Rag*-deficient mouse models [28]. Compound loss of FOXO1 and FOXO3 resulted in developmental block at the BLP stage and complete loss of B-lineage development, suggesting the CLP stage, where both *Foxo3* and *Foxo1* are expressed, was particularly vulnerable to the loss of both factors. The double knockout phenotype was associated with diminished expression of *Ebf1*, *Pax5* and multiple B-cell lineage-defining genes associated with pre-BCR signalling (*Igll1*, *VpreB1*, *Cd79a*, *Blk* and *Blnk*) as well as B-cell transcription factors and transcriptional activators (*Irf4*, *Ets1* and *Pou2af1*). Interestingly, motif analysis of chromatin-accessible regions in double knockout cells demonstrated decreased ATACseq signal in regions containing EBF and ETS consensus binding sites, suggesting potential FOXO1 and FOXO3 regulation of members of these transcription factor groups in CLPs: a role that had been proposed for FOXO1 in an EBF1 and E2A transcriptional network [32,38].

10. ZINC FINGER TRANSCRIPTION FACTORS

IKAROS, encoded by *Ikzf1*, is the original member [153] of a Krüppel-type zinc finger DNA-binding factor family that is essential for the development of all lymphoid lineages [154]. Like *Ikzf1*, other members HELIOS (*Ikzf2*), AIOLOS (*Ikzf3*), EOS (*Ikzf4*) and PEGASUS (*Ikzf5*) [155] possess a conserved N-terminal DNA-binding domain composed of four C2H2 zinc fingers that other than PEGASUS recognise the canonical 'GGGAA' motif, and a C-terminal two zinc finger dimerisation domain that allows homodimerisation or heterodimerisation with other family members [63]. IKAROS proteins form stable associations with members of the nucleosome-remodelling and histone-deacetylase (NuRD) and SWI-SNF chromatin remodelling complexes [64,156] that may contribute to the regulation of lineage specific genes.

IKAROS (*Ikzf1*) is essential for lymphoid development, with null mice failing to develop B, T or natural killer cells [154]. *Ikzf1* expression commences in the HSC compartment, with particularly strong expression in the LMPPs, a population lost in *Ikzf1* null mice. *Ikzf1* null HSCs were unable to undergo B-lineage differentiation or T-lineage expansion on in vitro culture, while myeloid differentiation potential was maintained including relative expansion of the megakaryocyte lineage [157]. Analysis of *Ikzf1* null bone marrow also demonstrated a virtual absence of IL7R$^+$ Flt3$^+$ CLPs [158]. Thus, Ikaros is critical determinant of early lymphoid lineage potential that is required for the formation of conventional LMPPs and CLPs. Examination of gene expression changes in *Ikzf1* null HSCs confirmed significant downregulation of *Flt3* and *Il7r* expression along with other lymphoid-lineage genes, as well as the upregulation of genes associated with inappropriate cell fates [159]. Ex vivo experiments in this latter study showed that downregulation of *Ikzf1* in sorted LMPPs resulted in significant reduction of B-lineage-forming potential and increased myeloid-forming potential, demonstrating Ikaros was also required for maintenance as well as establishment of lymphoid lineage potential.

Beyond the LMPP and CLP stages of early lymphoid development, analysis of models of an *Ikzf1* hypomorphic allele that was permissive for B-cell development demonstrated perturbed B-lineage development and a partial developmental block at the proB-cell stage [160]. This observation was confirmed in a B-cell-specific conditional C-terminal IKAROS dimerisation domain knockout model [161] that demonstrated the importance of IKAROS in proB to preB transition associated with a pre-BCR signalling defect. Multiple B-cell genes involved in surface receptor expression, signal transduction and transcriptional regulation were perturbed in defective proB cells including reduced expression of positive regulators of pre-BCR and

BCR signalling (*Slamf6*, *Lcp2*, *Lat*, *Grap2*, *Blk2* and *Pik3r5*) and de-repression of negative regulators of pre-BCR and BCR signalling (*Cd22*, *Cd72*, *Fcgr2b*, *Pdcd1*, *Ptpn6* and *Inppl1*) demonstrating an important role of IKAROS for appropriate signalling via calcium signalling, PI(3)K and MAPK pathways. Epigenetic analysis of H3K4me3 and H3K9ac marks in this model demonstrated induction or suppression of active chromatin associated with IKAROS-bound gene regulatory elements of activated or repressed target genes. Interestingly, this study noted that while significant colocalisation of the NuRD complex component Mi-2β was identified at IKAROS-bound proximal promoter regions, Mi-2β binding was independent of IKAROS binding in proB cells.

A separate B-cell-restricted conditional knockout model using a different *Ikzf1* conditional allele resulted in the failure of B-cell differentiation beyond the Hardy fraction C' 'large proliferative' proB stage, from which cell lines with a differentiation block could be derived [162]. This cell line model was used to explore the role of IKAROS in proB differentiation. Enforcing *Ikzf1* expression in *Ikzf1*-null proB cell lines rescued preB differentiation. Transcriptional analysis of this model suggested that IKAROS expression induced a transcriptional programme that was accentuated by IL7 withdrawal that included upregulation of genes that were enriched in the Hardy fraction D 'small quiescent' proB stage when compared to *Ikzf1* null proB fraction C′ cells. These included familiar B-lineage transcriptional regulators such as *Ebf1*, *Pax5* and *Spib*, as well as *Foxo1*, *Foxp1* and genes that enhanced FOXO function such as *Gadd45a*, *Pten* and *Ikzf3*. Several genes important to preB cells were also upregulated including *Irf4*, *Irf8*, as well as *Bcl2l1* required for preB-cell survival, *Fnip* that regulates the energy homoeostasis checkpoint in preB cells, as well as *Bach2* and *Bcl6*, the two important BCR checkpoint regulators in preB cells. There was also downregulation of components of the pre-BCR signalling complex *Igll1* and *VpreB1*, and upregulation of downstream BCR signalling components such as *Blnk* and *Syk*. Complimentary studies of IKAROS binding in separate mouse preB-cell lines at the *Igll1* locus [70] or genome wide with ChIPseq [66] identified direct binding of IKAROS to promoters and other gene regulatory elements of several of these genes. This includes IKAROS binding to the *Rag1/2* promoters and the *Erag* enhancer suggesting a potential cooperative role for IKAROS for inducing *Rag* gene expression in concert with FOXO proteins to facilitate *Igl* recombination in preB cells [60–62].

Notably, over 30 germline heterozygous mutations of *IKZF1* have been identified in humans that lead to immunological defects and various IKAROS-associated diseases that can manifest with immunodeficiency, autoimmunity and malignancy [68]. The known human mutations cluster in the DNA-binding domain (zinc fingers 2, 3) or dimerisation domain (zinc finger 6), with deletions, missense variants and point mutations resulting in mono-allelic loss of function. The type of mutation, such as those associated with residual haploinsufficient activity from the wild-type *IKZF1* allele, compared with dominant negative mutations resulting in complete loss of *IKZF1* function, can impact on disease manifestation and severity.

AIOLOS (*Ikzf3*), the closely related paralog to IKAROS (*Ikzf1*), is also expressed in the B-cell lineage, but its expression develops from later B-cell stages. Initiating from proB cells where it is a direct PAX5 transcriptional target [116,147], the expression of *Ikzf3* rapidly increases upon pre-BCR signalling until high expression levels are achieved in mature B cells [66,70]. As may be predicted from its expression pattern, germline loss of *Ikzf3* leads to defects in peripheral B cells that exist in a profoundly activated state, with mice prone to the development of autoimmunity and lymphoma [69]. During early B-cell development, *Ikzf3* null mice demonstrated an increase in the number of preB cells. AIOLOS and IKAROS can form hetero- and homodimers [70], and their function overlaps at common preB-cell gene targets [66], including silencing of the pre-BCR component genes (*Igll1* and *VpreB1*) in preB cells after pre-BCR signalling [70,163], as well as binding and regulating expression of genes involved B-cell transcription (*Foxo1*), BCR signalling (*Blnk*, *Syk*), cell metabolism and proliferation (*Myc*), cell cycle (*Ccnd2*, *Cdkn1a*) and *Igl* recombination (*Rag1*, *Rag2*, *Lig4*). Overexpression of *Ikzf1* and *Ikzf3* in B-cell lines resulted in termination of pre-BCR signalling and cell cycle exit with silencing of *Igll1* and *VpreB*, downregulation of c-*Myc* and *cyclin D3* [67].

A family with a human germline heterozygous missense *IKZF3* variant has been recently described resulting in an AIOLOS^{G159R} variant with reduced DNA-binding efficiency to canonical DNA motifs, but also showing aberrant binding to other genomic regions. This variant interfered with appropriate IKAROS binding and activity through AIOLOS^{G159R} IKAROS heterodimerisation, leading to a profound B-cell immunodeficiency syndrome with absent circulating B cells and reduced B-cell progenitors in the bone marrow [71]. This finding highlights the important role that IKAROS has across all stages of early B-cell development, and the close genomic and structural interaction with its AIOLOS paralog in B-cell development.

11. INTERFERON REGULATORY FACTORS

Interferon regulatory factor 4 (*Irf4*) and 8 (*Irf8*) are two of nine members of the interferon regulatory factor (IRF) family of transcriptional regulators. Members of this family possess an N-terminal helix–turn–helix DNA-binding domain that contains five conserved tryptophan repeats that resembles the DNA-binding domain of MYB. The IRF DNA-binding domain recognises the IFN-stimulated response element (ISRE) 'A/GNGAAANNGAAACT' motif [164].

IRF4 [54] (synonym *PU.1 interaction partner, Pip*) and **IRF8** [165] (synonym *interferon consensus sequence binding protein, ICSBP*) are two highly homologous members of the IRF family [166] that were originally identified through their binding to gene regulatory elements of immune genes, including a composite ETS:ISRE (EIRE) DNA motif located in the immunoglobulin Eκ3′ and λ gene enhancer regions that juxtaposed PU.1 and the IRF transcription factors [167]. Structural characterisation identified that IRF4 could form a ternary complex with PU.1 [168] which was required for IRF4 binding to the EIRE sequence [169].

In the *Irf4* and *Irf8* double knockout mouse, a preB development block is apparent with the failure of *Igl* recombination [72]. In this model, *Irf4* and *Irf8* were shown to be functionally redundant factors, either of which could rescue *Igκ* re-arrangement and preB differentiation in *Irf4* and *Irf8*-deficient preB cells in vitro [72]. Binding of the IRF4 and PU.1 complex could activate the Eκ3′ enhancer with *Igl* recombination in vitro [168] associated with H3/H4 acetylation and H3K4 trimethylation at this locus and the *Jkl* region [72]. Notably, *Spi1*$^{-/-}$*Spib*$^{-/-}$ double knockout mice demonstrated a differentiation block at the preB-cell stage with impaired *Igk* transcription [170], while *PU.1* and *Irf4* double knockouts have defects in preB-cell differentiation [171]. IRF4 has also been shown to enhance binding of other B-cell transcription factors such as the E2A E47 factor to a *Igκ 3′* enhancer E-box motif that orientates an IRF4 N-terminal and E47 HLH interaction, synergising transcriptional activation at the enhancer locus in vitro [49,172,173]. These findings demonstrate that other protein partners may complex with IRF4 to transactivate this locus.

In addition to loss of *Igl* rearrangement, the *Irf4*$^{-/-}$*Irf8*$^{-/-}$ Hardy C' 'proliferative' large proB cells demonstrate persistent cell cycling, surrogate light chain λ5 and VpreB expression, and pre-BCR signalling [174]. Retroviral transduction and conditional expression of IRF4 or IRF8 in these cells resulted in downregulation of the *Igll1* transcript and pre-BCR expression, while associated with upregulation of IKAROS and AIOLOS expression, which of themselves were sufficient to downregulate λ5 and the pre-BCR when expressed in these cells [73]. These results demonstrated IRF4 and IRF8 were required for preB-cell development, facilitating *Igl* rearrangement and transcription, and transitioning from the proliferative large preB cell to a quiescent small preB state. These features were associated with, and likely co-mediated by, an increase in IKAROS and AIOLOS, both of which have been shown to directly suppress *Myc* [67], a critical regulator of cell proliferation and metabolism.

12. MYC FAMILY PROTEINS

MYC family proteins, MYC (also known as c-MYC) and its closest relatives, N-MYC and L-MYC, are defined by a basic region helix−loop−helix leucine zipper domain (bHLH-ZIP) that binds to DNA at canonical E-box 'CACGTG' and at non-canonical variant motifs. MYC does not homodimerise or bind to DNA under physiologic circumstances and requires the related bHLH-ZIP factor, MAX for DNA binding, stabilisation and transactivation [77,175−177]. MYC also associates with WDR5, a WD40-repeat-containing protein, via a conserved IIIb MYC box. WDR5 assembles into several chromatin regulatory complexes including the MLL/SET methyltransferases that methylate H3K4 and the MOF/NSL histone acetyltransferases that acetylate histone H4. WDR5 and MYC demonstrate extensive colocalisation at proximal promoter and intergenic regions [78] with WDR5 appearing essential for MYC binding, stabilisation and activity in cell line models [178].

Myc is broadly expressed throughout haematopoiesis. Inactivation of *Myc* in early haematopoietic stem and progenitor cells results in profound reduction in lymphopoiesis and myelopoiesis due to a block in differentiation and lineage commitment [179]. B-lineage functions for *Myc* have been demonstrated using cre-driven deletion in early B-cell progenitors that results in a prominent defect in the proB to preB-cell transition. This included marked reduction of the large 'proliferative' Hardy C′ fraction that show increased markers of apoptosis, along with dramatically reduced and defective mature B-cell development [79] that was also observed with the deletion of *Myc* later in development using *CD19cre* [180]. In contrast, B-cell-specific inactivation of *Max* demonstrated a milder phenotype [177]; an observation that may relate to longer half-life of the MAX protein [77].

MYC regulates the expression of genes involved in a multitude of cell pathways, including biosynthetic processes such as ribosomal biogenesis and protein synthesis [181,182]. Additionally, MYC-deficient preB cells showed downregulation of *Tcf3*, *Ebf1*, *Pax5* and *Ikzf1* expression, along with PAX5 target genes *Cd19*, *Blnk*, *Cd79a*, *Cd79b* and de-repression of *Flt3* [79]. In this study, MYC bound to two conserved E-box motifs within the *Ebf1* α-promoter [79] with *Ebf1* transduction of *Myc*-deficient B cells able to promote B-cell differentiation associated with increased expression of *Pax5* and *Cd19*, but without restoration of normal proliferation. The transcriptional network relationship between EBF1 and MYC was reinforced with the recent functional delineation of six EBF1-responsive enhancer elements at the *Myc* gene locus and binding of

EBF1 to the distal *Myc* BENC enhancer region [183] promoted *Myc* expression for cell survival and proliferation [39]. Thus, MYC helps promote proB to preB transition with contribution to the regulation to B-lineage gene expression, but importantly facilitating cell proliferation and expansion of large proB cells in response to IL7 and pre-BCR signalling [152]. Thereafter, downregulation of *Myc* can occur via IKAROS and AIOLOS binding and inhibition at the *Myc* promoter, with subsequent de-repression of the cyclin inhibitor *Cdkn1b* and reduced expression of *Ccnd3* [184], facilitating transition to the small quiescent preB developmental stage [67].

13. CONCLUSIONS

Several decades on from the first studies into early B-cell development that occurred last century, our knowledge of the regulatory landscape required for B-cell lineage is growing at a remarkable pace. As observations from earlier studies continue to be refined through the application of contemporary approaches to investigating transcription factor function, the detailed insights afforded into these complex processes, particularly at the genomic level, continue to evolve our understanding of B-lineage development. We now have a detailed knowledge of all the key players involved in early lymphopoiesis, as well as a burgeoning model of the transcriptional network that powers the tightly controlled regulation of the B cell developmental pathway.

The study of traditional animal models can now also be complimented by the identification and characterisation of mutations of human diseases through genomic sequencing. The combination of these approaches will add to our growing knowledge of the basic biology of B-cell function, and importantly, help define how mutations and aberrant development translates into disease. To date, virtually all key regulators of B-cell development highlighted in this chapter have been shown to be the target of mutations in both primary immune deficiencies and in B-cell malignancies. Indeed, it can be anticipated that cross-fertilisation between this basic and translational knowledge will play an increasingly important role in our current era of genomic medicine and genetic therapies.

ACKNOWLEDGEMENTS

This work was supported by grants from the Cancer Council of Victoria Grant in Aid (1185830), National Stem Cell Foundation of Australia Metcalf Prize for Stem Cell Research and Alessandra's Fight Against Leukaemia (GoFundMe) to A.P.N. S.L.N. was supported by the National Health and Medical Research Council, Australia (1155243 and 2000461) and the Jacques and Margaret Miller Fellowship.

REFERENCES

[1] Murakami K, Tsai K-L, Kalisman N, et al. Structure of an RNA polymerase II preinitiation complex. Proc Natl Acad Sci USA 2015;112(44):13543–8.
[2] Aibara S, Schilbach S, Cramer P. Structures of mammalian RNA polymerase II pre-initiation complexes. Nature 2021;594(7861):124–8.
[3] Soufi A, Garcia MF, Jaroszewicz A, et al. Pioneer transcription factors target partial DNA motifs on nucleosomes to initiate reprogramming. Cell 2015;161(3):555–68.
[4] Geiman TM, Robertson KD. Chromatin remodeling, histone modifications, and DNA methylation—how does it all fit together? J Cell Biochem 2002;87(2):117–25.
[5] Martin BJE, Brind'Amour J, Kuzmin A, et al. Transcription shapes genome-wide histone acetylation patterns. Nat Commun 2021;12(1):210.
[6] Poon GMK, Kim HM. Signatures of DNA target selectivity by ETS transcription factors. Biochem Soc Symp 2017;8(3):193–203.
[7] Inukai S, Kock KH, Bulyk ML. Transcription factor–DNA binding: beyond binding site motifs. Curr Opin Genet Dev 2017;43:110–9.
[8] Suter DM. Transcription factors and DNA play hide and seek. Trends Cell Biol 2020;30(6):491–500.
[9] Sanyal A, Lajoie B, Jain G, Dekker J. The long-range interaction landscape of gene promoters. Nature 2012;489(7414):109–13.
[10] Szabo Q, Bantignies F, Cavalli G. Principles of genome folding into topologically associating domains. Sci Adv 2019;5(4):eaaw1668.
[11] Allen BL, Taatjes DJ. The Mediator complex: a central integrator of transcription. Nat Rev Mol Cell Biol 2015;16(3):155–66.
[12] Weintraub AS, Li CH, Zamudio AV, et al. YY1 is a structural regulator of enhancer-promoter loops. Cell 2017;171(7):1573–1588.e28.
[13] Nutt SL, Kee BL. The transcriptional regulation of B cell lineage commitment. Immunity 2007;26(6):715–25.
[14] Igarashi H, Gregory SC, Yokota T, Sakaguchi N, Kincade PW. Transcription from the RAG1 locus marks the earliest lymphocyte progenitors in bone marrow. Immunity 2002;17(2):117–30.
[15] Amann-Zalcenstein D, Tian L, Schreuder J, et al. A new lymphoid-primed progenitor marked by Dach1 downregulation identified with single cell multi-omics. Nat Immunol 2020;21(12):1574–84.
[16] Klein F, Roux J, Cvijetic G, et al. Dntt expression reveals developmental hierarchy and lineage specification of hematopoietic progenitors. Nat Immunol 2022;23(4):505–17.
[17] Sitnicka E, Bryder D, Theilgaard-Mönch K, et al. Key role of flt3 ligand in regulation of the common lymphoid progenitor but not in maintenance of the hematopoietic stem cell pool. Immunity 2002;17(4):463–72.
[18] Dias S, Silva H, Cumano A, Vieira P. Interleukin-7 is necessary to maintain the B cell potential in common lymphoid progenitors. J Exp Med 2005;201(6):971–9.
[19] Kondo M, Weissman IL, Akashi K. Identification of clonogenic common lymphoid progenitors in mouse bone marrow. Cell 1997;91(5):661–72.
[20] Inlay MA, Bhattacharya D, Sahoo D, et al. Ly6d marks the earliest stage of B-cell specification and identifies the branchpoint between B-cell and T-cell development. Gene Dev 2009;23(20):2376–81.
[21] Guo H, Barberi T, Suresh R, Friedman AD. Progression from the common lymphoid progenitor to B/myeloid preproB and proB

precursors during B lymphopoiesis requires C/EBPα. J Immunol 2018;201(6):1692–704.

[22] Hardy RR, Carmack CE, Shinton SA, Kemp JD, Hayakawa K. Resolution and characterization of pro-B and pre-pro-B cell stages in normal mouse bone marrow. J Exp Med 1991;173(5):1213–25.

[23] Rolink A, Grawunder U, Winkler TH, Karasuyama H, Melchers F. IL-2 receptor α chain (CD25JAC) expression defines a crucial stage in pre-B cell development. Int Immunol 1994;6(8):1257–64.

[24] O'Riordan M, Grosschedl R. Coordinate regulation of B cell differentiation by the transcription factors EBF and E2A. Immunity 1999;11(1):21–31.

[25] Reynaud D, Demarco IA, Reddy KL, et al. Regulation of B cell fate commitment and immunoglobulin heavy-chain gene rearrangements by Ikaros. Nat Immunol 2008;9(8):927–36.

[26] Pang SHM, Graaf CA de, Hilton DJ, et al. PU.1 is required for the developmental progression of multipotent progenitors to common lymphoid progenitors. Front Immunol 2018;9:1264.

[27] Medina KL, Pongubala JMR, Reddy KL, et al. Assembling a gene regulatory network for specification of the B cell fate. Dev Cell 2004;7(4):607–17.

[28] Peña-Pérez L, Kharazi S, Frengen N, et al. FOXO dictates initiation of B cell development and myeloid restriction in common lymphoid progenitors. Front Immunol 2022;13:880668.

[29] Stehling-Sun S, Dade J, Nutt SL, DeKoter RP, Camargo FD. Regulation of lymphoid versus myeloid fate "choice" by the transcription factor Mef2c. Nat Immunol 2009;10(3):289–96.

[30] Bain G, Maandag ECR, Izon DJ, et al. E2A proteins are required for proper B cell development and initiation of immunoglobulin gene rearrangements. Cell 1994;79(5):885–92.

[31] Greig KT, Graaf CA de, Murphy JM, et al. Critical roles for c-Myb in lymphoid priming and early B-cell development. Blood 2010;115(14):2796–805.

[32] Mansson R, Welinder E, Åhsberg J, et al. Positive intergenic feedback circuitry, involving EBF1 and FOXO1, orchestrates B-cell fate. Proc Natl Acad Sci USA 2012;109(51):21028–33.

[33] Lin H, Grosschedl R. Failure of B-cell differentiation in mice lacking the transcription factor EBF. Nature 1995;376(6537):263–7.

[34] Ng AP, Coughlan HD, Hediyeh-zadeh S, et al. An Erg-driven transcriptional program controls B cell lymphopoiesis. Nat Commun 2020;11(1):3013.

[35] Nutt SL, Heavey B, Rolink AG, Busslinger M. Commitment to the B-lymphoid lineage depends on the transcription factor Pax5. Nature 1999;402(Suppl. 6763):14–20.

[36] Li R, Cauchy P, Ramamoorthy S, et al. Dynamic EBF1 occupancy directs sequential epigenetic and transcriptional events in B-cell programming. Gene Dev 2018;32(2):96–111.

[37] Lenaerts A, Kucinski I, Deboutte W, et al. EBF1 primes B-lymphoid enhancers and limits the myeloid bias in murine multipotent progenitors. J Exp Med 2022;219(11):e20212437.

[38] Lin YC, Jhunjhunwala S, Benner C, et al. A global network of transcription factors, involving E2A, EBF1 and Foxo1, that orchestrates B cell fate. Nat Immunol 2010;11(7):635–43.

[39] Somasundaram R, Jensen CT, Tingvall-Gustafsson J, et al. EBF1 and PAX5 control pro-B cell expansion via opposing regulation of the Myc gene. Blood 2021;137(22):3037–49.

[40] Garvie CW, Hagman J, Wolberger C. Structural studies of Ets-1/Pax5 complex formation on DNA. Mol Cell 2001;8(6):1267–76.

[41] McManus S, Ebert A, Salvagiotto G, et al. The transcription factor Pax5 regulates its target genes by recruiting chromatin-modifying proteins in committed B cells. EMBO J 2011;30(12):2388–404.

[42] Fuxa M, Skok J, Souabni A, et al. Pax5 induces V-to-DJ rearrangements and locus contraction of the immunoglobulin heavy-chain gene. Gene Dev 2004;18(4):411–22.

[43] Johanson TM, Lun ATL, Coughlan HD, et al. Transcription-factor-mediated supervision of global genome architecture maintains B cell identity. Nat Immunol 2018;19(11):1257–64.

[44] Kaiser FMP, Gruenbacher S, Oyaga MR, et al. Biallelic PAX5 mutations cause hypogammaglobulinemia, sensorimotor deficits, and autism spectrum disorder. J Exp Med 2022;219(9):e20220498.

[45] Eckner R, Yao TP, Oldread E, Livingston DM. Interaction and functional collaboration of p300/CBP and bHLH proteins in muscle and B-cell differentiation. Gene Dev 1996;10(19):2478–90.

[46] Dias S, Månsson R, Gurbuxani S, Sigvardsson M, Kee BL. E2A proteins promote development of lymphoid-primed multipotent progenitors. Immunity 2008;29(2):217–27.

[47] Borghesi L, Aites J, Nelson S, et al. E47 is required for V(D)J recombinase activity in common lymphoid progenitors. J Exp Med 2005;202(12):1669–77.

[48] Miyazaki K, Watanabe H, Yoshikawa G, et al. The transcription factor E2A activates multiple enhancers that drive Rag expression in developing T and B cells. Sci Immunol 2020;5(51):eabb1455.

[49] Lazorchak AS, Schlissel MS, Zhuang Y. E2A and IRF-4/Pip promote chromatin modification and transcription of the immunoglobulin κ locus in pre-B cells. Mol Cell Biol 2006;26(3):810–21.

[50] Dobbs AK, Bosompem A, Coustan-Smith E, et al. Agammaglobulinemia associated with BCR− B cells and enhanced expression of CD19. Blood 2011;118(7):1828–37.

[51] Boisson B, Wang Y-D, Bosompem A, et al. A recurrent dominant negative E47 mutation causes agammaglobulinemia and BCR− B cells. J Clin Invest 2013;123(11):4781–5.

[52] Fahl SP, Crittenden RB, Allman D, Bender TP. c-Myb is required for pro-B cell differentiation. J Immunol 2009;183(9):5582–92.

[53] Minderjahn J, Schmidt A, Fuchs A, et al. Mechanisms governing the pioneering and redistribution capabilities of the non-classical pioneer PU.1. Nat Commun 2020;11(1):402.

[54] Eisenbeis CF, Singh H, Storb U. Pip, a novel IRF family member, is a lymphoid-specific, PU.1-dependent transcriptional activator. Gene Dev 1995;9(11):1377–87.

[55] Dakic A, Metcalf D, Rago LD, et al. PU.1 regulates the commitment of adult hematopoietic progenitors and restricts granulopoiesis. J Exp Med 2005;201(9):1487–502.

[56] Coz CL, Nguyen DN, Su C, et al. Constrained chromatin accessibility in PU.1-mutated agammaglobulinemia patients. J Exp Med 2021;218(7):e20201750.

[57] Garrett-Sinha LA, Su GH, Rao S, et al. PU.1 and Spi-B are required for normal B cell receptor−mediated signal transduction. Immunity 1999;10(4):399–408.

[58] Biggs WH, Meisenhelder J, Hunter T, Cavenee WK, Arden KC. Protein kinase B/Akt-mediated phosphorylation promotes nuclear exclusion of the winged helix transcription factor FKHR1. Proc Natl Acad Sci USA 1999;96(13):7421–6.

[59] Dengler HS, Baracho GV, Omori SA, et al. Distinct functions for the transcription factor Foxo1 at various stages of B cell differentiation. Nat Immunol 2008;9(12):1388–98.

[60] Amin RH, Schlissel MS. Foxo1 directly regulates the transcription of recombination-activating genes during B cell development. Nat Immunol 2008;9(6):613–22.
[61] Hu H, Wang B, Borde M, et al. Foxp1 is an essential transcriptional regulator of B cell development. Nat Immunol 2006;7(8):819–26.
[62] Herzog S, Hug E, Meixlsperger S, et al. SLP-65 regulates immunoglobulin light chain gene recombination through the PI(3)K-PKB-Foxo pathway. Nat Immunol 2008;9(6):623–31.
[63] John LB, Ward AC. The Ikaros gene family: transcriptional regulators of hematopoiesis and immunity. Mol Immunol 2011;48(9–10):1272–8.
[64] Kim J, Sif S, Jones B, et al. Ikaros DNA-binding proteins direct formation of chromatin remodeling complexes in lymphocytes. Immunity 1999;10(3):345–55.
[65] Arends T, Dege C, Bortnick A, et al. CHD4 is essential for transcriptional repression and lineage progression in B lymphopoiesis. Proc Natl Acad Sci USA 2019;116(22):10927–36.
[66] Ferreirós-Vidal I, Carroll T, Taylor B, et al. Genome-wide identification of Ikaros targets elucidates its contribution to mouse B-cell lineage specification and pre-B–cell differentiation. Blood 2013;121(10):1769–82.
[67] Ma S, Pathak S, Mandal M, et al. Ikaros and Aiolos inhibit pre-B-cell proliferation by directly suppressing c-Myc expression. Mol Cell Biol 2010;30(17):4149–58.
[68] Boast B, Nunes-Santos C de J, Kuehn HS, Rosenzweig SD. Ikaros-associated diseases: from mice to humans and back again. Front Pediatr 2021;9:705497.
[69] Wang J-H, Avitahl N, Cariappa A, et al. Aiolos regulates B cell activation and maturation to effector state. Immunity 1998;9(4):543–53.
[70] Thompson EC, Cobb BS, Sabbattini P, et al. Ikaros DNA-binding proteins as integral components of B cell developmental-stage-specific regulatory circuits. Immunity 2007;26(3):335–44.
[71] Yamashita M, Kuehn HS, Okuyama K, et al. A variant in human AIOLOS impairs adaptive immunity by interfering with IKAROS. Nat Immunol 2021;22(7):893–903.
[72] Ma S, Turetsky A, Trinh L, Lu R. IFN regulatory factor 4 and 8 promote Ig light chain κ locus activation in pre-B cell development. J Immunol 2006;177(11):7898–904.
[73] Ma S, Pathak S, Trinh L, Lu R. Interferon regulatory factors 4 and 8 induce the expression of Ikaros and Aiolos to down-regulate pre–B-cell receptor and promote cell-cycle withdrawal in pre–B-cell development. Blood 2008;111(3):1396–403.
[74] Swaminathan S, Duy C, Müschen M. BACH2–BCL6 balance regulates selection at the pre-B cell receptor checkpoint. Trends Immunol 2014;35(3):131–7.
[75] Duy C, Yu JJ, Nahar R, et al. BCL6 is critical for the development of a diverse primary B cell repertoire. J Exp Med 2010;207(6):1209–21.
[76] Swaminathan S, Huang C, Geng H, et al. BACH2 mediates negative selection and p53-dependent tumor suppression at the pre-B cell receptor checkpoint. Nat Med 2013;19(8):1014–22.
[77] Blackwood EM, Lüscher B, Eisenman RN. Myc and Max associate in vivo. Gene Dev 1992;6(1):71–80.
[78] Thomas LR, Wang Q, Grieb BC, et al. Interaction with WDR5 promotes target gene recognition and tumorigenesis by MYC. Mol Cell 2015;58(3):440–52.
[79] Vallespinós M, Fernández D, Rodríguez L, et al. B lymphocyte commitment program is driven by the proto-oncogene c-myc. J Immunol 2011;186(12):6726–36.
[80] Nechanitzky R, Akbas D, Scherer S, et al. Transcription factor EBF1 is essential for the maintenance of B cell identity and prevention of alternative fates in committed cells. Nat Immunol 2013;14(8):867–75.
[81] Zhuang Y, Soriano P, Weintraub H. The helix-loop-helix gene E2A is required for B cell formation. Cell 1994;79(5):875–84.
[82] Isbel L, Grand RS, Schübeler D. Generating specificity in genome regulation through transcription factor sensitivity to chromatin. Nat Rev Genet 2022;23(12):728–40.
[83] Wingender E, Schoeps T, Haubrock M, Krull M, Dönitz J. TFClass: expanding the classification of human transcription factors to their mammalian orthologs. Nucleic Acids Res 2017;46(Database issue):gkx987.
[84] Pratt HE, Andrews GR, Phalke N, et al. Factorbook: an updated catalog of transcription factor motifs and candidate regulatory motif sites. Nucleic Acids Res 2021;50(D1):D141–9.
[85] Hagman J, Travis A, Grosschedl R. A novel lineage-specific nuclear factor regulates mb-1 gene transcription at the early stages of B cell differentiation. EMBO J 1991;10(11):3409–17.
[86] Kozmik Z, Wang S, Dörfler P, Adams B, Busslinger M. The promoter of the CD19 gene is a target for the B-cell-specific transcription factor BSAP. Mol Cell Biol 1992;12(6):2662–72.
[87] Murre C, McCaw PS, Baltimore D. A new DNA binding and dimerization motif in immunoglobulin enhancer binding, daughterless, MyoD, and myc proteins. Cell 1989;56(5):777–83.
[88] Farnham PJ. Insights from genomic profiling of transcription factors. Nat Rev Genet 2009;10(9):605–16.
[89] Yoshida H, Lareau CA, Ramirez RN, et al. The cis-regulatory atlas of the mouse immune system. Cell 2019;176(4):897–912.e20.
[90] Buenrostro JD, Giresi PG, Zaba LC, Chang HY, Greenleaf WJ. Transposition of native chromatin for fast and sensitive epigenomic profiling of open chromatin, DNA-binding proteins and nucleosome position. Nat Methods 2013;10(12):1213–8.
[91] Project TIG, Aguilar SV, Aguilar O, et al. ImmGen at 15. Nat Immunol 2020;21(7):700–3.
[92] Castro-Mondragon JA, Riudavets-Puig R, Rauluseviciute I, et al. JASPAR 2022: the 9th release of the open-access database of transcription factor binding profiles. Nucleic Acids Res 2021;50(D1):gkab1113.
[93] Boller S, Grosschedl R. The regulatory network of B-cell differentiation: a focused view of early B-cell factor 1 function. Immunol Rev 2014;261(1):102–15.
[94] Treiber T, Mandel EM, Pott S, et al. Early B cell factor 1 regulates B cell gene networks by activation, repression, and transcription- independent poising of chromatin. Immunity 2010;32(5):714–25.
[95] Iwafuchi-Doi M, Zaret KS. Pioneer transcription factors in cell reprogramming. Gene Dev 2014;28(24):2679–92.
[96] Boller S, Ramamoorthy S, Akbas D, et al. Pioneering activity of the C-terminal domain of EBF1 shapes the chromatin landscape for B cell programming. Immunity 2016;44(3):527–41.
[97] Bossen C, Murre CS, Chang AN, et al. The chromatin remodeler Brg1 activates enhancer repertoires to establish B cell identity and modulate cell growth. Nat Immunol 2015;16(7):775–84.

[98] Zhang Z, Cotta CV, Stephan RP, deGuzman CG, Klug CA. Enforced expression of EBF in hematopoietic stem cells restricts lymphopoiesis to the B cell lineage. EMBO J 2003;22(18):4759–69.

[99] Pongubala JMR, Northrup DL, Lancki DW, et al. Transcription factor EBF restricts alternative lineage options and promotes B cell fate commitment independently of Pax5. Nat Immunol 2008;9(2):203–15.

[100] Seet CS, Brumbaugh RL, Kee BL. Early B cell factor promotes B lymphopoiesis with reduced interleukin 7 responsiveness in the absence of E2A. J Exp Med 2004;199(12):1689–700.

[101] Belle I, Zhuang Y. Chapter four E proteins in lymphocyte development and lymphoid diseases. Curr Top Dev Biol 2014;110(Blood116252010):153–87.

[102] Kwon K, Hutter C, Sun Q, et al. Instructive role of the transcription factor E2A in early B lymphopoiesis and germinal center B cell development. Immunity 2008;28(6):751–62.

[103] Thal MA, Carvalho TL, He T, et al. Ebf1-mediated down-regulation of Id2 and Id3 is essential for specification of the B cell lineage. Proc Natl Acad Sci USA 2009;106(2):552–7.

[104] Czerny T, Schaffner G, Busslinger M. DNA sequence recognition by Pax proteins: bipartite structure of the paired domain and its binding site. Gene Dev 1993;7(10):2048–61.

[105] Cobaleda C, Schebesta A, Delogu A, Busslinger M. Pax5: the guardian of B cell identity and function. Nat Immunol 2007;8(5):463–70.

[106] Hammal F, Langen P de, Bergon A, Lopez F, Ballester B. ReMap 2022: a database of Human, Mouse, Drosophila and Arabidopsis regulatory regions from an integrative analysis of DNA-binding sequencing experiments. Nucleic Acids Res 2021;50 (D1):D316–25.

[107] Bullerwell CE, Robichaud PP, Deprez PML, et al. EBF1 drives hallmark B cell gene expression by enabling the interaction of PAX5 with the MLL H3K4 methyltransferase complex. Sci Rep 2021;11(1):1537.

[108] Revilla-i-Domingo R, Bilic I, Vilagos B, et al. The B-cell identity factor Pax5 regulates distinct transcriptional programmes in early and late B lymphopoiesis. EMBO J 2012;31(14):3130–46.

[109] Nutt SL, Urbánek P, Rolink A, Busslinger M. Essential functions of Pax5 (BSAP) in pro-B cell development: difference between fetal and adult B lymphopoiesis and reduced V-to-DJ recombination at the IgH locus. Gene Dev 1997;11(4):476–91.

[110] Rolink AG, Nutt SL, Melchers F, Busslinger M. Long-term in vivo reconstitution of T-cell development by Pax5-deficient B-cell progenitors. Nature 1999;401(6753):603–6.

[111] Delogu A, Schebesta A, Sun Q, et al. Gene repression by Pax5 in B cells is essential for blood cell homeostasis and is reversed in plasma cells. Immunity 2006;24(3):269–81.

[112] Holmes ML, Carotta S, Corcoran LM, Nutt SL. Repression of Flt3 by Pax5 is crucial for B-cell lineage commitment. Gene Dev 2006;20(8):933–8.

[113] Holmes ML, Pridans C, Nutt SL. The regulation of the B-cell gene expression programme by Pax5. Immunol Cell Biol 2008;86(1):47–53.

[114] Souabni A, Cobaleda C, Schebesta M, Busslinger M. Pax5 promotes B lymphopoiesis and blocks T cell development by repressing Notch1. Immunity 2002;17(6):781–93.

[115] Tagoh H, Ingram R, Wilson N, et al. The mechanism of repression of the myeloid-specific c-fms gene by Pax5 during B lineage restriction. EMBO J 2006;25(5):1070–80.

[116] Pridans C, Holmes ML, Polli M, et al. Identification of Pax5 target genes in early B cell differentiation. J Immunol 2008;180(3):1719–28.

[117] Ikawa T, Kawamoto H, Wright LYT, Murre C. Long-Term cultured E2A-deficient hematopoietic progenitor cells are pluripotent. Immunity 2004;20(3):349–60.

[118] Ebert A, McManus S, Tagoh H, et al. The distal VH gene cluster of the Igh locus contains distinct regulatory elements with Pax5 transcription factor-dependent activity in pro-B cells. Immunity 2011;34(2):175–87.

[119] Hill L, Ebert A, Jaritz M, et al. Wapl repression by Pax5 promotes V gene recombination by Igh loop extrusion. Nature 2020;584(7819):142–7.

[120] Lieberman-Aiden E, Berkum NL van, Williams L, et al. Comprehensive mapping of long-range interactions reveals folding principles of the human genome. Science 2009;326 (5950):289–93.

[121] Oikawa T, Yamada T. Molecular biology of the Ets family of transcription factors. Gene 2003;303:11–34.

[122] Sharrocks AD. The ETS-domain transcription factor family. Nat Rev Mol Cell Biol 2001;2(11):827–37.

[123] Carrère S, Verger A, Flourens A, Stehelin D, Duterque-Coquillaud M. Erg proteins, transcription factors of the Ets family, form homo, heterodimers and ternary complexes via two distinct domains. Oncogene 1998;16(25):3261–8.

[124] Taoudi S, Bee T, Hilton A, et al. ERG dependence distinguishes developmental control of hematopoietic stem cell maintenance from hematopoietic specification. Gene Dev 2011;25(3):251–62.

[125] Loughran SJ, Kruse EA, Hacking DF, et al. The transcription factor Erg is essential for definitive hematopoiesis and the function of adult hematopoietic stem cells. Nat Immunol 2008;9(7):810–9.

[126] Ng AP, Loughran SJ, Metcalf D, et al. Erg is required for self-renewal of hematopoietic stem cells during stress hematopoiesis in mice. Blood 2011;118(9):2454–61.

[127] Ng AP, Hu Y, Metcalf D, et al. Early lineage priming by trisomy of Erg leads to myeloproliferation in a Down syndrome model. PLoS Genet 2015;11(5):e1005211.

[128] Bassuk AG, Leiden JM. The role of Ets transcription factors in the development and function of the mammalian immune system. Adv Immunol 1997;64:65–104.

[129] Gallant S, Gilkeson G. ETS transcription factors and regulation of immunity. Arch Immunol Ther Exp 2006;54(3):149–63.

[130] Vilagos B, Hoffmann M, Souabni A, et al. Essential role of EBF1 in the generation and function of distinct mature B cell types. J Exp Med 2012;209(4):775–92.

[131] Søndergaard E, Rauch A, Michaut M, et al. ERG controls B cell development by promoting Igh V-to-DJ recombination. Cell Rep 2019;29(9):2756–2769.e6.

[132] Dakic A, Wu L, Nutt SL. Is PU.1 a dosage-sensitive regulator of haemopoietic lineage commitment and leukaemogenesis? Trends Immunol 2007;28(3):108–14.

[133] McKercher SR, Torbett BE, Anderson KL, et al. Targeted disruption of the PU.1 gene results in multiple hematopoietic abnormalities. EMBO J 1996;15(20):5647–58.

[134] Scott EW, Simon MC, Anastasi J, Singh H. Requirement of transcription factor PU.1 in the development of multiple hematopoietic lineages. Science 1994;265(5178):1573–7.

[135] Nutt SL, Metcalf D, D'Amico A, Polli M, Wu L. Dynamic regulation of PU.1 expression in multipotent hematopoietic progenitors. J Exp Med 2005;201(2):221–31.

[136] DeKoter RP, Singh H. Regulation of B lymphocyte and macrophage development by graded expression of PU.1. Science 2000;288(5470):1439–41.

[137] Carotta S, Dakic A, D'Amico A, et al. The transcription factor PU.1 controls dendritic cell development and Flt3 cytokine receptor expression in a dose-dependent manner. Immunity 2010;32(5):628–41.

[138] DeKoter RP, Lee H-J, Singh H. PU.1 regulates expression of the interleukin-7 receptor in lymphoid progenitors. Immunity 2002;16(2):297–309.

[139] Sitnicka E, Buza-Vidas N, Ahlenius H, et al. Critical role of FLT3 ligand in IL-7 receptor–independent T lymphopoiesis and regulation of lymphoid-primed multipotent progenitors. Blood 2007;110(8):2955–64.

[140] Voßhenrich CAJ, Cumano A, Müller W, Santo JPD, Vieira P. Thymic stromal-derived lymphopoietin distinguishes fetal from adult B cell development. Nat Immunol 2003;4(8):773–9.

[141] Iwasaki H, Somoza C, Shigematsu H, et al. Distinctive and indispensable roles of PU.1 in maintenance of hematopoietic stem cells and their differentiation. Blood 2005;106(5):1590–600.

[142] Polli M, Dakic A, Light A, et al. The development of functional B lymphocytes in conditional PU.1 knock-out mice. Blood 2005;106(6):2083–90.

[143] Heinz S, Benner C, Spann N, et al. Simple combinations of lineage-determining transcription factors prime cis-regulatory elements required for macrophage and B cell identities. Mol Cell 2010;38(4):576–89.

[144] Ghisletti S, Barozzi I, Mietton F, et al. Identification and characterization of enhancers controlling the inflammatory gene expression program in macrophages. Immunity 2010;32(3):317–28.

[145] Su GH, Chen H, Muthusamy N, et al. Defective B cell receptor-mediated responses in mice lacking the Ets protein, Spi-B. EMBO J. 1997;16(23):7118–29.

[146] Dahl R, Ramirez-Bergeron DL, Rao S, Simon MC. Spi-B can functionally replace PU.1 in myeloid but not lymphoid development. EMBO J 2002;21(9):2220–30.

[147] Schebesta A, McManus S, Salvagiotto G, et al. Transcription factor Pax5 activates the chromatin of key genes involved in B cell signaling, adhesion, migration, and immune function. Immunity 2007;27(1):49–63.

[148] Sokalski KM, Li SKH, Welch I, et al. Deletion of genes encoding PU.1 and Spi-B in B cells impairs differentiation and induces pre-B cell acute lymphoblastic leukemia. Blood 2011;118(10):2801–8.

[149] Ushmorov A, Wirth T. FOXO in B-cell lymphopoiesis and B cell neoplasia. Semin Cancer Biol 2018;50:132–41.

[150] Tzivion G, Dobson M, Ramakrishnan G. FoxO transcription factors; Regulation by AKT and 14-3-3 proteins. Biochim Biophys Acta Mol Cell Res 2011;1813(11):1938–45.

[151] Calnan DR, Brunet A. The FoxO code. Oncogene 2008;27(16):2276–88.

[152] Herzog S, Reth M, Jumaa H. Regulation of B-cell proliferation and differentiation by pre-B-cell receptor signalling. Nat Rev Immunol 2009;9(3):195–205.

[153] Georgopoulos K, Moore DD, Derfler B. Ikaros, an early lymphoid-specific transcription factor and a putative mediator for T cell commitment. Science 1992;258(5083):808–12.

[154] Georgopoulos K, Bigby M, Wang J-H, et al. The ikaros gene is required for the development of all lymphoid lineages. Cell 1994;79(1):143–56.

[155] Heizmann B, Kastner P, Chan S. The Ikaros family in lymphocyte development. Curr Opin Immunol 2018;51:14–23.

[156] Georgopoulos K. Haematopoietic cell-fate decisions, chromatin regulation and ikaros. Nat Rev Immunol 2002;2(3):162–74.

[157] Yoshida T, Ng SY-M, Zuniga-Pflucker JC, Georgopoulos K. Early hematopoietic lineage restrictions directed by Ikaros. Nat Immunol 2006;7(4):382–91.

[158] Allman D, Sambandam A, Kim S, et al. Thymopoiesis independent of common lymphoid progenitors. Nat Immunol 2003;4(2):168–74.

[159] Ng SY-M, Yoshida T, Zhang J, Georgopoulos K. Genome-wide lineage-specific transcriptional networks underscore ikaros-dependent lymphoid priming in hematopoietic stem cells. Immunity 2009;30(4):493–507.

[160] Kirstetter P, Thomas M, Dierich A, Kastner P, Chan S. Ikaros is critical for B cell differentiation and function. Eur J Immunol 2002;32(3):720–30.

[161] Schwickert TA, Tagoh H, Gültekin S, et al. Stage-specific control of early B cell development by the transcription factor Ikaros. Nat Immunol 2014;15(3):283–93.

[162] Heizmann B, Kastner P, Chan S. Ikaros is absolutely required for pre-B cell differentiation by attenuating IL-7 signals. J Exp Med 2013;210(13):2823–32.

[163] Karnowski A, Cao C, Matthias G, et al. Silencing and nuclear repositioning of the λ5 gene locus at the pre-B cell stage requires Aiolos and OBF-1. PLoS One 2008;3(10):e3568.

[164] Tamura T, Yanai H, Savitsky D, Taniguchi T. The IRF family transcription factors in immunity and oncogenesis. Immunology 2008;26(1):535–84.

[165] Driggers PH, Ennist DL, Gleason SL, et al. An interferon gamma-regulated protein that binds the interferon-inducible enhancer element of major histocompatibility complex class I genes. Proc Natl Acad Sci USA 1990;87(10):3743–7.

[166] Lu R. Interferon regulatory factor 4 and 8 in B-cell development. Trends Immunol 2008;29(10):487–92.

[167] Eisenbeis CF, Singh H, Storb U. PU.1 is a component of a multiprotein complex which binds an essential site in the murine immunoglobulin lambda 2-4 enhancer. Mol Cell Biol 1993;13(10):6452–61.

[168] Brass AL, Zhu AQ, Singh H. Assembly requirements of PU.1–Pip (IRF-4) activator complexes: inhibiting function in vivo using fused dimers. EMBO J 1999;18(4):977–91.

[169] Brass AL, Kehrli E, Eisenbeis CF, Storb U, Singh H. Pip, a lymphoid-restricted IRF, contains a regulatory domain that is important for autoinhibition and ternary complex formation with the Ets factor PU.1. Gene Dev 1996;10(18):2335–47.

[170] Batista CR, Li SKH, Xu LS, Solomon LA, DeKoter RP. PU.1 regulates Ig light chain transcription and rearrangement in pre-B cells during B cell development. J Immunol 2017;198(4):1565–74.
[171] Pang SHM, Minnich M, Gangatirkar P, et al. PU.1 cooperates with IRF4 and IRF8 to suppress pre-B-cell leukemia. Leukemia 2016;30(6):1375–87.
[172] Nagulapalli S, Atchison ML. Transcription factor Pip can enhance DNA binding by E47, leading to transcriptional synergy involving multiple protein domains. Mol Cell Biol 1998;18(8):4639–50.
[173] Nagulapalli S, Goheer A, Pitt L, McIntosh LP, Atchison ML. Mechanism of E47-Pip interaction on DNA resulting in transcriptional synergy and activation of immunoglobulin germ line sterile transcripts. Mol Cell Biol 2002;22(20):7337–50.
[174] Lu R, Medina KL, Lancki DW, Singh H. IRF-4,8 orchestrate the pre-B-to-B transition in lymphocyte development. Gene Dev 2003;17(14):1703–8.
[175] Blackwood EM, Eisenman RN. Max: a helix-loop-helix zipper protein that forms a sequence-specific DNA-binding complex with myc. Science 1991;251(4998):1211–7.
[176] Mathsyaraja H, Freie B, Cheng P-F, et al. Max deletion destabilizes MYC protein and abrogates Eμ-Myc lymphomagenesis. Gene Dev 2019;33(17–18):1252–64.
[177] Pérez-Olivares M, Trento A, Rodriguez-Acebes S, et al. Functional interplay between c-Myc and Max in B lymphocyte differentiation. EMBO Rep 2018;19(10).
[178] Thomas LR, Adams CM, Wang J, et al. Interaction of the oncoprotein transcription factor MYC with its chromatin cofactor WDR5 is essential for tumor maintenance. Proc Natl Acad Sci USA 2019;116(50):25260–8.
[179] Wilson A, Murphy MJ, Oskarsson T, et al. c-Myc controls the balance between hematopoietic stem cell self-renewal and differentiation. Gene Dev 2004;18(22):2747–63.
[180] Alboran IM de, O'Hagan RC, Gärtner F, et al. Analysis of C-MYC function in normal cells via conditional gene-targeted mutation. Immunity 2001;14(1):45–55.
[181] Riggelen J van, Yetil A, Felsher DW. MYC as a regulator of ribosome biogenesis and protein synthesis. Nat Rev Cancer 2010;10(4):301–9.
[182] Muhar M, Ebert A, Neumann T, et al. SLAM-seq defines direct gene-regulatory functions of the BRD4-MYC axis. Science 2018;360(6390):800–5.
[183] Bahr C, Paleske L von, Uslu VV, et al. A Myc enhancer cluster regulates normal and leukaemic haematopoietic stem cell hierarchies. Nature 2018;553(7689):515–20.
[184] Bretones G, Delgado MD, León J. Myc and cell cycle control. Biochim Biophys Acta Gene Regul Mech 2015;1849(5):506–16.

Chapter 8

Relationship between B-Cell Populations, Development and Function of B-Cell Subsets

Eva Piano Mortari and Rita Carsetti

B Cell Unit, Immunology Research Area, Bambino Gesù Children's Hospital, Rome, Italy

1. INTRODUCTION

B cells maintain human health by exerting a wide range of effector functions including antibody secretion, antigen presentation, cytokine production, and memory responses. The intricate interplay among B-cell populations, their developmental trajectories, and multifaceted functionality are the subject of intense immunological research. Investigating the dynamic relationships that govern the emergence, maturation and specialised roles of various B-cell subsets offers insights into the orchestration of the whole immune response. This exploration not only enhances our comprehension of fundamental immunological processes but also holds the potential to unveil novel therapeutic avenues for immune-related disorders, infectious diseases and cancer.

The main role of B cells is to produce antibodies, molecules that are essential for protection against infectious diseases and are found in large amounts in the serum and mucosal secretion. The functional capability of different antibodies is determined by the isotype of the immunoglobulin heavy chain (IgH) and its specificity. B cells express IgM and IgD in the early stage of development and change to switched isotypes (IgG, IgA and IgE) later in differentiation. Antibodies are also expressed on the surface of B cells. Membrane-bound antibodies associate with the heterodimeric Igα and Igβ signal transduction machinery to form the B-cell receptor (BCR). The BCR is the tool for antigen recognition and the driving force of differentiation but is also indispensable for B-cell survival throughout development 1–3.

The importance of secreted antibodies is demonstrated by the observation that the number and severity of infections occurring in patients with primary B-cell immunodeficiencies are significantly reduced by the administration of immunoglobulin preparations purified from the serum of healthy donors (immunoglobulin replacement therapy, IgRT) 4,5. Chronic infections of the airways and gut, however, may persist despite appropriate treatment, suggesting that IgRT cannot wholly substitute the function of antibodies produced by B cells at mucosal sites [6].

Antibodies in the serum and at mucosal surfaces not only protect the individual from infection but also play a role in the survival of the species. Newborns have no immunological experience and lack all acquired immune defences. For this reason, they are particularly susceptible to communicable diseases, and mortality from infection is highest in the first year of life. To protect the neonate from infection, the mother transfers her own IgG antibodies to the child via the placenta during the second and third trimester of pregnancy [7,8]. After birth, the neonate continues to receive maternal antibodies through the colostrum and breastmilk, for protection of the respiratory and intestinal mucosal surfaces [9]. Making the appropriate antibody at the right time and place is a complex task that plays a fundamental role in the life of B cells.

Antibodies regulate the development of B cells and mediate their function. Several B-cell populations can be identified based on their phenotype, stage of development and capacity to express and secrete antibodies. These populations change with age, tissue localisation and

Molecular Biology of B Cells. https://doi.org/10.1016/B978-0-323-95895-0.00022-2

activation status and have distinct functions in health and disease and different body districts.

2. B-CELL POPULATIONS AND DEVELOPMENT

In the peripheral blood of the adult, B cells account for 3%−10% of the total lymphocytes. Different B-cell populations can be identified by the expression of surface markers [10−12], and, in each population, more subsets have been defined by deep phenotyping [13] and transcriptional profiling [14]. Several classifications have been proposed to identify well-characterised populations with different functions and roles in the normal immune response and pathologic conditions. For the sake of simplicity, we will describe the six largest populations of CD45+CD19+ B cells found in the blood of healthy adults, their developmental pathways, subsets and functions. In human peripheral blood, based on the expression of a few surface markers (CD19, CD24, CD27, CD38, IgM, IgG, IgA, IgD, CD21 and CD71), we can discriminate transitional, mature-naïve, resting memory B cells (MBCs), activated memory B cells (aMBCs), atypical B cells (AtBCs) and plasmablasts (PBs) (Figure 8.1 and Table 8.1). Whereas transitional and mature-naïve B cells have never encountered antigen, all the other populations are generated by the immune response.

The capacity to express antibodies, associated with a functional signalling complex, is the major requirement for the survival, proliferation and progression of B cells along their stepwise developmental pathway in the bone marrow. B cells leave the bone marrow at the transitional stage [15,16].

Transitional B cells ($CD24^{bright}CD27^{neg}CD38^{bright}$) are the most immature B-cell population in the peripheral blood (Figure 8.1A, in green). They are short-lived, express high amounts of membrane-bound IgM and IgD and include two subsets, $CD21^{dull}$ and $CD21^{bright}$ [11] (Figure 8.1A). $CD21^{dull}$ transitional 1 (T1) B cells are recent bone marrow emigrants and still express the marker CD10, which is present in all developing B cells in the bone marrow [11,17]. T1 B cells develop first into $CD21^{bright}$ T2 cells [18] and finally differentiate into long-lived mature-naïve B cells ($CD24^{pos}CD27^{neg}CD21^{pos}CD71^{neg}$) (Figure 8.1B, in light blue). Progression from the T1 to the mature-naïve stage results in the removal of autoreactive B cells [19] and requires the function of the BAFF-receptor (BAFFR) for survival [20].

Mature-naïve B cells form the primary follicles of lymphoid organs and recirculate with the blood throughout the body. This is the largest population in the peripheral blood and, similar to transitional B cells, expresses IgD and IgM. Mature-naïve B cells have a diverse antibody repertoire composed of useful and non-dangerous specificities 21 potentially able to recognise all pathogens and antigens [22].

Neither transitional B cells nor mature-naïve B cells produce antibodies unless they are induced to differentiate through the T-dependent or independent pathways [23]. If they are activated by the antigen, in collaboration with T cells and dendritic cells, mature-naive B cells become the major actors of the adaptive immune response (Figure 8.2). All the other populations of B cells detected in the peripheral blood are generated by immunological experience. They are interrelated and have different functions [14].

In the adult, 30%−50% of the B cells are resting $CD24^{pos}CD27^{pos}CD21^{pos}CD71^{neg}$ MBCs (Figure 8.1C, in dark blue): 40% of them express IgM and 60% switched isotypes (IgG and IgA) (Figure 8.1C).

Switched MBCs are the most refined, specific and effective product of the germinal centre (GC) reaction. Roughly 2/3 of the switched MBCs are IgG^{pos} and the rest are IgA^{pos} MBCs. In normal conditions, IgE MBCs are undetectable in the peripheral blood [24].

$CD27^{pos}$ B cells carry somatic mutations (SMs) in the immunoglobulin genes [25]. These are permanent genetic imprints left by previous activation events and due to the combined action of the activation-induced deaminase (AID), the uracil-DNA glycosylase (UNG) and several DNA repair enzymes and factors [26]. The same enzymes and factors are necessary for class-switch recombination (CSR), which leads to the generation of switched MBCs [27]. Mutated B cells compete for binding to the antigen that is entrapped in immune complexes on the surface of follicular dendritic cells. The ability to extract and present antigens to T follicular helper cells determines whether B cells are positively selected and, after several rounds of affinity maturation, differentiate into MBCs or antibody-secreting cells (ASCs) [28,29]. Thus, the GC reaction shapes the immunoglobulin to obtain an antibody that perfectly fits the antigen (Figure 8.2).

High-affinity switched MBCs and antibodies eliminate their target pathogen and remain in the organism to prevent reinfection [30−33]. High-affinity antibodies and MBCs are released from the GC two weeks after the primary stimulation, a very long time in case of infection. Besides switched MBCs, a second type of MBCs, expressing antibodies of the IgM isotype, play the function of first-line protection in the early phases after pathogen invasion [12]. This population, also called 'natural/natural effector', 'innate memory B cells' or 'marginal zone' B cells, can be found in patients lacking T cells or unable to build and carry on the GC reaction. For this reason, IgM MBCs are thought to develop through a T- and GC-independent pathway [34−36]. IgM MBCs can be generated from transitional B cells through an antigen-independent mechanism triggered by Toll-like receptor 9 (TLR9), which

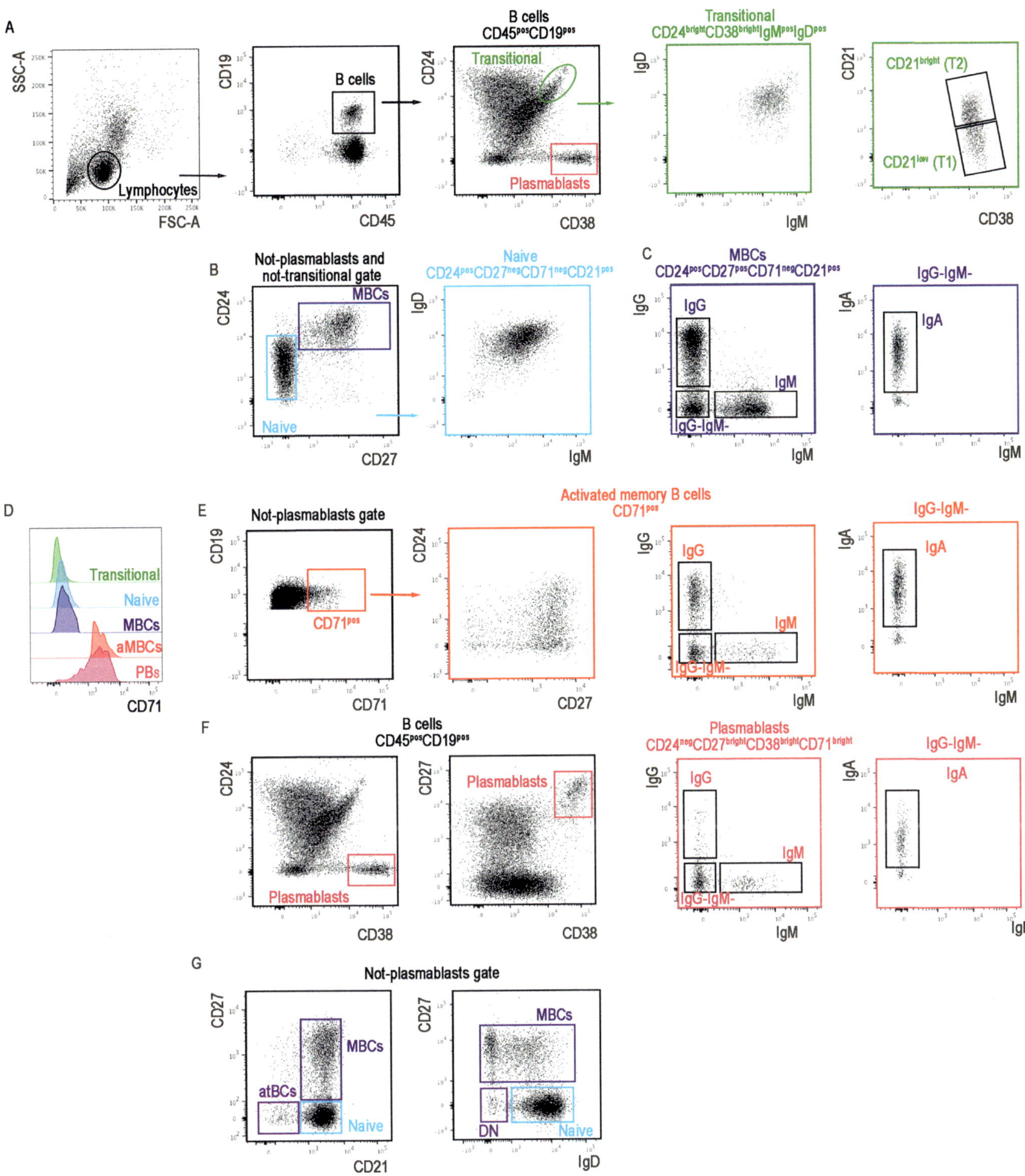

FIGURE 8.1 Identification of B-cell populations in the peripheral blood. (A) FACS plots show the staining of adult peripheral blood with antibodies against CD19, CD24, CD27, CD38, IgM, IgG, IgA, IgD, CD21 and CD71. Live cells were identified based on FSC/SSC and B cells were gated as $CD45^{pos}CD19^{pos}$. $CD24^{pos}CD38^{bright}$ transitional B cells (*green* frame plot) co-express IgM and IgD. Thanks to the level of CD21, it is possible to discriminate $CD21^{dull}$ T1 and $CD21^{bright}$ T2 B cells. Plasmablasts are $CD24^{neg}CD38^{bright}$. (B) After the exclusion of transitional B cells and plasmablasts (not-transitional and not-plasmablasts gate), we separate the mature-naïve ($CD24^{pos}CD27^{pos}$, in light *blue*) from the resting memory B-cell population (in dark *blue*). Whereas mature-naive B cells are $IgD^{pos}IgM^{pos}$, memory B cells express either IgM or switched isotypes (C). (D) Histogram depicts the staining of CD71 on B-cell populations: CD71 is expressed by activated memory B cells and plasmablasts (E) In the not-plasmablasts gate, all $CD71^{pos}$ B cells are activated. CD24 and CD27 expression and immunoglobulin isotype of activated memory B cells are shown. (F) The CD24, CD27 and CD38 markers identify plasmablasts (in dark *red*) as $CD24^{neg}CD38^{bright}$ and $CD27^{bright}$. Plasmablasts still express surface IgM, IgG or IgA. (G) Among the $CD71^{neg}$ population-based expression of CD21, CD27 and IgD, we were able to differentiate atypical memory ($CD21^{neg}CD27^{neg}$ or $CD27^{neg}IgD^{neg}$ in *purple*) from $CD21^{pos}CD27^{pos}$ memory B cells (both IgD^{pos} and IgD^{neg}) and mature-naïve B cells $CD21^{pos}CD27^{neg}$ IgD^{pos}.

TABLE 8.1 Surface Markers Expressed by B-Cell Populations

	Transitional B Cells	Mature-naïve B Cells	Memory B Cells	Activated Memory B Cells	Plasmablasts	Atypical B Cells
CD19	Pos	Pos	Pos	Pos	Pos	Pos
CD24	Bright	Pos	Pos	Pos and neg	Neg	Neg
CD27	Neg	Neg	Dull or bright	Pos	Bright	Neg
CD38	Bright	Pos	Neg	Pos/neg	Bright	Neg
CD21	Dull (T1) or bright (T2)	Pos	Pos	Pos/neg	Neg	Neg
CD71	Neg	Neg	Neg	Pos	Pos	Neg
IgM	Bright	Bright	Bright/neg	Pos (few)	Pos (few)	Neg
IgD	Bright	Neg	Pos/neg	Pos (few)	Pos (few)	Neg
IgG	Neg	Neg	Pos/neg	Pos	Pos/neg	Pos
IgA	Neg	Neg	Pos/neg	Pos	Pos/neg	Neg

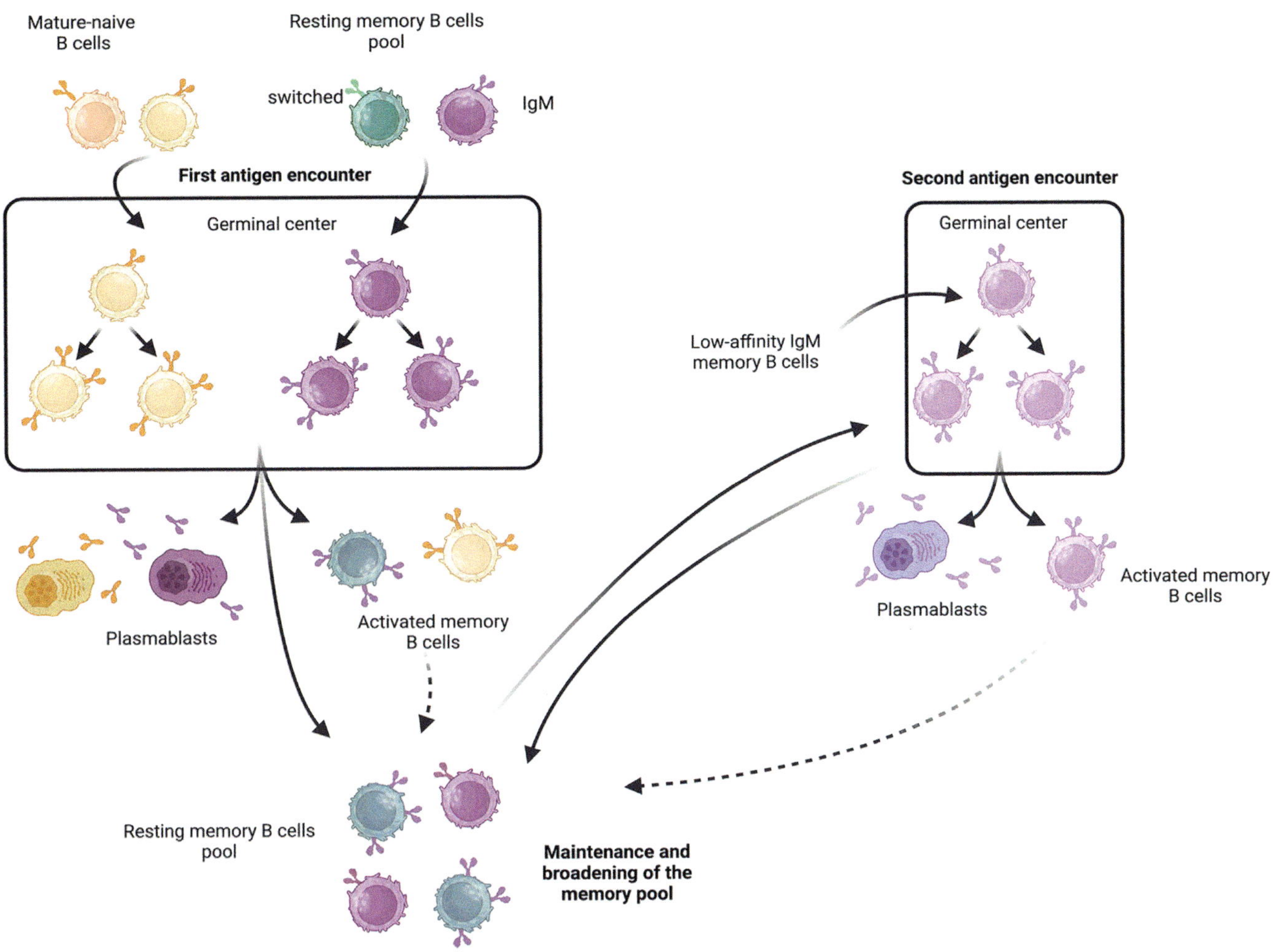

FIGURE 8.2 **Overview of the role of MBCs in primary and secondary immune reactions.** After primary infection or vaccination, both mature-naive and innate MBCs able to bind the antigen enter the GC. Clonal expansion and affinity maturation shape the antibody to increase its specificity. B cells with higher affinity for the antigen (here affinity is represented by the number of antibody molecules on the cell surface) exit from the GC as PBs or activated MBCs, mostly of switched isotypes. Low-affinity MBCs express IgM. These cells can re-enter the GC reaction upon restimulation. Resting MBCs ensure not only the stability and maintenance of the memory pool but also the possibility to broad the repertoire. The dotted line indicates a process for which we do not have certain data. It is possible that aMBCs re-enter the resting MBCs pool.

recognises unmethylated CpG (cytosine-guanosine dinucleotide) [35,37]. Following engagement of TLR9, transitional B cells proliferate, increase the expression of AID, acquire SMs and develop into IgM MBCs and ASCs that secrete IgM.

IgM MBCs play a role in the response to thymus-independent antigens, such as the polysaccharides of encapsulated bacteria, and prevent the permanent lung damage caused by chronic pneumococcal infections in patients with primary immunodeficiency [6]. Most of the functions of human IgM MBCs are attributed to B-1a B cells in the mouse [38]. Similarly, to human IgM MBCs, murine B-1a cells cooperate with the innate immune system to control early bacterial and viral growth [38–40], produce natural antibodies, predominantly of IgM isotype, without prior contact with cognate antigens and respond to thymus-independent antigens, such as the polysaccharides of encapsulated bacteria [40]. In mice, the congenital absence or removal of the spleen causes the depletion of B-1a B cells and the loss of all their functions [38]. B-1a B cells are also the precursors of IgA plasma cells (PCs) in the gut, as demonstrated by the observation that IgA PCs in the intestine are significantly reduced in asplenic and splenectomised mice and can be replenished by the transfer of B-1a B cells [41]. Thus, in the gut, B-1a B cells are necessary for the production of natural polyreactive IgA that rapidly appears in mice after weaning [42] and is generated by a T- and CD40L-independent mechanism 43–45. B cells of B-1a origin persist in the adult gut and revise the BCR in the local GCs (the self-sustaining layer of early life-origin B cells drives steady-state IgA responses in the adult gut). IgM MBCs play a similar role in the human intestine, where they differentiate into PCs that produce secretory IgA (SIgA) [46]. Removal or congenital absence of the spleen results in a significant reduction of circulating IgM MBCs because the marginal zone of the spleen functions as a reservoir for this population [34,36,47,48]. Consequently, asplenic or splenectomised patients have a decreased frequency of IgM MBCs in the bloodstream and a reduction of PCs producing SIgA in the gut [46]. The adaptive immune response is intact after splenectomy including the production of specific IgA in the GC of all lymph nodes including those of the gut, but the function of IgM MBCs cannot be replaced. Thus, IgM MBCs have a separate function from switched MBCs, representing a rapid first line of systemic and mucosal defence that protects the organism during the time necessary for the development and completion of the GC reaction.

In infants, IgM MBCs are unmutated or have very few SMs [49], similar to MBCs of patients with primary immunodeficiencies. In the adult, instead, IgM MBCs carry SMs and are clonally related to switched MBCs [50–52]. These contrasting observations are explained by the finding that IgM MBCs serve as substrates for the generation of high-affinity switched MBCs (Figures 8.2 and 8.3). Our MBCs compartment consists of large clones containing B cells with different numbers of SMs and different isotypes (Figure 8.3). Less mutated IgM MBCs are at the root of the clonal trees, whereas highly mutated, antigen-selected switched MBCs represent the final branches [50,52] (Figure 8.3A). The increase in the frequency of SMs is associated with improved affinity for the antigen. SMs localised in the complementary-determining regions (CDRs) of the immunoglobulin antigen-binding site and leading to amino acid replacement indicate that the B cell has been in the GC and has been selected to survive because of its high affinity to the antigen [53]. Switched MBCs contain a higher frequency of replacements compared to IgM MBCs (Figure 8.3B) [49,52]. Silent mutations that do not have any impact on the affinity for the antigen are equally distributed along the immunoglobulin in both IgM and switched MBCs (Figure 8.3B). The analysis of longitudinal samples collected at an interval of several years from donors confirmed that the memory repertoire contains large clones of MBCs that are stable over time and ensure the maintenance of antigen-specific memory [50]. The largest families of MBCs are primarily or exclusively of IgM isotype, followed by IgA, whereas IgG (IgG1, IgG3 and IgG4) families were mostly of small size (Figure 8.3C). Maintenance of B-cell memory, therefore, relies on the longevity of IgM memory clones that, when needed, switch to other isotypes [50].

In the mouse, MBC populations, with different abilities to seed the GC or generate ASCs were defined based on the expression of CD80 and PD-L2 [54]. Double-negative (DN) MBCs are mostly IgM^{pos} and can participate in the GC reaction, whereas the double-positive population expresses switched isotypes and gives rise to ASCs. We found that human MBCs with similar functions can be distinguished by the expression of CD27. CD27^{dull} MBCs are mostly of IgM isotype and carry few SMs. CD27^{bright} MBCs express highly mutated antibodies and are the precursors of ASCs [52]. MBCs of IgM isotype can re-enter the GC reaction [55] (Figure 8.2). Thanks to this mechanism, IgM MBCs may continue to improve their affinity or, instead, modify their antibodies before switching to other isotypes. Modified antibodies broaden the repertoire of the memory pool and, in case of viral infections, recognise not only the wild-type virus but also its variants [56–58] (Figure 8.2).

In response to vaccination or infection, resting MBCs develop into CD27^{pos}CD38^{pos}CD71^{pos} aMBCs (Figure 8.2) and express the proliferation marker Ki67 [59]. To gate aMBCs among CD19+ B cells, we exclude PBs because they express a high amount of CD71 (Figure 8.1D and E, in red). aMBCs are mostly switched with only 20% of IgM-expressing cells (Figure 8.1E). aMBCs become detectable in the first week after vaccination or infection and are clonally related to both CD71^{neg} resting MBCs and ASCs suggesting that B cells with the same specificity can adopt different fates [14] depending on the needs of the organism. Thus, antigenic stimulation drives resting MBCs to undergo a rapid clonal expansion and differentiate into ASCs, or aMBCs. We do not know whether aMBCs join the resting MBCs pool when the stimulation event is terminated (Figure 8.2).

ASCs, similar to aMBCs, increase in numbers in the first week after stimulation and rapidly decline thereafter. The term ASCs refers to both short-lived PBs (proliferating cells with a life span of 3–5 days) and long-lived PCs (non-proliferating cells with a life span of several months to a lifetime) [60]. ASCs express the *IRF4, XBP1 and PRDM1* genes that encode key transcription factors indispensable for their differentiation, survival and antibody secretion [61]. Blimp-1, encoded by *PRDM1*, is the transcriptional 'master regulator' indispensable for silencing the B-cell expression programme and activating the genes necessary for abundant and durable antibody secretion [61–63]. Long-lived PCs reside in specialised bone marrow niches and continuously produce high-affinity antibodies that maintain the memory of all past immune responses and prevent reinfection [61,64–68]. Long-lived PCs are generated by the GC reaction and are recruited in several waves, during primary and secondary immune responses [60,61,69]. Long-lived PCs generated in response to infections caused by enteric pathogens are also found in the intestine, where they persist for local protection [64,69–71]. In the blood, ASCs (CD24^{neg}CD27^{bright}CD38^{bright}CD21^{neg}CD71^{bright}) (Figure 8.1F, gated in dark red) are short-lived PBs. PBs are a heterogeneous population including ASCs formed both

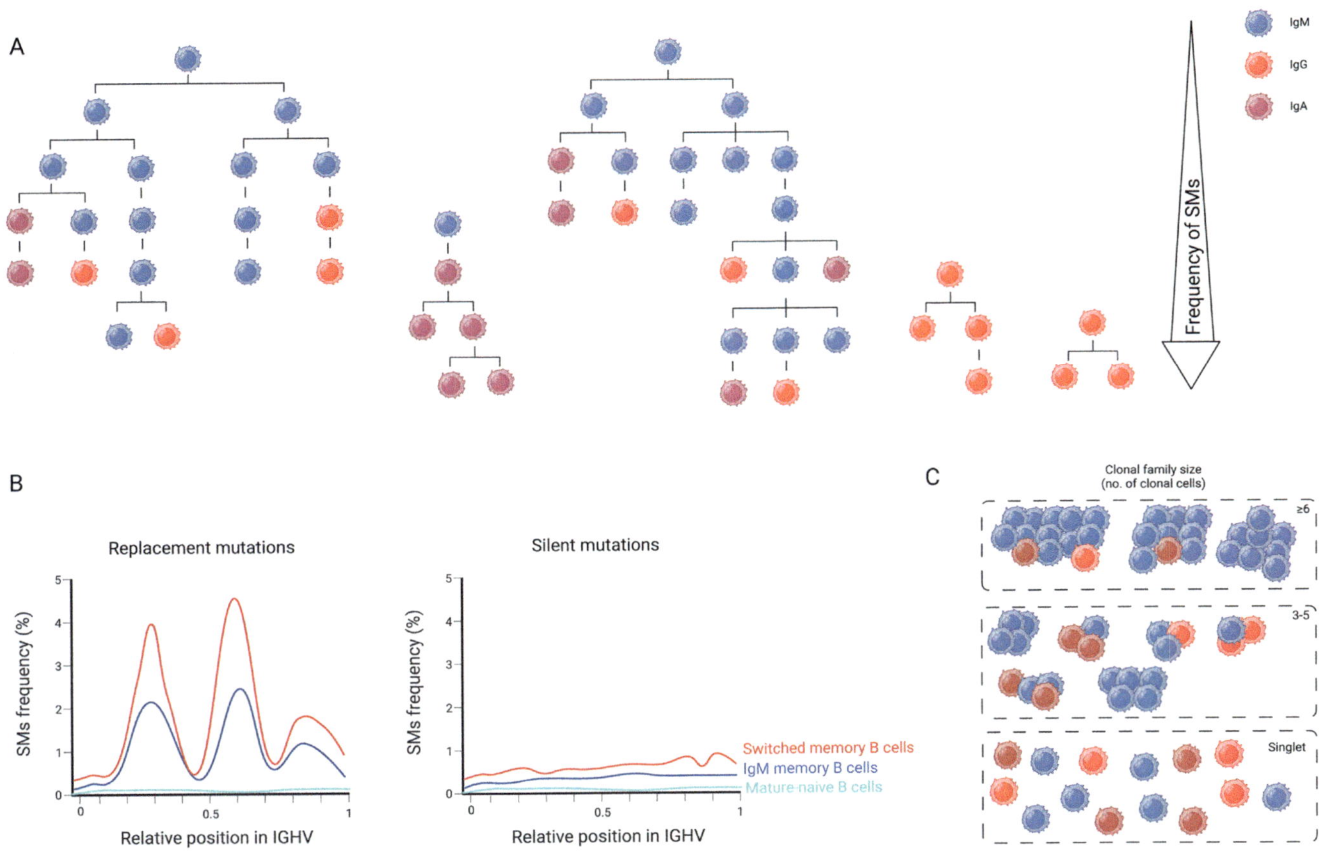

FIGURE 8.3 The clonal composition of the B-cell memory pool. (A) Representation of the human memory B-cell compartment. VH repertoire analysis demonstrates that MBCs form large clones. IgM MBCs carrying a low number of SMs are at the root of the clonal tree and highly mutated switched MBCs are at the extremities of the tree. The arrow indicates the frequency of somatic mutation that increases from the root to the final branches of the tree. (B) Frequency of replacement and silent SMs in mature-naive B cells, IgM and switched MBCs. Switched MBCs have a higher frequency of replacement mutations in the immunoglobulin portions determining the antigen binding. (C) The largest families of MBCs are primarily or exclusively of IgM isotype, whereas IgA and IgG families are of small size.

in the GC and at extra-follicular sites [60] (Figure 8.4). PBs that arise from the extra-follicular reaction are generally unmutated, whereas those generated by the GC reaction have a high frequency of SMs [72]. The presence of PBs in the peripheral blood indicates that B cells have been recently stimulated. In the GC, PBs differentiation is determined by their high affinity for the antigen and the strength of the CD40 signal [53]. Extrafollicular reactions occur immediately after antigen encounter and precede the GC response [72]. Naïve or innate IgM MBCs able to bind the antigen may initiate this type of response (Figure 8.4). Upon re-exposure to an already known antigen, pre-existing MBCs are recruited to extrafollicular reactions [72,73] (Figure 8.4).

In mice and in humans, extrafollicular reactions can be triggered by infections [74–76] and autoimmunity and lead to a persistent increase of PBs [72]. Both induction of SMs and class-switching can occur at extrafollicular sites, although it is unknown whether high-affinity switched MBCs can be selected and released in these conditions. The requirements and rules that dictate the recruitment of a few PBs into the long-lived PCs pool are still unknown [77].

Atypical MBCs were first described as $CD21^{neg}CD11c^{pos}$ in the murine spleen [78,79]. Over the years, these cells have been referred to by various names, often based on distinctive characteristics or the circumstances of their discovery: atypical memory B cells (atMBCs), age-associated B cells (ABCs), tissue-like MBCs, $CD11c^{pos}$ B cells, T-bet^{hi} B cells, $CD27^{neg}IgD^{neg}$ double-negative B cells and $CD27^{neg}CD21^{neg}$ B cells [80–86]. These populations do not completely overlap. The most studied type is usually identified as $CD27^{neg}CD21^{neg}$ and $Tbet^{pos}$. This population expresses different inhibitory receptors, including Fc receptor-like (FCRL) 5 [80,81,87,88]. We refer to this population as atBCs and gate them as $CD27^{neg}CD21^{neg}$ (Figure 8.1G, gating in purple, and we also show the gating strategy as double-negative $CD27^{neg}IgD^{neg}$) [11,89]. AtBCs are undetectable in infants and rare in normal adults [89], but

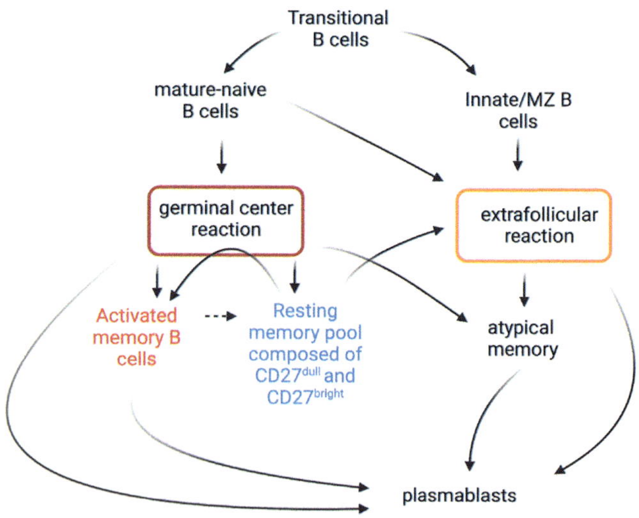

FIGURE 8.4 **The dynamic interplay among B-cell populations.** Transitional B cells differentiate either into mature-naive B cells or innate IgM MBCs. Both mature-naive B cells and innate IgM MBCs can initiate the immune response by either entering the GC or participating in extrafollicular reactions. The GC reaction generates aMBCs, PBs and atBCs. The resting memory pool includes innate and GC-derived MBCs. IgM MBCs from the resting memory pool can re-enter the GC reaction and further modify the BCR. It is not clear whether aMBCs join the pool of resting MBCs when the GC reaction is terminated. Mature-naive B cells, innate and resting MBCs may participate in extrafollicular reactions and develop into atBCs and PBs.

expand in conditions of chronic infections, inflammation, primary immunodeficiencies, autoimmune diseases and aging [80,89–92]. AtBCs were previously considered an anergic or exhausted population because of the expression of FcRL5 and the limited capacity to differentiate into PBs following BCR stimulation in vitro [80,81,88]. It has now been demonstrated that antigen-specific FCRL5^{pos} atBCs are detectable in every individual after vaccination or infection [93]. AtBCs are antigen-experienced and express switched isotypes. The lower frequency of SMs compared to CD27^{pos} MBCs suggests that they either derive from extrafollicular reactions [94] or have left the GC early, before the completion of the selection process [93,95]. Upon stimulation in vitro either with ligands of CD40 or TLR7, atBCs readily differentiate into PBs [96,97], thus contributing to both protection from infection and antibody-mediated autoimmunity. Antigen-specific atBCs, detected after immunisation, decrease over time indicating that they are not long-lived as typical MBCs. In the mouse, atBCs are not only found in the blood but also in tissues. Johnson and colleagues (2020) demonstrated that influenza-specific CD11c^{pos} Tbet^{hi} B cells could be found in the lungs and spleen of mice following influenza infection [98]. In humans, CD21^{neg}Tbet^{hi} B cells also reside in the spleen [98] and can be found bronchoalveolar space of patients with primary immunodeficiencies or the synovium of rheumatoid arthritis patients [99].

2.1 B Cells at Extreme Ages

Infectious diseases are more frequent and serious at extreme ages, in the very young and in the very old. Newborn children have no immunological memory. With time, vaccinations and infections stimulate the child's immune response and generate protective MBCs, PCs, and antibodies, but at birth, the only available protection is represented by maternal IgG transferred to the child during the second and third trimesters of pregnancy. Maternal antibodies of IgG isotype, protected by the neonatal Fc (FcRn) receptor and included in dedicated transport vesicles, cross the placenta to be released in the foetal blood. The system is very efficient, and, at birth, the concentration of maternal IgG antibodies is the same in the blood of the mother and her child [7,8]. These antibodies represent all maternal past and ongoing immune reactions and will protect the child from all infections she has experienced.

After birth, the immune system of the mother continues to protect the child with IgA and IgG antibodies contained in the colostrum and breast milk. IgA antibodies from PCs migrated to the lactating breast from respiratory and intestinal mucosal surfaces [100]. IgA is translocated through the ductal and alveolar epithelial cells by the polyIgR and released in the milk. IgG antibodies are a minor fraction of the antibodies in breast milk. Similarly, to the placenta, the lactating breast expresses the FcRn which transports IgG from the maternal serum to the milk. Breast milk IgA controls bacterial colonisation and the microbiota composition of the neonatal gut, and may also guide the establishment of tolerance to food antigens. IgG has a different role and defends the child from infections of the respiratory tract and gut [101].

Protected from severe infections by maternal antibodies, the child's immune response generates all the different B-cell populations. In cord blood and in the neonate, transitional B cells are the largest population that rapidly differentiate into mature-naïve B cells. IgM MBCs appear early in the first year of life, whereas switched MBCs increase with slower kinetics [10]. IgM and switched MBCs cells have a low frequency of SMs in infants [49] suggesting that either the GC reaction is not perfect or that repeated stimulations are necessary to create optimal MBCs at a young age. In agreement with this observation, repeated vaccine doses are required for optimal protection of young children. In 4–5-year-old children, the number of MBCs becomes more stable, and the frequency of SMs is similar to that of adult B cells. Thus, the B-cell compartment comes of age when children are 5 years old [49].

According with these findings, child mortality from infection is highest in the first year of life and is significantly reduced when children are 5 years old.

In contrast to other viral and bacterial infections, which are more severe in neonates and infants than in adults, SARS-CoV-2 has been less important and often asymptomatic in children [102−104]. The reason for the resistance of children to the acute respiratory distress syndrome that has caused many deaths in adult and elderly individuals may have different reasons. After birth, children have to face a world full of pathogens and commensals, all unknown to them. The child's immune system must be adapted to the challenge and able to react to multiple infections with a controlled inflammatory response. B cells may contribute to limiting excessive cytokine production and damage thanks to the increased frequency of B cells competent to secrete IL-10 [105]. On the other hand, IgM MBCs of children secrete natural antibodies able to neutralise SARS-CoV-2 and its variants. Potent cross-variant neutralising IgM monoclonal antibodies have been isolated from infected subjects. Most of them had few SMs or had almost germline sequencing confirming their origin from natural IgM MBCs [106]. Recognition of a novel pathogen might be more difficult in adults when most IgM MBCs produce antibodies shaped by repeated immune responses to other endemic pathogens.

More than 17% of the population in the USA and 20% in Europe is over 65 and suffers from age-associated diseases and increased susceptibility to infection. Avoiding hospitalisation, complications and death due to preventable diseases is mandatory to meet the challenge represented by an aging world. The SARS-CoV-2 pandemic has violently hit the elderly, showing their frailty and reduced ability to fight a pathogen never encountered before. The response to vaccination is impaired at an advanced age [107] advocating the development of tailored vaccines.

Large studies have demonstrated that individual disease history and genetic background are important determinants of immune health [108]. Immune age does not always coincide with chronological age [109]. Thus, establishing the individual immune age may identify at-risk subjects, who could be protected by appropriate immunotherapy and age-tailored vaccines. We cannot establish the B-cell immune age and health, because we do not know 'precisely' what changes in the B-cell compartment of elderly individuals, what makes them susceptible to severe and possibly lethal infections caused not only by novel viruses but also by common pathogens. In older individuals, a restriction of the B-cell repertoire has been observed, with a reduction of transitional and mature-naïve B cells. We have shown that the population of IgM MBCs is severely reduced, whereas switched MBCs increase [110]. The skewed distribution of several B-cell populations may partly explain why older individuals are not so efficient in reacting to novel pathogens, mutant viruses and T-independent pathogens, whereas they preserve the memory of long-past infections (Spanish flu) [111] [112].

The efficacy of the Herpes-Zoster [113] and respiratory syncytial virus (RSV) vaccines [114,115], both containing adjuvants, indicates that it is possible to design age-tailored vaccines against common pathogens. Vaccination schedules, including available and future adjuvanted vaccines, should be established and recommended for the elderly. Childhood vaccination prevents millions of deaths every year. It is now the right moment to establish the use of appropriate vaccines to avoid preventable hospitalisations and death in the aging population.

2.2 B Cells and Inflammation

The B-cell population landscape is complicated by the demonstration that B cells may have other important tasks, distinct from antibody production. The non-infectious complications of primary immune deficiencies cannot be prevented or cured by IgRT, thus demonstrating that B cells play an immune regulatory role and limit inflammation-driven organ damage [116]. On the contrary, B cells may sustain inflammation. The success of B-cell depletion therapies in the treatment of autoimmune conditions confirms that B cells have antibody-independent functions, as improvement of clinical conditions is not always associated with reducing levels of auto-antibodies in the serum [117]. The immunoregulatory and pathogenic functions have been attributed to different populations of B cells.

The term regulatory B (Breg) cell was coined to describe B cells that have an anti-inflammatory function and act to suppress immune responses [118]. Bregs contribute to the maintenance of tolerance, limiting ongoing immune responses and re-establishing immune homoeostasis. They produce different types of suppressive cytokines such as IL-10, TGF-β and IL-35 and express other suppressive molecules such as programmed death-ligand 1 (PD-L1), granzyme B, CD39 and CD73 and aryl hydrocarbon receptor (AhR). Considerable attempts have been made to discover a distinct molecule able to univocally identify Bregs without success. It has, instead, been demonstrated that different B-cell types can acquire a regulatory function in response to inflammation. Immature and mature B cells as well as PBs and PCs can act as Breg cells. Antigen stimulation and inflammatory cytokines induce the development of Breg cells. Other signals, associated with the termination of the inflammatory conditions, may allow the further evolution of Bregs into ASCs [118−121].

2.3 B-Cell Population Dynamics in the Response to SARS-CoV-2

The recent studies performed during the SARS-CoV-2 pandemic on large numbers of individuals after vaccination and/or infection have confirmed and expanded our knowledge of the human immune response and its reflection on the populations of B cells in the peripheral blood. Thanks to fluorescent probes recognising the spike protein and the receptor-binding domain (RBD) region, it has been possible to track antigen-specific B cells during the phases of induction, persistence and expansion of the immune response to a pathogen never encountered before. Most importantly, B cells were traced in the peripheral blood, the GC and at the sites of infection. At the same time, the concentrations of specific antibodies were measured in the serum and mucosal fluids [122–127]. Most of the studies have been performed on individuals receiving mRNA vaccines and/or infected.

In the peripheral blood, spike-specific MBCs become detectable one week after the first vaccine dose and increase after the second, thus confirming that the adaptation of B cells to the antigen requires time and is improved by re-stimulation. Spike-specific MBCs are of IgM and switched isotypes. Spike-specific antibodies are very low in the serum after the first dose and, similar to MBCs, increase after the second. However, while serum antibodies decline with time, MBCs persist and continue to increase, demonstrating that MBCs and antibodies have divergent kinetics [128–130]. The maintenance and expansion of spike-specific MBCs is explained by the longevity of vaccine-induced GCs, which persist for months after vaccination and remain active sites where the mechanism of antigen selection continues to create new specificities, thus broadening the B-cell repertoire [14,131]. Both antigens entrapped on follicular dendritic cells, and antigen-specific antibodies contribute to informing the selection process [57,132].

In non-vaccinated individuals with critical COVID-19, the extreme inflammation disrupted the GC reaction [133]. In these cases, the extrafollicular reaction generated atBCs [134] and PBs able to produce neutralising antibodies. In convalescent patients who recovered from severe or mild disease, the GC reaction was intact, and MBCs were produced, persisted and enlarged their repertoire [135]. Virus-generated MBCs also participated in the response induced by subsequent vaccine administration [135,136].

The findings on vaccinated or infected individuals show that human MBCs re-enter pre-existing GCs and undergo further diversification, thus becoming able to bind not only the original vaccine antigen but also its variants [129,136,137]. As only IgM, but not switched MBCs can re-enter the GCs [55], IgM MBCs may serve as a substrate for the generation of the broad repertoire against the SARS-CoV-2 virus and its variants. The advantage of this mechanism is the rapidity of the response that starts from pre-existing MBCs. Following a new encounter with the antigen, MBCs expand, differentiate into PBs and lead to the increase of serum antibodies in 3–4 days [130]. The negative aspect is that the administration of vaccines containing related, but not identical antigens, such as the spike protein of the omicron variants, stimulates and expands pre-existing spike-specific MBCs [135]. The omicron-only response, probably resulting from new GCs, is marginal [129,138,139]. A similar phenomenon, called original sin, was described in subjects repeatedly vaccinated against influenza [140,141]. It is debatable whether the original sin results in less effective protection or instead is a mechanism evolved to protect us from pathogen variants by ensuring non-precise but rapidly available protection. The number of deaths caused by our exposure to a pathogen never encountered before, such as SARS-CoV-2, demonstrates the risk of facing an unknown virus without any pre-formed protection. On the other hand, vaccines directed against a now-extinct wild-type virus prevent severe disease from distant viral variants suggesting that an immune memory, although imperfect, can ensure a fast and effective immune response.

Spike- and RBD-specific B cells can be found in all compartments of $CD27^{pos}$ B cells (resting and activated MBCs, atBCs, and PBs) at different time points after vaccination and/or infection. Resting and activated MBCs, atBCs and PBs are all clonally related [14], indicating that vaccination and/or infection creates a flexible pool of B cells, which are long-lived, responsive to antigen stimulation and adaptable to the need of the organism.

2.4 Memory B Cells as Therapeutic Tools

The ability of switched MBCs to produce high-affinity antibodies directed against pathogens has been exploited for therapeutic use. Antibodies cloned from switched MBCs can be administered for the prevention and cure of infectious diseases (RSV; SARS-CoV-2; Ebola; Zika virus and many others). This confirms the important function of B cells in the prevention of diseases and explains the effectiveness of vaccines.

The high affinity for a unique determinant has a side effect. Variation of the site against which the antibody is directed completely abolishes the efficacy of the antibody. At the beginning of the pandemic, monoclonal antibodies with high affinity for the RBD region of the spike protein of SARS-CoV-2 were obtained from MBCs of convalescent patients. The antibodies neutralized the virus in vitro and, when administered to patients, were effective as prophylaxis and treatment against COVID-19. However, the emergence of variants of concern (VOCs) has challenged the efficacy of antibody therapies. Monotherapies and cocktails of different monoclonal antibodies were ineffective against Omicron because of the high number of escape mutations in the RBD region [142–144]. Because of the rapid evolution of the virus and the continuous generation of new and genetically distant variants, the research is concentrated on identifying broadly neutralizing antibodies that recognise epitopes common to all coronavirus and are less prone to mutations.

The lesson learn from COVID-19 indicates that we should reconsider our strategy when designing monoclonal antibodies for the treatment of infectious diseases caused by mutating viruses.

A novel methodology proposes to use antibodies for protection and, at the same time, exploits the multifaceted functions of the different human B-cell populations. The first studies have been performed using HIV as a model pathogen. Very few patients produce neutralizing antibodies against HIV, and this occurs after a long time during chronic infections. Because of the rarity of human germline genes binding HIV, the antigen specificity of B cells was reprogrammed by introducing a rearranged anti-HIV antibody by genome editing. In mice, in vivo [145], and in humans, in vitro [146,147], this approach was successfully used. Engineered B cells were able to respond to stimulation. They developed into MBCs, accumulated SMs and produced neutralising antibodies. This method, still in development, allows the induction of SMs upon stimulation, thus avoiding the risks of a monoclonal repertoire. In addition, engineered B cells by differentiating into all B-cell populations should be able to exert their specific function at the required time and in the right tissue.

3. CONCLUSIONS

The B-cell compartment is fluid and in continuous evolution (Figure 8.4). Transitional and mature-naive B cells fuel all the other populations, which, in turn, are selected and shaped by the signals initiated by the BCR and modulated by cytokines. The sequence of the heavy and light chain combination and the modifications of the CDR3 fragments imposed by antigen experience identify each B cell. Thanks to variable heavy chain (VH) sequence analysis, it has been possible to unravel the mechanisms regulating B-cell development before and after antigen stimulation and follow the destiny of each B cell. We have learnt that B cells can assume different phenotypes and functions. Thus, MBCs, generated in response to vaccination or infection and expressing the same BCR, can turn into resting memory, aMBCs and also atMBCs. Then those populations can differentiate into PBs and then long-lived PCs. At all stages, B cells can act as regulatory B cells if they are exposed to inflammatory conditions. Each population expresses different sets of genes that define B-cell signatures and change their metabolism. Thanks to these changes, B cells acquire stage-specific functions. The evolution of B cells depends on epigenetic modifications that are transmissible to B cells of the same clone, but are also reversible [148–150]. We still do not know to what extent B cells move from one compartment to the other and which and when developmental choices become irreversible.

In humans, we study B cells isolated from the peripheral blood and have limited access to lymphoid organs and other tissues. B cells in the peripheral blood are in transit: they come from a lymphoid organ and are on their way to a destination, where they will exert their function. We do not know where all and each population will migrate and contribute to protective or pathologic immunity. Some PBs will be home to the bone marrow and have the possibility to join the pool of long-lived PCs, thus contributing to the constitutive defence against infection. We ignore the destination of the other PBs. We also do not know what is the fate of aMBCs and atBCs. We only know that some of them will differentiate into PBs.

Another open issue concerns the contribution of systemic immunity to local protection. SARS-CoV-2 vaccines generate a robust, broad and durable B and T cellular immunity, measured in the peripheral blood. Protection from infection is, however, more effective early after vaccination and correlates with high titres of specific and neutralising IgG in the serum [151]. Waning immunity is the term used to explain the progressive loss of protection. Specific immunity, however, does not wane at the cellular level [128–130]. In comparison to vaccination, infection-acquired immunity [152] remains effective for a longer time. SARS-CoV-2, like all respiratory viruses, invades the epithelial cells of the airways and here, at the virus entry site, immune protection is mostly needed. The question is whether and how systemic vaccination can be transformed into local protection. IgG antibodies, which transudate from the serum, can be measured in nasal secretion and saliva, but IgA is very low and increases only after breakthrough infections [130,153]. Perhaps, the recruitment of MBCs to the airways is addressed by inflammatory signals generated in response to the pathogens [130,154]. Which of the several populations identified in the blood migrates to the mucosae is unknown. We also ignore whether local factors may, then, imprint the character of tissue residency to migrated MBCs or, instead, resident MBCs derive from the immune response in the draining lymph nodes. A combination of systemic and mucosal memory, as in hybrid immunity or potentially after a mucosal booster dose, may ensure more effective protection than intra-muscular vaccination.

The use of new tools for the study of B cells at the granular level has confirmed the definition of different B-cell populations with distinct functions, based on phenotype, gene expression, VH usage and antigen specificity.

Further studies are necessary to identify the signals and mechanisms determining the developmental choices that address B cells from one compartment to the other, both in healthy subjects and patients. The recent pandemic has highlighted the need to understand the relationship between systemic and local immunity to develop better strategies for the protection of the population, from the very young to the very old. We need to codify markers that measure the level of immune health and B-cell function in each individual. Only with increased basic knowledge of markers, populations and mechanisms, we will develop methods, therapies and vaccines to overcome the failures of our immune system that predispose us to infection, cancer and autoimmunity.

REFERENCES

[1] Hobeika E, Levit-Zerdoun E, Anastasopoulou V, Pohlmeyer R, Altmeier A, et al. CD19 and BAFF-R can signal to promote B-cell survival in the absence of Syk. EMBO J 2015;34:925–39.

[2] Smith SH, Reth M. Perspectives on the nature of BCR-mediated survival signals. Mol Cell 2004;14(6):696–7.

[3] Kraus M, Alimzhanov MB, Rajewsky N, Rajewsky K. Survival of resting mature B lymphocytes depends on BCR signaling via the Igalpha/beta heterodimer. Cell 2004;117(6):787–800.

[4] Notarangelo LD, Fischer A, Geha RS, Chapel H, Conley ME, et al. Primary immunodeficiencies: 2009 update. J Allergy Clin Immunol 2009;124:1161–78.

[5] Perez EE, Orange JS, Bonilla F, Chinen J, Chinn IK, et al. Update on the use of immunoglobulin in human disease: a review of evidence. J Allergy Clin Immunol 2017;139(3S):S1–46.

[6] Carsetti R, Rosado MM, Donnanno S, Guazzi V, Soresina A, et al. The loss of IgM memory B cells correlates with clinical disease in common variable immunodeficiency. J Allergy Clin Immunol 2005;115:412–7.

[7] Zinkernagel RM. Maternal antibodies, childhood infections, and autoimmune diseases. N Engl J Med 2001;345(18):1331–5.

[8] Pou C, Nkulikiyimfura D, Henckel E, Olin A, Lakshmikanth T, et al. The repertoire of maternal anti-viral antibodies in human newborns. Nat Med 2019;25:591–6. https://doi.org/10.1038/s41591-019-0392-8.

[9] Brandtzaeg P. Mucosal immunity: integration between mother and the breast-fed infant. Vaccine 2003;21:3382–8.

[10] Blanco E, Pérez-Andrés M, Arriba-Méndez S, Contreras-Sanfeliciano T, Criado I, et al. Age-associated distribution of normal B-cell and plasma cell subsets in peripheral blood. J Allergy Clin Immunol 2018;141:2208–2219.e16.

[11] Carsetti R, Terreri S, Conti MG, Fernandez Salinas A, Corrente F, et al. Comprehensive phenotyping of human peripheral blood B lymphocytes in healthy conditions. Cytometry 2022;101:131–9.

[12] Carsetti R, Rosado MM, Wardmann H. Peripheral development of B cells in mouse and man. Immunol Rev 2004;197:179–91.

[13] Glass DR, Tsai AG, Oliveria JP, Hartmann FJ, Kimmey SC, et al. An integrated multi-omic single-cell atlas of human B cell identity. Immunity 2020;53:217–232.e5.

[14] Zurbuchen Y, Michler J, Taeschler P, Adamo S, Cervia C, et al. Human memory B cells show plasticity and adopt multiple fates upon recall response to SARS-CoV-2. Nat Immunol 2023;24:955–65.

[15] Carsetti R. The development of B cells in the bone marrow is controlled by the balance between cell-autonomous mechanisms and signals from the microenvironment. J Exp Med 2000;191(1):5–8.

[16] Melchers F. Checkpoints that control B cell development. J Clin Invest 2015;125(6):2203–10.

[17] Carrion C, Guérin E, Gachard N, le Guyader A, Giraut S, Feuillard J. Adult bone marrow three-dimensional phenotypic landscape of B-cell differentiation. Cytometry B Clin Cytom 2019;96:30–8.

[18] Suryani S, Fulcher DA, Santner-Nanan B, Nanan R, Wong M, et al. Differential expression of CD21 identifies developmentally and functionally distinct subsets of human transitional B cells. Blood 2010;115:519–29.

[19] Wardemann H, Nussenzweig MC. B-Cell self-tolerance in humans. Adv Immunol 2007;95:83–110. Academic Press.

[20] Smulski CR, Eibel H. BAFF and BAFF-receptor in B cell selection and survival. Front Immunol 2018;9.

[21] Meffre E, Wardemann H. B-cell tolerance checkpoints in health and autoimmunity. Curr Opin Immunol 2008;20(6):632–8.

[22] Meffre E, O'Connor KC. Impaired B-cell tolerance checkpoints promote the development of autoimmune diseases and pathogenic autoantibodies. Immunol Rev 2019;292(1):90–101.

[23] Marasco E, Farroni C, Cascioli S, Marcellini V, Scarsella M, et al. B-cell activation with CD40L or CpG measures the function of B-cell subsets and identifies specific defects in immunodeficient patients. Eur J Immunol 2017;47.

[24] Hoof I, Schulten V, Layhadi JA, Stranzl T, Christensen LH, et al. Allergen-specific IgG+ memory B cells are temporally linked to IgE memory responses. J Allergy Clin Immunol 2020;146:180–91.

[25] Klein U, Rajewsky K, Küppers R. Human immunoglobulin (Ig) M+IgD+ peripheral blood B cells expressing the CD27 cell surface antigen carry somatically mutated variable region genes: CD27 as a general marker for somatically mutated (memory) B cells. J Exp Med 1998;188:1679–89.

[26] Muramatsu M, Kinoshita K, Fagarasan S, Yamada S, Shinkai Y, Honjo T. Class switch recombination and hypermutation require activation-induced cytidine deaminase (AID), a potential RNA editing enzyme. Cell 2000;102:553–63.

[27] Okazaki I, Yoshikawa K, Kinoshita K, Muramatsu M, Nagaoka H, Honjo T. Activation-induced cytidine deaminase links class switch recombination and somatic hypermutation. Ann N Y Acad Sci 2003;987:1–8.

[28] Victora GD, Nussenzweig MC. Germinal centers. Annu Rev Immunol 2012;30:429–57.

[29] Victora GD, Schwickert TA, Fooksman DR, Kamphorst AO, Meyer-Hermann M, Dustin ML, et al. Germinal center dynamics revealed by multiphoton microscopy with a photoactivatable fluorescent reporter. Cell 2010;143:592–605.

[30] Kurosaki T, Kometani K, Ise W. Memory B cells. Nat Rev Immunol 2015;15:149–59.

[31] Seifert M, Küppers R. Human memory B cells. Leukemia 2016;30:2283–92.

[32] Tangye SG, Tarlinton DM. Memory B cells: effectors of long-lived immune responses. Eur J Immunol 2009;39:2065–75.

[33] Yoshida T, Mei H, Dörner T, Hiepe F, Radbruch A, Fillatreau S, et al. Memory B and memory plasma cells. Immunol Rev 2010;237:117–39.

[34] Weller S, Braun MC, Tan BK, Rosenwald A, Cordier C, et al. Human blood IgM "memory" B cells are circulating splenic marginal zone B cells harboring a prediversified immunoglobulin repertoire. Blood 2004;104:3647–54.

[35] Aranburu A, Ceccarelli S, Giorda E, Lasorella R, Ballatore G, Carsetti R. TLR ligation triggers somatic hypermutation in transitional B cells inducing the generation of IgM memory B cells. J Immunol 2010;185:7293–301.

[36] Kruetzmann S, Rosado MM, Weber H, Germing U, Tournilhac O, et al. Human immunoglobulin M memory B cells controlling streptococcus pneumoniae infections are generated in the spleen. J Exp Med 2003;197:939–45.

[37] Capolunghi F, Cascioli S, Giorda E, Rosado MM, Plebani A, et al. CpG drives human transitional B cells to terminal differentiation and production of natural antibodies. J Immunol 2008;180:800–8.

[38] Wardemann H, Boehm T, Dear N, Carsetti R. B-1a B cells that link the innate and adaptive immune responses are lacking in the absence of the spleen. J Exp Med 2002;195:771–80.

[39] Ochsenbein a F, Fehr T, Lutz C, Suter M, Brombacher R, Hengartner H, et al. Control of early viral and bacterial distribution and disease by natural antibodies. Science 1999;286:2156–9.

[40] Hayakawa K, Hardy RR. Development and function of B-1 cells: commentary. Curr Opin Immunol 2000;12:346–54.

[41] Rosado MM, Aranburu A, Capolunghi F, Giorda E, Cascioli S, et al. From the fetal liver to spleen and gut: the highway to natural antibody. Mucosal Immunol 2009;2:351–61.

[42] Bunker JJ, Erickson SA, Flynn TM, Henry C, Koval JC, et al. Natural polyreactive IgA antibodies coat the intestinal microbiota. Science 2017;358.

[43] Bergqvist P, Stensson A, Lycke NY, Bemark M. T cell-independent IgA class switch recombination is restricted to the GALT and occurs prior to manifest germinal center formation. J Immunol 2010.

[44] Bergqvist P, Gärdby E, Stensson A, Bemark M, Lycke NY. Gut IgA class switch recombination in the absence of CD40 does not occur in the lamina propria and is independent of germinal centers. J Immunol 2006;177(11):7772–83.

[45] Bemark M, Boysen P, Lycke NY. Induction of gut IgA production through T cell-dependent and T cell-independent pathways. Ann N Y Acad Sci 2012;1247:97–116.

[46] Carsetti R, Di Sabatino A, Rosado MM, Cascioli S, Piano Mortari E, et al. Lack of gut secretory immunoglobulin A in memory B-cell dysfunction-associated disorders: a possible gut-spleen Axis. Front Immunol 2020.

[47] Capolunghi F, Rosado MM, Sinibaldi M, Aranburu A, Carsetti R. Why do we need IgM memory B cells? Immunol Lett 2013;152.

[48] Weller S, Reynaud C-A, Weill J-C. Splenic marginal zone B cells in humans: where do they mutate their Ig receptor? Eur J Immunol 2005;35:2789–92.

[49] Aranburu A, Piano Mortari E, Baban A, Giorda E, Cascioli S, et al. Human B-cell memory is shaped by age- and tissue-specific T-independent and GC-dependent events. Eur J Immunol 2017;47:327–44.

[50] Phad GE, Pinto D, Foglierini M, Akhmedov M, Rossi RL, et al. Clonal structure, stability and dynamics of human memory B cells and circulating plasmablasts. Nat Immunol 2022;23:1–10.

[51] Seifert M, Kuppers R. Molecular footprints of a germinal center derivation of human IgM+(IgD+)CD27+ B cells and the dynamics of memory B cell generation. J Exp Med 2009;206:2659–69.

[52] Grimsholm O, Piano Mortari E, Davydov AN, Shugay M, Obraztsova AS, et al. The interplay between CD27dull and CD27bright B cells ensures the flexibility, stability, and resilience of human B cell memory. Cell Rep 2020;30:2963–2977.e6.

[53] Victora GD, Nussenzweig MC. Germinal centers. Annu Rev Immunol 2022;40. https://doi.org/10.1146/annurev-immunol-120419.

[54] Zuccarino-Catania GV, Sadanand S, Weisel FJ, Tomayko MM, Meng H, et al. CD80 and PD-L2 define functionally distinct memory B cell subsets that are independent of antibody isotype. Nat Immunol 2014;15:631–7.

[55] Dogan I, Bertocci B, Vilmont V, Delbos F, Mégret J, et al. Multiple layers of B cell memory with different effector functions. Nat Immunol 2009;10:1292–9.

[56] Quandt J, Muik A, Salisch N, Lui GB, Lutz S, et al. Omicron BA.1 breakthrough infection drives cross-variant neutralization and memory B cell formation against conserved epitopes. Sci Immunol 2023;7:eabq2427.

[57] Viant C, Weymar GHJ, Escolano A, Chen S, Hartweger H, et al. Antibody affinity shapes the choice between memory and germinal center B cell fates. Cell 2020;183:1298–1311.e11.

[58] Pierce SK. Spending the best but banking the rest. Cell 2020;183:1149–50.

[59] Ellebedy AH, Jackson KJL, Kissick HT, Nakaya HL, Davis CW, et al. Defining antigen-specific plasmablast and memory B cell subsets in human blood after viral infection or vaccination. Nat Immunol 2016;17:1226–34.

[60] Khodadadi L, Cheng Q, Radbruch A, Hiepe F. The maintenance of memory plasma cells. Front Immunol 2019;10.

[61] Riepperger TJ, Bhattacharya D. Transcriptional and metabolic control of memory B cells and plasma cells. Annu Rev Immunol 2021;39:345–68.

[62] Wöhner M, Pinter T, Bönelt P, Hagelkruys A, Kostanova-Poliakova D, et al. The Xbp1-regulated transcription factor Mist1 restricts antibody secretion by restraining Blimp1 expression in plasma cells. Front Immunol 2022;13. https://doi.org/10.3389/fimmu.2022.859598.

[63] Shapiro-Shelef M, Lin K-I, McHeyzer-Williams LJ, Liao J, McHeyzer-Williams MG, Calame K. Blimp-1 is required for the formation of immunoglobulin secreting plasma cells and pre-plasma memory B cells. Immunity 2003;19:607–20.

[64] Nutt SL, Hodgkin PD, Tarlinton DM, Corcoran LM. The generation of antibody-secreting plasma cells. Nat Rev Immunol 2015;15:160–71.

[65] Vijay R, Guthmiller JJ, Sturtz AJ, Surette FA, Rogers KJ, et al. Infection-induced plasmablasts are a nutrient sink that impairs humoral immunity to malaria. Nat Immunol 2020;790(801).

[66] Manz RA, Hauser AE, Hiepe F, Radbruch A. Maintenance of serum antibody levels. Ann Rev Immunol 2005;23:367–86.

[67] Amanna IJ, Carlson NE, Slifka MK. Duration of humoral immunity to common viral and vaccine antigens. N Engl J Med 2007;357(19):1903–15.

[68] Tarte K, Zhan F, De Vos J, Klein B, Shaughnessy Jr J. Gene expression profiling of plasma cells and plasmablasts: toward a better understanding of the late stages of B-cell differentiation. Blood 2003;102(2):592–600.

[69] Bortnick A, Allman D. What is and what should always have been: long-lived plasma cells induced by T cell—independent antigens. J Immunol 2013;190:5913—8.

[70] Landsverk OJB, Snir O, Casado RB, Richter L, Mold JE, et al. Antibody-secreting plasma cells persist for decades in human intestine. J Exp Med 2017;214:309—17.

[71] Wilmore JR, Gaudette BT, Gómez Atria D, Rosenthal RL, Kim Reiser S, et al. IgA plasma cells are long-lived residents of gut and bone marrow that express isotype- and tissue-specific gene expression patterns. Front Immunol 2021;12.

[72] Elsner RA, Shlomchik MJ. Germinal center and extrafollicular B cell responses in vaccination, immunity, and autoimmunity. Immunity 2020;53:1136—50.

[73] Fink K, Manjarrez-Orduño N, Schildknecht A, Weber J, Senn BM, Zinkernagel RM, Hengartner H. B cell activation state-governed formation of germinal centers following viral infection. J Immunol 2007;179:5877—85.

[74] Ryg-Cornejo V, Ioannidis LJ, Ly A, Chiu CY, Tellier J, et al. Severe malaria infections impair germinal center responses by inhibiting T follicular helper cell differentiation. Cell Rep 2016;14:68—81.

[75] Krishnamurty AT, Thouvenel CD, Portugal S, Keitany GJ, Kim KS, et al. Somatically hypermutated plasmodium-specific IgM+ memory B cells are rapid, plastic, early responders upon malaria rechallenge. Immunity 2016;45:402—14.

[76] Yam-Puc JC, Cedillo-Barrón L, Aguilar-Medina EM, Ramos-Payán R, Escobar-Gutiérrez A, Flores-Romo L. The cellular bases of antibody responses during dengue virus infection. Front Immunol 2016;7.

[77] Mashreghi MF, Ferreira-Gomes M, Durek P, Chen Y, Rincon-Arevalo H, et al. Recruitment of plasma cells to the bone marrow in primary and secondary immune reactions. Available at Research Square; 2023. Submitted for publication.

[78] Rubtsov AV, Rubtsova K, Fischer A, Meehan RT, Gillis JZ, Kappler JW, et al. Toll-like receptor 7 (TLR7)—driven accumulation of a novel CD11c+ B-cell population is important for the development of autoimmunity. Blood 2011;118:1305—15.

[79] Hao Y, O'Neill P, Naradikian MS, Scholz JL, Cancro MP. A B-cell subset uniquely responsive to innate stimuli accumulates in aged mice. Blood 2011;118:1294—304.

[80] Moir S, Ho J, Malaspina A, Wang W, DiPoto AC, et al. Evidence for HIV-associated B cell exhaustion in a dysfunctional memory B cell compartment in HIV-infected viremic individuals. J Exp Med 2008;205:1797—805.

[81] Portugal S, Tipton CM, Sohn H, Kone Y, Wang J, et al. Malaria-associated atypical memory B cells exhibit markedly reduced B cell receptor signaling and effector function. Elife 2015;4.

[82] Thorarinsdottir K, Camponeschi A, Cavallini N, Grimsholm O, Jacobsson L, Gjertsson I, et al. CD21—/low B cells in human blood are memory cells. Clin Exp Immunol 2016;185:252—62.

[83] Knox JJ, Myles A, Cancro MP. T-bet+ memory B cells: generation, function, and fate. Immunol Rev 2019.

[84] Karnell JL, Kumar V, Wang J, Wang S, Voynova E, Ettinger R. Role of CD11c+T-bet+B cells in human health and disease. Cell Immunol 2017. https://doi.org/10.1016/j.cellimm.2017.05.008.

[85] Li Y, Li Z, Hu F. Double-negative (DN) B cells: an under-recognized effector memory B cell subset in autoimmunity. Clin Exp Immunol 2021;205:119—27.

[86] Obeng-Adjei N, Portugal S, Holla P, Li S, Sohn H, et al. Malaria-induced interferon-γ drives the expansion of Tbethi atypical memory B cells. PLoS Pathog 2017;13. https://doi.org/10.1371/journal.ppat.1006576.

[87] Kim CC, Baccarella AM, Bayat A, Pepper M, Fontana MF. FCRL5^{+} memory B cells exhibit robust recall responses. Cell Rep 2019;27:1446—1460.e4.

[88] Sullivan RT, Kim CC, Fontana MF, Feeney ME, Jagannathan P, et al. FCRL5 delineates functionally impaired memory B cells associated with plasmodium falciparum exposure. PLoS Pathog 2015;11:e1004894.

[89] Corrente F, Terreri S, Palomba P, Capponi C, Mirabella M, Perno CF, Carsetti R. CD21— CD27— atypical B cells in a pediatric cohort study: an extensive single center flow cytometric analysis. Front Pediatr 2022;10.

[90] Weiss GE, Crompton PD, Li S, Walsh LA, Moir S, et al. Atypical memory B cells are greatly expanded in individuals living in a malaria-endemic area. J Immunol 2009;183:2176—82.

[91] Isnardi I, Ng Y-S, Menard L, Meyers G, Saadoun D, et al. Complement receptor 2/CD21- human naive B cells contain mostly autoreactive unresponsive clones. Blood 2010;115:5026—36.

[92] Wei C, Anolik J, Cappione A, Zheng B, Pugh-Bernard A, et al. A new population of cells lacking expression of CD27 represents a notable component of the B cell memory compartment in systemic lupus erythematosus. J Immunol 2007.

[93] Sutton HJ, Aye R, Idris AH, Vistein R, Nduati E, et al. Atypical B cells are part of an alternative lineage of B cells that participates in responses to vaccination and infection in humans. Cell Rep 2021;34.

[94] Sanz I, Wei C, Jenks SA, Cashman KS, Tipton C, et al. Challenges and opportunities for consistent classification of human B cell and plasma cell populations. Front Immunol 2019;10:2458.

[95] Cyster JG, Allen CDC. B cell responses: cell interaction dynamics and decisions. Cell 2019;177:524—40.

[96] Jenks SA, Cashman KS, Zumaquero E, Marigorta UM, Patel AV, et al. Distinct effector B cells induced by unregulated toll-like receptor 7 contribute to pathogenic responses in systemic lupus erythematosus. Immunity 2018;49:725—739.e6.

[97] Steuten J, Bos AV, Kuijper LH, Claireaux M, Olijhoek W, et al. Distinct dynamics of antigen-specific induction and differentiation of different CD11c+Tbet+ B-cell subsets. J Allergy Clin Immunol 2023.

[98] Johnson JL, Rosenthal RL, Knox JJ, Myles A, Naradikian MS, et al. The transcription factor T-bet resolves memory B cell subsets with distinct tissue distributions and antibody specificities in mice and humans. Immunity 2020;52:842—855.e6.

[99] Rakhmanov M, Keller B, Gutenberger S, Foerster C, Hoenig M, et al. Circulating CD21 low B cells in common variable immunodeficiency resemble tissue homing, innate-like B cells. Proc Natl Acad Sci U S A 2009;106(32):13451—6.

[100] Brandtzaeg P. The mucosal immune system and its integration with the mammary glands. J Pediatr 2010;156:S8—15.

[101] Ateyo C, Alter G. The multifaceted roles of breast milk antibodies. Cell 2021;184(6):1486—99.

[102] Carsetti R, Quintarelli C, Quinti I, Piano Mortari E, Zumla A, Ippolito G, Locatelli F. The immune system of children: the key to understanding SARS-CoV-2 susceptibility? Lancet Child Adolesc Health 2020;4:414—6.

[103] Cinicola BL, Piano Mortari E, Zicari AM, Agrati C, Bordoni V, et al. The BNT162b2 vaccine induces humoral and cellular immune memory to SARS-CoV-2 Wuhan strain and the Omicron variant in children 5 to 11 years of age. Front Immunol 2022;13.

[104] Gaythorpe KAM, Bhatia S, Mangal T, Unwin HJT, Imai N, et al. Children's role in the COVID-19 pandemic: a systematic review of early surveillance data on susceptibility, severity, and transmissibility. Sci Rep 2021;11:13903.

[105] Kalampokis I, Venturi GM, Poe JC, Dvergsten JA, Sleasman JW, Tedder TF. The regulatory B cell compartment expands transiently during childhood and is contracted in children with autoimmunity. Arthritis Rheumatol 2017;69:225–38.

[106] Hale M, Netland J, Chen Y, Thouvenel CD, Nabel Smith K, et al. IgM antibodies derived from memory B cells are potent cross-variant neutralizers of SARS-CoV-2. J Exp Med 2022;219.

[107] Wagner A, Garner-Spitzer E, Jasinska J, Kollaritsch H, Stiasny K, Kundi M, Wiedermann U. Age-related differences in humoral and cellular immune responses after primary immunisation: indications for stratified vaccination schedules. Sci Rep 2018;8:9825.

[108] Ahadi S, Zhou W, Schüssler-Fiorenza Rose SM, Sailani MR, Contrpois K, et al. Personal aging markers and ageotypes revealedby deep longitudinal profiling. Nat Med 2020;28:83–90.

[109] Alpert A, Pickman Y, Leipold M, Rosenberg-Hasson Y, Ji X, et al. A clinically meaningful metric of immune agederived from high-dimensional longitudinalmonitoring. Nat Med 2019;25:487–95.

[110] Ciocca M, Zaffina S, Fernandez Salinas A, Bocci C, Palomba P, et al. Evolution of human memory B cells from childhood to old age. Front Immunol 2021;12.

[111] Yu X, Tsibane T, McGraw PA, House FS, Keefer CJ, et al. Neutralizing antibodies derived from the B cells of 1918 influenza pandemic survivors. Nature 2008;455:532–6.

[112] Gagnon A, Miller MS, Hallman SA, Bourbeau R, Ann Herring D, Earn DJD, Madrenas J. Age-specific mortality during the 1918 influenza pandemic: unravelling the mystery of high young adult mortality. PLoS One 2013;8.

[113] Lal H, Cunningham AL, Godeaux O, Chlibek R, Diez-Domingo J, et al. Efficacy of an adjuvanted herpes zoster subunit vaccine in older adults. N Engl J Med 2015;372(22):2087–96.

[114] Papi A, Ison MG, Langley JM, Lee DG, Leroux-Roels I, et al. Respiratory syncytial virus prefusion F protein vaccine in older adults. N Engl J Med 2023;388:595–608.

[115] Walsh EE, Pérez Marc G, Zareba AM, Falsey AR, Jiang Q, et al. Efficacy and safety of a bivalent RSV prefusion F vaccine in older adults. N Engl J Med 2023;388:1465–77.

[116] Ho HE, Cunningham-Rundles C. Non-infectious complications of common variable immunodeficiency: updated clinical spectrum, sequelae, and insights to pathogenesis. Front Immunol 2020;11:149–63.

[117] Lee DSW, Rojas OL, Gommerman JL. B cell depletion therapies in autoimmune disease: advances and mechanistic insights. Nat Rev Drug Discov 2021;20(3):179–99.

[118] Rosser EC, Mauri C. Regulatory B cells: origin, phenotype, and function. Immunity 2015;42:607–12.

[119] Lenert P, Brummel R, Field EH, Ashman RF. TLR-9 activation of marginal zone B cells in lupus mice regulates immunity through increased IL-10 production. J Clin Immunol 2005;25:29–40.

[120] Duddy M, Niino M, Adatia F, Hebert S, Freedman M, et al. Distinct effector cytokine profiles of memory and naive human B cell subsets and implication in multiple sclerosis. J Immunol 2007;178:6092–9.

[121] Menon M, Blair PA, Isenberg DA, Mauri C. A regulatory feedback between plasmacytoid dendritic cells and regulatory B cells is aberrant in systemic lupus erythematosus. Immunity 2016;44:683–97.

[122] Sterlin D, Mathian A, Miyara M, Mohr A, Anna F, et al. IgA dominates the early neutralizing antibody response to SARS-CoV-2. Sci Transl Med 2021;13:eabd2223.

[123] Cervia C, Nilsson J, Zurbuchen Y, Valaperti A, Schreiner J, et al. Systemic and mucosal antibody responses specific to SARS-CoV-2 during mild versus severe COVID-19. J Allergy Clin Immunol 2021;147:545–557.e9.

[124] Isho B, Abe KT, Zuo M, Jamal AJ, Rathod B, et al. Persistence of serum and saliva antibody responses to SARS-CoV-2 spike antigens in COVID-19 patients. Sci Immunol 2020;5:eabe5511.

[125] Klingler J, Lambert GS, Itri V, Liu S, Bandres JC, et al. Detection of antibody responses against SARS-CoV-2 in plasma and saliva from vaccinated and infected individuals. Front Immunol 2021;12.

[126] Fox A, Marino J, Amanat F, Krammer F, Hann-Holbrook J, Zolla-Pazner S, Powell RL. Robust and specific secretory IgA against SARS-CoV-2 detected in human milk. iScience 2020;23.

[127] Pace RM, Williams JE, Järvinen KM, Belfort MB, Pace CDW, et al. Characterization of SARS-CoV-2 RNA, antibodies, and neutralizing capacity in milk produced by women with COVID-19. mBio 2021;12.

[128] Jernej P, Julia K, Karola M, Richter E, Zorn J, et al. Persistent maintenance of intermediate memory B cells following SARS-CoV-2 infection and vaccination recall response. J Virol 2022;96:e00760–22.

[129] Gaebler C, Wang Z, Lorenzi JCC, Muecksch F, Finkin S, et al. Evolution of antibody immunity to SARS-CoV-2. Nature 2021;591:639–44.

[130] Terreri S, Piano Mortari E, Vinci MR, Russo C, Alteri C, et al. Persistent B cell memory after SARS-CoV-2 vaccination is functional during breakthrough infections. Cell Host Microbe 2022;30:400–408.e4.

[131] Tong P, Gautam A, Windsor IW, Trevers M, Chen Y, et al. Memory B cell repertoire for recognition of evolving SARS-CoV-2 spike. Cell 2021;184:4969–4980.e15.

[132] Tas JMJ, Koo JH, Lin YC, Xie Z, Steichen JM, et al. Antibodies from primary humoral responses modulate the recruitment of naive B cells during secondary responses. Immunity 2022;55:1856–1871.e6.

[133] Kaneko N, Kuo HH, Boucau J, Farmer JR, Allard-Chamard H, et al. Loss of bcl-6-expressing T follicular helper cells and germinal centers in COVID-19. Cell 2020;183:143–157.e13.

[134] Woodruff MC, Ramonell RP, Nguyen DC, Cashman KS, Singh Saini A, et al. Extrafollicular B cell responses correlate with neutralizing antibodies and morbidity in COVID-19. Nat Immunol 2020;21:1506–16.

[135] Muecksch F, Wang Z, Cho A, Gaebler C, Tanfous TB, et al. Increased memory B cell potency and breadth after a SARS-CoV-2 mRNA boost. Nature 2022;607:128.

[136] Goel RR, Painter MM, Lundgreen KA, Apostolidis SA, Baxter AE, et al. Efficient recall of Omicron-reactive B cell memory after a

third dose of SARS-CoV-2 mRNA vaccine. Cell 2022;185:1875–1887.e8.
[137] Goel RR, Painter MM, Apostolidis SA, Mathew D, Meng W, et al. mRNA vaccines induce durable immune memory to SARS-CoV-2 and variants of concern. Science 2021;374.
[138] Kaku CI, Starr TN, Zhou P, Dugan HL, Khalifé P, et al. Evolution of antibody immunity following Omicron BA.1 breakthrough infection. Nat Commun 2023;14.
[139] Alsoussi WB, Malladi SK, Zhou JQ, Liu Z, Ying B, et al. SARS-CoV-2 Omicron boosting induces de novo B cell response in humans. Nature 2023;617:592–8.
[140] Horns F, Dekker CL, Quake SR. Memory B cell activation, broad anti-influenza antibodies, and bystander activation revealed by single-cell transcriptomics. Cell Rep 2020;30:905–913.e6.
[141] Henry C, Palm A-KE, Krammer F, Wilson PC. From original antigenic sin to the universal influenza virus vaccine. Trends Immunol 2018;39:70–9.
[142] Kudriavtsev AV, Vakhrusheva AV, Novoseletsky VN, Bozdaganyan ME, Shaitan KV, Kirpichnikov MP, Sokolova OS. Immune escape associated with RBD omicron mutations and SARS-CoV-2 evolution dynamics. Viruses 2022;14.
[143] Zhao Z, Zhou J, Tian M, Huang M, Liu S, et al. Omicron SARS-CoV-2 mutations stabilize spike up-RBD conformation and lead to a non-RBM-binding monoclonal antibody escape. Nat Commun 2022;13.
[144] Planas D, Saunders N, Maes P, Guivel-Benhassine F, Planchais C, et al. Considerable escape of SARS-CoV-2 Omicron to antibody neutralization. Nature 2022;602:671–5.
[145] Nahmad AD, Lazzarotto CR, Zelikson N, Kustin T, Tenuta M, et al. In vivo engineered B cells secrete high titers of broadly neutralizing anti-HIV antibodies in mice. Nat Biotechnol 2022;40:1241–9.
[146] Rogers GL, Cannon PM. Genome edited B cells: a new frontier in immune cell therapies. Mol Ther 2021;29:3192–204.
[147] Rogers GL, Huang C, Mathur A, Huang X, Chen H, et al. Reprogramming human B cells with custom heavy chain antibodies. bioRxiv 2023. https://doi.org/10.1101/2023.06.28.546944.
[148] Esteller M. Epigenetic drugs: more than meets the eye. Epigenetics 2017;12. 307–307.
[149] Obata Y, Furusawa Y, Hase K. Epigenetic modifications of the immune system in health and disease. Immunol Cell Biol 2015;93:226–32.
[150] Xiao F, Rui K, Shi X, Wu H, Cai X, et al. Epigenetic regulation of B cells and its role in autoimmune pathogenesis. Cell Mol Immunol 2022;19:1215–34.
[151] Regev-Yochay G, Lustig Y, Joseph G, Gilboa M, Barda N, et al. Correlates of protection against COVID-19 infection and intensity of symptomatic disease in vaccinated individuals exposed to SARS-CoV-2 in households in Israel (ICoFS): a prospective cohort study. Lancet Microbe 2023;4:e309–18.
[152] Hall V, Foulkes S, Insalata F, Kirwan P, Saei A, et al. Protection against SARS-CoV-2 after covid-19 vaccination and previous infection. N Engl J Med 2022;386:1207–20.
[153] Diem G, Lafon E, Bauer A, Lass-Flörl C, Reindl M, Wilflingseder D, Posch W. Salivary IgAs and their role in mucosal neutralization of SARS-CoV-2 variants of concern. J Clin Microbiol 2022;60.
[154] MacLean AJ, Richmond N, Koneva L, Attar M, Medina CAP, et al. Secondary influenza challenge triggers resident memory B cell migration and rapid relocation to boost antibody secretion at infected sites. Immunity 2022;55:718–733.e8.
[155] Rees AR. Understanding the human antibody repertoire. mAbs 2020;12:1729683.
[156] Briney BS, Willis JR, McKinney BA, Crowe Jr JE. High-throughput antibody sequencing reveals genetic evidence of global regulation of the naïve and memory repertoires that extends across individuals. Gene Immun 2012;13:469–73.
[157] Helguera-Repetto AC, Villegas-Mota I, Arredondo-Pulido GI, Cardona–Pérez JA, Moises LJ, et al. Cord blood SARS-CoV-2 IgG antibodies and their association with maternal immunity and neonatal outcomes. Front Pediatr 2022;10.

Chapter 9

B-Cell Development to Immunity and Tolerance

Fritz Melchers, Mir-Farzin Mashreghi and Peter K. Jani
German Rheumatism Research Centre Berlin (DRFZ), A Leibniz Institute, Berlin, Germany

1. INTRODUCTION

Two separate developmental pathways generate the peripheral pools of B cells during ontogeny in mice and in humans [1–3]; the first develops B1-a cells in the foetal liver, while the second follows with B2 cells in the bone marrow (BM) (Figure 9.1).

In B1-a and B2 developmental pathways, hematopoietic progenitor cells initially differentiate into B-lymphoid-committed precursor cells (initially CD19-negative, later CD19-positive) before finally becoming surface IgM-positive (sIgM+) B cells. They form their antigen-recognising repertoires by individual, successive variable (diversity) joining (V(D)J) rearrangements, first on the immunoglobulin (Ig) heavy (H) chain, then on the light (L) chain locus [4–6]. Over 98% of all B cells entering the mature peripheral B cell pools are Ig gene expression allelically excluded. Thus, they express Ig from only one uniquely and productively rearranged H-chain allele and one uniquely productively rearranged L-chain allele. The other H-chain allele remains either D_HJ_H –rearranged or is out of frame (non-productively $V_HD_HJ_H$-rearranged). Likewise, the other L-chain allele either remains in germline (non-rearranged) form or is in out-of-frame form, non-productively V_LJ_L-rearranged. Hence, one B cell makes one Ig with a given chemical affinity of their unique V_H/V_L-antigen-binding site. Importantly, terminal nucleotidyl transferase (TdT) is activated in common lymphoid progenitors (CLPs) during B-cell development and IgH- and L-chain gene rearrangements in BM, but not during B1-a development in the foetal liver [7]. Hence, the vast majority of rearranged V genes of the IgH and IgL loci of B1-a cells do not contain N-regions at their D_H to J_H, their V_H to D_HJ_H and their V_L to J_L-joints [8–12]. Consequently, B2-cell development generates Ig of higher diversity due to the insertion of N-region sequences at these joints of the V_H- and V_L-domains. IgM on the surface of B1-a and B2 cells functions as a receptor for antigen: the B-cell receptor (BcR).

2. B CELLS HAVE TWO MAJOR FUNCTIONS

A large part of all BcRs have been found to be structurally complementary to conformational, three-dimensionally folded determinants of antigens; these parts of an antigen are called haptens. The first major function of B cells results from binding of a fitting hapten of an antigen, for example, an infectious invader, to BcR, which then activates a positive cellular response inducing clonal B cell expansion by proliferation, differentiation from a naive to an antigen-experienced B cell and/or differentiation to Ig-secreting plasmablasts and plasma cells. In this way, a single antigen-binding B cell might expand a thousand-fold in numbers. Furthermore, it can increase the synthesis of Ig from 10^5 molecules/cell/day to a daily production and secretion of this specific Ig more than by one million-fold to bind and eliminate the invader by a so-called humoral immune response. B1-a cells can do so without the help of T cells, as they can be stimulated by T-independent antigens. They secrete complement (C3)-binding 19S pentameric IgM, allowing them to bind and lyse infectious bacteria. B2 cells organise their response to T-dependent antigens in cooperation with follicular helper T cells (Thf) in germinal centres.

Secondly, B cells can be antigen-presenting cells (APCs) [13], some properties of which they share with other major APCs, such as dendritic cells and macrophages. APCs can take up antigen by phagocytosis or endocytosis, process it by proteolysis into peptides, bind these peptides onto major histocompatibility complex (MHC) and present the MHC-peptide complexes back on the surface of the

Molecular Biology of B Cells. https://doi.org/10.1016/B978-0-323-95895-0.00009-X

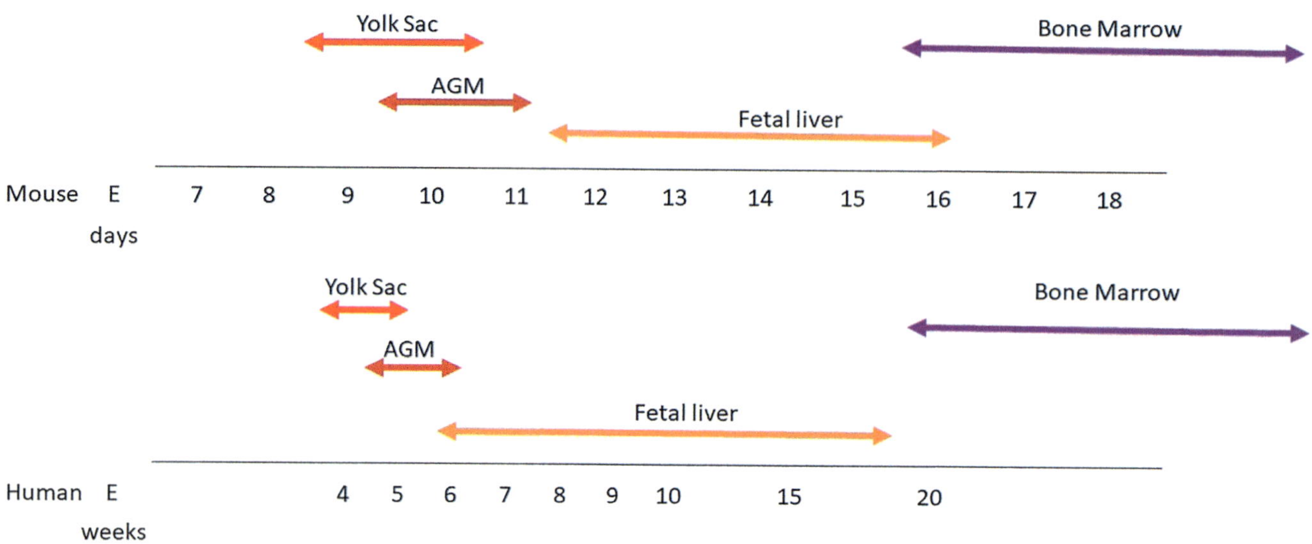

FIGURE 9.1 **Embryonic haematopoiesis in mice and humans.** This figure describes the B-cell development lineages in mice and humans. For details, see the text. *AGM*, aorta/gonad/mesonephros; *E* embryonic.

APC to T cells. These peptide parts of the antigen are called carrier determinants. However, the uptake affinity by phagocytosis or endocytosis is lower than the uptake affinity by an antigen-specific BcR [14]. Thus, antigen-specific BcRs on antigen-presenting B cells (APC-B) can 'sense' much lower concentrations of an antigen. If the BcR is hypermutated in its V_H/V_L-domains, then it can bind a hapten with even better affinity; such APC-B cells can therefore sense and effectively present even lower concentrations of antigen to T cells. Compared with the lower affinities of phagocytosis and endocytosis, the often hundred- to thousand-fold higher affinity of a hypermutated, specific BcR allows lower concentrations of antigen to increase specific immunity, not only to the hapten but also to the carrier. From the multiplicity of V-, D- and J-segments in the Ig loci of the genome, and from their combinatorial diversity of rearrangements, BcR repertoires generated during B cell development can be expected to be highly diverse – B2 cells even more so than B1-a cells due to their BcR containing N-sequences.

BcRs can also bind haptens of autoantigens, quasi-autoantigens of commensal microbiota, as well as of foreign antigens to produce secreted antibodies against them and to present their processed peptides on MHCII. How then can autoimmune attacks of B cells against autoantigens be avoided? In principle, this is achieved either by deleting or by suppressing the auto-aggressive B cells. This chapter reviews and discusses the different functions of B1-a and B2 cells in immunity and tolerance to self and non-self-antigens. B2 cells continue to delete, improving antibodies against foreign antigens by Ig class switching and hypermutation, while B1-a cells continue to suppress by presenting autoantigens for suppression.

3. SUCCESSIVE EMBRYONIC DEVELOPMENT OF B1-A CELLS IN FOETAL LIVER AND B2 CELLS IN FOETAL BONE MARROW

Embryonic erythroid-myeloid progenitors (EMPs) appear in the blood of the mouse soon after circulation has been established at embryonic day (E-day) 8.5 in the mouse and between embryonic week (E-week) 4 and 5 in humans (Figure 9.1). In the mouse, the yolk sac at E-day 8.5 (human E-week 4–5.5) is the first source of these progenitors, while the aorta/gonad/mesonephros (AGMs) area around E-day 9.5 (human E-week 4.5–6) is the second source [15–18]. EMPs develop erythrocytes, megakaryocytes, platelets, resident (long-lived) macrophages, epidermal Langerhans cells and brain microglial cells [19,20]. The yolk sac and AGM also develop myeloid-lymphoid progenitors, which express CSF-1R and IL-7Ra, induced by contact with the appropriate environment in foetal liver, which produces CSF-1 for myeloid and IL-7 for B-lymphoid development.

B1-a B cells are generated in one wave of embryonic and neonatal development, as soon as the main supportive site for this development – the foetal liver – has begun to develop at E-day 11.5 (human E-week 6 [21,22]). In this development, CSF-1R+ IL-7Ra+ EMPs migrate out of embryonic blood through the vascular endothelium of blood vessels into the developing stroma of the foetal liver, beginning at E-day 12.5 (human E-week 6). This migration through LYVE1+ vascular endothelium is guided by chemokine receptors, among them CCR7 and CXCR3, expressed on the progenitors, attracted by chemokines, such as CCL19 and CXCL10/11, respectively, produced by the vascular endothelium of blood vessels to transit into foetal liver stroma [23–27]. This first step of transmigration between E-day 12.5 and 14.5 is followed by the subsequent attraction of the progenitors by the chemokine CXCL12, produced by foetal liver ALCAM+ mesenchymal stroma. This stroma also produces either CSF-1 or IL-7, the latter to promote B1-a cell development. From E-day 13.5 to 16.5, IL-7Ra+ pro-pre-B cells and pre-B cells expand exponentially in contact with c-kit-ligand, IL-7-producing stroma, first as D_H- to J_H-, then as V_H- to D_HJ_H-rearranged B cell progenitors and precursors [21,27]. Within the next 3–4 days, surface IgM+ B1a cells appear. Thus, EMPs have the potential also to develop to B1-a lymphoid cells.

4. PRE-B1-A CELL LINES AND CLONES WITH EXTRAORDINARY PROLIFERATION CAPACITIES

D_HJ_H-D_HJ_H-rearranged, ckit+ CD19+ pre-B1-a cells from foetal liver at E15.5 until birth (E19.5) can be established as cell lines and clones, proliferating for dozens of cell divisions in serum-free media in the presence of IL-7 and immobilised kit-ligand [28–31]. These pre-B1-a cells can be induced in vitro by the removal of IL-7 to rearrange V_H- to D_HJ_H and V_L- to J_L-segments in- and out-of frame, generating – with productive rearrangements – sIgM+ B1-a cells. Transplantation of these pre-BI cells into immunodeficient SCID or RAG-deficient mice populates the spleen of the recipient mice with surface IgM+, MHC class II+ B1-a cells. Furthermore, they populate the lamina propria of the gut with sIgM+ and sIgA+ for longer than 25 weeks [28,32]. As the serum of the transplanted mice contains normal concentrations of IgM, less than 1% of normal levels of IgG and one-third of the normal level of IgA, this suggests that in the absence of T cells, B1-a cells repopulate their compartments in spleen and in the peritoneum. The BcR repertoires established in these pre-B-cell-derived splenic and peritoneal B1-a cells need to be compared with those established in the in vivo wave in foetal liver to evaluate the potency of the D_HJ_H-rearranged in in vitro cell lines as B1-a 'stem' cells.

These pre-B1-a cells can be stimulated by T-independent antigens and by polyclonal activators [33], such as lipopolysaccharide (LPS) [34] via TLR4 [35–37], and lipoprotein [38] via TLR2. At least one in three B1-a cells can be stimulated in vitro by either LPS or lipoprotein resulting in a limited number of cell divisions and the gradual development of IgM-secreting plasmablasts and plasma cells [39].

Foetal liver-derived EMP, but not BM-derived hematopoietic stem cells (HSCs), can generate B1-a cells upon transplantation. EMP can do so only when transplanted into recipients at early stages in embryonic/neonatal development. Foetal liver loses hematopoietic developmental capacities after birth, thus terminating B1-a generation capacities and therefore, the wave of B1-a development.

5. ESTABLISHMENT OF LIFE-LONG B2 CELL DEVELOPMENT IN THE BONE MARROW

B2 cell development follows B1-a development. From E15.5, when bone and BM begin to be formed, the BM attracts EMP from the blood to migrate into special niches. In these niches, the cells become quiescent, long-lived, erythroid/megakaryocytic/myeloid/lymphoid-multipotent HSC [40–50]. HSCs acquire and retain this capacity to be attracted into niches in the BM throughout life. This capacity is used in transplantation of HSC into hosts depleted of hematopoietic cells by lethal irradiation. Transplantation restores normal numbers of HSC in BM and, thus, restores normal numbers of progenitors and of differentiated hematopoietic cell lineages in the host [45].

B2 cells continuously develop in the BM throughout life by steady-state haematopoiesis from HSC, multipotent progenitors (MPPs) and CLPs. Steady-state haematopoiesis begins with the development of bone and BM at E15.5, hence later in embryonic development and at a different site, that is, not in foetal liver but in BM. EMP from the yolk sac and AGM, still circulating in embryonic blood, enter and seed the developing BM. Other HSC-like progenitors may also come from the developing foetal liver [40], but it is not clear whether HSC-like cells developing in foetal liver participate in a major way in BM haematopoiesis [41]. Importantly, EMP from embryonic blood transplanted into adult recipients will generate HSC in BM. Moreover, these HSCs generate B2-, not B1-a-type B cells in the host. This suggests that the environment of the host determines the type of B-cell development: foetal liver to B1-a, BM to B2.

HSCs in BM are heterogeneous in their functions. All are defined by their capacity to repopulate the long-lived CD150+CD48- compartment of approximately 5000 cells in the BM of lethally irradiated hosts upon transplantation. However, analyses of the differentiation capacities of single HSC have shown that HSCs can have different propensities for lineage differentiation fate decisions. Thus, while some of them are multipotent and develop erythrocytes, megakaryocytes, platelets, myeloid and lymphoid cells upon transplantation, others appear biased, or even restricted to megakaryocyte and platelet development, or to erythroid/myeloid and granulocyte differentiation, with a loss of lymphoid development [42,43]. Stress, ageing and/or deficiency in miR-221/222 expression increases this loss of multipotency [44].

In contrast to B1-a cell development, D_HJ_H-rearranged B2 pre-BI cells only proliferate for a limited number of divisions in vitro with stromal cells and IL-7. Furthermore, transplantation of these pre-B cells reconstitutes B2 cells in the BM and spleen only for a short time, since the sIgM+ B2 cells have no self-renewal capacities in vivo. Better reconstitution of B2 cells can be achieved by transplantation of CLP. CLP isolated from BM can be kept in tissue culture together with stromal cells in the presence of flt3-ligand and low concentrations of IL-7 for at least 2 months [51]. Transplantation of these cultured CLP into RAG-deficient recipients restores T and B cells in blood and spleen to normal levels after a month, but these B2 cells then disappear again, unless challenged by antigen. In fact, transplantation of HSC is the only way by which B2 cells in BM, blood and spleen can be restored in RAG-deficient hosts throughout life, as HSCs continuously generate MPPs, CLP and B2-lineage precursors in the BM to secure mature B2 cell pools in stable numbers.

A third group of B cells are primarily defined by their location: marginal zone (MZ) B cells, which are in part B1-a cells, but also B1-b cells [1–3] and some B2 cells, all of which are attracted by the debris of autoantigens and microbiota in the MZ. The B1, B2 and MZ B cells are each heterogeneous and can be further distinguished by different phenotypes, functions and anatomic locations.

6. SURROGATE L CHAIN GUIDES THE DEVELOPMENT OF PRE-B CELLS DURING THEIR V(D)J REARRANGEMENTS ON THE H-CHAIN LOCUS

During IL-7-induced B-cell development in the foetal liver and BM, the surrogate light (SL) chain, composed of Vpre-B and lambda 5 (λ5), is selectively expressed in progenitor and precursor B cells during the developmental stages of D_H to J_H and V_H to D_HJ_H rearrangements [7,52–55]. Most of the experimental evidence addressing the functions of SL chain has been obtained with B2 development in BM. C-kit+ CD19+ pre-BI cells, D_H-rearranged on both H-chain alleles, are induced by the removal of IL-7 to begin V_H rearrangement, inducing the generation of H-chain-allelically excluded pre-BII cells. Since quadruple-deficient λ5- VpreB1- VpreB2- as well as Vpre-B3-deficient mice have not yet been tested [56–60], it remains an open question whether SL is involved in H-chain allelic exclusion.

As soon as Ig μH chains become expressed in large pre-BII cells, the Ig-λL-like constant region domain of λ5 forms the classical IgH/L disulphide (SS) bonds with IgμH chain, which in turn allows Vpre-B1, Vpre-B2 and probably also Vpre-B3 [60] to bind non-covalently to the IgμH-λ5 complex and form a BcR-like structure: the pre-BcR. Inserted into the surface membrane, these pre-BcRs activate pre-B cells to expand by proliferation as large pre-BII cells. However, not all V_H can bind equally well to SL chain.

While these early stages of B cell development may function equally well in B1-a and B2 cell development, the following steps describe B2 cell development in the BM. Approximately half of all μH chains produced by V_H rearrangements cannot pair at all with SL chains and therefore do not form a pre-B-cell receptor (pre-BcR). Hence, it appears that a spectrum of affinities between Vpre-B and individual V_H domains generates pre-B2 cells with pre-BcR of different fitness which in turn determines the extent of the pre-B2-II-cell proliferation. SL chain deficiency, introduced by targeted deletion of the λ5, Vpre-BI and Vpre-BII loci, abolishes the proliferative expansion of pre-B2-II cells [56–59]. These deficiencies lead to a 40-fold reduction in the number of immature sIgM+ B2 cells in BM and a reduced development of mature B2 cells in the spleen. Thus, SL chain probes the emerging V_H repertoire of B2 cells for V_H/V_L fitness.

Once the pre-BcR is expressed in pre-BII cells, it signals for SL chain expression and RAG expression to be turned off. The IgH locus is closed for further V_H to D_HJ_H rearrangements. As pre-B2-II cells enter successive cell cycles, SL protein is diluted out, thus limiting pre-BCR formation. As a consequence, pre-B2-II cells stop proliferating, become small and resting and open the IgL chain loci to begin to rearrange V_L to J_L segments of the L-chain loci. Multiple rearrangements of the kappa L-chain alleles are often detected in wild-type small pre-B2-II cells [52,61,62]. Around 20% of the μH chain-expressing small pre-B2-II cells also express L-chains, but do not display the Ig on the surface. Hence, it is likely that not all L-chains originally generated in resting pre-B2-II cells can pair with the μH chain previously present in that cell. The best fitting ones are selected preferentially to generate sIg+ B cells.

When sIgM+ immature B2 cells have developed, they remain resting as G0-non-proliferating cells. Only 10% –20% of them are allowed to leave the BM, as the majority, possibly autoantigen-reactive cells, are deleted [52,53,63]. Immature, autoreactive B cells attempt to evade autoreactivity by upregulating RAG expression, which allows secondary L-chain rearrangements, a process termed "L-chain editing". The deletion of potentially autoantigen-reactive B cells continues as the cells exit the BM, migrate through arterial blood to the spleen and form follicles (hence also called follicular B cells), surrounded by T cells in the white pulp. B2-B cells are continuously generated throughout life. In fact, most naïve, newly

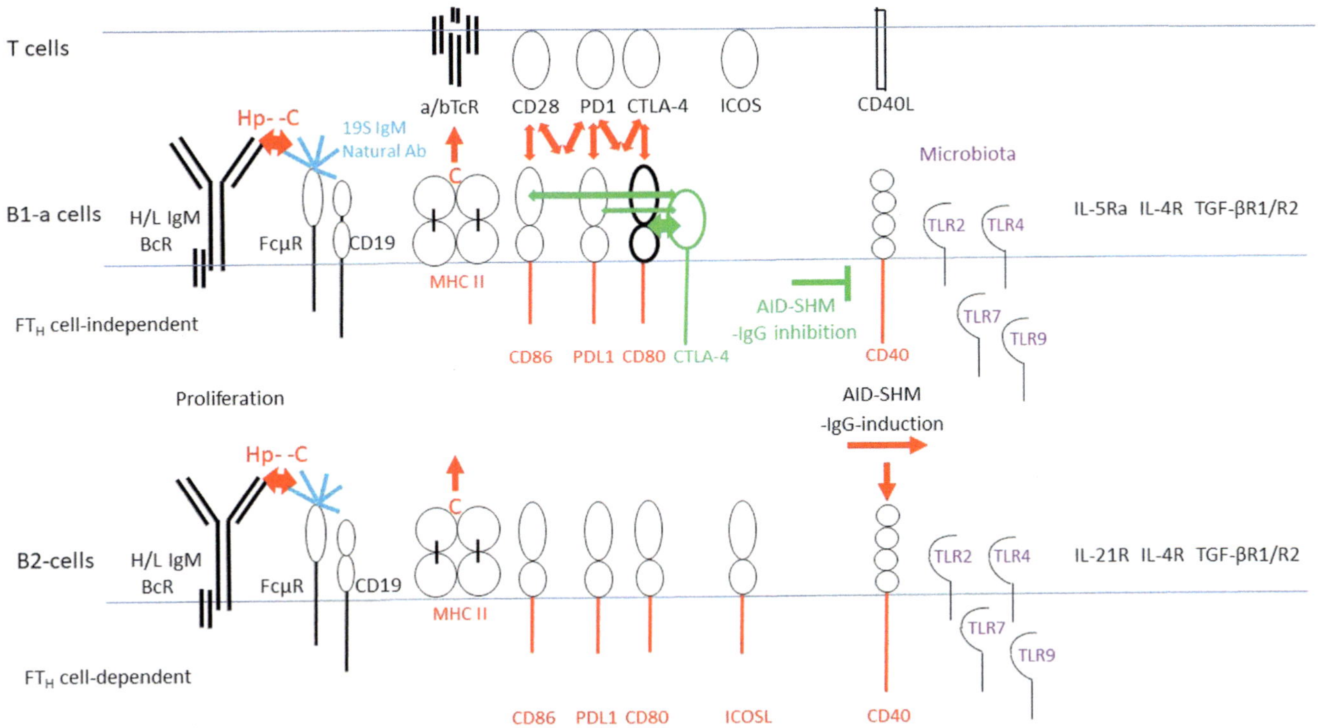

FIGURE 9.2 Antigen recognition and interactions of B1-a and B2 cells. 7S sIgM as BcR on B1-a and B2 cells bind hapten (H) of antigen (*red arrow left/right*) in complexes with pentameric 19S IgM natural antibodies, which link the BcR with Fcμ-receptor (FcμR), activating the resting cells. These complexes are taken up, processed to carrier peptides (C; in *red*) and presented, bound to MHCII molecules on the surface of the activated cells. T cells recognise (*red arrow up/down*) the presented carrier peptide-MHCII complexes and upregulate expression of CD28, PD-1 and CTLA-4. CD28 and CTLA-4 bind to CD80 and CD86 on both B1-a and B2 cells. The interaction between CTLA-4 and CD80 is stronger than with CD86 and stronger than CD28 with either CD80 or CD86. In B1-a, but not in B2 cells, CTLA-4 is expressed. Therefore, CD80 has the choice to react with CTLA-4, either on T cells or on B1-a cells. The presence or absence of CTLA-4 controls the development of CD40 ligand (CD40L)-CD40-induced expression of AID, hence Ig class switching and V-region-hypermutation in germinal centres. CTLA-4 interactions are depicted by *green arrows*. In B2 cells, T follicular helper cells (TF_H) induce this germinal centre reaction of Ig class switch and antibody affinity maturation, while CTLA-4 expressed in B1-a cells inhibits it. In addition, mature B1-a as well as B2 cells express and are responsive to TLR2, 4, 7, 9 and to IL-4 and TGF-β. B2 cells respond to IL-21, while B1-a cells respond to IL-5.

generated B2 cells are short-lived and are perpetually replaced by newly formed cells from the BM in steady-state haematopoiesis.

In the follicles, B2 cells are ready to enter Thf-dependent responses against foreign antigens by increasing antigen processing and carrier presentation on MHCII in germinal centres (Figure 9.2). B2 cells are stimulated to proliferate in the dark zone, induced by CD40-ligand on T cells interacting with CD40 on B2 cells, which results in the expression of activation-induced cytidine deaminase (AID). AID promotes Ig class switching and $V_{H/L}$-hypermutation. In the light zone, haptens of native foreign antigen displayed on follicular dendritic cells select mutated, better-fitting BcRs on the proliferating IgH-class-switched B2 cells. Antigen-experienced memory B2 cells and Ig-secreting plasma cells develop, and some of them are deposited in the BM as long-term resting cells, serving as memory to the invasion of haptens of foreign antigens.

B1-a cell development in SL-deficient foetal liver appears not to be reduced in SL-deficient mice [57–59], suggesting that pre-BcRs with SL-pairing V_H-μH chains are not needed to expand large pre-II cells, as they do B2 cells. V_H to D_H J_H rearrangements at the IgH chain loci and V_LJ_L rearrangements at the IgL chain loci occur simultaneously, apparently circumventing the need to control stepwise rearrangements of the IgH and IgL chain loci by pre-BcR [64]. In contrast to the immature B2 cells developing in the BM, which remain G0-resting cells, immature and then mature sIgM+ B1-a cells exhibit strong proliferative expansion, characteristic of self-renewing B1-a cells [7,53–59]. Pre-B1-a cells expressing SL-non-pairing IgH chains appear favoured to develop sIgM+ B1-a cells by self-renewal activities. In fact, in the foetal liver, instead of promoting pre-B cell proliferation, some SL-pairing pre-BcRs have been found to inhibit it [65]. The emerging BcR repertoires indicate that a larger part of all B1-a cells express SL-non-pairing V_H chains, which are disfavoured or even excluded in B2-cell development. It remains to be investigated in much greater detail how the original V(D)J-rearranged (auto- or non-autoreactive) SL-pairing or SL-non-pairing V_H repertoires develop in foetal liver B1-a cells and in BM B2 cells.

In contrast to adult steady-state haematopoiesis and B2 lymphopoiesis, the cellular stages of the embryonal wave of foetal haematopoiesis are less well defined. Nevertheless, progenitor phenotypes with similarities to BM counterparts [26,48–50] such as LSK CD150+CD48- HSC-like cells, LSK CD150+CD48+ multipotent progenitor-like MPPs, Lin-negative IL7Ra+ CSF1-R+ LMPP, Lin-negative IL7Ra+ CLP and C-kit+ CD19+ CD93+ SL+ pre-B1-a cells have been quantitated in foetal liver between E12.5 and E16.5 [4,5]. Pre-B1-a cells increase much more than pre-B2 cells, at least 100-fold between E13.5 and E16.5. This expansion needs a supportive stromal cell environment [21,24]. B1-a cells develop and expand more than another 100-fold between E16.5 and E19, suggesting strong capacities for proliferative expansion of pre-B1-a precursors and B1-a cells [28–30].

In the periphery, B1-a cells colonise multiple sites in the body, including being present in small numbers in the spleen and thymus, and in large numbers in pleural and in peritoneal cavities. They respond rapidly to T-independent antigens [66]. B1-a cells can be long-lived and can replenish and thus maintain their peripheral compartments in cell areas near the red pulp, called the MZ. The red pulp is enriched for erythrocytes and filters the blood of dying red cells and other cellular debris. Therefore, B1-a cells in the MZ are exposed to and possibly stimulated by autoantigens of dying cells. B1-a cells populate the MZ even in germ-free mice, whereas they are missing in the pleural and peritoneal cavities. This suggests that the colonisation of the pleural and peritoneal cavities by microbiota might attract B1-a cells to fill these cavities with microbiota-reactive B1-a cells. In the lamina propria of the gut, these B1-a cells are induced in T-independent responses to IL-5-stimulated B1-a proliferation. Additionally, TGF-β stimulates Ig class switching from IgM to IgA production and secretion [33,38,39,67–69]. It remains to be clarified whether, and if so how, the microbiota in the gastrointestinal (GI) tract affect/mediate this attraction and activation of B1-a cells inside the peritoneum. Nevertheless, IgA is transported into the GI tract to influence, contain and shape the composition of the microbiome.

7. AUTOANTIGEN-REACTIVE BCR REPERTOIRES OF B1-A AND B2 CELLS

It has long been recognised that B1-a and B2 BcR repertoires differ dramatically. During the first weeks of their generation in the foetal liver and establishment in diverse peripheral sites, the V_H/V_L-repertoires of B1-a cells have been found to contain less random, more repetitive sequences, even in germ-free mice [70]. B1-a cells are the source of 19S IgM natural antibodies [61,62], which are polyreactive, with specificities for multiple autoantigens and microbial antigens [71]. Importantly, by controlling the presence of the autoantigen Thy-1, Hayakawa et al. [72] showed that Thy-1-glycoprotein-specific B1-a cells and their production of Thy-1-specific autoantibodies required the presence of Thy-1, thereby providing direct evidence for positive selection of autoantigen-reactive B1-a cells during B-cell development in the foetal liver.

A large part of B1-a cells, but not of B2 cells, are reactive to phosphatidylcholine (PtC). PtC is a major component of eukaryotic cell membranes, hence an abundant potential autoantigen. In C57BL/6 mice, four of five PtC-binding B1-a BcRs express V_H 11-DSP-J_H1 Ig H-chains and Vk9 L-chains without N-region insertions [73]. This bias of V_H11/J_H1 usage is already visible in pre-B cells expressing V_H11/DSP/J_H1pre-BcRs. These H-chains are able to pair with Vpre-B/λ5 SL chain to form pre-BcRs [74,75]. The corresponding pre-BcRs induce strong pre-B1-cell proliferation and differentiation into V_H11-DSP-J_H1/Vk9-BcR expressing B1a-B cells. This suggests that the autoantigen PtC might bind to pre-BcRs and signal these pre-B1-a cells to expand by proliferation, and that SL pairing does not impede the capacity to bind the autoantigen PtC.

Another SL-pairing V_H-expressing pre-BcR B1-a cell has been found that cannot expand by proliferation, suggesting that SL pairing alone is not sufficient to induce B1-a proliferation [65]. Moreover, V_H11-expressing foetal liver-derived pre-B1-a BcR does not pair well with SL, but allows pre-B1-a proliferation. Thus, SL pairing is not required for PtC autoantigen-stimulated proliferation. It remains to be tested whether a yet undetected autoantigen, such as PtC, can bind to V_H11-DSP-J_H1/SL chain-pre-BcR expressing pre-B1-a cells and whether this binding stimulates proliferation, thereby positively selecting foetal liver-derived autoantigen-reactive pre-B1-a and B1-a cells. By contrast, during B2 cell development in the BM, PtC-binding pre-B2 or B2 cells, including the V_H11-DSP-J_H1-H chain/Vk9 L chain, do not expand. They are even difficult to detect as they are present in low or undetectable numbers in the follicular B2 cell pools. This might indicate that the autoantigen PtC might signal to these pre-B2 cells via their pre-BcR not to expand by proliferation.

The differences in V_H gene usage between foetal liver B1-a cells and BM B2 cells in B cell repertoires with V_H81X-rearranged H-chain loci are particularly striking. V_H81X, a V_H gene located within the most D-proximal V_H7183 family of genes of the mouse IgH chain locus, is heavily overutilised in early phases of B1-a and B2 cell ontogeny [76–79], but is rarely expressed in later phases of development in which B2 cells dominate the repertoires [80–82]. Analyses of the ratios of productive, in-frame and non-productive, out-of-frame V_H81X-rearrangements detected an accumulation of productive rearrangements in foetal liver-derived B1-a cells, while mature B2 cells had enriched non-productively V_H81X-rearranged H-chain loci [75,83–87]. The results of these analyses suggest that V_H81X is overutilised in the initiation of productive, as well as non-productive Ig H-chain rearrangement in foetal liver B1-a and in BM B2 development. As productively rearranged pre-BcR and BCR-expressing cells, V_H81X-expressing B1-a cells appear positively selected by yet unknown autoantigens. For B2 cell development, it suggests that these autoantigens might inhibit the development of pre-B2 and B2 cells, negatively selecting productively V_H81X-rearranged IgH chain alleles, but allowing the accumulation of non-productively V_H81X-rearranged H-chain alleles in B2 cells expressing other, non-autoreactive $V_HD_HJ_H$-rearrangements.

8. B1-A CELLS, BUT NOT B2 CELLS EXPRESS THE NEGATIVE REGULATOR OF T CELL FUNCTION: CTLA-4

Surprisingly cytotoxic T-lymphocyte-associated protein-4 (CTLA-4), a key regulator of inhibitory responses expressed in T cells [88], is expressed selectively in B1-a cells (Figure 9.2), with intensity of expression increasing in a step-wise manner from pre-B1-a cells in foetal liver, to B1-a cells in the neonatal spleen, to B1-a cells in pleural and peritoneal cavities, and finally natural antibody-IgM-secreting CD138+ plasma cells (Immunological Genome Project (ImmGen database)) [88]. It remains to be seen if CTLA-4 is expressed on all B-1a cells and whether other B-cell populations also express it. CTLA-4 is not expressed in B2 cells (ImmGen database) [88]. B1-a cells also express the two ligands of CTLA- 4: CD80 and CD86. CD80, the high-affinity ligand for CTLA-4, is most highly expressed in B1-a cells. Therefore, the actions of CTLA-4, induced by CD80 binding, may be executed autonomously in a B1-a cell. However, it cannot be excluded that other APCs, such as CD80/CD86-expressing dendritic cells, interact with CTLA-4 on B1-a cells [13,89].

To study the functions of CTLA-4 in B1-a cells, Yang et al. [88] created a $CTLA4^{fl/fl}$ $CD19^{CD19cre/+}$ mouse strain. In these mice, CTLA-4 is no longer expressed in B1-a cells, while regulatory T cells continue to express it. CTLA-4-deficient B1-a cells show increased replenishment in the spleen, as well as exhibiting activated signalling pathways downstream of the BCR, increased surface expression of MHCII, class switching to IgG1 and high levels of V_H/V_L domain hypermutation. Thus, a strong antigen-presenting programme in GC comparable to that of B2 cells stimulated by Thf is induced in CTLA-4-deficient B1-a cells – only that the response is not driven by BcR-mediated uptake, processing and presentation on MHCII of foreign antigens, but by autoantigens.

GC formation of the CTLA-4-B cell-deficient B1-a cells with Thf occurs in the spleen, near areas of follicular dendritic cells. Depletion of CD4 T cells, or blockade of CD40L-CD40-signalling by CD40-specific antibodies, abolishes class switching and somatic hypermutation. All these findings suggest that CTLA-4 does not inhibit the activation and replenishment of IgM natural antibody (19S) synthesis and secretion, thereby not abolishing the continued functioning of the first line of defence against infections and the potential use of these autoantibodies in the removal of debris from dying cells in the spleen. However, CTLA-4 safeguards B1-a cells from entering GC formation, Ig class switching and hypermutation – processes that B2 cells undergo when they recognise foreign antigen (Figure 9.2). CTLA-4 appears to do so by not allowing CD40-mediated AID expression and upregulation of APC functions. Hence, CTLA-4 inhibits B1-a cells producing low-level autoreactive IgM to switch to better fitting hypermutated IgG autoantibodies. In turn, CTLA-4 blockade by 'checkpoint' anti-CTLA-4 antibodies, which are used to activate immune responses against cancers by relieving the negative, inhibitory effects of regulatory T cells, should induce autoimmune antibody formation and autoimmune disease. Such CTLA-4-directed treatment might well allow lowly autoreactive B1-a cells to develop better-fitting autoantibodies and thus autoantigen-driven autoimmune disease, as it is frequently observed in these 'checkpoint' inhibitor cancer antibody treatments [90].

Furthermore, CTLA-4-expressing B1-a cells should be able to present antigen [91]. Since B1-a cells are positively selected for poly-autoreactivity and quasi-autoantigens of the microbiome, the uptake, processing and presentation of peptides of such autoantigens on MHCII should activate T cells expressing TcRs specific for MHCII-presenting carrier peptides of auto- and quasi-autoantigens. Thus, CTLA-4+ regulatory T cells are expected to collaborate with autoreactive B1-a cells to inhibit inflammatory helper T cells, such as Thf, from inducing auto- and quasi-autoantigen-directed inflammatory responses. Hence, the action of autoantigen-reactive B1-a cells appears to induce suppression and therefore tolerance to the autoantigens that they recognise.

Recently, antigen-presenting autoreactive sIgM+ sIgD+ CD72+ CCR7+ B cells have been described to present the autoantigen collagen type II to activate regulatory T cells and thereby suppress autoimmune arthritis in mice. Blockade of CD72 impedes this activation of regulatory T cells, suggesting a new suppressive mechanism that regulates autoantigen-specific T cell tolerisation, by which the development of autoantigen-specific inflammation is suppressed [92]. We suggest, that these B cells may, in fact, belong to the B1-a cell lineage, as a part of the autoreactive B cell repertoire developed from foetal liver during embryogenesis, with the life-long function of tolerising adaptive immune responses of B and T cells to autoantigens.

9. CONCLUSION

In conclusion, the two separate pathways of B cell development in mice and humans generate B1-a cells and B2 cells. B1-a cells are positively selected to have a BcR repertoire for the first line (19S) natural antibody-mediated defence against infections. Moreover, via autoantigen presentation, they induce T cell-mediated suppression of autoimmunity, and hence, tolerance. Furthermore, colonisation of the gut by microbiota attracts B1-a cells to the gut and induces T-independent, IL-5 and TGF-beta-stimulated switching from IgM to IgA autoantibodies, which cross-react with microbial antigens. The second developmental pathway generates B2 cells, which eliminates autoreactive cells in the BM to generate a BcR repertoire ready to recognise non-autoreactive, infectious agents. This recognition is improved by Ig class switching and V-hypermutation and results in long-lived memory to infections.

ACKNOWLEDGEMENTS

FM was a member of the Basel Institute for Immunology, founded and fully supported by F. Hoffmann-La Roche, Basel Switzerland from 1970 to 2001, and then a senior research group leader and Max Planck Fellow of the Max Planck Institute for Infection Biology, Berlin, partially supported by a DFG Kossellek Grant (ME-2764-2-1) from 2003 to 2016. Since 2017 FM has been a senior research group leader and Leibniz-Professor at the Deutsches Rheuma-Forschungszentrum Berlin, a Leibniz Institute, partially supported by a Leibniz Collaborative Excellence Grant (CHROQ).
MFM is in part supported by CHROQ. PKJ was a Humboldt Research Fellow from 2017 to 2018 and is a member of CHROQ since 2019. We would like to thank Mairi McGrath for her editing of the text.

REFERENCES

[1] Herzenberg LA, Herzenberg LA. Toward a layered immune system. Cell 1989;59(6):953–4.
[2] Hardy RR, Hayakawa K. A developmental switch in B lymphopoiesis. Proc Natl Acad Sci U S A 1991;88(24):11550–4.
[3] Rolink AG, Andersson J, Melchers F. Molecular mechanisms guiding late stages of B-cell development. Immunol Rev 2004;197:41–50.
[4] Alt FW, et al. Ordered rearrangement of immunoglobulin heavy chain variable region segments. EMBO J 1984;3(6):1209–19.
[5] Tonegawa S. Somatic generation of antibody diversity. Nature 1983;302(5909):575–81.
[6] ten Boekel E, Melchers F, Rolink AG. Changes in the V(H) gene repertoire of developing precursor B lymphocytes in mouse bone marrow mediated by the pre-B cell receptor. Immunity 1997;7(3):357–68.
[7] Li YS, Hayakawa K, Hardy RR. The regulated expression of B lineage associated genes during B cell differentiation in bone marrow and fetal liver. J Exp Med 1993;178(3):951–60.
[8] Feeney AJ. Lack of N regions in fetal and neonatal mouse immunoglobulin V-D-J junctional sequences. J Exp Med 1990;172(5):1377–90.
[9] Carlsson L, Holmberg D. Genetic basis of the neonatal antibody repertoire: germline V-gene expression and limited N-region diversity. Int Immunol 1990;2(7):639–43.
[10] Zemlin M, et al. Regulation and chance in the ontogeny of B and T cell antigen receptor repertoires. Immunol Res 2002;26(1–3):265–78.
[11] Gilfillan S, et al. Mice lacking TdT: mature animals with an immature lymphocyte repertoire. Science 1993;261(5125):1175–8.
[12] Komori T, et al. Lack of N regions in antigen receptor variable region genes of TdT-deficient lymphocytes. Science 1993;261(5125):1171–5.
[13] Popi AF, Longo-Maugeri IM, Mariano M. An overview of B-1 cells as antigen-presenting cells. Front Immunol 2016;7:138.
[14] Lanzavecchia A. Antigen-specific interaction between T and B cells. Nature 1985;314(6011):537–9.
[15] Medvinsky AL, et al. An early pre-liver intraembryonic source of CFU-S in the developing mouse. Nature 1993;364(6432):64–7.
[16] Godin IE, et al. Para-aortic splanchnopleura from early mouse embryos contains B1a cell progenitors. Nature 1993;364(6432):67–70.
[17] Medvinsky A, Dzierzak E. Definitive hematopoiesis is autonomously initiated by the AGM region. Cell 1996;86(6):897–906.
[18] Soares-da-Silva F, et al. Crosstalk between the hepatic and hematopoietic systems during embryonic development. Front Cell Dev Biol 2020;8:612.
[19] Gomez Perdiguero E, et al. Tissue-resident macrophages originate from yolk-sac-derived erythro-myeloid progenitors. Nature 2015;518(7540):547–51.
[20] Wu Y, Hirschi KK. Tissue-resident macrophage development and function. Front Cell Dev Biol 2020;8:617879.
[21] Strasser A, Rolink A, Melchers F. One synchronous wave of B cell development in mouse fetal liver changes at day 16 of gestation from dependence to independence of a stromal cell environment. J Exp Med 1989;170(6):1973–86.
[22] Kobayashi M, et al. Functional B-1 progenitor cells are present in the hematopoietic stem cell-deficient embryo and depend on Cbfbeta for their development. Proc Natl Acad Sci U S A 2014;111(33):12151–6.
[23] Kajikhina K, Melchers F, Tsuneto M. Chemokine polyreactivity of IL7Ralpha+CSF-1R+ lympho-myeloid progenitors in the developing fetal liver. Sci Rep 2015;5:12817.
[24] Kajikhina K, Tsuneto M, Melchers F. B-lymphopoiesis in fetal liver, guided by chemokines. Adv Immunol 2016;132:71–89.
[25] Alon R, Shulman Z. Chemokine triggered integrin activation and actin remodeling events guiding lymphocyte migration across vascular barriers. Exp Cell Res 2011;317(5):632–41.

[26] Tsuneto M, et al. Reprint of: Environments of B cell development. Immunol Lett 2014;160(2):109–12.
[27] Tsuneto M, et al. B-cell progenitors and precursors change their microenvironment in fetal liver during early development. Stem Cell 2013;31(12):2800–12.
[28] Rolink A, et al. Long-term proliferating early pre B cell lines and clones with the potential to develop to surface Ig-positive, mitogen reactive B cells in vitro and in vivo. EMBO J 1991;10(2):327–36.
[29] Kawano Y, et al. IL-7 and immobilized Kit-ligand stimulate serum- and stromal cell-free cultures of precursor B-cell lines and clones. Eur J Immunol 2017;47(1):206–12.
[30] Rolink AG, Schaniel C, Melchers F. Stability and plasticity of wild-type and Pax5-deficient precursor B cells. Immunol Rev 2002;187:87–95.
[31] Schaniel C, et al. Extensive in vivo self-renewal, long-term reconstitution capacity, and hematopoietic multipotency of Pax5-deficient precursor B-cell clones. Blood 2002;99(8):2760–6.
[32] Reininger L, et al. Development of autoimmune disease in SCID mice populated with long-term "in vitro" proliferating (NZB x NZW) F1 pre-B cells. J Exp Med 1992;176(5):1343–53.
[33] Andersson J, Coutinho A, Melchers F. Mitogen-activated B-cell blasts reactive to more than one mitogen. J Exp Med 1979;149(3):553–64.
[34] Andersson J, et al. The mitogenic effect of lipopolysaccharide on bone marrow-derived mouse lymphocytes. Lipid A as the mitogenic part of the molecule. J Exp Med 1973;137(4):943–53.
[35] Beutler B, Milsark IW, Cerami AC. Passive immunization against cachectin/tumor necrosis factor protects mice from lethal effect of endotoxin. Science 1985;229(4716):869–71.
[36] Poltorak A, et al. Defective LPS signaling in C3H/HeJ and C57BL/10ScCr mice: mutations in Tlr4 gene. Science 1998;282(5396):2085–8.
[37] Hoshino K, et al. Cutting edge: toll-like receptor 4 (TLR4)-deficient mice are hyporesponsive to lipopolysaccharide: evidence for TLR4 as the Lps gene product. J Immunol 1999;162(7):3749–52.
[38] Melchers F, Braun V, Galanos C. The lipoprotein of the outer membrane of *Escherichia coli*: a B-lymphocyte mitogen. J Exp Med 1975;142(2):473–82.
[39] Andersson J, et al. Clonal growth and maturation to immunoglobulin secretion in vitro of every growth-inducible B lymphocyte. Cell 1977;10(1):27–34.
[40] Ganuza M, et al. Murine foetal liver supports limited detectable expansion of life-long haematopoietic progenitors. Nat Cell Biol 2022;24(10):1475–86.
[41] Yokomizo T, et al. Independent origins of fetal liver haematopoietic stem and progenitor cells. Nature 2022;609(7928):779–84.
[42] Carrelha J, et al. Hierarchically related lineage-restricted fates of multipotent haematopoietic stem cells. Nature 2018;554(7690):106–11.
[43] Pei W, et al. Resolving fates and single-cell transcriptomes of hematopoietic stem cell clones by PolyloxExpress barcoding. Cell Stem Cell 2020;27(3):383–395 e8.
[44] Jani PK, et al. MicroRNA-221/222-expression in HSC and MPP safeguards their quiescence and multipotency by downregulating stress-independent and dependent expression of IEG and of several myelo/granulopoiesis-enhancing target genes. bioRxiv 2023. 2023.01.30.526397.
[45] Seita J, Weissman IL. Hematopoietic stem cell: self-renewal versus differentiation. Wiley Interdiscip Rev Syst Biol Med 2010;2(6):640–53.
[46] Morrison SJ, Scadden DT. The bone marrow niche for haematopoietic stem cells. Nature 2014;505(7483):327–34.
[47] Busch K, et al. Fundamental properties of unperturbed haematopoiesis from stem cells in vivo. Nature 2015;518(7540):542–6.
[48] Wilson A, et al. Hematopoietic stem cells reversibly switch from dormancy to self-renewal during homeostasis and repair. Cell 2008;135(6):1118–29.
[49] Kiel MJ, et al. SLAM family receptors distinguish hematopoietic stem and progenitor cells and reveal endothelial niches for stem cells. Cell 2005;121(7):1109–21.
[50] Cabezas-Wallscheid N, et al. Identification of regulatory networks in HSCs and their immediate progeny via integrated proteome, transcriptome, and DNA methylome analysis. Cell Stem Cell 2014;15(4):507–22.
[51] Kawano Y, et al. Stable lines and clones of long-term proliferating normal, genetically unmodified murine common lymphoid progenitors. Blood 2018;131(18):2026–35.
[52] Melchers F, et al. Repertoire selection by pre-B-cell receptors and B-cell receptors, and genetic control of B-cell development from immature to mature B cells. Immunol Rev 2000;175:33–46.
[53] Jani PK, Kubagawa H, Melchers F. A rheostat sets B-cell receptor repertoire selection to distinguish self from non-self. Curr Opin Immunol 2020;67:42–9.
[54] von Boehmer H, Melchers F. Checkpoints in lymphocyte development and autoimmune disease. Nat Immunol 2010;11(1):14–20.
[55] Melchers F. The pre-B-cell receptor: selector of fitting immunoglobulin heavy chains for the B-cell repertoire. Nat Rev Immunol 2005;5(7):578–84.
[56] Martensson IL, et al. The pre-B cell receptor and its role in proliferation and Ig heavy chain allelic exclusion. Semin Immunol 2002;14(5):335–42.
[57] Kitamura D, et al. A critical role of lambda 5 protein in B cell development. Cell 1992;69(5):823–31.
[58] Mundt C, et al. Loss of precursor B cell expansion but not allelic exclusion in VpreB1/VpreB2 double-deficient mice. J Exp Med 2001;193(4):435–45.
[59] Shimizu T, et al. VpreB1/VpreB2/lambda 5 triple-deficient mice show impaired B cell development but functional allelic exclusion of the IgH locus. J Immunol 2002;168(12):6286–93.
[60] Ohnishi K, Takemori T. Molecular components and assembly of mu.surrogate light chain complexes in pre-B cell lines. J Biol Chem 1994;269(45):28347–53.
[61] Melchers F, et al. Rules for the rearrangement events at the L chain gene loci of the mouse. Adv Exp Med Biol 2007;596:63–70.
[62] Yamagami T, et al. Frequencies of multiple IgL chain gene rearrangements in single normal or kappaL chain-deficient B lineage cells. Immunity 1999;11(3):317–27.
[63] Kubagawa H, et al. Functional roles of the IgM Fc receptor in the immune system. Front Immunol 2019;10:945.
[64] Wong JB, et al. B-1a cells acquire their unique characteristics by bypassing the pre-BCR selection stage. Nat Commun 2019;10(1):4768. https://doi.org/10.1038/s41467-019-12824-z.
[65] Wasserman R, et al. A novel mechanism for B cell repertoire maturation based on response by B cell precursors to pre-B receptor assembly. J Exp Med 1998;187(2):259–64.

[66] Martin F, Oliver AM, Kearney JF. Marginal zone and B1 B cells unite in the early response against T-independent blood-borne particulate antigens. Immunity 2001;14(5):617–29.
[67] Stavnezer J, Kang J. The surprising discovery that TGF beta specifically induces the IgA class switch. J Immunol 2009;182(1):5–7.
[68] Coffman RL, Lebman DA, Shrader B. Transforming growth factor beta specifically enhances IgA production by lipopolysaccharide-stimulated murine B lymphocytes. J Exp Med 1989;170(3):1039–44.
[69] Sonoda E, et al. Transforming growth factor beta induces IgA production and acts additively with interleukin 5 for IgA production. J Exp Med 1989;170(4):1415–20.
[70] Yang Y, et al. Distinct mechanisms define murine B cell lineage immunoglobulin heavy chain (IgH) repertoires. Elife 2015;4:e09083.
[71] Rollenske T, et al. Parallelism of intestinal secretory IgA shapes functional microbial fitness. Nature 2021;598(7882):657–61.
[72] Hayakawa K, et al. Positive selection of natural autoreactive B cells. Science 1999;285(5424):113–6.
[73] Seidl KJ, et al. Frequent occurrence of identical heavy and light chain Ig rearrangements. Int Immunol 1997;9(5):689–702.
[74] Yoshikawa S, et al. The skewed heavy-chain repertoire in peritoneal B-1 cells is predetermined by the selection via pre-B cell receptor during B cell ontogeny in the fetal liver. Int Immunol 2009;21(1):43–52.
[75] Kawano Y, et al. Selection of stereotyped VH81X-microH chains via pre-B cell receptor early in ontogeny and their conservation in adults by marginal zone B cells. Int Immunol 2005;17(7):857–67.
[76] Yancopoulos GD, et al. Preferential utilization of the most JH-proximal VH gene segments in pre-B-cell lines. Nature 1984;311(5988):727–33.
[77] Perlmutter RM, et al. Developmentally controlled expression of immunoglobulin VH genes. Science 1985;227(4694):1597–601.
[78] Yancopoulos GD, Malynn BA, Alt FW. Developmentally regulated and strain-specific expression of murine VH gene families. J Exp Med 1988;168(1):417–35.
[79] Schroeder Jr HW, Wang JY. Preferential utilization of conserved immunoglobulin heavy chain variable gene segments during human fetal life. Proc Natl Acad Sci U S A 1990;87(16):6146–50.
[80] Dildrop R, et al. VH-gene expression in murine lipopolysaccharide blasts distributes over the nine known VH-gene groups and may be random. Eur J Immunol 1985;15(11):1154–6.
[81] Wu GE, Paige CJ. VH gene family utilization in colonies derived from B and pre-B cells detected by the RNA colony blot assay. EMBO J 1986;5(13):3475–81.
[82] Sheehan KM, Brodeur PH. Molecular cloning of the primary IgH repertoire: a quantitative analysis of VH gene usage in adult mice. EMBO J 1989;8(8):2313–20.
[83] Decker DJ, Boyle NE, Klinman NR. Predominance of nonproductive rearrangements of VH81X gene segments evidences a dependence of B cell clonal maturation on the structure of nascent H chains. J Immunol 1991;147(4):1406–11.
[84] Carlsson L, Overmo C, Holmberg D. Developmentally controlled selection of antibody genes: characterization of individual VH7183 genes and evidence for stage-specific somatic diversification. Eur J Immunol 1992;22(1):71–8.
[85] Huetz F, et al. V-region directed selection in differentiating B lymphocytes. EMBO J 1993;12(5):1819–26.
[86] Marshall AJ, Wu GE, Paige GJ. Frequency of VH81x usage during B cell development: initial decline in usage is independent of Ig heavy chain cell surface expression. J Immunol 1996;156(6):2077–84.
[87] Decker DJ, et al. Heavy chain V gene-specific elimination of B cells during the pre-B cell to B cell transition. J Immunol 1995;154(10):4924–35.
[88] Yang Y, et al. CTLA-4 expression by B-1a B cells is essential for immune tolerance. Nat Commun 2021;12(1):525.
[89] Walker LS, Sansom DM. The emerging role of CTLA4 as a cell-extrinsic regulator of T cell responses. Nat Rev Immunol 2011;11(12):852–63.
[90] Khan S, Gerber DE. Autoimmunity, checkpoint inhibitor therapy and immune-related adverse events: a review. Semin Cancer Biol 2020;64:93–101.
[91] Chien CH, Chiang BL. Regulatory T cells induced by B cells: a novel subpopulation of regulatory T cells. J Biomed Sci 2017;24(1):86.
[92] Aoun M. Antigen-presenting autoreactive B cells activate regulatory T cells and suppress autoimmune arthritis in mice. J Exp Med 2023;220(11). https://doi.org/10.1084/jem.20230101.

Chapter 10

The Role of the BAFF and Lymphotoxin Pathways in B Cell Biology

Fabienne Mackay[1] and Jennifer L. Gommerman[2]

[1]Department of Immunology, Monash University, Clayton, VIC, Australia; [2]Department of Immunology, University of Toronto, Toronto, ON, Canada

1. BAFF/APRIL: IMPORTANT REGULATORS OF B CELL SURVIVAL, HOMOEOSTASIS, AND FUNCTION

1.1 Introduction

For many years, immunologists have believed that expression of a B cell receptor (BCR) was sufficient and essential to allow B cells to develop and survive in the periphery. This notion was later reconsidered after the discovery of the B cell activating factor belonging to the tumour necrosis factor (TNF) family (BAFF; also known as BLyS, TALL-1, zTNF-4, THANK, and TNFSF13b; reviewed in Refs. [1,2]), which emerged as an equally essential factor for B cell maturation and survival in the periphery. Not only was the discovery of BAFF a major paradigm shift in our understanding of B cell development and survival, a decade of studies on this factor led to the development of an inhibitor approved for use in the clinic as a novel B cell therapy treating a subset of patients with systemic lupus erythematosus (SLE) [3,4]. Apart from TNF, BAFF is the second ligand from the TNF superfamily that has led to bench-to-bedside outcomes.

1.2 The BAFF—APRIL System

1.2.1 The Ligands

BAFF and a related ligand, a proliferation-inducing ligand (APRIL; also known as TALL-2, TRDL-1, or TNFSF13a), are heterotrimeric ligand members of the TNF superfamily (reviewed in Refs. [1,2,5]); as such, they are typically produced as type II transmembrane proteins that are proteolytically cleaved at a furin protease site and released in a soluble form [6]. BAFF has also been described as membrane-bound ligand. In contrast, APRIL is cleaved in the Golgi before release and exists only in a soluble form. However, recently, APRIL-δ, a malignant isoform of APRIL that lacks a furin cleavage site, appears to remain membrane-bound on the surface of leukemia cell precursors [7]. The 285-amino-acid BAFF is cleaved by furin convertase between amino acids R133 and A134, which are situated at the end of a consensus furin cleavage motif (R-X-(R/K)-R) in the stalk region, and BAFF is released as a soluble 152-amino-acid, 17-kDa protein [8,9]. Likewise, 250-amino-acid APRIL is cleaved between R104 and A105 and is released as a 146-amino-acid, 17-kDa protein [10].

The human *baff* gene, encoded at 13q32-q34, contains a transmembrane domain and flanking regions in exon 1, a furin-processing site in exon 2, and a TNF homology domain (THD) in exons 3–6 [1]. The THD facilitates binding to the BAFF receptors (BAFF-R). The murine *baff* gene contains an additional exon encoding 30 amino acid residues between the furin processing site and the THD. Mouse and human APRIL contain a furin-processing site in exon 2 [1]. BAFF is an active ligand as a homotrimer [11]. The trimer contains two magnesium ions interacting with the side chains of Q234, N235, and N243 [12], and it is the main form of BAFF found in the circulation [13]. However, a 60-mer form of BAFF has been obtained at physiological pH conditions, and this form is also able to bind to its receptors [14]. BAFF 60-mers arise by the assembly of 20 BAFF trimers into a virus-like cluster. BAFF contains a "flap" region, which is absent in APRIL and is essential for the 60-mer assembly [1,14]. The "flap" region is also referred to as the DE loop because it is formed by five amino acids inserted between the β-strands D and E [15]. Interactions between BAFF trimers are extensive and complex because these involve hydrogen bonds, salt bridges, and hydrophobic contacts [14]. For this reason, the formation of 60-mer BAFF is not reversible at physiological pH, whereas dissociation of the 60-mer into trimers can occur at acidic pH. The strong dependency on a

Molecular Biology of B Cells. https://doi.org/10.1016/B978-0-323-95895-0.00018-0

physiological pH for the formation of the 60-mer suggests that BAFF 60-mer assembly is not due to random aggregation. Despite this argument, other studies have claimed that assembly of the BAFF 60-mer was merely an artefact involving fusion of recombinant N-terminal tags inserted to the BAFF proteins [11]. However, to address this criticism, another study was conducted using a construct of BAFF with no amino-terminal tag that was also able to form 60-mers in solution [16], hence ruling out the possibility that BAFF 60-mer assembly is an artefact caused by the fusion of protein tags. The function of BAFF 60-mer may involve the recruitment of numerous receptors; hence, clustering these receptors on the cell surface to increase the local concentration of receptor/ligand interaction thus enhancing signalling [14]. The BAFF 60-mer is the more active form of BAFF; therefore, it is likely to be physiologically relevant. In biological situations in which the local concentration of BAFF greatly increases, a corresponding increase in the formation of BAFF 60-mers might be expected, along with increased propensity for signalling through the BAFF/APRIL receptor transmembrane activator and cyclophilin ligand interactor (TACI; also known as TNFRSF13B), which requires the 60-mer form of BAFF to signal in some cases [13].

BAFF and APRIL are capable of forming heterotrimers when co-expressed [17]. These are biologically active and were detected in the serum of healthy humans and patients with systemic autoimmune rheumatic diseases [17]. Heterotrimers have been reported to be upregulated in a subset of patients with rheumatic diseases; however, it is unclear whether this may reflect a limitation in the anti-BAFF and anti-APRIL antibodies used for the capture and detection of heterotrimers [17]. However, purely based on structure prediction, it is also possible that other members of the TNF family are similarly capable of forming heterotrimers with BAFF and/or APRIL [17]. Indeed, a more recent study demonstrated that BAFF/APRIL complexes have biological activity and can modulate B1 B cell function and are frequently elevated in SLE but not rheumatoid arthritis (RA) patient sera [18].

ΔBAFF is a splice variant of BAFF that was identified in humans and mice [19,20]. The DNA sequence of ΔBAFF lacks the exon encoding the first β-sheet of the THD, which is exon 3 in humans and exon 4 in mice [19]. ΔBAFF is not released from the cell membrane, but it can form multimers with BAFF such that the release of BAFF from the cell is inhibited. As such, ΔBAFF appears to act as a negative regulator of BAFF [20]. Expression of ΔBAFF is less abundant than that of the full-length isoform of BAFF in lymphatic and central nervous system (CNS) tissues [21]. More recently, two new transcript variants of BAFF have been identified in peripheral blood mononuclear cells [22], but functional studies are required to determine the structural and biological significance of these new variants. Another study discovered yet another splice variant form of BAFF in mice and humans. This last variant form of BAFF was discovered using a new set of primers that generated a polymerase chain reaction band missing a 114-base-pair segment encoded by exon 4. It is interesting to note that early studies suggested that this new variant of BAFF, named Δ4BAFF, associated with DNA and possibly acted as a transcription factor for the full-length form of BAFF [23].

The APRIL locus is situated adjacent to the locus of TWEAK, another member of the TNF superfamily. A splice variation event at the APRIL/TWEAK locus leads to the production of an APRIL/TWEAK hybrid ligand, named TWE-PRIL. This ligand is a fusion of the APRIL extracellular domain with the TWEAK transmembrane portion; therefore, it is suspected to be biologically active similarly to APRIL [24]. However, the biological activity and regulation of this particular ligand remains unknown.

1.2.2 Sources of BAFF and APRIL Production

BAFF and APRIL are produced in various hematopoietic and nonhematopoietic cell lineages (Table 10.1) [2]. Myeloid cells in particular produce high amounts of BAFF, and cell types producing BAFF and APRIL include monocytes [25], macrophages, dendritic cells (DCs) [6,26], neutrophils [27–29], mast cells and eosinophils [27], and astrocytes [21]. These immune cells participate in innate immune responses, which act to defend the host from infection in a non-specific manner. They do not confer long-lasting protective immunity, but they have a more immediate action than cells of the adaptive immune system. However, many of these cells can also act as antigen-presenting cells, and production of BAFF by these cells can stimulate T cell activation and survival [27,30]. BAFF also influences the activation of innate cells; in particular, it induces monocyte survival, proinflammatory cytokine secretion, and co-stimulatory molecule expression [31]. In addition, BAFF production by innate cells is triggered upon stimulation by proinflammatory factors and co-stimulatory molecules. For example, BAFF production is upregulated by interferon-γ (IFN-γ), interleukin (IL)-10, and CD40L [32]. Pathogen-associated molecular patterns, such as lipopolysaccharide (LPS) or peptidoglycans, upregulate the expression of BAFF and APRIL in macrophages and DCs via toll-like receptors (TLRs) [33–35]. It is interesting to note that BAFF and APRIL expression is upregulated within B cells upon TLR activation [36]. TLRs are a means by which cells of the innate immune system recognise pathogenic material or pathogen-associated molecular patterns and initiate a T cell-independent immune response. TLR are especially important in marginal zone (MZ) B cells, which are situated adjacent to the splenic MZ sinus where the blood carrying pathogens enters the spleen. As such, MZ B cell activation via TLR is a strategic host

TABLE 10.1 Cellular Sources of B Cell Activating Factor of Tumour Necrosis Factor (BAFF) and a Proliferation-Inducing Ligand (APRIL) Production

	Hematopoietic producers	Nonhematopoietic producers
BAFF	Dendritic cells (DCs) Monocytes Macrophages Neutrophils T cells Natural killer cells B cells Leukaemic B cells Promyelocytic leukemia cells Bone marrow (BM) CD14$^+$ cells and osteoclasts	Salivary gland epithelial cells Fibroblast-like synoviocytes Cytotrophoblast cells and stromal cells in human placenta Astrocytes Renal cell carcinoma
APRIL	DCs Monocytes Macrophages Myeloid precursor cells Neutrophils B cells Leukaemic B cells T cells BM CD14$^+$ cells and osteoclasts	Intestinal epithelial cells Cytotrophoblast cells and stromal cells in human placenta Renal cell carcinoma Most tumour cells

response to blood-borne pathogens, and this may explain why BAFF plays an important role in upregulating TLR expression on the surface of these cells [33].

In the synovium of patients with RA, BAFF is produced by B and T cells, but also by synovial sublining cells [37]. Healthy fibroblast-like synoviocytes produce large amounts of BAFF in response to IFN-γ or TNF stimulation [38,39]. Synovial fibroblasts from RA patients constitutively express BAFF and APRIL, and they strongly upregulate the expression of these two factors upon stimulation of TLR3 and to a lesser extent TLR4 but not TLR2 [40].

APRIL expression in the bone marrow (BM) is critical for plasmablast survival because APRIL-deficient mice were unable to sustain a transferred population of plasmablasts [41]. BM-resident macrophages are an important source of this survival signal. In this context, APRIL binds preferentially to heparin sulphate proteoglycan (HSPG) on the surface of CD138$^+$ plasma cells (PCs), upregulating expression of the prosurvival factor Bcl-X_L [41]. However, more recently, it was shown that eosinophils in the BM are also a critical source of PC survival factors including APRIL [42]. Furthermore, it is now understood that BAFF and APRIL are mutually redundant sources of survival factors for BM PCs [43]. Tumour-associated macrophages are also a major source of APRIL in gastric mucosa-associated lymphoid tissue (MALT) lymphoma [44]. Activin A, a member of the transforming growth factor-β (TGF-β) family, strongly activates APRIL expression by mouse macrophages [45]. A mechanism for the role B cell maturation antigen (BCMA) in PC maintenance has recently been described. Indeed, BCMA signalling promotes the expression of Mcl1, a factor essential for PC survival [46].

BAFF and APRIL are produced by astrocytes as part of their active role in cerebral innate immunity [21,47]. Astrocytes in vitro produce high amounts of BAFF when stimulated with inflammatory cytokines [47], suggesting that they may be required to supply cerebral immune cells with survival signals. BAFF, together with other soluble glial-derived cytokines, may be responsible for triggering adaptive immunity in an inflamed CNS [47]. It is interesting to note that B cells may play an important pathogenic role in multiple sclerosis (MS) because patients responded to depletion of B cells using anti-CD20 (rituximab) with reduced inflammatory brain lesions and relapses [48]. BAFF production in the brain may contribute to the pathogenic effect of B cells in MS. Surprisingly and contrary to expectations, one attempt to block BAFF but also APRIL with atacicept (a soluble recombinant TACI protein fused with the Fc portion of a human antibody used as decoy) led to adverse effects and intensification of the inflammatory activity [49], suggesting a more complicated function of the BAFF/APRIL system beyond that of B cell survival.

Activated T cells produce APRIL [50,51] and TWE-PRIL [24], but BAFF expression from T cells has been somewhat controversial. BAFF mRNA has been detected within resting and stimulated T cells [26,27], stromal cells [52], and follicular dendritic cells (FDCs) [53]. A recent study has demonstrated the follicular helper T cells (T_{fh}) in the germinal centre (GC) are an important source of BAFF

[54]. Confirmation of BAFF expression at the protein level in these cell types in humans has been difficult to obtain [55]. Given that T cells co-express APRIL, it is possible that T cells produce BAFF/APRIL heterotrimers, which might complicate BAFF protein detection by anti-BAFF monoclonal antibodies [56].

BAFF and APRIL directly modulate T cell function by enhancing human T cell response to anti-CD3 activation [26,27] and T cell survival [51]. The addition of exogenous BAFF to activated splenic T cells increased anti-apoptotic Bcl-2 expression, suggesting that BAFF may have a positive influence on the survival of T cells [27]. Modification of the B cell compartment in BAFF Tg mice also indirectly affects the quality of T cell responses [33].

1.2.3 Receptors for BAFF and APRIL

BAFF and APRIL share two receptors: TACI and BCMA (also known as TNFRSF17; reviewed in Refs. [1,2,57]). APRIL and BAFF have different affinities for these receptors, with APRIL binding strongly to BCMA but moderately to TACI, and BAFF binding weakly to BCMA but strongly to TACI [1,58]. In addition, BAFF has a high affinity for the BAFF-R (also known as TNFRSF13C) [1,2]. BAFF binds preferentially to BAFF-R and TACI, and with relatively weaker affinity for BCMA (100-fold or 1000-fold weaker, respectively); APRIL binds comparably to TACI and BCMA [1,59]. Because TACI is the only receptor to bind relatively well to APRIL and BAFF, heterotrimers of these ligands are thought to be specific for TACI (Figure 10.1) [17].

The DxL motif on BAFF-R [60], TACI, and BCMA [61] is critical for binding to BAFF and APRIL. In the case of TACI, binding is favoured by an additional two amino acids for BAFF and APRIL. Four residues on BCMA (Y13, I22, Q25, and R27) have been implicated in this ligand specificity [62]. Interaction between BAFF/APRIL and their receptors occurs via conserved cysteine-rich domains (CRDs) in the extracellular regions of the receptors [61]. TNF receptor (TNFR) superfamily proteins typically contain three to six CRDs, TACI contains two CRDs, BCMA contains a single CRD, and BAFF-R contains a partial CRD [61]. The complete CRDs are each approximately 40 amino acid residues in length. A form of TACI exists in which the N-terminal CRD is removed by alternate splicing, and this form is fully capable of ligand-induced cell signalling [63]. This capability suggests that the second CRD of TACI alone mediates complete affinity for BAFF and APRIL [61]. This notion is supported by the extent of evolutionary conservation, which is higher in the second CRD than the first CRD, because highly conserved amino acid residues tend to be critical for protein function [61].

The expression pattern of these receptors on B cells differs depending on the type of B cells and their stage of maturation and activation. BAFF-R expression is not detectable on B cell precursors in the BM, but it appears on immature B cells as they progress towards their differentiation into splenic transitional type 1 peripheral B cells and upon acquiring a functional BCR [27,64,65]. BAFF-R is the dominant receptor expressed on mature B cells [27,65]. BAFF-R is essential for the survival and maturation of immature B cells [66,67]. BAFF-R expression is also upregulated on activated T cells [27] and constitutively expressed on regulatory T cells [68]. The role of BAFF-R on activated T cells appears to be to promote Bcl2 expression and T helper-1 (Th1) cytokine production in effector CD4 T cells [27]. The role of BAFF-R on regulatory T cells remains unclear.

Mouse TACI is expressed most highly on mature innate-like B cells such as MZ B cells and B-1 B cells, which are very important B cell subsets responding efficiently to type I and type II T-independent antigens [33,69]. It is interesting to note that T-independent type II and type I antibody responses are impaired in mice lacking TACI [70,71]. TACI is an important promoter of class switch recombination (CSR) in B cells, and this function involves direct interaction between TACI and the signalling adaptor myeloid differentiation primary response gene 88 [72]. Another feature of mice lacking TACI is a greater number of B cells, and some TACI-deficient mouse lines develop autoimmunity and B cell lymphoma [73]. Together, these observations suggest a negative regulation of B cell numbers by TACI, and some studies have suggested a role for TACI in driving B cell apoptosis [73]. However, TACI, unlike other receptors from the TNFR superfamily such as TNFR-1 or Fas [74], does not have a death domain in its intracellular portion, and currently described signalling elements downstream of TACI are not obviously linked to a death pathway [72,75]. A very recent study has shed some light on this aspect, demonstrating that the role of TACI in regulating B cells is indirect and involves indirect upregulation of the Fas and Fas ligand (FasL; a TNF–TNFR pair triggering apoptosis in lymphocytes) on the surface of LPS-activated MZ B cells [76]. $TACI^{-/-}$ MZ B cells fail to die after LPS activation and remain activated for long periods of time (Figure 10.2) [76]. TACI expression on B cells varies throughout life and is decreased in newborn humans and mice [77] and on GC B cells [27]. It is interesting to note that in TACI-deficient mice, numbers of T_{fh} and GC B cells are increased, suggesting a role for TACI in regulating these cells [78], although TACI seems to be more important in antibody-secreting cells (ASCs) in which it normally downregulates the proapoptotic molecule Bim, allowing ASC formation [78]. TACI signalling also supports Blimp-1 expression, a transcriptional regulator important for ASC differentiation [79].

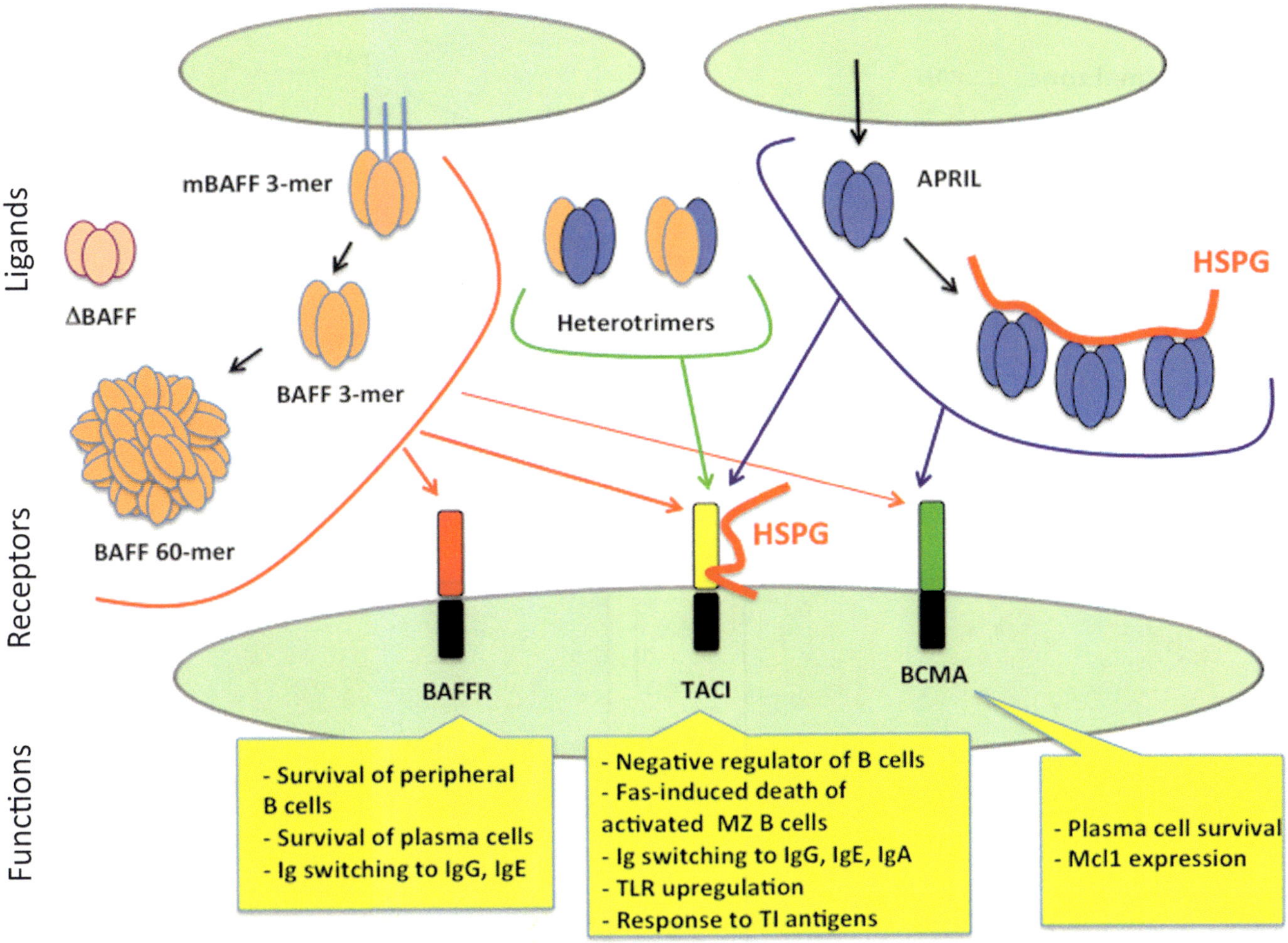

FIGURE 10.1 **Ligands and receptors of the BAFF/APRIL system and functions.** The upper part of the figure represents the various forms of the ligands. *Arrows* indicate the binding specificity of these different ligands to corresponding receptors (middle section of the figure). The *red line* binding APRIL ligands or TACI represents heparin sulphate proteoglycan (HSPG). Call boxes indicate the function of each of the receptors. Abbreviations: *APRIL*, a proliferation-inducing ligand; *BAFF*, B cell activating factor of tumour necrosis factor; *BCMA*, B cell maturation antigen; *mBAFF*, membrane-bound BAFF; *TACI*, transmembrane activator and cyclophilin ligand interactor; *TI*, thymus-independent; *TLR*, toll like receptor.

TACI is expressed on human memory B cells, which in healthy individuals also express higher levels of Fas, consistent with a general correlation seen between TACI and Fas expression [36,76,80]. TACI is also expressed by a subpopulation of activated CD27^{neg} B cells [36,80] and macrophages, and it influences their cell survival [31]. TACI mRNA is expressed by a subset of T cells [81]; however, polyclonal and monoclonal antibodies against TACI and gene profiling experiments failed to confirm TACI expression in T cells [27]. Many new T cell subsets have been identified in recent years, and it remains to be seen whether TACI can be expressed on these cells in some circumstances.

TACI expression is strongly upregulated on the surface of B cells in response to the activation of many TLRs, an aspect that is consistent with a key role for TACI in T-independent B cell responses [33,35,82,83]. TACI and APRIL interact with HSPG, forming a complex presumably designed to facilitate crosslinking of TACI by APRIL via stabilising ligand/receptor interaction and receptor cross-linking, thus allowing robust signalling (Figure 10.1) [84–86].

BCMA expression is limited to GC B cells, plasmablasts, and PCs [15,53,87]. BCMA is not expressed on splenic B cell subsets, but BCMA expression is high on BM PCs [88]. In fact, BCMA is important for the survival of long-lived PCs in the BM [89]. Recent work showed that BCMA signalling is critical for the expression of the pro-survival molecule Mcl-1 in PCs [46]. BCMA may play a role in antigen presentation by B cells [90] because, in contrast to TACI and BAFF-R, only BCMA can activate the JNK pathway, which leads to the induction of the antigen (Ag) presentation response [90]. However, further studies revealed that BAFF-induced MHCII upregulation

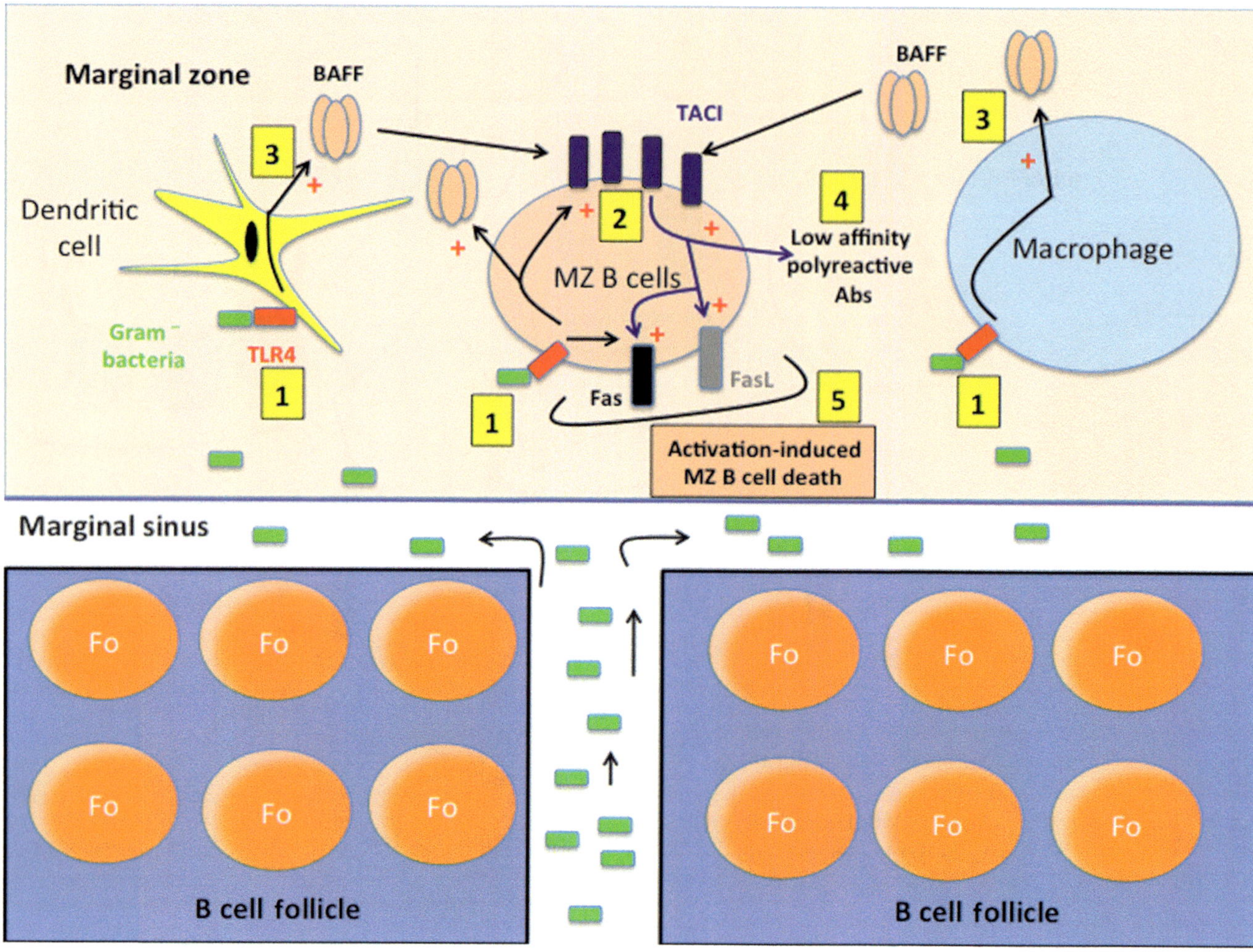

FIGURE 10.2 **Activation-induced B cell death controlled by TACI on marginal zone (MZ) B cells.** This figure depicts the anatomy of the mouse spleen with the outer part of the white pulp represented by the mouse splenic MZ (top in *pink*), the MZ sinus (middle T shape structure in *white*), and the B cell follicles (bottom with follicular B cell (Fo) in *orange*) in the inner part of the white pulp. In *green* are represented gram-negative bacteria entering the spleen via the MZ sinus. (1) LPS on the surface of gram-negative bacteria entering the MZ activate TLR4 on the surface of dendritic cells (DCs), MZ B cells, and MZ macrophage. (2) Activation of TLR4 in MZ B cells leads to the strong upregulation of TACI expression on the surface of these cells. (3) In response to TLR4 activation, DCs, MZ B cells, and macrophages produce BAFF, which binds to TACI on MZ B cells. (4) TACI activation is very important for the production of low-affinity polyreactive antibody in response to LPS activation of MZ B cells. (5) At the same time, LPS-activated MZ B cells upregulate Fas and Fas ligand (FasL) on their surface in a TLR4/TACI-dependent fashion, leading to activation-induced MZ B cell death. *BAFF*, B cell activating factor of tumour necrosis factor; *TACI*, transmembrane activator and cyclophilin ligand interactor; *TLR*, toll like receptor.

(an indicator of the Ag presentation response) was clearly dependent on BAFF-R and TACI [13].

A new receptor for BAFF, Nogo 66 receptor has been described, but it appears to be expressed in the CNS on neurites but not on B cells [91]. However, independent further confirmation of BAFF interaction with Nogo 66 receptor is needed.

1.3 The Role of the BAFF–APRIL System in B Cells in Health

1.3.1 B Cell Survival and Maturation

The primary function of BAFF is to ensure B cell survival during B cell maturation in the spleen [66,92,93]. B cells acquire BAFF-R expression at the immature stage and must undergo intermediate differentiation before reaching the mature B cells stage [15,67]. The intermediate differentiation stage that takes place in the spleen is called the immature transitional B cell stage, with several transitional stages described over the years from type 1, 2, and 3 [94,95]. Type 2 and 3 B cells are particularly dependent on BAFF signalling via BAFF-R for survival [2,66]. Mice lacking either BAFF or BAFF-R have no mature B cells apart from B-1 B cells and a small percentage of IgD^{hi}/IgM^{lo} B cells in the spleen [92]. Expression of Bcl2, a prosurvival oncogene, can rescue B cell maturation of all B cells in BAFF-deficient animals except MZ B cells, suggesting that the role of BAFF is not limited to the support of B cell survival but also plays a role in the differentiation of MZ B cells [96].

Activation of the survival receptor BAFF-R leads to the recruitment of TRAF2 and TRAF3, and ultimately NF-κB2 activation, and transcriptional activation of anti-apoptotic genes of the Bcl2 family [97,98]. BAFF also mediates B cell survival by blocking the BCR-induced upregulation of Bim (a proapoptotic molecule) and preventing the translocation of the proapoptotic protein kinase Cδ (PKCδ) to the nucleus [99,100]. Treatment of B cells with BAFF results in increases in cell size and metabolic activity of the B cell as well as upregulation of Akt [101]. Upstream of Akt, BAFF-mediated B cell survival requires the catalytic subunit of PI3K, p110δ [102]. More recent results suggest that BCR and BAFF signalling converge high on the signalling pathway, at the level of the Src-family tyrosine kinase [103,104]. Together, this work strongly suggests that signalling via the BCR is critical for optimal BAFF survival signals, although the exact regulation of BAFF-R signals by BCR signals still remains sketchy [105].

1.3.2 B Cell Activation

Initial work showed that BAFF also promotes the activation of B cells in culture [9]. However, how important this aspect of BAFF function is relative to other known B cell co-stimulators such as CD40L and IL-4 is not fully understood in vivo. B-1 B cells, as mentioned above, do not require BAFF for development or survival [92]; however, B-1 B cell numbers are increased in BAFF Tg mice, suggesting that the co-stimulatory effect of BAFF may have contributed to their expansion and proliferation post-activation [106]. The receptor TACI triggers IgG and IgA CSR [107,108]. It is interesting to note that the cytoplasmic portion of TACI interacts with the signalling adaptor myeloid differentiation primary response gene 88 [72]. Upon activation of TACI, the DNA-editing enzyme activation-induced cytidine deaminase (AID) is expressed, allowing CSR [72]. It is interesting to note that BAFF Tg mice lacking T cell help produce IgG isotypes typical of T-dependent B cell responses, suggesting that BAFF overexpression can act as a surrogate T cell-like co-stimulator of CSR via TACI [33]. BAFF Tg mice also produce high levels of IgA [33,109,110]; however, increased IgA levels in BAFF Tg mice are dependent on a T cell mechanism [33]. TACI also plays a role in controlling Blimp1 expression and as such contributes to generation of PCs [79]. BCMA plays a role at the end stage of B cell activation [111] and promotes the survival of a subset of PCs [89]. BCMA strongly promotes the expression of the pro-survival factor Mcl1 in the BM [46], which contributes to long-term survival of PCs in this compartment but not the spleen. APRIL expression in the BM is also critical for PC survival [5,41], especially in the murine setting, where BCMA exhibits poor affinity for BAFF [58].

1.4 The Role of the BAFF–APRIL System on B Cells in Disease

1.4.1 BAFF and B Cell Tolerance

The immune system generates lymphocytes daily [112,113]. Many newly generated lymphocytes are self-reactive and from the BM to the periphery, many immune checkpoints exist to eliminate or neutralise self-reactive and potentially harmful immune cells [114,115]. This process is called acquired immune tolerance and is driven by the recognition of self-antigen by self-reactive lymphocytes [114,116]. B cell immune tolerance takes place at various sites in the body, including the BM and the spleen [114–116]. Many different immune tolerance mechanisms have been described for B cells, from negative selection to receptor editing, which will allow the expression of an alternative BCR that is no longer self-reactive [114,115]. Another mechanism observed in the periphery is B cell anergy, in which self-reactive B cells become nonresponsive and fail to be activated upon BCR contact with self-antigens [114,115]. Exclusion of self-reactive B cells from important niches in the spleen, where survival factors are present, in particular when competing with non-self-reactive B cells, is another mechanism known to eliminate these potentially harmful B cells [117].

As mentioned above, BAFF is a B cell survival factor, and fine regulation of B cell survival is critical through the process of B cell maturation. Indeed, too much B cell survival can allow self-reactive B cells to survive and prevent negative selection [2]. Illustrating this point, BAFF Tg mice develop B cell-dependent autoimmune disorders [75,109]. How excessive BAFF breaks immune tolerance has been tested in vivo using an animal model in which all B cells recognise an experimental antigen, hen egg lysozyme (HEL), which has been intercrossed with mice expressing HEL as a transgene, hence generating progeny born with HEL as a self-antigen [118,119]. Experiments using this model have shown that self-reactive anti-HEL B cells have a greater dependence on BAFF for their survival [119]. It is interesting to note that HEL-specific B cells developing in an animal expressing HEL are generally more immature than HEL-specific B cells developing in the absence of the self-antigen [118]. Immature transitional B cells are known to be more dependent on BAFF for survival irrespective of their specificity [66,92]. The missing experiment would be the following: a transfer of transitional HEL-specific B cells (generated in wild-type (WT) or HEL-expressing animals) into HEL-expressing animals to check their BAFF requirement for survival. It is likely that in this case little difference will be seen.

Another study further clarified the role of BAFF in this model [118]. When HEL-specific B cells develop in a BAFF Tg mouse expressing HEL as a self-antigen, they expand in the periphery rather than being eliminated, when HEL is expressed as a soluble form (sHEL) [118]. However, when HEL is expressed as a membrane-bound form with strong affinity to the HEL-specific BCR, B cell negative selection in the BM occurs [120], and this is not affected by BAFF overexpression [118]. This result was expected because developing B cells in the BM do not express BAFF-R [27,64,65]. In contrast, the observed expansion of HEL-specific B cells in BAFF Tg mice expressing sHEL is an indication that excessive BAFF production corrupts peripheral B cell tolerance driven by sHEL [118]. However, the exceptionally high affinity of the HEL-specific BCR and the absence of competition with polyclonal non-self-reactive B cells are complicating factors in these experiments. To address these issues, mice were generated to express the HEL-specific heavy chain but not the light chain to generate high- and low-affinity HEL-specific B cells, and BM from these mice was used combined with WT BM to make chimaeric mice in which only 10% of the B cell repertoire was HEL-specific and the recipients expressed sHEL and BAFF [118]. In these chimaeric mice, high-affinity but not low-affinity HEL-specific B cells were eliminated despite the mice producing large amounts of BAFF; however, the number of low-affinity HEL-specific B cells, many of which exhibited an MZ phenotype, was greatly enhanced [118]. In the case of low-affinity HEL-specific B cells, unlike their high-affinity counterparts, these cells are allowed to mature to a stage when they express sufficient BAFF-R on the cell surface to be rescued by excessive BAFF production, survive, and expand [118]. In conclusion, a strong interaction between the HEL-specific BCR and HEL remains an effective way of negatively selecting self-reactive B cells that is not altered by excessive BAFF production. In contrast, excessive BAFF will alter the number of low-affinity self-reactive B cells in the periphery and increase their numbers.

As mentioned above, many of the low self-reactive B cells that are not eliminated because their affinity for the self-antigen is below the threshold that normally negatively selects self-reactive B cells populate the MZ B cell compartment [118]. MZ B cells are known to be innate-like B cells because they respond very efficiently to TLR activation, in particular TLR4 activation in response to LPS [121−123]. Moreover, MZ B cells are very good antigen-presenting cells to naïve B cells and as such form a bridge between the innate and the adaptive immune system [124]. However, considering that LPS can activate TLR4 on all B cells regardless of whether these cells are self-reactive or not, and MZ B cells are potentially self-reactive, there is a danger of generating activated autoreactive MZ B cells that can potentially access T cell help, increase their affinity in a GC, and eventually differentiate into antibody-forming cells producing high-affinity autoantibodies. However, most infections with gram-negative bacteria do not lead to an autoimmune disease, thereby raising the question as to how the activation of MZ B cells is regulated. Autoantibody production, such as anti-nuclear autoantibodies (ANAs), has been commonly observed in the clinic in the serum of individuals suffering from infections, but this production is usually transient and disappears when the infection is cleared [125]. In general, B cell response to type I antigens such as LPS is short-lived [126]. All of these observations point to a regulatory mechanism terminating the response of innate B cells such as MZ B cells to nonspecific activation with TLR ligands, but details of this mechanism are not fully understood.

Recent work has shed new light on one of such regulatory mechanism [76]. It is interesting to note that MZ B cells express much higher levels of TACI on the cell surface when compared with other B cell subsets in the spleen [33,69]. Moreover, exposure to LPS in vivo very strongly upregulates TACI on MZ B cells [76]. The reason for such a specific LPS-induced higher TACI expression on MZ B cells has remained unclear. It is interesting to note that LPS and BAFF cooperate to induce Fas expression on the surface of MZ B cells, a process dependent on TACI expression [76,82]. TACI expression on MZ B cells is essential for optimal induction of Fas and FasL expression on these cells [76]. Further work demonstrated that LPS leads to rapid activation of MZ B cells in vivo [123] but also induces MZ B cell apoptosis [76]. MZ B cell apoptosis was deficient in mice lacking TACI or the membrane form of FasL, which is the active form driving apoptosis via Fas [76].

TACI signalling has another key function contributing to the sensitivity of LPS-activated MZ B cells to apoptosis because it represses the expression of many inhibitors of the Fas signalling pathway, and in particular FLIP (p43) [76]. FLIP (p43) forms heterodimers with caspase-8, which interact directly or indirectly with other factors, such as TRAF-family adapters, rat insulin promoter (RIP) kinases, and possibly RAF-1 [127]. This results in the activation of NF-κB and certain other signalling pathways that promote cell survival and proliferation as well as in the suppression of RIPK3-mediated necroptosis [128,129]. Consistent with these findings, TACI-deficient MZ B cells but not WT MZ B cells are positive for the proliferation marker Ki67 [76]. Together, recent findings suggest that TACI signalling in MZ B cells serves two major purposes: (1) it upregulates Fas and FasL on the surface of these cells, and (2) it represses the expression of intracellular anti-apoptotic factors. TACI signalling sensitises MZ B cells to apoptosis. BAFF and APRIL can trigger this effect [76]. Moreover, TLR4 and the downstream signalling element Mal are very important for Fas and FasL upregulation on LPS-activated MZ B cells [76]. It is interesting to note that the MZ B cell compartment is enlarged in TLR4- and Mal-deficient animals [76], suggesting a role for these two molecules in regulating this particular B cell compartment. This effect is very specific for TLR4, TACI, and MZ B cells [76]. Other TLR can induce Fas on the surface of B cells, but this effect is TACI-independent [76]. B-1 B cells, which also express high levels of TACI, do not respond like MZ B cells to LPS in vivo. These differences suggest that possible BAFF-independent mechanisms exist to sensitise other B cells to apoptosis in response to innate activation.

TACI is the first receptor on B cells known to drive innate activation-induced B cell death, and this effect explains the short-lived nature of MZ B cell responses to type I independent antigens [126]. Upon contact with gram-negative bacteria entering the spleen via the marginal sinus, cells in the MZ such as macrophages, LPS-activated DC, and MZ B cells respond by producing BAFF and APRIL [2]. LPS also triggers TLR4 on the surface of MZ B cells, which in response strongly upregulates TACI on the cell surface [76]. LPS activation of MZ B cells leads to rapid and effective production of low-affinity nonspecific polyreactive antibodies that recognise conserved epitopes on bacteria [123,130]. These antibodies can opsonise bacteria and act as a first line of defence while the adaptive immune system prepares itself for an antigen-specific response. This first wave of low-affinity and nonspecific antibody production by LPS-activated MZ B cells is TACI dependent because these responses are deficient in TACI-deficient animals [71]. However, as soon as MZ B cells have been activated by LPS, these cells upregulate Fas and FasL on the cell surface and downregulate the expression of anti-Fas signalling factors such as XIAP and cFLIP, allowing these cells to die before they can migrate to the T cell zone, gain T cell help, and increase their affinity in a GC, a process that would not require TACI and is normal in TACI-deficient animals [71,131]. In conclusion, TACI plays a role in regulating and terminating activation of MZ B cells to innate activation.

The question is whether or not TACI is important for B cell tolerance as a whole. A recent study with common variable immune deficiency (CVID) patients carrying a mutation in the *TACI* gene confirmed that indeed B cell tolerance in these patients is defective [132], confirming in vivo the important role of TACI in not just regulating B cell responses to innate activation but for the maintenance of a healthy B cell repertoire in humans.

1.4.2 BAFF and B Cell Autoimmunity

In BAFF Tg mice, excessive BAFF production leads to the expansion of the B cell compartment and in particular MZ B cells [75,109,133−135]. In addition, BAFF Tg mice produce a wide range of autoantibodies, from rheumatoid factors (RFs), anti-double stranded and anti-single stranded DNA autoantibodies to ANAs, and anti-histone autoantibodies [75,109,133−135]. As BAFF Tg mice age, they develop autoimmune symptoms such as nephritis and sialadenitis, reminiscent of human SLE and Sjögren's syndrome (SS), respectively [75,109,133−135]. Mechanisms leading to tissue inflammation and destruction in BAFF Tg mice have been extensively studied [33,75,109,133−135]. Dysregulation of T and B lymphocyte functions contributes to development of autoimmune disorders such as SLE [136]. BAFF Tg mice have a greater proportion of activated T cells and spontaneously develop large GCs, which are T-dependent immune structures designed to select B cells with an improved affinity for their specific antigen or in this case potential autoantigens [33,109]. Moreover, BAFF is produced by activated T cells, which also upregulate BAFF-R after activation [27,56]. BAFF binds to BAFF-R on T cells, which induces the expression of prosurvival factors and the production of Th1 cytokines such as IFN-γ [27,56]. Th1 cells, a subset of IFN-γ-producing effector T cells, play a pathogenic role in autoimmunity, and, in BAFF Tg mice, it appeared that both of the activated T and B cell compartments are dysregulated, suggesting a strong role for T and B cell cooperation, leading to strong GC-dependent autoantibody production as the main underlying cause for autoimmune disorders. This model fitted the views that T cells provide help for B cells and B cells can serve as antigen-presenting cells to T cells, perpetuating a vicious cycle leading to inflammation and tissue destruction [136]. However, this model was challenged by the observation that BAFF Tg mice lacking GC develop autoimmune disorders undistinguishable to that of BAFF Tg mice developing GC spontaneously [106,135], ruling out a role for B cell affinity maturation as an important pathogenic event. Moreover, although the number of activated T cells is elevated in BAFF Tg mice [109], so is the number of Foxp3$^+$ regulatory T cells [33], questioning the exact role of T cells in driving disease in the BAFF Tg mouse model of SLE. It is surprizing to note that T cell-deficient BAFF Tg mice also develop an autoimmune disease undistinguishable from that of T cell-sufficient BAFF Tg mice [33], suggesting that T cells are dispensable for the autoimmune disorders developing in BAFF Tg mice. Disease in T cell-deficient BAFF Tg mice was characterised by the deposition of IgG1, 2b, and 2c autoantibodies in the kidneys of these animals and trapping of C3 complement, which triggers tissue inflammation and destruction [33]. IgG1, 2b, and 2c isotype production in the absence of T cells is unusual because normally isotype switching to these isotypes requires T cell help. As mentioned above, BAFF and APRIL are produced by activated T cells [56], and signalling via TACI is important for isotype class switching [72]. It is possible that excessive BAFF production in BAFF Tg mice has mimicked an overproduction of BAFF by T cells and led to switching to and production of IgG1, 2b, and 2c autoantibodies.

It is interesting to note that IgA levels are elevated in BAFF Tg mice [109,110,137], and this aspect is T cell-dependent because IgA levels become normalised in T cell-deficient BAFF Tg mice [33]. However, in T cell-sufficient BAFF Tg mice, elevated IgA levels play a significant role in the kidney pathology developing in BAFF Tg mice [110]. Indeed, in IgA-deficient BAFF Tg mice, the renal pathology was reduced [110]. In addition, commensal flora-reactive IgA were abnormally detected in the blood of BAFF Tg mice, suggesting an imbalance with the microflora and a defect of the normal mucosal-peripheral compartmentalisation in these animals [110]. IgA nephropathy in humans is a male-dominant disease and correlates with higher APRIL serum levels [110]. It is interesting to note that IgA-associated nephropathy in BAFF Tg mice is also more severe in male mice when compared with females [110], suggesting a possible gender-biased development of fatal nephropathy in BAFF Tg mice likely to be associated with different BAFF transgene expression in males and female mice. In conclusion, disease in BAFF Tg mice occurs in the absence of T cells; however, the nature of the renal disease is different than that of T cell-sufficient BAFF Tg mice. These findings shed some light on the contribution of T cells in kidney pathologies.

1.4.3 The Role of the BAFF/APRIL System in B Cell Immunodeficiencies

CVID is one of the most common immunodeficiency syndromes and is characterised by a reduction in antibody production, in particular IgG, IgA, and IgM, as well as poor responses to vaccinations and vulnerability to infections, in particular those of the respiratory track (reviewed in Ref. [138]). CVID is also characterised by autoimmune manifestations in approximately 20% of affected patients [138]. The aetiology of CVID remains unclear, but it appears to be linked to several genetic defects affecting immune functions, although the inheritance pattern of these genetic modifications does not always fit predictions [139]. Mutations or deletions of several immune genes affecting B cell function have been associated with CVID, such as the inducible co-stimulator (ICOS), CD19, CD81, and CD20 [139].

A recent advance in our understanding of CVID emerged with the discovery of a series of mutations in the *TACI* gene that were found in approximately 8%–10% of CVID patients [140–144]. Homozygosity for variants C104R, A181E, and S144X of the *TACI* gene was observed in several CVID patients [140,141]. Heterozygosity for variants C104R, A181E, S194X, and R202H has also been reported in CVID patients [140,141]. The most common variants are C104R affecting the extracellular domain and A181E located in the transmembrane domain, and for these, heterozygosity is more common than homozygosity [140,141]. One report described experiments establishing how the C104R mutation impairs TACI function [145]. It appears that this mutated form of TACI interferes with TACI signalling by associating with WT TACI in the absence of any ligand, and as such, forming signalling-defective oligomeric receptor complexes [145]. Later work looking at the equivalent TACI mutation in mice (C76R) demonstrated that B cell function was impaired in mice heterozygous for the mutation, suggesting that this mutation impairs TACI function in heterozygotes via haploinsufficiency [146].

Mice lacking TACI develop defects resembling that of CVID patients such as reduced IgA levels and impaired responses to T-independent antigens [71,131], a problem which appears to be linked in part to the role of TACI in maintaining Blimp-1 expression, essential for PC generation and persistence [79]. However, relatives of CVID patients can often carry the same TACI mutations without any immunological abnormalities [147,148]. It is interesting to note that both B cells from CVID patients and healthy relatives are defective in vitro, in particular in response to TLR9 activation and TACI stimulation [147]. In contrast, in vitro IgG and IgA production from healthy relative-derived B cells was normal when compared with B cells from CVID relatives [147]. Therefore, mutation in the *TACI* gene is not the sole contributing factor, but it may facilitate emergence of CVID in combination with other contributing factors.

BAFF-R is essential for B cell survival [67]; therefore, a dominant mutation in the *BAFF-R* gene is likely to lead to immunodeficiency. Indeed, BAFF-R variants, present at the heterozygous state, have been identified in CVID patients with low B cell numbers [149]. These variants appeared to be polymorphic and do not affect BAFF-R expression [149]. It is unclear whether these variants can signal [149]. Another report described siblings carrying a homozygous *BAFF-R* gene deletion preventing BAFF-R protein expression [150]. Although only one sibling had recurrent infections, the other remained relatively asymptomatic. Both siblings had reduced B cell numbers and reduced IgM and IgG levels, but IgA levels remained within a normal range [150]. TACI (which was not mutated in these patients) is critical for IgA production in mice [71,131], which may explain this difference. Therefore, similar to individuals treated with the B cell-depleting agent belimumab (anti-BAFF monoclonal antibody, also known as Benlystat), the incidence of infection in BAFF-R-null patients may vary from one individual to another depending on additional factors and/or exposure to pathogens [151]. For this reason, although defects in BAFF-R or TACI function will certainly increase an individual's vulnerability to immunodeficiency, additional factors are required to precipitate clear CVID clinical manifestations.

1.4.4 The Role of the BAFF/APRIL System in B Cell Malignancies

The classical picture of the BAFF system involved in B cell survival is slowly evolving into a more complex picture of interwoven regulatory function and downstream signalling designed to alter cellular survival, growth, and migration. Although the central role of BAFF in autoimmunity has now been well established, work is still ongoing trying to understand the contribution of the BAFF system to malignancies. The current conceptual picture suggests that BAFF and APRIL may be pathogenic players in microenvironments of haematological tumours of B cell origin such as chronic lymphocytic leukemia (CLL).

Indeed, there is strong evidence that the BAFF/APRIL system plays an important role in the progression of CLL, particularly promoting paracrine CLL survival because these cells produce BAFF and APRIL and express BAFF and APRIL receptors [152—154]. Other cells described in CLL, such as nurse-like cells, also produce BAFF and can contribute to CLL progression [155]. It is interesting to note that, unlike normal transitional B cells, BAFF and APRIL support the survival of CLL cells via activation of the classical but not the alternate NF-κB pathway, suggesting a different signalling pathway allowing persistence of CLL cells.

Overexpression of BAFF has been shown to enhance c-myc expression and promote CLL-like disease via activation of the canonical IKK in Myc/BAFF Tg mice [156]. BAFF production by stromal support cells, such as microvascular endothelial cells, has been shown to be triggered by CD40L aberrantly expressed on the surface of CLL B cells, further supporting the importance of support stroma for cancer cell survival and explaining observed ex vivo apoptosis [157]. BAFF Tg mice lacking TNF also develop B cell lymphoma and MALT lymphomas [106]. B lymphoma development has also been observed in TACI-deficient mice [73]. Moreover, APRIL transgenic mice develop a disease similar to CLL [158], and anti-APRIL blocking antibodies were effective at stopping the CLL-like pathology in these mice [159]. A picture is emerging in which too much BAFF or APRIL can allow inadequate survival of neoplastic B cells. This may occur via two mechanisms: one involving excessive survival via BAFF-R upon BAFF binding or defective TACI-mediated B cell regulation via B cell death.

Given these observations, measurement of serum BAFF levels were compared in relation to existing scoring systems of CLL [160]. Prognostic use of BAFF levels has already shown promise in CLL when combined with CD38, ZAP70 expression, and the mutational status [161] and in follicular lymphoma, in which expression of BAFF and BAFF-R are elevated and coincide with inferior progression-free survival, with some suggestion that overexpression of BAFF is secondary to elevated expression of BAFF-R and therefore increased sensitivity to BAFF [162].

It is interesting to note that soluble BCMA isoforms have been identified, and although their function is not yet elucidated, it suggests the exciting prospect of natural decoys produced to regulate levels of free BAFF or APRIL [7]. Most interestingly, overexpression of APRIL was found to coincide with overexpression of HSPGs [163], which are known co-receptors for APRIL when binding TACI. This has renewed speculation that HSPGs may act as primary receptors for APRIL in addition to co-receptor function. BAFF-R has been implicated in nuclear interactions with IKKβ and NF-κB/c-Rel in healthy and neoplastic B cells, including non-Hodgkin's lymphoma (NHL) B cells, which enhance histone H3 phosphorylation and binding of NF-κB on promoters for inducible genes, including Bcl-xl, CD40L, and most strikingly BAFF [164]. This complicates further the mechanism regulating the BAFF system, in which the receptor itself is a transcriptional regulator and exerts its effects via chromatin remodelling in addition to classical NF-κB activation at the membrane surface [164].

Mutations in BAFF-Rs have also emerged as critical factors in disease etiopathology and have emerging prognostic power. Substitution mutations in the *BAFF-R* gene at His159Tyr in patients with NHL increased recruitment of TRAF3, TRAF2, and, surprisingly, TRAF6, demonstrating for the first time that TRAF6 is an essential component of downstream BAFF-R signalling in NHL and in health [165].

Inhibition of BAFF or its receptors has been a strong focal point for therapeutic development. Merck's Phase Ib trial with the atacicept inhibitor of BAFF and APRIL has shown that the agent is tolerated at moderate dosage, but currently no data on the clinical activity in CLL trials are available [166]. A strategy using mutant BAFF fused with Pin2/TRF1-interacting protein X1, a telomerase reverse transcriptase inhibitor, has also been effective at selectively killing BAFF-R expressing Burkitt lymphoma lines [167]. The latter example further reinforces growing evidence that BAFF function is not confined to NF-κB activation and prosurvival transcriptional activity, but that BAFF also directs alteration at immunosenescence checkpoints of target cells at the chromatin level to alter proliferative capacity and ultimately survival. These developments will preface development of a new generation of cancer therapeutics.

Over a decade of research on the BAFF and APRIL system has clearly demonstrated how central this system is in B cell biology in health and disease states. This work led to the development of the second most important biologic treatment targeting B cells after rituximab. However, several questions remain, in particular differences in the function of these ligands and their receptors in mouse versus humans and how this may affect the development of targeting strategies in this system.

2. THE LYMPHOTOXIN PATHWAY: SHAPING B CELL ENVIRONMENTS

2.1 Introduction

Although BAFF and APRIL predominantly exert their functions by acting directly on B cells, other TNF superfamily members, in particular the lymphotoxin (LT) pathway, play a crucial role in shaping environments in which B cells are activated. Furthermore, expression of the LT-$\alpha\beta$ ligand by B cells plays an important role in the homoeostasis of several immune accessory cells in the steady state and during infection/inflammation. This section focuses on the contribution of the LT pathway to B cell biology.

2.2 Receptors and Ligands of the LT Pathway

LT-α was discovered as a soluble cytotoxic factor in 1968, approximately 7 years before the subsequent discovery of TNFα [168,169]. Homotrimers of LTα were found to induce cytotoxic activity, and these soluble homotrimers were later shown to bind to TNFRs I and II. Because LTα_3 and TNFα shared common receptors, for quite some time it was assumed that the biologies of these two cytokine pathways were mutually redundant. Indeed, after the discovery of TNFα, LTα was (temporarily) renamed TNFβ [170].

However, two major discoveries revealed that the LT pathway was a unique axis that governs biological processes distinct from TNFα/TNFR signalling. First, the discovery that LTα could form heteromers with another TNF superfamily molecule called LTβ to form a heterodimer (LT$\alpha_1\beta_2$) revealed greater complexity in this system [171]. Secondly, the discovery of the LTβ-receptor (LTβR) that could propagate signals, in particular the activation of the NF-κB pathway, revealed a TNFR-independent counter-receptor for LT$\alpha_1\beta_2$ [172]. In fact, LT$\alpha_1\beta_2$ does not bind to TNFR1/TNFRII.

Further cementing the concept that LT$\alpha_1\beta_2$/LTβR signalling provides unique biological function, genetic ablation of LTα, LTβ, and LTβR, but not TNFα, results in the absence of most lymph nodes (LNs). In addition to the LT$\alpha_1\beta_2$ heterotrimer, a second ligand for the LTβR was discovered called LIGHT (which stands for homologous to LTs, exhibits Inducible expression, and competes with HSV gD for Herpesvirus entry mediator, a receptor expressed by T lymphocytes) [173]. This ligand can bind to another receptor called HVEM, which itself can bind to other molecules. As can be imagined, one of the challenges in studying the LT pathway in the context of B cells has been to isolate the role of particular ligands in the context of specific cell types. This challenge can be addressed using chimaeric mice, tissue-specific deletion of the LTβR and its ligands, and adoptive transfer approaches.

2.2.1 Expression Pattern of LT Molecules

The LTβ moiety of the LT$\alpha_1\beta_2$ heterotrimer effectively anchors the LT$\alpha_1\beta_2$ complex to the cell surface. This provides a means for direct cell:cell communication between LT$\alpha_1\beta_2$-expressing cells and LTβR expressing cells [174]. In general, LT$\alpha_1\beta_2$ is not secreted, although it can be cleaved by matrix metalloproteinases in pathological environments such as in the RA synovium [175]. The expression of LT$\alpha_1\beta_2$ is largely limited to lymphocytes and NK cells. It is interesting to note that although expression of LT$\alpha_1\beta_2$ on T cells requires activation of the T cell (i.e., through the T cell receptor) [176], expression of LT$\alpha_1\beta_2$ on B cells is constitutive, albeit expressed at very low levels [177]. LT$\alpha_1\beta_2$ is further upregulated on B cells that have been activated by antigen and express GC B cell markers such as GL7 and FAS [178]. Some reports have also demonstrated the expression of LT$\alpha_1\beta_2$ on DCs [179,180]. LIGHT is also expressed on activated T cells as well as B cells, NK cells, platelets, and DC [181].

With respect to the LTβR, the expression pattern is quite different, being expressed primarily on "accessory" cells of the immune system including stromal cells within lymphoid tissues, macrophages, DCs, and FDCs. Given that the LTβR ligands are membrane-anchored and are upregulated on activated lymphocytes, this suggests dynamic crosstalk between LTβR-expressing accessory cells and lymphocytes recruited into immune responses. Indeed, the induced expression of LT$\alpha_1\beta_2$ on antigen-specific T cells is required for DC activation and cross-priming of CD8$^+$ T cell responses [182]. Furthermore, expression LT$\alpha_1\beta_2$ on innate lymphoid cells (ILCs) in the gut performs a similar crosstalk function by activating mucosal DCs to mediate host defence against intestinal pathogens [183,184]. Although expression of LT$\alpha_1\beta_2$ on T cells is important for optimising DC activation and cytokine secretion, as will be explained later, the role of LT$\alpha_1\beta_2$ on B cells in the immune response is less direct.

2.2.2 Genetic Targeting of the TNF/LT Locus

The *tnf*, *LTa*, and *LTb* genes are closely linked on mouse chromosome 17 (chromosome 6 in humans). This has caused difficulties in interpreting data derived from some of the original strains of TNF$^{-/-}$ mice, which in some cases had a concomitant disruption of *LTa*. Indeed, TNF null mice constructed using Cre-loxP technology have a very different Peyer's patch (PP) phenotype compared with conventionally generated $TNF^{-/-}$ mice [185]. Moreover, the *tnf*, *LTa*, and *LTb* genes are located within the H-2 locus. Thus, knockout mice generated using 129 ES cells will have a 129 MHC, potentially resulting in altered immune responses and incompatibilities due to minor mismatches. Such issues must be kept in mind when interpreting experiments that have used adoptive transfer or BM chimera models because the knockout cells could potentially be rejected by host immune responses.

Despite these caveats, consistent phenotypes have emerged in LT-deficient mice with respect to lymphoid tissue homoeostasis. As previously mentioned, deletion of *LTa* [186], *LTb* [187], or *LTbR* [188] genes results in the abnormal development of PPs and most LNs. Although LTβR and LTα-deficient mice lack all LNs, $LT\beta^{-/-}$ mice retain mesenteric LNs, which vary in size depending on the colony. Residual cervical LNs can also be observed in $LT\beta^{-/-}$ mice. The residual LNs in $LT\beta^{-/-}$ mice may be due to the compensatory triggering of LTβR signalling by LIGHT because mice that lack LTβ and LIGHT do not have mesenteric LNs [189]. However, LIGHT is not necessary for LN development because $LIGHT^{-/-}$ mice have a normal complement of LNs and PPs.

In addition to abnormal LN and PP development, several abnormalities in the finer structure of the spleen and LN are apparent. In the spleen, the MZ is located at the interface of the red pulp and the lymphocyte-rich white pulp. The MZ contains specialised MZ B cells as well as $CD169^+$ MZ metallophilic macrophages. These populations, as well as the MZ sinus itself, require LTβR signalling, and the MZ effectively collapses in LT-deficient mice or mice treated with lymphotoxin inhibitors [188,190]. In addition to effects on the MZ, stromal cells that support lymphocytes in the spleen are dysregulated in LT-deficient mice. In particular, a subset of stromal cells called FDCs, which are located in the B cell follicle, are not detectable based on their expression of markers such as CD35 and FDC-M1/MFG-E8. Corresponding to the reduction in FDC, the normally crisp separation between B cells in the follicle and T cells in the periarteriol lymphoid sheath in the spleen is blurred in LT-deficient mice [188]. This is likely due to a strong reduction in chemokine expression within the spleen, in particular CXCL13, CCL19, and CCL21, which are produced by LT-sensitive stromal cells [191]. Reduction in these chemokines results in suboptimal segregation of T cells and B cells into their physiological splenic niches.

In LT-deficient mice, the architecture and cellular composition of the LNs is also compromised. For example, the subcapsular sinus, which envelopes the LN, also hosts a population of $CD169^+$ macrophages that are absent in LT-deficient mice [192]. However, the segregation of B cells and T cells within the LN appears to be relatively preserved [193]. Within the T cell zone, the high-endothelial venules (HEVs), which are the portals of entry for naïve lymphocytes, do not appropriately display the adhesion molecule PNAd on the HEV lumen; thus, LNs in LT-deficient mice are hypocellular [194]. As in the spleen, FDCs are also absent. It is interesting to note that conditional ablation of the LTβR in another stromal cell subset called fibroblastic reticular cells (FRCs) results in some functional abnormalities of these cells, such as poor chemokine/cytokine production, although many physical attributes of FRCs, such as the ability to produce elaborate extracellular matrix networks, is intact [195].

Lastly, in LT-deficient mice, the gut-associated lymphoid tissue (GALT) organization is also affected. As mentioned, PPs are absent and mesenteric LNs are absent in LTα- and LTβR-deficient mice. Discrete follicles of lymphocytes, called isolated lymphoid follicles (ILFs), which depend on the host microbiota, are located in the lamina propria of the gut. ILFs host dense collections of B cells and ILCs. ILFs are absent in LT-deficient settings [196]. Because the ILF are more of an "inducible" nature, this finding has implications in considering ectopic "tertiary" follicles that can appear in the context of inflammation, which will be discussed later.

2.3 The Role of B Cell-Associated $LT\alpha_1\beta_2$ in Immune Cell Homoeostasis

2.3.1 *Expression of $LT\alpha_1\beta_2$ on B Cells and Maintenance of Lymphoid Tissue Accessory Cells*

LTβR signalling is well recognized for its important role in the maintenance of various stromal cell types. In the developing embryo, LTβR signalling on lymphoid tissue organising (LTo) cells is critical for the development of an LN anlage. In the case of embryonic development, the source of $LT\alpha_1\beta_2$ is derived from lymphoid tissue inducer (LTi) cells, and iterative interactions between LTo and LTi cells are required to induce LT-dependent chemokines that allow for the population of developing LN [197]. This same LT-dependent paradigm that plays out during development is recapitulated in the adult, with B cells being an important reservoir of $LT\alpha_1\beta_2$ for the constitutive LTβR signals that must take place in stromal cells for their maintenance. The most notable stromal cells that require constitutive LTβR signalling to be maintained in a mature state are FDCs [198], and the relevance of expression of $LT\alpha_1\beta_2$ on B cells for FDC differentiation will be discussed in the next section. Other stromal cells play a role in the peripheral immune response with variable dependency on the expression of $LT\alpha_1\beta_2$ on B cells. Notably, in responses against mouse cytomegalovirus, B cell-derived $LT\alpha_1\beta_2$ is critical for an early wave of type I IFN that is produced by stromal cells [199]. This early type I IFN response, which is not dependent on TLR signalling, is crucial for the early resolution of the virus. The nature of the stroma that produces this early wave of type I IFN has not been resolved. In addition, FRCs express chemokines and other cytokines that augment immunity. FRCs do not appear to be dependent on $LT\alpha_1\beta_2$ derived from B cells in the spleen, and in the LN, FRCs do not require LTβR signalling for the production of extracellular matrix. However, their ability to enhance an immune response through the production of IL-7 is indeed LT-dependent [195]. The cellular source of $LT\alpha_1\beta_2$ for FRC function has not been determined.

As mentioned, the integrity of the MZ and the subcapsular sinus of the spleen depends on LTβR signalling. It is unclear how LTβR signalling maintains the microarchitecture of the splenic MZ, which contains marginal reticular cells that are less well-characterised than the other stromal cell types [200]. Looking more closely at the MZ, the $MAdCAM^+$ $Sialodhesin^+$ sinus is ablated in the absence of LTβR signalling, and deletion of $LT\alpha_1\beta_2$ only on B cells results in a partial effect on the sinus [201], suggesting that $LT\alpha_1\beta_2$ on B cells plays a role in maintaining the MZ sinus, but other cell types expressing $LT\alpha_1\beta_2$ may partially compensate.

Particular macrophages within the splenic MZ and the subcapsular sinus are important for antigen capture because these zones are the first point of entry for antigen into the spleen and LN, respectively. $CD169^+$ macrophages require LTβR signalling for their persistence. These macrophages are largely, but not completely, absent in mice that lack $LT\alpha_1\beta_2$ only on B cells [201], suggesting that expression of $LT\alpha_1\beta_2$ on B cells plays an important role in maintaining these accessory cells. Given that the MZ is host to a large reservoir of MZ B cells, and the MZ and subcapsular sinus are adjacent to the B cell-rich follicle, it makes sense that B cells would provide an important source of $LT\alpha_1\beta_2$ for the maintenance of macrophages in these niches. Macrophage intrinsic LTβR signalling and LTβR signalling in radio-resistant cells appears to be required for macrophage maintenance because BM chimeras in which LTβR is deleted only in hematopoietic cells exhibit a partial reduction in $CD169^+$ macrophages whereas complete LTβR signalling blockade results in a more dramatic effect [202].

In addition to macrophages, DCs also rely on LTβR signalling; however, unlike macrophages, in which LTβR signalling must occur in macrophages as well as within the radio-resistant compartment, DC-intrinsic LTβR signalling is sufficient to maintain their homoeostasis by promoting DC proliferation [203,204]. The source of $LT\alpha_1\beta_2$ required to maintain splenic $CD11b^+$ DC is derived from B cells, suggesting an important cross-talk between these cell types [204]. Subsequent work has shown that LT-sensitive DCs have a particular phenotype in the spleen ($CD11b^+ESAM^{hi}$) and in the gut ($CD11b^+CD103^+$) [205]. The interaction between cells expressing $LT\alpha_1\beta_2$ and DCs in the mucosa will be further discussed in Section 2.5.

Collectively, these examples reveal that the expression of $LT\alpha_1\beta_2$ on B cells is important for the homoeostatic maintenance of various "accessory" cell types such as macrophages, DCs, and FDCs. The signalling mechanism by which LTβR maintains these cell types remains unclear.

2.4 The Role of B Cell-Associated $LT\alpha_1\beta_2$ and the GC Reaction

GCs are dynamic structures that facilitate the production of high-affinity antibodies and, as described elsewhere in this chapter, can be the source of autoreactive B cells in the context of excessive levels of BAFF. As mentioned, the expression of $LT\alpha_1\beta_2$ on B cells is necessary for the presence of FDCs in the LNs and the spleen. In the case of GCs, a very high level of $LT\alpha_1\beta_2$ is expressed on GC B cells, presumably to further aggregate FDCs into a tight GC structure that facilitates Ag trapping and presentation to GC B cells during the GC reaction [177,178]. This extra high level of $LT\alpha_1\beta_2$ on GC B cells is mediated via CD40 signalling in B cells, another TNF family member that is essential for B cell class switching to IgG and for T-dependent GC reactions. The source of CD40-ligand likely comes from T_{FH} cells [206]. ICOS is also involved in the GC response, and indeed LT-deficient and ICOS-deficient mice phenocopy in terms of their GC defects [178]. Thus, a multistep pathway involving ICOS and CD40L expression on T cells and subsequent T:B communication between T_{FH} and GC B cells inducing the expression of very high levels of $LT\alpha_1\beta_2$ is required to optimise the establishment of a GC niche.

Nevertheless, despite the induced expression of high levels of $LT\alpha_1\beta_2$ on GC B cells, the question of whether this is actually necessary for a GC response has been debated. Disruption of LTβR signalling certainly has profound effects on GC responses in the spleen. Significant evidence demonstrates the elimination of LTβR signalling and, as a consequence, mature $CD35^+FDC\text{-}M1^+$ FDCs result in a very poor GC response provoked by systemic antigens such as sheep red blood cells administered via the intraperitoneal route [190,207]. Furthermore, defective GC reactions in response to protein antigens such as the hapten nitrophenyl have been noted in LT-deficient mice, in which clusters of activated B cells expressing markers such as GL7, FAS, and peanut agglutinin in the spleen are not observed. Defects in GC responses in these mice also correspond with poor humoral immunity, with lower amounts of antibody (Ab) produced, and lower quality of

Ab due to poor affinity maturation, particularly at low doses of Ag [178,186].

On the other hand, LTβ deficient mice exhibit reasonable GC in the residual mesenteric LNs [208]. This begs the question of what is the role of LTβR signalling in the GC response in superficial peripheral LNs. To address this question, BM chimaeric mice have been examined in the context of a local subcutaneous administration of a model antigen phycoerytherin (PE). PE responses in the local draining LNs of LTβ chimaeric mice are surprisingly normal, with GC B cells upregulating AID and clustering in what appear to be GC niches within B cell follicles. Furthermore, sorting on PE-specific, AID-expressing GC B cells from either WT or LT-deficient LNs revealed that somatic hypermutation occurs normally, at least in the first 9 days of the immune response. However, by day 12 of the immune response, clusters of AID-expressing PE-specific GC B cells are no longer apparent in the LNs of LTβ chimaeric mice compared with WT chimaeric mice. The abrupt collapse of the GC is followed by a diminution in the affinity of the antibody response at late time points. Thus, it appears that the LT pathway, and by extension mature FDCs, are not required for the initiation of a GC response. Presumably other scaffold cells such as FRCs may substitute for the lack of FDCs and, indeed, FRCs are readily observed within FDC-less GCs in LTβ chimaeric mice [202]. Although FRCs do not express the receptors that would present Ag−Ab complexes (complement receptor CD35, Fc receptors), noncognate B cells could presumably step in and provide this function through their own Fc and complement receptors (Figure 10.3). However, as the GC response continues, presumably these non-FDC "replacement" elements become insufficient to continue to support an ongoing GC response. This is consistent with a study that showed that total ablation of FDCs using diptheria toxin at the peak of the immune response results in an abrupt collapse of the GC in a draining LN [209].

In FDC-less mice, it is unclear from where the source of CXCL13 required to draw B cells in the GC is derived. As mentioned, FDCs express the LTβR, and signalling through this receptor is required for stromal cells that secrete optimal levels of CXCL13, at least in the spleen. In that absence of LTβR signalling, FDCs that express CD35 and FDC-M1 are notably absent, and CXCL13 levels in the spleen are strongly reduced. However, despite the loss of LTβR signalling, B cell localization to splenic follicles is not completely disorganized, and in the LN it is actually quite normal. Therefore, there are likely other compensatory migratory cues and/or there are LT-independent alternative mechanisms for the induction of CXCL13, particularly in the LNs.

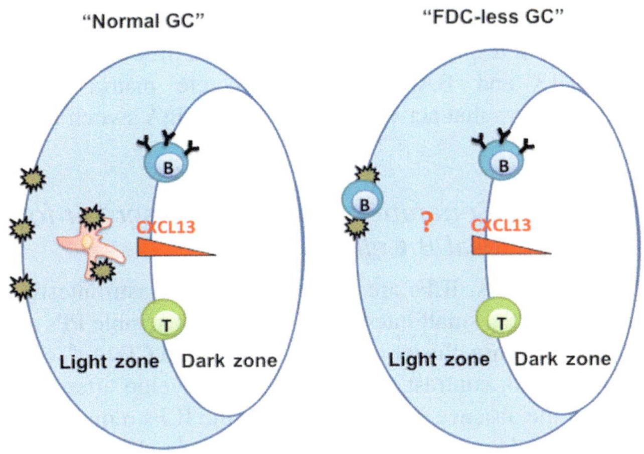

FIGURE 10.3 The key ingredients for a GC: B cells, T cells, and antigen (Ag). Following exposure to pathogen, Ag is transported from the SCS by SCS macrophages and noncognate B cells into the GC. Under normal circumstances, FDCs produce CXCL13 in the light zone of the GC and can display Ag on the surface of FDC extensions, thus providing a reservoir of Ag for the purposes of an affinity maturation template. T cells provide necessary help in the form of cytokines and CD40L. However, in an LN where LT βR signalling has been ablated, mature FDCs are not observed, but GC clusters appear. We hypothesise that an alternative reticular cell type, such as an FRC, may permit the display of CXCL13, but cannot retain Ag. On the other hand, noncognate B cells may provide a means for displaying Ag, albeit for limited periods of time. These minimal elements are sufficient for promoting a short-term GC response as well as somatic hypermutation of the BCR. *BCR*, B cell receptor; *FDC*, follicular dendritic cells; *FRC*, fibroblastic reticular cells; *GC*, germinal centre; *LN*, lymph nodes.

2.5 LT and Mucosal B Cell Biology

2.5.1 The Expression of LTαβ on Mucosal B Cells and PP Architecture

PPs are an important part of the mucosal immune system in which most priming of B cells occurs. In the absence of LTβR signalling, PPs do not develop, and if the LT pathway is blocked in the adult animal with LTβR-Ig, PPs tend to shrink in size [210]. It was later shown that $LT\alpha_1\beta_2$ expressed by B cells is required for the development of lymphoid follicles and optimal production of CXCL13 within follicle regions in PPs [211]. Because PPs are a rich source of B cells, this is a possible explanation for their smaller size in adult WT mice in which the LT pathway is blocked pharmacologically. It is interesting to note that the follicle-associated epithelium is intact in LT-deficient mice [211], suggesting that other pathways are important for maintaining this barrier (which includes antigen-transporting M cells). RANK, another TNF family member, is a likely candidate for this function [212]. Not unlike the LNs and spleen, the detection of mature FDCs in PPs is dependent on the expression of $LT\alpha_1\beta_2$ on B cells. FDCs in the PPs may play a particular role in shaping IgA responses because, in response to bacterial antigens, they can produce CXCL13 and BAFF and can secrete matrix metalloproteinases that activate TGFβ, a major IgA switch factor [213].

2.5.2 The Generation of ILFs—A Reservoir for Mucosal B Cells

Similar to PPs, ILFs are also found in the gastrointestinal mucosa of the small intestine. These ILFs resemble PPs and contain a dense cluster of $B220^+$ B cells and $CD11c^+$ cells. However, in contrast to PPs, ILFs can develop after birth, and in the absence of microbial Ags, the ILFs remain quite small. In the presence of luminal bacteria, ILFs become more organised and larger, taking on a more "mature" morphology. Expression of $LT\alpha_1\beta_2$ on B cells is sufficient to induce ILFs in LT-deficient mice [196]. However, in an LT-competent animal, ILCs are an important source of $LT\alpha_1\beta_2$ for inducing and maintaining ILFs. Thus, it is likely that B cells and ILCs may provide overlapping sources of $LT\alpha_1\beta_2$ for the generation of ILFs. Functions for ILFs and the colonic version of ILFs (colonic patches) in coordinating immune responses are emerging. For example, colonic patches orchestrate interactions between DCs and ILCs in an LT-dependent manner to provoke an innate response to *Citrobacter rodentium* [183,184]. In some cases, ILFs may also be an important site for IgA CSR [214], particularly for T-independent IgA CSR. The expression of AID in this location further argues that ILFs can support IgA CSR.

2.5.3 LT-Dependent Induction of IgA CSR

Significant amounts of IgA are produced in the GALT. B cells in PPs receive T cell help in GCs, and the cytokine milieu in this location promotes CSR to the IgA isotype [213]. IgA-producing B cells then leave the PPs as IgA-producing PCs, circulating through lymphatics to the thoracic duct and into the blood, ultimately entering the gut LP. Within the LP, IgA^+ PCs secrete copious amounts of IgA and can produce other molecules that may be involved in local defence mechanisms and/or homoeostasis of the IgA^+ PC pool [215].

LT-deficient mice have a profound reduction in IgA in their serum and faecal pellets, although the reason for this impairment remains unclear [216]. It is important to point out that although LT-deficient mice lack LNs and PPs, these structures are dispensable for the generation of IgA that can occur in the ILFs of the LPs [214]. Thus, unravelling the role of LTβR signalling requires an assessment of how IgA is made in the presence and absence of these organised lymphoid structures. In the absence of LNs and PPs, LTβR signalling is required for the expression of CXCL13 in radio-resistant cells to recruit B cells into the gut lamina propria and transplantation of LTβR-sufficient gut sections into LTβR-replete intestine is sufficient to induce a local IgA response [216]. On the other hand, B cells can still migrate, albeit in lower numbers, to the LTβR-deficient gut. Thus, there are likely other LT-dependent events that are required to optimise IgA CSR.

It is important to keep in mind that in the presence of PPs, LTβR signalling may actually lie downstream of other signals. Indeed, a recent study showed that the TNFR/LTα axis is required for IgA production. Specifically, the authors deleted LTα expression on ILCs and observed a profound reduction in B cell accumulation and IgA production in the gut. A role for LTβR/LTαβ signalling was also defined in T-independent IgA CSR [217]. Consistent with this observation, LTα-deficient mice have a strongly impaired immune response to *Rotavirus*, although curiously, the IgA response to *Rotavirus* eventually emerges with time [218]. It is possible that without $LT\alpha_1\beta_2$/LTβR and LTα/TNFR signalling, the LIGHT/LTβR axis may eventually compensate. Indeed, overexpression of LIGHT results in a hyper-IgA phenotype [219]. Thus, although a role for the LTα/TNFR signalling axis in IgA induction has been clearly demonstrated, unravelling the role of LTβR signalling in T-dependent (PP sufficient) and T-independent (PP deficient) settings remains a challenging question.

2.5.4 Expression of $LT\alpha_1\beta_2$ on Mucosal B Cells Is Required for Orchestrating Th2 Responses

In addition to its role in IgA responses, B cell-associated LT can also play a role in mucosal T cell responses. As mentioned, the interaction between $LT\alpha_1\beta_2$-expressing T cells and LTβR-expressing DCs is important for cross-priming a $CD8^+$ T cell response to protein antigen [182]. However, in the context of infection with a parasite, *Heligmosomoides polygyrus*, Lund and colleagues showed that expression of $LT\alpha_1\beta_2$ by B cells is important for B cell:DC interactions in the interfollicular regions of mesenteric LNs. These regions are typically not rich in B cells; however, upon infection with *H. polygyrus*, local production of CXCL13 in this region draws $CXCR5^+$ B cells and a subset of $CXCR5^+$ DCs into the interfollicular region. These interactions are critical for *H. polygyrus*-induced Th2 development. The Th2 programme induced by *H. polygyrus* infection is prevented by predepletion of follicular B cells with anti-CD20 antibodies, and it is the expression of $LT\alpha_1\beta_2$ on B cells that is required to orchestrate the encounters between $CXCR5^+$ DCs and T cells to drive the Th2 response. Therefore, in the context of an intestinal infection with *H. polygyrus*, expression of $LT\alpha_1\beta_2$ on B cells is critical for coordinating a Th2 response [220].

However, it is likely that other factors can also guide the movement of B cells to interfollicular regions. In particular, the G protein-coupled receptor Epstein–Barr virus-induced-2 (EBI2) is important for guiding B cells to the T/B interface during immune responses. Specialised stroma in these regions produce the enzymes required to convert cholesterol to 7α,25-dihydroxycholesterol, the ligand for EBI2 [221]. Such stromal cells are also present in the so-called bridging channels of the spleen, where DCs are localised. The localization of DCs in this region may be important for two reasons: (1) so that DCs may be positioned close to the MZ where Ag is first encountered upon systemic infection and (2) to position DCs close to B cells that express $LT\alpha_1\beta_2$ so that they may receive constitutive LTβR-mediated homoeostatic signals [222].

2.5.5 The Role of $LT\alpha_1\beta_2$ on B Cells During Pathological Gut Inflammation

Given its role in the mucosal immune system, it is not surprizing that the LT pathway can shape the inflammatory environment of the gut during disease. Inhibition of LTβR signalling has a beneficial effect in several models of mucosal inflammation. For example, the transfer of $CD45RB^{high}CD4^{+}$ T cells into lymphocyte-replete (SCID) mice results in colitis that is likewise inhibited by LTβR-Ig treatment, and similar results were obtained with a BM transplantation colitis model [223]. Th2-driven colitis initiated by the hapten TNBS can also be attenuated using LTβR pathway blockade [224,225]. Injection of dextran sodium sulphate (DSS) induces a chronic wasting disease and colitic inflammation that is often used as a mouse model for human colitis. It is interesting to note that although in one study LTβR-Ig treatment was shown to ameliorate DSS colitis and reduce inflammation concomitant with the reduction of the expression of homing marker MAdCAM-1 in the GALT [226], the opposite results were obtained in a second study in which LTβR-Ig treatment seemed to exacerbate colitis [227]. As of yet, no explanation has been provided for the discrepancy of the two studies. However, what was interesting about the second study was that the expression of $LT\alpha_1\beta_2$ on B cells was disease promoting whereas the expression of $LT\alpha_1\beta_2$ on T cells prevented disease. The authors speculate that interactions between of $LT\alpha_1\beta_2$ on T cells and LTβR on myeloid/DC may drive anti-inflammatory mediators whereas the expression of $LT\alpha_1\beta_2$ on B cells may promote inflammation by supporting the generation of tertiary structures in the colon (i.e., colonic patches).

Perhaps one way to resolve these contradictory data is to refer to another system of colonic inflammation: Infection with *C. rodentium* results in an inflammatory condition that recapitulates some aspects of colitis. Treatment of *C. rodentium* infected mice with LTβR-Ig increases disease-related mortality [228]. Specifically, weight loss is exacerbated, there are more intestinal bacterial abscesses, and these were associated with a higher burden of *C. rodentium* in the periphery (spleen, liver). Similar results are observed with $LT\beta R^{-/-}$ and $LT\beta^{-/-}$ mice. In addition, deletion of LTβR in DCs [183,184] can attenuate inflammation associated with *C. rodentium* infection, and this is also the case if LTβR is deleted in intestinal epithelial cells [229]. However, it is important to point out that in the case of *C. rodentium*, the generation of an antipathogen immune response is required to resolve the infection, and it would make sense that an unresolved infection would generate colonic inflammation. Indeed, $RAG^{-/-}$ mice are highly susceptible to *C. rodentium* infection [230]. Given that LTβR signalling is required for IgA responses and the T cell:DC collaboration, it is not surprizing that LT-deficient mice have problems clearing a *C. rodentium* infection. Thus, the inflammation observed in the *C. rodentium* setting in LT-deficient mice may be due to poor pathogen handling. Likewise, in the DSS colitis context in which host microbiota can penetrate the injured epithelial layer and cause inflammation, defective IgA responses or impaired T cell:DC communication could lead to local inflammation, as was observed in the aforementioned DSS studies performed in LT-deficient settings.

2.6 LT and Tertiary Lymphoid Structures

2.6.1 The LT Pathway and TLS Formation

The role of B cells in initiating tertiary lymphoid structures (TLS) has been well described using several different experimental systems, and the mechanisms of TLS formation and their role in disease are discussed in another chapter. The RIP has been used to drive ectopic expression of $LT\alpha_1\beta_2$ in the context of pancreatic tissue. Expression of LTα and LTβ is required to generate fully mature TLS in the pancreas with delineated T and B cell areas and fully mature HEVs [231]. Therefore, LTβR signalling on some kind of pancreas-resident cell type can be initiated by the RIP-induced $LT\alpha_1\beta_2$. However, to uncover whether B cells could be capable of providing an important source of $LT\alpha_1\beta_2$ to induce TLS in the pancreas, transgenic expression of CXCL13 by the RIP was also tested. RIP-CXCL13 mice also exhibited TLS in the pancreas, suggesting that B cells are an important initiating cell for TLS in non-lymphoid tissues [231]. Because the injection of LT inhibitors blocked follicle formation in RIP-CXCL13 mice, the main function for B cells in this context was via provision of $LT\alpha_1\beta_2$.

2.6.2 The LT Pathway and FDC Precursors

The implication of these studies is that during inflammation, immune cells, and particularly B cells that express $LT\alpha_1\beta_2$, can function to induce local LTβR signalling to promote the formation of TLS. However, this begs the question of what cell types express LTβR to initiate TLS formation. Recent studies by Aguzzi and colleagues have shown that TLS can be initiated by so-called "FDC precursor" cells (Figure 10.4). FDC precursors were revealed to originate from radio-resistant cell types that express plasma-derived growth factor receptor-β (PDGFRβ). Although mature FDCs are absent in $TNFR^{-/-}$ mice, such pre-FDC-expressing Mfge8 (also known as FDC-M1) and CXCL13 can be observed in the marginal sinus and white pulp areas. However, these pre-FDCs are greatly decreased and dispersed in mice that lack LTβR signalling or lack lymphoid cells ($RAG2^{-/-}$ $\gamma c^{-/-}$ mice). Introduction of B cells into $RAG2^{-/-}$ $\gamma c^{-/-}$ mice induced the formation of mature FDCs at perivascular locations in the spleen, suggesting that a precursor cell in this location could give rise to mature $CD35^+FcgR^+$ FDC. Such $Mfge8^+$ precursors were also found to be positive for PDGFRβ and were derived from cells (perhaps pericytes) that contact the vascular endothelium. It is important to note that by growing pericytes in a collagen-based matrix and transplanting into the kidney capsule of $LT\beta R^{-/-}$ mice, the authors were able to show that mature FDCs can be generated in the kidneys of recipient mice that supported the recruitment of lymphocytes. Therefore, FDC precursors are presumably sessile mural cell types that reside in perivascular locations throughout the body, and this provides a mechanism for why TLS may arise in a wide variety of tissues [232]. During inflammation, presumably the expression of $LT\alpha_1\beta_2$ on infiltrating activated immune cells (particularly B cells) is sufficient to trigger a maturation programme in these pericytes to establish TLS within the inflamed environment. Stimulation of LTβR on such perivascular cells induces the upregulation of CXCL13 and Mfge8, and further maturation of these cells induces the expression of Ag capture machinery.

However, TLS have also been observed to form in the absence of LTβR signalling. In the case of TLS in the lungs (so-called inducible bronchial lymphoid tissues (iBALT) structures), expression of CXCL13 can occur independent of the LT pathway, although in the absence of LTβR signalling, the iBALT were disorganized [233]. This implies that other pathways can compensate for the absence of LT in inducing CXCL13 for the recruitment of B cells.

Although LTβR signalling is dispensable for the formation of iBALT, Th17 cells (i.e., T cells that produce the inflammatory cytokines IL-17 and IL-22) are required to induce iBALT in the lungs [233]. Furthermore, it has been shown that TLS in the spinal cord that emerge during EAE can be induced by Th17 cells [233]. Th17 cells are a particularly abundant source of $LT\alpha_1\beta_2$ expression, and they can also produce TNFα. Thus, it makes sense that they may

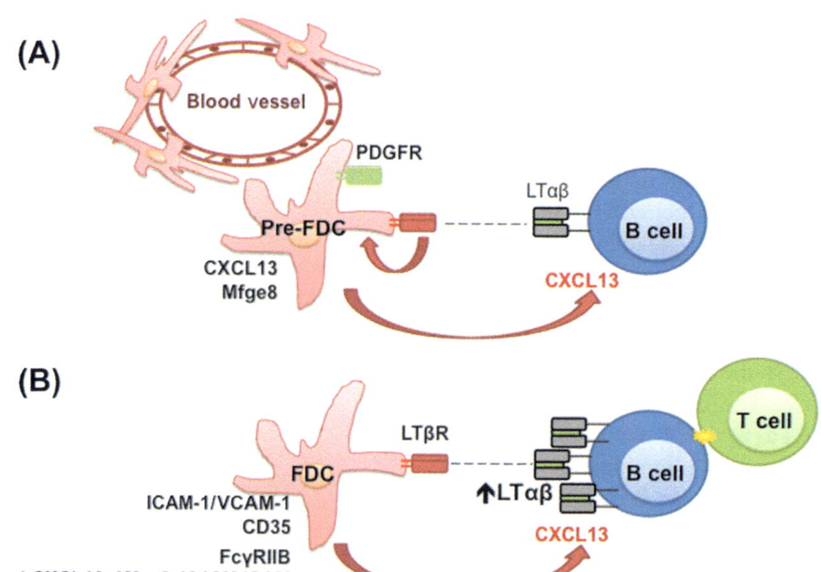

FIGURE 10.4 **Dynamic lymphotoxin (LT)-dependent maturation of follicular dendritic cells (FDC).** (A) FDCs are derived from sessile precursor mural cells associated with blood vessels. When these pre-FDCs make contact with LT $\alpha_1\beta_2$-expressing cells, particularly B cells, upon triggering of the LTβR the pre-FDC express markers such as Mfge8 and produce CXCL13. (B) Further differentiation of FDCs is required for them to reach a mature status. This can be achieved through TNFR1, or during a GC reaction exceptionally high levels of $LT\alpha_1\beta_2$ expression on GC B cells can also induce FDC maturation to express Mfge8 and high levels of CD35. *DGFR*, plasma-derived growth factor receptor.

stimulate local stromal cells through TNFR/LTβR to produce the necessary chemokines required to recruit additional lymphocytes to the site of inflammation, particularly B cells.

2.7 LT—Therapeutic Perspectives that Pertain to B Cells

2.7.1 Therapeutic Rationale for Targeting B Cells and the LT Pathway

The targeting of B cells for the treatment of disease has been a relatively new frontier for immunotherapeutics, with B cell depleting agents (anti-CD20) having efficacy in MS, RA, SLE, and lymphoma [234]. Anti-CD20 (rituximab/ocrelizumab) effectively couples with antibody-dependent cellular cytotoxicity mechanisms to deplete mature follicular B cells, sparing PCs and some subsets of B cells. However, ablation of mature B cells using anti-CD20 treatment can potentially leave patients vulnerable to infection, and not all B cells have pathogenic potential. In fact, anti-CD20 treatment depletes IL-10-producing B cells in mice, and this can worsen clinical outcomes in mouse models of MS [235].

Of course, an alternative strategy is to inhibit specific pathogenic functions attributable to B cells. As such, the LT inhibitor LTβR-Ig (baminercept) may have promise in disease settings in which B cells are playing a pathogenic role via their expression of $LT\alpha_1\beta_2$. As mentioned, significant evidence shows that $LT\alpha_1\beta_2$ expressed by B cells is not only important in FDC biology in the lymphoid tissues, and in some cases can cause a decay in the GC response, but expression of $LT\alpha_1\beta_2$ on naïve and Ag-activated B cells serves to stimulate LTβR on FDC precursor cells that exist in perivascular locations [232].

2.7.2 Targeting the LT Pathway in Autoimmunity

Inhibitors of the LT pathway have been tested in several preclinical animal models of disease [236]. Because of (1) the wide range of cells that express LTβR; (2) the complex role for LTβR in LN development, maintenance of lymphoid tissue architecture, DC/macrophage homoeostasis and function, and TLS formation; (3) the different ligands ($LT\alpha_1\beta_2$, LIGHT) that activate the LTβR; and (4) the interdigitating pathways (LTβR/$LT\alpha_1\beta_2$, TNFR/LTα, HVEM/LIGHT etc.), it has been very difficult to discern the mechanism of action of LTβR-Ig in disease, and indeed, different disease models likely invoke different mechanisms.

Some good examples of the effect of LTβR-Ig in disease settings where the B cell compartment has been affected are animal models of RA and SS. With respect to RA, prophylactic and therapeutic treatment of DBA-1 mice injected with collagen in complete Freund's adjuvant reduces disease burden. Efficacy in this collagen-induced arthritis (CIA) model correlates with a reduction in collagen titres in the periphery, rather than a decreased T cell response, suggesting that LTβR-Ig may affect the anti-collagen B cell response. Indeed, the FDC networks in the draining LNs were obliterated in this mouse model. Furthermore, the induction of RA by the adoptive transfer of complement containing immune complexes cannot be inhibited by LTβR-Ig treatment, suggesting that this therapeutic will not work in a setting in which the GC stage has been bypassed [237]. However, in a second study, LTβR-Ig treatment was not found to ameliorate CIA, but rather depletion of Th1/Th17 cells using an anti-LTα monoclonal antibody approach worked to reduce disease. Given its multiple functions in orchestrating immune responses, it is likely that the LT pathway can influence the CIA model by modulating T and B cell responses [238].

In the case of SS, the NOD mice develop a Sjögren's-like disease as they age. Treatment with LTβR-Ig at 9 weeks of age partially restores salivary flow in these mice, and this was associated with reduced FDC networks in the submandibulary glands [239]. More recent work has shown that the B cell content in the lacrimal glands of the Sjögren's model is greatly reduced with LTβR-Ig treatment, and this corresponds with reductions in HEV structures as well as chemokines such as CXCL13. A clinical trial testing the efficacy of LTβR-Ig (baminercept) in Sjögren's patients is ongoing (please see http://clinicaltrials.gov/show/NCT01552681).

Although these autoimmune disease examples suggest an involvement for B cells, the precise stage of B cell activation at which the LT pathway may interfere has not been determined. As mentioned, disruption of FDC by inhibition of the LT pathway can have variable effects on the GC response depending on the tissue in which the GC is taking place, with the spleen GC being especially LT-sensitive and the LN GC less so. Although FDC precursors are associated with perivascular areas throughout the body, the simulation of LTβR on these precursors is required for their differentiation into mature FDCs. Even in the presence of LTβR signalling, FDCs differ in terms of their capacity to express an antigen capture machine and to support local GC responses. So-called "primary" FDCs express less CD35 whereas "secondary" FDCs express high levels of CD35 and Fcγ receptors that aid in the capture of antibody/complement-coated antigen as well as in the opsonisation of apoptotic bodies [240]. Such secondary FDCs are typically associated with GC B cells that express AID, and indeed, the presence of secondary FDCs in ectopic lymphoid structures during autoimmunity is strongly associated with AID expression in B cells in the salivary glands of SS patients [241] and in the synovium of RA patients [242]. Perhaps the regulation of AID, although relatively intact in an LT-deficient LN, is much more sensitive to LTβR-dependent cell types in nonlymphoid tissues.

2.7.3 Targeting the LT Pathway in Cancer

The role of LT biology in cancer progression is complex because of divergent signalling outcomes of triggering LTβR in the tumour itself, versus the tumour microenvironment—and a full dissection of the LT pathway in tumourigenesis is beyond the scope of this chapter. Focussing on the tumour microenvironment, it is increasingly appreciated that chronic inflammation can have a negative outcome on tumour growth and metastasis, and this negative outcome is associated with local NF-κB activation within the tumour. Indeed, prostate cancer progression was shown to be associated with local NF-κB activation triggered by the LTβR, and the source of $LT\alpha_1\beta_2$in this scenario was shown to be derived from B cells [243]. However, in other cases, localization of B cells within tumour environments may also have beneficial outcomes, and the role of B cells in different tumours will likely be dictated by their effector or regulatory functions [244].

Follicular lymphomas, which can be supported by stromal cell networks within lymphoid tissues, may also depend on LTβR signalling. In vitro evidence using mesenchymal stem cells (MSCs) shows that the stimulation of LTβR on MSCs induces the adoption of an FRC phenotype [245], and, in vivo, expression of $LT\alpha_1\beta_2$on B cells allows for a coordinate interaction between lymphoma cells and the FRC network [246]. Lymphomas that are found in association with an FRC network are more aggressive with a significant survival advantage. Thus, one strategy to treat lymphoma may be to ablate LT-dependent networks within the lymphoid tissue to liberate lymphomas from microenvironments that are permissive to their growth.

3. CONCLUSIONS AND PERSPECTIVES

The coordinate interactions among TNF superfamily members shape immune responses. In this chapter, we have focused on two TNF superfamily pathways: the BAFF/APRIL pathway and the LT pathway. Given the role of the BAFF/APRIL and LT pathways in regulating B cell homoeostasis, the GC response, the fine balance between B cell tolerance and immunity, and innate B cell function and survival, it is reasonable to consider these pathways as important players in B cell biology. However, other TNF superfamily members, notably CD40, TNFα, and OX40, play important roles in the B cell response that are beyond the scope of this chapter to discuss. Nevertheless, targeting the TNF superfamily has been used to treat rheumatic B cell-mediated diseases as well as B cell malignancies, and it is expected that additional TNF superfamily therapies will emerge.

REFERENCES

[1] Bossen C, Schneider P. BAFF, APRIL and their receptors: structure, function and signaling. Semin Immunol 2006;18:263–75.
[2] Mackay F, Schneider P. Cracking the BAFF code. Nat Rev Immunol 2009;9:491–502.
[3] Wallace DJ, et al. A phase II, randomized, double-blind, placebo-controlled, dose-ranging study of belimumab in patients with active systemic lupus erythematosus. Arthritis Rheum 2009;61:1168–78.
[4] Sanz I, Yasothan U, Kirkpatrick P. Belimumab. Nat Rev Drug Discov 2011;10:335–6.
[5] Dillon SR, Gross JA, Ansell SM, Novak AJ. An APRIL to remember: novel TNF ligands as therapeutic targets. Nat Rev Drug Discov 2006;5:235–46.
[6] Nardelli B, et al. Synthesis and release of B-lymphocyte stimulator from myeloid cells. Blood 2001;97:198–204.
[7] Maia S, et al. Aberrant expression of functional baff-system receptors by malignant B-cell precursors impacts leukemia cell survival. PLoS One 2011;6:e20787.
[8] Moore PA, et al. BLyS: member of the tumor necrosis factor family and B lymphocyte stimulator. Science 1999;285:260–3.
[9] Schneider P, et al. BAFF, a novel ligand of the tumor necrosis factor family, stimulates B cell growth. J Exp Med 1999;189:1747–56.
[10] Lopez-Fraga M, Fernandez R, Albar JP, Hahne M. Biologically active APRIL is secreted following intracellular processing in the Golgi apparatus by furin convertase. EMBO Rep 2001;2:945–51.
[11] Zhukovsky EA, Lee JO, Villegas M, Chan C, Chu S, Mroske C. TNF ligands: is TALL-1 a trimer or a virus-like cluster? Nature 2004;427:413–4. discussion 414.
[12] Oren DA, et al. Structural basis of BLyS receptor recognition. Nat Struct Biol 2002;9:288–92.
[13] Bossen C, et al. TACI, unlike BAFF-R, is solely activated by oligomeric BAFF and APRIL to support survival of activated B cells and plasmablasts. Blood 2008;111:1004–12.
[14] Liu Y, Xu L, Opalka N, Kappler J, Shu HB, Zhang G. Crystal structure of sTALL-1 reveals a virus-like assembly of TNF family ligands. Cell 2002;108:383–94.
[15] Mackay F, Schneider P, Rennert P, Browning J. BAFF and APRIL: a tutorial on B cell survival. Annu Rev Immunol 2003;21:231–64.
[16] Cachero TG, et al. Formation of virus-like clusters is an intrinsic property of the tumor necrosis factor family member BAFF (B cell activating factor). Biochemistry 2006;45:2006–13.
[17] Roschke V, et al. BLyS and APRIL form biologically active heterotrimers that are expressed in patients with systemic immune-based rheumatic diseases. J Immunol 2002;169:4314–21.
[18] Stohl W. Systemic lupus erythematosus and its ABCs (APRIL/BLyS complexes). Arthritis Res Ther 2010;12:111.
[19] Gavin AL, Ait-Azzouzene D, Ware CF, Nemazee D. DeltaBAFF, an alternate splice isoform that regulates receptor binding and biopresentation of the B cell survival cytokine, BAFF. J Biol Chem 2003;278:38220–8.

[20] Gavin AL, et al. deltaBAFF, a splice isoform of BAFF, opposes full-length BAFF activity in vivo in transgenic mouse models. J Immunol 2005;175:319–28.
[21] Krumbholz M, et al. BAFF is produced by astrocytes and up-regulated in multiple sclerosis lesions and primary central nervous system lymphoma. J Exp Med 2005;201:195–200.
[22] Smirnova AS, Andrade-Oliveira V, Gerbase-DeLima M. Identification of new splice variants of the genes BAFF and BCMA. Mol Immunol 2008;45:1179–83.
[23] Youinou P, Pers JO. The late news on baff in autoimmune diseases. Autoimmun Rev 2010;9:804–6.
[24] Pradet-Balade B, et al. An endogenous hybrid mRNA encodes TWE-PRIL, a functional cell surface TWEAK-APRIL fusion protein. EMBO J 2002;21:5711–20.
[25] Lavie F, Miceli-Richard C, Ittah M, Sellam J, Gottenberg JE, Mariette X. B-cell activating factor of the tumour necrosis factor family expression in blood monocytes and T cells from patients with primary Sjogren's syndrome. Scand J Immunol 2008;67:185–92.
[26] Huard B, et al. BAFF production by antigen-presenting cells provides T cell co-stimulation. Int Immunol 2004;16:467–75.
[27] Ng LG, et al. B cell-activating factor belonging to the TNF family (BAFF)-R is the principal BAFF receptor facilitating BAFF costimulation of circulating T and B cells. J Immunol 2004;173:807–17.
[28] Scapini P, et al. G-CSF-stimulated neutrophils are a prominent source of functional BLyS. J Exp Med 2003;197:297–302.
[29] Scapini P, et al. Proinflammatory mediators elicit secretion of the intracellular B-lymphocyte stimulator pool (BLyS) that is stored in activated neutrophils: implications for inflammatory diseases. Blood 2005;105:830–7.
[30] Sutherland AP, et al. BAFF augments certain Th1-associated inflammatory responses. J Immunol 2005;174:5537–44.
[31] Chang SK, Arendt BK, Darce JR, Wu X, Jelinek DF. A role for BLyS in the activation of innate immune cells. Blood 2006;108:2687–94.
[32] Morimoto S, et al. Expression of B-cell activating factor of the tumour necrosis factor family (BAFF) in T cells in active systemic lupus erythematosus: the role of BAFF in T cell-dependent B cell pathogenic autoantibody production. Rheumatology 2007;46:1083–6.
[33] Groom JR, et al. BAFF and MyD88 signals promote a lupuslike disease independent of T cells. J Exp Med 2007;204:1959–71.
[34] Hardenberg G, et al. Specific TLR ligands regulate APRIL secretion by dendritic cells in a PKR-dependent manner. Eur J Immunol 2007;37:2900–11.
[35] Katsenelson N, Kanswal S, Puig M, Mostowski H, Verthelyi D, Akkoyunlu M. Synthetic CpG oligodeoxynucleotides augment BAFF- and APRIL-mediated immunoglobulin secretion. Eur J Immunol 2007;37:1785–95.
[36] Chu VT, Enghard P, Riemekasten G, Berek C. In vitro and in vivo activation induces BAFF and APRIL expression in B cells. J Immunol 2007;179:5947–57.
[37] Nakajima K, et al. Expression of BAFF and BAFF-R in the synovial tissue of patients with rheumatoid arthritis. Scand J Rheumatol 2007;36:365–72.
[38] Nagatani K, et al. Rheumatoid arthritis fibroblast-like synoviocytes express BCMA and are stimulated by APRIL. Arthritis Rheum 2007;56:3554–63.
[39] Ohata J, et al. Fibroblast-like synoviocytes of mesenchymal origin express functional B cell-activating factor of the TNF family in response to proinflammatory cytokines. J Immunol 2005;174:864–70.
[40] Bombardieri M, et al. A BAFF/APRIL-dependent TLR3-stimulated pathway enhances the capacity of rheumatoid synovial fibroblasts to induce AID expression and Ig class-switching in B cells. Ann Rheum Dis 2011;70:1857–65.
[41] Belnoue E, et al. APRIL is critical for plasmablast survival in the bone marrow and poorly expressed by early-life bone marrow stromal cells. Blood 2008;111:2755–64.
[42] Chu VT, et al. Eosinophils are required for the maintenance of plasma cells in the bone marrow. Nat Immunol 2011;12:151–9.
[43] Benson MJ, et al. Cutting edge: the dependence of plasma cells and independence of memory B cells on BAFF and APRIL. J Immunol 2008;180:3655–9.
[44] Munari F, et al. Tumor-associated macrophages as major source of APRIL in gastric MALT lymphoma. Blood 2011;117:6612–6.
[45] Lee H-J, Seo G-Y, Kim J-H, Lee M-R, Kim P-H. Activin A stimulates mouse macrophages to express APRIL via the Smad3 and ERK/CREB pathways. Immunol Lett 2011;140:92–6.
[46] Peperzak V, et al. Mcl-1 is essential for the survival of plasma cells. Nat Immunol 2013;14:290–7.
[47] Farina C, Aloisi F, Meinl E. Astrocytes are active players in cerebral innate immunity. Trends Immunol 2007;28:138–45.
[48] Hauser SL, et al. B-cell depletion with rituximab in relapsing-remitting multiple sclerosis. N Engl J Med 2008;358:676–88.
[49] Hartung H-P, Kieseier BC. Atacicept: targeting B cells in multiple sclerosis. Ther Adv Neurol Disord 2010;3:205–16.
[50] Hahne M, et al. APRIL, a new ligand of the tumor necrosis factor family, stimulates tumor cell growth. J Exp Med 1998;188:1185–90.
[51] Stein JV, et al. APRIL modulates B and T cell immunity. J Clin Invest 2002;109:1587–98.
[52] Gorelik L, Gilbride K, Dobles M, Kalled SL, Zandman D, Scott ML. Normal B cell homeostasis requires B cell activation factor production by radiation-resistant cells. J Exp Med 2003;198:937–45.
[53] Zhang M, et al. Expression and function of TNF family member B cell-activating factor in the development of autoimmune arthritis. Int Immunol 2005;17:1081–92.
[54] Goenka R, et al. Local BLyS production by T follicular cells mediates retention of high affinity B cells during affinity maturation. J Exp Med 2014;211:45–56.
[55] Tangye SG, Bryant VL, Cuss AK, Good KL. BAFF, APRIL and human B cell disorders. Semin Immunol 2006;18:305–17.
[56] Mackay F, Leung H. The role of the BAFF/APRIL system on T cell function. Semin Immunol 2006;18:284–9.
[57] Rennert P, et al. A soluble form of B cell maturation antigen, a receptor for the tumor necrosis factor family member APRIL, inhibits tumor cell growth. J Exp Med 2000;192:1677–84.
[58] Schneider P, et al. Maturation of marginal zone and follicular B cells requires B cell activating factor of the tumor necrosis factor family and is independent of B cell maturation antigen. J Exp Med 2001;194:1691–7.
[59] Day ES, et al. Selectivity of BAFF/BLyS and APRIL for binding to the TNF family receptors BAFFR/BR3 and BCMA. Biochemistry 2005;44:1919–31.
[60] Liu Y, et al. Ligand-receptor binding revealed by the TNF family member TALL-1. Nature 2003;423:49–56.

[61] Hymowitz SG, et al. Structures of APRIL-receptor complexes: like BCMA, TACI employs only a single cysteine-rich domain for high affinity ligand binding. J Biol Chem 2005;280:7218–27.

[62] Patel DR, et al. Engineering an APRIL-specific B cell maturation antigen. J Biol Chem 2004;279:16727–35.

[63] Wang R, et al. Expression and characterization of a variant of TACI (CRD2-shortTACIFc) in Pichia pastoris. Protein Pept Lett 2012;19:315–25.

[64] Cancro MP. The BLyS family of ligands and receptors: an archetype for niche-specific homeostatic regulation. Immunol Rev 2004;202:237–49.

[65] Zhang X, et al. BAFF supports human B cell differentiation in the lymphoid follicles through distinct receptors. Int Immunol 2005;17:779–88.

[66] Batten M, et al. BAFF mediates survival of peripheral immature B lymphocytes. J Exp Med 2000;192:1453–66.

[67] Thompson JS, et al. BAFF-R, a newly identified TNF receptor that specifically interacts with BAFF. Science 2001;293:2108–11.

[68] Ye Q, et al. BAFF binding to T cell-expressed BAFF-R costimulates T cell proliferation and alloresponses. Eur J Immunol 2004;34:2750–9.

[69] Ng LG, et al. BAFF costimulation of Toll-like receptor-activated B-1 cells. Eur J Immunol 2006;36:1837–46.

[70] Yan M, Marsters SA, Grewal IS, Wang H, Ashkenazi A, Dixit VM. Identification of a receptor for BLyS demonstrates a crucial role in humoral immunity. Nat Immunol 2000;1:37–41.

[71] von Bulow GU, van Deursen JM, Bram RJ. Regulation of the T-independent humoral response by TACI. Immunity 2001;14:573–82.

[72] He B, et al. The transmembrane activator TACI triggers immunoglobulin class switching by activating B cells through the adaptor MyD88. Nat Immunol 2010;11:836–45.

[73] Seshasayee D, Valdez P, Yan M, Dixit VM, Tumas D, Grewal IS. Loss of TACI causes fatal lymphoproliferation and autoimmunity, establishing TACI as an inhibitory BLyS receptor. Immunity 2003;18:279–88.

[74] Bodmer J-L, Schneider P, Tschopp J. The molecular architecture of the TNF superfamily. Trends Biochem Sci 2002;27:19–26.

[75] Gross JA, et al. TACI and BCMA are receptors for a TNF homologue implicated in B-cell autoimmune disease. Nature 2000;404:995–9.

[76] Figgett WA, et al. The TACI receptor regulates T-cell-independent marginal zone B cell responses through innate activation-induced cell death. Immunity 2013;39:573–83.

[77] Kanswal S, Katsenelson N, Selvapandiyan A, Bram RJ, Akkoyunlu M. Deficient TACI expression on B lymphocytes of newborn mice leads to defective Ig secretion in response to BAFF or APRIL. J Immunol 2008;181:976–90.

[78] Ou X, Xu S, Lam KP. Deficiency in TNFRSF13B (TACI) expands T-follicular helper and germinal center B cells via increased ICOS-ligand expression but impairs plasma cell survival. Proc Natl Acad Sci USA 2012;109:15401–6.

[79] Tsuji S, Cortesao C, Bram RJ, Platt JL, Cascalho M. TACI deficiency impairs sustained Blimp-1 expression in B cells decreasing long-lived plasma cells in the bone marrow. Blood 2011;118:5832–9.

[80] Darce JR, Arendt BK, Wu X, Jelinek DF. Regulated expression of BAFF-binding receptors during human B cell differentiation. J Immunol 2007;179:7276–86.

[81] von Bulow GU, Russell H, Copeland NG, Gilbert DJ, Jenkins NA, Bram RJ. Molecular cloning and functional characterization of murine transmembrane activator and CAML interactor (TACI) with chromosomal localization in human and mouse. Mamm Genome 2000;11:628–32.

[82] Acosta-Rodriguez EV, et al. BAFF and LPS cooperate to induce B cells to become susceptible to CD95/Fas-mediated cell death. Eur J Immunol 2007;37:990–1000.

[83] Treml LS, et al. TLR stimulation modifies BLyS receptor expression in follicular and marginal zone B cells. J Immunol 2007;178:7531–9.

[84] Hendriks J, et al. Heparan sulfate proteoglycan binding promotes APRIL-induced tumor cell proliferation. Cell Death Differ 2005;12:637–48.

[85] Ingold K, et al. Identification of proteoglycans as the APRIL-specific binding partners. J Exp Med 2005;201:1375–83.

[86] Sakurai D, Hase H, Kanno Y, Kojima H, Okumura K, Kobata T. TACI regulates IgA production by APRIL in collaboration with HSPG. Blood 2007;109:2961–7.

[87] Avery DT, et al. BAFF selectively enhances the survival of plasmablasts generated from human memory B cells. J Clin Invest 2003;112:286–97.

[88] Stadanlick JE, et al. Tonic B cell antigen receptor signals supply an NF-kappaB substrate for prosurvival BLyS signaling. Nat Immunol 2008;9:1379–87.

[89] O'Connor BP, et al. BCMA is essential for the survival of long-lived bone marrow plasma cells. J Exp Med 2004;199:91–8.

[90] Yang M, Hase H, Legarda-Addison D, Varughese L, Seed B, Ting AT. B cell maturation antigen, the receptor for a proliferation-inducing ligand and B cell-activating factor of the TNF family, induces antigen presentation in B cells. J Immunol 2005;175:2814–24.

[91] Zhang L, et al. Identification of BLyS (B lymphocyte stimulator), a non-myelin-associated protein, as a functional ligand for Nogo-66 receptor. J Neurosci 2009;29:6348–52.

[92] Gross JA, et al. TACI-Ig neutralizes molecules critical for B cell development and autoimmune disease. impaired B cell maturation in mice lacking BLyS. Immunity 2001;15:289–302.

[93] Schiemann B, et al. An essential role for BAFF in the normal development of B cells through a BCMA-independent pathway. Science 2001;293:2111–4.

[94] Allman D, Pillai S. Peripheral B cell subsets. Curr Opin Immunol 2008;20:149–57.

[95] Hayakawa K, Li YS, Wasserman R, Sauder S, Shinton S, Hardy RR. B lymphocyte developmental lineages. Ann N Y Acad Sci 1997;815:15–29.

[96] Tardivel A, et al. The anti-apoptotic factor Bcl-2 can functionally substitute for the B cell survival but not for the marginal zone B cell differentiation activity of BAFF. Eur J Immunol 2004;34:509–18.

[97] Kayagaki N, et al. BAFF/BLyS receptor 3 binds the B cell survival factor BAFF ligand through a discrete surface loop and promotes processing of NF-kappaB2. Immunity 2002;17:515–24.

[98] Gardam S, Sierro F, Basten A, Mackay F, Brink R. TRAF2 and TRAF3 signal adapters act cooperatively to control the maturation and survival signals delivered to B cells by the BAFF receptor. Immunity 2008;28:391–401.

[99] Craxton A, Draves KE, Gruppi A, Clark EA. BAFF regulates B cell survival by downregulating the BH3-only family member Bim via the ERK pathway. J Exp Med 2005;202:1363–74.

[100] Mecklenbrauker I, Kalled SL, Leitges M, Mackay F, Tarakhovsky A. Regulation of B-cell survival by BAFF-dependent PKCdelta-mediated nuclear signalling. Nature 2004;431:456–61.

[101] Patke A, Mecklenbrauker I, Erdjument-Bromage H, Tempst P, Tarakhovsky A. BAFF controls B cell metabolic fitness through a PKCβ- and Akt-dependent mechanism. J Exp Med 2006;203:2551–62.

[102] Henley T, Kovesdi D, Turner M. B-cell responses to B-cell activation factor of the TNF family (BAFF) are impaired in the absence of PI3K delta. Eur J Immunol 2008;38:3543–8.

[103] Schweighoffer E, et al. The BAFF receptor transduces survival signals by co-opting the B cell receptor signaling pathway. Immunity 2013;38:475–88.

[104] Infantino S, Tarlinton DM. Feeling a little SYK after mixing BAFF with BCR. Immunity 2013;38:406–8.

[105] Mackay F, Figgett WA, Saulep D, Lepage M, Hibbs ML. B-cell stage and context-dependent requirements for survival signals from BAFF and the B-cell receptor. Immunol Rev 2010;237:205–25.

[106] Batten M, et al. TNF deficiency fails to protect BAFF transgenic mice against autoimmunity and reveals a predisposition to B cell lymphoma. J Immunol 2004;172:812–22.

[107] Castigli E, et al. TACI is mutant in common variable immunodeficiency and IgA deficiency. Nat Genet 2005;37:829–34.

[108] Castigli E, et al. TACI and BAFF-R mediate isotype switching in B cells. J Exp Med 2005;201:35–9.

[109] Mackay F, et al. Mice transgenic for BAFF develop lymphocytic disorders along with autoimmune manifestations. J Exp Med 1999;190:1697–710.

[110] McCarthy DD, et al. Mice overexpressing BAFF develop a commensal flora-dependent, IgA-associated nephropathy. J Clin Invest 2011;121:3991–4002.

[111] Thompson JS, et al. BAFF binds to the tumor necrosis factor receptor-like molecule B cell maturation antigen and is important for maintaining the peripheral B cell population. J Exp Med 2000;192:129–35.

[112] Griesemer AD, Sorenson EC, Hardy MA. The role of the thymus in tolerance. Transplantation 2010;90:465–74.

[113] Hardy RR, Hayakawa K. B cell development pathways. Ann Rev Immunol 2001;19:595–621.

[114] Goodnow CC, et al. Self-tolerance checkpoints in B lymphocyte development. Adv Immunol 1995;59:279–368.

[115] Cornall RJ, Goodnow CC, Cyster JG. The regulation of self-reactive B cells. Curr Opin Immunol 1995;7:804–11.

[116] Glynne R, Ghandour G, Rayner J, Mack DH, Goodnow CC. B-lymphocyte quiescence, tolerance and activation as viewed by global gene expression profiling on microarrays. Immunol Rev 2000;176:216–46.

[117] Cyster JG, Hartley SB, Goodnow CC. Competition for follicular niches excludes self-reactive cells from the recirculating B-cell repertoire. Nature 1994;371:389–95.

[118] Thien M, et al. Excess BAFF rescues self-reactive B cells from peripheral deletion and allows them to enter forbidden follicular and marginal zone niches. Immunity 2004;20:785–98.

[119] Lesley R, et al. Reduced competitiveness of autoantigen-engaged B cells due to increased dependence on BAFF. Immunity 2004;20:441–53.

[120] Brink R, et al. Immunoglobulin M and D antigen receptors are both capable of mediating B lymphocyte activation, deletion, or anergy after interaction with specific antigen. J Exp Med 1992;176:991–1005.

[121] Kearney JF. Innate-like B cells. Springer Semin Immunopathol 2005;26:377–83.

[122] Martin F, Kearney JF. Marginal-zone B cells. Nat Rev Immunol 2002;2:323–35.

[123] Oliver AM, Martin F, Gartland GL, Carter RH, Kearney JF. Marginal zone B cells exhibit unique activation, proliferative and immunoglobulin secretory responses. Eur J Immunol 1997;27:2366–74.

[124] Attanavanich K, Kearney JF. Marginal zone, but not follicular B cells, are potent activators of naive CD4 T cells. J Immunol 2004;172:803–11. Baltimore, MD: 1950.

[125] Rose NR. Autoimmunity, infection and adjuvants. Lupus 2010;19:354–8.

[126] Liu YJ, Zhang J, Lane PJ, Chan EY, MacLennan IC. Sites of specific B cell activation in primary and secondary responses to T cell-dependent and T cell-independent antigens. Eur J Immunol 1991;21:2951–62.

[127] Hyer ML, Samuel T, Reed JC. The FLIP-side of Fas signaling. Clin Cancer Res 2006;12:5929–31.

[128] Kaiser WJ, et al. RIP3 mediates the embryonic lethality of caspase-8-deficient mice. Nature 2011;471:368–72.

[129] Oberst A, et al. Catalytic activity of the caspase-8-FLIP(L) complex inhibits RIPK3-dependent necrosis. Nature 2011;471:363–7.

[130] Chen X, Martin F, Forbush KA, Perlmutter RM, Kearney JF. Evidence for selection of a population of multi-reactive B cells into the splenic marginal zone. Int Immunol 1997;9:27–41.

[131] Yan M, et al. Activation and accumulation of B cells in TACI-deficient mice. Nat Immunol 2001;2:638–43.

[132] Romberg N, et al. CVID-associated TACI mutations affect autoreactive B cell selection and activation. J Clin Invest 2013;123(10):4283–93.

[133] Groom J, et al. Association of BAFF/BLyS overexpression and altered B cell differentiation with Sjogren's syndrome. J Clin Invest 2002;109:59–68.

[134] Fletcher CA, Groom JR, Woehl B, Leung H, Mackay C, Mackay F. Development of autoimmune nephritis in genetically asplenic and splenectomized BAFF transgenic mice. J Autoimmun 2011;36:125–34.

[135] Fletcher CA, et al. Development of nephritis but not sialadenitis in autoimmune-prone BAFF transgenic mice lacking marginal zone B cells. Eur J Immunol 2006;36:2504–14.

[136] Nagy G, Koncz A, Perl A. T- and B-cell abnormalities in systemic lupus erythematosus. Crit Rev Immunol 2005;25:123–40.

[137] McCarthy DD, Chiu S, Gao Y, Summers-deLuca LE, Gommerman JL. BAFF induces a hyper-IgA syndrome in the intestinal lamina propria concomitant with IgA deposition in the kidney independent of LIGHT. Cell Immunol 2006;241:85–94.

[138] Scharenberg AM, Hannibal MC, Torgerson T, Ochs HD, Rawlings DJ. Common variable immune deficiency overview. In: Pagon RA, Bird TD, Dolan CR, Stephens K, Adam MP, editors. GeneReviews. Seattle, WA: University of Washington; 1993–2013.

[139] Woo YJ, et al. Regulation of B cell activating factor (BAFF) receptor expression by NF-κB signaling in rheumatoid arthritis B cells. Exp Mol Med 2011;43:350–7.

[140] Salzer U, et al. Mutations in TNFRSF13B encoding TACI are associated with common variable immunodeficiency in humans. Nat Genet 2005;37:820–8.

[141] Zhang L, et al. Transmembrane activator and calcium-modulating cyclophilin ligand interactor mutations in common variable immunodeficiency: clinical and immunologic outcomes in heterozygotes. J Allergy Clin Immunol 2007;120:1178–85.

[142] Martinez-Pomar N, et al. Role of TNFRSF13B variants in patients with common variable immunodeficiency. Blood 2009;114:2846–8.

[143] Pan-Hammarstrom Q, et al. Reexamining the role of TACI coding variants in common variable immunodeficiency and selective IgA deficiency. Nat Genet 2007;39:429–30.

[144] Garibyan L, Lobito AA, Siegel RM, Call ME, Wucherpfennig KW, Geha RS. Dominant-negative effect of the heterozygous C104R TACI mutation in common variable immunodeficiency (CVID). J Clin Invest 2007;117:1550–7.

[145] Lee JJ, et al. The C104R mutant impairs the function of transmembrane activator and calcium modulator and cyclophilin ligand interactor (TACI) through haploinsufficiency. J Allergy Clin Immunol 2010;126:1234–41. e1232.

[146] Bacchelli C, et al. The C76R transmembrane activator and calcium modulator cyclophilin ligand interactor mutation disrupts antibody production and B-cell homeostasis in heterozygous and homozygous mice. J Allergy Clin Immunol 2011;127:1253–63.

[147] Martinez-Gallo M, Radigan L, Almejun MB, Martinez-Pomar N, Matamoros N, Cunningham-Rundles C. TACI mutations and impaired B-cell function in subjects with CVID and healthy heterozygotes. J Allergy Clin Immunol 2013;131:468–76.

[148] Park JH, Resnick ES, Cunningham-Rundles C. Perspectives on common variable immune deficiency. Ann N Y Acad Sci 2011;1246:41–9.

[149] Losi CG, et al. Mutational analysis of human BAFF receptor TNFRSF13C (BAFF-R) in patients with common variable immunodeficiency. J Clin Immunol 2005;25:496–502.

[150] Warnatz K, et al. B-cell activating factor receptor deficiency is associated with an adult-onset antibody deficiency syndrome in humans. Proc Natl Acad Sci USA 2009;106:13945–50.

[151] Jacobi AM, et al. Effect of long-term belimumab treatment on B cells in systemic lupus erythematosus: extension of a phase II, double-blind, placebo-controlled, dose-ranging study. Arthritis Rheum 2010;62:201–10.

[152] Sawicka-Powierza J, Jablonska E, Kloczko J, Piszcz J, Garley M, Ratajczk-Wrona W. Evaluation of TNF superfamily molecules release by neutrophils and B leukemic cells of patients with chronic B-cell lymphocytic leukemia. Neoplasma 2011;58:45–50.

[153] Haiat S, Billard C, Quiney C, Ajchenbaum-Cymbalista F, Kolb JP. Role of BAFF and APRIL in human B-cell chronic lymphocytic leukaemia. Immunology 2006;118:281–92.

[154] Kern C, et al. Involvement of BAFF and APRIL in the resistance to apoptosis of B-CLL through an autocrine pathway. Blood 2004;103:679–88.

[155] Nishio M, et al. Nurselike cells express BAFF and APRIL, which can promote survival of chronic lymphocytic leukemia cells via a paracrine pathway distinct from that of SDF-1α. Blood 2005;106:1012–20.

[156] Zhang W, et al. B-cell activating factor and v-Myc myelocytomatosis viral oncogene homolog (c-Myc) influence progression of chronic lymphocytic leukemia. Proc Natl Acad Sci USA 2010;107:18956–60.

[157] Cols M, et al. Stromal endothelial cells establish a bidirectional crosstalk with chronic lymphocytic leukemia cells through the TNF-related factors BAFF, APRIL, and CD40L. J Immunol 2012;188:6071–83.

[158] Planelles L, et al. APRIL promotes B-1 cell-associated neoplasm. Cancer Cell 2004;6:399–408.

[159] Guadagnoli M, et al. Development and characterization of APRIL antagonistic monoclonal antibodies for treatment of B-cell lymphomas. Blood 2011;117:6856–65.

[160] Ferrer G, et al. Combined analysis of levels of serum B-cell activating factor and a proliferation-inducing ligand as predictor of disease progression in patients with chronic lymphocytic leukemia. Leuk Lymphoma 2011;52:2064–8.

[161] Molica S, et al. Baff serum level predicts time to first treatment in early chronic lymphocytic leukemia. Eur J Haematol 2010;85:314–20.

[162] Li YJ, et al. Expression of BAFF and BAFF-R in follicular lymphoma: correlation with clinicopathologic characteristics and survival outcomes. PLoS One 2012;7:e50936.

[163] Moreaux J, et al. APRIL and TACI interact with syndecan-1 on the surface of multiple myeloma cells to form an essential survival loop. Eur J Haematol 2009;83:119–29.

[164] Fu L, Lin-Lee YC, Pham LV, Tamayo AT, Yoshimura LC, Ford RJ. BAFF-R promotes cell proliferation and survival through interaction with IKKβ and NF-κB/c-Rel in the nucleus of normal and neoplastic B-lymphoid cells. Blood 2009;113:4627–36.

[165] Hildebrand JM, et al. A BAFF-R mutation associated with non-Hodgkin lymphoma alters TRAF recruitment and reveals new insights into BAFF-R signaling. J Exp Med 2010;207:2569–79.

[166] Kofler DM, Gawlik BB, Elter T, Gianella-Borradori A, Wendtner CM, Hallek M. Phase 1b trial of atacicept, a recombinant protein binding BLyS and APRIL, in patients with chronic lymphocytic leukemia. Leukemia 2012;26:841–4.

[167] Zhang L, et al. Selective killing of Burkitt's lymphoma cells by mBAFF-targeted delivery of PinX1. Leukemia 2011;25:331–40.

[168] Granger GA, Williams TW. Lymphocyte cytotoxicity in vitro: activation and release of a cytotoxic factor. Nature 1968;218:1253–4.

[169] Ruddle NH, Waksman BH. Cytotoxicity mediated by soluble antigen and lymphocytes in delayed hypersensitivity. 3. Analysis of mechanism. J Exp Med 1968;128:1267–79.

[170] Ware CF, Van Arsdale TL, Crowe PD, Browning JL. The ligands and receptors of the lymphotoxin system. Curr Top Microbiol Immunol 1995;198:175–218.

[171] Browning JL, et al. Lymphotoxin β, a novel member of the TNF family that forms a heteromeric complex with lymphotoxin on the cell surface. Cell 1993;72:847–56.

[172] Crowe PD, et al. A lymphotoxin-beta-specific receptor. Science 1994;264:707–10.

[173] Mauri DN, et al. LIGHT, a new member of the TNF superfamily, and lymphotoxin α are ligands for herpesvirus entry mediator. Immunity 1998;8:21–30.

[174] Ware CF. Network communications: lymphotoxins, LIGHT, and TNF. Annu Rev Immunol 2005;23:787–819.

[175] Young J, et al. Lymphotoxin-αβ heterotrimers are cleaved by metalloproteinases and contribute to synovitis in rheumatoid arthritis. Cytokine 2010;51:78–86.

[176] Ware CF, Crowe PD, Grayson MH, Androlewicz MJ, Browning JL. Expression of surface lymphotoxin and tumor necrosis factor on activated T, B, and natural killer cells. J Immunol 1992;149:3881–8.

[177] Ansel KM, et al. A chemokine-driven positive feedback loop organizes lymphoid follicles. Nature 2000;406:309–14.

[178] Vu F, Dianzani U, Ware CF, Mak T, Gommerman JL. ICOS, CD40, and lymphotoxin β receptors signal sequentially and interdependently to initiate a germinal center reaction. J Immunol 2008;180:2284–93.

[179] GeurtsvanKessel CH, et al. Dendritic cells are crucial for maintenance of tertiary lymphoid structures in the lung of influenza virus-infected mice. J Exp Med 2009;206:2339–49.

[180] Moussion C, Girard JP. Dendritic cells control lymphocyte entry to lymph nodes through high endothelial venules. Nature 2011;479:542–6.

[181] del Rio ML, Lucas CL, Buhler L, Rayat G, Rodriguez-Barbosa JI. HVEM/LIGHT/BTLA/CD160 cosignaling pathways as targets for immune regulation. J Leukoc Biol 2010;87:223–35.

[182] Summers deLuca L, Ng D, Gao Y, Wortzman ME, Watts TH, Gommerman JL. LTβR signaling in dendritic cells induces a type I IFN response that is required for optimal clonal expansion of $CD8^+$ T cells. Proc Natl Acad Sci USA 2011;108:2046–51.

[183] Ota N, et al. IL-22 bridges the lymphotoxin pathway with the maintenance of colonic lymphoid structures during infection with *Citrobacter rodentium*. Nat Immunol 2011;12:941–8.

[184] Tumanov AV, et al. Lymphotoxin controls the IL-22 protection pathway in gut innate lymphoid cells during mucosal pathogen challenge. Cell Host Microbe 2011;10:44–53.

[185] Kuprash DV, et al. Novel tumor necrosis factor-knockout mice that lack Peyer's patches. Eur J Immunol 2005;35:1592–600.

[186] Matsumoto M, et al. Affinity maturation without germinal centres in lymphotoxin-α-deficient mice. Nature 1996;382:462–6.

[187] Koni PA, Sacca R, Lawton P, Browning JL, Ruddle NH, Flavell RA. Distinct roles in lymphoid organogenesis for lymphotoxins α and β revealed in lymphotoxin β-deficient mice. Immunity 1997;6:491–500.

[188] Futterer A, Mink K, Luz A, Kosco-Vilbois MH, Pfeffer K. The lymphotoxin β receptor controls organogenesis and affinity maturation in peripheral lymphoid tissues. Immunity 1998;9:59–70.

[189] Scheu S, Alferink J, Potzel T, Barchet W, Kalinke U, Pfeffer K. Targeted disruption of LIGHT causes defects in costimulatory T cell activation and reveals cooperation with lymphotoxin β in mesenteric lymph node genesis. J Exp Med 2002;195:1613–24.

[190] Mackay F, Majeau GR, Lawton P, Hochman PS, Browning JL. Lymphotoxin but not tumor necrosis factor functions to maintain splenic architecture and humoral responsiveness in adult mice. Eur J Immunol 1997;27:2033–42.

[191] Ngo VN, et al. Lymphotoxin α/β and tumor necrosis factor are required for stromal cell expression of homing chemokines in B and T cell areas of the spleen. J Exp Med 1999;189:403–12.

[192] Phan TG, Green JA, Gray EE, Xu Y, Cyster JG. Immune complex relay by subcapsular sinus macrophages and noncognate B cells drives antibody affinity maturation. Nat Immunol 2009;10:786–93.

[193] Fu YX, Huang G, Matsumoto M, Molina H, Chaplin DD. Independent signals regulate development of primary and secondary follicle structure in spleen and mesenteric lymph node. Proc Natl Acad Sci USA 1997;94:5739–43.

[194] Browning JL, et al. Lymphotoxin-β receptor signaling is required for the homeostatic control of HEV differentiation and function. Immunity 2005;23:539–50.

[195] Chai Q, et al. Maturation of lymph node fibroblastic reticular cells from myofibroblastic precursors is critical for antiviral immunity. Immunity 2013;38:1013–24.

[196] Lorenz RG, Chaplin DD, McDonald KG, McDonough JS, Newberry RD. Isolated lymphoid follicle formation is inducible and dependent upon lymphotoxin-sufficient B lymphocytes, lymphotoxin β receptor, and TNF receptor I function. J Immunol 2003;170:5475–82.

[197] van de Pavert SA, Mebius RE. New insights into the development of lymphoid tissues. Nat Rev Immunol 2010;10:664–74.

[198] Mackay F, Browning JL. Turning off follicular dendritic cells. Nature 1998;395:26–7.

[199] Schneider K, et al. Lymphotoxin-mediated crosstalk between B cells and splenic stroma promotes the initial type I interferon response to cytomegalovirus. Cell Host Microbe 2008;3:67–76.

[200] Katakai T. Marginal reticular cells: a stromal subset directly descended from the lymphoid tissue organizer. Front Immunol 2012;3:200.

[201] Tumanov A, et al. Distinct role of surface lymphotoxin expressed by B cells in the organization of secondary lymphoid tissues. Immunity 2002;17:239–50.

[202] Boulianne B, et al. AID-expressing germinal center B cells cluster normally within lymph node follicles in the absence of FDC-$M1^+$ $CD35^+$ follicular dendritic cells but dissipate prematurely. J Immunol 2013;191:4521–30.

[203] De Trez C, et al. The inhibitory HVEM-BTLA pathway counter regulates lymphotoxin receptor signaling to achieve homeostasis of dendritic cells. J Immunol 2008;180:238–48.

[204] Kabashima K, Banks TA, Ansel KM, Lu TT, Ware CF, Cyster JG. Intrinsic lymphotoxin-β receptor requirement for homeostasis of lymphoid tissue dendritic cells. Immunity 2005;22:439–50.

[205] Lewis KL, et al. Notch2 receptor signaling controls functional differentiation of dendritic cells in the spleen and intestine. Immunity 2011;35(5):780–91.

[206] Vinuesa CG, Linterman MA, Goodnow CC, Randall KL. T cells and follicular dendritic cells in germinal center B-cell formation and selection. Immunol Rev 2010;237:72–89.

[207] Fu YX, Molina H, Matsumoto M, Huang G, Min J, Chaplin DD. Lymphotoxin-α (LTα) supports development of splenic follicular structure that is required for IgG responses. J Exp Med 1997;185:2111–20.

[208] Koni PA, Flavell RA. Lymph node germinal centers form in the absence of follicular dendritic cell networks. J Exp Med 1999;189:855–64.

[209] Wang X, et al. Follicular dendritic cells help establish follicle identity and promote B cell retention in germinal centers. J Exp Med 2011;208:2497–510.

[210] Debard N, Sierro F, Browning J, Kraehenbuhl JP. Effect of mature lymphocytes and lymphotoxin on the development of the follicle-associated epithelium and M cells in mouse Peyer's patches. Gastroenterology 2001;120:1173–82.

[211] Tumanov AV, Kuprash DV, Mach JA, Nedospasov SA, Chervonsky AV. Lymphotoxin and TNF produced by B cells are

dispensable for maintenance of the follicle-associated epithelium but are required for development of lymphoid follicles in the Peyer's patches. J Immunol 2004;173:86–91.
[212] Knoop KA, et al. RANKL is necessary and sufficient to initiate development of antigen-sampling M cells in the intestinal epithelium. J Immunol 2009;183:5738–47.
[213] Suzuki K, et al. The sensing of environmental stimuli by follicular dendritic cells promotes immunoglobulin A generation in the gut. Immunity 2010;33:71–83.
[214] Tsuji M, et al. Requirement for lymphoid tissue-inducer cells in isolated follicle formation and T cell-independent immunoglobulin A generation in the gut. Immunity 2008;29:261–71.
[215] Fritz JH, et al. Acquisition of a multifunctional IgA^+ plasma cell phenotype in the gut. Nature 2012;481:199–203.
[216] Kang HS, et al. Signaling via LTβR on the lamina propria stromal cells of the gut is required for IgA production. Nat Immunol 2002;3:576–82.
[217] Kruglov AA, et al. Nonredundant function of soluble LTα3 produced by innate lymphoid cells in intestinal homeostasis. Science 2013;342:1243–6.
[218] Lopatin U, Blutt SE, Conner ME, Kelsall BL. Lymphotoxin alpha-deficient mice clear persistent rotavirus infection after local generation of mucosal IgA. J Virol 2013;87:524–30.
[219] Wang J, et al. Dysregulated LIGHT expression on T cells mediates intestinal inflammation and contributes to IgA nephropathy. J Clin Invest 2004;113:826–35.
[220] Leon B, Ballesteros-Tato A, Browning JL, Dunn R, Randall TD, Lund FE. Regulation of T(H)2 development by $CXCR5^+$ dendritic cells and lymphotoxin-expressing B cells. Nat Immunol 2012;13:681–90.
[221] Yi T, et al. Oxysterol gradient generation by lymphoid stromal cells guides activated B cell movement during humoral responses. Immunity 2012;37:535–48.
[222] Yi T, Cyster JG. EBI2-mediated bridging channel positioning supports splenic dendritic cell homeostasis and particulate antigen capture. Elife 2013;2:e00757.
[223] Mackay F, et al. Both the lymphotoxin and tumor necrosis factor pathways are involved in experimental murine models of colitis. Gastroenterology 1998;115:1464–75.
[224] An MM, et al. Lymphtoxin β receptor-Ig ameliorates TNBS-induced colitis via blocking LIGHT/HVEM signaling. Pharmacol Res 2005;52:234–44.
[225] Dohi T, et al. Elimination of colonic patches with lymphotoxin β receptor-Ig prevents Th2 cell-type colitis. J Immunol 2001;167:2781–90.
[226] Stopfer P, et al. Blocking lymphotoxin-β receptor activation diminishes inflammation via reduced mucosal addressin cell adhesion molecule-1 (MAdCAM-1) expression and leucocyte margination in chronic DSS-induced colitis. Clin Exp Immunol 2004;136:21–9.
[227] Jungbeck M, Stopfer P, Bataille F, Nedospasov SA, Mannel DN, Hehlgans T. Blocking lymphotoxin β receptor signalling exacerbates acute DSS-induced intestinal inflammation—opposite functions for surface lymphotoxin expressed by T and B lymphocytes. Mol Immunol 2008;45:34–41.
[228] Spahn TW, et al. The lymphotoxin-β receptor is critical for control of murine *Citrobacter rodentium*-induced colitis. Gastroenterology 2004;127:1463–73.
[229] Wang Y, et al. Lymphotoxin beta receptor signaling in intestinal epithelial cells orchestrates innate immune responses against mucosal bacterial infection. Immunity 2010;32:403–13.
[230] Bergstrom KS, et al. Modulation of intestinal goblet cell function during infection by an attaching and effacing bacterial pathogen. Infect Immun 2008;76:796–811.
[231] Luther SA, Lopez T, Bai W, Hanahan D, Cyster JG. BLC expression in pancreatic islets causes B cell recruitment and lymphotoxin-dependent lymphoid neogenesis. Immunity 2000;12:471–81.
[232] Krautler NJ, et al. Follicular dendritic cells emerge from ubiquitous perivascular precursors. Cell 2012;150:194–206.
[233] Rangel-Moreno J, et al. The development of inducible bronchus-associated lymphoid tissue depends on IL-17. Nat Immunol 2011;12:639–46.
[234] Browning JL. B cells move to centre stage: novel opportunities for autoimmune disease treatment. Nat Rev Drug Discov 2006;5:564–76.
[235] Matsushita T, Yanaba K, Bouaziz JD, Fujimoto M, Tedder TF. Regulatory B cells inhibit EAE initiation in mice while other B cells promote disease progression. J Clin Invest 2008;118:3420–30.
[236] Gommerman JL, Browning JL. Lymphotoxin/light, lymphoid microenvironments and autoimmune disease. Nat Rev Immunol 2003;3:642–55.
[237] Fava RA, et al. A role for the lymphotoxin/LIGHT axis in the pathogenesis of murine collagen-induced arthritis. J Immunol 2003;171:115–26.
[238] Chiang EY, et al. Targeted depletion of lymphotoxin-α-expressing TH1 and TH17 cells inhibits autoimmune disease. Nat Med 2009;15:766–73.
[239] Gatumu MK, Skarstein K, Papandile A, Browning JL, Fava RA, Bolstad AI. Blockade of lymphotoxin-beta receptor signaling reduces aspects of Sjogren's syndrome in salivary glands of non-obese diabetic mice. Arthritis Res Ther 2009;11:R24.
[240] El Shikh ME, Pitzalis C. Follicular dendritic cells in health and disease. Front Immunol 2012;3:292.
[241] Bombardieri M, et al. Activation-induced cytidine deaminase expression in follicular dendritic cell networks and interfollicular large B cells supports functionality of ectopic lymphoid neogenesis in autoimmune sialoadenitis and MALT lymphoma in Sjogren's syndrome. J Immunol 2007;179:4929–38.
[242] Humby F, et al. Ectopic lymphoid structures support ongoing production of class-switched autoantibodies in rheumatoid synovium. PLoS Med 2009;6:e1.
[243] Ammirante M, Luo JL, Grivennikov S, Nedospasov S, Karin M. B-cell-derived lymphotoxin promotes castration-resistant prostate cancer. Nature 2010;464:302–5.
[244] Nelson BH. $CD20^+$ B cells: the other tumor-infiltrating lymphocytes. J Immunol 2010;185:4977–82.
[245] Ame-Thomas P, et al. Human mesenchymal stem cells isolated from bone marrow and lymphoid organs support tumor B-cell growth: role of stromal cells in follicular lymphoma pathogenesis. Blood 2007;109:693–702.
[246] Rehm A, et al. Cooperative function of CCR7 and lymphotoxin in the formation of a lymphoma-permissive niche within murine secondary lymphoid organs. Blood 2011;118:1020–33.

Chapter 11

Mechanism and Regulation of Immunoglobulin Class Switch Recombination

Fei-Long Meng[1,2], Lekha Nair[3], Keith C. Fernandez[4], Uttiya Basu[3] and Jayanta Chaudhuri[4,5]

[1]State Key Laboratory of Molecular Biology, Shanghai Institute of Biochemistry and Cell Biology, Center for Excellence in Molecular Cell Science, Chinese Academy of Sciences, University of Chinese Academy of Sciences, Shanghai, China; [2]Shanghai Huashen Institute of Microbes and Infections, Shanghai, China; [3]Department of Microbiology and Immunology, Vagelos College of Physicians and Surgeons, Columbia University, New York, NY, United States; [4]Immunology and Microbial Pathogenesis Program, Weill Cornell Graduate School of Medical Sciences, New York, NY, United States; [5]Immunology Program, Sloan Kettering Institute, Memorial Sloan Kettering Cancer Center, New York, NY, United States

1. INTRODUCTION

Upon engaging antigens, mature IgM^+ B lymphocytes undergo immunoglobulin heavy chain (IgH) class switch recombination (CSR), wherein exons encoding the default Cμ constant gene segments are excised and replaced with a new constant gene segment (for example, Cγ, Cε or Cα). The B cell thereby changes from expressing IgM to producing IgG, IgE or IgA instead, with each antibody isotype having a different effector function during an immune reaction. CSR is a DNA deletional-recombination reaction that proceeds through the generation of DNA double-strand breaks (DSBs) in repetitive switch (S) sequences preceding each constant region gene segment (C_H gene) and is completed by the end-joining of donor Sμ and acceptor S regions. CSR requires an interplay of many factors and pathways including germline non-coding transcription, generation of non-canonical nucleic acid structures, AID-mediated DNA deamination, processing of the deaminated residues by DNA repair factors and finally, long-range end-joining of distal DSBs. Additionally, AID is a general mutator and deregulated AID activity leads to oncogenic mutations and translocations. Here, we review our current understanding of how these pathways and factors are choreographed during a reaction essential for humoral immunity.

2. OVERVIEW OF GENOMIC ALTERATIONS IN B CELLS

2.1 V(D)J Recombination

Developing B cells in the foetal liver and adult bone marrow undergo V(D)J recombination in which the variable region exons of the immunoglobulin (Ig) molecule are somatically assembled from component variable (V), diversity (D) and joining (J) gene segments (Figure 11.1). Initiated by the lymphocyte-specific RAG1/2 complex [1,2], the reaction proceeds via the generation of DNA DSBs at recombination signal sequences (RSSs) that flank the V, D and J gene segments [3]. Components of the generally expressed non-homologous end-joining (NHEJ) pathway joins the DSBs between the participating gene segments to complete V(D)J recombination [3]. At the *IgH* locus, the productive assembly of the V, D and J segments upstream of Cμ constant region exons generates the Igμ polypeptide that pairs with a similarly recombined Igκ or Igλ light chain, leading to the eventual development of a naïve, immature B cell expressing a fully assembled IgM and IgD receptor on its surface. The combinatorial usage of the V, D and J gene segments, untemplated addition of nucleotides by the lymphocyte-specific enzyme TdT and DNA end-modifications that occur via during the end-

Molecular Biology of B Cells. https://doi.org/10.1016/B978-0-323-95895-0.00021-0

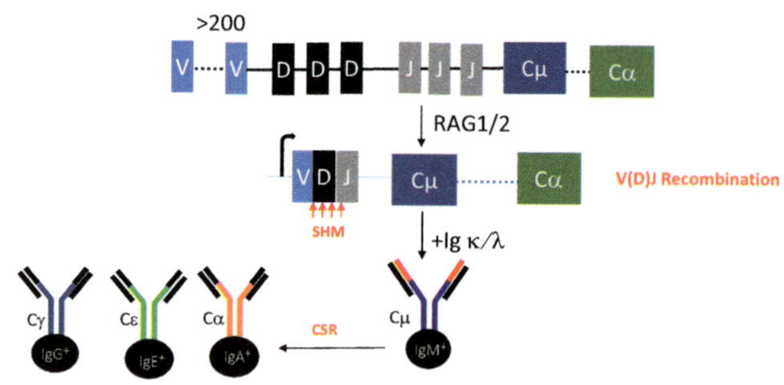

FIGURE 11.1 **Genomic alterations in B cells.** B cells undergo V(D)J recombination to assemble the variable region exons of the IgH and IgL chain genes. Subsequently, the B cells undergo somatic hypermutation (SHM) to mutate the variable region exons at a very high frequency and undergo class switch recombination (CSR) to change the isotype of the expressed antibody.

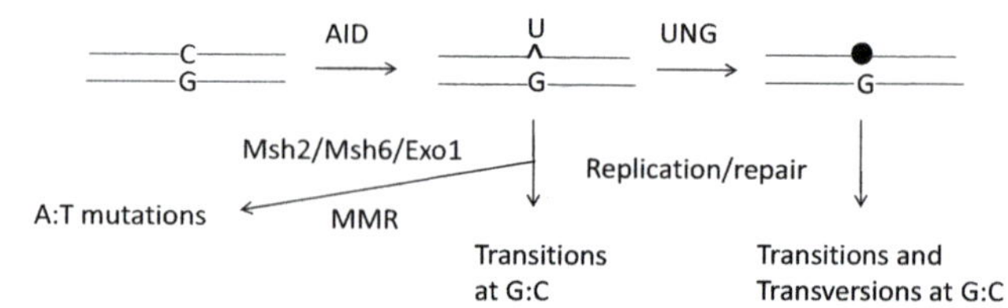

FIGURE 11.2 **Mechanism of SHM.** Components of the base excision repair (BER) and mismatch repair (MMR) pathways process the deaminated DNA, leading to transitions and transversions at G:C and A:T base pairs, respectively.

joining phase of V(D)J recombination generates B cells (and T cells) with a seemingly infinite repertoire of Ig (and T cell receptor) molecules [4]. The naïve $IgM^{+}IgD^{+}$ B cells subsequently migrate to the secondary lymphoid organs such as the spleen, lymph nodes and Peyer's patches where they encounter antigens and undergo two AID-initiated somatic alteration reactions: CSR and somatic hypermutation (SHM) [5] (Figure 11.1).

2.2 Somatic Hypermutation

During SHM, occurring primarily within microanatomical germinal centre (GC) structures that form within the lymphoid follicles, point mutations and sometimes deletions and insertions are introduced at a very high rate (10^{-2} to 10^{-3} per base pair per generation) into the variable region of both the *Igh* and *Igκ/λ* genes to ultimately produce B cells with increased receptor affinity [6]. Whilst mutations are found throughout the variable region exons, they are more frequently observed at RGYW/WRCY (R = A/G, Y=C/T, W = A/T) 'hot spot' motifs [7]. Typically, the mutations accumulate within the complementarity-determining region (CDR) loops that generally reside in the antigen-binding regions of an *Ig* protein. Recent studies have shown that variable region exons are enriched for SHM [8] and the sequences in the CDRs promote preferential AID deamination through enhancing DNA flexibility [9].

Transcription through the variable region exons is critical for SHM. Mutations are observed from 100 to 200 bp and 1.5 to 2 kb downstream of the promoter, thereby excluding the constant region gene segments from SHM [10–13]. In a simplified model for SHM (Figure 11.2), transcription through the variable region exons promotes the recruitment of AID, which deaminates deoxycytidines (dCs) to deoxyuridines (dUs) [14]. Replication across the dU:dG mismatches introduces transition mutations. On the other hand, removal of dU by the base excision repair (BER) enzyme uracil DNA glycosylase UNG prior to replication generates an abasic site, which, when replicated across by error-prone DNA polymerases, leads to the introduction of both transition and transversion mutations [14]. Furthermore, mismatch repair (MMR) proteins can recognise and process the DNA patch surrounding the dU:dG mismatches, after which it is filled in by error-prone DNA polymerases to give rise to point mutations at A:T residues and/or short insertions and deletions throughout the variable region [14].

2.3 Class Switch Recombination

The *IgH* locus is composed of a series of eight (mouse) or ten (human) constant region gene segments (C_H genes or exons) (Figure 11.3). CSR is a programmed DNA recombination event that exchanges the default Cμ exons for an alternative set of downstream *IgH* exons (Cγ, Cε, Cα) so that the B cell switches from expressing IgM to expressing a secondary Ig isotype such as IgG, IgE or IgA [15–17]. CSR is a deletional-recombination reaction that occurs between repetitive DNA elements called switch (S) regions that precede each of the C_H genes. CSR can also occur between Sμ and a poorly defined σδ region upstream of the Cδ exons, leading to class switching from IgM to IgD [18–22]. According to the prevailing model of CSR,

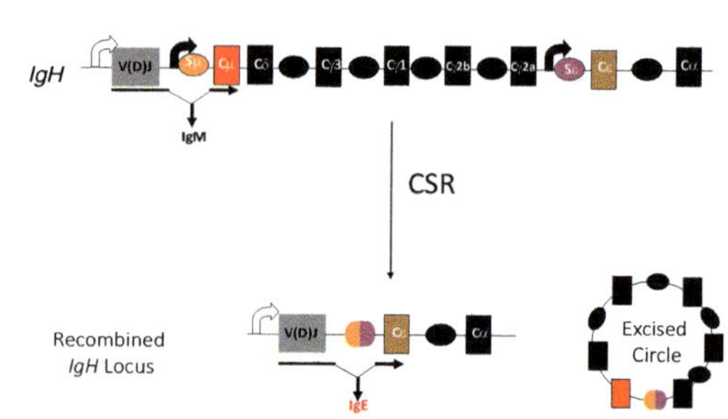

FIGURE 11.3 **Overview of CSR.** CSR is a deletional recombination reaction occurring at repetitive switch (S) region sequences that precede each of the constant region (C_H) gene segments. CSR leads to the juxtaposition of exons of different constant region gene downstream of the recombined VDJ exons and loss of the intervening sequence as an excised circle.

transcription through the S regions generates non-canonical nucleic acid structures such as RNA:DNA hybrids and G-quadruplexes that serve as targets for AID-mediated DNA deamination [23]. Subsequent processing by components of the BER and MMR machineries converts the deaminated residues into blunt DSBs or staggered single-stranded nicks that are processed into DSBs [5]. End-joining of DSBs between the acceptor and donor S regions by both NHEJ and a relatively poorly characterised alternative end-joining (alt-EJ) pathway juxtaposes a new set of C_H gene exons downstream of the VDJ exons and excises the intervening sequence as an extrachromosomal circle. Thus, CSR generates B cells expressing a new antibody with a distinct effector function but with the same affinity for the cognate antigen. For the remainder of this chapter, we will primarily discuss the roles of germline transcription, AID and DNA repair factors in CSR.

3. ACTIVATION-INDUCED CYTDINE DEAMINASE

3.1 Discovery of AID

AID was discovered in a subtractive hybridisation screen in the B lymphoma cell line CH12F3 [24,25]. AID, encoded by the *Aicda* locus, is a protein of 198 amino acids (24 kDa) that is barely detectable in unstimulated CH12F3 cells but whose expression increases by 10-fold in 6 h following stimulation of CH12F3 cells with a combination of anti-CD40, IL-4 and TGF-β [25]. The observations that AID-knockout mice were completely deficient in both CSR and SHM [26] and that AID mutations in patients with an autosomal recessive form of hyper-IgM syndrome (HIGM2) abrogate SHM and CSR [27] unequivocally established the essential role of AID in CSR and SHM.

3.2 AID Is a Single-Strand (ss) DNA-Specific Cytidine Deaminase

In vitro biochemical studies demonstrated that AID is a ssDNA-specific cytidine deaminase that converts dCs to dUs without observable activity on double-strand DNA, RNA or RNA:DNA hybrids [28–33]. Based on bacterial cytidine deaminases that are homologous to APOBEC1 and AID, the deamination reaction proceeds via a direct nucleophilic attack at position 4 of the pyrimidine ring of cytosine by a Zn^{2+} ion coordinated to AID. In vitro assays identified an 11 amino acid hotspot recognition motif (residues 113–123), distinct from the active site residues (56–58 and 87–90) that conferred specificity for the RGYW/WRCY hotspot motif. Replacing this recognition loop with one from another APOBEC family member (APOBEC3G or APOBEC3F) changed the mutational signature of AID towards that of the donor protein [34–36]. Finally, by solving crystal structures of AID bound to maltose-binding protein, AID was shown to possess a bifurcated substrate-binding surface that could capture two adjacent single-stranded overhangs simultaneously [37]. Furthermore, AID bound to nucleic acids that could fold into G4 structures induced cooperative AID oligomerisation. These studies implicated an intrinsic preference of AID for structured substrates and uncovered the importance of G4 recognition and oligomerisation of AID in CSR [37].

The ssDNA specificity of AID naturally leads to the question of how such substrates are generated during CSR. When transcribed in vitro in the physiological orientation, S regions form R-loops in which cytidines in the displaced G-rich non-template strand can be efficiently deaminated by AID [29,38–41] (Figure 11.4). These observations led to the prevailing model wherein transcription through mammalian S regions generates ssDNA R-loop substrates that permit AID-mediated deamination. The deaminated DNA is subsequently processed into DSBs through the activities of BER and MMR proteins (Figure 11.4). This R-loop-based model for AID access to S region DNA explains convincingly the essential requirements of S regions and S region transcription for CSR. However, A/T-rich S regions [42] and sequences from Xenopus inserted into the Ig loci [8] that at first approximation are not predisposed to forming R-loops can still be deaminated by AID, suggesting that non-R-loop-mediated mechanisms might also operate to facilitate CSR. In this regard, it is unclear how AID, an exclusively ssDNA-specific cytidine deaminase, accesses the variable region exons. One possibility is that transcription promotes negative supercoiling of DNA upstream of the elongating RNA polymerase, thus creating a DNA topology conducive for AID binding [43–45]. Another, though not necessarily mutually exclusive, proposal is that the ssDNA-binding protein replication protein A (RPA), which interacts with AID, stabilises ssDNA bubbles within transcribed variable region exons, allowing AID-mediated deamination to facilitate SHM [46]. Finally, the RNA-exosome complex can generate structures that are compatible with AID-mediated deamination on both DNA strands [47,48] (see below). Coordination of these factors can generate ssDNA tracks of >20 nt in length, the minimal length of ssDNA substrate required by AID to achieve CDR-preferential deamination [9,41].

4. GERMLINE TRANSCRIPTION

4.1 Germline Transcription Drives CSR

In addition to serving as exons of the mature IgH mRNA arising from the rearranged V_H promoter, each C_H region can also function as a distinct transcriptional unit. Each of the C_H genes (except Cδ) is organised as an independent

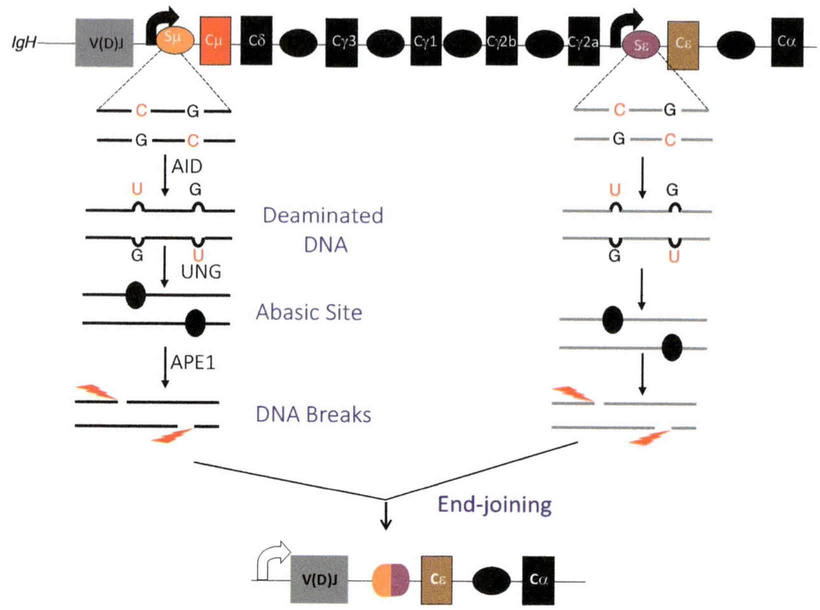

FIGURE 11.4 Mechanism of CSR. Activities of AID, and BER and MMR pathways lead to the generation of DSBs; end-joining of DSBs between acceptor and donor DSBs complete CSR.

transcription unit composed of a cytokine/activation-inducible promoter upstream of an intervening exon (I-exon), an intronic S region and C_H exons [49] (Figure 11.5). The primary transcript, initiated from the promoter upstream of the I-exon, proceeds through the S region and terminates downstream of the C_H exons. Splicing of the primary transcript removes the intronic S region and joins the I-exon to the C_H exons. The mature transcripts are polyadenylated but do not code for any proteins and are referred to as germline or sterile transcripts [50–53].

Transcription through individual C_H genes drives CSR to that particular isotype [54]. Ex vivo activation of mouse splenic B cells with bacterial LPS activates transcription through Iγ2b-Sγ2b-Cγ2b and Iγ3-Sγ3-Cγ3 transcription units and promotes CSR to IgG2b and IgG3, respectively. Conversely, addition of IL-4 inhibits transcription through γ2b and γ3 units and instead activates transcription through Iγ1-Sγ1-Cγ1 and Iε-Sε-Cε transcription units and CSR to IgG1 and IgE, respectively. Likewise, activation with TGF-β activates the Iα promoter and allows for CSR to IgA [55–57]. The μ promoter, on the other hand, drives constitutive transcription through Sμ even in unstimulated B cells [58]. Because of these stereotypical signal-specific CSR outcomes, ex vivo activation of mouse splenic B cells with distinct combinations of cytokines and activators has been a convenient tool to elucidate the mechanism of CSR.

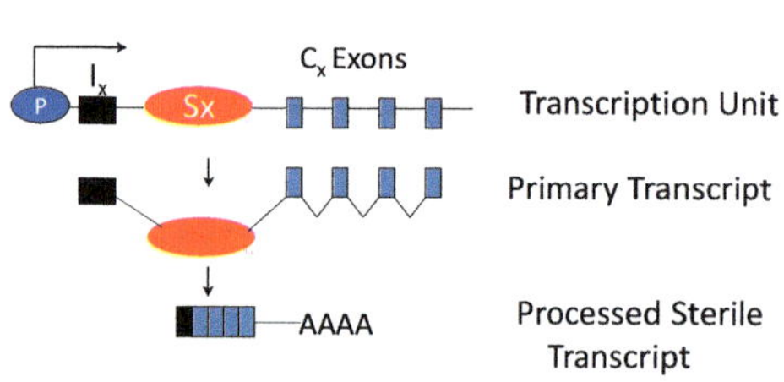

FIGURE 11.5 Germline transcription. Each of the C_H exons is a transcription unit comprising a cytokine-inducible promoter, an I-exon, the intronic S region sequence and the C_H exon. The primary transcript is spliced and polyadenylated, but the processed mature 'germline' transcript does not code for any protein.

The strict correlation between the induction of germline transcription and activation of CSR to particular isotypes was further validated through a series of mutational analyses. Deletion of I-exon promoters or regulatory regions (3′-RR) at the 3′-end of *Igh* abolished or severely reduced CSR, whilst replacing the inducible I-region promoters with heterologous constitutive promoters drove CSR in a cytokine-independent fashion [59–65]. These studies hence support the notion that germline transcription of particular C_H genes renders them 'accessible' for CSR [52,66,67]. We now know that the components vital for this accessibility include RNA polymerase II (Pol II)-dependent recruitment of AID to S regions [68], transcription-coupled modifications of histones at S regions [69,70], generation of RNA:DNA hybrid structures that provide ssDNA substrates of AID [71] and activity of the RNA exosome complex at the Ig loci [48,72].

4.2 Transcription Complex States and RNA Exosome in CSR

The functional association between SHM and transcription prompted experiments to determine whether AID physically interacts with components of the transcription machinery. Indeed, chromatin immunoprecipitation (ChIP) studies showed that AID occupies transcriptionally induced *IgH* loci during CSR [73], thus implicating AID at sites of RNA polymerase II (RNAP II) activity. Additionally, protein co-immunoprecipitation experiments revealed that AID and RNAP II physically interact within a complex [73]. In addition, Spt5, an evolutionarily conserved component of the transcription machinery, was found to interact with AID [68]. Depletion of Spt5 via shRNA knockdown reduced AID occupancy at the *IgH* locus and inhibited CSR in primary B cells [68]. Spt5 comprises one half of the heterodimeric DSIF complex [74]. DSIF inhibits transcription elongation and is critical for promoter proximal pausing of RNAP II [75,76]. Consequently, genomic occupancy of Spt5 is centred around transcription start sites (TSSs) [68,77]. Structural studies of DSIF/RNAP complexes have shown that Spt5 interacts with a region of RNAP close to the transcription bubble [78,79], indicating that Spt5 may tether AID onto RNAP II in close proximity to ssDNA [80]. Consistent with this hypothesis, genome-wide ChIP studies have found the genomic occupancy of AID to be concentrated near TSSs [81,82]. Co-localisation of RNAP II, Spt5 and AID near TSSs has led several groups to propose an RNAP II stalling model of AID function whereby stalled transcription complexes provide a platform for AID recruitment to ssDNA target sites [80,83–86]. Finally, stalled RNAP II recruits the AID cofactor RNA exosome for stalling resolution, providing a concomitant and independent evidence of AID function with transcriptional stalling.

IgH switch sequences interfere with transcription elongation due to their strong propensity to form R-loops [38–40]. RNAP II thus accumulates on *IgH* switch sequences during germline transcription of C_H genes in CSR, leading to RNA polymerase stalling [69,87] (Figure 11.6). Furthermore, R-loop formation and stability can hinder transcription elongation. This in turn creates a preponderance of premature transcription termination, leading to the recruitment of RNA exosome and subsequent resolution of arrested transcription complexes [72]. In fact, artificially inducing premature transcriptional arrest at an *Ig* locus through targeted insertion of a transcription termination element leads to enhanced hypermutation in the upstream region proximal to the site of transcription termination [44]. In addition, backtracked transcription complexes are also resolved through RNA exosome recognition of the extruding RNA 3′ end, resulting in degradation-coupled transcription termination [88]. Supporting the role of RNA exosome in CSR, the loss of RNA exosome-mediated RNA-decay activity in B cells leads to a significant decrease in CSR [72,89]. Moreover, both sense and antisense noncoding RNAs generated at switch sequences are processed by RNA exosome since loss of exosome activity leads to a robust accumulation of these noncoding RNAs [72,90,91]. Finally, RNA exosome complex and its cofactor complex NEXT biochemically interact with AID and promote the ability of AID to mutate both DNA strands of in vitro-transcribed AID substrates [47]. Taken together, RNA exosome recruitment to stalled RNAP II at switch regions stimulates AID activity to generate DNA DSBs and promote CSR.

4.3 G-Quadruplexes in CSR

A myriad of molecular factors affect AID targeting, including R-loops, the transcriptional and splicing machineries, the RNA exosome and G-quadruplex structures (G4s) [92]. G4s are non-canonical nucleic acid structures that can form in guanine-rich sequences via Hoogsteen base-pairing of planar guanine tetrads, generating a parallel four-stranded structure [93,94]. DNA G4s have been observed following in vitro transcription of S regions [95–97], which are composed of G-rich tandem repeats [98]. AID binding to transcription-induced G4s in vitro suggests that these structures may recruit AID in *cis* [99]. AID also binds to S region RNA G4s [37,100], and G4s bind to AID with 10-fold higher affinity than linear forms of the same sequence, perhaps accounting for the preferential deamination of dC residues adjacent to G4 structures [37].

G4 binding by AID contributes critically to effective CSR. Cooperative AID-G4 binding seeds the formation of large AID-G4 oligomers, and mutations that disrupt cooperativity and oligomer formation impair CSR without altering deamination activity [37]. Moreover, a Gly-to-Val mutation at residue 133 of AID (AID^{G133V}) disrupts AID-G4 RNA binding, abolishing AID targeting to S regions and CSR in ex vivo activated B cells [100]. Intriguingly, expression of S region transcripts that generate G4s, but not their antisense counterparts, rescues CSR in B cells with defective RNA splicing, suggesting that S region G4 RNAs ‘guide’ AID to the *Igh* locus, analogous to guide RNAs targeting Cas9 [100]. The identification of HIGM2 patients with homozygous G133V mutations in *AICDA* [101] suggests a conserved role for AID-G4 binding in CSR and underscores the importance of studying AID-G4 binding in vivo. An AID^{G133V} knock-in mouse provided in vivo evidence for the importance of G4 binding in AID targeting to the S regions [82]. Whilst AID^{G133V} and wild-type AID (AID^{WT}) have comparable DNA deaminase activity in vitro, AID^{G133V} mice completely lack CSR and SHM. We attribute the CSR defect to a failure of AID^{G133V} to localise to *Igh* S regions, as our ChIP-seq analysis found that AID^{G133V} exhibited a drastic defect in genome-wide chromatin localisation. Enforced localisation of AID^{G133V} to DNA by fusing it to a catalytically inactive AID protein markedly rescued CSR [82], strongly suggesting that AID^{G133V}, whilst unable to bind G4 structures, can mediate

(A)

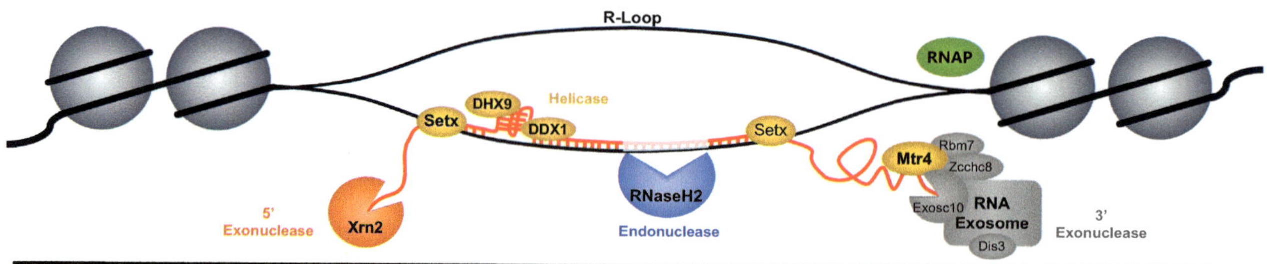

(B)

Process	Surveillance component	Main function (s)	Main subunits and Diseases
3'→ 5' Exonuclease	RNA exosome	R-loop processing	DIS3 (Multiple myeloma)[184] EXOSC2 (Retinitis pigmentosa)[170] EXOSC3 (Pontocerebellar hypoplasia type 1, Spinal motor neuron disease)[171–174] EXOSC8 (Cerebellar and corpus callosum hypoplasia, Spinal motor neuron disease)[169] EXOSC10
	PARN	Poly(A)-specific ribonuclease	Dyskeratosis congenita,[207–209] Familial pulmonary fibrosis[210]
	Regnase-1	Nonsense-mediated mRNA decay in immune system	
5'→ 3' Exonuclease	XRN1	Cytoplasmic RNA processing	
	XRN2	Transcription termination	
	DCP1/2	mRNA decapping enzyme	
Endonuclease	RNaseH1	Mitochondrial RNA processing/DNA replication	
	RNaseH2	Nuclear R-loop processing, ribonucleotide excision repair	Aicardi-Goutières syndrome[178,179]
Helicase	MTR4	R-loop processing	
	Senataxin (SETX)	R-loop processing, transcription termination	Ataxia with oculomotor apraxia 2,[58] Amyotrophic lateral sclerosis[59]
	DDX1	R-loop processing, G-quadruplex processing	
	DDX5	R-loop resolution with Xrn2	
	DDX19	RNA export, R-loop metabolism	
	DDX21	R-loop processing	
	DDX23	R-loop resolution	Adenoid cystic carcinoma[93]
	DHX9	R-loop processing, G-quadruplex processing	
	Aquarius (AQR)		
Cofactors	NEXT complex	Cofactor for RNA exosome, located in nucleoplasm	MTR4 RBM7 (Pontocerebellar hypoplasia type 1-like and spinal muscular atrophy)[175] ZCCHC8 (Familial pulmonary fibrosis)[211]

FIGURE 11.6 **Components of the RNA exosome machinery.** (A) The activities of the RNA exosome proteins at RNA in the context of R-loops. (B) Components of the RNA exosome complex.

CSR if properly targeted. Overall, the interaction between AID and G4s is critical in CSR; however, the relative contributions of AID:G4-RNA interactions versus a direct binding of AID to G4 DNA are yet to be fully resolved (Figure 11.7).

G4s and other highly ordered DNA structures may be extreme cases of substrates targeted by AID. In SHM, AID preferentially deaminates the Cs in the CDR region [8]. Recent work has shown that the flexibility of the ssDNA in the CDR region is key to this evolutionarily conserved phenotype [9]. In this context, the flexible ssDNA mimics the branched DNA structures seen in G4 DNA and transforms linear ssDNA into a shaped form [9,102]. Both the curly ssDNAs and G4 structures support efficient AID deamination through the salt-dependent AID−DNA interaction.

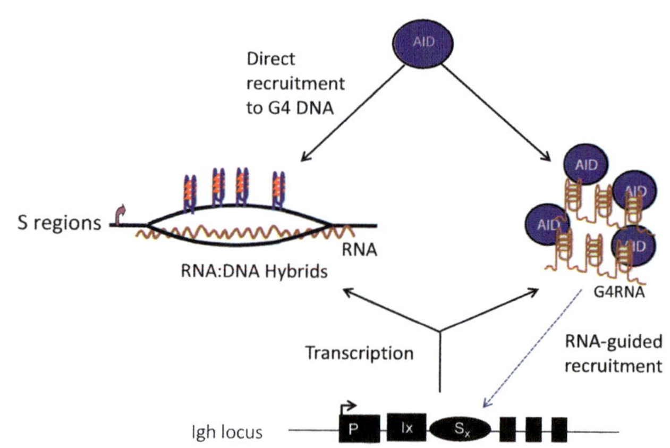

FIGURE 11.7 **Role of G-quadruplexes in CSR.** The S region RNA can bind AID and recruit it to DNA or G-quadruplexes formed at S regions can directly recruit AID.

5. FROM BASE LESIONS TO DSBs

5.1 Generation of DSB with BER and MMR Factors in CSR

Components of the BER pathway play a major role in CSR. Through BER, the AID-introduced dU in S region DNA is removed by UNG to generate an abasic site. The abasic site is recognised by the apurinic/apyrimidinic endonuclease APE1, generating a nick [14]. A closely spaced, similarly created nick on the opposite strand generates a staggered DSB. As predicted, mutations in UNG in humans and mice lead to impaired DSB formation at S regions and a severe block in CSR [103−108]. Additionally, mice heterozygous for APE1 deletion and CH12F3 cells with a homozygous deletion of APE1 have significantly impaired CSR [109−111]. Overall, these genetic data support the proposed AID/UNG/APE1 pathway in CSR. In addition to the well-established activities of UNG, UNG might have non-canonical properties that contribute to CSR, since a UNG mutant deficient in uracil removal activity can still promote CSR [112–114]; however, such non-canonical activity of UNG has yet to be defined. Besides UNG, SMUG1 was also shown to excise uracil from variable region genes, possibly serving as a backup pathway when UNG is unavailable [115].

The dU:dG mismatches can also be processed via the MMR pathway [116]. During this process, the MSH2-MSH6 heterodimer recognises dU:dG mismatches, triggering the recruitment of adaptors and other effectors such as the endonuclease heterodimer Pms2/Mlh1 and exonuclease 1 (Exo1) to generate nicks distal to the mismatch and subsequently convert the nicks into ssDNA gaps. Such activities occurring on both DNA strands could generate DSBs [15,117]. In mice, mutations in genes encoding the MMR proteins (Msh2, Msh6, Mlh1, Pms2 or Exo1) have significantly reduced CSR, providing further support for this model [103,118–121].

5.2 Role of AID Phosphorylation in Inducing DSBs at S Regions

Purification of AID from B cells undergoing CSR followed by mass spectrometric analyses demonstrated that AID is phosphorylated at the serine residue located at position 38 (S38) [122–126]. The S38 residue lies within a protein kinase A (PKA) consensus phosphorylation site and can be phosphorylated in vitro and in vivo by PKA [122–126]. The association of PKA with transcribed S regions undergoing CSR strongly suggests that PKA-mediated AID phosphorylation occurs at S regions [127].

B cells from mice with a S38A knock-in mutation (AID^{S38A}) are substantially impaired in mediating CSR, indicating that AID phosphorylated at S38 (pS38-AID) plays a critical role in CSR [125,126]. AID^{S38A} B cells were found to be impaired in their ability to generate DSBs at S regions, indicating that this specific phosphorylation event might be required at a step downstream of DNA deamination [128]. Indeed, immunoprecipitation experiments demonstrated that pS38-AID interacts with APE1, the enzyme critical for the generation of DNA breaks downstream of AID and UNG activities during CSR [128]. Furthermore, the phosphorylation-dependent interaction between AID and APE1 was significantly subdued in ATM-deficient cells, thereby unmasking a novel role of ATM in inducing DSBs [128]. Thus, a single protein, ATM, regulates both the generation and repair of DSBs during CSR, thereby coordinating the DNA damage response with DNA repair. Taken together, these observations provide strong evidence that phosphorylation of AID at S38 promotes its ability to interact with APE1 and to facilitate the formation of DSBs during CSR. The observation that DSBs were both dependent on and required for AID phosphorylation suggests that a positive feedback loop amplifies DNA breaks at S regions through AID phosphorylation. Envisioning a likely scenario wherein a high density of DSBs promotes end-joining between DSBs generated in two distal S regions necessitates a process of DNA break amplification for efficient CSR. A positive feedback loop provides a model whereby even after the assembly of AID and PKA at S regions [128], AID will not be efficiently phosphorylated until a DNA break is generated. Once a DNA break is formed, the rapid activation of AID phosphorylation and DSB formation will result in the synchronous activation of many molecules of AID bound to an S region. The high density of DSBs in S regions thus generates many broken DNA ends that promote the ligation of distal DSBs, thereby subverting normal DNA repair.

5.3 Different Outcomes of SHM and CSR

Whilst AID initiates both CSR and SHM [26,27], the outcomes of the two reactions are different. In CSR, AID-induced lesions at S regions are processed into DSBs, whilst in SHM, AID-induced lesions are propagated as point mutations (and sometimes deletions and insertions) without the requirement for a DSB intermediate. How AID-initiated DNA lesions are processed differentially during the two reactions remains poorly understood. Still, some clues emerge from examining amino acid substitutions in the AID protein itself [129]. The N-terminus of AID was found to be critical for SHM in a surrogate SHM assay [129], and an AID^{G23S} knock-in mouse line showed a dramatic SHM defect but with only a modest effect on CSR [130]. Likewise, the C-terminus of AID, which harbours a nuclear export sequence [131–134], plays a critical role in CSR but not in SHM [135–139]. The precise role of the C-terminus of AID in mediating CSR is yet to be fully resolved. AID lacking the C-terminus could still mutate the S regions, suggesting that AID targeting is not affected and that the defect in CSR lies downstream of AID-induced deamination [129,135]. Consistent with this idea, the C-terminus of AID was shown to recruit UNG and MSH2 proteins to S regions [140] and to alter the DNA damage response [141] for the efficient generation of DSBs [142]. In addition, the AID C-terminus deletion mutants have been reported in immunodeficient hyper-IgM syndrome (HIGM) type 2 patients and were found to exhibit a dominant negative (DN) effect [143]. The DN effect can be explained by the tendency of AID lacking the C-terminus to undergo phase separation, trapping the wild-type protein in inactive subnuclear compartments [139]. In this context, whether the *IgH* locus localises to different chromatin phase separation hubs during CSR and SHM requires further investigation.

5.4 The Error-Prone Choice in CSR

During CSR, BER and MMR enzymes process the base lesions into nicks that are then allowed to persist and be processed into DSBs, suggesting that these proteins participate in non-canonical pathways. For example, deficiency of XRCC1, a critical component of BER, has only a modest effect on CSR [144–146], which is dramatically different from the CSR defect observed in APE1-deficient B cells [109,111,147]. Although XRCC1 is also involved in alt-EJ as discussed in the next section, its role in sealing the BER nicks appears to be dispensible for CSR. Similarly, the role of MMR pathway in antibody diversification is considered to be non-canonical [148]. The regulation of error-prone choice also occurs at the early steps of BER. Genome-wide screenings of CSR factors have revealed that FAM72a can modulate a proper level of UNG to shift towards a mutagenic repair [149,150]. Similarly, HMCES, an abasic-site shielding enzyme [151], inhibits the generation of deletions during SHM [152], implicating it as a putative choice maker in CSR and SHM. The study of AID-dependent lesions has informed the development of base editors [153,154], and the dissection of pathway choice could inform the design of more precise or diverse base editors.

6. DNA END-JOINING DURING CSR

6.1 Distinct DNA Recombination Products

Productive *IgH* CSR is a long-range deletional recombination event [17] in which the DNA fragment between two DSBs separated by ~100 kb linear distance is deleted. The general property of cellular end-joining machinery ensures a high-frequency of intra-chromosomal recombination, as proven by a series of elegant genetic studies using yeast I-SceI meganuclease and other designer nucleases [155–157]. Independent of AID activity, I-SceI-introduced DSB at engineered *IgH* locus can initiate substantial levels of CSR [155], revealing a two-step model of CSR as AID first introduces DNA lesions in the S regions and the general DNA repair machinery ligates the broken ends. Besides the long-range deletional recombination, intra-S region deletion (ISD) represents a CSR byproduct, evidenced by examining the S region sequences from hybridoma lines [158]. DSB response factors were found to differentially regulate the formation of long-range CSR and short-range ISD [159–161], suggesting distinct underlying mechanisms.

At the *IgH* locus, the tandemly located S regions and S-like sequences further allow complex recombination products, including sequential CSR. The IgG/E class switch programme exemplifies sequential CSR, as a large portion of recombined junctions in IgE expressing cells harbour Sγ1 sequence [162,163]. Furthermore, AID targeting at S-like sequences in *IgH* 3′RR leads to locus suicide recombination [164], promoting the deletion of constant region genes and thus, the loss of that particular B cell. Overall, the physiological recombination products are not always byproducts and could modulate B-cell functions and fates.

6.2 Orientation-Specific End-Joining in Deletional CSR

In CSR, two breaks in the donor and acceptor S regions lead to four DSBs, and only the ligation of the DSBs between Sμ and the downstream S region DSB leads to a productive deletional class switch from IgM to IgG1. Whilst both deletional and inversional end-joining events can be visualised in B cells undergoing CSR [67,165,166], deletional recombination was considered to result from cellular selection for the productive *IgH* gene. So far, most chromosomal DSB ends are joined without orientation specificity, as shown by the joining of DSBs generated by designer enzymes or arising from endogenous damage [167–169]. However, the unbiased high-throughput genome-wide translocation sequencing (HTGTS) approach revealed an intrinsic orientation-specific end-joining during CSR [170]. In this context, the upstream and downstream DSBs are preferentially ligated to produce biased deletional recombination with a ~10:1 deletion/inversion ratio [170]. In addition, the orientation-specific joining in CSR can also be visualised by other semi-quantitative methods [170]. The predominantly deletional recombination event in CSR is reminiscent of RAG-mediated V(D)J recombination at the *IgH* locus [3], a feature that increases the probability of productive recombination.

The orientation-specific end-joining was further supported by several lines of genetic evidence. 53BP1-deficient B cells, which have extended DSB end resection at broken S regions, lose the biased end-joining and present a ~2:1 deletion/inversion ratio of CSR junctions [170]. DSB response factors including ATM and H2AX and 53BP1-downstream factors such as RIF1 and SHLD1 also participate in this process as B cells deficient in these factors partially lose orientation-specific end-joining [170,171]. Furthermore, a focussed CRISPR screening yielded a unique role of the end-joining factor ERCC6L2 in orientation-specific end-joining [172]. In the absence of ERCC6L2 protein or its putative helicase or remodeler activity, the orientation-specific end-joining was completely lost and resulted in a nearly 1:1 deletion/inversion ratio [172]. Overall, DSB response and end-joining factors work together with the cis-*IgH* chromatin and/or sequence features to ensure such deletional recombination.

6.3 The Chromatin Loop Extrusion Model

At the *IgH* locus, the participating switch regions could be as far apart as 100 kb and chromatin interactions mediated by transcription, AID and ATM-dependent DNA damage response proteins were proposed to mediate long-range S region synapsis during CSR [173–178]. However, experimental evidence provided by Alt and colleagues has led to the formulation of cohesion-dependent loop extrusion model to explain the dynamic nature of the *IgH* locus during CSR [179,180] (Figure 11.8). Chromosome conformation capture technologies revealed that the nuclear chromatin is organised into megabase-sized topological-associating domains (TADs) [181,182], with boundaries demarcated by CTCF-binding sites [181,183]. Cohesin was shown to play a central role in the extrusion of the chromatin loop [183–185] through its DNA compaction activity [186,187] and its ability to regulate DNA repair. The loop extrusion model satisfactorily explained not only the mechanism of S region synapsis during CSR but also the accessibility of RAG to the recombining gene segments during V(D)J recombination [188].

The chromatin loop extrusion model (Figure 11.8) reveals a clear mechanism of deletional CSR within the *IgH* domain [179,180]. According to this model, in naïve B cells, cohesin extrusion first generates a basal dynamic loop with the iEμ and 3′RR serving, respectively, as the upstream and downstream anchors of a CSR centre (CSRC). Upon cytokine stimulation, a particular I region promoter is brought proximal to the CSRC to initiate GLT, which further assists secondary cohesin loading to form a subloop. Deletional CSR is then achieved by the distinct kinetics of AID activity at the donor and acceptor S regions [189]. In this dynamic *IgH* loop domain, the orchestrated loop extrusion mechanism reels in the I promoters and S regions into the recombination centre with enhancer elements to induce GLT and AID targeting, underpinning the deletional CSR. The chromatin loop extrusion model explains CSR in the context of a 4D genome, in which both the three-dimension chromatin structure and the timing of AID activity play critical roles [91]. Supported by the early observation of sequential AID activity at S regions [189] and AID expression pattern [190], the model explains the orientation-specific joining in deletional CSR [179,180].

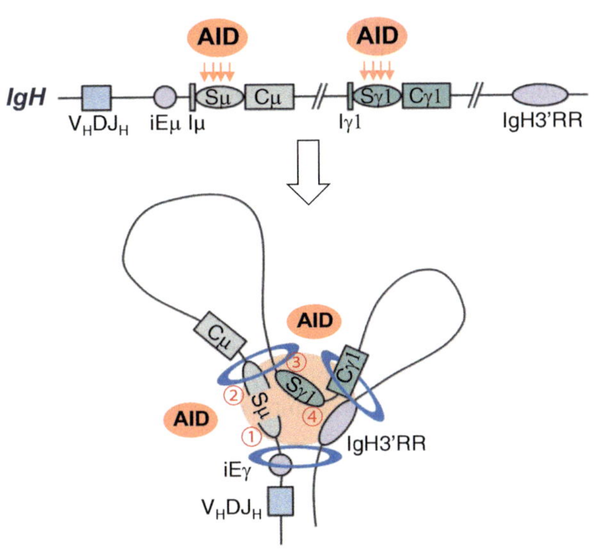

FIGURE 11.8 **The loop-extrusion model for CSR.** See text for details.

7. DSB DAMAGE RESPONSE AND END-JOINING

7.1 DSB Response and Broken end Protection

CSR requires the joining of DSBs in donor and acceptor S regions that may be separated by over 100 kb; however, some DSBs within an S region may be joined to another DSB within the same S region, resulting in an internal deletion rather than productive CSR [50,158,191]. In addition, DSBs in S regions can be ligated to a DSB on another chromosome to generate a chromosomal translocation [192]. The molecular mechanisms that promote the end-joining of DSBs in distal S regions rather than canonical DNA repair, internal deletions or chromosomal translocations are still being evaluated. During CSR, the *IgH* locus is re-organised into discrete transcriptionally active domains wherein I-promoters and regulatory enhancers (Eμ and Eα) are positioned close to one another to promote transcription, accessibility and synapsis of recombining S regions [83,173,179,180]. Productive CSR is observed in B cells that have S regions replaced by I-SceI restriction sites [155], suggesting that the general cellular DNA damage repair pathways, which function in the synapsis and long-range end-joining of S-region DSBs, are required for the resolution of DSBs in S regions and productive CSR. Following the introduction of DSBs in the S regions, ATM and its substrate 53BP1 are thought to promote S–S region synapsis and recombinogenic repair [160,193,194]. In the absence of ATM, DSBs at IgH and chromosomal translocations involving IgH are increased and CSR is decreased [160,193–195]. 53BP1 probably facilitates S–S synapsis, as Eα interactions with Eμ and the γ1 promoter are reduced in *53BP1* $^{-/-}$ B cells that are stimulated with LPS or LPS + IL4 [194].

ATM plays a key role in CSR not only in the stabilisation of S-region DSBs through the proposed synapsis and joining of S regions but also, as described earlier, in the generation of DSBs through the phosphorylation of AID and the subsequent interaction of AID with APE1 [128,160,193,195]. Activation of ATM kinase activity requires binding of the Mre11/RAD50/Nbs1 (MRN) complex to DSBs, which induces ATM-dependent phosphorylation of proteins mediating cell cycle checkpoints (for example, p53) and DNA repair, such as H2AX, MDC1, Nbs1 and 53BP1 [5,145]. The DNA damage response initiated by ATM promotes the assembly of macromolecular foci flanking DSBs and provides docking sites for DNA repair proteins to bind and stabilise DNA ends to promote recombinogenic repair during CSR. Null mutations in ATM substrates impair CSR and increase chromosomal abnormalities and translocations [160,193,195].

53BP1 deficiency leads to the most robust defect in CSR [196,197]. Mutation or deletion of 53BP1 results in a 90% defect in CSR with a significant proportion of chromosomal aberrations involving the *IgH* locus [195], as well as a high frequency of Sμ internal deletions in cells stimulated for CSR [161]. 53BP1 promotes the synapsis and long-range joining of S regions and protects DNA ends from end resection to direct DNA repair towards NHEJ [156,198]. Consistent with these roles for 53BP1, ATM-mediated phosphorylation of 53BP1 recruits Rap-1 interacting factor (Rif1) to sites of DNA damage to protect DNA ends from resection and to promote DNA repair [199]. Accordingly, Rif1-deficient B cells are significantly impaired in CSR [199].

REV7, a component of DNA polymerase zeta, was identified to protect the broken end from resection downstream of 53BP1-RIF1 [200,201]. As a multifunctional protein, REV7 is involved in many cellular DNA metabolism processes, and it also promotes efficient CSR by coordinating translesion synthesis through AP sites [202]. Additional biochemical studies have revealed the critical role of Shieldin complex in CSR, which contains REV7 and three uncharacterised protein subunits [203–208]. The Shieldin complex is recruited via 53BP1-RIF1 to protect the ssDNA ends, and it further recruits CTC1-STN1-TEN1 complex and polymerase α-primase complex to fill in the short ssDNA tracts [209]. Thus, the 53BP1-dependent role in CSR could function through both end protection and DNA synthesis [210]. Moreover, BRCT-repeat inhibitor of hTert expression (BRIT1) has been implicated as a novel effector of the DNA repair phase of CSR [211]. BRIT1 is a ubiquitously expressed protein that is rapidly recruited to radiation-induced DSBs through its C-terminal BRCT repeat domain, which is necessary for its interaction with γH2AX. In addition, the BRIT1-γH2AX pathway is further modulated by the interaction of γH2AX with MDC1 in CSR. Although BRIT1 or MDC1 deficiency alone leads to a moderate reduction in CSR, loss of both BRIT1 and MDC1 together markedly impairs CSR [211]. Thus, BRIT1 likely serves as a scaffold to recruit factors that resolve DSBs at S regions downstream of ATM.

7.2 Canonical NHEJ and A-EJ

Homologous recombination (HR) and NHEJ are the two major pathways for DSB repair in mammalian cells [17]. HR is restricted to the S/G_2 phase of the cell cycle and requires large stretches of homology, whereas NHEJ is active throughout the cell cycle and requires little or no homology. Since CSR-associated DSBs are observed primarily during the G_1 phase of the cell cycle and do not have extended stretches of homology, NHEJ is generally considered the major pathway in the joining of DSBs during CSR [17,145]. Consistent with this, mutations in the canonical NHEJ components Ku70/Ku80 heterodimer (Ku), XRCC4 and DNA ligase IV (Lig4) severely compromise CSR, whilst mutations in non-canonical NHEJ proteins such as DNA-dependent protein kinase catalytic subunit (DNA-PKcs), Artemis and XRCC4-like factor (XLF or Cernunnos) promote chromosomal translocations even though CSR is not severely impaired [17,145].

B cells lacking core NHEJ components are still capable of residual CSR mediated by the microhomology-biased A-EJ pathway. Although A-EJ is a poorly defined DNA repair mechanism guided by microhomology between two DSBs, factors from other DNA repair pathways, including XRCC1, Ligase III, Mre11, Parp1 and CtIP, have been shown to be necessary for A-EJ during CSR [212–215]. More recently, Rad52, an HR repair factor, was shown to facilitate microhomology-mediated A-EJ that favours intra-S region recombination and competes with Ku to mediate inter-S region DSB recombination [216]. Whether A-EJ is physiologically necessary to complete a productive class switch reaction is not clear at present.

Whilst C-NHEJ and A-EJ are the primary end-joining pathways ligating DSBs within IgH during CSR, HR has been proposed to repair AID-induced off-target DSBs. Deficiency of the Rad51 paralog XRCC2, a key component of HR-mediated repair, significantly enhances AID-dependent genome-wide DNA damage [217,218]. Notably, AID-expressing human chronic lymphocytic leukaemia cells are hypersensitive to HR inhibitors and this is possibly due to AID-dependent synthetic cytotoxicity from unrepaired DSBs at non-Ig loci [219]. Thus, HR is essential for repairing AID-generated DSBs, and dysregulated AID activity may provide a novel therapeutic approach to treat B-cell malignancies.

7.3 Switch Region RNA Coordinates AID Activity and End-Joining Pathways

G-rich ncRNA expressed from switch regions (e.g. Sμ GLT) is modified by N6 methyladenosine (m^6A) and recognised by the reader protein YTHDC1, which then recruits the RNA exosome complex via the adaptor protein MPP6 for processing of these transcripts (citation). Before m^6A-modified GLTs are recognised by RNA decay pathways, m^6A-recruited YTHDC1 transiently stabilises R-loops to initiate AID deamination on the non-template strand of switch regions. Following this step, recruitment of RNA exosome promotes the formation of AID-mediated DSBs, leading to robust CSR. Thus, the loss of m^6A modification on SmuGLT destabilises R-loops at switch sequences, impairing symmetrical generation of nicks on DNA strands and inhibiting robust DNA end-joining via NHEJ [90,91,220]. In conditions wherein DNA/RNA hybrids are stabilised and R-loops/G4 DNA persist, alt-EJ becomes the preferred mechanism of DNA repair during CSR [91,220,221]. The question arises: Why does m6A modification promote RNA processing of switch region transcripts? It has been shown that m^6A modification can alter RNA/RNA pairing and facilitate reader protein recognition of modified transcripts [222,223]. It is possible that m^6A modification inhibits RNA/RNA interactions and facilitates interaction between the RNA/DNA strand to stabilise the resulting R-loop and G4 DNA structures. In addition, this alteration of RNA secondary structure could facilitate m^6A reader protein binding (YTHDC1) to the ncRNA transcript. Similar to work done previously [222,223], m^6A modification could alter RNA/RNA interactions and RNA secondary structure to facilitate alternative nucleotide or reader protein binding.

7.4 Unique Players in CSR

The flexible nature of NHEJ has evolved to handle diverse cellular DSBs. During CSR, AID-initiated lesions are predominantly processed into staggered DSBs [224,225], with non-B DNA structures or the transcription machinery present in the vicinity. Thus, specific end-joining and end-processing factors could play unique roles in CSR. For example, 53BP1 deficiency severely reduces CSR, with the frequency of class switching only 5%–10% of normal levels [196,197]. Although the 53BP1-dependent end-protection pathway has been extensively dissected, deficiencies in none of the downstream factors cause such a dramatic defect. Whether this severe defect reflects an additive effect of known 53BP1 effectors or if there are unknown effectors downstream 53BP1 remains unclear. DSBs generated following the activities of AID and the MMR/BER proteins during CSR are distinct from the hairpin coding ends generated during V(D)J recombination [226]. Thus, the Artemis endonuclease, which is involved in hairpin opening and critical for V(D)J recombination [227,228], plays only a minor role in CSR [229]. On the contrary, ERCC6L2 plays a unique role in CSR [172]. Deleterious ERCC6L2 mutants caused bone marrow failure [230–236] similar to that observed upon deletion of canonical NHEJ factors [237–240]. ERCC6L2 is a canonical NHEJ factor whose deficiency leads to ionizing radiation (IR) sensitisation and defective CSR [172,241,242]. Strikingly, orientation-specific end-joining is completely lost in the absence of ERCC6L2, a phenotype distinct from the partial loss of end-joining bias observed in end-resection mutants [170,172]. The putative remodeler or helicase activity of ERCC6L2 is critical for the unique deletional recombination in CSR [172]. The dissection of unique CSR features could lead to new mechanisms of DSB response and end-joining.

8. CHROMOSOMAL TRANSLOCATIONS AND AID-INDUCED COLLATERAL DAMAGE

8.1 Identity of AID Off-Targets

Besides the *Ig* loci, AID occasionally targets other genome loci to generate deleterious outcomes. Recent single-cell whole-genome sequencing revealed that for every 1 mutation at *Ig* loci, a normal B cell accrues 18 off-target mutations [243]. Thus, the AID off-targets derived from aberrant SHM or CSR present a major source of chromosomal translocation partners in B-cell lymphomas [92,244]. Even before the discovery of AID, aberrant genomic alterations were observed at highly expressed genes such as *BCL6* in normal B cells [245,246]. Systemic sequencing analysis of mouse genes has yielded dozens of AID off-targets [247,248]. Finally, advances in sequencing and translocation-capture technologies have led to the genome-wide identification of AID off-targets during CSR [167,168,249,250]. Nonetheless, compared to the AID Ig on-targets, the mutation/translocation frequency at AID off-targets is orders of magnitude lower. Despite the low frequency, however, the absolute number of AID off-targets acts as a major source of genome instability in B cells ([243] #4303; [251]). Nonetheless, compared to actual AID targets, the mutation/translocation frequency at AID off-targets is orders of magnitude lower. Despite the low frequency, however, the absolute number of AID off-targets acts as a major source of genome instability in B cells [243]. The specific targeting mechanisms distinguishing *Ig* on-targets and off-targets is still an intriguing, unresolved question.

8.2 Preferential Targeting of AID at Proto-Oncogenes

Proto-oncogenes comprise a large fraction of non-Ig AID targets [92,244]. One could argue that the preference for proto-oncogenes could be due to cellular selection. However, the unbiased identification of AID off-targets with next-generation sequencing approaches in CSR-activated B cells without cellular selection [167,168] strongly suggests that the proto-oncogenes in B-cell lymphomas are indeed preferentially targeted by AID in which the super-enhancers (SEs) at these AID targets play a major role in this preferential targeting [72,89,249,250]. SEs are a group of enhancer clusters, usually a few hundred in a particular cell type, that are critical in cell identity establishment and tumourigenesis [252–255]. Through an unresolved mechanism, AID is recruited into SE hubs, where the general gene transcription machinery meets the enhancer elements and pervasive transcription generates ssDNA substrates for AID targeting [249,256]. In CSR-activated B cells, AID-dependent translocation junctions were found to be enriched at sites immediately downstream of transcriptional start sites [249]. Correlation analysis between AID targets and known chromatin features reveals strong enrichment of cryptic transcription [72,249] and epigenetic enhancer markers [249,250,257]. Thus, AID off-targets are a major source of genomic instability in lymphoma and leukaemia [92].

8.3 Deleterious Outcomes in Aberrant CSR

Recurrent chromosomal translocation between *Ig* loci and a proto-oncogene is a feature of B-cell lymphomas and in mouse models, *IgH*:*c-Myc* translocations have been demonstrated to be AID-dependent [192,258]. The factors that determine the formation of DSBs and the frequency of aberrant end-joining at the non-Ig translocation partners are poorly understood. At these non-Ig targets, AID-initiated lesions are likely differentially repaired, leading to distinct outcomes. Extensive studies in gene-editing strategies reported that different genomic loci could evoke diverse repair mechanisms following Cas9-generated DSBs [259,260]. The influence of the primary DNA sequence and epigenetic features of the off-target loci in directing error-prone versus error-free repair during AID-initiated chromosomal translocations is yet to be elucidated. DSBs arising at the *IgH* loci are restrained within defined chromatin domains to facilitate efficient CSR [180] and prevent aberrant end-joining. In this context, the recurrent translocations in tumours most likely represent a small fraction of DSB outcomes and reflect cellular selection. For example, the strong enhancer elements at *Ig* loci can enhance the activation of proto-oncogenes [261]. Additional non-exclusive possibilities that predispose certain loci to translocations include proximity [69,262], nuclear positioning of DSBs [263] and timing of DNA replication that could increase the chance of DSB ligation at different chromosomal loci [264]. Altogether, the chromatin context can contribute to the deleterious outcomes of aberrant CSR.

9. EVOLUTION OF CSR MECHANISM

9.1 Evolution of Cytidine Deaminase-Initiated Immune Diversification

Jawless vertebrates like lampreys have evolved a variable lymphocyte receptor (VLR) system composed of highly diverse leucine-rich repeats as an antigen receptor for adaptive immunity [265–267]. Despite the dramatic differences between immunoglobulins and the VLR systems, AID/APOBEC superfamily cytidine deaminases CDA1 and CDA2, with ssDNA specificity, have been proposed to initiate the diversification process in lampreys [268–270].

Diverse antibody isotypes, determined by the Ig heavy chain, can lead to different downstream effector functions; thus, CSR offers an efficient way to change the effector functions of an antibody without affecting its antigen-binding specificity. All jawed vertebrates undergo both V(D)J recombination and SHM; however, CSR was the last of the lymphocyte-specific DNA alteration reaction that emerged in evolution, appearing at the time of divergence of the amphibians. Whilst conserved in all tetrapods, CSR is absent in bony fish [271]. The emergence of CSR in evolution co-occurs with the S regions that are enriched in RGYW motif and has the tendency to form non-B secondary structures [272]. In this context, the mammalian and bird S regions show both RGYW-rich and GC-rich features [40,272]. Although many amphibian or reptile S regions share the feature of GC-richness, a few of them are relatively AT-rich [273–275] and lack the ability to form R-loop upon transcription [42,273]. A closer examination of the Sμ from *Xenopus laevis* identified repetitive GG motifs that can drive R-loop formation in a 2 kb S region DNA, suggesting that R-loop formation in CSR is evolutionarily conserved [276]. These evolved S regions may facilitate canonical CSR by attracting more AID activity [277] and possessing a higher tendency to break. Besides the features of S regions that promote destabilisation, the *IgH* architecture also contributes to CSR. With more than 11 kinds of isotypes documented, most of the *IgH* constant exons are transcribed in the same direction as the variable V(D)J exon, with a few exceptions [278]. In some birds, Cα exon is transcribed in the opposite direction, in which inversion CSR yields a productive recombination product. Similar inverted constant exons are also present in the reptile genome [278]. How inversion CSR is achieved with high efficiency is an intriguing question. Beside the canonical CSR, a CSR-like process has been described in shark [279]. Distinct from the translocon *IgH* architecture present in bony fishes and mammals, the shark *Ig* genes spread over hundreds of miniloci and contain only one set of V and constant exons [280]. Reciprocal exchange between IgM and IgW coding genes was documented [279], as well as exchange between two IgW-coding genes [281]. The molecular mechanism of shark CSR awaits further dissection.

Evolutionary analysis of DNA repair factors also reveals a correlation with CSR. For example, key subunits of Shieldin complex, including SHLD1-3, were identified in jawed vertebrates, and SHLD3 is only present in species that undergo CSR [205]. As the key effector of 53BP1-dependent end protection, Shieldin complex facilitates efficient end-joining in CSR [171,202–204,207].

Similarly, ERCC6L2, a homologue of CSB (ERCC6), emerges in vertebrates and promotes deletional recombination [172]. Thus, it is tempting to hypothesise that the burst of DSB response and end-joining factors underpins the evolution of CSR.

9.2 Physiological Role of AID Beyond the Ig Loci?

Several studies have shown that AID can localise to multiple MHCII genes [82,247,248]. AID expression was observed to correlate with decreased MHCII surface staining within DZ GC B cells [82]. Additionally, MHCII gene transcription is regulated by CIITA, and the gene encoding CIITA has been proposed to be a target of aberrant SHM [282–284]. These observations led to the hypothesis that AID may directly regulate MHCII gene expression in GC B cells, as DNA DSBs and deamination can regulate transcription [285–287]. Whilst non-Ig AID targets are generally considered to be collateral damage incurred during BCR secondary diversification, AID-dependent regulation of gene expression may instead be a programmed physiological adaption in B-cell responses. Alterations in MHCII expression would be expected to modulate T-cell help within the GC, a critical factor in B-cell expansion and affinity maturation. Intriguingly, another critical regulator of GC B-cell biology, *Bcl6*, is not only the most highly mutated non-Ig AID target, but is also thought to be selectively permissive to DNA damage [246,247,288].

AID-dependent gene regulation of MHCII and *Bcl6* could affect the probability of a B cell to leave the GC, acting as rheostat to prevent B cells from staying in an environment that is dangerously permissive to DNA damage. The observations that AID^{KO} GC B cells do not efficiently differentiate into plasma cells [289], and that low levels of BCL6 in LZ GC B cells are thought to favour GC exit and differentiation into plasma cells [290], suggest that AID may influence GC B-cell fates. The absence of this regulation could thus contribute to the GC B-cell hyperplasia observed in AID^{KO} mice. Importantly, AID-deficient mice and CRISPR-Cas9 technology will allow many of these ideas to transition into testable hypotheses, making AID-dependent regulation of gene expression an exciting avenue for future investigation in physiological and pathological B cells.

10. CONCLUDING REMARKS

B cells have evolved to perform CSR, a mechanistically counterintuitive process that involves a long-range deletional recombination event in which the deliberate generation of DSBs, one of the most toxic lesions that can happen in a cell, is obligatory. How this highly complex reaction is orchestrated to promote antibody gene diversification whilst minimising collateral genomic damage in a B cell is yet to be fully elucidated. Most research reviewed here has focussed on the positive regulation of CSR. But, one aspect that has to be borne in mind is that in *in vivo* contexts, CSR is only one of the multiple cell fates that a clonally expanding population of B cells undergoes. As such, there has to be a coordinated distribution of cell fates other than class-switched B cells including plasma cells and memory B cells. This could be achieved via the negative modulation of CSR by, for example, aryl hydrocarbon receptor [291]. There has to be active genetic and epigenetic modules that skew cell fate decisions towards one path versus another. Such fine-tuning would presumably be necessary to achieve a robust humoral response that fulfils all aspects of adaptive immunity (adaptability, specificity and memory). Future research is therefore warranted to understand the molecular mechanisms underlying the negative regulation of these modules of cell fate decision programmes in B cells during a GC response.

REFERENCES

[1] Schatz DG, Oettinger MA, Baltimore D. The V(D)J recombination activating gene, RAG-1. Cell 1989;59:1035–48.

[2] Oettinger MA, Schatz DG, Gorka C, Baltimore D. RAG-1 and RAG-2, adjacent genes that synergistically activate V(D)J recombination. Science 1990;248:1517–23.

[3] Schatz DG, Swanson PC. V(D)J recombination: mechanisms of initiation. Annu Rev Genet 2011;45:167–202.

[4] Jung D, Alt FW. Unraveling V(D)J recombination; insights into gene regulation. Cell 2004;116:299–311.

[5] Matthews AJ, Zheng S, DiMenna LJ, Chaudhuri J. Regulation of immunoglobulin class-switch recombination: choreography of noncoding transcription, targeted DNA deamination, and long-range DNA repair. Adv Immunol 2014;122:1–57.

[6] Mesin L, Ersching J, Victora GD. Germinal center B cell dynamics. Immunity 2016;45:471–82.

[7] Rogozin IB, Diaz M. Cutting edge: DGYW/WRCH is a better predictor of mutability at G:C bases in Ig hypermutation than the widely accepted RGYW/WRCY motif and probably reflects a two-step activation-induced cytidine deaminase-triggered process. J Immunol 2004;172:3382–4.

[8] Yeap LS, Hwang JK, Du Z, Meyers RM, Meng FL, et al. Sequence-intrinsic mechanisms that target AID mutational outcomes on antibody genes. Cell 2015;163:1124–37.

[9] Wang Y, Zhang S, Yang X, Hwang JK, Zhan C, et al. Mesoscale DNA feature in antibody-coding sequence facilitates somatic hypermutation. Cell 2023;186:2193–2207 e2119.

[10] Gearhart PJ, Bogenhagen DF. Clusters of point mutations are found exclusively around rearranged antibody variable genes. Proc Natl Acad Sci U S A 1983;80:3439–43.

[11] Lebecque SG, Gearhart PJ. Boundaries of somatic mutation in rearranged immunoglobulin genes: 5' boundary is near the promoter, and 3' boundary is approximately 1 kb from V(D)J gene. J Exp Med 1990;172:1717–27.

[12] Rada C, Gonzalez-Fernandez A, Jarvis JM, Milstein C. The 5' boundary of somatic hypermutation in a V kappa gene is in the leader intron. Eur J Immunol 1994;24:1453–7.

[13] Peters A, Storb U. Somatic hypermutation of immunoglobulin genes is linked to transcription initiation. Immunity 1996;4:57–65.

[14] Petersen-Mahrt SK, Harris RS, Neuberger MS. AID mutates *E. coli* suggesting a DNA deamination mechanism for antibody diversification. Nature 2002;418:99–103.

[15] Chaudhuri J, Alt FW. Class-switch recombination: interplay of transcription, DNA deamination and DNA repair. Nat Rev Immunol 2004;4:541–52.

[16] Vuong BQ, Chaudhuri J. Combinatorial mechanisms regulating AID-dependent DNA deamination: interacting proteins and post-translational modifications. Semin Immunol 2012;24:264–72.

[17] Alt FW, Zhang Y, Meng FL, Guo C, Schwer B. Mechanisms of programmed DNA lesions and genomic instability in the immune system. Cell 2013;152:417–29.

[18] Chen K, Xu W, Wilson M, He B, Miller NW, et al. Immunoglobulin D enhances immune surveillance by activating antimicrobial, proinflammatory and B cell-stimulating programs in basophils. Nat Immunol 2009;10:889–98.

[19] Rouaud P, Saintamand A, Saad F, Carrion C, Lecardeur S, Cogne M, et al. Elucidation of the enigmatic IgD class-switch recombination via germline deletion of the IgH 3' regulatory region. J Exp Med 2014;211:975–85.

[20] Choi JH, Wang KW, Zhang D, Zhan X, Wang T, et al. IgD class switching is initiated by microbiota and limited to mucosa-associated lymphoid tissue in mice. Proc Natl Acad Sci U S A 2017;114:E1196–204.

[21] Hussein I, Nour G, Francois B, Yves D, Alexis S. Deletion of the immunoglobulin heavy chain 3' regulatory region super-enhancer affects somatic hypermutation in B1 B cells. Cell Mol Immunol 2019;16:195–7.

[22] Xu Y, Zhou H, Post G, Zan H, Casali P. Rad52 mediates class-switch DNA recombination to IgD. Nat Commun 2022;13:980.

[23] Yewdell WT, Chaudhuri J. A transcriptional serenAID: the role of noncoding RNAs in class switch recombination. Int Immunol 2017;29:183–96.

[24] Nakamura M, Kondo S, Sugai M, Nazarea M, Imamura S, Honjo T. High frequency class switching of an IgM+ B lymphoma clone CH12F3 to IgA+ cells. Int Immunol 1996;8:193–201.

[25] Muramatsu M, Sankaranand VS, Anant S, Sugai M, Kinoshita K, Davidson NO, et al. Specific expression of activation-induced cytidine deaminase (AID), a novel member of the RNA-editing deaminase family in germinal center B cells. J Biol Chem 1999;274:18470–6.

[26] Muramatsu M, Kinoshita K, Fagarasan S, Yamada S, Shinkai Y, Honjo. Class switch recombination and hypermutation require activation-induced cytidine deaminase (AID), a potential RNA editing enzyme. Cell 2000;102.

[27] Revy P, Muto T, Levy Y, Geissmann F, Plebani A, et al. Activation-induced cytidine deaminase (AID) deficiency causes the autosomal recessive form of the Hyper-IgM syndrome (HIGM2). Cell 2000;102:565–75.

[28] Bransteitter R, Pham P, Scharff MD, Goodman MF. Activation-induced cytidine deaminase deaminates deoxycytidine on single-STRANDed DNA but requires the action of RNase. Proc Natl Acad Sci U S A 2003;100:4102–7.

[29] Chaudhuri J, Tian M, Khuong C, Chua K, Pinaud E, Alt FW. Transcription-targeted DNA deamination by the AID antibody diversification enzyme. Nature 2003;422:726–30.

[30] Dickerson SK, Market E, Besmer E, Papavasiliou FN. AID mediates hypermutation by deaminating single stranded DNA. J Exp Med 2003;197:1291–6.

[31] Ramiro AR, Stavropoulos P, Jankovic M, Nussenzweig MC. Transcription enhances AID-mediated cytidine deamination by exposing single-stranded DNA on the nontemplate strand. Nat Immunol 2003;4:452–6.

[32] Sohail A, Klapacz J, Samaranayake M, Ullah A, Bhagwat AS. Human activation-induced cytidine deaminase causes transcription-dependent, strand-biased C to U deaminations. Nucleic Acids Res 2003;31:2990–4.

[33] Nabel CS, Lee JW, Wang LC, Kohli RM. Nucleic acid determinants for selective deamination of DNA over RNA by activation-induced deaminase. Proc Natl Acad Sci U S A 2013;110:14225–30.

[34] Carpenter MA, Rajagurubandara E, Wijesinghe P, Bhagwat AS. Determinants of sequence-specificity within human AID and APOBEC3G. DNA Repair 2010;9:579–87.

[35] Wang M, Rada C, Neuberger MS. Altering the spectrum of immunoglobulin V gene somatic hypermutation by modifying the active site of AID. J Exp Med 2010;207:141–53.

[36] Pham P, Calabrese P, Park SJ, Goodman MF. Analysis of a single-stranded DNA-scanning process in which activation-induced deoxycytidine deaminase (AID) deaminates C to U haphazardly and inefficiently to ensure mutational diversity. J Biol Chem 2011;286:24931–42.

[37] Qiao Q, Wang L, Meng FL, Hwang JK, Alt FW, Wu H. AID recognizes structured DNA for class switch recombination. Mol Cell 2017;67:361–373 e364.

[38] Tian M, Alt FW. Transcription-induced cleavage of immunoglobulin switch regions by nucleotide excision repair nucleases in vitro. J Biol Chem 2000;275:24163–72.

[39] Shinkura R, Tian M, Smith M, Chua K, Fujiwara Y, Alt FW. The influence of transcriptional orientation on endogenous switch region function. Nat Immunol 2003;4:435–41.

[40] Yu K, Chédin F, Hsieh C-L, Wilson TE, Lieber MR. R-loops at immunoglobulin class switch regions in the chromosomes of stimulated B cells. Nat Immunol 2003;4:442–51.

[41] Yu K, Huang FT, Lieber MR. DNA substrate length and surrounding sequence affect the activation-induced deaminase activity at cytidine. J Biol Chem 2004;279:6496–500.

[42] Zarrin AA, Alt FW, Chaudhuri J, Stokes N, Kaushal D, Du Pasquier L, et al. An evolutionarily conserved target motif for immunoglobulin class-switch recombination. Nat Immunol 2004;5:1275–81.

[43] Shen HM, Storb U. Activation-induced cytidine deaminase (AID) can target both DNA strands when the DNA is supercoiled. Proc Natl Acad Sci U S A 2004;101:12997–3002.

[44] Kodgire P, Mukkawar P, North JA, Poirier MG, Storb U. Nucleosome stability dramatically impacts the targeting of somatic hypermutation. Mol Cell Biol 2012;32:2030–40.

[45] Parsa JY, Ramachandran S, Zaheen A, Nepal RM, Kapelnikov A, et al. Negative supercoiling creates single-stranded patches of DNA that are substrates for AID-mediated mutagenesis. PLoS Genet 2012;8:e1002518.

[46] Chaudhuri J, Khuong C, Alt FW. Replication protein A interacts with AID to promote deamination of somatic hypermutation targets. Nature 2004;430:992–8.
[47] Basu U, Meng F-l, Keim C, Grinstein V, Pefanis E, et al. The RNA exosome targets the AID cytidine deaminase to both strands of transcribed duplex DNA substrates. Cell 2011;144:353–63.
[48] Pefanis E, Basu U. RNA exosome regulates AID DNA mutator activity in the B cell genome. Adv Immunol 2015;127:257–308.
[49] Lennon GG, Perry RP. C mu-containing transcripts initiate heterogeneously within the IgH enhancer region and contain a novel 5'-nontranslatable exon. Nature 1985;318:475–8.
[50] Alt FW, Rosenberg N, Casanova RJ, Thomas E, Baltimore D. Immunoglobulin heavy-chain expression and class switching in a murine leukaemia cell line. Nature 1982;296:325–31.
[51] Alt FW, Rosenberg N, Enea V, Siden E, Baltimore D. Multiple immunoglobulin heavy-chain gene transcripts in Abelson murine leukemia virus-transformed lymphoid cell lines. Mol Cell Biol 1982;2:386–400.
[52] Stavnezer-Nordgren J, Sirlin S. Specificity of immunoglobulin heavy chain switch correlates with activity of germline heavy chain genes prior to switching. EMBO J 1986;5:95–102.
[53] Yancopoulos GD, Alt FW. Regulation of the assembly and expression of variable-region genes. Annu Rev Immunol 1986;4:339–68.
[54] Snapper CM, Finkelman FD, Paul WE. Differential regulation of IgG1 and IgE synthesis by interleukin 4. J Exp Med 1988;167:183–96.
[55] Lutzker S, Alt FW. Structure and expression of germ line immunoglobulin gamma 2b transcripts. Mol Cell Biol 1988;8:1849–52.
[56] Rothman P, Lutzker S, Cook W, Coffman R, Alt FW. Mitogen plus interleukin 4 induction of C epsilon transcripts in B lymphoid cells. J Exp Med 1988;168:2385–9.
[57] Stavnezer J, Radcliffe G, Lin YC, Nietupski J, Berggren L, Sitia R, et al. Immunoglobulin heavy-chain switching may be directed by prior induction of transcripts from constant-region genes. Proc Natl Acad Sci U S A 1988;85:7704–8.
[58] Li SC, Rothman PB, Zhang J, Chan C, Hirsh D, Alt FW. Expression of I mu-C gamma hybrid germline transcripts subsequent to immunoglobulin heavy chain class switching. Int Immunol 1994;6:491–7.
[59] Bottaro A, Lansford R, Xu L, Zhang J, Rothman P, Alt FW. S region transcription per se promotes basal IgE class switch recombination but additional factors regulate the efficiency of the process. EMBO J 1994;13:665–74.
[60] Cogne M, Lansford R, Bottaro A, Zhang J, Gorman J, et al. A class switch control region at the 3' end of the immunoglobulin heavy chain locus. Cell 1994;77:737–47.
[61] Manis JP, van der Stoep N, Tian M, Ferrini R, Davidson L, Bottaro A, et al. Class switching in B cells lacking 3' immunoglobulin heavy chain enhancers. J Exp Med 1998;188:1421–31.
[62] Qiu G, Stavnezer J. Overexpression of BSAP/Pax-5 inhibits switching to IgA and enhances switching to IgE in the I.29 mu B cell line. J Immunol 1998;161:2906–18.
[63] Seidl KJ, Manis JP, Bottaro A, Zhang J, Davidson L, et al. Position-dependent inhibition of class-switch recombination by PGK-neor cassettes inserted into the immunoglobulin heavy chain constant region locus. Proc Natl Acad Sci U S A 1999;96:3000–5.
[64] Kuzin II, Ugine GD, Wu D, Young F, Chen J, Bottaro A. Normal isotype switching in B cells lacking the I mu exon splice donor site: evidence for multiple I mu-like germline transcripts. J Immunol 2000;164:1451–7.
[65] Pinaud E, Khamlichi AA, Le Morvan C, Drouet M, Nalesso V, Le Bert M, et al. Localization of the 3' IgH locus elements that effect long-distance regulation of class switch recombination. Immunity 2001;15:187–99.
[66] Yancopoulos GD, Alt FW. Developmentally controlled and tissue-specific expression of unrearranged VH gene segments. Cell 1985;40:271–81.
[67] Yancopoulos GD, DePinho RA, Zimmerman KA, Lutzker SG, Rosenberg N, Alt FW. Secondary genomic rearrangement events in pre-B cells: VHDJH replacement by a LINE-1 sequence and directed class switching. EMBO J 1986;5:3259–66.
[68] Pavri R, Gazumyan A, Jankovic M, Di Virgilio M, Klein I, et al. Activation-induced cytidine deaminase targets DNA at sites of RNA polymerase II stalling by interaction with Spt5. Cell 2010;143:122–33.
[69] Wang L, Wuerffel R, Feldman S, Khamlichi AA, Kenter AL. S region sequence, RNA polymerase II, and histone modifications create chromatin accessibility during class switch recombination. J Exp Med 2009;206:1817–30.
[70] Jeevan-Raj BP, Robert I, Heyer V, Page A, Wang JH, et al. Epigenetic tethering of AID to the donor switch region during immunoglobulin class switch recombination. J Exp Med 2011;208:1649–60.
[71] Chaudhuri J, Basu U, Zarrin A, Yan C, Franco S, et al. Evolution of the immunoglobulin heavy chain class switch recombination mechanism. Adv Immunol 2007;94:157–214.
[72] Pefanis E, Wang J, Rothschild G, Lim J, Chao J, et al. Noncoding RNA transcription targets AID to divergently transcribed loci in B cells. Nature 2014;514:389–93.
[73] Nambu Y, Sugai M, Gonda H, Lee CG, Katakai T, et al. Transcription-coupled events associating with immunoglobulin switch region chromatin. Science 2003;302:2137–40.
[74] Wada T, Takagi T, Yamaguchi Y, Ferdous A, Imai T, et al. DSIF, a novel transcription elongation factor that regulates RNA polymerase II processivity, is composed of human Spt4 and Spt5 homologs. Gene Dev 1998;12:343–56.
[75] Yamaguchi Y, Takagi T, Wada T, Yano K, Furuya A, et al. NELF, a multisubunit complex containing RD, cooperates with DSIF to repress RNA polymerase II elongation. Cell 1999;97:41–51.
[76] Renner DB, Yamaguchi Y, Wada T, Handa H, Price DH. A highly purified RNA polymerase II elongation control system. J Biol Chem 2001;276:42601–9.
[77] Rahl PB, Lin CY, Seila AC, Flynn RA, McCuine S, et al. c-Myc regulates transcriptional pause release. Cell 2010;141:432–45.
[78] Klein BJ, Bose D, Baker KJ, Yusoff ZM, Zhang X, Murakami KS. RNA polymerase and transcription elongation factor Spt4/5 complex structure. Proc Natl Acad Sci USA 2011;108:546–50.
[79] Martinez-Rucobo FW, Sainsbury S, Cheung AC, Cramer P. Architecture of the RNA polymerase-Spt4/5 complex and basis of universal transcription processivity. EMBO J 2011;30:1302–10.
[80] Gazumyan A, Bothmer A, Klein IA, Nussenzweig MC, McBride KM. Activation-induced cytidine deaminase in antibody diversification and chromosome translocation. Adv Cancer Res 2012;113:167–90.

[81] Yamane A, Resch W, Kuo N, Kuchen S, Li Z, et al. Deep-sequencing identification of the genomic targets of the cytidine deaminase AID and its cofactor RPA in B lymphocytes. Nat Immunol 2011;12:62–9.
[82] Yewdell WT, Kim Y, Chowdhury P, Lau CM, Smolkin RM, et al. A hyper-IgM syndrome mutation in activation-induced cytidine deaminase disrupts G-quadruplex binding and genome-wide chromatin localization. Immunity 2020;53:952–970 e911.
[83] Kenter AL, Feldman S, Wuerffel R, Achour I, Wang L, Kumar S. Three-dimensional architecture of the IgH locus facilitates class switch recombination. Ann N Y Acad Sci 2012;1267:86–94.
[84] Sun J, Rothschild G, Pefanis E, Basu U. Transcriptional stalling in B-lymphocytes: a mechanism for antibody diversification and maintenance of genomic integrity. Transcription 2013;4:127–35.
[85] Storb U. Why does somatic hypermutation by AID require transcription of its target genes? Adv Immunol 2014;122:253–77.
[86] Methot SP, Litzler LC, Subramani PG, Eranki AK, Fifield H, et al. A licensing step links AID to transcription elongation for mutagenesis in B cells. Nat Commun 2018;9:1248.
[87] Rajagopal D, Maul RW, Ghosh A, Chakraborty T, Khamlichi AA, Sen R, et al. Immunoglobulin switch mu sequence causes RNA polymerase II accumulation and reduces dA hypermutation. J Exp Med 2009;206:1237–44.
[88] Lemay JF, Larochelle M, Marguerat S, Atkinson S, Bahler J, Bachand F. The RNA exosome promotes transcription termination of backtracked RNA polymerase II. Nat Struct Mol Biol 2014;21:919–26.
[89] Pefanis E, Wang J, Rothschild G, Lim J, Kazadi D, et al. RNA exosome-regulated long non-coding RNA transcription controls super-enhancer activity. Cell 2015;161:774–89.
[90] Lim J, Giri PK, Kazadi D, Laffleur B, Zhang W, et al. Nuclear proximity of Mtr4 to RNA exosome restricts DNA mutational asymmetry. Cell 2017;169:523–537 e515.
[91] Laffleur B, Lim J, Zhang W, Chen Y, Pefanis E, et al. Noncoding RNA processing by DIS3 regulates chromosomal architecture and somatic hypermutation in B cells. Nat Genet 2021;53:230–42.
[92] Casellas R, Basu U, Yewdell WT, Chaudhuri J, Robbiani DF, Di Noia JM. Mutations, kataegis and translocations in B cells: understanding AID promiscuous activity. Nat Rev Immunol 2016;16:164–76.
[93] Gellert M, Lipsett MN, Davies DR. Helix formation by guanylic acid. Proc Natl Acad Sci U S A 1962;48:2013–8.
[94] Sen D, Gilbert W. Formation of parallel four-stranded complexes by guanine-rich motifs in DNA and its implications for meiosis. Nature 1988;334:364–6.
[95] Duquette ML, Handa P, Vincent JA, Taylor AF, Maizels N. Intracellular transcription of G-rich DNAs induces formation of G-loops, novel structures containing G4 DNA. Gene Dev 2004;18:1618–29.
[96] Neaves KJ, Huppert JL, Henderson RM, Edwardson JM. Direct visualization of G-quadruplexes in DNA using atomic force microscopy. Nucleic Acids Res 2009;37:6269–75.
[97] Carrasco-Salas Y, Malapert A, Sulthana S, Molcrette B, Chazot-Franguiadakis L, et al. The extruded non-template strand determines the architecture of R-loops. Nucleic Acids Res 2019;47(13):6783–95.
[98] Dunnick W, Hertz GZ, Scappino L, Gritzmacher C. DNA sequences at immunoglobulin switch region recombination sites. Nucleic Acids Res 1993;21.
[99] Duquette ML, Pham P, Goodman MF, Maizels N. AID binds to transcription-induced structures in c-MYC that map to regions associated with translocation and hypermutation. Oncogene 2005;24:5791–8.
[100] Zheng S, Vuong BQ, Vaidyanathan B, Lin JY, Huang FT, Chaudhuri J. Non-coding RNA generated following lariat debranching mediates targeting of AID to DNA. Cell 2015;161:762–73.
[101] Mahdaviani SA, Hirbod-Mobarakeh A, Wang N, Aghamohammadi A, Hammarstrom L, et al. Novel mutation of the activation-induced cytidine deaminase gene in a Tajik family: special review on hyper-immunoglobulin M syndrome. Expet Rev Clin Immunol 2012;8:539–46.
[102] Yu K. AID function in somatic hypermutation and class switch recombination. Acta Biochim Biophys Sin 2022;54:759–66.
[103] Schrader CE, Edelmann W, Kucherlapati R, Stavnezer J. Reduced isotype switching in splenic B cells from mice deficient in mismatch repair enzymes. J Exp Med 1999;190:323–30.
[104] Di Noia J, Neuberger MS. Altering the pathway of immunoglobulin hypermutation by inhibiting uracil-DNA glycosylase. Nature 2002;419:43–8.
[105] Rada C, Williams GT, Nilsen H, Barnes DE, Lindahl T, Neuberger MS. Immunoglobulin isotype switching is inhibited and somatic hypermutation perturbed in UNG-deficient mice. Curr Biol 2002;12:1748–55.
[106] Imai K, Slupphaug G, Lee WI, Revy P, Nonoyama S, et al. Human uracil-DNA glycosylase deficiency associated with profoundly impaired immunoglobulin class-switch recombination. Nat Immunol 2003;4:1023–8.
[107] Schrader CE, Linehan EK, Mochegova SN, Woodland RT, Stavnezer J. Inducible DNA breaks in Ig S regions are dependent on AID and UNG. J Exp Med 2005;202:561–8.
[108] Schrader CE, Guikema JE, Wu X, Stavnezer J. The roles of APE1, APE2, DNA polymerase beta and mismatch repair in creating S region DNA breaks during antibody class switch. Philos Trans R Soc Lond B Biol Sci 2009;364:645–52.
[109] Guikema JE, Linehan EK, Tsuchimoto D, Nakabeppu Y, Strauss PR, Stavnezer J, et al. APE1- and APE2-dependent DNA breaks in immunoglobulin class switch recombination. J Exp Med 2007;204:3017–26.
[110] Guikema JE, Stavnezer J, Schrader CE. The role of Apex2 in class-switch recombination of immunoglobulin genes. Int Immunol 2010;22:213. author reply 213–214.
[111] Masani S, Han L, Yu K. Apurinic/apyrimidinic endonuclease 1 is the essential nuclease during immunoglobulin class switch recombination. Mol Cell Biol 2013;33:1468–73.
[112] Begum NA, Kinoshita K, Kakazu N, Muramatsu M, Nagaoka H, et al. Uracil DNA glycosylase activity is dispensable for immunoglobulin class switch. Science 2004;305:1160–3.
[113] Begum NA, Izumi N, Nishikori M, Nagaoka H, Shinkura R, Honjo T. Requirement of non-canonical activity of uracil DNA glycosylase for class switch recombination. J Biol Chem 2007;282:731–42.

[114] Begum NA, Stanlie A, Doi T, Sasaki Y, Jin HW, et al. Further evidence for involvement of a noncanonical function of uracil DNA glycosylase in class switch recombination. Proc Natl Acad Sci U S A 2009;106:2752–7.

[115] Pettersen HS, Sundheim O, Gilljam KM, Slupphaug G, Krokan HE, Kavli B. Uracil-DNA glycosylases SMUG1 and UNG2 coordinate the initial steps of base excision repair by distinct mechanisms. Nucleic Acids Res 2007;35:3879–92.

[116] Rada C, Di Noia JM, Neuberger MS. Mismatch recognition and uracil excision provide complementary paths to both Ig switching and the A/T-focused phase of somatic mutation. Mol Cell 2004;16:163–71.

[117] Stavnezer J, Guikema JE, Schrader CE. Mechanism and regulation of class switch recombination. Annu Rev Immunol 2008;26:261–92.

[118] Ehrenstein MR, Neuberger MS. Deficiency in Msh2 affects the efficiency and local sequence specificity of immunoglobulin class-switch recombination: parallels with somatic hypermutation. EMBO J 1999;18:3484–90.

[119] Schrader CE, Vardo J, Stavnezer J. Role for mismatch repair proteins Msh2, Mlh1, and Pms2 in immunoglobulin class switching shown by sequence analysis of recombination junctions. J Exp Med 2002;195:367–73.

[120] Bardwell PD, Woo CJ, Wei K, Li Z, Martin A, et al. Altered somatic hypermutation and reduced class-switch recombination in exonuclease 1-mutant mice. Nat Immunol 2004;5:224–9.

[121] Roa S, Li Z, Peled JU, Zhao C, Edelmann W, Scharff MD. MSH2/MSH6 complex promotes error-free repair of AID-induced dU:G mispairs as well as error-prone hypermutation of A:T sites. PLoS One 2010;5:e11182.

[122] Basu U, Chaudhuri J, Alpert C, Dutt S, Ranganath S, et al. The AID antibody diversification enzyme is regulated by protein kinase A phosphorylation. Nature 2005;438:508–11.

[123] McBride KM, Gazumyan A, Woo EM, Barreto VM, Robbiani DF, Chait BT, et al. Regulation of hypermutation by activation-induced cytidine deaminase phosphorylation. Proc Natl Acad Sci U S A 2006;103:8798–803.

[124] Pasqualucci L, Kitaura Y, Gu H, Dalla-Favera R. PKA-mediated phosphorylation regulates the function of activation-induced deaminase (AID) in B cells. Proc Natl Acad Sci U S A 2006;103:395–400.

[125] McBride KM, Gazumyan A, Woo EM, Schwickert TA, Chait BT, Nussenzweig MC. Regulation of class switch recombination and somatic mutation by AID phosphorylation. J Exp Med 2008;205:2585–94.

[126] Cheng HL, Vuong BQ, Basu U, Franklin A, Schwer B, et al. Integrity of the AID serine-38 phosphorylation site is critical for class switch recombination and somatic hypermutation in mice. Proc Natl Acad Sci U S A 2009;106:2717–22.

[127] Vuong BQ, Lee M, Kabir S, Irimia C, Macchiarulo S, McKnight GS, et al. Specific recruitment of protein kinase A to the immunoglobulin locus regulates class-switch recombination. Nat Immunol 2009;10:420–6.

[128] Vuong BQ, Herrick-Reynolds K, Vaidyanathan B, Pucella JN, Ucher AJ, et al. A DNA break- and phosphorylation-dependent positive feedback loop promotes immunoglobulin class-switch recombination. Nat Immunol 2013;14:1183–9.

[129] Shinkura R, Ito S, Begum NA, Nagaoka H, Muramatsu M, et al. Separate domains of AID are required for somatic hypermutation and class-switch recombination. Nat Immunol 2004;5:707–12.

[130] Wei M, Shinkura R, Doi Y, Maruya M, Fagarasan S, Honjo T. Mice carrying a knock-in mutation of Aicda resulting in a defect in somatic hypermutation have impaired gut homeostasis and compromised mucosal defense. Nat Immunol 2011;12:264–70.

[131] Ito S, Nagaoka H, Shinkura R, Begum N, Muramatsu M, Nakata M, et al. Activation-induced cytidine deaminase shuttles between nucleus and cytoplasm like apolipoprotein B mRNA editing catalytic polypeptide 1. Proc Natl Acad Sci U S A 2004;101:1975–80.

[132] McBride KM, Barreto V, Ramiro AR, Stavropoulos P, Nussenzweig MC. Somatic hypermutation is limited by CRM1-dependent nuclear export of activation-induced deaminase. J Exp Med 2004;199:1235–44.

[133] Brar SS, Sacho EJ, Tessmer I, Croteau DL, Erie DA, Diaz M. Activation-induced deaminase, AID, is catalytically active as a monomer on single-stranded DNA. DNA Repair 2008;7:77–87.

[134] Methot SP, Litzler LC, Trajtenberg F, Zahn A, Robert F, et al. Consecutive interactions with HSP90 and eEF1A underlie a functional maturation and storage pathway of AID in the cytoplasm. J Exp Med 2015;212:581–96.

[135] Barreto V, Reina-San-Martin B, Ramiro AR, McBride KM, Nussenzweig MC. C-terminal deletion of AID uncouples class switch recombination from somatic hypermutation and gene conversion. Mol Cell 2003;12:501–8.

[136] Ta VT, Nagaoka H, Catalan N, Durandy A, Fischer A, et al. AID mutant analyses indicate requirement for class-switch-specific cofactors. Nat Immunol 2003;4:843–8.

[137] Geisberger R, Rada C, Neuberger MS. The stability of AID and its function in class-switching are critically sensitive to the identity of its nuclear-export sequence. Proc Natl Acad Sci U S A 2009;106:6736–41.

[138] Zahn A, Daugan M, Safavi S, Godin D, Cheong C, Lamarre A, et al. Separation of function between isotype switching and affinity maturation in vivo during acute immune responses and circulating autoantibodies in UNG-deficient mice. J Immunol 2013;190:5949–60.

[139] Xie X, Gan T, Rao B, Zhang W, Panchakshari RA, et al. C-terminal deletion-induced condensation sequesters AID from IgH targets in immunodeficiency. EMBO J 2022;41:e109324.

[140] Ranjit S, Khair L, Linehan EK, Ucher AJ, Chakrabarti M, Schrader CE, et al. AID recruits UNG and Msh2 to Ig switch regions dependent upon the AID C terminus [corrected]. J Immunol 2011;187:2464–75.

[141] Zahn A, Eranki AK, Patenaude AM, Methot SP, Fifield H, et al. Activation induced deaminase C-terminal domain links DNA breaks to end protection and repair during class switch recombination. Proc Natl Acad Sci U S A 2014;111:E988–97.

[142] Sabouri S, Kobayashi M, Begum NA, Xu J, Hirota K, Honjo T. C-terminal region of activation-induced cytidine deaminase (AID) is required for efficient class switch recombination and gene conversion. Proc Natl Acad Sci U S A 2014;111:2253–8.

[143] Durandy A, Taubenheim N, Peron S, Fischer A. Pathophysiology of B-cell intrinsic immunoglobulin class switch recombination deficiencies. Adv Immunol 2007;94:275–306.

[144] Saribasak H, Maul RW, Cao Z, McClure RL, Yang W, et al. XRCC1 suppresses somatic hypermutation and promotes

alternative nonhomologous end joining in Igh genes. J Exp Med 2011;208:2209–16.
[145] Boboila C, Alt FW, Schwer B. Classical and alternative end-joining pathways for repair of lymphocyte-specific and general DNA double-strand breaks. Adv Immunol 2012;116:1–49.
[146] Han L, Mao W, Yu K. X-ray repair cross-complementing protein 1 (XRCC1) deficiency enhances class switch recombination and is permissive for alternative end joining. Proc Natl Acad Sci U S A 2012;109:4604–8.
[147] Xu J, Husain A, Hu W, Honjo T, Kobayashi M. APE1 is dispensable for S-region cleavage but required for its repair in class switch recombination. Proc Natl Acad Sci U S A 2014;111:17242–7.
[148] Pena-Diaz J, Bregenhorn S, Ghodgaonkar M, Follonier C, Artola-Boran M, et al. Noncanonical mismatch repair as a source of genomic instability in human cells. Mol Cell 2012;47:669–80.
[149] Feng Y, Li C, Stewart JA, Barbulescu P, Seija Desivo N, et al. FAM72A antagonizes UNG2 to promote mutagenic repair during antibody maturation. Nature 2021;600:324–8.
[150] Rogier M, Moritz J, Robert I, Lescale C, Heyer V, et al. Fam72a enforces error-prone DNA repair during antibody diversification. Nature 2021;600:329–33.
[151] Mohni KN, Wessel SR, Zhao R, Wojciechowski AC, Luzwick JW, et al. HMCES maintains genome integrity by shielding abasic sites in single-strand DNA. Cell 2019;176:144–153 e113.
[152] Wu L, Shukla V, Yadavalli AD, Dinesh RK, Xu D, Rao A, et al. HMCES protects immunoglobulin genes specifically from deletions during somatic hypermutation. Genes Dev 2022;36:433–50.
[153] Komor AC, Kim YB, Packer MS, Zuris JA, Liu DR. Programmable editing of a target base in genomic DNA without double-stranded DNA cleavage. Nature 2016;533:420–4.
[154] Gaudelli NM, Komor AC, Rees HA, Packer MS, Badran AH, Bryson DI, et al. Programmable base editing of A*T to G*C in genomic DNA without DNA cleavage. Nature 2017;551:464–71.
[155] Zarrin AA, Del Vecchio C, Tseng E, Gleason M, Zarin P, Tian M, et al. Antibody class switching mediated by yeast endonuclease-generated DNA breaks. Science 2007;315:377–81.
[156] Bothmer A, Robbiani DF, Feldhahn N, Gazumyan A, Nussenzweig A, Nussenzweig MC. 53BP1 regulates DNA resection and the choice between classical and alternative end joining during class switch recombination. J Exp Med 2010;207:855–65.
[157] Gostissa M, Schwer B, Chang A, Dong J, Meyers RM, et al. IgH class switching exploits a general property of two DNA breaks to be joined in cis over long chromosomal distances. Proc Natl Acad Sci U S A 2014;111:2644–9.
[158] Dudley DD, Manis JP, Zarrin AA, Kaylor L, Tian M, Alt FW. Internal IgH class switch region deletions are position-independent and enhanced by AID expression. Proc Natl Acad Sci U S A 2002;99:9984–9.
[159] Reina-San-Martin B, Difilippantonio S, Hanitsch L, Masilamani RF, Nussenzweig A, Nussenzweig MC. H2AX is required for recombination between immunoglobulin switch regions but not for intra-switch region recombination or somatic hypermutation. J Exp Med 2003;197:1767–78.
[160] Reina-San-Martin B, Chen HT, Nussenzweig A, Nussenzweig MC. ATM is required for efficient recombination between immunoglobulin switch regions. J Exp Med 2004;200:1103–10.
[161] Reina-San-Martin B, Chen J, Nussenzweig A, Nussenzweig MC. Enhanced intra-switch region recombination during immunoglobulin class switch recombination in 53BP1-/- B cells. Eur J Immunol 2007;37:235–9.
[162] Siebenkotten G, Esser C, Wabl M, Radbruch A. The murine IgG1/IgE class switch program. Eur J Immunol 1992;22:1827–34.
[163] Jabara HH, Loh R, Ramesh N, Vercelli D, Geha RS. Sequential switching from mu to epsilon via gamma 4 in human B cells stimulated with IL-4 and hydrocortisone. J Immunol 1993;151:4528–33.
[164] Peron S, Laffleur B, Denis-Lagache N, Cook-Moreau J, Tinguely A, et al. AID-driven deletion causes immunoglobulin heavy chain locus suicide recombination in B cells. Science 2012;336:931–4.
[165] Jack HM, McDowell M, Steinberg CM, Wabl M. Looping out and deletion mechanism for the immunoglobulin heavy-chain class switch. Proc Natl Acad Sci U S A 1988;85:1581–5.
[166] von Schwedler U, Jäck HM, Wabl M. Circular DNA is a product of the immunoglobulin class switch rearrangement. Nature 1990;345.
[167] Chiarle R, Zhang Y, Frock RL, Lewis SM, Molinie B, et al. Genome-wide translocation sequencing reveals mechanisms of chromosome breaks and rearrangements in B cells. Cell 2011;147:107–19.
[168] Klein IA, Resch W, Jankovic M, Oliveira T, Yamane A, et al. Translocation-capture sequencing reveals the extent and nature of chromosomal rearrangements in B lymphocytes. Cell 2011;147:95–106.
[169] Frock RL, Hu J, Meyers RM, Ho YJ, Kii E, Alt FW. Genome-wide detection of DNA double-stranded breaks induced by engineered nucleases. Nat Biotechnol 2015;33:179–86.
[170] Dong J, Panchakshari RA, Zhang T, Zhang Y, Hu J, et al. Orientation-specific joining of AID-initiated DNA breaks promotes antibody class switching. Nature 2015;525:134–9.
[171] Vincendeau E, Wei W, Zhang X, Planchais C, Yu W, et al. SHLD1 is dispensable for 53BP1-dependent V(D)J recombination but critical for productive class switch recombination. Nat Commun 2022;13:3707.
[172] Liu X, Liu T, Shang Y, Dai P, Zhang W, et al. ERCC6L2 promotes DNA orientation-specific recombination in mammalian cells. Cell Res 2020;30:732–44.
[173] Wuerffel R, Wang L, Grigera F, Manis J, Selsing E, et al. S-S synapsis during class switch recombination is promoted by distantly located transcriptional elements and activation-induced deaminase. Immunity 2007;27:711–22.
[174] Larson ED, Duquette ML, Cummings WJ, Streiff RJ, Maizels N. MutSalpha binds to and promotes synapsis of transcriptionally activated immunoglobulin switch regions. Curr Biol 2005;15:470–4.
[175] Rogakou EP, Boon C, Redon C, Bonner WM. Megabase chromatin domains involved in DNA double-strand breaks in vivo. J Cell Biol 1999;146:905–16.
[176] Bassing CH, Alt FW. H2AX may function as an anchor to hold broken chromosomal DNA ends in close proximity. Cell Cycle 2004;3:149–53.
[177] Stucki M, Jackson SP. gammaH2AX and MDC1: anchoring the DNA-damage-response machinery to broken chromosomes. DNA Repair 2006;5:534–43.

[178] Savic V, Yin B, Maas NL, Bredemeyer AL, Carpenter AC, et al. Formation of dynamic gamma-H2AX domains along broken DNA strands is distinctly regulated by ATM and MDC1 and dependent upon H2AX densities in chromatin. Mol Cell 2009;34:298–310.
[179] Zhang X, Zhang Y, Ba Z, Kyritsis N, Casellas R, Alt FW. Fundamental roles of chromatin loop extrusion in antibody class switching. Nature 2019;575:385–9.
[180] Zhang Y, Zhang X, Dai HQ, Hu H, Alt FW. The role of chromatin loop extrusion in antibody diversification. Nat Rev Immunol 2022;22:550–66.
[181] Dixon JR, Selvaraj S, Yue F, Kim A, Li Y, et al. Topological domains in mammalian genomes identified by analysis of chromatin interactions. Nature 2012;485:376–80.
[182] Nora EP, Lajoie BR, Schulz EG, Giorgetti L, Okamoto I, et al. Spatial partitioning of the regulatory landscape of the X-inactivation centre. Nature 2012;485:381–5.
[183] Rao SS, Huntley MH, Durand NC, Stamenova EK, Bochkov ID, et al. A 3D map of the human genome at kilobase resolution reveals principles of chromatin looping. Cell 2014;159:1665–80.
[184] Schwarzer W, Abdennur N, Goloborodko A, Pekowska A, Fudenberg G, et al. Two independent modes of chromatin organization revealed by cohesin removal. Nature 2017;551:51–6.
[185] Vian L, Pekowska A, Rao SSP, Kieffer-Kwon KR, Jung S, et al. The energetics and physiological impact of cohesin extrusion. Cell 2018;173:1165–1178 e1120.
[186] Davidson IF, Bauer B, Goetz D, Tang W, Wutz G, Peters JM. DNA loop extrusion by human cohesin. Science 2019;366:1338–45.
[187] Kim Y, Shi Z, Zhang H, Finkelstein IJ, Yu H. Human cohesin compacts DNA by loop extrusion. Science 2019;366:1345–9.
[188] Zhang Y, Zhang X, Ba Z, Liang Z, Dring EW, et al. The fundamental role of chromatin loop extrusion in physiological V(D)J recombination. Nature 2019;573:600–4.
[189] Schrader CE, Bradley SP, Vardo J, Mochegova SN, Flanagan E, Stavnezer J. Mutations occur in the Ig Smu region but rarely in Sgamma regions prior to class switch recombination. EMBO J 2003;22:5893–903.
[190] Horton MB, Cheon H, Duffy KR, Brown D, Naik SH, et al. Lineage tracing reveals B cell antibody class switching is stochastic, cell-autonomous, and tuneable. Immunity 2022;55:1843–1855 e1846.
[191] Gu H, Zou YR, Rajewsky K. Independent control of immunoglobulin switch recombination at individual switch regions evidenced through Cre-loxP-mediated gene targeting. Cell 1993;73:1155–64.
[192] Nussenzweig A, Nussenzweig MC. Origin of chromosomal translocations in lymphoid cancer. Cell 2010;141:27–38.
[193] Lumsden JM, McCarty T, Petiniot LK, Shen R, Barlow C, et al. Immunoglobulin class switch recombination is impaired in Atm-deficient mice. J Exp Med 2004;200:1111–21.
[194] Feldman S, Wuerffel R, Achour I, Wang L, Carpenter PB, Kenter AL. 53BP1 contributes to Igh locus chromatin topology during class switch recombination. J Immunol 2017;198:2434–44.
[195] Franco S, Gostissa M, Zha S, Lombard DB, Murphy MM, et al. H2AX prevents DNA breaks from progressing to chromosome breaks and translocations. Mol Cell 2006;21:201–14.
[196] Manis JP, Morales JC, Xia Z, Kutok JL, Alt FW, Carpenter PB. 53BP1 links DNA damage-response pathways to immunoglobulin heavy chain class-switch recombination. Nat Immunol 2004;5:481–7.
[197] Ward IM, Reina-San-Martin B, Olaru A, Minn K, Tamada K, et al. 53BP1 is required for class switch recombination. J Cell Biol 2004;165:459–64.
[198] Bothmer A, Robbiani DF, Di Virgilio M, Bunting SF, Klein IA, et al. Regulation of DNA end joining, resection, and immunoglobulin class switch recombination by 53BP1. Mol Cell 2011;42:319–29.
[199] Di Virgilio M, Callen E, Yamane A, Zhang W, Jankovic M, et al. Rif1 prevents resection of DNA breaks and promotes immunoglobulin class switching. Science 2013;339:711–5.
[200] Boersma V, Moatti N, Segura-Bayona S, Peuscher MH, van der Torre J, et al. MAD2L2 controls DNA repair at telomeres and DNA breaks by inhibiting 5' end resection. Nature 2015;521:537–40.
[201] Xu G, Chapman JR, Brandsma I, Yuan J, Mistrik M, et al. REV7 counteracts DNA double-strand break resection and affects PARP inhibition. Nature 2015;521:541–4.
[202] Yang D, Sun Y, Chen J, Zhang Y, Fan S, et al. REV7 is required for processing AID initiated DNA lesions in activated B cells. Nat Commun 2020;11:2812.
[203] Dev H, Chiang TW, Lescale C, de Krijger I, Martin AG, et al. Shieldin complex promotes DNA end-joining and counters homologous recombination in BRCA1-null cells. Nat Cell Biol 2018;20:954–65.
[204] Ghezraoui H, Oliveira C, Becker JR, Bilham K, Moralli D, et al. 53BP1 cooperation with the REV7-shieldin complex underpins DNA structure-specific NHEJ. Nature 2018;560:122–7.
[205] Gupta R, Somyajit K, Narita T, Maskey E, Stanlie A, et al. DNA repair network analysis reveals shieldin as a key regulator of NHEJ and PARP inhibitor sensitivity. Cell 2018;173:972–988 e923.
[206] Mirman Z, Lottersberger F, Takai H, Kibe T, Gong Y, et al. 53BP1-RIF1-shieldin counteracts DSB resection through CST- and Polalpha-dependent fill-in. Nature 2018;560:112–6.
[207] Noordermeer SM, Adam S, Setiaputra D, Barazas M, Pettitt SJ, et al. The shieldin complex mediates 53BP1-dependent DNA repair. Nature 2018;560:117–21.
[208] Tomida J, Takata KI, Bhetawal S, Person MD, Chao HP, Tang DG, et al. FAM35A associates with REV7 and modulates DNA damage responses of normal and BRCA1-defective cells. EMBO J 2018;37.
[209] Setiaputra D, Durocher D. Shieldin - the protector of DNA ends. EMBO Rep 2019;20.
[210] Paiano J, Zolnerowich N, Wu W, Pavani R, Wang C, et al. Role of 53BP1 in end protection and DNA synthesis at DNA breaks. Genes Dev 2021;35:1356–67.
[211] Yen WF, Chaudhry A, Vaidyanathan B, Yewdell WT, Pucella JN, et al. BRCT-domain protein BRIT1 influences class switch recombination. Proc Natl Acad Sci U S A 2017;114:8354–9.
[212] Dinkelmann M, Spehalski E, Stoneham T, Buis J, Wu Y, Sekiguchi JM, et al. Multiple functions of MRN in end-joining pathways during isotype class switching. Nat Struct Mol Biol 2009;16:808–13.
[213] Robert I, Dantzer F, Reina-San-Martin B. Parp1 facilitates alternative NHEJ, whereas Parp2 suppresses IgH/c-myc translocations during immunoglobulin class switch recombination. J Exp Med 2009;206:1047–56.

[214] Lee-Theilen M, Matthews AJ, Kelly D, Zheng S, Chaudhuri J. CtIP promotes microhomology-mediated alternative end joining during class-switch recombination. Nat Struct Mol Biol 2011;18:75–9.
[215] Robert I, Gaudot L, Rogier M, Heyer V, Noll A, Dantzer F, et al. Parp3 negatively regulates immunoglobulin class switch recombination. PLoS Genet 2015;11:e1005240.
[216] Zan H, Tat C, Qiu Z, Taylor JR, Guerrero JA, Shen T, et al. Rad52 competes with Ku70/Ku86 for binding to S-region DSB ends to modulate antibody class-switch DNA recombination. Nat Commun 2017;8:14244.
[217] Hasham MG, Donghia NM, Coffey E, Maynard J, Snow KJ, et al. Widespread genomic breaks generated by activation-induced cytidine deaminase are prevented by homologous recombination. Nat Immunol 2010;11:820–6.
[218] Hasham MG, Snow KJ, Donghia NM, Branca JA, Lessard MD, et al. Activation-induced cytidine deaminase-initiated off-target DNA breaks are detected and resolved during S phase. J Immunol 2012;189:2374–82.
[219] Lamont KR, Hasham MG, Donghia NM, Branca J, Chavaree M, et al. Attenuating homologous recombination stimulates an AID-induced antileukemic effect. J Exp Med 2013;210:1021–33.
[220] Nair L, Zhang W, Laffleur B, Jha MK, Lim J, et al. Mechanism of noncoding RNA-associated N(6)-methyladenosine recognition by an RNA processing complex during IgH DNA recombination. Mol Cell 2021;81:3949–64.
[221] Refaat AM, Nakata M, Husain A, Kosako H, Honjo T, Begum NA. HNRNPU facilitates antibody class-switch recombination through C-NHEJ promotion and R-loop suppression. Cell Rep 2023;42:112284.
[222] Liu N, Dai Q, Zheng G, He C, Parisien M, Pan T. N(6)-methyladenosine-dependent RNA structural switches regulate RNA-protein interactions. Nature 2015;518.
[223] Liu N, Zhou KI, Parisien M, Dai Q, Diatchenko L, Pan T. N6-methyladenosine alters RNA structure to regulate binding of a low-complexity protein. Nucleic Acids Res 2017;45.
[224] Rush JS, Fugmann SD, Schatz DG. Staggered AID-dependent DNA double strand breaks are the predominant DNA lesions targeted to S mu in Ig class switch recombination. Int Immunol 2004;16:549–57.
[225] Zan H, Casali P. AID- and Ung-dependent generation of staggered double-strand DNA breaks in immunoglobulin class switch DNA recombination: a post-cleavage role for AID. Mol Immunol 2008;46:45–61.
[226] Schatz DG, Ji Y. Recombination centres and the orchestration of V(D)J recombination. Nat Rev Immunol 2011;11:251–63.
[227] Moshous D, Callebaut I, de Chasseval R, Corneo B, Cavazzana-Calvo M, et al. Artemis, a novel DNA double-strand break repair/V(D)J recombination protein, is mutated in human severe combined immune deficiency. Cell 2001;105:177–86.
[228] Ma Y, Pannicke U, Schwarz K, Lieber MR. Hairpin opening and overhang processing by an Artemis/DNA-dependent protein kinase complex in nonhomologous end joining and V(D)J recombination. Cell 2002;108:781–94.
[229] Rooney S, Alt FW, Sekiguchi J, Manis JP. Artemis-independent functions of DNA-dependent protein kinase in Ig heavy chain class switch recombination and development. Proc Natl Acad Sci U S A 2005;102:2471–5.
[230] Tummala H, Kirwan M, Walne AJ, Hossain U, Jackson N, et al. ERCC6L2 mutations link a distinct bone-marrow-failure syndrome to DNA repair and mitochondrial function. Am J Hum Genet 2014;94:246–56.
[231] Zhang S, Pondarre C, Pennarun G, Labussiere-Wallet H, Vera G, et al. A nonsense mutation in the DNA repair factor Hebo causes mild bone marrow failure and microcephaly. J Exp Med 2016;213:1011–28.
[232] Jarviaho T, Halt K, Hirvikoski P, Moilanen J, Mottonen M, Niinimaki R. Bone marrow failure syndrome caused by homozygous frameshift mutation in the ERCC6L2 gene. Clin Genet 2018;93:392–5.
[233] Shabanova I, Cohen E, Cada M, Vincent A, Cohn RD, Dror Y. ERCC6L2-associated inherited bone marrow failure syndrome. Mol Genet Genomic Med 2018;6:463–8.
[234] Tummala H, Dokal AD, Walne A, Ellison A, Cardoso S, et al. Genome instability is a consequence of transcription deficiency in patients with bone marrow failure harboring biallelic ERCC6L2 variants. Proc Natl Acad Sci U S A 2018;115:7777–82.
[235] Dardiotis E, Karampinis E, Siokas V, Aloizou AM, Rikos D, et al. ERCC6L2 rs591486 polymorphism and risk for amyotrophic lateral sclerosis in Greek population. Neurol Sci 2019;40:1237–44.
[236] Douglas SPM, Siipola P, Kovanen PE, Pyorala M, Kakko S, et al. ERCC6L2 defines a novel entity within inherited acute myeloid leukemia. Blood 2019;133:2724–8.
[237] Nijnik A, Woodbine L, Marchetti C, Dawson S, Lambe T, et al. DNA repair is limiting for haematopoietic stem cells during ageing. Nature 2007;447:686–90.
[238] Rossi DJ, Bryder D, Seita J, Nussenzweig A, Hoeijmakers J, Weissman IL. Deficiencies in DNA damage repair limit the function of haematopoietic stem cells with age. Nature 2007;447:725–9.
[239] Avagyan S, Churchill M, Yamamoto K, Crowe JL, Li C, et al. Hematopoietic stem cell dysfunction underlies the progressive lymphocytopenia in XLF/Cernunnos deficiency. Blood 2014;124:1622–5.
[240] Bluteau O, Sebert M, Leblanc T, Peffault de Latour R, Quentin S, et al. A landscape of germ line mutations in a cohort of inherited bone marrow failure patients. Blood 2018;131:717–32.
[241] Francica P, Mutlu M, Blomen VA, Oliveira C, Nowicka Z, et al. Functional radiogenetic profiling implicates ERCC6L2 in non-homologous end joining. Cell Rep 2020;32:108068.
[242] Olivieri M, Cho T, Alvarez-Quilon A, Li K, Schellenberg MJ, et al. A genetic map of the response to DNA damage in human cells. Cell 2020;182:481–96.
[243] Machado HE, Mitchell E, Obro NF, Kubler K, Davies M, et al. Diverse mutational landscapes in human lymphocytes. Nature 2022;608:724–32.
[244] Kuppers R, Dalla-Favera R. Mechanisms of chromosomal translocations in B cell lymphomas. Oncogene 2001;20:5580–94.
[245] Pasqualucci L, Migliazza A, Fracchiolla N, William C, Neri A, et al. BCL-6 mutations in normal germinal center B cells: evidence of somatic hypermutation acting outside Ig loci. Proc Natl Acad Sci U S A 1998;95:11816–21.
[246] Shen HM, Peters A, Baron B, Zhu X, Storb U. Mutation of BCL-6 gene in normal B cells by the process of somatic hypermutation of Ig genes. Science 1998;280:1750–2.

[247] Liu M, Duke JL, Richter DJ, Vinuesa CG, Goodnow CC, Kleinstein SH, et al. Two levels of protection for the B cell genome during somatic hypermutation. Nature 2008;451:841–5.

[248] Alvarez-Prado AF, Perez-Duran P, Perez-Garcia A, Benguria A, Torroja C, de Yebenes VG, et al. A broad atlas of somatic hypermutation allows prediction of activation-induced deaminase targets. J Exp Med 2018;215:761–71.

[249] Meng FL, Du Z, Federation A, Hu J, Wang Q, et al. Convergent transcription at intragenic super-enhancers targets AID-initiated genomic instability. Cell 2014;159:1538–48.

[250] Qian J, Wang Q, Dose M, Pruett N, Kieffer-Kwon KR, et al. B cell super-enhancers and regulatory clusters recruit AID tumorigenic activity. Cell 2014;159:1524–37.

[251] Compagno M, Wang Q, Pighi C, Cheong TC, Meng FL, et al. Phosphatidylinositol 3-kinase delta blockade increases genomic instability in B cells. Nature 2017;542:489–93.

[252] Chapuy B, McKeown MR, Lin CY, Monti S, Roemer MG, et al. Discovery and characterization of super-enhancer-associated dependencies in diffuse large B cell lymphoma. Cancer Cell 2013;24:777–90.

[253] Hnisz D, Abraham BJ, Lee TI, Lau A, Saint-Andre V, et al. Super-enhancers in the control of cell identity and disease. Cell 2013;155:934–47.

[254] Loven J, Hoke HA, Lin CY, Lau A, Orlando DA, et al. Selective inhibition of tumor oncogenes by disruption of super-enhancers. Cell 2013;153:320–34.

[255] Whyte WA, Orlando DA, Hnisz D, Abraham BJ, Lin CY, et al. Master transcription factors and mediator establish super-enhancers at key cell identity genes. Cell 2013;153:307–19.

[256] Alinikula J, Schatz DG. Super-enhancer transcription converges on AID. Cell 2014;159:1490–2.

[257] Wang Q, Oliveira T, Jankovic M, Silva IT, Hakim O, et al. Epigenetic targeting of activation-induced cytidine deaminase. Proc Natl Acad Sci U S A 2014;111:18667–72.

[258] Gostissa M, Alt FW, Chiarle R. Mechanisms that promote and suppress chromosomal translocations in lymphocytes. Annu Rev Immunol 2011;29:319–50.

[259] van Overbeek M, Capurso D, Carter MM, Thompson MS, Frias E, et al. DNA repair profiling reveals nonrandom outcomes at Cas9-mediated breaks. Mol Cell 2016;63:633–46.

[260] Schep R, Brinkman EK, Leemans C, Vergara X, van der Weide RH, et al. Impact of chromatin context on Cas9-induced DNA double-strand break repair pathway balance. Mol Cell 2021;81:2216–2230 e2210.

[261] Gostissa M, Yan CT, Bianco JM, Cogne M, Pinaud E, Alt FW. Long-range oncogenic activation of Igh-c-myc translocations by the Igh 3' regulatory region. Nature 2009;462:803–7.

[262] Lin C, Yang L, Tanasa B, Hutt K, Ju BG, et al. Nuclear receptor-induced chromosomal proximity and DNA breaks underlie specific translocations in cancer. Cell 2009;139:1069–83.

[263] Marnef A, Legube G. Organizing DNA repair in the nucleus: DSBs hit the road. Curr Opin Cell Biol 2017;46:1–8.

[264] Peycheva M, Neumann T, Malzl D, Nazarova M, Schoeberl UE, Pavri R. DNA replication timing directly regulates the frequency of oncogenic chromosomal translocations. Science 2022;377:eabj5502.

[265] Pancer Z, Amemiya CT, Ehrhardt GR, Ceitlin J, Gartland GL, Cooper MD. Somatic diversification of variable lymphocyte receptors in the agnathan sea lamprey. Nature 2004;430:174–80.

[266] Alder MN, Rogozin IB, Iyer LM, Glazko GV, Cooper MD, Pancer Z. Diversity and function of adaptive immune receptors in a jawless vertebrate. Science 2005;310:1970–3.

[267] Pancer Z, Saha NR, Kasamatsu J, Suzuki T, Amemiya CT, Kasahara M, et al. Variable lymphocyte receptors in hagfish. Proc Natl Acad Sci U S A 2005;102:9224–9.

[268] Rogozin IB, Iyer LM, Liang L, Glazko GV, Liston VG, et al. Evolution and diversification of lamprey antigen receptors: evidence for involvement of an AID-APOBEC family cytosine deaminase. Nat Immunol 2007;8:647–56.

[269] Quinlan EM, King JJ, Amemiya CT, Hsu E, Larijani M. Biochemical regulatory features of activation-induced cytidine deaminase remain conserved from lampreys to humans. Mol Cell Biol 2017;37.

[270] Holland SJ, Berghuis LM, King JJ, Iyer LM, Sikora K, et al. Expansions, diversification, and interindividual copy number variations of AID/APOBEC family cytidine deaminase genes in lampreys. Proc Natl Acad Sci U S A 2018;115:E3211–20.

[271] Barreto VM, Pan-Hammarstrom Q, Zhao Y, Hammarstrom L, Misulovin Z, Nussenzweig MC. AID from bony fish catalyzes class switch recombination. J Exp Med 2005;202:733–8.

[272] Hackney JA, Misaghi S, Senger K, Garris C, Sun Y, Lorenzo MN, et al. DNA targets of AID evolutionary link between antibody somatic hypermutation and class switch recombination. Adv Immunol 2009;101:163–89.

[273] Mussmann R, Courtet M, Schwager J, Du Pasquier L. Microsites for immunoglobulin switch recombination breakpoints from Xenopus to mammals. Eur J Immunol 1997;27:2610–9.

[274] Yu K, Lieber MR. Nucleic acid structures and enzymes in the immunoglobulin class switch recombination mechanism. DNA Repair 2003;2:1163–74.

[275] Cheng G, Gao Y, Wang T, Sun Y, Wei Z, et al. Extensive diversification of IgH subclass-encoding genes and IgM subclass switching in crocodilians. Nat Commun 2013;4:1337.

[276] Zhang ZZ, Pannunzio NR, Lu Z, Hsu E, Yu K, Lieber MR. The repetitive portion of the Xenopus IgH Mu switch region mediates orientation-dependent class switch recombination. Mol Immunol 2015;67:524–31.

[277] Yeap LS, Meng FL. Cis- and trans-factors affecting AID targeting and mutagenic outcomes in antibody diversification. Adv Immunol 2019;141:51–103.

[278] Sun Y, Huang T, Hammarstrom L, Zhao Y. The immunoglobulins: new insights, implications, and applications. Annu Rev Anim Biosci 2020;8:145–69.

[279] Zhu C, Lee V, Finn A, Senger K, Zarrin AA, Du Pasquier L, et al. Origin of immunoglobulin isotype switching. Curr Biol 2012;22:872–80.

[280] Hsu E. Assembly and expression of shark Ig genes. J Immunol 2016;196:3517–23.

[281] Zhang C, Du Pasquier L, Hsu E. Shark IgW C region diversification through RNA processing and isotype switching. J Immunol 2013;191:3410–8.

[282] Steidl C, Shah SP, Woolcock BW, Rui L, Kawahara M, et al. MHC class II transactivator CIITA is a recurrent gene fusion partner in lymphoid cancers. Nature 2011;471:377–81.

[283] Khodabakhshi AH, Morin RD, Fejes AP, Mungall AJ, Mungall KL, et al. Recurrent targets of aberrant somatic hypermutation in lymphoma. Oncotarget 2012;3:1308–19.

[284] Mottok A, Woolcock B, Chan FC, Tong KM, Chong L, et al. Genomic alterations in CIITA are frequent in primary mediastinal large B cell lymphoma and are associated with diminished MHC class II expression. Cell Rep 2015;13:1418–31.

[285] Shanbhag NM, Rafalska-Metcalf IU, Balane-Bolivar C, Janicki SM, Greenberg RA. ATM-dependent chromatin changes silence transcription in cis to DNA double-strand breaks. Cell 2010;141:970–81.

[286] Pankotai T, Bonhomme C, Chen D, Soutoglou E. DNAPKcs-dependent arrest of RNA polymerase II transcription in the presence of DNA breaks. Nat Struct Mol Biol 2012;19:276–82.

[287] Kim J, Sturgill D, Tran AD, Sinclair DA, Oberdoerffer P. Controlled DNA double-strand break induction in mice reveals post-damage transcriptome stability. Nucleic Acids Res 2016;44:e64.

[288] Ye BH, Lista F, Lo Coco F, Knowles DM, Offit K, Chaganti RS, et al. Alterations of a zinc finger-encoding gene, BCL-6, in diffuse large-cell lymphoma. Science 1993;262:747–50.

[289] Boulianne B, Rojas OL, Haddad D, Zaheen A, Kapelnikov A, et al. AID and caspase 8 shape the germinal center response through apoptosis. J Immunol 2013;191:5840–7.

[290] Ise W, Fujii K, Shiroguchi K, Ito A, Kometani K, et al. T follicular helper cell-germinal center B cell interaction strength regulates entry into plasma cell or recycling germinal center cell fate. Immunity 2018;48:702–715 e704.

[291] Vaidyanathan B, Chaudhry A, Yewdell WT, Angeletti D, Yen WF, et al. The aryl hydrocarbon receptor controls cell-fate decisions in B cells. J Exp Med 2017;214:197–208.

Chapter 12

Somatic Hypermutation

Jana Ridani[1,2,a], Philip Barbulescu[3,a], Alberto Martin[3] and Javier M. Di Noia[1,2,4]

[1]*Institut de Recherches Cliniques de Montréal, Montréal, QC, Canada;* [2]*Division of Experimental Medicine, Department of Medicine, McGill University, Montréal, QC, Canada;* [3]*Department of Immunology, University of Toronto, Toronto, ON, Canada;* [4]*Department of Medicine, University of Montreal, Montréal, QC, Canada*

1. INTRODUCTION

The adaptive immune system has evolved to produce a large repertoire of B cells, each expressing a different antibody specificity. Overall, this antibody repertoire allows the recognition of a vast range of antigens and virtually any pathogen. Further diversification of this primary antibody repertoire after antigen exposure by the process of somatic hypermutation (SHM) fosters the generation of B cells that produce antibodies displaying high-affinity binding to the cognate antigen. SHM also shapes the memory B cell compartment, producing a large range of affinities and specificities to the immunising antigen that protects from future infections by the same pathogen, but also potential variants [1,2].

B cells develop in the bone marrow and undergo stepwise V(D)J recombination (see Chapter 3) to assemble functional immunoglobulin heavy and light antibody genes (*IgH* and *IgL*). Each B cell generates a unique antibody molecule to produce a primary antibody repertoire estimated to be in the order of 3.5×10^{10} distinct specificities or clonotypes in the human population [3]. The newly formed B cells migrate to and recirculate through the secondary lymphoid organs, where they can be activated upon first encounter with cognate antigen. This event triggers an additional round of modifications of the Ig genes. Activated B cells first undergo class switch recombination (CSR), which causes the B cell to change the isotype of the antibody from IgM to another isotype (i.e. IgG, IgE, IgA) (see Chapter 11) [4]. Some of the activated B cells can initiate or join the germinal centre structures [5] (see Chapter 16), where the exons encoding for the variable regions (IgV) of the antibody heavy and light chains accumulate point mutations by the mechanism of SHM. Mutations improving the antibody affinity can be positively selected in the germinal centre, underpinning the affinity maturation of the antibody response, with highly mutated B cells dominating the high-affinity primary and memory responses to foreign antigens [5]. Despite earlier studies that proposed that both CSR and SHM occurred in the germinal centre, recent studies in mice and humans suggest that the bulk of CSR takes place at a pre-germinal centre B-cell stage [6,7], while SHM occurs during the germinal centre reaction. CSR is most often a unique event, with the notable exception of sequential switching from IgG1 to IgE, while SHM is an iterative process. The germinal centre is divided into two distinct compartments: the dark zone and the light zone, with SHM occurring when the B cells are inside the dark zone B [5,8]. However, despite CSR and SHM occurring independently and having different molecular outcomes, they are mechanistically related. In this chapter, we will focus on the basic principles of and recent advances in understanding the mechanism of SHM.

1.1 What Does Somatic Hypermutation Look Like?

Well before the elucidation of the molecular mechanism, the patterns and frequency of SHM were extensively characterised. SHM involves single nucleotide substitutions at frequencies of 10^{-3} mutations per base pair per generation, which is a million times higher than the average spontaneous mutation rate in vertebrate genomes [9]. Pioneer studies determined that SHM affects both DNA strands of the *IgV* [10], with mutations accumulating in the exon made by the rearranged VDJ (*IgH*) and VJ (*IgL*) regions. Mutations typically start ~100 bp downstream of the *IgV* promoter and extend over a ~2 kb region into the adjacent intron [11–14]. Mathematical models showed that the mutability within this window decreased exponentially

[a] Shared First author

Molecular Biology of B Cells. https://doi.org/10.1016/B978-0-323-95895-0.00012-X

from the promoter [14]. However, this pattern is interrupted by the presence of hotspots, a few positions that are preferentially mutated, and that tend to cluster within the three regions of the *IgV* region that encode for the protein loops that make direct contact with the antigen. Mutational hotspot sequence motifs were characterised, the first one defined as RGYW DNA motifs (where W = A/T, R = A/G, Y = C/T and the G was the mutated base) [15,16]. Subsequent studies showed that the complementary WRCY notation is the preferentially targeted motif, but nonetheless, all four bases can be mutated to any other base, which maximises the potential for amino acid changes.

The analysis of the SHM pattern in mice deficient in DNA repair enzymes then suggested that there were two mutation phases, one preferentially targeting C and G bases with preference for WRCY, and another one for A and T bases with preference for AT motifs that sometimes are also referred to as hotspots [17]. The advent of next-generation sequencing and new computational methods of analysis has further refined the understanding of mutational hotspots in mice and human V(D)J sequences, uncovering high mutability motifs such as WWRCT or the association between WRCY and AT hotspots [18–20]. These preferences can be explained by the preferences and activity of the enzymes underlying the mechanism, which are reviewed below.

The regions of the *IgV* encoding the antigen-binding surface of the antibody are denominated complementarity-determining regions (CDR), and there are three in each *IgH* and *IgL*. These regions are where most of the diversity between V regions resides. The CDR1 and 2 are encoded in the germline, while CDR3 is formed by the junction between V and DJ or V and J segments [21]. During the germinal centre reaction, the CDRs accumulate SHM more frequently than their surroundings [22]. Interestingly, the accumulation of mutations in the CDRs is not solely due to selection, since *IgV* sequences in non-productive alleles also accumulate more mutation at the CDRs [23,24]. The IgV sequence appears to have acquired this feature during evolution by enriching the CDRs with WRCY hotspots motifs and in some contexts AT hotspot motifs, partly by biasing the choice of the degenerated codons encoding certain amino acids [25–27]. Recently, the sequence upstream of these WRCY motifs has been shown to underlie the SHM bias for those hotspots at the CDR regions. Adjacent sequence with pyrimidine–pyrimidine motifs are the favoured substrates for mutations as they confer a higher single-strand DNA flexibility due to a weak base stacking strength [28]. Moreover, overlapping hotspots within CDR1 and CDR2 may act as preferential entry points for the SHM machinery, as suggested by the correlation between their mutated state and overall SHM accumulation [29].

Aside from the localised enrichment in mutational hotspots, the V(D)J exon sequence appears to award no additional benefit, as replacing it with unrelated sequences resulted in a similar frequency and pattern of mutations [23,24,30]. The accumulation of SHM exclusively downstream from the *IgV* promoter suggested the importance of transcription in this process. Indeed, inserting a promoter proximally upstream of the *Ig* exons encoding the constant region, which is usually spared from SHM, resulted in mutations in the constant exons with a similar frequency than at the *IgV*, revealing the essential role of transcription in SHM [31].

2. A MUTAGENIC ENZYME INITIATES SHM

The discovery of the enzyme activation-induced deaminase (AID) in the murine B lymphoma cell line CH12F3, which can switch from IgM to IgA, enabled the elucidation of the mechanisms of SHM and CSR [32]. AID-deficient mice are completely defective in SHM and CSR [33]. Consistently, immunocompromised patients with hyper-IgM syndrome exhibit inactivating AID mutations and lack SHM and CSR [34]. These results confirmed previous proposals that the processes of SHM and CSR shared at least part of their mechanism [35,36].

AID, encoded by *Aicda* in mice and *AICDA* in humans, is expressed almost exclusively in activated B cells and in GC B cells [32,37]. The expression of AID in a small proportion of B cell progenitors has been confirmed and helps establishing central B-cell tolerance [38,39]. The underlying mechanisms behind this finding are not fully understood. Whether SHM is involved in this mechanism is unclear; the mechanism might be related to the ability of AID to create DNA damage and trigger apoptosis [40] but will not be further reviewed here.

AID belongs to a superfamily of zinc-dependent deaminases and was initially proposed to be an RNA editing enzyme owing to its sequence homology with its paralog, APOBEC-1, a bona fide RNA-editing enzyme [32]. However, while APOBEC1 plays a role in lipid metabolism by editing the C base at position 6666 of the apolipoprotein B mRNA [41], this might not be its ancestral activity. This is suggested by APOBEC1 being largely a DNA-editing enzyme in reptiles, unable to perform RNA editing for lack of a C-terminal domain acquired later in evolution [42]. The ability of AID to mutate *Escherichia coli* and subsequent biochemical studies have established that AID most probably deaminates deoxycytidine (dC) bases exclusively in the context of single-stranded DNA (ssDNA) [43,44]. The deaminated dC becomes a dU, which has the coding equivalence to thymidine and therefore leads to a mutation if unrepaired. No evidence of RNA-editing capacity has been found so far for AID, albeit it binds RNA in multiple ways, which can inhibit its catalytic activity and limit SHM [44,45]. The DNA-editing activity of AID seems to be an ancestral trait [46], which was passed on to its paralogs, the APOBEC enzymes, including APOBEC1 and the APOBEC3s that mediate cellular innate immunity by deaminating dC to dU in viral DNA, although some of them can also deaminate viral RNA [33,47]. As such it cannot be ruled out that AID may deaminate RNA in some context, but there has not been any direct evidence of this so far.

Forced expression of AID in hybridomas and other cell types is sufficient to induce SHM, predominantly at dC:dG pairs [48,49], indicating that AID is the only B-cell-specific factor required for SHM. The dU created by AID at the *Ig* loci can then be processed by DNA repair enzymes to extend the spectrum of SHM, which is reviewed below. Despite the capacity of the APOBECs to bind genomic DNA and mutate it [50], none of them can induce SHM or CSR at the Ig locus, as shown by the non-redundant role of AID in B cells and by overexpression assays in vitro [33,51].

AID shares with the members of the cytidine deaminase superfamily a conserved catalytic deaminase domain [50,52]. This domain consists of a histidine and two cysteines coordinating a Zn^{2+} atom, within an H[AV]E-x [24−36]-PCxxC motif [50,52]. Deamination occurs when nucleophilic water attacks carbon-4 of the cytidine ring and the neighbouring proton donor in the glutamate residue [41,53]. Biochemically, AID has a predilection for deaminating C residues within the consensus WRCY motif, explaining the SHM hotspots described above. The abundance of WRCY motifs at the CDRs suggests that the *Ig* loci have evolved to maximise AID activity at these regions [15,26,46]. This DNA motif preference is dictated by a putative substrate recognition loop made of amino acids 113−123. Accordingly, substituting this loop with the corresponding region from an APOBEC3 paralog changes the preference to that of the APOBEC3, which is reflected in an altered SHM spectrum [54,55]. In addition, a positively charged 'assistant patch' in AID rich in arginine residues located in alpha helix 6 (R171, R174, R178) was proposed to recognise the region immediately 5′ of the WRCY motifs at the CDRs to facilitate their deamination [28]. Nonetheless, in vitro experiments showed that AID exhibits a very low catalytic efficiency, with each AID molecule performing a single deamination event every 4 min [56]. This might be considered a mechanism to limit AID's mutagenic activity across the genome. On the other hand, AID can catalyse several deamination events per DNA binding event, at least in vitro [57,58], by moving bidirectionally via sliding and hopping, according to experiments performed on *ssDNA* in vitro [59]. This capacity may increase the chance of a fraction of the events surviving faithful DNA repair and being fixed as mutations. These biochemical properties apply to AID incubated with purified ssDNA so presumably reflect the activity of AID after it gains access to ssDNA at the Ig genes, which represents another level of complexity in SHM.

Transcription exposes ssDNA at the Ig in various ways, as is discussed below, providing the biochemical substrate of AID [44], explaining why transcription is necessary for SHM [31]. AID moves in the same direction as the transcription machinery on transcribed DNA [59], most likely through the association with the RNA Pol II complex [60]. Consistent with the SHM profile, the activity of AID is restricted to the IgV and does not mutate farther than a few kb downstream of the IgV promoter. The mechanistic basis of this limited window of mutagenesis has not been elucidated, but it was most likely selected to avoid mutating the exons encoding the antibody constant regions. While DNA replication is another potential source of ssDNA, AID has not shown any preference for ssDNA at the replication fork, at least in *E. coli* [61].

3. TRANSCRIPTIONAL AND EPIGENETIC FEATURES THAT FAVOUR SHM AT *IG* GENES

The mechanisms that afford preferential access and activity of AID at the *Ig* loci compared to other genomic loci are complex and still incompletely understood. Neither the preference of AID for WRCY motifs nor transcription can explain the selective activity of AID at *Ig* loci over other genomic regions [62,63]. In fact, ChIP-seq, super-resolution microscopy and chromatin fractionation have suggested that AID associates with a relatively large fraction of the genome [64,65]. Nonetheless, the *Ig* loci have some distinguishing features that are required for AID activity and therefore SHM, and it is conceivable that the combination of multiple different features creates an environment unlike any other at the *IgH* and *IgL* loci, to explain the uniquely high mutation rates at *IgV* regions.

3.1 Cis-Acting Elements in SHM

Early research focussed on the role of transcription-regulatory elements present at the *Ig* loci in recruiting SHM activity. Transcription at the *Ig* genes starts at the promoter upstream of the VJ and VDJ exons and proceeds past the downstream exons that encode the constant regions to produce a full-length mRNA encoding each antibody chain. Although the *IgH* has additional internal promoters that play a role in CSR (see Chapter 11), the 5′ boundary of SHM at the *IgV* is defined by the main transcription start site (TSS) upstream of the small exon encoding the leader sequence of the antibody chain [11]. This 5′ boundary reflects the lack of AID-catalysed deamination in the first 100 bp after the IgV promoter [66]. This promoter is essential for SHM largely because of the need for transcription, rather than harbouring some specific mutation-targeting feature. Indeed, replacing it with the β-globulin gene promoter in transgenic mice, and other promoters in cell lines, induced similar levels of SHM in V-regions [24,67–69].

In addition to the promoter, the *Ig* loci harbour enhancer sequences that regulate transcription. The *IgH* contains an intronic enhancer 5′Eμ and a downstream 3′ regulatory region (3′RR) (Figure 12.1A) [70]. Both the 5′Eμ and 3′RR are part of a super-enhancer [71]. The 30 kb *Igh* 3′RR contains four transcriptional enhancers: hs3a, hs1,2, hs3b and hs4 [72] flanked by palindromic sequences. The 3′RR seems to play a role in AID targeting since SHM is fully abolished in mice lacking the 3′RR despite having only a partial effect on transcription [73]. The critical role of the 3′RR in SHM is dependent on the palindromic arrangement of this region, and, to a lesser extent, on the individual enhancer elements [74,75]. However, this is not the case for CSR, which appears to be less dependent on the palindromic arrangement of the 3′RR [76]. The mechanism by which the 3′RR enables SHM is unclear, but the need for a palindromic configuration suggests a contribution of the

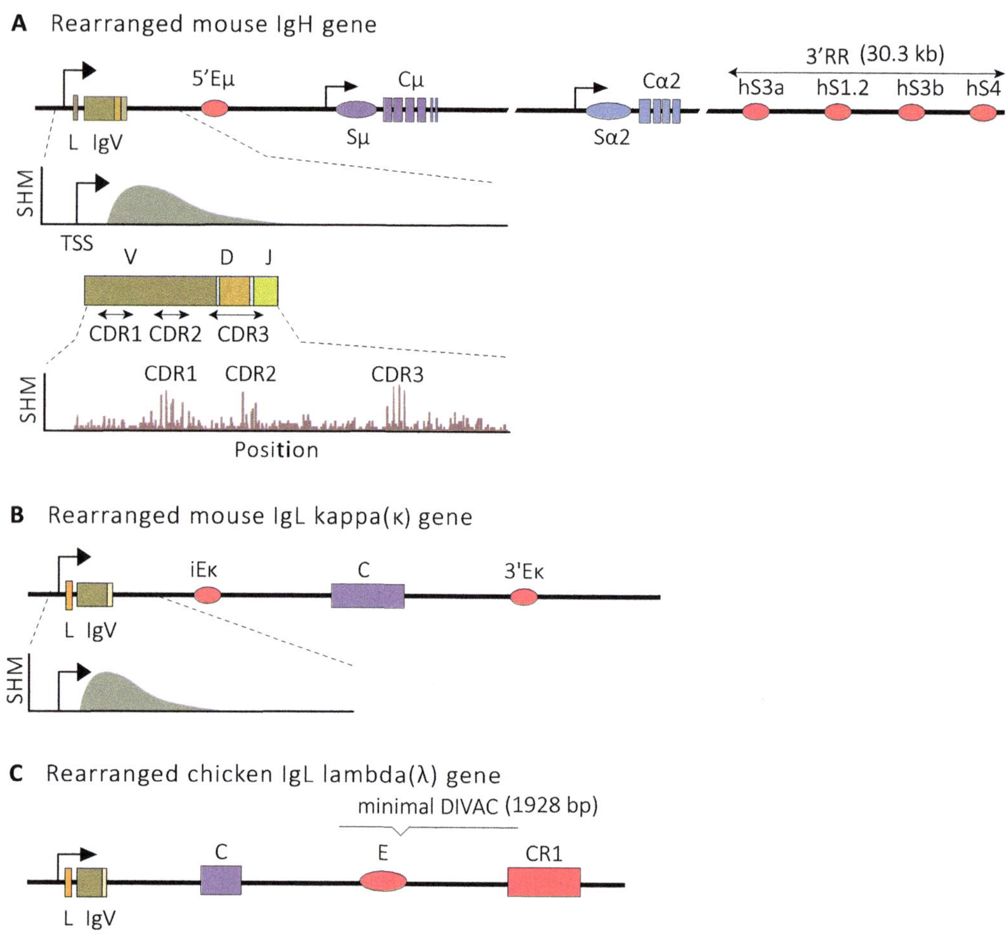

FIGURE 12.1 **SHM and cis elements in Ig genes.** (A) **Immunoglobulin heavy chain** showing the IgV made of recombined variable (V), diversity (D), joining (J) gene segments preceded by the main IgH promoter. Transcription at this site drives SHM with mutations starting to accumulate 100 bp after the transcription start site (TSS). The VDJ possesses three intrinsic mutational hotspots called complementarity-determining region (CDR). CDR1 and CDR2 are fully encoded in the germline V fragment, while CDR3 is formed during the rearrangement by the VDJ junction. Mutation frequency decreases as transcription progresses through the gene body, up until ~2 Kb from the TSS. An intronic 5′ enhancer (5′Eμ) can be found in the intron between the VDJ and first exon encoding the constant region (Cμ). The human immunoglobulin heavy chain contains 9 C units encoding the different isotypes. Each C is preceded by a switch region (S). For simplicity, only first m Cμ (IgM) and last Cα2 (IgA_2) are shown, without the intermediate isotypes. The 3′ regulatory region (3′RR) is located downstream from IgA_2. The 3′RR contains four transcriptional enhancers: DNase I hypersensitive sites (hs)3a, hs1,2, hs3b and hs4. (B) **Immunoglobulin kappa light chain:** The light chain IgV lacks a D segment and is made of a rearranged VJ exon, with the unique upstream promoter driving transcription of the whole gene. The intron enhancer (iEκ) and a downstream enhancer (3′Eκ), as well as SHM distribution are shown. Human and mice also have a IgH lambda that is not depicted by SHM distribution is like the IgL kappa. (C) **Chicken immunoglobulin light (IgL) lambda chain** and VJ segment followed by a C region transcribed under the same promoter. An IgL enhancer and chicken repeat region 1(CR1) can be found downstream. The minimal diversification activator (DIVAC) enhancer element that can direct SHM is shown. (A–C) A leader exon (L) encoding the endoplasmic reticulum signal peptide for the secretory pathway is present at the 5′ of all IgV.

DNA topology or chromatin conformation. The 3′RR harbours binding sites for multiple transcription factors [77], which could contribute to stabilise an optimal chromatin organisation for recruiting AID. Indeed, AID was shown to occupy the 3′RR region and could deaminate the core enhancers as well as flanking DNA [78]. Therefore, the 3′RR might impact AID targeting to the *IgH* IgV. In the same vein, the chromatin reader ZMYND8, which binds at the 3′RR enhancers, is required for SHM [79], further suggesting physical connection between the IgH transcriptional regulatory regions during SHM. ZMYND8 represses the 3′RR transcription by reducing RNA pol II occupancy at 3′RR. Therefore, ZMYND8 could be favouring RNA pol II loading at the IgV promoter by eliminating the competition with the 3′RR and thus providing a substrate for AID targeting to IgV [79].

The *Igk* gene contains an intron enhancer (iEκ) and a downstream enhancer (3′Eκ) (Figure 12.1B). Deleting these elements in the context of transgenes in mice was found to be either dispensable [80] or essential [67] for SHM at the IgVκ exon, albeit the latter also noticed their importance for transcription. Later work in mice with a deletion in the endogenous Igκ locus found that the 3′Eκ was required for optimal SHM at the IgVκ, most likely because it was necessary for optimal *Igκ* transcription, while the iEκ$^{-/-}$ mice showed normal SHM levels [81]. The *IgLλ* enhancer is less studied but, in a transgene, it supported some SHM, albeit at lower frequency than in the *IgLλ* locus [82].

The chicken B cell lymphoma DT40 cell line, which undergoes constitutive V(D)J diversification by gene conversion, has provided additional insight into cis-acting elements controlling SHM. Gene conversion is a templated homologous recombination-based mechanism used by avian species [83]. However, it shares the initial steps including AID targeting and deamination with SHM, as shown by eliminating homology repair components or the pseudogenes that are used as templates, which leads to diversification by C:G-focussed SHM [84,85]. Dissection of the chicken *IgLλ* identified a gene minimal region named diversification activator element (DIVAC) as essential for SHM (Figure 12.1C) [86]. Surprisingly, the DIVAC region proved to be sufficient to confer SHM to other loci when ectopically inserted close to another promoter, an activity that was maintained in a human B cell line [87] indicating that it satisfies most of the requirements for SHM. Like the *IgH* 3′RR, the DIVAC elements contain multiple transcription factor-binding sites. Mutation of some of these transcription factors-binding sites, such as the E-box, NF-KB or Ets motifs, reduced SHM of sequences next to the DIVAC, indicating that its activity is dependent on the combined activity of these transcription factors [88]. DIVAC elements might be evolutionarily conserved, as mouse and human *IgH* and *IgLλ* enhancer sequences that contain sequences with similarity to the DIVAC elements could induce some SHM in DT40 cells [88]. On the other hand, the ability of the *IgLκ* enhancer to attract SHM in the same setting was quite low, despite enhancing transcription [89]. A recent study, using a GFP reporter for SHM in DT40 and human Ramos B cell lymphoma line, showed that the chicken DIVAC SHM activity correlated with an increase in RNA pol II phosphorylation at serine 2 and 5, indicative of transcriptional stalling [90], which could potentially facilitate AID activity by exposing ssDNA at adjacent sequences.

A side effect of SHM is that AID can target at least a few hundred other genes in the genome, a phenomenon known as off-target activity that is relevant for the aetiology and progression of B cell neoplasms [62,91]. DIVAC-like elements have been proposed as a factor that contributes to AID activity outside of the *Ig* loci [87]. The analysis of AID off-target activity, as well as experiments with the DIVAC, further indicates that 3D chromatin conformation contributes to SHM. The *Ig* as well as the other loci that are mutated by AID in a cell type tend to be clustered within chromatin loops denominated topologically associating domain (TAD) containing super-enhancer elements [92]. Super-enhancers are highly active clusters of nearby enhancers engaged in long-range interactions via chromatin loops, rich in transcription factor-binding sites [93]. The DIVAC shares some of these features [87], which may partly explain its activities. Inserting a DIVAC element into an otherwise non-targeted TAD was sufficient to induce SHM in several genes within the TAD, while the regions flanking the TAD were not mutated [87]. This suggests that TADs act as boundaries for the spread of SHM. Accordingly, AID is predominately targeted to genes associated to super-enhancers in normal mouse B cells [71,92]. The role of chromatin conformation has been largely studied in the context of CSR [94], but it is likely to contribute to SHM and warrants further investigation. Another feature of super-enhancers is the production of non-coding enhancer RNA. Furthermore, super-enhancers generate robust antisense transcription and it has been suggested that collision between the sense and antisense RNA Pol II leads to transcriptional stalling and the generation of ssDNA, which may facilitate AID activity [71]. Processing of the enhancer RNA and other non-coding RNAs generated at the *Ig* loci by the RNA exosome and helicase activities contribute to AID recruitment or activity (reviewed in Ref. [95] and see below).

3.2 Trans-Acting Elements in SHM

The role of cis-acting elements in SHM is likely underpinned by their recruiting of trans-acting factors (Figure 12.2). The role of transcription factors has been noted above and includes E2A, YY1 as well as E-box, MEF2, Ets and Ikaros family members, likely acting in a

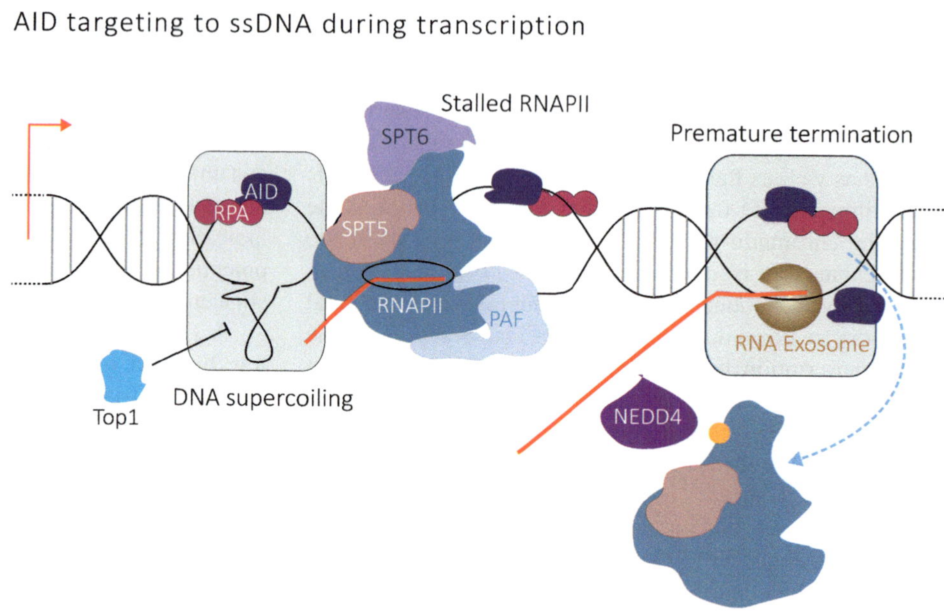

FIGURE 12.2 AID interplay and access to ssDNA during transcription for SHM. AID is associated to the RNA pol II, which within the gene body is associated with factors that promote transcription elongation like, SPT5, SPT6 and PAF. Transcription of the IgV can produce the required ssDNA substrate for AID by three mechanisms. First, negative supercoiling behind the RNA Pol II. Top1 can counteract negative supercoiling therefore limiting SHM. Second, the transcriptional machinery can be stalled by either the DNA structure or antisense transcription, which may provide time for ssDNA to be exposed. Third, premature transcriptional termination caused by SPT5 or monoubiquitination and eviction of RNA Pol II by NEDD4 that provides access to the RNA exosome and degradation of the transcript. AID and RPA interact, presumably at ssDNA in any of these settings.

combinatorial fashion [96]. The exact relative contribution of each transcription factor to SHM via regulating transcription, chromatin conformation and/or AID recruitment remains to be clarified.

The RNA Pol II was one of the first protein complexes to be discovered associated to AID [60]. Consistent with the distribution of SHM, AID occupies the IgV promoter proximal region around the TSS with RNA Pol II and AID occupancy rapidly decays over the first few kilobases of the gene, albeit the latter observation is based on genome-wide ChIP-seq data rather than a detailed analysis of the *Ig* V(D) J region [65,97]. Transcription can be divided into several phases and AID activity and thus SHM has been linked to transcriptional pausing, elongation and termination.

The association of AID to the RNA pol II requires the co-factor suppressor of Ty5 homologue (SPT5), which binds the RNA Pol II at the TSS and remains associated during transcription, regulating pausing and elongation [98]. SPT5 is not only necessary but sufficient to recruit AID to chromatin [64], suggesting it physically links AID to the RNA pol II, although we ignore as this is a direct interaction. Accordingly, the AID−SPT5 interaction is independent of DNA or RNA [98]. Together with SPT4, SPT5 forms the DSIF complex, which initially pauses the RNA Pol II some 50 bp downstream from the TSS. Phosphorylation of SPT5 by CDK9 and other kinases transforms DSIF into an elongation complex [99]. Elongating RNA pol II can stop inside the gene, a phenomenon called stalling, and probably mediated by SPT5, whereby SPT5 may favour the access of AID to the exposed ssDNA for mutagenesis [98]. Another model proposed envisages SPT5 causing premature transcriptional termination at the IgV, which would lead to ssDNA exposure by the degradation of the transcript by the RNA exosome [100]. The RNA exosome is a complex made of 11 subunits. The core is composed of: (1). a catalytically inactive EXO9 that forms a barrel shape; (2). on top of which there are three subunits that contain S1 and the KH RNA-binding domain: RRP4, CSL4 and EXOSC3 and (3). six subunits that form a ring complex with RNase activity. These subunits associate with DIS3 that possesses 3−5′ exo- and endoribonuclease activities [101]. The RNA exosome facilitates deamination of a transcribed dsDNA template in vitro [102]. Depleting the core subunit EXOSC3 in mice reduced SHM at IgV [103]. In addition, EXOSC3 deletion leads to increased persistence of divergent RNA at the TSS and of anti-sense RNA over the gene body, which has been suggested to facilitate AID substrate access [71,103]. Eliminating DISC3 caused a change in mutation patterns on both DNA strands [104]. Collectively, the evidence would suggest that the main role of the RNA exosome during SHM is to expose ssDNA on both strands, explaining how AID targets both strands to approximately similar extents during transcription.

Another transcription elongation complex associated with RNA pol II is the RNAPII-associated factor (PAF). PAF has also been proposed to help recruit AID, based on the association observed by coimmunoprecipitation and proteomics, and the negative effect on SHM and CSR of depleting several of the PAF subunits [105]. Transcription termination is facilitated through RNA Pol II ubiquitination at K63 by E3 ubiquitin-protein ligase NEDD4 [106]. Knocking down NEDD4 in the mouse lymphoma cell line

CH12 significantly reduced RNA Pol II ubiquitination as well as AID and EXOCS3 occupancy at the switch region (Sμ) [106]. RNA-seq in *Nedd4*$^{-/-}$ primary B cells showed an enrichment of long intergenic non-coding RNAs (LincRNA) [106]. This supports a role for NEDD4 in giving access to the RNA exosome to degrade ncRNA, allowing AID activity during CSR, but remains to be studied in the context of SHM.

While SHM takes place in germinal centre B cells in vivo, primary B cells stimulated ex vivo show very low levels of SHM in their IgV region, detectable only by high-throughput sequencing [24], despite displaying higher AID expression than germinal centre B cells and robust deamination of the S-regions and CSR [107]. A much reduced (undetectable by ChIP) occupancy of SPT5 and AID at the IgV in B cells activated ex vivo likely underlies the difference in SHM compared to germinal centre B cells. Interestingly, SPT5 and AID are readily detected at the *IgH* S-regions in the ex vivo-activated B cells, which undergo robust CSR, but not at the VDJ region of the same locus, despite being equivalently transcribed and showing RNA pol II accumulation (potentially stalled enzyme) [97]. Given the crucial role of SPT5 in transcription, the lack of SPT5 detection probably reflects more the detection limit of chromatin immunoprecipitation than the real absence, as supported by the verification of SHM using more sensitive methods [24]. Nonetheless, these results illustrate that transcription quality, rather than quantity, defines AID accessibility, and that SHM and CSR can happen independently, consistent with the proposal that CSR precedes SHM in GC B cells [6].

The nuclear DNA is wrapped around nucleosomes to form chromatin. During transcription elongation, two histone chaperones act to facilitate the progression of the RNA pol II through the gene length [108]. Ablation either one of these histone chaperones, SPT6 (encoded by *SUPT6H*) and FAcilitates Chromatin Transcription (FACT), results in reduced AID occupancy and activity [109,110]. Nucleosome eviction is necessary for AID activity on transcribed DNA [111] and is likely one function of these histone chaperones [111]. It is unclear if they directly participate in recruiting AID, or if their absence renders the gene body inaccessible by affecting transcription. Other histone chaperones are involved in the deposition of histone variants and will be discussed below.

A licensing step for AID has been proposed based on the observation that certain AID mutants that retain catalytic activity and can associate at the *Ig* TSS fail to occupy the gene body and show no SHM, CSR or even off-target activity [64]. These AID variants have mutations in one of three Arginine residues present in the alpha helix 6 of the AID structure, which is part of the "assistant patch". These residues were implicated in recognising structured DNA, which is deemed important during CSR but has not been tested for SHM [112]. Licencing is supported by systems in which AID can be found at the IgV, but no CSR is detectable, or vice versa, cells undergo CSR, but no IgV SHM is observed [113–115]. Either SHM and all AID activity entails binding to structured DNA or this AID helix 6 mediates two distinct functions and the mechanism for AID licencing remains to be elucidated.

It has been proposed that AID gains access to ssDNA by interacting with ssDNA-binding replication protein A (RPA) [116]. However, RPA deposition at AID target regions is AID-dependent [65], arguing against a main targeting role. Nonetheless, the regulated nature of the interaction between AID and RPA further suggests a functional relationship. The interaction with RPA requires phosphorylation of AID serine-38 by protein kinase A (PKA) [117,118]. This AID post-translational modification is required for SHM in vitro and in vivo, as mutating serine-38 to alanine in AID to prevent phosphorylation reduces SHM [119–121]. The implication of RPA during SHM would make sense given the requirement of ssDNA for AID activity. RPA would stabilise the ssDNA conformation, giving AID time to deaminate. However, this raises the question of whether AID can compete with RPA for accessing ssDNA, an issue that has so far only been studied for some of the APOBECs [122].

Splicing is an RNA-processing mechanism that proceeds co-transcriptionally and thus many splicing factors are associated to the RNA pol II complex [123]. Splicing has been implicated in CSR [124], notably by suggesting that G4 structures formed in the debranched S-region intron can bind to and target AID back to the S-region DNA [124]. This mechanism would not apply to SHM, and there is little evidence of whether splicing is implicated in SHM. At least in Ramos B cells, deletion of the intron downstream from the VDJ did not affect SHM [125]. Nonetheless, factors associated to the splicing machinery have been implicated in SHM. AID localises to nuclear speckles enriched in splicing factors [126] and has been shown to interact with major spliceosome subunits snRNPs [126], as well as individual splicing factors U2AF65 [126], SRSF1-3 [127], GANP [128], PTBP2 [129] and HNRNP proteins [130]. However, splicing has seldom been analysed in these works, and the effects observed are difficult to reconcile with a splicing effect.

Heterogeneous nuclear ribonucleoproteins (HNRNPs) are a family of RNA-binding proteins that regulate alternative splicing [131]. HNRNPs form a complex with AID in an RNA-dependent manner [130]. Depletion of HNRNP U, HNRNPI (PTBP1), HNRNP, HNRNPK and HNRNPL in CH12 cell line reduced CSR to IgA, with HNRNPL proposed to be involved in S–S synapse and HNRNP U in post-break repair [130,132]. The involvement of most HNRNPs in SHM has not been studied, except for HNRNPK, which was proposed to promote SHM by favouring a putative DNA cleavage step, and HNRNPL, which limits SHM in a human B-cell line BL2 reconstituted with a highly active nuclear AID variant [132].

An extensively studied interaction partner of AID has been CTNNBL1 [133], a factor reported to activate pre-mRNA splicing [134]. AID and CTNNBL1 colocalised at nuclear speckles in the HeLa cell line [126] and at the chromatin in DT40 cells [133]. CTNNBL1 deletion reduced SHM in DT40 cells [133]. Recently, it was shown that patients carrying a CTNNBL1 mutation that reduced its interaction with AID displayed reduced SHM [135]. The CTNNBL1 mutant cells displayed reduced levels of AID in the nucleus [135]. In fact, CTNNBL1 can bind nuclear localisation signals and has armadillo repeat structures typical of the karyopherin proteins that are the main mediators of protein nuclear import [136,137]. Thus, the data are more consistent with an AID nuclear import than splicing role for CTNNBL1 in SHM. While the structural nuclear localisation signal of AID mediates its interaction with importins in vitro [138], CTNNBL1 may represent a specialised pathway for AID nuclear import in vivo.

SRSF1-3 is another splicing factor that was first implicated in SHM in DT40 cells [127], but the mechanism remains ill-defined and there is little evidence that it involves splicing, except for the observation that it inhibits *Ig* RNA splicing in an AID-dependent manner [127]. Several indirect effects were proposed including regulating TOP1 topoisomerase via direct interaction [139], upregulating the chromatin chaperone histone cell cycle regulator (HIRA)/UBN1 that deposits histone variant H3.3 which promotes SHM [140,141] and increasing *Ig* transcription [142]. Moreover, SRSF1-3 also resembles CTNNBL1 in that it contributes to AID nuclear localisation, albeit indirectly [143].

The histone H3.3 variant [110,140] and its chaperone HIRA [141] have also been implicated in SHM, as knockout of HIRA reduces mutational frequency and changes the IgV mutational pattern in Ramos cells, *IgH* transcription, and increases chromatin accessibility. Additionally, the H3.3G34V mutant incapable of subsequent methylation was observed to reduce recruitment of elongation factors including ZMYND11 [79]. This has led to the identification of HIRA impacting chromatin structure and accessibility during SHM through H3.3K36me3 modifications.

The polypyrimidine tract-binding proteins (PTBPs), RNA-binding proteins involved in alternative splicing [144], have also been implicated in AID biology, but their contribution to SHM remains to be understood. PTBP2 coimmunoprecipitates with AID, and its depletion reduces AID occupancy at the switch regions and CSR in CH12 cell line [129]. While CH12 cells do not normally undergo SHM or display AID at the IgV, *Ptbp2*$^{-/-}$ CH12 cells showed AID occupancy at the IgV region but no detectable SHM [115]. This supports the view that AID targeting is not enough to induce mutations and would be consistent with a licensing step mediated by as yet unknown factors. Another unexplained observation that highlights the need for additional factors or specific conditions to enable SHM is that overexpressing AID in AID-deficient CH12 cells and even in *Aicda*$^{-/-}$ B cells activated ex vivo, induced detectable SHM, while the same strategy failed to achieve SHM in wt AID-sufficient B cells [113]. It is possible that endogenous AID is titrating a limiting factor but further investigation will be needed to understand this observation. Nonetheless, the role of PTBP2 in SHM is unlikely after the discovery that it is not expressed in germinal centre B cells [145]. However, the very similar proteins PTBP1 and PTBP3 may be required. *Ptbp3*$^{-/-}$ mice show strong CSR defect in vitro and reduced SHM in Peyer's patches germinal centre B cells [146]. The mechanism suggested is that PTBP3 mediates genome-wide recruitment of AID to the chromatin via nascent RNAs [146]. There must be some redundancy, perhaps with PTBP1 because there was residual CSR and SHM in the mutant cells. The link with splicing was not addressed.

The germinal centre-associated nuclear protein (GANP) is a factor linked to mRNA export post-splicing by homology with a yeast protein that seems to facilitate SHM by helping connect AID to SPT5 and RNA pol II, and GANP deletion reduces, and overexpression increases SHM in vivo [128]. Thus, while several factors associated with RNA processing and splicing have been implicated in SHM or CSR, it is not clear that splicing itself is implicated rather than other activities of those factors. Moreover, there seems to be functional redundancy, for AID targeting as well as AID nuclear import, that needs to be understood and put into a comprehensive model.

4. B-CELL INTRINSIC REGULATION OF SHM FREQUENCY BY AID REGULATION

The SHM process needs to be restricted to the *Ig* genes by tight regulation of AID. Regulating the subcellular location of AID is one efficient mechanism to minimise AID-induced mutations at off-target sites. AID shuttles between the cytoplasm and nucleus with the majority of AID protein being in the cytoplasm in steady state [147,148]. This is the result of several mechanisms. AID is actively imported into the nucleus by the interaction of its nuclear localisation signal in the N-terminus [138]. The karyopherin family was indirectly implicated by using treatments that inhibit karyopherin-mediated nuclear import and blocked AID nuclear access, and by pull down of AID with karyopherins that depended on the mapped AID nuclear localisation signal [138]. As reviewed above, other proteins have also been implicated in AID nuclear access. As a counter mechanism, AID is constantly exported out of the nucleus as it possesses a nuclear export signal (NES) [121] on the C-terminus that can be recognised by the chromosomal maintenance 1 (CRM1) exportin complex [147]. Mutating the NES or treatment with the CRM1 inhibitor leptomycin B results in the accumulation of AID in the nucleus of B cells [147]. Because of its small 24 kDa size, AID could passively diffuse through the nuclear pores into the nucleus [138]. This is prevented by a cytoplasmic retention mechanism through the interaction of AID with elongation factor 1α (EAF1A) [149,150]. Mutations disrupting this interaction, as well as the drugs that stabilise the interaction of EEF1A with tRNA, which is mutually exclusive with AID interaction, displace the equilibrium and facilitate AID nuclear import [149,150]. Regulating AID protein levels and stability is another mechanism for regulating SHM in B cells. Following B cell activation and *Aicda* gene expression, the 3′UTR of *Aicda* mRNA is recognised by mi-155 [151] and mi-181b [152] miRNA, which represses its translation. Cytoplasmic AID is stabilised by the heat shock protein 90 (HSP90) chaperone complex [153]. Nuclear AID is instead destabilised by proteasome activator γ (REG-γ) in a ubiquitin-independent fashion [154] and by polyubiquitination by the CULLIN7 (CUL7) E3 ligase [155]. In contrast, AID deubiquitination by ubiquitin-specific peptidase 10 (USP10) stabilises nuclear AID [156]. The transcription factor YY1 has also been shown to stabilise AID within the nucleus by blocking its ubiquitination [157,158]. Hence, ubiquitination and deubiquitination play important roles in maintaining the levels of AID for SHM and CSR in B cells. The net result of all the mechanisms regulating AID subcellular localisation and stability lead to only ~10% of cellular AID being in the nucleus at steady state [159].

5. AID SUBSTRATES

AID activity is limited to substrates with a minimum of five bases in an accessible ssDNA state to begin catalysing the deamination of dC residues [44,56]. The bifurcated substrate-binding domain of AID is dispensable for deaminating ssDNA but allows for structured DNA binding of higher affinity in scenarios where adjacent ssDNA tracks, including branched and G4 substrates are available [112]. Double-stranded DNA and DNA:RNA hybrids are unsuitable substrates for AID as the complementary base pairings shield dC from deamination [160,161]. Typically, ssDNA is scarce and occurs transiently and is highly protected to prevent spontaneous or enzymatic base modifications. Active transcription remains the primary mechanism generating ssDNA in cells for AID, as reviewed above.

Several scenarios of stable ssDNA intermediates may appear during gene transcription including DNA:RNA hybrids, branched DNA G-quadruplex (G4) structures and unorganised sections of supercoiled DNA [162,163]. In addition, there are intermediates of DNA repair or replication that lead to ssDNA that could theoretically be targeted by AID but are unlikely to contribute to SHM. G4 and R-loop structures transiently appearing during transcription and the stalling of the transcriptional machinery prolong the time of ssDNA substrate availability [112]. While AID possesses high activity on G4 and branched DNA substrates as observed through biochemical assays [112], and these structures are likely important in CSR. Bioinformatic analysis has identified potential for G4 formation in certain human IgVH families [164], but whether this is a general enough phenomenon to explain SHM remains to be studied. An AID mutant that cannot bind G4 structures was shown to be unable to associate with the chromatin altogether [165]. Thus, the role of G4 in SHM requires further investigation. G4 quadruplex structures may contribute to AID off-targeting as telomer transcripts are capable of forming transient G4 structures in B cells and have been observed to recruit AID [166,167].

The co-transcriptional RNA:DNA hybrid structures – referred to as R-loops – have also been described as AID substrates but would be expected to cause AID mutations to the non-hybridised DNA strand during transcription [44,168,169]. The discovery of R-loops within *IgH* S-regions was first detected in bacteria through transcriptional studies where the presence of stable R-loops was observed following the transcription of a plasmid that harbours a human S-region [170]. These R-loops occur during active transcription of S-regions due to the particularly high levels of dG clusters that positively influence the frequency of R-loop formation [160,161,169,171]. The dG-rich nature of the non-template S-region strand enriches for R-loop structure formation and can promote transcriptional stalling that is favorable for CSR as it prolongs ssDNA accessibility for AID targeting [169]. Studies using *Ung*$^{-/-}$*Msh2*$^{-/-}$ mice observed few R-loops in the IgV region due to a lack of G-clustering, suggesting R-loops do not contribute to SHM [172,173]. Enrichment of short ssDNA has instead been observed in IgV regions undergoing SHM in both primary B cells and human B cell lines suggesting assembly of ssDNA-protein complex that may promote ssDNA accessibility for AID targeting [172].

6. DNA REPAIR PATHWAYS DOWNSTREAM OF AID

Mutations induced by AID within IgV regions initiate networks of repair pathways involved in the resolution of dU:dG lesions that result in the full spectrum of SHM. Indeed, instead of repairing these lesions, these DNA repair pathways promote mutation downstream from AID [174]. The base excision repair (BER) factor, UNG2, and the mismatch repair (MMR) factors MSH2/MSH6 are key enzymes involved in dU recognition and processing, both competing for the same substrate, although possibly at different stages of the cell cycle. There is sufficient evidence to suggest that there is crosstalk between these two DNA repair pathways that generate intermediates which demand error-prone translesional polymerases to achieve resolution, ultimately contributing to the mutational-prone SHM outcome (Figure 12.3) (see below).

6.1 Base Excision Repair During SHM

BER serves as an important cellular mechanism for the preservation and repair of DNA following endogenous or exogenous sources of genotoxic damage. This pathway is highly conserved and ubiquitous among both prokaryotes and eukaryotes [176,177], resolving a wide range of base modifications including alkylation, deamination, depurination and single-stranded nicks [178]. Enzymes involved in BER were initially purified from *E. coli*, where the presence of endonucleases capable of recognising and catalysing the cleavage of apurinic/apyrimidinic sites (AP-sites) was first shown [179,180]. These initial findings supported the concept of DNA repair machinery dedicated to the recognition of small-scale DNA damage and founded the interest towards understanding the mechanism behind BER.

The process of BER can be categorised into several distinct phases: recognition of damaged bases, AP endonuclease processing, DNA resynthesis and finally DNA ligation to complete the repair. Resolution of base lesions requires two key enzymes, DNA N-glycosylases for recognition as well as base cleavage and AP endonucleases – APE1 and APE2 – for subsequent processing of the ensuing abasic (AP) site. The DNA N-glycosylase family is extensive, with each member specialised in recognising specific base lesions that arise through exogenous mutagens or spontaneous endogenous errors during cellular

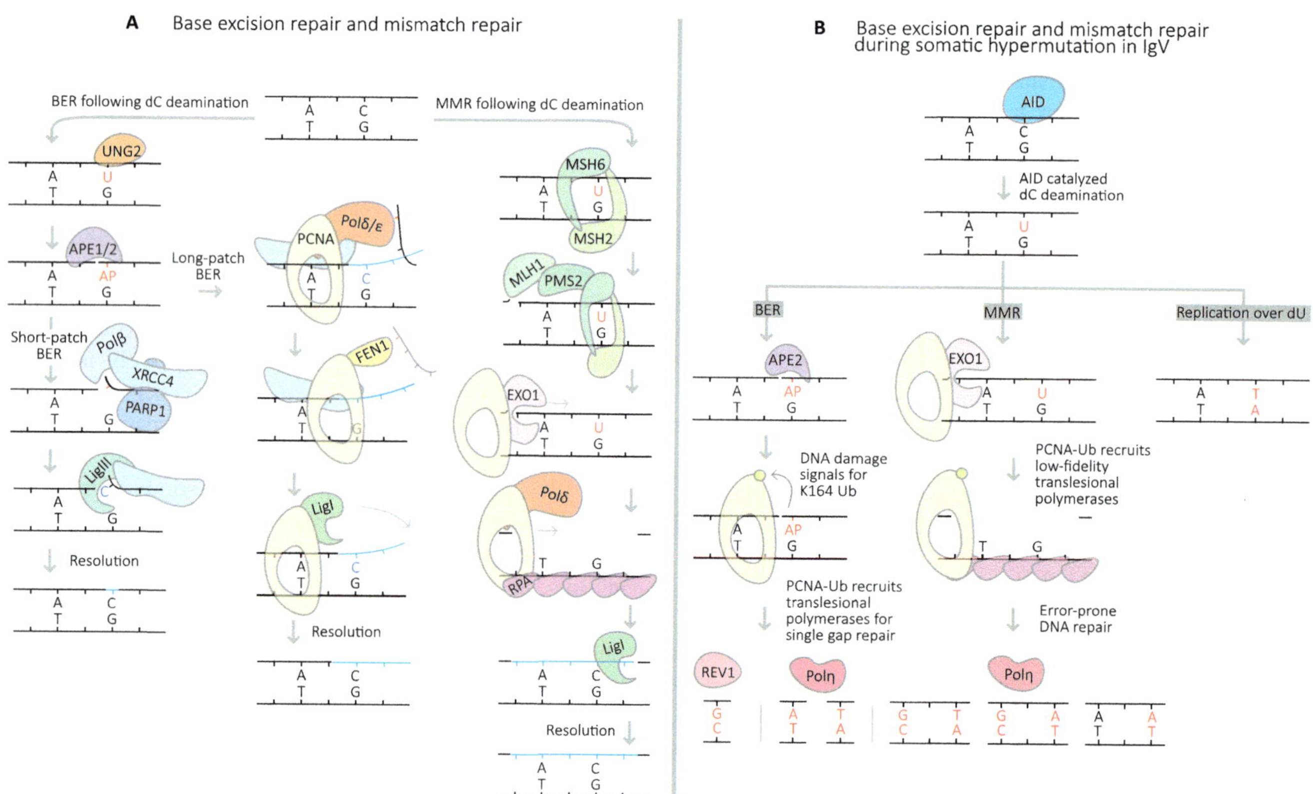

FIGURE 12.3 **Base excision and mismatch repair of dU:dG lesions for error-free and error-promoting resolution during somatic hypermutation.** (A) Nucleotide mismatches are recognised and processed through base excision repair (BER) or mismatch repair (MMR) to achieve faithful resolution. BER is initiated through the recognition of dU by uracil DNA glycosylase 2 (UNG2), which is responsible for generating an apyrimidinic (AP) site. AP endonuclease 1 or 2 (APE1/2) catalyse a nick 5′ of the AP site which is processed by the BER machinery for short-patch or long-patch repair. In short-patch BER, PARP1 is the first responder to the DNA damage signal, recruiting the XRCC1 scaffold and high-fidelity polymerase β (Pol β). The AP lyase activity of Pol β allows for the excision of the AP nucleotide before proceeding with DNA synthesis over the nucleotide gap and ligation through Ligase III (LigIII). For long-patch BER, proliferating cell nuclear antigen (PCNA) responds to the site of DNA damage, recruiting Pol δ and Pol ε to synthesise DNA and displace the original DNA strand. FEN1 catalyses the removal the displaced DNA strand, followed by ligation through Lig I. MMR is initiated by dU recognition through the MutSα heterodimer (MSH2-MSH6), recruiting the MutLα heterodimer (PMS2-MLH1) possessing endonuclease activity to nick DNA 5′ upstream of the mismatch. PCNA is recruited to the site of damage, signalling for exonuclease 1 (EXO1) to process 5′–3′ and produce a ssDNA track [175]. Replication protein A (RPA) binds to and stabilises the complementary ssDNA, while Pol δ synthesises over the gap and Lig I completes the resolution. (B) During SHM, BER and MMR function in a similar manner as above, but instead utilise translesional polymerases recruited by monoubiquitinated PCNA at K164. These translesional polymerases – REV1 and Polη for BER; Polη for MMR – contribute to the mutational spectrum. The dU lesions which fail to be repaired during this process act as templates during DNA replication to generate dC:dG to dT:dA transition mutations.

metabolism. Four N-glycosylase family members capable of recognising uracil misincorporation exist in mammalian cells, all varying in cellular localisation and substrate specificity – uracil N-glycosylase (UNG or UDG), single-stranded selective monofunctional UDG1 (SMUG1), T:G mismatch thymidine DNA glycosylase (TDG) and methyl-CpG-binding endonuclease (MBD4). During the initial recognition of modified bases, N-glycosylases flip damaged bases out of the DNA helix before catalysing the cleavage of the β-N-glycosidic bond to generate an AP site [181,182]. If BER fails to resolve the lesion, AP sites can persist and be fixed as mutations following replication because they lack the template information for DNA polymerases.

During SHM, AID-induced uracil lesions act as the principal substrate for the BER machinery. While TDG and MBD4 function solely on double-stranded DNA, UNG and SMUG1 can function on ssDNA substrate, like AID. Genomic uracil is predominantly excised through the nuclear isoform of UNG, uracil DNA glycosylase 2 (UNG2) and to a lesser extent, SMUG1 [183–186]. This is also the case for AID-catalysed uracil in SHM and CSR [187–191] Unlike UNG, SMUG1 does not have a significant natural role in antibody diversification as its expression levels are low throughout B-cell activation [191,192]. However, SMUG1 can act as a surrogate during SHM and CSR in UNG-deficient models when overexpressed [192,193]. A study in chicken DT40 B cells showed disruptions in the catalytic region of MBD4 resulted in increased SHM frequency, suggesting it may play a minor role in SHM through its ability to recognise AID-induced uracil, at least in chicken B cells [194]. Deficiency in UNG alters the mutational spectrum of SHM, limiting the number of dC:dG transversion mutations in V-regions [187–189]. However, mutations at dA:dT are not severely impacted by UNG deficiency [187], suggesting that the MMR machinery is largely responsible for these mutations.

Following N-glycosylase activity, the subsequent AP sites are substrates for the APE1/APE2 endonucleases, that cleave the DNA backbone to set up the canonical scenario for processing through factors such as AP lyase, polymerase and ligase to resolve the lesion [195,196]. During SHM, however, only APE1 endonuclease has been shown to be necessary throughout BER of dU lesions [197]. APE1/APE2 signal through PARP1 activation to recruit DNA repair factors including XRCC1, DNA ligases and polymerases to the nicked site. DNA polymerase β (Pol β) and to a lesser extent Pol λ are generally involved in high-fidelity DNA synthesis. However, monoubiquitinated PCNA (PCNA-Ub) at lysine 164 (K164) typically associated to long-patch BER may instead recruit the translesional error-prone Pol η [198]. The synthesis phase of BER is initiated once AP endonucleases catalyse a nick adjacent to the damaged base and reveals the 3′-hydroxyl group that behaves as a substrate for DNA polymerases. Opposing 5′-deoxyribose phosphates (dRPs) are cleaved through enzymes possessing dRP lyase domains, including several polymerases, pol β, pol θ and REV1 [199–202]. REV1 is multifaceted, catalysing the incorporation of dCMP, functioning as a scaffold, and might recruit other translesional polymerases, and possibly contributes to the stabilisation of UNG [203–205]. *Rev1*$^{-/-}$ B cells have reduced CSR, which might be due to reduced UNG protein stability that is caused by the loss of REV1–UNG interaction [204]. Interestingly, catalytically inactive REV1 mutant rescues CSR in REV1-deficient cells, showing that the scaffolding role for UNG stability outweighs the value of its catalytic activity [204], with the caveat that UNG is required for REV1 activity to take place. Given that REV1 is responsible for most transition mutations at dC:dG in the DT40 model [206] and in mice [205], its catalytic activity is likely involved in SHM.

DNA nicks generated following APE1 activity are traditionally catalysed through canonical BER (cBER) which recruits high-fidelity DNA pol β or pol λ to resynthesise single nucleotide errors through short-patch BER. In a less frequent scenario, long-patch BER recruits the PCNA processivity factor along with a separate set of polymerases, pol δ and ε to handle scenarios involving tandem mutations where several close proximity base mutations may be resolved concurrently. The ligases recruited to short- or long-patch repair also differ, with Ligase I being involved in short-patch single nucleotide lesions while Ligase III is dedicated to long-batch BER resolution [207]. The decision between short and long-patch BER is believed to be influenced by the base modification, the state of cell differentiation and the cell cycle [208]. Short-patch BER is likely a minimal contributor to SHM due to the high-fidelity nature of this DNA repair process [208]. In the context of SHM requiring an enrichment of mutation, short-patch BER does not give opportunity for mutational incorporation and is less relevant to generating the mutational spectrum at IgV regions as opposed to long-patch BER that involves error-prone polymerases [209].

HMCES, the 5-hydroxymethylcytosine binding, embryonic stem cell-specific protein is implicated in SHM through its ability to form covalent cross-links with abasic sites generated by UNG2 [210]. HMCES protects AP sites from breakage and therefore from larger genomic deletions [210]. Interestingly, this function of HMCES in SHM is separate from its role in DNA end-joining that occurs during CSR [211].

6.2 Mismatch Repair During SHM

MMR typically corrects erroneous base misincorporations, insertions or deletions that occur following errors in DNA replication and recombination. Despite this function, the MMR system instead expands the mutational spectrum during SHM to dA/dT bases that are not substrates of AID, generating variability within the IgV regions that prioritises antibody affinity maturation over maintenance of genomic integrity. The MMR pathway is highly conserved among prokaryotes and eukaryotes [212–216]. Initial studies of this DNA repair pathway in *E. coli* identified that inactivating MMR factors such as the Mut protein family (MutS, MutL and MutH) led to a ~100-fold increase in mutational rates [217,218]. The importance of MMR in maintaining genomic integrity was later confirmed in eukaryotic models, where it was observed that MMR deficiency resulted in up to 1000-fold increases in mutation frequencies [219,220].

The mechanism of how eukaryotes differentiate between the parent and error-containing daughter strands during the MMR process is still not clear compared to the bacterial mechanism which utilises hemi-methylation for strand differentiation [221]. Previous work suggested that the misincorporations of ribonucleotides into DNA may act as a differentiation signal during MMR in eukaryotes, while another study suggests that the asymmetrical loading of the replicative clamp, PCNA, positions the MutSα endonuclease along the daughter strand relative to the nick [222–224]. Under normal circumstances, MMR is highly effective at maintaining genomic integrity. Initiation of MMR begins with the recognition of mismatched bases through the partially redundant MutSα (MSH2-MSH6) or MutSβ (MSH2-MSH3) heterodimers, the latter having little to no roles during SHM and CSR [216,225]. MutSα further recruits the MutLα complex (MLH1-PMS2) to nick the DNA backbone adjacent to the mismatched base through the endonuclease activity of PMS2 [226,227] to permit the 5′–3′ EXO1 exonuclease to catalyse the excision of DNA harbouring the misincorporated base [228]. Surprisingly, deficiency in catalytically functional EXO1 did not affect SHM or CSR mice [229]. MMR and EXO1 can produce either short-patch or long-patch excisions. Short-patch MMR efficiently resolves single nucleotide mismatches and excises small patches of 2–10 nucleotides in both non-proliferating and proliferating cells, while long-patch repair occurs in proliferating cells and can create kilobase-wide tract excisions [213,216]. The exposed ssDNA template allows for high-fidelity replicative polymerases such as pol δ to resolve the initial mismatch. However, during SHM, the MMR system is the driver of mutation and instead leads to the recruitment of error-prone translesional DNA polymerases such as Pol η through PCNA-Ub to resynthesise ssDNA gaps [230–233]. MSH2, MSH6 and EXO1-deficient B cells exhibit dramatic changes in SHM mutational spectrum with a reduced accumulation of dA:dT mutations [17,228,234,235]. As the function of the MMR system during SHM was opposite to its canonical function, it was termed non-canonical MMR (ncMMR) or mutagenic MMR.

Mutagenic MMR is governed through the direct association of MutSα heterodimers to PCNA. PCNA monoubiquitination (PCNA-Ub) at K164 through RAD6/RAD18 following its association to MutSα is a key post-translational modification – along with its deubiquitination by enzymes such as USP1 – that results in the activation of the translesional synthesis pathway and preferential recruitment of error-prone polymerases such as pol η, biasing towards mutagenic MMR [236–238]. Human Ramos B cells expressing a fusion of Ub-PCNA show that translesional Pol η is recruited to promote dA:dT mutations during SHM [239].

6.3 Translesional DNA Synthesis

Following the generation of mismatched bases, mutations produced by the MMR pathway depend on translesion DNA synthesis (TLS) to overcome replication blocks. PCNA-Ub at lysine 164 occurs in response to DNA damage and initiates the interaction and recruitment of low-fidelity translesion polymerases including Pol η, Pol θ, Pol ζ and REV1 which are responsible for the dA and dT mutational spectrum observed in SHM [230,240–244]. Pol η acts as the primary TLS polymerase involved in generating dA:dT mutations during SHM, as Pol η-deficient mice experience much greater losses of the dA:dT mutations when compared to other TLS polymerase deficiency that generate more subtle contributions to the mutation spectrum [230,241–244].

The primase-polymerase hybrid, PrimPol, is implicated in TLS, and a previous study found it to have a minor influence on the SHM spectrum [245]. In particular, PrimPol was observed to specifically prevent dC to dG transversions in the leading strand during SHM presumably through its ability to maintain transcription fork progression by priming downstream of AP sites in the leading replication strand. AP sites primed downstream by PrimPol are no longer processing by TLS, instead using homology directed error-free repair [245,246].

7. CELL CYCLE REGULATION DURING SHM

Most AID-induced dU lesions of the Ig locus predominantly occur within the G_1 phase of the cell cycle [247]. Additionally, AID may be more stable in the G_1 phase in comparison to S and G_2 phases, which might be influenced by ubiquitin-dependent proteolysis [248]. Recently, restricted expression of AID at different stages of the cell cycle has allowed for the study of G_1- or S/G_2-specific events of antibody diversification in B-cell lines, observing that C/G mutagenesis occurs throughout the cell cycle, while A/T mutagenesis and CSR require G1-restricted AID activity [249]. Collectively, these works indicate that AID activity in G1 underlies SHM.

The cell cycle stage is also a major determinant of the expression levels of various DNA repair factors and influences which DNA repair pathway is used during SHM (Figure 12.4). MMR is believed to be functional throughout all stages of the cell cycle but is thought to reach its highest

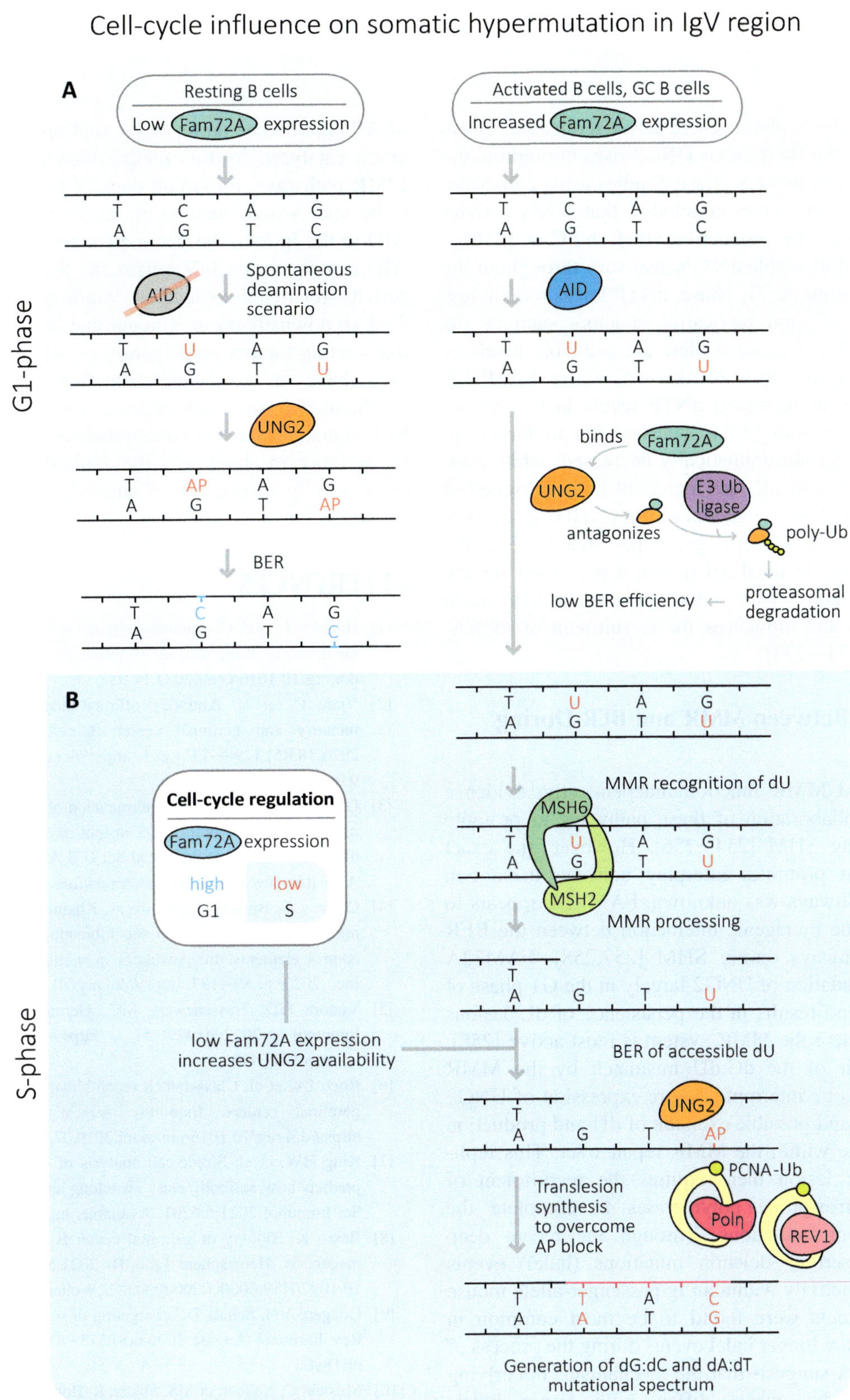

FIGURE 12.4 Influence of the cell cycle on repair of AID-induced mutations during somatic hypermutation in activated mouse B cells. (A) Following B-cell activation, induction of activation-induced cytidine deaminase (AID) and FAM72A promotes mutagenic repair. Expression of AID within the G_1 phase of the cell cycle overlaps with elevated expression of FAM72A. FAM72A interacts with the key BER-initiating factor, uracil DNA glycosylase 2 (UNG2), leading to proteasomal degradation of UNG2 and a reduced repair efficiency of dU lesions in G_1, which results in the dU lesion persisting into S-phase. Cells with low FAM72A expression, including resting B cells, have normal UNG2 levels and can therefore repair dU efficiently in an error-free manner. (B) Persistence of dU lesions in S-phase is processed through mismatch repair (MMR), initiated by MutSα heterodimer (MSH2-MSH6) mismatch recognition. EXOI excises the dU-containing strand or the opposite DNA strand. In the scenario that MMR excises the non-dU containing DNA strand or if other dU lesions are present in the exposed ssDNA patch, UNG2 can generate an apyrimidinic (AP) site that prevents resolution of MMR through standard high-fidelity polymerases, polymerase δ or ε. The AP site instead signals for the monoubiquitination of proliferating cell nuclear antigen (PCNA) at K164, initiating the recruitment of translesional polymerases including Polη and REV1 to complete repair. Of note, REV1 is specialised to incorporate dC opposite of AP sites and generate dG to dC transversion mutations post-replication. The error-permitting property of these translesional polymerases generates the mutational spectrum observed during SHM.

activity during the S-phase where DNA replication errors arise. Likewise, the BER factor UNG2 rises throughout the G_1 phase and peaks in the S-phase. Studies using cell cycle-dependent UNG inhibitors concluded that UNG activity was required in G_1 for productive SHM and CSR [250].

The levels of available dNTPs also vary throughout the cell cycle, as during the G_1 phase, dNTP levels remain low to reduce risk of viral replication. Factors such as the SAMHD1 dNTP reductase deplete the available levels of dNTPs by catalysing their breakdown, where SAMHD1 deficiency leads to increased dNTP levels in G_1 and an altered SHM spectrum [251]. Transversion mutations to dA:dT and dC:dG are dramatically decreased, while transition mutations at dC:dG are increased in the absence of SAMHD1 [251]. The consequence of SAMHD1 deficiency was observed to phenocopy genotypes that prevent the recruitment of translesional pol η, as it is presumed that the depleted dNTP pool in G_1 stalls conventional long-patch DNA synthesis and influences the recruitment of PCNA-Ub and TLS [251–253].

7.1 Synergy Between MMR and BER During SHM

While BER and MMR function independently, evidence supports the collaboration of these pathways to promote mutations during SHM [254–256]. However, the exact mechanism that promotes interplay between these two DNA repair pathways was unknown. FAM72A appears to be the key to the mutagenic interaction between the BER and MMR pathways during SHM [257,258]. FAM72A causes the degradation of UNG2 largely in the G1 phase of the cell cycle and results in the persistence of dU lesions into S-phase where the MMR system is most active [258]. However, repair of the dG:dU mismatch by the MMR system seems to be interrupted by re-expression of UNG2 in the S-phase, and possible excision of dU and production of an abasic site within the MMR repair tract. This replicative blocking lesion then requires the recruitment of translesional error-prone polymerases to complete the repair, promoting mutations. Through the recent deep profiling of insertion–deletion mutations (indel) events following AID activity within an Ig passenger-allele mouse model, 1 bp indels were found to be most common in comparison to rare longer indel events during the process of SHM. This work suggests that the mechanisms underlying various-sized indel events differ, with larger indels requiring additional DNA-processing factors [259].

8. CONCLUDING REMARKS

The advances in understanding the molecular mechanisms of SHM in the past 20 years have been able to explain many of the characteristics of this process by the properties of AID preference and activity, and processing of genomic uracil catalysed by this unique enzyme by the BER and MMR pathways. Important parts of the mechanism remain to be understood, such as the exact mechanisms targeting AID to the Ig loci, the parameters and factors that regulate AID activity at the IgV versus the S-regions, how is AID activity restricted to a short window after the TSS and 2–3 kb downstream or a model that integrates the multiple trans-acting factors participating in SHM. Nonetheless, we now have a good understanding of how a single biochemical base transformation can be transformed into a large mutation spectrum that translates into a large potential for amino acid changes at the antibody variable region to drive affinity maturation of antibodies against almost any antigen.

REFERENCES

[1] Hägglöf T, et al. Continuous germinal center invasion contributes to the diversity of the immune response. Cell December 2022. https://doi.org/10.1016/j.cell.2022.11.032.

[2] Viant C, et al. Antibody affinity shapes the choice between memory and germinal center B cell fates. Cell November 2020;183(5):1298–1311.e11. https://doi.org/10.1016/j.cell.2020.09.063.

[3] Glanville J, et al. Precise determination of the diversity of a combinatorial antibody library gives insight into the human immunoglobulin repertoire. Proc Natl Acad Sci U S A 2009;106(48):20216–21. Available: www.pnas.org/cgi/content/full/.

[4] Oudinet C, Braikia FZ, Dauba A, Khamlichi AA. Mechanism and regulation of class switch recombination by IgH transcriptional control elements. In: Advances in immunology. Academic Press Inc.; 2020. p. 89–137. https://doi.org/10.1016/bs.ai.2020.06.003.

[5] Victora GD, Nussenzweig MC. Germinal centers. Annu Rev Immunol 2022;30:429–57. https://doi.org/10.1146/annurev-immunol-120419.

[6] Roco JA, et al. Class-switch recombination occurs infrequently in germinal centers. Immunity August 2019;51(2):337–350.e7. https://doi.org/10.1016/j.immuni.2019.07.001.

[7] King HW, et al. Single-cell analysis of human B cell maturation predicts how antibody class switching shapes selection dynamics. Sci Immunol 2021;6:6291. Available: https://www.science.org.

[8] Basso K. Biology of germinal center B cells relating to lymphomagenesis. Hemasphere June 01, 2021;5(6):E582. https://doi.org/10.1097/HS9.0000000000000582. Wolters Kluwer Health.

[9] Odegard VH, Schatz DG. Targeting of somatic hypermutation. Nat Rev Immunol August 2006;6(8):573–83. https://doi.org/10.1038/nri1896.

[10] Milstein C, Neuberger MS, Staden R. Both DNA strands of antibody genes are hypermutation targets. 1998. Available: www.pnas.org.

[11] Lebecque SG, Gearhart PJ. Boundaries of somatic mutation in rearranged immunoglobulin genes: 5' boundary is near the promoter, and 3' boundary is rut kb from V(D)J gene. J Exp Med 1990;172(6):1717–27.

[12] Both GW, et al. Distribution of mutations around rearranged heavy-chain antibody variable-region genes. Mol Cell Biol 1990;10(10):5187–96.

[13] Gearhart PJ, Bogenhagent DF. Clusters of point mutations are found exclusively around rearranged antibody variable genes (somatic mutation/nucleotide substitutions/error-prone repair). Proc Natl Acad Sci U S A 1983;80(11):3439–43. Available: https://www.pnas.org.

[14] Rada C, César M. The intrinsic hypermutability of antibody heavy and light chain genes decays exponentially. EMBO J 2001;20(16). https://doi.org/10.1093/emboj/20.16.4570.

[15] Dörner T, Brezinschek H-P, 1 Brezinschek R, Foster SJ, Domiati-Saad R, Lipsky3 PE. Analysis of the frequency and pattern of somatic mutations within nonproductively rearranged human variable heavy chain genes'. J Immunol 1997. Available: http://www.peter.lipskyQemaiI.swmed.edu.

[16] Betz AG, Neuberger MS, Milstein C. Discriminating intrinsic and antigen-selected mutational hotspots in immunoglobulin V genes. Immunol Today 1993;14(8):405–11.

[17] Rada C, Ehrenstein MR, Neuberger MS, Milstein C. Hot spot focusing of somatic hypermutation in MSH2-deficient mice suggests two stages of mutational targeting. Immunity July 1998;9(1):135–41. https://doi.org/10.1016/s1074-7613(00)80595-6.

[18] Pérez-Durán P, Belver L, de Yébenes VG, Delgado P, Pisano DG, Ramiro AR. UNG shapes the specificity of AID-induced somatic hypermutation. J Exp Med July 2012;209(7):1379–89. https://doi.org/10.1084/jem.20112253.

[19] Álvarez-Prado ÁF, et al. A broad atlas of somatic hypermutation allows prediction of activation-induced deaminase targets. J Exp Med March 2018;215(3):761–71. https://doi.org/10.1084/jem.20171738.

[20] Tang C, Krantsevich A, MacCarthy T. Deep learning model of somatic hypermutation reveals importance of sequence context beyond hotspot targeting. iScience January 2022;25(1). https://doi.org/10.1016/j.isci.2021.103668.

[21] Hwang JK, Alt FW, Yeap L-S. Related mechanisms of antibody somatic hypermutation and class switch recombination. Microbiol Spectr February 2015;3(1). https://doi.org/10.1128/microbiolspec.mdna3-0037-2014.

[22] Di Noia JM, Neuberger MS. Molecular mechanisms of antibody somatic hypermutation. Annu Rev Biochem 2007;76:1–22. https://doi.org/10.1146/annurev.biochem.76.061705.090740.

[23] Betz AG, Rada C, Pannell R, Milstein C, Neuberger MS. Passenger transgenes reveal intrinsic specificity of the antibody hypermutation mechanism: clustering, polarity, and specific hot spots. Proc Natl Acad Sci U S A 1993;90:2385–8.

[24] Yeap LS, et al. Sequence-intrinsic mechanisms that target AID mutational outcomes on antibody genes. Cell November 2015;163(5):1124–37. https://doi.org/10.1016/j.cell.2015.10.042.

[25] Tang C, Bagnara D, Chiorazzi N, Scharff MD, MacCarthy T. AID overlapping and Polη hotspots are key features of evolutionary variation within the human antibody heavy chain (IGHV) genes. Front Immunol April 2020;11. https://doi.org/10.3389/fimmu.2020.00788.

[26] Wagner SD, Milstein C, Neuberger MS. Codon bias targets mutation. Nature August 1995;376(6543):732. https://doi.org/10.1038/376732a0.

[27] Jolly CJ, Wagner SD, Rada C, Klix N, Milstein C, Neuberger MS. The targeting of somatic hypermutation. Semin Immunol 1996;8:159–68.

[28] Wang Y, et al. Mesoscale DNA feature in antibody-coding sequence facilitates somatic hypermutation. Cell May 2023;186(10):2193–2207.e19.

[29] Wei L, et al. Overlapping hotspots in CDRs are critical sites for V region diversification. Proc Natl Acad Sci U S A February 2015;112(7):E728–37. https://doi.org/10.1073/pnas.1500788112.

[30] Azuma T, Motoyama N, Fields LE, Loh DY. Mutations of the chloramphenicol acetyl transferase transgene driven by the immunoglobulin promoter and intron enhancer. Int Immunol February 1993;5(2):121–30. https://doi.org/10.1093/intimm/5.2.121.

[31] Peters A, Storb U. Somatic hypermutation of immunoglobulin genes is linked to transcription initiation. Immunity January 1996;4(1):57–65. https://doi.org/10.1016/s1074-7613(00)80298-8.

[32] Muramatsu M, et al. Specific expression of activation-induced cytidine deaminase (AID), a novel member of the RNA-editing deaminase family in germinal center B cells. J Biol Chem 1999. Available: http://www.jbc.org.

[33] Muramatsu M, Kinoshita K, Fagarasan S, Yamada S, Shinkai Y, Honjo T. Class switch recombination and hypermutation require activation-induced cytidine deaminase (AID), a potential RNA editing enzyme. Cell 2000;102(5):553–63.

[34] Revy P, et al. Activation-induced cytidine deaminase (AID) deficiency causes the autosomal recessive form of the hyper-IgM syndrome (HIGM2). Cell 2000;102:565–75.

[35] Ehrenstein MR, Neuberger MS. Deficiency in Msh2 affects the efficiency and local sequence specificity of immunoglobulin class-switch recombination: parallels with somatic hypermutation. EMBO J June 1999;18(12):3484–90. https://doi.org/10.1093/emboj/18.12.3484.

[36] Ehrenstein MR, Rada C, Jones A-M, Sar Milstein C, Neuberger MS. Switch junction sequences in PMS2-deficient mice reveal a microhomology-mediated mechanism of Ig class switch recombination. Proc Natl Acad Sci U S A 2001;4:14553–8. Available: www.pnas.orgcgidoi10.1073pnas.241525998.

[37] Crouch EE, et al. Regulation of AID expression in the immune response. J Exp Med May 2007;204(5):1145–56. https://doi.org/10.1084/jem.20061952.

[38] Cantaert T, et al. Activation-induced cytidine deaminase expression in human B cell precursors is essential for central B cell tolerance. Immunity November 2015;43(5):884–95. https://doi.org/10.1016/j.immuni.2015.10.002.

[39] Kuraoka M, McWilliams L, Kelsoe G. AID expression during B-cell development: searching for answers. Immunol Res April 2011;49(1–3):3–13. https://doi.org/10.1007/s12026-010-8185-7.

[40] Zaheen A, Boulianne B, Parsa JY, Ramachandran S, Gommerman JL, Martin A. AID constrains germinal center size by rendering B cells susceptible to apoptosis. Blood 2009;114(3):547–54. https://doi.org/10.1182/blood-2009-03-211763.

[41] Navaratnam N, Bhattacharya S, Fujino T, Patel D, Jarmuz AL, Scott J. Evolutionary origins of apoB mRNA editing: catalysis by a cytidine deaminase that has acquired a novel RNA-binding motif at its active site. Cell 1995;81(2):187–95.

[42] Severi F, Chicca A, Conticello SG. Analysis of reptilian APOBEC1 suggests that RNA editing may not be its ancestral function. Mol Biol Evol March 2011;28(3):1125–9. https://doi.org/10.1093/molbev/msq338.

[43] Petersen-Mahrt SK, Harris RS, Neuberger MS. AID mutates *E. coli* suggesting a DNA deamination mechanism for antibody diversification. Nature July 2002;418(6893):99–103. https://doi.org/10.1038/nature00862.

[44] Bransteitter R, Pham P, Scharfft MD, Goodman MF, Scharff MD, Goodman MF. Activation-induced cytidine deaminase deaminates deoxycytidine on single-stranded DNA but requires the action of RNase. Proc Natl Acad Sci U S A April 2003;100(7):4102–7. Available: www.pnas.orgcgidoi10.1073pnas.0730835100. [Accessed 1 January 2023].

[45] King J, Larijani M. Structural plasticity of substrate selection by activation-induced cytidine deaminase as a regulator of its genome-wide mutagenic activity. FEBS Letters January 1, 2021;595(1):3–13. https://doi.org/10.1002/1873-3468.13962. Wiley Blackwell.

[46] Conticello SG, Thomas CJF, Petersen-Mahrt SK, Neuberger MS. Evolution of the AID/APOBEC family of polynucleotide (deoxy) cytidine deaminases. Mol Biol Evol February 2005;22(2):367–77. https://doi.org/10.1093/molbev/msi026.

[47] Pecori R, Di Giorgio S, Paulo Lorenzo J, Nina Papavasiliou F. Functions and consequences of AID/APOBEC-mediated DNA and RNA deamination. Nat Rev Genet August 2022;23(8):505–18. https://doi.org/10.1038/s41576-022-00459-8.

[48] Martin A, Bardwell PD, Woo CJ, Fan M, Shulman MJ, Scharff MD. Activation-induced cytidine deaminase turns on somatic hypermutation in hybridomas. Nature February 2002;415(6873):802–6. https://doi.org/10.1038/nature714.

[49] Martin A, Scharff MD. Somatic hypermutation of the AID transgene in B and non-B cells. 2002. Available: www.pnas.orgcgidoi10.1073pnas.192442899.

[50] Conticello SG. The AID/APOBEC family of nucleic acid mutators. Genome Biol June 17, 2008;9(6). https://doi.org/10.1186/gb-2008-9-6-229.

[51] Fugmann SD, Rush JS, Schatz DG. Non-redundancy of cytidine deaminases in class switch recombination. Eur J Immunol March 2004;34(3):844–9. https://doi.org/10.1002/eji.200324418.

[52] Conticello SG, Langlois MA, Yang Z, Neuberger MS. DNA deamination in immunity: AID in the context of its APOBEC relatives. Adv Immunol 2007;94:37–73. https://doi.org/10.1016/S0065-2776(06)94002-4.

[53] Scott J, Navaratnam N, Carter C. Molecular modelling of the biosynthesis of the Rna-editing enzyme Apobec-1, responsible for generating the alternative forms of apolipoprotein B. Exp Physiol July 1999;84(4):791–800. https://doi.org/10.1111/j.1469-445x.1999.01805.x.

[54] Wang M, Rada C, Neuberger MS. Altering the spectrum of immunoglobulin V gene somatic hypermutation by modifying the active site of AID. J Exp Med January 2010;207(1):141–53. https://doi.org/10.1084/jem.20092238.

[55] Kohli RM, Abrams SR, Gajula KS, Maul RW, Gearhart PJ, Stivers JT. A portable hot spot recognition loop transfers sequence preferences from APOBEC family members to activation-induced cytidine deaminase. J Biol Chem August 2009;284(34):22898–904. https://doi.org/10.1074/jbc.M109.025536.

[56] Larijani M, Martin A. Single-stranded DNA structure and positional context of the target cytidine determine the enzymatic efficiency of AID. Mol Cell Biol December 2007;27(23). https://doi.org/10.1128/MCB.01046-07.

[57] Pham P, Bransteitter R, Petruska J, Goodman MF. Processive AID-catalysed cytosine deamination on single-stranded DNA simulates somatic hypermutation. Nature 2003;424:103–7.

[58] Pham P, Malik S, Mak C, Calabrese PC, Roeder RG, Goodman MF. AID-RNA polymerase II transcription-dependent deamination of IgV DNA. Nucleic Acids Res November 2019;47(20):10815–29. https://doi.org/10.1093/nar/gkz821.

[59] Senavirathne G, et al. Activation-induced deoxycytidine deaminase (AID) co-transcriptional scanning at single-molecule resolution. Nat Commun December 2015;6. https://doi.org/10.1038/ncomms10209.

[60] Nambu Y, et al. Transcription-coupled events associating with immunoglobulin switch region chromatin. Science (1979) December 2003;302(5653):2137–40. https://doi.org/10.1126/science.1092481.

[61] Sakhtemani R, Perera MLW, Hübschmann D, Siebert R, Lawrence MS, Bhagwat AS. Human activation-induced deaminase lacks strong replicative strand bias or preference for cytosines in hairpin loops. Nucleic Acids Res May 2022;50(9):5145–57. https://doi.org/10.1093/nar/gkac296.

[62] Feng Y, Seija N, Di Noia JM, Martin A. AID in antibody diversification: there and back again. Trends Immunol July 1, 2020;41(7):586–600. https://doi.org/10.1016/j.it.2020.04.009. Elsevier Ltd.

[63] Yu K. AID function in somatic hypermutation and class switch recombination. Acta Biochim Biophys Sin 2022;54(6):759–66. https://doi.org/10.3724/abbs.2022070. Science Press.

[64] Methot SP, et al. A licensing step links AID to transcription elongation for mutagenesis in B cells. Nat Commun December 2018;9(1). https://doi.org/10.1038/s41467-018-03387-6.

[65] Yamane A, et al. Deep-sequencing identification of the genomic targets of the cytidine deaminase AID and its cofactor RPA in B lymphocytes. Nat Immunol January 2011;12(1):62–9. https://doi.org/10.1038/ni.1964.

[66] Longerich S, Tanaka A, Bozek G, Nicolae D, Storb U. The very 5′ end and the constant region of Ig genes are spared from somatic mutation because AID does not access these regions. J Exp Med November 2005;202(10):1443–54. https://doi.org/10.1084/jem.20051604.

[67] Betz AG, Milstein C, Gonzblez-Fernbndez A, Pannell R, Larson T, Neuberger MS. Elements regulating somatic hypermutation of an immunoglobulin K gene: critical role for the lntron enhancer/matrix attachment region. Cell 1994;77(2):239–48.

[68] Tumas-Brundage K, Manser T. The transcriptional promoter regulates hypermutation of the antibody heavy chain locus. J Exp Med 1997;185(2):239–50.

[69] Fukita Y, Jacobs H, Rajewsky K. Somatic hypermutation in the heavy chain locus correlates with transcription. Immunity July 1998;9(1):105–14. https://doi.org/10.1016/S1074-7613(00)80592-0.

[70] Matthias P, Baltimore D. The immunoglobulin heavy chain locus contains another B-cell-specific 3' enhancer close to the ot constant region. 1993.

[71] Meng FL, et al. Convergent transcription at intragenic super-enhancers targets AID-initiated genomic instability. Cell December 2014;159(7):1538–48. https://doi.org/10.1016/j.cell.2014.11.014.

[72] Madisen L, Groudine M. Identification of a locus control region in the immunoglobulin heavy-chain locus that deregulates c-myc expression in plasmacytoma and Burkitt's lymphoma cells. Genes Dev 1994;15;8(18):2212–6.

[73] Rouaud P, et al. The IgH 3' regulatory region controls somatic hypermutation in germinal center B cells. J Exp Med 2013;210(8):1501–7. https://doi.org/10.1084/jem.20130072.

[74] Bruzeau C, Moreau J, le Noir S, Pinaud E. Panorama of stepwise involvement of the IgH 3′ regulatory region in murine B cells. In: Advances in immunology. Academic Press Inc.; 2021. p. 95–114. https://doi.org/10.1016/bs.ai.2021.03.004.

[75] le Noir S, et al. Functional anatomy of the immunoglobulin heavy chain 3' super-enhancer needs not only core enhancer elements but also their unique DNA context. Nucleic Acids Res June 2017;45(10):5829–37. https://doi.org/10.1093/nar/gkx203.

[76] Saintamand A, et al. Deciphering the importance of the palindromic architecture of the immunoglobulin heavy-chain 3' regulatory region. Nat Commun February 2016;7. https://doi.org/10.1038/ncomms10730.

[77] Pinaud E, et al. The IgH locus 3' regulatory region: pulling the strings from behind. In: Advances in immunology. Academic Press Inc.; 2011. p. 27–70. https://doi.org/10.1016/B978-0-12-387663-8.00002-8.

[78] Péron S, et al. AID-driven deletion causes immunoglobulin heavy chain locus suicide recombination in B cells. Science (1979) May 2012;336(6083):931–4. https://doi.org/10.1126/science.1218692.

[79] Delgado-Benito V, et al. The chromatin reader ZMYND8 regulates Igh enhancers to promote immunoglobulin class switch recombination. Mol Cell November 2018;72(4):636–649.e8. https://doi.org/10.1016/j.molcel.2018.08.042.

[80] van der Stoep N, Gorman JR, Alt FW. Reevaluation of 3E function in stage-and lineage-specific rearrangement and somatic hypermutation. 1998.

[81] Inlay MA, Gao HH, Odegard VH, Lin T, Schatz DG, Xu Y. Roles of the Ig κ light chain intronic and 3′ enhancers in Igk somatic hypermutation. J Immunol July. 2006;177(2):1146–51. https://doi.org/10.4049/jimmunol.177.2.1146.

[82] Klotz EL, Storb U. Somatic hypermutation of a lambda 2 transgene under the control of the lambda enhancer or the heavy chain intron enhancer. J Immunol November 1996;157(10):4458–63.

[83] Sale JE. Immunoglobulin diversification in DT40: a model for vertebrate DNA damage tolerance. DNA Repair July 2, 2004;3(7):693–702. https://doi.org/10.1016/j.dnarep.2004.03.042.

[84] Sale JE, Calandrini DM, Takata M, Takeda S, Neuberger MS. Ablation of XRCC2/3 transforms immunoglobulin V gene conversion into somatic hypermutation. Nature August 2001;412(6850):921–6. https://doi.org/10.1038/35091100.

[85] Arakawa H, Saribasak H, Buerstedde JM. Activation-induced cytidine deaminase initiates immunoglobulin gene conversion and hypermutation by a common intermediate. PLoS Biol 2004;2(7). https://doi.org/10.1371/journal.pbio.0020179.

[86] Blagodatski A, et al. A cis-acting diversification activator both necessary and sufficient for AID-mediated hypermutation. PLoS Genet January 2009;5(1). https://doi.org/10.1371/journal.pgen.1000332.

[87] Senigl F, et al. Topologically associated domains delineate susceptibility to somatic hypermutation. Cell Rep December 2019;29(12):3902–3915.e8. https://doi.org/10.1016/j.celrep.2019.11.039.

[88] Buerstedde JM, Alinikula J, Arakawa H, McDonald JJ, Schatz DG. Targeting of somatic hypermutation by immunoglobulin enhancer and enhancer-like sequences. PLoS Biol 2014;12(4). https://doi.org/10.1371/journal.pbio.1001831.

[89] Kothapalli NR, Norton DD, Fugmann SD. Classical Mus musculus Igκ enhancers support transcription but not high level somatic hypermutation from a V-Lambda promoter in chicken DT40 cells. PLoS One 2011;6(4). https://doi.org/10.1371/journal.pone.0018955.

[90] Tarsalainen A, et al. Ig enhancers increase RNA polymerase II stalling at somatic hypermutation target sequences. J Immunol January 2022;208(1):143–54. https://doi.org/10.4049/jimmunol.2100923.

[91] Swaminathan S, et al. Mechanisms of clonal evolution in childhood acute lymphoblastic leukemia. Nat Immunol July 2015;16(7):766–74. https://doi.org/10.1038/ni.3160.

[92] Qian J, et al. B cell super-enhancers and regulatory clusters recruit AID tumorigenic activity. Cell December 2014;159(7):1524–37. https://doi.org/10.1016/j.cell.2014.11.013.

[93] Zhang J, Yue W, Zhou Y, Liao M, Chen X, Hua J. Super enhancers—functional cores under the 3D genome. Cell Prolif Febraury 1, 2021;54(2). https://doi.org/10.1111/cpr.12970. Blackwell Publishing Ltd.

[94] Zhang Y, Zhang X, Dai HQ, Hu H, Alt FW. The role of chromatin loop extrusion in antibody diversification. Nat Rev Immunol September 1, 2022;22(9):550–66. https://doi.org/10.1038/s41577-022-00679-3. Nature Research.

[95] Laffleur B, Basu U, Lim J. RNA exosome and non-coding RNA-coupled mechanisms in AID-mediated genomic alterations. J Mol Biol October 27, 2017;429(21):3230–41. https://doi.org/10.1016/j.jmb.2016.12.021. Academic Press.

[96] Michael N, Shen HM, Longerich S, Kim N, Longacre A, Storb U. The E box motif CAGGTG enhances somatic hypermutation without enhancing transcription. Immunity August 2003;19(2):235–42. https://doi.org/10.1016/s1074-7613(03)00204-8.

[97] Maul RW, et al. Spt5 accumulation at variable genes distinguishes somatic hypermutation in germinal center B cells from ex vivo-activated cells. J Exp Med 2014;211(11):2297–306. https://doi.org/10.1084/jem.20131512.

[98] Pavri R, et al. Activation-induced cytidine deaminase targets DNA at sites of RNA polymerase II stalling by interaction with Spt5. Cell 2010;143(1):122–33. https://doi.org/10.1016/j.cell.2010.09.017.

[99] Singh HR, Ostwal YB. Post-translational modification, phase separation, and robust gene transcription. Trends Genet February 2019;35(2):89–92. https://doi.org/10.1016/j.tig.2018.11.002.

[100] Wang X, Fan M, Kalis S, Wei L, Scharff MD. A source of the single-stranded DNA substrate for activation-induced deaminase during somatic hypermutation. Nat Commun June 2014;5. https://doi.org/10.1038/ncomms5137.

[101] Kilchert C, Wittmann S, Vasiljeva L. The regulation and functions of the nuclear RNA exosome complex. Nat Rev Mol Cell Biol April 01, 2016;17(4):227–39. https://doi.org/10.1038/nrm.2015.15. Nature Publishing Group.

[102] Basu U, et al. The RNA exosome targets the AID cytidine deaminase to both strands of transcribed duplex DNA substrates. Cell February 2011;144(3):353–63. https://doi.org/10.1016/j.cell.2011.01.001.

[103] Pefanis E, et al. Noncoding RNA transcription targets AID to divergently transcribed loci in B cells. Nature October 2014;514(7522):389–93. https://doi.org/10.1038/nature13580.

[104] Laffleur B, et al. Noncoding RNA processing by DIS3 regulates chromosomal architecture and somatic hypermutation in B cells. Nat Genet February 2021;53(2):230–42. https://doi.org/10.1038/s41588-020-00772-0.

[105] Willmann KL, et al. A role for the RNA pol II-associated PAF complex in AID-induced immune diversification. J Exp Med October 2012;209(11):2099–111. https://doi.org/10.1084/jem.20112145.
[106] Sun J, et al. E3-ubiquitin ligase Nedd4 determines the fate of AID-associated RNA polymerase II in B cells. Genes Dev August 2013;27(16):1821–33. https://doi.org/10.1101/gad.210211.112.
[107] Nagaoka H, Muramatsu M, Yamamura N, Kinoshita K, Honjo T. Brief definitive report activation-induced deaminase (AID)-directed hypermutation in the immunoglobulin S region: implication of AID involvement in a common step of class switch recombination and somatic hypermutation. J Exp Med 2002;195(4):529–34. Available: http://www.jem.org/cgi/content/full/195/4/529.
[108] Kwak H, Lis JT. Control of transcriptional elongation. Ann Rev Genet 2013;47:483–508. https://doi.org/10.1146/annurev-genet-110711-155440. Annual Reviews Inc.
[109] Begum NA, Stanlie A, Nakata M, Akiyama H, Honjo T. The histone chaperone Spt6 is required for activation-induced cytidine deaminase target determination through H3K4me3 regulation. J Biol Chem September 2012;287(39):32415–29. https://doi.org/10.1074/jbc.M112.351569.
[110] Aida M, Hamad N, Stanlie A, Begum NA, Honjo T. Accumulation of the FACT complex, as well as histone H3.3, serves as a target marker for somatic hypermutation. Proc Natl Acad Sci USA May 2013;110(19):7784–9. https://doi.org/10.1073/pnas.1305859110.
[111] Shen HM, et al. The activation-induced cytidine deaminase (AID) efficiently targets DNA in nucleosomes but only during transcription. J Exp Med May 2009;206(5):1057–71. https://doi.org/10.1084/jem.20082678.
[112] Qiao Q, Wang L, Meng FL, Hwang JK, Alt FW, Wu H. AID recognizes structured DNA for class switch recombination. Mol Cell August 2017;67(3):361–373.e4. Available: https://pubmed.ncbi.nlm.nih.gov/28757211/. [Accessed 1 January 2023].
[113] Liu J, et al. Efficient induction of Ig gene hypermutation in ex vivo–activated primary B cells. J Immunol November 2017;199(9):3023–30. https://doi.org/10.4049/jimmunol.1700868.
[114] Bonaud A, Lechouane F, Le Noir S, Monestier O, Cogne M, Sirac C. Efficient AID targeting of switch regions is not sufficient for optimal class switch recombination. Nat Commun July 2015;6. https://doi.org/10.1038/ncomms8613.
[115] Matthews AJ, Husain S, Chaudhuri J. Binding of AID to DNA does not correlate with mutator activity. J Immunol July 2014;193(1):252–7. https://doi.org/10.4049/jimmunol.1400433.
[116] Chaudhuri J, Khuong C, Alt FW. Replication protein A interacts with AID to promote deamination of somatic hypermutation targets. Nature 2004. Available: www.nature.com/nature.
[117] Pasqualucci L, Kitaura Y, Gu H, Dalla-Favera R. PKA-mediated phosphorylation regulates the function of activation-induced deaminase (AID) in B cells. Proc Natl Acad Sci U S A 2005. Available: www.pnas.orgcgidoi10.1073pnas.0509969103.
[118] Basu U, et al. The AID antibody diversification enzyme is regulated by protein kinase a phosphorylation. Nature November 2005;438(7067):508–11. https://doi.org/10.1038/nature04255.
[119] Cheng H-L, et al. Integrity of the AID serine-38 phosphorylation site is critical for class switch recombination and somatic hypermutation in mice. Proc Natl Acad Sci U S A February 2009;106(8):2717–22. https://doi.org/10.1073/pnas.0812304106.
[120] Chatterji M, Unniraman S, Mcbride KM, Schatz DG. Role of activation-induced deaminase protein kinase A phosphorylation sites in Ig gene conversion and somatic hypermutation. J Immunol 2007. Available: www.jimmunol.org.
[121] McBride KM, Barreto V, Ramiro AR, Stavropoulos P, Nussenzweig MC. Somatic hypermutation is limited by CRM1-dependent nuclear export of activation-induced deaminase. J Exp Med May 2004;199(9):1235–44. https://doi.org/10.1084/jem.20040373.
[122] Wong L, Sami A, Chelico L. Competition for DNA binding between the genome protector replication protein A and the genome modifying APOBEC3 single-stranded DNA deaminases. Nucleic Acids Res November 2022;50(21):12039–57. https://doi.org/10.1093/nar/gkac1121.
[123] Merkhofer EC, Hu P, Johnson TL. Introduction to cotranscriptional RNA splicing. Methods Mol Biol 2014;1126:83–96. https://doi.org/10.1007/978-1-62703-980-2_6. Humana Press Inc.
[124] Zheng S, Vuong BQ, Vaidyanathan B, Lin JY, Huang FT, Chaudhuri J. Non-coding RNA generated following lariat debranching mediates targeting of AID to DNA. Cell May 2015;161(4):762–73. https://doi.org/10.1016/j.cell.2015.03.020.
[125] Castiblanco DP, Norton DD, Maul RW, Gearhart PJ. JH6 downstream intronic sequence is dispensable for RNA polymerase II accumulation and somatic hypermutation of the variable gene in Ramos cells. Mol Immunol May 2018;97:101–8. https://doi.org/10.1016/j.molimm.2018.03.031.
[126] Hu Y, Ericsson I, Doseth B, Liabakk NB, Krokan HE, Kavli B. Activation-induced cytidine deaminase (AID) is localized to subnuclear domains enriched in splicing factors. Exp Cell Res March 2014;322(1):178–92. https://doi.org/10.1016/j.yexcr.2014.01.004.
[127] Kanehiro Y, et al. Activation-induced cytidine deaminase (AID)-dependent somatic hypermutation requires a splice isoform of the serine/arginine-rich (SR) protein SRSF1. Proc Natl Acad Sci U S A January 2012;109(4):1216–21. https://doi.org/10.1073/pnas.1120368109.
[128] Singh SK, et al. GANP regulates recruitment of AID to immunoglobulin variable regions by modulating transcription and nucleosome occupancy. Nat Commun May 2013;4(1):1–12. Available: https://www.nature.com/articles/ncomms2823. [Accessed 1 January 2023].
[129] Nowak U, Matthews AJ, Zheng S, Chaudhuri J. The splicing regulator PTBP2 interacts with the cytidine deaminase AID and promotes binding of AID to switch-region DNA. Nat Immunol February 2011;12(2):160–6. https://doi.org/10.1038/ni.1977.
[130] Mondal S, Begum NA, Hu W, Honjo T. Functional requirements of AID's higher order structures and their interaction with RNA-binding proteins. Proc Natl Acad Sci U S A March 2016;113(11):E1545–54. https://doi.org/10.1073/pnas.1601678113.
[131] Geuens T, Bouhy D, Timmerman V. The hnRNP family: insights into their role in health and disease. Hum Genet August 1, 2016;135(8):851–67. https://doi.org/10.1007/s00439-016-1683-5. Springer Verlag.
[132] Hu W, Begum NA, Mondal S, Stanlie A, Honjo T. Identification of DNA cleavage- and recombination-specific hnRNP cofactors for activation-induced cytidine deaminase. Proc Natl Acad Sci U S A May 2015;112(18):5791–6. https://doi.org/10.1073/pnas.1506167112.
[133] Conticello SG, Ganesh K, Xue K, Lu M, Rada C, Neuberger MS. Interaction between antibody-diversification enzyme AID and spliceosome-associated factor CTNNBL1. Mol Cell August 2008;31(4):474–84. https://doi.org/10.1016/j.molcel.2008.07.009.
[134] van Maldegem F, et al. CTNNBL1 facilitates the association of CWC15 with CDC5L and is required to maintain the abundance of

the Prp19 spliceosomal complex. Nucleic Acids Res August 2015;43(14):7058–69. https://doi.org/10.1093/nar/gkv643.

[135] Kuhny M, et al. Disease-associated CTNNBL1 mutation impairs somatic hypermutation by decreasing nuclear AID. J Clin Invest August 2020;140(8):4411–22. https://doi.org/10.1172/JCI131297.

[136] Ganesh K, Adam S, Taylor B, Simpson P, Rada C, Neuberger M. CTNNBL1 is a novel nuclear localization sequence-binding protein that recognizes RNA-splicing factors CDC5L and Prp31. J Biol Chem May 2011;286(19):17091–102. https://doi.org/10.1074/jbc.M110.208769.

[137] Huang X, Wang G, Wu Y, Du Z. The structure of full-length human CTNNBL1 reveals a distinct member of the armadillo-repeat protein family. Acta Crystallogr D Biol Crystallogr August 2013;69(8):1598–608. https://doi.org/10.1107/S0907444913011360.

[138] Patenaude AM, et al. Active nuclear import and cytoplasmic retention of activation-induced deaminase. Nat Struct Mol Biol May 2009;16(5):517–27. https://doi.org/10.1038/nsmb.1598.

[139] Kumar Singh A, et al. Splicing regulator SRSF1-3 that controls somatic hypermutation of IgV genes interacts with topoisomerase 1 and AID. Mol Immunol December 2019;116:63–72. https://doi.org/10.1016/j.molimm.2019.10.002.

[140] Romanello M, Schiavone D, Frey A, Sale JE. Histone H3.3 promotes IgV gene diversification by enhancing formation of AID-accessible single-stranded DNA. EMBO J July 2016;35(13):1452–64. https://doi.org/10.15252/EMBJ.201693958.

[141] Yu G, et al. The role of HIRA-dependent H3.3 deposition and its modifications in the somatic hypermutation of immunoglobulin variable regions. Proc Natl Acad Sci U S A December 2021;118(50):2114743118. https://doi.org/10.1073/pnas.2114743118.

[142] Singh AK, Tamrakar A, Jaiswal A, Kanayama N, Kodgire P. SRSF1-3, a splicing and somatic hypermutation regulator, controls transcription of IgV genes via chromatin regulators SATB2, UBN1 and histone variant H3.3. Mol Immunol March 2020;119:69–82. https://doi.org/10.1016/j.molimm.2020.01.005.

[143] Kawaguchi Y, et al. SRSF1-3 contributes to diversification of the immunoglobulin variable region gene by promoting accumulation of AID in the nucleus. Biochem Biophys Res Commun April 2017;485(2):261–6. https://doi.org/10.1016/j.bbrc.2017.02.097.

[144] Hu J, Qian H, Xue Y, Fu X-D. PTB/nPTB: master regulators of neuronal fate in mammals. Biophys Rep August 2018;4(4):204–14. https://doi.org/10.1007/s41048-018-0066-y.

[145] Monzón-Casanova E, et al. The RNA-binding protein PTBP1 is necessary for B cell selection in germinal centers article. Nat Immunol March 2018;19(3):267–78. https://doi.org/10.1038/s41590-017-0035-5.

[146] Chen J, et al. The RNA-binding protein ROD1/PTBP3 cotranscriptionally defines AID-loading sites to mediate antibody class switch in mammalian genomes. Cell Res October 2018;28(10):981–95. https://doi.org/10.1038/s41422-018-0076-9.

[147] Ito S, et al. Activation-induced cytidine deaminase shuttles between nucleus and cytoplasm like apolipoprotein B mRNA editing catalytic polypeptide 1. Proc Natl Acad Sci U S A 2004;101(7):1975–80. Available: www.pnas.orgcgidoi10.1073pnas.0307335101.

[148] Patenaude A-M, Di Noia JM. The mechanisms regulating the subcellular localization of AID. Nucleus 2010;1(4):325–31. https://doi.org/10.4161/nucl.1.4.12107.

[149] Häsler J, Rada C, Neuberger MS. Cytoplasmic activation-induced cytidine deaminase (AID) exists in stoichiometric complex with translation elongation factor 1α (eEF1A). Proc Natl Acad Sci U S A November 2011;108(45):18366–71. https://doi.org/10.1073/pnas.1106729108.

[150] Methot SP, et al. Consecutive interactions with HSP90 and eEF1A underlie a functional maturation and storage pathway of AID in the cytoplasm. J Exp Med April 2015;212(4):581–96. https://doi.org/10.1084/jem.20141157.

[151] Dorsett Y, et al. MicroRNA-155 suppresses activation-induced cytidine deaminase-mediated myc-Igh translocation. Immunity May 2008;28(5):630–8. https://doi.org/10.1016/j.immuni.2008.04.002.

[152] De Yébenes, et al. miR-181b negatively regulates activation-induced cytidine deaminase in B cells. J Exp Med September 2008;205(10):2199–206. https://doi.org/10.1084/jem.20080579.

[153] Orthwein A, Patenaude AM, Affar EB, Lamarre A, Young JC, di Noia JM. Regulation of activation-induced deaminase stability and antibody gene diversification by Hsp90. J Exp Med November 2010;207(12):2751–65. https://doi.org/10.1084/jem.20101321.

[154] Uchimura Y, Barton LF, Rada C, Neuberger MS. Reg-γ associates with and modulates the abundance of nuclear activation-induced deaminase. J Exp Med November 2011;208(12):2385–91. https://doi.org/10.1084/jem.20110856.

[155] Luo Y, et al. CUL7 E3 ubiquitin ligase mediates the degradation of activation-induced cytidine deaminase and regulates the Ig class switch recombination in B lymphocytes. J Immunol July 2019;203(1):269–81. https://doi.org/10.4049/jimmunol.1900125.

[156] Luo Y, et al. USP10 regulates B cell response to SARS-CoV-2 or HIV-1 nanoparticle vaccines through deubiquitinating AID. Signal Transduct Target Ther December 2022;7(1). https://doi.org/10.1038/s41392-021-00858-z.

[157] Zaprazna K, et al. Transcription factor YY1 can control AID-mediated mutagenesis in mice. Eur J Immunol February 2018;48(2):273–82. https://doi.org/10.1002/eji.201747065.

[158] Zaprazna K, Atchison ML. YY1 controls immunoglobulin class switch recombination and nuclear activation-induced deaminase levels. Mol Cell Biol April 2012;32(8):1542–54. https://doi.org/10.1128/mcb.05989-11.

[159] Pasqualucci L, et al. Expression of the AID protein in normal and neoplastic B cells. Blood November 2004;104(10):3318–25. https://doi.org/10.1182/blood-2004-04-1558.

[160] Peled JU, et al. The biochemistry of somatic hypermutation. Annu Rev Immunol 2008;26:481–511. https://doi.org/10.1146/ANNUREV.IMMUNOL.26.021607.090236.

[161] Larijani M, Martin A. The biochemistry of activation-induced deaminase and its physiological functions. Semin Immunol August 2012;24(4):255–63. Available: https://pubmed.ncbi.nlm.nih.gov/22695318/. [Accessed 1 January 2023].

[162] Shen HM, Storb U. Activation-induced cytidine deaminase (AID) can target both DNA strands when the DNA is supercoiled. Proc Natl Acad Sci U S A August 2004;35. https://doi.org/10.1073/PNAS.0404974101.

[163] Parsa JY, et al. Negative supercoiling creates single-stranded patches of DNA that are substrates for AID–mediated mutagenesis. PLoS Genet February 2012;8(2):e1002518. https://doi.org/10.1371/JOURNAL.PGEN.1002518.

[164] Tang C, MacCarthy T. Characterization of DNA G-quadruplex structures in human immunoglobulin heavy variable (IGHV) genes. Front Immunol May 2021;12. https://doi.org/10.3389/fimmu.2021.671944.

[165] Yewdell WT, et al. A hyper-IgM syndrome mutation in activation-induced cytidine deaminase disrupts G-quadruplex binding and genome-wide chromatin localization. Immunity November 2020;53(5):952–970.e11. https://doi.org/10.1016/j.immuni.2020.10.003.

[166] Cortizas EM, et al. UNG protects B cells from AID-induced telomere loss. J Exp Med October 2016;213(11):2459–72. https://doi.org/10.1084/JEM.20160635.

[167] Rhodes D, Giraldo R. Telomere structure and function. Curr Opin Struct Biol 1995;5(3):311–22. https://doi.org/10.1016/0959-440X(95)80092-1.

[168] Hamperl S, Cimprich KA. The contribution of co-transcriptional RNA:DNA hybrid structures to DNA damage and genome instability. DNA Repair 2014;19:84–94. Available: https://pubmed.ncbi.nlm.nih.gov/24746923/. [Accessed 1 January 2023].

[169] Yu K, Chedin F, Hsieh CL, Wilson TE, Lieber MR. R-loops at immunoglobulin class switch regions in the chromosomes of stimulated B cells. Nat Immunol 2003;4(5):442–51. https://doi.org/10.1038/NI919.

[170] Reaban ME, Griffin JA. Induction of RNA-stabilized DMA conformers by transcription of an immunoglobulin switch region. Nature 1990;348(6299):342–4. https://doi.org/10.1038/348342a0.

[171] Rajagopal D, et al. Immunoglobulin switch mu sequence causes RNA polymerase II accumulation and reduces dA hypermutation. J Exp Med June 2009;206(6):1237–44. https://doi.org/10.1084/JEM.20082514.

[172] Ronai D, Iglesias-Ussel MD, Fan M, Li Z, Martin A, Scharff MD. Detection of chromatin-associated single-stranded DNA in regions targeted for somatic hypermutation. J Exp Med January 2007;204(1):181–90. https://doi.org/10.1084/JEM.20062032.

[173] Xue K, Rada C, Neuberger MS. The in vivo pattern of AID targeting to immunoglobulin switch regions deduced from mutation spectra in msh2-/- ung-/- mice. J Exp Med September 2006;203(9). https://doi.org/10.1084/JEM.20061067.

[174] Cascalho M, Wong J, Steinberg C, Wabl M. Mismatch repair co-opted by hypermutation. Science (1979) February 1998;279(5354):1207–10. https://doi.org/10.1126/SCIENCE.279.5354.1207.

[175] Chen X, Paudyal SC, Chin R, You Z. PCNA promotes processive DNA end resection by Exo1. Nucleic Acids Res 2013;41(20):9325–38. https://doi.org/10.1093/nar/gkt672.

[176] Friedberg EC, et al. DNA repair: from molecular mechanism to human disease. DNA Repair August 2006;5(8):986–96. https://doi.org/10.1016/J.DNAREP.2006.05.005.

[177] Lindahl T. An N-glycosidase from *Escherichia coli* that releases free uracil from DNA containing deaminated cytosine residues. Proc Natl Acad Sci U S A 1974;71(9):3649. https://doi.org/10.1073/PNAS.71.9.3649.

[178] Ljungquist S, Lindahl T. A mammalian endonuclease specific for apurinic sites in double-stranded deoxyribonucleic acid. I. Purification and general properties. J Biol Chem March 1974;249(5):1530–5. https://doi.org/10.1016/S0021-9258(19)42915-3.

[179] Lindahl T, Ljungquist S. Apurinic and apyrimidinic sites in DNA. Basic Life Sci 1975;5A:31–8. https://doi.org/10.1007/978-1-4684-2895-7_5.

[180] Ljungquist S. A new endonuclease from *Escherichia coli* acting at apurinic sites in DNA. J Biol Chem 1977;252(9):2808–14. https://doi.org/10.1016/s0021-9258(17)40434-0.

[181] David SS, O'Shea VL, Kundu S. Base-excision repair of oxidative DNA damage. Nature June 2007;447:941–50. https://doi.org/10.1038/nature05978.

[182] Zharkov DO. Base excision DNA repair. Cell Mol Life Sci May 2008;65(10):1544–65. https://doi.org/10.1007/S00018-008-7543-2.

[183] Kavli B, Otterlei M, Slupphaug G, Krokan HE. Uracil in DNA–general mutagen, but normal intermediate in acquired immunity. DNA Repair April 2007;6(4):505–16. https://doi.org/10.1016/J.DNAREP.2006.10.014.

[184] Kavli B, et al. hUNG2 is the major repair enzyme for removal of uracil from U:A matches, U:G mismatches, and U in single-stranded DNA, with hSMUG1 as a broad specificity backup. J Biol Chem October 2002;277(42):39926–36. https://doi.org/10.1074/JBC.M207107200.

[185] Kavli B, et al. B cells from hyper-IgM patients carrying UNG mutations lack ability to remove uracil from ssDNA and have elevated genomic uracil. J Exp Med June 2005;201(12):2011. https://doi.org/10.1084/JEM.20050042.

[186] Shen HM, Tanaka A, Bozek G, Nicolae D, Storb U. Somatic hypermutation and class switch recombination in Msh6(-/-)Ung(-/-) double-knockout mice. J Immunol October 2006;177(8):5386–92. https://doi.org/10.4049/JIMMUNOL.177.8.5386.

[187] Rada C, Williams GT, Nilsen H, Barnes DE, Lindahl T, Neuberger MS. Immunoglobulin isotype switching is inhibited and somatic hypermutation perturbed in UNG-deficient mice. Curr Biol October 2002;12(20):1748–55. https://doi.org/10.1016/S0960-9822(02)01215-0.

[188] Di Noia JM, Neuberger MS. Altering the pathway of immunoglobulin hypermutation by inhibiting uracil-DNA glycosylase. Nature July 2002;419(6902):43–8. https://doi.org/10.1038/nature00981.

[189] Imai K, et al. Human uracil–DNA glycosylase deficiency associated with profoundly impaired immunoglobulin class-switch recombination. Nat Immunol September 2003;4(10):1023–8. https://doi.org/10.1038/ni974.

[190] Di Noia JM, Williams GT, Chan DTY, Buerstedde J-M, Baldwin GS, Neuberger MS. Dependence of antibody gene diversification on uracil excision. J Exp Med December 2007;204(13):3209–19. https://doi.org/10.1084/jem.20071768.

[191] Rada C, Di Noia JM, Neuberger MS. Mismatch recognition and uracil excision provide complementary paths to both Ig switching and the A/T-focused phase of somatic mutation. Mol Cell October 2004;16(2):163–71. https://doi.org/10.1016/J.MOLCEL.2004.10.011.

[192] Di Noia JM, Rada C, Neuberger MS. SMUG1 is able to excise uracil from immunoglobulin genes: insight into mutation versus repair. EMBO J February 2006;25(3):585–95. https://doi.org/10.1038/SJ.EMBOJ.7600939.

[193] Kemmerich K, Dingler FA, Rada C, Neuberger MS. Germline ablation of SMUG1 DNA glycosylase causes loss of 5-hydroxymethyluracil- and UNG-backup uracil-excision activities and increases cancer predisposition of Ung-/-Msh2-/- mice. Nucleic Acids Res July 2012;40(13):6016–25. https://doi.org/10.1093/NAR/GKS259.

[194] Costello R, Cantillo JF, Kenter AL. Chicken MBD4 regulates immunoglobulin diversification by somatic hypermutation. Front Immunol November 2019;10. Available: https://pubmed.ncbi.nlm.nih.gov/31736964/. [Accessed 1 January 2023].

[195] Schrader CE, Guikema JEJ, Wu X, Stavnezer J. The roles of APE1, APE2, DNA polymerase β and mismatch repair in creating S region DNA breaks during antibody class switch. Phil Trans Biol Sci March 2009;364(1517):645. https://doi.org/10.1098/RSTB.2008.0200.

[196] Guikema JEJ, et al. APE1- and APE2-dependent DNA breaks in immunoglobulin class switch recombination. J Exp Med November 2007;204(12):3017. https://doi.org/10.1084/JEM.20071289.

[197] Masani S, Han L, Yu K. Apurinic/apyrimidinic endonuclease 1 is the essential nuclease during immunoglobulin class switch recombination. Mol Cell Biol April 2013;33(7):1468–73. https://doi.org/10.1128/MCB.00026-13.

[198] Bennett RAO, Wilson DM, Wong D, Demple B. Interaction of human apurinic endonuclease and DNA polymerase beta in the base excision repair pathway. Proc Natl Acad Sci U S A July 1997;94(14):7166–9. https://doi.org/10.1073/PNAS.94.14.7166.

[199] Prasad R, Poltoratsky V, Hou EW, Wilson SH. Rev1 is a base excision repair enzyme with 5'-deoxyribose phosphate lyase activity. Nucleic Acids Res December 2016;44(22):10824–33. https://doi.org/10.1093/NAR/GKW869.

[200] Daskalova SM, Bai X, Hecht SM. Study of the lyase activity of human DNA polymerase β using analogues of the intermediate Schiff base complex. Biochemistry May 2018;57(18):2711–22. https://doi.org/10.1021/ACS.BIOCHEM.8B00308.

[201] Prasad R, Longley MJ, Sharief FS, Hou EW, Copeland WC, Wilson SH. Human DNA polymerase theta possesses 5'-dRP lyase activity and functions in single-nucleotide base excision repair in vitro. Nucleic Acids Res 2009;37(6):1868–77. https://doi.org/10.1093/NAR/GKP035.

[202] Allinson SL, Dianova II, Dianov GL. DNA polymerase beta is the major dRP lyase involved in repair of oxidative base lesions in DNA by mammalian cell extracts. EMBO J December 2001;20(23):6919–26. https://doi.org/10.1093/EMBOJ/20.23.6919.

[203] Guo C, et al. Mouse Rev1 protein interacts with multiple DNA polymerases involved in translesion DNA synthesis. EMBO J December 2003;22(24):6621–30. https://doi.org/10.1093/EMBOJ/CDG626.

[204] Zan H, et al. Rev1 recruits ung to switch regions and enhances dU glycosylation for immunoglobulin class switch DNA recombination. Cell Rep November 2012;2(5):1220–32. https://doi.org/10.1016/j.celrep.2012.09.029.

[205] Krijger PH, Tsaalbi-Shtylik A, Wit N, van den Berk PCM, de Wind N, Jacobs H. Rev1 is essential in generating G to C transversions downstream of the Ung2 pathway but not the Msh2+Ung2 hybrid pathway. Eur J Immunol October 2013;43(10):2765–70. https://doi.org/10.1002/EJI.201243191.

[206] Simpson LJ, Sale JE. Rev1 is essential for DNA damage tolerance and non-templated immunoglobulin gene mutation in a vertebrate cell line. EMBO J April 2003;22(7):1654–64. https://doi.org/10.1093/EMBOJ/CDG161.

[207] Sattler U, Frit P, Salles B, Calsou P. Long-patch DNA repair synthesis during base excision repair in mammalian cells. EMBO Rep April 2003;4(4):363. https://doi.org/10.1038/SJ.EMBOR.EMBOR796.

[208] Fortini P, Dogliotti E. Base damage and single-strand break repair: mechanisms and functional significance of short- and long-patch repair subpathways. DNA Repair April 2007;6(4):398–409. https://doi.org/10.1016/J.DNAREP.2006.10.008.

[209] Schrader CE, Linehan EK, Ucher AJ, Bertocci B, Stavnezer J. DNA polymerases β and λ do not directly affect Ig variable region somatic hypermutation although their absence reduces the frequency of mutations. DNA Repair December 2013;12(12):1087. https://doi.org/10.1016/J.DNAREP.2013.09.002.

[210] Wu L, et al. HMCES protects immunoglobulin genes specifically from deletions during somatic hypermutation. Genes Dev April 2022;36(7):433–50. https://doi.org/10.1101/gad.349438.122.

[211] Shukla V, et al. HMCES functions in the alternative end-joining pathway of the DNA DSB repair during class switch recombination in B cells. Mol Cell January 2020;77(2):384–394.e4. https://doi.org/10.1016/j.molcel.2019.10.031.

[212] Claverys JP, Mejean V, Gasc AM, Sicard AM. Mismatch repair in Streptococcus pneumoniae: relationship between base mismatches and transformation efficiencies. Proc Natl Acad Sci U S A October 1983;80(19):5956–60. https://doi.org/10.1073/PNAS.80.19.5956.

[213] Lyer RR, Pluciennik A, Burdett V, Modrich PL. DNA mismatch repair: functions and mechanisms. Chem Rev February 2006;106(2):302–23. https://doi.org/10.1021/CR0404794.

[214] Hsieh P. DNA mismatch repair: Dr. Jekyll and Mr. Hyde? Mol Cell September 2012;47(5):665. https://doi.org/10.1016/J.MOLCEL.2012.08.020.

[215] Shin-San S, Modrich P. *Escherichia coli* mutS-encoded protein binds to mismatched DNA base pairs. Proc Natl Acad Sci U S A July 1986;83(14):5057–61. https://doi.org/10.1073/PNAS.83.14.5057.

[216] Jiricny J. Postreplicative mismatch repair. Cold Spring Harbor Perspect Biol April 2013;5(4):1–23. https://doi.org/10.1101/CSHPERSPECT.A012633.

[217] Lee H, Popodi E, Tang H, Foster PL. Rate and molecular spectrum of spontaneous mutations in the bacterium *Escherichia coli* as determined by whole-genome sequencing. Proc Natl Acad Sci U S A October 2012;109(41):E2774. https://doi.org/10.1073/PNAS.1210309109/-/DCSUPPLEMENTAL.

[218] Marinus MG. DNA methylation and mutator genes in *Escherichia coli* K-12. Mutat Res October 2010;705(2):71. https://doi.org/10.1016/J.MRREV.2010.05.001.

[219] Eshleman JR, Lang EZ, Bowerfind GK, Parsons R, Vogelstein B, Willson JK, et al. Increased mutation rate at the hprt locus accompanies microsatellite instability in colon cancer. Oncogene 1995;10(1):33–7.

[220] Bhattacharyya NP, Skandalis A, Ganesh A, Grodent J, Meuth M. Mutator phenotypes in human colorectal carcinoma cell lines. 1994. Available: https://www.pnas.org.

[221] Harfe BD, Jinks-Robertson S. DNA polymerase zeta introduces multiple mutations when bypassing spontaneous DNA damage in *Saccharomyces cerevisiae*. Mol Cell 2000;6(6):1491–9. https://doi.org/10.1016/S1097-2765(00)00145-3.

[222] Putnam CD. Strand discrimination in DNA mismatch repair. DNA Repair September 2021;105:103161. https://doi.org/10.1016/J.DNAREP.2021.103161.

[223] Ghodgaonkar MM, et al. Ribonucleotides misincorporated into DNA act as strand-discrimination signals in eukaryotic mismatch repair. Mol Cell May 2013;50(3):323–32. https://doi.org/10.1016/J.MOLCEL.2013.03.019.

[224] Lujan SA, Williams JS, Clausen AR, Clark AB, Kunkel TA. Ribonucleotides are signals for mismatch repair of leading-strand replication errors. Mol Cell May 2013;50(3):437–43. https://doi.org/10.1016/J.MOLCEL.2013.03.017.

[225] Li Z, et al. Examination of Msh6- and Msh3-deficient mice in class switching reveals overlapping and distinct roles of MutS homologues in antibody diversification. J Exp Med July 2004;200(1):47–59. https://doi.org/10.1084/JEM.20040355.

[226] Kadyrov FA, Dzantiev L, Constantin N, Modrich P. Endonucleolytic function of MutLα in human mismatch repair. Cell July 2006;126(2):297–308. https://doi.org/10.1016/J.CELL.2006.05.039.

[227] Kadyrov FA, et al. *Saccharomyces cerevisiae* MutLα is a mismatch repair endonuclease. J Biol Chem December 2007;282(51):37181–90. https://doi.org/10.1074/JBC.M707617200.

[228] Bardwell PD, et al. Altered somatic hypermutation and reduced class-switch recombination in exonuclease 1–mutant mice. Nat Immunol January 2004;5(2):224–9. https://doi.org/10.1038/ni1031.

[229] Schaetzlein S, et al. Mammalian Exo1 encodes both structural and catalytic functions that play distinct roles in essential biological processes. Proc Natl Acad Sci U S A July 2013;110(27):E2470–9. https://doi.org/10.1073/PNAS.1308512110/SUPPL_FILE/PNAS.201308512SI.PDF.

[230] Zeng X, Winter DB, Kasmer C, Kraemer KH, Lehmann AR, Gearhart PJ. DNA polymerase η is an A-T mutator in somatic hypermutation of immunoglobulin variable genes. Nat Immunol June 2001;2:537–41. https://doi.org/10.1038/88740.

[231] Rogozin IB, Pavlov YI, Bebenek K, Matsuda T, Kunkel TA. Somatic mutation hotspots correlate with DNA polymerase eta error spectrum. Nat Immunol June 2001;2(6):530–6. https://doi.org/10.1038/88732.

[232] Pavlov YI, et al. Correlation of somatic hypermutation specificity and A-T base pair substitution errors by DNA polymerase eta during copying of a mouse immunoglobulin kappa light chain transgene. Proc Natl Acad Sci U S A July 2002;99(15):9954–9. https://doi.org/10.1073/PNAS.152126799.

[233] Faili A, et al. DNA polymerase eta is involved in hypermutation occurring during immunoglobulin class switch recombination. J Exp Med January 2004;199(2):265–70. https://doi.org/10.1084/JEM.20031831.

[234] Wiesendanger M, Kneitz B, Edelmann W, Scharff MD. Somatic hypermutation in muts homologue (Msh)3-, Msh6-, and Msh3/Msh6-deficient mice reveals a role for the msh2–Msh6 heterodimer in modulating the base substitution pattern. J Exp Med February 2000;191(3):579. https://doi.org/10.1084/JEM.191.3.579.

[235] Kim N, Bozek G, Lo JC, Storb U. Different mismatch repair deficiencies all have the same effects on somatic hypermutation: intact primary mechanism accompanied by secondary modifications. J Exp Med July 1999;190(1):21. https://doi.org/10.1084/JEM.190.1.21.

[236] Kannouche PL, Wing J, Lehmann AR. Interaction of human DNA polymerase η with monoubiquitinated PCNA: a possible mechanism for the polymerase switch in response to DNA damage. Mol Cell May 2004;14(4):491–500. https://doi.org/10.1016/S1097-2765(04)00259-X.

[237] Ulrich HD. Deubiquitinating PCNA: a downside to DNA damage tolerance. Nat Cell Biol 2006;8:303–5.

[238] Langerak P, Nygren AOH, Krijger PHL, Van Den Berk PCM, Jacobs H. A/T mutagenesis in hypermutated immunoglobulin genes strongly depends on PCNAK164 modification. J Exp Med August 2007;204(8):1989–98. https://doi.org/10.1084/JEM.20070902.

[239] Lerner LK, et al. Expression of constitutive fusion of ubiquitin to PCNA restores the level of immunoglobulin A/T mutations during somatic hypermutation in the Ramos cell line. Front Immunol April 2022;13:1429. Available: https://pubmed.ncbi.nlm.nih.gov/35432321/. [Accessed 19 December 2022].

[240] Zan H, et al. The translesion DNA polymerase θ plays a dominant role in immunoglobulin gene somatic hypermutation. EMBO J November 2005;24(21):3757. https://doi.org/10.1038/SJ.EMBOJ.7600833.

[241] Delbos F, Aoufouchi S, Faili A, Weill JC, Reynaud CA. DNA polymerase η is the sole contributor of A/T modifications during immunoglobulin gene hypermutation in the mouse. J Exp Med January 2007;204(1):17–23. https://doi.org/10.1084/jem.20062131.

[242] Delbos F, De Smet A, Faili A, Aoufouchi S, Weill JC, Reynaud CA. Contribution of DNA polymerase η to immunoglobulin gene hypermutation in the mouse. J Exp Med April. 2005;201(8):1191–6. https://doi.org/10.1084/JEM.20050292.

[243] Faili A, et al. A backup role of DNA polymerase κ in Ig gene hypermutation only takes place in the complete absence of DNA polymerase η. J Immunol May 2009;182(10):6353–9. https://doi.org/10.4049/JIMMUNOL.0900177.

[244] Saribasak H, et al. DNA polymerase ζ generates tandem mutations in immunoglobulin variable regions. J Exp Med June 2012;209(6):1075–81. https://doi.org/10.1084/JEM.20112234.

[245] Pilzecker B, et al. PrimPol prevents APOBEC/AID family mediated DNA mutagenesis. Nucleic Acids Res June 2016;44(10):4734–44. https://doi.org/10.1093/NAR/GKW123.

[246] Guilliam TA, Doherty AJ. PrimPol—prime time to reprime. Genes January 2017;8(1):20. https://doi.org/10.3390/GENES8010020.

[247] Wang Q, et al. The cell cycle restricts activation-induced cytidine deaminase activity to early G1. J Exp Med 2017;214(1). Available: https://pubmed.ncbi.nlm.nih.gov/27998928/. [Accessed 1 January 2023].

[248] Le Q, Maizels N. Cell cycle regulates nuclear stability of AID and determines the cellular response to AID. PLoS Genet 2015;11(9):e1005411. Available: https://journals.plos.org/plosgenetics/article?id=10.1371/journal.pgen.1005411. [Accessed 1 January 2023].

[249] Bello A, et al. Cell cycle–mediated regulation of secondary Ig diversification. J Immunol March 2023;31. https://doi.org/10.4049/JIMMUNOL.2100880.

[250] Sharbeen G, Yee CWY, Smith AL, Jolly CJ. Ectopic restriction of DNA repair reveals that UNG2 excises AID-induced uracils predominantly or exclusively during G1 phase. J Exp Med May 2012;209(5):965. https://doi.org/10.1084/JEM.20112379.

[251] Thientosapol ES, et al. SAMHD1 enhances immunoglobulin hypermutation by promoting transversion mutation. Proc Natl Acad Sci U S A May 2018;115(19):4921–6. https://doi.org/10.1073/PNAS.1719771115/SUPPL_FILE/PNAS.201719771SI.PDF.

[252] Chabes A, Stillman B. Constitutively high dNTP concentration inhibits cell cycle progression and the DNA damage checkpoint in yeast *Saccharomyces cerevisiae*. Proc Natl Acad Sci U S A January 2007;104(4):1183–8. https://doi.org/10.1073/PNAS.0610585104/SUPPL_FILE/10585FIG6.PDF.

[253] Bonifati S, et al. SAMHD1 controls cell cycle status, apoptosis and HIV-1 infection in monocytic THP-1 cells. Virology August 2016;495:92–100. https://doi.org/10.1016/J.VIROL.2016.05.002.

[254] Frieder D, Larijani M, Collins C, Shulman M, Martin A. The concerted action of Msh2 and UNG stimulates somatic hypermutation at A:T base pairs. Mol Cell Biol September 2009;29(18):5148–57. https://doi.org/10.1128/MCB.00647-09.

[255] Krijger PHL, Langerak P, Van Den Berk PCM, Jacobs H. Dependence of nucleotide substitutions on Ung2, Msh2, and PCNA-Ub during somatic hypermutation. J Exp Med November 2009;206(12):2603. https://doi.org/10.1084/JEM.20091707.

[256] Schanz S, Castor D, Fischer F, Jiricny J. Interference of mismatch and base excision repair during the processing of adjacent U/G mispairs may play a key role in somatic hypermutation. Proc Natl Acad Sci U S A April 2009;106(14):5593–8. https://doi.org/10.1073/PNAS.0901726106.

[257] Rogier M, et al. Fam72a enforces error-prone DNA repair during antibody diversification. Nature 2021;600:329–33. https://doi.org/10.1038/s41586-021-04093-y.

[258] Feng Y, et al. FAM72A antagonizes UNG2 to promote mutagenic repair during antibody maturation. Nature November 2021;600(7888):324–8. https://doi.org/10.1038/S41586-021-04144-4.

[259] Hao Q, et al. DNA repair mechanisms that promote insertion-deletion events during immunoglobulin gene diversification. Sci Immunol March 2023;8(81). https://doi.org/10.1126/SCIIMMUNOL.ADE1167.

FURTHER READING

[1] Wilson TM, et al. MSH2–MSH6 stimulates DNA polymerase η, suggesting a role for A:T mutations in antibody genes. J Exp Med February 2005;201(4):637–45. https://doi.org/10.1084/JEM.20042066.

Chapter 13

Molecular Mechanism of Activation-Induced Cytidine Deaminase

Nasim A. Begum[1], Maki Kobayashi[1], Hitoshi Nagaoka[2] and Tasuku Honjo[1]

[1]Department of Immunology and Genomic Medicine, Center for Cancer Immunology and Immunobiology, Kyoto University Graduate School Medicine, Kyoto, Japan; [2]Department of Molecular Pathobiochemistry, Graduate School of Medicine, Gifu University, Gifu, Japan

1. INTRODUCTION

'How is antibody memory generated?' is one of the core questions about the mechanisms of acquired immunity. Antigen-induced antibody memory is formed by the diversification of *immunoglobulin* (Ig) genes, which is achieved by two main genetic recombination events: somatic hypermutation (SHM) and class switch recombination (CSR). The master regulator of SHM and CSR is activation-induced cytidine deaminase (AID), which mainly occurs in the variable (V) and switch (S) regions of the Ig gene, respectively.

In brief, the processes of SHM and CSR are as shown in Figure 13.1. The *Ig* gene is embedded in high-order chromatin structure or topologically associating domains [1—4], which are presumably prepared during development [5,6] and reorganised upon activation [7]. Additionally, recombined DNA regions must be transcribed for cleavage and repair to occur.

Based on this environment, AID performs DNA cleavage and repair in its N-terminus- and C-terminus-dependent manners, respectively. Because SHM is assumed to be the product of error-prone DNA repair after a single-strand break (SSB) [8,9], the molecular mechanism of SHM is divided into two phases: the target-specific SSB and the subsequent error-prone DNA synthesis that is mediated by translesion synthesis (TLS) polymerases. Efficient SHM, including the putative direct deamination of cytosine (C) to uridine (U) on the DNA, requires the suppression of error-free repair at the DNA damage site.

On the other hand, CSR requires two additional steps that depends on AID's C-terminus. The first step is the AID-dependent S—S synapsis formation, which pulls the two *S* regions together. These regions are typically located kilobases to megabases apart. The second step involves processing the initial DNA cleavage (SSB form [10]) to nearly blunt-type double-strand breaks (DSBs), which are typical substrates for non-homologous end joining (NHEJ) repair.

Before the discovery of AID, many researchers basically predicted that the enzymes mediating SHM were distinct from those required for CSR based on the mechanistic considerations [11]. However, elucidating AID's function revealed its unexpected importance in SHM and CSR [12,13]. As is usually the case with scientific progress, this discovery gave rise to more questions than answers. For instance, whether AID deaminates C in DNA or RNA

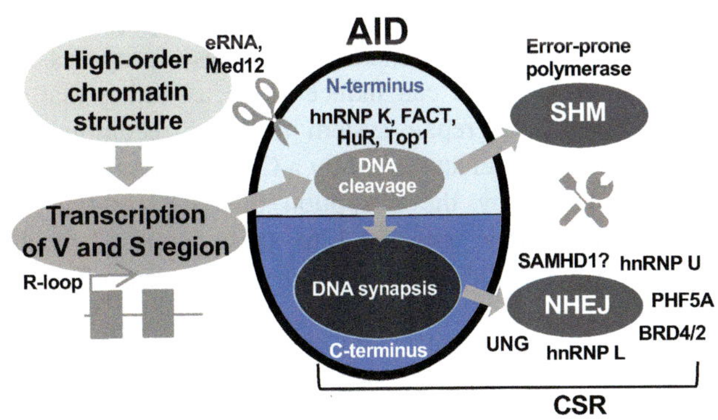

FIGURE 13.1 **AID orchestrates IgH recombination by regulating DNA cleavage and DNA repair through its N-terminus and C-terminus, respectively.** High-order chromatin structure is formed around IgH gene which undergoes class switch recombination (CSR) and somatic hypermutation (SHM). CSR and SHM require transcription of the non-coding RNA of the recombined regions. Based on these environments, AID makes DNA cleavage, DNA synapsis and processing of SSB to DSB with many collaborative molecules. SHM is the result of repair of DNA cleavage by error-prone DNA polymerases. Final sealing step of CSR is completed by mainly non-homologous end joining (NHEJ) and partially, alternative end joining. The factors particularly mentioned in this chapter is shown in this illustration.

Molecular Biology of B Cells. https://doi.org/10.1016/B978-0-323-95895-0.00001-5

is still debated. Two hypotheses on this subject, the 'DNA deamination model' and the 'RNA editing model', are still extensively debated [12,14–16]. This chapter will summarise these hypotheses and the evidence for and against them. Another mystery is how AID, which has 198 residues, can select specific cleavage targets in the whole genome. AID is not absolutely specific to the *Ig* genes, nor is it wholly promiscuous [17–19]. The number of genes outside the *Ig* locus that are mutated or translocated after AID activation is limited, and several hypotheses have been proposed to explain how AID limits its targets [18,20,21]. The two hypotheses of C's deamination in DNA versus RNA propose different mechanisms by which AID determines its specific targets. This chapter will review experimental data that bear on these critical questions, from the two different views of AID's function.

Distinct molecular mechanisms accomplish CSR and SHM because CSR requires not only DNA cleavage but also its recombination. A series of genetic findings indicated that the AID protein molecule itself, although relatively small, mediates these two reactions in CSR [12,22]. Natural and artificial AID mutagenesis revealed that C-terminal mutations specifically abolish CSR without affecting SHM [23]. Some C-terminally truncated AID mutants enhance SHM but almost completely lose CSR, which suggests that AID's C-terminus is dispensable for its DNA-cleavage activity [24]. On the other hand, N-terminal mutations affect SHM more severely than CSR, because AID's ability to cleave DNA is compromised [25]. These findings were interpreted by the RNA-editing model to indicate that AID may interact with at least two different RNA-cofactor complexes, generating two or more edited RNA products that mediate DNA cleavage and recombination separately [26,27]. In contrast, the DNA deamination hypothesis proposes that DNA cleavage alone can induce CSR, assuming that the endogenous DNA repair system mediates recombination. Although a CSR defect caused by a C-terminal mutation has been explained by the disruption of AID's stability or nuclear-cytoplasmic transport [28], these proposed mechanisms are not sufficient to explain the different effects of AID mutations on SHM and CSR.

Finally, there is the critical question of how AID expression is restricted to activated B cells. AID expression must be strictly regulated, because its target specificity is not absolute, and its aberrant expression can induce wide-ranging genome instability. In fact, dysregulated or ectopic AID expression can cause various types of tumours [27,29].

While some of the questions that emerged with AID's discovery 23 years ago have been answered, many others are still a matter of extensive debate. This chapter summarises the current evidence supporting the two differing views of AID's basic function: whether it deaminates C in DNA or in RNA.

2. FEATURES AND PROPERTIES OF AID

2.1 AID's Primary Structure and Functional Domains

The AID gene encodes a 26 kDa cytidine deaminase (CD) protein with a nuclear import signal at the N-terminus and an export signal at the C-terminus [12,13]. The protein contains a catalytic domain, a linker and an APOBEC-like domain between the N- and C-terminal region (Figure 13.2). Genetic analysis of patients with hyper-IgM (HIGM)-syndrome type II and their families first provided clues about AID's structure and function, indicating that it is responsible for Ig-isotype deficiencies [13,30,31]. In these patients, point mutations were found throughout the AID coding region, including the nuclear localisation signal (NLS) and nuclear export signal (NES) motifs [13,23,32]. The wide distribution of loss-of-function mutations outside the catalytic domain indicates that AID's structural integrity is necessary for its physiological function. AID mutants can be classified into two categories: N-terminal mutants that are defective in both CSR and SHM and C-terminal mutants that are only defective in CSR (Figure 13.2).

2.2 AID Requires N-Terminus for DNA Cleavage

Although the AID point mutations found in French and Japanese HIGM-II cases provided many clues to the consequences of those mutations [13,23], detailed functional analyses required a model system. Initial efforts to establish a suitable model system included generating a library of mouse AID mutants and their functional assays in $AID^{-/-}$ primary B cells. Shinkura et al. [25] reported several N-terminal mutants (Y13H, V18R, V18SR19V, W20K and G23S) with AID's catalytic activity was not disturbed that displayed intact CSR but decreased SHM. Shivarov et al. [33] conducted an extensive AID mutant analysis by comparing CSR and SHM with CD activity and found a clear dissociation between the two parameters. The same study also found that most N-terminal mutants are defective in both CSR and SHM. Another study reported that an N-terminal AID mutant was defective in introducing DNA cleavage in the variable (V) and S regions, a prerequisite function for AID-induced SHM and CSR [24]. Because no AID mutants with defective SHM but intact CSR have been identified, AID-induced DNA cleavage for V and S regions is considered common in both. The disassociation of AID's functions in CSR and SHM was demonstrated more clearly by in vivo G23S-mutation studies. In AID-G23S knock-in mice, CSR was fully intact, but SHM was severely defective [34]. Impaired DNA cleavage and AID–AID dimerisation defect are two prominent features of the AID-G23S mutant.

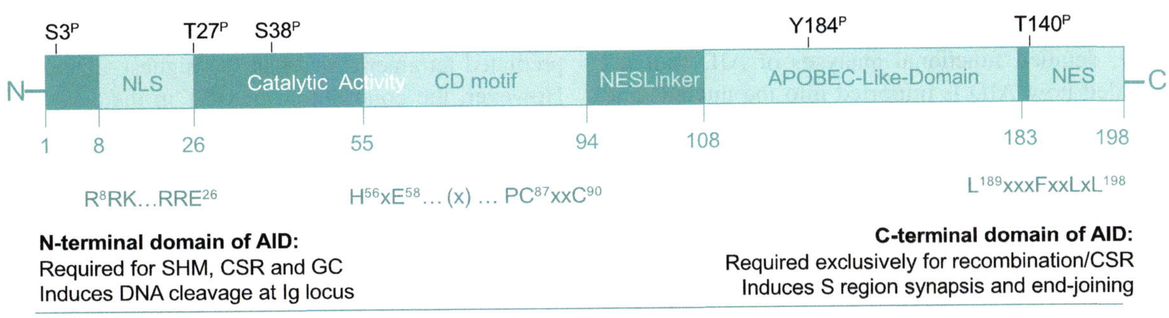

Mutations with different phenotypes

SHM(-)CSR(+):	SHM(-)CSR(-):	SHM(+)CSR(-):
G23S,V18R,W20K, H48A	M6T, R24W, D45A/F46A, C55A, H56Y, W80R, C87R/S, L106P, A111E, R112C/H, M139, R174S	JP8Bdel(L183X), JP41(R190X), P20 (P182ins), L198A, L196A, F193A

FIGURE 13.2 **General structural features and functionally important mutants of AID.** The schematic diagram shows the hAID structure with various functional domains and motifs. The numbers below the scheme indicate the amino acid residues, and the potential phospho-Ser and -Thr residues are indicated above the scheme. Consensus sequence motifs for NLS, CD and NES are shown below their respective locations. Selected AID mutations that significantly contributed to understanding AID's CSR and SHM functions are grouped into three categories and shown under the scheme.

2.3 AID Requires C-Terminus for Recombination

The analysis of several C-terminal AID mutants identified in Japanese patients revealed that any truncation, frameshift or defect in AID's C-terminus causes CSR to fail but leaves SHM intact [23]. When expressed in AID-deficient splenic B cells, the AID mutants P20 (hAID with a 34-aa insertion at residue 182), JP41 (hAID190X), JP8B (hAID frameshift at 183) and JP8Bdel (hAID183X) showed severe CSR defects, whereas SHM was intact or enhanced. Because a C-terminal AID mutant (Δ189–198) retained its full CD activity in *ung-1 Escherichia coli* cells, it is intriguing that C-terminal-defective AID mutants cannot induce CSR in vivo [35].

Another interesting observation is that mutants of AID's NES motif are defective in CSR and accumulate in the nucleus [32,36]. However, in the AID P20 mutant, which has a 34-residue insertion that did not destroy the NES motif, the CSR function was found to be completely defective [23,24]. All C-terminal AID mutants tested in these studies showed a normal or augmented ability to cleave DNA in both the V and S regions [24]. These results suggest that DNA cleavage alone is insufficient for CSR and that the C-terminal region may be responsible for an as-yet-unknown mechanism required for CSR.

In agreement with this idea, the wild-type (WT) but not the P20 mutant supported Sμ-Sα synapse during IgM-to-IgA switching in CH12F3-2A cells [37]. Furthermore, the CSR junctions in residual switched clones were predominantly repaired by the alternative end-joining (A-EJ) pathway because classical NHEJ (C-NHEJ) recruitment, such as KU80, was reduced in the S region [38,39]. Conversely, accumulation of the single-stranded gap (SSG) repair-binding protein PARP1 was enhanced, suggesting abundant SSBs that are not processed appropriately to undergo C-NHEJ-mediated DSB repair. Therefore, it is likely that the DNA end-processing required for the efficient generation and subsequent recombination of DSB is impaired in C-terminal AID mutants. However, another C-terminal mutant, JB8Bdel, showed intact DSB at the S region but was defective in dimerisation [40]. Other studies also suggest that the C-terminal region possibly helps recruit DNA repair proteins to facilitate C-NHEJ [41]. Recently, the crystal structure of maltose-binding protein (MBP)-fused AID has been solved, suggesting that AID binds to the structured substrate through oligomerisation [42]. The 'assistant patch' region (residues 170–178) at the C-terminus contributes to substrate binding by enhancing the affinity of AID for the substrate. Consistently, mutation of the basic residues within the assistant patch region also inhibited CSR.

Taken together, it is clear that AID has two distinct roles: (1) it induces DNA cleavage in the *Ig heavy-chain* (*IgH*) locus (V and S regions), for which AID's N-terminal region is responsible, and (2) it promotes CSR by influencing S–S synapse and C-NHEJ-mediated repair, although the precise mechanism has yet to be explored.

2.4 AID Shuttles Between Nucleus and Cytoplasm

One of AID's puzzling properties is its preferential localisation in the cytoplasm, even though AID-induced DNA cleavage and recombination are restricted to the nucleus. Although it has not been experimentally demonstrated whether AID's migration to the nucleus is mandatory, detailed functional analyses of AID mutants have revealed how AID is imported into the nucleus and exported to the cytoplasm [32,36]. AID has the unique signatures of a shuttling protein, namely bipartite NLS and NES motifs, which reside at the AID's N-terminal (residues 8–26) and C-terminal (residues 189–198) regions, respectively. Deleting the C-terminal region or modifying the NES motif causes AID to accumulate in the nucleus. Because leptomycin-B (LMB) [43] treatment causes AID–GFP fusion proteins to accumulate in the nucleus [32,36,44], AID's import into the nucleus is likely active rather than occurring by passive diffusion [36,44]. Consistent with this idea, importin-α/β and other nuclear import factors can be found in association with both WT and constitutively nuclear forms of AID. Not only AID's NLS but also additional N-terminal residues located between residues 19 and 54 have been shown to interact with the nuclear import complex [44]. Therefore, in addition to the bipartite NLS motif, other structural features of the entire N-terminal region may contribute to AID's active transport into the nucleus. A putative nuclear localisation motif overlapping with the NLS has also been reported [45].

AID's NES resides at the very end of the C-terminus and consists of more than 10 residues [24,25,46]. Most AID C-terminal mutants accumulate in the nucleus; in other words, they fail to diffuse out of the nucleus. In addition, it is clear that the NES-mediated nuclear export activity is not the sole requirement for CSR because a heterologous NES that preserves interaction with the export factor CRM1 does not support CSR [24,46]. Similarly, a mutation in the NES motif that abolishes the CSR function does not abrogate AID's export function [46]. This finding is further supported by the fact that some AID C-terminal mutants with an intact NES are CSR defective. In fact, the P20 mutant, which has an intact WT NES, has severe defects in CSR, probably because the insertion perturbs the C-terminal-specific structural integrity required for CSR by separating the NES from the rest of AID through an insertion [23].

2.5 AID Forms Dimer and Multimer

To understand the mechanisms of AID's functions, it is important to know the active form of AID and how it behaves in the physiological environment inside the cell. Although CD enzymes are generally monomeric, dimeric and tetrameric forms of CD enzymes, including members of the AID-APOBEC family, are also present [47–49].

Early studies showed that purified N-terminally streptavidin-tagged AID forms a tetramer, but its functionality remained unclear [50]. Later analyses of GST- or His-tagged AID activity, purified by size-exclusion chromatography, showed that AID fractions containing the predicted tetrameric size had the highest CD activity [51]. However, the contribution of GST in the AID–AID multimer formation cannot be excluded from the study, as GST inherently forms a dimer. In contrast, an atomic force microscopy study of AID indicated that AID could be monomeric in solution and that the monomeric form is catalytically active [52]. Nevertheless, recent crystal structures of N-terminally MBP-fused AID support the dimeric and tetrameric structures [42].

When AID proteins with two different tags were coexpressed, immunoprecipitation (IP) by either tag pulled down AID proteins with the other tag [23], indicating an AID–AID association in the cells. However, this result neither excluded the possibility of large complex formation nor distinguished the dimer from other multimers. Bimolecular fluorescence complementation (BiFC) assays [53], which enable a fluorescent reporter protein to be reconstituted through dimerisation with fused bait proteins, have been used to explore AID's interaction properties in living cells. This study revealed that truncating AID at either the N- or C-terminus causes defects in AID–AID interaction [54]. Therefore, it is possible that AID can form head-to-tail dimers or multimers in vivo. Orientation-dependent monomeric and multimeric states might unmask and mask the NES/NLS motifs, as has been observed for other nucleocytoplasmic shuttling proteins. Interestingly, the P20 mutant is also unable to show BiFC-based dimerisation in live cells, which supports the idea that AID's C-terminus has functional aspects other than CRM1-mediated export by the NES motif.

Using semi-purified AID from HEK293T cells, it was shown that AID proteins with CD activity are present in a high-molecular weight (HMW) complex of 500 kDa [55]. Glycerol-gradient sedimentation also revealed that cytoplasmic AID protein, expressed as a GFP-fusion protein in CH12F3-2A cells, migrated as a high molecular mass complex of 11S [56]. Another sedimentation and migration study identified a similar cytoplasmic complex of 300 kDa of flag-AID expressed from the endogenous locus of DT40 cells [57]. The HMW complex is highly sensitive to high salt treatment [55,56], suggesting that it may include many proteins loosely or indirectly associated with AID. Interestingly, an AID-chromatin complex isolated from DT40 cells also appears to exist as an HMW complex [58]. Due to the high abundance of positively charged residues, AID has an intrinsic affinity for RNA, which is likely responsible for forming RNA-protein (RNP) complexes of monomeric and/or oligomeric AID. AID fused with monomeric Kusabira Green (mKG) split fragments, used for AID–AID interaction analysis using BiFC assays, also showed that high salt or RNase A treatment disrupts the homomeric interaction-induced higher-order structures in the glycerol gradient [54].

2.6 AID Interacts with Numerous Proteins (Table 13.1)

An early co-immunoprecipitation (co-IP) study revealed that AID interacts with RNA polymerase II (RNAPII), suggesting that AID interacts functionally with the transcriptional complex [59]. Later, other AID co-IP approaches, including IP from B-cell chromatin fractions, identified RNAPII-associated transcription elongation factors such as SPT5, SPT6, TRIM28/KAP1 and the PAF and FACT complexes [56,58,60,61]. Co-immunoprecipitation analyses suggested that AID's N-terminus is critical for interacting with SPT6, whereas its C-terminus interacts with TRIM28 [58,60]. Depleting SPT6 or TRIM28 reduced CSR and affected AID gene expression in CH12F3-2A cells. AID expression was also severely reduced in mb1-cre-Trim28-CKO mice [56]. However, using CD19-cre-Trim28 CKO mice, it was concluded that Trim28 regulates CSR but not SHM [60]. Two DNA-binding proteins, RPA and 14-3-3γ, are reported to interact with AID's N-terminus (phospho-S38) and C-terminus, respectively [62–64]. Whereas RPA is known to bind single-stranded DNA (ssDNA) nonspecifically, 14-3-3γ recognises a sequence-specific DNA motif [65].

Yeast two-hybrid screening using AID as bait identified CTNNBL1, a splicing-associated factor that interacts with AID through its NLS [66,67]. Although knockdown (KD) of CTNNBL1 decreased Ig-gene conversion in DT40 cells, no CSR defect was observed in CTNNBL1-null spleen B or CH12F3-2A cells [68]. By applying an in-cell protein-biotinylation strategy followed by AID-IP, Nowak et al. [69] later identified the splicing factor PTBP2 as another interactor of AID. This study reported that PTBP2 depletion in CH12F3-2A cells reduced CSR and proposed that PTBP2 acts as an AID-targeting factor to the S region for CSR. Basu et al. [70] reported that 11 subunits of the RNA-exosome complex associate with AID and proposed that these components enhance CSR by increasing DNA deamination on both DNA strands. Several other RNA processing and splicing-associated proteins and helicases, such as DDX1, have also been found in AID-IP complexes isolated from chromatin fractions of DT40 cells or by other methods [58,71,72].

CRM1 and proteins from the karyopherin family play central roles in nucleocytoplasmic transport [73,74]. AID's interaction with the exportin CRM1 was confirmed by leptomycin B sensitivity tests and by mutagenesis of the NES motif [24,32,36]. The EEF1 protein, a translation elongation factor, was proposed to be involved in AID's cytoplasmic retention [57]. EEF1 was found to interact directly with AID's C-terminus with a 1:1 stoichiometry. Besides bona fide subcellular trafficking factors, CTNNBL1 and germinal centre nuclear protein (GANP) were reported to influence AID's intracellular localisation. The GANP overexpression caused AID to relocate from the cytoplasm to the nucleus [32]. The AID residue D143 was significant for GANP interaction, both in vitro and in vivo [75]. Conversely, AID-mediated transport regulation of other proteins was reported. AID overexpression was observed to relocalise nuclear Tet-family enzymes to the cytoplasm [76].

AID associates with the cytoplasm's molecular chaperones of the heat-shock protein family (HSP40, HSP70 and HSP90). AID's interaction with the HSP90 system protects it from polyubiquitination-mediated proteasomal degradation [77,78]. A deficiency of HSP40 DnaJa1 destabilises and decreases AID protein, resulting in proportional decreases in CSR and SHM. The degradation of nuclear AID is faster than that of cytoplasmic AID and occurs through both ubiquitin-dependent and -independent mechanisms. Although the pathway is not fully understood, an interaction with nuclear REG has been shown to promote AID's nuclear degradation [28]. AID also interacts with two E3 ubiquitin ligases, MDM2 and RNF126, through its residues 189–198 and 88–166, respectively [79,80]. However, MDM2 deficiency did not lead to significant defects in Ig-gene conversion in DT40 cells. Similarly, RNF126's functional significance has yet to be determined. A screening based on AID protein solubility identified 127 proteins interacting with AID, including RNF126, Rad51, FEN1 and RRM2 [79]. Previously, co-IP with AID identified DNA-PKcs, a critical CSR-specific DNA repair factor [81,82]. Moreover, DNA damage signals in CSR induced the interaction of APE1, a base excision repair (BER) enzyme, with a phosphorylated form of AID (p-Ser38) [83].

As AID has an intrinsic affinity for RNA, protein complexes related to RNA processing may interact with AID in co-IP experiments or during affinity purification. Additionally, SPT5- and SPT6-like proteins can form a macromolecular complex by associating with exosomes, RNAPII and other elongation-associated factors. AID may also form multiprotein RNP complexes in nuclear and cytoplasmic sub-compartments. Innovative approaches will be required to resolve the complexity and functional significance of forming nuclear, chromatin and cytoplasmic AID complexes. Overall, the exact functions of many AID-binding proteins have not been clarified for either the DNA deamination model or the RNA-editing model. As it is unlikely that most of these diverse proteins are relevant to AID function, it is unclear why AID interacts with many different proteins.

2.7 AID-Interacting Proteins and DNA Deamination

Based on AID's association with RNAPII, an early study proposed that AID was functionally associated with the

TABLE 13.1 AID Interacting Proteins

Interacting Proteins	Interaction Region	General Function/Property	Associated with			Specific Functions/Proposed Mechanisms	Refs
HSP40	AID internal	Protein chaperone	CSR	SHM	GC	Stabilises AID protein in the cytoplasm	[78]
HSP90	AID (19-84 aa)	Protein chaperone	CSR	SHM	GC	Stabilises AID protein in the cytoplasm	[77]
REG-γ	N-terminal	Proteasome activator subunit	CSR	nd	nd	Reduces nuclear resident time of AID via its degradation	[28]
YY1	N-terminal	Transcription factor	CSR	nd	nd	Regulates stability of AID in the nucleus	[471]
EEF1A	N and C-termini	Translation elongation complex component	CSR	nd	nd	Involved in AID retention in the cytoplasm	[57]
Importin-α3	N-terminal/NLS	Nuclear import of proteins with NLS	nd	nd	nd	May be involved in nuclear import of AID	[44]
CTNNBL1	N-terminal/NLS	Pre-mRNA processing associated factor	–	nd	GC	Interact with the NLS any may have redundant role in trafficking	[66–68]
CRM1	C-terminal/NES	Transport of NES containing proteins	CSR	SHM	nd	Exports AID out from the nucleus to the cytoplasm	[36,46]
GANP	D143 position	Transcription/mRNP export	–	SHM	nd	Targeting AID to V(D)J; Transports AID into the nucleus	[32,252]
TRIM28	C-terminal	Transcriptional regulator	CSR	–	nd	Targeting AID to H3K9me3 enriched S region	[56,60]
HP1	nd	Transcriptional corepressor	CSR	nd	nd	Targeting AID to H3K9me3 enriched S region	[60]
RNAP II	nd	Transcription	nd	nd	nd	Targeting of AID in the transcribe locus	[58,59]
RPA	AID p-S38	ssDNA-binding protein	CSR	SHM	nd	Targeting AID p-S38 to the transcribed DNA	[62,63]
SPT5	nd	Transcription elongation factor	CSR	SHM	nd	Targeting AID to the transcriptionally paused regions; target chromatin regulation and DNA repair	[61,257]
SPT6	N-terminal	Histone chaperone, Histone-PTM regulator	CSR	SHM	nd	Target chromatin regulation; present in AID-associated complex	[56,58]
FACT complex	nd	Histone chaperone, Histone-PTM regulator	CSR	SHM	nd	Target chromatin regulation; present in AID and Top1-associated chromatin complex	[58,239,257]
PAF complex	N-terminal	Transcription elongation	CSR	nd	nd	Targeting AID in the transcription elongation complex	[58,257]
14-3-3 proteins	C-terminal	Adapter/scaffold protein; signal transduction	CSR	nd	nd	Targeting AID to S-region AGCT motif and H3K4me3.-K9ac.S10P mark	[64,248,249]
RNA exosomes	nd	RNA processing and degradation	CSR	SHM	nd	Targeting AID to both strands of the transcribed DNA	[70]
PTBP2	nd	Pre-mRNA splicing associated factor	CSR	nd	nd	Targeting AID to S region through GLT	[69]

PKA	N-terminal	Site specific protein phosphorylation	CSR	SHM	GC	Phosphorylation of AID at S38 and promotes RPA association; tethers to the S region	[62,83,86]
DNA-PKcs	C-term, CDD	DSB repair in NHEJ/ Recombination	CSR	–	GC	Promotes repair and protects from AID-induced apoptosis	[60,82]
APE1	AID p-S38	Repair and transcription	CSR	–	nd	Abasic site cleavage	[83]
MDM2	C-terminus	E3 ubiquitin protein ligase	nd	–	–	Dispensable for AID-associated function	[80]
CIB1	nd	Calcium and integrin-binding protein	–	nd	–	Dispensable for AID-associated function	[472]
SMN; U2AF65	nd	Splicing-associated factor	nd	nd	nd	Unknown	[71]
RNF126	AID (88–166)	E3 ubiquitin protein ligase	nd	nd	nd	Unknown	[79]
Other proteins identified in AID co-Ips							**Refs**
TOP2B, RAD50, SMCA5, GCN1L1, TCP1, TBB3, Ac–Co carboxylase, Q3U2N8, Niban, Symplekin, Dynine, Q3TML0, Mre11, ZF protein, DDX47, GALK1, RL4, Trip12, AK129105, DNL1, SMRC2, DPOD1, IRF8, NF-kB inhibitor, N-pac homologue, PCD4, ARP3, RBBP4, AMPM1,γ-Actin like, ORC5, PDK3							[69]
DDX17, DDX20, HNRNPA1, IGF2BP1, PABPC1, REG-γ/PSME3, AFF4							[56]
SPT5, SF3A, CTR9, RNAPII subunits, PAF1, SPT6, Lamin, TOP1, DDX3Y, IL3IF, LEO1, MCM5, SMU1, YBOX1, PRP6, PRP4, SPT16, SSRP1, DDX1, C17orf85, SF3B, DDX15							[58]
HRNP-M, HRNP-A0, DDX21, DDX19, LGTN/Sui1, CPSF73, XRN2, EEF1A1, RBM39, NIP30, NOL11, CTNNBL1, NUP93, importin β1, RAD51, RRM2B, IER5, FEN1, Actin, RDH1, PGAM1, IMPDH2, UbcH7, PPP2R1A, RPA1, TCP1, HSPA5, HERPUD2, C14orf94, C22orf28, CXorf9, ZNF44, DAZAP2, ZMIZ22							[79]
DNAJA1-2-3, HSPA8, HSP90, AHSA1, BAG2, TCP1, PSMD1,2,6, PSMC2							[77,78]
MSH2, MSH6, DNA-PKcs, CRM1, Ikaros							[60]
HRNP-L, HRNP-K, HNRNP-U, HRNP-C, HRNP-I, HRNP-M, PCBP1, PABP, SERBP1							[54,178]

transcription machinery [59]. Later, the ssDNA-binding protein RPA was suggested to be a cofactor that interacts specifically with PKA-phosphorylated AID (p-S38) [62,63]. In an in vitro transcription-coupled system using T7 polymerase, RPA enhances dC deamination by AID, leading to the proposal that RPA stabilises AID (p-S38) on the transcribed S region to promote AID-induced DNA deamination. Supporting this view, CSR is impaired by pharmacological inhibition of PKA, and CSR and SHM are reduced in AID-S38A mice [62,84]. However, the S38A AID mutant was proficient in dC deamination on ssDNA and in a mutagenic catalytic activity assay in *E. coli* [62,85]. Subsequent studies showed that RPA deposition in the S region is AID-dependent, as RPA coats ssDNA generated after CSR-associated DSBs [83,86]. Biphasic DNA deamination in the S region was also proposed. Initially, dC is deaminated at a low level independently of AID phosphorylation, producing a DNA damage signal that induces AID phosphorylation and RPA recruitment in the S regions; this leads to a second, elevated phase of dC deamination that enhances DSBs [83].

Basu et al. [70] proposed a mechanism by which AID can deaminate dCs in the template strand, which remains inaccessible due to RNA–DNA hybrid formation during transcription. The presence of the RNA-exosome complex was shown to be important for AID targeting both strands of SHM substrates transcribed by T7 polymerase in vitro. Sun and Basu [70,87] proposed that a macromolecular association exists among the RNA exosome SPT5, AID and NEDD4 in the stalling RNAPII complex, in which germline transcripts (GLTs) undergo exosome-mediated degradation. The RNA degradation process may release the complex from pausing to initiate AID-induced dC deamination on both strands of the DNA. It has been proposed that AID interacts with SPT5, an RNAPII-stalling factor, enhancing AID's interaction with the stalled RNAPII [61,88]. Therefore, AID's association with the poised RNAPII gives rise to the hypothesis that as AID moves along with the elongating RNAPII complex, repeated pausing may allow sufficient time for deamination in the transcribed track [89,90]. AID is reported to interact with other factors associated with the elongating RNAPII complexes, such as PAF, FACT, SPT5 and SPT6 [56,58,61]. All the subunits of the PAF complex have been detected in AID-IP experiments. Knockdown of LEO1, one of the PAF complex components, reduces AID recruitment in the Ig locus [58].

Although AID-induced mutations are preferentially restricted to the regions 1–2 kb downstream of the V- and I-promoters, core S regions also undergo extensive DNA breaks [91]. In fact, RNAPII stalls throughout the S regions, and accumulations of the stalling factor SPT5, as well as AID, can be detected from the S regions all the way to the *C* regions [61,92]. The 14-3-3 series of adapter proteins were reported to bind the 5′-AGCT-3′ sequence motif and AID. As S regions are enriched in AGCT motifs, it has been suggested that 14-3-3 family proteins promote S region deamination by recruiting AID [65]. The chromatin marks H3K9acH3S10p and H3K9me3 are also implicated in recruiting AID through 14-3-3 and TRIM28/KAP1, respectively [60,93]. Both proteins cannot interact with a CSR-defective but SHM-proficient AID mutant with a C-terminal truncation. Interestingly, C-terminally truncated AID fused to 14-3-3 completely rescued the defective CSR in AID-deficient B cells [94]. These authors [64] also concluded that the HIV accessory protein Vpr, which is reported to interact with uracil-DNA glycosylase (UNG) and inhibit CSR [95], disrupts the S region macromolecular complex containing AID, RPA and PKA, along with other molecules.

Nowak et al. [69] suggested that antisense transcripts produced from the S region bind PTBP2, a cofactor that interacts with AID, and thus RNA-mediated AID targeting occurs in the S region for CSR. A similar idea was proposed regarding the AID-cofactor GANP, which binds to the V transcript and thereby recruits AID specifically to the V region. This explains why mice lacking GANP are defective in SHM but not in CSR [96]. More recently, factors like DDX1 and YTHDC1, binding to switch GLT, have also been proposed to recruit AID to the S region [72,97]. Direct interaction of AID with G-quadruplex structures in S region-derived RNA and DNA has also been reported [42,98]. It is unclear why AID requires binding to GLT when it can directly access S region ssDNA via DNA G-quadruplex structures. Similarly, the question arises regarding the requirement of AID-interacting proteins that are proposed to facilitate AID targeting. Further study will be necessary to understand the *S*-region targeting mechanism and the role of various interacting proteins, especially in the context of AID-induced DNA deamination in vivo.

2.8 AID's Similarity with APOBEC Family: Evolutionary Perspective

AID is a member of the large apolipoprotein B mRNA-editing catalytic-component (APOBEC) protein family, which carries a zinc-dependent CD domain [47,99]. The active site of AID's CD consists of two cysteines, a histidine residue for zinc coordination and a glutamic acid that serves as a proton donor during rC/dC deamination. Currently, 11 APOBEC family members are known in humans, including APOBEC1, APOBEC2, APOBEC3A–3H, APOBEC4 and AID. These members possess either a single or double catalytic active site for CD activity and an as-yet-uncharacterised domain called the APOBEC domain. While AID contains one active CD domain, some APOBEC3 proteins, which evolved through duplication and fusion in primates, possess two domains, one of which is often catalytically inert [100,101].

Despite their structural similarities, the physiological functions of the APOBEC family members are diverse and distinct, and other family members cannot replace AID. While bacterial and yeast CDs convert free cytidine nucleotide to uridine, APOBEC1 converts cytosine to uracil in ApoB mRNA with the assistance of the RNA-binding protein cofactor ACF, causing the generation of a truncated protein with a different function [95,96]. Later, APOBEC1 was found to convert dC to dU in DNA in vitro and *E. coli*, demonstrating that AID-APOBEC family proteins can edit both RNA and DNA substrates [102,103]. Although APOBEC3G proteins act on the human immunodeficiency virus genome and APOBEC3A acts nonspecifically on the host genome [104,105], we do not know the physiological substrates for the deamination activity of APOBEC family members other than APOBEC1. Interestingly, the APOBEC3 family members can restrict viral and retrotransposon activity independently of their CD function, although the mechanism is poorly understood [106,107]. As most of these studies were conducted by protein overexpression, the potential for genome-wide mutagenesis through the CD activity of the APOBEC family proteins, including AID, in a physiological context remains to be determined.

AID-like genes have been identified in bony fish and cartilaginous fish, the earliest jawed vertebrates, based on sequence conservation [47]. This finding correlates well with the origin of the antibody diversification process that appeared in jawed vertebrates [26,108,109]. The lamprey, a non-jawed vertebrate, has two AID-like genes, pmcda1 and pmcda2 [110]. The diversification of immune receptors in lampreys occurs via a gene conversion-like process reminiscent of the Ig-gene conversion process in chicken B cells, which is AID-dependent. Phylogenetic tree analysis suggests that the ancestor of the AID-APOBEC family of CDs may have appeared first in non-jawed vertebrates. The APOBEC3 family members are believed to have expanded through the duplication and amplification of an ancestral AID-like gene during primate evolution [111,112]. It has been proposed that this massive expansion took place to combat the evolving nature of viral pathogens.

3. MOLECULAR MECHANISM FOR DNA CLEAVAGE AND RECOMBINATION BY AID

3.1 Properties of AID-Induced DNA Cleavage

Before delving into AID, it is essential to introduce some characteristics of the DNA that AID targets. We will use the *IgH* gene as a representative example, which undergoes SHM and CSR by AID.

3.1.1 AID-Initiated DNA Cleavage Requires Transcription of the Locus

The *Ig* heavy-chain (H) locus contains the multiple units of non-coding RNA in addition to the ordinary *Ig* promoter, located 5′ to the V region, to produce Ig protein. Each constant (C) gene, except for Cδ, contains an I promoter that is located at the 5′ region [113–115] (Figure 13.3A). Non-coding transcripts from I promoters are known as GLTs. These GLTs start with an I promoter, followed by an intron switch (S) region containing a heavily repeated core region, and end with a C region (Figure 13.3A and B).

The Iμ promoter is constitutively active, and another I promoter downstream of the *IgM* locus is activated by specific cytokines in a germinal centre B cell. Each GLT promoter is induced by a specific cytokine. For example, *IgHγ1* and *IgHε* GLTs are induced by IL-4, their promoter sequences contain STAT6-binding sites, which are induced by IL-4. Similarly, the *IgHγ2b* and *IgHα* promoters are induced by TGF-β, which activates small mothers against decapentaplegic (SMAD) and Runt-related transcription factor (RUNX) [116]. Other factors known to regulate GLTs include p50 of NF-κB [117], a mediator complex factor MED1 [118], MED12 [118,119] or a transcription elongation factor SPT5 [120].

Transcription from the Iμ promoter and the other promoter allows AID-induced DNA breaks in the Sμ and the other transcribed S regions, resulting in CSR between these two S regions. Each S region contains its characteristic repeat structure [121–127]. For example, mouse Sμ and Sα regions harbour pentameric repeats containing highly frequent G or T, typically AGCTG and GGCTG. Mouse Sγ1 is formed by a more complex repeat structure, with 49 bp repeated 45 times with small variations (Figure 13.3B). These repeats are assumed to take an unusual non-B DNA structure when they are transcribed [128].

Clear evidence shows that transcription efficiency is correlated with CSR efficiency, as demonstrated by artificial switch constructs [129]. Transcription of the V region gene is also required for SHM, and the level of transcriptional activity regulates the level of SHM [130–132].

Interestingly, it seems that the transcriptional activity of the locus, rather than the transcript itself, is essential for DNA cleavage, because the replacement of the promoter and GLT sequences does not abolish SHM or CSR, as shown in several knock-in mouse experiments [133–136]. For instance, knock-in experiments clearly showed that the RNA polymerase I (RNAPI) promoter could substitute for the RNAPII promoter to induce SHM in the IgV region, indicating that the RNA polymerase protein per se is not critical for the initiation of DNA cleavage [134]. Moreover, a series of in vivo mutational analyses further confirmed that the GLTs are replaceable, meaning that no specific DNA sequence is required to initiate DNA cleavage. These

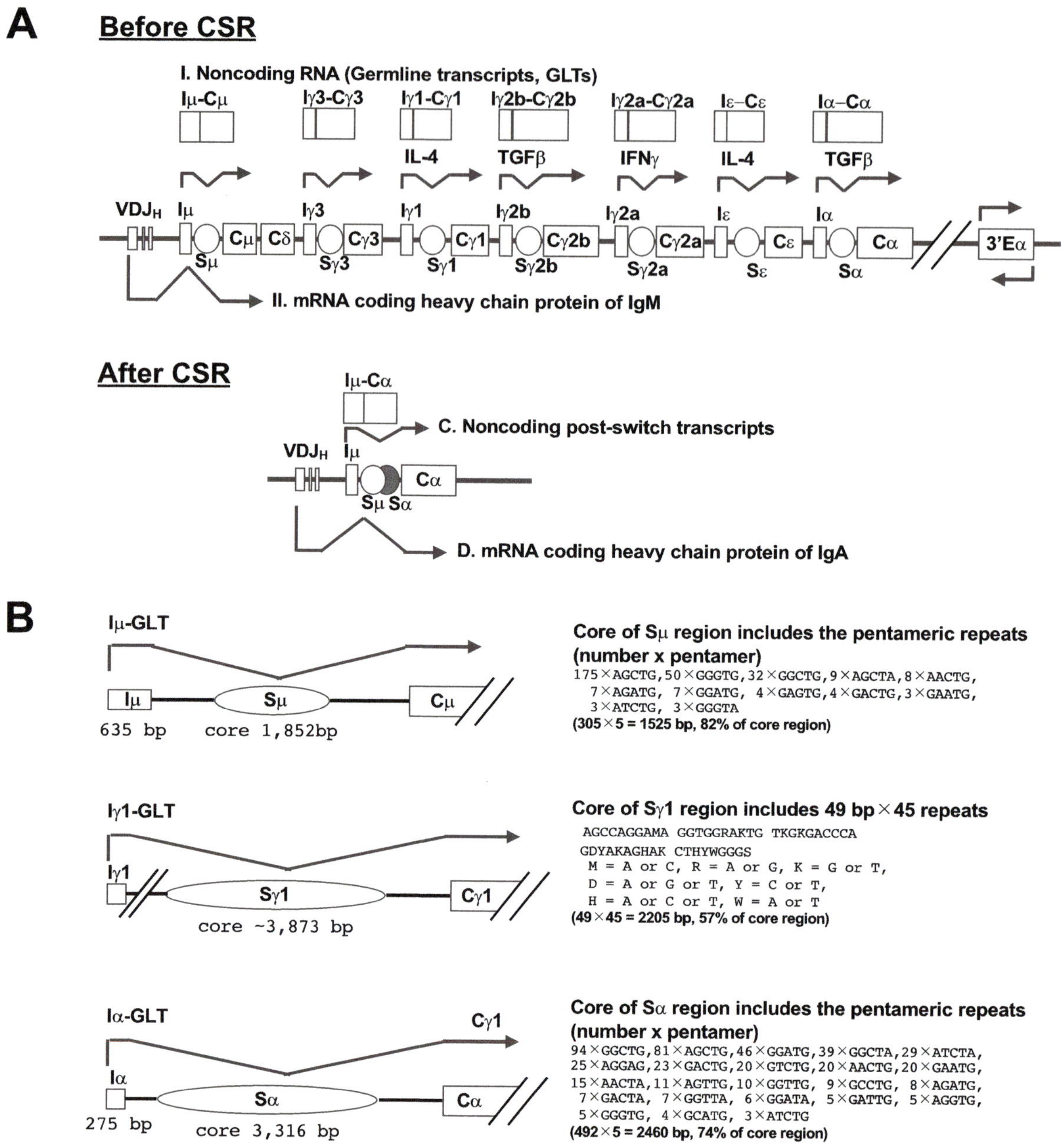

FIGURE 13.3 **Non-coding RNA transcription units required for DNA cleavage of S regions in the IgH gene.** A. Each non-coding transcription unit consists by I promoter, S region and constant (C) exon. One of the I promoters is activated in a splenic B cell, by the particular cytokines described above the arrows. An I promoter and the constant (C) region are fused after splicing and S region is spliced out. On the contrary, mRNA coding heavy chain protein is transcribed from 5′ promoter of VDJ exon to 3′end of a constant region. A far downstream enhancer of the IgH gene is bidirectionally transcribed. B. Repeat structure of the core of the representative switch (S) regions in mouse IgH gene. Each S regions contains their proprietary repeat structures. Sμ and Sα regions heavily consisted by the pentameric repeats which includes minor variations. S regions in noncoding GLTs are spliced out.

results suggest that transcriptional activity probably induces a chromatin structure that becomes accessible to DNA-cleaving enzymes [27,116,137], as transcription of the S regions was shown to associate with the sensitivity to the restriction enzyme of the stimulated B cells' nuclei [92]. Transcription in the DNA break sites reportedly induces the accumulation of RNAPII or transcription pausing, which may be required for AID's targeting of S regions [61,92].

Although the sequence of GLTs is not absolutely essential for DNA cleavage, the characteristic repeat sequences in the S regions facilitate the efficiency of CSR. This was revealed by deletion experiments of the Sμ core region in mice. Mice with a deletion of 2.7 kb Sμ core region showed in vitro switching efficiency ranging from 20% to 60% of WT, and the switched antibodies in their serum were approximately 20% to 50% of the WT [138]. Mice that lacked the larger Sμ region but retained the shorter Iμ-Cμ GLT demonstrated more severe CSR defects [139]. These results suggest that longer repeat regions enhance non-B DNA structure and increase the frequency of DNA cleavage. In cases where topoisomerase 1 (TOP1)-mediated SSBs [140] occur in S regions, longer S regions with the repeat sequences are necessary for efficient formation of non-B DNA structure and further DNA cleavage [141,142].

Although the importance of GLT products has not been demonstrated for many years, a function of GLT has recently been proposed. It has been reported that Sμ-RNA promotes DNA cleavage by bringing cytoplasmic AID into the nucleus, Sμ-RNA trap AID with the G-quadruplex structure contained in Sμ and further interact with S regions. This mechanism is similar to the Cas9 enzyme in the CRISPR/Cas9 system [98,143]. DDX1 and PTBP3 support this GLT-mediated AID recruitment through their helicase activity or RNA-binding property [72,144].

Furthermore, transcriptional activity (and/or GLT-RNA) from I promoters is involved not only in DNA cleavage but also in S—S synapsis formation under the control of the 3′ enhancer of *IgH* (3′RR). Depletion of SPT5 decreases transcription of the 3′RR and *IgHγ1* GLT, whereas forced transcription of 3′RR restores *IgHγ1* GLT and S—S synapses in primary mouse B cells [120]. The function of GLT transcription in S—S synapsis formation is not fully elucidated and will be analysed in the future.

Transcription of the locus is necessary; however, replication is not absolutely necessary for AID-dependent DNA cleavage and repair in the *IgH* gene. Tamoxifen-inducible AID-ER, a fusion protein of AID and ER (the ligand-binding domain of an oestrogen receptor mutant, ER-T2) [24], enables CSR even within three hours after activation, as detected by digestion-circularisation-PCR (DC-PCR) [141]. As three hours are not long enough to complete a cell cycle, replication is dispensable for CSR.

3.1.2 The Initial Cleavage Induced by AID Appears to be SSBs

AID-dependent CSR heavily relies on C-NHEJ to ligate the blunt-type DSBs generated from SSBs by exo/endonucleases [116,145]. C-NHEJ is a DNA repair system that joins DSB ends through the concerted action of specific repair proteins, such as 53BP1, DNA-PKcs, XRCC4, Ku70/80, Ligase IV and others (Table 13.2). A hallmark of C-NHEJ is observed at the joint sequences between the two S regions, which show almost no homologous ends, at most less than 2—3 bp. In contrast, A-EJ repairs staggered-type DNA break ends that retain ssDNA overhang, with the help of PARP1 and others. In the case of A-EJ, the joint sequence harbours a longer microhomology.

While CSR is mainly repaired by C-NHEJ, AID-dependent DNA cleavage appears to be SSBs. This was first indicated by an experiment using a unique artificial switch substrate in CH12F3-2 cells [10,146]. This substrate undergoes AID-dependent DNA cleavage and repair in CH12 cells, and the analysed junction sequences between Sμ and Sα regions demonstrated remarkably frequent duplication, insertion and deletion in the junctional ends. If the AID-dependent DNA cleavage was exactly the blunt and double-stranded, the junctional sequences should be simpler and have few insertions or deletions.

This finding that AID initially generates SSBs is further confirmed by analysing CSR, C-NHEJ and A-EJ repair pathways. When the blunting mechanism from SSBs to DSBs is impaired, CSR efficiency decreases, and A-EJ supports the remaining CSR, suggesting that initial DNA cleavage is SSBs. For example, APE1 is an endo/exonuclease that senses SSBs to initiate 3′ to 5′ SSB resection [147]. Deletion of the *Ape1* gene in B lymphocytic cells reduced CSR [148] (Table 13.2), while it did not affect DNA cleavage and SHM [149,150]. This was because CSR was blocked at the DSB formation from SSBs [151]. Blunt-type DSBs were detected by the double-stranded linker-ligation-mediated PCR (LM-PCR) method, which does not include an in vitro blunting step, in *Ape1*-heterozygous mouse B cells [151]. Similarly, *Ung* deletions resulted in decreased CSR caused by a reduced frequency of blunt-type DSB formation, detected by the same LM-PCR method [152], although *Ung* deletions had normal DNA cleavage in S regions and even increased SHM [153,154]. These results of UNG and APE1 suggest that BER proteins are involved in processing SSBs to DSBs.

Other proteins have been implicated in the processing of SSBs to DSBs. Endo/exonuclease APE2 is specifically necessary for SHM [155]. The MRN complex, which consists of MRE11, RAD50 and NBS1, is recruited by γH2AX, processes DNA damage and it is necessary for CSR [156—159]. The MRN complex harbours 3′ to 5′ exonuclease activity [160] and endonuclease activity at the DNA hairpin structure [161]. Exonuclease I (EXO1), belonging to the 5′ to 3′ structure-specific exonuclease family, also contributes to CSR largely, whereas its effect on SHM is marginal [162] (Table 13.2). However, these mismatch repair (MMR)-related nucleases' function in CSR has remained elusive, because the function of exo/endonuclease of MRE11 and EXO1 is generally reported in homologous end joining repair, to trim and further broaden the ssDNA region from the nick in the double-stranded DNA (dsDNA) [163,164]. Alternatively, their structure-specific exo/endonuclease functions [165,166] presumably resolve the staggered-type DSBs formed in non-B DNA structure and process into blunt-type DSBs.

An AID knockdown experiment also provides supportive evidence for processing SSBs into blunt-type DSBs. AID knockdown cells are supposed to have fewer SSBs, and the processing of SSB ends into blunt-type DSB

TABLE 13.2 DNA Repair Factors and Their Enrolment in Immunoglobulin Diversification

Repair Pathway	Gene	V/JH4-SHM	Sμ-SHM	CSR	c-Myc Translocation	GC[a]
Classical NHEJ (C-NHEJ)	53BP1$^{-/-}$ mouse	No change [473]	No change [474]	<10% [347,473,475] with enhanced intra-switch recombination [474]	No change [347]	120% 53BP1$^{-/-}$ embryonic cells [b] [475]
	Ku70$^{-/-}$ mouse	ND	ND	~10% (in Ku70$^{-/-}$ mouse) [476] with increased intra-switch deletion [477] increased microhomology [167]	250% in Ku70$^{-/-}$ mouse ES cells [478]	~600% (Ku70$^{-/-}$ DT40) [479]
	Ku80$^{-/-}$ mouse	ND	~50% (in Ku80$^{-/-}$ mouse) [480]	~10% (in Ku80$^{-/-}$ mouse) with reduced intra-switch deletion [480]	ND	ND
	H2AX$^{-/-}$ mouse	No change [480]	~60% [480]	13%–30% (4, 10, 11) [347,480,481]	~120% [480]	ND
	DNA PKcs mutant mouse, inhibitor and/ or knockout	No change in scid/ scid mouse [482]	ND	70% in DNA-PKcs$^{-/-}$ HL B cells [483], ~100% by inhibitor, PKi NU7026 [484]	180%[c] ATM$^{-/-}$ PKi mouse splenic B cells [483]	1000% (DNA-PKcs$^{-/-}$DT40) [479]
	DNA PKcs deficient patients		Reduced G/C mutation in patients [212]	SCID phenotype with increased microhomology [212]		
	Ligase4 P/Δ CH12 cells	ND	ND	~40% [485]	ND	ND
	Ligase4 syndrome			Longer microhomology [213]		
	XRCC4 $^{\Delta/\Delta}$ mouse	ND	ND	40%–80% [486]		400%–700% [d] (in CHO cells) [487]
	CD21-*cre* XRCC4$^{c/-}$ mouse			25%–50% with longer microhomology and increased chromosomal breaks [39]		
	Cernunnos (XLF)-deficient patient	ND	40%–50% in human [214]	~Decreased CSR with longer microhomology in human [214],	ND	ND
	XLF$^{-/-}$ mouse	ND	ND	30%–50% in XLF$^{-/-}$ mouse [484,488]	Increased chromosome/chromatic breaks [484]	
	Shieldin2$^{-/-}$ mouse or	ND	ND	10%–50% with longer microhomology [215]	ND	ND
	Shieldin2$^{-/-/-}$ CH12 cells			~25% [356]		
	Rif1$^{\Delta/\Delta}$ mouse or U2OS cells	ND	ND	~20% [489]	150% [489]	50%[e] in U2OS cells [490]

Alternative EJ (A-EJ)	PARP1$^{-/-}$ mouse and/or inhibitor	~150% in PARP1$^{-/-}$ GC B cells [491]	ND	No change in PARP1$^{-/-}$ mouse B cells [492]	~300%^{f} in PARP1$^{-/-}$ mouse with KU0058948 [492]	~10% (PARP1$^{-/-}$ DT40) [493]
	CtIP shRNA in CH12 cells	ND	ND	~60% [494] with 60% reduced AID expression	~30% by I-SceI and pCr15 translocation reporter [495]	ND
	ERCC1$^{-/-}$ mouse	ND	No change [496]	No change [497,498] or 60–80% [496]	ND	ND
	ERCC1$^{-/-}$Lig4$^{-/-}$ CH12 cells			~30% of Lig4$^{-/-}$ CH12 cells [498]	ND	ND
	XPF(ERCC4)$^{-/-}$ CH12 cells	ND	ND	No change [498]	ND	ND
	XPF$^{-/-}$Lig4$^{-/-}$ CH12 cells	ND	ND	~50% of Lig4$^{-/-}$ CH12 cells [498]	ND	ND
Shared between C-NHEJ and A-EJ	Mre11-deficient patient	ND	No change in ATLD patients [156]	Less efficient in ATLD patients [156]	ND	ND
	Mre11$^{-/-}$ mouse	ND	ND	~20% in MRE11$^{-/-}$ mouse [157]		
	NBS1-deficient patient	No change [499]	ND	ND	ND	ND
	NBS1$^{\Delta/-}$ mouse	ND	~20% [158]	40%–50% [158]	~500% [347]	ND
	Over-expressed NBS1	290% (in Ramos cells) [500]	ND	ND	ND	300% in DT40 [500]
	ATM$^{-/-}$ mouse	~80% [501]	~60% [501]	~30% [347,501,502], longer micro-homology [502]	~700% [347]	~10% (ATM$^{-/-}$ DT40) ^{g} [503]
	ATM patients	Normal [504]	ND	Decreased ~ normal with longer microhomology [504]		
	MDC1−/− mouse	ND	ND	~50% [505]	ND	ND

Continued

TABLE 13.2 DNA Repair Factors and Their Enrolment in Immunoglobulin Diversification—cont'd

Repair Pathway	Gene	V/JH4-SHM	Sμ-SHM	CSR	c-Myc Translocation	GC[a]
BER	$UNG^{-/-}$	~200% [204]	200%–600% [154,225,506]	5%–25% [154,346,507]	~0% [347]	<5% (UNG $^{-/-}$DT40) [508]
	Overexpression of UNG in CH12 cells	ND	~50% [203]	~40% [203,205]	ND	
	Overexpression of UNG in the Primary B cells	ND	~20% [154]			
	$PARP2^{-/-}$ mouse	ND	ND	No change [492]		ND
	$PARP2^{-/-}$ mouse with PARP inhibitor, KU0058948	ND	ND	~50% [492]	~300% [492]	
	XRCC1+/−mouse	150%–200% [509]	ND	No change [509], less micro-homology	~50% [509]	ND
	$APE1^{-/-}$ BL2 cells	No change [150]	No change [150]	ND	ND	ND
	$APE1^{-/-/-}$ CH12 cells	ND	No change [149]	~20% [148,149]	No change [149]	
	$APE2^{-/y}$ mouse	30%–40% [155]	No change [155]	No change [155]	ND	ND
	Pol beta deficient cells	~700% in pol beta deficient BL2 cell clones [201]	ND	ND	ND	ND
	Pol beta $^{-/-}$ mouse	ND	~200% in pol beta $^{-/-}$ B cells [202]	No change or increased in pol $beta^{-/-}$ B cells [202]		
	$FEN1^{E160D/E160}$ mouse [h]	No change [510]	ND	No change [510]		~30% (FEN1 $^{-/-}$DT40) [511]
MMR	$MSH2^{-/-}$ mouse	~20% (AT↓) [206]	ND	10%–40% [512]	ND	ND
	$MSH2^{-/-}UNG^{-/-}$ mouse	ND	Increased [513]		ND	ND
	$MSH6^{-/-}$ mouse	~15% (AT↓) [514]	No change [514]	~30% [514]	ND	No change in $Msh6^{-/-}$ DT40 [515]
	PMS2-deficient patients	ND	ND	Decreased [187]	ND	No change in $PMS2^{-/-}$DT-40 [515]
	PMS2 $^{-/-}$ mouse	No change [186]	Not changed frequency [184,186]	30%–80% with increased microhomology [186]		
	$MLH1^{-/-}$ mouse	ND	Unchanged frequency [185]	20%–50% with increased microhomology [185]		
	$Exo1^{-/-}$ mouse	~80% (GC↑) [162]	No change [162]	10%–20% [162]	ND	ND

	$PCNA^{-/-}tg^{K164R}$ mouse [i]	ND	Sμ-SHM ~ 15% [516] Sγ3-SHM ~ 30% [516]	~50%–70% of wild-type [516]	ND	ND
TLS (Y-family)	Pol eta-deficiency patients	No change (AT↓) in XPV patients [198,207]	No change (AT↓) in XPV patients [198]		ND	~50% (pol $eta^{-/-}$ DT40) [517]
	Pol $eta^{-/-}$ mouse	No change ~ 130% (AT↓) [208,518]	~208% (AT↓GC↑) [208]	No change [518]		
	Pol iota null BL2 cells	13%–27% [9]	ND		ND	ND
	Pol iota −/− mouse	No change [518]	ND	No change [518]		
	Pol $kappa^{-/-}$ mouse	No change [519]	ND	No change [519]	ND	ND
	REV1AA mutant [j] mouse	65% in AA mutant [199]		No change in AA mutant [199]	ND	30% in $REV1^{-/-}$ DT40 [520]
	$REV1^{-/-}$ mouse		130% (AT↑) [190]	40%–50% ($Rev1^{-/-}$ mouse) [190]		
(A-family)	Pol theta polymerase (−) helicase (+) mouse	~80% (GC↓) [200]	ND	Modest decrease [200]	ND	No change in pol $theta^{-/-}$ DT 40 [521]
	Pol $nu^{\Delta Ex2/\Delta Ex2}$ mouse	ND	No change [522]	No change [522]	ND	No change in pol $nu^{-/-}$ DT 40 [521]
	Pol $theta^{-/-}$ pol $n^{-/-}$ DT40	ND	ND	ND	ND	~20% [521]
(B-family)	REV3 (pol zeta) Δ/Δ mouse	~5% [191,192]	ND	35%–40% [191,192]	ND	No change ($REV3^{-/-}$ DT40) [520]
	REV3 L2610F mutant mouse[k]	~170% [192]	ND	No change [192]	ND	ND

ND, not done; Δ, conditional gene deletion by cre-loxP system
[a]*The evaluation system of GC is described for each gene.*
[b]*I-SceI-mediated HR substrate.*
[c]*$ATM^{-/-}$ DNA-PKcs inhibitor vs. $ATM^{-/-}$.*
[d]*I-SceI-mediated gene-conversion/homologous recombination substrate.*
[e]*Homology directed repair system, by siRNA.*
[f]*(c-myc-IgH translocation in $PARP1^{-/-}$ spleen B cells with PARP inhibitor KU0058948).*
[g]*I-SceI-mediated HR substrate in HeLa cells. siATM + ATM inhibitor KU55933.*
[h]*$FEN1^{E160D/E160D}$ is a catalytic-deficient mutant.*
[i]*PCNA knockout mouse with transgene of K164R mutant, an ubiquitination mutant.*
[j]*Catalytically inactive REV1 D568AE569A mutant.*
[k]*Knock-in mutation of REV3 L2610F, a hyperactive mutant.*

was insufficient, leading to more frequent repair by A-EJ [167]. In control cells, more SSBs accumulated in the limited region than in the AID knockdown cells, and excision of the overhanging nucleotides and formation of blunt-type DSBs would be easier and repaired by C-NHEJ.

Because AID is unlikely to employ a different cleavage mechanism for CSR and SHM, SHM is probably initiated by SSBs, followed by gap generation, after which the gap is filled with error-prone DNA polymerases and then sealed with DNA ligases [8]. It should be noted that SHM would have to be accompanied by as-yet-unknown mechanisms to suppress the error-free repair because most DNA damage in non-B-cells is correctly repaired [168].

3.2 Evidence for AID's Cytidine Deamination Activity on DNA or RNA

There is a clear consensus about the deamination activity of AID. First, purified AID has been found to exhibit in vitro deamination activity of cytidine (C) in DNA [55,169]. Interestingly, this in vitro DNA deamination activity is highly specific to ssDNA [55,169] and is drastically reduced in dsDNA, especially in the form of chromatin [170,171]. No evidence supports AID cytidine deamination activity on naked RNA as a substrate in vitro.

Second, the deamination activity of AID has been observed in eukaryotic cells. Several genome-editing systems have been developed to purposefully induce C to U editing of genomic DNA in mammalian cells to correct pathogenic mutations. These systems utilise CDs such as lamprey pmCDA1, human AID, rat APOBEC1 or APOBEC3 in combination with dCas9 and specific gRNAs [172−175]. In this case, the gRNA forms a short R-loop with its target DNA sequence, creating a region of ssDNA. Interestingly, APOBEC 1, a bona fide RNA-editing enzyme has also been shown to possess DNA deamination activity in vitro, in *E. coli* [102] and even in mammalian cells [174]. These results suggest that CDs, including AID and APOBEC enzymes, do not distinguish between DNA and RNA when their targets are encoded in single-stranded nucleic acids. In fact, AID has been shown to exhibit deamination activity on both RNA and DNA when encapsulated in hepatitis B virus [176]. Collectively, AID can deaminate C on single-stranded nucleic acids regardless of whether they are DNA or RNA.

3.3 Two Separate Functions of AID

As described in the second section, mutations in AID genes identified in hyper IgM type II syndrome patients are located throughout the molecule. Among these mutations, those in the C-terminus region, such as P20 and JP8B, greatly reduce CSR but not SHM activity [23,35]. These results indicate that CSR is not solely completed by AID's DNA cleavage activity but also requires AID's C-terminus-dependent DNA repair function. It has been reported that the formation of S−S DNA synapsis requires the C-terminus of AID [37], as well as UNG [154], APE1 [149] and other factors, suggesting that these factors form a complex to support S−S synapsis. Another report has shown that AID's C-terminus, UNG and MSH2 collaboratively bind to S regions [177] for CSR activity. AID's C-terminus-dependent complex formation has been directly shown by glycerol gradient to be a heavy ribonucleoprotein (RNP) complex including heterogenous nuclear ribonucleoprotein L (HNRNPL), SREBP1 and heterogenous nuclear ribonucleoprotein U (HNRNPU), all of which are required for CSR [54,178,179]. Hypothetically, AID may edit some putative RNA contained in this RNP complex, which could facilitate DNA synapsis formation. Alternatively, AID's C-terminus-dependent protein−protein interaction may be required for the repair step of CSR, by recruiting the proteins for C-NHEJ and A-EJ.

Other C-terminus-dependent functions of AID have been proposed, including the processing of SSBs to DSBs for the utilisation of C-NHEJ. This is supported by the fact that deletion of the C-terminus of AID increases the frequency of A-EJ with longer microhomology. Furthermore, C-terminus deletion changes the repair factors that are recruited to the S regions from C-NHEJ-type to A-EJ-type. Specifically, C-terminus deletion mutants decrease the accumulation of Ku80, XRCC4, DNA-PKcs, 53BP1 and UNG but increase PARP1 and MSH2 recruitment in S regions [37]. Zahn et al. has reported that a different mutant of AID's C-terminus also recruits RPA, EXO1 and RAD51, which are responsible for A-EJ [41].

In contrast, mutations in the N-terminus of AID generally abolishes both CSR and SHM, as do mutations in the catalytic domain, since the N-terminus mutations abolish DNA cleavage step which is essential for both CSR and SHM [23]. G23S of AID is a rare mutation because G23S knock-in mice exhibit sufficient in vivo CSR ability comparable to that of WT mice, while losing SHM in the V region [34]. This discrepancy between CSR and SHM may be due to the decreased DNA cleavage capacity in G23S mutant, as shown by a tamoxifen-inducible fusion protein, G23S-ER, tested in CH12 cells [24]. Taken together, AID appears to have two separate functions: DNA cleavage, which requires an intact N-terminus in addition to the catalytic domain, and CSR-specific activity, located in AID's C-terminal region (Figure 13.1).

3.4 Requirement of the Other Factors in CSR and SHM

3.4.1 MMR in CSR

MMR proteins, such as MSH2 and MSH6, play a substantial role in CSR because the deletion of MutSα proteins

(MSH2 or MSH6) decreases not only SHM but also CSR frequency. Deleting MutL proteins (MLH1 and PMS2) and Exo1 does not change the SHM frequency but decreases CSR efficiency, indicating that they are unnecessary or redundant in SHM but uniquely required for CSR.

Considering the role of MMR proteins in CSR, MMR-mediated instability of repetitive DNA may be important, as recently revealed in studies of trinucleotide repeat (TNR) diseases [180]. MMR proteins are expected to maintain genome integrity during mitosis and meiosis. However, they sometimes fail and cause further genome instability. MMR proteins recognise, bind and stabilise unusual DNA structures in TNR and contribute to their expansion or contraction as shown by several knockout/knockdown studies. When *Mlh1* or *Exo1* is deleted in mice, CSR decreases, and intra-switch recombination in the Sμ region increases, leading to CSR failure [181]. This result is reasonable, because *Mlh1* deficiency leads to the contraction of TNR, which is a failure to protect repetitive DNA sequences [180]. Because S regions contain heavily repetitive sequences, non-B DNA structures formed in these regions may require MMR proteins to support complex DNA ends, as is the case with TNR DNA.

These MMR proteins also bind to NHEJ factors. MSH6 has been reported to interact with Ku70 and is recruited to irradiation-induced DSB foci in colocalisation with γ-H2AX in HeLa cells [182]. MSH6 deletion impairs NHEJ, indicating that MSH6 is necessary not only for recognising unusual DNA but also for recruiting other repair proteins. Both MLH1 and PMS2 also bind to DNA-PKcs [183]. This seems reasonable because these three factors are involved in CSR but are not essential for SHM (Table 13.2). *Pms2* or *Mlh1* deletion decreases CSR, and the remaining switch products have longer microhomology, indicating that both PMS2 and MLH1 contribute to C-NHEJ [184–187].

MMR proteins may also support A-EJ. When MMR proteins, *Msh2, Mlh1* and *Exo1*, are deleted in a conditional knockout mouse of XRCC4, a C-NHEJ factor, CSR is further decreased compared to the single deletion of XRCC4 or the others. This suggests that these MMR proteins support A-EJ [181].

3.4.2 Translesion Synthesis Polymerases in CSR and SHM

Another mechanism for processing of SSBs into DSBs than exo/endonucleases depends on DNA synthesis from the 3′ OH end of the SSB. The 5′ single-strand overhang will be filled by TLS using error-prone polymerases. Generally, TLS polymerases are responsible for post-replication repair (PRR) by synthesising the daughter strand from the damaged template DNA during replication using its 5′ to 3′ gap-filling activity or template switching [188]. By this mechanism, TLS polymerases avoid genetic instability at the expense of sequence fidelity. TLS polymerases work on single-stranded templates in replication forks. When DNA damage occurs in these replication forks, a single-strand gap is formed, inducing mono-ubiquitination of proliferating cell nuclear antigen (PCNA) and further recruiting TLS polymerases to repair the damage [188]. TLS polymerases are important not only for PRR in the damaged DNA sequences but also for PRR in the undamaged but difficult-to-replicate DNA sequences, such as repetitive DNA forming non-B structure and common fragile sequences (CFSs) [189].

Actually, most error-prone DNA polymerases have no effect on CSR when they are solely deleted, although they are involved in V region-SHM (V-SHM) (Table 13.2). This suggests that these polymerases either have no contribution to CSR or are functionally redundant with each other. However, REV1 or REV3, included in the polymerase (pol) zeta complex, is essential for CSR because knocking out each of them decreases CSR efficiency [190–192]. Specifically, REV1 is important for genomic stability of G4 or telomere sequences [193], and REV3 is necessary for the genomic maintenance of CFS [194] and heterochromatin regions [195]. Deleting keratin5-specific Rev3L in mice revealed higher sensitivity to ultraviolet B (UVB) irradiation and increased frequency of chromosomal gaps, DNA breaks and 53BP1 and γH2AX foci formation, showing the clear involvement of polymerase zeta in DSB repair in vivo [196].

TLS polymerases are thought to be involved in CSR through a function other than gap-filling. For instance, pol eta promotes NHEJ in a catalytic domain-independent manner [197], indicating its role in DNA repair as a scaffolding protein.

TLS polymerases, which are crucial for SHM [9,192,198–200], are counteracted by BER system, consisting of the error-free polymerase beta, UNG and other factors. Mice with a knockout of pol beta display increased S-SHM and V-SHM, with only a modest increase in IgG2a CSR [201,202]. Since pol beta localises closely with UNG, UNG overexpression exhibits a competing effect against error-prone DNA polymerases, decreasing SHM [154,203]. In fact, in *Ung*-deleted mice, affinity maturation is normal [204]. Recently FAM72a was identified as a negative regulator of UNG. Cytokines induce it and degrade UNG in proteasome-dependent manner. Deletion of *Fam72a* results in excessive amounts of UNG protein and a decrease in SHM [203,205].

3.4.3 Linkage of MMR Proteins with TLS Polymerases for Promotion of SHM and CSR

There are some functional linkages between MMR proteins and TLS polymerases based on their repairing abilities for unusual DNA regions. Among the MMR proteins, MSH2, MSH6 and EXO1 are known to be necessary in SHM (Table 13.2). In particular, MSH2 seems to have a strong interaction with pol eta. B cells deficient in MSH2 exhibit decreased SHM with reduced AT mutations [206], which is similar to the effects of pol eta deficiency [207,208] (Table 13.2). It has been reported that MSH2 facilitates TLS focus after UV irradiation through PCNA monoubiquitination [209]. Since MSH2/MSH6 stimulates pol eta, deficiency of *Msh2/Msh6* decreases V-SHM to approximately 20%, with a particular lack of mutation at A or T bases [210].

3.4.4 The Final Step of CSR Mediated by Classical NHEJ and A-EJ

When C-NHEJ is suppressed, A-EJ compensates for the loss, as suggested by the deletion experiments of C-NHEJ factors such as 53BP1 [211], DNA PKcs [212], Ku70/80 [167], Ligase IV [213], XRCC4-like factor (XLF/cernunnos) [214], Shieldin2 [215] and XRCC4 [39] (Table 13.2). While C-NHEJ preferentially joins blunt-type DSBs, A-EJ selectively joins a staggered-type DNA end that contains a homologous sequence with another DNA end with the support of homologous end-joining factors such as XLF and XRCC4.

In addition to the typical NHEJ factors, UNG appears to have a role in end-joining. UNG-deficient patients exhibit longer microhomology than controls, suggesting its involvement in C-NHEJ [216]. Another study using *Ung*-knockout mice also demonstrated that UNG is necessary to recruit DNA-PKcs and 53BP1, both essential C-NHEJ factors [154]. However, *Ung*-deficient mouse B cells showed increased recruitment of Ku80 and XRCC4 in S regions and fewer microhomologies at the junctions, suggesting a role of UNG in A-EJ [154].

3.4.5 Difference Between V and S Regions

Extensive SHM accumulations appear in the S regions of the heavy-chain locus, which are the primary cleavage target for CSR [217]. Deletion or mutation of C-terminus of AID fails to induce CSR; however, hypermutation in S regions or in an artificial hypermutation substrate has been observed [23,35]. These results indicate that not all AID-dependent nicks and SSGs induced in the S region are processed into DSBs. So far, the mechanisms of DNA cleavage or SHM in the V and S regions are indistinguishable [24,137].

However, DNA synapsis only occurs between Sμ and the downstream S regions and never involves the V region. Some differences in property between the V and S regions may explain why DNA cleavage in the V region is not used for DNA synapsis.

First, the level of DNA breaks or SHM frequency in the V region seems to be lower than that in the S region, as observed in BL2 cells [24,142], germinal centre B cells [218] or an artificial V region allele [219]. This is partially due to the difference of DNA sequences. As shown in Figure 13.3, the S regions, which are cleaved during CSR, include repetitive sequences, whereas the V region does not harbour such remarkable repeat structures. G-quadruplex (G4), which forms structurally altered non-B DNA and is bound by many repair proteins [220], is a hotspot for DNA cleavage and is bound by AID when it is embedded in both RNA and DNA [42,98]. Analysis using a pre-trained deep learning model suggests that potential G4s are more abundantly distributed in S region than V region [221].

The importance of DNA sequences on SHM has been further demonstrated. Yeap et al. developed a unique and convenient tool for analysing V-SHM [219] using *Rag-2*-deficient blastocysts injected with knocked-in embryonic stem (ES) cells. In these mice, all B cells are derived from the knocked-in ES cells. Their results showed that the sequence of the V region after the leader (promoter) sequence is exchangeable with the *β-globin* or *E. coli*-derived genes, without affecting the SHM frequency. When it is replaced with partial fragment of Sμ sequence, V-SHM is increased because of the increased number of AID's hotspot motif sequences, demonstrating the strong SHM-inducing ability of AID's hotspot motifs. Furthermore, this Sμ embedded in V region shows still lower SHM frequency than that of the original one in S region, probably because of the difference of the promoter activity [219]. Therefore, V region SHM is controlled by both of its DNA sequence motifs susceptible to AID-dependent DNA cleavage and its promoter activity.

Second, S−S synapsis is regulated by the 3′RR enhancer and higher-order chromatin structure, as described in detail in the later part of this chapter. Higher-order chromatin structure is developmentally regulated and plays a role in determining cell fate [5,6]. It can also be reorganised by activation and other processes [7]. Therefore, B cells may be predisposed to form S−S synapsis even without the V region.

Third, the formation S−S synapsis requires the recruitment of several factors to the S regions, which may have a different pattern of the chromatin complex, long-noncoding RNAs and other molecules than the V regions. As shown in Table 13.2, several repair factors are involved in SHM and/or CSR. Among them, the deletion of H2AX does not affect V-SHM, but slightly decreases S region SHM, possibly because H2AX may not be phosphorylated in V region. NBS1 also only affects S-SHM and not V-SHM. In B cells from *Ung*-deficient mice and CH12 cells *lacking Ape1*, the formation of Sμ-Sγ1 and Sμ-Sα synapsis, respectively, is reduced, suggesting that recruitments of UNG or APE1 may be more prominent in S regions than in V region.

Together, the weaker DNA cleavage in the V region compared to S regions, higher-order chromatin structure and recruitment of repair factors may determine the fate of V and S regions. Further studies are needed to elucidate the differences between V and S regions.

3.5 DNA Deamination Hypothesis

3.5.1 The Model (Figure 13.4)

In the DNA deamination model, the recruitment of AID to S regions of the *IgH* gene is mediated by GLTs, particularly in G4 in GLTs [98,143]. This recruitment process is supported by PTBP3 and DDX1 [72,144]. Deoxyuridine, converted from deoxycytidine by AID, is repaired by BER, MMR or passed on to daughter cells without repair via replication. AID has been proposed to have DNA-deamination activity based on its in vitro cytidine deamination activity on ssDNA and G/C-specific mutations when overexpressed in *E. coli* [16,222]. The DNA deamination model postulates that AID's deaminase activity generates U from C on DNA, creating U/G mismatched base-pairs. Uracil DNA glycosylase (UNG) is proposed to remove the Us and create abasic sites, after which APE1 cleaves the phosphodiester bond to produce nicking cleavage. Generating a DSB requires additional nicking on the opposite strand, exo-nucleolytic or endo-nucleolytic digestion or helicase or DNA synthesis, as discussed previously. It has also been suggested that U/G mismatches are recognised and cleaved by the MMR system of the MSH2/MSH6 complex. SHM has been explained by either accumulation of U/G or incorrect repairs of Ape1 nicking by error-prone DNA polymerases. The DNA deamination hypothesis postulates that AID is only responsible for DNA cleavage, and that the end-processing and recombination in CSR are mediated by ubiquitous DNA repair systems such as NHEJ.

3.5.2 Evidence for and Against the DNA Deamination Model

The DNA deamination activity of AID is supported by several in vitro experiments, as described previously [50,55,169,223]. However, the in vitro deamination activities of various AID mutants do not necessarily correlate with in vivo SHM and CSR activity [33]. Notably, the AID N51A mutation lost in vitro DNA deamination activity but retained CSR activity. The mutation at the homologous position N57 in APOBEC1 also abolished in vitro DNA deamination activity but retained RNA-editing activity.

Genomic uracil base accumulation in the Ig locus has actually been reported [224]. Intact DNA fragments were evaluated after genomic DNA purification, in vitro treatment with UNG and APE1 and amplification of the *Igλ* or control gene using quantitative PCR (qPCR). VJλ region in stimulated UNG-deficient DT40 chicken B cells showed a decrease in qPCR signals in an AID-dependent manner, suggesting genomic uracil accumulation by AID.

However, it appears that proteins involved in the BER pathway do not contribute to CSR and SHM as initially expected. Deficiency of UNG reduces CSR, but not to background (zero) level and does not impair AID-dependent DNA cleavage in both SHM and CSR. Instead, UNG suppresses SHM through an error-free pol beta-containing BER complex and competes with TLS polymerases. Besides its impact on SHM, UNG supports CSR through its multiple positive functions, including processing from SSBs to DSBs, S–S synapsis formation and recruitment of NHEJ. Notably, loss-of-catalysis UNG mutations had little effect on CSR efficiency as well as U-removal activity and suppression of SHM using rescue experiments of *Ung*-deficient splenic B cells [153,154,225].

As UNG is a multifunctional protein, the following is a unique feature of UNG: CSR is reduced in both UNG-deficient and UNG-excessive situations. Excessive UNG resulting from the deletion of FAM72a also decreases CSR, probably because excessive UNG over-repairs U:G mismatches and suppresses SSBs and DSBs, which are the substrates of NHEJ in CSR if AID can deaminate cytosines in DNA.

In addition, *Ape1* deficiency in CH12F3-2 cells reduces CSR but does not affect SHM and DNA cleavage levels. Like UNG, *Ape1*-deficient cells show decreases in DSB formation from SSB, S–S synapsis and recruitment of Ku80 [148,149]. Furthermore, DNA cleavage by APE1 at abasic DNA sites produced by UNG leaves 3′-OH (hydroxyl) group, as observed in typical BER-induced SSBs [140,226]. However, AID-induced DNA break ends harbour a 3′-phosphate, which is evidence of TOP1-induced SSBs, as shown by the recently improved DNA break assay [227]. Therefore, APE1 is not likely to be the DNA-cleaving enzyme in CSR.

3.6 RNA-Editing Hypothesis

3.6.1 The Model (Figure 13.4)

This alternative model suggests that AID deaminates C in RNA. However, since AID does not deaminate RNA in vitro, AID is believed to form various complexes with the several RNA-binding cofactors and target RNAs to achieve this. In addition, AID performs two separate functions. Therefore, AID's N-terminus and C-terminus form two or more distinct complexes to achieve DNA cleavage and DNA repair, respectively [54,178,228]. For instance, heterologous nuclear ribonucleoprotein K (HNRNPK) interacts with AID's N-terminus in its RNA-binding domain-dependent manner and is specifically required for the DNA cleavage step [178,228]. On the other hand, HNRNPL, HNRNPU and other RNA-binding proteins are required for the DNA repair step [54,178,179].

This model is also based on the distinctive features of the *IgH* gene mentioned in the preceding part of this

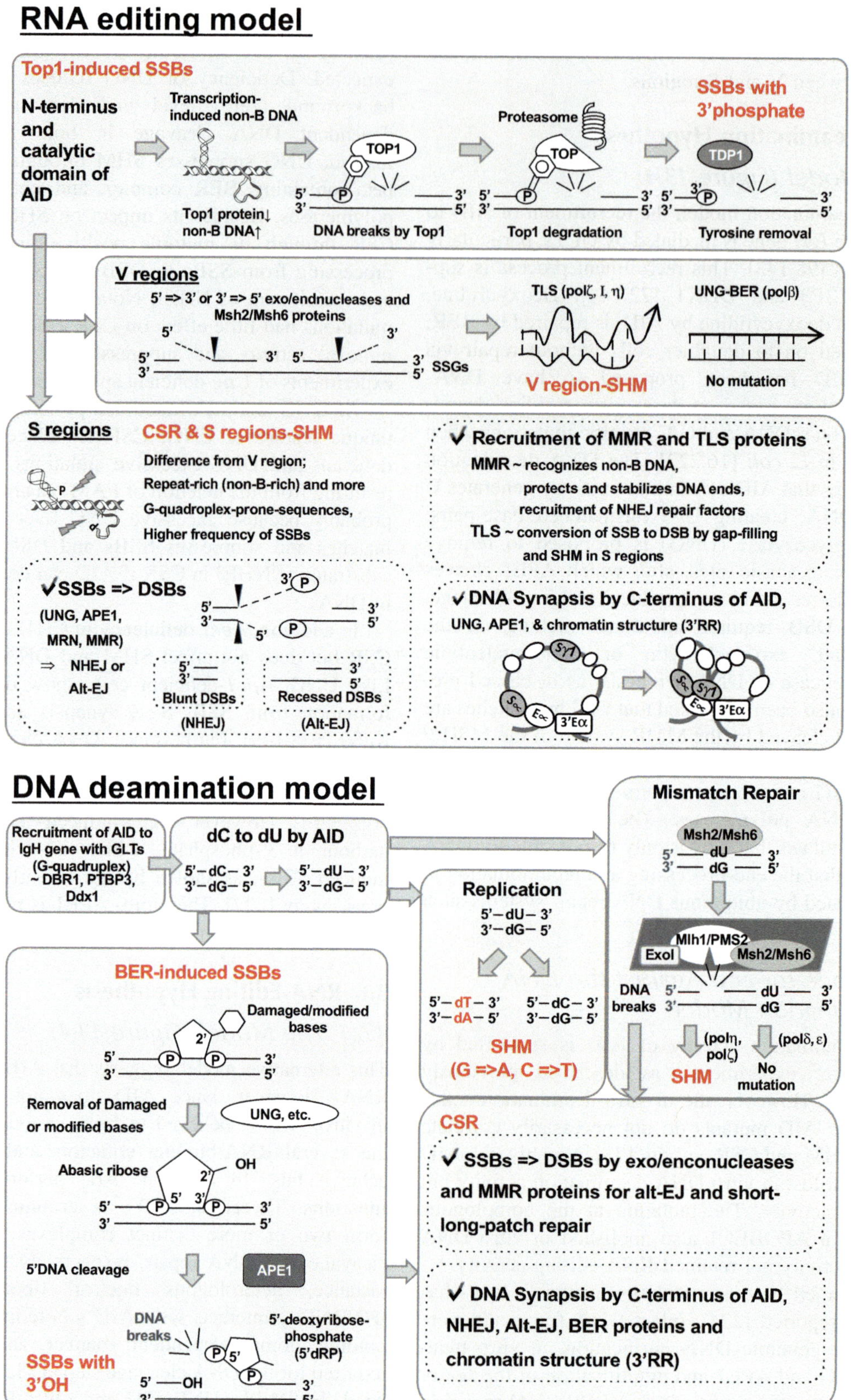

FIGURE 13.4 Two hypotheses on AID's function in CSR and SHM. In the RNA-editing model, TOP1-induced single-strand DNA breaks (SSBs) occur first and SSBs will be repaired by several pathways. In V region, SSBs are processed into single-strand gaps (SSGs) and repaired by translesion synthesis (TLS) by error-prone DNA polymerases. In S regions, CSR and S-region SHM occur. On the basis of the repeat-rich and non-B-rich DNA sequences upon transcription and high frequency of SSBs, recruitment of MMR and TLS proteins, formation of double-strand breaks (DSBs) from SSB by exo/endonucleases, DNA synapsis dependent of C-terminus of AID are necessary for achievement of CSR. Blunt-type DSBs are repaired by NHEJ with 0–2 bp microhomology and recessed-type DSBs are repaired by alternative end joining with longer microhomology. In the DNA editing model, recruitment of AID to IgH gene is helped by germline transcripts which work as guideRNAs in case of CRISPR/Cas9 system. Deoxy uracil (dU) converted from dC in DNA by AID is processed by base excision repair (BER), mismatch repair (MMR) or just replicated without modification. BER-induced SSBs and MMR-processed SSBs are further processed in to DSB by short-patch or long-patch repair, and joined by DNA synapsis, NHEJ and alt-EJ.

section. One feature is the transcription occurring in the DNA break sites [113–115], and the other is the DNA sequences in the core part of S regions that are prone to form non-B DNA structures when transcribed [121–127]. AID-dependent DNA break sites in the *IgH* gene, namely the V and S regions, contain AID's hotspot motifs, G-rich and repetitive sequences [221,229]. During transcription, a non-B DNA structure is induced, which becomes the target of TOP1-induced SSBs.

AID induces the accumulation of helical stress in the above-mentioned DNA regions susceptible to TOP1-induced SSBs because TOP1, a DNA helical stress regulator, is decreased by AID [141,142,227]. Notably, although AID decreases TOP1, the decrease is not negligible; almost half of the TOP1 remains after 24 h and so on [141,142,227]. The small amount of TOP1 remaining can form the irreversible TOP1-DNA cleavage complex (TOP1-cc). This TOP1-cc or TOP1-induced SSB is the failure to seal the two DNA break ends after resolving DNA helical stress. Supporting this notion, knockdown of TOP1 increases CSR, SHM frequency and DNA cleavage frequency, indicating that non-B DNA is induced when the TOP1 protein decreases. In contrast, TOP1 inhibition by camptothecin (CPT) decreases CSR and SHM frequency because CPT blocks the processing step of TOP1-tethered DNA ends after the formation of TOP1-cc [230]. For CSR or SHM, the TOP1-cc is removed by the proteasome and TDP1 enzyme, and 3′ phosphates remain at the 3′ DNA ends [140].

TOP1-induced SSBs are processed differently in the V and S regions. In V regions, SSBs are processed into SSGs by exo/endonucleases, such as APE2, and repaired by TLS polymerases. In contrast, SSBs in the two S regions are processed into blunt-type or recessed-type DSBs. As the frequency of SSBs is higher in S regions than in V region, and the DNA structures in S regions are more G-, hotspot-, and repeat-rich than in V region, the processing of SSBs in S regions is more complex and supported by more proteins than in the V region. The two S regions are united by S–S synapsis, which is induced by cytokine stimulation and regulated by GLTs, 3′RR enhancer and processing proteins such as UNG and APE1. The processed DSBs in the two separate S regions are sealed by C-NHEJ or A-EJ supported by the numerous repair proteins (Figure 13.4 and Table 13.2).

3.6.2 Evidence for and Against the RNA Editing Model

The AID-dependent, TOP1-mediated DNA cleavage in the *IgH* gene is conserved among other genomic instability systems in eukaryotes (Figure 13.5A). As transcription increases, positive and negative supercoils in DNA accumulate in the front and rear of transcription machinery [231], respectively, producing non-B DNA structures [128]. This accumulation of helical stress is enhanced by a decrease in the amount of TOP1 protein, such as *Top1* knockdown or AID-dependent regulation [141]. In contrast, the deletion of *Top1* in yeast abolishes transcription-coupled DNA cleavage or instability, which is known as transcription-associated mutagenesis (TAM) [232,233]. Additionally, triplet-repeat contraction and expansion cause genetic disorders, such as Huntington's disease, depending on TOP1 activity [234]. These findings indicate that TOP1 introduces irreversible cleavage at the repetitive sequence region, where non-B DNA structures likely form during extensive transcription. Consistent with this possibility, the Ig S and V regions, where DNA cleavage frequently occurs, are enriched in inverted or tandem repeats that can form non-B structures [18,235].

Additionally, non-B DNA structure formation in DNA cleavage sites reasonably explains the contribution of the MMR and TLS proteins as discussed earlier in this section.

In particular, AID utilises RNA editing to regulate efficient DNA cleavage, as editing of a micro-RNA (miRNA) that can regulate TOP1 protein synthesis has been proposed (Figure 13.5B) [141,142,227]. Recently, it was revealed that AID-dependent TOP1 protein decrease is mediated by the *Top1* 3′UTR, which is bound by AGO2 at particular sites in the 3′UTR in an AID-dependent manner [227]. Moreover, miR-92a-3p positively affects DNA cleavage, CSR and SHM. Knockdown of miR-92a-3p cancels AID-dependent TOP1 protein decrease. As AGO2 interacts with AID and colocalises with AID in RNA-binding fractions during sucrose gradient separation, it is postulated that AID-AGO2-miR-92a-3p form a complex and bind to the *Top1* 3′UTR to regulate TOP1 protein levels in the soluble fraction. Ultimately, the soluble TOP1 protein level controls the efficiencies of AID-dependent DNA breaks in S regions, CSR and SHM.

Regarding the AID-dependent TOP1 decrease, the RNA-binding protein Hu antigen R (HuR)/ELAVL1 also contributes to the AID-dependent decrease of TOP1 protein [236]. Conditional deletion of HuR in mouse spleen B cells revealed increased oxidative stress and decreased CSR, which was recovered by scavengers eliminating reactive oxidative species (ROS) [237]. However, the *HuR*-knockout mouse B lymphoma CH12 cell line indicated that (1) CSR decrease is not due to ROS, (2) HuR binds to the *Top1* 3′UTR, (3) AID-dependent TOP1 decrease is cancelled in *HuR*-knockout cells, and (4) *Top1* knockdown enables catch-up of CSR defect observed in *HuR*-knockout cells [236]. Since the CH12 cell line may have obtained some resistance mechanism to ROS, the function of HuR in *Top1* regulation could be revealed in *HuR*-knockout CH12 cells.

This TOP1-mediated DNA cleavage is further supported by the analysis of SMARCA4/BRG1 [238], as SMARCA4 supports the recruitment of TOP1 to the *IgH*

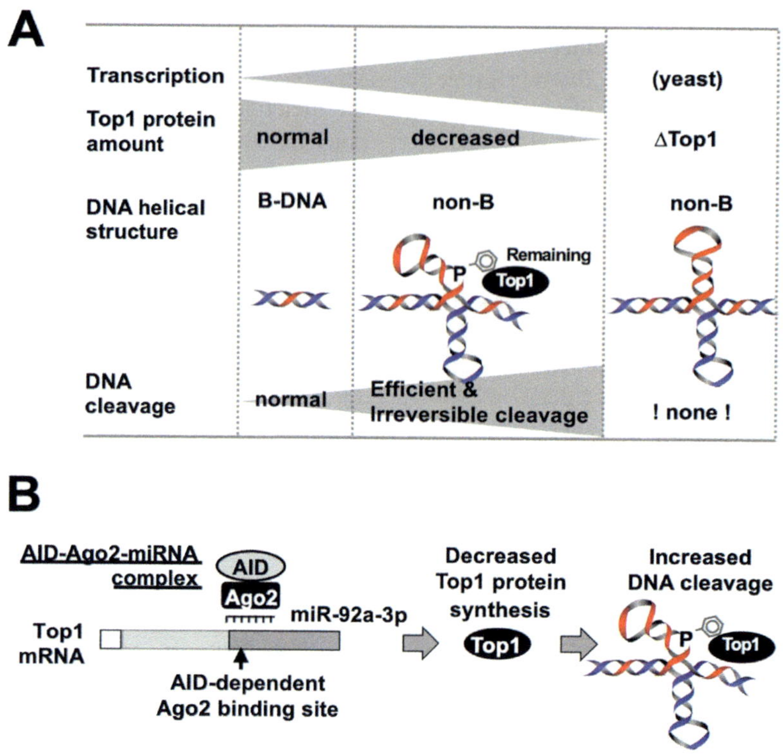

FIGURE 13.5 **AID decreases TOP1 protein synthesis with a miRNA and Ago2 through *Top1* 3′UTR.** A. Relation of transcription, TOP1 protein amount, DNA helical structure and cleavage efficiencies. B. AID-Ago2-miRNA complex binds to *Top1* 3′UTR to decrease TOP1 protein amount. DNA cleavage by remaining TOP1 is increased in AID-dependent manner [227].

gene locus and knockdown of SMARCA4 resembles the knockdown of *Top1*, leading to genomic instability. Furthermore, FACT complex, consisting of SSRP1 and SUPT16, is necessary for binding of TOP1 to tri-methylated H3K4, an important chromatin modification for AID-dependent DNA cleavage in S regions [239].

Although evidence accumulates in favour of AID-mediated RNA editing, some questions remain unresolved. For example, RNA editing by AID for CSR or SHM has not been directly proven yet. It is unclear whether miR-92a-3p or its precursor is genuinely edited by AID. Since the putative alignment of miR-92a-3p to the *Top1* 3′UTR does not follow the classical 'seed' rule, editing of the 3′ side of miR-92a-3p by AID may increase its binding affinity. Additionally, the TOP1 decrease is insufficient to induce DNA cleavage, as siTop1 experiments did not show an increased DNA cleavage level [141,142]. Therefore, another function of AID to initiate DNA cleavage will be necessary. When *Top1* knockdown is continued for weeks to months in non-B lymphocytic cells, DNA instability can be induced even without AID in genomic DNA or introduced artificial constructs [240–242]. These results suggest that another function of AID to initiate DNA cleavage within 24 h will be replaced by accumulating other types of DNA damage.

The TOP1-cc is processed by the proteasome [149] and then TDP1 [140]. However, the processing mechanism of the remaining 3′ phosphate at the DNA cleavage ends remains elusive. It should be determined which molecules remove 3′ phosphate in V and S regions.

AID's C-terminus-dependent complex, which is specifically required for S–S synapsis and processing of SSB to DSB, will include not only repair proteins but also some putative RNAs that may be edited by AID. RNA editing by AID of these unknown RNAs may support the repairing steps, processing of SSBs to DSBs, S–S synapsis and final NHEJ.

To summarise, the recently revealed function of UNG and APE1 is not reasonably explained by the classical DNA deamination hypothesis (Figure 13.4). On the other hand, evidence for the RNA-editing hypothesis has been strengthened by detailed data about TOP1 activity and AID's C-terminal function. Nonetheless, the RNA-editing hypothesis has not been directly proven by identifying the miRNA or mRNA used as AID targets.

4. REGULATION OF DNA CLEAVAGE, REPAIR AND RECOMBINATION

4.1 Characteristics of Target DNA

CSR occurs between a universal donor S region (Sμ) and one of the acceptor switch regions (Sγ/Sα/Sε) located upstream of each constant region [116,243]. The I promoter immediately upstream of each S region drives active transcription through the recombining S region. The S regions are 2–12 kb long, with highly repetitive units of 25–80 bp arranged in tandem and containing motifs such as TGGGG, GGGGT, GGGCT, GAGCT and AGCT [127,244].

Various assay systems were developed by introducing artificial switch substrates into CH12F3-2A B cells or through genetic manipulation of switch regions in mice [245–247] to evaluate the impact of repetitive sequence motifs. In the earliest effort, the Sμ and Sα regions and their derivatives were used to construct artificial switch substrates. The results showed that S regions from different species and isotypes, and S regions with an inverted orientation, all showed efficient CSR. Surprisingly, artificial multiple linker sequences could also serve as a CSR target.

Computational secondary structure analysis showed that S regions have the potential to form non-B DNA structures, including stem-loop structures. Although the Xenopus S region is AT-rich, it harbours AGCT motifs and secondary structures with inverted repeats [248]. Replacing the mouse S region with the Xenopus S region does not prevent efficient CSR [170]; however, CSR efficiency is reduced if the region is replaced with a mouse S region with an altered orientation [245,247]. More recently, a similar replacement study in the CH12F3-2A cell line demonstrated that transcriptional orientation and the presence of successive AGCT motifs are critical for CSR [249].

The transcription-coupled DNA structural changes in the S region sequences are considered essential for AID targeting, as CSR is abolished when the S regions or their promoters are deleted. Studies using bisulfite sequencing also demonstrated that the transcription of repetitive GC-rich sequences in vivo and in vitro generates non-B structures, such as R-loops and G-quadruplexes (G4) [235,250–253]. Altering sequence motifs, such as changing from AGCT to ACGT, can dramatically reduce GLT and CSR [249], suggesting that individual S region repeats are naturally set for optimal transcription and secondary structure formation to initiate DNA cleavage and recombination.

R-loops can be generated in an S region when the template strand DNA hybridises with the GLT, causing the nontemplate DNA strand to loop out. The nontemplate strand in an R-loop provides a long stretch of exposed ssDNA, which is proposed to be an ideal substrate for CSR DNA cleavage. Although AID deaminates the non-template strand of DNA in vitro and in bacteria and yeast, AID-induced SHM occurs on both strands in B cells in vivo [254,255]. AID has also been reported to deaminate dCs on both strands in supercoiled but not relaxed plasmid DNA [256,257]. Interestingly, overexpression of RNaseH1, which resolves the RNA/DNA hybrid and reduces R-loop formation, did not affect CSR in CH12F3-2A cells [129].

Emerging studies suggest that R-loops can contribute to DNA repair pathway choice, such as homologous recombination (HR) and NHEJ [258,259]. DSBs themselves can act as a trigger for small non-coding RNA transcription in the vicinity to form R-loops; the RNAs at the site can also support C-NHEJ through RNP complex formation and/or serving as a template for DNA repair [260,261]. Although these possibilities remain to be explored in the context of CSR, recent studies suggest that GLT processing/surveillance by the exosome, epitranscriptomic modification such as m6A and secondary structures such as G4 can significantly impact the R-loop, and consequently DNA break and repair in CSR.

A defect in exosome-mediated GLT processing caused strand-biased R-loop accumulation, which was suggested to cause asymmetric DNA breaks with ssDNA overhangs and induction of A-EJ-mediated repair [262]. In contrast, the defect of m6A modification in GLT reduced R-loop formation and symmetric breaks, leading to A-EJ-mediated repair [97]. The m6A modification on switch GLT facilitates G4 resolution and R-loop formation, in turn, and AID recruitment via the m6A reader protein YTHDC1. In the absence of RNA helicase DDX1, which resolves G4-RNA structures, reduced R-loop formation and AID recruitment also elevated A-EJ-mediated repair [72]. In contrast, loss of HNRNPU, a G4 and DNA/RNA-binding protein [263,264] increased S region R-loop and decreased C-NHEJ-mediated repair [179].

Although bidirectional transcription-induced S region R-loops are mainly implicated in the DNA breaks on both strands, their role in S–S synapsis has also been suggested [55,265,266]. Like S regions, the IgH super-enhancers or 3′regulatory regions (3′RR) harbour repetitive sequences and abundant G4 structures [267–269]. Super-enhancer activation induces enhancer RNA (eRNA) transcription and R-loop formation, and the repetitive regions are found to be a target of AID-induced DNA breaks and A-EJ-mediated joining with the S regions [267,269].

Like CSR, SHM is tightly coupled with transcription. In SHM, however, the transcribed V region does not form an R-loop but generates an abundant ssDNA structure [256,270]. Although the V region lacks the prototype-repetitive sequence of the S region, inverted repeats and triplet-like motifs (AGC, TAC, GCT and GTA) were found in the V region [271,272]. Although these triplets are not consecutively arranged as a triplet track, their abundant clusters are frequent mutation targets in the V region.

Moreover, SHM is enhanced by introducing the transcription factor E2A-binding sequence CAGCTC (E-box motif), which may be prone to forming secondary non-B DNA structures in the V region or the GFP gene [273–275]. The transcribed GFP transgene is often used as an efficient SHM substrate owing to its high GC content and abundant AGCT motifs [132]. Intriguingly, CAGCTC motifs are frequently found in the S region.

A genome-wide approach to identifying AID-induced DNA breakpoints in BL2, a human Burkitt lymphoma line, identified several known and novel non-Ig AID target loci, including *MYC*, *MALAT1* and *SNHG3* [18]. Like the IgH locus, all identified non-Ig loci showed frequent DNA breaks and characteristic SHM near the breakpoints. These loci are highly transcribed, and the break regions harbour abundant repetitive sequences, including AGCT sequences [18,20]. These findings support the concept that transcription-induced R-loop and non-B structure formation in a repetitive sequence is critical for inducing breaks at both the Ig and non-Ig loci during AID-induced genomic instability [20,26].

Although purified and semi-purified AID showed binding to G4-RNA and G4-DNA structures [42,98], no dC deamination was observed in the G4-structured region of DNA. It is also unclear why AID-induced DNA deamination requires G4-RNA or cofactors like RPA, PTBP1, 14-3-3-γ, YTHDC1 or DDX1 when it can directly bind to the switch G4-DNA.

4.2 Characteristics of Target Chromatin

4.2.1 Nucleosome Assembly of the Target Chromatin

Chromatin consists of repeating units of nucleosomes, which are wrapped with a defined length of DNA and many proteins packaged into a compact structure. Four histones, H2A, H2B, H3 and H4, are the building blocks of every nucleosome in which the core H3−H4 tetramer is flanked on both sides by H2A-H2B dimers [276]. Since DNA is tightly packaged into chromatin, accessibility to DNA is essential during transcription, replication and recombination. Although nucleosomes can be displaced by RNAPII, in many transcribed genes, they are dynamically restored by various mechanisms that are not fully understood [277−279].

During SHM and CSR, AID possibly accesses the DNA after RNAPII migration destabilises the nucleosome core [278,280]. AID's ability to deaminate nucleosomal DNA or chromatinised substrate has been investigated in a plasmid DNA with a strong nucleosome positioning sequence (NPS) [170,171]. AID-mediated dC deamination on nucleosomal DNA was observed only when the plasmid was transcribed. When introduced into the V region in DT40 cells, the same NPS sequence severely reduced SHM efficiency [171]. The introduced NPS may have failed to establish the native chromatin architecture and the associated epigenetic state.

There is little information on nucleosomal organisation or behaviour in the context of the natural V or S region. In CH12F3-2A cells, Stanlie et al. [239] mapped the position of 2.5 nucleosomes at the 5′end of the core Sμ, a region that contains hotspots for AID-induced mutations. The target chromatin appeared to undergo rapid nucleosome assembly, which indicates that the integrity of the locus relies on specialised histone chaperones [239,281]. Nucleosome positioning and stability are known to be greatly influenced by the nature of the target DNA sequence. Because the strong NPS did not favour AID-induced mutations, the Ig locus may naturally contain flexible nucleosomes as a result of the repetitious nature of the S region. Furthermore, histone variants like H3.3 and H2A.Z were detected in the transcribing V and S regions [282,283], suggesting dynamic histone exchange that can facilitate accessibility to DNA during cleavage and repair. The presence of H3.3 also promotes recovery from the transient transcription shutdowns that frequently occur when a transcribed locus encounters DNA breaks [284].

Despite active transcription of the Ig locus, the rapid nucleosome reassembly mechanism presumably preserves its epigenetic integrity, which is required for recruiting various effector proteins involved in DNA cleavage and repair. On the other hand, the execution of complex DNA transactions requires adequate access to the target locus, which can be achieved through dynamic nucleosomal reorganisation and/or histone variant exchange in the target chromatin.

4.2.2 Histone Epigenetics of the Target Chromatin

AID-induced DNA cleavage is always restricted to the promoter (V and I)-proximal 1−2 kb region and to the repetitive S regions (1−10 kb) [244,246,285]. Notably, the S regions, but not the C regions, are the substrate of AID-induced DNA breaks, although the C regions are enriched with AID-interacting proteins, SPT5 and RNAPII (p-Ser5), and are transcribed along the same transcriptional unit [61,137]. The characteristics of the local DNA and the transcribed chromatin may contribute to the restricted DNA cleavage. For instance, transcription induces non-B DNA structures in repetitive DNA sequences that are present in the S regions but not in the C regions. Another decisive factor for the cleavage target could be the distribution of the open-chromatin-associated histone epigenetics, which marks transcribed chromatin differentially, such as to distinguish the promoter-proximal and promoter-distal regions [286−288]. Several studies also align with a strong correlation between non-B structure−associated DNA breaks and epigenetic marks on the breakpoint chromatin [141,239,281,282]. In general, the promoter-proximal sites (1 kb from the transcriptional start site) of actively transcribed loci are enriched with trimethyl histone H3 lysine 4 (H3K4me3), hyperacetylated H3 and H4 and RNAPII (p-Ser5). In contrast, the distal coding regions are enriched with H3K36me3 and RNAPII (p-Ser2). In the context of the Ig locus, H3K4me3, H3Ac and H4Ac are elevated downstream of the active I promoters and throughout the transcribed S regions, whereas C regions have more H3K36me3 and H4K20me1 [92,289]. Active histone marks are generally confined to the promoter-proximal region of the transcriptionally active loci. Strikingly, however, the transcribed S regions show unusually extended H3K4me3 tracks, whereas the C regions are completely devoid of H3K4me3 [92,239,281]. These results strongly suggest that H3K4me3 may be one of the epigenetic marks associated with AID-induced DNA cleavage.

IgM-expressing B cells express Sμ constitutively, which is enriched with active histone marks and RNAPII even before CSR is induced [92,239,281,290]. In contrast, downstream acceptor S regions accumulate H3K4me3, H3K36me3 and RNAPII only after isotype-specific CSR stimulation and transcription [290]. Extensive H3K4me3 accumulation, spreading up to 7 kb, is observed in the activated Sγ3 region in response to lipopolysaccharide (LPS) stimulation. Interestingly, when the Sμ tandem repeat sequences were deleted in mice, H3K4me3 and H3 acetylation marks accumulated upstream of Sμ, and AID-induced cleavage shifted accordingly [291,292]. PAX interaction with transcription-activation domain protein-1-deficient mice did not accumulate H3K4me3, H3-acetylation or RNAPII in Sγ3, Sγ2b and Sγ1, and transcription from these loci was completely inhibited [290]. A ChIP-seq study also shows that AID's localisation significantly correlates with active histone marks in the Ig locus, including H3K4me3 and RNAPII (p-Ser5) [88]. H3K9acS10p has also been proposed to serve as an additional histone code on H3K4me3-marked chromatin, which may provide additional specificity [93]. This concept is supported by the finding that H3K9acS10p can be found in both Sμ and Sγ3, but not Sγ1, in response to LPS activation, which promotes switching to Sγ3. Similarly, LPS and IL4 induction, which are required for IgG1 switching, induce H3K9acS10p specifically in Sμ and Sγ1 but not in Sγ3. Furthermore, the presence of H3K9me3, a repressive chromatin mark, was reported specifically in Sμ [60]. Consequently, CSR is defective in mice lacking the H3K9 methyltransferase SUV39H1 [293]. As none of the chromatin marks discussed are exclusive to the actively transcribed S regions [294], additional criteria, such as DNA structure and regulatory factors, including proteins and RNA, are needed to help assemble the DNA cleavage and repair machinery.

4.3 Factors Influencing DNA Cleavage and Recombination

Understanding how AID induces targeted DNA breaks in the IgH locus is essential for gaining insight into this process. For example, it is still unclear how AID targets the widely separated V and S regions or how it selects non-random regions outside the IgH locus. However, chromatin accessibility and epigenetic features are crucial for targeted DNA breakage and post-break DNA repair and recombination, particularly in CSR. CSR requires additional steps such as S—S synapse and NHEJ-mediated DNA repair compared to SHM. Histone post-translational modifications (PTMs) also play a crucial role in creating an accessible chromatin environment before and after AID-induced DNA breakage. This presumably alters the chromatin structure and/or generates new recognition sites for chromatin regulatory proteins.

The highly dynamic environment involving transcription, splicing, DNA breakage and repair requires an organised regulatory cascade to execute the recombination process appropriately. This chapter discusses identifying and characterising several molecules and cofactors that regulate the recombination process underlying Ig-isotype switching. Various strategies, such as screening a large set of candidate genes against siRNAs in a widely used CSR-inducible mouse B-cell line (CH12F3-2A), analysis of co-immunoprecipitated proteins with AID and TOP1 and isolation of IgH locus-specific binding proteins, have identified various regulatory factors. These factors include histone chaperones, transcription elongation factors, chromatin remodelers, spliceosome components, RNA-binding proteins and regulators of cellular nucleotide pools. Diverse regulatory cofactors show distinctive effects on DNA cleavage, repair or locus-specific conformation, indicating that CSR is intricate and sensitive to the physiological milieu. Therefore, functional perturbation or defects in any of these regulatory molecules can cause catastrophic DNA lesions and aberrant genomic rearrangements, leading to genomic instability and the development of B-cell lymphomas.

4.3.1 Histone Chaperone FACT: SSRP1 and SPT16

The histone chaperone FACT, a heterodimeric protein complex composed of SSRP1 and SPT16, is known to assist in transcription elongation by RNAPII on a chromatin template [295,296]. Knockdown of either SSRP1 or SPT16 severely inhibited IgA class switching in CH12F3-2A cells [239]. The requirement of SSRP1 and SPT16 in CSR as part of a heterodimeric transcription elongation complex was further confirmed by introducing disruptive mutations at the interaction surface. Thus, the FACT complex was found to be an essential chromatin regulator for CSR. A mechanistic study revealed that SSRP1 depletion significantly reduced the AID-induced DNA breaks and γH2AX accumulation in both donor and acceptor S regions.

Interestingly, no transcriptional defects were observed for the essential CSR-associated transcripts such as μGLT, αGLT or AID. This phenomenon was unexpected, considering that the FACT complex is known to promote RNAPII transcription through chromatin [297,298]. However, in FACT-depleted cells, the chromatin landscape of the S regions showed two distinct changes: defects in histone stability and reduced histone PTMs, both associated with the transcribed locus. The dramatic loss of all four core histones was especially observed in Sμ, whereas active histone epigenetic marks were downregulated in both Sμ and Sα. Under FACT-depleted conditions, the H3K4me3 modification was significantly reduced in both Sμ and Sα, which correlated well with the impaired DNA cleavage observed at the respective locus [239].

Intriguingly, depleting H3K4me3 by knocking down WDR5 and ASH2, the core components of H3K4me3 histone methyltransferase (HMTase), recapitulated the results of FACT depletion – intact GLT and AID transcription, but a strong blockade of S region DNA cleavage [239]. SHM is also significantly correlated with FACT and H3K4me3 enrichment in the V region [282]. In contrast to Sμ but similar to Sα, the V regions show altered histone PTMs with no histone loss. However, FACT-enriched V and other non-Ig SHM target loci accumulate the H3.3 histone variant [282,299], which is present in highly transcribed regions in a replication-independent manner [300].

Although H3K4me3 is generally associated with an actively transcribing locus, it also has a critical role in marking meiotic recombination hotspots to be cleaved by Spo11 [301,302]. The PHD domain of RAG2 recognises the same epigenetic mark and promotes RAG-induced DNA cleavage in V(D)J recombination [303]. Therefore, the specific histone epigenetic mark H3K4me3 appears to be critical in all three recombination systems. In the context of CSR and SHM, FACT-mediated nucleosome assembly, H3K4me3 regulation and H3.3 exchange appear to be the critical parameters in accessible DNA-cleavage complex formation [239,282]. Sequential ChIP analysis suggested that H3K4me3 is present in both H3 and H3.3 in V and S region chromatin. TOP1 binds to these histones and coimmunoprecipitates with FACT and H3K4me3. FACT was also enriched in the IgH locus in the absence of TOP1 [238], a condition that elevates negative supercoiling and non-B structures, the preferred sites of AID-induced DNA cleavage.

4.3.2 Histone Chaperone SPT6: RNAPII Elongation Complex

Among the transcription elongation-associated histone chaperones examined, the Ty6 homologue (SPT6) suppressor was another essential histone chaperone involved in CSR [56,281]. SPT6, which was originally identified in *Saccharomyces cerevisiae*, is required for transcription elongation and chromatin structure maintenance. SPT6 knockdown in CH12F3-2A cells reduced the αGLT and AID transcription to some extent but strongly inhibited CSR. This inhibition was not caused by altered AID transcription because the CSR was inhibited even when AID was overexpressed [56]. Global gene expression analysis showed that only ~5% of genes were affected, none of which were known to be required for CSR. As with FACT knockdown, SPT6 depletion elevated the μGLT transcription and Sμ-cryptic transcripts. Consistent with these findings, there was no alteration in the amount of elongating RNAPII in the S regions. Deregulated histone PTMs such as H3K4me3 and H3K36me3 were evident in Sμ and Sα, and histone loss was observed in Sα. As SPT6 depletion hampered H3K4me3 formation in the S and V regions [281], both CSR and SHM were reduced, which correlated with impaired DNA cleavage at the target loci. SPT6 depletion also reduced SHM frequency in two non-coding genes, MALAT1 and SNHG3, with a concomitant decrease in H3K4me3 at these loci [281]. FACT and SPT6 are also involved in S region H3K36me3 regulation, but the knockdown of individual H3K36me3 histone methyltransferases (SETD2, WHSC1 and NSD1) did not dramatically affect CSR.

Interestingly, SPT6 was involved in histone epigenetic regulation (H3K4me3 and H3K36me3) at the AID locus, but FACT was not [281]. AID gene expression correlated well with SPT6-mediated H3K36me3 formation but not with H3K4me3 modulation. Thus, SPT6 appears to have dual functions in CSR: regulation of AID expression and S region DNA cleavage by preserving the histone epigenetic integrity of their respective genomic loci.

To understand how SPT6 regulates the H3K4me3 status at the IgH locus, various SPT6 mutants were analysed based on their ability to restore H3K4me3 upon SPT6 depletion. The C-terminal tandem SH2 domains of SPT6 were found to be necessary for recruiting SET1A H3K4me3-HMTase in both universal donor Sμ and acceptor Sα regions. Consistent with this observation, depleting the SET1-specific cofactor CXXC1 strongly inhibited CSR [281,304]. Later studies also showed the involvement of SPT6 in the positive regulation of H3 modification during transcription and the inhibition of the negative histone epigenetic mark H3K27me3 [305,306].

It is worth noting that H3K4me3 is just one of several histone modifications that have been implicated in CSR, and the exact mechanisms by which these modifications contribute to the process are still an active area of research. Intriguingly, H3K4me3 has also been implicated in RAG-mediated V(D)J recombination, which generates a diverse repertoire of antigen receptor genes [303,307]. Studies have shown that H3K4me3 is enriched at recombination signal sequences (RSSs) flanking antigen receptor gene segments, and this modification is important for the efficient recruitment of the RAG complex. Specifically, H3K4me3 is thought to create a permissive chromatin environment at the RSSs that facilitates RAG binding and cleavage. Although the RAG proteins are directly involved in DNA cleavage and bind to both putative and cryptic RSSs, meaningful RAG-mediated cleavage can occur only at some, and most likely not all, of these sequences [308–311]. Thus, CSR likely requires a yet unknown combinatorial epigenetic signature and *cis*- and *trans*-acting factors for target specificity [93,94,312].

4.3.3 SMARCA4: ATP-Dependent Chromatin Remodeler

SMARCA4, also known as BRG1, is a chromatin-associated ATPase and a component of the BAFF chromatin remodelling complex [313,314]. It is known to modify chromatin structure by reorganising the nucleosomes and their movement in an ATP-dependent manner [315]. TOP1 protein interactome analysis identified SMARCA4 as a regulatory cofactor for both CSR and SHM [238]. Regarding CSR, it was found that SMARCA4 depletion elevates Sμ DSB but inhibits αGLT expression, presumably the main cause of IgM-to-IgA class switch inhibition in CH12F3-2A cells. On the other hand, SMARCA4's effect on SHM has not been associated with V or Sμ region-specific transcription but with TOP1. Previous studies have shown that Top1 deficiency elevates SHM in B-cell lines and Peyer's patch B cells from mice, indicating an inverse correlation between TOP1 expression and SHM. Consistently, *Top1*-deficient p388/CPT45 cells show elevated SHM and its suppression upon ectopic expression of TOP1 [141,142,238]. However, the depletion of SMARCA4 counteracted this phenomenon and augmented the SHM, similar to TOP1 deficiency, emphasising the requirement of SMARCA4 in TOP1-mediated SHM suppression. To understand the mechanism, the authors performed switch region-specific ChIP analysis and *Smarca4* mutagenesis, which revealed that SMARCA4 with intact ATPase activity is essential for TOP1 recruitment in the S region chromatin.

TOP1 is involved in releasing superhelical tension in DNA that can arise during transcription or replication. In the absence of TOP1, the accumulation of negative supercoils increases the propensity of DNA secondary structures, potentially leading to genomic instability. To better understand the molecular mechanisms underlying TOP1-mediated genomic instability, the authors also examined the effect of SMARCA4 depletion, which mimics TOP1 deficiency, on AID-induced non-IgH locus mutagenesis [238]. Similar to TOP1 deficiency, SMARCA4 depletion in CH12F3-2A cells led to an accumulation of negative supercoils and reduced TOP1 occupancy in a number of non-IgH loci, including c-MYC, BCL-2, BCL-6 and FGFR. Recurrent chromosomal translocations involving these proto-oncogenes and Ig loci are observed in B-cell malignancies such as Burkitt's lymphoma, follicular lymphoma, diffuse large-cell lymphoma and multiple myeloma. A significant increase in chromosomal translocation frequency involving the IgH and c-Myc loci was observed upon SMARCA4 or TOP1 depletion. These findings demonstrate that in the absence of SMARCA4, impaired TOP1 recruitment to oncogenic gene loci can potentially drive AID-induced genomic instability further.

The study also showed that FACT could associate with the TOP1 complex involving H3K4me3, as TOP1 does not show interaction with H3K4me3 in FACT-depleted cells [238]. However, TOP1 can interact with H3 and H3.3, suggesting that FACT likely mediates TOP1 association with H3K4me3-enriched chromatin. In the absence of FACT, although there was also less SMARCA4-TOP1 association, it did not affect TOP1 recruitment on chromatin, which largely depends on SMARCA4. It is not mutually exclusive that FACT might contribute to TOP1 recruitment in a fraction of chromatin or chromatin subdomain-specific manner. In conclusion, the identification of SMARCA4 as a TOP1-interacting partner helps to reveal the dominant mechanism by which TOP1 gets recruited to the IgH and non-IgH loci chromatin, which prevents the accumulation of non-B DNA structures, potential targets of AID/TOP1-induced DNA breaks and causes genomic instability in B cells (Figure 13.6).

4.3.4 MED12: IgH Super-Enhancer Activation and eRNA Transcription

RNAi screening in CH12F3-2A cells identified the requirement of MED12 in CSR. Depletion of MED12 strongly impairs the ability of cells to switch from IgM to IgA in response to CIT [119]. MED12 depletion also reduces AID-induced IgH/c-Myc translocations, while TOP1 deficiency elicits supra-IgH/c-Myc translocation.

In activated B cells, the IgH locus remains in a specific conformation, with a long-range interaction between Eμ and 3′RR, establishing a loop structure between the IgH promoter and enhancer. In response to CSR activation, the locus undergoes another conformational transition that brings the donor and acceptor S regions into proximity to form an S–S synapse [316,317].

The chromosome conformation capture (3C) assay reveals that depletion of MED12 markedly diminishes the interaction between Eμ and 3′RR and between Sμ and Sα, which are crucial for CSR. The appropriate locus conformation promotes germline transcription and assembly of the DNA repair-recombination machinery for NHEJ, which joins the DSB ends between the donor and acceptor S regions. In an NHEJ-mediated DNA repair assay, which uses

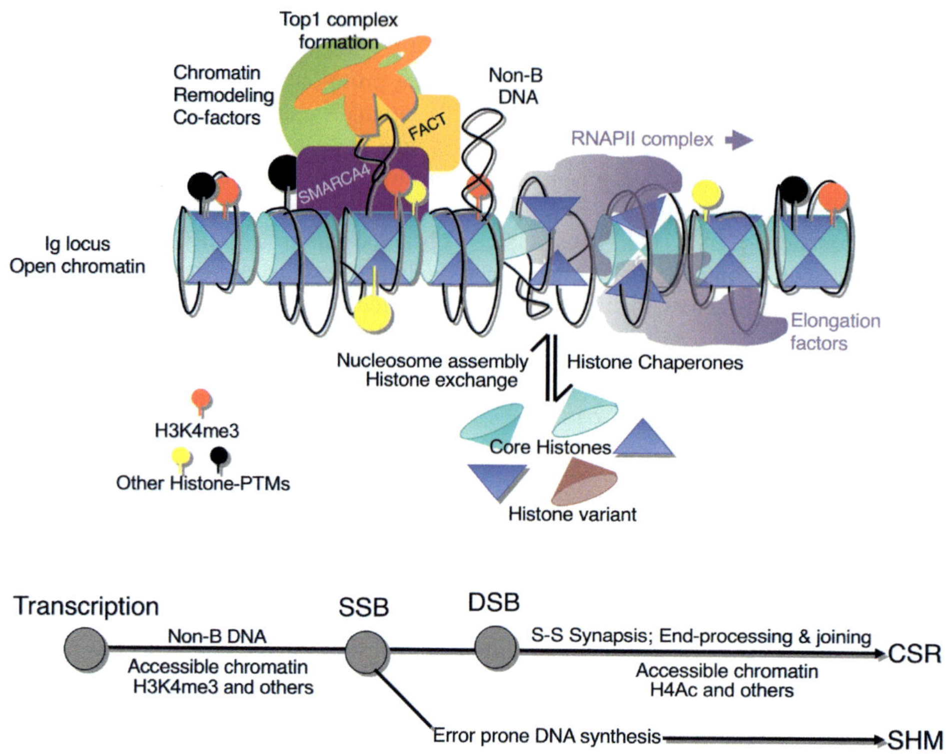

FIGURE 13.6 **A proposed model of Top1 and non-B DNA cleavage complex formation in the IgH locus.** This model proposes how topoisomerase 1 (Top1) and non-B DNA cleavage complex form in the IgH locus. The actively transcribed chromatin in this locus is enriched with various histone post-translational modifications (PTMs), including H3K4me3, H4Ac and others. Nucleosome assembly and histone exchange by histone chaperones help maintain the dynamic and accessible state of the chromatin for both DNA breakage and repair. Combinatorial histone PTMs and chromatin remodelers like SMARCA4 and FACT can facilitate site-specific Top1 complex formation and/or DNA cleavage. Negative supercoils generated behind the RNA polymerase II (RNAP II) can produce non-B DNA structures, which can be substrates of single-strand DNA breaks (SSBs) that lead to double-strand DNA breaks (DSBs). The scheme at the bottom of the figure shows the key events associated with activation-induced cytidine deaminase (AID)-induced somatic hypermutation (SHM) and class switch recombination (CSR) in B cells.

I-Sce-I meganuclease-induced DSB and GFP reporter as a readout, depletion of MED12 results in a fourfold decrease in NHEJ efficiency.

Thus, several sequential and interconnected mechanistic requirements for CSR are tightly linked with MED12. Mutational analysis has revealed that the LS and PQL domains of MED12 are responsible for AID-induced DNA DSB and S–S synapsis, respectively. MED12 can regulate gene expression positively and negatively by influencing RNAPII activity through its N-terminal kinase module, which interacts with the core mediator [318,319]. The kinase module, consisting of MED12, CyclinC, CDK8/CDK19 and MED13, is reversibly associated with the mediator. While MED13 is required for the association between the kinase module and the core, the MED12-CyclinC interaction is necessary to stimulate kinase activity in the module. Specifically, the L domain of MED12 is essential to stimulate the kinase activity, and disruptive point mutations, such as L36R, Q43P or G44S on the L domain, abolish the kinase module-dependent functions [320]. Notably, neither the point mutations nor the depletion of MED13, which dissociates the kinase module from the core mediator, impaired CSR. Additional experiments, including pharmacological inhibition of kinase activity or CDK8/CDK19 knockdown, suggest that MED12 functions in CSR independent of its kinase module and core mediator.

A report suggests that MED12 likely possesses a kinase module-independent function in hematopoietic stem cell-specific super-enhancer regulation [321]. Haque et al. found that MED12 was essential for depositing H3K27ac, a hallmark of enhancer activation, at the IgH 3′RR super-enhancer. Depletion of either MED12 or p300 HAT markedly reduced H3K27ac deposition across the IgH 3′enhancer clusters (hs3a, hs1.2, hs3b, hs4, hs5, hs6 and hs7) [322,323]. Moreover, MED12 and p300 HAT are also crucial for both S region chromatin remodelling and long-range S–S synapse formation. To confirm the requirement of 3′RR activation for CSR, the authors showed that CSR could be restored by enforced enhancer activation using a CRISPR–dCas9-p300c targeted H3K27ac formation at the 3′RR following MED12 deficiency, and S–S synapse formation. Intriguingly, this also restored the H3K4me3 epigenetic signature in the S region, essential to

DNA DSB induction [239,281]. Additional studies revealed that MED12 primarily exerts its function through eRNAs, which are short transcripts transcribed unidirectionally or bidirectionally from enhancers. To further support this, the authors depleted 3′RR eRNAs by antisense oligos targeting the hs4 enhancer. Along with additional studies, the authors showed that eRNAs are required at multiple steps of CSR, including 3′RR and S region epigenomics, S–S synapsis and AID recruitment.

Although the requirement of the 3′RR super-enhancer in CSR has long been known [317,324], the mechanism of 3′RR activation and its implication in AID-induced CSR remained unclear. An early study on MED1 showed its association with 3′RR and the requirement for switch germline transcription [118]. Recent advancements in the field have revealed that RNAPII loading and its pause release at the 3′RR super-enhancer can be influenced by ZYMND8 and SPT5, respectively [120,325]. Upstream, MED12 activates the super-enhancer, and its interaction with p300 HAT and Jmjd6/Carm1 helps stabilise the complexes, promoting super-enhancer and S region-specific histone epigenetics, eRNA synthesis and subsequent steps necessary for CSR. Loss of MED12 leads to both epigenomic and architectural dysregulation of the locus, abrogating AID-induced DSB and post-break recombination/repair essential to Ig-isotype switching. Further studies will be required to fully understand the mechanism of 3′RR activation, recombination and associated locus-specific nuclear condensate formation, which appear to be linked to MED12 and eRNAs [119].

4.4 Factors Influencing DNA Repair and Recombination (Figure 13.7)

4.4.1 DSIF Complex: SPT4 and SPT5

Screening with siRNA also revealed that the DSIF components SPT5 and SPT4 are critical for CSR but not SHM [326]. SPT5 and SPT4 constitute the large and small subunits of the DSIF complex, respectively. Because DSIF is required for transcription elongation, KD of either subunit was expected to cause CSR defects due to the disruption of the complex. However, contrary to this expectation, SPT4 and SPT5 have distinct yet novel mechanistic roles in CSR [326]. The first indication of their dissimilarity came from global transcription analysis in CH12F3-2A cells without either SPT4 or SPT5. Although SPT4 or SPT5 depletion affected only a small number of CSR-irrelevant genes, most of the affected transcripts did not correlate well between the SPT4 and SPT5 KD samples. In addition, cryptic switch transcripts derived from the intronic region of Sμ were dramatically enhanced in SPT4 but not SPT5 KD cells, indicating the difference in transcriptional regulation. Interestingly, SPT5 but not SPT4 KD slightly reduced the amount of AID protein without affecting its transcripts, suggesting that SPT5 is involved in AID's protein stability.

While SPT4 depletion did not affect the H3K4me3 signal in Sμ and Sα, SPT5 depletion blocked H3K4me3 formation specifically in Sα. Consistent with this observation, SPT5 deficiency reduced Sα DNA cleavage, whereas SPT4 did not. Although it is not known why SPT5 regulates Sα chromatin more strongly than Sμ chromatin, it could be attributed to differences in the inducible or constitutive promoter activity, nucleosome-free regions (NFRs), or a non-NFR arrangement, or the specific role of the H2B ubiquitin ligase BRE1 in the histone trans-modification cascade and thus in H3K4me3 at Sα [239,327,328]. As αGLT transcription is more sensitive to DRB inhibition than μGLT transcription, DSIF may affect the poised state in the Iα promoter more robustly than in the constitutively active Iμ promoter.

Other studies also indicate that SPT5 has additional roles in the IgH locus, such as AID recruitment and 3′ enhancer regulation [120,329], which can significantly influence the CSR outcome. Intriguingly, SPT4 and SPT5 also have novel functions in the DNA repair phase of CSR because both the NHEJ and HR pathways were deregulated in the absence of either of the DSIF components [326]. Earlier studies in yeast showed that SPT4 or SPT5 mutants resulted in methanesulfonate (MMS) sensitivity [330], indicating defective DNA repair and recombination. As proteins involved in NHEJ or HR, such as Ku70/80, DNA-PKcs and RAD51, associate with the RNAPII complex [331], DNA breaks appear to be repaired through transcription-coupled processes at some loci. Indeed, DSBs take longer to repair when located in heterochromatin, where transcription is silenced [332]. Therefore, DSIF subunits are involved in both H3K4me3 and DNA repair regulation in CSR and can function independently or alongside their bona fide role in transcription elongation. In activated primary B cells, the deficiency of PTIP/PAXIP1, which stably associates with the MLL3-MLL4 H3K4me3 methyltransferase complex, impairs CSR [290]. PTIP was found to accumulate at AID-induced DSBs and suggested to play a role in CSR-DSB repair independent of its involvement in transcriptional regulation [333].

4.4.2 PHF5A: Histone Variant Remodelling

To explore the role of H3K4me3 in AID-induced DNA cleavage, a systematic screening of proteins containing the PHD zinc finger domain [334,335], known as H3K4me3 readers, was carried out. Among the candidate genes whose knockdown impaired CSR, PHF5A, a small 14 kDa protein also known as SF3B14b, one of the two small subunits of the U2SnRNP major spliceosome complex, stood out as its depletion drastically inhibited CSR in CH12F3-2A cells [283]. Although CSR has been associated with germline

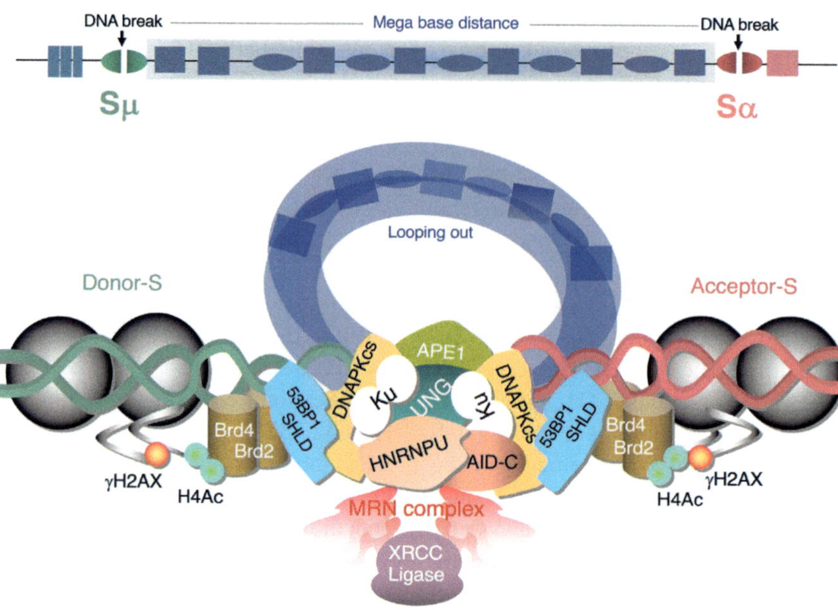

FIGURE 13.7 **The recombination step of CSR requires DSB-end joining in a switch synapse.** During class switch recombination (CSR), the repair of AID-induced double-strand breaks (DSBs) through non-homologous end joining (NHEJ) requires adequate proximity between donor and acceptor S regions, which are typically far apart. The AID C-terminus has been proposed to play an essential role during the S–S synapse and DNA repair phase, although the precise mechanism is unclear. The NHEJ during CSR is a complex interplay of various DNA repair enzymes, scaffold proteins and RNA-binding proteins. Many chromatin and non-chromatin proteins participate directly and/or indirectly, promoting the synapse and DNA break-end joining. The efficient repair-recombination also depends on target chromatin-associated histone variants, histone epigenetics, R-loops, non-coding RNAs and liquid–liquid phase separation, which remain to be fully elucidated.

transcription and splicing, the implication of the latter is poorly understood. Therefore, the identification of PHF5A raised intriguing possibilities that warranted further investigation.

The loss of PHF5A had a significant impact not only on CSR but also on AID-induced IgH/c-Myc translocation, including the hyper-translocation induced by depleting TOP1 or BPTF [283]. Interestingly, PHF5A depletion had minimal effects on SHM in the V region, suggesting its dispensability for AID-induced DNA breaks and possible involvement in the DNA repair phase of CSR. Since DNA repair by NHEJ is crucial for CSR, the requirement of PHF5A was examined in a reporter experiment that evaluates NHEJ-mediated joining in DSBs introduced with I-SceI endonuclease [336]. The results showed an increased frequency of microhomology, insertion and deletion in both CSR and I-SceI-cleaved recombination junctions, representing the signature of A-EJ. Further investigation revealed that PHF5A affects NHEJ by regulating variant H2A deposition and other repair proteins at the IgH locus.

Consistent with the requirement of H2A.Z, depletion of the H2A.Z depositing histone chaperone p400 affected DNA damage response, resulting in a recruitment defect of crucial DSB end-protecting factors such as KU80 and 53BP1 [283,337]. This depletion could allow DNA end resection through the CtIP and MRN/DNA2 complex, as the S-region DSBs are no longer flanked by sufficient H2A.Z. In response to AID-induced DSBs in the S region, increased recruitment of DDR factors such as γH2AX and KU80 is evident, but there was no elevation of H2AX or H2A.Z. Since histone variants are essential to promote DNA repair events, the S regions are either preloaded or continuously replenished with H2A variants through the joint action of PHF5A and p400. Knockdown of PHF5A or p400 also affects the deposition of two additional H2A variants, H2A.Bbd and macroH2A, indicating that the PHF5A-p400 axis plays a central role in regulating H2A variant stoichiometry in the S region.

Despite its interaction with the SF3B splicing complex, PHF5A's function in CSR is found to be unrelated to mRNA splicing of numerous CSR-associated genes, including GLTs [283,338,339]. Depletion of PHF5A did not affect the pre-mRNA splicing of several CSR-associated genes, such as μGLT, aGLT, AID, p400 or H2A.Z. On the other hand, depletion of SF3B1 or treatment with splicing inhibitors had adverse effects on CSR-associated gene expression and CH12F3-2A proliferation. Interestingly, while depletion of PHF5A did not affect the pre-mRNA splicing of dozens of additional genes, an increase in the protein expression of specific histones and their PTMs was observed. Furthermore, the unique structure of PHF5A allows it to interact dynamically with different protein complexes in various physiological settings, including histone variant regulation, which may influence gene-specific promoter remodelling activity. Unique transcriptional regulation and subset-specific gene regulation have also been reported in pluripotent and glioblastoma cells, respectively.

The unique structure of PHF5A, which contains zinc finger motifs, allows it to interact with various large protein complexes, such as the p400 histone chaperone complex and the SF3B complex. These interactions promote stabilisation of the p400 histone chaperone complex in the S region and are important for PHF5A's role in CSR [283]. Studies have shown that the SF3B subunits, along with PHF5A, H2A.Z and H2A.X, likely form a large complex at the site of DSBs on chromatin, especially when PHF5A binds to H3K4me3 [340]. After DSB formation, ATM-dependent γH2AX formation activates DNA repair cascades. Downstream of H2AX phosphorylation, H2A.Z facilitates the recruitment of KU80 and, subsequently, DNA-PKcs and the four XRCC4-XLF-DNA ligase complexes. The H2A.Z flanking the DSBs creates an upstream and downstream boundary, protecting the DSB ends from excessive resection to promote NHEJ. H2A.Z-containing nucleosomes can also promote the chromatin accessibility of DNA-processing enzymes. Thus, the regulation of histone variants such as H2AX and H2A.Z significantly impacts CSR and genome stability. PHF5A, in cooperation with p400, regulates the H2AX and H2A.Z balance in the S region, which is essential for an optimal DDR and efficient DNA repair by NHEJ [283].

4.4.3 BETs: Bromodomain Histone Acetyl Readers

Similar to regulating DNA cleavage by various chromatin proteins, novel regulators of CSR at the step of DNA repair and recombination have been predicted. Proteins that can recognise acetylated histones (H4Ac) were screened to identify these novel players because this particular histone PTM is implicated in various DNA repair systems [341,342]. RNAi screening for bromodomain (BRD)-containing acetylated histone readers has identified BRD4 as a novel CSR regulator [154,343]. Depletion of BRD4 in the S region chromatin using either siRNA or JQ1, a chemical inhibitor [344], significantly reduced CSR in CH12F3-2A cells and primary B cells [345]. Interestingly, genome-wide transcription, DSB formation in the S regions and SHM frequency at the AID target loci were largely unperturbed in BRD4-depleted cells, suggesting its function downstream of AID-induced DNA cleavage or, in other words, at the DNA repair-recombination phase of CSR. Indeed, in the absence of BRD4, increased microhomology usage at the CSR junctions impaired IgH/c-myc translocation, and reduced occupancy of 53BP1 and MSH2/MLH1 in the S region were observed, implying that DNA repair was indeed defective. As BRD4 reduction negatively affected UNG occupancy in the S regions, many of these phenomena overlapped with UNG deficiency [154,225,346,347]. UNG, through its non-canonical and/or enzymatic activity-independent function [348–350], has been implicated in mediating DNA repair and CSR after DNA cleavage [95,225].

BRD4's involvement in DNA repair was directly demonstrated using the I-SceI-introduced DSB repair assay for NHEJ, the dominant DNA repair pathway during CSR [38,336]. BRD4 knockdown reduced NHEJ efficiency and 53BP1 recruitment in I-SceI-cleaved DSBs. These results delineate the importance of BRD4 during NHEJ repair in CSR [345]. Sequential chromatin immunoprecipitation (ChIP) assays further demonstrated that BRD4 tethers 53BP1 and UNG to the recombining S region chromatin through acetylated histones following γH2AX formation. Thus, BRD4 binding to H4Ac creates a platform for recruiting CSR repair components.

This finding raised an intriguing question of whether other members of the BRD protein family are also involved in CSR regulation, as BRD4 is a member of the Bromodomain and Extra Terminal motif (BET) family, which includes BRD2, BRD3, BRD4 and BRDT [351]. Although BET family proteins share structural similarities, including two conserved BRDs and one ET domain, only depletion of BRD2 and BRD4 affected CSR. Knockdown of BRD3 did not affect CSR, and no detectable expression of BRDT was observed in CH12F3-2A cells (Gothwal et al., unpublished). This raises the critical question of whether BRD2 and BRD4 function similarly in CSR. However, depletion of BRD2 did not impair AID-induced DSBs in the S regions or germline transcription through the S regions, yet it significantly impaired CSR by impairing NHEJ. Since the depletion of each protein independently impaired CSR through DNA repair defects, they are not functionally complementary in the context of CSR.

Similar to a report in T cells, BRD2 but not BRD4 predominantly interacted with NIPBL in CH12F3-2A cells [351,352]. Additionally, a report showed that recombination junctions in B cells from NIPBL-defective Cornelia de Lange syndrome patients show microhomology-mediated repair, indicating impaired NHEJ during CSR [353]. Intriguingly, a comparison of Sμ-Sα junctions in CH12F3-2A cells revealed that BRD2 and NIPBL depletion led to elevated microhomology-mediated DNA repair, a typical feature of A-EJ, which overrides the repair pathway when C-NHEJ is impaired (Gothwal et al., unpublished). A high incidence of long insertions at the CSR junctions was evident in BRD2 and NIPBL deficiency, presumably due to elevated ssDNA, as a free 3′ DNA break can opt for homology-directed repair via ectopic sequence capture from other genomic loci. A significant decrease in replication protein A (RPA) phosphorylation was observed in BRD2 and NIPBL deficiency, presumably driving the pathway towards error-prone DNA repair via the microhomology-mediated end-joining pathway [354].

Surprisingly, BRD2 depletion reduced NIPBL deposition in the S region, suggesting that BRD2 contributes to stabilising NIPBL at the IgH locus. In contrast, NIPBL depletion elevated BRD4 but not BRD2, implying intricate regulation underlying the balance between BRD2 and BRD4 in the S region, which involves NIPBL. Moreover, the R-loop in the S region was elevated, particularly in the absence of BRD4 but not BRD2, further emphasising their differential contributions to the IgH locus and CSR (Gothwal et al., unpublished). Although the R-loop has long been implicated as a requirement for AID-induced DNA breaks, its role in DNA repair has become increasingly evident [355]. Overall, this study reveals that although BRD2 and BRD4 distinctly affect S-region R-loop and DNA damage response, the outcome converges into NHEJ defect and CSR impairment.

4.4.4 HNRNPU: A G-quadruplex RNA/DNA-Binding Protein

HNRNPU was identified as a cofactor associated with CSR by two independent approaches: analysis of co-immunoprecipitated proteins with AID in 293T cells and RNAi screening of HNRNP family genes in CH12F3-2A cells [54]. HNRNPU depletion decreased the switch from IgM to IgA and IgG1 isotypes in CH12F3-2A cells and primary B cells, respectively. Although germline transcription or AID-induced S-region DSBs were not compromised in HNRNPU knockdown, S—S recombination junctions showed impaired C-NHEJ. Consistently, HNRNPU loss caused a recruitment defect of C-NHEJ factors, including 53BP1 and the Shieldin complex essential for DSB end protection. Moreover, HNRNPU deficiency elevated R-loop and ssDNA in the S region, which promoted persistent DNA breaks, end resection, and homology-directed repair. Co-immunoprecipitation analysis showed that HNRNPU interacts with R-loop and C-NHEJ-associated regulatory factors, suggesting a dual role of HNRNPU in CSR [179].

As HNRNPU showed an interaction with the 53BP1-Shieldin complex [343,356], the CSR junctions of HNRNPU-deficient cells were compared with 53BP1 and Shieldin deficiencies in CH12F3-2A cells. Depletion of SHLD1/2 reduced the C-NHEJ-mediated S—S joining, resulting in increased frequencies of longer microhomologies and insertions, as observed in HNRNPU deficiency. In contrast, SHLD3 and 53BP1 depletions modestly reduced C-NHEJ-mediated repair with increased longer microhomologies at the CSR junctions. Interestingly, 53BP1 but not SHLD3 deficiency increased the insertion frequency, suggesting subtle functional differences among the end-protective DNA repair factors. The HNRNPU-deficiency-associated S—S joining defect generally showed a high similarity to the SHLD1 and SHLD2 deficiencies in CH12F3-2A cells [179]. As expected, co-depletion of HNRNPU along with SHLD1/SHLD2 did not further reduce C-NHEJ-mediated repair frequency, implying that the C-NHEJ defect in HNRNPU deficiency was due to impaired recruitment of the 53BP1-Shieldin complex at the break site.

There was a striking increase in RPA occupancy at the S regions in the absence of HNRNPU, suggesting excessive ssDNA formation as observed in Shieldin deficiency. In the absence of the 53BP1-Shieldin complex, the DSB ends are excessively resected, resulting in free ssDNA. Since SHLD2 binds directly to ssDNA, in its absence, RPA accumulation occurs on the exposed ssDNA [357]. Furthermore, the XPG family 3′ DNA endonuclease ASTE1 is associated with SHLD2 to facilitate C-NHEJ by processing the 3′ overhangs at DSBs [358]. Notably, another XPG family nuclease, FEN1, which possesses 5′-flap endonuclease and 5′—3′ exonuclease activities, associates with both HNRNPU and SHLD1. Knockdown of FEN1 also reduces IgM-to-IgA switching efficiency and C-NHEJ-mediated joining during CSR in CH12F3-2A cells. It is possible that ASTE1 and FEN1, associated with SHLD2 and SHLD1, respectively, may facilitate DNA overhang processing at S-region DSBs to promote C-NHEJ. As aberrant DNA rearrangements also deploy the C-NHEJ pathway, AID-induced *IgH/c-Myc* translocation frequency was significantly decreased after depletion of HNRNPU, SHLD1 and SHLD2, confirming the requirement of HNRNPU in B cells for CSR and genome stability [179].

Refaat et al. showed that the interaction of HNRNPU with the C-NHEJ factors is through its RGG RNA-binding domain and is sensitive to RNase A treatment [179]. Many C-NHEJ factors, including 53BP1, KU80 and DNA-PKcs, are known to bind long non-coding RNA and DNA damage-specific small non-coding RNAs. At the DNA damage sites, these RNA—protein interactions promote 53BP1/KU80 foci formation, which can be disrupted in the presence of RNase A. Long non-coding RNA, *LINP1*, has been reported to stabilise KU80 and DNA-PKc in the C-NHEJ synaptic complex [359]. LINP1 and switch GLTs can be easily detected in HNRNPU or SHLD1 immunoprecipitates, suggesting possible RNA-associated C-NHEJ complex formation. RNA-dependent DNA repair complex formation is further supported by the fact that occupancy of HNRNPU and 53BP1, which are upstream of the Shieldin complex recruitment, is sensitive to inhibitors of transcription or liquid—liquid phase separation (LLPS).

The C-terminal RGG/RNA-binding domain is embedded within a low-complexity or intrinsically disordered region (IDR), which also has the characteristic of a prion-like domain (PLD) [360]. Proteins harbouring these domains are prone to form a polymer-like 'mesh' involving RNA and DNA. Several RNA-binding proteins with such characteristics, such as FUS, are recruited to the DNA damage site to promote the formation of LLPS and PAR to facilitate DNA repair [361]. HNRNPU has been reported to be recruited to the laser-induced damage site in a PAR-dependent manner through its C-terminus [362]. Increasing RNA and PAR formation at the DNA damage site promotes multivalent RNP and/or protein–protein non-covalent interactions, leading to LLPS or condensate formation. Interestingly, 53BP1 also possesses IDR and RGG-like domains in its C-terminus and undergoes RNA-dependent oligomerisation. Intriguingly, in response to AID-induced DSB, 53BP1 recruitment to the S region can be completely abolished by LLPS perturbation induced by chemical inhibitors or osmotic stress, suggesting that CSR repair foci undergo condensate formation [179,363].

Several studies suggest that the S region R-loop plays a crucial role in AID-induced DNA break formation [72,364]. Conversely, emerging studies indicate that DSB site-specific R-loop formation promotes DNA repair in both HR and NHEJ [258,259,365]. The loss of HNRNPU leads to elevated R-loop and unperturbed AID-induced DNA breaks. HNRNPU occupancy in the GLT likely affects the S region R-loop formation [179], as the binding of RNA-binding proteins to the nascent transcript can prevent its snapback to the template DNA strand, thereby reducing the R-loop. The winding/unwinding of the G4 structures in RNA also influences the S region R-loop, including AID recruitment, as seen in the case of G4-helicase DDX1 [72]. Although it is unclear whether HNRNPU can contribute to RNA-G4 structure modulation, such as on GLTs, it is known to facilitate DNA-G4 formation at the telomere to protect the ssDNA end from resection [263]. It is possible that, in the presence of HNRNPU, the ssDNA end at the CSR-DSB adopts an analogous G4 structure. At the repair phase, FEN1, associated with SHLD1-HNRNPU, can help remove such a G4 obstruction before C-NHEJ. Interestingly, several CSR-associated DNA repair factors, such as EXO1, BLM, and RIF1, are G4-interacting proteins, but their functions in CSR are unclear from a G4 perspective [366,367]. In meiotic recombination, DNA-G4 structures are suggested to contribute to DNA strand pairing and recombination.

In response to AID-induced DSBs, HNRNPU is upregulated at the S regions, stabilising the C-NHEJ complex through various RNA scaffolds. HNRNPU's presence in the S region also limits the accumulation of ssDNA and the resection of the DNA end through the R-loop and/or DNA-G4 regulation. RNAs with G4 structures, as observed in switch GLTs, are known to facilitate LLPS/condensate formation. It is not mutually exclusive that GLTs, along with IDR-containing proteins such as HNRNPU, participate in forming a repair/recombination condensate [179].

4.4.5 SAMHD1: A Nucleotide Reductase and 3′→5′ Exonuclease

The insertional chromatin immunoprecipitation (iChIP) technique was utilised to identify proteins that bind to AID-induced DNA break sites in the S region of CH12F3-2A cells [368,369]. The authors of this study found that the sterile alpha motif and histidine-aspartic acid domain-containing protein 1 (SAMHD1) is a novel regulator of CSR. SAMHD1 is a dNTP triphosphohydrolase that hydrolyses nucleotide triphosphates into triphosphate and nucleoside [370,371] and is crucial for maintaining an optimal balance of the dNTP pool in mammalian cells. The CH12F3-2A mouse B-cell line depleted of SAMHD1 and spleen B cells generated from *Samhd1* knockout animals showed increased dNTPs, particularly purines. This elevated dNTP imbalance reduced CSR and aberrant recombination between the IgH and c-Myc loci. However, SAMHD1 depletion did not affect AID-induced DNA breaks or the S–S synapsis during CSR.

The recombination junctions of CSR and IgH/c-Myc exhibited a high frequency of long insertions, a distinct characteristic of SAMHD1 deficiency, indicating that nucleotide accumulation in the cell affects both *cis* and *trans* recombination. The impact of nucleotides was directly demonstrated by supplementing the cell culture medium with purine nucleosides, which impaired CSR and IgH/c-Myc translocation in CH12F3-2A cells. Consequently, SAMHD1 depletion also reduced the CSR induced by CRISPR/Cas9 nickase, which generated staggered DSBs in the donor and acceptor S regions [372]. In B cells, AID induces staggered DNA breaks, necessitating effective end-processing prior to repair by C-NHEJ/A-EJ.

An increase in the cellular dNTP pool led to significant increases in the frequency and length of insertions at the recombination junctions of CRISPR/Cas9-nickase-induced CSR and I-SceI-cleavage-induced NHEJ in plasmid substrates [369]. These findings support the notion that the cellular dNTP pool plays a significant role in regulating the outcome of DNA repair. Consistently, SAMHD1 localises at DNA break sites, which may decrease the local dNTP pool during DSB repair. SAMHD1 has also been shown to facilitate HR by interacting with and recruiting the DNA end-processing factor CtIP to the DSB site [373]. Employing SAMHD1 mutants, the authors confirmed that the effects of SAMHD1 depletion are entirely dependent on its dNTPase activity [369]. The elevated dNTP pool in germinal centre B cells caused by *Samhd1* inactivation affected SHM and increased C/G to A/T transition mutations while decreasing transversion mutations [374]. A reduction in the SHM frequency was also observed in the S regions of primary B cells from *Samhd1*$^{-/-}$ mice [369]. The imbalance in the dNTP pool in SAMHD1 deficiency hampered SHM by influencing the recruitment or activity of the translesion (TLS) DNA polymerases responsible for SHM.

A recent finding indicates that DNA polymerase theta (POLQ) introduces nucleotide insertions at the DNA break site [375,376]. As expected, POLQ deficiency decreased insertions at the CSR junctions and increased the frequency of IgH/c-Myc translocation, implying that the function of DNA polymerases affecting end-joining is modulated by cellular dNTP pool imbalance. Intriguingly, the ectopic expression of terminal deoxynucleotidyl transferase (TdT) in CH12F3-2A cells, which do not express TdT, elevated base insertion and decreased CSR with increasing purine nucleotide pool [369]. Since TdT plays a crucial role in V(D)J recombination by N-nucleotide insertion [377], loss of SAMHD1 may also impact V(D)J joining. Both V(D)J recombination and CSR require NHEJ, and NHEJ relies on effective DNA end-processing. NHEJ activity is highest during the G1 phase of the cell cycle, where dNTP concentrations are lowest due to low ribonucleotide reductase (RNR) and high SAMHD1 activities [369].

Importantly, V(D)J recombination, SHM and CSR are mainly limited to the G1 phase of the cell cycle, indicating that antibody gene diversification has evolved to function at low dNTP concentrations [378]. In conclusion, cellular dNTP balance is essential, and its dysregulation can affect the activity of various dNTP-sensitive enzymes involved in DNA repair and recombination, which are crucial for antibody diversity and chromosomal stability.

5. THE REGULATION OF AID EXPRESSION

5.1 AID's Expression Profile

5.1.1 AID Expression in Normal Tissues

AID is expressed in activated B cells in GCs and in the extrafollicular regions of the secondary lymphoid organ where SHM and CSR take place [12,22,379]. In resting mature B cells, AID expression is virtually undetectable, but low levels can be detected in immature and pro-/pre- B cells [12,380–384]. The expression was observed in the autoreactive cellular fraction of immature cells, and it is involved in removal of these cells by apoptosis, thereby contributing to central tolerance [385].

AID expression is not absolutely restricted to B cells. It has been observed in T lymphocytes, and even in non-lymphoid tissues, such as in the ovary and testis [382,386,387]. AID's physiological roles in non-B cells are unknown. It has been proposed to be involved in epigenetic gene regulation through its ability to deaminate 5 methyl-cytosine on the DNA [388,389]. However, recent reports have not supported the notion that AID has a major role in demethylation of DNA in normal tissues [390,391].

5.1.2 AID Expression in Pathological Conditions

The majority of GC-derived human B-cell lymphomas, such as diffuse large B-cell lymphoma (DLBCL), follicular B-cell lymphoma and Burkitt lymphoma, constitutively express AID [22,392–396], and some of non-GC-derived tumours also express it [397–401]. AID expression in B-cell tumours is required for ongoing SHM in the *IgV* gene or in oncogenes, although the levels of AID in these systems do not always correlate with SHM [396,399,401,402]. The gross mutation efficiency may be influenced by other factors, such as the transcription efficiency of the target gene. The actual impact of AID expression in human lymphoid tumours has been strongly suggested by extensive mutation analysis with deep sequencing. The feature of mutations, namely mutation signatures, in 10 human DLBCL primary tumours have been well matched with those by AID-dependent mutagenesis. Especially, a cluster of mutations, called kataegis, in DLBCL are likely caused by AID because the features of mutations are well matched to those of AID-mediated mutations, such as the preference of target gene selection, the requirement of ongoing transcription and base substitution patterns [1]. Since AID expression in tumours is related to the mutator phenotype, it is conceivable to expect a correlation with the malignant potential of tumours. In fact, some studies have reported that AID expression correlated with a poor prognosis in several human B-cell lymphomas and leukaemias [396,399,401–407]. In this context, the BCR-ABL1 kinase encoded by the Philadelphia chromosome, which is a hallmark of acute lymphatic leukaemia (ALL), can induce AID expression, and the prognosis is worse for those cases of ALL that express AID [408–410].

The AID expression has also been reported in non-lymphoid tumours, such as epithelial breast cancer, hepatoma, squamous cell carcinoma, renal cell carcinoma and bladder cell carcinoma [411–415]. Involvement of AID in occurrence and/or progression of such non-lymphoid malignancies has also been proposed.

Some pathogens induce AID expression in B cells and non-B cells. The Epstein–Barr virus (EBV) infection induces AID expression by two ways: direct binding of EBV nuclear antigen 3C (EBNA3C) to the regulatory region of the AID gene (*Aicda*) and mimicking of continuous CD40 signalling by another EBV protein, latent membrane protein 1 [416,417]. Infection with the hepatitis C virus as well as transfection of the viral genome induces AID in B lymphocytes and hepatocytes, respectively [418,419]. AID is also expressed in some adult T-cell leukaemia cells after HTLV-1 infection [420]. Infection with mouse retroviruses, such as Abelson murine leukaemia virus and Moloney murine leukaemia virus, can induce AID expression in mouse B cells [421,422]. Bacteria can also induce AID expression; for example, *Helicobacter pylori* induces AID expression in gastric epithelium [423]. Induction of AID expression can be due to the direct effect of a product made by a pathogen, or it can be an indirect effect mediated by cytokines, such as TNF-α produced during inflammation [419]. The induction of AID expression by pathogens has been discussed in two contexts: its antiviral activity and involvement in cancer development. Many pathogens that induce AID expression are accompanied by tumours, leading to the hypothesis that AID plays a substantial role in tumourigenesis. However, an experiment with animal models of HCV or HPV oncogenesis has not supported this idea, as they revealed that AID is unnecessary for virus-mediated liver and skin tumour development [424]. Inflammatory tissue infiltrating activated B cells are likely to be the major AID-expressing cells in the tumour tissues, and the actual contribution of AID to the inflammation-associated tumourigenesis should be carefully evaluated.

5.2 *Aicda* Regulation

Because of its potential genotoxic activity, AID expression must be tightly regulated to maintain genomic integrity in both B cells and non-B cells [27]. It is especially critical to regulate AID's expression under the control of B cell and activation-specific promoters, to avoid its misfiring by accidental signalling.

5.2.1 Signals and Transcription Factors Involved in AID Expression

The expression of the AID gene (*Aicda*) depends on proper outside-in signals introduced upon the activation of B cells. Although BCR signalling is essential for specific immune responses in which CSR and SHM play important roles, the signal alone cannot induce appreciable AID expression. The BCR signal cooperatively induces *Aicda* expression with other stimuli from a receptor, such as a member of the TNF-receptor superfamily and TLRs that are featuring NF-κB activation [152,425]. Both canonical (p65) and non-canonical (p52) NF-κB pathways are critical for normal *Aicda* expression [425–427].

Signals from CD40 and TACI/BAFFR, members of the TNF-receptor superfamily whose ligands are CD154 and BAFF/APRIL, respectively, activate NF-kB pathways. They synergise with TLR and BCR signals, thereby enhancing *Aicda* expression [428,429]. The mechanism by which these signals cooperate may not be straightforward because BCR engagement can induce multiple signalling pathways, including an NF-kB pathway. One possibility is that the BCR may not activate enough NF-kB, either quantitively or qualitatively, for AID induction. An alternative possibility has been proposed from knockout experiments of TNF-alpha receptor-associated factor (TRAF) 2 and TRAF3, which are signalling adaptor proteins of TNF-receptor family. Interestingly, TRAF2 and/or TRAF3-deficient B cells can induce CSR by the engagement of their BCR without co-stimulation, suggesting inhibitory role of these TRAF proteins to the signal from BCR towards *Aicda* expression [430]. The degradation of TRAF proteins upon co-stimulation has been reported, suggesting that the inhibitory effects are removed, thereby allowing AID induction [431].

TLRs are pattern-recognition receptors that detect and bind microbial components [432]. The engagement of TLRs with their ligands elicits signal transduction mediated by adaptor proteins, including MYD88, TIRAP, TRIF and TRAM, thereby activating TFs such as NF-κB, p38 and IRFs. TLR signalling can induce *Aicda* expression, but it is synergistically enhanced when BCR or cytokine signals are provided [425]. Because many TLR ligands are constituents of the bacterial outer cell wall and membrane, B-cell clones whose BCRs are reactive to bacteria should frequently receive signals mediated by both BCR and TLR signals.

Basic leucine zipper transcription factors, ATF-like (BATF), E2A and interferon regulatory factor 4 (IRF4) are critical for *Aicda* expression [433–435]. These TFs have crucial roles in the development of certain lymphocyte subsets as well as the normal response during immune response [436–438]. Because these TFs involve in many phases of lymphocyte development, it is difficult to determine the signals that lead to the activation of BATF, E2A or IRF4 for *Aicda* induction from surface receptors. It is known that IRF4 expression is upregulated by NF-κB signalling [439], and that E2F activity depends on the balance between E2A and its antagonists, ID proteins [440].

Cytokines are another important component in regulating AID expression. Interleukin-4 (IL-4) directs B cells to switch their Ig class to IgG4 and IgE in humans, and to IgG1 and IgE in the mouse. IL-4R signalling efficiently induces *Aicda* in combination with TLR- or CD40-signalling [152,441,442]. The binding of IL-4 to its receptor activates multiple signal transduction molecules, including STAT6, IRS-2, SHC, SHP-1 and SHIP. STAT6 is important in activating IL-4-induced gene expression, including *Aicda* expression [443,444]. TGF-β critically contributes to inducing IgA in mucosae. It also induces AID expression in B-cell and non-B-cell tumour cell lines [441]. The transcription factor, SMAD is involved in TGF-β signalling [445]. Oestrogen is another humoral factor that enhances *Aicda* expression. It induces *Aicda* indirectly by upregulating HOXC4, and the oestrogen receptor is also reported to bind directly to the *Aicda* promoter [446,447].

5.2.2 Cis-Regulatory Elements of Aicda

Recent advances in massive sequencing technology have provided knowledge of the genome-wide cis-regulatory network of genes. The *Aicda* gene region has also been extensively examined by next-generation sequencer-based strategies. The gene regulatory domain that *Aicda* belongs to spans approximately 90 Kb of the mouse genome, as determined by chromatin interaction analysis using paired-end-tag sequencing (ChIA-PET), which enables the detection of RNA polII-mediated associations between promoters and enhancers [448]. The domain includes three genes (*Aicda*, *Apobec1* and a part of *Mfap5* genes), and six intergenic/intronic DNA regions, designated as E1−E6, which were defined by the ChIA-PET study [448]. These DNA regions found by the new technology include four major cis-regulatory regions that had been defined as evolutionarily conserved previously [449]. The relative positions of these elements are shown in Figure 13.8. Deposition of H3K27Ac, an epigenetic marker of transcription enhancer, to E1−E6 regions is also observed, strongly supporting the idea that these DNA regions involve in *Aicda* regulation [1]. The contribution of these elements to *Aicda* regulation has been strongly supported by evidence at different levels, except for E6. Deletion experiments of E1−E5 regions have been performed either in the endogenous *Aicda* locus of CSR-competent mouse lymphoma cell line CH12F3, (for E1 and E2) or in a bacterial artificial chromosome (BAC) carrying whole mouse *Aicda* locus, which was used to generate transgenics mice (for E3, E4 and E5), and every study supported the importance of these regions [381,448,450]. Besides, in terms of E3, E4 and E5, sequence conservation of inter-species, clustering of transcription factor motifs, as well as direct assay of classical enhancer activities have provided solid evidence that these elements are important for normal *Aicda* regulation [426]. In contrast, the involvement of E6, which exists in the intron of next neighbour gene *Apocec1*, located approximately 40 kb downstream of the *Aicda* gene locus, has not been clearly demonstrated. In the human genome, the relative position of *Aicda* and *Apobec1* is not syntenic with that in the mouse; it is almost 1 Mbp away, presumably due to an inversion event that happened after the segregation of mouse and human. Therefore, even if E6 would have a role in *Aicda* regulation, it could be a mouse-specific phenomenon.

5.2.3 Feature of Aicda Promoter

Luciferase assays in both B and non-B cells revealed that the *Aicda* promoter alone exhibits basic activity [426,451]. This promoter contains well-conserved binding motifs for OCT1 and/or OCT2 (5′-ATTTGAAT-3′), which include the HOXC4-binding motif (5′-ATTT-3′). Additionally, it has NF-κB (non-canonical) and SP1/SP3-binding motifs [427]. HOXC4 is a homeodomain transcription factor that is expressed in B cells and is upregulated by CD40 signalling. AID expression is impaired in HOXC4-knockout mice, indicating that HOXC4 upregulation contributes to AID's induction in activated B cells [427]. HOXC4 was reported to play a role in oestrogen-mediated AID expression enhancement [446,447]. The promoter region per se is not sufficient to achieve physiological regulation of *Aicda*.

5.2.4 Binding of TFs to each Enhancer

E3, E4 and E5 were originally designated as region-4, -2 and -3, respectively, and had been characterised by

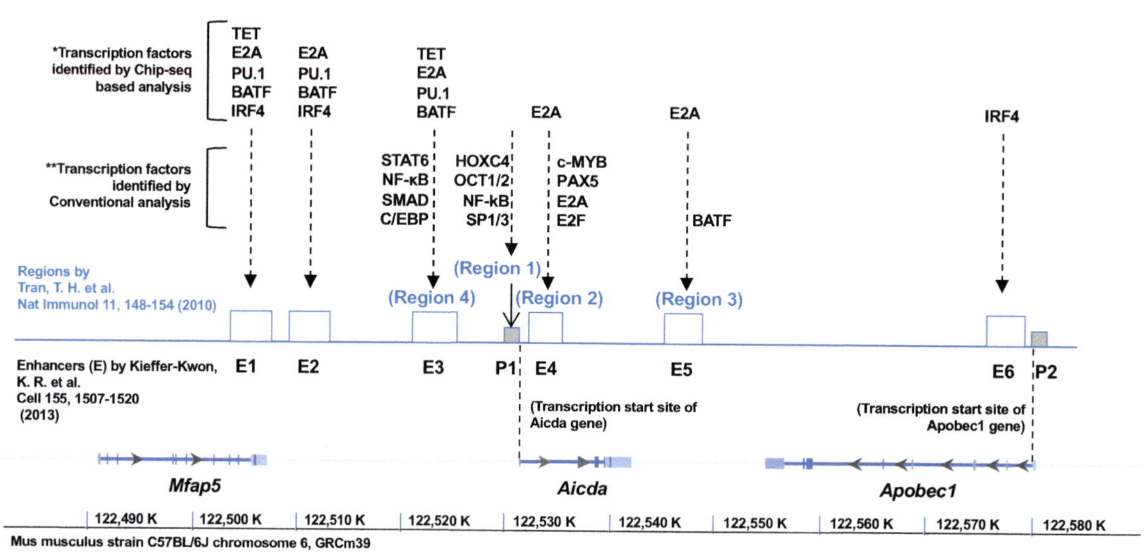

FIGURE 13.8 **Regulatory elements of *Aicda* gene and the transcription factors binding to them.** * binding peak found in Chip-seq analysis. Lio et al. [453]; Ochiai et al. [470]; Wöhner et al. [452] ** binding motifs found in conventional analysis. Yadav et al. [451]; Tran et al. [426]; Park et al. [427]; Huong et al. [450]; Ise et al. [433]; Gonda et al. [454]; Sayegh et al. [434].

conventional strategies such as reporter assay or electrophoretic mobility shift assays. The latest strategies such as ChIP-seq have updated our knowledge on this issue significantly. Apparently, many TFs bind to multiple enhancers that are separated from each other and associate with a topological structure formation (Figure 13.8). Remarkably, E2A appears to bind from E1 through E5 [452,453]. E-proteins are class I bHLH proteins that consist of E2A, E2-2 and HEB [436]. E2A and E2-2 positively regulate *Aicda* expression, and ID2 or ID3 substantially represses *Aicda* expression [452,454]. Both BATF and IRF4 play critical roles in Aicda expression and have been shown to bind to multiple positions, specifically E1, E2 and E3 for BATF and E1 and E2 for IRF4. As IRF4 forms heterocomplex with BATF, it is plausible to expect the IRF4-BATF complex formation on these elements and directly regulates *Aicda* expression. PU.1, an ETS family transcription factor, activates many genes during B-lymphoid and myeloid cell development and can form a heterodimer with IRF4 to regulate a different set of genes. Thus, co-localisation of these TFs would be suggestive, but despite its potential importance, the major contribution of PU.1 protein in *Aicda* regulation has not been clearly demonstrated so far.

The binding motifs of TFs in E3 and E4 have been well characterised by conventional reporter assay both in vivo and in vitro [426,450]. In addition to the protein bindings detected by ChIP-seqs, as described above, E3 contains binding motifs for STAT6, NF-κB and SMAD proteins, which are downstream regulators for IL-4, CD40L and TGF-β stimulation, respectively. The element shows an enhancer activity reactive to these stimulations in reporter assays. In addition, tandem C/EBP-binding sites found in this element are required for the enhancer activity [426,450]. E4 contains binding sites for MYB, Pax5, E-proteins and E2F. The E2F and c-MYB-binding motifs in E4 show silencer activity to the promoter [426,450]. Although significance of the silencing activity is still unclear, it may contribute to avoid ectopic expression.

Other TFs, including BACH2, BCL6 and Blimp1, have also been reported to play important roles in the normal regulation of AID expression. However, factors seem to regulate *Aicda* via indirect mechanisms [455–458].

5.2.5 Epigenetic Regulation of Aicda Gene

The CpG sites in the promoter region of *Aicda* in immature/naïve B cells are heavily methylated. Upon stimulation that induces AID expression, a demethylation process of the promoter takes place [459]. The initial step of the demethylation is catalysed by ten-eleven-translocation (TET) proteins, which oxidates 5-methylcytosine and a DNA repair process replaces it with non-methylated cytosine. TET2 and TET3 are recruited to E1 and E3 regions of *Aicda* to initiate this process, and the recruitment is BATF protein-dependent. AID expression is significantly impaired in *Tet2/3* double knockout B cells, indicates pivotal role of TETs in AID induction [453].

The histone modifications are also associated with *Aicda* regulation. In activated B cells, a cluster of H3K27Ac deposition appears in and around the *Aicda*, spanning from E1 through E6, which is referred to as the B cell super-enhancer. H3K36me3 histone modification on the gene body indicates ongoing active transcription. Depletion of the major H3K36me3 methyltransferase significantly down-regulates AID expression, and the histone chaperone SPT6 is involved in the regulation [281]. These epigenetic mechanisms may serve as a safeguard against AID, a genotoxic molecule.

5.3 Post-transcriptional and Translational Regulations

5.3.1 MicroRNAs

MicroRNAs (miRs) are short, single-stranded noncoding RNAs that suppress the gene expression of a specific target. Several microRNAs, including miR-155, miR-181b, miR-361, miR-93, miR125a and miR-26a, regulate AID expression by binding to evolutionarily conserved sites in the 3′UTR of AID mRNA [455,460–464]. The miR-mediated translation suppression of AID has been reported in both activated B and resting B cells. The expression of miR155 is very low in resting B cells but increases along with AID expression in mouse B cells stimulated with LPS and IL-4; miR155 appears to counteract increased AID activity in these cells [460]. In contrast, miR181b is expressed in resting B cells, but its expression is decreased by LPS and IL-4 stimulation [462]. BCL6, a transcription repressor essential for forming GC B cells, represses miR-155 and miR-361 expression, and thereby contributes to inducing AID in GC B cells [455]. MicroRNA also mediates fine tuning of immune system relating with gut flora. Short-chain fatty acids that are metabolite of gut microbiota inhibit histone deacetylases to induce the expression of various genes, including miR-155, miR-26a and miR-125a, resulting in down-regulation of AID [464].

5.3.2 Phosphorylation

AID is phosphorylated at serine 3, threonine 27, serine 38, threonine 140 and tyrosine 184 [63,85,465–467]. The roles of these phosphorylations are not fully understood, although many of them influence AID's activity. The phosphorylation of S38 and T140 increases AID's activity [84,85]. Phosphorylation at S38 is thought to facilitate

AID's interaction with chromatin, possibly by binding to RPA [62,85]. In contrast, S3 phosphorylation is reported to reduce AID's activity [467]. Both protein kinase C and cyclic AMP-dependent protein kinase A can phosphorylate AID [62,85,466].

5.3.3 *Regulation of Protein Stability*

AID is a nucleocytoplasmic shuttling protein [32,36]. Nuclear AID undergoes rapid degradation, and two degradation mechanisms have been reported: the polyubiquitination and proteasome pathway, and a ubiquitin- and ATP-independent but proteasome-dependent degradation pathway [28,468]. The ubiquitin-dependent degradation is mediated by the RING-domain E3 enzyme, CUL7. In fact, AID expression increases in CUL7 knockout B cells, and the efficiency of CSR goes up [469]. The ubiquitin-independent pathway relies on the nuclear protein RDG-γ. Class switching upon stimulation with LPS and IL-4 was increased in REG-γ-deficient B cells [28]. Given the potential genotoxic activity of nuclear AID, these dual rapid-degradation mechanisms may represent another level of control of AID activity and could be a good target for manipulation of AID activity.

6. CONCLUSIONS

6.1 Concluding Remark

Although the exact molecular mechanisms of AID's functions have been extensively debated, there are several points of consensus regarding experimental observations that have yet to be explained by any model of AID function. The first point, which is critical, is that AID has at least two functions, one for both SHM and CSR, and the other only for CSR. Since there are no AID mutants that abolish SHM without affecting CSR, the common AID function in SHM and CSR is probably DNA cleavage, and it is unlikely that the mechanism of that cleavage is distinct for SHM and CSR. AID's CSR-specific function is probably associated with recombination.

The second important point is that AID cannot itself specify the target to be cleaved. There are several biochemical events that may be involved in specifying the AID target, including transcription of the locus and unique DNA structures such as R-loops, non-B DNA structures and epigenetic marks. To recognise these modifications of the locus, AID obviously requires cofactors. The DNA deamination model proposes that cofactors recruit AID to the marked loci, meaning that a complex of cofactors must interact with AID, locus-specific transcripts, DNA structure and/or epigenetic marks. In contrast, the RNA-editing model proposes that cofactors capture target RNA, which is edited by AID or by part of the RNA-editosome complex. The edited RNA is assumed to generate conditions that facilitate TOP1's action in SHM and CSR. The critical difference between AID and TOP1 targeting is that AID must be recruited to the target locus by cofactors, while TOP1 is already present, because it is required for general purposes such as transcription and splicing. Given that global TOP1 reduction by siRNA elevates SHM and CSR, minimal TOP1 must ensure target-specific DNA cleavage, especially when the conditions trigger excessive substrate availability (negative supercoil-induced non-B structures). It can be speculated that only a fraction of the TOP1 available is normally involved in the context of CSR or SHM. For instance, TOP1 may associate with the FACT complex or an as-yet unknown modulator to gain localised and controlled access to target DNA at the H3K4me3-marked chromatin. It is likely that multiple chromatin factors are involved in forming global or local TOP1 cleavage complexes, for general-purpose DNA transactions as well as specialised CSR or SHM functions. In addition, in cooperation with histone chaperones, TOP1 has a tremendous influence on the density of the DNA supercoil along the transcribed target.

Third, SHM mostly depends on error-prone DNA synthesis. In this context, there must be a mechanism to suppress the error-free mechanisms in B cells, since most DNA damages are correctly repaired in non-B cells. A recent report that UNG suppresses TLP [154] provided a clue toward solving this puzzle, at least in part.

Finally, it is important to recognise that biological mechanisms must evolve from related mechanisms. So far, there are no other established DNA deamination systems with important physiological functions. On the other hand, the RNA-editing mechanism is well established by cytidine deamination by APOBEC1 family members, and by adenosine deamination by ADAR I and II. An evolutionary consideration of the molecular mechanism of AID may be critical in resolving the controversies surrounding this important molecule.

REFERENCES

[1] Qian J, et al. B cell super-enhancers and regulatory clusters recruit AID tumorigenic activity. Cell 2014;159(7):1524–37.

[2] Montefiori L, et al. Extremely long-range chromatin loops link topological domains to facilitate a diverse antibody repertoire. Cell Rep 2016;14(4):896–906.

[3] Senigl F, et al. Topologically associated domains delineate susceptibility to somatic hypermutation. Cell Rep 2019;29 (12):3902–15.

[4] Rothschild G, et al. Noncoding RNA transcription alters chromosomal topology to promote isotype-specific class switch recombination. Sci Immunol 2020;5(44):eaay5864.

[5] Boya R, et al. Developmentally regulated higher-order chromatin interactions orchestrate B cell fate commitment. Nucleic Acids Res 2017;45(19):11070–87.

[6] Perino M, Veenstra GJC. Chromatin control of developmental dynamics and plasticity. Dev Cell 2016;38(6):610–20.
[7] Kurotaki D, et al. Chromatin structure undergoes global and local reorganization during murine dendritic cell development and activation. Proc Natl Acad Sci USA 2022;119(34).
[8] Kong Q, Maizels N. DNA breaks in hypermutating immunoglobulin genes: evidence for a break-and-repair pathway of somatic hypermutation. Genetics 2001;158(1):369–78.
[9] Faili A, et al. Induction of somatic hypermutation in immunoglobulin genes is dependent on DNA polymerase iota. Nature 2002;419(6910):944–7.
[10] Kinoshita K, et al. Target specificity of immunoglobulin class switch recombination is not determined by nucleotide sequences of S regions. Immunity 1998;9(6):849–58.
[11] Weill JC, Reynaud CA. Rearrangement/hypermutation/gene conversion: when, where and why? Immunol Today 1996;17(2):92–7.
[12] Muramatsu M, et al. Class switch recombination and hypermutation require activation-induced cytidine deaminase (AID), a potential RNA editing enzyme. Cell 2000;102(5):553–63.
[13] Revy P, et al. Activation-induced cytidine deaminase (AID) deficiency causes the autosomal recessive form of the Hyper-IgM syndrome (HIGM2). Cell 2000;102(5):565–75.
[14] Honjo T, Muramatsu M, Fagarasan S. AID: how does it aid antibody diversity? Immunity 2004;20(6):659–68.
[15] Di Noia J, Neuberger MS. Altering the pathway of immunoglobulin hypermutation by inhibiting uracil-DNA glycosylase. Nature 2002;419(6902):43–8.
[16] Petersen-Mahrt SK, Harris RS, Neuberger MS. AID mutates *E. coli* suggesting a DNA deamination mechanism for antibody diversification. Nature 2002;418(6893):99–104.
[17] Liu M, et al. Two levels of protection for the B cell genome during somatic hypermutation. Nature 2008;451(7180):841–5.
[18] Kato L, et al. Nonimmunoglobulin target loci of activation-induced cytidine deaminase (AID) share unique features with immunoglobulin genes. Proc Natl Acad Sci USA 2012;109(7):2479–84.
[19] Hasham MG, et al. Activation-induced cytidine deaminase-initiated off-target DNA breaks are detected and resolved during S phase. J Immunol 2012;189(5):2374–82.
[20] Begum NA, Honjo T. Evolutionary comparison of the mechanism of DNA cleavage with respect to immune diversity and genomic instability. Biochemistry 2012;51(26):5243–56.
[21] Alt FW, et al. Mechanisms of programmed DNA lesions and genomic instability in the immune system. Cell 2013;152(3):417–29.
[22] Muto T, et al. Isolation, tissue distribution, and chromosomal localization of the human activation-induced cytidine deaminase (AID) gene. Genomics 2000;68(1):85–8.
[23] Ta VT, et al. AID mutant analyses indicate requirement for class-switch-specific cofactors. Nat Immunol 2003;4(9):843–8.
[24] Doi T, et al. The C-terminal region of activation-induced cytidine deaminase is responsible for a recombination function other than DNA cleavage in class switch recombination. Proc Natl Acad Sci USA 2009;106(8):2758–63.
[25] Shinkura R, et al. Separate domains of AID are required for somatic hypermutation and class-switch recombination. Nat Immunol 2004;5(7):707–12.
[26] Kato L, et al. An evolutionary view of the mechanism for immune and genome diversity. J Immunol 2012;188(8):3559–66.
[27] Honjo T, et al. The AID dilemma: infection, or cancer? Adv Cancer Res 2012;113:1–44.
[28] Uchimura Y, et al. REG-γ associates with and modulates the abundance of nuclear activation-induced deaminase. J Exp Med 2011;208(12):2385–91.
[29] Robbiani DF, Nussenzweig MC. Chromosome translocation, B cell lymphoma, and activation-induced cytidine deaminase. Annu Rev Pathol 2013;8:79–103.
[30] Durandy A. Hyper-IgM syndromes: a model for studying the regulation of class switch recombination and somatic hypermutation generation. Biochem Soc Trans 2002;30(4):815–8.
[31] Durandy A, et al. Activation-induced cytidine deaminase: structure-function relationship as based on the study of mutants. Hum Mutat 2006;27(12):1185–91.
[32] Ito S, et al. Activation-induced cytidine deaminase shuttles between nucleus and cytoplasm like apolipoprotein B mRNA editing catalytic polypeptide 1. Proc Natl Acad Sci USA 2004;101(7):1975–80.
[33] Shivarov V, Shinkura R, Honjo T. Dissociation of in vitro DNA deamination activity and physiological functions of AID mutants. Proc Natl Acad Sci USA 2008;105(41):15866–71.
[34] Wei M, et al. Mice carrying a knock-in mutation of Aicda resulting in a defect in somatic hypermutation have impaired gut homeostasis and compromised mucosal defense. Nat Immunol 2011;12(3):264–70.
[35] Barreto V, et al. C-terminal deletion of AID uncouples class switch recombination from somatic hypermutation and gene conversion. Mol Cell 2003;12(2):501–8.
[36] McBride KM, et al. Somatic hypermutation is limited by CRM1-dependent nuclear export of activation-induced deaminase. J Exp Med 2004;199(9):1235–44.
[37] Sabouri S, et al. C-terminal region of activation-induced cytidine deaminase (AID) is required for efficient class switch recombination and gene conversion. Proc Natl Acad Sci USA 2014;111(6):2253–8.
[38] Boboila C, Alt FW, Schwer B. Classical and alternative end-joining pathways for repair of lymphocyte-specific and general DNA double-strand breaks. Adv Immunol 2012;116:1–49.
[39] Yan CT, et al. IgH class switching and translocations use a robust non-classical end-joining pathway. Nature 2007;449(7161):478–82.
[40] Al Ismail A, et al. Depletion of recombination-specific cofactors by the C-terminal mutant of the activation-induced cytidine deaminase causes the dominant negative effect on class switch recombination. Int Immunol 2017;29(11):525–37.
[41] Zahn A, et al. Activation induced deaminase C-terminal domain links DNA breaks to end protection and repair during class switch recombination. Proc Natl Acad Sci USA 2014;111(11):E988–97.
[42] Qiao Q, et al. AID recognizes structured DNA for class switch AID recognizes structured DNA for class switch recombination. Mol Cell 2017;67(3):361–73.
[43] Kudo N, et al. Leptomycin B inhibition of signal-mediated nuclear export by direct binding to CRM1. Exp Cell Res 1998;242(2):540–7.
[44] Patenaude AM, et al. Active nuclear import and cytoplasmic retention of activation-induced deaminase. Nat Struct Mol Biol 2009;16(5):517–27.

[45] Hu Y, et al. A combined nuclear and nucleolar localization motif in activation-induced cytidine deaminase (AID) controls immunoglobulin class switching. J Mol Biol 2013;425(2):424–43.
[46] Ellyard JI, et al. The dependence of Ig class-switching on the nuclear export sequence of AID likely reflects interaction with factors additional to Crm1 exportin. Eur J Immunol 2011;41(2):485–90.
[47] Conticello SG, et al. Evolution of the AID/APOBEC family of polynucleotide (deoxy)cytidine deaminases. Mol Biol Evol 2005;22(2):367–77.
[48] Costanzi S, et al. Human cytidine deaminase: a three-dimensional homology model of a tetrameric metallo-enzyme inferred from the crystal structure of a distantly related dimeric homologue. J Mol Graph Model 2006;25(1):10–6.
[49] Johansson E, et al. Crystal structure of the tetrameric cytidine deaminase from Bacillus subtilis at 2.0 A resolution. Biochemistry 2002;41(8):2563–70.
[50] Dickerson SK, et al. AID mediates hypermutation by deaminating single stranded DNA. J Exp Med 2003;197(10):1291–6.
[51] Larijani M, et al. AID associates with single-stranded DNA with high affinity and a long complex half-life in a sequence-independent manner. Mol Cell Biol 2007;27(1):20–30.
[52] Brar SS, et al. Activation-induced deaminase, AID, is catalytically active as a monomer on single-stranded DNA. DNA Repair 2008;7(1):77–87.
[53] Kerppola TK. Bimolecular fluorescence complementation (BiFC) analysis as a probe of protein interactions in living cells. Annu Rev Biophys 2008;37:465–87.
[54] Mondal S, et al. Functional requirements of AID's higher order structures and their interaction with RNA-binding proteins. Proc Natl Acad Sci USA 2016;113(11):E1545–54.
[55] Chaudhuri J, et al. Transcription-targeted DNA deamination by the AID antibody diversification enzyme. Nature 2003;422(6933):726–30.
[56] Okazaki IM, et al. Histone chaperone Spt6 is required for class switch recombination but not somatic hypermutation. Proc Natl Acad Sci USA 2011;108(19):7920–5.
[57] Hasler J, Rada C, Neuberger MS. Cytoplasmic activation-induced cytidine deaminase (AID) exists in stoichiometric complex with translation elongation factor 1alpha (eEF1A). Proc Natl Acad Sci USA 2011;108(45):18366–71.
[58] Willmann KL, et al. A role for the RNA pol II-associated PAF complex in AID-induced immune diversification. J Exp Med 2012;209(11):2099–111.
[59] Nambu Y, et al. Transcription-coupled events associating with immunoglobulin switch region chromatin. Science 2003;302 (5653):2137–40.
[60] Jeevan-Raj BP, et al. Epigenetic tethering of AID to the donor switch region during immunoglobulin class switch recombination. J Exp Med 2011;208(8):1649–60.
[61] Pavri R, et al. Activation-induced cytidine deaminase targets DNA at sites of RNA polymerase II stalling by interaction with Spt5. Cell 2010;143(1):122–33.
[62] Basu U, et al. The AID antibody diversification enzyme is regulated by protein kinase A phosphorylation. Nature 2005;438 (7067):508–11.
[63] Chaudhuri J, Khuong C, Alt FW. Replication protein A interacts with AID to promote deamination of somatic hypermutation targets. Nature 2004;430(7003):992–8.
[64] Lam T, et al. Scaffold functions of 14-3-3 adaptors in B cell immunoglobulin class switch DNA recombination. PLoS One 2013;8(11):e80414.
[65] Xu Z, et al. 14-3-3 adaptor proteins recruit AID to 5'-AGCT-3'-rich switch regions for class switch recombination. Nat Struct Mol Biol 2010;17(9):1124–35.
[66] Conticello SG, et al. Interaction between antibody-diversification enzyme AID and spliceosome-associated factor CTNNBL1. Mol Cell 2008;31(4):474–84.
[67] Ganesh K, et al. CTNNBL1 is a novel nuclear localization sequence-binding protein that recognizes RNA-splicing factors CDC5L and Prp31. J Biol Chem 2011;286(19):17091–102.
[68] Han L, Masani S, Yu K. Cutting edge: CTNNBL1 is dispensable for Ig class switch recombination. J Immunol 2010;185(3):1379–81.
[69] Nowak U, et al. The splicing regulator PTBP2 interacts with the cytidine deaminase AID and promotes binding of AID to switch-region DNA. Nat Immunol 2011;12(2):160–6.
[70] Basu U, et al. The RNA exosome targets the AID cytidine deaminase to both strands of transcribed duplex DNA substrates. Cell 2011;144(3):353–63.
[71] Hu Y, et al. Activation-induced cytidine deaminase (AID) is localized to subnuclear domains enriched in splicing factors. Exp Cell Res 2014;322:178–92.
[72] Ribeiro de Almeida C, et al. RNA helicase DDX1 converts RNA G-quadruplex structures into R-loops to promote IgH class switch recombination. Mol Cell 2018;70(4):650–62.
[73] Dong J, et al. Orientation-specific joining of AID-initiated DNA breaks promotes antibody class switching. Nature 2015;525(7567):134–9.
[74] Fukuda M, et al. CRM1 is responsible for intracellular transport mediated by the nuclear export signal. Nature 1997;390 (6657):308–11.
[75] Maeda K, et al. GANP interacts with APOBEC3G and facilitates its encapsidation into the virions to reduce HIV-1 infectivity. J Immunol 2013;191(12):6030–9.
[76] Arioka Y, et al. Activation-induced cytidine deaminase alters the subcellular localization of Tet family proteins. PLoS One 2012;7(9):e45031.
[77] Orthwein A, et al. Regulation of activation-induced deaminase stability and antibody gene diversification by Hsp90. J Exp Med 2010;207(12):2751–65.
[78] Orthwein A, et al. Optimal functional levels of activation-induced deaminase specifically require the Hsp40 DnaJa1. EMBO J 2012;31(3):679–91.
[79] Delker RK, et al. Solubility-based genetic screen identifies RING finger protein 126 as an E3 ligase for activation-induced cytidine deaminase. Proc Natl Acad Sci USA 2013;110(3):1029–34.
[80] MacDuff DA, Neuberger MS, Harris RS. MDM2 can interact with the C-terminus of AID but it is inessential for antibody diversification in DT40 B cells. Mol Immunol 2006;43(8):1099–108.
[81] Cook AJ, et al. Reduced switching in SCID B cells is associated with altered somatic mutation of recombined S regions. J Immunol 2003;171(12):6556–64.
[82] Wu X, et al. The double-edged sword of activation-induced cytidine deaminase. J Immunol 2005;174(2):934–41.
[83] Vuong BQ, et al. A DNA break- and phosphorylation-dependent positive feedback loop promotes immunoglobulin class-switch recombination. Nat Immunol 2013;14(11):1183–9.

[84] Cheng HL, et al. Integrity of the AID serine-38 phosphorylation site is critical for class switch recombination and somatic hypermutation in mice. Proc Natl Acad Sci USA 2009;106(8):2717–22.
[85] McBride KM, et al. Regulation of class switch recombination and somatic mutation by AID phosphorylation. J Exp Med 2008;205(11):2585–94.
[86] Vuong BQ, et al. Specific recruitment of protein kinase A to the immunoglobulin locus regulates class-switch recombination. Nat Immunol 2009;10(4):420–6.
[87] Sun J, et al. E3-ubiquitin ligase Nedd4 determines the fate of AID-associated RNA polymerase II in B cells. Genes Dev 2013;27(16):1821–33.
[88] Yamane A, et al. Deep-sequencing identification of the genomic targets of the cytidine deaminase AID and its cofactor RPA in B lymphocytes. Nat Immunol 2011;12(1):62–9.
[89] Besmer E, Market E, Papavasiliou FN. The transcription elongation complex directs activation-induced cytidine deaminase-mediated DNA deamination. Mol Cell Biol 2006;26(11):4378–85.
[90] Kodgire P, et al. Changes in RNA polymerase II progression influence somatic hypermutation of Ig-related genes by AID. J Exp Med 2013;210(7):1481–92.
[91] Rush JS, Fugmann SD, Schatz DG. Staggered AID-dependent DNA double strand breaks are the predominant DNA lesions targeted to S mu in Ig class switch recombination. Int Immunol 2004;16(4):549–57.
[92] Wang L, et al. S region sequence, RNA polymerase II, and histone modifications create chromatin accessibility during class switch recombination. J Exp Med 2009;206(8):1817–30.
[93] Li G, et al. Combinatorial H3K9acS10ph histone modification in IgH locus S regions targets 14-3-3 adaptors and AID to specify antibody class-switch DNA recombination. Cell Rep 2013;5(3):702–14.
[94] Xu Z, et al. Immunoglobulin class-switch DNA recombination: induction, targeting and beyond. Nat Rev Immunol 2012;12(7):517–31.
[95] Begum NA, et al. Requirement of non-canonical activity of uracil DNA glycosylase for class switch recombination. J Biol Chem 2007;282(1):731–42.
[96] Singh SK, et al. GANP regulates recruitment of AID to immunoglobulin variable regions by modulating transcription and nucleosome occupancy. Nat Commun 2013;4:1830.
[97] Nair L, et al. Mechanism of noncoding RNA-associated N(6)-methyladenosine recognition by an RNA processing complex during IgH DNA recombination. Mol Cell 2021;81(19):3949–64.
[98] Zheng S, et al. Non-coding RNA generated following lariat debranching mediates targeting of AID to DNA. Cell 2015;161(4):762–73.
[99] Conticello SG, et al. DNA deamination in immunity: AID in the context of its APOBEC relatives. Adv Immunol 2007;94:37–73.
[100] Li X, et al. Functional analysis of the two cytidine deaminase domains in APOBEC3G. Virology 2011;414(2):130–6.
[101] Navarro F, et al. Complementary function of the two catalytic domains of APOBEC3G. Virology 2005;333(2):374–86.
[102] Harris RS, Petersen-Mahrt SK, Neuberger MS. RNA editing enzyme APOBEC1 and some of its homologs can act as DNA mutators. Mol Cell 2002;10(5):1247–53.
[103] Petersen-Mahrt SK, Neuberger MS. In vitro deamination of cytosine to uracil in single-stranded DNA by apolipoprotein B editing complex catalytic subunit 1 (APOBEC1). J Biol Chem 2003;278(22):19583–6.
[104] Harris RS, Liddament MT. Retroviral restriction by APOBEC proteins. Nat Rev Immunol 2004;4(11):868–77.
[105] Mussil B, et al. Human APOBEC3A isoforms translocate to the nucleus and induce DNA double strand breaks leading to cell stress and death. PLoS One 2013;8(8):e73641.
[106] Belanger K, et al. Binding of RNA by APOBEC3G controls deamination-independent restriction of retroviruses. Nucleic Acids Res 2013;41(15):7438–52.
[107] Ikeda T, et al. Intrinsic restriction activity by apolipoprotein B mRNA editing enzyme APOBEC1 against the mobility of autonomous retrotransposons. Nucleic Acids Res 2011;39(13):5538–54.
[108] Alder MN, et al. Diversity and function of adaptive immune receptors in a jawless vertebrate. Science 2005;310(5756):1970–3.
[109] Rogozin IB, et al. Evolution and diversification of lamprey antigen receptors: evidence for involvement of an AID-APOBEC family cytosine deaminase. Nat Immunol 2007;8(6):647–56.
[110] Pancer Z, et al. Somatic diversification of variable lymphocyte receptors in the agnathan sea lamprey. Nature 2004;430(6996):174–80.
[111] Conticello SG. The AID/APOBEC family of nucleic acid mutators. Genome Biol 2008;9(6):229.
[112] Harris RS, et al. DNA deamination mediates innate immunity to retroviral infection. Cell 2003;113(6):803–9.
[113] Stavnezer-Nordgren J, Sirlin S. Specificity of immunoglobulin heavy chain switch correlates with activity of germline heavy chain genes prior to switching. EMBO J 1986;5(1):95–102.
[114] Lutzker S, Alt FW. Structure and expression of germ line immunoglobulin gamma 2b transcripts. Mol Cell Biol 1988;8(4):1849–52.
[115] Berton MT, Uhr JW, Vitetta ES. Synthesis of germ-line gamma 1 immunoglobulin heavy-chain transcripts in resting B cells: induction by interleukin 4 and inhibition by interferon gamma. Proc Natl Acad Sci USA 1989;86(8):2829–33.
[116] Stavnezer J, Guikema JEJ, Schrader CE. Mechanism and regulation of class switch recombination. Annu Rev Immunol 2008;26:261–92.
[117] Kenter AL, et al. Mapping of a functional recombination motif that defines isotype specificity for mu→gamma3 switch recombination implicates NF-kappaB p50 as the isotype-specific switching factor. J Exp Med 2004;199(5):617–27.
[118] Thomas-Claudepierre AS, et al. Mediator facilitates transcriptional activation and dynamic long-range contacts at the IgH locus during class switch recombination. J Exp Med 2016;213(3):303–12.
[119] Haque F, Honjo T, Begum NA. XLID syndrome gene Med12 promotes Ig isotype switching through chromatin modification and enhancer RNA regulation. Sci Adv 2022;8(47):eadd1466.
[120] Fitz J, et al. Spt5-mediated enhancer transcription directly couples enhancer activation with physical promoter interaction. Nat Genet 2020;52(5):505–15.
[121] Sakano H, et al. Two types of somatic recombination are necessary for the generation of complete immunoglobulin heavy-chain genes. Nature 1980;286(5774):676–83.
[122] Dunnick W, Rabbitts TH, Milstein C. An immunoglobulin deletion mutant with implications for the heavy-chain switch and RNA splicing. Nature 1980;286(5774):669–75.

[123] Davis MM, Kim SK, Hood LE. DNA sequences at immunoglobulin switch region recombination sites. Science 1980;209(4463):1360–5.

[124] Kataoka T, Miyata T, Honjo T. Repetitive sequences in class-switch recombination regions of immunoglobulin heavy chain genes. Cell 1981;23(2):357–68.

[125] Nikaido T, Yamawaki-Kataoka Y, Honjo T. Nucleotide sequences of switch regions of immunoglobulin C epsilon and C gamma genes and their comparison. J Biol Chem 1982;257(13):7322–9.

[126] Stanton LW, Marcu KB. Nucleotide sequence and properties of the murine gamma 3 immunoglobulin heavy chain gene switch region: implications for successive C gamma gene switching. Nucleic Acids Res 1982;10(19):5993–6006.

[127] Dunnick W, et al. DNA sequences at immunoglobulin switch region recombination sites. Nucleic Acids Res 1993;21(3):365–72.

[128] Bacolla A, Wells RD. Non-B DNA conformations, genomic rearrangements, and human disease. J Biol Chem 2004;279(46):47411–4.

[129] Lee CG, et al. Quantitative regulation of class switch recombination by switch region transcription. J Exp Med 2001;194(3):365–73.

[130] Peters A, Storb U. Somatic hypermutation of immunoglobulin genes is linked to transcription initiation. Immunity 1996;4(1):57–65.

[131] Bachl J, et al. Increased transcription levels induce higher mutation rates in a hypermutating cell line. J Immunol 2001;166(8):5051–7.

[132] Yoshikawa K, et al. AID enzyme-induced hypermutation in an actively transcribed gene in fibroblasts. Science 2002;296(5575):2033–6.

[133] Harriman GR, et al. IgA class switch in I alpha exon-deficient mice. Role of germline transcription in class switch recombination. J Clin Invest 1996;97(2):477–85.

[134] Fukita Y, Jacobs H, Rajewsky K. Somatic hypermutation in the heavy chain locus correlates with transcription. Immunity 1998;9(1):105–14.

[135] Hein K, et al. Processing of switch transcripts is required for targeting of antibody class switch recombination. J Exp Med 1998;188(12):2369–74.

[136] Jacobs H, et al. Tuning somatic hypermutation by transcription. Curr Top Microbiol Immunol 1999;246:149–58.

[137] Pavri R, Nussenzweig MC. AID targeting in antibody diversity. Adv Immunol 2011;110:1–26.

[138] Luby TM, et al. The mu switch region tandem repeats are important, but not required, for antibody class switch recombination. J Exp Med 2001;193(2):159–68.

[139] Khamlichi AA, et al. Immunoglobulin class-switch recombination in mice devoid of any S mu tandem repeat. Blood 2004;103(10):3828–36.

[140] Caldecott KW. DNA single-strand break repair and human genetic disease. Trends Cell Biol 2022;32(9):733–45.

[141] Kobayashi M, et al. AID-induced decrease in topoisomerase 1 induces DNA structural alteration and DNA cleavage for class switch recombination. Proc Natl Acad Sci USA 2009;106(52).

[142] Kobayashi M, et al. Decrease in topoisomerase I is responsible for activation-induced cytidine deaminase (AID)-dependent somatic hypermutation. Proc Natl Acad Sci USA 2011;108(48).

[143] Yewdell WT, et al. A hyper-IgM syndrome mutation in activation-induced cytidine deaminase disrupts G-quadruplex binding and genome-wide chromatin localization. Immunity 2020;53(5):952–70.

[144] Chen J, et al. The RNA-binding protein ROD1/PTBP3 cotranscriptionally defines AID-loading sites to mediate antibody class switch in mammalian genomes. Cell Res 2018;28(10):981–95.

[145] Kotnis A, et al. Non-homologous end joining in class switch recombination: the beginning of the end. Philos Trans R Soc Lond B Biol Sci 2009;364(1517):653–65.

[146] Chen X, Kinoshita K, Honjo T. Variable deletion and duplication at recombination junction ends: implication for staggered double-strand cleavage in class-switch recombination. Proc Natl Acad Sci USA 2001;98(24):13860–5.

[147] Lin Y, et al. APE1 senses DNA single-strand breaks for repair and signaling. Nucleic Acids Res 2020;48(4):1925–40.

[148] Masani S, Han L, Yu K. Apurinic/apyrimidinic endonuclease 1 is the essential nuclease during immunoglobulin class switch recombination. Mol Cell Biol 2013;33(7):1468–73.

[149] Xu J, et al. Ape1 is dispensable for s-region cleavage but required for its repair in class switch recombination. Proc Natl Acad Sci USA 2014;111(48).

[150] Islam H, Kobayashi M, Honjo T. Apurinic/apyrimidinic endonuclease 1 (APE1) is dispensable for activation-induced cytidine deaminase (AID)-dependent somatic hypermutation in the immunoglobulin gene. Int Immunol 2019;31(8):543–54.

[151] Guikema JE, et al. APE1- and APE2-dependent DNA breaks in immunoglobulin class switch recombination. J Exp Med 2007;204(12):3017–26.

[152] Schrader CE, et al. Inducible DNA breaks in Ig S regions are dependent on AID and UNG. J Exp Med 2005;202(4):561–8.

[153] Begum NA, et al. Further evidence for involvement of a noncanonical function of uracil DNA glycosylase in class switch recombination. Proc Natl Acad Sci USA 2009;106(8):2752–7.

[154] Yousif AS, et al. Differential regulation of S-region hypermutation and class-switch recombination by noncanonical functions of uracil DNA glycosylase. Proc Natl Acad Sci USA 2014;111(11):E1016–24.

[155] Sabouri Z, et al. Apex2 is required for efficient somatic hypermutation but not for class switch recombination of immunoglobulin genes. Int Immunol 2009;21(8):947–55.

[156] Lähdesmäki A, et al. Delineation of the role of the Mre11 complex in class switch recombination. J Biol Chem 2004;279(16):16479–87.

[157] Dinkelmann M, et al. Multiple functions of MRN in end-joining pathways during isotype class switching. Nat Struct Mol Biol 2009;16(8):808–13.

[158] Reina-San-Martin B, et al. Genomic instability, endoreduplication, and diminished Ig class-switch recombination in B cells lacking Nbs1. Proc Natl Acad Sci USA 2005;102(5):1590–5.

[159] Bian L, et al. MRE11-RAD50-NBS1 complex alterations and DNA damage response: implications for cancer treatment. Mol Cancer 2019;18(1):169.

[160] Paull TT, Gellert M. The 3' to 5' exonuclease activity of Mre 11 facilitates repair of DNA double-strand breaks. Mol Cell 1998;1(7):969–79.

[161] Paull TT, Gellert M. Nbs1 potentiates ATP-driven DNA unwinding and endonuclease cleavage by the Mre11/Rad50 complex. Genes Dev 1999;13(10):1276–88.

[162] Bardwell PD, et al. Altered somatic hypermutation and reduced class-switch recombination in exonuclease 1-mutant mice. Nat Immunol 2004;5(2):224–9.

[163] Shibata A, et al. DNA double-strand break repair pathway choice is directed by distinct MRE11 nuclease activities. Mol Cell 2014;53(1):7–18.

[164] Yang SH, et al. The SOSS1 single-stranded DNA binding complex promotes DNA end resection in concert with Exo1. EMBO J 2013;32(1):126–39.

[165] Trujillo KM, Sung P. DNA structure-specific nuclease activities in the *Saccharomyces cerevisiae* Rad50*Mre11 complex. J Biol Chem 2001;276(38):35458–64.

[166] Sertic S, et al. EXO1: a tightly regulated nuclease. DNA Repair 2020;93:102929.

[167] Cortizas EM, et al. Alternative end-joining and classical nonhomologous end-joining pathways repair different types of double-strand breaks during class-switch recombination. J Immunol 2013;191(11):5751–63.

[168] Kumar S, Subramanian S. Mutation rates in mammalian genomes. Proc Natl Acad Sci USA 2002;99(2):803–8.

[169] Bransteitter R, et al. Activation-induced cytidine deaminase deaminates deoxycytidine on single-stranded DNA but requires the action of RNase. Proc Natl Acad Sci USA 2003;100(7):4102–7.

[170] Shen HM, et al. The activation-induced cytidine deaminase (AID) efficiently targets DNA in nucleosomes but only during transcription. J Exp Med 2009;206(5):1057–71.

[171] Kodgire P, et al. Nucleosome stability dramatically impacts the targeting of somatic hypermutation. Mol Cell Biol 2012;32(10):2030–40.

[172] Nishida K, et al. Targeted nucleotide editing using hybrid prokaryotic and vertebrate adaptive immune systems. Science 2016;(6305):353.

[173] Ma Y, et al. Targeted AID-mediated mutagenesis (TAM) enables efficient genomic diversification in mammalian cells. Nat Methods 2016;13(12):1029–35.

[174] Komor AC, et al. Programmable editing of a target base in genomic DNA without double-stranded DNA cleavage. Nature 2016;533(7603):420–4.

[175] Komor AC, et al. Improved base excision repair inhibition and bacteriophage Mu Gam protein yields C:G-to-T:A base editors with higher efficiency and product purity. Sci Adv 2017;3(8):eaao4774.

[176] Liang G, et al. RNA editing of hepatitis B virus transcripts by activation-induced cytidine deaminase. Proc Natl Acad Sci USA 2013;110(6):2246–51.

[177] Ranjit S, et al. AID recruits UNG and Msh2 to Ig switch regions dependent upon the AID C terminus. J Immunol 2011;187(5):2464–75.

[178] Hu W, et al. Identification of DNA cleavage- and recombination-specific hnRNP cofactors for activation-induced cytidine deaminase. Proc Natl Acad Sci USA 2015;112(18):5791–6.

[179] Refaat AM, et al. HNRNPU facilitates antibody class-switch recombination through C-NHEJ promotion and R-loop suppression. Cell Rep 2023;42(3):112284.

[180] Schmidt MHM, Pearson CE. Disease-associated repeat instability and mismatch repair. DNA Repair 2016;38:117–26.

[181] Eccleston J, et al. Mismatch repair proteins MSH2, MLH1, and EXO1 are important for class-switch recombination events occurring in B cells that lack nonhomologous end joining. J Immunol 2011;186(4):2336–43.

[182] Shahi A, et al. Mismatch-repair protein MSH6 is associated with Ku70 and regulates DNA double-strand break repair. Nucleic Acids Res 2011;39(6):2130–43.

[183] Cannavo E, et al. Characterization of the interactome of the human MutL homologues MLH1, PMS1, and PMS2. J Biol Chem 2007;282(5):2976–86.

[184] Schrader CE, Vardo J, Stavnezer J. Role for mismatch repair proteins Msh2, Mlh1, and Pms2 in immunoglobulin class switching shown by sequence analysis of recombination junctions. J Exp Med 2002;195(3):367–73.

[185] Schrader CE, Vardo J, Stavnezer J. Mlh1 can function in antibody class switch recombination independently of Msh2. J Exp Med 2003;197(10):1377–83.

[186] Ehrenstein MR, et al. Switch junction sequences in PMS2-deficient mice reveal a microhomology-mediated mechanism of Ig class switch recombination. Proc Natl Acad Sci USA 2001;98(25):14553–8.

[187] Péron S, et al. Human PMS2 deficiency is associated with impaired immunoglobulin class switch recombination. J Exp Med 2008;205(11):2465–72.

[188] Gao Y, et al. Mechanisms of post-replication DNA repair. Genes 2017;8(2).

[189] Gallo D, Brown GW. Post-replication repair: Rad5/HLTF regulation, activity on undamaged templates, and relationship to cancer. Crit Rev Biochem Mol Biol 2019;54(3):301–32.

[190] Zan H, et al. Rev1 recruits ung to switch regions and enhances dU glycosylation for immunoglobulin class switch DNA recombination. Cell Rep 2012;2(5):1220–32.

[191] Schenten D, et al. Polζ ablation in B cells impairs the germinal center reaction, class switch recombination, DNA break repair, and genome stability. J Exp Med 2009;206(2):477–90.

[192] Daly J, et al. Altered Ig hypermutation pattern and frequency in complementary mouse models of DNA polymerase ζ activity. J Immunol 2012;188(11):5528–37.

[193] Sarkies P, et al. Epigenetic instability due to defective replication of structured DNA. Mol Cell 2010;40(5):703–13.

[194] Bhat A, et al. Rev3, the catalytic subunit of Polzeta, is required for maintaining fragile site stability in human cells. Nucleic Acids Res 2013;41(4):2328–39.

[195] Ben Yamin B, et al. DNA polymerase zeta contributes to heterochromatin replication to prevent genome instability. EMBO J 2021;40(21):e104543.

[196] Lange SS, et al. Dual role for mammalian DNA polymerase ζ in maintaining genome stability and proliferative responses. Proc Natl Acad Sci USA 2013;110(8):E687–96.

[197] Ma X, et al. DNA polymerase η promotes nonhomologous end joining upon etoposide exposure dependent on the scaffolding protein Kap1. J Biol Chem 2022;298(5):101861.

[198] Faili A, et al. DNA polymerase eta is involved in hypermutation occurring during immunoglobulin class switch recombination. J Exp Med 2004;199(2):265–70.

[199] Masuda K, et al. A critical role for REV1 in regulating the induction of C: G transitions and A: T mutations during Ig gene hypermutation. J Immunol 2009;183(3):1846–50.

[200] Masuda K, et al. DNA polymerase θ contributes to the generation of C/G mutations during somatic hypermutation of Ig genes. PNAS USA 2005;102(39):13986–91.

[201] Poltoratsky V, et al. Down-regulation of DNA polymerase beta accompanies somatic hypermutation in human BL2 cell lines. DNA Repair 2007;6(2):244–53.

[202] Wu X, Stavnezer J. DNA polymerase beta is able to repair breaks in switch regions and plays an inhibitory role during immunoglobulin class switch recombination. J Exp Med 2007;204(7):1677–89.

[203] Feng Y, et al. FAM72A antagonizes UNG2 to promote mutagenic repair during antibody maturation. Nature 2021;600(7888):324–8.

[204] Zahn A, et al. Separation of function between isotype switching and affinity maturation in vivo during acute immune responses and circulating autoantibodies in UNG-deficient mice. J Immunol 2013;190(12):5949–60.

[205] Rogier M, et al. Fam72a enforces error-prone DNA repair during antibody diversification. Nature 2021;600(7888):329–33.

[206] Rada C, et al. Hot spot focusing of somatic hypermutation in MSH2-deficient mice suggests two stages of mutational targeting. Immunity 1998;9(1):135–41.

[207] Zeng X, et al. DNA polymerase eta is an A-T mutator in somatic hypermutation of immunoglobulin variable genes. Nat Immunol 2001;2(6):537–41.

[208] Martomo SA, et al. Different mutation signatures in DNA polymerase eta- and MSH6-deficient mice suggest separate roles in antibody diversification. Proc Natl Acad Sci USA 2005;102(24):8656–61.

[209] Lv L, et al. Mismatch repair protein MSH2 regulates translesion DNA synthesis following exposure of cells to UV radiation. Nucleic Acids Res 2013;41(22):10312–22.

[210] Wilson TM, et al. MSH2-MSH6 stimulates DNA polymerase eta, suggesting a role for A: T mutations in antibody genes. J Exp Med 2005;201(4):637–45.

[211] Bothmer A, et al. 53BP1 regulates DNA resection and the choice between classical and alternative end joining during class switch recombination. J Exp Med 2010;207(4):855–65.

[212] Björkman A, et al. DNA-PKcs is involved in Ig class switch recombination in human B cells. J Immunol 2015;195(12):5608–15.

[213] Pan-Hammarström Q, et al. Impact of DNA ligase IV on nonhomologous end joining pathways during class switch recombination in human cells. J Exp Med 2005;201(2):189–94.

[214] Du L, et al. Cernunnos influences human immunoglobulin class switch recombination and may be associated with B cell lymphomagenesis. J Exp Med 2012;209(2):291–305.

[215] Ling AK, et al. SHLD2 promotes class switch recombination by preventing inactivating deletions within the Igh locus. EMBO Rep 2020;21(8):e49823.

[216] Kracker S, et al. Impaired induction of DNA lesions during immunoglobulin class-switch recombination in humans influences end-joining repair. Proc Natl Acad Sci USA 2010;107(51):22225–30.

[217] Nagaoka H, et al. Activation-induced deaminase (AID)-directed hypermutation in the immunoglobulin Smu region: implication of AID involvement in a common step of class switch recombination and somatic hypermutation. J Exp Med 2002;195(4):529–34.

[218] Odegard VH, et al. Histone modifications associated with somatic hypermutation. Immunity 2005;23(1):101–10.

[219] Yeap L-S, et al. Sequence-intrinsic mechanisms that target AID mutational outcomes on antibody genes. Cell 2015;163(5):1124–37.

[220] Pavlova AV, et al. Impact of G-quadruplexes on the regulation of genome integrity, DNA Damage and Repair. Biomolecules 2021;11(9).

[221] Tang C, MacCarthy T. Characterization of DNA G-quadruplex structures in human immunoglobulin heavy variable (IGHV) genes. Front Immunol 2021;12:671944.

[222] Neuberger MS, et al. Immunity through DNA deamination. Trends Biochem Sci 2003;28(6):305–12.

[223] Yu K, Huang FT, Lieber MR. DNA substrate length and surrounding sequence affect the activation-induced deaminase activity at cytidine. J Biol Chem 2004;279(8):6496–500.

[224] Maul RW, et al. Uracil residues dependent on the deaminase AID in immunoglobulin gene variable and switch regions. Nat Immunol 2011;12(1):70–6.

[225] Begum NA, et al. Uracil DNA glycosylase activity is dispensable for immunoglobulin class switch. Science 2004;305(5687):1160–3.

[226] Boiteux S, Guillet M. Abasic sites in DNA: repair and biological consequences in *Saccharomyces cerevisiae*. DNA Repair 2004;3(1):1–12.

[227] Kobayashi M, et al. Ago2 and a miRNA reduce Topoisomerase 1 for enhancing DNA cleavage in antibody diversification by activation-induced cytidine deaminase. PNAS USA 2023;120.

[228] Yin Z, et al. RNA-binding motifs of hnRNP K are critical for induction of antibody diversification by activation-induced cytidine deaminase. Proc Natl Acad Sci USA 2020;117(21):11624–35.

[229] Rogozin IB, Diaz M. Cutting edge: DGYW/WRCH is a better predictor of mutability at G: C bases in Ig hypermutation than the widely accepted RGYW/WRCY motif and probably reflects a two-step activation-induced cytidine deaminase-triggered process. J Immunol 2004;172(6):3382–4.

[230] Pommier Y. Topoisomerase I inhibitors: camptothecins and beyond. Nat Rev Cancer 2006;6(10):789–802.

[231] Liu LF, Wang JC. Supercoiling of the DNA template during transcription. Proc Natl Acad Sci USA 1987;84(20):7024–7.

[232] Lippert MJ, et al. Role for topoisomerase 1 in transcription-associated mutagenesis in yeast. Proc Natl Acad Sci USA 2011;108(2):698–703.

[233] Takahashi DT, et al. Topoisomerase 1 provokes the formation of short deletions in repeated sequences upon high transcription in *Saccharomyces cerevisiae*. Proc Natl Acad Sci USA 2011;108(2):692–7.

[234] Hubert Jr L, et al. Topoisomerase 1 and single-strand break repair modulate transcription-induced CAG repeat contraction in human cells. Mol Cell Biol 2011;31(15):3105–12.

[235] Tashiro J, Kinoshita K, Honjo T. Palindromic but not G-rich sequences are targets of class switch recombination. Int Immunol 2001;13(4):495–505.

[236] Amin W, et al. Necessity of HuR/ELAVL1 for activation-induced cytidine deaminase-dependent decrease in topoisomerase 1 in antibody diversification. Int Immunol 2023;35:361–75.

[237] Diaz-Munoz MD, et al. The RNA-binding protein HuR is essential for the B cell antibody response. Nat Immunol 2015;16(4):415–25.

[238] Husain A, et al. Chromatin remodeller SMARCA4 recruits topoisomerase 1 and suppresses transcription-associated genomic instability. Nat Commun 2016;7:10549.

[239] Stanlie A, et al. Histone3 lysine4 trimethylation regulated by the facilitates chromatin transcription complex is critical for DNA cleavage in class switch recombination. Proc Natl Acad Sci USA 2010;107(51):22190–5.

[240] Miao ZH, et al. Nonclassic functions of human topoisomerase I: genome-wide and pharmacologic analyses. Cancer Res 2007;67(18):8752–61.

[241] Tuduri S, et al. Topoisomerase I suppresses genomic instability by preventing interference between replication and transcription. Nat Cell Biol 2009;11(11):1315–24.

[242] Nakatani R, et al. Large expansion of CTG*CAG repeats is exacerbated by MutSbeta in human cells. Sci Rep 2015;5:11020.

[243] Honjo T, Kinoshita K, Muramatsu M. Molecular mechanism of class switch recombination: linkage with somatic hypermutation. Annu Rev Immunol 2002;20:165–96.

[244] Nikaido T, Nakai S, Honjo T. Switch region of immunoglobulin Cmu gene is composed of simple tandem repetitive sequences. Nature 1981;292(5826):845–8.

[245] Haddad D, et al. Sense transcription through the S region is essential for immunoglobulin class switch recombination. EMBO J 2011;30(8):1608–20.

[246] Kinoshita K, Honjo T. Linking class-switch recombination with somatic hypermutation. Nat Rev Mol Cell Biol 2001;2(7):493–503.

[247] Shinkura R, et al. The influence of transcriptional orientation on endogenous switch region function. Nat Immunol 2003;4(5):435–41.

[248] Mussmann R, et al. Microsites for immunoglobulin switch recombination breakpoints from Xenopus to mammals. Eur J Immunol 1997;27(10):2610–9.

[249] Han L, Masani S, Yu K. Overlapping activation-induced cytidine deaminase hotspot motifs in Ig class-switch recombination. Proc Natl Acad Sci USA 2011;108(28):11584–9.

[250] Dempsey LA, Hanakahi LA, Maizels N. A specific isoform of hnRNP D interacts with DNA in the LR1 heterodimer: canonical RNA binding motifs in a sequence-specific duplex DNA binding protein. J Biol Chem 1998;273(44):29224–9.

[251] Duquette ML, et al. Intracellular transcription of G-rich DNAs induces formation of G-loops, novel structures containing G4 DNA. Genes Dev 2004;18(13):1618–29.

[252] Huang FT, et al. Sequence dependence of chromosomal R-loops at the immunoglobulin heavy-chain Smu class switch region. Mol Cell Biol 2007;27(16):5921–32.

[253] Huang FT, et al. Downstream boundary of chromosomal R-loops at murine switch regions: implications for the mechanism of class switch recombination. Proc Natl Acad Sci USA 2006;103(13):5030–5.

[254] Li Z, et al. The generation of antibody diversity through somatic hypermutation and class switch recombination. Genes Dev 2004;18(1):1–11.

[255] Milstein C, Neuberger MS, Staden R. Both DNA strands of antibody genes are hypermutation targets. Proc Natl Acad Sci USA 1998;95(15):8791–4.

[256] Parsa JY, et al. Negative supercoiling creates single-stranded patches of DNA that are substrates for AID-mediated mutagenesis. PLoS Genet 2012;8(2):e1002518.

[257] Shen HM, Storb U. Activation-induced cytidine deaminase (AID) can target both DNA strands when the DNA is supercoiled. Proc Natl Acad Sci USA 2004;101(35):12997–3002.

[258] Liu S, et al. RNA polymerase III is required for the repair of DNA double-strand breaks by homologous recombination. Cell 2021;184(5):1314–29.

[259] Ohle C, et al. Transient RNA-DNA hybrids are required for efficient double-strand break repair. Cell 2016;167(4):1001–1013 e7.

[260] Michelini F, et al. Damage-induced lncRNAs control the DNA damage response through interaction with DDRNAs at individual double-strand breaks. Nat Cell Biol 2017;19(12):1400–11.

[261] Sharma S, et al. MRE11-RAD50-NBS1 complex is sufficient to promote transcription by RNA polymerase II at double-strand breaks by melting DNA ends. Cell Rep 2021;34(1):108565.

[262] Laffleur B, et al. Noncoding RNA processing by DIS3 regulates chromosomal architecture and somatic hypermutation in B cells. Nat Genet 2021;53(2):230–42.

[263] Izumi H, Funa K. Telomere function and the G-quadruplex formation are regulated by hnRNP U. Cells 2019;8(5).

[264] Kiledjian M, Dreyfuss G. Primary structure and binding activity of the hnRNP U protein: binding RNA through RGG box. EMBO J 1992;11(7):2655–64.

[265] Perlot T, Li G, Alt FW. Antisense transcripts from immunoglobulin heavy-chain locus V(D)J and switch regions. Proc Natl Acad Sci USA 2008;105(10):3843–8.

[266] Wiedemann EM, Peycheva M, Pavri R. DNA replication origins in immunoglobulin switch regions regulate class switch recombination in an R-loop-dependent manner. Cell Rep 2016;17(11):2927–42.

[267] Dalloul I, et al. Locus suicide recombination actively occurs on the functionally rearranged IgH allele in B-cells from inflamed human lymphoid tissues. PLoS Genet 2019;15(6):e1007721.

[268] Deze O, Laffleur B, Cogne M. Roles of G4-DNA and G4-RNA in class switch recombination and additional regulations in B-lymphocytes. Molecules 2023;28(3).

[269] Peron S, et al. AID-driven deletion causes immunoglobulin heavy chain locus suicide recombination in B cells. Science 2012;336(6083):931–4.

[270] Ronai D, et al. Detection of chromatin-associated single-stranded DNA in regions targeted for somatic hypermutation. J Exp Med 2007;204(1):181–90.

[271] Shapiro GS, et al. Predicting regional mutability in antibody V genes based solely on di- and trinucleotide sequence composition. J Immunol 1999;163(1):259–68.

[272] Smith DS, et al. Di- and trinucleotide target preferences of somatic mutagenesis in normal and autoreactive B cells. J Immunol 1996;156(7):2642–52.

[273] McDonald JJ, et al. A critical context-dependent role for E boxes in the targeting of somatic hypermutation. J Immunol 2013;191(4):1556–66.

[274] Michael N, et al. The E box motif CAGGTG enhances somatic hypermutation without enhancing transcription. Immunity 2003;19(2):235–42.

[275] Tanaka A, et al. Attracting AID to targets of somatic hypermutation. J Exp Med 2010;207(2):405–15.

[276] Kornberg RD, Lorch Y. Twenty-five years of the nucleosome, fundamental particle of the eukaryote chromosome. Cell 1999;98(3):285–94.

[277] Hsieh FK, et al. Histone chaperone FACT action during transcription through chromatin by RNA polymerase II. Proc Natl Acad Sci USA 2013;110(19):7654–9.

[278] Kulaeva OI, et al. Mechanism of chromatin remodeling and recovery during passage of RNA polymerase II. Nat Struct Mol Biol 2009;16(12):1272–8.

[279] Petesch SJ, Lis JT. Overcoming the nucleosome barrier during transcript elongation. Trends Genet 2012;28(6):285–94.

[280] Sims 3rd RJ, Belotserkovskaya R, Reinberg D. Elongation by RNA polymerase II: the short and long of it. Genes Dev 2004;18(20):2437–68.
[281] Begum NA, et al. The histone chaperone Spt6 is required for activation-induced cytidine deaminase target determination through H3K4me3 regulation. J Biol Chem 2012;287(39):32415–29.
[282] Aida M, et al. Accumulation of the FACT complex, as well as histone H3.3, serves as a target marker for somatic hypermutation. Proc Natl Acad Sci USA 2013;110(19):7784–9.
[283] Begum NA, et al. Phf5a regulates DNA repair in class switch recombination via p400 and histone H2A variant deposition. EMBO J 2021;40(12):e106393.
[284] Adam S, Polo SE, Almouzni G. Transcription recovery after DNA damage requires chromatin priming by the H3.3 histone chaperone HIRA. Cell 2013;155(1):94–106.
[285] Stavnezer J. Immunoglobulin class switching. Curr Opin Immunol 1996;8(2):199–205.
[286] Berger SL. The complex language of chromatin regulation during transcription. Nature 2007;447(7143):407–12.
[287] Li B, Carey M, Workman JL. The role of chromatin during transcription. Cell 2007;128(4):707–19.
[288] Shilatifard A. Chromatin modifications by methylation and ubiquitination: implications in the regulation of gene expression. Annu Rev Biochem 2006;75:243–69.
[289] Wang L, et al. AID-dependent histone acetylation is detected in immunoglobulin S regions. J Exp Med 2006;203(1):215–26.
[290] Daniel JA, et al. PTIP promotes chromatin changes critical for immunoglobulin class switch recombination. Science 2010;329(5994):917–23.
[291] Balter BB, et al. Mice lacking Smu tandem repeats maintain RNA polymerase patterns but exhibit histone modification pattern shifts linked to class switch site locations. Mol Immunol 2012;52(1):1–8.
[292] Min IM, et al. Shifts in targeting of class switch recombination sites in mice that lack mu switch region tandem repeats or Msh2. J Exp Med 2005;201(12):1885–90.
[293] Bradley SP, et al. The histone methyltransferase suv39h1 increases class switch recombination specifically to IgA. J Immunol 2006;177(2):1179–88.
[294] Zippo A, et al. Histone crosstalk between H3S10ph and H4K16ac generates a histone code that mediates transcription elongation. Cell 2009;138(6):1122–36.
[295] Belotserkovskaya R, et al. FACT facilitates transcription-dependent nucleosome alteration. Science 2003;301(5636):1090–3.
[296] Reinberg D, Sims 3rd RJ. De FACTo nucleosome dynamics. J Biol Chem 2006;281(33):23297–301.
[297] Orphanides G, et al. FACT, a factor that facilitates transcript elongation through nucleosomes. Cell 1998;92(1):105–16.
[298] Orphanides G, et al. The chromatin-specific transcription elongation factor FACT comprises human SPT16 and SSRP1 proteins. Nature 1999;400(6741):284–8.
[299] Aida M, Honjo T. FACT and H3.3: new markers for the somatic hypermutation. Cell Cycle 2013;12(18):2923–4.
[300] Tagami H, et al. Histone H3.1 and H3.3 complexes mediate nucleosome assembly pathways dependent or independent of DNA synthesis. Cell 2004;116(1):51–61.
[301] Borde V, et al. Histone H3 lysine 4 trimethylation marks meiotic recombination initiation sites. EMBO J 2009;28(2):99–111.
[302] Smagulova F, et al. Genome-wide analysis reveals novel molecular features of mouse recombination hotspots. Nature 2011;472(7343):375–8.
[303] Matthews AG, et al. RAG2 PHD finger couples histone H3 lysine 4 trimethylation with V(D)J recombination. Nature 2007;450(7172):1106–10.
[304] Lee JH, Skalnik DG. CpG-binding protein (CXXC finger protein 1) is a component of the mammalian Set1 histone H3-Lys4 methyltransferase complex, the analogue of the yeast Set1/COMPASS complex. J Biol Chem 2005;280(50):41725–31.
[305] Chen S, et al. The histone H3 Lys 27 demethylase JMJD3 regulates gene expression by impacting transcriptional elongation. Genes Dev 2012;26(12):1364–75.
[306] Kato H, et al. Spt6 prevents transcription-coupled loss of post-translationally modified histone H3. Sci Rep 2013;3:2186.
[307] Shimazaki N, Tsai AG, Lieber MR. H3K4me3 stimulates the V(D)J RAG complex for both nicking and hairpinning in trans in addition to tethering in cis: implications for translocations. Mol Cell 2009;34(5):535–44.
[308] Marculescu R, et al. V(D)J-mediated translocations in lymphoid neoplasms: a functional assessment of genomic instability by cryptic sites. J Exp Med 2002;195(1):85–98.
[309] Merelli I, et al. RSSsite: a reference database and prediction tool for the identification of cryptic Recombination Signal Sequences in human and murine genomes. Nucleic Acids Res 2010;38:W262–7.
[310] Tsai AG, et al. Human chromosomal translocations at CpG sites and a theoretical basis for their lineage and stage specificity. Cell 2008;135(6):1130–42.
[311] Zhang M, Swanson PC. V(D)J recombinase binding and cleavage of cryptic recombination signal sequences identified from lymphoid malignancies. J Biol Chem 2008;283(11):6717–27.
[312] Vaidyanathan B, Chaudhuri J. Epigenetic codes programing class switch recombination. Front Immunol 2015;6:405.
[313] Bultman S, et al. A Brg1 null mutation in the mouse reveals functional differences among mammalian SWI/SNF complexes. Mol Cell 2000;6(6):1287–95.
[314] Singhal N, et al. Chromatin-remodeling components of the BAF complex facilitate reprogramming. Cell 2010;141(6):943–55.
[315] Wang X, et al. Oncogenesis caused by loss of the SNF5 tumor suppressor is dependent on activity of BRG1, the ATPase of the SWI/SNF chromatin remodeling complex. Cancer Res 2009;69(20):8094–101.
[316] Dong X, Biswas A, Chook YM. Structural basis for assembly and disassembly of the CRM1 nuclear export complex. Nat Struct Mol Biol 2009;16(5):558–60.
[317] Wuerffel R, et al. S-S synapsis during class switch recombination is promoted by distantly located transcriptional elements and activation-induced deaminase. Immunity 2007;27(5):711–22.
[318] Knuesel MT, et al. The human CDK8 subcomplex is a molecular switch that controls Mediator coactivator function. Genes Dev 2009;23(4):439–51.
[319] Soutourina J. Transcription regulation by the Mediator complex. Nat Rev Mol Cell Biol 2018;19(4):262–74.
[320] Turunen M, et al. Uterine leiomyoma-linked MED12 mutations disrupt mediator-associated CDK activity. Cell Rep 2014;7(3):654–60.

[321] Aranda-Orgilles B, et al. MED12 regulates HSC-specific enhancers independently of mediator kinase activity to control hematopoiesis. Cell Stem Cell 2016;19(6):784−99.
[322] Pinaud E, et al. Localization of the 3' IgH locus elements that effect long-distance regulation of class switch recombination. Immunity 2001;15(2):187−99.
[323] Vincent-Fabert C, et al. Genomic deletion of the whole IgH 3' regulatory region (hs3a, hs1,2, hs3b, and hs4) dramatically affects class switch recombination and Ig secretion to all isotypes. Blood 2010;116(11):1895−8.
[324] Kenter AL, et al. Three-dimensional architecture of the IgH locus facilitates class switch recombination. Ann N Y Acad Sci 2012;1267(1):86−94.
[325] Delgado-Benito V, et al. The chromatin reader ZMYND8 regulates igh enhancers to promote immunoglobulin class switch recombination. Mol Cell 2018;72(4):636−49.
[326] Stanlie A, et al. The DSIF subunits Spt4 and Spt5 have distinct roles at various phases of immunoglobulin class switch recombination. PLoS Genet 2012;8. https://doi.org/10.1371/journal.pgen.1002675.
[327] Fingerman IM, Du HN, Briggs SD. Controlling histone methylation via trans-histone pathways. Epigenetics 2008;3(5):237−42.
[328] Zhou K, et al. Control of transcriptional elongation and cotranscriptional histone modification by the yeast BUR kinase substrate Spt5. Proc Natl Acad Sci USA 2009;106(17):6956−61.
[329] Saintamand A, et al. Elucidation of IgH 3' region regulatory role during class switch recombination via germline deletion. Nat Commun 2015;6:7084.
[330] Winston F, et al. Mutations affecting Ty-mediated expression of the HIS4 gene of *Saccharomyces cerevisiae*. Genetics 1984;107(2):179−97.
[331] Maldonado E, et al. A human RNA polymerase II complex associated with SRB and DNA-repair proteins. Nature 1996;381(6577):86−9.
[332] Goodarzi AA, et al. ATM signaling facilitates repair of DNA double-strand breaks associated with heterochromatin. Mol Cell 2008;31(2):167−77.
[333] Escribano-Diaz C, Durocher D. DNA repair pathway choice−a PTIP of the hat to 53BP1. EMBO Rep 2013;14(8):665−6.
[334] Bienz M. The PHD finger, a nuclear protein-interaction domain. Trends Biochem Sci 2006;31(1):35−40.
[335] Sanchez R, Zhou MM. The PHD finger: a versatile epigenome reader. Trends Biochem Sci 2011;36(7):364−72.
[336] Ogiwara H, et al. Histone acetylation by CBP and p300 at double-strand break sites facilitates SWI/SNF chromatin remodeling and the recruitment of non-homologous end joining factors. Oncogene 2011;30(18):2135−46.
[337] Mizuguchi G, et al. ATP-driven exchange of histone H2AZ variant catalyzed by SWR1 chromatin remodeling complex. Science 2004;303(5656):343−8.
[338] van Roon AM, et al. Solution structure of the U2 snRNP protein Rds3p reveals a knotted zinc-finger motif. Proc Natl Acad Sci USA 2008;105(28):9621−6.
[339] Marchalot A, et al. Uncoupling splicing from transcription using antisense oligonucleotides reveals a dual role for I exon donor splice sites in antibody class switching. Front Immunol 2020;11:780.
[340] Zheng YZ, et al. PHF5A epigenetically inhibits apoptosis to promote breast cancer progression. Cancer Res 2018;78(12):3190−206.
[341] Gong F, Miller KM. Mammalian DNA repair: HATs and HDACs make their mark through histone acetylation. Mutat Res 2013;750(1−2):23−30.
[342] Polo SE, Jackson SP. Dynamics of DNA damage response proteins at DNA breaks: a focus on protein modifications. Genes Dev 2011;25(5):409−33.
[343] Dev H, et al. Shieldin complex promotes DNA end-joining and counters homologous recombination in BRCA1-null cells. Nat Cell Biol 2018;20(8):954−65.
[344] Filippakopoulos P, et al. Selective inhibition of BET bromodomains. Nature 2010;468(7327):1067−73.
[345] Stanlie A, et al. Acetylated histone reader Brd4 functions as repair adaptor in immunoglobulin class switch recombination. Mol Cell 2014;55:97−110.
[346] Rada C, et al. Immunoglobulin isotype switching is inhibited and somatic hypermutation perturbed in UNG-deficient mice. Curr Biol 2002;12(20):1748−55.
[347] Ramiro AR, et al. Role of genomic instability and p53 in AID-induced c-myc-Igh translocations. Nature 2006;440(7080):105−9.
[348] De Silva FS, Moss B. Vaccinia virus uracil DNA glycosylase has an essential role in DNA synthesis that is independent of its glycosylase activity: catalytic site mutations reduce virulence but not virus replication in cultured cells. J Virol 2003;77(1):159−66.
[349] Dong Q, et al. Combinatorial loss of the enzymatic activities of viral uracil-DNA glycosylase and viral dUTPase impairs murine gammaherpesvirus pathogenesis and leads to increased recombination-based deletion in the viral genome. mBio 2018;9(5).
[350] Su MT, et al. Uracil DNA glycosylase BKRF3 contributes to Epstein-Barr virus DNA replication through physical interactions with proteins in viral DNA replication complex. J Virol 2014;88(16):8883−99.
[351] Garcia-Gutierrez P, Garcia-Dominguez M. BETting on a transcriptional deficit as the main cause for Cornelia de Lange syndrome. Front Mol Biosci 2021;8:709232.
[352] Cheung KL, et al. Distinct roles of Brd2 and Brd4 in potentiating the transcriptional program for Th17 cell differentiation. Mol Cell 2017;65(6):1068−80.
[353] Enervald E, et al. A regulatory role for the cohesin loader NIPBL in nonhomologous end joining during immunoglobulin class switch recombination. J Exp Med 2013;210(12):2503−13.
[354] Guirouilh-Barbat J, et al. 53BP1 protects against CtIP-dependent capture of ectopic chromosomal sequences at the junction of distant double-strand breaks. PLoS Genet 2016;12(10):e1006230.
[355] Lam FC, et al. BRD4 prevents the accumulation of R-loops and protects against transcription-replication collision events and DNA damage. Nat Commun 2020;11(1):4083.
[356] Noordermeer SM, et al. The shieldin complex mediates 53BP1-dependent DNA repair. Nature 2018;560(7716):117−21.
[357] Ghezraoui H, et al. 53BP1 cooperation with the REV7-shieldin complex underpins DNA structure-specific NHEJ. Nature 2018;560(7716):122−7.
[358] Zhao F, et al. ASTE1 promotes shieldin-complex-mediated DNA repair by attenuating end resection. Nat Cell Biol 2021;23(8):894−904.
[359] Zhang Y, et al. Long noncoding RNA LINP1 regulates repair of DNA double-strand breaks in triple-negative breast cancer. Nat Struct Mol Biol 2016;23(6):522−30.
[360] Nozawa RS, Gilbert N. RNA: nuclear glue for folding the genome. Trends Cell Biol 2019;29(3):201−11.

[361] Altmeyer M, et al. Liquid demixing of intrinsically disordered proteins is seeded by poly(ADP-ribose). Nat Commun 2015;6:8088.

[362] Britton S, et al. DNA damage triggers SAF-A and RNA biogenesis factors exclusion from chromatin coupled to R-loops removal. Nucleic Acids Res 2014;42(14):9047–62.

[363] Pessina F, et al. Functional transcription promoters at DNA double-strand breaks mediate RNA-driven phase separation of damage-response factors. Nat Cell Biol 2019;21(10):1286–99.

[364] Roy D, Lieber MR. G clustering is important for the initiation of transcription-induced R-loops in vitro, whereas high G density without clustering is sufficient thereafter. Mol Cell Biol 2009;29(11):3124–33.

[365] Lu WT, et al. Drosha drives the formation of DNA: RNA hybrids around DNA break sites to facilitate DNA repair. Nat Commun 2018;9(1):532.

[366] Masai H, et al. Rif1 promotes association of G-quadruplex (G4) by its specific G4 binding and oligomerization activities. Sci Rep 2019;9(1):8618.

[367] Moriyama K, Yoshizawa-Sugata N, Masai H. Oligomer formation and G-quadruplex binding by purified murine Rif1 protein, a key organizer of higher-order chromatin architecture. J Biol Chem 2018;293(10):3607–24.

[368] Fujita T, Fujii H. Efficient isolation of specific genomic regions retaining molecular interactions by the iChIP system using recombinant exogenous DNA-binding proteins. BMC Mol Biol 2014;15:26.

[369] Husain A, et al. SAMHD1-mediated dNTP degradation is required for efficient DNA repair during antibody class switch recombination. EMBO J 2020;39(15):e102931.

[370] Clifford R, et al. SAMHD1 is mutated recurrently in chronic lymphocytic leukemia and is involved in response to DNA damage. Blood 2014;123(7):1021–31.

[371] Franzolin E, et al. The deoxynucleotide triphosphohydrolase SAMHD1 is a major regulator of DNA precursor pools in mammalian cells. Proc Natl Acad Sci USA 2013;110(35):14272–7.

[372] Ling AK, et al. Double-stranded DNA break polarity skews repair pathway choice during intrachromosomal and interchromosomal recombination. Proc Natl Acad Sci USA 2018;115(11):2800–5.

[373] Daddacha W, et al. SAMHD1 promotes DNA end resection to facilitate DNA repair by homologous recombination. Cell Rep 2017;20(8):1921–35.

[374] Thientosapol ES, et al. SAMHD1 enhances immunoglobulin hypermutation by promoting transversion mutation. Proc Natl Acad Sci USA 2018;115(19):4921–6.

[375] Yousefzadeh MJ, et al. Mechanism of suppression of chromosomal instability by DNA polymerase POLQ. PLoS Genet 2014;10(10):e1004654.

[376] Zelensky AN, et al. Inactivation of Pol theta and C-NHEJ eliminates off-target integration of exogenous DNA. Nat Commun 2017;8(1):66.

[377] Komori T, et al. Lack of N regions in antigen receptor variable region genes of TdT-deficient lymphocytes. Science 1993;261(5125):1171–5.

[378] Azhar A, Begum NA, Husain A. Nucleotide pool imbalance and antibody gene diversification. Vaccines (Basel) 2021;9(10).

[379] Cattoretti G, et al. Nuclear and cytoplasmic AID in extrafollicular and germinal center B cells. Blood 2006;107(10):3967–75.

[380] Mao C, et al. T cell-independent somatic hypermutation in murine B cells with an immature phenotype. Immunity 2004;20(2):133–44.

[381] Crouch EE, et al. Regulation of AID expression in the immune response. J Exp Med 2007;204(5):1145–56.

[382] Qin H, et al. Activation-induced cytidine deaminase expression in CD4+ T cells is associated with a unique IL-10-producing subset that increases with age. PLoS One 2011;6(12):e29141.

[383] Kuraoka M, et al. Activation-induced cytidine deaminase expression and activity in the absence of germinal centers: insights into hyper-IgM syndrome. J Immunol 2009;183(5):3237–48.

[384] Ueda Y, et al. T-independent activation-induced cytidine deaminase expression, class-switch recombination, and antibody production by immature/transitional 1 B cells. J Immunol 2007;178(6):3593–601.

[385] Cantaert T, et al. Activation-induced cytidine deaminase expression in human B cell precursors is essential for central B cell tolerance. Immunity 2015;43(5):884–95.

[386] Morgan HD, et al. Activation-induced cytidine deaminase deaminates 5-methylcytosine in DNA and is expressed in pluripotent tissues: implications for epigenetic reprogramming. J Biol Chem 2004;279(50):52353–60.

[387] Schreck S, et al. Activation-induced cytidine deaminase (AID) is expressed in normal spermatogenesis but only infrequently in testicular germ cell tumours. J Pathol 2006;210(1):26–31.

[388] Pastor WA, Aravind L, Rao A. TETonic shift: biological roles of TET proteins in DNA demethylation and transcription. Nat Rev Mol Cell Biol 2013;14(6):341–56.

[389] Fritz EL, Papavasiliou FN. Cytidine deaminases: AIDing DNA demethylation? Genes Dev 2010;24(19):2107–14.

[390] Català-Moll F, et al. Activation-induced deaminase is critical for the establishment of DNA methylation patterns prior to the germinal center reaction. Nucleic Acids Res 2021;49(9):5057–73.

[391] Fritz EL, et al. A comprehensive analysis of the effects of the deaminase AID on the transcriptome and methylome of activated B cells. Nat Immunol 2013;14(7):749–55.

[392] Faili A, et al. AID-dependent somatic hypermutation occurs as a DNA single-strand event in the BL2 cell line. Nat Immunol 2002;3(9):815–21.

[393] Greeve J, et al. Expression of activation-induced cytidine deaminase in human B-cell non-Hodgkin lymphomas. Blood 2003;101(9):3574–80.

[394] Hardianti MS, et al. Activation-induced cytidine deaminase expression in follicular lymphoma: association between AID expression and ongoing mutation in FL. Leukemia 2004;18(4):826–31.

[395] Pasqualucci L, et al. Hypermutation of multiple proto-oncogenes in B-cell diffuse large-cell lymphomas. Nature 2001;412(6844):341–6.

[396] Smit LA, et al. Expression of activation-induced cytidine deaminase is confined to B-cell non-Hodgkin's lymphomas of germinal-center phenotype. Cancer Res 2003;63(14):3894–8.

[397] Albesiano E, et al. Activation-induced cytidine deaminase in chronic lymphocytic leukemia B cells: expression as multiple forms in a dynamic, variably sized fraction of the clone. Blood 2003;102(9):3333–9.

[398] Cerutti A, et al. Ongoing in vivo immunoglobulin class switch DNA recombination in chronic lymphocytic leukemia B cells. J Immunol 2002;169(11):6594–603.

[399] Heintel D, et al. High expression of activation-induced cytidine deaminase (AID) mRNA is associated with unmutated IGVH gene status and unfavourable cytogenetic aberrations in patients with chronic lymphocytic leukaemia. Leukemia 2004;18(4):756–62.

[400] Oppezzo P, et al. Chronic lymphocytic leukemia B cells expressing AID display dissociation between class switch recombination and somatic hypermutation. Blood 2003;101(10):4029–32.

[401] McCarthy H, et al. High expression of activation-induced cytidine deaminase (AID) and splice variants is a distinctive feature of poor-prognosis chronic lymphocytic leukemia. Blood 2003;101(12):4903–8.

[402] Pasqualucci L, et al. Expression of the AID protein in normal and neoplastic B cells. Blood 2004;104(10):3318–25.

[403] Palacios F, et al. High expression of AID and active class switch recombination might account for a more aggressive disease in unmutated CLL patients: link with an activated microenvironment in CLL disease. Blood 2010;115(22):4488–96.

[404] Reiniger L, et al. Richter's and prolymphocytic transformation of chronic lymphocytic leukemia are associated with high mRNA expression of activation-induced cytidine deaminase and aberrant somatic hypermutation. Leukemia 2006;20(6):1089–95.

[405] Lossos IS, Levy R, Alizadeh AA. AID is expressed in germinal center B-cell-like and activated B-cell-like diffuse large-cell lymphomas and is not correlated with intraclonal heterogeneity. Leukemia 2004;18(11):1775–9.

[406] Dijkman R, et al. Primary cutaneous follicle center lymphoma and primary cutaneous large B-cell lymphoma, leg type, are both targeted by aberrant somatic hypermutation but demonstrate differential expression of AID. Blood 2006;107(12):4926–9.

[407] Kawamura K, et al. Expression of activation-induced cytidine deaminase is associated with a poor prognosis of diffuse large B cell lymphoma patients treated with CHOP-based chemotherapy. J Cancer Res Clin Oncol 2016;142(1):27–36.

[408] Feldhahn N, et al. Activation-induced cytidine deaminase acts as a mutator in BCR-ABL1-transformed acute lymphoblastic leukemia cells. J Exp Med 2007;204(5):1157–66.

[409] Iacobucci I, et al. Different isoforms of the B-cell mutator activation-induced cytidine deaminase are aberrantly expressed in BCR-ABL1-positive acute lymphoblastic leukemia patients. Leukemia 2010;24(1):66–73.

[410] Shi Y, et al. Aberrant activation-induced cytidine deaminase expression in Philadelphia chromosome-positive B-cell acute lymphoblastic leukemia. Hum Pathol 2016;52:173–8.

[411] Komori J, et al. Activation-induced cytidine deaminase links bile duct inflammation to human cholangiocarcinoma. Hepatology 2008;47(3):888–96.

[412] Kou T, et al. Expression of activation-induced cytidine deaminase in human hepatocytes during hepatocarcinogenesis. Int J Cancer 2007;120(3):469–76.

[413] Nakanishi Y, et al. Role of activation-induced cytidine deaminase in the development of oral squamous cell carcinoma. PLoS One 2013;8(4):e62066.

[414] Li H, et al. AID modulates carcinogenesis network via DNA demethylation in bladder urothelial cell carcinoma. Cell Death Dis 2019;10(4):251.

[415] Che Z, et al. Activation-induced cytidine deaminase expression facilitates the malignant phenotype and epithelial-to-mesenchymal transition in clear cell renal cell carcinoma. DNA Cell Biol 2020;39(7):1299–312.

[416] He B, et al. EBV-encoded latent membrane protein 1 cooperates with BAFF/BLyS and APRIL to induce T cell-independent Ig heavy chain class switching. J Immunol 2003;171(10):5215–24.

[417] Kalchschmidt JS, et al. Epstein-Barr virus nuclear protein EBNA3C directly induces expression of AID and somatic mutations in B cells. J Exp Med 2016;213(6):921–8.

[418] Machida K, et al. Hepatitis C virus induces a mutator phenotype: enhanced mutations of immunoglobulin and protooncogenes. Proc Natl Acad Sci USA 2004;101(12):4262–7.

[419] Endo Y, et al. Expression of activation-induced cytidine deaminase in human hepatocytes via NF-kappaB signaling. Oncogene 2007;26(38):5587–95.

[420] Ishikawa C, et al. Activation of AID by human T-cell leukemia virus Tax oncoprotein and the possible role of its constitutive expression in ATL genesis. Carcinogenesis 2011;32(1):110–9.

[421] Gourzi P, Leonova T, Papavasiliou FN. A role for activation-induced cytidine deaminase in the host response against a transforming retrovirus. Immunity 2006;24(6):779–86.

[422] Gourzi P, Leonova T, Papavasiliou FN. Viral induction of AID is independent of the interferon and the Toll-like receptor signaling pathways but requires NF-kappaB. J Exp Med 2007;204(2):259–65.

[423] Matsumoto Y, et al. *Helicobacter pylori* infection triggers aberrant expression of activation-induced cytidine deaminase in gastric epithelium. Nat Med 2007;13(4):470–6.

[424] Nguyen T, et al. Activation-induced cytidine deaminase is dispensable for virus-mediated liver and skin tumor development in mouse models. Int Immunol 2014;26(7):397–406.

[425] Pone EJ, et al. B cell TLRs and induction of immunoglobulin class-switch DNA recombination. Front Biosci 2012;17(7):2594–615.

[426] Tran TH, et al. B cell-specific and stimulation-responsive enhancers derepress Aicda by overcoming the effects of silencers. Nat Immunol 2010;11(2):148–54.

[427] Park SR, et al. HoxC4 binds to the promoter of the cytidine deaminase AID gene to induce AID expression, class-switch DNA recombination and somatic hypermutation. Nat Immunol 2009;10(5):540–50.

[428] Bossen C, Schneider P. BAFF, APRIL and their receptors: structure, function and signaling. Semin Immunol 2006;18(5):263–75.

[429] He B, et al. The transmembrane activator TACI triggers immunoglobulin class switching by activating B cells through the adaptor MyD88. Nat Immunol 2010;11(9):836–45.

[430] Chen Z, et al. TRAF3 acts as a checkpoint of B cell receptor signaling to control antibody class switch recombination and anergy. J Immunol 2020;205(3):830–41.

[431] Chen Z, Wang JH. How the signaling crosstalk of B cell receptor (BCR) and Co-receptors regulates antibody class switch recombination: a new perspective of checkpoints of BCR signaling. Front Immunol 2021;12:663443.

[432] Kawai T, Akira S. The role of pattern-recognition receptors in innate immunity: update on Toll-like receptors. Nat Immunol 2010;11(5):373–84.

[433] Ise W, et al. The transcription factor BATF controls the global regulators of class-switch recombination in both B cells and T cells. Nat Immunol 2011;12(6):536–43.

[434] Sayegh CE, et al. E-proteins directly regulate expression of activation-induced deaminase in mature B cells. Nat Immunol 2003;4(6):586–93.

[435] Sciammas R, et al. Graded expression of interferon regulatory factor-4 coordinates isotype switching with plasma cell differentiation. Immunity 2006;25(2):225–36.

[436] Murre C. Helix-loop-helix proteins and lymphocyte development. Nat Immunol 2005;6(11):1079–86.

[437] Murphy TL, Tussiwand R, Murphy KM. Specificity through cooperation: BATF-IRF interactions control immune-regulatory networks. Nat Rev Immunol 2013;13(7):499–509.

[438] Nutt SL, Tarlinton DM. Germinal center B and follicular helper T cells: siblings, cousins or just good friends? Nat Immunol 2011;12(6):472–7.

[439] Grumont RJ, Gerondakis S. Rel induces interferon regulatory factor 4 (IRF-4) expression in lymphocytes: modulation of interferon-regulated gene expression by rel/nuclear factor kappaB. J Exp Med 2000;191(8):1281–92.

[440] Gloury R, et al. Dynamic changes in Id3 and E-protein activity orchestrate germinal center and plasma cell development. J Exp Med 2016;213(6):1095–111.

[441] Muramatsu M, et al. Specific expression of activation-induced cytidine deaminase (AID), a novel member of the RNA-editing deaminase family in germinal center B cells. J Biol Chem 1999;274(26):18470–6.

[442] Jabara H, et al. The binding site for TRAF2 and TRAF3 but not for TRAF6 is essential for CD40-mediated immunoglobulin class switching. Immunity 2002;17(3):265–76.

[443] Dedeoglu F, et al. Induction of activation-induced cytidine deaminase gene expression by IL-4 and CD40 ligation is dependent on STAT6 and NFkappaB. Int Immunol 2004;16(3):395–404.

[444] Nelms K, et al. The IL-4 receptor: signaling mechanisms and biologic functions. Annu Rev Immunol 1999;17:701–38.

[445] Heldin CH, Miyazono K, ten Dijke P. TGF-beta signalling from cell membrane to nucleus through SMAD proteins. Nature 1997;390(6659):465–71.

[446] Pauklin S, et al. Estrogen directly activates AID transcription and function. J Exp Med 2009;206(1):99–111.

[447] Mai T, et al. Estrogen receptors bind to and activate the HOXC4/HoxC4 promoter to potentiate HoxC4-mediated activation-induced cytosine deaminase induction, immunoglobulin class switch DNA recombination, and somatic hypermutation. J Biol Chem 2010;285(48):37797–810.

[448] Kieffer-Kwon KR, et al. Interactome maps of mouse gene regulatory domains reveal basic principles of transcriptional regulation. Cell 2013;155(7):1507–20.

[449] Nagaoka H, et al. Preventing AID, a physiological mutator, from deleterious activation: regulation of the genomic instability that is associated with antibody diversity. Int Immunol 2010;22(4):227–35.

[450] Huong IT, et al. In vivo analysis of Aicda gene regulation: a critical balance between upstream enhancers and intronic silencers governs appropriate expression. PLoS One 2013;8(4):e61433.

[451] Yadav A, et al. Identification of a ubiquitously active promoter of the murine activation-induced cytidine deaminase (AICDA) gene. Mol Immunol 2006;43(6):529–41.

[452] Wohner M, et al. Molecular functions of the transcription factors E2A and E2-2 in controlling germinal center B cell and plasma cell development. J Exp Med 2016;213(7):1201–21.

[453] Lio CJ, et al. TET enzymes augment activation-induced deaminase (AID) expression via 5-hydroxymethylcytosine modifications at the. Sci Immunol 2019;4(34).

[454] Gonda H, et al. The balance between Pax5 and Id2 activities is the key to AID gene expression. J Exp Med 2003;198(9):1427–37.

[455] Basso K, et al. BCL6 positively regulates AID and germinal center gene expression via repression of miR-155. J Exp Med 2012;209(13):2455–65.

[456] Shaffer AL, et al. Blimp-1 orchestrates plasma cell differentiation by extinguishing the mature B cell gene expression program. Immunity 2002;17(1):51–62.

[457] Igarashi K, et al. Orchestration of plasma cell differentiation by Bach2 and its gene regulatory network. Immunol Rev 2014;261(1):116–25.

[458] Xu Z, et al. Regulation of aicda expression and AID activity: relevance to somatic hypermutation and class switch DNA recombination. Crit Rev Immunol 2007;27(4):367–97.

[459] Fujimura S, et al. Germinal center B-cell-associated DNA hypomethylation at transcriptional regions of the AID gene. Mol Immunol 2008;45(6):1712–9.

[460] Teng G, et al. MicroRNA-155 is a negative regulator of activation-induced cytidine deaminase. Immunity 2008;28(5):621–9.

[461] Dorsett Y, et al. MicroRNA-155 suppresses activation-induced cytidine deaminase-mediated Myc-Igh translocation. Immunity 2008;28(5):630–8.

[462] de Yébenes VG, et al. miR-181b negatively regulates activation-induced cytidine deaminase in B cells. J Exp Med 2008;205(10):2199–206.

[463] Borchert GM, Holton NW, Larson ED. Repression of human activation induced cytidine deaminase by miR-93 and miR-155. BMC Cancer 2011;11:347.

[464] Sanchez HN, et al. B cell-intrinsic epigenetic modulation of antibody responses by dietary fiber-derived short-chain fatty acids. Nat Commun 2020;11(1):60.

[465] McBride KM, et al. Regulation of hypermutation by activation-induced cytidine deaminase phosphorylation. Proc Natl Acad Sci USA 2006;103(23):8798–803.

[466] Pasqualucci L, et al. PKA-mediated phosphorylation regulates the function of activation-induced deaminase (AID) in B cells. Proc Natl Acad Sci USA 2006;103(2):395–400.

[467] Gazumyan A, et al. Amino-terminal phosphorylation of activation-induced cytidine deaminase suppresses c-myc/IgH translocation. Mol Cell Biol 2011;31(3):442–9.

[468] Aoufouchi S, et al. Proteasomal degradation restricts the nuclear lifespan of AID. J Exp Med 2008;205(6):1357–68.

[469] Luo Y, et al. CUL7 E3 ubiquitin ligase mediates the degradation of activation-induced cytidine deaminase and regulates the Ig class switch recombination in B lymphocytes. J Immunol 2019;203(1):269–81.

[470] Ochiai K, et al. Zinc finger-IRF composite elements bound by Ikaros/IRF4 complexes function as gene repression in plasma cell. Blood Adv 2018;2(8):883–94.

[471] Zaprazna K, Atchison ML. YY1 controls immunoglobulin class switch recombination and nuclear activation-induced deaminase levels. Mol Cell Biol 2012;32(8):1542–54.

[472] Demorest ZL, et al. The interaction between AID and CIB1 is nonessential for antibody gene diversification by gene conversion or class switch recombination. PLoS One 2010;5(7):e11660.
[473] Manis JP, et al. 53BP1 links DNA damage-response pathways to immunoglobulin heavy chain class-switch recombination. Nat Immunol 2004;5(5):481–7.
[474] Reina-San-Martin B, et al. Enhanced intra-switch region recombination during immunoglobulin class switch recombination in 53BP1-/- B cells. Eur J Immunol 2007;37(1):235–9.
[475] Ward IM, et al. 53BP1 is required for class switch recombination. J Cell Biol 2004;165(4):459–64.
[476] Casellas R, et al. Ku80 is required for immunoglobulin isotype switching. EMBO J 1998;17(8):2404–11.
[477] Boboila C, et al. Alternative end-joining catalyzes robust IgH locus deletions and translocations in the combined absence of ligase 4 and Ku70. Proc Natl Acad Sci USA 2010;107(7):3034–9.
[478] Weinstock DM, Brunet E, Jasin M. Formation of NHEJ-derived reciprocal chromosomal translocations does not require Ku70. Nat Cell Biol 2007;9(8):978–81.
[479] Tang ES, Martin A. NHEJ-deficient DT40 cells have increased levels of immunoglobulin gene conversion: evidence for a double strand break intermediate. Nucleic Acids Res 2006;34(21):6345–51.
[480] Reina-San-Martin B, et al. H2AX is required for recombination between immunoglobulin switch regions but not for intra-switch region recombination or somatic hypermutation. J Exp Med 2003;197(12):1767–78.
[481] Petersen S, et al. AID is required to initiate Nbs1/gamma-H2AX focus formation and mutations at sites of class switching. Nature 2001;414(6864):660–5.
[482] Bemark M, et al. Somatic hypermutation in the absence of DNA-dependent protein kinase catalytic subunit (DNA-PKcs) or recombination-activating gene (rag)1 activity. J Exp Med 2000;192(10):1509–14.
[483] Callén E, et al. Essential role for DNA-PKcs in DNA double-strand break repair and apoptosis in ATM-deficient lymphocytes. Mol Cell 2009;34(3):285–97.
[484] Oksenych V, et al. Functional redundancy between the XLF and DNA-PKcs DNA repair factors in V(D)J recombination and nonhomologous DNA end joining. Proc Natl Acad Sci USA 2013;110(6):2234–9.
[485] Han L, Yu K. Altered kinetics of nonhomologous end joining and class switch recombination in ligase IV-deficient B cells. J Exp Med 2008;205(12):2745–53.
[486] Soulas-Sprauel P, et al. Role for DNA repair factor XRCC4 in immunoglobulin class switch recombination. J Exp Med 2007;204(7):1717–27.
[487] Delacote F, et al. An xrcc4 defect or Wortmannin stimulates homologous recombination specifically induced by double-strand breaks in mammalian cells. Nucleic Acids Res 2002;30(15):3454–63.
[488] Li G, et al. Lymphocyte-specific compensation for XLF/cernunnos end-joining functions in V(D)J recombination. Mol Cell 2008;31(5):631–40.
[489] Di Virgilio M, et al. Rif1 prevents resection of DNA breaks and promotes immunoglobulin class switching. Science 2013;339(6120):711–5.
[490] Buonomo SBC, et al. Mammalian Rif1 contributes to replication stress survival and homology-directed repair. J Cell Biol 2009;187(3):385–98.
[491] Tepper S, et al. Restriction of AID activity and somatic hypermutation by PARP-1. Nucleic Acids Res 2019;47(14):7418–29.
[492] Robert I, Dantzer F, Reina-San-Martin B. Parp1 facilitates alternative NHEJ, whereas Parp2 suppresses IgH/c-myc translocations during immunoglobulin class switch recombination. J Exp Med 2009;206(5):1047–56.
[493] Paddock MN, et al. The BRCT domain of PARP-1 is required for immunoglobulin gene conversion. PLoS Biol 2010;8(7):e1000428.
[494] Lee-Theilen M, et al. CtIP promotes microhomology-mediated alternative end joining during class-switch recombination. Nat Struct Mol Biol 2011;18(1):75–9.
[495] Zhang Y, Jasin M. An essential role for CtIP in chromosomal translocation formation through an alternative end-joining pathway. Nat Struct Mol Biol 2011;18(1):80–4.
[496] Schrader CE, et al. Deletion of the nucleotide excision repair gene Ercc1 reduces immunoglobulin class switching and alters mutations near switch recombination junctions. J Exp Med 2004;200(3):321–30.
[497] Winter AG, et al. The repair and recombination enzyme ERCC1 is not required for immunoglobulin class switching. DNA Repair 2003;2(5):561–9.
[498] Bai W, et al. The 3′-flap endonuclease XPF-ERCC1 promotes alternative end joining and chromosomal translocation during B cell class switching. Cell Rep 2021;36(13).
[499] Du L, et al. A regulatory role for NBS1 in strand-specific mutagenesis during somatic hypermutation. PLoS One 2008;3(6):e2482.
[500] Yabuki M, Fujii MM, Maizels N. The MRE11-RAD50-NBS1 complex accelerates somatic hypermutation and gene conversion of immunoglobulin variable regions. Nat Immunol 2005;6(7):730–6.
[501] Reina-San-Martin B, et al. ATM is required for efficient recombination between immunoglobulin switch regions. J Exp Med 2004;200(9):1103–10.
[502] Lumsden JM, et al. Immunoglobulin class switch recombination is impaired in Atm-deficient mice. J Exp Med 2004;200(9):1111–21.
[503] Köcher S, et al. Radiation-induced double-strand breaks require ATM but not Artemis for homologous recombination during S-phase. Nucleic Acids Res 2012;40(17):8336–47.
[504] Pan-Hammarström Q, et al. Disparate roles of ATR and ATM in immunoglobulin class switch recombination and somatic hypermutation. J Exp Med 2006;203(1):99–110.
[505] Lou Z, et al. MDC1 maintains genomic stability by participating in the amplification of ATM-dependent DNA damage signals. Mol Cell 2006;21(2):187–200.
[506] Pérez-Durán P, et al. UNG shapes the specificity of AID-induced somatic hypermutation. J Exp Med 2012;209(7):1379–89.
[507] Imai K, et al. Human uracil-DNA glycosylase deficiency associated with profoundly impaired immunoglobulin class-switch recombination. Nat Immunol 2003;4(10):1023–8.
[508] Saribasak H, et al. Uracil DNA glycosylase disruption blocks Ig gene conversion and induces transition Mutations1. J Immunol 2006;176(1):365–71.
[509] Saribasak H, et al. XRCC1 suppresses somatic hypermutation and promotes alternative nonhomologous end joining in Igh genes. J Exp Med 2011;208(11):2209–16.
[510] Larsen E, et al. Early-onset lymphoma and extensive embryonic apoptosis in two domain-specific Fen1 mice mutants. Cancer Res 2008;68(12):4571–9.

[511] Kikuchi K, et al. Fen-1 facilitates homologous recombination by removing divergent sequences at DNA break ends. Mol Cell Biol 2005;25(16):6948–55.

[512] Ehrenstein MR, Neuberger MS. Deficiency in Msh2 affects the efficiency and local sequence specificity of immunoglobulin class-switch recombination: parallels with somatic hypermutation. EMBO J 1999;18(12):3484–90.

[513] Xue K, Rada C, Neuberger MS. The in vivo pattern of AID targeting to immunoglobulin switch regions deduced from mutation spectra in msh2−/− ung−/− mice. J Exp Med 2006;203(9):2085–94.

[514] Martomo SA, Yang WW, Gearhart PJ. A role for Msh6 but not Msh3 in somatic hypermutation and class switch recombination. J Exp Med 2004;200(1):61–8.

[515] Campo VA, et al. MSH6- or PMS2-deficiency causes re-replication in DT40 B cells, but it has little effect on immunoglobulin gene conversion or on repair of AID-generated uracils. Nucleic Acids Res 2013;41(5):3032–46.

[516] Roa S, et al. Ubiquitylated PCNA plays a role in somatic hypermutation and class-switch recombination and is required for meiotic progression. Proc Natl Acad Sci USA 2008;105(42):16248–53.

[517] Hirota K, et al. Simultaneous disruption of two DNA polymerases, Polη and Polζ, in Avian DT40 cells unmasks the role of Polη in cellular response to various DNA lesions. PLoS Genet 2010;6(10).

[518] Martomo SA, et al. Normal hypermutation in antibody genes from congenic mice defective for DNA polymerase iota. DNA Repair 2006;5(3):392–8.

[519] Schenten D, et al. DNA polymerase kappa deficiency does not affect somatic hypermutation in mice. Eur J Immunol 2002;32(11):3152–60.

[520] Okada T, et al. Multiple roles of vertebrate REV genes in DNA repair and recombination. Mol Cell Biol 2005;25(14):6103–11.

[521] Kohzaki M, et al. DNA polymerases nu and theta are required for efficient immunoglobulin V gene diversification in chicken. J Cell Biol 2010;189(7):1117–27.

[522] Takata K-I, et al. Analysis of DNA polymerase ν function in meiotic recombination, immunoglobulin class-switching, and DNA damage tolerance. PLoS Genet 2017;13(6):e1006818.

Chapter 14

Molecular Pathogenesis of B-Cell Lymphomas

Laura Pasqualucci[1,2,3], Pantaleo De Simone[1] and Riccardo Dalla-Favera[1,2,3,4,5]

[1]*Institute for Cancer Genetics, Columbia University, New York, NY, United States;* [2]*Department of Pathology & Cell Biology, Columbia University, New York, NY, United States;* [3]*Herbert Irving Comprehensive Cancer Center, Vagelos College of Physicians and Surgeons, Columbia University, New York, NY, United States;* [4]*Department of Genetics & Development, Columbia University, New York, NY, United States;* [5]*Department of Microbiology & Immunology, Columbia University, New York, NY, United States*

1. INTRODUCTION

B-cell lymphomas (BCLs) represent a heterogeneous group of biologically and clinically distinct neoplasms, including over 40 different subtypes derived from the malignant transformation of mature B cells that, in most cases, have experienced the germinal centre (GC) reaction. This chapter will focus on the molecular pathogenesis of the most common and well-characterised BCLs, including mantle cell lymphoma (MCL), Burkitt lymphoma (BL), follicular lymphoma (FL), diffuse large B-cell lymphoma (DLBCL), Hodgkin lymphoma (HL), and chronic lymphocytic leukaemia (CLL), which also derives from mature B cells. Emphasis will be given to the mechanisms of genetic lesion and the nature of the involved genes in relationship to the normal biology of GC B lymphocytes.

2. THE CELL OF ORIGIN OF LYMPHOMAS

2.1 The Biology of the Germinal Centre Reaction

With the exception of most MCLs and a subset of CLLs, the majority of BCLs derive from the malignant expansion of B cells reflecting various stages of GC transit. While we refer to other chapters in this book for a detailed description of the GC, we summarise here aspects that are relevant to BCL pathogenesis.

GCs are specialised structures that form in peripheral lymphoid organs when mature, naive B cells encounter a foreign antigen, in the context of signals delivered by CD4+ T cells and antigen-presenting cells (Figure 14.1) [1,2]. They represent a highly dynamic microenvironment characterised by two histologically well-defined zones: the dark zone (DZ), which consists of rapidly proliferating centroblasts (doubling time 6–12 h), and the light zone (LZ), which consists of more quiescent cells termed centrocytes, amidst a network of resident accessory cells (e.g. follicular dendritic cells, FDC and T follicular helper cells, Tfh) [3–5]. The DZ is the site where GC B cells modify the variable region of their immunoglobulin (*IG*) genes by the process of somatic hypermutation (SHM), which introduces mostly single nucleotide substitutions in order to change their affinity for the antigen [1,6–8]. Conversely, the LZ is the site of selection based on affinity to the antigen [2]. Here, B cells encounter the antigen again, presented by FDCs, and compete for survival signals delivered by a limited number of CD4+ Tfh cells [2]. Only the few LZ B cells expressing a BCR with high affinity for the antigen are selected for survival and differentiation into memory B cells or plasma cells, while the majority of cells will be eliminated by apoptosis [4]; a subset of cells will re-enter the DZ for additional rounds of proliferation, SHM and affinity selection, following stimulation by a variety of different signals that lead to the upregulation of MYC expression [9–11]. The LZ is also the site where B cells undergo class-switch recombination (CSR) [12], a DNA remodelling event that confers distinct effector functions to antibodies with identical specificities. However, studies in mice have shown that CSR may occur immediately after Ag encounter, prior to GC entry and SHM [13]. More recently, the advent of single-cell transcriptomic technologies has provided a more granular picture of this

Molecular Biology of B Cells. https://doi.org/10.1016/B978-0-323-95895-0.00020-9

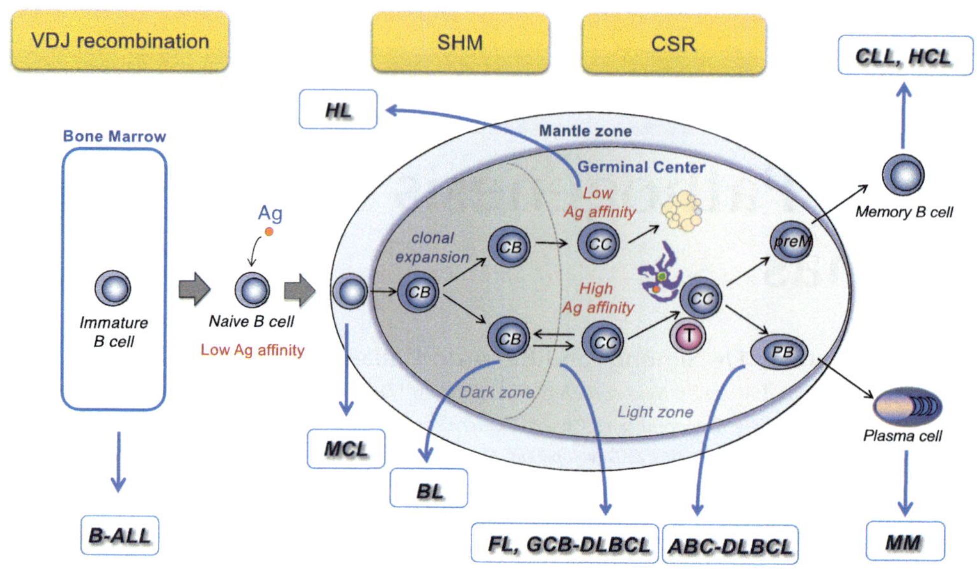

FIGURE 14.1 **The germinal centre reaction and lymphomagenesis.** Schematic representation of a lymphoid follicle illustrating the germinal centre (GC) and the mantle zone (MZ). B cells that have successfully rearranged their *IG* genes in the bone marrow through the process of V(D)J recombination move to peripheral lymphoid organs as naive B cells. Upon encounter with a T-cell-dependent antigen, B cells become proliferating centroblasts (CBs) in the dark zone of the GC and subsequently transition into centrocytes (CCs) in the light zone; these cells shuttle continuously between the dark and light zone while undergoing iterative rounds of SHM and affinity selection. Only GC B cells with high affinity for the antigen are positively selected to leave the GC and terminally differentiate into plasma cells or memory B cells; low-affinity clones are eliminated by apoptosis. Thick, blue arrows link various lymphoid malignancies to their postulated normal counterpart, as identified based on the expression of phenotypic markers and the presence of somatically mutated *IG* genes. *B-ALL*, B cell acute lymphoblastic leukemia; *BL*, Burkitt lymphoma; *CLL*, chronic lymphocytic leukaemia; *DLBCL*, diffuse large B-cell lymphoma (GCB, germinal centre B cell-like; ABC, activated B cell-like); *FL*, follicular lymphoma; *HCL*, hairy cell leukaemia; *HL*, Hodgkin lymphoma; *MCL*, mantle cell lymphoma; *MM*, multiple myeloma.

heterogeneous population in both human and murine GCs, revealing multiple clusters that correspond to transitional states in cells recirculating between the DZ and LZ, with a significant proportion of 'intermediate' B cells, as well as the precursors of memory B cells and plasma cells [14,15].

SHM and CSR represent B cell-specific functions that are executed by the activation-induced cytidine deaminase (AID) enzyme and involve single- or double-strand breaks [16–18], a notion that will become important in the understanding of the mechanisms generating genetic alterations in BCL.

This oversimplified description of the GC reaction is useful to introduce three basic concepts that led to the current understanding of BCL pathogenesis as a GC-derived malignancy. First, the presence of somatically mutated *IG variable region* (*IGV*) genes unequivocally demonstrated that most BCL types derive from the clonal expansion of GC-experienced B cells [19]. Second, two common mechanisms of genetic lesion associated with BCL – namely chromosomal translocations and aberrant somatic hypermutation (ASHM [20]) – are due to mistakes in the AID-dependent machinery that normally diversifies the *IG* genes during B lymphocyte differentiation. Third, the DZ and LZ phenotypes can be recognised in the transcriptional profile of different lymphoma subtypes, further supporting their origin from B cells in distinct states of the GC reaction. In particular, BL and DLBCL with a double-hit signature (DHSig) share significant similarities with the highly proliferative DZ B-cell population [21,22], whereas FL and the so-called germinal centre B-cell-like (GCB)-DLBCL (discussed in the DLBCL section) appear more closely related to LZ B cells [21,23], and activated B-cell-like (ABC)-DLBCL is presumably derived from a committed plasma blast [23], although recent analyses support a pre-memory B cell as its putative normal counterpart in a subset of cases [14,24] (Figure 14.1).

2.2 Transcriptional Regulation of the Germinal Centre

The unique biology of the GC reaction is under the control of an intricate network of transcription factors (TFs), epigenetic modifiers and cytokine receptors/signal transduction molecules whose coordinated regulation guarantees the ordered execution of distinct biological functions associated with GC seeding, DZ:LZ recirculation, and GC exit (Figure 14.2).

MYC is a nuclear phosphoprotein that functions as a sequence-specific DNA-binding transcriptional regulator controlling the transcription of target genes with diverse roles in regulating cell growth by affecting DNA replication, energy metabolism, protein synthesis and telomere

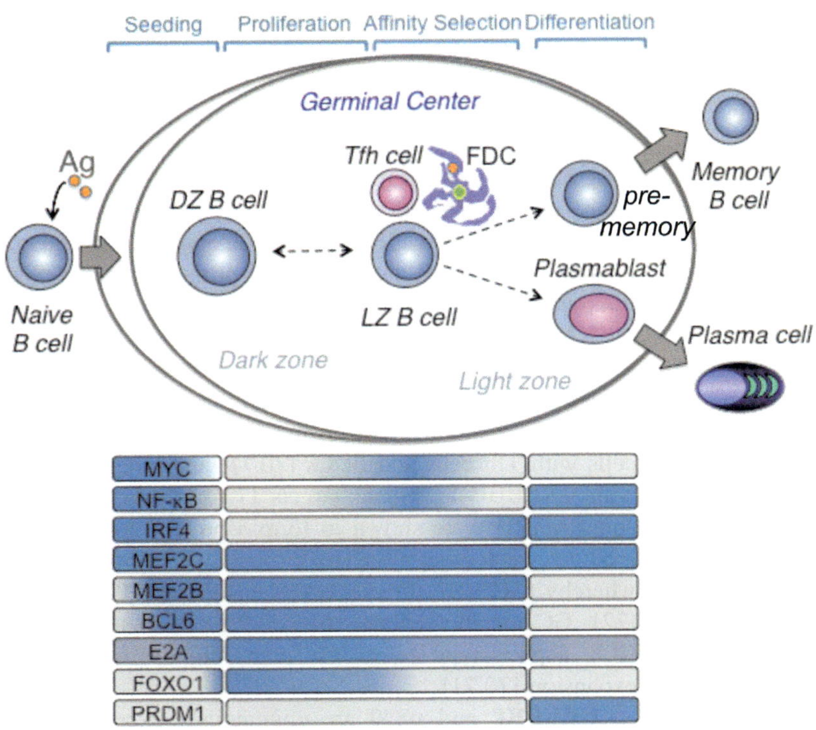

FIGURE 14.2 **Major transcription factors implicated in the control of the GC reaction.** The expression pattern of major TFs implicated in various phases of the GC reaction is illustrated by colour shades (blue, high expression levels or nuclear localisation (FOXO1); grey, no expression). *Reproduced with permission from Pasqualucci et al. [25].*

elongation [26–28]. It is a major regulator of both GC initiation and DZ re-entry, whose protein expression is induced in GC-founding B cells but is downregulated in DZ B cells, partly due to direct transcriptional repression by BCL6, and is then re-induced in a small subset of LZ B cells undergoing affinity-based selection, priming them for re-entry into the DZ [11]. Consistently, GC formation is abrogated in vivo when *MYC* is conditionally deleted early upon antigenic stimulation, whereas its deletion in the GC LZ leads to the dissolution of established GCs [10,11].

An absolute requirement for GC formation is the transcriptional repressor **BCL6** [29–31]. BCL6 is selectively upregulated in the GC [32] by signals that are still incompletely understood but include IRF4 [33], IRF8 [34] and MEF2B [35]. BCL6 binds to specific sequences in the promoters and enhancers of hundreds of target genes and negatively modulates their expression to control signals emanating from the B-cell receptor (BCR) and CD40 receptor [36,37], T-cell-mediated B-cell activation [36], the induction of apoptosis (via suppression of BCL2) [36,38], the sensing and response to DNA damage (via suppression of p53 and ATR) [39–42], various cytokine and chemokine signalling pathways (e.g. those triggered by interferon and TGFβ) [36,38] and terminal differentiation, via direct suppression of *BLIMP1* [43–46]. This transcriptional programme suggests that BCL6 is critical to establish the proliferative status of DZ B cells and to allow the execution of antigen-specific DNA modification processes (SHM and CSR) without eliciting DNA damage responses. In addition, BCL6 prevents the engagement of a variety of signalling pathways that could lead to premature activation and differentiation of cells prior to affinity selection. Once these processes are completed, engagement of the BCR by the antigen and activation of the CD40 receptor by the CD40 ligand on CD4+ T-cells, among other signals, downregulate BCL6 at both translational and transcriptional level, relieving the expression of its targets and restoring the capabilities of the B cell to differentiate. Another important player in opposing the activity of BCL6 is the epigenetic writer CREBBP, which counteracts its repressive function both directly, by acetylating and inactivating the BCL6 protein [47,48], and indirectly, by depositing activation marks on an overlapping set of target enhancers/super-enhancers (SEs), thus priming them for rapid firing in response to LZ-associated signals [49,50].

Myocyte-specific Enhancer Factor 2B (**MEF2B**) and 2C (**MEF2C**) [51] regulate the GC reaction through slightly different but partially redundant functions, as demonstrated by the requirement of both genes' deletion to completely abrogate the GC response in vivo [52]. In particular, MEF2B expression is confined to GC B cells, where it instructs the DZ programme by modulating multiple cis-regulatory elements that include, among others, the full set of GC-specific SEs [35,52].

FOXO1, a member of the Fox-O family of Forkhead TFs [53], is activated in the DZ and in a fraction of LZ B cells destined to re-enter the DZ, where its antagonistic player PI3K-AKT is not expressed [54,55]. FOXO1 sustains the DZ B cell phenotype and maintains GC polarity by regulating the expression of multiple genes involved in DZ-associated functions (e.g. the chemokine receptor CXCR4, cell proliferation and negative modulation of DNA repair), while silencing activation and differentiation pathways characteristic of the LZ programme, partially in cooperation with BCL6 [54,55]. This dual activity is reflected in the phenotype of GC-specific *Foxo1*-null mouse models, which form apparently normal GCs entirely composed of LZ B cells and devoid of DZ gene programms.

E2A, encoded by the *TCF3* gene, is dispensable during the initial phases of antigen activation [56] but is required for the proliferative expansion of GC B cells, in part via upregulation of CCND3 [57], and consistent with its higher expression levels in the DZ [21].

Finally, termination of the mature GC programme and differentiation into effector plasma cells require various signalling cascades emanating from the BCR, the CD40 receptor or the endolysosomal Toll-like receptor (TLR) 9, culminating in the activation of **NF-κB** [58]. Among its targets, **IRF4** shows a biphasic expression pattern, contributing first to the initiation of the GC and then being re-expressed in a subset of LZ B cells to ensure terminal differentiation by suppressing BCL6 [59–61]. The downregulation of BCL6 relieves the transcription of **BLIMP1**, a zinc finger transcriptional repressor that is also required for terminal B-cell differentiation [46,62] and, in turn, maintains the plasma cell programme through a negative feedback loop with BCL6 [45].

2.3 Epigenetic Regulators Relevant to the GC

GC B cells implement these phenotypic changes by rapidly reprogramming their epigenome, modulating the accessibility of the underlying DNA to specific TFs through massive chromatin organisation and the post-translational modification of histone tails [63]. This interplay involves the opposing functions of macromolecular complexes that function as transcriptional co-activators (e.g. the acetyltransferases CREBBP/p300 and the histone methyltransferase KMT2D) or repressors (e.g. the methyltransferase EZH2), as well as enzymes implicated in the control of DNA methylation. Notably, mutations in histone modifier genes represent a hallmark of GC-derived lymphomas found in virtually all FLs and >50% of GCB-DLBCL, indicating a prominent role for epigenetic perturbations in driving malignant transformation.

EZH2 is a methyltransferase that is specifically upregulated in GC B cells to establish bivalent promoters at genes involved in proliferation checkpoints (*CDKN1A*), antigen presentation, the immune-synapse and terminal B-cell differentiation [64,65]. EZH2 poises gene transcription by trimethylating the lysine 27 of histone H3, which in turn recruits the Polycomb-like BCOR complex in combination with BCL6 [66,67], enabling the proliferative phenotype of GC B cells at the expenses of differentiation [68]. *Ezh2* knock-out mice are unable to form GCs, while expression of the most common lymphoma-associated mutant $EZH2^{Y646}$ allele promotes GC hyperplasia and cooperates with BCL2 deregulation to promote overt FL-like tumours [64,65].

KMT2D primes active transcription by monomethylating H3K4 at enhancer chromatin [69]. It occupies a broad array of GC-specific E/SEs that are involved in signalling pathways triggered by the interaction of B cells with Tfh cells and FDCs [70,71] and largely overlap with CREBBP-dependent targets [72]. Indeed, CREBBP and KMT2D were shown to form a protein complex and modulate each other's activity through a feed forward loop, whereby CREBBP directly acetylates the KMT2D protein [72]. GC-specific deletion of KMT2D leads to enlarged GCs, a phenotype commonly preceding overt neoplastic expansion [70,71].

CREBBP and **p300** are ubiquitously expressed acetyltransferases that deposit H3K27Ac/H3K18Ac marks on both histone and non-histone proteins [73]. CREBBP is involved in the decision-making of B cells that exit the GC, by counteracting the activity of BCL6 and facilitating the reactivation of its target genes [49,50]. The CREBBP-dependent transcriptional network is preferentially enriched in GC LZ signature genes, including those implicated in antigen presentation, BCR- and CD40 signalling [49,50,74]. In contrast, the p300-controlled enhanceosome is more closely related to the DZ programme (e.g. DNA damage response and replication genes), although p300 can partially compensate for CREBBP loss in vivo, revealing a synthetic lethal interaction that could be exploited for therapeutic targeting in CREBBP-mutant lymphomas [75].

3. MECHANISMS OF GENETIC LESION IN LYMPHOMA

3.1 Chromosomal Translocations

Chromosomal translocations are generated through the reciprocal and balanced recombination of two chromosomes. They are often recurrently associated with a given tumour type, where they are clonally represented in each tumour sample. In B-cell malignancies, chromosomal translocations occur at least in part as a consequence of mistakes in *IG* gene remodelling mechanisms (Figure 14.3) and, based on the characteristics of the chromosomal breakpoint, can be broadly divided into three groups: (1) translocations derived from mistakes of the RAG-mediated

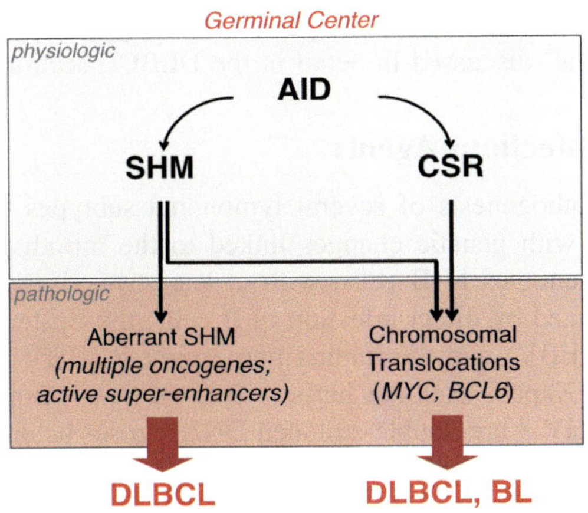

FIGURE 14.3 Model for the initiation of chromosomal translocations and ASHM during lymphomagenesis. BCL-associated genetic lesions are initiated by mistakes occurring during the physiologic SHM and CSR processes in the highly proliferative environment of the GC (top). These events lead to chromosomal translocations, which in most cases juxtapose the *IG* genes to one of several proto-oncogenes (e.g. *BCL2,MYC, BCL6*), and ASHM, which targets multiple genes specifically at active super-enhancers leading to dysregulated gene expression, thus contributing to the pathogenesis of lymphoma.

V(D)J recombination process, as it is the case for translocations involving *IGH* and *CCND1* in MCL or *IGH* and *BCL2* in FL [76–78]; (2) translocations mediated by errors in the AID-dependent CSR process, such as those involving the *IG* genes and *MYC* in sporadic BL (sBL) [76] and (3) translocations occurring as by-products of the AID-mediated SHM mechanism, which also generates DNA breaks, such as those joining the *IG* loci and the *MYC* locus in endemic BL (eBL) or the *BCL6* locus in FL and DLBCL [76]. In lymphoma-prone mouse models, removal of AID was sufficient to abrogate the generation of *MYC-IGH* translocations in normal B cells undergoing CSR [79,80] and to prevent the development of GC-derived BCLs [81,82]. These studies provided experimental evidence for the requirement of *IG* gene remodelling mechanisms in the pathogenesis of BCL in mice, suggesting that the same mechanism may be responsible for the human tumours.

In all BCL-associated chromosomal translocations, the recombination site occurs in variable genomic proximity of a proto-oncogene; however, and in contrast with acute leukaemias, it generally does not affect the coding domain of the oncogene, and no gene fusion is generated (an exception being represented by the BIRC3/MALT1 fusion protein in extranodal marginal zone lymphomas [83,84]). Instead, heterologous regulatory sequences derived from the partner chromosome are juxtaposed in front of the oncogene, deregulating its pattern of expression (proto-oncogene deregulation). This process is defined as homotopic when a proto-oncogene whose expression is tightly regulated in the normal counterpart of the tumour cell becomes constitutively expressed, and heterotopic when the proto-oncogene is not expressed in the putative normal counterpart of the tumour cell and undergoes ectopic expression in the lymphoma. The heterologous regulatory sequences responsible for proto-oncogene deregulation are often derived from antigen receptor loci, which are expressed at high levels in the target tissue [76]. However, in certain translocations, such as the ones involving *BCL6* in DLBCL, promoter regions from distinct chromosomal sites can be found juxtaposed to the proto-oncogene in different tumour cases, leading to so-called 'promiscuous' translocations [85–92]. The molecular cloning of the genetic loci involved in most recurrent translocations has led to the identification of several proto-oncogenes involved in lymphomagenesis.

3.2 Copy Number Gains and Amplifications

In addition to chromosomal translocations, the structure of proto-oncogenes and their pattern of expression can be altered by copy number gains and amplifications, which lead to overexpression of an intact protein. Compared to epithelial cancers, relatively few genes have been identified so far as specific targets of amplification in BCL, including *REL* and *BCL2* in DLBCL [93–96], and the genes encoding for PD-1 ligands in primary mediastinal B-cell lymphoma (PMBCL) and classical Hodgkin lymphoma (cHL) [97].

3.3 Gain-of-Function Mutations

The biological properties of a proto-oncogene can be altered by somatic point mutations in its coding sequences, which may lead to stabilisation or constitutive activation of its protein product by several mechanisms. Over the past decade, the use of massively parallel sequencing technologies has revolutionised the field by allowing the identification of numerous previously unappreciated targets of somatic mutations in cancer, including lymphomas. These genes will be discussed in individual disease sections. Of note, mutations of the *RAS* genes, a very frequent proto-oncogene alteration in human neoplasia, are rare in lymphomas [98,99].

3.4 Inactivating Mutations and Deletions

Inactivation of the *TP53* tumour suppressor gene, possibly the most common target of genetic alterations in human cancer [100], is found at relatively lower frequencies in BCL, where it is mostly restricted to specific disease subtypes, such as BL and DLBCL derived from the transformation of FL or CLL [101,102]. The mechanism of *TP53* inactivation in BCL is analogous to the one observed in human neoplasia in general, entailing point mutation of one allele with chromosomal deletion or mutation of the second allele, or heterozygous missense mutations leading to dominant negative proteins. Additional candidate tumour suppressor genes that are lost in BCL through specific chromosomal deletions and/or deleterious mutations include, among others, the *PRDM1* gene on 6q21, which encodes BLIMP1 and is biallelically inactivated in ~25% of ABC-DLBCL cases [103–105], and the gene encoding for the negative NF-κB regulator A20 on chromosome 6q23, which is inactivated in ABC-DLBCL, PMBCL and subtypes of marginal zone and HL [106–109]. Monoallelic inactivating mutations and deletions affect the acetyltransferase genes *CREBBP* and *EP300* in a significant proportion of DLBCL and FL, consistent with a role as haploinsufficient tumour suppressors [48]. These two lymphoma types also frequently harbour monoallelic or biallelic truncating mutations of *KMT2D*. In over 50% of CLL cases, genetic losses affect the *DLEU2/miR15-a/16.1* cluster on chromosome 13q14.3[110], while the *CDKN2A/B* (p16/INK4a) locus is inactivated by focal homozygous deletions in a substantial fraction of transformed FL (tFL), Richter syndrome (RS) and ABC-DLBCL cases, and more rarely by epigenetic transcriptional silencing in various BCL [110].

3.5 Aberrant Somatic Hypermutation

The term ASHM was originally coined to indicate a malfunction of the physiologic SHM process, leading to the mutational targeting of multiple non-*IG* genes, most commonly in DLBCL [111]. Most recently, the unbiased analysis of the non-coding genome of DLBCL, integrated with chromatin studies, revealed that this pervasive AID-dependent hypermutation activity specifically targets regulatory domains, in particular SEs, in over 90% of all DLBCLs, establishing a major mechanism of genetic damage in this disease [112].

In normal GC B cells, the process of SHM is tightly regulated to introduce mutations only in the rearranged *IGV* genes [6] as well as in the 5′ region of a few other genes, including *BCL6* and the *CD79* components of the BCR [113–115], although the functional role of mutations found in non-*IG* genes remains obscure. Conversely, most DLBCL cases and, at lower frequencies, few other lymphoma types were found to exhibit multiple mutational events attributable to this process in numerous loci, including well-known proto-oncogenes such as *PIM1*, *PAX5* and *MYC* (Figure 14.3) [111,116–121]. These mutations are typically distributed within ~2 Kb from the transcription initiation site (the hypermutable domain in the *IG* locus) affect regions corresponding to active intragenic SEs, although promoter-distal E/SEs are also recurrently hit [112]. Depending on the genomic configuration of the target gene, mutations may affect non-coding as well as coding regions, with the potential of altering the regulation and/or function of critical genes activating their oncogenic potential (discussed in detail in the DLBCL section).

3.6 Infectious Agents

The pathogenesis of several lymphoma subtypes is associated with genetic changes linked to the introduction of viral genomes in B cells, with at least two viruses being implicated by direct infection of B cells: the Epstein–Barr virus (EBV) and the human herpesvirus-8 (HHV-8), also called Kaposi sarcoma herpes virus (KSHV). Both EBV and HHV-8 are double-stranded DNA viruses belonging to the *Gammaherpesvirinae* subfamily. Other infectious agents, including the human immunodeficiency virus (HIV), hepatitis C virus (HCV), *Helicobacter pylori* (*H. pylori*) and *Chlamydophila psittaci* (*C. psittaci*) have an indirect role in lymphoma pathogenesis by either impairing the immune system and/or by providing chronic antigenic stimulation, and will not be discussed in this chapter [122–124].

EBV was initially identified in cases of endemic African BL [125] and subsequently detected also in a fraction of sBL, HIV-related lymphomas, and primary effusion lymphomas (PELs), a rare neoplasm that arises in the context of severe immunosuppression (e.g. HIV infection, organ transplantation or physiological immunosenescence) [126–131]. Multiple evidences support a pathogenic role for this virus in BCL: first, EBV is able to significantly alter the growth of B cells [132,133]; and second, EBV-infected lymphomas usually display a single form of the otherwise highly heterogeneous fused EBV genome termini, suggesting that the lymphoma cell population derived from the clonal expansion of a single infected cell [126,134]. Nonetheless, the role of EBV in lymphomagenesis remains unclear. In particular, despite EBV gene expression related to different latency programmes can be identified in most of EBV-associated BCLs, this pattern is highly heterogeneous, with EBV-positive BL cases often lacking the expression of essential transforming genes [133,135]. Indeed, a ‘hit and run’ model has been recently proposed, by which the virus leads to B-cell transformation followed by the acquisition of oncogenic aberrations that make the clones independent from the initial EBV gene programme [136,137].

HHV-8 was first identified in tissues of HIV-related Kaposi’s sarcoma [138] and subsequently found in PEL cells as well as in a large proportion of multicentric Castleman’s disease cases [139–142]. Like other gamma herpesviruses, HHV-8 is also lymphotropic and can infect lymphocytes both in vitro and in vivo, establishing latency in its host and undergoing periods of lytic replications upon immune suppression [138,141,142]. Lymphoma cells naturally infected by HHV-8 harbour the viral genome in episomal configuration and display a marked restriction of viral gene expression, suggesting a pattern of latent infection [143].

4. MOLECULAR PATHOGENESIS OF MOST COMMON LYMPHOMA TYPES

4.1 Mantle Cell Lymphoma

MCL comprises 3%–10% of all mature BCL diagnoses and is generally regarded as an incurable disease [110,144]. Two molecular subtypes of MCL are recognised, based on *IGHV* mutational status, genetic aberrations and clinical behaviour [145,146]. The conventional MCL (cMCL) is characterised by absent/low *IGHV* mutation load, genetic instability and high expression of the TF SOX11 [147]; the leukaemic non-nodal subtype (nnMCL) is defined by the presence of hypermutated *IGHV* genes, non-complex karyotype and low SOX11 expression [148]. Based on *IGHV* mutation status and related gene-expression profile, cMCL and nnMCL seem to derive from a naive-like and memory-like B cell, respectively [149].

MCL is typically associated with the t(11; 14) (q13; q32) translocation that juxtaposes the *IGH* gene to a region containing the *CCND1* gene on chromosome 11q13 [150,151]. In a minority of cases, the rearrangement involves the *IGL* or *IGK* genes; this includes rare cases of t(11; 14) negative MCLs where cryptic rearrangements have been described using WGS, leading to the juxtaposition of the enhancer elements of the *IG* light chain genes [152].

These translocations cause homotopic deregulation and overexpression of cyclin D1, a member of the D-type G_1 cyclins that regulates the early phases of the cell cycle and is normally not expressed in resting B cells [153–155]. In addition to t(11; 14), ~10% of MCLs over-express aberrant or shorter cyclin D1 transcripts, as a consequence of secondary rearrangements, microdeletions or somatic point mutations in the gene 3′ untranslated region (at least 40% of cases) [156–159]. These alterations lead to cyclin D1 overexpression through the removal of destabilising sequences and the consequent increase in the mRNA half-life, and are more commonly observed in cases characterised by high proliferative activity and a more aggressive clinical course [160]. The pathogenic role of cyclin D1 deregulation in human neoplasia is suggested by the ability of the overexpressed protein to transform cells in vitro and to promote B-cell lymphomagenesis in transgenic mice, although only when combined to other oncogenic events [161,162]. Nonetheless, an animal model that faithfully recapitulates the features of the human MCL is still lacking. Importantly, the frequency and specificity of the ectopic expression of Cyclin D1 in the tumour cells provides an excellent immunohistochemical marker for MCL diagnosis, at least in those cases lacking mutations in the C-terminal domain [144]. A minority of MCL cases that not present *CCND1* aberrations are characterised by the overexpression of cyclin D2 and translocations that involve the *CCND2* gene with a light chain *IG* gene [163]. The oncogenic role of *CCND2* in MCL has been recently described in an MCL-like (B1a-derived) conditional mouse model, where the overexpression of Cyclin D2 drives lymphomagenesis with an aggressive behaviour when associated with *TP53* loss [164].

Besides the canonical t(11; 14), almost 90% of MCL cases display additional chromosomal aberrations, both numerical and structural [160,165–168]. Recurrent genetic alterations involved in MCL include genomic deletions and mutations inactivating the *ATM* gene (50% of SOX11-positive cases), loss of *TP53* (20%–25% of patients, typically associated with adverse prognosis) and inactivation of the *CDKN2A* cell cycle regulator gene ($p16^{INK4a}$) by deletions, point mutations or promoter hypermethylation [159,169–173]. In a small number of cases, *BMI1* is amplified and/or overexpressed, possibly as an alternative mechanism to the loss of *CDKN2A* [159,174]. Also associated with aggressive tumours are mutations activating the Notch signalling pathway, including *NOTCH1* (12% of clinical samples) and NOTCH2 (5% of samples) [170–172]. These alterations are mutually exclusive and mostly consist of truncating events that remove the PEST sequences required for protein degradation, leading to protein stabilisation. Recently, novel recurrent noncoding mutations were found in the sequences surrounding the *HNRNPH1* gene, leading to altered splicing and deregulated protein expression [172]. Finally, less common yet recurrent and presumably functionally significant mutations involve genes encoding the anti-apoptotic protein BIRC3, the Toll-like receptor 2 (TLR2), the chromatin modifiers WHSC1 and KMT2D and the MEF2B TF [170]. Analysis of the mutational landscape in the two different MCL subtypes showed that most of these mutations are enriched in cMCL, while mutations in *CCND1*, *TP53* and *TERT* are more represented in nnMCL [159].

4.2 Burkitt Lymphoma

BL is an aggressive lymphoma deriving from GC B cells and resembling transformed DZ centroblasts, as documented by the presence of highly mutated *IGV* sequences [175–178] and a distinct transcriptional signature [21,179,180]. BL has been historically divided in three epidemiological variants: sBL, eBL and immunodeficiency-associated BL (iBL), which is often diagnosed as the initial manifestation of AIDS [145,146]. Thanks to recent advances in the comprehension of the molecular background of the disease, it is now clear that the established clinical variants do not recapitulate the biological diversity of the disease. Conversely, EBV status associates with distinct mutational landscapes [181–184], leading the latest revision of the WHO classification of lymphoid neoplasms recommending the distinction between EBV-positive and EBV-negative BLs [145].

The genetic hallmark of BL is the presence of chromosomal translocations involving the *MYC* gene on region 8q24 and one of the *IG* loci on the partner chromosome [185,186], which drives oncogene activation by enhancer hijacking. In ~80% of cases, this is represented by the *IGH* locus, leading to the characteristic t(8; 14) (q24; q32), while in the remaining 20% of cases, either *IGK* (2p12) or *IGL* (22q11) are involved [185–187]. While fairly homogeneous at the microscopic level, these translocations show a high degree of molecular heterogeneity, since the breakpoints are located 5′ and centromeric to *MYC* in t(8; 14), but map 3′ to *MYC* in t(2; 8) and t(8; 22) [188]. Further molecular heterogeneity derives from the exact breakpoint sites observed on chromosomes 8 and 14 in t(8; 14): translocations of eBL tend to involve sequences at an undefined distance (>1000 Kb) 5′ to *MYC* on chromosome 8 (Class III breakpoint) and sequences within or in proximity to the *IGHJ* region on chromosome 14 [189–191]. In sBL, t(8; 14) preferentially involves sequences within (Class I breakpoint) [190] or immediately 5′ to *MYC* (<3 Kb; Class II breakpoint) on chromosome 8 and within the *IG* switch regions on chromosome 14 [189–191]. The diverse molecular architecture of these translocations is thought to reflect the distinct mechanism of *IG* gene remodelling involved in their generation, that is, CSR in sBL and iBL and SHM in eBL [76].

The common consequence of t(8; 14), t(2; 8) and t(8; 22) is the ectopic and constitutive overexpression of the MYC proto-oncogene [192–194], which is normally absent in the majority of proliferating GC B cells, in part due to BCL6-mediated transcriptional repression [11,195]. Oncogenic activation of MYC in BL is mediated by at least three distinct mechanisms: (i) juxtaposition of the *MYC* coding sequences to heterologous enhancers derived from the *IG* loci [192–194]; (ii) structural alterations of the gene 5′ regulatory sequences, which alter the responsiveness to cellular factors controlling its expression (in particular, the *MYC* exon 1/intron 1 junction encompasses critical regulatory elements that are either decapitated by the translocation or mutated in the translocated alleles, such as the suppressive BCL6 element that is altered in class I/II breakpoints) [196,197]; and (iii) amino acid substitutions within the gene exon 2, encoding for the protein transactivation domain [198,199]; these mutations mainly cluster in proximity of known MYC phosphorylation sites and can abolish the ability of *p107*, a nuclear protein related to *RB1*, to suppress *MYC* activity [200], or increase protein stability [201,202]. Somatic mutations in the *MYC* gene can be identified in 40%–70% of BL cases and reflect ASHM activity [111,197,203,204].

Besides disrupting the normal DZ:LZ dynamics of the GC, deregulated expression of MYC is thought to cause genomic instability, and thus contribute to tumour progression by facilitating the occurrence of additional genetic lesions [205]. Dysregulation of MYC expression in a number of transgenic mouse models leads to the development of aggressive BCLs with high penetrance and short latency, although not fully recapitulating the biology of human BL [202,206,207]. In contrast, a mouse model with features highly reminiscent of human BL was obtained by combining GC-specific deregulated expression of MYC in GC B cells with aberrant PI3 kinase signalling [208].

Integrative genomic and transcriptomic studies have recently revealed additional oncogenic mechanisms that cooperate with MYC in the development of this aggressive lymphoma. These include mutations in several genes associated with BCR signalling (*TCF3*, *ID3*), cell cycle and genomic stability (*CCDND3*, *TP53*) and chromatin remodelling (*SMARCA4*, *ARID1A*) [203,204]. Mutations of the TF TCF3 (10%–25%) and its negative regulator ID3 (35%–58%) are highly recurrent in all three subtypes of BL, where they promote tonic BCR signalling and sustain survival of the tumour cell by engaging the PI3K pathway (Figure 14.4) [204,209]. Inactivating mutations of *ID3* are mostly biallelic and affect the HLH domain of the protein, impairing the negative regulation of TCF3 and TCF4. On the other hand, recurrent somatic mutations in *TCF3* are often monoallelic and have been identified only in the b-HLH DNA binding and dimerisation domain of the splicing isoform E47. These mutations are associated with higher expression of the protein and are, thus, gain-of-function [204]. In addition, TCF3 can promote cell-cycle progression by transactivating *CCND3*, itself a target of gain-of-function mutations in 38% of sBL. These events affect conserved residues in the carboxyl terminus of this D-type cyclin implicated in the control of protein stability, leading to higher expression levels. Other common genetic lesions include loss of *TP53* by mutation and/or deletion (35% of both sBL and eBL cases) [101], inactivation of *CDKN2B* by deletion or hypermethylation (17% of samples) [210] and deletions of 6q, detected in ~30% of cases, independent of the clinical variant [211]. Mutations in the SWI-SNF (SWItch/Sucrose Non-Fermentable) complex family members *SMARCA4* and *ARID1A* have been identified in at least 25% of cases of BL [182,184,203]. Alterations in those genes have been described as mutually exclusive, with mutations in *SMARCA4* mainly affecting the DEAD-box or the helicase region of the protein [183].

Finally, one contributing factor to the development of BL is monoclonal EBV infection, present in virtually all cases of eBL and in ~30% of sBL and iBL [125,134,212,213]. The pathogenetic role of EBV in BL has been a topic of intense debate, as tumour cells lack the expression of both EBV transforming antigens LMP1 and

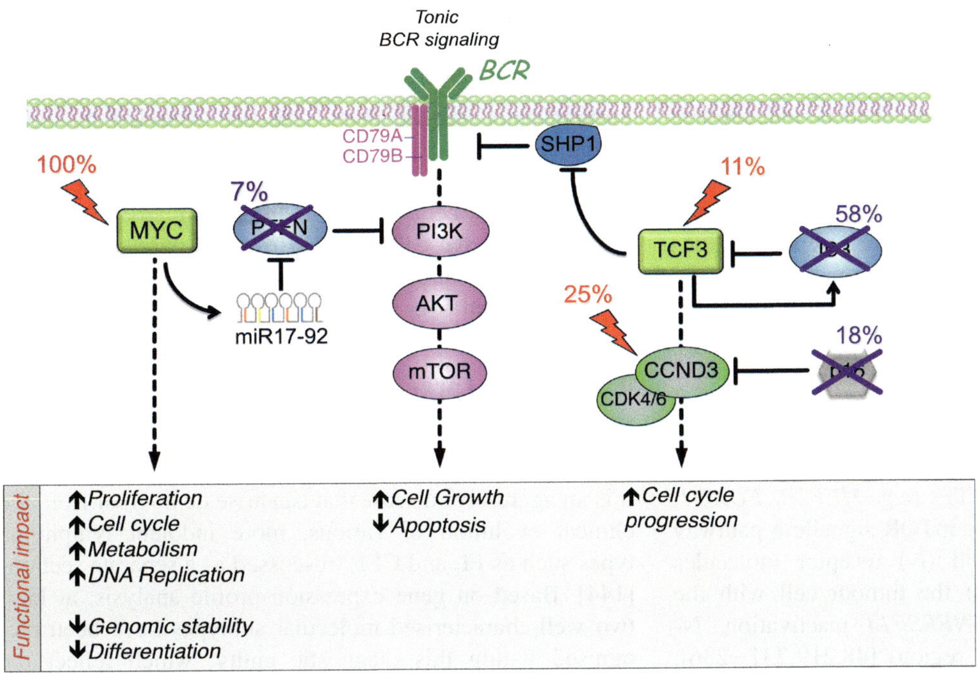

FIGURE 14.4 **Molecular basis of BL pathogenesis.** Most common genetic lesions associated with BL. Bolts indicate genes targeted by gain-of-function mutations or chromosomal translocations (MYC) and crosses denote genes targeted by inactivating events.

EBNA2 [214]. Nonetheless, recent studies integrating gene expression and mutational profile of BL cases have shown that EBV-positive cases are associated with a distinct genetic profile compared to the negative cases [181,182,184,215]. In particular, EBV+ cases are associated with higher mutational burden, an enrichment in noncoding mutations associated with ASHM, and higher expression of genes related to the AKT/mTOR pathway; on the contrary, EBV cases are enriched in mutations of known BL driver genes such as *ID3*, *CCND3*, *TCF3* and *SMARCA4* [181,182,215].

4.3 Follicular Lymphoma

FL, the second most common lymphoma type (~20% of diagnoses), is an indolent, but largely incurable disease, characterised by a continuous pattern of progression and relapses that can culminate, in 20%–30% of cases, in its histologic transformation to an aggressive lymphoma (typically, a DLBCL) [216,217].

The normal counterpart of FL is a GC B cell, as supported by the morphologic appearance of these tumours, the expression of specific GC B-cell markers (e.g. BCL6 and CD10) and the evidence of ongoing SHM in their clonally rearranged *IG* genes [144], although a desynchronisation of GC-specific gene expression programmes was recently revealed by single-cell studies of FL samples [218]. Nonetheless, the genesis of FL is a complex process that starts well before this final clonal expansion in the GC. The currently accepted model of FL ontogenesis infers that, in the bone marrow, precursor B cells acquire a chromosomal translocation juxtaposing the *IG* enhancer to the *BCL2* locus (the genetic hallmark of FL) due to mistakes in V(D)J recombination. These cells can continue their maturation process into naive B cells and, if challenged by an antigen, are engaged in the GC reaction. Due to the ectopic expression of BCL2, t(14; 18)+ B cells will resist the default apoptosis programme, even in the presence of low-affinity antigen receptors, remaining in circulation for decades as memory B cells. These cells can thus undergo repeated GC transits, being exposed to the risk of accumulating additional genetic hits that ultimately drive their malignant expansion [219]. Consistent with this model, t(14; 18)+ B cells can be found at extremely low levels in the peripheral blood of ~70% of healthy subjects [220], but most of these individuals will never develop an FL, whereas FL samples display many additional genetic alterations, of which some (i.e. *CREBBP* mutations) can be detected years prior to FL diagnosis [221].

The t(14; 18) translocation, detected in 80%–90% of FL cases, joins the 3′ untranslated region of *BCL2* to an *IGHJ* segment, resulting in the ectopic expression of BCL2 in GC B cells [222–227], where its transcription is normally repressed by BCL6 [38,228]. Approximately 70% of the breakpoints on chromosome 18 cluster within the major breakpoint region, while the remaining 5%–25% map to the more distant minor cluster region, located ~20 kb downstream of *BCL2* [222–224,229]. Rearrangements involving the 5′ flanking region of *BCL2* have been described in a minority of cases [230]. The *BCL2* gene encodes a 26-kD integral membrane protein that controls the cell apoptotic threshold by preventing programed cell death and may thus contribute to lymphomagenesis by inducing apoptosis resistance in tumour cells, independent of antigen selection.

In addition to *BCL2* translocations, the pathogenesis of FL involves a complex network of genetic, epigenetic and microenvironmental factors driven by recurrent mutations in multiple genes encoding (i) histone/chromatin modifiers, including the methyltransferase KMT2D (80%−90% of cases), the acetyltransferases CREBBP and EP300 (40%−60% of cases), the polycomb-group oncogene EZH2 (20% of cases), linker histones (40% of cases) and ARID1A (10%−15% of cases), which all contribute to transformation by remodelling the epigenetic landscape of the precursor tumour cell; (ii) TFs (e.g. *MEF2B, FOXO1, STAT6*); (iii) components of the mTOR signalling pathway (e.g. *RRAGC, ATP6V1B2*) and (iv) receptor molecules implicated in the cross-talk of the tumour cell with the immune environment (e.g. *TNFRSF14* inactivation, N-glycosylation sites in the *IGV* region) [48,219,231−236]. Among these lesions, inactivating mutations of the CREBBP acetyltransferase and the KMT2D methyltransferase represent early events that are introduced in a putative common ancestral clone prior to its final clonal expansion [237−239]; indeed, *CREBBP* mutations can be found in the circulating tumor DNA of apparently healthy individuals as early as six years prior to FL diagnosis, but are rare in subjects that will not develop the disease [221]. Loss of *Crebbp* or *Kmt2d* was shown to favour tumour development in vivo by reprogramming the expression of genes implicated in signalling pathways that are typically activated when B cells are primed to exit the GC [49,50,70,71]. Moreover, lesions of *CREBBP/EP300* alter the balance between the activity of the BCL6 oncogene, which is typically inactivated by acetylation, and the tumour suppressor p53, which requires acetylation for its function [48]. Interestingly, in vivo combined haploinsufficiency of *Crebbp* and *Kmt2d* leads to GC hyperplasia and accelerates BCL2-driven lymphomagenesis, indicating a cooperative role [72,240]. Consistently, the transcriptional changes induced by dual *Crebbp/Kmt2d* loss in GC B cells are significantly higher than those observed upon individual gene deletion [72]. The identification of the FL common precursor cell has important therapeutic implications, as it could pave the basis for the development of more effective therapeutic approaches aimed to eradicate not only the dominant tumour clone but also the reservoir of initiating cells, thereby preventing transformation and relapse.

In addition to genetic alterations, a major role in the pathogenesis of FL has been attributed to chronic antigen stimulation [241]. In particular, N-glycosylation sites are recurrently introduced by SHM in the antigen-binding pocket of the BCR. Since such sites are specifically enriched in FL, as compared to normal GC B cells, it has been suggested that they are positively selected by the tumour. Interestingly, these sites are occupied by oligomannose glycans, potentially subverting the binding of conventional antigens towards mannose-binding lectins on the surface of cells from the tumour microenvironment (FDCs or stromal cells) [242]. In this scenario, BCR signalling might acquire a complete autonomy from the initial cognate antigen over time.

4.4 Diffuse Large B-Cell Lymphoma

DLBCL is the most common B-cell malignancy, accounting for 30%−40% of all lymphoma diagnoses in adulthood. It is an aggressive disease that can arise de novo or from the clinical evolution of various, more indolent lymphoma types such as FL and CLL (discussed as a separate section) [144]. Based on gene expression profile analysis, at least two well-characterised molecular subtypes have been recognised within this diagnostic entity, which reflect the derivation from different stages of GC B-cell development: GCB-like DLBCL shares significant similarities with the phenotype of cells reflecting intermediate stages between DZ and LZ B cells; ABC-like DLBCL shows a transcriptional signature related to BCR-activated B cells or to B cells committed to plasmablastic differentiation, although more recent studies suggest that a fraction of cases may be more related to memory B cells [21,23]. The remaining 15%−30% of cases remain unclassified and reflect intermediate stages between GCB- and ABC-DLBCL [243]; however, the enrichment of specific genetic lesions in this group suggests it may represent a discrete entity (see genetic classification). Stratification according to gene expression profiles has prognostic value, since patients diagnosed with GCB-DLBCL display better overall survival compared to those with ABC-DLBCL upon standard immune-chemotherapy [96]. In addition, a double-hit signature has been identified in tumours with concurrent BCL2 and MYC translocations, which is closely related to the phenotype of DZ B cells and defines a group of aggressive lymphomas within the GCB-DLBCL [22].

The heterogeneity of DLBCL is reflected by the broad set of genetic lesions that are associated with its pathogenesis, as revealed by the application of high-throughput genomic technologies. Compared to other B-cell malignancies, DLBCL shows a higher degree of genomic complexity, with significant diversity across patients [20,231,244,245]. These genomic alterations are not uniformly distributed across the different phenotypic subtypes, indicating that GCB- and ABC-DLBCL utilise distinct oncogenic pathways. Additional programmes are disrupted in both molecular subtypes of the disease, although through distinct mechanisms, consistent with a more general role during transformation. Recently, the analysis of large patient cohorts using advanced computational methods [244–246] revealed further granularity within this picture based on the pattern of concurrent genetic alterations, paving the basis for two novel genetic classifications of the disease (Figure 14.5). In the so-called LymphGen classification, seven groups were termed based on the most recurrent defining lesions: EZB (for *EZH2* and *BCL2* translocations) and ST2 (*SGK1* and *TET2* mutations) include mostly GCB-DLBCL cases; MCD (*MYD88*^{L265P} and *CD79B* mutations) and the rare N1 subgroup (*NOTCH1* mutations) are primarily composed of ABC-DLBCL; BN2 (*BCL6* translocations and gain-of-function mutations in *NOTCH2*) includes a significant number of 'unclassified' DLBCLs along with ABC-DLBCL; and A53 (aneuploidy and *TP53* alterations) is a heterogeneous group mostly unrelated to cell-of-origin (COO) [244,245,247] phenotype. A separate study utilised a different approach to define five distinct subsets of cluster-associated genes, including the so-called *Cluster 1*, which shares features with both MCD and BN2 DLBCL; *Cluster 2*, defined by TP53 gene alterations, loss of 9p21.3/*CDKN2A* and associated genomic instability; *Cluster 3*, largely overlapping with EZB-DLBCL; *Cluster 4*, characterised by mutations in histone genes along with multiple immune evasion molecules, and also primarily displaying a GCB-like phenotype, thus likely corresponding to the LymphGen ST2 class; and *Cluster 0*, which included a few cases devoid of any defining genetic drivers, but enriched in T cell/histiocyte rich large cell lymphomas [244]. While additional studies will be necessary to confirm the clinical applicability of these genetic classifications, their clinical relevance is supported by the association of individual subgroups with discrete clinical outcomes following conventional therapies [244,245,247,248]. As such, the new genetic subtypes may assist in identifying novel therapeutic targets for precision medicine approaches and stratifying patients that are more likely to respond. The next paragraphs summarise the most common genetic alterations, as related to the COO subtypes and to more recent genetic taxonomies.

4.4.1 Programmes Disrupted in Multiple DLBCL Subtypes

Three broad biological programmes are frequently disrupted in DLBCL independent of subtype, although the specific target genes and/or the underlying mechanisms may differ in distinct phenotypic or genetic clusters: (i) epigenetic remodelling; (ii) regulation of BCL6 and (iii) escape from immune surveillance mechanisms.

Epigenetic dysregulation is driven by genetic alterations in multiple genes encoding for histone modifiers, histones themselves and, at lower frequency, chromatin remodelers. Among them, the most common are inactivating mutations of *CREBBP/EP300* (35% of cases) and *KMT2D* (~30% of all DLBCL), together with gain-of-function mutations of the EZH2 trimethyltransferase [20,48,231,232]. These lesions are enriched in GCB-DLBCL belonging to the EZB/C3 genetic cluster [244,245]. In contrast, mutations of the *TET2* DNA hydroxymethyltransferase and of linker histones are more frequent in the ST2/C4 genetic cluster of GCB-DLBCL [244,245,249].

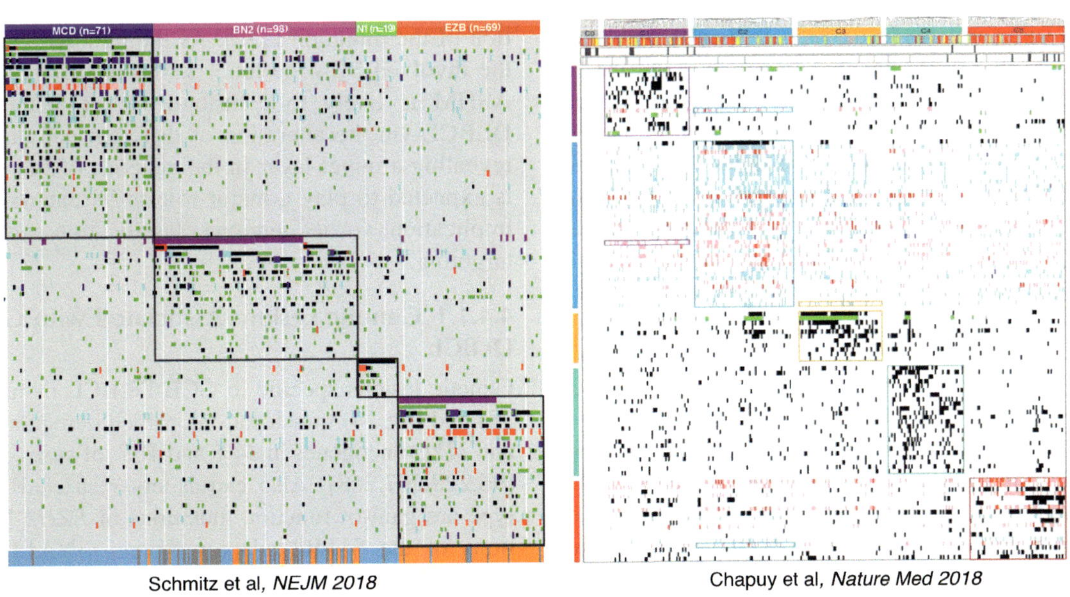

FIGURE 14.5 Genetic classifications of DLBCL. Distribution of coordinated genetic lesions associated with distinct genetic subtypes of DLBCL. In the figure, rows are distinct genetic alterations and columns are DLBCL samples. *Reproduced with permission from Refs. [244,245].*

Deregulated activity of the BCL6 oncoprotein is a major contributor to the pathogenesis of both GCB and ABC-DLBCL, due to a multitude of genetic lesions that target this gene either directly or indirectly. Chromosomal rearrangements involving band 3q27 are observed in up to 35% of cases overall [250,251], although with a twofold higher frequency in the ABC-DLBCL subtype [252]. These rearrangements juxtapose the intact coding domain of *BCL6* downstream and in the same transcriptional orientation to heterologous sequences derived from the partner chromosome, including *IGH* (14q23), *IGK* (2p12), *IGL* (22q11) and at least 20 other chromosomal sites unrelated to the *IG* loci [85–92]. The majority of these translocations result in a fusion transcript in which the promoter region and the first non-coding exon of BCL6 are replaced by sequences derived from the partner gene [86,253]. Since the common denominator of these promoters is a broader spectrum of activity throughout B-cell development, including expression in the post-GC differentiation stage, the translocation is thought to prevent the downregulation of BCL6 expression that is normally associated with GC exit. In addition, AID-mediated point mutations affecting its intragenic SE are detected in up to 75% of DLBCL cases [114,254,255], where they reflect the activity of the physiologic SHM mechanism that operates in GC B cells [114,115]. However, tumours display evidence of strong selective pressure for the accumulation of specific functionally relevant mutations, including recurrent hotspots targeting an autoregulatory domain within the *BCL6* 5′ untranslated exon 1 [256,257], and two binding motifs used by BLIMP1 and IRF4 within the *BCL6* first intron [61,112]. These mutations are most common in GCB-DLBCL and deregulate BCL6 transcription by either disrupting the ability of the BCL6 protein to control its own expression levels [256,257] or by preventing CD40-induced BCL6 downregulation [61,112]. As the full extent of mutations deregulating BCL6 expression has not been functionally characterised, it is likely that the fraction of DLBCL cases with BCL6 abnormalities is higher than currently estimated. Deregulated expression of an intact *BCL6* gene product is also sustained by a variety of indirect mechanisms, including gain-of-function mutations in its positive regulator MEF2B (~11% of cases) [35], inactivating mutations/deletions of *CREBBP/EP300* [48] and mutations/deletions of *FBXO11* (~5%), which encodes a ubiquitin ligase involved in the control of BCL6 protein degradation [258]. The above lesions are also mostly observed in GCB-DLBCL and promote lymphomagenesis by enforcing the proliferative phenotype typical of GC B cells, as well as by suppressing proper DNA damage responses and terminal differentiation. Accordingly, a mouse model enforcing deregulated BCL6 expression in GC B cells develops DLBCL [259], and mice expressing the MEF2B hotspot D83V point mutation or lacking *FBXO11* display preneoplastic GC expansion [52,260].

Escape from immune surveillance, including both cytotoxic T cell (CTL)-mediated cytotoxicity and Natural Killer (NK) cell-mediated death, is a recurrent feature that DLBCLs acquire through the loss of surface MHC-I and CD58 expression (~75% of cases) [261]. This is explained in part by the genetic disruption of *B2M/HLA-I* and *CD58*, respectively [261,262]. Additionally, signalling through the MHC-II receptor molecule is frequently impaired, either by direct mechanisms (mutations in the MHC-II transactivator CIITA, amplification of the genes encoding for the immunomodulatory proteins PDL1/PDL2) [263,264] or because of *CREBBP* inactivation and *EZH2* mutations [49,50,74,265,266].

Aberrant SE hypermutation: Over 90% of DLBCL cases exhibit the footprint of an aberrant hypermutation activity that specifically targets chromatin domains with features of active SEs and, less frequently, typical enhancers [111,112]. These regions are mostly linked to genes encoding important regulators of B-cell development and GC biology, including well-known lymphoma oncogenes, and can be individually affected in as many as 60% of all DLBCLs. Proof of concept for the functional consequences of these lesions has been provided for representative targets, namely the intragenic SEs of *BCL2* and *CXCR4*, besides *BCL6* [112]. Interestingly, both *BCL2* and *CXCR4* mutational hotspots abrogate the binding of the glucocorticoid receptor, which was identified as a novel negative regulator of their transcription and a potential new player in the pathogenesis of this disease. Consistently, expression of BCL2 and CXCR4 was deregulated in isogenic cell lines carrying these changes. Notably, tumour cells were addicted to the SE mutation, as indicated by the counterselection of clones upon correction of the mutation [112]. As the number and identity of the involved SE-gene loops varies in different cases, SE-ASHM is likely to contribute to the heterogeneity of DLBCL via the alteration of different cellular pathways, revealing a novel layer in the genetics of this disease that is expected to play a major role in its pathogenesis, with implications for its genomic classification and therapeutic targeting.

4.4.1.1 Genetic Lesions Associated with GCB-DLBCL

Genetic lesions specific to GCB-DLBCL include the t(14; 18) and t(8; 14) translocations, which deregulate the *BCL2* and *MYC* oncogenes in 34% and 10% of cases, respectively [96,228,267,268]. Also exquisitely restricted to this subtype are gain-of-function mutations of *EZH2* [232], mutations of the S1PR2 adaptor protein GNA13, which are involved in the ability of DLBCL to spread [269]; deletions of the tumour suppressor *PTEN* [270,271] and mutations/deletions of *TNFRSF14*, encoding for the HVEM receptor [235,272] (Figure 14.6).

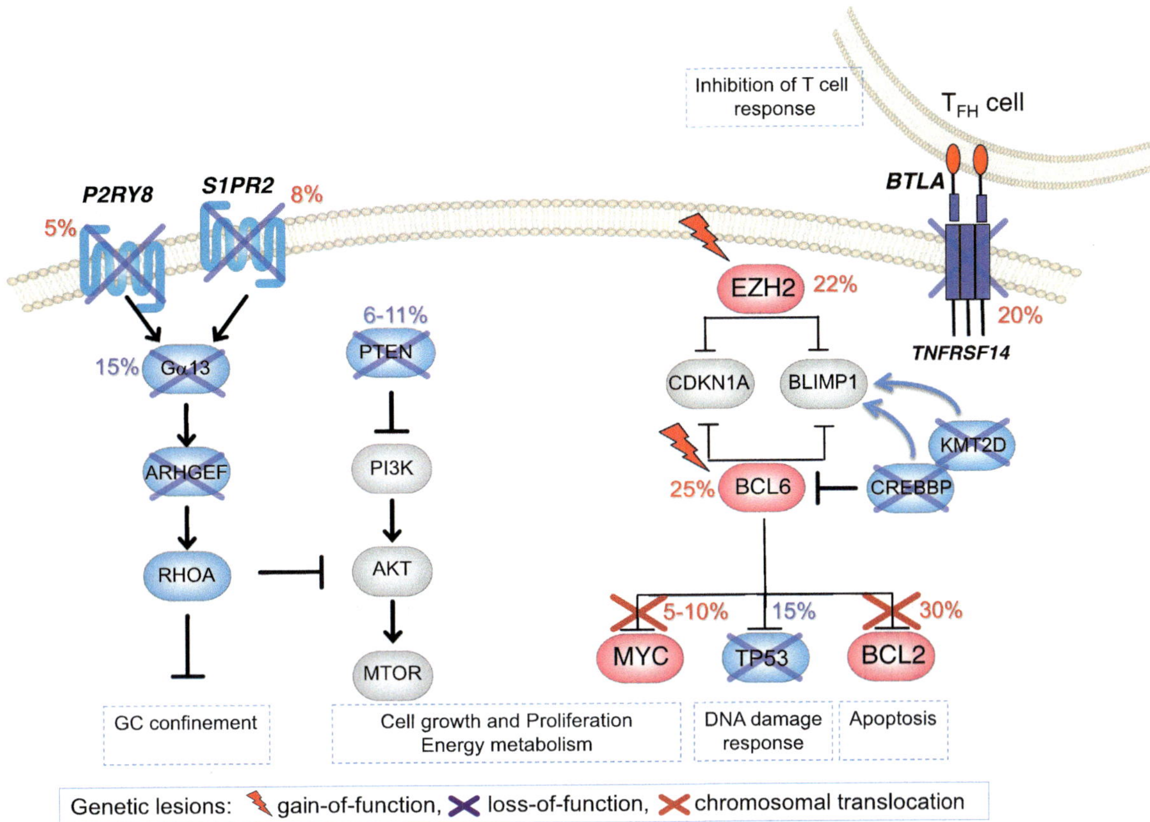

FIGURE 14.6 **Pathway lesions in GCB-DLBCL.** Schematic representation of signalling pathways and transcription factors recurrently targeted by mutations in GCB-DLBCL. *Reproduced with permission from Pasqualucci et al. [25].*

4.4.1.2 Genetic Lesions Associated with ABC-DLBCL

A predominant feature of ABC-DLBCL is the constitutive activation of the NF-κB signalling pathway, initially evidenced by a transcriptional signature enriched in NF-κB target genes and by the requirement of NF-κB for the proliferation and survival of ABC-DLBCL cell lines [273]. Constitutive NF-κB activation is caused by a variety of alterations affecting positive and negative regulators of this pathway, downstream of an active BCR (Figure 14.7). In ~10% of cases, somatic mutations affect the immunoreceptor tyrosine-based activation motif (ITAM) signalling modules of CD79B and CD79A [274]. These mutations sustain oncogenic BCR signalling in a multiprotein supercomplex formed by MYD88, TLR9 and the BCR (termed the My-T-BCR supercomplex), which co-localises with mTOR on endolysosomes and drives pro-survival NF-κB and mTOR signalling [275]. The *MYD88* gene is itself commonly mutated in 30% of ABC-DLBCL cases due to a hotspot L265P substitution in the intracellular Toll/interleukin-1 receptor domain that activates NF-κB by inducing IRAK4 phosphorylation, as well as JAK/STAT3 transcriptional responses [276]. An additional ~10% of ABC-DLBCL carry oncogenic mutations clustering in the CARD11 coiled-coil domain and enhancing its ability to transactivate NF-κB target genes [277]. Finally, the *TNFAIP3* gene, encoding for the negative regulator A20, is biallelically inactivated by mutations and/or deletions in up to 30% of ABC-DLBCL, preventing termination of NF-κB responses [106,107]. The tumour suppressor role of A20 was documented by the observation that reconstitution of A20 knock-out cell lines with a wild-type protein induces apoptosis and blocks proliferation, in part due to the suppression of NF-κB activity [106,107]. The BCR signalling plays an equally critical role in cases lacking direct mutations, where silencing of several BCR proximal and distal subunits showed selective toxicity. In these cases, activation of NF-κB is driven by autoantigens through a kinase signalling cascade of SYK, BTK and PKCβ that triggers the assembly of the CARD11-BCL10-MALT1 adaptor complex, activating the IκB kinase [278]. Collectively, these findings provided strong genetic support to the development of therapies that interfere with BCR signalling, such as the Bruton tyrosine kinase (BTK) inhibitor Ibrutinib [279,280]. This small molecule compound was especially toxic to ABC-DLBCL, with tumours carrying mutations in both *CD79B* and *MYD88* showing exceptional responses [279], explained by the role of My-T-BCR supercomplexes in coordinating drug activity [275].

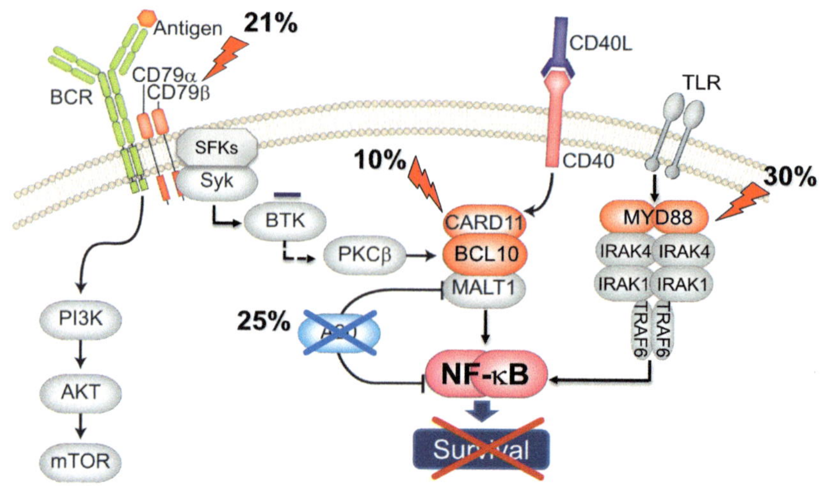

FIGURE 14.7 **Pathway lesions in ABC-DLBCL.** Schematic representation of signalling pathways that are induced in late GC B cells following engagement of the BCR by the antigen, CD40-CD40L interaction, and activation of TLR (note that TLR9 is localized in the endosomal compartment). These signals converge on activation of the NF-κB pathway and lead to upregulation of multiple target genes including IRF4 and A20. IRF4, in turn, downregulates BCL6 expression, allowing the release of BLIMP1 expression as required for terminal differentiation. This cascade is disrupted at multiple levels due to genetic lesions that favour the anti-apoptotic and pro-proliferative function of NF-κB, induce chronic active BCR and JAK/STAT3 signalling, and block terminal B cell differentiation through mutually exclusive deregulation of BCL6 and inactivation of BLIMP1. *Reproduced with permission from Pasqualucci et al. [25].*

The second oncogenic programme implicated in ABC-DLBCL is blockade of terminal B cell differentiation, due to biallelic loss of the *PRDM1* gene (truncating and missense mutations, genomic deletions) or to transcriptional repression of its protein product BLIMP1 by constitutively active *BCL6* alleles involved in chromosomal translocations [103–105]. Consistent with this model, *BCL6* translocations and *PRDM1* inactivation are generally mutually exclusive and *Prdm1* loss causes the development of lymphomas when recapitulated in vivo in GC-specific mouse models [103,281,282].

4.5 DLBCL Derived from CLL and FL Transformation

Insights into the molecular events that are specifically acquired during the histologic transformation of indolent lymphomas to DLBCL have been obtained by integrating exome sequencing and copy number analysis of sequential biopsies from CLL/RS or FL/tFL patients. These studies allowed to reconstruct the evolutionary history of the dominant tumour clone during transformation, revealing that FL and tFL arise from a common mutated precursor clone by divergent evolution [238,239], while RS derives from the dominant CLL clone through a predominantly linear pattern involving the maintenance of the CLL-associated lesions [283,284]. tFL-specific lesions include inactivation of *CDKN2A/B* through deletion, mutation and hypermethylation (one-third of patients) [169,285], rearrangements and amplifications of *MYC* [239,286], *TP53* mutations/deletions (25%–30% of cases) [102,287], loss of chromosome 6q (20%) [211], ASHM [239] and biallelic inactivation of the *B2M* gene causing the loss of cell surface MHC-I to favour escape from immune surveillance [239]. Conversely, RS-acquired alterations include *NOTCH1* mutations, *CDKN2A/B* loss, *TP53* loss and *MYC* translocations [283]. Comparison with de novo DLBCL showed that, despite their morphologic resemblance, the genomic landscapes of RS and tFL are largely unique, being characterised by distinct combinations of alterations otherwise not commonly observed in de novo DLBCL [239,283]. This distinction may have important future implications for the development of targeted therapies.

4.6 Primary Mediastinal B-Cell Lymphoma

PMBCL is a tumour postulated to derive from post-GC thymic B cells in the mediastinum, which displays a unique transcriptional profile highly similar to that of cHL [288,289]. A genetic hallmark of both PMBCL and cHL is the amplification of chromosomal region 9q24, detected in nearly 50% of patients [97,270]. This relatively large interval encompasses multiple genes of possible pathogenetic significance, including *JAK2* and *PDL1/PDL2*, which encode for inhibitors of T-cell responses [97,270,290] and have been linked to impaired anti-tumour immune responses in several cancers. Other lesions affecting regulators of immune responses in PMBCL include genomic breakpoints and mutations of the MHC class II transactivator gene *CIITA*, which may reduce tumour cell immunogenicity by downregulating surface MHC-II expression [264,291]. Besides contributing to immune escape, elevated expression levels of these genes may in part explain the unique features of these lymphoma types, which are characterised by a significant inflammatory infiltrate. PMBCL also shares with cHL the presence of genetic lesions affecting the NF-κB pathway and the deregulated expression of receptor tyrosine kinases [292,347,348]. In particular, mutations of the TF *STAT6* and inactivating mutations of its negative regulator *SOCS1* are highly recurrent in PMBCL [291,292]. Together with amplifications/overexpression of *JAK2*, which promote

STAT6 activation via IL-3/IL-4, these lesions reveal a pervasive role for deregulation of the JAK-STAT signalling pathway as an important contributor to PMBCL pathogenesis [292,347].

4.7 Hodgkin Lymphoma

HL is a histologically unique group of malignancies characterised by the presence of rare (<5%) scattered large atypical cells – the mononucleated Hodgkin cells and the multinucleated Reed–Sternberg cells (HRS) – residing in a complex admixture of inflammatory cells [293]. The morphology and phenotype of the tumour cells, together with the composition of the infiltrate, allow to distinguish two major subtypes of HL: the most common cHL, comprising the nodular sclerosis, mixed cellularity, lymphocyte-depletion and lymphocyte-rich variants, and nodular lymphocyte-predominant HL (NLPHL) (~5% of cases). Both HL types represent clonal populations of mature GC-experienced B cells, as they harbour clonally rearranged and somatically mutated *IG* genes [294,295]. In about 25% of cHL cases, nonsense mutations disrupt originally in-frame *IGHV* gene rearrangements (crippling mutations), preventing antigen selection and suggesting that HRS cells have escaped apoptosis through a mechanism not linked to antigen stimulation [295]. Recently, the evidence of divergent biological and clinical aspects between cHL and NLPHL has prompted the redefinition of the latter as 'nodular predominant B cell lymphoma', which is now considered a discrete histopathological entity [145,146].

Giving the rarity of the HRS cells in the tumour milieu and the technical difficulties to isolate them, knowledge of the mutational landscape of the disease was confined for decades to targeted-gene studies. Only recently, a more comprehensive molecular picture of cHL has been revealed by whole exome sequencing (WES) and whole genome sequencing (WGS) efforts [296–299]. A known feature of HL is the marked genomic instability, with high recurrence of aneuploidy and about 60% of cases presenting whole-genome duplication [299]. Copy number alterations, structural variants and somatic mutations mostly affect genes involved in the NF-κB and JAK/STAT signalling pathways as well as in immune surveillance. In particular, recurrent mutations leading to constitutive activation of NF-κB are found in more than 50% of patients [297] and include the frequent amplification of *REL* (50%–70% of cases), associated with increased protein expression levels [299–301], and/or inactivating mutations in genes coding for negative regulators of NF-κB (*NFKBIA*, 20% of cases; *NFKBIE*, 15%–25%, and *TNFAIP3*, 25%–40% of cases, among others) [109,298,302,303]. Amplification of *JAK2*, mutations of *STAT6* and inactivating mutations of *SOCS1* and *PTPN1* collectively account for over 85% of cHL cases [348,297–299,304]. Finally, a distinctive feature of HL, analogous to PMBCL, is represented by recurrent genomic gains of *PD-L1*(CD274) and *PD-L2*(CD273) at 9p24.1 and translocations of *CIITA* [97,264,290]. Inactivating mutations of *B2M* were reported in one-third of cases where they associate with impaired expression of surface β2M and MHC-I [296]. Other frequent genetic alterations affect *XPO1* (26% of cases), with consequences on the nuclear transport of several TFs, *GNA13* (26%) and *ARID1A* (26%). An important pathogenic cofactor in cHL is monoclonal EBV infection, observed in about 40% of cHL and up to 90% of HIV-related HL, suggesting that infection precedes clonal expansion [293]. Of the viral proteins encoded by the EBV genome, infected HRS cells most commonly express LMP1 and LMP2A, which lead to constitutive BCR and CD40 signalling and may thus contribute to HL pathogenesis [293,297,299].

4.8 Chronic Lymphocytic Leukaemia

CLL is an indolent malignancy whose precise cellular origin remains a subject of debate. The discovery of two different disease subgroups carrying mutated (M-CLL) or unmutated (U-CLL) *IGHV* genes [305,306] and associated with distinct clinical outcome [307,308] initially suggested their origin from GC-experienced and naive B cells, respectively [305,306]. However, gene expression profile analysis of CLL samples revealed a homogeneous signature that was independent of *IGHV* mutation status and appeared related to that of CD27+ memory B cells, suggesting both subtypes derive from antigen-experienced memory B cells, either post-GC or GC-independent [309,310]. Supporting a role for antigen receptor signalling in CLL pathogenesis, analysis of the *IG* gene repertoire in CLL patients indicates the 'stereotypic' use of a recurrent restricted subset of V regions, with more than 200 *IG* stereotypes identified until now [311–316]. Subsequent transcriptome analysis of normal and CLL samples identified a $CD5^+CD27^+$ post-GC B-cell subset as the possible normal counterpart of M-CLL, while the U-CLL cases seem to cluster with a $CD5^+CD27^-$ pre-GC naive B-cell population [317]. Consistently, DNA methylation studies have identified three subsets of CLL that display distinct signatures (memory-like, naïve-like and intermediate), likely reflecting their putative normal counterpart. These epigenetic subtypes differ in *IGHV* mutation status, *IG* stereotypes and prognosis [318].

Of note, a few studies suggested the initial genetic/epigenetic events leading to CLL are acquired early in the hematopoietic stem cells (HSCs) or in early progenitors, as demonstrated in xenotransplant models in which the infusion of HSCs from CLL patients, but not from healthy donors, led to the appearance of clonal B cells with a CLL-like phenotype [319]. Moreover, mutations affecting known driver genes in CLL (i.e. *NOTCH1*, *SF3B1*) have been identified in HSCs isolated from CLL patients [320,321].

In contrast with other BCLs, and consistent with the derivation from a post-GC or GC-independent B cell in which *IG* gene remodelling mechanisms are no longer active, balanced, reciprocal chromosomal translocations are exceedingly rare [322,323,324]. Conversely, CLL is recurrently associated with several numerical abnormalities, including trisomy 12 and monoallelic or biallelic deletion/inactivation of chromosomal regions 17p, 11q and the most common 13q14 [322,323]. The minimal deleted region (MDR) of 13q14 deletions encompasses a long noncoding RNA (*DLEU2*) and two microRNAs expressed as a cluster, namely miR-15a and miR-16-1 [325–327]. Deletion of this locus in mice leads to the development of clonal lymphoproliferative diseases with features of MBL, CLL and DLBCL at 25%–40% penetrance [328], documenting its pathogenetic role in CLL. Trisomy 12 is found in approximately 16% of patients and correlates with poor survival, but no specific gene/s have been identified [329–331]. Deletions of chromosomal region 11q22-23 (18% of cases) almost invariably encompass the *ATM* gene and may promote genomic instability [332–334]. These lesions can be observed in the patient germline, and may thus account, at least in part, for the familial form of the disease. Deletions of 17p13, which include the *TP53* tumour suppressor and are frequently accompanied by mutation of the second allele [101,335], are observed in ~4%–10% of CLL at diagnosis [336–338] but are enriched in cases that underwent transformation to RS, a highly aggressive lymphoma with poor clinical outcome [101,283].

More recently, WES and WGS studies have revealed the complexity and interpatient heterogeneity of the disease [336–342]. These studies showed a generally lower mutational burden compared to epithelial tumours and other lymphoid malignancies, with M-CLL cases harbouring more somatic substitutions than U-CLL cases, but the latter being characterised by a higher incidence of driver mutations [343]. Of the many potential driver mutations detected in the disease, only a few are recurrent in at least 5% of patients, including *SF3B1*, *NOTCH1*, *ATM, BIRC3, CHD2* and *TP53* [337,338]. Commonly affected programmes comprise cell signalling (*NOTCH1*, *FBXW7*, *MYD88*, *EGR2*), DNA damage response (*TP53*, *ATM*), RNA splicing (*SF3B1*, *XPO1*, *U1*) and chromatin modification (*IKZF3*, *ASXL1, CHD2*). Interestingly, some driver mutations seem to be associated to specific CLL features and clonal/subclonal evolution [284,340]. Gain-of-function mutations of *NOTCH1* and mutations of *SF3B1*, found in 5%–10% of diagnostic CLL samples, are preferentially enriched in RS (30% of cases) and chemoresistant cases (25% of cases), respectively, portending negative prognostic value [283,339,341,344,345,346]. More recently, mutations affecting the 3′ untranslated region of the *NOTCH1* gene were identified in approximately 20% of CLL cases, where they cause an alternative splicing event leading to a truncated protein with increased stability [343].

REFERENCES

[1] MacLennan IC. Germinal centers. Annu Rev Immunol 1994;12:117–39.

[2] Rajewsky K. Clonal selection and learning in the antibody system. Nature 1996;381:751–8.

[3] Allen CD, Okada T, Cyster JG. Germinal-center organization and cellular dynamics. Immunity 2007;27:190–202.

[4] De Silva NS, Klein U. Dynamics of B cells in germinal centres. Nat Rev Immunol 2015;15:137–48.

[5] Mesin L, Ersching J, Victora GD. Germinal center B cell dynamics. Immunity 2016;45:471–82.

[6] Neuberger MS. Antibody diversification by somatic mutation: from burnet onwards. Immunol Cell Biol 2008;86:124–32.

[7] Kuppers R, Zhao M, Hansmann ML, Rajewsky K. Tracing B cell development in human germinal centres by molecular analysis of single cells picked from histological sections. EMBO J 1993;12:4955–67.

[8] Jacob J, Kassir R, Kelsoe G. In situ studies of the primary immune response to (4-hydroxy-3-nitrophenyl)acetyl. I. The architecture and dynamics of responding cell populations. J Exp Med 1991;173:1165–75.

[9] Victora GD, Nussenzweig MC. Germinal centers. Annu Rev Immunol 2022;40:413–42.

[10] Calado DP, et al. The cell-cycle regulator c-Myc is essential for the formation and maintenance of germinal centers. Nat Immunol 2012;13:1092–100.

[11] Dominguez-Sola D, et al. The proto-oncogene MYC is required for selection in the germinal center and cyclic reentry. Nat Immunol 2012;13:1083–91.

[12] Liu YJ, et al. Within germinal centers, isotype switching of immunoglobulin genes occurs after the onset of somatic mutation. Immunity 1996;4:241–50.

[13] Roco JA, et al. Class-switch recombination occurs infrequently in germinal centers. Immunity 2019;51:337–350 e7.

[14] Holmes AB, et al. Single-cell analysis of germinal-center B cells informs on lymphoma cell of origin and outcome. J Exp Med 2020;217.

[15] Kennedy DE, et al. Novel specialized cell state and spatial compartments within the germinal center. Nat Immunol 2020;21:660–70.

[16] Longerich S, Basu U, Alt F, Storb U. AID in somatic hypermutation and class switch recombination. Curr Opin Immunol 2006;18:164–74.

[17] Muramatsu M, et al. Class switch recombination and hypermutation require activation-induced cytidine deaminase (AID), a potential RNA editing enzyme. Cell 2000;102:553–63.

[18] Revy P, et al. Activation-induced cytidine deaminase (AID) deficiency causes the autosomal recessive form of the Hyper-IgM syndrome (HIGM2). Cell 2000;102:565–75.

[19] Kuppers R, Klein U, Hansmann ML, Rajewsky K. Cellular origin of human B-cell lymphomas. N Engl J Med 1999;341:1520–9.

[20] Pasqualucci L, et al. Analysis of the coding genome of diffuse large B-cell lymphoma. Nat Genet 2011;43:830–7.

[21] Victora GD, et al. Identification of human germinal center light and dark zone cells and their relationship to human B-cell lymphomas. Blood 2012;120:2240–8.
[22] Ennishi D, et al. Double-hit gene expression signature defines a distinct subgroup of germinal center B-cell-like diffuse large B-cell lymphoma. J Clin Oncol 2019;37:190–201.
[23] Alizadeh AA, et al. Distinct types of diffuse large B-cell lymphoma identified by gene expression profiling. Nature 2000;403:503–11.
[24] Venturutti L, Melnick AM. The dangers of deja vu: memory B cells as the cells of origin of ABC-DLBCLs. Blood 2020;136:2263–74.
[25] Pasqualucci L. Molecular pathogenesis of germinal center-derived B cell lymphomas. Immunol Rev 2019;288:240–61.
[26] Dang CV, et al. The c-Myc target gene network. Semin Cancer Biol 2006;16:253–64.
[27] Grandori C, Cowley SM, James LP, Eisenman RN. The Myc/Max/Mad network and the transcriptional control of cell behavior. Annu Rev Cell Dev Biol 2000;16:653–99.
[28] Meyer N, Penn LZ. Reflecting on 25 years with MYC. Nat Rev Cancer 2008;8:976–90.
[29] Chang CC, Ye BH, Chaganti RS, Dalla-Favera R. BCL-6, a POZ/zinc-finger protein, is a sequence-specific transcriptional repressor. Proc Natl Acad Sci U S A 1996;93:6947–52.
[30] Dent AL, Shaffer AL, Yu X, Allman D, Staudt LM. Control of inflammation, cytokine expression, and germinal center formation by BCL-6. Science 1997;276:589–92.
[31] Ye BH, et al. The BCL-6 proto-oncogene controls germinal-centre formation and Th2-type inflammation. Nat Genet 1997;16:161–70.
[32] Cattoretti G, et al. BCL-6 protein is expressed in germinal-center B cells. Blood 1995;86:45–53.
[33] Ochiai K, et al. Transcriptional regulation of germinal center B and plasma cell fates by dynamical control of IRF4. Immunity 2013;38:918–29.
[34] Lee CH, et al. Regulation of the germinal center gene program by interferon (IFN) regulatory factor 8/IFN consensus sequence-binding protein. J Exp Med 2006;203:63–72.
[35] Ying CY, et al. MEF2B mutations lead to deregulated expression of the oncogene BCL6 in diffuse large B cell lymphoma. Nat Immunol 2013;14:1084–92.
[36] Basso K, et al. Integrated biochemical and computational approach identifies BCL6 direct target genes controlling multiple pathways in normal germinal center B cells. Blood 2010;115:975–84.
[37] Niu H, Ye BH, Dalla-Favera R. Antigen receptor signaling induces MAP kinase-mediated phosphorylation and degradation of the BCL-6 transcription factor. Genes Dev 1998;12:1953–61.
[38] Ci W, et al. The BCL6 transcriptional program features repression of multiple oncogenes in primary B cells and is deregulated in DLBCL. Blood 2009;113:5536–48.
[39] Phan RT, Dalla-Favera R. The BCL6 proto-oncogene suppresses p53 expression in germinal-centre B cells. Nature 2004;432:635–9.
[40] Phan RT, Saito M, Basso K, Niu H, Dalla-Favera R. BCL6 interacts with the transcription factor Miz-1 to suppress the cyclin-dependent kinase inhibitor p21 and cell cycle arrest in germinal center B cells. Nat Immunol 2005;6:1054–60.
[41] Ranuncolo SM, et al. Bcl-6 mediates the germinal center B cell phenotype and lymphomagenesis through transcriptional repression of the DNA-damage sensor ATR. Nat Immunol 2007;8:705–14.
[42] Ranuncolo SM, Polo JM, Melnick A. BCL6 represses CHEK1 and suppresses DNA damage pathways in normal and malignant B-cells. Blood Cells Mol Dis 2008;41:95–9.
[43] Keller AD, Maniatis T. Identification and characterization of a novel repressor of beta-interferon gene expression. Genes Dev 1991;5:868–79.
[44] Turner Jr CA, Mack DH, Davis MM. Blimp-1, a novel zinc finger-containing protein that can drive the maturation of B lymphocytes into immunoglobulin-secreting cells. Cell 1994;77:297–306.
[45] Shaffer AL, et al. BCL-6 represses genes that function in lymphocyte differentiation, inflammation, and cell cycle control. Immunity 2000;13:199–212.
[46] Tunyaplin C, et al. Direct repression of prdm1 by Bcl-6 inhibits plasmacytic differentiation. J Immunol 2004;173:1158–65.
[47] Bereshchenko OR, Gu W, Dalla-Favera R. Acetylation inactivates the transcriptional repressor BCL6. Nat Genet 2002;32:606–13.
[48] Pasqualucci L, et al. Inactivating mutations of acetyltransferase genes in B-cell lymphoma. Nature 2011;471:189–95.
[49] Zhang J, et al. The CREBBP acetyltransferase is a haploinsufficient tumor suppressor in B-cell lymphoma. Cancer Discov 2017;7:322–37.
[50] Jiang Y, et al. CREBBP inactivation promotes the development of HDAC3-dependent lymphomas. Cancer Discov 2017;7:38–53.
[51] Potthoff MJ, Olson EN. MEF2: a central regulator of diverse developmental programs. Development 2007;134:4131–40.
[52] Brescia P, et al. MEF2B instructs germinal center development and acts as an oncogene in B cell lymphomagenesis. Cancer Cell 2018;34:453–65.
[53] Eijkelenboom A, Burgering BM. FOXOs: signalling integrators for homeostasis maintenance. Nat Rev Mol Cell Biol 2013;14:83–97.
[54] Dominguez-Sola D, et al. The FOXO1 transcription factor instructs the germinal center dark zone program. Immunity 2015;43:1064–74.
[55] Sander S, et al. PI3 kinase and FOXO1 transcription factor activity differentially control B cells in the germinal center light and dark zones. Immunity 2015;43:1075–86.
[56] Kwon K, et al. Instructive role of the transcription factor E2A in early B lymphopoiesis and germinal center B cell development. Immunity 2008;28:751–62.
[57] Cato MH, Chintalapati SK, Yau IW, Omori SA, Rickert RC. Cyclin D3 is selectively required for proliferative expansion of germinal center B cells. Mol Cell Biol 2011;31:127–37.
[58] Gerondakis S, Siebenlist U. Roles of the NF-kappaB pathway in lymphocyte development and function. Cold Spring Harbor Perspect Biol 2010;2:a000182.
[59] Klein U, et al. Transcription factor IRF4 controls plasma cell differentiation and class-switch recombination. Nat Immunol 2006;7:773–82.
[60] Sciammas R, et al. Graded expression of interferon regulatory factor-4 coordinates isotype switching with plasma cell differentiation. Immunity 2006;25:225–36.
[61] Saito M, et al. A signaling pathway mediating downregulation of BCL6 in germinal center B cells is blocked by BCL6 gene alterations in B cell lymphoma. Cancer Cell 2007;12:280–92.
[62] Shapiro-Shelef M, et al. Blimp-1 is required for the formation of immunoglobulin secreting plasma cells and pre-plasma memory B cells. Immunity 2003;19:607–20.
[63] Goldberg AD, Allis CD, Bernstein E. Epigenetics: a landscape takes shape. Cell 2007;128:635–8.
[64] Beguelin W, et al. EZH2 is required for germinal center formation and somatic EZH2 mutations promote lymphoid transformation. Cancer Cell 2013;23:677–92.

[65] Caganova M, et al. Germinal center dysregulation by histone methyltransferase EZH2 promotes lymphomagenesis. J Clin Invest 2013;123:5009–22.
[66] Czermin B, et al. Drosophila enhancer of Zeste/ESC complexes have a histone H3 methyltransferase activity that marks chromosomal Polycomb sites. Cell 2002;111:185–96.
[67] Beguelin W, et al. EZH2 and BCL6 cooperate to assemble CBX8-BCOR complex to repress bivalent promoters, mediate germinal center formation and lymphomagenesis. Cancer Cell 2016;30:197–213.
[68] Beguelin W, et al. EZH2 enables germinal centre formation through epigenetic silencing of CDKN1A and an Rb-E2F1 feedback loop. Nat Commun 2017;8:877.
[69] Shilatifard A. The COMPASS family of histone H3K4 methylases: mechanisms of regulation in development and disease pathogenesis. Annu Rev Biochem 2012;81:65–95.
[70] Ortega-Molina A, et al. The histone lysine methyltransferase KMT2D sustains a gene expression program that represses B cell lymphoma development. Nat Med 2015;21:1199–208.
[71] Zhang J, et al. Disruption of KMT2D perturbs germinal center B cell development and promotes lymphomagenesis. Nat Med 2015;21:1190–8.
[72] Vlasevska S, et al. KMT2D acetylation by CREBBP reveals a cooperative functional interaction at enhancers in normal and malignant germinal center B cells. Proc Natl Acad Sci U S A 2023;120:e2218330120.
[73] Goodman RH, Smolik S. CBP/p300 in cell growth, transformation, and development. Genes Dev 2000;14:1553–77.
[74] Hashwah H, et al. Inactivation of CREBBP expands the germinal center B cell compartment, down-regulates MHCII expression and promotes DLBCL growth. Proc Natl Acad Sci U S A 2017;114:9701–6.
[75] Meyer SN, et al. Unique and shared epigenetic programs of the CREBBP and EP300 acetyltransferases in germinal center B cells reveal targetable dependencies in lymphoma. Immunity 2019;51:535.
[76] Kuppers R, Dalla-Favera R. Mechanisms of chromosomal translocations in B cell lymphomas. Oncogene 2001;20:5580–94.
[77] Tsujimoto Y, Gorham J, Cossman J, Jaffe E, Croce CM. The t(14;18) chromosome translocations involved in B-cell neoplasms result from mistakes in VDJ joining. Science 1985;229:1390–3.
[78] Tsujimoto Y, Louie E, Bashir MM, Croce CM. The reciprocal partners of both the t(14; 18) and the t(11; 14) translocations involved in B-cell neoplasms are rearranged by the same mechanism. Oncogene 1988;2:347–51.
[79] Ramiro AR, et al. AID is required for c-myc/IgH chromosome translocations in vivo. Cell 2004;118:431–8.
[80] Robbiani DF, et al. AID is required for the chromosomal breaks in c-myc that lead to c-myc/IgH translocations. Cell 2008;135:1028–38.
[81] Pasqualucci L, et al. AID is required for germinal center-derived lymphomagenesis. Nat Genet 2008;40:108–12.
[82] Takizawa M, et al. AID expression levels determine the extent of cMyc oncogenic translocations and the incidence of B cell tumor development. J Exp Med 2008;205:1949–57.
[83] Akagi T, et al. A novel gene, MALT1 at 18q21, is involved in t(11;18) (q21;q21) found in low-grade B-cell lymphoma of mucosa-associated lymphoid tissue. Oncogene 1999;18:5785–94.
[84] Baens M, Steyls A, Dierlamm J, De Wolf-Peeters C, Marynen P. Structure of the MLT gene and molecular characterization of the genomic breakpoint junctions in the t(11;18)(q21;q21) of marginal zone B-cell lymphomas of MALT type. Genes Chromosomes Cancer 2000;29:281–91.
[85] Akasaka H, et al. Molecular anatomy of BCL6 translocations revealed by long-distance polymerase chain reaction-based assays. Cancer Res 2000;60:2335–41.
[86] Chen W, Iida S, Louie DC, Dalla-Favera R, Chaganti RS. Heterologous promoters fused to BCL6 by chromosomal translocations affecting band 3q27 cause its deregulated expression during B-cell differentiation. Blood 1998;91:603–7.
[87] Ye BH, et al. Alterations of a zinc finger-encoding gene, BCL-6, in diffuse large-cell lymphoma. Science 1993;262:747–50.
[88] Ye BH, Rao PH, Chaganti RS, Dalla-Favera R. Cloning of bcl-6, the locus involved in chromosome translocations affecting band 3q27 in B-cell lymphoma. Cancer Res 1993;53:2732–5.
[89] Yoshida S, et al. Identification of heterologous translocation partner genes fused to the BCL6 gene in diffuse large B-cell lymphomas: 5'-RACE and LA - PCR analyses of biopsy samples. Oncogene 1999;18:7994–9.
[90] Baron BW, et al. Identification of the gene associated with the recurring chromosomal translocations t(3;14)(q27;q32) and t(3;22)(q27;q11) in B-cell lymphomas. Proc Natl Acad Sci U S A 1993;90:5262–6.
[91] Kerckaert JP, et al. Laz3, a novel zinc-finger encoding gene, is disrupted by recurring chromosome 3q27 translocations in human lymphomas. Nat Genet 1993;5:66–70.
[92] Miki T, et al. Molecular cloning of the breakpoint for 3q27 translocation in B-cell lymphomas and leukemias. Blood 1994;83:217–22.
[93] Houldsworth J, et al. REL proto-oncogene is frequently amplified in extranodal diffuse large cell lymphoma. Blood 1996;87:25–9.
[94] Houldsworth J, et al. Relationship between REL amplification, REL function, and clinical and biologic features in diffuse large B-cell lymphomas. Blood 2004;103:1862–8.
[95] Rao PH, et al. Chromosomal and gene amplification in diffuse large B-cell lymphoma. Blood 1998;92:234–40.
[96] Rosenwald A, et al. The use of molecular profiling to predict survival after chemotherapy for diffuse large-B-cell lymphoma. N Engl J Med 2002;346:1937–47.
[97] Green MR, et al. Integrative analysis reveals selective 9p24.1 amplification, increased PD-1 ligand expression, and further induction via JAK2 in nodular sclerosing Hodgkin lymphoma and primary mediastinal large B-cell lymphoma. Blood 2010;116:3268–77.
[98] Varano G, et al. The B-cell receptor controls fitness of MYC-driven lymphoma cells via GSK3beta inhibition. Nature 2017;546:302–6.
[99] Neri A, Knowles DM, Greco A, McCormick F, Dalla-Favera R. Analysis of RAS oncogene mutations in human lymphoid malignancies. Proc Natl Acad Sci U S A 1988:9268–72.
[100] Hollstein M, Sidransky D, Vogelstein B, Harris CC. p53 mutations in human cancers. Science 1991;253:49–53.
[101] Gaidano G, et al. p53 mutations in human lymphoid malignancies: association with Burkitt lymphoma and chronic lymphocytic leukemia. Proc Natl Acad Sci U S A 1991;88:5413–7.
[102] Lo Coco F, et al. p53 mutations are associated with histologic transformation of follicular lymphoma. Blood 1993;82:2289–95.

[103] Mandelbaum J, et al. BLIMP1 is a tumor suppressor gene frequently disrupted in activated B cell-like diffuse large B cell lymphoma. Cancer Cell 2010;18:568–79.
[104] Pasqualucci L, et al. Inactivation of the PRDM1/BLIMP1 gene in diffuse large B cell lymphoma. J Exp Med 2006;203:311–7.
[105] Tam W, et al. Mutational analysis of PRDM1 indicates a tumor-suppressor role in diffuse large B-cell lymphomas. Blood 2006;107:4090–100.
[106] Compagno M, et al. Mutations of multiple genes cause deregulation of NF-kappaB in diffuse large B-cell lymphoma. Nature 2009;459:717–21.
[107] Kato M, et al. Frequent inactivation of A20 in B-cell lymphomas. Nature 2009;459:712–6.
[108] Novak U, et al. The NF-{kappa}B negative regulator TNFAIP3 (A20) is inactivated by somatic mutations and genomic deletions in marginal zone lymphomas. Blood 2009;113:4918–21.
[109] Schmitz R, et al. TNFAIP3 (A20) is a tumor suppressor gene in Hodgkin lymphoma and primary mediastinal B cell lymphoma. J Exp Med 2009;206:981–9.
[110] Jares P, Colomer D, Campo E. Genetic and molecular pathogenesis of mantle cell lymphoma: perspectives for new targeted therapeutics. Nat Rev Cancer 2007;7:750–62.
[111] Pasqualucci L, et al. Hypermutation of multiple proto-oncogenes in B-cell diffuse large-cell lymphomas. Nature 2001;412:341–6.
[112] Bal E, et al. Super-enhancer hypermutation alters oncogene expression in B cell lymphoma. Nature 2022;607:808–15.
[113] Gordon MS, Kanegai CM, Doerr JR, Wall R. Somatic hypermutation of the B cell receptor genes B29 (Igbeta, CD79b) and mb1 (Igalpha, CD79a). Proc Natl Acad Sci U S A 2003;100:4126–31.
[114] Pasqualucci L, et al. BCL-6 mutations in normal germinal center B cells: evidence of somatic hypermutation acting outside Ig loci. Proc Natl Acad Sci U S A 1998;95:11816–21.
[115] Shen HM, Peters A, Baron B, Zhu X, Storb U. Mutation of BCL-6 gene in normal B cells by the process of somatic hypermutation of Ig genes. Science 1998;280:1750–2.
[116] Cerri M, et al. Aberrant somatic hypermutation in post-transplant lymphoproliferative disorders. Br J Haematol 2004;127:362–4.
[117] Deutsch AJ, et al. MALT lymphoma and extranodal diffuse large B-cell lymphoma are targeted by aberrant somatic hypermutation. Blood 2007;109:3500–4.
[118] Gaidano G, et al. Aberrant somatic hypermutation in multiple subtypes of AIDS-associated non-Hodgkin lymphoma. Blood 2003;102:1833–41.
[119] Montesinos-Rongen M, Van Roost D, Schaller C, Wiestler OD, Deckert M. Primary diffuse large B-cell lymphomas of the central nervous system are targeted by aberrant somatic hypermutation. Blood 2004;103:1869–75.
[120] Vakiani E, et al. Genetic and phenotypic analysis of B-cell post-transplant lymphoproliferative disorders provides insights into disease biology. Hematol Oncol 2008;26:199–211.
[121] Khodabakhshi AH, et al. Recurrent targets of aberrant somatic hypermutation in lymphoma. Oncotarget 2012;3:1308–19.
[122] Brunnberg U, Hentrich M, Hoffmann C, Wolf T, Hubel K. HIV-associated malignant lymphoma. Oncol Res Treat 2017;40:82–7.
[123] Pozzato G, et al. Low-grade malignant lymphoma, hepatitis C virus infection, and mixed cryoglobulinemia. Blood 1994;84:3047–53.
[124] Zucca E, Bertoni F. The spectrum of MALT lymphoma at different sites: biological and therapeutic relevance. Blood 2016;127:2082–92.
[125] zur Hausen H, et al. EBV DNA in biopsies of Burkitt tumours and anaplastic carcinomas of the nasopharynx. Nature 1970;228:1056–8.
[126] Ballerini P, et al. Multiple genetic lesions in acquired immunodeficiency syndrome-related non-Hodgkin's lymphoma. Blood 1993;81:166–76.
[127] Hamilton-Dutoit SJ, Pallesen G. A survey of Epstein-Barr virus gene expression in sporadic non-Hodgkin's lymphomas. Detection of Epstein-Barr virus in a subset of peripheral T-cell lymphomas. Am J Pathol 1992;140:1315–25.
[128] Horenstein MG, et al. Epstein-Barr virus latent gene expression in primary effusion lymphomas containing Kaposi's sarcoma-associated herpesvirus/human herpesvirus-8. Blood 1997;90:1186–91.
[129] Cingolani A, et al. Epstein-Barr virus infection is predictive of CNS involvement in systemic AIDS-related non-Hodgkin's lymphomas. J Clin Oncol 2000;18:3325–30.
[130] Fassone L, et al. Molecular profile of Epstein-Barr virus infection in HHV-8-positive primary effusion lymphoma. Leukemia 2000;14:271–7.
[131] Larocca LM, et al. The molecular and phenotypic profile of primary central nervous system lymphoma identifies distinct categories of the disease and is consistent with histogenetic derivation from germinal center-related B cells. Blood 1998;92:1011–9.
[132] Kieff E, Leibowitz D. Oncogenesis by herpesvirus. NY: Cold Spring Harbor Laboratory Press, Cold Spring Harbor; 1989.
[133] Farrell PJ. Epstein–barr virus and cancer. Annu Rev Pathol 2019;14:29–53.
[134] Neri A, et al. Epstein-Barr virus infection precedes clonal expansion in Burkitt's and acquired immunodeficiency syndrome-associated lymphoma. Blood 1991;77:1092–5.
[135] Hutcheson RL, Chakravorty A, Sugden B. Burkitt lymphomas evolve to escape dependencies on epstein-barr virus. Front Cell Infect Microbiol 2021;10.
[136] Vereide D, Sugden B. Proof for EBV's sustaining role in Burkitt's lymphomas. Semin Cancer Biol 2009;19:389–93.
[137] Mundo L, et al. Frequent traces of EBV infection in Hodgkin and non-Hodgkin lymphomas classified as EBV-negative by routine methods: expanding the landscape of EBV-related lymphomas. Mod Pathol 2020;33:2407–21.
[138] Chang Y, et al. Identification of herpesvirus-like DNA sequences in AIDS-associated Kaposi's sarcoma. Science 1994;266:1865–9.
[139] Soulier J, et al. Kaposi's sarcoma-associated herpesvirus-like DNA sequences in multicentric Castleman's disease. Blood 1995;86:1276–80.
[140] Carbone A, Gaidano G. HHV-8-positive body-cavity-based lymphoma: a novel lymphoma entity. Br J Haematol 1997;97:515–22.
[141] Cesarman E, Chang Y, Moore PS, Said JW, Knowles DM. Kaposi's sarcoma-associated herpesvirus-like DNA sequences in AIDS-related body-cavity-based lymphomas. N Engl J Med 1995;332:1186–91.
[142] Gaidano G, et al. Distribution of human herpesvirus-8 sequences throughout the spectrum of AIDS-related neoplasia. AIDS 1996;10:941–9.
[143] Moore PS, et al. Primary characterization of a herpesvirus agent associated with Kaposi's sarcomae. J Virol 1996;70:549–58.
[144] Swerdlow SH, C E, Harris NL, Jaffe ES, Pileri SA, Stein H, et al. WHO classification of haematopoietic and lymphoid tissues. Revised 4th ed. *Lyon*: IARC; 2017.
[145] Alaggio R, et al. The 5th edition of the world health organization classification of haematolymphoid tumours: lymphoid neoplasms. Leukemia 2022;36:1720–48.
[146] Campo E, et al. The international consensus classification of mature lymphoid neoplasms: a report from the clinical advisory committee. Blood 2022;140:1229–53.

[147] Bea S, Amador V. Role of SOX11 and genetic events cooperating with cyclin D1 in mantle cell lymphoma. Curr Oncol Rep 2017;19:43.

[148] Fernàndez V, et al. Genomic and gene expression profiling defines indolent forms of mantle cell lymphoma. Cancer Res 2010;70:1408–18.

[149] Navarro A, et al. Molecular subsets of mantle cell lymphoma defined by the IGHV mutational status and SOX11 expression have distinct biologic and clinical features. Cancer Res 2012;72:5307–16.

[150] Tsujimoto Y, et al. Clustering of breakpoints on chromosome 11 in human B-cell neoplasms with the t(11;14) chromosome translocation. Nature 1985;315:340–3.

[151] Tsujimoto Y, et al. Molecular cloning of the chromosomal breakpoint of B-cell lymphomas and leukemias with the t(11;14) chromosome translocation. Science 1984;224:1403–6.

[152] Fuster C, et al. Cryptic insertions of the immunoglobulin light chain enhancer region near CCND1 in t(11;14)-negative mantle cell lymphoma. Haematologica 2020;105:e408–11.

[153] Motokura T, et al. A novel cyclin encoded by a bcl1-linked candidate oncogene. Nature 1991;350:512–5.

[154] Rosenberg CL, et al. PRAD1, a candidate BCL1 oncogene: mapping and expression in centrocytic lymphoma. Proc Natl Acad Sci U S A 1991;88:9638–42.

[155] Withers DA, et al. Characterization of a candidate bcl-1 gene. Mol Cell Biol 1991;11:4846–53.

[156] Seto M, et al. Gene rearrangement and overexpression of PRAD1 in lymphoid malignancy with t(11;14)(q13;q32) translocation. Oncogene 1992;7:1401–6.

[157] Komatsu H, et al. A variant chromosome translocation at 11q13 identifying PRAD1/cyclin D1 as the BCL-1 gene. Blood 1994;84:1226–31.

[158] Wiestner A, et al. Point mutations and genomic deletions in CCND1 create stable truncated cyclin D1 mRNAs that are associated with increased proliferation rate and shorter survival. Blood 2007;109:4599–606.

[159] Nadeu F, et al. Genomic and epigenomic insights into the origin, pathogenesis, and clinical behavior of mantle cell lymphoma subtypes. Blood 2020;136:1419–32.

[160] Bea S, et al. Uniparental disomies, homozygous deletions, amplifications, and target genes in mantle cell lymphoma revealed by integrative high-resolution whole-genome profiling. Blood 2009;113:3059–69.

[161] Bodrug SE, et al. Cyclin D1 transgene impedes lymphocyte maturation and collaborates in lymphomagenesis with the myc gene. EMBO J 1994;13:2124–30.

[162] Lovec H, Grzeschiczek A, Kowalski MB, Moroy T. Cyclin D1/bcl-1 cooperates with myc genes in the generation of B-cell lymphoma in transgenic mice. EMBO J 1994;13:3487–95.

[163] Salaverria I, et al. CCND2 rearrangements are the most frequent genetic events in cyclin D1– mantle cell lymphoma. Blood 2013;121:1394–402.

[164] Pieters T, et al. Cyclin D2 overexpression drives B1a-derived MCL-like lymphoma in mice. J Exp Med 2021;218.

[165] Salaverria I, et al. Specific secondary genetic alterations in mantle cell lymphoma provide prognostic information independent of the gene expression–based proliferation signature. J Clin Oncol 2007;25:1216–22.

[166] Wlodarska I, Pittaluga S, Hagemeijer A, De Wolf-Peeters C, Van Den Berghe H. Secondary chromosome changes in mantle cell lymphoma. Haematologica 1999;84:594–9.

[167] Allen JE, et al. Identification of novel regions of amplification and deletion within mantle cell lymphoma DNA by comparative genomic hybridization. Br J Haematol 2002;116:291–8.

[168] Schraders M, et al. Novel chromosomal imbalances in mantle cell lymphoma detected by genome-wide array-based comparative genomic hybridization. Blood 2005;105:1686–93.

[169] Pinyol M, et al. Deletions and loss of expression of p16INK4a and p21Waf1 genes are associated with aggressive variants of mantle cell lymphomas. Blood 1997;89:272–80.

[170] Bea S, et al. Landscape of somatic mutations and clonal evolution in mantle cell lymphoma. Proc Natl Acad Sci U S A 2013;110:18250–5.

[171] Kridel R, et al. Whole transcriptome sequencing reveals recurrent NOTCH1 mutations in mantle cell lymphoma. Blood 2012;119:1963–71.

[172] Pararajalingam P, et al. Coding and noncoding drivers of mantle cell lymphoma identified through exome and genome sequencing. Blood 2020;136:572–84.

[173] Schaffner C, Idler I, Stilgenbauer S, Dohner H, Lichter P. Mantle cell lymphoma is characterized by inactivation of the ATM gene. Proc Natl Acad Sci U S A 2000;97:2773–8.

[174] Bea S, et al. BMI-1 gene amplification and overexpression in hematological malignancies occur mainly in mantle cell lymphomas. Cancer Res 2001;61:2409–12.

[175] Chapman CJ, Mockridge CI, Rowe M, Rickinson AB, Stevenson FK. Analysis of VH genes used by neoplastic B cells in endemic Burkitt's lymphoma shows somatic hypermutation and intraclonal heterogeneity. Blood 1995;85:2176–81.

[176] Chapman CJ, Zhou JX, Gregory C, Rickinson AB, Stevenson FK. VH and VL gene analysis in sporadic Burkitt's lymphoma shows somatic hypermutation, intraclonal heterogeneity, and a role for antigen selection. Blood 1996;88:3562–8.

[177] Jain R, et al. A potential role for antigen selection in the clonal evolution of Burkitt's lymphoma. J Immunol 1994;153:45–52.

[178] Tamaru J, et al. Burkitt's lymphomas express VH genes with a moderate number of antigen-selected somatic mutations. Am J Pathol 1995;147:1398–407.

[179] Dave SS, et al. Molecular diagnosis of Burkitt's lymphoma. N Engl J Med 2006;354:2431–42.

[180] Hummel M, et al. A biologic definition of Burkitt's lymphoma from transcriptional and genomic profiling. N Engl J Med 2006;354:2419–30.

[181] Kaymaz Y, et al. Comprehensive transcriptome and mutational profiling of endemic Burkitt lymphoma reveals EBV type-specific differences. Mol Cancer Res 2017;15:563–76.

[182] Grande BM, et al. Genome-wide discovery of somatic coding and noncoding mutations in pediatric endemic and sporadic Burkitt lymphoma. Blood 2019;133:1313–24.

[183] Burkhardt B, et al. Clinical relevance of molecular characteristics in Burkitt lymphoma differs according to age. Nat Commun 2022;13:3881.

[184] Thomas N, et al. Genetic subgroups inform on pathobiology in adult and pediatric Burkitt lymphoma. Blood 2023;141:904–16.

[185] Dalla-Favera R, et al. Human c-myc onc gene is located on the region of chromosome 8 that is translocated in Burkitt lymphoma cells. Proc Natl Acad Sci U S A 1982;79:7824–7.
[186] Dalla-Favera R, Martinotti S, Gallo RC, Erikson J, Croce CM. Translocation and rearrangements of the c-myc oncogene locus in human undifferentiated B-cell lymphomas. Science 1983;219:963–7.
[187] Taub R, et al. Translocation of the c-myc gene into the immunoglobulin heavy chain locus in human Burkitt lymphoma and murine plasmacytoma cells. Proc Natl Acad Sci U S A 1982;79:7837–41.
[188] Davis M, Malcolm S, Rabbitts TH. Chromosome translocation can occur on either side of the c-myc oncogene in Burkitt lymphoma cells. Nature 1984;308:286–8.
[189] Neri A, Barriga F, Knowles DM, Magrath IT, Dalla-Favera R. Different regions of the immunoglobulin heavy-chain locus are involved in chromosomal translocations in distinct pathogenetic forms of Burkitt lymphoma. Proc Natl Acad Sci U S A 1988;85:2748–52.
[190] Boxer LM, Dang CV. Translocations involving c-myc and c-myc function. Oncogene 2001;20:5595–610.
[191] Pelicci PG, Knowles 2nd DM, Magrath I, Dalla-Favera R. Chromosomal breakpoints and structural alterations of the c-myc locus differ in endemic and sporadic forms of Burkitt lymphoma. Proc Natl Acad Sci U S A 1986;83:2984–8.
[192] ar-Rushdi A, et al. Differential expression of the translocated and the untranslocated c-myc oncogene in Burkitt lymphoma. Science 1983;222:390–3.
[193] Hayday AC, et al. Activation of a translocated human c-myc gene by an enhancer in the immunoglobulin heavy-chain locus. Nature 1984;307:334–40.
[194] Rabbitts TH, Forster A, Baer R, Hamlyn PH. Transcription enhancer identified near the human C mu immunoglobulin heavy chain gene is unavailable to the translocated c-myc gene in a Burkitt lymphoma. Nature 1983;306:806–9.
[195] Klein U, et al. Transcriptional analysis of the B cell germinal center reaction. Proc Natl Acad Sci U S A 2003;100:2639–44.
[196] Cesarman E, Dalla-Favera R, Bentley D, Groudine M. Mutations in the first exon are associated with altered transcription of c-myc in Burkitt lymphoma. Science 1987;238:1272–5.
[197] Lopez C, et al. Genomic and transcriptomic changes complement each other in the pathogenesis of sporadic Burkitt lymphoma. Nat Commun 2019;10:1459.
[198] Bhatia K, et al. Point mutations in the c-Myc transactivation domain are common in Burkitt's lymphoma and mouse plasmacytomas. Nat Genet 1993;5:56–61.
[199] Bhatia K, et al. Mutations in the coding region of c-myc occur frequently in acquired immunodeficiency syndrome-associated lymphomas. Blood 1994;84:883–8.
[200] Gu W, Bhatia K, Magrath IT, Dang CV, Dalla-Favera R. Binding and suppression of the Myc transcriptional activation domain by p107. Science 1994;264:251–4.
[201] Gregory MA, Hann SR. c-Myc proteolysis by the ubiquitin-proteasome pathway: stabilization of c-Myc in Burkitt's lymphoma cells. Mol Cell Biol 2000;20:2423–35.
[202] Hemann MT, et al. Evasion of the p53 tumour surveillance network by tumour-derived MYC mutants. Nature 2005;436:807–11.
[203] Love C, et al. The genetic landscape of mutations in Burkitt lymphoma. Nat Genet 2012;44:1321–5.
[204] Schmitz R, et al. Burkitt lymphoma pathogenesis and therapeutic targets from structural and functional genomics. Nature 2012;490:116–20.
[205] Felsher DW, Bishop JM. Transient excess of MYC activity can elicit genomic instability and tumorigenesis. Proc Natl Acad Sci U S A 1999;96:3940–4.
[206] Adams JM, et al. The c-myc oncogene driven by immunoglobulin enhancers induces lymphoid malignancy in transgenic mice. Nature 1985;318:533–8.
[207] Kovalchuk AL, et al. Burkitt lymphoma in the mouse. J Exp Med 2000;192:1183–90.
[208] Sander S, et al. Synergy between PI3K signaling and MYC in Burkitt lymphomagenesis. Cancer Cell 2012;22:167–79.
[209] Richter J, et al. Recurrent mutation of the ID3 gene in Burkitt lymphoma identified by integrated genome, exome and transcriptome sequencing. Nat Genet 2012;44:1316–20.
[210] Martinez-Delgado B, et al. Hypermethylation of p15/ink4b/MTS2 gene is differentially implicated among non-Hodgkin's lymphomas. Leukemia 1998;12:937–41.
[211] Gaidano G, et al. Deletions involving two distinct regions of 6q in B-cell non-Hodgkin lymphoma. Blood 1992;80:1781–7.
[212] Lombardi L, Newcomb EW, Dalla-Favera R. Pathogenesis of Burkitt lymphoma: expression of an activated c-myc oncogene causes the tumorigenic conversion of EBV-infected human B lymphoblasts. Cell 1987;49:161–70.
[213] Prevot S, et al. Analysis of African Burkitt's and high-grade B cell non-Burkitt's lymphoma for Epstein-Barr virus genomes using in situ hybridization. Br J Haematol 1992;80:27–32.
[214] Thorley-Lawson DA, Allday MJ. The curious case of the tumour virus: 50 years of Burkitt's lymphoma. Nat Rev Microbiol 2008;6:913–24.
[215] Panea RI, et al. The whole-genome landscape of Burkitt lymphoma subtypes. Blood 2019;134:1598–607.
[216] Kridel R, Sehn LH, Gascoyne RD. Pathogenesis of follicular lymphoma. J Clin Invest 2012;122:3424–31.
[217] Montoto S, Fitzgibbon J. Transformation of indolent B-cell lymphomas. J Clin Oncol 2011;29:1827–34.
[218] Milpied P, et al. Human germinal center transcriptional programs are de-synchronized in B cell lymphoma. Nat Immunol 2018;19:1013–24.
[219] Carbone A, et al. Follicular lymphoma. Nat Rev Dis Prim 2019;5:83.
[220] Roulland S, et al. Follicular lymphoma-like B cells in healthy individuals: a novel intermediate step in early lymphomagenesis. J Exp Med 2006;203:2425–31.
[221] Schroers-Martin JG, et al. Tracing founder mutations in circulating and tissue-resident follicular lymphoma precursors. Cancer Discov 2023.
[222] Bakhshi A, et al. Cloning the chromosomal breakpoint of t(14;18) human lymphomas: clustering around JH on chromosome 14 and near a transcriptional unit on 18. Cell 1985;41:899–906.
[223] Tsujimoto Y, Finger LR, Yunis J, Nowell PC, Croce CM. Cloning of the chromosome breakpoint of neoplastic B cells with the t(14;18) chromosome translocation. Science 1984;226:1097–9.
[224] Cleary ML, Sklar J. Nucleotide sequence of a t(14;18) chromosomal breakpoint in follicular lymphoma and demonstration of a breakpoint-cluster region near a transcriptionally active locus on chromosome 18. Proc Natl Acad Sci U S A 1985;82:7439–43.

[225] Graninger WB, Seto M, Boutain B, Goldman P, Korsmeyer SJ. Expression of Bcl-2 and Bcl-2-Ig fusion transcripts in normal and neoplastic cells. J Clin Invest 1987;80:1512–5.
[226] Ngan BY, Chen-Levy Z, Weiss LM, Warnke RA, Cleary ML. Expression in non-Hodgkin's lymphoma of the bcl-2 protein associated with the t(14;18) chromosomal translocation. N Engl J Med 1988;318:1638–44.
[227] Petrovic AS, et al. The Ig heavy chain 3' end confers a post-transcriptional processing advantage to Bcl-2-IgH fusion RNA in t(14;18) lymphoma. Blood 1998;91:3952–61.
[228] Saito M, et al. BCL6 suppression of BCL2 via Miz1 and its disruption in diffuse large B cell lymphoma. Proc Natl Acad Sci U S A 2009;106:11294–9.
[229] Cleary ML, Galili N, Sklar J. Detection of a second t(14;18) breakpoint cluster region in human follicular lymphomas. J Exp Med 1986;164:315–20.
[230] Buchonnet G, et al. Distribution of BCL2 breakpoints in follicular lymphoma and correlation with clinical features: specific subtypes or same disease? Leukemia 2002;16:1852–6.
[231] Morin RD, et al. Frequent mutation of histone-modifying genes in non-Hodgkin lymphoma. Nature 2011;476:298–303.
[232] Morin RD, et al. Somatic mutations altering EZH2 (Tyr641) in follicular and diffuse large B-cell lymphomas of germinal-center origin. Nat Genet 2010;42:181–5.
[233] Okosun J, et al. Recurrent mTORC1-activating RRAGC mutations in follicular lymphoma. Nat Genet 2016;48:183–8.
[234] Cheung KJ, et al. Acquired TNFRSF14 mutations in follicular lymphoma are associated with worse prognosis. Cancer Res 2010;70:9166–74.
[235] Boice M, et al. Loss of the HVEM tumor suppressor in lymphoma and restoration by modified CAR-T cells. Cell 2016;167:405–18.
[236] Zhu D, et al. Acquisition of potential N-glycosylation sites in the immunoglobulin variable region by somatic mutation is a distinctive feature of follicular lymphoma. Blood 2002;99:2562–8.
[237] Green MR, et al. Mutations in early follicular lymphoma progenitors are associated with suppressed antigen presentation. Proc Natl Acad Sci U S A 2015;112:E1116–25.
[238] Okosun J, et al. Integrated genomic analysis identifies recurrent mutations and evolution patterns driving the initiation and progression of follicular lymphoma. Nat Genet 2014;46:176–81.
[239] Pasqualucci L, et al. Genetics of follicular lymphoma transformation. Cell Rep 2014;6:130–40.
[240] Li J, et al. Cooperative super-enhancer inactivation caused by heterozygous loss of CREBBP and KMT2D skews B cell fate decisions and yields T cell-depleted lymphomas. bioRxiv 2023:6770–4.
[241] Bahler DW, Levy R. Clonal evolution of a follicular lymphoma: evidence for antigen selection. Proc Natl Acad Sci U S A 1992;89:6770–4.
[242] Radcliffe CM, et al. Human follicular lymphoma cells contáin oligomannose glycans in the antigen-binding site of the B-cell receptor. J Biol Chem 2007;282:7405–15.
[243] Wright G, et al. A gene expression-based method to diagnose clinically distinct subgroups of diffuse large B cell lymphoma. Proc Natl Acad Sci U S A 2003;100:9991–6.
[244] Chapuy B, et al. Molecular subtypes of diffuse large B cell lymphoma are associated with distinct pathogenic mechanisms and outcomes. Nat Med 2018;24:679–90.
[245] Schmitz R, et al. Genetics and pathogenesis of diffuse large B-cell lymphoma. N Engl J Med 2018;378:1396–407.
[246] Lacy SE, et al. Targeted sequencing in DLBCL, molecular subtypes, and outcomes: a Haematological Malignancy Research Network report. Blood 2020;135:1759–71.
[247] Wright GW, et al. A probabilistic classification tool for genetic subtypes of diffuse large B cell lymphoma with therapeutic implications. Cancer Cell 2020;37:551–568 e14.
[248] Wilson WH, et al. Effect of ibrutinib with R-CHOP chemotherapy in genetic subtypes of DLBCL. Cancer Cell 2021;39:1643–1653 e3.
[249] Dominguez PM, et al. TET2 deficiency causes germinal center hyperplasia, impairs plasma cell differentiation, and promotes B-cell lymphomagenesis. Cancer Discov 2018;8:1632–53.
[250] Lo Coco F, et al. Rearrangements of the BCL6 gene in diffuse large cell non-Hodgkin's lymphoma. Blood 1994;83:1757–9.
[251] Offit K, et al. t(3;22)(q27;q11): a novel translocation associated with diffuse non-Hodgkin's lymphoma. Blood 1989;74:1876–9.
[252] Iqbal J, et al. Distinctive patterns of BCL6 molecular alterations and their functional consequences in different subgroups of diffuse large B-cell lymphoma. Leukemia 2007;21:2332–43.
[253] Ye BH, et al. Chromosomal translocations cause deregulated BCL6 expression by promoter substitution in B cell lymphoma. EMBO J 1995;14:6209–17.
[254] Capello D, et al. Distribution and pattern of BCL-6 mutations throughout the spectrum of B-cell neoplasia. Blood 2000;95:651–9.
[255] Migliazza A, et al. Frequent somatic hypermutation of the 5' noncoding region of the BCL6 gene in B-cell lymphoma. Proc Natl Acad Sci U S A 1995;92:12520–4.
[256] Pasqualucci L, et al. Mutations of the BCL6 proto-oncogene disrupt its negative autoregulation in diffuse large B-cell lymphoma. Blood 2003;101:2914–23.
[257] Wang X, Li Z, Naganuma A, Ye BH. Negative autoregulation of BCL-6 is bypassed by genetic alterations in diffuse large B cell lymphomas. Proc Natl Acad Sci U S A 2002;99:15018–23.
[258] Duan S, et al. FBXO11 targets BCL6 for degradation and is inactivated in diffuse large B-cell lymphomas. Nature 2012;481:90–3.
[259] Cattoretti G, et al. PRDM1/Blimp-1 is expressed in human B-lymphocytes committed to the plasma cell lineage. J Pathol 2005;206:76–86.
[260] Schneider C, et al. FBXO11 inactivation leads to abnormal germinal-center formation and lymphoproliferative disease. Blood 2016;128:660–6.
[261] Challa-Malladi M, et al. Combined genetic inactivation of beta2-Microglobulin and CD58 reveals frequent escape from immune recognition in diffuse large B cell lymphoma. Cancer Cell 2011;20:728–40.
[262] Fangazio M, et al. Genetic mechanisms of immune escape in diffuse large B cell lymphoma. Blood 2014;124.
[263] Cycon KA, Rimsza LM, Murphy SP. Alterations in CIITA constitute a common mechanism accounting for downregulation of MHC class II expression in diffuse large B-cell lymphoma (DLBCL). Exp Hematol 2009;37:184–94.
[264] Steidl C, et al. MHC class II transactivator CIITA is a recurrent gene fusion partner in lymphoid cancers. Nature 2011;471:377–81.
[265] Ennishi D, et al. Molecular and genetic characterization of MHC deficiency identifies EZH2 as therapeutic target for enhancing immune recognition. Cancer Discov 2019;9:546–63.
[266] Mondello P, et al. Selective inhibition of HDAC3 targets synthetic vulnerabilities and activates immune surveillance in lymphoma. Cancer Discov 2020;10(3):440–59.

[267] Iqbal J, et al. BCL2 translocation defines a unique tumor subset within the germinal center B-cell-like diffuse large B-cell lymphoma. Am J Pathol 2004;165:159–66.

[268] Ladanyi M, Offit K, Jhanwar SC, Filippa DA, Chaganti RS. MYC rearrangement and translocations involving band 8q24 in diffuse large cell lymphomas. Blood 1991;77:1057–63.

[269] Muppidi JR, et al. Loss of signalling via Galpha13 in germinal centre B-cell-derived lymphoma. Nature 2014;516(7530):254–8.

[270] Lenz G, et al. Molecular subtypes of diffuse large B-cell lymphoma arise by distinct genetic pathways. Proc Natl Acad Sci U S A 2008;105:13520–5.

[271] Pfeifer M, et al. PTEN loss defines a PI3K/AKT pathway-dependent germinal center subtype of diffuse large B-cell lymphoma. Proc Natl Acad Sci U S A 2013;110:12420–5.

[272] Steinberg MW, Cheung TC, Ware CF. The signaling networks of the herpesvirus entry mediator (TNFRSF14) in immune regulation. Immunol Rev 2011;244:169–87.

[273] Davis RE, Brown KD, Siebenlist U, Staudt LM. Constitutive nuclear factor kappaB activity is required for survival of activated B cell-like diffuse large B cell lymphoma cells. J Exp Med 2001;194:1861–74.

[274] Davis RE, et al. Chronic active B-cell-receptor signalling in diffuse large B-cell lymphoma. Nature 2010;463:88–92.

[275] Phelan JD, et al. A multiprotein supercomplex controlling oncogenic signalling in lymphoma. Nature 2018;560:387–91.

[276] Ngo VN, et al. Oncogenically active MYD88 mutations in human lymphoma. Nature 2011;470:115–9.

[277] Lenz G, et al. Oncogenic CARD11 mutations in human diffuse large B cell lymphoma. Science 2008;319:1676–9.

[278] Young RM, et al. Survival of human lymphoma cells requires B-cell receptor engagement by self-antigens. Proc Natl Acad Sci U S A 2015;112:13447–54.

[279] Wilson WH, et al. Targeting B cell receptor signaling with ibrutinib in diffuse large B cell lymphoma. Nat Med 2015;21:922–6.

[280] Roschewski M, Staudt LM, Wilson WH. Diffuse large B-cell lymphoma-treatment approaches in the molecular era. Nat Rev Clin Oncol 2014;11:12–23.

[281] Cattoretti G, et al. Deregulated BCL6 expression recapitulates the pathogenesis of human diffuse large B cell lymphomas in mice. Cancer Cell 2005;7:445–55.

[282] Calado DP, et al. Constitutive canonical NF-kappaB activation cooperates with disruption of BLIMP1 in the pathogenesis of activated B cell-like diffuse large cell lymphoma. Cancer Cell 2010;18:580–9.

[283] Fabbri G, et al. Genetic lesions associated with chronic lymphocytic leukemia transformation to Richter syndrome. J Exp Med 2013;210:2273–88.

[284] Landau DA, et al. Evolution and impact of subclonal mutations in chronic lymphocytic leukemia. Cell 2013;152:714–26.

[285] Elenitoba-Johnson KS, et al. Homozygous deletions at chromosome 9p21 involving p16 and p15 are associated with histologic progression in follicle center lymphoma. Blood 1998;91:4677–85.

[286] Yano T, Jaffe ES, Longo DL, Raffeld M. MYC rearrangements in histologically progressed follicular lymphomas. Blood 1992;80:758–67.

[287] Sander CA, et al. p53 mutation is associated with progression in follicular lymphomas. Blood 1993;82:1994–2004.

[288] Rosenwald A, et al. Molecular diagnosis of primary mediastinal B cell lymphoma identifies a clinically favorable subgroup of diffuse large B cell lymphoma related to Hodgkin lymphoma 2003;198:851–62.

[289] Savage KJ, et al. The molecular signature of mediastinal large B-cell lymphoma differs from that of other diffuse large B-cell lymphomas and shares features with classical Hodgkin lymphoma. Blood 2003;102:3871–9.

[290] Rui L, et al. Cooperative epigenetic modulation by cancer amplicon genes. Cancer Cell 2010;18:590–605.

[291] Mottok A, Woolcock B, Chan FC, Tong KM, Chong L, Farinha P, et al. Genomic alterations in CIITA are frequent in primary mediastinal large B cell lymphoma and are associated with diminished MHC class II expression. Cell Rep 2015;13(7):1418–31. https://doi.org/10.1016/j.celrep.2015.10.008. Epub 2015 Nov 5.

[292] Mottok A, Hung SS, Chavez EA, Woolcock B, Telenius A, Chong LC, et al. Integrative genomic analysis identifies key pathogenic mechanisms in primary mediastinal large B-cell lymphoma. Blood 2019;134(10):802–13. https://doi.org/10.1182/blood.2019001126. Epub 2019 Jul 10.

[293] Kuppers R. The biology of Hodgkin's lymphoma. Nat Rev Cancer 2009;9:15–27.

[294] Kuppers R, et al. Hodgkin disease: Hodgkin and Reed-Sternberg cells picked from histological sections show clonal immunoglobulin gene rearrangements and appear to be derived from B cells at various stages of development. Proc Natl Acad Sci U S A 1994;91:10962–6.

[295] Kanzler H, Kuppers R, Hansmann ML, Rajewsky K. Hodgkin and Reed-Sternberg cells in Hodgkin's disease represent the outgrowth of a dominant tumor clone derived from (crippled) germinal center B cells. J Exp Med 1996;184:1495–505.

[296] Reichel J, et al. Flow sorting and exome sequencing reveal the oncogenome of primary Hodgkin and Reed-Sternberg cells. Blood 2015;125:1061–72.

[297] Tiacci E, et al. Pervasive mutations of JAK-STAT pathway genes in classical Hodgkin lymphoma. Blood 2018;131:2454–65.

[298] Wienand K, et al. Genomic analyses of flow-sorted Hodgkin Reed-Sternberg cells reveal complementary mechanisms of immune evasion. Blood Adv 2019;3:4065–80.

[299] Maura F, et al. Molecular evolution of classic Hodgkin lymphoma revealed through whole-genome sequencing of Hodgkin and reed sternberg cells. Blood Cancer Discov 2023;4:208–27.

[300] Martin-Subero JI, et al. Recurrent involvement of the REL and BCL11A loci in classical Hodgkin lymphoma. Blood 2002;99:1474–7.

[301] Barth TF, et al. Gains of 2p involving the REL locus correlate with nuclear c-Rel protein accumulation in neoplastic cells of classical Hodgkin lymphoma. Blood 2003;101:3681–6.

[302] Jungnickel B, et al. Clonal deleterious mutations in the Ikappa-Balpha gene in the malignant cells in Hodgkin's lymphoma. J Exp Med 2000;191:395–402.

[303] Emmerich F, et al. Overexpression of I kappa B alpha without inhibition of NF-kappaB activity and mutations in the I kappa B alpha gene in Reed-Sternberg cells. Blood 1999;94:3129–34.

[304] Joos S, et al. Genomic imbalances including amplification of the tyrosine kinase gene JAK2 in CD30+ Hodgkin cells. Cancer Res 2000;60:549–52.

[305] Fais F, et al. Chronic lymphocytic leukemia B cells express restricted sets of mutated and unmutated antigen receptors. J Clin Invest 1998;102:1515–25.
[306] Oscier DG, Thompsett A, Zhu D, Stevenson FK. Differential rates of somatic hypermutation in V(H) genes among subsets of chronic lymphocytic leukemia defined by chromosomal abnormalities. Blood 1997;89:4153–60.
[307] Damle RN, et al. Ig V gene mutation status and CD38 expression as novel prognostic indicators in chronic lymphocytic leukemia. Blood 1999;94:1840–7.
[308] Hamblin TJ, Davis Z, Gardiner A, Oscier DG, Stevenson FK. Unmutated Ig V(H) genes are associated with a more aggressive form of chronic lymphocytic leukemia. Blood 1999;94:1848–54.
[309] Klein U, et al. Gene expression profiling of B cell chronic lymphocytic leukemia reveals a homogeneous phenotype related to memory B cells. J Exp Med 2001;194:1625–38.
[310] Rosenwald A, et al. Relation of gene expression phenotype to immunoglobulin mutation genotype in B cell chronic lymphocytic leukemia. J Exp Med 2001;194:1639–47.
[311] Widhopf GF, et al. Chronic lymphocytic leukemia B cells of more than 1% of patients express virtually identical immunoglobulins. Blood 2004;104:2499–504.
[312] Messmer BT, et al. Multiple distinct sets of stereotyped antigen receptors indicate a role for antigen in promoting chronic lymphocytic leukemia. J Exp Med 2004;200:519–25.
[313] Tobin G, et al. Subsets with restricted immunoglobulin gene rearrangement features indicate a role for antigen selection in the development of chronic lymphocytic leukemia. Blood 2004;104:2879–85.
[314] Murray F, et al. Stereotyped patterns of somatic hypermutation in subsets of patients with chronic lymphocytic leukemia: implications for the role of antigen selection in leukemogenesis. Blood 2008;111:1524–33.
[315] Stamatopoulos K, et al. Over 20% of patients with chronic lymphocytic leukemia carry stereotyped receptors: pathogenetic implications and clinical correlations. Blood 2007;109:259–70.
[316] Tobin G, et al. Chronic lymphocytic leukemias utilizing the VH3-21 gene display highly restricted Vlambda2-14 gene use and homologous CDR3s: implicating recognition of a common antigen epitope. Blood 2003;101:4952–7.
[317] Seifert M, et al. Cellular origin and pathophysiology of chronic lymphocytic leukemia. J Exp Med 2012;209:2183–98.
[318] Queiros AC, et al. A B-cell epigenetic signature defines three biologic subgroups of chronic lymphocytic leukemia with clinical impact. Leukemia 2015;29:598–605.
[319] Kikushige Y, et al. Self-renewing hematopoietic stem cell is the primary target in pathogenesis of human chronic lymphocytic leukemia. Cancer Cell 2011;20:246–59.
[320] Damm F, et al. Acquired initiating mutations in early hematopoietic cells of CLL patients. Cancer Discov 2014;4:1088–101.
[321] Marsilio S, et al. Somatic CLL mutations occur at multiple distinct hematopoietic maturation stages: documentation and cautionary note regarding cell fraction purity. Leukemia 2018;32:1041–4.
[322] Dohner H, et al. Genomic aberrations and survival in chronic lymphocytic leukemia. N Engl J Med 2000;343:1910–6.
[323] Dohner H, Stilgenbauer S, Dohner K, Bentz M, Lichter P. Chromosome aberrations in B-cell chronic lymphocytic leukemia: reassessment based on molecular cytogenetic analysis. J Mol Med (Berl) 1999;77:266–81.
[324] Haferlach C, Dicker F, Schnittger S, Kern W, Haferlach T. Comprehensive genetic characterization of CLL: a study on 506 cases analysed with chromosome banding analysis, interphase FISH, IgV(H) status and immunophenotyping. Leukemia 2007;21:2442–51.
[325] Calin GA, et al. Frequent deletions and down-regulation of micro-RNA genes miR15 and miR16 at 13q14 in chronic lymphocytic leukemia. Proc Natl Acad Sci U S A 2002;99:15524–9.
[326] Kalachikov S, et al. Cloning and gene mapping of the chromosome 13q14 region deleted in chronic lymphocytic leukemia. Genomics 1997;42:369–77.
[327] Migliazza A, et al. Nucleotide sequence, transcription map, and mutation analysis of the 13q14 chromosomal region deleted in B-cell chronic lymphocytic leukemia. Blood 2001;97:2098–104.
[328] Klein U, et al. The DLEU2/miR-15a/16-1 cluster controls B cell proliferation and its deletion leads to chronic lymphocytic leukemia. Cancer Cell 2010;17:28–40.
[329] Anastasi J, et al. Detection of trisomy 12 in chronic lymphocytic leukemia by fluorescence in situ hybridization to interphase cells: a simple and sensitive method. Blood 1992;79:1796–801.
[330] Hjalmar V, et al. Trisomy 12 and lymphoplasmacytoid lymphocytes in chronic leukemic B-cell disorders. Haematologica 1998;83:602–9.
[331] Juliusson G, et al. Prognostic subgroups in B-cell chronic lymphocytic leukemia defined by specific chromosomal abnormalities. N Engl J Med 1990;323:720–4.
[332] Bullrich F, et al. ATM mutations in B-cell chronic lymphocytic leukemia. Cancer Res 1999;59:24–7.
[333] Stankovic T, et al. Inactivation of ataxia telangiectasia mutated gene in B-cell chronic lymphocytic leukaemia. Lancet 1999;353:26–9.
[334] Starostik P, et al. Deficiency of the ATM protein expression defines an aggressive subgroup of B-cell chronic lymphocytic leukemia. Cancer Res 1998;58:4552–7.
[335] Gaidano G, et al. Analysis of alterations of oncogenes and tumor suppressor genes in chronic lymphocytic leukemia. Am J Pathol 1994;144:1312–9.
[336] Burns A, et al. Whole-genome sequencing of chronic lymphocytic leukaemia reveals distinct differences in the mutational landscape between IgHV(mut) and IgHV(unmut) subgroups. Leukemia 2018;32:332–42.
[337] Robbe P, et al. Whole-genome sequencing of chronic lymphocytic leukemia identifies subgroups with distinct biological and clinical features. Nat Genet 2022;54:1675–89.
[338] Knisbacher BA, et al. Molecular map of chronic lymphocytic leukemia and its impact on outcome. Nat Genet 2022;54:1664–74.
[339] Puente XS, et al. Whole-genome sequencing identifies recurrent mutations in chronic lymphocytic leukaemia. Nature 2011;475:101–5.
[340] Landau DA, et al. Mutations driving CLL and their evolution in progression and relapse. Nature 2015;526:525–30.
[341] Fabbri G, et al. Analysis of the chronic lymphocytic leukemia coding genome: role of NOTCH1 mutational activation. J Exp Med 2011;208:1389–401.
[342] Gill H, et al. Two tales of two lymphomas. J Clin Oncol 2011;29:e610–2.
[343] Puente XS, et al. Non-coding recurrent mutations in chronic lymphocytic leukaemia. Nature 2015;526:519–24.
[344] Wang L, et al. SF3B1 and other novel cancer genes in chronic lymphocytic leukemia. N Engl J Med 2011;365:2497–506.
[345] Quesada V, et al. Exome sequencing identifies recurrent mutations of the splicing factor SF3B1 gene in chronic lymphocytic leukemia. Nat Genet 2012;44:47–52.

[346] Rossi D, et al. Mutations of the SF3B1 splicing factor in chronic lymphocytic leukemia: association with progression and fludarabine-refractoriness. Blood 2011;118:6904–8.

[347] Chapuy B, Stewart C, Dunford AJ, Kim J, Wienand K, Kamburov A, et al. Genomic analyses of PMBL reveal new drivers and mechanisms of sensitivity to PD-1 blockade. Blood 2019;134(26):2369–82. https://doi.org/10.1182/blood.2019002067. PMID: 31697821. Free PMC article.

[348] Gunawardana J, Chan FC, Telenius A, Woolcock B, Kridel R, Tan KL, et al. Recurrent somatic mutations of PTPN1 in primary mediastinal B cell lymphoma and Hodgkin lymphoma. Nat Genet 2014;46(4):329–35. https://doi.org/10.1038/ng.2900. Epub 2014 Feb 16. PMID: 24531327.

Chapter 15

Human Immunodeficiencies Caused by Inborn Errors of B-Cell Development or Function

Stuart G. Tangye[1,2], Jean-Laurent Casanova[3,4,5,6] and Cindy S. Ma[1,2]

[1]Garvan Institute of Medical Research, Darlinghurst, NSW, Australia; [2]School of Clinical Medicine, Faculty of Medicine and Health, UNSW Sydney, Sydney, NSW, Australia; [3]Laboratory of Human Genetics of Infectious Diseases, Paris, France; [4]University of Paris Cité, Imagine Institute, Paris, France; [5]St. Giles Laboratory of Human Genetics of Infectious Diseases, The Rockefeller University, New York, NY, United States; [6]Howard Hughes Medical Institute, New York, NY, United States

ABBREVIATIONS

AID activation-induced cytidine deaminase
APDS activated PI3K delta syndrome
ARA autosomal recessive agammaglobulinemia
BTK Bruton's tyrosine kinase
CID combined immunodeficiency
CSR class switch recombination
CVID common variable immunodeficiency
DN dominant negative
GC germinal centre
GOF gain of function
HIES hyper IgE syndrome
HIGM hyper IgM syndrome
IEI inborn errors of immunity
LOF loss of function
PC plasma cell
SCID severe combined immunodeficiency
SHM somatic hypermutation
Tfh T follicular helper cell
X-SCID X-linked severe combined immunodeficiency
XL X-linked
XL-EDA-ID XL anhidrotic ectodermal dysplasia with immunodeficiency
XLA X-linked agammaglobulinemia

1. INTRODUCTION

Mature B cells develop from hematopoietic stem cells in the bone marrow. Once generated, B cells serve multiple roles in immune regulation and host defence. However, their most important function is producing neutralising antibodies (Abs) that efficiently clear invading and recurring pathogens. This is achieved by naïve B cells differentiating into memory B cells that rapidly respond to subsequent exposure to the initial Ag, and plasma cells (PCs) that continually secrete Ab. These B cell subsets cooperate to maintain humoral immunity and host protection against recurrent infections for extended periods of time. Thus, the generation of antigen (Ag)-specific memory B cells and PCs underlies long-lived serological immunity, explaining the rarity of secondary disease and contributing to the success of most vaccines. Our understanding of immunity is often derived from animal models. However, analysis of individuals with rare or common monogenic defects that disrupt immune function of leucocytes and non-hematopoietic cells are unprecedented models to link genotypes to clinical phenotypes, establish mechanisms of disease pathogenesis, and elucidate non-redundant pathways for immune cell development, differentiation and host defence. Here, we review the fundamental breakthroughs in unravelling the complexities of humoral immunity that have come from the discovery and characterisation of inborn errors disrupting humoral immunity in humans.

2. B-CELL DEVELOPMENT AND FUNCTION

B cells develop from multipotent hematopoietic stem cells (HSCs) present in the bone marrow (BM). This involves a series of iterative steps whereby HSCs progressively develop into a common lymphoid progenitor (CLP), then

Molecular Biology of B Cells. https://doi.org/10.1016/B978-0-323-95895-0.00013-1

into a progenitor (pro-B cells), precursor (large and small pre-B cells) and then an immature B cell [1–3] (Figure 15.1). Each of these stages is characterised by the sequential and temporal rearrangement of genes encoding variable (V), diversity (D) and joining (J) elements of the immunoglobulin heavy (IgH) chain, acquisition of expression of a pre-B-cell receptor (BCR) complex comprising IgH chain and a surrogate light (L) chain on large pre-B cells, followed by recombination of genes encoding an Ig L chain in small pre-B cells and then finally expression of a functional BCR on immature B cells. Rearrangement of genes encoding the V, D and J elements of the IgH (VDJ) and Ig L (VJ) chains during B-cell development is regulated by the coordinated induction and activity of the recombination activating genes *RAG1* and *RAG2* in pro-(VDJ) and pre-B cells (VJ). Immature B cells are then exported from the BM to the periphery as transitional B cells, where they undergo final maturation steps to yield immunocompetent naïve B cells. This process generates a pool of Ag-inexperienced B cells expressing a functional BCR capable of transmitting survival, proliferative and differentiative signals in response to a potentially unlimited number and array of foreign antigens (Ags) (Figure 15.1) [1–3].

B cells have multiple roles in immune homoeostasis and host defence against infections. These include capture and presentation of specific cognate Ag, $CD4^{+}$ T cell stimulation, cytokine production, wound healing and lymphoid tissue neogenesis [4]. However, their most important and fundamental function is unequivocally the production of neutralising antibodies (Abs) that efficiently clear invading pathogens from the host. Depending on the pathogen, Abs contribute to protection from primary and/or secondary infection. Indeed, the ability of naïve B cells to differentiate into plasma cells (PCs) that continually secrete Ab, as well as into memory B cells that promptly generate PCs following subsequent exposure to the initial Ag/pathogen, enable these B-cell subsets to maintain humoral immunity and host protection against recurrent infections for extended periods of time [5–7]. Indeed, the generation of Ag-specific memory B cells and PCs underlies the cellular basis of long-lived serological immunity and underpins the rarity of recurrent infectious disease and the success of most current vaccines [7].

3. NAÏVE B CELLS DIFFERENTIATE INTO MEMORY AND PLASMA CELLS DURING GERMINAL CENTRE REACTIONS

When naïve B cells traffic through secondary lymphoid tissues and encounter foreign Ag, specific $CD4^{+}$ T cells and accessory cells, they can differentiate into various fates depending on signals provided by the cellular and cytokine milieu of the microenvironment. Activated B cells can undergo Ig class switch recombination (CSR) to express IgG, IgA or IgE, and seed germinal centres (GCs), which form transiently in lymphoid tissues following cognate interactions with Ag-specific $CD4^{+}$ T cells and are the

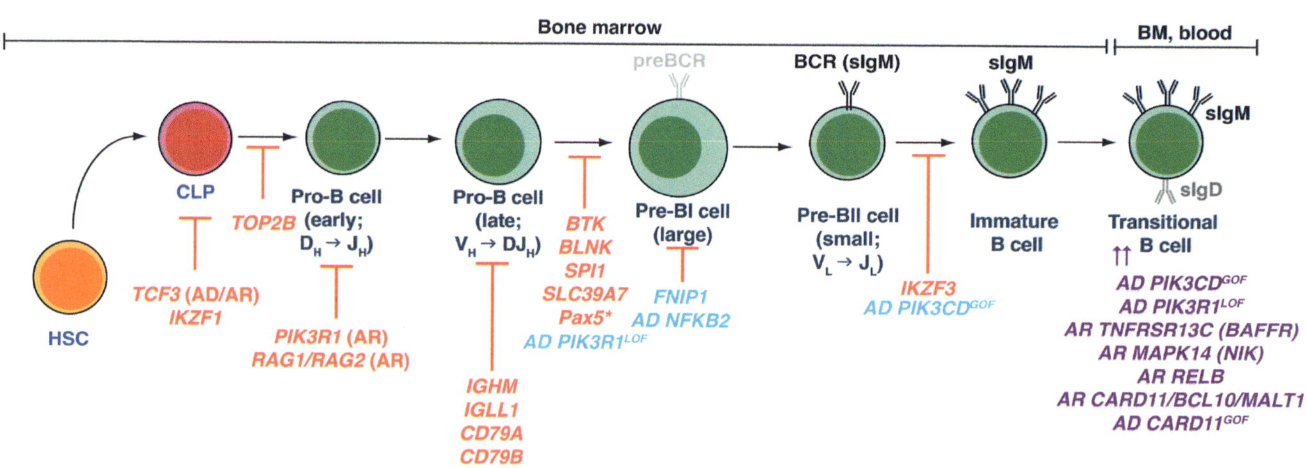

FIGURE 15.1 **Intrinsic molecular requirements for human B-cell development as revealed by inborn errors of immunity.** B-cell development occurs in the bone marrow and involves the sequential progression of HSCs into CLP which then give rise to progenitor B cells committed to a B-cell fate. B-cell development requires assembly and expression of a functional BCR. The initial stages of Ig gene rearrangement occur at the early and late pro-B cell stages. Pro-B cells that successfully express cytoplasmic Igμ chains develop into pre-BI cells that express a preBCR; rearrangement of Ig L chain genes occurs in pre-BII cells, which then express a functional sIgM molecule. Pre-B cells develop into immature B cells which then give rise into transitional B cells which egress from the BM and enter the peripheral circulation. Genetic variants causing severe disruption to B-cell development are shown in *red*; those having a milder effect on B-cell development are shown in *blue*. Some of these latter variants also result in an accumulation of transitional B cells (shown in *purple*). *Data for impact of Pax5 deficiency on B-cell development is inferred for studies of a mouse model expressing the human mutations.

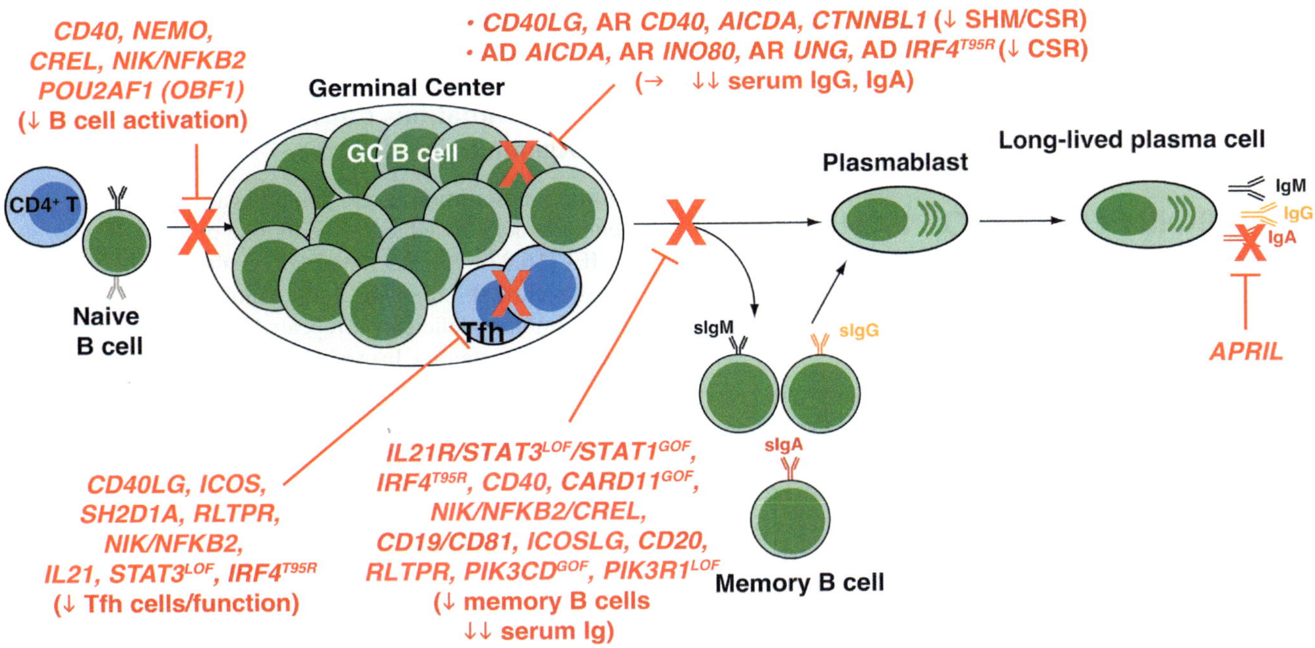

FIGURE 15.2 **Inborn errors of immunity can disrupt human B-cell differentiation by intrinsic and extrinsic mechanisms.** In secondary lymphoid tissues, Ag-specific naïve B cells interact with cognate CD4$^+$ T cells and seed a germinal centre. Here, B cells undergo intense proliferation (clonal expansion) and somatic hypermutation. GC B cells with highest affinity for Ag compete for survival signals provided by Tfh cells, and then differentiate into memory or plasma cells that produce high affinity neutralising Ig (IgM, IgG, IgA). IEIs affecting various stages of T-dependent B-cell differentiation are shown in *red*.

crucible for generating memory B cells and PCs. Within GCs, Ag-specific B cells undergo clonal expansion and somatic hypermutation (SHM) of IgV genes. GC B cells that acquire highest affinity for specific Ag then compete for limited signals provided by CD4$^+$ T follicular helper (Tfh) cells – CD4$^+$ T cells specialised to support B-cell differentiation – dendritic cells (DCs) and follicular DCs to survive this competitive microenvironment to differentiate into memory B cells or PCs (Figure 15.2) [8].

Once generated, memory B cells and PCs migrate from GCs to various sites. PCs can localise to specific regions within their tissue of origin, such as the splenic red pulp, medullary cords of lymph nodes (LNs) or mucosal-associated lymphoid tissues of the gut [9,10]. Alternatively, PCs exit these lymphoid tissues, traverse the circulation and enter the BM. In these tissues, PCs are maintained within specialised survival niches which enables constitutive production of protective Abs [9,10]. Similarly, memory B cells can become tethered in the splenic marginal zone or tonsillar epithelium, or egress from their tissue of origin, recirculate through the peripheral blood and/or traffic to distal sites including other lymphoid tissues or the BM [6,9–14] (Figures 15.1 and 15.2). The motility and anatomical distribution of PCs and memory B cells thereby enables the immune system to provide basal protection against infection by producing Abs that clear invading pathogens and maximises the potential for the host's immune system to respond rapidly following re-encounter with specific pathogens, thereby mitigating infectious diseases [6,9,10].

4. INBORN ERRORS OF IMMUNITY

Inborn errors of immunity (IEIs) affect at least 1/500 individuals and usually result from germline monogenic genetic variants [1–5]. IEIs are characterised by defects in immune cell development, and/or impaired innate and adaptive immune function of hematopoietic and non-hematopoietic cells. Due to this immune-deficient state, affected individuals are highly susceptible to severe, recurrent and potentially fatal infections [4,6]. Vaccine efficacy can also be compromised in IEI, rendering even vaccinated individuals vulnerable to many opportunistic infectious diseases [9]. Currently, approximately 500 genes have been identified that, when mutated, can cause an IEI. The clinical phenotypes can be autosomal dominant (heterozygous), autosomal recessive (AR; homozygous/compound heterozygous), X-linked recessive (XR), X-linked dominant (XD) or autosomal co-dominant (ACD) (when heterozygotes have a milder phenotype). The genetic lesions can lead to loss of expression or not, complete or partial loss of function (LOF) and gain of function (GOF; hypermorphic). In heterozygous individuals, the mechanism of dominance can involve haploinsufficiency, negative dominance or GOF. IEIs have elucidated key roles of individual genes, proteins, signalling pathways and cell types in immune cell development, immune regulation and host defence against infectious diseases.

5. INBORN ERRORS OF IMMUNITY REVEAL THE FUNDAMENTAL ROLES OF B CELLS IN HOST DEFENCE

The critical and non-redundant roles of B cells and secreted Ig in protection against infection, as well as the importance of immune regulation in preventing immune pathologies such as autoimmunity, malignancy and allergy, are evident from the discoveries of IEI that disrupt B-cell development and/or function [3,15]. These breakthroughs have provided not only insights into the molecular and cellular requirements for generating effective humoral immune responses and established mechanisms of disease pathogenesis but also enabled implementation of treatments not just for individuals with specific IEI but also for people affected by similar maladies.

A classic example is the description by Bruton in 1952 of the first case of agammaglobulinemia in an 8-year-old boy who experienced 19 episodes of severe bacterial infection; importantly, infusion of donor immune serum prevented further infections [16]. These observations established that agammaglobulinemia caused recurrent infections, and serum gamma globulins contained Ab capable of preventing infection. It cannot be emphasised enough that these observations in a single patient not only pre-dated the discovery of B cells by >10 years [17] but also led to the findings that:

- X-linked agammaglobulinemia (XLA) is caused by mutations in the *BTK* gene [18,19];
- Bruton's tyrosine kinase (BTK) links the BCR to many intracellular signalling pathways and is critical for human B-cell development and function;
- Ab deficiency can be treated by Ig replacement therapy irrespective of genetic cause; and
- pharmacological targeting of BTK would revolutionise treatment for B-cell malignancies and some autoimmune conditions decades later [20,21].

Such insights, often gleaned from single patients [22], kindreds or small numbers of affected individuals, are frequently repeated such that our understanding of human B-cell biology has been greatly enriched by seminal discoveries of IEI that impact B cells (Figure 15.3).

6. AGAMMAGLOBULINEMIA AND B-CELL DEFICIENCY

6.1 BTK Mutations Cause X-Linked Agammaglobulinemia

Bruton's description of congenital agammaglobulinemia is generally acknowledged as being one of the first reports of an IEI [16,23–25]. As noted above, the index patient suffered recurrent infections [16]. Clinical features of early onset immunodeficiency, repeated hospitalisation for severe life-threatening bacterial infections and the presence of <1% peripheral blood B cells largely defined patients with agammaglobulinemia. The first genetic aetiology of agammaglobulinemia was identified in 1993 as variants in *BTK*, an X-linked gene [18,19] (Figure 15.3). Thus, BTK deficiency causes XLA and accounts for ~85% of cases of agammaglobulinemia [23]. BTK is expressed in numerous cell types including B cells, myeloid cells and platelets and functions downstream of several cell surface receptors including cytokine, chemokine and Toll-like receptors. However, its most fundamental role is unequivocally in enabling BCR signalling in B cells.

Following BCR engagement, a myriad of intracellular signalling intermediates and pathways are activated, including protein tyrosine kinases (SYK, LYN, BTK), phospholipase C γ2 (PLCγ2), protein kinase C (PKC) and phosphatidylinositol-3 kinase (PI3K)/Akt/mTOR, Ras/MAP kinase/JNK, NFAT and NFKB pathways (Figure 15.4) [26]. In the absence of BTK-mediated signalling, B-cell development is arrested at the pro- to pre-B-cell stage (Figure 15.1). Consequently, up to 90% of B cells in BM of individuals with XLA correspond to pro-B cells, compared to 10%–25% in healthy individuals. This leads to 100–1000-fold fewer mature recirculating B cells than in healthy donors and a lack of all serum Ig isotypes [23,24,27]. The few B cells that can be detected in peripheral blood of XLA patients exhibit a phenotype of immature transitional B cells, characterised by elevated expression of IgM, reduced expression of CD19 and CD21 and BCRs enriched for autoreactive specificities [27–30]. Female carriers of *BTK* mutations are healthy. Furthermore, their B cells exhibit skewed inactivation of the X chromosome harbouring the mutant *BTK* allele [31]. This indicates that impaired or absent BTK function causes an intrinsic survival disadvantage in developing B cells. Thus, BTK is required for the generation of a pool of mature B cells expressing an immunocompetent non-self-reactive BCR capable of responding to foreign Ags to provide Ab-mediated host defence against infection pathogens.

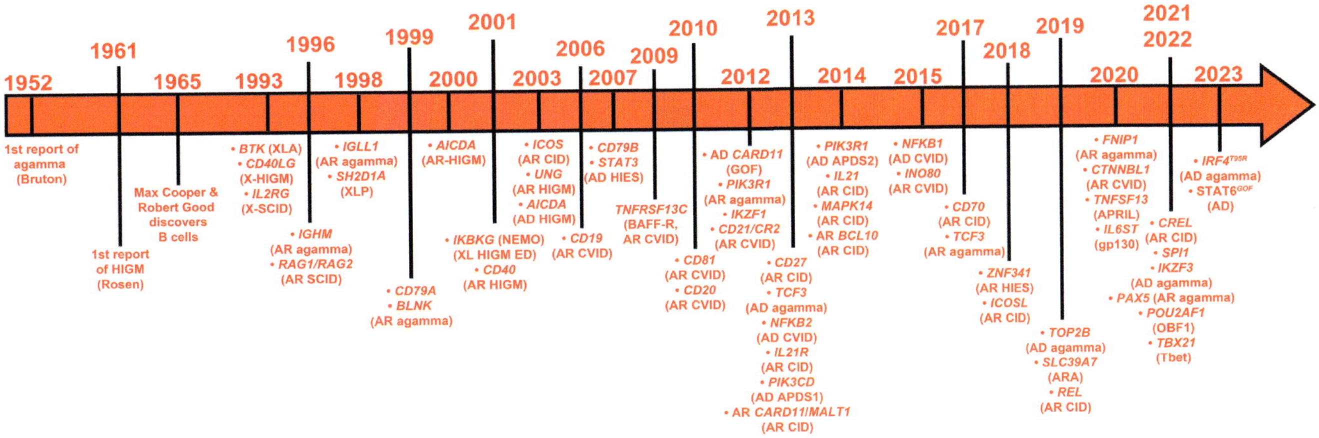

FIGURE 15.3 **Timeline of key discoveries on IEI and human B cells.** *agamma*, agammaglobuliemia; *APDS*, activated PI3K delta syndrome; *AR*, autosomal recessive; *AD*, autosomal dominant; *CID*, combined immunodeficiency; *CVID*, common variable immunodeficiency; *HIES*, hyper IgE syndrome; *HIGM*, hyper IgM syndrome; *XL-HIGM ED*, X-linked hyper IgM syndrome with ectodermal dysplasia; *XLA*, X-linked agammaglobuliemia; *X-SCID*, X-linked severe combined immunodeficiency; *XLP*, X-linked lymphoproliferative disease. Key discoveries over the last 70 years that have advanced the field of human B-cell biology.

6.2 Phenocopies of XLA: Elucidation of Molecular Causes of Autosomal Recessive Agammaglobulinemia

Subsequent studies of male and female patients who had a clinical presentation resembling XLA (early onset infections, profound hypogammaglobulinemia, few/no peripheral blood B cells) identified pathogenic biallelic variants in genes encoding key components of the BCR complex or associated signalling pathways: *IGHM* (Ig μ chain) [32], *IGLL1* (λ5 surrogate light chain) [33], *CD79A* (Igα) [34], *CD79B* (Igβ) [35,36], *BLNK* [37], *PIK3R1* [38,39] or *PIK3CD* [40] (Figures 15.3 and 15.4). Interestingly, there were some subtle differences regarding the stage of B-cell development impacted by these genetic lesions. Variants affecting expression of the BCR – that is, *IGHM*, *IGLL1*, *CD79A* and *CD79B* – usually arrest B-cell development earlier, and also present clinically at a younger age, than patients with BTK or BLNK deficiency [23,27].

Igμ, λ5, Igα and Igβ comprise the pre-BCR expressed during B-cell development. *PIK3R1* and *PIK3CD* encode the p85 regulatory and p110δ catalytic subunits of PI3K, which is activated following BCR ligation to generate phosphatidylinositol 3,4,5-triphosphate (PIP3). PIP3 binds to and recruits BTK to the cell membrane to undergo autophosphorylation-mediated activation, interacts with BLNK and then phosphorylates specific substrates (Figure 15.4) [23,26,41]. Collectively, these cases further highlighted the fundamental requirement for intact BCR signalling in human B-cell development and revealed critical, complementary, coordinated and integrated roles of BTK, BLNK and PI3K in this process (Figures 15.1 and 15.4). Interestingly, individuals heterozygous for *BLNK*, *IGHM*, *IGLL1*, *CD79A* or *CD79B* variants are healthy, indicating that haploinsufficiency or dominant negative (DN) functions of the encoded proteins are not pathogenic. Thus, B-cell deficiency and agammaglobulinemia result from loss of expression or LOF of encoded proteins. As *BTK*, *BLNK*, *IGHM*, *IGLL1*, *CD79A* and *CD79B* are prominently – or exclusively – expressed in B cells, T-cell development and differentiation were largely unaffected in individuals with mutations in these genes.

B-cell development is also severely disrupted by biallelic variants in either *RAG1* or *RAG2* (Figure 15.1) [42]. These variants reduce the activity of RAG1 and RAG2 proteins, thus preventing rearrangement and recombination of IgH and IgL chains and subsequent assembly of a functional BCR complex (Figure 15.1) [42]. However, unlike other IEI affecting B-cell development resulting in agammaglobulinemia, variants in *RAG1* or *RAG2* cause severe combined immunodeficiency (SCID) because recombination of T-cell receptor chains during thymic development is also compromised.

6.3 Transcription Factors Required for B-Cell Development

IEI also identified key transcription factors necessary for B-cell lineage commitment. PU.1 is encoded by *SPI1*, belongs to the Ets-family of transcription factors and regulates haematopoiesis by interacting with a multitude of target genes. Complete germline deletion of PU.1 in mice is embryonically lethal and significantly impairs the generation of myeloid and lymphoid cells [43]. Subsequent studies established that PU.1 functions at the early stages of specification of HSC to the B-cell lineage [43]. *TCF3* encodes the E2A transcription factors E12 and E47 that are critical for

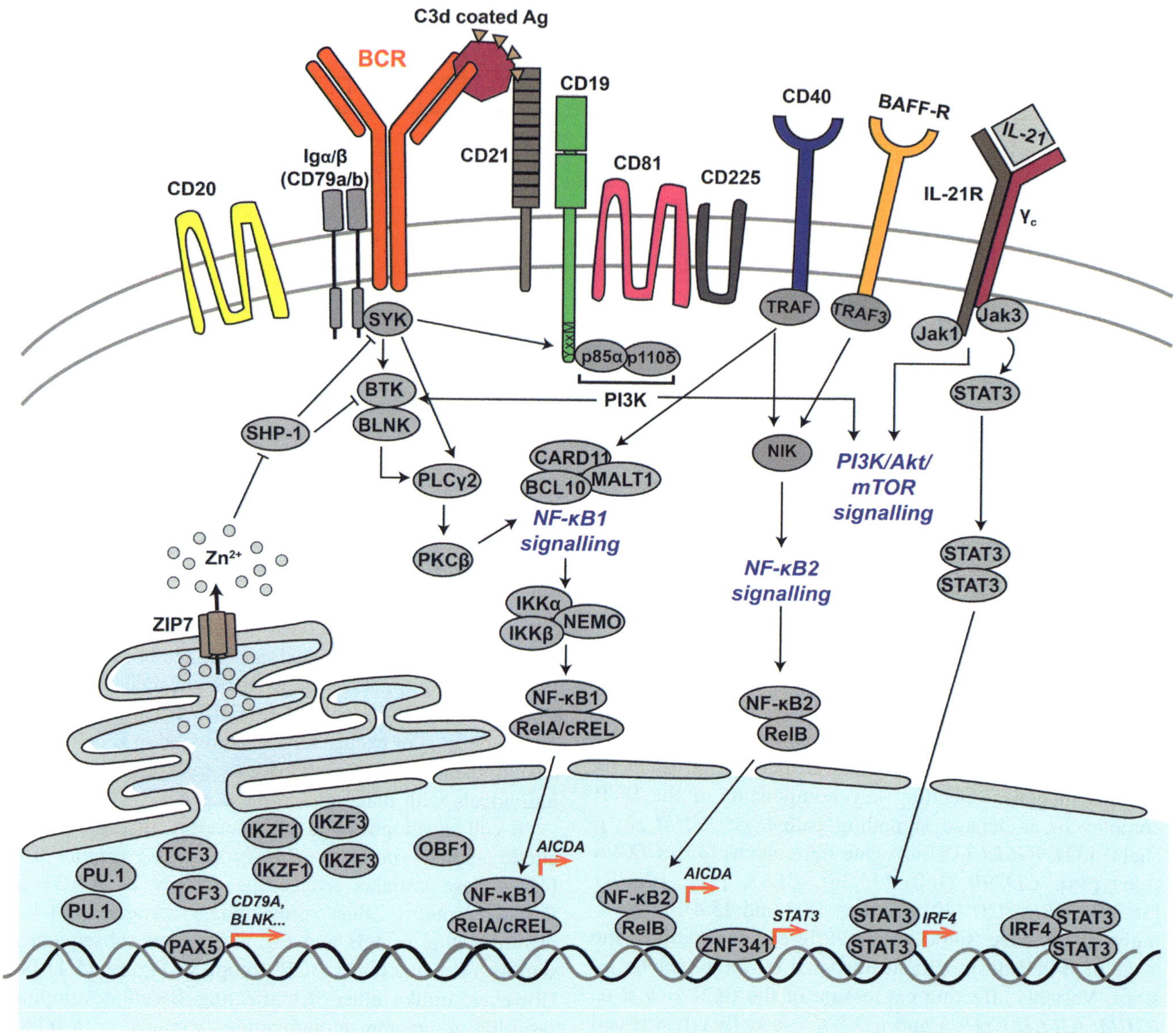

FIGURE 15.4 **Intracellular signalling pathways and transcriptional networks necessary for human B-cell development and differentiation.** Diagrammatic representation of the key receptor signalling pathways, intermediates and transcription factors that co-operatively underpin human B-cell development, differentiation and effector function.

lymphoid progenitors to commit to the B-lineage. This is in part regulated by inducing the transcription factor EBF1, which contributes to B-cell specification, as well as regulating Ig light chain gene rearrangement [43].

Within hematopoietic cells, PAX5 is exclusively and stably expressed in B cells and is required for commitment of progenitor cells to the B-lineage and progression of developing B cells beyond the pro-B cell stage. PAX5 initiates and maintains B-cell commitment by concomitantly repressing genes that influence commitment of HSC to other lineages and promoting expression of genes crucial for B-cell development, including components of the BCR signalling pathway *CD79A* and *BLNK* [43]. The Ikaros zinc finger family of transcription factors – IKAROS (*IKZF1*), HELIOS (*IKZF2*) and AILIOS (*IKZF3*) – are highly expressed in leucocytes and form homo- and hetero-dimers to regulate gene expression and haematopoiesis [44]. Recently, individuals with B-lymphopenia, agammaglobulinemia and recurrent sinopulmonary infections due to mutations in *TCF3*, *SPI1*, *PAX5*, *IKZF1* or *IKZF3* have been reported (Figure 15.3).

TCF3: Five unrelated individuals with <1% blood B cells, and dramatic reductions in B-lineage BM cells due to a loss of CLPs, had heterozygous DN mutations in *TCF3* [27,45,46]. Biallelic nonsense *TCF3* variants were also identified in two individuals with B-cell deficiency and hypo/agammaglobulinemia [47]. In this latter setting, heterozygous carriers of nonsense *TCF3* variants were healthy, indicating E2A haploinsufficiency does not impact B-cell development (Figure 15.1 and 15.3).

SPI1: Six unrelated individuals have been identified with pathogenic heterozygous variants in *SPI1*. The *SPI1* variants destabilised PU.1 expression and compromised the ability of PU.1 to localise to the nucleus and bind target genes. Affected individuals had 0%–0.2% B cells in their peripheral blood. Pro-B cells were the only B cells detected in their BM [48], consistent with PU.1 expression increasing as B cells transit through the pro → pre stages of development [43] (Figure 15.1). Molecular characterisation of pro-B cells from healthy donors and patients revealed dramatically reduced expression of key genes required for B-cell development and function, including genes that can disrupt B-cell development and cause agammaglobulinemia (*IGLL5*, *IGHM*, *CD79B*, *BTK*), in PU-1-deficient versus PU.1-sufficient pro-B cells [48]. Notably, mutant PU.1 did not affect WT PU.1 function, indicating the mechanism of pathogenicity is PU.1-haploinsufficiency. Thus, human B-cell development is strictly dependent on PU.1 dose, as one WT *SPI1* allele could not support B-cell development in affected individuals [48].

PAX5: A single individual with compound heterozygous *PAX5* variants had reduced levels of all serum Ig isotypes and peripheral B-lymphopenia (~fivefold lower than healthy donors) [49]. Whilst B-cell development in the BM of this patient was not reported, analysis of a gene-edited mouse strain engineered to express the patient-specific *PAX5* mutant alleles ($Pax5^{R31Q/E242*}$) revealed an accumulation of pro-B cells and 5–10-fold fewer pre-B and immature B cells in BM compared to WT mice (Figure 15.1). Consistent with the phenotype of the PAX5-deficient human, $Pax5^{R31Q/E242*}$ mice also had significantly reduced peripheral B cells and levels of serum IgG (10–20-fold), IgA (~fivefold) and IgM (twofold). Mechanistically, $Pax5^{R31Q}$ encoded a hypomorphic protein, impaired in regulating 10%–20% of PAX5 target genes [49]. Thus, human PAX5 deficiency blocks B-cell development resulting in B-cell cytopenia, hypogammaglobulinemia and impaired specific Ab responses following vaccination.

IKZF1: The first case of a possible pathogenic variant in *IKZF1* was reported in 2012 in a premature infant suffering from pancytopenia, including a near absence of B cells (Figure 15.3) [50]. Since then, heterozygous *IKZF1* mutations have been reported in ~100 individuals who generally present with impressive B-cell lymphopenia, hypogammaglobulinemia and recurrent infections. IKAROS deficiency also predisposes to immune dysregulation and B- or T-cell malignancy in 25%–50% and ~10%, respectively, of affected individuals [44,51]. Several mechanisms of disease pathogenesis due to *IKZF1* variants have been proposed, with each depending in the nature of the variant. Thus, some cause IKAROS haploinsufficiency, whilst others are disease-causing due to negative dominance or affecting homo- or hetero-dimerisation [44,51,52].

Despite these distinct mechanisms of disease causation, B-cell lymphopenia and hypogammaglobulinemia are highly penetrant clinical features of most patients with *IKZF1* variants [44,51,52]. IKAROS deficiency disrupts B-cell development by impairing the generation of lymphoid progenitors (Figure 15.1), evidenced by a >10-fold reduction in numbers of CLP and B-lineage cells in the BM of individuals with *IKZF1* variants compared to healthy donors [44,51–53]. Notably, despite the overall reduction in BM B cells, proportions of developing B-cell subsets in IKAROS deficiency were largely within normal ranges, indicating that the block in B-cell development is quantitative rather than qualitative [52,53]. T-cell development per se is intact in IKAROS-deficient individuals; however, differentiation defects are apparent in peripheral $CD4^+$ and $CD8^+$ T cells [44,51–53]. Thus, whilst *IKZF1* variants affect lymphoid progenitors, this predominantly impacts the trajectory of CLP to the B-cell, rather than T-cell, lineage even though functional consequences are evident in other lymphocyte populations (Figure 15.1).

IKZF3: Three individuals from one family experienced recurrent sinopulmonary infections and had undetectable peripheral blood B cells and very low numbers of all B-lineage subsets in their BM (Figure 15.3). Genetic investigation identified a heterozygous variant in *IKZF3*, encoding AIOLOS ($AIOLOS^{G159R}$) [54]. Unlike IKAROS, AIOLOS is absent from CLP and early B-cell progenitors but is highly expressed in pre-B cells [54]. Thus, the *IKZF3* variant identified in this family likely causes B-cell lymphopenia by disrupting progression of pre-B cells to immature B cells and beyond, rather than impacting the ability of CLP to give rise to B-lineage precursor cells. This was confirmed by generating mice with the patient-specific allele ($Ikzf3^{+/G159R}$). This revealed a block in B-cell development from the pre-B-cell stage (Figure 15.1) [54]. Importantly, $Ikzf3^{+/G159R}$ intrinsically affected murine B-cell development [54]. Thus, *IKZF3* variants cause B-lymphopenia by disrupting progression of pre-B to immature B cells, rather than impacting the progression of CLP into B-lineage precursors.

Biochemical and molecular analysis of the $AIOLOS^{G159R}$ variant provided fascinating insights into mechanisms of AIOLOS function and of disease. $AIOLOS^{G159R}$ was LOF for binding to canonical DNA sequences defined for WT AIOLOS; it was also DN, impairing WT AIOLOS from binding to its target sequences. However, $AIOLOS^{G159R}$ was also neomorphic, binding DNA motifs not previously recognised as targets of WT AIOLOS. Furthermore, whilst $AIOLOS^{G159R}$ retained its ability to interact with IKAROS, resultant AIOLOS^{G159R}/IKAROS heterodimers were dysfunctional, exhibiting altered chromatin localisation and reduced binding to DNA. Thus, AIOLOS^{G159R} compromised B-cell development and humoral immunity not only by mechanisms of LOF and negative dominance of WT AIOLOS but also by impeding the function of IKAROS via a mechanism termed heterodimeric interference [44,54]. This provides a clear explanation for the similarity between clinical features of IKAROS and AIOLOS deficiency.

Combined, novel IEI that intrinsically disrupt human B-cell development identified key transcriptional regulators with non-redundant roles in commitment of CLP to a B-cell fate or regulating expression of key genes required for B-cell development (Figures 15.1 and 15.3).

6.4 Additional Genetic Causes of Agammaglobulinemia and B-Cell Deficiency

Whilst the studies summarised above underscored the necessity of specific transcriptional regulators and BCR expression, function and signalling in B-cell development, several other novel IEI identified additional requirements for this process. Interestingly, several presented as syndromic conditions together with B-cell deficiency and agammaglobulinemia.

SLC39A7: The obligate role of divalent cations in signalling pathways is well established [55]. This has been further highlighted by identifying individuals with mutations in genes that disrupt mobilisation and release of Mg^{2+} (*MAGT1*) or Ca^{2+} (*STIM1*, *ORAI1*) ions following engagement of the BCR, T-cell receptors and Fc receptors on B cells, T cells and myeloid and lymphoid cells, respectively [55]. Affected individuals have a combined immune deficiency characterised by increased susceptibility to infectious diseases, as well as some non-immune features [55]. Recently, the importance of Zn^{2+} in B-cell development has been revealed by the discovery in 2019 of six individuals from five unrelated families with biallelic hypomorphic variants in *SLC39A7*, encoding the zinc transporter ZIP7 [56] and a seventh unrelated patient in 2022 [57] (Figure 15.3).

All patients had recurrent bacterial infections, agammaglobulinemia, very few blood B cells and a severe block at the pro → pre-stage of B-cell development (Figure 15.1) [56,57]. By generating mice carrying patient hypomorphic *Slc39a7* alleles, Anzilotti et al. (2019) elegantly determined the causes of the selective B-cell deficiency due to impaired ZIP7 function. ZIP7$^{P198A/P198A}$ mice recapitulated the severe reductions in BM and peripheral B cells, low levels of serum Ig and an intrinsic block at the pro-B cell stage of development noted in affected humans (Figure 15.1). Total phosphatase activity was increased in ZIP7$^{P198A/P198A}$ B cells, dampening overall BCR signalling. Thus, consistent with zinc-inhibiting phosphatases [55], hypomorphic ZIP7 reduces cytoplasmic zinc concentrations thereby enabling increased phosphatase activity which sufficiently attenuates BCR signalling to disrupt B-cell development [56]. Notably, biallelic *SLC39A7* variants had no effect on T-cell signalling, development or function [56]. Thus, similar to mutations in genes encoding the BCR complex (*IGHM*, *IGLL1*, *CD79A*, *CD79B*), downstream mediators of BCR signalling (*BTK*, *BLNK*, *PIK3R1*, *PIK3CD*), or transcription factors that regulate some of these key genes (*SPIB*, *PAX5*), ZIP7 deficiency abolishes early B-cell development by impairing BCR function, causing severe B-cell deficiency and agammaglobulinemia (Figure 15.1).

FNIP1: Folliculin-interacting protein 1, encoded by *FNIP1*, interacts with folliculin to modulate the activity of AMP kinase, a key sensor of energy consumption and regulator of cellular metabolism. Earlier studies in mice demonstrated that Fnip1 deficiency disrupted B-cell development resulting in agammaglobulinemia. This appeared to be caused by increases in B-cell intrinsic mTORC1 activity, mitochondrial biogenesis and autophagy [58]. Interestingly, *Fnip1*-deficient mice also developed hypertropic cardiomyopathy [58]. To date, six patients from five unrelated families have been identified with biallelic damaging variants in *FNIP1* [59,60]. All patients presented with frank B-cell deficiency or B-cell lymphopenia, agammaglobulinemia and recurrent respiratory infections. Several patients also had additional clinical features typically not observed in other cases of genetic agammaglobulinemia, such as neutropenia and – similar to *Fnip1*-deficient mice – cardiomyopathy. Limited analysis of BM samples suggested FNIP1 deficiency affects early B-cell development (Figure 15.1) [59,60]. Whilst the molecular mechanism underlying this selective effect on B cells is unclear, increased mTOR and AMPK activity in cells from these patients indicates perturbations to cellular energy homoeostasis compromise B-cell development and/or survival.

TOP2B: Topoisomerase 2B induces transient double-stranded DNA breaks to modulate topological chromatin conformation and maintain genomic integrity during replication; it also regulates transcription of particularly long genes [61]. Eleven individuals with BILU (B-cell immunodeficiency, limb anomalies and urogenital malformations) or Hoffman's syndrome were found to have heterozygous *TOP2B* variants [62–64] (Figure 15.3). Despite phenotypic variability in the impact of these variants on facial appearance and skeletal and neurological development, a unifying trait of all patients was greatly reduced peripheral B cells, hypogammaglobulinemia and recurrent respiratory infections [62–64]. Analysis of BM from patients and mice harbouring a patient mutation (*Top2b*$^{EE587E/WT}$) identified a block at the CLP → pro-B-cell stage of development (Figure 15.1), consistent with intact generation of T cells in the patients [62,63].

The *TOP2B* variants were LOF as well as DN [62,63]. Interestingly, Top2b is more highly expressed in murine B cells than T cells, and CLPs isolated from *Top2b*$^{\mathrm{EE587E/WT}}$ mice expressed significantly lower levels of *Pax5* and *Foxo1* mRNA – transcription factors with well-established roles during B-cell development [43] – than *Top2b*$^{\mathrm{WT/WT}}$ CLP's [62]. These observations, together with *TOP2B* variants affecting early B cells but not CLPs, provide an explanation for why immune cell lineages other than B cells were unaffected by *TOP2B* variants [62–64]. Whilst the molecular mechanism underlying the selective loss of B cells due to TOP2B negative dominance remains incompletely defined, B-cell development is highly sensitive to impaired TOP2B function, affecting genomic integrity and regulation of gene expression.

Overall, these studies have demonstrated that whilst individuals can survive without functioning B cells, or B cells at all, their mortality is greatly compromised by an absence of specific Ab, thereby requiring life-long Ig replacement therapy [23].

7. INBORN ERRORS AFFECTING B-CELL DIFFERENTIATION AND HUMORAL IMMUNITY

In the 1960s and 1970s, pioneering clinical immunologists were harnessing emerging techniques (of the times!) to characterise different classes of serum gamma globulins or lymphocytes in distinct immune deficiencies. These studies followed the seminal discoveries of XLA by Bruton in 1952 [16] and of B cells by Max Cooper and Robert Good in 1965 [17] (Figure 15.3). Thus, it became possible to define immune-deficient states according to immune phenotype and serology. Rosen et al. described two males with recurrent bacterial infections and increased levels of 19S (i.e. modern day IgM) but lacked 7S (i.e. IgG, IgA) gamma globulins [65]. These probably represent the first reports of X-linked hyper-IgM syndrome (XL-HIGM) (Figure 15.3). Cooper et al. and Siegal et al. extended Rosen's observation by reporting additional cases of 'XR immunodeficiency with HIGM' and likely cases of AR HIGM, as well as examples of 'agammaglobulinemia with B cells' [24,66]. Over the past 30 years, the genetic cause of many IEI presenting as severe infections despite intact B-cell development has been identified (Figure 15.3). These molecular lesions disrupt signalling pathways initiated following interactions between B cells, specific Ag and cognate CD4$^+$ T cells, resulting in impaired CSR, GC formation, SHM, memory B cell and PC generation, hypogammaglobulinemia and/or functional Ab deficiency (Figure 15.2).

7.1 CD40L/CD40 Signalling: A Fundamental Requirement for B Cell Ig Class Switching and Somatic Hypermutation

A key component of an effective B-cell response is initiation of CSR in naïve B cells to express IgG, IgA or IgE [67,68]. This is frequently coupled with subsequent induction of SHM during GC reactions, resulting in the selection of B cells with the highest affinities for foreign Ag, thus yielding memory B cells, PC and secreted Ig with appropriate affinity, dictated by Ag binding and effector function mediated by the Ig H chain isotype. CD40 is a member of the TNFR superfamily and is expressed by professional Ag-presenting cells, including B cells, monocytes, macrophages and DCs [69]. The counter structure of CD40 – CD40 ligand (CD40L) – belongs to the TNF superfamily and is predominantly and transiently expressed on activated CD4$^+$ T cells [69]. Engagement of CD40 on B cells activates a series of intracellular signalling pathways, including the canonical (NFκB1) and non-canonical (NFκB2) NFκB pathways, as well as JNK/MAPK and PI3K pathways (Figure 15.4). CD40L/CD40 interactions are critical for inducing many aspects of B-cell biology including migration, activation, adhesion, survival, proliferation, CSR and differentiation [69].

Many IEIs have been identified that impact CD40 signalling and function [68]. As noted above, the first cases of XR-HIGM were reported by Rosen et al. in 1961 [65]. XR-HIGM is a combined immunodeficiency characterised by severe and recurrent bacterial, sinopulmonary or opportunistic infections (*Pneumocystis jiroveci*, *Cryptosporidium parvum*), dysgammaglobulinemia (increased IgM, low/absent IgG, IgA), diminished specific Ab responses and reduced proportions of class-switched memory B cells [67,68,70]. Due to severely depressed humoral immunity, it was long assumed that XR-HIGM resulted from a B-cell intrinsic defect. However, studies in the 1980s indicated that B cells from XR-HIGM patients are intrinsically functional, at least in the context of assessing in vitro B-cell responses [71]. In 1993, XR-HIGM was found to be caused by hemizygous variants in *CD40LG* [72–76]. These variants were loss of expression as well as LOF, with the latter observation highlighting the need to assess CD40L function to obtain accurate diagnoses. This was because some mutant CD40L proteins continued to be expressed on activated CD4$^+$ T cells and could be detected by anti-CD40L mAbs; however, the ability of mutant CD40L to bind soluble CD40 was dramatically reduced or abolished [72–76]. The fact that CD40L is predominantly expressed by CD4$^+$ T cells indeed confirmed that impaired B-cell differentiation and humoral immunity in XL HIGM results from B-cell extrinsic mechanisms, namely in the provision of help from Tfh cells for inducing CD4$^+$ T-cell-dependent B-cell responses. Thus, even though signalling through the BCR and other surface receptors is intact in patients with XL-HIGM due to *CD40L* deficiency, cognate CD40/CD40L-mediated interactions with Ag-specific $CD4^+$ T cells are fundamental for B-cell differentiation, CSR and function (Figure 15.4).

Similar to BTK and XLA, there were cases of HIGM reported in females, as well as males with WT *CD40LG*, indicating there were likely AR causes of this condition. Identifying hemizygous variants in *CD40LG* provided a significant clue as to what other genes may result in AR HIGM. Indeed, since 2001, five patients have been identified with biallelic mutations in *CD40*, leading to CD40 deficiency [69,77,78], which essentially clinically phenocopies XL-HIGM due to *CD40LG* mutations. Importantly, CD40/CD40L deficiency also compromises $CD4^+$ T cell-mediated activation of CD40-expressing myeloid cells, which contributes to the severe opportunistic infections observed in these individuals which extend beyond defects in Ab class switching and GC formation [68].

7.2 IEI Causing B-Cell Intrinsic Impairments in CSR and SHM

The defining features of HIGM – low-to-absent serum IgG and IgA, diminished Ag-specific Ab responses following infection or vaccination, reduced proportions of class-switched memory B cells – point to defects in GC reactions and outputs [70]. This was highlighted by the finding that patients with XL-HIGM had very few or only poorly formed GCs in their lymphoid tissues [79]. A breakthrough in understanding molecular processes driving SHM and CSR was provided by Honjo and colleagues when they discovered the novel DNA-editing enzyme activation-induced cytidine deaminase (AID), encoded by the *AICDA* gene [80]. AID was cloned from a murine B-cell line induced to undergo Ig class switching and was then found to be exclusively expressed in situ in GC B cells [80].

Definitive proof of a fundamental role for AID in SHM and CSR came in 2000 with the co-publication of studies by the Honjo lab describing the phenotype of *Aicda*$^{-/-}$ mice [81], and the first description by Durandy and colleagues of individuals with bi-allelic mutations in *AICDA* [82] (Figure 15.3). The phenotypes of AID-deficient humans and mice were strikingly similar: hyper IgM, complete lack of Ig CSR in vivo and in vitro, reduced/absent SHM in Ig V genes, low serum IgG and IgA and poor Ab responses to immunisation [81,82]. In contrast to CD40/CD40L deficiency, AID deficiency resulted in lymphoid tissue hyperplasia due to the formation of giant – but ineffectual – GCs, as well as autoimmunity [81,82]. Thus, GCs could form in the absence of AID, due to intact signalling via the BCR, CD40 and cytokine receptors; however, GC outputs were severely compromised due to the inability of GC B cells to undergo AID-mediated SHM, and subsequent selection and differentiation into high-affinity memory B cells and PC. Similarly, abolished AID-mediated CSR prevented production of switched Ig isotypes, rendering affected individuals vulnerable to recurrent infections. Thus, the phenotype of AID deficiency established the absolute requirement for AID in several fundamental steps of terminal differentiation of B cells necessary for efficient Ab responses [67,68].

Subsequently, patients were identified with biallelic mutations in other genes involved in DNA mismatch repair which underpins CSR and SHM. These include IEI due to biallelic mutations in *UNG* [83], *MSH6* [84], *PMS2* [85], *INO80* [86] or *CTNNBL1* [87] (Figure 15.3). Interestingly, patients lacking UNG or CTNNBL1 also exhibit follicular hypoplasia and autoimmunity [67,68,83,87], indicating important roles for DNA repair mediated by AID, UNG and CTNNBL1 in maintaining and/or establishing B-cell tolerance.

Interestingly, monoallelic *AICDA* variants that truncate the last 10–12 amino acids of AID protein have been identified in 18 patients from 7 kindreds with a hyper-IgM-type phenotype [88–92] (Figure 15.3). These patients presented with recurrent infection, increased levels of serum IgM, reduced serum IgG and IgA, and impaired Ab responses to vaccines and infection. However, whilst CSR was abolished in vivo and in vitro, SHM was unaffected in these patients [88–92] (Figure 15.2). Patients also did not develop autoimmune features characteristic of AR AID deficiency [92]. Thus, heterozygous *AICDA* variants affecting the C-terminal domain of AID can disrupt Ig CSR resulting in humoral immunodeficiency. The mechanism(s) of pathogenicity of these C-terminal truncating variants are unresolved but may reflect effects on nuclear export or dimerisation of AID, or a DN effect of the truncated AID protein, rather than haploinsufficiency, because carriers of heterozygous *AICDA* variants – including null mutation – that are pathogenic in the homozygous state are healthy [90,92]. Regardless of the mechanism, heterozygous *AICDA* variants cause an AD form of HIGM; however, only CSR and not SHM are impacted in these individuals [88–92].

CD40 signalling in B cells, in combination with inputs provided by cytokines, initiates SHM and CSR via induction of AID. Thus, it is not surprising that these processes are abolished by AR mutations in *CD40* or *AICDA*, or XL *CD40LG* mutations [67,68,78]. Interestingly, whilst variants in *UNG*, *MSH6*, *PMS2*, *INO80* or *CTNNBL1* all impact CSR, they do not universally affect SHM. Thus, *UNG* mutations alter the pattern but not the overall level of SHM [67,68,83], the *CTNNBL1* variant modestly reduced but certainly did not ablate SHM [87], whilst deficiency of PMS2, MSH2 or INO80 had little effect on SHM [67,68,84–86]. Combined, these findings suggest that the effector function of Ig acquired following CSR is critical for effective host defence even when SHM and affinity maturation of B cells is intact.

OBF-1 (also known as OCA-B or BOB1; encoded by *POU2AF1*) is a B-cell-specific transcription factor originally considered to be important for regulating Ig transcripts during B-cell development [93]. The generation of gene-targeted mice revealed that Obf-1 does have a modest role during B-cell development, evidenced by two to fourfold reductions in numbers of splenic B cells. However, the more striking function of OBF-1 was in governing the formation of GCs and establishing humoral immunity in murine models of viral infection or immunisation [93]. Recently, the first case of human OBF-1 deficiency, due to homozygous variants in *POU2AF1*, was reported [94] (Figure 15.3). Despite normal numbers of peripheral B cells, the OBF-1-deficient patient presented with recurrent respiratory infection, agammaglobulinemia and an absence of class-switched memory B cells [94]. Furthermore, OBF-1-deficient B cells exhibited impaired biochemical responses and reduced differentiation into IgG-expressing or plasmablast-type cells following in vitro stimulation with CD40L, TLR or BCR agonists. Importantly, reconstitution of patient-derived OBF-1-deficient B cells with wild-type OBF-1 restored these functional defects, indicating that the *POU2AF1* variants identified are likely pathogenic [94]. These findings highlight a critical role for OBF-1 in human B-cell differentiation. As CD40 stimulation induces OBF-1 expression in B cells [93], OBF-1 deficiency probably disrupts aspects of CD40-induced human B cell activation (Figures 15.2 and 15.4).

7.3 B-Cell Defects Due to Inborn Errors of NFKB Signalling

The NFκB family of transcription factors regulate the expression of >500 genes involved in cell development, proliferation, survival and inflammation [95]. The canonical pathway (NF-κB1) is activated by many receptors, including the BCR, Toll-like receptors and CD40, whilst the non-canonical (NF-κB2) pathway is restricted to a few receptors which are typically expressed by B cells such as the TNFR superfamily members CD40, BAFF-R and lymphotoxin-β receptor (LT-βR) (Figure 15.4) [95]. When activated, NF-κB1 (p105) and NFκB2 (p100) are phosphorylated by a kinase complex and then cleaved to their respective active subunits (p50, p52) that dimerise with other NF-κB members (RelA/p65, RelB, c-Rel) and translocate to the nucleus. NFκB signalling is restrained by inhibitory proteins (e.g. inhibitor of κB [IκBα]) that retain NFκB1/2 in the cytoplasm, preventing their activation [95]. Here, we discuss some IEIs due to germline mutations in genes encoding canonical and non-canonical NF-κB pathway proteins that contribute to diseases by disrupting B-cell function. Remarkably, most of these gene defects have little, if any, effect on B-cell development in the BM, indicating that NFKB signalling is largely redundant for B-cell development.

Impaired NF-κB1 signalling: Heterozygous mutations in *NFKB1*, resulting in NF-κB1/p105 haploinsufficiency, was initially discovered in 13 individuals from 3 kindreds with a diagnosis of combined variable immune deficiency (CVID) [96] (Figure 15.3). Since then, >200 patients with *NFKB1* mutations have been identified that disrupt NF-κB1 expression, stability, phosphorylation and/or nuclear transport [97–101]. Most affected individuals present with late onset hypogammaglobulinemia, recurrent respiratory tract infections, lymphoproliferation, autoimmunity, B-cell lymphopenia and reductions in class-switched memory B cells [97–101]. Impaired differentiation of B cells in vitro demonstrates that pathogenic *NFKB1* variants can contribute to disease by intrinsically affecting B-cell function [101]. Consistent with this, most studies have found relatively intact T-cell differentiation and function in affected individuals. Thus, NF-κB1 LOF compromises human B-cell survival and/or differentiation, via a mechanism involves haploinsufficiency [98] resulting in immune dysregulation (Figure 15.4).

IKKγ (NEMO, *NF-κB* essential modulator) is encoded by the X-linked gene *IKBKG* and is the regulatory subunit of the kinase complex that phosphorylates IκBα leading to its degradation and subsequent NF-κB1 activation and nuclear translocation [95]. Hemizygous hypomorphic *IKBKG* mutations cause XR anhidrotic ectodermal dysplasia with immunodeficiency (XL-EDA-ID) [102,103]. More than 90% of patients suffer from severe recurrent infections [78,104–106]. This is consistent with ~60%–80% of patients exhibiting hypogammaglobulinemia, impaired functional Ab responses and reduced memory B cells [78,104–107]. The heterogeneity in clinical presentation of NEMO deficiency reflects the extreme variability in NEMO function due to different hypomorphic *IKBKG* variants [78,104–106,108]. LOF/null variants in *IKBKG* are lethal in utero in male foetuses and underlie incontinentia pigmenti in heterozygous women [108], whose XD condition has not been immunologically characterised thoroughly yet but underlies the production of auto-Abs to type I IFNs [109].

There is considerable overlap in the clinical features of NEMO deficiency and *CD40LG* or *CD40* deficiency [67,68,78]. This reflects a key role of NEMO in CD40/NF-κB1 signalling (Figure 15.4). Indeed, CD40L-induced cRel activation, proliferation and Ig CSR were dramatically disrupted in B cells with hypomorphic *IKBKB* mutations [103,104,110], indicating that B cell-intrinsic defects contribute to disease aetiology due to NEMO deficiency. Furthermore, complete cRel deficiency prevented the generation of Ag-specific Abs and reduced memory B-cell formation and Ig class switching in vivo, as well intrinsically abolished CD40L-induced B-cell differentiation in vitro [111,112]. Thus, CD40L-mediated activation of the NFκB1 pathway, involving NEMO, NFκB1 and cRel, is critical for inducing B-cell differentiation and establishing humoral immunity.

CARD11/BCL10/MALT1 (CBM) complex: The BCR is functionally linked to the NFκB1 pathway by a multimeric complex comprising CARD11, BCL10 and MALT1 [113,114] (Figure 15.4). CARD11 is recruited to the BCR following ligation, where it is phosphorylated and undergoes conformational change enabling recruitment of BCL10 and MALT1, which constitutively associate with one another. The CBM complex then regulates activation of the canonical NF-κB1 pathway [113]. Recessive mutations in *CARD11* (8 patients) [115–118], *MALT1* (~20 patients) [119–125] or *BCL10* (4 patients) [126–129] all result in a CID due to profound defects in lymphocyte differentiation [113,114]. Deficiencies of these key components of BCR signalling did not disrupt B-cell development per se, as circulating B cells were present in the patients in normal numbers. However, there was a marked defect in peripheral B-cell development and differentiation, evidenced by an accumulation of immature transitional B cells, a complete absence of memory B cells and hypogammaglobulinemia (Figure 15.1). Thus, canonical NFκB1 signalling mediated via the CBM complex is redundant for B-cell development but critical for the generation of a pool of naïve B cells [113,114]. This likely reflects a B-cell intrinsic requirement for CARD11, BCL10 and MALT1 during the final stages of B-cell development that occur in peripheral lymphoid organs, whilst the paucity of memory B cells may also arise from impaired $CD4^+$ T cell help as these proteins are also expressed in T cells and have similar functions in TCR signalling [113,114].

Monoallelic activating variants in *CARD11* have also been identified in individuals with childhood-onset polyclonal B-cell lymphocytosis. The first clinical description of this condition was actually reported in 1971 [130], and it was solved genetically in 2012 when the two daughters of the index case also presented with these clinical features and were investigated by next-generation sequencing [131] (Figure 15.3). Since then, ~30 cases of *CARD11* GOF have been described, with clinical features including B-cell lymphocytosis, splenomegaly, lymphadenopathy and recurrent bacterial (100% of cases) and viral (30%–45% cases) infections [113,132]. Laboratory investigations revealed that the vast increase in peripheral blood B-cell numbers was due to an expansion of transitional and naive B cells, but a paucity of total and Ig class-switched memory B cells [133] (Figures 15.1 and 15.2). Consistent with this block in peripheral B-cell maturation, and recurrent infections, serum Ig levels and Ab responses to polysaccharide or protein vaccines were reduced in many patients with *CARD11* GOF [113,114]. *CARD11* GOF B cells were unable to differentiate into PCs following in vitro stimulation, establishing that aberrant CARD11 function disrupts B-cell differentiation in a cell-intrinsic manner [133]. Mechanistically, GOF variants result in spontaneous activation of CARD11 and CBM formation, and subsequent constitutive NF-κB1 signalling without requiring BCR engagement [113,114]. Thus, elevated basal NF-κB1 signalling leads to B-cell lymphoproliferation and impaired humoral immunity.

Impaired NF-κB2 signalling: Heterozygous mutations in *NFKB2* have been identified in individuals who typically present with clinical features including early onset hypogammaglobulinaemia, recurrent respiratory infections, poor Ab responses to vaccines, autoimmunity and adrenocorticotropic deficiency (Figure 15.3) [134–138]. Patients with *NFKB2* deficiency exhibit normal/low B-cell numbers or even complete B-cell deficiency, possibly due to progressive B-cell lymphopenia [134–138]. Analysis of BM from one patient revealed a block in B-cell development at the pre-B stage [139]. When peripheral B cells are present, severe reductions in memory B cells has been observed [134–138]. Notably, the T cell compartment is generally less affected by *NFKB2* variants than B cells [138]. Thus, NF-κB2 signalling, likely downstream of CD40L/CD40 and BAFF/BAFF-R interactions, is critical for intact human B-cell development, differentiation, and function (Figure 15.4).

The importance of non-canonical NF-κB2 signalling is further highlighted by the clinical phenotypes of individuals with biallelic variants in either *MAP3K14* (encoding NF-κB-inducing kinase [NIK]) [140] which acts upstream of NF-κB2, or *RELB* [141] which dimerises with NF-κB2 (Figure 15.4). Deficiency of NIK or RelB impacts peripheral B-cell maturation, evidenced by an accumulation of transitional B cells, hypogammaglobulinemia and dramatic reductions in circulating memory B cells, as well as impaired CSR and Ab responses to infections and vaccines [140,141]. B cells from these patients exhibit poor responses to stimulation with BAFF or CD40L in vitro, including reduced NFκB2 signalling, survival, upregulation of activation markers and Ig class switching [140,141], indicating B cell-intrinsic defects contribute to the clinical features of NIK deficiency or RelB deficiency. NIK-deficient B cells also express reduced levels of ICOS-L and fail to further upregulate its expression in response to CD40L signalling [140]. Consistent with this, and similar to individuals with mutations in *CD40LG*, *ICOS*, *ICOSLG* or *NFKB2*, cTfh cells are reduced in NIK-deficient individuals [140,141] (Figure 15.2). Whilst the Tfh deficit may be secondary to defective B-cell function, bi-allelic mutations in *MAP3K14* or *RELB* do significantly affect T cells, indicating that the clinical features of these IEI likely result from impaired effector functions of many immune lineages, rather than being disease-causing by predominantly impacting B cells [140,141].

Impaired signalling via BAFF and APRIL: Two siblings have been reported with homozygous mutations in *TNFRSF13C*, encoding BAFF-R which signals predominantly via the NFκB2 pathway [142]. Laboratory findings of these individuals partially overlapped with NIK deficiency or RelB deficiency, including B-cell lymphopenia, reduced serum IgM and IgG (but normal levels of IgA), poor responses to pneumococcal polysaccharides vaccines, an accumulation of transitional B cells and corresponding paucity of total and class switched memory B cells in peripheral blood. However, B-cell development in the BM, SHM in peripheral blood B cells, Ab responses to tetanus vaccination, as well as distribution of T cell subsets were unaffected by BAFF-R deficiency [142]. Thus, impaired BAFF-R signalling results in a selective block in B-cell maturation at the transitional → naïve stages (Figure 15.4), thereby significantly impacting the final stages of B-cell development that occur in the periphery and manifesting as B-cell deficiency and impaired T-independent immune responses. This is consistent with BAFF functioning as a survival factor for human and murine B cells [143]. Interestingly, the index case was diagnosed with CVID late in life (aged 57 years) following a history of recurrent infections [142], whilst recessive *TNFRSF13C* mutations were clinically silent in the sibling, suggesting compensatory pathways may overcome BAFF-R deficiency. Despite this, the discovery of individuals with BAFF-R deficiency has shed light on mechanisms underlying different IEIs affecting NFκB. Thus, as BAFF-R almost exclusively signals via the NFκB2 pathway [143], impaired responsiveness to BAFF likely explains defective peripheral B-cell maturation in individuals with NIK or RelB deficiency [140,141], whilst an inability to respond to CD40L and other TNFR ligands would contribute to the more severe clinical phenotype of these patients compared to those with BAFF-R deficiency.

BAFF can also bind the TNFR-SF members TACI and BCMA; furthermore, the related TNF ligand APRIL binds TACI and BCMA, but not BAFF-R [3]. Heterozygous variants have been identified in *TNFRSF13B*, encoding TACI, in 5%–10% of CVID patients [3,144]. However, the same *TNFRSF13B* variants are present in 1%–2% of healthy donors, and the penetrance of a CVID phenotype due to *TNFRSF13B* variants is incredibly variable [144]. Thus, *TNFRSF13B* variants are likely to be disease-modifying alleles rather than pathogenic. Recently, the first case of APRIL deficiency due to homozygous nonsense variants in *TNFSF13*, encoding APRIL, was reported in an individual diagnosed with adult-onset CVID [145] (Figure 15.3). The patient presented with hypogammaglobulinemia (low IgM, IgG; very low IgA) and mild infections that were controlled with Ab replacement. Whilst numbers of peripheral B cells, including IgG^+ and IgA^+ memory B cells, the BCR repertoire and levels of SHM were similar to healthy donors, circulating PCs were reduced 10-fold in the absence of APRIL [145] (Figure 15.3). Although it is difficult to draw too many conclusions from a single case, these findings confirm a likely role for APRIL in maintaining PC survival and serum Ig levels. However, like BAFF-R deficiency, the clinical phenotype of APRIL deficiency was late onset and relatively mild, suggesting modest, albeit important, roles for the BAFF/APRIL system in maintaining humoral immunity in humans. Indeed, the findings that IgA levels are particularly low in APRIL deficiency [145], but IgA^+CD27^- B cells are present at normal numbers in CD40L deficiency [146], suggests APRIL predominantly regulates IgA class switching and production by human PC.

7.4 Defects in the BCR/CD19 Co-receptor Complex

The BCR interacts with a complex comprising CD19, CD21, CD81 and CD225 (Figure 15.4) [147]. CD21 is the receptor for complement component C3d, and CD19 transduces signals following engagement of CD21; CD81 is required for CD19 expression. Ag-C3d aggregates the BCR with the CD19/CD21/CD81 complex, amplifying and sustaining BCR signalling. This also decreases the threshold for B cells to respond to limited Ag and bridges the adaptive and innate immune systems [147].

Biallelic mutations in *CD19* ($n = 11$) [147–152] or *CD81* ($n = 2$) [153,154] cause severe recurrent respiratory tract infections, hypogammaglobulinemia (reduced IgG in all; low IgM and/or IgA in ~50%), impaired vaccine responses and reduced memory B cells; however, B-cell development is intact (Figure 15.2). Although GCs were detected in LNs of CD19-deficient patients [148], SHM in Ig V genes of patient B cells was reduced compared to healthy donors [146,148,149,153,155]. Thus, CD19 or CD81 deficiency impedes affinity maturation. Mutations in *CD81* abolished CD19 expression on patients' B cells, indicating the requirement for CD81 in stabilising CD19 expression, and explaining similarities in clinical features of these IEI [153,154]. CD19 or CD81 deficiency reduced BCR-mediated signalling, proliferation and differentiation in B cells from affected patients [148,153,154]. *CD19* mutations also reduced CD21 expression on B cells [148–150], potentially further compromising function of the BCR/CD19 complex. Heterozygous carriers of *CD19* or *CD81* mutations are asymptomatic, establishing haploinsufficiency of these genes is not pathogenic.

In contrast to CD19 or CD81 deficiency, biallelic *CD21* (*CR2*) mutations cause a milder immunodeficiency: hypogammaglobulinemia, normal/mildly reduced serum IgM or

IgA, fewer memory B cells, variable respiratory infections and low-normal responses to protein and polysaccharide Ags ($n = 4$) [155–157]. Responses of CD21-deficient B cells to an Ag-C3 complex were reduced, but direct BCR engagement induced intact B-cell activation [157]. Furthermore, SHM in CD21-deficient memory B cells was only modestly reduced and significantly greater than CD19-deficient B cells [155]. Thus, although CD19, CD81 and CD21 form a multimeric signalling complex with the BCR, CD19/CD81 and CD21 have unique roles in regulating human B-cell responses. Specifically, CD21 deficiency can be compensated by signalling through receptors whose expression and function does not require CD21. CD19 or CD81 deficiency may also reflect effects of mutations on expression/function of additional co-receptors – CD21 – together with loss of CD19. These findings also highlight that CD19 has key roles in B-cell responses beyond it functioning exclusively as the signal transduction partner of CD21.

The tetraspanin transmembrane molecule CD20 is another well-defined receptor expressed on B cells (Figure 15.4) – indeed, it is typically used to identify and enumerate human B cells and is the target of the biological therapeutic rituximab used to treatment human B-cell malignancies and some autoimmune conditions [4,158]. Despite its ubiquitous expression during B-cell development and maturation, the specific function of CD20 during humoral immune responses remains enigmatic. Some insights have been gleaned from the discovery of one individual with a homozygous *CD20* mutation [159]. Clinically, the patient experienced intermittent respiratory infections and persistently low levels of serum IgG and was diagnosed with CVID during her first decade of life. Numbers and proportions of B cells, titres and affinity of tetanus-specific IgG, levels of SHM, as well as an in vitro proliferation and calcium flux in response to BCR ligation and T-independent and T-dependent stimuli were intact. However, proportions of class-switched memory B cells, Ab responses to pneumococcal vaccination and in vitro production of IgG by activated B cells were reduced (Figure 15.3) [159]. It should be noted though that as the functional assays relied on responses of total B cells, then any reduction in IgG secretion noted in vitro could reflect the presence of fewer IgG^+ memory B cells in the patient ex vivo compared to healthy donors. Despite this caveat, whilst CD20 appears to be dispensable for B-cell development and responses to T-dependent Ag, it appears to have a key role in regulating humoral immunity to T-independent Ags.

7.5 B-Cell Extrinsic Defects: Impaired Generation and Function of Tfh Cells

The generation of robust and effective humoral memory is strictly dependent on the provision of $CD4^+$ T cell help provided by Tfh cells at distinct stages of a GC reaction. This ensures stringent selection of high-affinity B cells, subsequent differentiation of activated B cells into memory cells and PC and the quantity and quality of the Ab response [160–162] (Figure 15.5). Tfh cells were discovered in human tonsils as $CD4^+$ T cells expressing the B-cell zone homing chemokine receptor CXCR5, but lacking the T-cell zone chemokine receptor CCR7 [160–164]. Additional canonical features of Tfh cells include expression of the transcriptional repressor BCL-6, the cytokine IL-21, the chemokine CXCL13 and a suite of surface receptors – CD40L, ICOS, SLAM receptors, PD-1 – and signalling proteins (SLAM-associated protein [SAP]) involved in T-cell/B-cell interactions [160–163] (Figure 15.5). It is the co-ordinated and spatiotemporal function and regulation of chemokines, chemokine receptors, cytokines and cognate receptor/ligand pairs that allow Tfh cells to be precisely positioned within B-cell follicles and GCs to provide requisite survival and helper signals for B-cell differentiation (Figure 15.5).

As exemplified by XR HIGM, caused by hemizygous variants in *CD40LG* [72–76], defective humoral immunity can result from B-cell extrinsic mechanisms due to mutations in genes predominantly expressed in T cells [15,165]. Additional examples include:

- ICOS deficiency, resulting in recurrent bacterial, viral and opportunistic infections, immune dysregulation and malignancy [166,167];
- a single case of IL-21 deficiency presenting with recurrent sinopulmonary infections and early onset colitis [168]; and
- X-linked lymphoproliferative disease (XLP1) due to *SH2D1A* mutations (encoding SAP) resulting in recurrent infections and impaired vaccine responses in ~50% of affected patients [169].

These IEI all share hypogammaglobulinemia, reduced class-switched memory B cells, few or poorly formed GCs and diminished specific Ab responses [68,78,166,168,169]. Importantly, in vitro function of B cells from these patients is intact [71,110,168,169], consistent with low or absent expression of *CD40L*, *IL21*, *ICOS* or *SH2D1A* by human B cells. These findings indicated that impaired B-cell differentiation results from compromised $CD4^+$ T-cell function, specifically Tfh cells. CD40L, ICOS and IL-21 expressed by Tfh cells interact, respectively, with CD40, ICOS-ligand (ICOS-L) and IL-21R on B cells to mediate memory cell and PC generation [165], whilst SAP is also highly expressed by Tfh cells and signals downstream of SLAM family receptors on T cells [169]. IEI due to mutations in *CD40* [69,77], *ICOSL* [170] or *IL21R* [171,172] phenocopy CD40L-, ICOS- or IL-21-deficiencies. Patients with impaired CD40L/CD40, ICOS/ICOS-L or IL-21/IL21R signalling also have reduced proportions of circulating

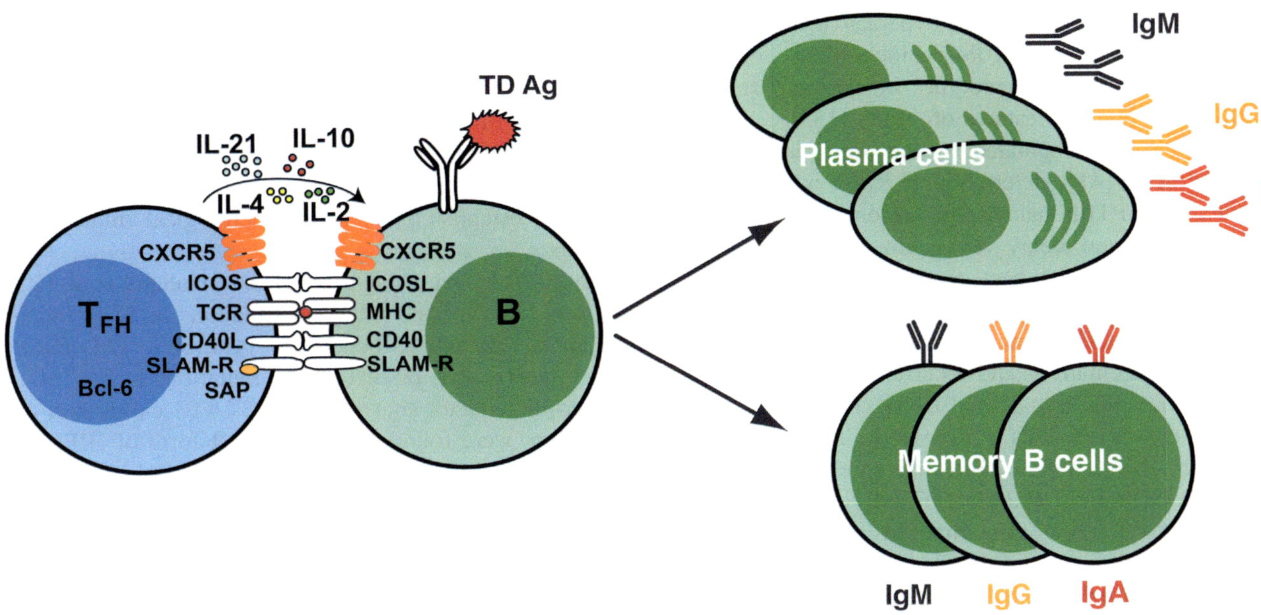

FIGURE 15.5 **Tfh cell-mediated B-cell differentiation.** Naive $CD4^+$ T cells differentiate into Tfh cells, which can be delineated from other subsets of effector $CD4^+$ T cells by a distinct phenotype, cytokine profile and expression of the transcription factor Bcl-6. Tfh cells promote the differentiation of Ag-specific B cells into memory cells or PCs. This effect is largely mediated by cognate interactions between Tfh cells and GC B cells, as well as the production of cytokines, particularly IL-21.

Tfh cells, whilst SAP deficiency compromises Tfh cell function (Figures 15.3 and 15.5) [165].

Tfh cells are also reduced or functionally impaired in many other IEI characterised by hypogammaglobulinemia, poor humoral/Ag-specific immune responses and a paucity of memory B cells. These include immune deficiencies due heterozygous mutations in *STAT3*, *NFKB2* or *IRF4*, or biallelic mutations in *MAPK14* (NIK) or *CARMIL2* (encoding RLTPR) [140,165,173,174] (Figure 15.5). As these genes are also expressed by and have key functions in other cell types, including B cells, the humoral defects in affected individuals are likely to result from defects in both B cells and $CD4^+$ T cells (Figure 15.5). Interestingly, RLTPR functions downstream of CD28, enabling CD4+ T-cell activation and differentiation [174]. Strikingly though, despite studies in mice reporting CD28 signalling in T cells is necessary for TD B-cell differentiation [175,176], the recent description of the first CD28-deficient human noted intact proportions of Tfh cells, total and class-switched memory B cells, as well as intact levels of serum Ig, and Abs against numerous pathogens [177]. Thus, this finding reveals that CD28 is dispensable for generating long-lived humoral immunity in humans and indicates fundamental roles for RLTPR in this process that are CD28-independent.

These findings underscore the critical role of Tfh cells in mediating B-cell differentiation during natural infection and vaccination, even when B cells are intrinsically functional, and highlight how IEI can identify key Tfh-dependent mechanisms regulating humoral immunity.

7.6 Dysregulated PI3K Signalling Causes Immune Dysregulation

Biallelic mutations in *PIK3R1* or *PIK3CD* established the fundamental requirement for PI3K signalling in B-cell development (Figure 15.1) [38–40]. The discovery of heterozygous GOF *PIK3CD* (p110δ catalytic subunit) [178,179] or LOF *PIK3R1* (p85 regulatory subunit) mutations [180,181] – which both increase PI3K activation – revealed a key role for balanced PI3K signalling in B-cell development and differentiation (Figure 15.3). Affected individuals present with recurrent respiratory infections, poor immune responses to vaccines, hypogammaglobulinaemia with occasional hyper-IgM, susceptibility to viral infections, lymphadenopathy, autoimmunity and B-cell malignancy [41,178–181]. These conditions are termed activated PI3K-δ syndrome (APDS1: *PIK3CD*; APDS2: *PIK3R1*).

PIK3CD GOF and *PIK3R1* LOF variants cause a partial block at the pro/pre-B cell stage, resulting in few mature B cells in the BM (Figure 15.1) [182–184]. APDS patients also have low numbers of peripheral B cells, significantly increased frequencies of transitional cells and reductions in naïve and memory cells (Figures 15.1 and 15.2) [182–184]. In vitro assessment revealed intrinsic differentiation defects, evidenced by impaired CSR and Ig secretion by *PIK3CD* GOF/*PIK3R1* LOF B cells [183,184], which likely contribute to poor humoral immune responses in APDS. Importantly, some in vitro B-cell defects could be improved by pharmacological attenuation of p110δ

[183,184], thus establishing that hyperactive PI3K causes this disorder. Indeed, the p110δ-specific inhibitor leniolisib corrects aberrant B-cell phenotypes in vivo and substantially improves clinical features of APDS [185].

PI3K is directly regulated by the lipid phosphatases PTEN and SHIP [41]. Heterozygous *PTEN* mutations are associated with PTEN hamartoma tumour syndromes and neurodevelopmental delay [81]. However, PTEN deficiency has recently also been associated with an APDS-like clinical phenotype, including autoimmunity, lymphoid hyperplasia, hypogammaglobulinemia, reduced responses to vaccinations and recurrent infections, $CD4^+$ T cell lymphopenia, transitional B-cell accumulation and reduced memory B cells [186–189]. The underlying biochemical defect in these patients is likely to be similar to *PIK3CD* GOF and *PIK3R1* LOF individuals, as PTEN mutations also result in enhanced PI3K signalling [186,187]. However, more detailed analysis of immune defects in individuals with *PTEN* mutations is required to determine whether the severity and penetrance of their clinical feature are as dramatic as those observed in individuals with *PIK3CD* GOF or *PIK3R1* LOF mutations.

7.7 IL21R/STAT3 Signalling in B Cells Is Required for Generating Long-Lived Humoral Immunity and Memory

Whilst inputs via the BCR and co-stimulatory molecules (e.g. CD40) from cognate $CD4^+$ T cells are important for inducing B-cell activation and proliferation, cytokines also play critical roles in humoral immunity by mediating the differentiation of B cells into memory cells or PC [190,191]. To date, >60 cytokines have been identified, including interleukins (IL-1 to IL-40), interferons, growth factors (TGF, CSF) and members of the TNF superfamily [192–196]. The biological effects of most cytokines are mediated by JAK (Janus activating kinases)/STAT (signal transducer and activator of transcription) pathways [197–199]. Four JAKs (JAK1, JAK2, JAK3, Tyk2) and seven STATs (STAT1, 2, 3, 4, 5a, 5b, 6) have been identified in mammalian genomes [197,198]. JAKs associate with the cytoplasmic domains of cytokine receptors and, following engagement by specific ligands, phosphorylate key tyrosine residues to provide docking sites for STATs. Receptor-associated STATs undergo JAK-mediated phosphorylation, resulting in the formation of multimers that translocate to the nucleus, bind specific DNA sequences and regulate expression of target genes [197–199].

Over the past four decades, studies have identified many cytokines that induce human B-cell differentiation in vitro [190]. However, the non-redundant functions of specific cytokines and associated JAK/STAT pathways in B-cell function in humans have been revealed by the identification of individuals with inborn errors in cytokine signalling that cause immunodeficient and/or autoimmune states [15,199,200]. Since 2007, we have witnessed the discovery of myriad IEIs, and analyses of which have converged to reveal the fundamental requirement for B-cell intrinsic signalling via the IL21R/γc complex in response to IL-21 stimulation in mediating the generation of memory B cells and establishing long-lived serological immunity in humans.

In 2007, Minegishi et al. [201] and Holland et al. [202] reported that DN heterozygous mutations in *STAT3* cause autosomal dominant hyper-IgE syndrome (AD-HIES). Eleven years later, bi-allelic LOF mutations in *ZNF341*, a transcription factor that regulates STAT3 expression and activation, were found to cause an AR form of HIES that phenocopies *STAT3* DN variants [203,204]. Whilst individuals with germline inactivating variants in *STAT3* or *ZNF341* predominantly suffer from recurrent bacterial and fungal infections and atopic conditions [205,206], they also have impaired functional Ab responses following infection or vaccination [203,204,207–210]. Consistent with defective humoral immunity, *STAT3* DN and AR *ZNF341* LOF patients have few circulating memory B cells [107,203–205,207,211], and their B cells are impaired in differentiating in vitro into Ab-secreting cells in response to STAT3-activating cytokines IL-10 and IL-21 [110,203,207,211,212] (Figure 15.2). Another AD aetiology of HIES is due to DN variants in *IL6ST* (encoding gp130) [213,214]. Interestingly, patients with *STAT1* GOF mutations share many features of patients with *STAT3* DN variants [215], such as infectious susceptibility, and impaired Ab responses to protein Ags or vaccines in some patients [210,215,216], and reduced memory B cells and poor naïve B cell differentiation in vitro [107,110] (Figure 15.2). Mechanistically, clinical and cellular similarities in patients with *STAT1* GOF or *STAT3* DN variants likely result from increased expression of total and phospho-STAT1 in lymphocytes from patients with either genotype and/or impaired induction of STAT3-target genes in the presence of hyper-active STAT1 function [217–219]. Indeed, cellular defects in lymphocytes from these patients can be restored by JAK inhibitors, at least in vitro [217,219].

In addition to IL-10 and IL-21, STAT3 is activated by many cytokines – IL-6, IL-12, IL-23, IL-27 – found to modulate human B cell function, at least in vitro [190,206,220]. Thus, it was initially difficult to ascribe a key requirement for a specific STAT3-activating cytokine in human memory B-cell formation and humoral immunity [207,211]. However, subsequent discoveries of individuals with germline variants in *IL21R* [171,172,221,222], *IL21* [168], *IL6R* [223,224], *IL6ST* [213,214,225,226], *IL12RB2* [227], *IL23R* [227,228] and earlier findings of IEI due to mutations in *IL7R* [229,230], *IL10/IL10R* [231,232], *IFNG/IFNGR1/IFNGR2* [233,234] or *IL12B/IL12RB1* [235] provided mechanistic insights into the requirements of these individual cytokines in human B-cell responses.

Memory B cells and/or Ag-specific Abs induced following natural infection or routine vaccination are unaffected by IL-12R-, IL-23R-, IFNα1/2 or IFNγ/IFNγR-deficiency [107,110,227,228,234,235]. Consistent with this, memory B-cell formation and serum Ig levels are within or exceed normal ranges in patients with mutations in *STAT1*, which is activated by several of these cytokines [107,207,211]. Thus, IL-12- or IL-23-induced STAT3 activation is unlikely to contribute to STAT3-dependent humoral immunity. Unlike mice, B cells develop in humans with combined immunodeficiency ($T^-NK^+B^+$) due to biallelic variants in *IL7RA* [229,230]. Thus, although affected individuals exhibit hypogammaglobulinemia and require HSCT due to a severe T-cell deficit, these observations suggest human B-cell development is independent of IL-7. However, several studies have established that IL-7R is expressed on human B-cell precursors in the BM, IL-7 activates STAT5 in these cells and IL-7 promotes expansion of human pro-B cells in vitro [236,237]. Thus, it remains possible that whilst human B cells can develop in the absence of IL-7R, IL-7/IL-7R signalling may have a nuanced role in human B-cell maturation. This could be revealed by detailed analysis of B-cell subsets in the BM of IL7RA-deficient patients.

In contrast to the IEI described above, bi-allelic variants in *IL21* or *IL21R* cause a combined immunodeficiency characterised by recurrent infections, reduced specific Ab titres, hypogammaglobulinemia and dramatic reductions in total and class-switched memory B cells [107,110,168,171,172,211,221,222] (Figures 15.2 and 15.5). Similarly, HSCT of patients with *IL2RG* variants (causing XR SCID), who retained their own γc-deficient B cells but engrafted donor (i.e. *IL2RG* WT) T cells post-transplant, did not rescue memory B-cell formation, humoral immunity or hypogammaglobulinemia [238]. This revealed that a γc-activating cytokine – likely IL-21 – that functioned intrinsically in B cells was required for these processes. Further evidence for a B-cell intrinsic requirement for IL-21/STAT3 signalling was provided by the finding of an individual who was somatic for a pathogenic *STAT3* variant that caused a mild form of HIES [239]. In this individual, B cells were the only leucocyte population carrying the heterozygous *STAT3* variant; importantly, memory B cells were greatly diminished, despite the availability of help from *STAT3* WT $CD4^+$ T cells [239].

Not surprisingly, pathogenic variants in *IL21R* or *IL2RG* prevent responses of naïve B cells to the stimulatory effects of IL-21 [110,171,172,211,212,222,238]. Compared to IL-21R deficiency, impaired IL6R/gp130 signaling only modestly impacts humoral immunity, as shown by partial reductions in total memory B cells, reduced serum Ig levels and lower specific Ab responses in some but not all affected individuals [107,206,213,214,223–226]. In contrast, IL-10- and IL-10R-deficient patients have normal-to-elevated levels of serum IgM, IgG and IgA, and generate specific Ab following vaccination [231,232]. Combined, these data establish that B-cell-intrinsic STAT3-dependent signals downstream of IL-21/IL-21R are predominantly responsible for inducing and maintaining B-cell-mediated humoral immune responses (memory B cell formation, class switching, specific Ig production) in humans (Figure 15.2), whilst signals provided by IL-6/IL-6R may complement the dominant role of IL-21. These findings explain the shared clinical features of impaired humoral immunity and memory B-cell deficiency resulting from pathogenic *STAT3* DN variants or biallelic variants in *IL21*/*IL21R* or *ZNF341*. Mechanistically, IL-21 induces all the intrinsic molecular requirements necessary for GC formation and B-cell differentiation: proliferation, *AICDA* and *IRF4* for CSR (and SHM), *BCL6* to form GCs, and *PRDM1* (Blimp-1), *XBP1* and *IRF4* which all play key roles in PC differentiation [191,207,211,240,241].

A novel IEI was recently reported in seven individuals from six families due to a recurrent heterozygous and DN *IRF4* variant ($IRF4^{T95R}$) [173]. These patients had opportunistic infections, agammaglobulinemia and a severe deficiency of memory B cells and PC. Patient naïve B cells were unable to undergo CSR and PC differentiation in vitro, establishing that B cell-intrinsic effects of $IRF4^{T95R}$ contributed to their humoral defects [173]. Patients with a LOF but not DN *IRF4* allele reported earlier did not have detectable B cell anomalies but Whipple's disease [242]. Another type of AD IRF4 disorder was more recently reported [243]. IL-21 induces IRF4 expression in activated B cells [211,244], and IRF4 upregulates AID and promotes CSR and PC generation in vivo in an IL-21-dependent manner [244,245]. Thus, the overlap in clinical features of patients with IL21R-deficiency or $IRF4^{T95R}$ may reflect the IL-21/IRF4 axis being a major regulator of B-cell differentiation (Figure 15.4).

8. INSIGHTS INTO REGULATION OF IGE PRODUCTION FROM IEI AFFECTING CYTOKINE SIGNALING

IgE production is stringently regulated: IgE concentrations in human serum typically range from 50 to 200 ng/mL in healthy donors, compared to 1–10 mg/mL for other Ig isotypes. Thus, IgE comprises <0.05% of all serum Ig [246,247]. Similarly, serum half-life of human IgE is ~2.5 days, compared to ~7 days for IgA and ~21–25 days for IgG. The need for such regulation is highlighted by the pathognomonic role of IgE in atopic and allergic diseases [246,247].

8.1 IL-6/STAT3 Signalling

Inborn errors affecting cytokine signalling that result in HIES have shed significant light on the complexities and redundancies of regulating IgE in humans. HIES due to heterozygous *STAT3* and biallelic *ZNF341* LOF variants reveal an unequivocal and critical regulatory role for STAT3/ZNF341 function in constraining human IgE production. The hyper-IgE phenotype, along with clinical features of atopic disease (dermatitis, food allergy, eosinophilia), of STAT3/ZNF341 deficiency are phenocopied by pathogenic variants in *IL6R* or *IL6ST* [206,213,223–226]. Interestingly, some IL-21R-deficient patients also have elevated levels of IgE [172]. This parallels studies showing *Il21* and *Il21r* gene-targeted mice have elevated levels of serum IgE and IgE plasmablasts, and that IL-21 repressed production of Igε germline transcripts, thereby preventing IgE CSR in murine B cells [191]. However, the increase in serum IgE levels in IL21R-deficient humans is relatively modest (200–4000 IU/mL) [171,172,221,222], compared to the extremely elevated levels (5000–100 000 IU/mL) resulting from AD *STAT3* DN or AR *ZNF341* LOF variants [205]. IgE levels in patients with AR IL-10- or IL-10R-deficiency are within normal reference ranges [231,232], indicating that impaired IL-10/STAT3 signalling does not influence CSR to IgE. Thus, IL-6 appears to be the major STAT3-activating cytokine that suppresses IgE production by human B cells. Interestingly, the individual with B-cell-restricted mosaic expression of a pathogenic *STAT3* variant did exhibit high levels of serum IgE; however, these levels (500–5000 IU/mL) were generally lower than typically observed for AD *STAT3* HIES patients [239]. This suggests IL-6 likely constrains IgE production via direct effects on B cells and indirect effect on $CD4^+$ T cells.

8.2 STAT6 GOF

STAT6 regulates the generation of IL-4, IL-5 and IL-13-secreting Th2 cells, as well as promoting IL-4- and IL-13-induced IgE CSR in B cells [246,247]. Consequently, STAT6 has a prominent role in promoting protection against parasitic and helminth infections, but also in allergy and atopic diseases [246,247]. Recently, AD heterozygous GOF variants in *STAT6* were identified in 19 individuals from 13 unrelated kindreds with severe atopic disease [248–251]. Common clinical presentations associated with *STAT6* GOF include early onset, severe allergic immune dysregulation exemplified by often treatment-resistant atopic dermatitis, multiple allergies with serial anaphylaxis, hypereosinophilia with eosinophilic oesophagitis, gastritis and duodenitis, high levels of serum IgE (>5000 U/mL) and skin infections probably secondary to eczema; two reported patients died (Table 15.1 and Figure 15.1C). Additional features in some patients included gastroenteritis with protein-losing enteropathy, dental enamel hypoplasia, cerebral vascular anomalies and follicular lymphoma which has also been associated with somatic mutations in the DNA-binding domain of *STAT6* [248–251].

Pathogenic variants resulted in increased basal or IL-4-induced STAT6 phosphorylation and/or enhanced DNA binding and greater transcription of STAT6 target genes. Overexpression studies suggested a putative therapeutic role for JAK inhibition, evidenced by reductions in both baseline and IL-4 stimulated STAT6 phosphorylation. However, the utility of JAK inhibition in these patients may depend on the specific *STAT6* GOF mutation and mechanism of action of GOF STAT6 protein. The variable clinical and molecular findings in these recent reports highlight the complex biology of STAT6 in regulating gene expression, which will be important to determine the most appropriate precision medicine treatment approaches for this novel IEI. Regardless of these uncertainties, the discovery of pathogenic *STAT6* GOF variants establishes that STAT6 activation, downstream of IL-4 and IL-13, is the central driver of IgE production by human B cells [248,249,250,251](Figure 15.1C).

8.3 Other Inborn Errors of Cytokine Signalling Affecting IgE Production

Variable increases in serum IgE levels have also been found in some individuals with pathogenic variants in *IFNGR1* (300–3000 IU/mL) [252], AR *STAT5B* deficiency (200–1000 IU/mL) [253–255], DN *STAT5B* variants (100–400 IU/mL) [256] or heterozygous variants in *TGFBR1* or *TGFBR2* causing Loeys–Dietz syndrome (400–5000 IU/mL) [257–260]. The prevalence of Th2-type diseases such as eczema, allergies, rhinitis and asthma was also greatly increased in individuals with increased IgE due to these pathogenic variants compared to the general population [252–258,260]. These observations are consistent with data from mice and humans showing IFNγ and TGFβ can inhibit IgE production, as well as Th2 differentiation, in vivo and in vitro [190]. However, it is worth emphasising that the mechanisms by which variants in *TGFBR1* or *TGFBR2* disrupt TGFβ signalling are enigmatic. Depending on the readout measured, the cell type examined or the kinetics of analysis, these variants can be either GOF or LOF [257,259]. Consequently, it is unclear whether increased IgE levels in Loeys–Dietz syndrome results from increased or decreased TGFβ signalling, and whether this is predominantly occurring in B cells or $CD4^+$ T cells [257,259]. Irrespective of this, increased serum IgE due to variants in *IFNGR1*, *STAT5B*, *TGFBR1* or *TGFBR2* is only observed for some patients and is generally less than that observed in individuals with variants in *IL6R/IL6ST*, *STAT3* or *ZNF341* [205,206,252–258]. Thus, whilst signalling via IFNγ, TGFβ and STAT5 may contribute to the general regulation of IgE production by B cells, the predominant pathway that represses IgE CSR is clearly IL-6/IL6R-gp130/STAT3/ZNF341.

9. MOLECULAR REQUIREMENTS FOR GENERATING HUMAN $CD21^{LO}CD11C^{HI}T\text{-}BET^{+}$ B CELLS

A B-cell population that has attracted increasing interest over the past decade is the subset of $CD21^{lo}CD11c^{+}T\text{-}bet^{+}$ B cells (also termed 'atypical memory', 'tissue-like memory' or 'age-associated B cells' [ABCs]). $CD21^{lo}CD11c^{+}T\text{-}bet^{+}$ B cells comprise a minor subset of peripheral blood B cells in healthy individuals [261,262]. However, they are substantially increased in various disease states such as autoimmunity (SLE, scleroderma, rheumatoid arthritis) [261], acute SARS-CoV-2 infection [263,264] or chronic infection (malaria, HIV, hepatitis C) [261,262,265], as well as following some vaccinations [266,267]. Several of these settings are associated with B-cell dysfunction, such as production of autoAbs [261–264], exhaustion or reduced participation in GC reactions resulting in poor responses to infectious pathogens [262,265]. Whilst some studies have identified an IFNγ-gene signature in human $T\text{-}bet^{+}$ B cells ex vivo [268], and demonstrated that stimulation of naïve B cells via the BCR, TLRs and IFNγ-R can give rise to T-bet-expressing cells in vitro [269], the exact requirements for generating and maintaining $CD21^{lo}CD11c^{+}T\text{-}bet^{+}$ B cells in vivo are largely unknown.

This has been recently addressed by quantifying $CD21^{lo}T\text{-}bet^{+}$ B cells in peripheral blood of patients with a wide array of IEI. Proportions of circulating $CD21^{lo}T\text{-}bet^{+}$ B cells were unaffected by LOF/DN variants in *MYD88*, *IRAK4*, *CD19*, *NFKB2*, *IL2RG* (γc), *IFNGR1*, *IL12RB1*, *IL27R*, *IL6ST*, *STAT1* or *STAT3* [270,271]. In contrast, $CD21^{lo}T\text{-}bet^{+}$ B cells were reduced in patients with variants that compromise provision of $CD4^{+}$ T cell help (*CD40LG*, *IL21R*) [270], as well as the only known individual with AR T-bet deficiency [271] (Table 15.1 and Figure 15.1B). Further analysis revealed that whilst $CD21^{lo}T\text{-}bet^{+}$ B cells could be generated in the absence of IFNγ/STAT1 signalling and were reduced but not abolished by T-bet deficiency, the subset expressing canonical features of $CD21^{lo}T\text{-}bet^{+}$ B cells – elevated expression of CD11c, CXCR3 and FCRL5 – was significantly diminished by pathogenic variants in *IFNGR1*, *STAT1* or *TBX21* [271] (Table 15.1 and Figure 15.1B). Thus, whilst the initial differentiation of B cells to a $CD21^{lo}T\text{-}bet^{+}$ B cell fate is independent of STAT1/T-bet signalling, their subsequent progression into $CD21^{lo}T\text{-}bet^{+}CXCR3^{+}CD11c^{hi}FCRL5^{+}$ B cells is strictly dependent on STAT1/T-bet downstream of IFNγ, and possibly IL-27 (Table 15.1 and Figure 15.1B). Interestingly, STAT1 is redundant for generating human memory B cells and maintaining serological memory [107,110,207,211,212]. The study of $CD21^{lo}T\text{-}bet^{+}$ B cells has now established a non-redundant role for STAT1 in human B-cell physiology.

$CD21^{lo}T\text{-}bet^{+}$ B cells were found to be increased in peripheral blood of patients with monogenic immune dysregulation and autoimmunity resulting from LOF variants in *CTLA4*, *LRBA*, *FAS*, *NFKB1*, or *STAT5B*, or GOF variants in *STAT1* or *STAT3* [270,272] (Table 15.1 and Figure 15.1B). Whilst excessive basal or IFNγ/IL-27-induced signalling via STAT1 may promote accumulation of these cells in individuals with *STAT1* GOF variants, the mechanism(s) underlying expansion of $CD21^{lo}T\text{-}bet^{+}$ B cells in other conditions is unknown. It is possible that *STAT3* GOF variants render B cells hyper-responsive to cytokines (IFNγ/IL-27) that ordinarily do not require STAT3 for cytokine-induced generation of $CD21^{lo}T\text{-}bet^{+}$ B cells. However, this remains to be examined experimentally.

10. B CELLS ARE KEY ANTIGEN-PRESENTING CELLS TO INDUCE EFFECTIVE CELL-MEDIATED IMMUNITY AGAINST EBV

Whilst the best-studied and most well-defined function of B cells is the production of neutralising Abs that protect against infection, B cells are also capable of acting as Ag-presenting cells [4,273]. This is particularly true for defence against pathogens that preferentially infect B cells, such as the human herpes virus Epstein–Barr virus (EBV). During primary EBV infection, oral epithelial cells and some infiltrating B cells at mucosal surfaces are sites for primary viral replication. However, EBV subsequently switches to latent infection of B cells, which is critical for colonisation of the host [274,275]. Viral persistence is achieved largely through silent infection of memory B cells, which are the exclusive host of EBV in healthy donors. Viral reactivation likely occurs when these cells enter oropharyngeal lymphoid tissues, resulting in viral replication and viral shedding [274,275].

Clinical manifestations of primary EBV infection depend on the age and immunocompetent state of the host. EBV infection in childhood is generally asymptomatic. Acquisition in healthy adolescents is also mostly benign; however, EBV infection can result in infectious mononucleosis (IM) – a febrile disease characterised by sore throat, swollen LNs, body aches and fatigue – in 10% of infected adolescents. EBV can also cause induce lymphoproliferation, lymphoma and hemophagocytic lymphohistiocytosis (HLH) in immunocompromised individuals, suggesting continuous immune surveillance is critical for virus-host homoeostasis. IEI that selectively disrupt the ability of Ag-presenting B cells to interact with and activate EBV-specific $CD8^+$ T cells are characterised by severe EBV-induced pathologies [276]. These experiments of nature have identified critical non-redundant pathways for anti-EBV immunity and highlight the fundamental role played by B cells in eliciting cytotoxic lymphocyte-mediated EBV-specific immune responses [276].

10.1 X-Linked Lymphoproliferative Disease (XLP1): Inactivating Mutations in *SH2D1A*

XLP was first reported in the 1970s [277–279] and is the classic example of an IEI with a selective inability to generate effective immunity against EBV. The defining features of XLP are severe EBV-induced IM, HLH, B-cell lymphoma and hypogammaglobulinemia [276,280–283]. As discussed above, XLP is caused by mutations in *SH2D1A*, encoding the cytoplasmic adaptor protein SAP [284–286] which binds to the cytoplasmic domains of the SLAM family of surface receptors (SLAM/CD150, 2B4/CD244, CD84, NTB-A, Ly9). In doing so, SAP enables intracellular signalling pathways induced following engagement of these receptors on hematopoietic cells by cognate ligands to regulate immune cell function [169,287,288].

The exquisite sensitivity of XLP patients to disease caused by EBV, but not other pathogens including other herpes viruses, raised the question of how SAP deficiency selectively cripples host defence to a single pathogen, which generally causes only mild sequelae in most healthy individuals. This dichotomy was resolved by the seminal discovery that induction of $CD8^+$ T-cell activation by distinct types of APCs exhibited differential dependence on SAP. Thus, whilst killing of DCs, monocytes or fibroblasts was unaffected by SAP deficiency, the ability of $CD8^+$ T cells to respond to Ag-presenting B cells was strictly dependent on SAP signalling downstream of 2B4 and NTB-A [289,290] (Figure 15.6). Consequently, XLP patients are unable to control EBV infection due to defective immunological synapse formation and subsequent interactions between SAP-deficient cytotoxic cells and Ag-presenting/virally infected B cells, thereby specifically impairing B-cell-mediated activation of these cells (Figure 15.6). As EBV predominantly maintains latent infection in human B cells [291–293], the selective inability of SAP-deficient cytotoxic cells to be activated by B cells explained the unique susceptibility of XLP patients to EBV. Thus, when infected with viruses other than EBV, alternate pathways exist for the activation of cytotoxic T-cell responses by non-B-cell APCs, enabling the host to generate protective immune responses in the presence of *SH2D1A* mutations.

10.2 MAGT1 Deficiency Impairs NKG2D Expression and B-Cell-Mediated Cytotoxic Lymphocyte Activation

MAGT1 encodes a magnesium transporter that acts as a second messenger in TCR-mediated activation [294]. Hemizygous mutations were first reported in 2011 in three males presenting with recurrent sinopulmonary infections, EBV viraemia, EBV-induced lymphoproliferative disease and/or EBV^+ B-lymphoma [294]. More than 40 MAGT1-deficient patients have now been described [294–305]. A common feature of MAGT1-deficient patients was markedly reduced expression of the activating receptor NKG2D on MAGT1-deficient cytotoxic lymphocytes [296,298,302–305] (Figure 15.6). Reduced NKG2D expression was sufficient to compromise the cytotoxic function of MAGT1-deficient NK and $CD8^+$ T cells against EBV^+ B cells [296], which express high levels of NKG2D ligands [306] (Figure 15.6). Thus, MAGT1 deficiency established the critical role of interactions between NKG2D on cytolytic $CD8^+$ T and NK cells and its ligands on Ag-presenting B cells to induce effective cell-mediated anti-EBV immunity [276,305].

10.3 Signalling via TNFR Superfamily Members Is Critical for the Initiation of B-Cell-Mediated EBV-Specific $CD8^+$ T-Cell Responses

The study of XLP1 and MAGT1 deficiency elucidated fundamental requirements for interactions between activating receptors on cytotoxic cells and corresponding ligands on cognate B cells in controlling EBV infection [282,297,307]. Further analyses of patients with EBV-associated IEI have identified additional receptor/ligand pairs with non-redundant functions in cell-mediated anti-EBV immunity.

CD27 and CD137 (4-1BB) are members of the TNFRSF of cell surface receptors. CD27 is constitutively expressed on naïve T cells, some memory T cells and NK cells, as well as GC B cells, memory B cells and PC. Unlike CD27, CD137 is not expressed by resting T cells; rather it is induced following activation [308]. Both molecules function as co-stimulatory receptors on T cells, promoting proliferation, cytokine secretion, survival and cytotoxicity [308] (Figure 15.6). This likely reflects the overlapping expression patterns of their ligands, the TNFR superfamily members CD70 and CD137 ligand (CD137L/4-1BBL) on activated B cells, EBV-infected B cells in IM patients, myeloid cells (monocytes, DCs) and some T cells

[308,309]. LOF bi-allelic mutations in *CD27* [309–312], its counter-structure *CD70* [309,313,314], have been identified in ~50 individuals. EBV-induced disease including chronic EBV viraemia, EBV-associated lymphoproliferation, EBV^+ lymphoma and/or HLH was evident in >90% and fatal in 15% of patients. Homozygous null mutations in *TNFRSF9* (4-1BB/CD137) or *TNFSF9* (4-1BBL/CD137L) have also been identified in patients with chronic EBV viraemia, lymphoproliferation, lymphadenopathy, lymphoma, HLH or EBV^+ smooth muscle tumours [315–317] (Figure 15.6).

Mechanistically, $CD8^+$ T cells from CD27-, CD70-, CD137- or CD137L-deficient patients exhibited impaired proliferation towards autologous EBV-infected B cells (Figure 15.6), compromising their ability to lyse these target cells [309,313,315–317]. CD27/CD70 and CD137/CD137L interactions were critical for initial priming and expansion of EBV-specific $CD8^+$ T cells, rather than cytotoxicity [309] (Figure 15.6). This potentially explains reduced proportions of EBV-specific $CD8^+$ T cells in patients with *CD70* [313] or *CD27* mutations. Consistent with CD27 and CD137 functioning as a co-stimulatory receptor on T cells, as well as with high expression of CD70 and CD137L on activated (and EBV-infected) B cells [308], CD70 and CD137L on B cells were required to engage CD27 and CD137 on cognate $CD8^+$ T cells to elicit a response [309]. These observations suggest CD27 and CD137 have similar but non-redundant roles in EBV immunity and probably modulate T-cell functions at distinct stages of an immune response.

Combined, these findings establish that poor effector function of $CD8^+$ T cells in IEIs characterised by severe EBV-induced disease results from:

(1) impaired proliferation of EBV-specific $CD8^+$ T cells (*MAGT1*, *CD27/CD70*, *TNFRSF9/TNFSF9* mutations) induced by Ag-presenting B cells, and/or

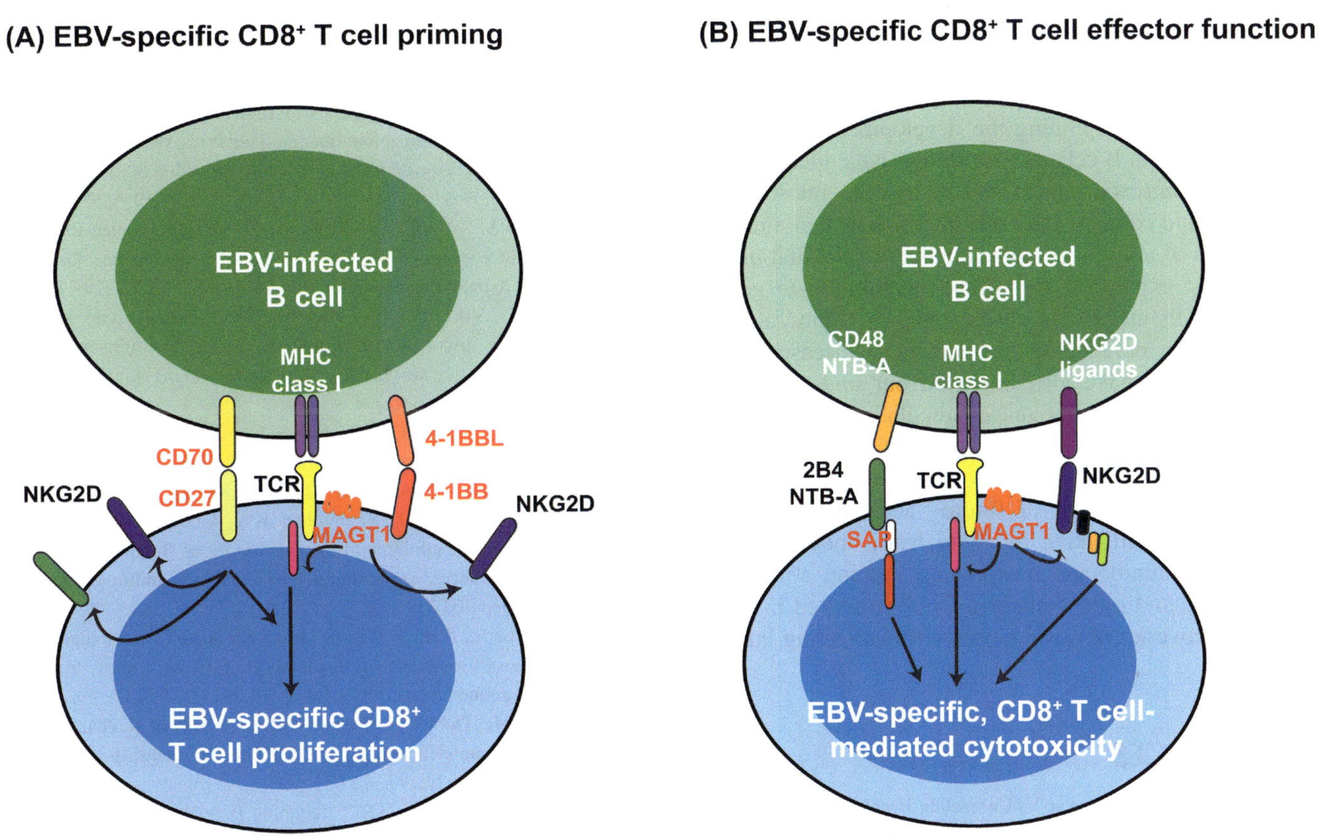

FIGURE 15.6 **Molecular and biochemical requirements for inducing B cell-dependent $CD8^+$ T cell-mediated EBV-specific immunity.** The scheme depicts the key cell surface receptors required for generating $CD8^+$ T cell-mediated immunity against EBV-infected B cells, and how specific inborn errors of immunity (*SH12D1A* [SAP], *MAGT1*, *CD27*, *CD70*, *TNFRSF9* [4-1BB], depicted in *red*) compromise this process. (A) Initial interactions between EBV-infected or Ag-presenting B cells and $CD8^+$ T cells involve CD27/CD70, 4-1BB/4-1BBL and MHC class I/TCR. Signals elicited downstream of these receptors induce proliferation of EBV-specific $CD8^+$ T cells. Signals via CD27/CD70 and MAGT1 contribute to inducing or maintaining expression of activating receptors such as 2B4 and NKG2D. (B) Following expansion of EBV-specific $CD8^+$ T cells, engagement of the activating SLAM family receptors 2B4 and NTB-A (via SAP), and the glycoprotein NKG2D, by their ligands highly expressed on EBV-infected B cells, induce cytolytic effector function resulting in $CD8^+$ T cell-mediated killing of EBV^+ target B cells.

(2) impaired formation of lytic synapses between EBV-infected B cells and cytolytic lymphocytes resulting in compromised activation of cytotoxic pathways in response to Ag-presenting B cells (*SH2D1A*, *MAGT1*, *CD27/CD70*, *TNSRSF9* mutations) [276] (Figure 15.6).

These mechanisms have broad implications, not only restricted to controlling EBV infection but also in tumour surveillance of B-cell neoplasms that arise due to the oncogenicity, or even independently, of EBV. This is highlighted by the finding of somatic mutations in *CD70* and *TNFSF9* in EBV^{+} and EBV^{-} human B-cell lymphomas [318,319] which would potentially cripple $CD8^{+}$ T-cell-mediated immune surveillance of malignant B cells.

11. CONCLUSION

It has been over 70 years since Bruton first reported agammaglobulinemia, and 30 years since the discovery that *BTK* mutations cause XLA. By studying inborn errors affecting humoral immunity, the past three decades have seen an exponential increase in our understanding of fundamental molecular, biochemical and cellular requirements underpinning the development and differentiation of human B cells. These studies have identified non-redundant intrinsic signalling pathways, transcriptional networks and cognate cell—cell interactions that enable the generation of immunocompetent naive B cells that undergo Ag-specific expansion and affinity maturation to generate memory cells and PC to ensure long-live humoral immunity and host defence against infectious disease. Discoveries from IEI have also impacted our understanding of mechanisms causing immune dysregulation (autoimmunity, allergy, malignancy), enabled therapeutic interventions for B-cell-mediated immunopathologies and have the potential to improve the design of future vaccinations strategies. However, we still have much to learn about long-lived humoral immunity and maximising vaccine efficacy. No doubt these and other unknowns will be solved in part by future discoveries of novel inborn errors affecting human B cells.

REFERENCES

[1] Morgan D, Tergaonkar V. Unraveling B cell trajectories at single cell resolution. Trends Immunol 2022;43:210—29.

[2] Uckun FM. Regulation of human B-cell ontogeny. Blood 1990;76:1908—23.

[3] Pieper K, Grimbacher B, Eibel H. B-cell biology and development. J Allergy Clin Immunol 2013;131:959—71.

[4] LeBien TW, Tedder TF. B lymphocytes: how they develop and function. Blood 2008;112:1570—80.

[5] Amanna IJ, Carlson NE, Slifka MK. Duration of humoral immunity to common viral and vaccine antigens. N Engl J Med 2007;357:1903—15.

[6] Tangye SG, Tarlinton DM. Memory B cells: effectors of long-lived immune responses. Eur J Immunol 2009;39:2065—75.

[7] Rajewsky K. Clonal selection and learning in the antibody system. Nature 1996;381:751—8.

[8] Victora GD, Nussenzweig MC. Germinal centers. Annu Rev Immunol 2022;40:413—42.

[9] Tangye SG. Staying alive: regulation of plasma cell survival. Trends Immunol 2011;32:595—602.

[10] Robinson MJ, Webster RH, Tarlinton DM. How intrinsic and extrinsic regulators of plasma cell survival might intersect for durable humoral immunity. Immunol Rev 2020;296:87—103.

[11] Allen CD, Okada T, Cyster JG. Germinal-center organization and cellular dynamics. Immunity 2007;27:190—202.

[12] Elgueta R, de Vries VC, Noelle RJ. The immortality of humoral immunity. Immunol Rev 2010;236:139—50.

[13] Liu YJ, Banchereau J. The paths and molecular controls of peripheral B-cell development. Immunologist 1996;4:55—66.

[14] Kunkel EJ, Butcher EC. Plasma-cell homing. Nat Rev Immunol 2003;3:822—9.

[15] Tangye SG, Al-Herz W, Bousfiha A, Cunningham-Rundles C, Franco JL, et al. Human inborn errors of immunity: 2022 update on the classification from the International Union of Immunological Societies Expert Committee. J Clin Immunol 2022;42(7):1473—507.

[16] Bruton OC. Agammaglobulinemia. Pediatrics 1952;9:722—8.

[17] Cooper MD, Peterson RD, Good RA. Delineation of the thymic and bursal lymphoid systems in the chicken. Nature 1965;205:143—6.

[18] Tsukada S, Saffran DC, Rawlings DJ, Parolini O, Allen RC, et al. Deficient expression of a B cell cytoplasmic tyrosine kinase in human X-linked agammaglobulinemia. Cell 1993;72:279—90.

[19] Vetrie D, Vorechovsky I, Sideras P, Holland J, Davies A, et al. The gene involved in X-linked agammaglobulinaemia is a member of the src family of protein-tyrosine kinases. Nature 1993;361:226—33.

[20] Gayko U, Fung M, Clow F, Sun S, Faust E, et al. Development of the Bruton's tyrosine kinase inhibitor ibrutinib for B cell malignancies. Ann N Y Acad Sci 2015;1358:82—94.

[21] Garg N, Padron EJ, Rammohan KW, Goodman CF. Bruton's tyrosine kinase inhibitors: the next frontier of B-cell-targeted therapies for cancer, autoimmune disorders, and multiple sclerosis. J Clin Med 2022;11.

[22] Casanova JL, Conley ME, Seligman SJ, Abel L, Notarangelo LD. Guidelines for genetic studies in single patients: lessons from primary immunodeficiencies. J Exp Med 2014;211:2137—49.

[23] Conley ME, Dobbs AK, Farmer DM, Kilic S, Paris K, et al. Primary B cell immunodeficiencies: comparisons and contrasts. Annu Rev Immunol 2009;27:199—227.

[24] Cooper MD, Lawton AR. Circulating B-cells in patients with immunodeficiency. Am J Pathol 1972;69:513—28.

[25] Casanova JL, Hammarstrom L. Foreword to the English translation of Kostmann's memoirs. J Clin Immunol 2023;43:671—4.

[26] Niiro H, Clark EA. Regulation of B-cell fate by antigen-receptor signals. Nat Rev Immunol 2002;2:945—56.

[27] Dobbs AK, Bosompem A, Coustan-Smith E, Tyerman G, Saulsbury FT, Conley ME. Agammaglobulinemia associated with

BCR(-) B cells and enhanced expression of CD19. Blood 2011;118:1828–37.
[28] Conley ME. B cells in patients with X-linked agammaglobulinemia. J Immunol 1985;134:3070–4.
[29] Suryani S, Fulcher DA, Santner-Nanan B, Nanan R, Wong M, et al. Differential expression of CD21 identifies developmentally and functionally distinct subsets of human transitional B cells. Blood 2010;115:519–29.
[30] Ng YS, Wardemann H, Chelnis J, Cunningham-Rundles C, Meffre E. Bruton's tyrosine kinase is essential for human B cell tolerance. J Exp Med 2004;200:927–34.
[31] Conley ME, Brown P, Pickard AR, Buckley RH, Miller DS, et al. Expression of the gene defect in X-linked agammaglobulinemia. N Engl J Med 1986;315:564–7.
[32] Yel L, Minegishi Y, Coustan-Smith E, Buckley RH, Trubel H, et al. Mutations in the mu heavy-chain gene in patients with agammaglobulinemia. N Engl J Med 1996;335:1486–93.
[33] Minegishi Y, Coustan-Smith E, Wang YH, Cooper MD, Campana D, Conley ME. Mutations in the human lambda5/14.1 gene result in B cell deficiency and agammaglobulinemia. J Exp Med 1998;187:71–7.
[34] Minegishi Y, Coustan-Smith E, Rapalus L, Ersoy F, Campana D, Conley ME. Mutations in Igalpha (CD79a) result in a complete block in B-cell development. J Clin Invest 1999;104:1115–21.
[35] Ferrari S, Lougaris V, Caraffi S, Zuntini R, Yang J, et al. Mutations of the Igbeta gene cause agammaglobulinemia in man. J Exp Med 2007;204:2047–51.
[36] Dobbs AK, Yang T, Farmer D, Kager L, Parolini O, Conley ME. Cutting edge: a hypomorphic mutation in Igbeta (CD79b) in a patient with immunodeficiency and a leaky defect in B cell development. J Immunol 2007;179:2055–9.
[37] Minegishi Y, Rohrer J, Coustan-Smith E, Lederman HM, Pappu R, et al. An essential role for BLNK in human B cell development. Science 1999;286:1954–7.
[38] Conley ME, Dobbs AK, Quintana AM, Bosompem A, Wang YD, et al. Agammaglobulinemia and absent B lineage cells in a patient lacking the p85alpha subunit of PI3K. J Exp Med 2012;209:463–70.
[39] Tang P, Upton JEM, Barton-Forbes MA, Salvadori MI, Clynick MP, Price AK, et al. Autosomal recessive agammaglobulinemia due to a homozygous mutation in PIK3R1. J Clin Immunol 2018;38(1):88–95.
[40] Sogkas G, Fedchenko M, Dhingra A, Jablonka A, Schmidt RE, Atschekzei F. Primary immunodeficiency disorder caused by phosphoinositide 3-kinase delta deficiency. J Allergy Clin Immunol 2018;142:1650–1653 e1652.
[41] Tangye SG, Bier J, Lau A, Nguyen T, Uzel G, Deenick EK. Immune dysregulation and disease pathogenesis due to activating mutations in PIK3CD-the Goldilocks' effect. J Clin Immunol 2019;39:148–58.
[42] Schwarz K, Gauss GH, Ludwig L, Pannicke U, Li Z, et al. RAG mutations in human B cell-negative SCID. Science 1996;274:97–9.
[43] Nutt SL, Kee BL. The transcriptional regulation of B cell lineage commitment. Immunity 2007;26:715–25.
[44] Yamashita M, Morio T. Inborn errors of IKAROS and AIOLOS. Curr Opin Immunol 2021;72:239–48.
[45] Al Sheikh E, Arkwright PD, Herwadkar A, Hussell T, Briggs TA. TCF3 dominant negative variant causes an early block in B-lymphopoiesis and agammaglobulinemia. J Clin Immunol 2021;41:1391–4.
[46] Boisson B, Wang YD, Bosompem A, Ma CS, Lim A, et al. A recurrent dominant negative E47 mutation causes agammaglobulinemia and BCR(-) B cells. J Clin Invest 2013;123:4781–5.
[47] Ben-Ali M, Yang J, Chan KW, Ben-Mustapha I, Mekki N, et al. Homozygous transcription factor 3 gene (TCF3) mutation is associated with severe hypogammaglobulinemia and B-cell acute lymphoblastic leukemia. J Allergy Clin Immunol 2017;140:1191–1194 e1194.
[48] Le Coz C, Nguyen DN, Su C, Nolan BE, Albrecht AV, et al. Constrained chromatin accessibility in PU.1-mutated agammaglobulinemia patients. J Exp Med 2021;218.
[49] Kaiser FMP, Gruenbacher S, Oyaga MR, Nio E, Jaritz M, et al. Biallelic PAX5 mutations cause hypogammaglobulinemia, sensorimotor deficits, and autism spectrum disorder. J Exp Med 2022;219.
[50] Goldman FD, Gurel Z, Al-Zubeidi D, Fried AJ, Icardi M, Song C, et al. Congenital pancytopenia and absence of B lymphocytes in a neonate with a mutation in the Ikaros gene. Pediatr Blood Cancer 2012;58:591–7.
[51] Kuehn HS, Boast B, Rosenzweig SD. Inborn errors of human IKAROS: LOF and GOF variants associated with primary immunodeficiency. Clin Exp Immunol 2023;212(2):129–36.
[52] Hoshino A, Okada S, Yoshida K, Nishida N, Okuno Y, et al. Abnormal hematopoiesis and autoimmunity in human subjects with germline IKZF1 mutations. J Allergy Clin Immunol 2017;140:223–31.
[53] Kuehn HS, Boisson B, Cunningham-Rundles C, Reichenbach J, Stray-Pedersen A, et al. Loss of B cells in patients with heterozygous mutations in IKAROS. N Engl J Med 2016;374:1032–43.
[54] Yamashita M, Kuehn HS, Okuyama K, Okada S, Inoue Y, et al. A variant in human AIOLOS impairs adaptive immunity by interfering with IKAROS. Nat Immunol 2021;22:893–903.
[55] Vaeth M, Feske S. Ion channelopathies of the immune system. Curr Opin Immunol 2018;52:39–50.
[56] Anzilotti C, Swan DJ, Boisson B, Deobagkar-Lele M, Oliveira C, et al. An essential role for the Zn(2+) transporter ZIP7 in B cell development. Nat Immunol 2019;20:350–61.
[57] Erdos M, Mironska K, Kareva L, Stavric K, Hasani A, et al. A novel mutation in SLC39A7 identified in a patient with autosomal recessive agammaglobulinemia: the impact of the J Project. Pediatr Allergy Immunol 2022;33:e13805.
[58] Siggs OM, Stockenhuber A, Deobagkar-Lele M, Bull KR, Crockford TL, et al. Mutation of Fnip1 is associated with B-cell deficiency, cardiomyopathy, and elevated AMPK activity. Proc Natl Acad Sci U S A 2016;113:E3706–15.
[59] Niehues T, Ozgur TT, Bickes M, Waldmann R, Schoning J, et al. Mutations of the gene FNIP1 associated with a syndromic autosomal recessive immunodeficiency with cardiomyopathy and pre-excitation syndrome. Eur J Immunol 2020;50:1078–80.
[60] Saettini F, Poli C, Vengoechea J, Bonanomi S, Orellana JC, et al. Absent B cells, agammaglobulinemia, and hypertrophic cardiomyopathy in folliculin-interacting protein 1 deficiency. Blood 2021;137:493–9.
[61] Austin CA, Cowell IG, Khazeem MM, Lok D, Ng HT. TOP2B's contributions to transcription. Biochem Soc Trans 2021;49:2483–93.

[62] Broderick L, Yost S, Li D, McGeough MD, Booshehri LM, et al. Mutations in topoisomerase IIbeta result in a B cell immunodeficiency. Nat Commun 2019;10:3644.
[63] Papapietro O, Chandra A, Eletto D, Inglott S, Plagnol V, et al. Topoisomerase 2beta mutation impairs early B-cell development. Blood 2020;135:1497–501.
[64] Erdos M, Lanyi A, Balazs G, Casanova JL, Boisson B, Marodi L. Inherited TOP2B mutation: possible confirmation of mutational hotspots in the TOPRIM domain. J Clin Immunol 2021;41:817–9.
[65] Rosen FS, Kevy SV, Merler E, Janeway CA, Gitlin D. Recurrent bacterial infections and dysgamma-globulinemia: deficiency of 7S gamma-globulins in the presence of elevated 19S gamma-globulins. Report of two cases. Pediatrics 1961;28:182–95.
[66] Siegal FP, Pernis B, Kunkel HG. Lymphocytes in human immunodeficiency states: a study of membrane-associated immunoglobulins. Eur J Immunol 1971;1:482–6.
[67] Notarangelo LD, Lanzi G, Peron S, Durandy A. Defects of class-switch recombination. J Allergy Clin Immunol 2006;117:855–64.
[68] de la Morena MT. Clinical phenotypes of hyper-IgM syndromes. J Allergy Clin Immunol Pract 2016;4:1023–36.
[69] Lougaris V, Badolato R, Ferrari S, Plebani A. Hyper immunoglobulin M syndrome due to CD40 deficiency: clinical, molecular, and immunological features. Immunol Rev 2005;203:48–66.
[70] Agematsu K, Nagumo H, Shinozaki K, Hokibara S, Yasui K, et al. Absence of IgD-CD27(+) memory B cell population in X-linked hyper-IgM syndrome. J Clin Invest 1998;102:853–60.
[71] Mayer L, Kwan SP, Thompson C, Ko HS, Chiorazzi N, Waldmann T, et al. Evidence for a defect in "switch" T cells in patients with immunodeficiency and hyperimmunoglobulinemia M. N Engl J Med 1986;314:409–13.
[72] Allen RC, Armitage RJ, Conley ME, Rosenblatt H, Jenkins NA, et al. CD40 ligand gene defects responsible for X-linked hyper-IgM syndrome. Science 1993;259:990–3.
[73] Aruffo A, Farrington M, Hollenbaugh D, Li X, Milatovich A, et al. The CD40 ligand, gp39, is defective in activated T cells from patients with X-linked hyper-IgM syndrome. Cell 1993;72:291–300.
[74] DiSanto JP, Bonnefoy JY, Gauchat JF, Fischer A, de Saint Basile G. CD40 ligand mutations in x-linked immunodeficiency with hyper-IgM. Nature 1993;361:541–3.
[75] Korthauer U, Graf D, Mages HW, Briere F, Padayachee M, et al. Defective expression of T-cell CD40 ligand causes X-linked immunodeficiency with hyper-IgM. Nature 1993;361:539–41.
[76] Ramesh N, Fuleihan R, Ramesh V, Lederman S, Yellin MJ, et al. Deletions in the ligand for CD40 in X-linked immunoglobulin deficiency with normal or elevated IgM (HIGMX-1). Int Immunol 1993;5:769–73.
[77] Ferrari S, Giliani S, Insalaco A, Al-Ghonaium A, Soresina AR, et al. Mutations of CD40 gene cause an autosomal recessive form of immunodeficiency with hyper IgM. Proc Natl Acad Sci U S A 2001;98:12614–9.
[78] Renner ED, Kratz CE, Orange JS, Hagl B, Rylaarsdam S, et al. Class switch recombination defects: impact on B cell maturation and antibody responses. Clin Immunol 2021;222:108638.
[79] Facchetti F, Appiani C, Salvi L, Levy J, Notarangelo LD. Immunohistologic analysis of ineffective CD40-CD40 ligand interaction in lymphoid tissues from patients with X-linked immunodeficiency with hyper-IgM. Abortive germinal center cell reaction and severe depletion of follicular dendritic cells. J Immunol 1995;154:6624–33.
[80] Muramatsu M, Sankaranand VS, Anant S, Sugai M, Kinoshita K, Davidson NO, et al. Specific expression of activation-induced cytidine deaminase (AID), a novel member of the RNA-editing deaminase family in germinal center B cells. J Biol Chem 1999;274:18470–6.
[81] Muramatsu M, Kinoshita K, Fagarasan S, Yamada S, Shinkai Y, Honjo T. Class switch recombination and hypermutation require activation-induced cytidine deaminase (AID), a potential RNA editing enzyme. Cell 2000;102:553–63.
[82] Revy P, Muto T, Levy Y, Geissmann F, Plebani A, et al. Activation-induced cytidine deaminase (AID) deficiency causes the autosomal recessive form of the Hyper-IgM syndrome (HIGM2). Cell 2000;102:565–75.
[83] Imai K, Slupphaug G, Lee WI, Revy P, Nonoyama S, et al. Human uracil-DNA glycosylase deficiency associated with profoundly impaired immunoglobulin class-switch recombination. Nat Immunol 2003;4:1023–8.
[84] Gardes P, Forveille M, Alyanakian MA, Aucouturier P, Ilencikova D, et al. Human MSH6 deficiency is associated with impaired antibody maturation. J Immunol 2012;188:2023–9.
[85] Peron S, Metin A, Gardes P, Alyanakian MA, Sheridan E, et al. Human PMS2 deficiency is associated with impaired immunoglobulin class switch recombination. J Exp Med 2008;205:2465–72.
[86] Kracker S, Di Virgilio M, Schwartzentruber J, Cuenin C, Forveille M, et al. An inherited immunoglobulin class-switch recombination deficiency associated with a defect in the INO80 chromatin remodeling complex. J Allergy Clin Immunol 2015;135:998–1007 e1006.
[87] Kuhny M, Forbes LR, Cakan E, Vega-Loza A, Kostiuk V, et al. Disease-associated CTNNBL1 mutation impairs somatic hypermutation by decreasing nuclear AID. J Clin Invest 2020;130:4411–22.
[88] Kasahara Y, Kaneko H, Fukao T, Terada T, Asano T, Kasahara K, et al. Hyper-IgM syndrome with putative dominant negative mutation in activation-induced cytidine deaminase. J Allergy Clin Immunol 2003;112:755–60.
[89] Imai K, Zhu Y, Revy P, Morio T, Mizutani S, et al. Analysis of class switch recombination and somatic hypermutation in patients affected with autosomal dominant hyper-IgM syndrome type 2. Clin Immunol 2005;115:277–85.
[90] Fadlallah J, Chentout L, Boisson B, Pouliet A, Masson C, et al. From dysgammaglobulinemia to autosomal-dominant activation-induced cytidine deaminase deficiency: unraveling an inherited immunodeficiency after 50 years. J Pediatr 2020;223:207–211 e201.
[91] Kermode W, De Santis D, Truong L, Della Mina E, Salman S, et al. A novel targeted amplicon next-generation sequencing gene panel for the diagnosis of common variable immunodeficiency has a high diagnostic yield: results from the Perth CVID Cohort Study. J Mol Diagn 2022;24:586–99.
[92] Durandy A, Taubenheim N, Peron S, Fischer A. Pathophysiology of B-cell intrinsic immunoglobulin class switch recombination deficiencies. Adv Immunol 2007;94:275–306.
[93] Teitell MA. OCA-B regulation of B-cell development and function. Trends Immunol 2003;24:546–53.
[94] Kury P, Staniek J, Wegehaupt O, Janowska I, Eckenweiler M, et al. Agammaglobulinemia with normal B-cell numbers in a patient lacking Bob1. J Allergy Clin Immunol 2021;147:1977–80.

[95] Zhang Q, Lenardo MJ, Baltimore D. 30 years of NF-kappaB: a blossoming of relevance to human pathobiology. Cell 2017;168:37−57.

[96] Fliegauf M, Bryant VL, Frede N, Slade C, Woon ST, et al. Haploinsufficiency of the NF-kappaB1 subunit p50 in common variable immunodeficiency. Am J Hum Genet 2015;97:389−403.

[97] Lorenzini T, Fliegauf M, Klammer N, Frede N, Proietti M, et al. Characterization of the clinical and immunologic phenotype and management of 157 individuals with 56 distinct heterozygous NFKB1 mutations. J Allergy Clin Immunol 2020;146:901−11.

[98] Li J, Lei WT, Zhang P, Rapaport F, Seeleuthner Y, et al. Biochemically deleterious human NFKB1 variants underlie an autosomal dominant form of common variable immunodeficiency. J Exp Med 2021;218:e20210566.

[99] Kaustio M, Haapaniemi E, Goos H, Hautala T, Park G, et al. Damaging heterozygous mutations in NFKB1 lead to diverse immunologic phenotypes. J Allergy Clin Immunol 2017;140:782−96.

[100] Tuovinen EA, Kuismin O, Soikkonen L, Martelius T, Kaustio M, et al. Long-term follow up of families with pathogenic NFKB1 variants reveals incomplete penetrance and frequent inflammatory sequelae. Clin Immunol 2023;246:109181.

[101] Tuijnenburg P, Lango Allen H, Burns SO, Greene D, Jansen MH, et al. Loss-of-function nuclear factor kappaB subunit 1 (NFKB1) variants are the most common monogenic cause of common variable immunodeficiency in Europeans. J Allergy Clin Immunol 2018;142:1285−96.

[102] Doffinger R, Smahi A, Bessia C, Geissmann F, Feinberg J, et al. X-linked anhidrotic ectodermal dysplasia with immunodeficiency is caused by impaired NF-kappaB signaling. Nat Genet 2001;27:277−85.

[103] Jain A, Ma CA, Liu S, Brown M, Cohen J, Strober W. Specific missense mutations in NEMO result in hyper-IgM syndrome with hypohydrotic ectodermal dysplasia. Nat Immunol 2001;2:223−8.

[104] Jain A, Ma CA, Lopez-Granados E, Means G, Brady W, et al. Specific NEMO mutations impair CD40-mediated c-Rel activation and B cell terminal differentiation. J Clin Invest 2004;114:1593−602.

[105] Hanson EP, Monaco-Shawver L, Solt LA, Madge LA, Banerjee PP, May MJ, et al. Hypomorphic nuclear factor-kappaB essential modulator mutation database and reconstitution system identifies phenotypic and immunologic diversity. J Allergy Clin Immunol 2008;122:1169−1177 e1116.

[106] Orange JS, Jain A, Ballas ZK, Schneider LC, Geha RS, Bonilla FA. The presentation and natural history of immunodeficiency caused by nuclear factor kappaB essential modulator mutation. J Allergy Clin Immunol 2004;113:725−33.

[107] Ma CS, Wong N, Rao G, Avery DT, Torpy J, et al. Monogenic mutations differentially affect the quantity and quality of T follicular helper cells in patients with human primary immunodeficiencies. J Allergy Clin Immunol 2015;136:993−1006 e1001.

[108] How KN, Leong HJY, Pramono ZAD, Leong KF, Lai ZW, Yap WH. Uncovering incontinentia pigmenti: from DNA sequence to pathophysiology. Front Pediatr 2022;10:900606.

[109] Bastard P, Rosen LB, Zhang Q, Michailidis E, Hoffmann HH, et al. Autoantibodies against type I IFNs in patients with life-threatening COVID-19. Science 2020;370.

[110] Ma CS, Wong N, Rao G, Nguyen A, Avery DT, et al. Unique and shared signaling pathways cooperate to regulate the differentiation of human CD4+ T cells into distinct effector subsets. J Exp Med 2016;213:1589−608.

[111] Levy R, Langlais D, Beziat V, Rapaport F, Rao G, et al. Inherited human c-Rel deficiency disrupts myeloid and lymphoid immunity to multiple infectious agents. J Clin Invest 2021;131.

[112] Beaussant-Cohen S, Jaber F, Massaad MJ, Weeks S, Jones J, et al. Combined immunodeficiency in a patient with c-Rel deficiency. J Allergy Clin Immunol 2019;144:606−608 e604.

[113] Lu HY, Bauman BM, Arjunaraja S, Dorjbal B, Milner JD, Snow AL, et al. The CBM-opathies-A rapidly expanding spectrum of human inborn errors of immunity caused by mutations in the CARD11-BCL10-MALT1 complex. Front Immunol 2018;9:2078.

[114] Lu HY, Biggs CM, Blanchard-Rohner G, Fung SY, Sharma M, Turvey SE. Germline CBM-opathies: from immunodeficiency to atopy. J Allergy Clin Immunol 2019;143:1661−73.

[115] Greil J, Rausch T, Giese T, Bandapalli OR, Daniel V, et al. Whole-exome sequencing links caspase recruitment domain 11 (CARD11) inactivation to severe combined immunodeficiency. J Allergy Clin Immunol 2013;131:1376−1383 e1373.

[116] Stepensky P, Keller B, Buchta M, Kienzler AK, Elpeleg O, et al. Deficiency of caspase recruitment domain family, member 11 (CARD11), causes profound combined immunodeficiency in human subjects. J Allergy Clin Immunol 2013;131:477−485 e471.

[117] Nguyen A, Lu HY, Consortium U-N-K, Turvey SE, Snow AL. Autosomal recessive inflammatory skin disease caused by a novel biallelic loss-of-function variant in CARD11. J Clin Immunol 2023;43:709−13.

[118] Lu HY, Sharma M, Sharma AA, Lacson A, Szpurko A, et al. Mechanistic understanding of the combined immunodeficiency in complete human CARD11 deficiency. J Allergy Clin Immunol 2021;148:1559−1574 e1513.

[119] Jabara HH, Ohsumi T, Chou J, Massaad MJ, Benson H, et al. A homozygous mucosa-associated lymphoid tissue 1 (MALT1) mutation in a family with combined immunodeficiency. J Allergy Clin Immunol 2013;132:151−8.

[120] McKinnon ML, Rozmus J, Fung SY, Hirschfeld AF, Del Bel KL, et al. Combined immunodeficiency associated with homozygous MALT1 mutations. J Allergy Clin Immunol 2014;133:1458−62. 1462 e1451−1457.

[121] Punwani D, Wang H, Chan AY, Cowan MJ, Mallott J, et al. Combined immunodeficiency due to MALT1 mutations, treated by hematopoietic cell transplantation. J Clin Immunol 2015;35:135−46.

[122] Frizinsky S, Rechavi E, Barel O, Najeeb RH, Greenberger S, et al. Novel MALT1 mutation linked to immunodeficiency, immune dysregulation, and an abnormal T cell receptor repertoire. J Clin Immunol 2019;39:401−13.

[123] Kutukculer N, Seeholzer T, O'Neill TJ, Grass C, Aykut A, et al. Human immune disorder associated with homozygous hypomorphic mutation affecting MALT1B splice variant. J Allergy Clin Immunol 2021;147:775−778 e778.

[124] Sonoda M, Ishimura M, Eguchi K, Yada Y, Lenhartova N, et al. Progressive B cell depletion in human MALT1 deficiency. Clin Exp Immunol 2021;206:237−47.

[125] Charbit-Henrion F, Jeverica AK, Begue B, Markelj G, Parlato M, et al. Deficiency in mucosa-associated lymphoid tissue lymphoma translocation 1: a novel cause of IPEX-like syndrome. J Pediatr Gastroenterol Nutr 2017;64:378−84.

[126] Garcia-Solis B, Van Den Rym A, Perez-Caraballo JJ, Al-Ayoubi A, Alazami AM, et al. Clinical and immunological features of human BCL10 deficiency. Front Immunol 2021;12:786572.

[127] Torres JM, Martinez-Barricarte R, Garcia-Gomez S, Mazariegos MS, Itan Y, et al. Inherited BCL10 deficiency impairs hematopoietic and nonhematopoietic immunity. J Clin Invest 2014;124:5239–48.

[128] Van Den Rym A, Taur P, Martinez-Barricarte R, Lorenzo L, Puel A, et al. Human BCL10 deficiency due to homozygosity for a rare allele. J Clin Immunol 2020;40:388–98.

[129] Al-Tamemi S, Alhinai Z, Al-Rahbi N, Al-Abdawani R, Al-Yazidi L, et al. BCL10 loss-of-function novel mutation leading to atypical severe combined immunodeficiency. Clin Immunol 2022;241:109067.

[130] Darte JM, McClure PD, Saunders EF, Weber JL, Donohue WL. Congenital lymphoid hyperplasia with persistent hyper-lymphocytosis. N Engl J Med 1971;284:431–2.

[131] Snow AL, Xiao W, Stinson JR, Lu W, Chaigne-Delalande B, et al. Congenital B cell lymphocytosis explained by novel germline CARD11 mutations. J Exp Med 2012;209:2247–61.

[132] Shields AM, Bauman BM, Hargreaves CE, Pollard AJ, Snow AL, Patel SY. A novel, heterozygous three base-pair deletion in CARD11 results in B cell expansion with NF-kappaB and T cell anergy disease. J Clin Immunol 2020;40:406–11.

[133] Arjunaraja S, Nose BD, Sukumar G, Lott NM, Dalgard CL, Snow AL. Intrinsic plasma cell differentiation defects in B cell expansion with NF-kappaB and T cell anergy patient B cells. Front Immunol 2017;8:913.

[134] Chen K, Coonrod EM, Kumanovics A, Franks ZF, Durtschi JD, et al. Germline mutations in NFKB2 implicate the noncanonical NF-kappaB pathway in the pathogenesis of common variable immunodeficiency. Am J Hum Genet 2013;93:812–24.

[135] Liu Y, Hanson S, Gurugama P, Jones A, Clark B, Ibrahim MA. Novel NFKB2 mutation in early-onset CVID. J Clin Immunol 2014;34:686–90.

[136] Brue T, Quentien MH, Khetchoumian K, Bensa M, Capo-Chichi JM, et al. Mutations in NFKB2 and potential genetic heterogeneity in patients with DAVID syndrome, having variable endocrine and immune deficiencies. BMC Med Genet 2014;15:139.

[137] Lee CE, Fulcher DA, Whittle B, Chand R, Fewings N, et al. Autosomal-dominant B-cell deficiency with alopecia due to a mutation in NFKB2 that results in nonprocessable p100. Blood 2014;124:2964–72.

[138] Klemann C, Camacho-Ordonez N, Yang L, Eskandarian Z, Rojas-Restrepo JL, et al. Clinical and immunological phenotype of patients with primary immunodeficiency due to damaging mutations in NFKB2. Front Immunol 2019;10:297.

[139] Lougaris V, Moratto D, Baronio M, Lorenzini T, Rossi S, et al. Early B cell developmental impairment with progressive B cell deficiency in NFKB2 mutated CVID disease without autoimmunity. Clin Immunol 2019;205:153–5.

[140] Willmann KL, Klaver S, Dogu F, Santos-Valente E, Garncarz W, et al. Biallelic loss-of-function mutation in NIK causes a primary immunodeficiency with multifaceted aberrant lymphoid immunity. Nat Commun 2014;5:5360.

[141] Sharfe N, Merico D, Karanxha A, Macdonald C, Dadi H, et al. The effects of RelB deficiency on lymphocyte development and function. J Autoimmun 2015;65:90–100.

[142] Warnatz K, Salzer U, Rizzi M, Fischer B, Gutenberger S, et al. B-cell activating factor receptor deficiency is associated with an adult-onset antibody deficiency syndrome in humans. Proc Natl Acad Sci U S A 2009;106:13945–50.

[143] Tangye SG, Bryant VL, Cuss AK, Good KL. BAFF, APRIL and human B cell disorders. Semin Immunol 2006;18:305–17.

[144] Bogaert DJ, Dullaers M, Lambrecht BN, Vermaelen KY, De Baere E, Haerynck F. Genes associated with common variable immunodeficiency: one diagnosis to rule them all? J Med Genet 2016;53:575–90.

[145] Yeh TW, Okano T, Naruto T, Yamashita M, Okamura M, et al. APRIL-dependent lifelong plasmacyte maintenance and immunoglobulin production in humans. J Allergy Clin Immunol 2020;146:1109–1120 e1104.

[146] van Zelm MC, Bartol SJ, Driessen GJ, Mascart F, Reisli I, et al. Human CD19 and CD40L deficiencies impair antibody selection and differentially affect somatic hypermutation. J Allergy Clin Immunol 2014;134:135–44.

[147] Wentink MWJ, van Zelm MC, van Dongen JJM, Warnatz K, van der Burg M. Deficiencies in the CD19 complex. Clin Immunol 2018;195:82–7.

[148] van Zelm MC, Reisli I, van der Burg M, Castano D, van Noesel CJ, et al. An antibody-deficiency syndrome due to mutations in the CD19 gene. N Engl J Med 2006;354:1901–12.

[149] Kanegane H, Agematsu K, Futatani T, Sira MM, Suga K, et al. Novel mutations in a Japanese patient with CD19 deficiency. Gene Immun 2007;8:663–70.

[150] Vince N, Boutboul D, Mouillot G, Just N, Peralta M, et al. Defects in the CD19 complex predispose to glomerulonephritis, as well as IgG1 subclass deficiency. J Allergy Clin Immunol 2011;127:538–41. e531-535.

[151] Skendros P, Rondeau S, Chateil JF, Bui S, Bocly V, et al. Misdiagnosed CD19 deficiency leads to severe lung disease. Pediatr Allergy Immunol 2014;25:603–6.

[152] Walker K, Mistry A, Watson CM, Nadat F, O'Callaghan E, et al. Inherited CD19 deficiency does not impair plasma cell formation or response to CXCL12. J Clin Immunol 2023;43(7):1543–56.

[153] van Zelm MC, Smet J, Adams B, Mascart F, Schandene L, et al. CD81 gene defect in humans disrupts CD19 complex formation and leads to antibody deficiency. J Clin Invest 2010;120:1265–74.

[154] Yang L, Liu P, Du H, Chen R, Zhou B, et al. Novel CD81 mutations in a Chinese patient led to IgA nephropathy and impaired BCR signaling. J Clin Immunol 2022;42:1672–84.

[155] Wentink MW, Lambeck AJ, van Zelm MC, Simons E, van Dongen JJ, et al. CD21 and CD19 deficiency: two defects in the same complex leading to different disease modalities. Clin Immunol 2015;161:120–7.

[156] Rosain J, Miot C, Lambert N, Rousselet MC, Pellier I, Picard C. CD21 deficiency in 2 siblings with recurrent respiratory infections and hypogammaglobulinemia. J Allergy Clin Immunol Pract 2017;5:1765–1767 e1763.

[157] Thiel J, Kimmig L, Salzer U, Grudzien M, Lebrecht D, et al. Genetic CD21 deficiency is associated with hypogammaglobulinemia. J Allergy Clin Immunol 2012;129:801–810 e806.

[158] Paley MA, Strand V, Kim AH. From mechanism to therapies in systemic lupus erythematosus. Curr Opin Rheumatol 2017;29:178–86.

[159] Kuijpers TW, Bende RJ, Baars PA, Grummels A, Derks IA, et al. CD20 deficiency in humans results in impaired T cell-independent antibody responses. J Clin Invest 2010;120:214–22.

[160] Crotty S. Follicular helper CD4 T cells (TFH). Annu Rev Immunol 2011;29:621–63.

[161] Tangye SG, Ma CS, Brink R, Deenick EK. The good, the bad and the ugly – TFH cells in human health and disease. Nat Rev Immunol 2013;13:412–26.

[162] Crotty S. T follicular helper cell biology: a decade of discovery and diseases. Immunity 2019;50:1132–48.

[163] Ma CS, Deenick EK, Batten M, Tangye SG. The origins, function, and regulation of T follicular helper cells. J Exp Med 2012;209:1241–53.

[164] Ma CS, Uzel G, Tangye SG. Human T follicular helper cells in primary immunodeficiencies. Curr Opin Pediatr 2014;26:720–6.

[165] Tangye SG, Ma CS. Molecular regulation and dysregulation of T follicular helper cells – learning from inborn errors of immunity. Curr Opin Immunol 2021;72:249–61.

[166] Schepp J, Chou J, Skrabl-Baumgartner A, Arkwright PD, Engelhardt KR, et al. 14 years after discovery: clinical follow-up on 15 patients with inducible co-stimulator deficiency. Front Immunol 2017;8:964.

[167] Grimbacher B, Hutloff A, Schlesier M, Glocker E, Warnatz K, et al. Homozygous loss of ICOS is associated with adult-onset common variable immunodeficiency. Nat Immunol 2003;4:261–8.

[168] Salzer E, Kansu A, Sic H, Majek P, Ikinciogullari A, et al. Early-onset inflammatory bowel disease and common variable immunodeficiency-like disease caused by IL-21 deficiency. J Allergy Clin Immunol 2014;133:1651–1659 e1612.

[169] Ma CS, Nichols KE, Tangye SG. Regulation of cellular and humoral immune responses by the SLAM and SAP families of molecules. Annu Rev Immunol 2007;25:337–79.

[170] Roussel L, Landekic M, Golizeh M, Gavino C, Zhong MC, et al. Loss of human ICOSL results in combined immunodeficiency. J Exp Med 2018;215:3151–64.

[171] Kotlarz D, Zietara N, Uzel G, Weidemann T, Braun CJ, et al. Loss-of-function mutations in the IL-21 receptor gene cause a primary immunodeficiency syndrome. J Exp Med 2013;210:433–43.

[172] Cagdas D, Mayr D, Baris S, Worley L, Langley DB, et al. Genomic spectrum and phenotypic heterogeneity of human IL-21 receptor deficiency. J Clin Immunol 2021;41:1272–90.

[173] International Consortium I, Fornes O, Jia A, Kuehn HS, Min Q, et al. A multimorphic mutation in IRF4 causes human autosomal dominant combined immunodeficiency. Sci Immunol 2023;8:eade7953.

[174] Wang Y, Ma CS, Ling Y, Bousfiha A, Camcioglu Y, et al. Dual T cell- and B cell-intrinsic deficiency in humans with biallelic RLTPR mutations. J Exp Med 2016;213:2413–35.

[175] Akiba H, Takeda K, Kojima Y, Usui Y, Harada N, et al. The role of ICOS in the CXCR5+ follicular B helper T cell maintenance in vivo. J Immunol 2005;175:2340–8.

[176] Ferguson SE, Han S, Kelsoe G, Thompson CB. CD28 is required for germinal center formation. J Immunol 1996;156:4576–81.

[177] Béziat V, et al. Inherited CD28 deficiency in otherwise healthy patients with disseminated warts and giant horns. Cell 2021;184(14):3812–3828.e30.

[178] Angulo I, Vadas O, Garcon F, Banham-Hall E, Plagnol V, et al. Phosphoinositide 3-kinase delta gene mutation predisposes to respiratory infection and airway damage. Science 2013;342:866–71.

[179] Lucas CL, Kuehn HS, Zhao F, Niemela JE, Deenick EK, et al. Dominant-activating germline mutations in the gene encoding the PI(3)K catalytic subunit p110delta result in T cell senescence and human immunodeficiency. Nat Immunol 2014;15:88–97.

[180] Deau MC, Heurtier L, Frange P, Suarez F, Bole-Feysot C, et al. A human immunodeficiency caused by mutations in the PIK3R1 gene. J Clin Invest 2014;124:3923–8.

[181] Lucas CL, Zhang Y, Venida A, Wang Y, Hughes J, et al. Heterozygous splice mutation in PIK3R1 causes human immunodeficiency with lymphoproliferation due to dominant activation of PI3K. J Exp Med 2014;211:2537–47.

[182] Dulau Florea AE, Braylan RC, Schafernak KT, Williams KW, Daub J, et al. Abnormal B-cell maturation in the bone marrow of patients with germline mutations in PIK3CD. J Allergy Clin Immunol 2017;139:1032–1035 e1036.

[183] Avery DT, Kane A, Nguyen T, Lau A, Nguyen A, et al. Germline-activating mutations in PIK3CD compromise B cell development and function. J Exp Med 2018;215:2073–95.

[184] Nguyen T. Aberrant lymphocyte differentiation due to human PIK3R1 mutations underlies activated PI3Kd syndrome 2. J Exp Med 2023 (in press).

[185] Rao VK, Webster S, Sediva A, Plebani A, Schuetz C, et al. Randomized, placebo-controlled, phase 3 trial of PI3Kdelta inhibitor leniolisib for activated PI3Kdelta syndrome. Blood 2023;141(9):971–83.

[186] Tsujita Y, Mitsui-Sekinaka K, Imai K, Yeh TW, Mitsuiki N, et al. Phosphatase and tensin homolog (PTEN) mutation can cause activated phosphatidylinositol 3-kinase delta syndrome-like immunodeficiency. J Allergy Clin Immunol 2016;138:1672–1680 e1610.

[187] Browning MJ, Chandra A, Carbonaro V, Okkenhaug K, Barwell J. Cowden's syndrome with immunodeficiency. J Med Genet 2015;52:856–9.

[188] Driessen GJ, IJspeert H, Wentink M, Yntema HG, van Hagen PM, et al. Increased PI3K/Akt activity and deregulated humoral immune response in human PTEN deficiency. J Allergy Clin Immunol 2016;138:1744–1747 e1745.

[189] Chen HH, Handel N, Ngeow J, Muller J, Huhn M, et al. Immune dysregulation in patients with PTEN hamartoma tumor syndrome: analysis of FOXP3 regulatory T cells. J Allergy Clin Immunol 2017;139:607–620 e615.

[190] Moens L, Tangye SG. Cytokine-mediated regulation of plasma cell generation: IL-21 takes center stage. Front Immunol 2014;5:65.

[191] Tangye SG, Ma CS. Regulation of the germinal center and humoral immunity by interleukin-21. J Exp Med 2020;217.

[192] Akdis M, Aab A, Altunbulakli C, Azkur K, Costa RA, et al. Interleukins (from IL-1 to IL-38), interferons, transforming growth factor beta, and TNF-alpha: receptors, functions, and roles in diseases. J Allergy Clin Immunol 2016;138:984–1010.

[193] Catalan-Dibene J, McIntyre LL, Zlotnik A. Interleukin 30 to interleukin 40. J Interferon Cytokine Res 2018;38:423–39.

[194] Crow MK, Olferiev M, Kirou KA. Type I interferons in autoimmune disease. Annu Rev Pathol 2019;14:369–93.

[195] Locksley RM, Killeen N, Lenardo MJ. The TNF and TNF receptor superfamilies: integrating mammalian biology. Cell 2001;104:487–501.

[196] Metcalf D. Hematopoietic cytokines. Blood 2008;111:485–91.

[197] Levy DE, Darnell Jr JE. Stats: transcriptional control and biological impact. Nat Rev Mol Cell Biol 2002;3:651–62.

[198] Shuai K, Liu B. Regulation of JAK-STAT signalling in the immune system. Nat Rev Immunol 2003;3:900–11.

[199] O'Shea JJ, Holland SM, Staudt LM. JAKs and STATs in immunity, immunodeficiency, and cancer. N Engl J Med 2013;368:161–70.

[200] Casanova JL, Holland SM, Notarangelo LD. Inborn errors of human JAKs and STATs. Immunity 2012;36:515–28.

[201] Minegishi Y, Saito M, Tsuchiya S, Tsuge I, Takada H, et al. Dominant-negative mutations in the DNA-binding domain of STAT3 cause hyper-IgE syndrome. Nature 2007;448:1058–62.

[202] Holland SM, DeLeo FR, Elloumi HZ, Hsu AP, Uzel G, et al. STAT3 mutations in the hyper-IgE syndrome. N Engl J Med 2007;357:1608–19.

[203] Beziat V, Li J, Lin JX, Ma CS, Li P, et al. A recessive form of hyper-IgE syndrome by disruption of ZNF341-dependent STAT3 transcription and activity. Sci Immunol 2018;3.

[204] Frey-Jakobs S, Hartberger JM, Fliegauf M, Bossen C, Wehmeyer ML, et al. ZNF341 controls STAT3 expression and thereby immunocompetence. Sci Immunol 2018;3.

[205] Chandesris MO, Melki I, Natividad A, Puel A, Fieschi C, et al. Autosomal dominant STAT3 deficiency and hyper-IgE syndrome: molecular, cellular, and clinical features from a French national survey. Medicine (Baltim) 2012;91:e1–19.

[206] Chen YH, Spencer S, Laurence A, Thaventhiran JE, Uhlig HH. Inborn errors of IL-6 family cytokine responses. Curr Opin Immunol 2021;72:135–45.

[207] Avery DT, Deenick EK, Ma CS, Suryani S, Simpson N, et al. B cell-intrinsic signaling through IL-21 receptor and STAT3 is required for establishing long-lived antibody responses in humans. J Exp Med 2010;207:155–71.

[208] Leung DY, Ambrosino DM, Arbeit RD, Newton JL, Geha RS. Impaired antibody responses in the hyperimmunoglobulin E syndrome. J Allergy Clin Immunol 1988;81:1082–7.

[209] Sheerin KA, Buckley RH. Antibody responses to protein, polysaccharide, and phi X174 antigens in the hyperimmunoglobulinemia E (hyper-IgE) syndrome. J Allergy Clin Immunol 1991;87:803–11.

[210] Delmonte OM, Bergerson JRE, Burbelo PD, Durkee-Shock JR, Dobbs K, et al. Antibody responses to the SARS-CoV-2 vaccine in individuals with various inborn errors of immunity. J Allergy Clin Immunol 2021;148:1192–7.

[211] Deenick EK, Avery DT, Chan A, Berglund LJ, Ives ML, et al. Naive and memory human B cells have distinct requirements for STAT3 activation to differentiate into antibody-secreting plasma cells. J Exp Med 2013;210:2739–53.

[212] Berglund LJ, Avery DT, Ma CS, Moens L, Deenick EK, et al. IL-21 signalling via STAT3 primes human naive B cells to respond to IL-2 to enhance their differentiation into plasmablasts. Blood 2013;122:3940–50.

[213] Beziat V, Tavernier SJ, Chen YH, Ma CS, Materna M, et al. Dominant-negative mutations in human IL6ST underlie hyper-IgE syndrome. J Exp Med 2020;217.

[214] Arlabosse T, Materna M, Riccio O, Schnider C, Angelini F, et al. New dominant-negative IL6ST variants expand the immunological and clinical spectrum of GP130-dependent hyper-IgE syndrome. J Clin Immunol 2023;43(7):1566–80.

[215] Toubiana J, Okada S, Hiller J, Oleastro M, Lagos Gomez M, et al. Heterozygous STAT1 gain-of-function mutations underlie an unexpectedly broad clinical phenotype. Blood 2016;127:3154–64.

[216] Bloomfield M, Parackova Z, Hanzlikova J, Lastovicka J, Sediva A. Immunogenicity and safety of COVID-19 mRNA vaccine in STAT1 GOF patients. J Clin Immunol 2022;42:266–9.

[217] Zheng J, van de Veerdonk FL, Crossland KL, Smeekens SP, Chan CM, et al. Gain-of-function STAT1 mutations impair STAT3 activity in patients with chronic mucocutaneous candidiasis (CMC). Eur J Immunol 2015;45:2834–46.

[218] Zhang Y, Ma CA, Lawrence MG, Break TJ, O'Connell MP, et al. PD-L1 up-regulation restrains Th17 cell differentiation in STAT3 loss- and STAT1 gain-of-function patients. J Exp Med 2017;214:2523–33.

[219] Lobo PB, Guisado-Hernandez P, Villaoslada I, de Felipe B, Carreras C, et al. Ex vivo effect of JAK inhibition on JAK-STAT1 pathway hyperactivation in patients with dominant-negative STAT3 mutations. J Clin Immunol 2022;42:1193–204.

[220] Tangye SG, Pelham SJ, Deenick EK, Ma CS. Cytokine-mediated regulation of human lymphocyte development and function: insights from primary immunodeficiencies. J Immunol 2017;199:1949–58.

[221] Erman B, Bilic I, Hirschmugl T, Salzer E, Cagdas D, et al. Combined immunodeficiency with CD4 lymphopenia and sclerosing cholangitis caused by a novel loss-of-function mutation affecting IL21R. Haematologica 2015;100:e216–9.

[222] Stepensky P, Keller B, Abuzaitoun O, Shaag A, Yaacov B, et al. Extending the clinical and immunological phenotype of human Interleukin-21 receptor deficiency. Haematologica 2015;100(2):e72–6.

[223] Spencer S, Kostel Bal S, Egner W, Lango Allen H, Raza SI, et al. Loss of the interleukin-6 receptor causes immunodeficiency, atopy, and abnormal inflammatory responses. J Exp Med 2019;216:1986–98.

[224] Nahum A, Sharfe N, Broides A, Dadi H, Naghdi Z, et al. Defining the biological responses of IL-6 by the study of a novel IL-6 receptor chain immunodeficiency. J Allergy Clin Immunol 2020;145:1011–1015 e1016.

[225] Schwerd T, Twigg SRF, Aschenbrenner D, Manrique S, Miller KA, et al. A biallelic mutation in IL6ST encoding the GP130 co-receptor causes immunodeficiency and craniosynostosis. J Exp Med 2017;214:2547–62.

[226] Shahin T, Aschenbrenner D, Cagdas D, Bal SK, Conde CD, et al. Selective loss of function variants in IL6ST cause Hyper-IgE syndrome with distinct impairments of T-cell phenotype and function. Haematologica 2019;104:609–21.

[227] Martinez-Barricarte R, Markle JG, Ma CS, Deenick EK, Ramirez-Alejo N, et al. Human IFN-gamma immunity to mycobacteria is governed by both IL-12 and IL-23. Sci Immunol 2018;3.

[228] Philippot Q, Ogishi M, Bohlen J, Puchan J, Arias AA, et al. Human IL-23 is essential for IFN-gamma-dependent immunity to mycobacteria. Sci Immunol 2023;8:eabq5204.

[229] Puel A, Leonard WJ. Mutations in the gene for the IL-7 receptor result in T(-)B(+)NK(+) severe combined immunodeficiency disease. Curr Opin Immunol 2000;12:468–73.

[230] Giliani S, Mori L, de Saint Basile G, Le Deist F, Rodriguez-Perez C, et al. Interleukin-7 receptor alpha (IL-7Ralpha) deficiency: cellular and molecular bases. Analysis of clinical,

immunological, and molecular features in 16 novel patients. Immunol Rev 2005;203:110–26.
[231] Kotlarz D, Beier R, Murugan D, Diestelhorst J, Jensen O, et al. Loss of interleukin-10 signaling and infantile inflammatory bowel disease: implications for diagnosis and therapy. Gastroenterology 2012;143:347–55.
[232] Sharifinejad N, Zaki-Dizaji M, Sepahvandi R, Fayyaz F, Dos Santos Vilela MM, et al. The clinical, molecular, and therapeutic features of patients with IL10/IL10R deficiency: a systematic review. Clin Exp Immunol 2022;208:281–91.
[233] Dorman SE, Picard C, Lammas D, Heyne K, van Dissel JT, et al. Clinical features of dominant and recessive interferon gamma receptor 1 deficiencies. Lancet 2004;364:2113–21.
[234] Kerner G, Rosain J, Guerin A, Al-Khabaz A, Oleaga-Quintas C, et al. Inherited human IFN-gamma deficiency underlies mycobacterial disease. J Clin Invest 2020;130:3158–71.
[235] de Beaucoudrey L, Samarina A, Bustamante J, Cobat A, Boisson-Dupuis S, et al. Revisiting human IL-12Rbeta1 deficiency: a survey of 141 patients from 30 countries. Medicine (Baltim) 2010;89:381–402.
[236] Johnson SE, Shah N, Panoskaltsis-Mortari A, LeBien TW. Murine and human IL-7 activate STAT5 and induce proliferation of normal human pro-B cells. J Immunol 2005;175:7325–31.
[237] Parrish YK, Baez I, Milford TA, Benitez A, Galloway N, et al. IL-7 dependence in human B lymphopoiesis increases during progression of ontogeny from cord blood to bone marrow. J Immunol 2009;182:4255–66.
[238] Recher M, Berglund LJ, Avery DT, Cowan MJ, Gennery AR, et al. IL-21 is the primary common gamma chain-binding cytokine required for human B-cell differentiation in vivo. Blood 2011;118:6824–35.
[239] Alcantara-Montiel JC, Staines-Boone T, Lopez-Herrera G, Berron-Ruiz L, Borrego-Montoya CR, Santos-Argumedo L. Somatic mosaicism in B cells of a patient with autosomal dominant hyper IgE syndrome. Eur J Immunol 2016;46:2438–43.
[240] Dvorscek AR, McKenzie CI, Robinson MJ, Ding Z, Pitt C, et al. IL-21 has a critical role in establishing germinal centers by amplifying early B cell proliferation. EMBO Rep 2022;23:e54677.
[241] Quast I, Dvorscek AR, Pattaroni C, Steiner TM, McKenzie CI, et al. Interleukin-21, acting beyond the immunological synapse, independently controls T follicular helper and germinal center B cells. Immunity 2022;55:1414–1430 e1415.
[242] Guerin A, Kerner G, Marr N, Markle JG, Fenollar F, et al. IRF4 haploinsufficiency in a family with Whipple's disease. Elife 2018;7.
[243] Thouenon R, Chentout L, Moreno-Corona N, Poggi L, Lombardi EP, et al. A neomorphic mutation in the interferon activation domain of IRF4 causes a dominant primary immunodeficiency. J Exp Med 2023;220.
[244] Luo W, Conter L, Elsner RA, Smita S, Weisel F, et al. IL-21R signal reprogramming cooperates with CD40 and BCR signals to select and differentiate germinal center B cells. Sci Immunol 2023;8:eadd1823.
[245] Klein U, Casola S, Cattoretti G, Shen Q, Lia M, et al. Transcription factor IRF4 controls plasma cell differentiation and class-switch recombination. Nat Immunol 2006;7:773–82.
[246] Haase P, Voehringer D. Regulation of the humoral type 2 immune response against allergens and helminths. Eur J Immunol 2021;51:273–9.
[247] Junttila IS. Tuning the cytokine responses: an update on interleukin (IL)-4 and IL-13 receptor complexes. Front Immunol 2018;9:888.
[248] Sharma M, Leung D, Momenilandi M, Jones LCW, Pacillo L, et al. Human germline heterozygous gain-of-function STAT6 variants cause severe allergic disease. J Exp Med 2023;220.
[249] Suratannon N, Ittiwut C, Dik WA, Ittiwut R, Meesilpavikkai K, et al. A germline STAT6 gain-of-function variant is associated with early-onset allergies. J Allergy Clin Immunol 2023;151(2):565–571.e9.
[250] Takeuchi I, Yanagi K, Takada S, Uchiyama T, Igarashi A, et al. STAT6 gain-of-function variant exacerbates multiple allergic symptoms. J Allergy Clin Immunol 2023;151(5):1402–1409.e6.
[251] Baris S, Benamar M, Chen Q, Catak MC, Martinez-Blanco M, et al. Severe allergic dysregulation due to a gain of function mutation in the transcription factor STAT6. J Allergy Clin Immunol 2023;152(1):182–194.e7.
[252] Wood PM, Fieschi C, Picard C, Ottenhoff TH, Casanova JL, Kumararatne DS. Inherited defects in the interferon-gamma receptor or interleukin-12 signalling pathways are not sufficient to cause allergic disease in children. Eur J Pediatr 2005;164:741–7.
[253] Bezrodnik L, Di Giovanni D, Caldirola MS, Azcoiti ME, Torgerson T, Gaillard MI. Long-term follow-up of STAT5B deficiency in three argentinian patients: clinical and immunological features. J Clin Immunol 2015;35:264–72.
[254] Foley CL, Al Remeithi SS, Towe CT, Dauber A, Backeljauw PF, et al. Developmental adaptive immune defects associated with STAT5B deficiency in three young siblings. J Clin Immunol 2021;41:136–46.
[255] Nadeau K, Hwa V, Rosenfeld RG. STAT5b deficiency: an unsuspected cause of growth failure, immunodeficiency, and severe pulmonary disease. J Pediatr 2011;158:701–8.
[256] Klammt J, Neumann D, Gevers EF, Andrew SF, Schwartz ID, et al. Dominant-negative STAT5B mutations cause growth hormone insensitivity with short stature and mild immune dysregulation. Nat Commun 2018;9:2105.
[257] Frischmeyer-Guerrerio PA, Guerrerio AL, Oswald G, Chichester K, Myers L, et al. TGFbeta receptor mutations impose a strong predisposition for human allergic disease. Sci Transl Med 2013;5:195ra194.
[258] Felgentreff K, Siepe M, Kotthoff S, von Kodolitsch Y, Schachtrup K, et al. Severe eczema and Hyper-IgE in Loeys-Dietz-syndrome – contribution to new findings of immune dysregulation in connective tissue disorders. Clin Immunol 2014;150:43–50.
[259] Loeys BL, Chen J, Neptune ER, Judge DP, Podowski M, et al. A syndrome of altered cardiovascular, craniofacial, neurocognitive and skeletal development caused by mutations in TGFBR1 or TGFBR2. Nat Genet 2005;37:275–81.
[260] Ma CS. T-helper-2 cells and atopic disease: lessons learnt from inborn errors of immunity. Curr Opin Immunol 2023;81:102298.
[261] Cancro MP. Age-associated B cells. Annu Rev Immunol 2020;38:315–40.
[262] Ambegaonkar AA, Holla P, Dizon BL, Sohn H, Pierce SK. Atypical B cells in chronic infectious diseases and systemic autoimmunity: puzzles with many missing pieces. Curr Opin Immunol 2022;77:102227.
[263] Woodruff MC, Ramonell RP, Nguyen DC, Cashman KS, Saini AS, et al. Extrafollicular B cell responses correlate with neutralizing

antibodies and morbidity in COVID-19. Nat Immunol 2020;21:1506–16.
[264] Woodruff MC, Ramonell RP, Haddad NS, Anam FA, Rudolph ME, et al. Dysregulated naive B cells and de novo autoreactivity in severe COVID-19. Nature 2022;611(7934):139–47.
[265] Austin JW, Buckner CM, Kardava L, Wang W, Zhang X, et al. Overexpression of T-bet in HIV infection is associated with accumulation of B cells outside germinal centers and poor affinity maturation. Sci Transl Med 2019;11.
[266] Lau D, Lan LY, Andrews SF, Henry C, Rojas KT, et al. Low CD21 expression defines a population of recent germinal center graduates primed for plasma cell differentiation. Sci Immunol 2017;2.
[267] Koutsakos M, Wheatley AK, Loh L, Clemens EB, Sant S, et al. Circulating TFH cells, serological memory, and tissue compartmentalization shape human influenza-specific B cell immunity. Sci Transl Med 2018;10.
[268] Holla P, Dizon B, Ambegaonkar AA, Rogel N, Goldschmidt E, et al. Shared transcriptional profiles of atypical B cells suggest common drivers of expansion and function in malaria, HIV, and autoimmunity. Sci Adv 2021;7.
[269] Ambegaonkar AA, Nagata S, Pierce SK, Sohn H. The differentiation in vitro of human tonsil B cells with the phenotypic and functional characteristics of T-bet+ atypical memory B cells in malaria. Front Immunol 2019;10:852.
[270] Keller B, Strohmeier V, Harder I, Unger S, Payne KJ, et al. The expansion of human T-bet(high)CD21(low) B cells is T cell dependent. Sci Immunol 2021;6:eabh0891.
[271] Yang R, Avery DT, Jackson KJL, Ogishi M, Benhsaien I, et al. Human T-bet governs the generation of a distinct subset of CD11c(high)CD21(low) B cells. Sci Immunol 2022;7:eabq3277.
[272] Pelham SJ, Caldirola MS, Avery DT, Mackie J, Rao G, et al. STAT5B restrains human B-cell differentiation to maintain humoral immune homeostasis. J Allergy Clin Immunol 2022;150:931–46.
[273] Lanzavecchia A. Antigen-specific interaction between T and B cells. Nature 1985;314:537–9.
[274] Thorley-Lawson DA. EBV persistence–introducing the virus. Curr Top Microbiol Immunol 2015;390:151–209.
[275] Kuppers R. B cells under influence: transformation of B cells by Epstein-Barr virus. Nat Rev Immunol 2003;3:801–12.
[276] Tangye SG, Latour S. Primary immunodeficiencies reveal the molecular requirements for effective host defense against EBV infection. Blood 2020;135:644–55.
[277] Bar RS, DeLor CJ, Clausen KP, Hurtubise P, Henle W, Hewetson JF. Fatal infectious mononucleosis in a family. N Engl J Med 1974;290:363–7.
[278] Purtilo DT, Cassel CK, Yang JP, Harper R. X-linked recessive progressive combined variable immunodeficiency (Duncan's disease). Lancet 1975;1:935–40.
[279] Provisor AJ, Iacuone JJ, Chilcote RR, Neiburger RG, Crussi FG. Acquired agammaglobulinemia after a life-threatening illness with clinical and laboratory features of infectious mononucleosis in three related male children. N Engl J Med 1975;293:62–5.
[280] Sumegi J, Huang D, Lanyi A, Davis JD, Seemayer TA, et al. Correlation of mutations of the SH2D1A gene and epstein-barr virus infection with clinical phenotype and outcome in X-linked lymphoproliferative disease. Blood 2000;96:3118–25.
[281] Booth C, Gilmour KC, Veys P, Gennery AR, Slatter MA, et al. X-linked lymphoproliferative disease due to SAP/SH2D1A deficiency: a multicenter study on the manifestations, management and outcome of the disease. Blood 2011;117:53–62.
[282] Tangye SG. XLP: clinical features and molecular etiology due to mutations in SH2D1A encoding SAP. J Clin Immunol 2014;34:772–9.
[283] Pachlopnik Schmid J, Canioni D, Moshous D, Touzot F, Mahlaoui N, et al. Clinical similarities and differences of patients with X-linked lymphoproliferative syndrome type 1 (XLP-1/SAP deficiency) versus type 2 (XLP-2/XIAP deficiency). Blood 2011;117:1522–9.
[284] Coffey AJ, Brooksbank RA, Brandau O, Oohashi T, Howell GR, et al. Host response to EBV infection in X-linked lymphoproliferative disease results from mutations in an SH2-domain encoding gene. Nat Genet 1998;20:129–35.
[285] Nichols KE, Harkin DP, Levitz S, Krainer M, Kolquist KA, et al. Inactivating mutations in an SH2 domain-encoding gene in X-linked lymphoproliferative syndrome. Proc Natl Acad Sci U S A 1998;95:13765–70.
[286] Sayos J, Wu C, Morra M, Wang N, Zhang X, et al. The X-linked lymphoproliferative-disease gene product SAP regulates signals induced through the co-receptor SLAM. Nature 1998;395:462–9.
[287] Cannons JL, Tangye SG, Schwartzberg PL. SLAM family receptors and SAP adaptors in immunity. Annu Rev Immunol 2011;29:665–705.
[288] Latour S, Veillette A. Molecular and immunological basis of X-linked lymphoproliferative disease. Immunol Rev 2003;192:212–24.
[289] Hislop AD, Palendira U, Leese AM, Arkwright PD, Rohrlich PS, et al. Impaired Epstein-Barr virus-specific CD8+ T-cell function in X-linked lymphoproliferative disease is restricted to SLAM family-positive B-cell targets. Blood 2010;116:3249–57.
[290] Palendira U, Low C, Chan A, Hislop AD, Ho E, et al. Molecular pathogenesis of EBV susceptibility in XLP as revealed by analysis of female carriers with heterozygous expression of SAP. PLoS Biol 2011;9:e1001187.
[291] Thorley-Lawson DA, Gross A. Persistence of the Epstein-Barr virus and the origins of associated lymphomas. N Engl J Med 2004;350:1328–37.
[292] Rickinson AB, Long HM, Palendira U, Munz C, Hislop AD. Cellular immune controls over Epstein-Barr virus infection: new lessons from the clinic and the laboratory. Trends Immunol 2014;35:159–69.
[293] Young LS, Rickinson AB. Epstein-Barr virus: 40 years on. Nat Rev Cancer 2004;4:757–68.
[294] Li FY, Chaigne-Delalande B, Kanellopoulou C, Davis JC, Matthews HF, et al. Second messenger role for Mg2+ revealed by human T-cell immunodeficiency. Nature 2011;475:471–6.
[295] Brigida I, Chiriaco M, Di Cesare S, Cittaro D, Di Matteo G, et al. Large deletion of MAGT1 gene in a patient with classic Kaposi sarcoma, CD4 lymphopenia, and EBV infection. J Clin Immunol 2017;37:32–5.
[296] Chaigne-Delalande B, Li FY, O'Connor GM, Lukacs MJ, Jiang P, et al. Mg2+ regulates cytotoxic functions of NK and CD8 T cells in chronic EBV infection through NKG2D. Science 2013;341:186–91.
[297] Li FY, Chaigne-Delalande B, Su H, Uzel G, Matthews H, Lenardo MJ. XMEN disease: a new primary immunodeficiency affecting Mg2+ regulation of immunity against Epstein-Barr virus. Blood 2014;123:2148–52.

[298] Dhalla F, Murray S, Sadler R, Chaigne-Delalande B, Sadaoka T, et al. Identification of a novel mutation in MAGT1 and progressive multifocal leucoencephalopathy in a 58-year-old man with XMEN disease. J Clin Immunol 2015;35:112–8.

[299] Patiroglu T, Haluk Akar H, Gilmour K, Unal E, Akif Ozdemir M, et al. A case of XMEN syndrome presented with severe autoimmune disorders mimicking autoimmune lymphoproliferative disease. Clin Immunol 2015;159:58–62.

[300] Dimitrova D, Rose JJ, Uzel G, Cohen JI, Rao KV, et al. Successful bone marrow transplantation for XMEN: hemorrhagic risk uncovered. J Clin Immunol 2019;39:1–3.

[301] Blommaert E, Peanne R, Cherepanova NA, Rymen D, Staels F, et al. Mutations in MAGT1 lead to a glycosylation disorder with a variable phenotype. Proc Natl Acad Sci USA 2019;116:9865–70.

[302] Ravell JC, Matsuda-Lennikov M, Chauvin SD, Zou J, Biancalana M, et al. Defective glycosylation and multisystem abnormalities characterize the primary immunodeficiency XMEN disease. J Clin Invest 2020;130:507–22.

[303] Hoyos-Bachiloglu R, Concha S, Sepulveda P, Campos R, Perez-Mateluna G, King A, et al. The many faces of XMEN disease, report of two patients with novel mutations. J Clin Immunol 2020;40(2):415–7.

[304] Klinken EM, Gray PE, Pillay B, Worley L, Edwards ESJ, et al. Diversity of XMEN disease: description of 2 novel variants and analysis of the lymphocyte phenotype. J Clin Immunol 2020;40(2):299–309.

[305] Ravell JC, Chauvin SD, He T, Lenardo M. An update on XMEN disease. J Clin Immunol 2020;40:671–81.

[306] Lanier LL. NKG2D receptor and its ligands in host defense. Cancer Immunol Res 2015;3:575–82.

[307] Tangye SG, Palendira U, Edwards ES. Human immunity against EBV-lessons from the clinic. J Exp Med 2017;214:269–83.

[308] Nolte MA, van Olffen RW, van Gisbergen KP, van Lier RA. Timing and tuning of CD27-CD70 interactions: the impact of signal strength in setting the balance between adaptive responses and immunopathology. Immunol Rev 2009;229:216–31.

[309] Izawa K, Martin E, Soudais C, Bruneau J, Boutboul D, et al. Inherited CD70 deficiency in humans reveals a critical role for the CD70-CD27 pathway in immunity to Epstein-Barr virus infection. J Exp Med 2017;214:73–89.

[310] van Montfrans JM, Hoepelman AI, Otto S, van Gijn M, van de Corput L, et al. CD27 deficiency is associated with combined immunodeficiency and persistent symptomatic EBV viremia. J Allergy Clin Immunol 2012;129:787–793 e786.

[311] Salzer E, Daschkey S, Choo S, Gombert M, Santos-Valente E, et al. Combined immunodeficiency with life-threatening EBV-associated lymphoproliferative disorder in patients lacking functional CD27. Haematologica 2013;98:473–8.

[312] Alkhairy OK, Perez-Becker R, Driessen GJ, Abolhassani H, van Montfrans J, et al. Novel mutations in TNFRSF7/CD27: clinical, immunologic, and genetic characterization of human CD27 deficiency. J Allergy Clin Immunol 2015;136:703–712 e710.

[313] Abolhassani H, Edwards ES, Ikinciogullari A, Jing H, Borte S, et al. Combined immunodeficiency and Epstein-Barr virus-induced B cell malignancy in humans with inherited CD70 deficiency. J Exp Med 2017;214:91–106.

[314] Caorsi R, Rusmini M, Volpi S, Chiesa S, Pastorino C, et al. CD70 deficiency due to a novel mutation in a patient with severe chronic EBV infection presenting as a periodic fever. Front Immunol 2017;8:2015.

[315] Alosaimi MF, Hoenig M, Jaber F, Platt CD, Jones J, et al. Immunodeficiency and EBV-induced lymphoproliferation caused by 4-1BB deficiency. J Allergy Clin Immunol 2019;144:574–583 e575.

[316] Somekh I, Thian M, Medgyesi D, Gulez N, Magg T, et al. CD137 deficiency causes immune dysregulation with predisposition to lymphomagenesis. Blood 2019;134(18):1510–6.

[317] Rodriguez R, Fournier B, Cordeiro DJ, Winter S, Izawa K, et al. Concomitant PIK3CD and TNFRSF9 deficiencies cause chronic active Epstein-Barr virus infection of T cells. J Exp Med 2019;216(12):2800–18.

[318] de Miranda NF, Georgiou K, Chen L, Wu C, Gao Z, et al. Exome sequencing reveals novel mutation targets in diffuse large B-cell lymphomas derived from Chinese patients. Blood 2014;124:2544–53.

[319] Scholtysik R, Nagel I, Kreuz M, Vater I, Giefing M, et al. Recurrent deletions of the TNFSF7 and TNFSF9 genes in 19p13.3 in diffuse large B-cell and Burkitt lymphomas. Int. J. Cancer [Journal international du cancer] 2012;131:E830–5.

Chapter 16

Memory B Cells and Plasma Cells

Mairi Anne McGrath[1], Florian Weisel[2], Andreas Radbruch[1,3] and Mark Shlomchik[2]

[1]German Rheumatism Research Centre Berlin (DRFZ), A Leibniz Institute, Berlin, Germany; [2]Department of Immunology, School of Medicine, University of Pittsburgh, Pittsburgh, PA, United States; [3]Charité – Universitätsmedizin Berlin, Berlin, Germany

1. INTRODUCTION

Immunological memory provided by memory B cells (MBCs), memory T cells and long-lived (memory) plasma cells (PCs) yields long-lasting, effective cellular and humoral immune protection to previously encountered pathogens and vaccines. MBCs provide a fast and potent anamnestic antibody response following a reencounter with antigen, thereby providing an extra arm of immunity to pathogens that are not cleared by pre-existing antibodies. In addition, they qualify as potent antigen-presenting and immunoregulatory cells in secondary immune reactions. We are only beginning to understand the heterogeneity of MBCs and their compartmentalisation. PCs persist, potentially for a lifetime, in specialised niches in the tissues, especially in bone marrow and gut, but also in inflamed tissue. The molecular mechanisms governing the long-term persistence of memory B cells and PCs, in the apparent absence of antigen, are only now beginning to be unravelled. While long-lived plasma cells (LLPCs) provide essential protection against pathogens, their longevity and resistance to irradiation, immunosuppression and therapies targeting B cells can be problematic for the treatment of chronic antibody-mediated diseases.

1.1 Generation of Memory B Cells and Plasma Cells

The immune state consists of antibodies (Ab, in serum and at mucosal surfaces), the cells that make Ab—PCs and plasmablasts (PBs) – and MBCs, the dynamic cellular component of the humoral immune state [1]. Preformed Ab will intercept pathogens prior to infection and can create 'sterilising' immunity. But, unless produced by relatively rarer LLPCs, Ab titres can wane after an initial infection. In addition, many pathogens change their antigenic profiles over time, potentially evading pre-existing Ab.

When and where do MBC, PB and LLPC emerge during an immune response? It has long been known that PBs are created rather early during various types of immune responses [2,3]. These PBs are from the early extrafollicular response and are short-lived, with a half-life of only a few days. They are generally IgM, but can undergo isotype switching, even as early as day 2–3 of the immune response [4]. The extent of switching depends on the nature of the immunogen and type of immune response. Infection of mice with *Salmonella* yields a very robust extrafollicular response with large numbers of PB, many of which begin to express IgG within the first and second week of the response [5,6]. These PBs can also be found circulating in blood and in humans, a peak PB response to secondary immunisation (as in a vaccine response) is often seen around a week after exposure [7,8]. From a variety of studies, it has become apparent that MBCs are also made during this time, in a germinal centre-(GC) independent manner [9–15].

Perhaps most convincing regarding the timing of MBC generation are in vivo 5-bromo-2′-deoxyuridine (BrdU) pulse-labelling experiments performed in a synchronised transfer immunisation scheme, monitoring the response to the hapten 4-hydroxy-3-nitrophenyl acetyl (NP), delivered conjugated to chicken gamma globulin formulated in alum adjuvant [12]. In this response, nearly half of all MBCs are generated in the first 5 days of the response, with about one quarter formed even as early as the first 2 days. In most T-dependent contexts, the immune response gradually shifts to the GC, with early precursors detected anywhere from day 4 to day 7 [16–19] after initial exposure to an immunogen. MBCs continue to be generated during this time, with a second discernible peak of MBC formation in the above-described transfer system just before the GC peaks in size.

The response also shifts from generation of IgM MBC, which predominate early, to a mixture of IgM and IgG

Molecular Biology of B Cells. https://doi.org/10.1016/B978-0-323-95895-0.00011-8

MBC, with most IgG MBC being made around when the GC is the dominant response. Notably, MBC subsets defined by CD80 and programmed cell death-ligand 2 (PD-L2) expression also arise in temporal order, with CD80/PD-L2 double negative (DN) MBC being made very early, followed by a wave of CD80⁻ PD-L2$^+$ single positive (SP) MBC and then CD80$^+$ PD-L2$^+$ double positive (DP) MBC, which comprise the majority of MBC made at later time points. The GC also begins to generate antibody-forming cells (AFCs), although most, if not all, of the early AFCs are likely to be short-lived PB that populate spleen, blood and migrate to the bone marrow. At a much later time, as the GC is involuting, the output switches to PC, some of which are destined to be long-lived [12]. Many of these will migrate to the bone marrow where, if they find the right niches, will assume an oxidative phosphorylation metabolic state [20] and continue to secrete Ab for an indefinite period. Most data indicate that all, or nearly all, of the LLPC that are generated in a T-dependent response derive from GC precursors [12,21]. Consistent with this, eliminating the GC using anti-CD40L at a time before most LLPC had formed had a drastic effect on LLPC numbers [12,22], but it could not be excluded that the treatment also eliminated a cryptic non-GC site that also contributed. Similarly, when B-cell lymphoma 6 (Bcl6) is deleted, thus precluding GC formation, LLPCs are also depleted [23,24]. T-independent (TI) responses can also generate both MBC and LLPC [24–26]. Although some TI-responses can generate transient and early appearing GC responses [24,27,28], it is less clear whether the MBCs generated from such responses are GC-dependent, -independent or both [24,29]. The subset quality of MBC requires further study, although it appears that some TI-derived MBCs have IgG isotypes [29]. Similarly, in both mice and humans, responses to TI antigens, such as polysaccharides, can lead to long-lived AFC and antibodies [25,26,29,30]; however, a large proportion of the initial response is composed of short-lived PB.

2. HETEROGENEITY OF MEMORY B CELLS

In the study of MBC, it is important to be able to distinguish them from their naïve B cell (NBC) counterparts. Resting antigen-experienced, that is, 'memory' B cells can be distinguished from NBC if they have switched their BCR isotype from IgM/IgD to any of the downstream isotypes, such as IgG or IgA. In early studies [31] and to this day [32], this has been a major useful method of isolating or studying a population of MBC. However, this approach failed to identify IgM MBC, which can be a prevalent population in both mice and humans [13,15,23,33–40]. A more desirable approach is to identify a surface marker, or combination of surface markers, that correlates with MBC identity. In 1998, both CD27 and CD148 were identified as marking most, if not all, MBC in humans [34,41]. CD27, in particular, has been extensively used as a marker for human MBC. In mice, surface markers that are expressed on all or most MBC have been elusive. Tomayko et al. published that CD73, PD-L2 and CD80 mark a substantial proportion of, but by no means all, murine MBC [37]; however, these markers have been used for partial tracking of the MBC compartment in many subsequent studies [13–15,24,38,39,42,43]. More recently, the laboratory of Mark Shlomchik published a compendium of both murine and human MBC surface phenotypes that includes both known and novel markers of MBC, including some markers that appear to be expressed on the great majority of MBC [40]. The knowledge of such markers should greatly facilitate future research into the function and nature of human and murine MBC in many settings.

Gene expression and flow cytometry-based phenotyping of MBC has revealed a surprising amount of heterogeneity among MBC in both humans and mice [40,44–48]. These observations raise the questions of how these phenotypically identified MBC 'subsets' differ in their function, origin and even location. One possible correlate of differential function is BCR isotype. In mice, both IgM and isotype-switched MBC have been identified. Variable (V) regions of IgM MBC may harbour mutations, but in general tend to be less mutated than those from isotype-switched MBC [13,33,49]. Dogan et al. showed that IgM MBCs are more prone to reenter GC upon reimmunisation compared to IgG MBC; they proposed therefore that BCR isotype may correlate with or determine function [50]. MBC elicited by a variety of immunisations and infections express CD73, PD-L2, CD80 or a combination of these [37,38,42,43]. Shlomchik and colleagues proposed the use of PD-L2 and CD80 to define up to four MBC subsets, acknowledging that the inclusion of CD73 would imply the existence of even more distinct subsets [37]. In immune responses to NP hapten-derivatised proteins given in alum adjuvant, three subsets predominate: PD-L2^{neg}CD80^{neg} 'double negative' (DN), PD-L2$^+$CD80^{neg} 'single positive' (SP) and PD-L2$^+$CD80$^+$ 'double positive' (DP). IgM$^+$ DN MBCs are prone to reenter GC, whereas IgM$^+$ DP MBCs do not generate a GC response upon reimmunisation; rather, DP are adept at creating a rapid and robust isotype-switched PB response. IgM SP MBCs are capable of both early PB generation as well as GC seeding [49]. Thus, MBC isotype by itself does not define functional capacity. DP MBCs are, however, enriched for IgG MBC, while DN MBCs are almost exclusively IgM, so there is some correlation with the data correlating IgM expression with GC repopulation capacity [49]. More recently, a population of murine MBC that expresses CD80 only has been appreciated [30,39,51,52]. This population seems to emerge under more inflammatory conditions, such as in infection and autoimmunity. The functional significance of this additional MBC subset is not yet clear.

CD73 is expressed on a large proportion of MBC. It has been proposed as a marker of GC-derived MBC, and indeed, CD73 MBCs tend to have V-region mutations [13]; but clearly MBCs that are formed at very early time points after immunisation, prior to GC formation, can express CD73. Conversely, not all MBCs formed at later time points express CD73. Hence, it is not a reliable marker for the origin of a particular MBC [12].

It is further notable that, although PD-L2 and CD80 are B7-family members that can interact with important CD28-family members expressed on T cells, such as programmed cell death protein-1 (PD-1), cytotoxic T-lymphocyte associated protein-4 (CTLA-4) and CD28, the functions of CD80 and PD-L2 on established MBC have not been determined. CD80 and PD-L2 expression do play a role in GC outcomes [53,54]. CD73 is important for extracellular adenosine production; it is notable that genes encoding multiple proteins in this pathway are differentially expressed in MBC vs NBC [45]. However, the function of CD73 is not known in the context of MBC, although CD73 expression on either GC B cells or follicular T-helper cells (TFH) can modulate the GC response output [55].

While the discovery and exploration of these markers has been of great interest, none of them can fully mark all, or nearly all, murine MBC in the way that CD27 marks most MBC in humans. Via a flow cytometry-based proteomics screen that monitored expression of 255 surface markers on murine MBC and 332 surface markers on human MBC, a number of novel surface proteins that are differentially expressed on murine MBC compared to NBC were identified. Some of these were expressed only on some MBC and may either be alternative subset markers or reveal yet more heterogeneity. However, some (such as CD180 and CD205) appeared to mark nearly all MBC and may function in the mouse system in a way similar to CD27 [40].

Human MBCs also display heterogeneity [48,56–58]. Like the mouse, both IgM and switched MBC have been identified, with IgM MBC being remarkably prominent, comprising the majority of $CD27^+$ MBC in some people. IgA MBCs are, as expected, most prominently found in gut mucosal lamina propria [48,59]. Whether there are functional differences between IgM and IgG MBC is not clear. So far, a clear association has not been made between murine MBC subsets and human MBC subsets.

Additional human MBC subsets, delineated after a comprehensive flow-based proteomics screen [40], revealed that CD45RB and CD69 expression could divide splenic $CD27^+$ B cells into four compartments [48]. CD45RB, the expression of which reflects a particular O-linked glycoepitope, had previously been identified as a surface marker expressed on a substantial proportion of $CD27^+$ cells in blood and tonsil [60]. CD69 is thought to mark tissue-residency and early activation among T cells. Among MBC, CD69 was expressed only on splenic, lymph node (LN)-derived and gut-derived MBC, but not on blood-derived or bone marrow-derived cells. While most tissues expressed all of the possible subsets, gut B cells were almost exclusively $CD27^+CD45RB^+CD69^+$ [48]. The functional significance and relationships among these subsets remain to be determined.

Human $IgM^+IgD^+CD27^+$ cells have been the subject of controversy as to whether they are human counterparts of mouse marginal zone B cells or else authentic IgM MBC [56,61–64]. An emerging consensus is that these cells are most likely authentic MBC. Much of the confusion may have stemmed from the preconceived notion that MBC must be GC-derived and highly mutated, which is in fact not the case.

In the last 10 years, a type of MBC generally known as age-associated B cells (ABCs) or 'atypical MBC' has become appreciated and increasingly studied. ABCs are a subset of B cells that were initially found in older female B6 mice and then subsequently associated with certain autoimmune states, like lupus, as well as certain infections, including Ehrlichia, SARS-CoV-2 and influenza [52,65–74]. These cells can best be identified in mice as $CD11b^+/CD11c^+$ [52]; they had in some cases been identified as being CD21 and CD23 negative [75], but this definition also includes other cell types. Many of these cells express the transcription factor T-bet, encoded by the gene Tbx21 and are thus considered ABC-like. The role of T-bet expression in these cells is not clear, as it appears in mice that they can be generated in the absence of T-bet and can function if T-bet is deleted in them [68,76], although T-bet is required for IgG2a/c switching and possibly for maintenance of some Abs [69,77]. This is an area of active research.

In humans, ABC-like cells are also $CD11b^+/CD11c^+$ and appear to lack the otherwise-canonical MBC marker CD27 [71,72]. Human B cells that are isotype-switched and thus Ag-experienced, but lack both CD27 and IgD (and thus are considered Ag-experienced by virtue of having switched isotype), have also been termed DN2 (indicated DN for both markers) [71]. In some contexts, human ABC-like cells express Fc-related inhibitory type receptors such as FCRL4 and FCRL5 [78]; they have sometimes been called 'exhausted', perhaps for this reason [79,80]. Although these cells are indeed not highly responsive to BCR stimuli, they are very sensitive to Toll-like receptor (TLR) stimulation [52,72,81,82], rendering this concept questionable in our view. They have also been termed 'atypical' MBC [83,84], and the circulating cells have even been called 'tissue-like' MBC, perhaps because similar cells were initially identified in the tonsil [78].

It is clear that these cells share many properties with MBC, including being Ag-experienced and harbouring V region mutations. From this perspective, they have been deemed to be MBC [65,69,70,84,85]. They are notable for responding relatively poorly to BCR signals but well to TLR signals [75]. The linkage of the accumulation of these cells to autoimmune and chronic infection states, however, calls into question whether all ABCs are truly resting MBC or whether they are an intermediate, memory-like cell that has been recently activated or is destined to soon be reactivated. The exact status of these cells is likely to be context-dependent, with ABC constituting a type of true MBC in post-infection states [39,68,69] while turning over more rapidly and mainly not constituting a resting MBC subset in the face of active autoimmunity. Indeed, we have recently found that ABCs in lupus-prone mice are a highly dynamic, proliferating population [52].

In lupus mice, it appears that ABC-like cells can be found among all of the MBC subsets that are defined by the expression of CD80 and PD-L2, although there is a preference for the DP and CD80 SP subsets among the CD11b/CD11c-positive ABC [52]. Hence, it appears that, at least in this context, the ABC programme is another layer of complexity or heterogeneity that coexists with the subset identities defined by the expression of B7 family members. While in vitro exposure to the cytokines interferon-γ (IFN-γ) and IL-21, concurrently with TLR stimulation, can cause differentiation that appears to favour ABC identity [81], the exact in vivo signals that trigger this programme remain to be defined. Overall, though, it is clear that pro-inflammatory conditions tend to favour ABC-like MBC development.

2.1 Functional and Molecular Correlates of Subset Identity

One of the most fundamental questions in B cell memory is how, mechanistically, do MBCs differ from NBCs. With the elucidation of MBC subsets with different functional responses, this question is extended to the comparison between these MBC subsets. In principle, this question can be addressed at the phenotypic, functional, signal-reprogramming, transcriptional and epigenetic levels.

In general, it is thought that MBCs are more responsive to stimuli compared to NBCs. Human MBCs have mainly been studied in vitro and for the most part, subset identity other than IgM vs switched has not been considered in these studies. Human MBCs are notable for entering the cell cycle more quickly than NBC upon stimulation. They are also capable of more overall proliferation and better survival after stimulation with TLR ligands [86,87]. Depending on their isotype expression, they show distinct patterns of migration to chemokines in vitro [88]. Human MBCs may be more prone to terminal differentiation compared to NBCs [89].

In mice, subsets defined by B7-family members PD-L2 and CD80 have been investigated mostly in vivo for the types of immune responses they support. Strikingly, although DP MBC made very rapid PB responses, they were unable to seed subsequent GC responses. In contrast, DN MBCs were not as able to support quick PB responses (though they still did better than NBCs), but could reseed GC responses and did so nearly as well as NBC. SP MBC had intermediate properties, showing some ability to generate both PB and GC progeny upon reimmunisation in vivo. When tested whether the cells expressed IgM or IgG, the results showed no detectable differences in their functional capabilities [49].

At the level of signalling, MBC can alter their responses to BCR, innate immune receptor and cytokine signals, among others [90–92]. The response to BCR stimulus of many MBC is that IgG will be altered by virtue of unique signalling properties of the IgG cytoplasmic tail, which can engage Igα/β-independent pathways via growth factor receptor-bound protein 2 (Grb2) [93–97]. Gene expression analysis suggests that MBC could also reprogramme signal responses via the protein kinase A pathway [45], which is downstream of many G-protein-coupled receptors. TLR signals, including notably TLR7 and TLR9, are also reprogrammed to be more sensitive in MBC vs NBC, in part by increased receptor expression [98].

MBCs and NBCs differ both in their propensity to secrete certain cytokines, as well as their response to them. As an example, MBCs are less likely to secrete IL-10, compared to NBCs [90]. Conversely, they are more apt to secrete inflammatory cytokines tumour necrosis factor-α (TNF-α) and lymphotoxin-α [90]. Similarly, a subset of human MBC, termed 'innate reactive activator' [99] or IRA B cells, is endowed with the unique capacity to secrete granulocyte-macrophage colony-stimulating factor (GM-CSF). Similarly, MBCs may modulate their expression of cytokine receptors; for example, human MBCs express higher levels of the IL-2Ra and IL-6R [46,88].

A wealth of information is now available about the differing transcriptomes expressed in MBC vs NBC. The earliest studies compared human MBC [46] and subsequently more rare mouse MBC [44,45] using gene array methods. Subsequently, multiple studies produced microarray and RNA-seq datasets of both murine and human MBC, including the specific MBC subsets described above [14,15,47–49,100,101]. These studies revealed several themes, including expression of stem cell-related programmes, greater expression of innate immune receptors, increased sensitivity to BCR stimulation, a 'heme' signature and adhesion molecule and migration differences [100,101]. Moreover, work on subsets of MBC revealed distinct transcriptional profiles [102], with DN MBC of mice having a cell cycle signature [49]. In particular, Riedel and colleagues, when comparing the transcriptomes and repertoires of isotype-switched murine MBC from spleen and bone marrow, have seen a remarkable heterogeneity and segregation, with marginal zone like MBC exclusively in the spleen and a quiescent MBC population exclusive for the bone marrow. Exclusive repertoires and gene expression signatures suggest that more than 80% of the MBC of spleen and bone marrow are permanent residents of the respective tissue [32].

Much less is currently known about the epigenetic state of MBC, although this is particularly intriguing, as presumably there is an epigenetic basis for imprinting of memory. Epigenetic changes can be measured at several levels. DNA methylation can be inherited from cell to cell and in some cases even across generations of individuals in a species. Chromatin marks, based on specific histone modifications, are epigenetic modifiers that can control gene accessibility as well as transcription. A few studies in humans have documented changes in DNA methylation, as well as open chromatin regions, confirming the distinctive nature of MBC [101,103,104]. This work is a start towards explaining how MBC function differs from that of NBC, but much more work needs to be done to link epigenetic changes to either MBC origin or function in humans. Even less is known about murine MBC epigenetic changes, probably due to the relatively small numbers of MBC that can routinely be isolated from mice. Nonetheless, Boss, Scharer and colleagues have been making recent progress on this front, characterising open chromatin regions as well as transcriptomes of murine MBC [100].

3. HETEROGENEITY OF PLASMA CELLS

3.1 A Brief History of Plasma Cells

While the term 'plasma cell' was first coined in 1875 by Wilhelm Waldeyer, an anatomist from the Charité University Hospital in Germany [105], PC as we know them today were actually identified by several other eminent scientists of the time. In 1890, Ramon y Cajal described them as cells with an eccentric nucleus [106], in 1891, Paul Unna showed that they could be stained by basic dyes [107] and in 1896, Tamas Marschalko pointed to their perinuclear 'Hof' and their condensed chromatin [108]. This period coincided with the discovery of 'antitoxins', that is, serum antibodies, by Kitasato and Behring in 1890 [109] and the recognition of antibodies as the mediators of specific protective and transferable immunity. More than 50 years passed until Astrid Fagreaus from the Karolinska Institute in Stockholm described a striking correlation between the presence of PC in the spleen of immunised rabbits and antibody production in vivo and ex vivo [110]. A further major discovery came from Max D. Cooper and colleagues, who showed that avian lymphocytes of the bursa gland, that is, B lymphocytes, were the precursors of antibody-secreting PC [111,112].

Following these initial discoveries, PCs were considered to be short-lived cells at the end stage of B-cell differentiation due to two observations, namely (1) that they disappeared from secondary lymphoid organs after immune reactions and (2) that they could not be cultured in vitro. This led to the perception that PC must be short-lived and constantly generated de novo from activated B cells [113,114]. This interpretation of the maintenance of PC persisted for many years due to a limited interest in understanding their lifestyle. It was perhaps for this reason that an important observation made by Benner and colleagues in 1974 was largely neglected at the time [115]. They showed that while PC did indeed disappear from the spleen within 2–3 weeks after the onset of an immune reaction, IgG secreting PC persisted in the bone marrow for more than 4 weeks, suggesting that PC had the potential to live longer than previously thought. Evidence for distinct short- and long-lived PC was provided in 1986 by Ho and colleagues [116]. The definite proof rebutting the short-lived/regeneration cell theory came more than 20 years later from Manz and colleagues who demonstrated that in fact, murine bone marrow PCs were maintained in constant numbers as non-dividing LLPC for more than 120 days after the onset of an immune response [117]. This work was closely followed by a study from Slifka and colleagues using a different murine model system [118]. Both groups showed that in the long run, specific serum antibodies were maintained by LLPC. Moreover, this long-term persistence of PC was independent of antigen, as shown by adoptive transfer of bone marrow PC into naive hosts [119]. The concept that LLPCs serve as a correlate of humoral immune protection has been further corroborated by clinical observations of patients treated with the therapeutic antibody rituximab. This CD20-targeting antibody eliminates CD20-expressing B cells, while sparing CD20-negative PC. Treatment with rituximab has not been found to alter serum antibody titres against antigens like tetanus toxoid in patients [120,121], adding to the proof that LLPCs provide persistent humoral protection and constitute a compartment of immunological memory, independent of (memory) B cells (summarised in Figure 16.1).

3.2 Plasma Cell Tissue Tropism and Heterogeneity

Activation of B cells, their interaction with T cells and their maturation into antibody-secreting cells occurs in secondary lymphoid organs, specifically in the follicular GCs of the spleen and LNs. In humans, a wave of specific PB is observed in the blood approximately 1 week after the onset

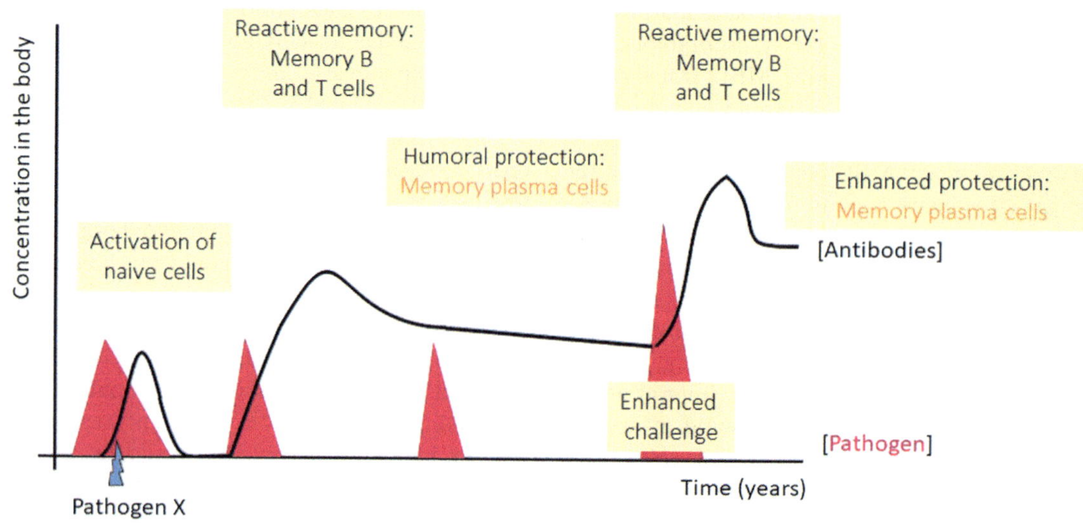

FIGURE 16.1 **Immunological memory adaptation.** Following an initial encounter with a pathogen (exposure represented by *pink triangles*), naive cells of the immune system are activated, resulting in the generation of antibodies (*black line*), the levels of which decline over time. Should the same pathogen be reencountered, memory B cells and T cells generated in the primary response are reactivated, resulting in an increased production of antibodies. Long-lived memory plasma cells in the bone marrow continue to secrete antibodies at a higher level than in the primary response. Enhanced protection can also occur following an enhanced challenge by the same pathogen, resulting in further reactivation of memory B- and T- and plasma cells and a further increase in the concentration of antibody levels.

of a secondary immune reaction [122–124]. Thereafter, these cells disappear from the blood. This corresponds with a decline in the number of splenic antibody-secreting cells and an analogous increase of such cells in the bone marrow. These cells then remain in the bone marrow in constant numbers (at approximately 10% of the maximum number generated at the peak of the immune reaction) as non-proliferating LLPC.

While the translocation process of antibody-secreting cells from secondary lymphoid organs to the bone marrow remains to be elucidated in detail, it is known that there is an intricate and dynamic regulation of chemokine receptor expression that helps to guide PC from the spleen into the bone marrow. This process involves downregulation of receptors that help to keep B and T cells in close contact to one another in the spleen (namely CXCR5 and CCR7), and upregulation of CXCR4 and sphingosine-1-phosphate receptor (S1PR1) [125–128]. The high levels of sphingosine-1-phosphate (S1P) in blood are sensed by S1PR1, luring the cells out of the tissue and into the blood [129]. The importance of this pathway in mediating cellular emigration from secondary lymphoid organs has been demonstrated by blocking the S1P-S1PR1 interaction using the drug FTY720, resulting in the inhibition of PC egress from the spleen and LNs and failure to accumulate in the bone marrow [126]. CXCR4, on the other hand, appears to be necessary to attract PC to tissues such as the bone marrow, where stromal cells are known to be a major source of its ligand, CXCL12 [130–132]. Expression of CXCR4 is upregulated on activated B cells when they differentiate into PB, and further, when they become PC [133]. Furthermore, homing of newly generated PB to the bone marrow has been shown to be inhibited by AMD3100, a CXCR4 antagonist [134], with similar results also shown using CXCR4-deficient cells [125], highlighting the importance of this receptor in bone marrow tropism. Importantly, expression of chemokine receptors on PC may not always equate to the ability of the cell to migrate in response to a particular chemokine. For example, although PCs are known to express CXCR4, they do not migrate in response to CXCL12 [127,128], while their precursors do. This immobility of bone marrow PC is actually a key requirement for clearing PC from inflamed and regenerated tissues, for their long-term survival, and probably also for their competition with newly generated PB for habitation of survival niches.

Although the bone marrow appears as the place where memory PCs are maintained rather than generated, this does not preclude maturation of immigrating PB, or bone marrow-resident MBC into LLPC in the bone marrow. Indeed, a considerable phenotypic heterogeneity of PC has been determined by cytometry, with high and low expressors of CD45, CD56, CD19, CD95 and CD9 [135,136]. The role of $CD19^{low}$ versus $CD19^{high}$ PC in particular has been controversial: Is downregulation of CD19 a hallmark of maturation of PB into LLPC or does it indicate derivation from distinct types of immune reactions [137,138]? More recently, single-cell transcriptomic technology has provided a global picture of the diversity of antibody-secreting cells in the bone marrow. Transcriptomes of PC clusters in the bone marrow indicate their origin in various types of immune reactions, their instruction by interferons,

TGF-ß and other cytokines and their extrafollicular or follicular origin, with their follicular activation marked by downregulation of CD19 [139–141].

In healthy individuals, the bone marrow is the major reservoir of LLPC, although considerable numbers of PC are also found in other tissues. While mucosal tissues were originally only considered as hosting short-lived cells [142], several studies have documented the existence of long-lived mucosal PC in the meantime [116,143–147]. Interestingly, 80% of PB and PC analysed in human blood during steady-state conditions were found to express IgA, pointing to the fact that they originated from a mucosal immune response [123]. Expression of the mucosal chemokine receptors CCR9 and CCR10 facilitate the ability of PC to home to mucosal tissues [148–150], although intriguingly, some long-lived bone marrow PCs have also been shown to express CCR10 [123] and/or IgA [143]. In these cases, CXCR4-mediated homing to the bone marrow is thought to dominate over CCR10 homing to the mucosa [123]. In non-steady-state conditions, that is, during a chronic inflammatory response, expression of CXCR3 is induced on PC, presumably mediating their entry into inflamed tissues [127,133]. Induction of expression of CXCR3 is dependent on IFN-γ present at the time of activation of B cells, and this expression is then maintained by MBCs and PC [133]. The expression of the ligands for CXCR3, namely CXCL9, -10 and -11 in inflamed tissue, is similarly dependent on IFN-γ [151]. Interestingly, PB can home to inflamed tissue and persist there as PC independent of their specificity, as long as the tissue is inflamed [152]. Small numbers of PC also appear to persist in the secondary lymphoid organs, although their lifespan remains to be analysed.

4. MAINTENANCE OF MEMORY B CELLS AND LONG-LIVED PLASMA CELLS

In order not to lose acquired immunity, MBC must be maintained long-term. Pulse chase-labelling and time-course experiments revealed that follicular and marginal zone B cells have a half-life of 13–22 weeks, decay rates for murine MBC revealed cellular half-lives greater than the 2-year lifespan of the mouse [153,154] (and F.W. and M.J.S. unpublished observations). MBCs are quiescent and can exceed the lifetime of mice [154]. Little is known about how humoral memory is maintained over long periods of time. Several factors are thought to be *not* required for long-term persistence. IgG1 MBC persistence is independent of B-cell lymphoma 2 (Bcl-2) family members [155,156] and continued exposure to Ag (Ag deposition on follicular dendritic cells as reservoir for continued Ag exposure via immune-complexes) [31,153]. T cells are also not required [157]. It is reported that TNF-family cytokine B-cell activating factor/B lymphocyte stimulator (BAFF/BLyS) were not essential for the survival of MBCs, in contrast to NBC [158,159], though more recent literature challenges the requirement for BAFF and BAFF-R in MBC survival [160].

On the other hand, Syk and phospholipase Cγ2 (PLCγ2) are necessary for MBC persistence [161,162]. BCR and the immunoreceptor tyrosine-based inhibitory motif (ITIM) domain of the signalling subunit CD79A are involved in MBC maintenance. MBC survival is also strongly dependent on NF-κB signalling via IKK2 [160]. It has been suggested that differential expression of survival genes of the Bcl-2 family system, such as Puma, could be responsible for differential lifespan of MBC compared to NBC [163]. MBCs upregulate a stem cell-like programme and bone morphogenetic protein receptor 1a, which could render them to rely on cytokines for stemness, such as leukaemia inhibitory factor [44,45,164]. It is tempting to speculate that there is a stem-like MBC subpopulation that underpins the overall antigen-specific memory population. Such a subset would be able to replenish the entire humoral immune compartment and therefore serves as a long-lived reservoir to prevent acquired memory from depletion over time. This idea was originally suggested by Fearon and colleagues in 2001 [165]. It is also tempting to speculate that bone marrow-resident MBC, which inhabit the same kind of niches as bone marrow-resident PC [32] may require phosphoinositide 3-kinase (PI3K) signalling for their survival, induced by (integrin-mediated) contact to stromal cells, alike the PC [166].

In line with this notion, as discussed above, MBC subsets are generated in a temporal order in a GC-dependent and independent way and have differential immediate effector functions in a recall response. These data suggest that maintenance and regeneration of the MBC compartment is dependent on which subset of MBCs gets activated during recall responses. The fact that not all MBC subsets are equally capable of generating secondary MBC subsets – only DN MBCs can generate all – supports the concept of differential 'stemness' of MBCs with respect to their ability to replenish and maintain long-lived humoral immunity, and probably also different mechanisms for their maintenance. We integrate these data along with some of our emerging observations in a model that explains how MBC subsets are generated and then their potential to contribute to secondary immune responses and even secondary memory compartments that ultimately would maintain memory in the face of repeated challenges (Figure 16.2).

LLPCs have been identified as cells residing in the bone marrow and in chronically inflamed tissues during and after the conclusion of immune responses [117,118,127,167]. In contrast, some MBCs circulate in the blood and lymph for

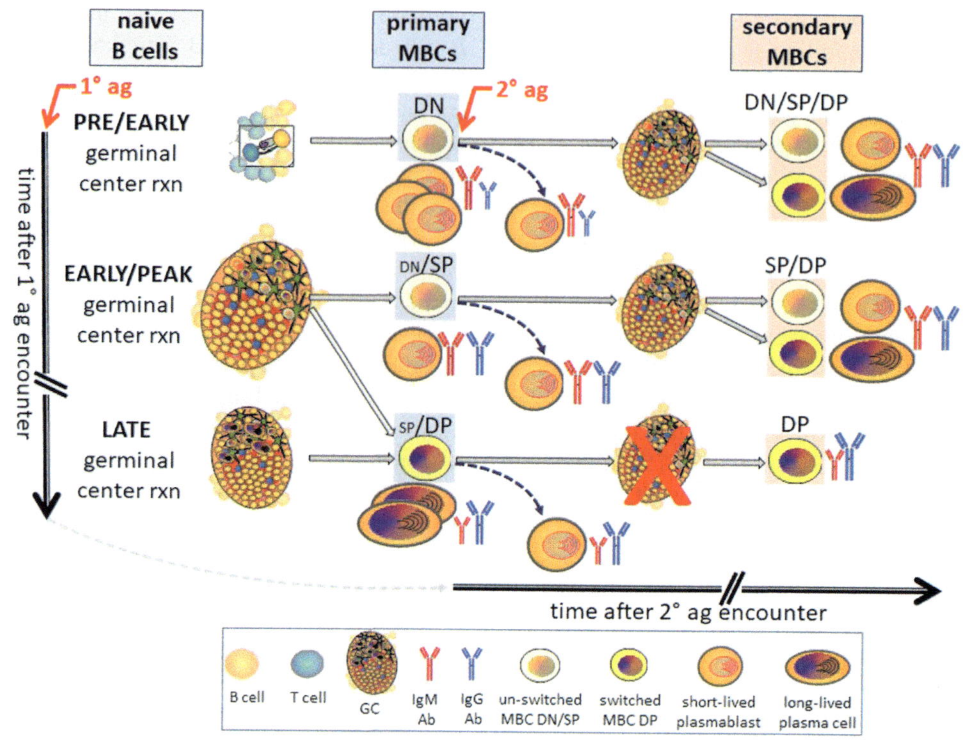

FIGURE 16.2 **Primary and secondary B-cell responses.** Initial antigen encounter and successful cognate B–T interaction leads to an early extra-follicular B-cell response with strong proliferation and appearance of short-lived plasmablasts and the generation of mainly un-switched DN ('double negative'; $CD80^{neg}$/PD-$L2^{neg}$) MBCs. Within 3–5 days, germinal centres (GCs) are formed. 'Early' GC reactions mainly give raise to DN and SP ('single positive'; $CD80^{neg}$ PD-$L2^{pos}$) MBCs, whereas DP ('double positive'; $CD80^{pos}$ PD-$L2^{pos}$) are mainly formed during late GC reaction. Somatically mutated and isotype-switched LLPCs are mainly generated during 'late' GC reaction (*left panel*). MBCs are activated upon second antigen exposure. The nature of this recall response depends on the subset of MBCs. DN MBCs form secondary GCs and can give rise to all secondary MBC subsets. DN are also able to facilitate delayed plasmablast responses. The antigen recall response of DP MBCs is characterised by a fast differentiation into plasma blasts which lead to increased antibody titres. DP MBCs do not induce secondary GC reactions, and DP MBCs are limited in their ability to only generate secondary DP MBC. SP MBCs can obtain both phenotypes of DN and DP MBCs and can give rise to secondary SP and DP MBCs but not secondary DN MBCs (*right panel*).

extended periods, while others reside in spleen, lymph nodes and bone marrow. Resident populations of distinct MBC have been identified, in the spleen [168–171] and in the bone marrow [32,172] as well as the lung and gut [48,51,173,174]. In retrospect, this is not too surprising given the eminent work of James Gowans and colleagues in the 1960s who demonstrated that, while primary immune responses are dependent on circulating cells, secondary antibody responses are not, showing that circulating lymphocytes are not required to mount secondary immune responses [175]. While the lifestyle and survival mechanisms of circulating MBC remains enigmatic, bone marrow memory B and memory T cells apparently reside individually in niches resembling those of LLPC, that is, in contact to bone marrow stromal cells [176].

In stark contrast to mobile PB that can compete for survival niches by changing their location, immobile bone marrow PCs [177] are completely dependent on their environment to provide the survival signals they require. Indeed, if the environment changes or PCs are dislocated from the survival niche they inhabit, they will die [178]. For bone marrow PC, stromal cell contact is vital for their survival. This contact is thought to be mediated by integrin signalling, as interfering with the integrin-mediated linkage between PC and stromal cells with very late antigen-4 (VLA-4; α4β1) and lymphocyte function-associated antigen-1 (LFA-1; αLβ2) specific antibodies in vivo resulted in the (temporary) elimination of PC from the bone marrow [179], as did joint blockade of VLA-4 and CXCR4 [180]. Moreover, PC isolated from bone marrow die within days in cell culture, emphasising their need for a survival niche [181]. Artificial cell culture conditions have been described that allow the maintenance of PC for several weeks in vitro [182–184]. As the bone marrow is hypoxic, physiological adaptation to low oxygen levels may play an important role in the survival of long-lived PC [166,182,185]. Survival has also been reported to be prolonged by cytokines and ligands of PC receptors, including hyaluronic acid, (a ligand of CD44 on PC), IL-6, TNF-alpha, CXCL12, BAFF and A proliferation-inducing ligand (APRIL) [134,159,166,181,183,184,186]. For BAFF and APRIL, both of which signal via the same PC receptor, B-cell maturation antigen (BCMA), it has been shown that they deliver a critical and non-redundant signal for the survival of the PC in vivo [186]. However, other signals, such as IL-6, are apparently not involved, or are redundant in vivo, since their genetic ablation in mice does not interfere with the persistence of PC in the bone marrow [181]. Although BAFF or APRIL are critical mediators for the persistence of PC in the bone marrow, they alone are not sufficient to ensure the long-term survival of PC. A second pro-survival signal is required to prevent apoptosis of PC, which is apparently provided by contact of the PC with the stromal cell, most likely the integrin-mediated contact, as is suggested by the work of DiLillo and colleagues [179]. Contact between stromal cells and PC is known to induce PI3K signalling in the PC, which is vital for their survival both ex vivo and in vivo [166]. Stromal cell contact provides the resilience against metabolic stress by inducing downregulation of Forkhead box protein O1/3 (FoxO1/3) and prevention of the activation of pro-apoptotic caspases 3 and 7. BCMA signalling activates the NF-kB pathway and prevents activation of caspase 12, thus making the PC resilient against stress related to protein synthesis (summarised in Figure 16.3). This complementation is necessary and sufficient to prevent apoptosis of PC ex vivo as efficiently as a pan-caspase inhibitor.

Additional signals may be required to secure long-term persistence of PC in the bone marrow, such as Sec22b [187], or CXCL12 from stromal cells, addressing CXCR4 on the PC. Other signals may be required by only certain subpopulations of PC, such as laminin-ß1 of bone marrow stromal cells for IgG secreting PC, which have several potential receptors for it. The relevance of laminin-ß1 for persistence of IgG-secreting PC has been shown to be exploited by *Salmonella* resulting in a significant reduction of IgG-secreting PC in the bone marrow during infection [188]. Whether *Salmonella* primarily targets bone marrow PC or MBC remains to be shown as IgG-expressing MBC of the bone marrow inhabit the same niches as IgG-secreting PC, docking on to laminin-ß1 expressing vascular cell adhesion protein-1+ (VCAM-1) and fibronectin+ stromal cells [32].

5. THE INTERPLAY OF HUMORAL AND REACTIVE MEMORY IN ADAPTIVE IMMUNITY

Reactive memory B and T cells and LLPC secreting protective antibodies provide effective and specific immunity and protect against environmental pathogens and toxins. Collaboration of activated T and B cells is the starting point for the induction of long-lasting immunity, that is, immunological memory. Upon repeated challenge with the same or a similar antigen, more bone marrow PCs are generated from reactivated MBC [140] and serum antibody titres increase [189]. However, serum antibodies are also known to 'mask' the antigen and thus suppress the reactivation of specific memory B and T cells [190], a type of antibody feedback regulation that had already been observed in the very early days of antibody discovery [191]. Once sufficient bone marrow PCs have been generated, protective immunity by secreted antibodies prevents further immune

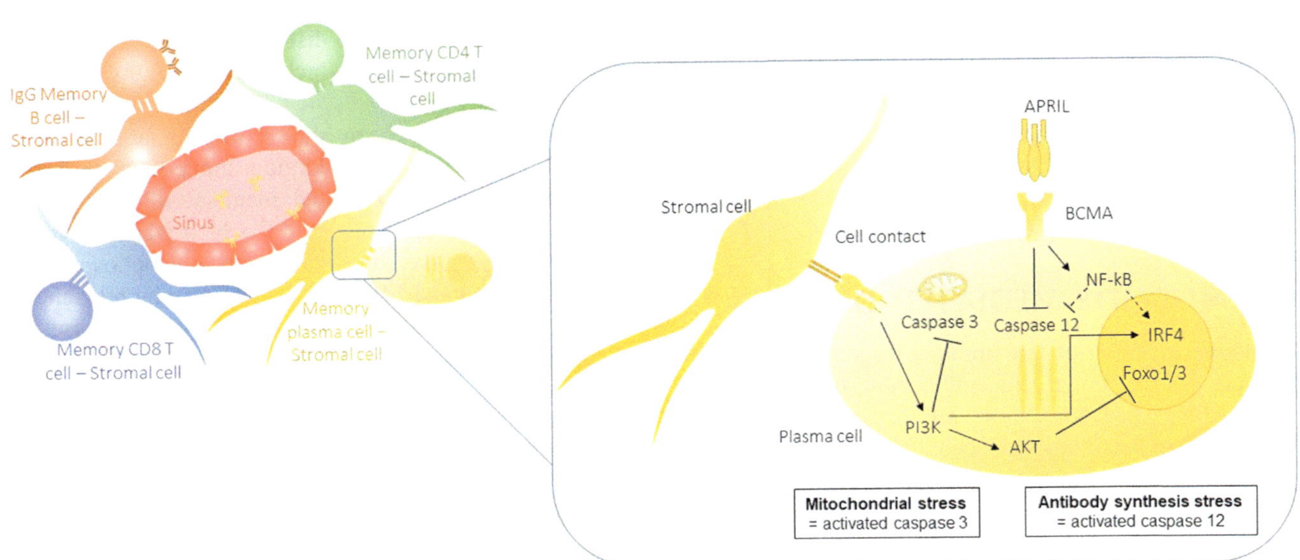

FIGURE 16.3 **Survival niche of long-lived plasma cells in the bone marrow.** The bone marrow hosts memory CD4 and CD8 T cells, memory B cells and long-lived plasma cells in stromal cell survival niches. The two main survival pathways used by long-lived bone marrow plasma cells include PI3K, activated by cell–cell contact, and NF-κB, activated by the cytokine APRIL signalling via its receptor BCMA. This synergistic signalling inhibits the activation of caspases 3 and 12, thus preventing apoptosis via either mitochondrial or antibody-synthesis stress, respectively.

reactions to that specific antigen, and reinfections will go unnoticed, at least on a systemic level [192–195]. If the environmental pathogen or toxin changes in quantity or quality and 'escapes' protection mediated by secreted antibodies, new PC can be generated from MBC. 'Escape' in this sense means escape from antibody feedback regulation and is not related to 'neutralising' capacity of the antibodies. While the molecular mechanism of antibody feedback regulation is still not clear, it should be noted that it is epitope-specific, suggesting that it works at the level of antigen presentation [190].

Adaptive immunity to environmental pathogens and toxins is also conservative and imprinted by the original instructions. This has been termed 'original antigenic sin' [196], or nowadays is often considered as 'imprinted immunity', referring to reactivation of MBC from previous immune reactions without further affinity maturation and adaptation to variation of the reactive challenge, as exemplified for repeated challenges of SARS-CoV-2 and influenza and their variants [197,198].

An important note may be that feedback regulation can occur on two different levels: (a) at mucosal surfaces and (b) systemically. Commensals and pathogens at mucosal surfaces will interact with mucosal antibodies in the first place, agglutinating them, and in case of infectious agents, eventually preventing infection by 'neutralising' antibodies. The presence of antibodies at mucosal surfaces requires their active transport through the epithelial cell layer, by means of Fc receptors, in particular the neonatal Fc receptor and the poly-Ig Fc receptor [199–201]. Little is known today how activity of these Fc receptors is regulated, but often mucosal antibody titres wane quickly, while systemic antibody titres remain stable, implying that systemic protection from severe disease is persistent, while protection from infection is not [202–206]. In the case of systemic protection, it may be not so relevant whether antibodies are neutralising or just aggregating, opsonising, complement activating or ADCC-mediating antibodies. In other words, neutralising antibodies have a role in preventing infection, but less so in preventing severe disease.

6. ANTIBODY-MEDIATED DISEASES

While the intention of any immune response is to protect the host from a foreign antigenic assault, despite layers of safeguarding measures, there are occasions when self-harming immune responses occur. In such a 'pathogenic immune reaction', PC secreting antibodies significantly contribute to antibody-mediated diseases, such as in autoimmune diseases and allergies. In this sense, these antibodies embody pathogenicity. Beyond chronic immune reactions, memory PCs provide a level of chronicity on their own, persisting in niches of the bone marrow and inflamed tissues, where they are refractory to therapies targeting chronic immune reactions [207]. The relevance of LLPC-secreting pathogenic antibodies, as an independent pharmacological target, is increasingly realised, but strategies for their targeted elimination are in their infancy.

The key role of LLPC in perpetuating chronic autoimmune diseases was first demonstrated in NZB/W mice, a murine model for the human disease systemic lupus erythematosus (SLE). Feeding of NZB/W mice with BrdU marked proliferating from non-proliferating cells in the spleen and in particular, in the bone marrow, revealing a substantial proportion of LLPC [208]. These LLPCs were resistant to immunosuppression with cyclophosphamide and glucocorticoids [208,209] as well as irradiation [118]. While in lupus mouse models, as well as in humans with SLE or rheumatoid arthritis (RA), many autoantibodies – including anti-dsDNA, anti-chromatin and rheumatoid factor—are secreted by short-lived PC (SLPC), it remains to be determined to what extent they are also secreted by LLPC, in particular in those patients who have 30% of their IgG autoantibodies refractory to cyclophosphamide and rituximab [210]. Therapy-free remission could be induced in such patients by immunoablation and autologous stem cell transplantation [211]. The recent experience using CAR-19 T cells in several SLE patients shows that, at least in these patients, autoantibody titres were so significantly reduced that the disease went into remission [212]. Whether this was due to ablation of pathogenic B cells with APC and cytokine secreting function, autoantibody-secreting short-lived PC or long-lived $CD19^{+}$ PC [137,138] is not clear at present. In general, even partial reduction of autoantibody production following treatment with anti-proliferative drugs, like cyclophosphamide and drugs depleting the precursors of PC, like rituximab, in patients with autoantibody-driven diseases suggests a (critical) contribution of LLPC, which are refractory to these treatments.

Likewise, it has been shown that in an ovalbumin-induced murine model of allergic asthma, inhalation of an aerosolised allergen generated memory, allergen-specific IgG1-, IgA- and IgE-secreting PC which were resistant to an anti-proliferative treatment with cyclophosphamide, qualifying them as LLPC [213]. The potential relevance of pathogenic LLPC for the etiopathology of disease was again demonstrated in the NZB/W model of lupus nephritis following transfer of PB and PC from lupus prone NZB/W into immunodeficient $Rag1^{-/-}$ mice, resulting in the development of long-lived memory cells in bone marrow and spleen. Over a time period of 21 weeks, the host mice developed lupus nephritis indistinguishable from that of NZB/W mice [214]. The contribution of pathogenic PC to disease pathology may vary between diseases, but their relevance may also have been underestimated. While the case is clear for IgE-mediated allergies, autoantibodies may not only cause classical autoimmune diseases but also may be of broader clinical relevance, for example, causing adaptive immunodeficiencies, like autoantibodies directed against interferons and their receptors, and impairing effective anti-viral immune responses [215–217].

Whether or not PC secrete pathogenic antibodies in a given situation can be concluded from the efficacy of therapies aiming to deplete B cells or activated B cells, that is, their precursors. As CD20 is not expressed by PC, B-cell depletion (or more precisely, depletion of circulating B cells) with the anti-CD20 antibody rituximab has been informative to distinguish between antibodies secreted by SLPC versus LLPC. Studies in mice confirm that mature and MBCs are not required for maintaining splenic and bone marrow PC populations [179,218]. Moreover, protective antibody titres are known to remain stable for many years [219] and do not change after rituximab therapy [220,221]. By contrast, the responses of autoantibody levels are much more heterogeneous: some remain unchanged by rituximab, others are ablated and some decrease but do not disappear completely. This indicates that autoantibodies that decrease or disappear after B-cell depletion therapy must have been secreted by short-lived PB and PC, while any persisting autoantibodies will have been secreted by memory PC. A more efficient way to ablate B cells may be anti-CD19 CAR-T cells, which supposedly also target resident CD19 B cells and CD19 expressing PC. Reportedly, anti-CD19 CAR-T cells in SLE did not affect PC secreting protective antibodies in the patients analysed, but significantly diminished the titres of autoantibodies, in line with the notion that a significant proportion of autoantibodies in these patients is derived from SLPC or LLPC expressing CD19 [212].

IgG4-related disease is another autoimmune disease with apparent contributions of both SLPC and LLPCs. IgG4-related disease is a fibroinflammatory disorder characterised by tumefactive lesions, a dense lymphoplasmacytic infiltrate rich in IgG4-positive PC, storiform fibrosis and, often but not always, elevated serum IgG4 concentrations [222,223]. The disease was first described in Japan [224,225] and can affect almost any organ in the body. Commonly, the disease responds well to treatment with glucocorticoids or B-cell depletion therapy with rituximab, but therapy-free remission is not attainable. The rapid and selective rituximab-induced decline in IgG4 levels suggests that the bulk of IgG4 in this disease originates from SLPC [222]. However, it is likely that the IgG4+ PC in the affected organs are LLPCs, persisting as long as the organs are inflamed. It seems possible that the main effect of both glucocorticoids and rituximab is due to the resolution of inflammation which terminates these IgG4+ memory PC [178].

Less is known about the components of the PC niche in inflamed tissues; however, it is interesting that PCs have also been found – like in bone marrow – to be in contact with cells expressing CXCL12. For example, in patients with Sjögren's syndrome, PCs are located in close proximity to ductal and acinar epithelium as well as mononuclear cells within the focal infiltrate expressing CXCL12 and IL-6 [226]. CXCL12 is also overexpressed in inflamed kidneys [227], whereas the CXCL10–CXCR3 interaction seems not to be crucial for PC homing, as infiltration of PC lacking CXCR3 in inflamed kidneys was not blocked [228]. In autoimmune thrombocytopaenia, the spleen provides niches for autoreactive LLPC [229]. Finally, PC themselves also can contribute to their survival in autoimmune diseases, as it is known that PC in the nephritic kidneys derived from SLE patients express the cytokines APRIL and BAFF [230], providing them with resilience to anabolic stress [166].

Therapeutic targeting of pathogenic PC is a medical need in antibody-mediated diseases, since they are refractory to therapeutic strategies aiming to suppress (chronic) immune responses or to ablate pathogenic T- and B-cells. Several generic approaches targeting PC as such have been developed, based on what we know about their lifestyle. Such options include the targeting of PC precursors, the inhibition of migration of newly generated PB into the bone marrow or inflamed tissues, the targeting of the survival signals and the direct targeting of PC. Preventing the establishment of pathogenic PC is not a feasible option in most diseases, since this already happens in the preclinical phase of disease, as demonstrated for SLE and RA [231,232].

Generic ablation of PC may be possible by targeting one of their essential, non-redundant survival signals. For example, VLA-4-mediated contact to stromal cells or BAFF and APRIL-induced signalling via BCMA can therapeutically targeted using natalizumab, belimumab, tabalumab or atacicept [179,186,233,234]. Targeting the unfolded protein response of PC using the proteasome inhibitor bortezomib has also been successful in a number of preclinical and clinical settings, including in NZB/W mice and in a murine model of myasthenia gravis [235,236], in SLE patients [237] and patients with autoantibody-mediated neuroinflammatory diseases [238]. There are case reports describing the induction of clinical remission in patients with refractory autoantibody-mediated diseases such as thrombotic thrombocytopenic purpura (TTP) [239,240] and autoimmune haemolytic anaemia [241,242] by bortezomib, which was accompanied by a striking reduction of autoantibody levels. PC depletion with bortezomib has also emerged as a therapeutic approach in transplantation medicine where alloantibodies induce acute rejection [243–245]. There is one case report of a patient with an IgG4-related disease refractory to glucocorticoids who responded to PC depletion with bortezomib [242]. In the SLE patients who had been refractory to glucocorticoids and immunosuppressive drugs such as cyclophosphamide or mycophenolate mofetil before treatment, bortezomib not only significantly ameliorated disease activity but also broke refraction. This is in line with the notion that LLPCs are the key roadblock to regeneration of tolerance, and that they make the patients refractory to conventional immunosuppressive treatments.

Recently, therapeutic antibodies targeting CD38 (daratumumab) have shown efficacy in SLE [246]. CD38 is highly expressed by PC, but also expressed by activated B- and T cells. To what extent the therapeutic effect was due to PC depletion remains to be shown.

Generic ablation of PC may be effective, but it inevitably also leads to depletion of PC secreting protective antibodies, that is, ablation of humoral immunity, which then has to be restored by vaccination, if vaccines are available. Therapeutic strategies for the targeted selective ablation of pathogenic PC have not been described so far. One such strategy may be the ablation of PC according to the specificity of the antibodies they secrete. Since IgG expressing PCs do not express their antibodies on the cell surface, they do not reveal the specificity of their antibodies directly. A therapeutic conjugate of an antigen-of-relevance with a PC-binding antibody has been used successfully to selectively ablate PC secreting antibodies of a defined specificity, in a murine model system, providing proof-of-principle that such a treatment works in vivo [247]. This approach might not only provide a unique therapeutic option but also could help to identify relevant antigens in antibody-mediated diseases.

REFERENCES

[1] Shlomchik MJ, Weisel F. B cell primary immune responses. Immunol Rev March 2019;288(1):5–9.

[2] MacLennan IC, Johnson GD, Liu YJ, Gordon J. The heterogeneity of follicular reactions. Res Immunol 1991;142(3):253–7.

[3] MacLennan IC, Toellner KM, Cunningham AF, Serre K, Sze DM, Zuniga E, et al. Extrafollicular antibody responses. Immunol Rev August 2003;194:8–18.

[4] Marshall JL, Zhang Y, Pallan L, Hsu MC, Khan M, Cunningham AF, et al. Early B blasts acquire a capacity for Ig class switch recombination that is lost as they become plasmablasts. Eur J Immunol December 2011;41(12):3506–12.

[5] Cunningham AF, Gaspal F, Serre K, Mohr E, Henderson IR, Scott-Tucker A, et al. Salmonella induces a switched antibody response without germinal centers that impedes the extracellular spread of infection. J Immunol May 15, 2007;178(10):6200–7.

[6] Di Niro R, Lee SJ, Vander Heiden JA, Elsner RA, Trivedi N, Bannock JM, et al. Salmonella infection drives promiscuous B cell activation followed by extrafollicular affinity maturation. Immunity July 21, 2015;43(1):120–31.

[7] Wrammert J, Smith K, Miller J, Langley WA, Kokko K, Larsen C, et al. Rapid cloning of high-affinity human monoclonal antibodies against influenza virus. Nature May 29, 2008;453(7195):667–71.

[8] Wrammert J, Onlamoon N, Akondy RS, Perng GC, Polsrila K, Chandele A, et al. Rapid and massive virus-specific plasmablast responses during acute dengue virus infection in humans. J Virol March 2012;86(6):2911–8.

[9] Inamine A, Takahashi Y, Baba N, Miyake K, Tokuhisa T, Takemori T, et al. Two waves of memory B-cell generation in the primary immune response. Int Immunol May 2005;17(5):581–9.

[10] Takemori T, Kaji T, Takahashi Y, Shimoda M, Rajewsky K. Generation of memory B cells inside and outside germinal centers. Eur J Immunol 2014;44(5):1258–64.

[11] Shlomchik MJ, Weisel F. Germinal center selection and the development of memory B and plasma cells. Immunol Rev 2012;247(1):52–63.

[12] Weisel FJ, Zuccarino-Catania GV, Chikina M, Shlomchik MJ. A temporal switch in the germinal center determines differential output of memory B and plasma cells. Immunity January 19, 2016;44(1):116–30.

[13] Taylor JJ, Pape KA, Jenkins MK. A germinal center-independent pathway generates unswitched memory B cells early in the primary response. J Exp Med March 12, 2012;209(3):597–606.

[14] Viant C, Wirthmiller T, ElTanbouly MA, Chen ST, Cipolla M, Ramos V, et al. Germinal center-dependent and -independent memory B cells produced throughout the immune response. J Exp Med August 2, 2021;218(8).

[15] Glaros V, Rauschmeier R, Artemov AV, Reinhardt A, Ols S, Emmanouilidi A, et al. Limited access to antigen drives generation of early B cell memory while restraining the plasmablast response. Immunity September 14, 2021;54(9):2005–23. e2010.

[16] Kitano M, Moriyama S, Ando Y, Hikida M, Mori Y, Kurosaki T, et al. Bcl6 protein expression shapes pre-germinal center B cell dynamics and follicular helper T cell heterogeneity. Immunity June 24, 2011;34(6):961–72.

[17] Kerfoot SM, Yaari G, Patel JR, Johnson KL, Gonzalez DG, Kleinstein SH, et al. Germinal center B cell and T follicular helper cell development initiates in the interfollicular zone. Immunity June 24, 2011;34(6):947–60.

[18] Zhang T-t, Gonzalez DG, Cote CM, Kerfoot SM, Deng S, Cheng Y, et al. Germinal center B cell development has distinctly regulated stages completed by disengagement from T cell help. Elife May 12, 2017;6:e19552.

[19] Schwickert TA, Victora GD, Fooksman DR, Kamphorst AO, Mugnier MR, Gitlin AD, et al. A dynamic T cell-limited checkpoint regulates affinity-dependent B cell entry into the germinal center. J Exp Med June 6, 2011;208(6):1243–52.

[20] Lam WY, Jash A, Yao CH, D'Souza L, Wong R, Nunley RM, et al. Metabolic and transcriptional modules independently diversify plasma cell lifespan and function. Cell Rep August 28, 2018;24(9):2479–2492 e2476.

[21] Nutt SL, Hodgkin PD, Tarlinton DM, Corcoran LM. The generation of antibody-secreting plasma cells. Nat Rev Immunol 2015;15(3):160–71.

[22] Takahashi Y, Ohta H, Takemori T. Fas is required for clonal selection in germinal centers and the subsequent establishment of the memory B cell repertoire. Immunity February 2001;14(2):181–92.

[23] Toyama H, Okada S, Hatano M, Takahashi Y, Takeda N, Ichii H, et al. Memory B cells without somatic hypermutation are generated from Bcl6-deficient B cells. Immunity September 2002;17(3):329.

[24] Liu X, Zhao Y, Qi H. T-independent antigen induces humoral memory through germinal centers. J Exp Med March 7, 2022;219(3).

[25] Bortnick A, Chernova I, Quinn 3rd WJ, Mugnier M, Cancro MP, Allman D. Long-lived bone marrow plasma cells are induced early in response to T cell-independent or T cell-dependent antigens. J Immunol June 1, 2012;188(11):5389–96.

[26] Taillardet M, Haffar G, Mondi√®re P, Asensio M-J, Gheit H, Burdin N, et al. The thymus-independent immunity conferred by a pneumococcal polysaccharide is mediated by long-lived plasma cells. Blood November 12, 2009;114(20):4432–40.

[27] de Vinuesa CG, Cook MC, Ball J, Drew M, Sunners Y, Cascalho M, et al. Germinal centers without T cells. J Exp Med February 7, 2000;191(3):485–94.

[28] Toellner KM, Jenkinson WE, Taylor DR, Khan M, Sze DM, Sansom DM, et al. Low-level hypermutation in T cell-independent germinal centers compared with high mutation rates associated with T cell-dependent germinal centers. J Exp Med 2002;195(3):383–9.

[29] Obukhanych TV, Nussenzweig MC. T-independent type II immune responses generate memory B cells. J Exp Med February 20, 2006;203(2):305–10.

[30] Daly CA, Spurrier MA, Jennings-Gee JE, Haas KM. B cell subsets differentially contribute to the T cell–independent memory pool. J Immunol 2020;205(9):2362–74.

[31] Maruyama M, Lam KP, Rajewsky K. Memory B-cell persistence is independent of persisting immunizing antigen. Nature October 5, 2000;407(6804):636–42.

[32] Riedel R, Addo R, Ferreira-Gomes M, Heinz GA, Heinrich F, Kummer J, et al. Discrete populations of isotype-switched memory B lymphocytes are maintained in murine spleen and bone marrow. Nat Commun May 22, 2020;11(1):2570.

[33] Klein U, Kuppers R, Rajewsky K. Evidence for a large compartment of IgM-expressing memory B cells in humans. Blood 1997;89:1288–98.

[34] Klein U, Rajewsky K, Kuppers R. Human immunoglobulin (Ig) M+IgD+ peripheral blood B cells expressing the CD27 cell surface antigen carry somatically mutated variable region genes: CD27 as a general marker for somatically mutated (memory) B cells. J Exp Med November 2, 1998;188(9):1679–89.

[35] Tierens A, Delabie J, Michiels L, Vandenberghe P, De Wolf-Peeters C. Marginal-zone B cells in the human lymph node and spleen show somatic hypermutations and display clonal expansion. Blood 1999;93(1):226–34.

[36] White H, Gray D. Analysis of immunoglobulin (Ig) isotype diversity and Igm/D memory in the response to phenyl-oxazolone. J Exp Med June 19, 2000;191(12):2209–20.

[37] Tomayko MM, Steinel NC, Anderson SM, Shlomchik MJ. Cutting edge: hierarchy of maturity of murine memory B cell subsets. J Immunol December 15, 2010;185(12):7146–50.

[38] Krishnamurty AT, Thouvenel CD, Portugal S, Keitany GJ, Kim KS, Holder A, et al. Somatically hypermutated plasmodium-specific IgM(+) memory B cells are rapid, plastic, early responders upon malaria rechallenge. Immunity August 16, 2016;45(2):402–14.

[39] Kenderes KJ, Levack RC, Papillion AM, Cabrera-Martinez B, Dishaw LM, Winslow GM. T-Bet(+) IgM memory cells generate multi-lineage effector B cells. Cell Rep July 24, 2018;24(4):824–837 e823.

[40] Weisel NM, Joachim SM, Smita S, Callahan D, Elsner RA, Conter LJ, et al. Surface phenotypes of naive and memory B cells in mouse and human tissues. Nat Immunol January 2022;23(1):135–45.

[41] Tangye SG, Liu YJ, Aversa G, Phillips JH, de Vries JE. Identification of functional human splenic memory B cells by expression of CD148 and CD27. J Exp Med November 2, 1998;188(9):1691–703.

[42] Bemark M, Hazanov H, Stromberg A, Komban R, Holmqvist J, Koster S, et al. Limited clonal relatedness between gut IgA plasma cells and memory B cells after oral immunization. Nat Commun September 6, 2016;7:12698.

[43] Onodera T, Hosono A, Odagiri T, Tashiro M, Kaminogawa S, Okuno Y, et al. Whole-virion influenza vaccine recalls an early burst of high-affinity memory B cell response through TLR signaling. J Immunol May 15, 2016;196(10):4172–84.

[44] Bhattacharya D, Cheah MT, Franco CB, Hosen N, Pin CL, Sha WC, et al. Transcriptional profiling of antigen-dependent murine B cell differentiation and memory formation. J Immunol November 15, 2007;179(10):6808–19.

[45] Tomayko MM, Anderson SM, Brayton CE, Sadanand S, Steinel NC, Behrens TW, et al. Systematic comparison of gene expression between murine memory and naive B cells demonstrates that memory B cells have unique signaling capabilities. J Immunol July 1, 2008;181(1):27–38.

[46] Klein U, Tu Y, Stolovitzky GA, Keller JL, Haddad Jr J, Miljkovic V, et al. Transcriptional analysis of the B cell germinal center reaction. Proc Natl Acad Sci U S A February 25, 2003;100(5):2639–44.

[47] Glass DR, Tsai AG, Oliveria JP, Hartmann FJ, Kimmey SC, Calderon AA, et al. An integrated multi-omic single-cell atlas of human B cell identity. Immunity July 14, 2020;53(1):217–232 e215.

[48] Weisel NM, Weisel FJ, Farber DL, Borghesi LA, Shen Y, Ma W, et al. Comprehensive analyses of B-cell compartments across the human body reveal novel subsets and a gut-resident memory phenotype. Blood December 10, 2020;136(24):2774–85.

[49] Zuccarino-Catania GV, Sadanand S, Weisel FJ, Tomayko MM, Meng HL, Kleinstein SH, et al. CD80 and PD-L2 define functionally distinct memory B cell subsets that are independent of antibody isotype. Nat Immunol July 2014;15(7):631–7.

[50] Dogan I, Bertocci B, Vilmont V, Delbos F, Megret J, Storck S, et al. Multiple layers of B cell memory with different effector functions. Nat Immunol December 2009;10(12):1292–9.

[51] Allie SR, Bradley JE, Mudunuru U, Schultz MD, Graf BA, Lund FE, et al. The establishment of resident memory B cells in the lung requires local antigen encounter. Nat Immunol January 2019;20(1):97–108.

[52] Nickerson KM, Smita S, Hoehn KB, Marinov AD, Thomas KB, Kos JT, et al. Age-associated B cells are heterogeneous and dynamic drivers of autoimmunity in mice. J Exp Med May 1, 2023;220(5).

[53] Good-Jacobson K, Szumilas C, Chen L, Sharpe A, Tomayko M, Shlomchik M. PD-1 regulates germinal center B cell survival and the formation and affinity of long-lived plasma cells. Nat Immunol 2010;11(6):535–42.

[54] Good-Jacobson KL, Song E, Anderson S, Sharpe AH, Shlomchik MJ. CD80 expression on B cells regulates murine T follicular helper development, germinal center B cell survival, and plasma cell generation. J Immunol May 1, 2012;188(9):4217–25.

[55] Conter LJ, Song E, Shlomchik MJ, Tomayko MM. CD73 expression is dynamically regulated in the germinal center and bone marrow plasma cells are diminished in its absence. PLoS One 2014;9(3):e92009.

[56] Budeus B, Schweigle de Reynoso S, Przekopowitz M, Hoffmann D, Seifert M, Kuppers R. Complexity of the human memory B-cell compartment is determined by the versatility of clonal diversification in germinal centers. Proc Natl Acad Sci U S A September 22, 2015;112(38):E5281–9.

[57] Harms Pritchard G, Pepper M. Memory B cell heterogeneity: remembrance of things past. J Leukoc Biol 2018;103(2):269–74.

[58] Sanz I, Wei C, Lee FE-H, Anolik J. Phenotypic and functional heterogeneity of human memory B cells. Semin Immunol 2008;20(1):67–82.

[59] Magri G, Comerma L, Pybus M, Sintes J, Lligé D, Segura-Garzón D, et al. Human secretory IgM emerges from plasma cells clonally related to gut memory B cells and targets highly diverse commensals. Immunity July 18, 2017;47(1):118–134.e118.

[60] Koethe S, Zander L, Koster S, Annan A, Ebenfelt A, Spencer J, et al. Pivotal advance: CD45RB glycosylation is specifically regulated during human peripheral B cell differentiation. J Leukoc Biol July 2011;90(1):5–19.

[61] Tangye SG, Good KL. Human IgM+CD27+ B cells: memory B cells or "memory" B cells? J Immunol July 1, 2007;179(1):13–9.

[62] Reynaud CA, Descatoire M, Dogan I, Huetz F, Weller S, Weill JC. IgM memory B cells: a mouse/human paradox. Cell Mol Life Sci May 2012;69(10):1625–34.

[63] Weill J, Weller S, Reynaud C. Human marginal zone B cells. Annu Rev Immunol 2009;27.

[64] Weller S, Mamani-Matsuda M, Picard C, Cordier C, Lecoeuche D, Gauthier F, et al. Somatic diversification in the absence of antigen-driven responses is the hallmark of the IgM+ IgD+ CD27+ B cell repertoire in infants. J Exp Med 2008;205.

[65] Aranburu A, Höök N, Gerasimcik N, Corleis B, Ren W, Camponeschi A, et al. Age-associated B cells expanded in autoimmune mice are memory cells sharing H-CDR3-selected repertoires. Eur J Immunol 2018;48(3):509–21.

[66] Russell Knode LM, Naradikian MS, Myles A, Scholz JL, Hao Y, Liu D, et al. Age-associated B cells express a diverse repertoire of VH and Vkappa genes with somatic hypermutation. J Immunol March 1, 2017;198(5):1921–7.

[67] Du SW, Arkatkar T, Al Qureshah F, Jacobs HM, Thouvenel CD, Chiang K, et al. Functional characterization of CD11c^{+} age-associated B cells as memory B cells. J Immunol 2019:ji1900404.

[68] Trivedi N, Weisel F, Smita S, Joachim S, Kader M, Radhakrishnan A, et al. Liver is a generative site for the B cell response to Ehrlichia muris. Immunity December 17, 2019;51(6):1088–101. e1085.

[69] Johnson JL, Rosenthal RL, Knox JJ, Myles A, Naradikian MS, Madej J, et al. The transcription factor T-bet resolves memory B cell subsets with distinct tissue distributions and antibody specificities in mice and humans. Immunity May 19, 2020;52(5):842–55. e846.

[70] Winslow GM, Papillion AM, Kenderes KJ, Levack RC. CD11c+ T-bet+ memory B cells: immune maintenance during chronic infection and inflammation? Cell Immunol November 2017;321:8–17.

[71] Jenks SA, Cashman KS, Woodruff MC, Lee FE, Sanz I. Extrafollicular responses in humans and SLE. Immunol Rev March 2019;288(1):136–48.

[72] Jenks SA, Cashman KS, Zumaquero E, Marigorta UM, Patel AV, Wang X, et al. Distinct effector B cells induced by unregulated toll-like receptor 7 contribute to pathogenic responses in systemic lupus erythematosus. Immunity October 16, 2018;49(4):725–739 e726.

[73] Woodruff MC, Ramonell RP, Nguyen DC, Cashman KS, Saini AS, Haddad NS, et al. Extrafollicular B cell responses correlate with neutralizing antibodies and morbidity in COVID-19. Nat Immunol December 2020;21(12):1506–16.

[74] Wang S, Wang J, Kumar V, Karnell JL, Naiman B, Gross PS, et al. IL-21 drives expansion and plasma cell differentiation of autoreactive CD11c(hi)T-bet(+) B cells in SLE. Nat Commun May 1, 2018;9(1):1758.

[75] Hao Y, O'Neill P, Naradikian MS, Scholz JL, Cancro MP. A B-cell subset uniquely responsive to innate stimuli accumulates in aged mice. Blood 2011;118.

[76] Du SW, Arkatkar T, Jacobs HM, Rawlings DJ, Jackson SW. Generation of functional murine CD11c(+) age-associated B cells in the absence of B cell T-bet expression. Eur J Immunol January 2019;49(1):170–8.

[77] Peng SL, Szabo SJ, Glimcher LH. T-bet regulates IgG class switching and pathogenic autoantibody production. Proc Natl Acad Sci U S A April 16, 2002;99(8):5545–50.

[78] Ehrhardt GR, Hsu JT, Gartland L, Leu CM, Zhang S, Davis RS, et al. Expression of the immunoregulatory molecule FcRH4 defines a distinctive tissue-based population of memory B cells. J Exp Med 2005;202.

[79] Portugal S, Tipton CM, Sohn H, Kone Y, Wang J, Li S, et al. Malaria-associated atypical memory B cells exhibit markedly reduced B cell receptor signaling and effector function. Elife 2015;4.

[80] Sullivan RT, Kim CC, Fontana MF, Feeney ME, Jagannathan P, Boyle MJ, et al. FCRL5 delineates functionally impaired memory B cells associated with *Plasmodium falciparum* exposure. PLoS Pathog 2015;11(5):e1004894.

[81] Naradikian MS, Myles A, Beiting DP, Roberts KJ, Dawson L, Herati RS, et al. Cutting edge: IL-4, IL-21, and IFN-gamma interact to govern T-bet and CD11c expression in TLR-activated B cells. J Immunol August 15, 2016;197(4):1023–8.

[82] Leibler C, John S, Elsner RA, Thomas KB, Smita S, Joachim S, et al. Genetic dissection of TLR9 reveals complex regulatory and cryptic proinflammatory roles in mouse lupus. Nat Immunol October 2022;23(10):1457–69.

[83] Weiss GE, Crompton PD, Li S, Walsh LA, Moir S, Traore B, et al. Atypical memory B cells are greatly expanded in individuals living in a malaria-endemic area. J Immunol August 1, 2009;183(3):2176–82.

[84] Sutton HJ, Aye R, Idris AH, Vistein R, Nduati E, Kai O, et al. Atypical B cells are part of an alternative lineage of B cells that participates in responses to vaccination and infection in humans. Cell Rep February 9, 2021;34(6):108684.

[85] Knox JJ, Myles A, Cancro MP. T-bet(+) memory B cells: generation, function, and fate. Immunol Rev March 2019;288(1):149–60.

[86] Good KL, Avery DT, Tangye SG. Resting human memory B cells are intrinsically programmed for enhanced survival and responsiveness to diverse stimuli compared to naive B cells. J Immunol January 15, 2009;182(2):890–901.

[87] Good KL, Bryant VL, Tangye SG. Kinetics of human B cell behavior and amplification of proliferative responses following stimulation with IL-21. J Immunol October 15, 2006;177(8):5236–47.

[88] Seifert M, Przekopowitz M, Taudien S, Lollies A, Ronge V, Drees B, et al. Functional capacities of human IgM memory B cells in early inflammatory responses and secondary germinal center

reactions. Proc Natl Acad Sci U S A February 10, 2015;112(6):E546–55.

[89] Arpin C, Dechanet J, Van Kooten C, Merville P, Grouard G, Briere F, et al. Generation of memory B cells and plasma cells in vitro. Science 1995;268:720–2.

[90] Duddy M, Niino M, Adatia F, Hebert S, Freedman M, Atkins H, et al. Distinct effector cytokine profiles of memory and naive human B cell subsets and implication in multiple sclerosis. J Immunol May 15, 2007;178(10):6092–9.

[91] Tabrizi SJ, Niiro H, Masui M, Yoshimoto G, Iino T, Kikushige Y, et al. T cell leukemia/lymphoma 1 and galectin-1 regulate survival/cell death pathways in human naive and IgM+ memory B cells through altering balances in Bcl-2 family proteins1. J Immunol 2009;182(3):1490–9.

[92] Moens L, Kane A, Tangye SG. Naïve and memory B cells exhibit distinct biochemical responses following BCR engagement. Immunol Cell Biol September 2016;94(8):774–86.

[93] Davey AM, Pierce SK. Intrinsic differences in the initiation of B cell receptor signaling favor responses of human IgG(+) memory B cells over IgM(+) naive B cells. J Immunol April 1, 2012;188(7):3332–41.

[94] Engels N, Konig LM, Heemann C, Lutz J, Tsubata T, Griep S, et al. Recruitment of the cytoplasmic adaptor Grb2 to surface IgG and IgE provides antigen receptor-intrinsic costimulation to class-switched B cells. Nat Immunol September 2009;10(9):1018–25.

[95] Engels N, Konig LM, Schulze W, Radtke D, Vanshylla K, Lutz J, et al. The immunoglobulin tail tyrosine motif upgrades memory-type BCRs by incorporating a Grb2-Btk signalling module. Nat Commun November 21, 2014;5:5456.

[96] Vanshylla K, Bartsch C, Hitzing C, Krumpelmann L, Wienands J, Engels N. Grb2 and GRAP connect the B cell antigen receptor to Erk MAP kinase activation in human B cells. Sci Rep March 9, 2018;8(1):4244.

[97] Ackermann JA, Radtke D, Maurberger A, Winkler TH, Nitschke L. Grb2 regulates B-cell maturation, B-cell memory responses and inhibits B-cell Ca2+ signalling. EMBO J April 20, 2011;30(8):1621–33.

[98] Bernasconi NL, Onai N, Lanzavecchia A. A role for Toll-like receptors in acquired immunity: up-regulation of TLR9 by BCR triggering in naive B cells and constitutive expression in memory B cells. Blood June 1, 2003;101(11):4500–4.

[99] Rauch PJ, Chudnovskiy A, Robbins CS, Weber GF, Etzrodt M, Hilgendorf I, et al. Innate response activator B cells protect against microbial sepsis. Science 2012;335(6068):597–601.

[100] Price MJ, Scharer CD, Kania AK, Randall TD, Boss JM. Conserved epigenetic programming and enhanced heme metabolism drive memory B cell reactivation. J Immunol April 1, 2021;206(7):1493–504.

[101] Moroney JB, Vasudev A, Pertsemlidis A, Zan H, Casali P. Integrative transcriptome and chromatin landscape analysis reveals distinct epigenetic regulations in human memory B cells. Nat Commun October 28, 2020;11(1):5435.

[102] Ehrhardt GR, Hijikata A, Kitamura H, Ohara O, Wang JY, Cooper MD. Discriminating gene expression profiles of memory B cell subpopulations. J Exp Med 2008;205.

[103] Lai AY, Mav D, Shah R, Grimm SA, Phadke D, Hatzi K, et al. DNA methylation profiling in human B cells reveals immune regulatory elements and epigenetic plasticity at Alu elements during B-cell activation. Genome Res December 2013;23(12):2030–41.

[104] Scharer CD, Blalock EL, Mi T, Barwick BG, Jenks SA, Deguchi T, et al. Epigenetic programming underpins B cell dysfunction in human SLE. Nat Immunol August 1, 2019;20(8):1071–82.

[105] Waldeyer W. Ueber Bindegewebszellen. Archiv f mikrosk Anat 1875;11:176–94.

[106] Cajal R. Manual de Anatomía Patológica General y Fundamentos de Bacteriología. 1890.

[107] Unna P. Über Plasmazellen, insbesondere beim Lupus. Monatshefte für praktische Dermatologie. 1891.

[108] Marschalko TV. Ueber die sogenannten Plasmazellen, ein Beitrag zur Kenntniss der Herkunft der entzündlichen Infiltrationszellen. T Arch Dermatol u Syphilis 1985;30(3).

[109] Von Behring EKS. Ueber das Zustandekommen der Diphtherie-Immunitat and der Tetanus-Immunitat bei Thieren. Dtsch Med Wochenschr 1890;16:1113–4.

[110] Fagraeus A. Plasma cellular reaction and its relation to the formation of antibodies in vitro. Nature April 12, 1947;159(4041):499.

[111] Cooper MD, Raymond DA, Peterson RD, South MA, Good RA. The functions of the thymus system and the bursa system in the chicken. J Exp Med January 1, 1966;123(1):75–102.

[112] Cooper MD, Peterson RD, Good RA. Delineation of the thymic and bursal lymphoid systems in the chicken. Nature January 9, 1965;205:143–6.

[113] Mäkelä O, Nossal GJV. Autoradiographic studies on the immune response: II. DNA synthesis amongst single antibody-producing cells. J Exp Med 1962;115(1):231–44.

[114] Nossal GJV, Mäkelä O. Autoradiographic studies on the immune response: I. The kinetics of plasma cell proliferation. J Exp Med 1962;115(1):209–30.

[115] Benner R, Meima F, van der Meulen GM, van Muiswinkel WB. Antibody formation in mouse bone marrow. I. Evidence for the development of plaque-forming cells in situ. Immunology February 1974;26(2):247–55.

[116] Ho F, Lortan JE, MacLennan IC, Khan M. Distinct short-lived and long-lived antibody-producing cell populations. Eur J Immunol October 1986;16(10):1297–301.

[117] Manz RA, Thiel A, Radbruch A. Lifetime of plasma cells in the bone marrow. Nature July 10, 1997;388(6638):133–4.

[118] Slifka MK, Antia R, Whitmire JK, Ahmed R. Humoral immunity due to long-lived plasma cells. Immunity March 1998;8(3):363–72.

[119] Manz RA, Lohning M, Cassese G, Thiel A, Radbruch A. Survival of long-lived plasma cells is independent of antigen. Int Immunol November 1998;10(11):1703–11.

[120] Pestronk A, Florence J, Miller T, Choksi R, Al-Lozi MT, Levine TD. Treatment of IgM antibody associated polyneuropathies using rituximab. J Neurol Neurosurg Psychiatry April 2003;74(4):485–9.

[121] Edwards JC, Szczepanski L, Szechinski J, Filipowicz-Sosnowska A, Emery P, Close DR, et al. Efficacy of B-cell-targeted therapy with rituximab in patients with rheumatoid arthritis. N Engl J Med June 17, 2004;350(25):2572–81.

[122] Odendahl M, Mei H, Hoyer BF, Jacobi AM, Hansen A, Muehlinghaus G, et al. Generation of migratory antigen-specific plasma blasts and mobilization of resident plasma cells in a secondary immune response. Blood February 15, 2005;105(4):1614–21.

[123] Mei HE, Yoshida T, Sime W, Hiepe F, Thiele K, Manz RA, et al. Blood-borne human plasma cells in steady state are derived from mucosal immune responses. Blood March 12, 2009;113(11):2461–9.

[124] Bernasconi NL, Traggiai E, Lanzavecchia A. Maintenance of serological memory by polyclonal activation of human memory B cells. Science December 13, 2002;298(5601):2199–202.

[125] Hargreaves DC, Hyman PL, Lu TT, Ngo VN, Bidgol A, Suzuki G, et al. A coordinated change in chemokine responsiveness guides plasma cell movements. J Exp Med July 2, 2001;194(1):45–56.

[126] Kabashima K, Haynes NM, Xu Y, Nutt SL, Allende ML, Proia RL, et al. Plasma cell S1P1 expression determines secondary lymphoid organ retention versus bone marrow tropism. J Exp Med November 27, 2006;203(12):2683–90.

[127] Hauser AE, Debes GF, Arce S, Cassese G, Hamann A, Radbruch A, et al. Chemotactic responsiveness toward ligands for CXCR3 and CXCR4 is regulated on plasma blasts during the time course of a memory immune response. J Immunol August 1, 2002;169(3):1277–82.

[128] Wehrli N, Legler DF, Finke D, Toellner KM, Loetscher P, Baggiolini M, et al. Changing responsiveness to chemokines allows medullary plasmablasts to leave lymph nodes. Eur J Immunol February 2001;31(2):609–16.

[129] Pappu R, Schwab SR, Cornelissen I, Pereira JP, Regard JB, Xu Y, et al. Promotion of lymphocyte egress into blood and lymph by distinct sources of sphingosine-1-phosphate. Science April 13, 2007;316(5822):295–8.

[130] Tashiro K, Tada H, Heilker R, Shirozu M, Nakano T, Honjo T. Signal sequence trap: a cloning strategy for secreted proteins and type I membrane proteins. Science July 30, 1993;261(5121):600–3.

[131] Nagasawa T, Kikutani H, Kishimoto T. Molecular cloning and structure of a pre-B-cell growth-stimulating factor. Proc Natl Acad Sci U S A March 15, 1994;91(6):2305–9.

[132] Bleul CC, Fuhlbrigge RC, Casasnovas JM, Aiuti A, Springer TA. A highly efficacious lymphocyte chemoattractant, stromal cell-derived factor 1 (SDF-1). J Exp Med 1996;184(3):1101–9.

[133] Muehlinghaus G, Cigliano L, Huehn S, Peddinghaus A, Leyendeckers H, Hauser AE, et al. Regulation of CXCR3 and CXCR4 expression during terminal differentiation of memory B cells into plasma cells. Blood May 15, 2005;105(10):3965–71.

[134] Cheng Q, Khodadadi L, Taddeo A, Klotsche J, Hoyer BF, Radbruch A, et al. CXCR4-CXCL12 interaction is important for plasma cell homing and survival in NZB/W mice. Eur J Immunol June 2018;48(6):1020–9.

[135] Arce S, Luger E, Muehlinghaus G, Cassese G, Hauser A, Horst A, et al. CD38 low IgG-secreting cells are precursors of various CD38 high-expressing plasma cell populations. J Leukoc Biol June 2004;75(6):1022–8.

[136] Medina F, Segundo C, Campos-Caro A, Gonzalez-Garcia I, Brieva JA. The heterogeneity shown by human plasma cells from tonsil, blood, and bone marrow reveals graded stages of increasing maturity, but local profiles of adhesion molecule expression. Blood March 15, 2002;99(6):2154–61.

[137] Mei HE, Wirries I, Frolich D, Brisslert M, Giesecke C, Grun JR, et al. A unique population of IgG-expressing plasma cells lacking CD19 is enriched in human bone marrow. Blood March 12, 2015;125(11):1739–48.

[138] Halliley JL, Tipton CM, Liesveld J, Rosenberg AF, Darce J, Gregoretti IV, et al. Long-lived plasma cells are contained within the CD19(-)CD38(hi)CD138(+) subset in human bone marrow. Immunity July 21, 2015;43(1):132–45.

[139] Duan M, Nguyen DC, Joyner CJ, Saney CL, Tipton CM, Andrews J, et al. Human bone marrow plasma cell atlas: maturation and survival pathways unraveled by single cell analyses. bioRxiv January 20, 2023.

[140] Mir-Farzin Mashreghi MF-G, Durek P, et al. Recruitment of plasma cells to the bone marrow in primary and secondary immune reactions. Research Square 2023. 10.21203/rs.3.rs-2378630/v1.

[141] Delaloy C, Schuh W, Jack HM, Bonaud A, Espeli M. Single-cell resolution of plasma cell fate programming in health and disease. Eur J Immunol January 2022;52(1):10–23.

[142] Mattioli CA, Tomasi Jr TB. The life span of IgA plasma cells from the mouse intestine. J Exp Med August 1, 1973;138(2):452–60.

[143] Wilmore JR, Gaudette BT, Gómez Atria D, Rosenthal RL, Reiser SK, Meng W, et al. IgA plasma cells are long-lived residents of gut and bone marrow that express isotype- and tissue-specific gene expression patterns. Front Immunol December 24, 2021;12.

[144] Mesin L, Di Niro R, Thompson KM, Lundin KE, Sollid LM. Long-lived plasma cells from human small intestine biopsies secrete immunoglobulins for many weeks in vitro. J Immunol September 15, 2011;187(6):2867–74.

[145] Lemke A, Kraft M, Roth K, Riedel R, Lammerding D, Hauser AE. Long-lived plasma cells are generated in mucosal immune responses and contribute to the bone marrow plasma cell pool in mice. Mucosal Immunol January 2016;9(1):83–97.

[146] Landsverk OJ, Snir O, Casado RB, Richter L, Mold JE, Reu P, et al. Antibody-secreting plasma cells persist for decades in human intestine. J Exp Med February 2017;214(2):309–17.

[147] Hapfelmeier S, Lawson MA, Slack E, Kirundi JK, Stoel M, Heikenwalder M, et al. Reversible microbial colonization of germ-free mice reveals the dynamics of IgA immune responses. Science June 25, 2010;328(5986):1705–9.

[148] Bowman EP, Kuklin NA, Youngman KR, Lazarus NH, Kunkel EJ, Pan J, et al. The intestinal chemokine thymus-expressed chemokine (CCL25) attracts IgA antibody-secreting cells. J Exp Med January 21, 2002;195(2):269–75.

[149] Kunkel EJ, Kim CH, Lazarus NH, Vierra MA, Soler D, Bowman EP, et al. CCR10 expression is a common feature of circulating and mucosal epithelial tissue IgA Ab-secreting cells. J Clin Invest April 2003;111(7):1001–10.

[150] Pabst O, Ohl L, Wendland M, Wurbel MA, Kremmer E, Malissen B, et al. Chemokine receptor CCR9 contributes to the localization of plasma cells to the small intestine. J Exp Med February 2, 2004;199(3):411–6.

[151] Baggiolini M. Chemokines and leukocyte traffic. Nature April 9, 1998;392(6676):565–8.

[152] Cassese G, Lindenau S, de Boer B, Arce S, Hauser A, Riemekasten G, et al. Inflamed kidneys of NZB/W mice are a major site for the homeostasis of plasma cells. Eur J Immunol September 2001;31(9):2726–32.

[153] Anderson SM, Hannum LG, Shlomchik MJ. Cutting Edge: memory B cell survival and function in the absence of secreted antibody and immune complexes on follicular dendritic cells. J Immunol April 15, 2006;176(8):4515–9.

[154] Jones DD, Wilmore JR, Allman D. Cellular dynamics of memory B cell populations: IgM+ and IgG+ memory B cells persist indefinitely as quiescent cells. J Immunol November 15, 2015;195(10):4753–9.
[155] Vikstrom I, Carotta S, Luthje K, Peperzak V, Jost PJ, Glaser S, et al. Mcl-1 is essential for germinal center formation and B cell memory. Science November 19, 2010;330(6007):1095–9.
[156] Carrington EM, Vikstrom IB, Light A, Sutherland RM, Londrigan SL, Mason KD, et al. BH3 mimetics antagonizing restricted prosurvival Bcl-2 proteins represent another class of selective immune modulatory drugs. Proc Natl Acad Sci U S A June 15, 2010;107(24):10967–71.
[157] Vieira P, Rajewsky K. Persistence of memory B cells in mice deprived of T cell help. Int Immunol 1990;2(6):487–94.
[158] Scholz JL, Crowley JE, Tomayko MM, Steinel N, O'Neill PJ, Quinn 3rd WJ, et al. BLyS inhibition eliminates primary B cells but leaves natural and acquired humoral immunity intact. Proc Natl Acad Sci U S A October 7, 2008;105(40):15517–22.
[159] Benson MJ, Dillon SR, Castigli E, Geha RS, Xu S, Lam K-P, et al. Cutting edge: the dependence of plasma cells and independence of memory B cells on BAFF and APRIL. J Immunol March 15, 2008;180(6):3655–9. 2008.
[160] Müller-Winkler J, Mitter R, Rappe JCF, Vanes L, Schweighoffer E, Mohammadi H, et al. Critical requirement for BCR, BAFF, and BAFFR in memory B cell survival. J Exp Med 2020;218(2).
[161] Ackermann JA, Nys J, Schweighoffer E, McCleary S, Smithers N, Tybulewicz VL. Syk tyrosine kinase is critical for B cell antibody responses and memory B cell survival. J Immunol May 15, 2015;194(10):4650–6.
[162] Hikida M, Casola S, Takahashi N, Kaji T, Takemori T, Rajewsky K, et al. PLC-γ2 is essential for formation and maintenance of memory B cells. J Exp Med 2009;206(3):681–9.
[163] Clybouw C, Fischer S, Auffredou MT, Hugues P, Alexia C, Bouillet P, et al. Regulation of memory B-cell survival by the BH3-only protein Puma. Blood October 13, 2011;118(15):4120–8.
[164] Luckey CJ, Bhattacharya D, Goldrath AW, Weissman IL, Benoist C, Mathis D. Memory T and memory B cells share a transcriptional program of self-renewal with long-term hematopoietic stem cells. Proc Natl Acad Sci U S A February 28, 2006;103(9):3304–9.
[165] Fearon DT, Manders P, Wagner SD. Arrested differentiation, the self-renewing memory lymphocyte, and vaccination. Science July 13, 2001;293(5528):248–50.
[166] Cornelis R, Hahne S, Taddeo A, Petkau G, Malko D, Durek P, et al. Stromal cell-contact dependent PI3K and APRIL induced NF-kappaB signaling prevent mitochondrial- and ER stress induced death of memory plasma cells. Cell Rep August 4, 2020;32(5):107982.
[167] Lindquist RL, Niesner RA, Hauser AE. In the right place, at the right time: spatiotemporal conditions determining plasma cell survival and function. Front Immunol 2019;10:788.
[168] Mamani-Matsuda M, Cosma A, Weller S, Faili A, Staib C, Garcon L, et al. The human spleen is a major reservoir for long-lived vaccinia virus-specific memory B cells. Blood May 1, 2008;111(9):4653–9.
[169] Chappert P, Huetz F, Espinasse MA, Chatonnet F, Pannetier L, Da Silva L, et al. Human anti-smallpox long-lived memory B cells are defined by dynamic interactions in the splenic niche and long-lasting germinal center imprinting. Immunity October 11, 2022;55(10):1872–90. e1879.
[170] Kibler A, Budeus B, Homp E, Bronischewski K, Berg V, Sellmann L, et al. Systematic memory B cell archiving and random display shape the human splenic marginal zone throughout life. J Exp Med April 5, 2021;218(4).
[171] Anderson SM, Tomayko MM, Ahuja A, Haberman AM, Shlomchik MJ. New markers for murine memory B cells that define mutated and unmutated subsets. J Exp Med September 3, 2007;204(9):2103–14.
[172] Paramithiotis E, Cooper MD. Memory B lymphocytes migrate to bone marrow in humans. Proc Natl Acad Sci U S A January 7, 1997;94(1):208–12.
[173] Barker KA, Etesami NS, Shenoy AT, Arafa EI, Lyon de Ana C, Smith NM, et al. Lung-resident memory B cells protect against bacterial pneumonia. J Clin Invest June 1, 2021;131(11).
[174] Onodera T, Takahashi Y, Yokoi Y, Ato M, Kodama Y, Hachimura S, et al. Memory B cells in the lung participate in protective humoral immune responses to pulmonary influenza virus reinfection. Proc Natl Acad Sci U S A February 14, 2012;109(7):2485–90.
[175] McGregor DD, Gowans JL. The antibody response of rats depleted of lymphocytes by chronic drainage from the thoracic duct. J Exp Med January 31, 1963;117(2):303–20.
[176] Chang HD, Radbruch A. Maintenance of quiescent immune memory in the bone marrow. Eur J Immunol July 2021;51(7):1592–601.
[177] Zehentmeier S, Roth K, Cseresnyes Z, Sercan O, Horn K, Niesner RA, et al. Static and dynamic components synergize to form a stable survival niche for bone marrow plasma cells. Eur J Immunol August 2014;44(8):2306–17.
[178] Radbruch A, Muehlinghaus G, Luger EO, Inamine A, Smith KG, Dorner T, et al. Competence and competition: the challenge of becoming a long-lived plasma cell. Nat Rev Immunol October 2006;6(10):741–50.
[179] DiLillo DJ, Hamaguchi Y, Ueda Y, Yang K, Uchida J, Haas KM, et al. Maintenance of long-lived plasma cells and serological memory despite mature and memory B cell depletion during CD20 immunotherapy in mice. J Immunol January 1, 2008;180(1):361–71.
[180] Benet Z, Jing Z, Fooksman DR. Plasma cell dynamics in the bone marrow niche. Cell Rep February 9, 2021;34(6):108733.
[181] Cassese G, Arce S, Hauser AE, Lehnert K, Moewes B, Mostarac M, et al. Plasma cell survival is mediated by synergistic effects of cytokines and adhesion-dependent signals. J Immunol August 15, 2003;171(4):1684–90.
[182] Nguyen DC, Garimalla S, Xiao H, Kyu S, Albizua I, Galipeau J, et al. Factors of the bone marrow microniche that support human plasma cell survival and immunoglobulin secretion. Nat Commun September 12, 2018;9(1):3698.
[183] Jourdan M, Cren M, Robert N, Bollore K, Fest T, Duperray C, et al. IL-6 supports the generation of human long-lived plasma cells in combination with either APRIL or stromal cell-soluble factors. Leukemia August 2014;28(8):1647–56.
[184] Minges Wols HA, Underhill GH, Kansas GS, Witte PL. The role of bone marrow-derived stromal cells in the maintenance of plasma cell longevity. J Immunol October 15, 2002;169(8):4213–21.
[185] Spencer JA, Ferraro F, Roussakis E, Klein A, Wu J, Runnels JM, et al. Direct measurement of local oxygen concentration in the bone marrow of live animals. Nature April 10, 2014;508(7495):269–73.

[186] O'Connor BP, Raman VS, Erickson LD, Cook WJ, Weaver LK, Ahonen C, et al. BCMA is essential for the survival of long-lived bone marrow plasma cells. J Exp Med January 5, 2004;199(1):91–8.
[187] Bonaud A, Gargowitsch L, Gilbert SM, Rajan E, Canales-Herrerias P, Stockholm D, et al. Sec22b is a critical and nonredundant regulator of plasma cell maintenance. Proc Natl Acad Sci U S A January 10, 2023;120(2):e2213056120.
[188] Manne C, Takaya A, Yamasaki Y, Mursell M, Hojyo S, Wu TY, et al. Salmonella SiiE prevents an efficient humoral immune memory by interfering with IgG(+) plasma cell persistence in the bone marrow. Proc Natl Acad Sci U S A April 9, 2019;116(15):7425–30.
[189] Canetti M, Barda N, Gilboa M, Indenbaum V, Asraf K, Gonen T, et al. Six-month follow-up after a fourth BNT162b2 vaccine dose. N Engl J Med December 1, 2022;387(22):2092–4.
[190] Schaefer-Babajew D, Wang Z, Muecksch F, Cho A, Loewe M, Cipolla M, et al. Antibody feedback regulates immune memory after SARS-CoV-2 mRNA vaccination. Nature January 2023;613(7945):735–42.
[191] Smith T. Active immunity produced by so called balanced or neutral mixtures of diphtheria toxin and antitoxin. J Exp Med March 1, 1909;11(2):241–56.
[192] Munro APS, Feng S, Janani L, Cornelius V, Aley PK, Babbage G, et al. Safety, immunogenicity, and reactogenicity of BNT162b2 and mRNA-1273 COVID-19 vaccines given as fourth-dose boosters following two doses of ChAdOx1 nCoV-19 or BNT162b2 and a third dose of BNT162b2 (COV-BOOST): a multicentre, blinded, phase 2, randomised trial. Lancet Infect Dis August 2022;22(8):1131–41.
[193] Sanyal M, Holmes TH, Maecker HT, Albrecht RA, Dekker CL, He XS, et al. Diminished B-cell response after repeat influenza vaccination. J Infect Dis April 19, 2019;219(10):1586–95.
[194] Ellebedy AH, Nachbagauer R, Jackson KJL, Dai YN, Han J, Alsoussi WB, et al. Adjuvanted H5N1 influenza vaccine enhances both cross-reactive memory B cell and strain-specific naive B cell responses in humans. Proc Natl Acad Sci U S A July 28, 2020;117(30):17957–64.
[195] Zarnitsyna VI, Ellebedy AH, Davis C, Jacob J, Ahmed R, Antia R. Masking of antigenic epitopes by antibodies shapes the humoral immune response to influenza. Philos Trans R Soc Lond B Biol Sci September 5, 2015;370(1676).
[196] Francis T. On the doctrine of original antigenic sin. Proc Am Phil Soc 1960;104(6):572–8.
[197] Reynolds CJ, Gibbons JM, Pade C, Lin KM, Sandoval DM, Pieper F, et al. Heterologous infection and vaccination shapes immunity against SARS-CoV-2 variants. Science January 14, 2022;375(6577):183–92.
[198] Arevalo CP, Le Sage V, Bolton MJ, Eilola T, Jones JE, Kormuth KA, et al. Original antigenic sin priming of influenza virus hemagglutinin stalk antibodies. Proc Natl Acad Sci U S A July 21, 2020;117(29):17221–7.
[199] Aaen KH, Anthi AK, Sandlie I, Nilsen J, Mester S, Andersen JT. The neonatal Fc receptor in mucosal immune regulation. Scand J Immunol February 2021;93(2):e13017.
[200] Spiekermann GM, Finn PW, Ward ES, Dumont J, Dickinson BL, Blumberg RS, et al. Receptor-mediated immunoglobulin G transport across mucosal barriers in adult life: functional expression of FcRn in the mammalian lung. J Exp Med August 5, 2002;196(3):303–10.
[201] Turula H, Wobus CE. The role of the polymeric immunoglobulin receptor and secretory immunoglobulins during mucosal infection and immunity. Viruses May 3, 2018;10(5).
[202] Chan RWY, Liu S, Cheung JY, Tsun JGS, Chan KC, Chan KYY, et al. The mucosal and serological immune responses to the novel coronavirus (SARS-CoV-2) vaccines. Front Immunol 2021;12:744887.
[203] Hall V, Foulkes S, Insalata F, Kirwan P, Saei A, Atti A, et al. Protection against SARS-CoV-2 after Covid-19 vaccination and previous infection. N Engl J Med March 31, 2022;386(13):1207–20.
[204] Sano K, Bhavsar D, Singh G, Floda D, Srivastava K, Gleason C, et al. SARS-CoV-2 vaccination induces mucosal antibody responses in previously infected individuals. Nat Commun September 1, 2022;13(1):5135.
[205] Turner JS, O'Halloran JA, Kalaidina E, Kim W, Schmitz AJ, Zhou JQ, et al. SARS-CoV-2 mRNA vaccines induce persistent human germinal centre responses. Nature August 2021;596(7870):109–13.
[206] Kim W, Zhou JQ, Horvath SC, Schmitz AJ, Sturtz AJ, Lei T, et al. Germinal centre-driven maturation of B cell response to mRNA vaccination. Nature April 2022;604(7904):141–5.
[207] Maschmeyer P, Chang HD, Cheng Q, Mashreghi MF, Hiepe F, Alexander T, et al. Immunological memory in rheumatic inflammation – A roadblock to tolerance induction. Nat Rev Rheumatol May 2021;17(5):291–305.
[208] Hoyer B, Moser K, Hauser A, Peddinghaus A, Voigt C, Eilat D, et al. Short-lived plasmablasts and long-lived plasma cells contribute to chronic humoral autoimmunity in NZB/W mice. J Exp Med June 7, 2004;199(11):1577–84.
[209] Mumtaz IM, Hoyer BF, Panne D, Moser K, Winter O, Cheng QY, et al. Bone marrow of NZB/W mice is the major site for plasma cells resistant to dexamethasone and cyclophosphamide: implications for the treatment of autoimmunity. J Autoimmun September 2012;39(3):180–8.
[210] Jonsdottir T, Gunnarsson I, Risselada A, Henriksson EW, Klareskog L, van Vollenhoven RF. Treatment of refractory SLE with rituximab plus cyclophosphamide: clinical effects, serological changes, and predictors of response. Ann Rheum Dis March 2008;67(3):330–4.
[211] Alexander T, Thiel A, Rosen O, Massenkeil G, Sattler A, Kohler S, et al. Depletion of autoreactive immunologic memory followed by autologous hematopoietic stem cell transplantation in patients with refractory SLE induces long-term remission through de novo generation of a juvenile and tolerant immune system. Blood January 1, 2009;113(1):214–23.
[212] Mackensen A, Muller F, Mougiakakos D, Boltz S, Wilhelm A, Aigner M, et al. Anti-CD19 CAR T cell therapy for refractory systemic lupus erythematosus. Nat Med October 2022;28(10):2124–32.
[213] Luger EO, Fokuhl V, Wegmann M, Abram M, Tillack K, Achatz G, et al. Induction of long-lived allergen-specific plasma cells by mucosal allergen challenge. J Allergy Clin Immunol October 2009;124(4):819–26.
[214] Cheng Q, Mumtaz IM, Khodadadi L, Radbruch A, Hoyer BF, Hiepe F. Autoantibodies from long-lived 'memory' plasma cells of NZB/W mice drive immune complex nephritis. Ann Rheum Dis October 10, 2013.

[215] Bastard P, Gervais A, Le Voyer T, Rosain J, Philippot Q, Manry J, et al. Autoantibodies neutralizing type I IFNs are present in ~4% of uninfected individuals over 70 years old and account for ~20% of COVID-19 deaths. Sci Immunol August 19, 2021;6(62).

[216] Bastard P, Rosen LB, Zhang Q, Michailidis E, Hoffmann HH, Zhang Y, et al. Autoantibodies against type I IFNs in patients with life-threatening COVID-19. Science October 23, 2020;370(6515).

[217] Puel A, Bastard P, Bustamante J, Casanova JL. Human autoantibodies underlying infectious diseases. J Exp Med April 4, 2022;219(4).

[218] Ahuja A, Anderson SM, Khalil A, Shlomchik MJ. Maintenance of the plasma cell pool is independent of memory B cells. Proc Natl Acad Sci U S A March 25, 2008;105(12):4802–7.

[219] Amanna IJ, Carlson NE, Slifka MK. Duration of humoral immunity to common viral and vaccine antigens. N Engl J Med November 8, 2007;357(19):1903–15.

[220] Cambridge G, Leandro MJ, Edwards JC, Ehrenstein MR, Salden M, Bodman-Smith M, et al. Serologic changes following B lymphocyte depletion therapy for rheumatoid arthritis. Arthritis Rheum August 2003;48(8):2146–54.

[221] Ferraro AJ, Drayson MT, Savage CO, MacLennan IC. Levels of autoantibodies, unlike antibodies to all extrinsic antigen groups, fall following B cell depletion with Rituximab. Eur J Immunol January 2008;38(1):292–8.

[222] Mahajan VS, Mattoo H, Deshpande V, Pillai SS, Stone JH. IgG4-related disease. Ann Rev Pathol 2014;9:315–47.

[223] Stone JH, Zen Y, Deshpande V. IgG4-related disease. N Engl J Med February 9, 2012;366(6):539–51.

[224] Hamano H, Kawa S, Horiuchi A, Unno H, Furuya N, Akamatsu T, et al. High serum IgG4 concentrations in patients with sclerosing pancreatitis. N Engl J Med March 8, 2001;344(10):732–8.

[225] Kamisawa T, Funata N, Hayashi Y, Eishi Y, Koike M, Tsuruta K, et al. A new clinicopathological entity of IgG4-related autoimmune disease. J Gastroenterol 2003;38(10):982–4.

[226] Szyszko EA, Brokstad KA, Oijordsbakken G, Jonsson MV, Jonsson R, Skarstein K. Salivary glands of primary Sjogren's syndrome patients express factors vital for plasma cell survival. Arthritis Res Ther January 7, 2011;13(1):R2.

[227] Balabanian K, Couderc J, Bouchet-Delbos L, Amara A, Berrebi D, Foussat A, et al. Role of the chemokine stromal cell-derived factor 1 in autoantibody production and nephritis in murine lupus. J Immunol March 15, 2003;170(6):3392–400.

[228] Moser K, Kalies K, Szyska M, Humrich JY, Amann K, Manz RA. CXCR3 promotes the production of IgG1 autoantibodies but is not essential for the development of lupus nephritis in NZB/NZW mice. Arthritis Rheum April 2012;64(4):1237–46.

[229] Mahevas M, Patin P, Huetz F, Descatoire M, Cagnard N, Bole-Feysot C, et al. B cell depletion in immune thrombocytopenia reveals splenic long-lived plasma cells. J Clin Invest January 2, 2013;123(1):432–42.

[230] Chu VT, Enghard P, Schurer S, Steinhauser G, Rudolph B, Riemekasten G, et al. Systemic activation of the immune system induces aberrant BAFF and APRIL expression in B cells in patients with systemic lupus erythematosus. Arthritis Rheum July 2009;60(7):2083–93.

[231] Choi MY, Costenbader KH. Understanding the concept of pre-clinical autoimmunity: prediction and prevention of systemic lupus erythematosus: identifying risk factors and developing strategies against disease development. Front Immunol 2022;13:890522.

[232] Romao VC, Fonseca JE. Disease mechanisms in preclinical rheumatoid arthritis: a narrative review. Front Med 2022;9:689711.

[233] Hofmann K, Clauder AK, Manz RA. Targeting B cells and plasma cells in autoimmune diseases. Front Immunol 2018;9:835.

[234] Rudick R, Polman C, Clifford D, Miller D, Steinman L. Natalizumab: bench to bedside and beyond. JAMA Neurol February 2013;70(2):172–82.

[235] Neubert K, Meister S, Moser K, Weisel F, Maseda D, Amann K, et al. The proteasome inhibitor bortezomib depletes plasma cells and protects mice with lupus-like disease from nephritis. Nat Med July 2008;14(7):748–55.

[236] Gomez AM, Vrolix K, Martinez-Martinez P, Molenaar PC, Phernambucq M, van der Esch E, et al. Proteasome inhibition with bortezomib depletes plasma cells and autoantibodies in experimental autoimmune myasthenia gravis. J Immunol February 15, 2011;186(4):2503–13.

[237] Alexander T, Sarfert R, Klotsche J, Kuhl AA, Rubbert-Roth A, Lorenz HM, et al. The proteasome inhibitior bortezomib depletes plasma cells and ameliorates clinical manifestations of refractory systemic lupus erythematosus. Ann Rheum Dis July 2015;74(7):1474–8.

[238] Pruss H. Autoantibodies in neurological disease. Nat Rev Immunol December 2021;21(12):798–813.

[239] Mazepa MA, Raval JS, Moll S, Ma A, Park YA. Bortezomib induces clinical remission and reduction of ADAMTS13 inhibitory antibodies in relapsed refractory idiopathic thrombotic thrombocytopenic purpura. Br J Haematol March 2014;164(6):900–2.

[240] van Balen T, Schreuder MF, de Jong H, van de Kar NC. Refractory thrombotic thrombocytopenic purpura in a 16-year-old girl: successful treatment with bortezomib. Eur J Haematol January 2014;92(1):80–2.

[241] Carson KR, Beckwith LG, Mehta J. Successful treatment of IgM-mediated autoimmune hemolytic anemia with bortezomib. Blood January 28, 2010;115(4):915.

[242] Khan ML, Colby TV, Viggiano RW, Fonseca R. Treatment with bortezomib of a patient having hyper IgG4 disease. Clin Lymphoma, Myeloma & Leukemia June 2010;10(3):217–9.

[243] Ejaz NS, Alloway RR, Halleck F, Durr M, Budde K, Woodle ES. Review of bortezomib treatment of antibody-mediated rejection in renal transplantation. Antioxidants Redox Signal March 18, 2014.

[244] Walsh RC, Alloway RR, Girnita AL, Woodle ES. Proteasome inhibitor-based therapy for antibody-mediated rejection. Kidney Int June 2012;81(11):1067–74.

[245] Woodle ES, Alloway RR, Girnita A. Proteasome inhibitor treatment of antibody-mediated allograft rejection. Curr Opin Organ Transplant August 2011;16(4):434–8.

[246] Ostendorf L, Burns M, Durek P, Heinz GA, Heinrich F, Garantziotis P, et al. Targeting CD38 with daratumumab in refractory systemic lupus erythematosus. N Engl J Med September 17, 2020;383(12):1149–55.

[247] Cheng Q, Pelz A, Taddeo A, Khodadadi L, Klotsche J, Hoyer BF, et al. Selective depletion of plasma cells in vivo based on the specificity of their secreted antibodies. Eur J Immunol February 2020;50(2):284–91.

Chapter 17

Homoeostatic Versus Pathogenic Autoantibodies: Origin, Structure and Effector Functions

Yemil Atisha-Fregoso[1], Rita Pozovskiy[1,2], Shabirul Haque[1], Manami Watanabe[1], Yong-Rui Zou[1] and Betty Diamond[1]

[1] *Center for Autoimmune Musculoskeletal and Hematopoietic Diseases, The Feinstein Institutes for Medical Research, Northwell Health, Manhasset, NY, United States;* [2] *Zucker School of Medicine at Hofstra/Northwell, Hempstead, NY, United States*

1. INTRODUCTION

A consequence of generating immunoglobulin (Ig) molecules by random rearrangement of variable, diversity and joining gene (V(D)J) segments and random association of heavy and light chains (LCs) (described in Chapter 2) is that many B cells express germline-encoded Ig molecules that recognise self-components. While many of these autoreactive B cells are censored during the process of B-cell maturation, all subsets of mature B cells include some cells with specificity for self [1,2]. This phenomenon does not have an intrinsic negative consequence, as IgM autoreactive antibodies play a protective role in maintaining physiologic homoeostasis [3]. In contrast, high titres of IgG autoantibodies are characteristic of a breach of immune tolerance checkpoints and are often harmful, contributing to the development and pathogenesis of autoimmune diseases. These antibodies can either cause direct cellular dysfunction or tissue injury or can form immune complexes with self-antigen and generate an inflammatory milieu.

The B-cell lineage consists of several subsets which diverge early in development. All of them are able to differentiate into antibody-secreting plasma cells (PCs). B1 cells and a subtype of B2 cells, marginal zone B cells (MZB cells), are enriched in autoreactive clones. Many of the IgM autoantibodies produced by these subtypes belong to the repertoire of 'natural antibodies', which have antimicrobial properties and function in the non-inflammatory clearance of apoptotic debris. Autoantibodies that are involved in autoimmune diseases primarily have IgG constant regions and are produced by B2 cells with T cell help. Many have acquired autoreactivity by somatic hypermutation. Moreover, as Ig molecules are expressed on B-cell membranes (B-cell receptors; BCRs), B cells with an autoreactive BCR may function as antigen-presenting cells (APCs) to autoreactive T cells.

In this chapter, we will first discuss the relevance of natural autoantibodies to homoeostasis and defence against pathogens. We will describe how B1 and MZB are selected for homoeostatic autoreactivity, meanwhile follicular B2 cells are selected against organ-damaging autoreactivity. We will briefly discuss post-translational modifications that make an antibody immunosuppressive or proinflammatory. We will address mechanisms by which pathogenic autoantibodies are generated in autoimmune diseases, and their mechanisms for inducing damage. In this regard, we will use systemic lupus erythematosus (SLE) as an example of a disease in which abnormal B-cell activation is central for pathogenesis. Indeed, most of our current knowledge of how multiple mechanisms are combined to induce disease come from studies in SLE. Finally, we will discuss therapeutic strategies that target B cells and pathogenic antibodies in SLE.

2. NATURAL AUTOANTIBODIES AND THEIR BIOLOGICAL EFFECTS

Natural antibodies are germline-encoded Ig molecules produced by B cells even in early ontogeny. Studies conducted in germ-free, antigen-free and conventional or specific pathogen-free mice describe maintenance of a normal serum level of IgM with a profile of specificity for endogenous antigen [4]; human newborns possess an analogous IgM repertoire [5]. Natural antibodies provide a

Molecular Biology of B Cells. https://doi.org/10.1016/B978-0-323-95895-0.00006-4

first line of defence before adaptive immunity becomes fully mature or activated. In addition to direct neutralisation of invading bacteria or viruses, natural antibodies can also opsonise pathogens for phagocytosis by innate immune cells or for complement-mediated lysis [6]. These antibodies can also scavenge molecules of microbial origin (pathogen-associated molecular patterns: PAMPs) and prevent them from binding and activating pattern recognition receptors (PRRs) [7].

Natural antibodies are often described as 'polyreactive' due to their recognition of a broad range of structural motifs ubiquitously expressed on both microbial and self-components. For example, healthy mice harbour natural antibodies with a dominant usage of the T15 idiotype which binds phosphorylcholine, a determinant on the cell wall of *Streptococcus pneumoniae* and several other microbes. While these naturally occurring T15 antibodies provide an innate-like immune response against virulent pneumococcal infections, they also recognise the phosphorylcholine head group of oxidised phospholipids present on oxidised low-density lipoprotein (OxLDL) and on apoptotic cells [8]. Thus, natural antibodies not only function to clear pathogens, they also have a protective role to prevent the development of autoimmune diseases by clearing apoptotic debris in a non-immunogenic fashion [9]. They neutralise and remove self-molecules that are damage-associated molecular patterns (DAMPs) that otherwise would bind PRR. These include phospholipids, other reactive lipoproteins and DNA molecules bearing CpG motifs, especially oxidised DNA [10,11]. Natural autoantibodies also recognise insulin and thus are involved in the regulation of physiological blood glucose levels [12]. The extent to which natural antibodies regulate homoeostasis remains only partially appreciated, as our understanding of the repertoire and function of natural autoantibodies is continuously expanding.

Support of the protective role of natural antibodies in preventing autoimmunity comes from murine studies in which infusion of monoclonal or polyclonal anti-double-stranded (ds)DNA IgM delays onset of disease in lupus-prone mice [13,14]. Similarly, studies of twins discordant for SLE show that the healthy twin has a higher level of natural anti-phosphorylcholine IgM than the affected twin, suggesting that natural IgM antibodies may reduce risk for SLE in those that are genetically predisposed [15]. Higher levels of IgM anti-phosphorylcholine or anti−dsDNA antibodies in SLE patients also correlate with less organ damage [16]. Natural antibodies can also recognise modifications present on damaged cells and so help regulate homoeostasis. For example, senescent or injured red blood cells display phosphatidylcholine and I/i carbohydrate epitopes, self-determinants recognised by natural antibodies. The autoantibodies binding these motifs help in the clearance of senescent and damaged cells. Moreover, roughly one-third of natural antibodies are able to bind oxidation-specific epitopes [17] and mitigate the effects of oxidative stress caused by ageing, inflammation and tissue injury.

There are, however, certain inflammatory conditions in which natural antibodies exhibit pathogenic properties. In ischaemia-reperfusion injury, the binding of IgM natural antibodies to neo-epitopes (such as annexin IV) exposed on ischaemic cells with subsequent activation of the complement cascade leads to tissue injury [18,19]. Similarly, IgM antibodies to blood-borne microbial antigens which cross react with erythrocytes and platelets can opsonise healthy cells causing their destruction or elimination by the reticuloendothelial system and, hence, induce the pathology of autoimmune diseases (autoimmune anaemia or thrombocytopaenia) [20,21]. SLE is especially instructive as numerous perturbation that alter B-cell activation lead to a lupus-like serology in mice, demonstrating clearly that checkpoints on natural autoantibodies are essential to prevent pathologic autoreactivity.

3. IMMUNO-SUPPRESSIVE VERSUS IMMUNO-ACTIVATING ANTIBODIES: IGM, IGG AND THEIR RECEPTORS

Natural autoantibodies that are protective for autoimmunity are predominantly of the IgM isotype, while pathogenic antibodies are usually of the IgG isotype. It is important to remind that in general both isotypes are protective against pathogens; however, there are relevant differences in their receptors and pathways that are activated upon their binding that help to explain their differential function in autoimmune disease generation.

Binding of IgM antibodies to apoptotic cells recruits C1q and enhances phagocytosis, supporting an important pathway for clearing apoptotic debris. Antibodies to apoptotic cells can also inhibit the inflammatory response mediated by macrophages activated by oxidised phospholipid by blocking its binding to scavenger receptors [8]. Apoptotic debris bound by IgM autoantibodies and C1q binds LAIR-1, an inhibitory receptor on monocytes and DCs [22]. LAIR1 is also expressed on the surface of T cells and human, but not mouse, B cells, although its immunomodulatory function in these cells is still to be established. It is notable that LAIR1 expression is low on B cells in SLE [23].

There are three Fc receptors for IgM: Fcα/μ receptor (Fcα/μR), polymeric immunoglobulin receptor (pIgR) and Fcμ receptor (FcμR). All of them are type I transmembrane proteins. Fcα/μR is necessary for toll-like receptor 4 (TLR4) response to LPS in mouse MZB in sepsis [24]. pIgR is expressed on the basolateral surface of ciliated epithelial cell in the mucosal epithelium and transports dimeric IgA and polymeric IgM from the lamina propria across the epithelial barrier to the lumenal mucosal surfaces [25]. The FcμR was recently discovered, initially in humans and later in mice. The function of FcμR has not been fully elucidated [26], but it is the only IgM receptor for which there is evidence that it may modulate autoimmunity. B6/lpr mice deficient in FcμR have fewer marginal zone and follicular B cells and develop autoantibodies earlier than B6/lpr mice competent for this receptor [27]. In contrast, B6 mice deficient for FcμR are protected from severe experimental autoimmune encephalomyelitis (EAE); this effect is associated with a diminished response of dendritic cells (DCs) to TLR stimulation and to an increased number of regulatory T cells (Tregs) [28]. Thus, the effect of loss of FcμR depends on both, genetic background and disease model. There are some relevant differences in the characteristics of this receptor across species. The receptor is only expressed on B cells in mice, and in humans is expressed also on T cells and on natural killer (NK) cells. The binding of IgM to the human FcμR is stronger than the binding to mouse FcμR [29]. How this difference affects the effector function of IgM antibodies, and how FcμR impedes or enhances the development of autoimmunity are unanswered questions.

IgG autoantibodies form immune complexes that are capable of engaging activating FcRs on myeloid lineage cells. This activation sets up an amplification loop whereby the activated myeloid cells establish a pro-inflammatory milieu, promoting presentation of self-antigen and rescue of autoreactive B cells that would otherwise be censored. In humans, there are four subclasses of IgG, IgG1 to 4. They differ from each other in their constant region, and their ability to bind Fc receptors and C1q, with the consequent activation of the complement cascade. IgG3 and IgG1 are the most potent activators of the complement cascade [30], IgG4 is a less potent activator of complement and IgG2 only activates complement when antigen is present at a high density, allowing the formation of hexamers of Ig [31]. In a similar way, IgG1 and IgG3 bind more efficiently to FcγRs compared to IgG2 and IgG4. IgG2 only binds to FcγRIIa, the most widely expressed FcγR on myeloid cells. Binding to the inhibitory receptor FcγRIIb is relatively low for all IgG subclasses, and similar for IgG4 compared to IgG1/3. This can explain a lower proinflammatory profile of IgG4 [32].

The neonatal Fc receptor (FcRn) is expressed on myeloid and epithelial cells, and it is responsible of the long half-life of IgG (approximately 3 weeks). This receptor is expressed in early development and is responsible for the transport of IgG antibodies from the mother to the foetus [33], explaining also the presence of some autoantibody mediated pathology, such as thyroid disease or neonatal lupus congenital cardiac block associated with SSa antibodies in the foetus [34].

Besides functional changes in complement activation or FcγR activation, changes in the constant region of an Ig chain can influence antigen binding [35]. Thus, an antibody with identical variable (VH and VL) domains, but different constant region can bind antigen with different affinity [36,37] or change their specificity [38]. Class switching of a monoclonal antibody directed against DNA (PL9-11, originally an IgG3 antibody) to IgG1, IgG2a/b or IgM reduced its affinity for DNA and chromatin [39]. When hybridoma cells producing these antibodies were transferred to a mouse lacking B and T cells (SCID mice), different isotypes of antibodies showed differential binding to the glomeruli, with increased renal pathology and decreased survival in mice administered antibodies with the highest affinity (IgG3) [39]. Of note, not all antibodies are equally susceptible to alterations in their binding properties after changes to their constant region, and this phenomenon is probably associated with specific properties of different V regions [35].

4. GLYCOSYLATION OF ANTIBODIES REGULATES THEIR PRO- OR ANTI-INFLAMMATORY EFFECT

Post-translational modifications of antibodies can render them more immunosuppressive or more proinflammatory. Every molecule of IgG (IgG 1,2,3,4 in humans) has a single N-linked glycan attached at asparagine 297 (Asn297) on the heavy chain (HC). This glycan is located near the 'hinge region' which contains the disulfide bonds and provides flexibility to the antibody. Modifications in this glycan can affect antibody conformation and modify potential interactions with FcRs, cell surface lectins and complement components [32]. Each IgG molecule may also have additional glycans in the antigen-binding sites [40]. The functional significance of Fab glycosylation is incompletely understood. During Ig translation, a preformed lipid-linked glycan is attached to Asn297. This glycan is composed of two N-acetylglucosamines followed by nine mannose and three glucose (Glc) residues. The Ig is then transferred from the endoplasmic reticulum to the Golgi, where glycosyl-hydrolases and -transferases can modify the glycan structure leading to diverse glycans.

Most circulating IgG is fucosylated [41]. Fucosylation reduces IgG affinity for the activating receptor FcγRIII and diminishes cellular cytotoxicity [42] and phagocytosis [43]. Clinical trials with non-fucosylated antibodies have shown improvement in target cell depletion [44]. Approximately 40% of glycans have one galactose, but they can be bigalactosylated or non-galactosylated. The galactosylation profile is age- and gender-dependent [45]. Galactose provides the basis for the addition of sialic acid; sialylated IgG binds DC-SIGN on DCs to suppress their function. Thus, the effect of changes in galactosylation may be explained by secondary changes in sialylation [46,47]. Some studies show a decreased level of galactosylation with RA and SLE, and others show decreased sialylation [48,49].

Around 5%–10% of glycans on IgG molecules carry sialic acid (neuraminic acid, Neu5Ac) on one arm, and approximately 1% are bi-sialylated [50]. Decreased levels of sialylation have been associated with chronic inflammatory diseases [51]. Intravenous immunoglobulin (IVIG), which produces a potent anti-inflammatory response in some autoimmune diseases, loses its anti-inflammatory effect upon removal of sialic acid with neuraminidase [46]. In mouse models, IVIG increases its anti-inflammatory activity when enriched for sialic acid residues [47,52,53]. Engagement of DC-SIGN induces IL-33 production by DCs and production of IL-4 by basophils, and upregulation of the inhibitory FcγRIIb on macrophages [54]. In humans, Fc sialylation also impairs complement-dependent cytotoxicity by reducing the binding of the IgG CH2 domain to C1q [55].

Abnormal glycosylation of surface proteins in B cells has also been associated with autoimmunity [56]. Interestingly, a traditional way to detect germinal centre (GC) B cells is by testing their reactivity against the plant lectin peanut agglutinin (PNA), which binds serine/threonine (O-linked) glycans containing an asialylated disaccharide. The binding of PNA is associated with downregulation of a specific sialyltransferase, ST3Gal1, which causes O-glycosylation of CD45, a tyrosine phosphatase [57]. Specific antibodies to CD45 epitopes modified by its O-glycosylation status, such as MEM-55, which binds to sialylated O-glycans, preferentially bind antigen-experienced B cells over naïve B cells [58]. Additionally, the GL-7 monoclonal antibody that is specific for GC B cells recognises the sialic acid glycan moiety Neu5AC, suggesting that protein glycosylation is a characteristic of the B-cell differentiation state.

5. B-CELL SUBSETS ENRICHED FOR HOMOEOSTATIC AUTOREACTIVITY

5.1 B1 Cells

B1 cells were first found in mice as a population of B cells expressing CD5, high levels of IgM and low levels of IgD, often residing in the pleural and peritoneal cavities [59]. Up to 80% of the circulating IgM in mice is produced by B1 cells [60]. B-cell precursors in the foetal liver give rise predominantly to B1 cells [61], while B2 cells are not produced until lymphopoiesis shifts from the foetal liver to the bone marrow (BM). The BM hematopoietic stem cells constantly produce B2 cells but do not support B1 cell generation. Along with the drastic decline of foetal liver haematopoiesis after birth, the B1 population in adults is maintained through self-renewal [62,63].

During early B-cell development, there is limited V(D)J diversity in Ig molecules with biased usage of D proximal V and J gene segments. Extremely low Tdt activity in the foetal liver B-cell progenitors results in a short CDR3 of the Ig HC [64,65].

Autoreactive BCRs are positively selected within the B1 population. For example, $V_{H6\text{-}1}$, the most D_H proximal human V_H gene, is overrepresented in human foetal B cells, but rare in the adult B2 cell repertoire [66]. $V_{H6\text{-}1}$ is thought to be involved in the clearance of DNA-containing cellular breakdown products due to its reactivity with single-stranded DNA and cardiolipin [67]. In mice, $V_{HS107.1}$, a D proximal V gene, responsible for encoding phosphorylcholine-binding T15 natural antibodies, is enriched in the B1 cell compartment [68]. The need for antigen in the bias to an autoreactive BCR repertoire was demonstrated in a study of B1 cells specific for CD90/Thy-1. These are present in wildtype but not CD90 knockout mice [69]. That reactivity to self-antigen drives B1 lineage determination, and cell survival was further demonstrated in a study showing that enforced expression of V_{H12}, a V_H gene specific for phosphatidylcholine on senescent erythrocytes, leads to an expansion of the B1 population in transgenic mice [70]. In contrast, co-expression of V_{H12} with $V_{H1\text{-}8}$, a V_H gene expressed by B2 cells, promotes precursor B cells to differentiate into B2 cells in the BM, presumably because the presence of $V_{H1\text{-}8}$ dilutes the density of the self-reactive V_{H12} on the cell surface and weakens the BCR signalling needed for positive selection into the B1 cell compartment [71]. Consistent with this interpretation, genetically modified mice with diminished positive regulators of BCR signalling, such as CD19 and BTK, have impaired B1 development. Reciprocally, genetically modified mice with enhanced BCR signalling have an increased B1 population [72–74]. Moreover, enhanced BCR signalling can prolong B1 survival.

It is not entirely clear how autoreactive B1 cells avoid the tolerance mechanisms that are present in B2 cells (described below). Siglec G is a negative regulator of BCR signalling in B1 cells [75]. Siglec G-deficient mice have an enlarged pool of B1 cells that exhibit upregulation of NFATc1, promoting resistance to apoptosis, and prolonged B1-cell survival [76,77]. NFATc1 knockout in foetal liver cells renders them incapable of populating the B1-cell compartment [78]. Vitamin A deficiency in mice leads to a downregulation of NFATc1 in B cells and results in a reduction in the B1-cell population and in the level of natural antibodies [79]. Thus, both B-cell intrinsic and extrinsic signals act to regulate the B1-cell pool.

Overproduction of natural antibodies and an expanded B1 population can contribute to the development of autoimmunity. Motheaten mice with a deficiency in SHP-1 have a relative expansion of B1 cells and tissue injury secondary to autoantibodies [80]. Lupus-prone NZB/W mice also have an expanded B1 population [59]; however, this feature is not present in all lupus-prone strains and how important it is to the autoimmune phenotype in NZB/W mice is not clear. Of note, SLE triple congenic mice (TC) harbour chromosomal loci from NZB/W mice which render them autoreactive with a lupus-like phenotype. The SLE2 locus leads to an expansion of B1 cells, but only to autoantibody production when present in conjunction with the SLE1 locus [81]. Patients with CD5+ B cell chronic lymphocytic leukaemia develop haemolytic anaemia due to an autoreactive BCR on the malignant B cells [82]. Thus, while many studies show that autoreactive B1 cells produce natural autoantibodies that contribute to protection against certain microbes and help dispose of apoptotic debris through a non-inflammatory process, they nonetheless may also contribute to pathogenic autoreactivity and to autoimmunity. It is important to note that B1 cells do not require T-cell help for activation [83].

5.2 Marginal Zone B Cells

MZB cells are innate-like B cells named after their microanatomical localisation in the marginal zone of the spleen in mice [84,85] where they are strategically positioned at the meeting place of the vascular and the immune system and where the concentration of blood borne antigen is high. In humans, MZB cells are not exclusive to the spleen and have been found in other secondary lymphoid organs such as tonsillar crypts, Peyer's patches and the subcapsular sinus of lymph nodes [86–88]. It has even been hypothesised that human blood IgM memory B cells are recirculating MZB [89].

MZB cells arise in the BM and are selected on the basis of functional BCR assembly and specificity. The early stages of B-cell maturation in the BM involve negative selection of B cells with high-affinity autoreactive BCRs, but the final stage of maturation of MZB involves selection for self-reactivity. The mechanism of repertoire selection for polyreactive BCRs of MZB leads to restricted V-gene usage and limited junctional diversity. Akin to B1 cells, recognition of evolutionarily conserved structural motifs and reactivity to self is involved in the survival of MZB. The repertoire of MZB cells is not complete in humans at birth, taking around 2 years to be fully developed [90], and there is evidence from patients deficient for CD40 ligand that the diversification of the repertoire of MZB is independent of T-cell help [91]. High expression of BAFF favours an expansion of MZB [92] as does a persistent high level of oestradiol [93]. Thus, the commitment to become an MZB depends on both cell intrinsic and extrinsic factors.

Like B1 cells, MZB cells produce natural IgM and serve a housekeeping function as well as first responders to blood-borne pathogens. To perform their principal functions, MZB cells produce low-affinity antibodies with reactivity to apoptotic debris and to microbial antigens. As their localisation allows them to have constant exposure to antigens, they have a recognised function to present antigens. They contribute to an autoimmune response in part by their function as APCs. MZB cells present self-antigen to T cells in a model of collagen-induced arthritis, causing the activation and proliferation of pathogenic antigen-specific T cells [94,95], and in a mouse model of diabetes [96]. In triple congenic lupus-prone mice, but not in WT mice, MZB cells, identified as B220+ CD1d high cells, have been observed to abnormally localise in the follicles and interact with CD4+ T cells [97], further suggesting a role for MZB as APCs in SLE.

MZB cells have other unique features. They are entirely dependent on Notch signalling for development and survival. They are highly responsive to TLR signalling which increases expression of the complement receptor CD21. Thus, they are easily activated by microbial particles opsonised with antibody and complement. MZB activation is enhanced by T-cell help, but there is no requirement for cognate T-cell help [91]. MZB cells are activated as part of the extrafollicular response. The extrafollicular reaction provides an initial wave of secreted antibodies early after initiation of the immune response. Differentiation of extrafollicular PCs is driven by T cell–B cell interactions with a specific subset of Th cells that resembles T follicular helper (TFH) cells but located in extrafollicular areas [98]. Although recent evidence suggests that some characteristics traditionally attributed to GC responses such as the formation of memory B cells, class switch recombination (CSR), somatic hypermutation (SHM) and induction of long-lived (LL) PCs can also occur in extrafollicular responses [99], most of the PCs generated in extrafollicular responses are short-lived [100]. The preferential generation of short-lived PC by MZB, together with limited IgG CSR, is protective against the generation of autoimmune responses.

6. B CELLS DEVELOPMENT, SELECTION AND TOLERANCE

Although B1 cells and MZB can undergo CSR, high-affinity IgG autoantibodies are produced mostly by follicular B2 cells, and their activation requires cognate T-cell help. B2 cells continuously differentiate from precursors in the BM and populate peripheral lymphoid organs after birth. Autoreactive B cells arise frequently in the BM as a consequence of the random gene rearrangements that generate the BCR. Due to Tdt upregulation and a broader use of V genes, the BCR repertoire of B2 cells is much more diverse than that of B1 cells and even the repertoire of MZB. As opposed to B1 cells and MZB in which BCR development involves positive selection for autoreactivity,

B2 cells undergo negative selection. This actually begins in early stages of B-cell development even before MZB and B2 lineage bifurcation, which in humans occurs at the T2 stage of transitional B cells [101]. Approximately 75% of immature BM B cells carry self-reactive BCRs in both healthy individuals and mice [2].

Successful assembly of BCR on the cell surface allows newly generated B cells to migrate to the periphery as transitional B cells, where they further mature into phenotypically distinct B cells that localise in the MZ or the B-cell follicles of the spleen and other peripheral lymphoid tissues. The immune system employs tolerance mechanisms to remove much self-reactivity at each stage of B-cell development. B-cell intrinsic properties triggered by BCR signalling are capable of activating these tolerance mechanisms. The objective of each checkpoint is to decrease the frequency and affinity of autoreactive B cells in the mature B-cell repertoire.

Central tolerance mechanisms begin at early stages of B-cell development in the BM. Strength of BCR engagement with self-antigen decides whether central tolerance processes are imposed. The BCR transmits a tonic signal, a weak signalling emanating from a receptor in the absence of its ligand, necessary for survival and maturation [102]. A high-affinity interaction of the BCR with self-antigen will trigger BCR signalling above tonic levels, and the activation of tolerance mechanisms (receptor editing/clonal deletion explained below) [103].

Central tolerance has a key role in reducing the frequency of autoreactive cells in the naïve B-cell repertoire. Receptor editing and clonal deletion are employed in both immature B cells and the next stage of maturation, transitional B cells. Receptor editing is a process by which a novel LC is generated to pair with the HC to create a non-autoreactive BCR, and it is the preferred mechanism of central tolerance following strong BCR signalling in immature B cells. By permitting ongoing LC Ig gene recombination to modify the specificity of B cells carrying autoreactive BCRs, receptor editing contributes to immune diversity by promoting the generation of BCRs with a single HC paired potentially with multiple distinct LCs.

If receptor editing fails to produce a functional LC or one that can pair with the existing HC and generate a non-autoreactive BCR, the B cell undergoes clonal deletion through apoptosis. BCR ligation in immature BM B cells results in calcium mobilisation. The intensity of the response correlates with an increased likelihood of apoptosis. Apoptosis is driven by the proapoptotic protein BIM. BIM can be inhibited by enforced expression of its anti-apoptotic binding partners, one of which is BCL-2 [104–107]. Enforced BCL-2 expression results in prolonged receptor editing, supporting the notion that receptor editing is limited by BIM-dependent death [108]. It is worth noting that conditions which diminish receptor editing lead to an increased frequency of MZB, suggesting that clonal deletion is more operative in B2 cells than in MZB.

Central tolerance is incomplete. Not all self-antigens are accessible in their native form to B cells in the BM. Moreover, some polyreactive BCRs fail to induce receptor editing or deletion. While it is unclear how they avoid negative selection, certain self-antigens may engage activating or inhibitory receptors on B cells such as Siglecs or complement receptors that may help regulate central tolerance to certain classes of antigens [109–111].

Tolerance mechanisms continue to operate in the transitional B-cell population that are new emigrants from the BM. B-cell anergy is a state of relative non-responsiveness to antigen due in part to low surface expression of the BCR and is the predominant tolerance mechanism induced in mature B cells [112]. While anergy is also induced in transitional B cells, it is unclear if it occurs in the immature stages of development [113]. Anergy is induced by chronic exposure to antigen [114]. Anergic B cells have a shorter lifespan, when compared with other mature B cells [115]. Defective anergy has been proposed as a potential mechanism of disease in SLE [1]. Many self-reactive B2 cells in the periphery exist in an anergic state. Anergic B cells have a diminished response after BCR activation and produce fewer PC after activation [116]. A higher percentage of mature naïve anergic B cells produce anti-nuclear antibodies (ANAs) in humans [1] than naïve non-anergic B cells. It is not totally clear how or why some autoreactive B cells are deleted and others become anergic, but there is evidence that anergic cells can be rescued from this state and can be involved in subsequent activation [117]. In BAFF-transgenic mice, self-reactive B cells are rescued from deletion and from anergy [118], suggesting that BAFF has a critical role rescuing cells from becoming anergic. Excess BAFF has been reported in patients with SLE after B-cell depletion therapies, and adding a belimumab, an anti-BAFF antibody, enhances the negative selection of mature autoreactive B cells [119], probably because transitional B cells, that are enriched in autoreactivity compared to mature naïve B2 cells, are more dependent on BAFF.

7. B CELL DIFFERENTIATION

7.1 Germinal Centre Reaction

The GC is a highly dynamic and open structure containing multiple lineages of cells. It is a specialised location for B cells to undergo SHM leading to the selection of B cells with BCRs with higher affinity for antigen (affinity maturation). Formation of GCs requires B-cell interaction with cognate TFH cells and follicular dendritic cells (FDCs). T cells are primed in the T-cell zone where antigen-bearing DCs are localised; activated T cells that upregulate CXCR5 migrate to the T-B border. Once at the T-B border, CXCR5 and Bcl-6 expressing CD4+ T cells that are able to maintain an extended interaction with antigen-presenting B cells differentiate into functional TFH and produce IL-21 [120–122]. The early GC is an FDC-rich microenvironment where B cells proliferate. Within a few days, the 'mature' GC develops containing a dark zone (DZ) full of highly proliferating centroblasts, and a light zone (LZ), where centrocytes co-localise with FDCs and TFH [113] (described in Chapter 16).

FDCs mature from perivascular mesenchymal cells in both lymphoid and non-lymphoid organs. Their ubiquitous presence, enabled by lymphotoxin provided by infiltrating lymphocytes, potentially contributes to the de novo development of ectopic lymphoid follicles during inflammation and autoimmune disease [113]. Since FDCs are not phagocytic and do not express MHCII, they are not able to present antigen to T cells [123]. FDCs do, however, express high levels of CD35 and FcγRIIb, allowing them to trap and retain immune complexes for prolonged periods [124,125]. Immune complexes bound to complement receptors or FcγRIIb on FDCs are termed icosomes and concentrate antigen for B cells to engulf and present to T cells. Because B cells in the GC downregulate expression of the BCR, antigen-bearing FDCs select for high-affinity B cells in the GC LZ. B cells with higher affinity BCRs engulf more antigen and present it to T cells; the interaction with T cells allows B cells to upregulate c-Myc in B cells [126,127]. These positively selected GC B cells in the LZ upregulate CXCR4 and move to the DZ, where B cells divide and undergo SHM. B cells bearing mutated BCR move back to the LZ in a chemokine determined fashion. The cyclic migration of GC B cells between LZ and DZ allows cells that have undergone SHM and gained higher affinity BCR to be selected for clonal expansion, thus achieving affinity maturation. After several rounds of proliferation and hypermutation, GC B cells can differentiate into LL memory B cells or antibody-secreting PCs [128,129].

The B cell-specific enzyme activation-induced cytidine deaminase (AID) is crucial for both CSR and SHM of Ig genes during the GC response [130–132]. AID-deficient B cells generate only low-affinity IgM antibodies and are not capable of CSR or of differentiating into memory B cells [133]. Altered AID expression appears to have a role in the development of autoimmunity, but it is not monolithic. In certain mouse models, AID deficiency promotes autoimmunity [134]; interestingly, however, AID expression is increased in circulating and synovium infiltrating B cells of patients with RA [135].

7.1.1 *Regulating Autoreactivity of GC Reaction*

While affinity maturation is critical for the generation of effective antibodies to control pathogens, high-affinity autoantibody production is a hallmark of autoimmune diseases. There are several mechanisms to prevent the autoreactivity of GC reactions. One study in mice suggests that autoreactive B cells are excluded from entrance into the GC. In humans, B cells expressing VH4-34, which are autoreactive, are excluded from the GC, except in patients with SLE [136]. Thus, abnormal entrance of autoreactive B cells into the GC has been proposed as a potential pathogenic mechanism in SLE [137].

During GC responses, acquiring autoreactivity is inevitable, as SHM can lead to the generation of novel BCRs that bind self-components. Analysis of the BCRs of autoreactive B cells from patients with SLE shows that the germline-encoded BCRs of many of the cloned autoantibodies are not autoreactive, supporting the notion that SHM is responsible for the generation of novel autoreactivity [138,139]. The role of SHM in generating autoreactivity was first demonstrated in a mouse myeloma cell line. A single base substitution in HC V-gene of an anti-phosphorylcholine antibody protective against pneumococcal infection spawned new reactivity to DNA [140]. When an effort was made to determine if this routinely occurred in vivo, B-cell hybridomas from immunised mice were made using a fusion partner engineered to overexpress BCL-2, which prevents apoptosis of autoreactive cells. Approximately 40% of hybridomas from spleen cells of mice immunised with phosphorylcholine coupled to a protein carrier are cross-reactive with DNA. This demonstrated the high frequency of autoreactivity generated within the GC [141].

Both B cell intrinsic and extrinsic factors are involved in eliminating autoreactive B cells in the GC or preventing entry of post-GC B cells into the memory B cell or LL PC compartments [142]. The main mechanism that prevents the propagation of autoreactive B cells in the GC has traditionally been considered the absence of T cell help and FDC-derived survival factors for autoreactive B cells. B cells reactive only with autoantigen will not have antigen to bind on icosomes and will not form cognate interactions with T cells.

Another mechanism for the deletion of autoreactive B cells in the GC is the presence of soluble antigen. The evidence supporting this mechanism comes mostly from experiments in which an excess of soluble antigen was administered to mice that were previously immunised against the same antigen. The presence of soluble antigen led a majority of the GC B cells to undergo apoptosis. The mechanism for this effect is not clear; however, it is possible that an excess of soluble antigen might interfere with the interaction between B cells and FDC, precluding B cells from obtaining other signals that are necessary for their survival [143].

Alternatively, tolerance mechanisms may operate as B cells exit the GC. Some data suggest B cells exiting the

GC are similar to immature B cells in that they can respond to antigen by re-expressing RAG and initiating receptor editing. RAG expression in early B-cell development is dependent on IL-7, and the IL-7 receptor is expressed on GC emigrants. IL-6 is required for the expression of the IL-7 receptor in early lymphocyte development, and IL-6 is also required for the expression of the IL-7 receptor in GC mature B cells [144–146]. Non-autoimmune mice immunised with a peptide mimetope of DNA develop peptide- and DNA- cross-reactive antibodies. As these autoreactive B cells move out of the GC, many express RAG protein and appear to undergo receptor editing or apoptosis. RAG expression has been reported in ectopic GCs in synovium of patients with RA [147] and circulating B cells of SLE patients. The presence and relevance of receptor editing as a mechanism of peripheral tolerance is still to be confirmed.

While many autoreactive B cells generated by SHM in the GC will die due to lack of positive selection by antigen on FDCs and cognate TFH, some may gain the necessary survival signals and differentiate into memory cells or LL PCs. This is especially true if the B cells bind both the eliciting antigen present on FDCs and self-antigen. There is evidence in mouse models that a single autoreactive B-cell clone can promote autoreactive GCs, in which autoreactive B cells against other autoantigens can proliferate. When WT mice are reconstituted with mixed BM cells from transgenic mice expressing an autoreactive BCR targeting ribonuclear complexes and with WT BM cells, they spontaneously form GCs which contain autoreactive B cells predominantly originating from the WT B cells [148]. Moreover, autoreactive B cells arising in the chimaeric mice target autoantigens other than ribonucleoproteins, suggesting that the GC has a mechanism, such as epitope spreading, for amplifying autoreactivity.

The lower frequency of autoreactive B cells in BM PCs compared to circulating blood memory cells is suggestive of a tolerance checkpoint before differentiation to LL PCs. Alternatively, engagement of Fc receptors by immune complexes on LL PCs may reduce the numbers of autoreactive PCs [149].

Spontaneous germinal centres (SpGCs) are GCs that form in the absence of immunisation or infection [150]. Similar to conventional GCs, SpGCs are characterised by the presence of B-cell follicles that contain proliferating B cells, T cells and FDCs. Increased BAFF expression was one of the first factors to be associated with the formation of SpGCs [151], but the formation of SpGCs has been linked to alterations in innate immune response, BCR signalling and levels of expression or costimulatory signals [152]. There are experimental data that show that affinity maturation and epitope spreading occur in SpGCs (explained later in this chapter) and may lead to autoreactivity [148].

Ectopic GCs allow for the activation and differentiation of B cells in inflamed tissues. Tissue-resident fibroblasts can support the formation of ectopic aggregates of lymphoid and myeloid cells termed tertiary lymphoid structures (TLSs) [153]. TLSs provide alternative sites for immune activation in response to pathogens or cancer cells. Indeed, in mice without secondary lymphoid organs, there can be an effective immune response that is dependent on TLS [154]. Moreover, affinity maturation and memory B cells can be generated in the context of TLS [154]. TLSs have variable degrees of organisation, ranging from B-T aggregates to organised structures similar to GC. Because part of the regulation described in GC is dependent on a highly organised anatomical microstructure, these mechanisms may be altered in TLS, potentially allowing for the activation of autoreactive B cells. Supporting their relevance in autoimmunity, TLSs are present in multiple autoimmune diseases (thyroiditis, diabetes, multiple sclerosis (MS), Sjögrens syndrome, rheumatoid arthritis [RA]) [155], and some studies showed an association of TLS to disease progression in RA [156].

7.1.2 Tregs and Autoantibodies

Tregs play an essential role in suppressing autoimmunity. Tregs consist of a heterogeneous population of T cells with different suppressive functions. Traditional Tregs are CD25+ Foxp3+ CD4+; these cells characteristically inhibit T-dependent autoantibody production through suppression of both autoreactive T cells and B cells. CD4+ CD25+ Tregs have multiple effector mechanisms; they can directly kill antigen-presenting B cells through a Fas or a granzyme-perforin dependent pathway. They can also activate a PD1 pathway, which inhibits BCR signalling by recruiting SRC homology 2-domain-containing tyrosine phosphatase 2 to phosphotyrosine-dependent kinases in the BCR signalling pathway [157].

There is a population of Tregs, T follicular regulatory (TFR) cells that regulate the GC response. These cells are CD4+, Foxp3+ and express the TFH markers CXCR5 and PD-1. In the absence of TFR, there are increased numbers of TFH and GC B cells. TFR directly acts on TFH, and there is evidence that they can also directly modulate B cells within the GC through the production of neuritin [158]. Neuritin is taken up by B cells and leads to downregulated BLIMP-1, upregulated BCL6 and decreased differentiation of mouse and human GC B cells into PCs [158].

A population of CD8+ T cells expressing CD122 has also been found to be immunosuppressive. These cells suppress GC reactions and the production of IFN-γ by conventional T cells [159]. They have been associated with a number of autoimmune diseases in mice, such as EAE and autoimmune diabetes [159]. In humans, their equivalent are CD8+, CXCR3+ T cells [160]. A subpopulation of CD8+, CD122+ Tregs is Qa-1 restricted. This T-cell subset directly inhibits TFH survival through perforin-dependent killing [113]. Qa-1 is a nonclassical MHC class I molecule present in mice that is homologous to HLA-E in humans. In mice, a point mutation in Qa-1 (D227K) disrupts binding of Qa-1 to TCRs on CD8+ T cells, resulting in a fivefold expansion in the TFH population as well as spontaneous GC formation and SLE-like disease. Notably, GC reactions in D227K mutant mice are not contracted in the late stage of the immune response and continue to generate autoantibodies [161].

7.1.3 FDCs and Autoantibodies

The formation of FDC-GC B cell clusters is mediated by the integrins ICAM-1 and VCAM-1. ICAM-1 and VCAM-1 are upregulated in mature FDCs through an NF-κB-dependent pathway [162,163]. In mice, FDCs express TLR4, which modulates their maturation; TLR4 signalling plays a role in the ability of FDCs to mediate GC formation and SHM [164]. Increased TLR4 signalling may contribute to FDC function in aberrant GC reactions during autoimmunity. Interference with FDC development attenuates GC reactions and disease activity in K/BxN mice that develop autoantibody-mediated spontaneous arthritis. This diminished disease phenotype is attributed to the impaired generation of GC and to a reduced recruitment of arthritogenic TFH [165].

7.1.4 TFH Cells and Autoantibodies

TFH cells in the GC provide co-stimulatory signals through ICOS, CD40-ligand and IL-4, IL-21 and IFNγ. The combination of co-stimulatory signals and antigen stimulation results in vigorous expansion of GC B cells. CD40 signalling stabilises c-FLIP and prevents Fas-mediated apoptosis of GC B cells. IL-21 acts in autocrine fashion to promote TFH differentiation and in a paracrine fashion to increase expression of Bcl-6 in B cells. An increase in IL-21 has been observed in SLE [166], and polymorphisms near the IL-21 gene are associated with autoimmunity [167,168], although it has not been shown that they result in increased expression of IL21.

Involvement of TFH cells in autoantibody formation has been demonstrated in 'Sanroque' mice. Sanroque mice have a M119R amino acid substitution in E3 ubiquitin ligase Roquin (Rc3h1) [169]. Homozygous mice bearing the 'san' allele of Roquin have increased numbers of TFH cells. Adoptive transfer of TFH cells from sanroque mice induces spontaneous GC formation and SLE-like autoantibody production in recipient mice [170,171].

The mechanism of dysregulation of TFH cells in sanroque mice is in part attributed to lack of ICOS repression, as roquin represses ICOS [172,173]. Deletion of roquin results in robust upregulation of ICOS and an increase in spontaneous GC formation; however, this is not sufficient for autoantibody development [174]. IFNγ is also overexpressed in sanroque mice, due to delayed degradation of IFNγ m-RNA [175]. Genetic ablation of SAP in the sanroque model prevents renal pathology [171]. TFH involvement in autoantibody responses has been also shown in other models. Mice with a deletion of Blimp1 in DCs have an increased number of TFH and increased number of GCs [176]. Blimp1-deficient DCs phenocopy DCs with the Blimp1 risk allele for SLE.

7.2 Memory B Cells

The origin of memory B cells has been traditionally linked to the GC response, where they are routinely generated. Some evidence suggests they can also be generated as part of an extrafollicular response and can derive from MZB as well as B2 cells. As their name suggests, these cells are critical for adaptive immunological memory. After re-exposure to antigen, memory B cells initiate the secondary immune response. The secondary immune response is faster and produces higher titres of IgG antibodies with greater specificity and increased affinity for the antigen. These cells are often linked to pathogenic responses in patients with autoimmune diseases. In patients with SLE, a delayed recovery in memory B cell number after B cell depletion therapy has been associated with better clinical responses [177].

7.3 Age- or Autoimmunity-Associated B Cells

Age- or autoimmunity-associated B cells (ABCs) are a relatively recently identified B-cell subset that is increased in frequency in older individuals and individuals with autoimmune disease or chronic viral infection. The origin of ABCs is disputed in humans with evidence that they can originate directly from naïve B cells or from memory B cells. Some studies suggest they are a major source of autoantibodies and that they are precursors of PCs. An antigen-presenting function has also been proposed [178]. ABCs are present in the kidney in patients with lupus nephritis [179]; however, their specific relevance and function in disease pathogenesis is still to be defined.

8. MECHANISMS THAT BREACH B-CELL TOLERANCE – WHY PATHOGENIC AUTOANTIBODIES ARISE IN INFLAMMATORY CONDITIONS?

Almost every autoimmune disorder has a complex aetiology involving both environmental and genetic influences.

Both cell intrinsic and extrinsic factors can breach B-cell tolerance at various checkpoints through modulating B-cell activation or regulating survival of autoreactive B cells, culminating in systemic inflammation and cellular or tissue damage.

8.1 Environmental Factors

The importance of environmental or stochastic factors is emphasised by relatively low concordance for autoimmune disorders between monozygotic twins, which is generally in the range of 15%–30%. The incidence of human autoimmune diseases increases with modern industrial development that affects our diet, medication, environmental pathogens, microbiome and stress. Non-infectious environmental factors, such as physical trauma, stress, drugs, and smoking or vitamin D deficiency can predispose individuals to autoantibody-mediated autoimmunity [180]. They may break the barriers that sequester some self-antigens from the immune system. Alternatively, they can induce innate and adaptive immune cells to express pro-inflammatory genes that can impair normal tolerance mechanisms [181]. They may also alter post-translational modifications on proteins thereby allowing autoantibodies to be made to neoantigens [182]. Finally, some drugs may act as haptens, enabling self-proteins to become immunogenic and leading through epitope spreading to autoreactivity [183,184]. In RA, long-term smoking induces the activation of PAD4 in the lung, with subsequent citrullination of proteins. In the context of the HLA-DRB1 shared epitope, T cells are activated, by citrullinated peptides and induce activation of antigen-specific B cells, and production of anti-citrullinated protein/peptide antibodies that are considered pathogenic [185].

8.2 Molecular Mimicry

The concept of molecular mimicry refers to the structural similarity between an exogenous antigen and a self-antigen. Thus, B cells that are activated due to reactivity with a pathogen can potentially cause tissue and organ damage. In most scenarios, even in the presence of molecular mimicry, self-tolerance mechanisms will prevail [186]. However, there are some autoimmune diseases with a clearly defined infectious aetiology. Rheumatic fever and Guillain–Barré syndrome (GBS) are classical examples of autoimmune diseases initiated by acute infection with defined pathogens. Rheumatic fever is triggered by infection with group A streptococcus, the streptococcal M protein and the carbohydrate epitope N-acetyl-glucosamine (GlcNAc) share homology with proteins such as myosin, tropomyosin and laminin. In mice and humans, anti-M protein and anti-GlcNAc antibodies broadly cross-react with these self-antigens and cause damage to multiple organs, including the heart, joints and kidney. In some individuals, anti-GlcNAc antibodies penetrate the blood–brain barrier, bind the D2 dopamine receptor and cause symptoms of Sydenham's chorea. GBS is characterised by lymphocytic infiltration and demyelination in the peripheral nervous system (PNS). *Campylobacter jejuni* is the principal infectious organism causing GBS. Antibodies produced during Campylobacter infection can bind to gangliosides on the myelin sheath of peripheral nerves. Mice immunised with *C. jejuni* and lipopolysaccharide (LPS) produce antibodies against gangliosides and develop neural pathology analogous to human GBS.

8.2.1 Cryptic Epitopes and Epitope Spreading

Initial exposure to self-antigen can occur when a pathogen or trauma causes the death of host cells or when inflammation induced by trauma or a pathogen causes tissue damage. Novel or cryptic epitopes exposed by these processes are presented to B and T cells, causing an immune response not only against the inciting pathogen but also to self-antigen. Epitope spreading happens when reactivity to a single epitope on an antigen or antigen complex leads to reactivity to multiple components of the complex. In models of experimental myasthenia gravis, animals that have been immunised with peptide from the α-subunit of the acetylcholine receptor (AchR) develop autoantibodies not only specific to the immunogen but also to the other subunits of the AchR [187]. Patients with SLE who initially have anti-histone antibodies develop anti-DNA antibodies and those with anti-Sm autoantibodies may later develop antibodies to ribonuclear protein (RNP), another component of the splicesomal complex [188]. B cells acting as APCs can activate T cells by presenting peptides of the autoantigen in class II MHC generating a peptide MHC complex that was not present during T-cell development and, therefore, did not participate in negative selection [189].

8.2.2 Non-specific Activation of B Cells

Inflammation can trigger nonspecific activation of autoreactive B cells and promote their expansion. This process is known as bystander activation. Pathogens can activate both innate and adaptive cells through the release of PAMPs which activate PRRs. It is interesting to note that many experimental autoimmune diseases are induced by immunising animals with autoantigens emulsified with complete Freund's adjuvant (CFA), which contains killed mycobacterium tuberculosis and numerous PAMPs. For example, administration of autoantigen AchR and CFA in C57BL/6 mice induces symptoms similar to those of human myasthenia gravis. It has been found that NK cells activated by CFA in this experimental model are required for polarisation of autoreactive Th1 cells, production of anti-AchR autoantibodies and disease development [190]. This observation reveals that NK cells and other innate immune cells can participate in the development of destructive autoimmunity. There are several families of PRRs through which innate and adaptive immune cells detect DAMPs or PAMPs. TLRs detect PAMPs on the cell surface or in endosomes. The cytosolic PRRs, including NOD-like receptors (NLRs) and RIG-I like receptors (RLRs) sense intracellular bacterial products and nucleic acids, respectively. Activation of PRRs has been linked with chronic inflammation and autoimmune diseases. In particular, murine studies show that TLR7 contributes to the generation of autoantibodies against RNA-containing antigens such as Smith (Sm), and TLR9 for spontaneous generation of anti-DNA autoantibodies in the presence of the SLE1 risk allele or in the MRL/lpr mouse model [191–193]. These studies clearly demonstrate that infection might lead to aberrant activation of innate immune cells and to activation of autoreactive lymphocytes, and thus contribute to inflammatory responses and autoimmunity. Superantigens are a class of unconventional antigens present on microbial pathogens which bind BCRs outside their usual recognition sites, thereby allowing the activation of a vast array of B cells, resulting in excessive activation of the immune system [110,111] and production of numerous antibodies.

Innate immune cells can augment autoimmunity by enhancing processing of autoantigens to activate autoreactive B cells, and by producing inflammatory cytokines and chemokines to overcome autoreactive B-cell tolerance and to promote autoreactive B-cell survival and expansion. For example, there is an expanded population of neutrophils in the blood of SLE patients. Unlike those in healthy individuals, circulating neutrophils in SLE patients are of low density and bear an immature phenotype [194]. Activated immature neutrophils undergo a specific form of cell death called NETosis and form neutrophil extracellular traps (NETs) containing self-DNA and nuclear proteins that are principal autoantigens of SLE [195]. Through NETosis, neutrophils also release inflammatory cytokines and trigger plasmacytoid DCs to produce Type I interferon (IFN) [196]. Upregulation of Type I IFN and its target genes is recognised to contribute to the pathogenesis of several systemic autoimmune diseases, including SLE, in part through activation of PC differentiation through an extrafollicular pathway [197]. Accordingly, neutrophils have been shown to be particularly important in the activation of MZB [198].

8.2.3 Apoptotic Debris

B cells encounter extracellular antigens in secondary lymphoid organs. Many autoantigens are located intracellularly and therefore are not routinely available to autoreactive B cells. Injury, inflammation and infections can cause intracellular autoantigens to become extracellular. Decreased clearance and impaired degradation of NETs and other forms of cellular debris increase access to autoantigens. Macrophages are responsible for phagocytosis of apoptotic cells. Ablation of Mer, a receptor tyrosine kinase that mediates clearance of apoptotic debris in mice leads to SLE-like autoimmunity [199,200], and an increase of apoptotic debris has been noted in GCs in SLE [201]. This will expose B cells to usually sequestered self-antigen.

8.3 Why Are Pathogenic Autoantibodies More Prevalent in Women?

Most autoimmune disease affects women more commonly than men. The presence of two X chromosomes contributes to the female predominance of autoimmunity. For example, TLR7 and CD40 ligand (CD40L) are both encoded on the X chromosome and overexpressed in lymphocytes of women with SLE. There are also data demonstrating that hormones contribute to the female predominance of autoimmune disease. Female castration reduces the incidence and severity of disease in NZB/W lupus-prone mice, whereas castration increases the incidence of lupus in males [202]. Oestrogen has been shown to alter the function of numerous cells within the immune system. In particular, it has been shown to allow DNA-reactive transitional B cells to pass an early tolerance checkpoint and mature into MZB [93]. The molecular mechanisms for this are not clear, but oestrogen increases expression of Bcl-2, an antiapoptotic molecule, and CD22, an inhibitory coreceptor of the BCR [203]. Oestrogen synergises with Type I IFN to increase expression of TLRs and to decrease expression of another inhibitory coreceptor of the BCR, FcγRIIb in B cells [204,205]. The effects of oestrogen on B cells are mediated through ERα; absence of ERα prevents an oestrogen-mediated breach of B-cell tolerance [206]. Oestrogen also augments AID expression and function which leads to increased CSR in GC B cells [207]. Additionally, ERα directly binds to several regulatory regions within the Ig heavy-chain gene sequence, such as Sμ and Eμ, and through binding to these sites, promotes CSR [208]. Surprisingly, a study of oestrogen receptor α (ERα)-deficient NZB/W mice suggests that the absence of ERα diminishes renal diseases but does not alter autoantibody titre [209]. Oestrogen also regulates non-coding microRNAs which contribute to the development and progression of autoimmunity. This effect might be caused by the binding of oestradiol to the promoter elements of the microRNAs [210–212]. Hyperprolactinemia has been reported in patients with several autoimmune diseases [213]. One study shows that prolactin spares transitional B cells from tolerance induction by increasing CD40L on T cells. B cells in a hyperprolactinemic milieu develop as follicular B cells.

There are also differences in Tregs based on sex. Tregs from young men have significantly increased suppressive capacity in vitro compared to those from women and a distinct transcriptomic signature enriched for genes in the PI3K–AKT signalling pathway. Gender-affirming sex hormones in transgender men and transgender women induce changes in the Treg-cell transcriptome; many of these changes are in the same pathways that differ between cis-gender men and women, suggesting a hormonal influence on the differential Treg-cell function [214].

8.4 Genetics and Pathogenic Autoantibodies

The genetic composition of an individual has a very significant role in the development of pathogenic autoantibodies. Many autoimmune diseases present with family aggregation. Genome-wide association studies (GWASs) have identified multiple genetic risk factors that might be involved in the pathogenesis of autoantibodies by modulating the inflammatory response, such as genes involved in clearance of apoptotic debris (C1q and C4, Mfge8, DNase I, Trex1), as well as those that may predispose to autoantibody production through B-cell pathway signalling (Lyn, FcgRIIB, PTPN22, BANK10, BAFF-R, CD40, IL-6, Fas) [215]. The genetic risk factors for autoimmune diseases are both distinct to a particular disease and overlapping.

The predominant genetic factor affecting the susceptibility to almost all autoimmune diseases is HLA class II. Polymorphisms in HLA class II are a high-risk factor for Graves' disease, myasthenia gravis, RA, SLE and many other autoimmune diseases [216]. In SLE patients, HLA-DR3 haplotypes are related to the production of autoantibodies against Ro/SSA or La/SSB, HLA-DR2 to the production of anti-DNA and anti-Sm antibodies and HLA-DR4 to the production of antibodies specific for the small nuclear ribonucleoproteins [217,218]. The fact that MHC class II haplotypes shape the specificity of autoantibodies suggests that T cells help drive the activation and expansion of autoreactive B cells in a cognate interaction.

8.4.1 Genes that Affect Apoptotic Cell Clearance

Damaged tissues present autoantigens and neoantigens to the immune system and are major triggers of inflammation and autoimmunity. The rapid clearance of autoantigens and immune complexes is important to prevent autoimmune responses [219]. Impaired clearance of apoptotic cells by DCs and macrophages in mice deficient for milk fat globule epidermal growth factor 8 (Mfge8) causes spontaneous GC reactions and production of lupus-like autoantibodies [200]. Efficient removal of immune complexes and opsonised cell debris is dependent on complement components and Fc receptors (FcR) on phagocytic cells. C1q and C4 deficiencies in humans predispose to SLE. FcγRIIIA and FcγRIIIB polymorphisms have also been reported to be linked to SLE [220]. Polymorphisms in DNase I and Trex1 are responsible for the removal of extracellular or cytosolic nucleic acids, respectively, are also linked to SLE [221,222].

8.4.2 Genes that Affect B-Cell Activation and Induce a Lupus-Like Serology

BCR signalling directly affects B-cell activation, and changes in BCR activation can influence the fate of a B cell. BANK1, a positive regulator of the BCR signalling pathway, is a risk allele for SLE. BANK1 induces sustained BCR signalling and B-cell hyperactivity, leading to the proliferation and expansion of autoreactive B cells [223,224]. Many negative regulators of BCR signalling have also been linked with autoimmunity, including PTPN22, CSK, LYN, CD22 and FcγRIIB. PTPN22 is a phosphatase that associates with Csk to regulate phosphorylation of the Src family kinases in both B and T cells. PTPN22 has a strong association in GWAS with many autoimmune diseases, including RA, Type 1 diabetes and SLE. Lyn deficiency compromises pathways that diminish B-cell activation and so contributes to B cell–mediated autoimmunity. Expression of LYN is decreased in B cells of SLE patients [225,226]. In mice, Lyn inactivation leads to a severe lupus-like disease. LYN phosphorylates CD22, a negative regulator of BCR signalling. Phosphorylated CD22 then recruits the tyrosine phosphatase SHP-1 to the CD22/BCR complex and dampens BCR signalling [227]. LYN also negatively controls BCR signalling by phosphorylating FcγRIIB and recruiting SHIP, an inositol phosphatase. FcγRIIB is the only FcR on B cells and plays a critical role in maintaining B-cell tolerance. Non-spontaneously autoimmune mice carrying a transgenic BCR specific for DNA produce autoreactive IgM but do not develop autoimmune nephritis; introducing FcγRIIB deficiency promotes class switch of nonpathogenic autoreactive IgM to pathogenic IgG2a and IgG2b through a TLR9/MyD88-dependent pathway, with resulting kidney damage [228].

8.4.3 Genes that Affect B-Cell Survival

Apoptosis is critical not only for the removal of B cells with high-affinity autoreactive BCRs but also for maintaining B-cell homoeostasis. A family of proteins that plays an important role in supporting B-cell survival is the TNF family, including BAFF and APRIL [229]. BAFF is needed for the survival of transitional B cells and MZB and has a role in the survival of GC B cells. APRIL has been demonstrated to be a critical survival factor for PCs. Elevated BAFF expression can rescue autoreactive B cells from apoptosis and direct them to the MZB subset. BAFF and APRIL expression is reported to be augmented in many autoimmune diseases, such as SLE and RA [230]. The direct evidence that increased BAFF levels lead to the development of autoimmune disease comes from murine studies in which mice constitutively expressing excessive amounts of BAFF develop a lupus-like disease [151].

Polymorphisms within the CD40 gene have been demonstrated to be linked with SLE, MS and RA [231,232]; they are thought to increase the expression of CD40 molecules on the surface of B cells. Co-stimulation of the BCR together with CD40 drives B cells to proliferate vigorously, rescues immature and transitional autoreactive B cells from tolerance mechanisms and prevents spontaneous apoptosis of GC B cells. CD40 signalling also induces B cells and DCs to produce IL-6, which promotes CSR and helps PC survival [233].

The death receptor Fas has a critical role in the maintenance of self-tolerance in the GC. B cells receiving a survival signal from FDCs, but lacking a positive signal provided by interaction with cognate TFH are destroyed through Fas-dependent apoptosis. The autoimmune lymphoproliferative syndrome (ALPS) is caused by mutations in Fas that alter its function. Individuals with ALPS present with autoimmune anaemia, thrombocytopaenia and neutropenia, and can develop glomerulonephritis. Fas deficiency leads to the accumulation of GC B cells with a high rate with SHM, potentially because these cells survive despite losing reactivity to antigen presented on FDCs [234]. This appears to be a cell-intrinsic property of Fas-deficient B cells.

Intrinsic characteristics of the B cells that affect the BCR response can also affect tolerance in the GC. Studies of the GC response in SLE1 mice have shown that the Sle1 locus controls tolerance in GC-matured B cells. GC B cells with the SLE1 locus have high levels of Bcl-2 and c-FLIP, which protect autoreactive B cells from normal tolerance mechanisms [235,236]. Overexpression of Bcl-2 or Bcl-XL can lead to the entry of autoreactive B cells into the memory or PC compartments, depending on the model system studied [237–240]. Polymorphisms in BCR signalling pathways also regulate autoreactive B cells in the GC response. For instance, a *CSK* allele is a risk factor for human autoimmunity. In BALB/c mice, a *Csk* allele leading to increased protein expression is permissive for anti-DNA antibody production in an immunisation model [241].

9. MECHANISMS OF PATHOGENICITY OF AUTOREACTIVE ANTIBODIES

There are multiple autoantibodies that appear in the context of autoimmune diseases; however, an autoantibody, notwithstanding its specificity for any condition, is not equivalent to a pathogenic antibody. Autoantibodies utilise the same mechanisms as antibodies that function in a protective response. Autoantibody containing immune complexes activate Fc receptors on myeloid cells or initiate the complement cascade inciting inflammation and causing tissue damage. Autoantibodies can also act as either antagonists or agonists of cellular receptors to cause derangement of cellular functions without inducing cell death or target organ inflammation. Similarly, antibodies to soluble factors, such as insulin and thyroid hormone, can also alter physiology without causing inflammation.

9.1 Non-inflammatory Injury from Pathogenic Autoantibodies

A few examples of autoantibodies that mediate non-inflammatory pathogenicity follow. Patients with pernicious anaemia have autoantibodies against gastric intrinsic factor, a glycoprotein produced by gastric parietal cells that is essential for vitamin B12 absorption by ileal enterocytes. By neutralising the B12-binding site of intrinsic factor (type I) or blocking the interaction between the intrinsic factor—B12 complex and its receptor on ileal enterocytes (type II), intrinsic factor—specific autoantibodies prevent vitamin B12 uptake and consequently lead to anaemia [242].

Neutralising autoantibodies are also found in patients with myasthenia gravis and Lambert—Eaton myasthenic syndrome. In both cases, muscle weakness is due to autoantibodies against proteins in the neuromuscular junction. Acetylcholine binds to its receptor, AchR, on the postsynaptic membrane of skeletal muscle fibres to trigger muscle contraction. Lambert—Eaton myasthenic syndrome is caused by autoantibodies specific to the presynaptic voltage-gated calcium channels and myasthenia gravis by autoantibodies against postsynaptic AchR [243]. Binding of autoantibodies to these proteins in the neuromuscular junction antagonises the ion channels with subsequent disturbance of the proper neuromuscular function. Ultimately, there is also complement-mediated destruction of the neuromuscular junction.

Graves' disease is an organ-specific autoimmune disorder that specifically affects the thyroid gland. Normally, hormone production by the thyroid is induced by thyroid-stimulating hormone (TSH) from the anterior pituitary, which itself is negatively regulated by the thyroid hormone levels in a feedback loop. TSH receptor (TSHR), located on the basal surface of thyroid follicular cells, is the prime autoantigen in Graves' disease. In patients with Graves' disease, autoantibodies bind to the TSHR and mimic the action of TSH [244]. Therefore, the gland produces an excessive amount of thyroid hormone causing hyperthyroidism. Graves' disease is the most common cause of hyperthyroidism, often affecting women. Mothers with this disease can pass anti-TSHR IgG to the foetus, and the baby will be born with hyperthyroidism.

9.2 Inflammatory Effects of Pathogenic Antibodies

The most widely recognised destructive effects of autoantibodies are antibody-dependent cell-mediated cytotoxicity (ADCC), and complement mediated cytotoxicity, but activation of proinflammatory pathways is increasingly recognised. These mechanisms can operate independently, but in fact, in most scenarios, they operate together to mediate inflammation and tissue damage.

9.2.1 Antibody-Dependent Cell-Mediated Cytotoxicity

ADCC is a lytic mechanism that can be mediated by autoreactive IgG. Autoreactive IgG recruits effector cells to the target cells with the Fab portion of the IgG binding to the target cell and the Fc region associating with FcγR on effector cells. Crosslinking FcγR on effector cells activates the secretion of cytotoxic molecules. Classical ADCC is mediated by NK cells that express CD16 (FcγRIII). Upon FcγR activation, NK cells release cytotoxic granules containing perforin and granzymes to kill opsonised target cells. In patients with Hashimoto's thyroiditis, a condition leading to hypothyroidism, anti-thyroperoxidase (TPO) antibodies bind to the basement membrane of the thyroid follicles, after which ADCC destroys the thyroid follicles. Effector cells, including macrophages and neutrophils, may lyse autoantibody opsonised target cells through production of reactive oxygen species (ROS). A rare disease is granulomatosis with polyangiitis (GPA), an autoimmune vasculitis characterised by antineutrophil cytoplasmic antibodies (ANCAs) that affects small and medium-size vessels in many organs [245]. Many ANCAs bind proteinase 3, which is normally present in the cytoplasm of neutrophils. Activated neutrophils express proteinase 3 on their plasma membrane. Upon ANCA binding, neutrophils are immobilised on vessel walls and trigger inflammation by releasing chemokines that induce further neutrophil influx and by undergoing a respiratory burst, leading to vascular necrosis. Approximately 70% of GPA patients develop ANCA-associated glomerulonephritis with no evidence of immune complex deposition (known as pauci-immune), the antibodies that activate neutrophils rather than form deposits on glomerular cells.

9.2.2 Complement-Mediated Cytotoxicity

The complement system plays a primary pathogenic role in autoantibody-induced inflammation and cytotoxicity [246]. Both IgM and IgG autoantibodies activate the classic complement pathway by binding to C1q through the Fc portion, which leads to the activation of the autocatalytic enzyme activity of the C1 complex and initiates the complement cascade that eventually leads to the assembly of the membrane-attack complex, a structure with a hydrophilic internal channel and a hydrophobic external portion that allows it to insert into the lipid bilayer of a cell membrane. The free passage of electrolytes and water through this transmembrane channel results in osmotic lysis of cells. Complement-mediated cytolysis represents a major pathogenic mechanism underlying autoimmune haemolytic anaemia and autoimmune thrombocytopenic purpura. Autoantibodies specific to the cell surface molecules on red blood cells or to glycoprotein IIb/IIIa (the fibrinogen receptor) on platelets activate the classic complement pathway and lead to intravascular lysis of circulating red blood cells and platelets.

In addition to cytolysis, complement also incites inflammation. The autoantibody-activated complement cascade yields the small cleavage products C3a, C4a and C5a, which are powerful inflammatory mediators that bind receptors on phagocytes and activate them to secrete inflammatory cytokines and chemokines. C5a is also a potent chemoattractant that induces homing of macrophages and neutrophils into the target organs. Tissue-bound antibody can activate the complement cascade and generate complement components that recruit and activate phagocytes, which release cytokines and ROS, to cause inflammatory injury. Moreover, C5a binds to and activates endothelial cells. Indeed, C5a binding to brain microvascular endothelial cells potently impairs the integrity of the blood–brain barrier.

9.2.3 Phagocytosis

Autoreactive IgM and IgG are able to promote phagocytosis of antibody-sensitised cells. In patients with autoimmune haemolytic anaemia, IgG antibodies recognise protein antigens on erythrocyte surface at physiological temperature (warm agglutinins), whereas IgM antibodies react with polysaccharide antigens on erythrocytes at a cold temperature of 0–4°C (cold agglutinins) [247]. IgG-coated erythrocytes activate phagocytic cells through FcγR crosslinking and trigger phagocytosis. Erythrocytes sensitised with IgG are cleared from the circulation by macrophages in the splenic sinusoids. IgM activates the complement cascade that produces C3b. C3b and C4b can act as opsonins, which mediate the attachment of IgM-sensitised erythrocytes to phagocytic cells through complement receptor 1 (CR1). In general, phagocytosis of IgM and C3b opsonised erythrocytes takes place in the liver by Kupffer cells. Thus, IgM-associated erythrocyte destruction occurs both intravascularly through the complement system and extra-vascularly through hepatic phagocytosis. Importantly, this phagocytic mechanism has been co-opted for its therapeutic potential. Many therapeutic antibodies used to deplete certain cell subsets do so by opsonising target cells, thereby priming them for removal by the reticuloendothelial system.

9.2.4 Immune Complex-Mediated Inflammation and Tissue Damage

Immune complexes might be formed when antibodies bind to soluble autoantigens. Normally, immune complexes do not impose significant harm, as they are rapidly cleared from the bloodstream by phagocytic cells that express both complement receptors and FcγRs. Under certain circumstances when either clearance mechanisms fail or an excessive amount of complex is produced, immune complexes persist in the body. Circulating immune complexes may be progressively trapped in tissues, especially synovium and kidney glomeruli, due in part to increased vascular permeability at those sites. Immune complex deposition in tissues activates the complement cascade and FcγRs on inflammatory immune cells such as neutrophils, monocytes, macrophages and DCs. Arthritis, glomerulonephritis and vasculitis are often manifestations of autoimmune diseases caused by immune complexes.

Autoantibodies can also bind antigen in tissues and form local noncirculating immune complexes activate complement and trigger the activation of inflammatory immune cells in situ. SLE is characterised by tissue-bound immune complexes. Antinuclear autoantibodies that lead to multiorgan damage are the hallmark of the disease. Nucleoproteins and dsDNA released from dead cells can activate autoreactive B cells. In lupus patients with active disease, an excessive amount of DNA can be found in the plasma. Nuclear autoantigens containing nucleic acids RNA or DNA form immune complexes that activate TLRs in both myeloid and lymphoid cells after they are transported into cells by FcRs. Thus, with continuous stimulation by autoantigens, innate and adaptive immune cells are activated, resulting in inflammation and an expansion of B cells producing autoreactive IgG. Autoantibodies in SLE patients can also deposit in the skin, the kidney and in other organs. By activating the complement cascade and promoting ADCC and CDC, immune complexes kill cells in tissues. Dead cells, in turn, release more intracellular autoantigens continuing to drive the autoimmune response.

10. THERAPEUTIC STRATEGIES TARGETING PATHOGENIC B CELL-MEDIATED AUTOREACTIVITY

Most of the B cell-targeting therapies so far have shown reduction in levels of circulating autoantibodies; however, this is not the only mechanism of action involved. B cell-targeting therapeutic molecules are described in a schematic diagram in Figure 17.1.

Targeting CD20 to deplete B cells has been remarkably successful in some diseases and surprisingly unsuccessful in others. Patients with neuromyelitis optical (NMO) lose their pathogenic anti-aquaporin 4 antibodies with B-cell depletion. Patients with MS also do extremely well with ant-CD20 antibody therapy, presumably because of loss of the APC function of B cells and a resulting decrease in T-cell activation. Surprisingly, it has been a less successful therapy for patients with SLE in whom it leads to a significant but relatively small decrease in autoantibody titres. This observation has led to the suggestion that the autoantibodies in SLE patients are made by LL PCs, a population that does not need continuous replenishments from the B-cell pool.

To improve autoantibody reduction, chimaeric antigen receptor T cells (CAR-T) targeting both CD20 and CD19 have been used in SLE patients and in a small case series have shown dramatic benefit. This has been attributed to the depletion of PCs as well as B cells. Because titres of anti-tetanus antibodies are reduced far less than titres of autoantibodies, the suggestion has been made that anti-tetanus antibodies are made by CD19-LL PCs and autoantibodies in SLE are predominantly made by CD19+ short-lived PCs. These therapeutic trials highlight current ignorance regarding the origin of the autoantibodies in SLE. This uncertainty extends to many diseases characterised by the presence of autoantibodies.

Proteasome inhibitors have also been proposed as a treatment for autoantibody-mediated diseases as they effectively target PCs, but they have multiple toxicities that have precluded their adoption in the clinic. Daratumumab, an anti-CD38 antibody that is used in patients with multiple myeloma and depletes PCs, was used successfully in two patients with refractory SLE [248].

Belimumab is an IgG1 human monoclonal antibody directed against soluble BAFF. It interferes with the binding of BAFF with its three receptors, BCMA, TACI and BAFF-R. Based on the efficacy shown in clinical trials, belimumab was approved by the FDA for the treatment of both adult and paediatric SLE. The use of belimumab has been associated with reduction in the levels of circulating pathogenic autoantibodies [249,250].

Other therapeutic modality that has been widely explored is blockade of molecules associated with B-cell activation or costimulation. Therapy against interferon, with a human monoclonal antibody against type I IFN receptor subunit 1, anifrolumab, was also recently approved for the treatment of moderate to severe SLE. However, only limited reduction in the titres of circulating autoantibodies was observed after treatment with anifrolumab [251,252]. It is also possible to target B cells by targeting T cell help. To

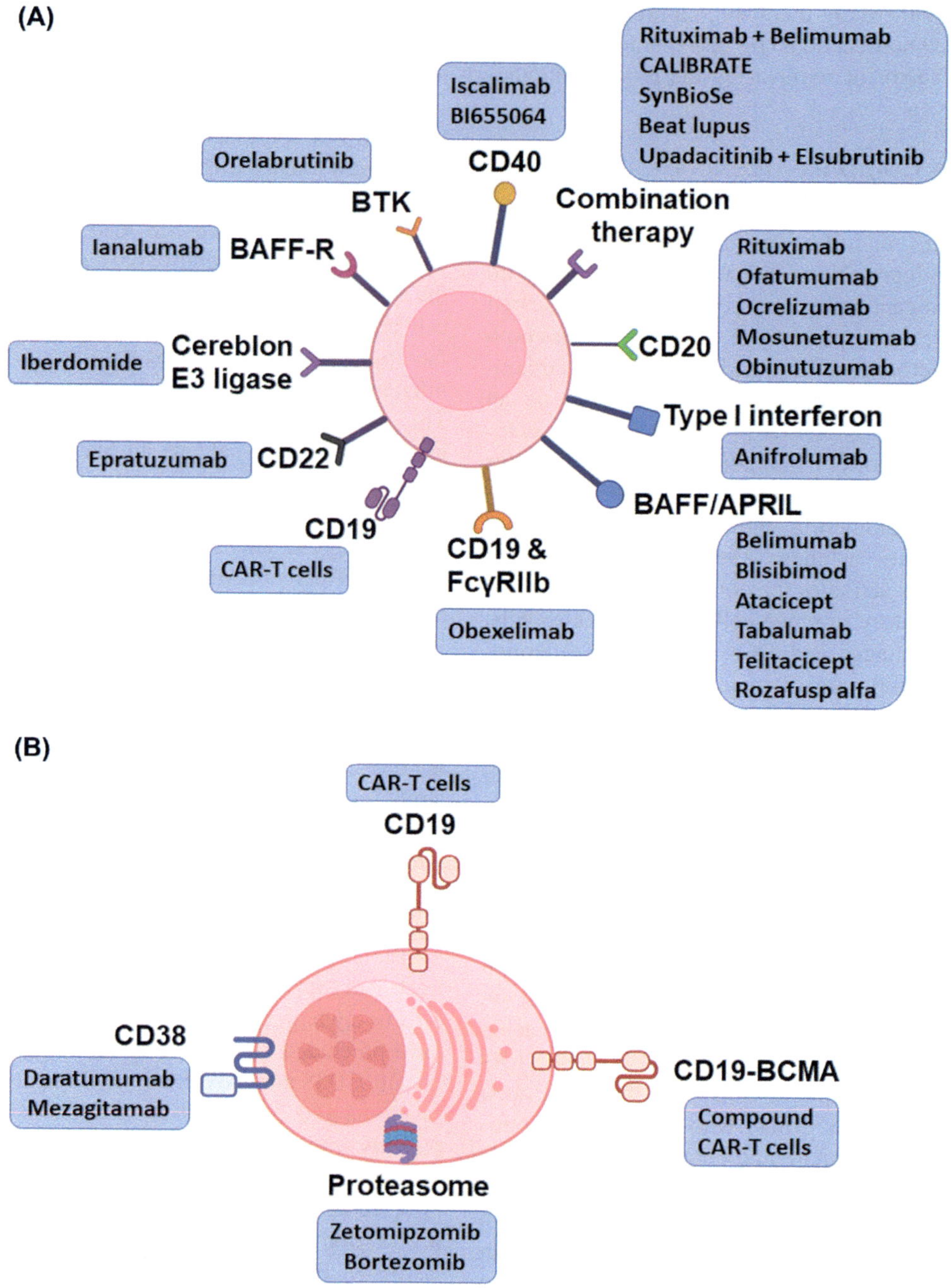

FIGURE 17.1 (A) B cell targeting therapeutic molecules. (B) Plasma B cell targeting therapeutic molecules.

date, there has been minimal effect on autoantibodies by targeting co-stimulation pathways.

There are multiple molecules that have been targeted or are currently under study for the treatment of SLE focusing on B cells (Table 17.1) or PCs (Table 17.2). Some of these therapeutic approaches potentially can affect the generation of pathogenic autoantibodies; however, clinical and mechanistic data will be required to confirm their efficacy and their effect on the generation of autoantibodies. A thorough analysis of the effects on autoantibodies of new medications, considering the specific characteristics of pathogenic antibodies, is warranted in order to properly understand their mechanism of action and potential biomarkers for efficacy. It is worth noting that, at present, the only therapy that has specificity for autoreactive B cells is BAFF blockade.

TABLE 17.1 Relevant B Cell Targeting Therapeutic Therapies

Target molecule	Therapeutic molecule	Mechanism of action	Relevant studies
CD20	Rituximab	Chimaeric monoclonal antibody. Depletes CD20+ B cells.	[253–256]
	Ofatumumab	Human monoclonal antibody. Depletes CD20+ B cells.	[257–259]
	Ocrelizumab	Humanised monoclonal antibody. Depletes CD20+ B cells.	[254,260]
	Mosunetuzumab	Bispecific antibody targets CD20 and CD3 (redirects T cells to eliminate B cells).	[253,261]
	Obinutuzumab	Humanised antibody with superior B cell depleting capacity than rituximab.	[253,262–264]
BAFF-R	Ianalumab	Humanised monoclonal antibody.	[265]
BTK	Orelabrutinib	BTK inhibitors. BTK is necessary for development, survival and activation of B cells.	[266,267]
CD40	Iscalimab	Anti-CD40 humanised monoclonal antibody. Inhibits CD154-induced activation of human leucocyte activation.	[268]
	BI655064	Humanised antagonistic anti-CD40 monoclonal antibody.	[269,270]
Cereblon E3 ligase	Iberdomide	Cereblon modulator binds to the cullin-RING E3 ubiquitin ligase complex, promoting ubiquitination and proteasomal degradation of Ikaros and Aiolos.	[271,272]
CD22	Epratuzumab	Monoclonal antibody. Enhances the normal inhibitory role of CD22 on the BCR, diminishing activation of B cells.	[273,274]
CD19	CAR-T cells	Chimaeric antigen receptor (CAR) T cells. CD19 CAR-T cells treatment depletes CD19+ B cells, including CD19+ plasmablasts.	[275–280]
CD19 + FcγRIIb	Obexelimab	Bispecific antibody for CD19 and FcγRIIb. Inhibits B cell activation.	[281–284]
BAFF/APRIL	Belimumab	Monoclonal antibody against BAFF. Approved for the treatment of SLE in children and adults.	[285–287]
	Blisibimod	BAFF-inhibiting peptibody, composed of both a peptide and antibody.	[288–290]
	Atacicept	Human recombinant TACI receptor targets both BAFF and APRIL.	[291]
	Tabalumab	Monoclonal antibody targeted for soluble and membrane-bound BAFF.	[292]
	Telitacicept	Recombinant fusion protein composed of the TACI receptor and the Fc component of human IgG (TACI-Fc).	[293–295]
	Rozafusp alfa	Bispecific monoclonal antibody-peptide conjugate which inhibits both Inducible T-cell costimulator Ligand (ICOSL) and BAFF.	[296,297]
Type I interferon	Anifrolumab	Type I IFN inhibitor.	[251,252]
Combination therapy	Rituximab + belimumab	Co-administration of rituximab and belimumab. Different studies have evaluated this combination, with differences in the sequence of both treatments. The administration of belimumab before rituximab may increase the mobilisation of B cells enhancing B-cell depletion. The administration of belimumab after rituximab may reduce the emergence of autoreactive B cells after B-cell depletion.	
	CALIBRATE	CALIBRATE compared treatment responses between LN patients who received rituximab with or without subsequent belimumab.	[119]
	SynBioSe	Belimumab followed by rituximab in SLE (SynBioSe-1) and in LN patients (SynBioSe-2).	[298,299]
	Beat lupus	Belimumab following rituximab in SLE patients.	[300,301]
	Upadacitinib + Elsubrutinib	Elsubrutinib, a BTK inhibitor, co-administered with upadacitinib, a janus kinase inhibitor.	[302]

TABLE 17.2 Therapeutic Molecules Targeting Plasma Cells

Target molecule	Therapeutic molecule	Mechanism of action	Relevant studies
CD38	Daratumumab	Monoclonal antibody targets CD38.	[248,303]
	Mezagitamab	Fully human IgG1 monoclonal antibody targets CD38.	[304]
Proteasome	Zetomipzomib	Selective proteasome inhibitor, targets the immunoproteasome, whose distribution is restricted to inflammatory cells.	[305,306]
	Bortezomib	Constitutive proteasome inhibitor.	[307,308]
CD19-BCMA	Compound CAR-T cells	CD19-BCMA compound CAR (cCAR) designed to target CD19 on B cells and BCMA on plasma cells. Case report in a patient with stage IV diffuse large B-cell lymphoma (DLBCL) and 20-year history of SLE.	[309]

11. CONCLUSION

Autoreactive B cells and autoreactive antibodies are part of the normal repertoire of healthy individuals, and they can be not only non-pathogenic, but part of the mechanisms that prevent infection and autoimmune diseases and regulate physiologic homoeostasis. The generation of pathogenic antibodies is related to intrinsic alterations in B cells, post-transcriptional changes in antibodies, or extrinsic factors and environmental factors. Therefore, the mechanisms that prevent pathogenic autoreactivity in B cells are designed to modulate B-cell survival, activation and differentiation in a lineage and circumstance dependent fashion. Understanding the multiple components that are involved in the pathogenesis of autoimmune diseases associated with autoantibodies is critical to the development of therapies that focus on the pathogenic B cells or antibodies, sparing those that are relevant to maintain homoeostasis.

REFERENCES

[1] Malkiel S, et al. Checkpoints for autoreactive B cells in the peripheral blood of lupus patients assessed by flow cytometry. Arthritis Rheumatol 2016;68(9):2210–20.

[2] Wardemann H, et al. Predominant autoantibody production by early human B cell precursors. Science 2003;301(5638):1374–7.

[3] Grönwall C, Vas J, Silverman GJ. Protective roles of natural IgM antibodies. Front Immunol 2012;3:66.

[4] Bos NA, et al. Serum immunoglobulin levels and naturally occurring antibodies against carbohydrate antigens in germ-free BALB/c mice fed chemically defined ultrafiltered diet. Eur J Immunol 1989;19(12):2335–9.

[5] Merbl Y, et al. Newborn humans manifest autoantibodies to defined self molecules detected by antigen microarray informatics. J Clin Invest 2007;117(3):712–8.

[6] Panda S, Ding JL. Natural antibodies bridge innate and adaptive immunity. J Immunol 2015;194(1):13–20.

[7] Lobo PI. Role of natural autoantibodies and natural IgM anti-leucocyte autoantibodies in health and disease. Front Immunol 2016;7:198.

[8] Shaw PX, et al. Natural antibodies with the T15 idiotype may act in atherosclerosis, apoptotic clearance, and protective immunity. J Clin Invest 2000;105(12):1731–40.

[9] Silverman GJ, et al. Natural autoantibodies to apoptotic cell membranes regulate fundamental innate immune functions and suppress inflammation. Discov Med 2009;8(42):151–6.

[10] Hörkkö S, et al. Monoclonal autoantibodies specific for oxidized phospholipids or oxidized phospholipid-protein adducts inhibit macrophage uptake of oxidized low-density lipoproteins. J Clin Invest 1999;103(1):117–28.

[11] Uchida K. Natural antibodies as a sensor of electronegative damage-associated molecular patterns (DAMPs). Free Radic Biol Med 2014;72:156–61.

[12] Amendt T, et al. Autoreactive antibodies control blood glucose by regulating insulin homeostasis. Proc Natl Acad Sci U S A 2022;119(6).

[13] Werwitzke S, et al. Inhibition of lupus disease by anti-double-stranded DNA antibodies of the IgM isotype in the (NZB x NZW) F1 mouse. Arthritis Rheum 2005;52(11):3629–38.

[14] Jiang C, et al. Activation-induced deaminase-deficient MRL/lpr mice secrete high levels of protective antibodies against lupus nephritis. Arthritis Rheum 2011;63(4):1086–96.

[15] Silverman GJ, et al. Genetic imprinting of autoantibody repertoires in systemic lupus erythematosus patients. Clin Exp Immunol 2008;153(1):102–16.

[16] Grönwall C, et al. IgM autoantibodies to distinct apoptosis-associated antigens correlate with protection from cardiovascular events and renal disease in patients with SLE. Clin Immunol 2012;142(3):390–8.

[17] Chou MY, et al. Oxidation-specific epitopes are dominant targets of innate natural antibodies in mice and humans. J Clin Invest 2009;119(5):1335–49.

[18] Zhang M, et al. Identification of a specific self-reactive IgM antibody that initiates intestinal ischemia/reperfusion injury. Proc Natl Acad Sci U S A 2004;101(11):3886–91.

[19] Kulik L, et al. Pathogenic natural antibodies recognizing annexin IV are required to develop intestinal ischemia-reperfusion injury. J Immunol 2009;182(9):5363–73.

[20] Barcellini W. New insights in the pathogenesis of autoimmune hemolytic anemia. Transfus Med Hemother 2015;42(5):287–93.

[21] Wright JF, et al. Characterization of platelet-reactive antibodies in children with varicella-associated acute immune thrombocytopenic purpura (ITP). Br J Haematol 1996;95(1):145–52.

[22] Son M, et al. C1q limits dendritic cell differentiation and activation by engaging LAIR-1. Proc Natl Acad Sci U S A 2012;109(46):E3160–7.

[23] Colombo BM, et al. Defective expression and function of the leukocyte associated Ig-like receptor 1 in B lymphocytes from systemic lupus erythematosus patients. PLoS One 2012;7(2):e31903.

[24] Honda S, et al. Marginal zone B cells exacerbate endotoxic shock via interleukin-6 secretion induced by Fcα/μR-coupled TLR4 signalling. Nat Commun 2016;7:11498.

[25] Kaetzel CS, et al. The polymeric immunoglobulin receptor (secretory component) mediates transport of immune complexes across epithelial cells: a local defense function for IgA. Proc Natl Acad Sci U S A 1991;88(19):8796–800.

[26] Liu J, et al. Role of the IgM Fc receptor in immunity and tolerance. Front Immunol 2019;10:529.

[27] Honjo K, et al. Enhanced auto-antibody production and Mott cell formation in FcμR-deficient autoimmune mice. Int Immunol 2014;26(12):659–72.

[28] Brenner D, et al. Toso controls encephalitogenic immune responses by dendritic cells and regulatory T cells. Proc Natl Acad Sci U S A 2014;111(3):1060–5.

[29] Kubagawa H, et al. Differences between human and mouse IgM Fc receptor (FcμR). Int J Mol Sci 2021;22(13).

[30] Bindon CI, et al. Human monoclonal IgG isotypes differ in complement activating function at the level of C4 as well as C1q. J Exp Med 1988;168(1):127–42.

[31] Diebolder CA, et al. Complement is activated by IgG hexamers assembled at the cell surface. Science 2014;343(6176):1260–3.

[32] Vidarsson G, Dekkers G, Rispens T. IgG subclasses and allotypes: from structure to effector functions. Front Immunol 2014;5:520.

[33] Israel EJ, et al. Requirement for a beta 2-microglobulin-associated Fc receptor for acquisition of maternal IgG by fetal and neonatal mice. J Immunol 1995;154(12):6246–51.

[34] Eftekhari P, et al. Anti-SSA/Ro52 autoantibodies blocking the cardiac 5-HT4 serotoninergic receptor could explain neonatal lupus congenital heart block. Eur J Immunol 2000;30(10):2782–90.

[35] Janda A, et al. Ig constant region effects on variable region structure and function. Front Microbiol 2016;7:22.

[36] Pritsch O, et al. Can immunoglobulin C(H)1 constant region domain modulate antigen binding affinity of antibodies? J Clin Invest 1996;98(10):2235–43.

[37] Pritsch O, et al. Can isotype switch modulate antigen-binding affinity and influence clonal selection? Eur J Immunol 2000;30(12):3387–95.

[38] McLean GR, et al. Isotype can affect the fine specificity of an antibody for a polysaccharide antigen. J Immunol 2002;169(3):1379–86.

[39] Xia Y, et al. The constant region contributes to the antigenic specificity and renal pathogenicity of murine anti-DNA antibodies. J Autoimmun 2012;39(4):398–411.

[40] Quast I, Lunemann JD. Fc glycan-modulated immunoglobulin G effector functions. J Clin Immunol 2014;34(Suppl. 1):S51–5.

[41] Mizuochi T, et al. Structural and numerical variations of the carbohydrate moiety of immunoglobulin G. J Immunol 1982;129(5):2016–20.

[42] Shields RL, et al. Lack of fucose on human IgG1 N-linked oligosaccharide improves binding to human Fcgamma RIII and antibody-dependent cellular toxicity. J Biol Chem 2002;277(30):26733–40.

[43] Herter S, et al. Glycoengineering of therapeutic antibodies enhances monocyte/macrophage-mediated phagocytosis and cytotoxicity. J Immunol 2014;192(5):2252–60.

[44] Gasdaska JR, et al. An afucosylated anti-CD20 monoclonal antibody with greater antibody-dependent cellular cytotoxicity and B-cell depletion and lower complement-dependent cytotoxicity than rituximab. Mol Immunol 2012;50(3):134–41.

[45] Zauner G, et al. Glycoproteomic analysis of antibodies. Mol Cell Proteomics 2013;12(4):856–65.

[46] Kaneko Y, Nimmerjahn F, Ravetch JV. Anti-inflammatory activity of immunoglobulin G resulting from Fc sialylation. Science 2006;313(5787):670–3.

[47] Anthony RM, et al. Recapitulation of IVIG anti-inflammatory activity with a recombinant IgG Fc. Science 2008;320(5874):373–6.

[48] Parekh RB, et al. Association of rheumatoid arthritis and primary osteoarthritis with changes in the glycosylation pattern of total serum IgG. Nature 1985;316(6027):452–7.

[49] Vučković F, et al. Association of systemic lupus erythematosus with decreased immunosuppressive potential of the IgG glycome. Arthritis Rheumatol 2015;67(11):2978–89.

[50] Wuhrer M, et al. Glycosylation profiling of immunoglobulin G (IgG) subclasses from human serum. Proteomics 2007;7(22):4070–81.

[51] Scherer HU, et al. Glycan profiling of anti-citrullinated protein antibodies isolated from human serum and synovial fluid. Arthritis Rheum 2010;62(6):1620–9.

[52] Washburn N, et al. Controlled tetra-Fc sialylation of IVIg results in a drug candidate with consistent enhanced anti-inflammatory activity. Proc Natl Acad Sci U S A 2015;112(11):E1297–306.

[53] Schwab I, et al. Broad requirement for terminal sialic acid residues and FcgammaRIIB for the preventive and therapeutic activity of intravenous immunoglobulins in vivo. Eur J Immunol 2014;44(5):1444–53.

[54] Anthony RM, et al. Intravenous gammaglobulin suppresses inflammation through a novel T(H)2 pathway. Nature 2011;475(7354):110–3.

[55] Quast I, et al. Sialylation of IgG Fc domain impairs complement-dependent cytotoxicity. J Clin Invest 2015;125(11):4160–70.

[56] Morel M, et al. Abnormal B cell glycosylation in autoimmunity: a new potential treatment strategy. Front Immunol 2022;13:975963.

[57] Giovannone N, et al. Human B cell differentiation is characterized by progressive remodeling of O-linked glycans. Front Immunol 2018;9:2857.

[58] Koethe S, et al. Pivotal advance: CD45RB glycosylation is specifically regulated during human peripheral B cell differentiation. J Leukoc Biol 2011;90(1):5–19.

[59] Hayakawa K, et al. The "Ly-1 B" cell subpopulation in normal immunodefective, and autoimmune mice. J Exp Med 1983;157(1):202–18.

[60] Baumgarth N, et al. Innate and acquired humoral immunities to influenza virus are mediated by distinct arms of the immune system. Proc Natl Acad Sci U S A 1999;96(5):2250–5.

[61] Montecino-Rodriguez E, Leathers H, Dorshkind K. Identification of a B-1 B cell-specified progenitor. Nat Immunol 2006;7(3):293–301.

[62] Hardy RR. B-1 B cell development. J Immunol 2006;177(5):2749–54.

[63] Baumgarth N. The double life of a B-1 cell: self-reactivity selects for protective effector functions. Nat Rev Immunol 2011;11(1):34–46.
[64] Souto-Carneiro MM, et al. Developmental changes in the human heavy chain CDR3. J Immunol 2005;175(11):7425–36.
[65] Gu H, Forster I, Rajewsky K. Sequence homologies, N sequence insertion and JH gene utilization in VHDJH joining: implications for the joining mechanism and the ontogenetic timing of Ly1 B cell and B-CLL progenitor generation. EMBO J 1990;9(7):2133–40.
[66] Schroeder Jr HW, et al. Developmental regulation of the human antibody repertoire. Ann N Y Acad Sci 1995;764:242–60.
[67] Logtenberg T, et al. Autoantibodies encoded by the most Jh-proximal human immunoglobulin heavy chain variable region gene. J Exp Med 1989;170(4):1347–55.
[68] Crews S, et al. A single VH gene segment encodes the immune response to phosphorylcholine: somatic mutation is correlated with the class of the antibody. Cell 1981;25(1):59–66.
[69] Hayakawa K, et al. Positive selection of natural autoreactive B cells. Science 1999;285(5424):113–6.
[70] Arnold LW, et al. Development of B-1 cells: segregation of phosphatidyl choline-specific B cells to the B-1 population occurs after immunoglobulin gene expression. J Exp Med 1994;179(5):1585–95.
[71] Lam KP, Rajewsky K. B cell antigen receptor specificity and surface density together determine B-1 versus B-2 cell development. J Exp Med 1999;190(4):471–7.
[72] Rickert RC, Rajewsky K, Roes J. Impairment of T-cell-dependent B-cell responses and B-1 cell development in CD19-deficient mice. Nature 1995;376(6538):352–5.
[73] Engel P, et al. Abnormal B lymphocyte development, activation, and differentiation in mice that lack or overexpress the CD19 signal transduction molecule. Immunity 1995;3(1):39–50.
[74] Satterthwaite AB, Li Z, Witte ON. Btk function in B cell development and response. Semin Immunol 1998;10(4):309–16.
[75] Hoffmann A, et al. Siglec-G is a B1 cell-inhibitory receptor that controls expansion and calcium signaling of the B1 cell population. Nat Immunol 2007;8(7):695–704.
[76] Jellusova J, et al. Siglec-G regulates B1 cell survival and selection. J Immunol 2010;185(6):3277–84.
[77] Ueffing N, et al. Up-regulation of c-FLIP short by NFAT contributes to apoptosis resistance of short-term activated T cells. Blood 2008;112(3):690–8.
[78] Berland R, Wortis HH. Normal B-1a cell development requires B cell-intrinsic NFATc1 activity. Proc Natl Acad Sci U S A 2003;100(23):13459–64.
[79] Maruya M, et al. Vitamin A-dependent transcriptional activation of the nuclear factor of activated T cells c1 (NFATc1) is critical for the development and survival of B1 cells. Proc Natl Acad Sci U S A 2011;108(2):722–7.
[80] Pao LI, et al. B cell-specific deletion of protein-tyrosine phosphatase Shp1 promotes B-1a cell development and causes systemic autoimmunity. Immunity 2007;27(1):35–48.
[81] Morel L, et al. Functional dissection of systemic lupus erythematosus using congenic mouse strains. J Immunol 1997;158(12):6019–28.
[82] Ghia P, et al. From normal to clonal B cells: chronic lymphocytic leukemia (CLL) at the crossroad between neoplasia and autoimmunity. Autoimmun Rev 2007;7(2):127–31.
[83] Berland R, Wortis HH. Origins and functions of B-1 cells with notes on the role of CD5. Annu Rev Immunol 2002;20:253–300.
[84] Bendelac A, Bonneville M, Kearney JF. Autoreactivity by design: innate B and T lymphocytes. Nat Rev Immunol 2001;1(3):177–86.
[85] Zhou Z, et al. Autoreactive marginal zone B cells enter the follicles and interact with CD4+ T cells in lupus-prone mice. BMC Immunol 2011;12:7.
[86] Dono M, et al. Heterogeneity of tonsillar subepithelial B lymphocytes, the splenic marginal zone equivalents. J Immunol 2000;164(11):5596–604.
[87] Garraud O, et al. Revisiting the B-cell compartment in mouse and humans: more than one B-cell subset exists in the marginal zone and beyond. BMC Immunol 2012;13:63.
[88] Hendricks J, Bos NA, Kroese FGM. Heterogeneity of memory marginal zone B cells. Crit Rev Immunol 2018;38(2):145–58.
[89] Weller S, et al. Human blood IgM "memory" B cells are circulating splenic marginal zone B cells harboring a prediversified immunoglobulin repertoire. Blood 2004;104(12):3647–54.
[90] Weller S, et al. Somatic diversification in the absence of antigen-driven responses is the hallmark of the IgM+ IgD+ CD27+ B cell repertoire in infants. J Exp Med 2008;205(6):1331–42.
[91] Weller S, et al. CD40-CD40L independent Ig gene hypermutation suggests a second B cell diversification pathway in humans. Proc Natl Acad Sci U S A 2001;98(3):1166–70.
[92] Jia W, et al. BAFF promotes heightened BCR responsiveness and manifestations of chronic GVHD after allogeneic stem cell transplantation. Blood 2021;137(18):2544–57.
[93] Grimaldi CM, Michael DJ, Diamond B. Cutting edge: expansion and activation of a population of autoreactive marginal zone B cells in a model of estrogen-induced lupus. J Immunol 2001;167(4):1886–90.
[94] Palm AK, et al. Function and regulation of self-reactive marginal zone B cells in autoimmune arthritis. Cell Mol Immunol 2015;12(4):493–504.
[95] Palm AK, Friedrich HC, Kleinau S. Nodal marginal zone B cells in mice: a novel subset with dormant self-reactivity. Sci Rep 2016;6:27687.
[96] Mariño E, et al. Marginal-zone B-cells of nonobese diabetic mice expand with diabetes onset, invade the pancreatic lymph nodes, and present autoantigen to diabetogenic T-cells. Diabetes 2008;57(2):395–404.
[97] Duan B, et al. Intrafollicular location of marginal zone/CD1d(hi) B cells is associated with autoimmune pathology in a mouse model of lupus. Lab Invest 2008;88(9):1008–20.
[98] Lee SK, et al. B cell priming for extrafollicular antibody responses requires Bcl-6 expression by T cells. J Exp Med 2011;208(7):1377–88.
[99] Malkiel S, et al. Plasma cell differentiation pathways in systemic lupus erythematosus. Front Immunol 2018;9:427.
[100] Chan TD, et al. Antigen affinity controls rapid T-dependent antibody production by driving the expansion rather than the differentiation or extrafollicular migration of early plasmablasts. J Immunol 2009;183(5):3139–49.
[101] Tull TJ, et al. Human marginal zone B cell development from early T2 progenitors. J Exp Med 2021;218(4).
[102] Monroe JG. ITAM-mediated tonic signalling through pre-BCR and BCR complexes. Nat Rev Immunol 2006;6(4):283–94.

[103] Nemazee D. Receptor editing in lymphocyte development and central tolerance. Nat Rev Immunol 2006;6(10):728–40.
[104] Melamed D, et al. Developmental regulation of B lymphocyte immune tolerance compartmentalizes clonal selection from receptor selection. Cell 1998;92(2):173–82.
[105] Carsetti R, Kohler G, Lamers MC. Transitional B cells are the target of negative selection in the B cell compartment. J Exp Med 1995;181(6):2129–40.
[106] Enders A, et al. Loss of the pro-apoptotic BH3-only Bcl-2 family member Bim inhibits BCR stimulation-induced apoptosis and deletion of autoreactive B cells. J Exp Med 2003;198(7):1119–26.
[107] Fang W, et al. Self-reactive B lymphocytes overexpressing Bcl-xL escape negative selection and are tolerized by clonal anergy and receptor editing. Immunity 1998;9(1):35–45.
[108] Lang J, et al. Enforced Bcl-2 expression inhibits antigen-mediated clonal elimination of peripheral B cells in an antigen dose-dependent manner and promotes receptor editing in autoreactive, immature B cells. J Exp Med 1997;186(9):1513–22.
[109] Dempsey PW, et al. C3d of complement as a molecular adjuvant: bridging innate and acquired immunity. Science 1996;271(5247):348–50.
[110] Macauley MS, et al. Antigenic liposomes displaying CD22 ligands induce antigen-specific B cell apoptosis. J Clin Invest 2013;123(7):3074–83.
[111] Nemazee D. Mechanisms of central tolerance for B cells. Nat Rev Immunol 2017;17(5):281–94.
[112] Zikherman J, Parameswaran R, Weiss A. Endogenous antigen tunes the responsiveness of naive B cells but not T cells. Nature 2012;489(7414):160–4.
[113] Zou YR, Diamond B. Fate determination of mature autoreactive B cells. Adv Immunol 2013;118:1–36.
[114] Cooke MP, et al. Immunoglobulin signal transduction guides the specificity of B cell-T cell interactions and is blocked in tolerant self-reactive B cells. J Exp Med 1994;179(2):425–38.
[115] Fulcher DA, Basten A. Reduced life span of anergic self-reactive B cells in a double-transgenic model. J Exp Med 1994;179(1):125–34.
[116] Quách TD, et al. Anergic responses characterize a large fraction of human autoreactive naive B cells expressing low levels of surface IgM. J Immunol 2011;186(8):4640–8.
[117] Sabouri Z, et al. Redemption of autoantibodies on anergic B cells by variable-region glycosylation and mutation away from self-reactivity. Proc Natl Acad Sci U S A 2014;111(25):E2567–75.
[118] Thien M, et al. Excess BAFF rescues self-reactive B cells from peripheral deletion and allows them to enter forbidden follicular and marginal zone niches. Immunity 2004;20(6):785–98.
[119] Atisha-Fregoso Y, et al. Phase II randomized trial of rituximab plus cyclophosphamide followed by belimumab for the treatment of lupus nephritis. Arthritis Rheumatol 2021;73(1):121–31.
[120] Ansel KM, et al. In vivo-activated CD4 T cells upregulate CXC chemokine receptor 5 and reprogram their response to lymphoid chemokines. J Exp Med 1999;190(8):1123–34.
[121] Haynes NM, et al. Role of CXCR5 and CCR7 in follicular Th cell positioning and appearance of a programmed cell death gene-1high germinal center-associated subpopulation. J Immunol 2007;179(8):5099–108.
[122] Kerfoot SM, et al. Germinal center B cell and T follicular helper cell development initiates in the interfollicular zone. Immunity 2011;34(6):947–60.
[123] Schnizlein CT, et al. Follicular dendritic cells in suspension: identification, enrichment, and initial characterization indicating immune complex trapping and lack of adherence and phagocytic activity. J Immunol 1985;134(3):1360–8.
[124] Allen CD, Cyster JG. Follicular dendritic cell networks of primary follicles and germinal centers: phenotype and function. Semin Immunol 2008;20(1):14–25.
[125] Gray D, Kosco M, Stockinger B. Novel pathways of antigen presentation for the maintenance of memory. Int Immunol 1991;3(2):141–8.
[126] Allen CD, Okada T, Cyster JG. Germinal-center organization and cellular dynamics. Immunity 2007;27(2):190–202.
[127] Calado DP, et al. The cell-cycle regulator c-Myc is essential for the formation and maintenance of germinal centers. Nat Immunol 2012;13(11):1092–100.
[128] Berek C, Berger A, Apel M. Maturation of the immune response in germinal centers. Cell 1991;67(6):1121–9.
[129] Chan TD, Brink R. Affinity-based selection and the germinal center response. Immunol Rev 2012;247(1):11–23.
[130] Hase K, et al. Activation-induced cytidine deaminase deficiency causes organ-specific autoimmune disease. PLoS One 2008;3(8):e3033.
[131] Muramatsu M, et al. Class switch recombination and hypermutation require activation-induced cytidine deaminase (AID), a potential RNA editing enzyme. Cell 2000;102(5):553–63.
[132] Zaheen A, et al. AID constrains germinal center size by rendering B cells susceptible to apoptosis. Blood 2009;114(3):547–54.
[133] Revy P, et al. Activation-induced cytidine deaminase (AID) deficiency causes the autosomal recessive form of the Hyper-IgM syndrome (HIGM2). Cell 2000;102(5):565–75.
[134] Tan Q, et al. Activation-induced cytidine deaminase deficiency accelerates autoimmune diabetes in NOD mice. JCI Insight 2018;3(1).
[135] Xu X, et al. Increased expression of activation-induced cytidine deaminase is associated with anti-CCP and rheumatoid factor in rheumatoid arthritis. Scand J Immunol 2009;70(3):309–16.
[136] Pugh-Bernard AE, et al. Regulation of inherently autoreactive VH4-34 B cells in the maintenance of human B cell tolerance. J Clin Invest 2001;108(7):1061–70.
[137] Cappione 3rd A, et al. Germinal center exclusion of autoreactive B cells is defective in human systemic lupus erythematosus. J Clin Invest 2005;115(11):3205–16.
[138] Wellmann U, et al. The evolution of human anti-double-stranded DNA autoantibodies. Proc Natl Acad Sci U S A 2005;102(26):9258–63.
[139] Mietzner B, et al. Autoreactive IgG memory antibodies in patients with systemic lupus erythematosus arise from nonreactive and polyreactive precursors. Proc Natl Acad Sci U S A 2008;105(28):9727–32.
[140] Diamond B, Scharff MD. Somatic mutation of the T15 heavy chain gives rise to an antibody with autoantibody specificity. Proc Natl Acad Sci U S A 1984;81(18):5841–4.
[141] Ray SK, Putterman C, Diamond B. Pathogenic autoantibodies are routinely generated during the response to foreign antigen: a

paradigm for autoimmune disease. Proc Natl Acad Sci U S A 1996;93(5):2019–24.
[142] Sweet RA, Lee SK, Vinuesa CG. Developing connections amongst key cytokines and dysregulated germinal centers in autoimmunity. Curr Opin Immunol 2012;24(6):658–64.
[143] Brink R, Phan TG. Self-reactive B cells in the germinal center reaction. Annu Rev Immunol 2018;36:339–57.
[144] Billips LG, et al. Immunoglobulin recombinase gene activity is modulated reciprocally by interleukin 7 and CD19 in B cell progenitors. J Exp Med 1995;182(4):973–82.
[145] Hikida M, et al. Expression of recombination activating genes in germinal center B cells: involvement of interleukin 7 (IL-7) and the IL-7 receptor. J Exp Med 1998;188(2):365–72.
[146] Huang J, Durum SK, Muegge K. Cutting edge: histone acetylation and recombination at the TCR gamma locus follows IL-7 induction. J Immunol 2001;167(11):6073–7.
[147] Zhang Z, et al. Expression of recombination-activating genes and terminal deoxynucleotidyl transferase and secondary rearrangement of immunoglobulin kappa light chains in rheumatoid arthritis synovial tissue. Arthritis Rheum 2001;44(10):2275–84.
[148] Degn SE, et al. Clonal evolution of autoreactive germinal centers. Cell 2017;170(5):913–926.e19.
[149] Fukuyama H, Nimmerjahn F, Ravetch JV. The inhibitory Fcgamma receptor modulates autoimmunity by limiting the accumulation of immunoglobulin G+ anti-DNA plasma cells. Nat Immunol 2005;6(1):99–106.
[150] Luzina IG, et al. Spontaneous formation of germinal centers in autoimmune mice. J Leukoc Biol 2001;70(4):578–84.
[151] Mackay F, et al. Mice transgenic for BAFF develop lymphocytic disorders along with autoimmune manifestations. J Exp Med 1999;190(11):1697–710.
[152] Domeier PP, Schell SL, Rahman ZS. Spontaneous germinal centers and autoimmunity. Autoimmunity 2017;50(1):4–18.
[153] Gago da Graça C, van Baarsen LGM, Mebius RE. Tertiary lymphoid structures: diversity in their development, composition, and role. J Immunol 2021;206(2):273–81.
[154] Moyron-Quiroz JE, et al. Persistence and responsiveness of immunologic memory in the absence of secondary lymphoid organs. Immunity 2006;25(4):643–54.
[155] Pitzalis C, et al. Ectopic lymphoid-like structures in infection, cancer and autoimmunity. Nat Rev Immunol 2014;14(7):447–62.
[156] Humby F, et al. Synovial cellular and molecular signatures stratify clinical response to csDMARD therapy and predict radiographic progression in early rheumatoid arthritis patients. Ann Rheum Dis 2019;78(6):761–72.
[157] Okazaki T, et al. PD-1 immunoreceptor inhibits B cell receptor-mediated signaling by recruiting src homology 2-domain-containing tyrosine phosphatase 2 to phosphotyrosine. Proc Natl Acad Sci U S A 2001;98(24):13866–71.
[158] Gonzalez-Figueroa P, et al. Follicular regulatory T cells produce neuritin to regulate B cells. Cell 2021;184(7):1775–1789.e19.
[159] Niederlova V, et al. CD8(+) Tregs revisited: a heterogeneous population with different phenotypes and properties. Eur J Immunol 2021;51(3):512–30.
[160] Shi Z, et al. Human CD8+CXCR3+ T cells have the same function as murine CD8+CD122+ Treg. Eur J Immunol 2009;39(8):2106–19.
[161] Kim HJ, et al. Inhibition of follicular T-helper cells by CD8(+) regulatory T cells is essential for self tolerance. Nature 2010;467(7313):328–32.
[162] Koopman G, et al. Adhesion through the LFA-1 (CD11a/CD18)-ICAM-1 (CD54) and the VLA-4 (CD49d)-VCAM-1 (CD106) pathways prevents apoptosis of germinal center B cells. J Immunol 1994;152(8):3760–7.
[163] Victoratos P, et al. FDC-specific functions of p55TNFR and IKK2 in the development of FDC networks and of antibody responses. Immunity 2006;24(1):65–77.
[164] Garin A, et al. Toll-like receptor 4 signaling by follicular dendritic cells is pivotal for germinal center onset and affinity maturation. Immunity 2010;33(1):84–95.
[165] Victoratos P, Kollias G. Induction of autoantibody-mediated spontaneous arthritis critically depends on follicular dendritic cells. Immunity 2009;30(1):130–42.
[166] Sarra M, Monteleone G. Interleukin-21: a new mediator of inflammation in systemic lupus erythematosus. J Biomed Biotechnol 2010;2010:294582.
[167] Hughes T, et al. Fine-mapping and transethnic genotyping establish IL2/IL21 genetic association with lupus and localize this genetic effect to IL21. Arthritis Rheum 2011;63(6):1689–97.
[168] Sawalha AH, et al. Genetic association of interleukin-21 polymorphisms with systemic lupus erythematosus. Ann Rheum Dis 2008;67(4):458–61.
[169] Vinuesa CG, et al. A RING-type ubiquitin ligase family member required to repress follicular helper T cells and autoimmunity. Nature 2005;435(7041):452–8.
[170] Linterman MA, et al. Roquin differentiates the specialized functions of duplicated T cell costimulatory receptor genes CD28 and ICOS. Immunity 2009;30(2):228–41.
[171] Linterman MA, et al. Follicular helper T cells are required for systemic autoimmunity. J Exp Med 2009;206(3):561–76.
[172] Athanasopoulos V, et al. The ROQUIN family of proteins localizes to stress granules via the ROQ domain and binds target mRNAs. FEBS J 2010;277(9):2109–27.
[173] Glasmacher E, et al. Roquin binds inducible costimulator mRNA and effectors of mRNA decay to induce microRNA-independent post-transcriptional repression. Nat Immunol 2010;11(8):725–33.
[174] Bertossi A, et al. Loss of Roquin induces early death and immune deregulation but not autoimmunity. J Exp Med 2011;208(9):1749–56.
[175] Lee SK, et al. Interferon-gamma excess leads to pathogenic accumulation of follicular helper T cells and germinal centers. Immunity 2012;37(5):880–92.
[176] Kim SJ, et al. Tolerogenic function of Blimp-1 in dendritic cells. J Exp Med 2011;208(11):2193–9.
[177] Anolik JH, et al. Delayed memory B cell recovery in peripheral blood and lymphoid tissue in systemic lupus erythematosus after B cell depletion therapy. Arthritis Rheum 2007;56(9):3044–56.
[178] Zhang W, et al. Excessive CD11c(+)Tbet(+) B cells promote aberrant TFH differentiation and affinity-based germinal center selection in lupus. Proc Natl Acad Sci U S A 2019;116(37):18550–60.
[179] Arazi A, et al. The immune cell landscape in kidneys of patients with lupus nephritis. Nat Immunol 2019;20(7):902–14.

[180] Barbhaiya M, Costenbader KH. Environmental exposures and the development of systemic lupus erythematosus. Curr Opin Rheumatol 2016;28(5):497–505.

[181] Parks CG, et al. Understanding the role of environmental factors in the development of systemic lupus erythematosus. Best Pract Res Clin Rheumatol 2017;31(3):306–20.

[182] Baka Z, Buzás E, Nagy G. Rheumatoid arthritis and smoking: putting the pieces together. Arthritis Res Ther 2009;11(4):238.

[183] Sakamoto E, et al. Chemical- and drug-induced allergic, inflammatory, and autoimmune diseases via haptenation. Biology 2023;12(1).

[184] Craft J, Fatenejad S. Self antigens and epitope spreading in systemic autoimmunity. Arthritis Rheum 1997;40(8):1374–82.

[185] Klareskog L, et al. Smoking, citrullination and genetic variability in the immunopathogenesis of rheumatoid arthritis. Semin Immunol 2011;23(2):92–8.

[186] Chan TD, et al. Elimination of germinal-center-derived self-reactive B cells is governed by the location and concentration of self-antigen. Immunity 2012;37(5):893–904.

[187] Agius MA, Twaddle GM, Fairclough RH. Epitope spreading in experimental autoimmune myasthenia gravis. Ann N Y Acad Sci 1998;841:365–7.

[188] Deshmukh US, et al. Mechanisms of autoantibody diversification to SLE-related autoantigens. Ann N Y Acad Sci 2003;987:91–8.

[189] Mamula MJ, Janeway Jr CA. Do B cells drive the diversification of immune responses? Immunol Today 1993;14(4):151–2. discussion 153–4.

[190] Shi FD, et al. Natural killer cells determine the outcome of B cell-mediated autoimmunity. Nat Immunol 2000;1(3):245–51.

[191] Hwang SH, et al. B cell TLR7 expression drives anti-RNA autoantibody production and exacerbates disease in systemic lupus erythematosus-prone mice. J Immunol 2012;189(12):5786–96.

[192] Lartigue A, et al. Role of TLR9 in anti-nucleosome and anti-DNA antibody production in lpr mutation-induced murine lupus. J Immunol 2006;177(2):1349–54.

[193] Christensen SR, et al. Toll-like receptor 9 controls anti-DNA autoantibody production in murine lupus. J Exp Med 2005;202(2):321–31.

[194] Knight JS, Kaplan MJ. Lupus neutrophils: 'NET' gain in understanding lupus pathogenesis. Curr Opin Rheumatol 2012;24(5):441–50.

[195] Villanueva E, et al. Netting neutrophils induce endothelial damage, infiltrate tissues, and expose immunostimulatory molecules in systemic lupus erythematosus. J Immunol 2011;187(1):538–52.

[196] Garcia-Romo GS, et al. Netting neutrophils are major inducers of type I IFN production in pediatric systemic lupus erythematosus. Sci Transl Med 2011;3(73):73ra20.

[197] Ronnblom L. The type I interferon system in the etiopathogenesis of autoimmune diseases. Ups J Med Sci 2011;116(4):227–37.

[198] Puga I, et al. B cell-helper neutrophils stimulate the diversification and production of immunoglobulin in the marginal zone of the spleen. Nat Immunol 2011;13(2):170–80.

[199] Cohen PL, et al. Delayed apoptotic cell clearance and lupus-like autoimmunity in mice lacking the c-mer membrane tyrosine kinase. J Exp Med 2002;196(1):135–40.

[200] Hanayama R, et al. Autoimmune disease and impaired uptake of apoptotic cells in MFG-E8-deficient mice. Science 2004;304(5674):1147–50.

[201] Rahman ZS, et al. Impaired apoptotic cell clearance in the germinal center by Mer-deficient tingible body macrophages leads to enhanced antibody-forming cell and germinal center responses. J Immunol 2010;185(10):5859–68.

[202] Roubinian JR, et al. Effect of castration and sex hormone treatment on survival, anti-nucleic acid antibodies, and glomerulonephritis in NZB/NZW F1 mice. J Exp Med 1978;147(6):1568–83.

[203] Grimaldi CM, et al. Estrogen alters thresholds for B cell apoptosis and activation. J Clin Invest 2002;109(12):1625–33.

[204] Calippe B, et al. 17Beta-estradiol promotes TLR4-triggered proinflammatory mediator production through direct estrogen receptor alpha signaling in macrophages in vivo. J Immunol 2010;185(2):1169–76.

[205] Seillet C, et al. The TLR-mediated response of plasmacytoid dendritic cells is positively regulated by estradiol in vivo through cell-intrinsic estrogen receptor α signaling. Blood 2012;119(2):454–64.

[206] Hill L, et al. Differential roles of estrogen receptors α and β in control of B-cell maturation and selection. Mol Med 2011;17(3–4):211–20.

[207] Kojetin DJ, Burris TP. REV-ERB and ROR nuclear receptors as drug targets. Nat Rev Drug Discov 2014;13(3):197–216.

[208] Devarajan P, Chen Z. Autoimmune effector memory T cells: the bad and the good. Immunol Res 2013;57(1–3):12–22.

[209] Tabor DE, Gould KA. Estrogen receptor alpha promotes lupus in (NZB×NZW)F1 mice in a B cell intrinsic manner. Clin Immunol 2017;174:41–52.

[210] Gregory RI, et al. The Microprocessor complex mediates the genesis of microRNAs. Nature 2004;432(7014):235–40.

[211] Dai R, et al. Sex differences in the expression of lupus-associated miRNAs in splenocytes from lupus-prone NZB/WF1 mice. Biol Sex Differ 2013;4(1):19.

[212] Macias S, Michlewski G, Cáceres JF. Hormonal regulation of microRNA biogenesis. Mol Cell 2009;36(2):172–3.

[213] Shelly S, Boaz M, Orbach H. Prolactin and autoimmunity. Autoimmun Rev 2012;11(6–7):A465–70.

[214] Robinson GA, et al. Investigating sex differences in T regulatory cells from cisgender and transgender healthy individuals and patients with autoimmune inflammatory disease: a cross-sectional study. Lancet Rheumatol 2022;4(10):e710–24.

[215] Zou Y-R, Diamond B. B cells producing pathogenic autoantibodies. In: Frederick TH, Alt W, Radbruch A, Reth M, editors. Molecular biology of B cells. Academic Press; 2015. p. 417–39.

[216] Gough SC, Simmonds MJ. The HLA region and autoimmune disease: associations and mechanisms of action. Curr Genomics 2007;8(7):453–65.

[217] Hamilton RG, et al. Two Ro (SS-A) autoantibody responses in systemic lupus erythematosus. Correlation of HLA-DR/DQ specificities with quantitative expression of Ro (SS-A) autoantibody. Arthritis Rheum 1988;31(4):496–505.

[218] Schur PH. Genetics of systemic lupus erythematosus. Lupus 1995;4(6):425–37.

[219] Elliott MR, Ravichandran KS. Clearance of apoptotic cells: implications in health and disease. J Cell Biol 2010;189(7):1059–70.

[220] Li X, et al. Fcgamma receptors: structure, function and role as genetic risk factors in SLE. Genes Immun 2009;10(5):380–9.

[221] Yasutomo K, et al. Mutation of DNASE1 in people with systemic lupus erythematosus. Nat Genet 2001;28(4):313–4.

[222] Namjou B, et al. Evaluation of the TREX1 gene in a large multi-ancestral lupus cohort. Genes Immun 2011;12(4):270–9.
[223] Kozyrev SV, et al. Functional variants in the B-cell gene BANK1 are associated with systemic lupus erythematosus. Nat Genet 2008;40(2):211–6.
[224] Castillejo-López C, et al. Genetic and physical interaction of the B-cell systemic lupus erythematosus-associated genes BANK1 and BLK. Ann Rheum Dis 2012;71(1):136–42.
[225] Flores-Borja F, et al. Decreased Lyn expression and translocation to lipid raft signaling domains in B lymphocytes from patients with systemic lupus erythematosus. Arthritis Rheum 2005;52(12):3955–65.
[226] Liossis SN, et al. B-cell kinase lyn deficiency in patients with systemic lupus erythematosus. J Invest Med 2001;49(2):157–65.
[227] Cornall RJ, et al. Polygenic autoimmune traits: lyn, CD22, and SHP-1 are limiting elements of a biochemical pathway regulating BCR signaling and selection. Immunity 1998;8(4):497–508.
[228] Ehlers M, et al. TLR9/MyD88 signaling is required for class switching to pathogenic IgG2a and 2b autoantibodies in SLE. J Exp Med 2006;203(3):553–61.
[229] Vincent FB, et al. The BAFF/APRIL system: emerging functions beyond B cell biology and autoimmunity. Cytokine Growth Factor Rev 2013;24(3):203–15.
[230] Moisini I, Davidson A. BAFF: a local and systemic target in autoimmune diseases. Clin Exp Immunol 2009;158(2):155–63.
[231] Peters AL, Stunz LL, Bishop GA. CD40 and autoimmunity: the dark side of a great activator. Semin Immunol 2009;21(5):293–300.
[232] Li G, et al. Human genetics in rheumatoid arthritis guides a high-throughput drug screen of the CD40 signaling pathway. PLoS Genet 2013;9(5):e1003487.
[233] Grammer AC, et al. Abnormal germinal center reactions in systemic lupus erythematosus demonstrated by blockade of CD154-CD40 interactions. J Clin Invest 2003;112(10):1506–20.
[234] Butt D, et al. FAS inactivation releases unconventional germinal center B cells that escape antigen control and drive IgE and auto-antibody production. Immunity 2015;42(5):890–902.
[235] Kumar KR, et al. Regulation of B cell tolerance by the lupus susceptibility gene Ly108. Science 2006;312(5780):1665–9.
[236] Liu Y, et al. Lupus susceptibility genes may breach tolerance to DNA by impairing receptor editing of nuclear antigen-reactive B cells. J Immunol 2007;179(2):1340–52.
[237] Hande S, Notidis E, Manser T. Bcl-2 obstructs negative selection of autoreactive, hypermutated antibody V regions during memory B cell development. Immunity 1998;8(2):189–98.
[238] Nisitani S, et al. The bcl-2 gene product inhibits clonal deletion of self-reactive B lymphocytes in the periphery but not in the bone marrow. J Exp Med 1993;178(4):1247–54.
[239] Takahashi Y, et al. Relaxed negative selection in germinal centers and impaired affinity maturation in bcl-xL transgenic mice. J Exp Med 1999;190(3):399–410.
[240] Kuo P, et al. Bcl-2 leads to expression of anti-DNA B cells but no nephritis: a model for a clinical subset. Eur J Immunol 1999;29(10):3168–78.
[241] Manjarrez-Orduno N, et al. CSK regulatory polymorphism is associated with systemic lupus erythematosus and influences B-cell signaling and activation. Nat Genet 2012;44(11):1227–30.
[242] Schade SG, Abels J, Schilling RF. Studies on antibody to intrinsic factor. J Clin Invest 1967;46(4):615–20.
[243] Takamori M. Lambert-Eaton myasthenic syndrome: search for alternative autoimmune targets and possible compensatory mechanisms based on presynaptic calcium homeostasis. J Neuroimmunol 2008;201–202:145–52.
[244] Smith BR, Sanders J, Furmaniak J. TSH receptor antibodies. Thyroid 2007;17(10):923–38.
[245] Falk RJ, Jennette JC. ANCA disease: where is this field heading? J Am Soc Nephrol 2010;21(5):745–52.
[246] Markiewski MM, Lambris JD. The role of complement in inflammatory diseases from behind the scenes into the spotlight. Am J Pathol 2007;171(3):715–27.
[247] Gehrs BC, Friedberg RC. Autoimmune hemolytic anemia. Am J Hematol 2002;69(4):258–71.
[248] Ostendorf L, et al. Targeting CD38 with daratumumab in refractory systemic lupus erythematosus. N Engl J Med 2020;383(12):1149–55.
[249] Parodis I, et al. Autoantibody and cytokine profiles during treatment with belimumab in patients with systemic lupus erythematosus. Int J Mol Sci 2020;21(10).
[250] Stohl W, et al. Belimumab reduces autoantibodies, normalizes low complement levels, and reduces select B cell populations in patients with systemic lupus erythematosus. Arthritis Rheum 2012;64(7):2328–37.
[251] Furie RA, et al. Type I interferon inhibitor anifrolumab in active systemic lupus erythematosus (TULIP-1): a randomised, controlled, phase 3 trial. Lancet Rheumatol 2019;1(4):e208–19.
[252] Morand EF, et al. Trial of anifrolumab in active systemic lupus erythematosus. N Engl J Med 2020;382(3):211–21.
[253] Boross P, Leusen JH. Mechanisms of action of CD20 antibodies. Am J Cancer Res 2012;2(6):676–90.
[254] Oflazoglu E, Audoly LP. Evolution of anti-CD20 monoclonal antibody therapeutics in oncology. mAbs 2010;2(1):14–9.
[255] Merrill JT, et al. Efficacy and safety of rituximab in moderately-to-severely active systemic lupus erythematosus: the randomized, double-blind, phase II/III systemic lupus erythematosus evaluation of rituximab trial. Arthritis Rheum 2010;62(1):222–33.
[256] Rovin BH, et al. Efficacy and safety of rituximab in patients with active proliferative lupus nephritis: the Lupus Nephritis Assessment with Rituximab study. Arthritis Rheum 2012;64(4):1215–26.
[257] Montillo M, et al. Ibrutinib in previously treated chronic lymphocytic leukemia patients with autoimmune cytopenias in the RESONATE study. Blood Cancer J 2017;7:e524. United States.
[258] Masoud S, et al. Ofatumumab for B cell depletion in patients with systemic lupus erythematosus who are allergic to rituximab. Rheumatology 2018;57(7):1156–61.
[259] Zhang B. Ofatumumab. MAbs 2009;1(4):326–31.
[260] Mysler EF, et al. Efficacy and safety of ocrelizumab in active proliferative lupus nephritis: results from a randomized, double-blind, phase III study. Arthritis Rheum 2013;65(9):2368–79.
[261] Roche H-L. A study to evaluate the safety, tolerability, pharmacokinetics, and pharmacodynamics of subcutaneously administered mosunetuzumab to participants with systemic lupus erythematosus. 2022. https://ClinicalTrials.gov/show/NCT05155345.
[262] Reddy V, et al. Obinutuzumab induces superior B-cell cytotoxicity to rituximab in rheumatoid arthritis and systemic lupus erythematosus patient samples. Rheumatology 2017;56(7):1227–37.

[263] Clark EA, Ledbetter JA. How does B cell depletion therapy work, and how can it be improved? Ann Rheum Dis 2005;64(Suppl. 4):iv77–80.
[264] Chen DR, Cohen PL. Living life without B cells: is repeated B-cell depletion a safe and effective long-term treatment plan for rheumatoid arthritis? Int J Clin Rheumtol 2012;7(2):159–66.
[265] Novartis Pharmaceuticals. Study the efficacy and safety of VAY736 and CFZ533 in SLE patients. 2018. https://ClinicalTrials.gov/show/NCT03656562.
[266] Isenberg D, et al. Efficacy, safety, and pharmacodynamic effects of the Bruton's tyrosine kinase inhibitor fenebrutinib (GDC-0853) in systemic lupus erythematosus: results of a phase II, randomized, double-blind, placebo-controlled trial. Arthritis Rheumatol 2021;73(10):1835–46.
[267] Wallace DJ, et al. Efficacy and safety of evobrutinib (M2951) in adult patients with systemic lupus erythematosus who received standard of care therapy: a phase II, randomized, double-blind, placebo-controlled dose ranging study. 2020 [cited 2022 December 28]; Available from: https://acrabstracts.org/abstract/efficacy-and-safety-of-evobrutinib-m2951-in-adult-patients-with-systemic-lupus-erythematosus-who-received-standard-of-care-therapy-a-phase-ii-randomized-double-blind-placebo-controlled-dose-rang/.
[268] Novartis Pharmaceuticals. Safety, pharmacokinetics and preliminary efficacy study of CFZ533 in patients with lupus nephritis. 2018. https://ClinicalTrials.gov/show/NCT03610516.
[269] Schwabe C, et al. Safety, pharmacokinetics, and pharmacodynamics of multiple rising doses of BI 655064, an antagonistic anti-CD40 antibody, in healthy subjects: a potential novel treatment for autoimmune diseases. J Clin Pharmacol 2018;58(12):1566–77.
[270] Jayne D, et al. POS0687 a randomised dose ranging, placebo-controlled, phase II study assessing the efficacy and safety of BI 655064, an antagonistic anti-cd40 antibody. In: Patients with lupus nephritis. BMJ Publishing Group Ltd; 2021.
[271] Schafer PH, et al. Cereblon modulator iberdomide induces degradation of the transcription factors Ikaros and Aiolos: immunomodulation in healthy volunteers and relevance to systemic lupus erythematosus. Ann Rheum Dis 2018;77(10):1516–23.
[272] Lipsky PE, et al. Biological impact of iberdomide in patients with active systemic lupus erythematosus. Ann Rheum Dis 2022;81(8):1136–42.
[273] Clowse ME, et al. Efficacy and safety of epratuzumab in moderately to severely active systemic lupus erythematosus: results from two phase III randomized, double-blind, placebo-controlled trials. Arthritis Rheumatol 2017;69(2):362–75.
[274] Wallace DJ, et al. Efficacy and safety of epratuzumab in patients with moderate/severe active systemic lupus erythematosus: results from EMBLEM, a phase IIb, randomised, double-blind, placebo-controlled, multicentre study. Ann Rheum Dis 2014;73(1):183–90.
[275] Kuwana Y, et al. Expression of chimeric receptor composed of immunoglobulin-derived V regions and T-cell receptor-derived C regions. Biochem Biophys Res Commun 1987;149(3):960–8.
[276] Gross G, Waks T, Eshhar Z. Expression of immunoglobulin-T-cell receptor chimeric molecules as functional receptors with antibody-type specificity. Proc Natl Acad Sci U S A 1989;86(24):10024–8.
[277] Kansal R, et al. Sustained B cell depletion by CD19-targeted CAR T cells is a highly effective treatment for murine lupus. Sci Transl Med 2019;11(482).
[278] Mougiakakos D, et al. CD19-Targeted CAR T cells in refractory systemic lupus erythematosus. N Engl J Med 2021;385(6):567–9.
[279] Schett G, et al. OP0279 CAR-T cell treatment of refractory systemic lupus erythematosus-safety and preliminary efficacy data from the first four patients. BMJ Publishing Group Ltd; 2022.
[280] Mackensen A, et al. Anti-CD19 CAR T cell therapy for refractory systemic lupus erythematosus. Nat Med 2022;28(10):2124–32.
[281] Merrill JT, Guthridge J, Zack D, et al. SAT0187 discrimination of systemic lupus (SLE) patients with clinical response to obexelimab (XMAB®5871) based on a pattern of immunologic markers. Ann Rheum Dis 2020;79:1035–6.
[282] Chu SY, et al. Inhibition of B cell activation following in vivo co-engagement of B cell antigen receptor and Fcγ receptor IIb in non-autoimmune-prone and SLE-prone mice. J Transl Autoimmun 2021;4:100075.
[283] Merrill J, et al. FRI0176 phase 2, double-blind, randomized, placebo-controlled study of a reversible b cell inhibitor, XMAB® 5871. In: Systemic lupus erythematosus (SLE). BMJ Publishing Group Ltd; 2019.
[284] Stone JH, et al. FRI0588 Interim results of a phase 2 study of XMAB® 5871, a reversible B cell inhibitor. In: IGG4-related disease. BMJ Publishing Group Ltd; 2017.
[285] Navarra SV, et al. Efficacy and safety of belimumab in patients with active systemic lupus erythematosus: a randomised, placebo-controlled, phase 3 trial. Lancet 2011;377(9767):721–31.
[286] Furie R, et al. A phase III, randomized, placebo-controlled study of belimumab, a monoclonal antibody that inhibits B lymphocyte stimulator, in patients with systemic lupus erythematosus. Arthritis Rheum 2011;63(12):3918–30.
[287] Stohl W, et al. Efficacy and safety of subcutaneous belimumab in systemic lupus erythematosus: a fifty-two-week randomized, double-blind, placebo-controlled study. Arthritis Rheumatol 2017;69(5):1016–27.
[288] Lenert A, Niewold TB, Lenert P. Spotlight on blisibimod and its potential in the treatment of systemic lupus erythematosus: evidence to date. Drug Des Dev Ther 2017;11:747–57.
[289] Furie RA, et al. A phase 2, randomised, placebo-controlled clinical trial of blisibimod, an inhibitor of B cell activating factor, in patients with moderate-to-severe systemic lupus erythematosus, the PEARL-SC study. Ann Rheum Dis 2015;74(9):1667–75.
[290] Merrill JT, et al. Phase III trial results with blisibimod, a selective inhibitor of B-cell activating factor, in subjects with systemic lupus erythematosus (SLE): results from a randomised, double-blind, placebo-controlled trial. Ann Rheum Dis 2018;77(6):883–9.
[291] Kaegi C, et al. Systematic review of safety and efficacy of atacicept in treating immune-mediated disorders. Front Immunol 2020;11:433.
[292] Isenberg DA, et al. Efficacy and safety of subcutaneous tabalumab in patients with systemic lupus erythematosus: results from ILLUMINATE-1, a 52-week, phase III, multicentre, randomised, double-blind, placebo-controlled study. Ann Rheum Dis 2016;75(2):323–31.
[293] Fan Y, Gao D, Zhang Z. Telitacicept, a novel humanized, recombinant TACI-Fc fusion protein, for the treatment of systemic lupus erythematosus. Drugs Today 2022;58(1):23–32.
[294] Dhillon S. Telitacicept: first approval. Drugs 2021;81(14):1671–5.

[295] RemeGen Co., Ltd.. A study of telitacicept for the treatment of moderately to severely active systemic lupus erythematosus. 2022. https://ClinicalTrials.gov/show/NCT05306574.

[296] Abuqayyas L, et al. Safety and biological activity of rozibafusp alfa, a bispecific inhibitor of inducible costimulator ligand and B cell activating factor, in patients with rheumatoid arthritis: results of a phase 1b, randomized, double-blind, placebo-controlled, multiple ascending dose study. ACR Open Rheumatol 2022;4(10):903−11.

[297] Amgen. Efficacy and safety of AMG 570 in subjects with active systemic lupus erythematosus (SLE). 2020. https://ClinicalTrials.gov/show/NCT04058028.

[298] Kraaij T, et al. The NET-effect of combining rituximab with belimumab in severe systemic lupus erythematosus. J Autoimmun 2018;91:45−54.

[299] Leiden University Medical Center, Dutch Kidney Foundation, and GlaxoSmithKline. Synergetic B-cell immunomodulation in SLE - 2nd study. 2018. https://ClinicalTrials.gov/show/NCT03747159.

[300] Jones A, et al. Belimumab after B cell depletion therapy in patients with systemic lupus erythematosus (BEAT Lupus) protocol: a prospective multicentre, double-blind, randomised, placebo-controlled, 52-week phase II clinical trial. BMJ Open 2019;9(12):e032569.

[301] Shipa M, et al. Effectiveness of belimumab after rituximab in systemic lupus erythematosus: a randomized controlled trial. Ann Intern Med 2021;174(12):1647−57.

[302] AbbVie Advances. Study to investigate the safety and efficacy of elsubrutinib and upadacitinib given alone or in combination in participants with moderately to severely active systemic lupus erythematosus (SLE). 2019. https://ClinicalTrials.gov/show/NCT03978520.

[303] Charite University, Berlin, Germany, et al.. An open label study to evaluate daratumumab in participants with moderate to severe systemic lupus erythematosus. 2021. https://ClinicalTrials.gov/show/NCT04810754.

[304] Takeda (Millennium Pharmaceuticals, Inc.). A study to evaluate the safety, pharmacokinetics (PK), and pharmacodynamics (PD) of TAK-079 in combination with standard background therapy in participants with moderate to severe systemic lupus erythematosus (SLE). 2018. https://ClinicalTrials.gov/show/NCT03724916.

[305] Furie R, et al. POS0695 KZR-616, a selective immunoproteasome inhibitor for the treatment of systemic lupus erythematosus: results from the completed dose escalation phase 1B portion of the mission study. BMJ Publishing Group Ltd; 2021.

[306] Fan A, et al. POS0715 treatment of SLE patients with zetomipzomib (KZR-616), a selective inhibitor of the immunoproteasome, results in circulating gene expression, protein level, and immune cell phenotypic changes with potential correlations to clinical response. Ann Rheum Dis 2022;81(Suppl. 1):639−40.

[307] Alexander T, et al. The proteasome inhibitior bortezomib depletes plasma cells and ameliorates clinical manifestations of refractory systemic lupus erythematosus. Ann Rheum Dis 2015;74(7):1474−8.

[308] Ishii T, et al. Multicenter double-blind randomized controlled trial to evaluate the effectiveness and safety of bortezomib as a treatment for refractory systemic lupus erythematosus. Mod Rheumatol 2018;28(6):986−92.

[309] Zhang W, et al. Treatment of systemic lupus erythematosus using BCMA-CD19 compound CAR. Stem Cell Rev Rep 2021;17:2120−3. United States.

Chapter 18

Anti-Tumour Necrosis Factor and New Paradigms for Therapies Using Antibodies

Sergei A. Nedospasov[1,2,3], Marina S. Drutskaya[1,2] and Andrey A. Kruglov[3,4]
[1]Engelhardt Institute of Molecular Biology, Russian Academy of Sciences, Moscow, Russia; [2]Sirius University of Science and Technology, Federal Territory Sirius, Russia; [3]Belozersky Institute of Physico-Chemical Biology and Biological Faculty, Lomonosov Moscow State University, Moscow, Russia; [4]German Rheumatism Research Center (DRFZ), A Leibniz Institute, Berlin, Germany

1. INTRODUCTION

The original concept of 'cytokine network' does not claim that one particular cytokine may be at the apex of cytokine regulation [1]. Rather, multiple cytokines were viewed as 'equal' in the sense that they affect the expression of other cytokines involved in the network. Nevertheless, a number of cytokines, including several proinflammatory mediators, appeared to have *non-redundant* functions making them potential therapeutic targets. This notion has been corroborated by the stunning success of anti-tumour necrosis factor (TNF) therapy in rheumatoid arthritis (RA) [2,3]. It was also supported by studies in mouse models which demonstrated that overexpression of a single cytokine, TNF, led to spontaneous arthritis [4–6]. This may seem surprizing due to a complex immunobiology of TNF with its several known non-redundant protective functions that would be also blocked. The mechanisms of anti-TNF therapeutic activity were extensively studied over the last 25 years. For example, anti-TNF antibodies may bind to the membrane-associated TNF expressed by monocytes and T cells and promote their depletion via antibody-induced cytotoxicity. In the inflamed tissue, TNF may induce production of chemokines that mediate recruitment of proinflammatory myeloid cells, as well as various adhesion molecules directing the homing of cells to the inflammation site. It was also shown that TNF regulates programed death of various cells that may result in extensive tissue damage and organ function failure, and that administration of anti-TNF agents may block these processes. Additionally, TNF interferes with homoeostatic tissue repair processes via regulation of IL-22 activity [7]. In line with this, TNF may amplify production of metalloproteases that destruct cartilage tissue and induce impairment in joint mobility [7,8]. Apart from its role in boosting inflammation at the given site, TNF is crucial for the organisation of secondary lymphoid organ microstructure, including ectopic lymphoid follicles (ELFs) and follicular dendritic cell (FDC) formation in B-cell follicles. As the FDC network is important for the induction of germinal centres (GCs) and subsequent generation of specific antibody responses characterised by high affinity and specificity, anti-TNF therapy via disruption of these cell subsets may affect the generation of new autoreactive B cells [9]. Whether this effect contributes to the overall therapeutic activity of anti-TNF biologics remains a matter of debate. On the other hand, preclinical and clinical evidence indicate that continuous TNF blockade may reduce efficiency of antibody responses during vaccination [10–13] and predispose patients to increased risk of tuberculosis and other granulomatous conditions associated with latent infection.

2. BRIEF OVERVIEW OF TNF IMMUNOBIOLOGY

TNF was discovered due to its potent LPS-induced anti-tumour activity in mice [14]. The *TNF* gene was cloned and expressed in 1984–1985 [15–17] and evaluated as an anti-tumour drug (reviewed in Ref. [18]). However, within

Molecular Biology of B Cells. https://doi.org/10.1016/B978-0-323-95895-0.00008-8

several years, it was recognised that systemic application of recombinant human TNF is too toxic due to a so called 'dark side of TNF' [17,19]. In reality, TNF is a pleiotropic cytokine induced by many receptors of innate immunity and produced by various cell types, including myeloid and lymphoid cells, stromal cells, muscle cells and neurons, initially in the form of a membrane-bound trimer that can be cleaved off by ADAM17 (TACE) protease [20,21]. This trimeric cytokine interacts with two types of cognate receptors and can transmit directly or indirectly multiple signals: activation, proliferation, differentiation, two types of regulated cell death and even migration (through activation of chemokine expression) (reviewed in Ref. [18]). Since the membrane-associated form is an obligatory intermediate in TNF biosynthesis, all TNF-producing cells contain certain amount of this form on their surface (Figure 18.1). Unlike soluble trimeric TNF that interacts with both TNFR1 (primarily) and TNFR2, the membrane-associated form preferentially signals through TNFR2 [22]. TNFR1 employs distinct set of adaptor molecules to induce several biochemical pathways leading to the activation of transcription factors, most notably, NFκB. Although most cells are protected from TNF-induced regulated cell death (apoptosis or necroptosis) due to the expression of protective proteins, under certain conditions, two types of cytoplasmic complexes can be formed that induce either caspase-8-mediated apoptosis or RIP3-MLKL-mediated necroptosis (reviewed in Ref. [23]). The switch between two extreme modes of cellular signalling – activation or death – is regulated by a complex pathway of reversible ubiquitination (reviewed in Ref. [24]). 'The dark side' of TNF manifests itself when this cytokine is produced systemically at a high level leading to hyper-activation of multiple cell types bearing TNFR1. One of the early examples has been provided by a mouse model of septic shock in which anti-TNF antibodies protected animals from lethal toxicity [25].

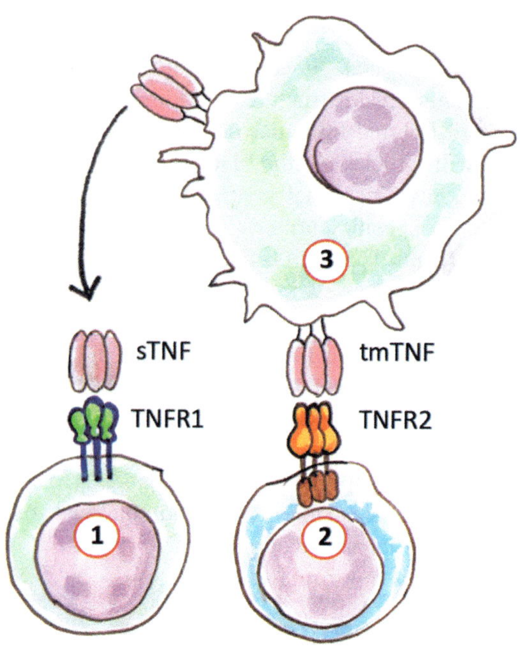

FIGURE 18.1 **Three types of intracellular signalling cascades can be induced upon TNF interaction with TNFRs.** Multiple functions of TNF are mediated by two distinct receptors, which can transmit directly or indirectly multiple types of signals ranging from activation and proliferation to regulated cell death. Since the membrane-associated form is an obligatory intermediate in TNF biosynthesis, all TNF-producing cells contain certain amount of such a form on the surface. sTNF interacts primarily with TNFR1 (signal 1), while tmTNF preferentially signals through TNFR2 (signal 2). Upon interaction of tmTNF with TNFR2, a so called 'reverse signalling' cascade can be activated (signal 3).

3. OVERVIEW OF ANTI-TNF BIOLOGICS

There are five major anti-TNF agents approved in many countries for clinical use as treatment of several autoimmune conditions (Figure 18.2 and Table 18.1). Infliximab (IFX) was the very first therapeutic human TNF inhibitor represented by a chimaeric antibody consisting of a murine part (cloned from a hybridoma A2 initially characterised in J. Vilcek's laboratory) and a much larger human part that included constant regions and Fc from human IgG1 [30]. For their pioneering work on therapeutic application of IFX in RA, M. Feldmann and R. Maini received the Lasker Award in 2003.

Another early biologic, etanercept (ETN), represents a soluble part of human TNFR2 in a dimeric form fused to human Fc of the IgG1 antibody. It can also bind soluble lymphotoxin alpha (LTα), the closest TNF relative, albeit similar efficiency of ETN and IFX in RA. This suggests that LTα does not play the major pathogenic role in this autoimmune condition. Interestingly, the known side effect – reactivation of tuberculosis – is less frequent in the case of ETN therapy as compared to other anti-TNF biologics [31,32,33,34]. A similar blocker, Lenercept, was developed based on soluble TNFR1, but its failure in sepsis and MS clinical trials [35,36] apparently prevented the evaluation of this otherwise potent TNF inhibitor for other human diseases.

More recently, two other biologics, adalimumab and golimumab, were developed in a form of fully human antibodies that can bind and neutralise human TNF. The generation of adalimumab involved the phage display technology and is therefore directly related to 2018 Nobel Prize in Chemistry awarded to G. Winter [37]. Golimumab was originally isolated from a hybridoma clone produced by HuMab transgenic mice that were immunised with human TNF. These mice contained non-rearranged human Ig loci allowing generation of fully human antibodies. The golimumab-producing clone was selected on the basis of its

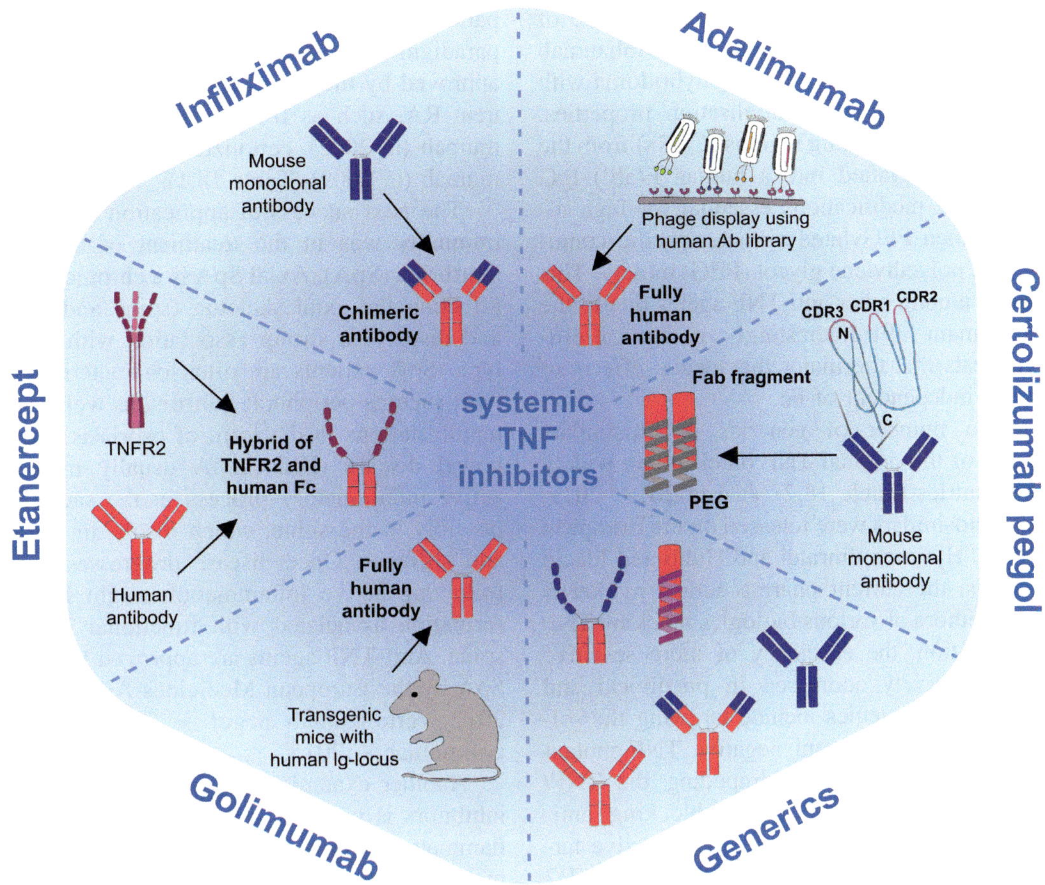

FIGURE 18.2 **Systemic TNF inhibitors.** Etanercept, Infliximab, Adalimumab, Golimumab and Certolizumab pegol are in clinical use for a number of autoimmune diseases. Some biosimilars and non-originator biologics have been approved, while others are still undergoing clinical evaluation.

TABLE 18.1 Anti-TNF Biologics Approved for Clinical Use

Biologic Agent (Trade name)	Antibody Structure	Approved for Treatment	Company	References
Infliximab (Remicade)	Chimaeric human-mouse IgG1 monoclonal antibody	RA, PsA, AS, PP, CD and UC	Janssen Biotech, Inc	[3]
Etanercept (Enbrel)	Fusion protein of soluble TNFR2 and human IgG1 (hinge and Fc regions) antibody	RA, PsA, AS, PP and JIA	Amgen	[26]
Certolizumab pegol (Cimzia)	Pegylated F(ab′) fragment of humanised monoclonal antibody	RA, PsA, AS and CD	UCB Pharma	[27]
Golimumab (Simponi)	Fully human IgG1 monoclonal antibody	RA, PsA, AS, JIA and UC	Janssen Biotech, Inc	[28]
Adalimumab (Humira)	Fully human IgG1 monoclonal antibody	RA, PsA, AS, PP, JIA, CD, UC, uveitis and hidradenitis suppurativa	AbbVie	[29]

high-affinity binding to human TNF and the presence of human light and heavy chains [38]. Finally, certolizumab pegol, as IFX, was derived from a murine hybridoma with good human TNF-binding and neutralisation properties. The complementarity-determining regions (CDRs) from the murine antibody were grafted into a human F(ab') IgG along with additional modifications essential for high affinity. F(ab') was then PEGylated via site-specific conjugation of a 40 kDa polyethylene glycol (PEG) moiety. This biologic is unique among other anti-TNF agents due to the lack of the Fc domain [39]. Interestingly, its clinical efficacy in RA suggests that the major therapeutic effects of TNF blockers are independent of Fc.

Furthermore, a number of generics or biosimilars related to this list of the original TNF blockers are widely used in many countries (Table 18.2). Just in 2018–2019, five adalimumab biosimilars were released on the European market: Amgevita, Hyrimoz, Imraldi and Hulio and Idacio (Table 18.2). Thus, the current pharmaceutical market is saturated with a plethora of various biologics blocking TNF in humans. In addition, the efficiency of more selective TNF targeting is actively addressed in preclinical and clinical studies. These strategies include targeting the soluble form of TNF by a dominant negative TNF mutant disrupting its activity (ЧPro1595), impeding the TNF/TNFR1 signalling cascade with TNFR1-blocking antibodies (ATROSAB and TROS), as well as selective targeting of TNF/TNFR2 pathway with agonistic TNFR2 antibodies in chronic inflammatory conditions, or antagonistic TNFR2 antibodies as cancer therapy for TNFR2-overexpressing tumours (reviewed in Ref. [40]).

4. SUCCESSES AND FAILURES OF ANTI-TNF THERAPY

Immune system attack on host tissues, that is, autoimmune disease manifestation, is characterised by local or systemic activation of both innate and adaptive immune responses. This often coincides with the elevated levels of proinflammatory cytokines, such as TNF, IL-6 and IL-1β, that further amplifies the extent of tissue damage.

Many autoimmune disorders can be successfully treated with anti-TNF drugs. The most known autoimmune condition in which anti-TNF therapy was implemented is RA – a chronic, inflammatory joint disease that is characterised by autoantibodies to IgG collectively referred to as a rheumatoid factor (RF), as well as the appearance of various antibodies to citrullinated proteins (ACPAs) and several other autoantigens. RA is a heterogeneous disease and although autoantibodies remain an important characteristic of RA, some individuals appear to be negative for such autoantibodies and are classified as suffering from seronegative RA. TNF production is increased in RA patients and TNF blockade emerged as a novel therapeutic paradigm at a time [2,3]. Multiple TNF inhibitors were approved by the Food and Drug Administration (FDA) to treat RA, such as IFX (in 1998), ETN (in 1998), adalimumab (in 2002), certolizumab pegol (in 2008) and golimumab (in 2009) (Table 18.1).

The next successful application of anti-TNF in autoimmunity was in the treatment of ankylosing spondyloarthritis (SpA). Axial SpA is a chronic rheumatic disease affecting the axial skeleton (spine and sacroiliac joints) and displays a strong association with HLA-B27 haplotype. SpA patients are often characterised by inflammation, such as peripheral arthritis, as well as extra-articular manifestations in the form of psoriasis and inflammatory bowel disease (IBD). SpA usually manifests itself by active inflammatory process in the sacroiliac joints and possibly in the spine, which results in chronic back pain and stiffness. Once disease progresses, repair processes following active inflammation might lead to new bone formation associated with functional impairment of the spine. Anti-TNF agents are approved for both subtypes of SpA by the European Medicines Agency: adalimumab in 2012, certolizumab pegol – 2013, ETN – 2014 and golimumab – 2015.

Another example of a disease that is treated by TNF inhibitors is psoriatic arthritis (PsA), a heterogeneous inflammatory disease with musculoskeletal involvement, including arthritis, enthesitis, dactylitis and axial involvement, as well as potential skin and nail ailments. PsA is diagnosed in 30%–50% of patients with multiple affected joints, while psoriasis and nail abnormalities (pitting, onycholysis, etc.) are present in more than half of cases. PsA patients have increased TNF, IL-23 and IL-17A production, but are usually RF and ACPA negative, in contrast to RA. Multiple anti-TNF agents were approved for clinical use, such as ENT (in 2004), IFX (in 2006), adalimumab (in 2008), as well as drugs targeting Th17 axis, like anti-IL12p40 antibody, ustekinumab (in 2009) and anti-IL-17A antibody, secukinumab (in 2015).

Finally, anti-TNF drugs are proven effective for the treatment of autoimmune inflammation in the intestinal tract IBD. IBD is mainly divided into ulcerative colitis (UC) and Crohn's disease (CD). UC is localised to the large intestine, where the inflammation affects either the entire organ or its part. CD can affect any part of the GI tract, but most commonly involves both the large and small intestines. Both subtypes are characterised by increased TNF production, and several anti-TNF agents were approved by the FDA for the treatment of this autoimmune condition: IFX (in 1998), adalimumab (in 2007) and certolizumab pegol (2008).

There are several examples when anti-TNF therapy is not effective, even though the disease is accompanied by elevated TNF levels. The most known case is multiple

TABLE 18.2 Approved Anti-TNF Biosimilars and Non-Originator Biologicals

Original Drug	Product name	Company name, Country	Approval Status
Infliximab	Remsima/Inflectra/Flammegis (CT-P13)	Celltrion/Hospira (Pfizer), South Korea/USA	EU (2013); Canada (2014); Australia (2015); Russia (2015); US (2016); China (2017)
	Flixabi/Renflexis (SB2)	Samsung Bioepis (Biogen/Samsung)/Merck (MSD), South Korea/USA	EU (2016); South Korea (2015); Australia (2016); US (2017)
	Ixifi/Zessly (PF-06438179)	Pfizer/Sandoz, USA/Switzerland	USA (2017); EU (2018)
	Avsola (ABP 710)	Amgen, USA	US (2019); Canada (2020)
	Infliximab BS/NI-071	Nippon Kayaku/Nichi-Iko Pharmaceutical, Japan	Japan (2014/2017)
	BOW015*/Infimab*	Ranbaxy Laboratories/Epirus Biopharmaceuticals, India/USA	India (2014)
	BCD-055*	Biocad, Russia	Russia (2018)
Adalimumab	Imraldi/Hadlima (SB5)	Samsung Bioepis (Biogen/Samsung)/Merck, South Korea/USA	EU (2018); South Korea (2017); Australia (2018); Canada (2018); US (2019)
	Amjevita/Amgevita/Solymbic (ABP 501)	Amgen, USA	US (2016); EU (2018); Columbia (2020)
	Hulio (FKB327)	Fujifilm/Kyowa Hakko Kirin (Fujifilm Kyowa Kirin Biologics)/Mylan, Japan/USA	EU (2018); Japan (2020); US (2020)
	Cyltezo (BI 695501)	Boehringer Ingelheim, Germany	US (2017); EU (2018)
	Idacio/Kromeya (MSB11022)	Fresenius Kabi, Germany	EU (2019); US (2022)
	Abrilada/Amsparity (PF-06410293)	Pfizer, USA	US (2019); EU (2019)
	Halimatoz/Hefiya/Hyrimoz (GP2017)	Sandoz, Switzerland	EU (2018); US (2018)
	Exemptia*(ZRC3197)/Adfrar*/Mabura*	Zydus Cadila/Torrent Pharmaceuticals / Hetero Drugs, India	India (2014/2016/2018)
	Sulinno* (IBI-303)	Innovent Biologics, China	China (2020)
Etanercept	Brenzys/Benepali/Eticovo (SB4)	Samsung Bioepis (Biogen/Samsung)/Merck, South Korea/USA	South Korea (2015); EU (2016); Australia (2016); Canada (2016); US (2019)
	Erelzi (GP2015)	Sandoz, Switzerland	USA (2016); EU (2017); Canada (2017); Australia (2017)
	Eucept (LBEC0101)	LG Life Sciences (LG Chem)/Mochida Pharmaceutical, South Korea/Japan	Japan (2018); South Korea (2018)
	Nepexto (YLB113)*	YL Biologics (Lupin/Yoshida), India/Japan; Mylan/Lupin, USA/India	Japan (2019); EU (2020)
	Davictrel (HD203)	Hanwha Chemical, South Korea	South Korea (2014)
	AnBaiNuo*	Hisun Pharmaceuticals, China	China (2015)
	Intacept*	Intas Pharmaceuticals, India	India (2015)

*, Non-originator biological.
Adopted from Generics and biosimilars Initiative at GaBI-online: https://www.gabionline.net.

sclerosis [36]. Also, several small clinical studies showed that TNF inhibition may be effective as a short-term treatment for systemic lupus erythematosus (SLE) although anti-TNF therapy was associated with multiple adverse effects if SLE patients were treated for a longer time period (reviewed in Ref. [41]). Of note, therapeutic antibodies against CD20 show efficacy in multiple sclerosis (see Chapter 20).

Despite the tremendous success of anti-TNF therapy, significant cohorts of patients in each of the aforementioned disease category fail to respond to this type of therapy. Lack of clinical response to anti-TNF drugs may be subdivided into primary and secondary ones. Primary anti-TNF therapy-resistant patients may elicit inflammation that is not principally driven by TNF overexpression or simply characterised by fewer TNF-producing cells in the affected organ [42]. In contrast, secondary, or acquired, unresponsiveness can be caused by changes in inflammatory milieu during the treatment [43,44] or by the development of antibodies against anti-TNF biologics.

Taking into account that autoimmune diseases are characterised by heterogeneity despite sharing common features, such as the presence of autoantibodies, autoreactive T cells and comparable cytokine profiles, it is plausible that TNF blockade may work only in a subset of autoimmune patients. It became evident from multiple histological studies of the synovial tissues that autoimmune processes in the tissue can be dominated by various distinct pathotypes: diffuse myeloid subtype, which is characterised by the increased myeloid cell infiltrates to the tissue; lympho-myeloid subtype, featuring abundance of both myeloid cells and T and B lymphocytes and fibroid/pauci-immune pathotype, in which case inflamed tissue does not contain significant numbers of leucocytes and disease is driven by activated fibroblasts [45]. In addition, chronic inflammation often results in the formation of ELF in the targeted tissue where local autoimmune responses may take place. Interestingly, destruction of joints driven by pauci-immune pathotype is refractory to many current biological treatments, such as anti-TNF, anti-IL-6 or anti-CD20 [45]. Notably, synovial myeloid, but not lymphoid, gene signature expression correlated with a good clinical response to anti-TNF therapy. Furthermore, sICAM1^{high}/CXCL13^{low} patients revealed the highest response rate to anti-TNF blockers as compared to patients with high sICAM1^{low}/CXCL13^{high} expression [43]. These data concur with other studies showing that responsiveness to anti-TNF therapy in RA is dependent on the initial high inflammatory status of disease and the presence of myeloid cells in the synovium [46–49]. These observations are further supported by studies implicating the pivotal role of TNF produced by myeloid cells in the pathogenesis of experimental arthritis and colitis [7,50].

Interestingly, anti-TNF therapy in RA does not induce apoptosis of myeloid cells, but rather blocks their tissue infiltration/retention [49], whereas the response to anti-TNF during IBD is associated with apoptosis of T cells in the gut [51]. Also, the IBD response to anti-TNF therapies correlates with the levels of TNF-producing cells in the gut and in circulation [42,52]. Failure to respond to TNF inhibition in IBD is associated with the specific cell-type composition at the inflamed site, such as the abundance of activated fibroblasts or increased amount of IgG plasma cells, activated T cells and inflammatory macrophages [44]. Thus, variability of response to anti-TNF therapy is likely due to the heterogeneous nature of inflammation and depends on the composition of cellular infiltrates promoting the disease. These conclusions suggest that further stricter stratification of patients into distinct subgroups may improve clinical efficiency of anti-TNF therapy.

5. IMPORTANCE OF MOUSE MODELS

Mice with transgenic overexpression of human TNF unexpectedly turned into predictive animal model of arthritis [41]. These and similar mice can be used as preclinical models for evaluating clinically used TNF blockers in both preventive and therapeutic settings. Other useful models included mice 'humanised' for TNF and its receptors [53,54], as well as mice with conditional inactivation of TNF gene in specific cell types [55]. It was the latter mouse model that allowed to document opposite effects of TNF produced by distinct types of immunocytes in experimental arthritis [50]. Additionally, mice carrying humanised TNF and/or TNF receptors enabled efficacy evaluation of agonistic and antagonistic antibodies targeting both TNF receptors in mouse models of autoimmunity [56].

Numerous experiments involving mice with conditional TNF deficiency [55] uncovered distinct TNF activities associated with specific types of TNF-producing cells. Specifically, non-redundant homoeostatic functions of TNF produced by mast cells, basophils and smooth muscle cells were established [57–60]. Deleterious effects of TNF produced by myeloid cells were demonstrated in various disease models including arthritis [50], EAE [61], IBD [7], diabetic kidney injury [62], spinal cord injury [63] and acute response to endotoxin [55,64]. On the contrary, TNF produced by T cells appears to be protective in experimental arthritis [50], mycobacterial infection [65], endotoxin challenge [64] and renal injury in experimental nephritis [66].

Contrasting functions of TNF may be partially explained by signalling via distinct receptors (TNFR1 vs.

TNFR2) suggesting novel therapeutic opportunities using antibodies selective against these two TNF receptors. Another likely explanation is the role of tissue microenvironment in which TNF-producing cells cross-talk by activating the TNF-TNFR signalling axis. Recent advances in antibody engineering brought about various types of bispecific antibodies, both containing and lacking Fc. Of particular interest are bispecific mini-antibodies based either on scFv or VHH antigen-binding modules. For example, Bite [67] represents a new type of immune activators, called engagers, that may guide the cytotoxic T-cell response directly to malignant cells. Bispecific antibody simultaneously targeting TNF and IL-21 was evaluated in preclinical model [68], whereas bispecific antibody that binds and neutralises both TNF and IL-17A was evaluated in a clinical study for the treatment of plaque psoriasis [69]. Additionally, an emerging paradigm in the field of TNF-targeting therapeutics pursues a possibility to selectively neutralise only those TNF molecules that are secreted by a specific type of cell, such as myeloid cells. So far, however, this concept has only been evaluated in mouse models using a bi-specific mini-antibody called myeloid cell-specific TNF inhibitor (MYSTI) (see below).

6. B CELLS AND ANTI-TNF THERAPY

B cells are key players in the induction of autoimmune diseases via production of autoantibodies as well as by secretion of proinflammatory cytokines (reviewed in Ref. [70]). The B-cell compartment is modified in many autoimmune diseases both in the periphery and at the sites of inflammation [71,72]. For instance, RA patients contain in their synovium a significant fraction of naive and memory B cells, age-associated B cells and plasmablasts. Moreover, naive, memory and age-associated B cells from synovium were found to produce significant levels of TNF and IL-6 mRNA, thus mediating the deleterious functions via production of these molecules. Additionally, the expression of RANKL was reported on memory and age-associated B cells [72], and TNF can itself increase the expression of RANKL by B cells in the presence of IL-1β [73].

Both TNF and RANKL are critical factors for the induction of osteoclastogenesis and subsequent bone erosion. TNF production by peripheral B cells is increased with ageing restraining B-cell lymphopoiesis in the bone marrow via regulation of insulin-like growth factor 1 (IGF-1) [74]. Indeed, several studies showed a significant increase in the number of circulating B lymphocytes (and concomitantly in the subset of IgM memory B lymphocytes), similar to observations made in healthy control individuals and in patients with IBD following anti-TNF therapy. Restoration of B-cell numbers correlated with the biological response to therapy [75,76]. Thus, B cells may exert their pathogenic functions not only through autoantibody production but also by induction of pro-inflammatory cytokines.

Many autoimmune diseases are characterised by the development ELFs at the site of inflammation. Such ELFs consist of B-cell follicle with FDC network and scattered T cells. Interestingly, ELFs contain B cells resembling GC B cells by high expression of NR4A transcription factors and enhanced expression of IL-6, as well as LTα and LTβ [74]. Such ELFs may further contribute to the in situ development of pathogenic autoantibodies. Notably, the efficacy of anti-TNF therapy in autoimmunity conditions correlates with the disappearance of these structures in PsA [75] and RA [76]. Moreover, TNF blockade by ETN resulted in reduction of FDC and GC structures [7]. As a result, the number of CD38+ GC B cells and peripheral blood memory B cells were decreased in RA patients undergoing anti-TNF treatment. Consistent with these findings, TNF blockade suppressed the response to vaccination against hepatitis B, influenza and SARS-CoV-2, whereas the T cell-independent response against pneumococcal polysaccharides was only marginally affected [8,77,78].

7. CONCLUSIONS/PERSPECTIVES

Thirty years of the exciting anti-TNF saga helped to introduce and critically evaluate several novel concepts that may have a broader significance for immune regulation and for improved therapeutic strategies:

1. Mice with regulated overexpression of a single cytokine as predictive model of human disease.
2. Systemic blockade of a single 'pathogenic' cytokine as a therapy.
3. A particular cytokine may have simultaneously deleterious and protective effects.
4. 'Pathogenic' cytokine activity may be linked to a limited subset of producing cells.

As emphasised above, an emerging paradigm of more selective and therapeutically beneficial TNF inhibition is based on a possibility to neutralise only those TNF molecules that are secreted by a specific cell type. This concept was realised in a bi-specific mini-antibody MYSTI (Figure 18.3) that limits the secretion of human TNF in primary bone marrow-derived macrophage cultures obtained from humanised mice [77,78]. Importantly, MYSTI shows activity in vivo — in models of lethal hepatotoxicity and experimental arthritis [79]. Further studies are needed to demonstrate the advantage of selective inhibitors over traditional systemic TNF blockers.

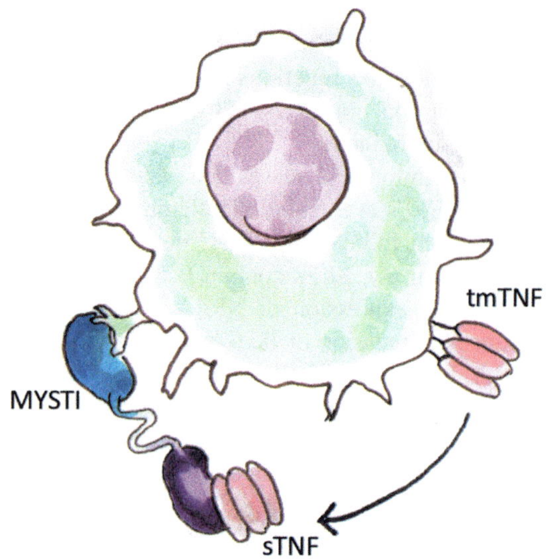

FIGURE 18.3 **Myeloid cell type-specific TNF inhibitor.** MYSTI is a prototype of several bispecific antibodies, which bind to the surface of a myeloid cell and can capture and retain sTNF secreted from this cell, thus limiting its systemic bioavailability.

GLOSSARY

ACPA Anti-citrullinated peptide antibodies
CD Crohn's disease
EAE Experimental autoimmune encephalomyelitis
ELF Ectopic lymphoid follicles
EMA European Medicines Agency
ETN Etanercept
FDA US Food and Drug Administration
FDC Follicular dendritic cells
GC Germinal centres
IBD Inflammatory bowel disease
IFX Infliximab
JIA Juvenile idiopathic arthritis
LT Lymphotoxin
MYSTI Myeloid cell-specific TNF inhibitor
PP Plaque psoriasis
PsA Psoriatic arthritis
RA Rheumatoid arthritis
RF Rheumatoid factor
scFv Single-chain mini-antibody
SpA Spondyloarthritis
TNF Tumour necrosis factor
UC Ulcerative colitis
VHH Single variable domain of a heavy chain only antibody

ACKNOWLEDGEMENTS

We are indebted to many colleagues who over the years helped to develop mouse strains for studying TNF immunobiology and for their numerous insights. We thank Thomas Dörner for useful discussion, Serguei Kozlov for critical reading the manuscript. Ekaterina Gubernatorova, Ekaterina Bulekova and Svetlana Bozrova for technical assistance with figure preparation. Ekaterina Gorshkova for her artistic view on MYSTI and TNF signaling. This work was supported by the Russian Science Foundation grant #19-75-30032.

REFERENCES

[1] Balkwill FR, Burke F. The cytokine network. Immunol Today 1989;10(9):299–304. https://doi.org/10.1016/0167-5699(89)90085-6.
[2] Maini RN, Elliott M, Brennan FM, Williams RO, Feldmann M. Targeting TNF alpha for the therapy of rheumatoid arthritis. Clin Exp Rheumatol 1994;12(Suppl. 11):S63–6. Epub 1994/11/01. PubMed PMID: 7768056.
[3] Elliott MJ, Maini RN, Feldmann M, Kalden JR, Antoni C, Smolen JS, et al. Randomised double-blind comparison of chimeric monoclonal antibody to tumour necrosis factor alpha (cA2) versus placebo in rheumatoid arthritis. Lancet 1994;344(8930):1105–10. https://doi.org/10.1016/s0140-6736(94)90628-9. PubMed PMID: 7934491.
[4] Keffer J, Probert L, Cazlaris H, Georgopoulos S, Kaslaris E, Kioussis D, et al. Transgenic mice expressing human tumour necrosis factor: a predictive genetic model of arthritis. EMBO J 1991;10(13):4025–31. https://doi.org/10.1002/j.1460-2075.1991.tb04978.x. PubMed PMID: 1721867; PubMed Central PMCID: PMCPMC453150.
[5] Butler DM, Malfait AM, Mason LJ, Warden PJ, Kollias G, Maini RN, et al. DBA/1 mice expressing the human TNF-alpha transgene develop a severe, erosive arthritis: characterization of the cytokine cascade and cellular composition. J Immunol 1997;159(6):2867–76. Epub 1997/09/23. PubMed PMID: 9300710.
[6] Williams RO, Feldmann M, Maini RN. Anti-tumor necrosis factor ameliorates joint disease in murine collagen-induced arthritis. Proc Natl Acad Sci U S A 1992;89(20):9784–8. https://doi.org/10.1073/pnas.89.20.9784. PubMed PMID: 1409699; PubMed Central PMCID: PMCPMC50217.
[7] Ninnemann J, Winsauer C, Bondareva M, Kuhl AA, Lozza L, Durek P, et al. TNF hampers intestinal tissue repair in colitis by restricting IL-22 bioavailability. Mucosal Immunol 2022;15(4):698–716. https://doi.org/10.1038/s41385-022-00506-x. PubMed PMID: 35383266; PubMed Central PMCID: PMCPMC9259490.
[8] Armaka M, Apostolaki M, Jacques P, Kontoyiannis DL, Elewaut D, Kollias G. Mesenchymal cell targeting by TNF as a common pathogenic principle in chronic inflammatory joint and intestinal diseases. J Exp Med 2008;205(2):331–7. https://doi.org/10.1084/jem.20070906. PubMed PMID: 18250193; PubMed Central PMCID: PMCPMC2271010.
[9] Anolik JH, Ravikumar R, Barnard J, Owen T, Almudevar A, Milner EC, et al. Cutting edge: anti-tumor necrosis factor therapy in rheumatoid arthritis inhibits memory B lymphocytes via effects on lymphoid germinal centers and follicular dendritic cell networks. J Immunol 2008;180(2):688–92. https://doi.org/10.4049/jimmunol.180.2.688.
[10] Kobie JJ, Zheng B, Bryk P, Barnes M, Ritchlin CT, Tabechian DA, et al. Decreased influenza-specific B cell responses in rheumatoid arthritis patients treated with anti-tumor necrosis factor. Arthritis Res Ther 2011;13(6):R209. https://doi.org/10.1186/ar3542. PubMed PMID: 22177419; PubMed Central PMCID: PMCPMC3334662.

[11] Melmed GY, Agarwal N, Frenck RW, Ippoliti AF, Ibanez P, Papadakis KA, et al. Immunosuppression impairs response to pneumococcal polysaccharide vaccination in patients with inflammatory bowel disease. Am J Gastroenterol 2010;105(1):148–54. https://doi.org/10.1038/ajg.2009.523.

[12] Caldera F, Hillman L, Saha S, Wald A, Grimes I, Zhang Y, et al. Immunogenicity of high dose influenza vaccine for patients with inflammatory bowel disease on anti-TNF monotherapy: a randomized clinical trial. Inflamm Bowel Dis 2020;26(4):593–602. https://doi.org/10.1093/ibd/izz164.

[13] Lin S, Kennedy NA, Saifuddin A, Sandoval DM, Reynolds CJ, Seoane RC, et al. Antibody decay, T cell immunity and breakthrough infections following two SARS-CoV-2 vaccine doses in inflammatory bowel disease patients treated with infliximab and vedolizumab. Nat Commun 2022;13(1):1379. https://doi.org/10.1038/s41467-022-28517-z. PubMed PMID: 35296643; PubMed Central PMCID: PMCPMC8927425.

[14] Carswell EA, Old LJ, Kassel RL, Green S, Fiore N, Williamson B. An endotoxin-induced serum factor that causes necrosis of tumors. Proc Natl Acad Sci U S A 1975;72(9):3666–70. https://doi.org/10.1073/pnas.72.9.3666. PubMed PMID: 1103152; PubMed Central PMCID: PMCPMC433057.

[15] Pennica D, Nedwin GE, Hayflick JS, Seeburg PH, Derynck R, Palladino MA, et al. Human tumour necrosis factor: precursor structure, expression and homology to lymphotoxin. Nature 1984;312(5996):724–9. https://doi.org/10.1038/312724a0.

[16] Fransen L, Muller R, Marmenout A, Tavernier J, Van der Heyden J, Kawashima E, et al. Molecular cloning of mouse tumour necrosis factor cDNA and its eukaryotic expression. Nucleic Acids Res 1985;13(12):4417–29. https://doi.org/10.1093/nar/13.12.4417. PubMed PMID: 2989794; PubMed Central PMCID: PMCPMC321797.

[17] Beutler B, Greenwald D, Hulmes JD, Chang M, Pan YC, Mathison J, et al. Identity of tumour necrosis factor and the macrophage-secreted factor cachectin. Nature 1985;316(6028):552–4. https://doi.org/10.1038/316552a0.

[18] Balkwill F. Tumour necrosis factor and cancer. Nat Rev Cancer 2009;9(5):361–71. https://doi.org/10.1038/nrc2628.

[19] Beutler B, Cerami A. Cachectin and tumour necrosis factor as two sides of the same biological coin. Nature 1986;320(6063):584–8. https://doi.org/10.1038/320584a0.

[20] Black RA, Rauch CT, Kozlosky CJ, Peschon JJ, Slack JL, Wolfson MF, et al. A metalloproteinase disintegrin that releases tumour-necrosis factor-alpha from cells. Nature 1997;385(6618):729–33. https://doi.org/10.1038/385729a0.

[21] Moss ML, Jin SL, Milla ME, Bickett DM, Burkhart W, Carter HL, et al. Cloning of a disintegrin metalloproteinase that processes precursor tumour-necrosis factor-alpha. Nature 1997;385(6618):733–6. https://doi.org/10.1038/385733a0.

[22] Grell M, Douni E, Wajant H, Lohden M, Clauss M, Maxeiner B, et al. The transmembrane form of tumor necrosis factor is the prime activating ligand of the 80 kDa tumor necrosis factor receptor. Cell 1995;83(5):793–802. https://doi.org/10.1016/0092-8674(95)90192-2.

[23] van Loo G, Bertrand MJM. Death by TNF: a road to inflammation. Nat Rev Immunol 2022:1–15. https://doi.org/10.1038/s41577-022-00792-3. PubMed PMID: 36380021; PubMed Central PMCID: PMCPMC9665039.

[24] Lork M, Verhelst K, Beyaert R. CYLD, A20 and OTULIN deubiquitinases in NF-kappaB signaling and cell death: so similar, yet so different. Cell Death Differ 2017;24(7):1172–83. https://doi.org/10.1038/cdd.2017.46. PubMed PMID: 28362430; PubMed Central PMCID: PMCPMC5520167.

[25] Beutler B, Milsark IW, Cerami AC. Passive immunization against cachectin/tumor necrosis factor protects mice from lethal effect of endotoxin. Science 1985;229(4716):869–71. https://doi.org/10.1126/science.3895437.

[26] Moreland LW, Baumgartner SW, Schiff MH, Tindall EA, Fleischmann RM, Weaver AL, et al. Treatment of rheumatoid arthritis with a recombinant human tumor necrosis factor receptor (p75)-Fc fusion protein. N Engl J Med 1997;337(3):141–7. https://doi.org/10.1056/NEJM199707173370301.

[27] Schreiber S, Rutgeerts P, Fedorak RN, Khaliq-Kareemi M, Kamm MA, Boivin M, et al. A randomized, placebo-controlled trial of certolizumab pegol (CDP870) for treatment of Crohn's disease. Gastroenterology 2005;129(3):807–18. https://doi.org/10.1053/j.gastro.2005.06.064.

[28] Inman RD, Davis Jr JC, Heijde D, Diekman L, Sieper J, Kim SI, et al. Efficacy and safety of golimumab in patients with ankylosing spondylitis: results of a randomized, double-blind, placebo-controlled, phase III trial. Arthritis Rheum 2008;58(11):3402–12. https://doi.org/10.1002/art.23969.

[29] van de Putte LB, Rau R, Breedveld FC, Kalden JR, Malaise MG, van Riel PL, et al. Efficacy and safety of the fully human anti-tumour necrosis factor alpha monoclonal antibody adalimumab (D2E7) in DMARD refractory patients with rheumatoid arthritis: a 12 week, phase II study. Ann Rheum Dis 2003;62(12):1168–77. https://doi.org/10.1136/ard.2003.009563. PubMed PMID: 14644854; PubMed Central PMCID: PMCPMC1754401.

[30] Knight DM, Trinh H, Le J, Siegel S, Shealy D, McDonough M, et al. Construction and initial characterization of a mouse-human chimeric anti-TNF antibody. Mol Immunol 1993;30(16):1443–53. https://doi.org/10.1016/0161-5890(93)90106-l. PubMed PMID: 8232330.

[31] Keane J, Gershon S, Wise RP, Mirabile-Levens E, Kasznica J, Schwieterman WD, et al. Tuberculosis associated with infliximab, a tumor necrosis factor alpha-neutralizing agent. N Engl J Med 2001;345(15):1098–104. https://doi.org/10.1056/NEJMoa011110.

[32] Tubach F, Salmon D, Ravaud P, Allanore Y, Goupille P, Breban M, et al. Risk of tuberculosis is higher with anti-tumor necrosis factor monoclonal antibody therapy than with soluble tumor necrosis factor receptor therapy: the three-year prospective French Research Axed on Tolerance of Biotherapies registry. Arthritis Rheum 2009;60(7):1884–94. https://doi.org/10.1002/art.24632. PubMed PMID: 19565495; PubMed Central PMCID: PMCPMC2921546.

[33] Dixon WG, Hyrich KL, Watson KD, Lunt M, Galloway J, Ustianowski A, et al. Drug-specific risk of tuberculosis in patients with rheumatoid arthritis treated with anti-TNF therapy: results from the British Society for Rheumatology Biologics Register (BSRBR). Ann Rheum Dis 2010;69(3):522–8. https://doi.org/10.1136/ard.2009.118935. PubMed PMID: 19854715; PubMed Central PMCID: PMCPMC2927681.

[34] Yamauchi PS, Bissonnette R, Teixeira HD, Valdecantos WC. Systematic review of efficacy of anti-tumor necrosis factor (TNF) therapy in patients with psoriasis previously treated with a different anti-TNF agent. J Am Acad Dermatol 2016;75(3):612–618 e6. https://doi.org/10.1016/j.jaad.2016.02.1221.

[35] Abraham E, Laterre PF, Garbino J, Pingleton S, Butler T, Dugernier T, et al. Lenercept (p55 tumor necrosis factor receptor fusion protein) in severe sepsis and early septic shock: a randomized, double-blind, placebo-controlled, multicenter phase III trial with 1,342 patients. Crit Care Med 2001;29(3):503−10. https://doi.org/10.1097/00003246-200103000-00006.

[36] TNF neutralization in MS: results of a randomized, placebo-controlled multicenter study. The Lenercept multiple sclerosis study group and the university of British columbia MS/MRI analysis group. Neurology 1999;53(3):457−65.

[37] Jespers LS, Roberts A, Mahler SM, Winter G, Hoogenboom HR. Guiding the selection of human antibodies from phage display repertoires to a single epitope of an antigen. Biotechnology 1994;12(9):899−903. https://doi.org/10.1038/nbt0994-899. PubMed PMID: 7521646.

[38] Martin PL, Cornacoff J, Prabhakar U, Lohr T, Treacy G, Sutherland JE, et al. Reviews preclinical safety and immune-modulating effects of therapeutic monoclonal antibodies to interleukin-6 and tumor necrosis factor-alpha in cynomolgus macaques. J Immunot 2005;1(3):131−9. https://doi.org/10.1080/15476910490894904.

[39] Nesbitt A, Fossati G, Bergin M, Stephens P, Stephens S, Foulkes R, et al. Mechanism of action of certolizumab pegol (CDP870): in vitro comparison with other anti-tumor necrosis factor alpha agents. Inflamm Bowel Dis 2007;13(11):1323−32. https://doi.org/10.1002/ibd.20225.

[40] Fischer R, Kontermann RE, Pfizenmaier K. Selective targeting of TNF receptors as a novel therapeutic approach. Front Cell Dev Biol 2020;8:401. https://doi.org/10.3389/fcell.2020.00401. PubMed PMID: 32528961; PubMed Central PMCID: PMCPMC7264106.

[41] Lorenzo-Vizcaya A, Isenberg DA. The use of anti-TNF-alpha therapies for patients with systemic lupus erythematosus. Where are we now? Expet Opin Biol Ther 2021;21(5):639−47. https://doi.org/10.1080/14712598.2021.1853096.

[42] Atreya R, Neumann H, Neufert C, Waldner MJ, Billmeier U, Zopf Y, et al. In vivo imaging using fluorescent antibodies to tumor necrosis factor predicts therapeutic response in Crohn's disease. Nat Med 2014;20(3):313−8. https://doi.org/10.1038/nm.3462. PubMed PMID: 24562382; PubMed Central PMCID: PMCPMC4479137.

[43] Dennis Jr G, Holweg CT, Kummerfeld SK, Choy DF, Setiadi AF, Hackney JA, et al. Synovial phenotypes in rheumatoid arthritis correlate with response to biologic therapeutics. Arthritis Res Ther 2014;16(2):R90. https://doi.org/10.1186/ar4555. PubMed PMID: 25167216; PubMed Central PMCID: PMCPMC4060385.

[44] Martin JC, Chang C, Boschetti G, Ungaro R, Giri M, Grout JA, et al. Single-cell analysis of crohn's disease lesions identifies a pathogenic cellular module associated with resistance to anti-TNF therapy. Cell 2019;178(6):1493−508. https://doi.org/10.1016/j.cell.2019.08.008. PubMed PMID: 31474370; PubMed Central PMCID: PMCPMC7060942.

[45] Rivellese F, Surace AEA, Goldmann K, Sciacca E, Cubuk C, Giorli G, et al. Rituximab versus tocilizumab in rheumatoid arthritis: synovial biopsy-based biomarker analysis of the phase 4 R4RA randomized trial. Nat Med 2022;28(6):1256−68. https://doi.org/10.1038/s41591-022-01789-0. PubMed PMID: 35589854; PubMed Central PMCID: PMCPMC9205785.

[46] van der Pouw Kraan TC, Wijbrandts CA, van Baarsen LG, Rustenburg F, Baggen JM, Verweij CL, et al. Responsiveness to anti-tumour necrosis factor alpha therapy is related to pre-treatment tissue inflammation levels in rheumatoid arthritis patients. Ann Rheum Dis 2008;67(4):563−6. https://doi.org/10.1136/ard.2007.081950.

[47] Choi IY, Gerlag DM, Herenius MJ, Thurlings RM, Wijbrandts CA, Foell D, et al. MRP8/14 serum levels as a strong predictor of response to biological treatments in patients with rheumatoid arthritis. Ann Rheum Dis 2015;74(3):499−505. https://doi.org/10.1136/annrheumdis-2013-203923.

[48] Klaasen R, Thurlings RM, Wijbrandts CA, van Kuijk AW, Baeten D, Gerlag DM, et al. The relationship between synovial lymphocyte aggregates and the clinical response to infliximab in rheumatoid arthritis: a prospective study. Arthritis Rheum 2009;60(11):3217−24. https://doi.org/10.1002/art.24913.

[49] Wijbrandts CA, Remans PH, Klarenbeek PL, Wouters D, van den Bergh Weerman MA, Smeets TJ, et al. Analysis of apoptosis in peripheral blood and synovial tissue very early after initiation of infliximab treatment in rheumatoid arthritis patients. Arthritis Rheum 2008;58(11):3330−9. https://doi.org/10.1002/art.23989.

[50] Kruglov A, Drutskaya M, Schlienz D, Gorshkova E, Kurz K, Morawietz L, et al. Contrasting contributions of TNF from distinct cellular sources in arthritis. Ann Rheum Dis 2020;79(11):1453−9. https://doi.org/10.1136/annrheumdis-2019-216068. PubMed PMID: 32796044; PubMed Central PMCID: PMCPMC7569389.

[51] Atreya R, Zimmer M, Bartsch B, Waldner MJ, Atreya I, Neumann H, et al. Antibodies against tumor necrosis factor (TNF) induce T-cell apoptosis in patients with inflammatory bowel diseases via TNF receptor 2 and intestinal CD14(+) macrophages. Gastroenterology 2011;141(6):2026−38. https://doi.org/10.1053/j.gastro.2011.08.032.

[52] Jessen B, Rodriguez-Sillke Y, Sonnenberg E, Schumann M, Kruglov A, Freise I, et al. Level of tumor necrosis factor production by stimulated blood mononuclear cells can Be used to predict response of patients with inflammatory bowel diseases to infliximab. Clin Gastroenterol Hepatol 2021;19(4):721−31. https://doi.org/10.1016/j.cgh.2020.03.066.

[53] Olleros ML, Chavez-Galan L, Segueni N, Bourigault ML, Vesin D, Kruglov AA, et al. Control of mycobacterial infections in mice expressing human tumor necrosis factor (TNF) but not mouse TNF. Infect Immun 2015;83(9):3612−23. https://doi.org/10.1128/IAI.00743-15. PubMed PMID: 26123801; PubMed Central PMCID: PMCPMC4534677.

[54] Atretkhany KN, Mufazalov IA, Dunst J, Kuchmiy A, Gogoleva VS, Andruszewski D, et al. Intrinsic TNFR2 signaling in T regulatory cells provides protection in CNS autoimmunity. Proc Natl Acad Sci U S A 2018;115(51):13051−6. https://doi.org/10.1073/pnas.1807499115. PubMed PMID: 30498033; PubMed Central PMCID: PMCPMC6304938.

[55] Grivennikov SI, Tumanov AV, Liepinsh DJ, Kruglov AA, Marakusha BI, Shakhov AN, et al. Distinct and nonredundant in vivo functions of TNF produced by t cells and macrophages/neutrophils: protective and deleterious effects. Immunity 2005;22(1):93−104. https://doi.org/10.1016/j.immuni.2004.11.016.

[56] Dong Y, Fischer R, Naude PJ, Maier O, Nyakas C, Duffey M, et al. Essential protective role of tumor necrosis factor receptor 2 in neurodegeneration. Proc Natl Acad Sci U S A 2016;113(43):12304−9. https://doi.org/10.1073/pnas.1605195113. PubMed PMID: 27791020; PubMed Central PMCID: PMCPMC5087045.

[57] Dudeck J, Kotrba J, Immler R, Hoffmann A, Voss M, Alexaki VI, et al. Directional mast cell degranulation of tumor necrosis factor into blood vessels primes neutrophil extravasation. Immunity 2021;54(3):468–83. https://doi.org/10.1016/j.immuni.2020.12.017.

[58] Dudeck J, Ghouse SM, Lehmann CH, Hoppe A, Schubert N, Nedospasov SA, et al. Mast-cell-derived TNF amplifies CD8(+) dendritic cell functionality and CD8(+) T cell priming. Cell Rep 2015;13(2):399–411. https://doi.org/10.1016/j.celrep.2015.08.078.

[59] Piliponsky AM, Shubin NJ, Lahiri AK, Truong P, Clauson M, Niino K, et al. Basophil-derived tumor necrosis factor can enhance survival in a sepsis model in mice. Nat Immunol 2019;20(2):129–40. https://doi.org/10.1038/s41590-018-0288-7. PubMed PMID: 30664762; PubMed Central PMCID: PMCPMC6352314.

[60] Kroetsch JT, Levy AS, Zhang H, Aschar-Sobbi R, Lidington D, Offermanns S, et al. Constitutive smooth muscle tumour necrosis factor regulates microvascular myogenic responsiveness and systemic blood pressure. Nat Commun 2017;8:14805. https://doi.org/10.1038/ncomms14805. PubMed PMID: 28378814; PubMed Central PMCID: PMCPMC5382284.

[61] Kruglov AA, Lampropoulou V, Fillatreau S, Nedospasov SA. Pathogenic and protective functions of TNF in neuroinflammation are defined by its expression in T lymphocytes and myeloid cells. J Immunol 2011;187(11):5660–70. https://doi.org/10.4049/jimmunol.1100663.

[62] Awad AS, You H, Gao T, Cooper TK, Nedospasov SA, Vacher J, et al. Macrophage-derived tumor necrosis factor-alpha mediates diabetic renal injury. Kidney Int 2015;88(4):722–33. https://doi.org/10.1038/ki.2015.162. PubMed PMID: 26061548; PubMed Central PMCID: PMCPMC4589442.

[63] Ellman DG, Lund MC, Nissen M, Nielsen PS, Sorensen C, Lester EB, et al. Conditional ablation of myeloid TNF improves functional outcome and decreases lesion size after spinal cord injury in mice. Cells 2020;9(11). https://doi.org/10.3390/cells9112407. PubMed PMID: 33153044; PubMed Central PMCID: PMCPMC7692197.

[64] Togbe D, Grivennikov SI, Noulin N, Couillin I, Maillet I, Jacobs M, et al. T cell-derived TNF down-regulates acute airway response to endotoxin. Eur J Immunol 2007;37(3):768–79. https://doi.org/10.1002/eji.200636371.

[65] Allie N, Grivennikov SI, Keeton R, Hsu NJ, Bourigault ML, Court N, et al. Prominent role for T cell-derived tumour necrosis factor for sustained control of *Mycobacterium tuberculosis* infection. Sci Rep 2013;3:1809. https://doi.org/10.1038/srep01809. PubMed PMID: 23657146; PubMed Central PMCID: PMCPMC3648802.

[66] Wen Y, Lu X, Ren J, Privratsky JR, Yang B, Rudemiller NP, et al. KLF4 in macrophages attenuates TNFalpha-mediated kidney injury and fibrosis. J Am Soc Nephrol 2019;30(10):1925–38. https://doi.org/10.1681/ASN.2019020111. PubMed PMID: 31337692; PubMed Central PMCID: PMCPMC6779357.

[67] Nagorsen D, Baeuerle PA. Immunomodulatory therapy of cancer with T cell-engaging BiTE antibody blinatumomab. Exp Cell Res 2011;317(9):1255–60. https://doi.org/10.1016/j.yexcr.2011.03.010.

[68] Park JS, Yoo SH, Lim MA, Ryu JG, Oh HJ, Hwang SH, et al. A bispecific soluble receptor fusion protein that targets TNF-alpha and IL-21 for synergistic therapy in inflammatory arthritis. FASEB J 2020;34(1):248–62. https://doi.org/10.1096/fj.201900816RR.

[69] Silacci M, Lembke W, Woods R, Attinger-Toller I, Baenziger-Tobler N, Batey S, et al. Discovery and characterization of COVA322, a clinical-stage bispecific TNF/IL-17A inhibitor for the treatment of inflammatory diseases. mAbs 2016;8(1):141–9. https://doi.org/10.1080/19420862.2015.1093266. PubMed PMID: 26390837; PubMed Central PMCID: PMCPMC4966518.

[70] Merino-Vico A, Frazzei G, van Hamburg JP, Tas SW. Targeting B cells and plasma cells in autoimmune diseases: from established treatments to novel therapeutic approaches. Eur J Immunol 2022. https://doi.org/10.1002/eji.202149675.

[71] Elsner RA, Shlomchik MJ. Germinal center and extrafollicular B cell responses in vaccination, immunity, and autoimmunity. Immunity 2020;53(6):1136–50. https://doi.org/10.1016/j.immuni.2020.11.006. PubMed PMID: 33326765; PubMed Central PMCID: PMCPMC7748291.

[72] Zhang F, Wei K, Slowikowski K, Fonseka CY, Rao DA, Kelly S, et al. Defining inflammatory cell states in rheumatoid arthritis joint synovial tissues by integrating single-cell transcriptomics and mass cytometry. Nat Immunol 2019;20(7):928–42. https://doi.org/10.1038/s41590-019-0378-1. PubMed PMID: 31061532; PubMed Central PMCID: PMCPMC6602051.

[73] Banko Z, Pozsgay J, Szili D, Toth M, Gati T, Nagy G, et al. Induction and differentiation of IL-10-producing regulatory B cells from healthy blood donors and rheumatoid arthritis patients. J Immunol 2017;198(4):1512–20. https://doi.org/10.4049/jimmunol.1600218.

[74] Dowery R, Benhamou D, Benchetrit E, Harel O, Nevelsky A, Zisman-Rozen S, et al. Peripheral B cells repress B-cell regeneration in aging through a TNF-alpha/IGFBP-1/IGF-1 immune-endocrine axis. Blood 2021;138(19):1817–29. https://doi.org/10.1182/blood.2021012428. PubMed PMID: 34297797; PubMed Central PMCID: PMCPMC9642783.

[75] Li Z, Vermeire S, Bullens D, Ferrante M, Van Steen K, Noman M, et al. Anti-tumor necrosis factor therapy restores peripheral blood B-cell subsets and CD40 expression in inflammatory bowel diseases. Inflamm Bowel Dis 2015;21(12):2787–96. https://doi.org/10.1097/MIB.0000000000000554.

[76] Defendenti C, Atzeni F, Malandrin S, Ardizzone S, Almasio PL, Saibeni S, et al. Anti-tumour necrosis factor-alpha antibodies and B cell homeostasis in human inflammatory bowel diseases. Int Immunopharm 2018;54:329–35. https://doi.org/10.1016/j.intimp.2017.11.016.

[77] Efimov GA, Kruglov AA, Khlopchatnikova ZV, Rozov FN, Mokhonov VV, Rose-John S, et al. Cell-type-restricted anti-cytokine therapy: TNF inhibition from one pathogenic source. Proc Natl Acad Sci U S A 2016;113(11):3006–11. https://doi.org/10.1073/pnas.1520175113. PubMed PMID: 26936954; PubMed Central PMCID: PMCPMC4801281.

[78] Nosenko MA, Atretkhany KN, Mokhonov VV, Efimov GA, Kruglov AA, Tillib SV, et al. VHH-based bispecific antibodies targeting cytokine production. Front Immunol 2017;8:1073. https://doi.org/10.3389/fimmu.2017.01073. PubMed PMID: 28919896; PubMed Central PMCID: PMCPMC5585155.

[79] Drutskaya MS, Nosenko MA, Gorshkova EA, Mokhonov VV, Zvartsev RV, Polinova AI, et al. Effects of myeloid cell-restricted TNF inhibitors in vitro and in vivo. J Leukoc Biol 2020;107(6):933–9. https://doi.org/10.1002/JLB.3AB0120-532R.

Chapter 19

Therapeutic Targeting of B Cells and Plasma Cells with a Focus on Multiple Sclerosis and Other Autoimmune Conditions

Andreia C. Lino[1,2] and Thomas Dörner[1,2]

[1]*Department of Rheumatology and Clinical Immunology, Charité Universitätsmedizin Berlin, Berlin, Germany;* [2]*German Rheumatism Research Center Berlin-Leibniz Institute, Berlin, Germany*

1. INTRODUCTION

Multiple sclerosis (MS) is a complex, chronic, disabling neurological disease with a wide range of potential symptoms and physical impairments. The disease is related to an autoimmune reaction that leads to inflammatory damage of the myelin sheaths that protect the nerve fibres in the brain and spinal cord. The exact cause of MS is unknown, but it is believed to involve genetic, environmental and immunological factors. There are several types of MS, each with a distinct B-cell signature and varying levels of severity. Relapsing-remitting MS (RRMS) is the most common form, occurring in about 80%–85% of patients. RRMS characterised by B cell-mediated inflammation in the brain and spinal cord and leading to episodes of flares with new or worsening symptoms (relapses). The symptoms are followed by periods of recovery or remissions. Secondary progressive MS (SPMS) is a more severe form of MS, occurring in about 15% of patients. It begins as RRMS and then gradually progresses over time, with fewer and less severe relapses and more gradual disability accumulation. B cells are still involved in this type of MS especially during early stages, but the exact role they play is still unclear. Primary progressive MS (PPMS) is a rarer form of MS, affecting only about 5% of patients. It is characterised by gradual onset of disability, with no periods of remission. Its association with B cell-mediated inflammation is less clear, thus B cell-targeted therapies were less successful [1–3], even though ocrelizumab (anti-human CD20) is approved for the treatment of PPMS [4]. Here, we refer preferentially to RRMS as it is the most common MS subset including well-documented pathogenic role of B cells.

B cells play a central role in the pathogenesis of several autoimmune diseases, such as systemic lupus erythematosus (SLE), rheumatoid arthritis (RA) and Sjogren's syndrome (SS). While it was clear that B cells played a role in these diseases due to their unique capacity of differentiating into autoreactive B cells producing autoantibodies, the involvement of B cells in MS, a disease that has been considered for decades uniquely T cell driven, was less clear. Therapeutic anti-CD20 interventions in MS patients led to disease improvement. Thus, these results led to new concepts and several paradigm changes regarding the pathophysiology of MS and treatment approaches. In this context, B-cell depletion therapies were effective in the treatment of autoimmunity in humans, but such effect could not be explained entirely by the elimination of cells producing autoantibodies. Thus, it became clear that B cells contributed to autoimmunity by antibody independent functions, such as cytokine production and antigen presentation.

Currently, there are two disease-modifying agents (biologicals – ocrelizumab and ofatumumab), which target CD20 (type I) on the surface of B cells approved for MS treatment. But several other agents have been or are presently in clinical trials for MS, such as BTK inhibitors. Notably, the BAFF inhibitor belimumab approved for SLE is currently undergoing clinical trial for RRMS (NCT04767698), while the trial for atacicept, which inhibits BAFF as well as APRIL with an additional impact on plasma cell (PC) survival, was terminated early due to increased MS lesions and relapse rate during treatment [5].

Molecular Biology of B Cells. https://doi.org/10.1016/B978-0-323-95895-0.00007-6

Atacicept treatment of SLE patients gave rise to increased number infections, and the two trials were terminated early due to safety issues [6,7]. Another receptor fusion protein inhibiting BAFF and APRIL, telitacicept is currently being evaluated in clinical trials of SLE [8].

Anti-B-cell treatment (directly through cellular targeting or indirectly by blocking cytokines such as BAFF) provided further proof for substantial B-cell involvement in RA and SLE. From all these trials, we gained knowledge about the role of B cells beyond autoantibodies produced by PCs in these autoimmune conditions. Here, we review the current knowledge on therapeutic targeting of B-cell lineage cells in MS and other autoimmune diseases from a translational and reverse translational perspective. In particular, we aim to develop take-home messages from trials targeting B-lineage cells that failed and discuss future approaches and opportunities for B-cell targeting, including the targeting of certain B-cell subsets as well as selective depletion of autoantigen-specific B-lineage cells only.

A key lesson from all these observations is that different stages of B-cell development (conventional naïve and memory B cells) may show abnormalities reflected by the occurrence of atypical memory B cells and distinct contributions by PCs. The latter compartment evolves as an interesting area as we are just witnessing the deconvolution of this rather heterogeneous compartment in the human bone marrow [9–11]. With the deconvolution of pathogenic relevant B-lineage subsets (B cells and PCs), their distinct role among various autoimmune diseases and certain T-cell subset interaction needed for their induction may become apparent. In addition, the survival of pathogenic B-lineage subsets may be critical during chronic autoimmunity and likely depends on the composition of a niche composed by stromal cells. However, knowledge about the latter is very limited in MS and therefore not further addressed here.

2. THE ROLES OF B-CELL LINEAGE CELLS IN MS AND OTHER AUTOIMMUNE CONDITIONS

It is now widely accepted that B-lineage cells play a significant role in MS and other autoimmune conditions though production of antibodies by plasmablasts and PCs, but also through antibody-independent mechanisms. Antibody-independent roles played by B cell include: production of soluble factors such as proinflammatory cytokines and chemokines; antigen presentation to T cells and formation of tertiary lymphoid structures.

2.1 Autoantibody-Producing Cells

Regarding the role of B cells as antibody producers, it is known that immunoglobulins are found in abnormal quantities in the cerebrospinal fluid (CSF) of MS patients, with more than 90% of patients being positive for IgG oligoclonal bands in protein gel electrophoresis of CSF – a useful diagnostic biomarker [12,13], but not essentially required by the current MacDonald criteria [14]. The antigens recognized by these IgG antibodies remained elusive and might be patient-specific [15]. Therefore, target strategies to eliminate antigen-specific PC in MS are currently difficult to imagine. However, this observation is critical for our clinical understanding.

On the other hand, autoantigen-specific B-cell targeting could be a possible strategy to treat neuromyelitis optica spectrum disorder (NMOSD) that represent a subentity of neuroinflammatory diseases with defined autoantigens, such as aquaporin-4 (AQP4) or myelin oligodendrocyte glycoprotein (MOG). NMOSD patients respond to anti-CD20 (rituximab) treatment, even though CD20 is not expressed by PCs and therefore is not able to deplete these cells. NMOSD activity of patients treated with rituximab correlated with B-cell counts, but not with the levels of anti-AQP4 antibodies [16] indicating that PC precursors are of pathogenic relevance in contrast to PCs. Another study of patients with repeated rituximab courses concluded that the frequency of reemerging $CD27^{+}$ memory B cells in circulation [17] are a surrogate for subsequent anti-CD20 treatment in order to achieve or maintain remission consistent with observations in RA [18] and SLE [19]. These results suggest that memory B cells are the main precursors of pathogenic PC in NMOSD.

Tocilizumab is a humanised monoclonal antibody against IL-6 receptor that also showed efficacy in the treatment of NMOSD [20,21] by causing decreased serum levels of anti-AQP4 antibodies [20]. A subsequent study in which B-cell subsets were analysed before and after tocilizumab showed that after tocilizumab treatment, the frequencies of $CD27^{hi}CD38^{hi}$ PC as well as differentiated B-cell subsets such as switched memory $CD27^{+}IgD^{-}$ B cells or $CD27^{-}IgD^{-}$ double negative (DN) cells were decreased in NMOSD patients. In addition, the levels of PD-L1 that are on the surface of switched memory $CD27^{+}IgD^{-}$ B cells and PCs were also decreased after treatment [22]. IL-6 is an essential factor for PC survival [23] but also for B-cell differentiation. Thus, tocilizumab might contribute to the amelioration of NMOSD though the elimination of pathogenic PC survival niches or simply blockade of B-cell differentiation into PC, which has also been observed in RA [24,25].

It has long been known that SLE is characterised by a marked peripheral lymphopenia, which affects $CD27^{-}$ naive B cells more severely and results in higher frequencies of $CD27^{+}$ memory B cells and $CD27^{hi}$ plasmablasts/PCs in the peripheral blood [26] that correlate with disease activity [27,28], highlighting the pathogenic role of these cells in SLE. A more detailed description about the role of B cells, including PCs, in SLE can be found elsewhere [29–31]. A recent report identified a population of antigen-experienced $CXCR5^{-}CD19^{low}$ B cells as precursors of plasmablasts. This population is expanded in SLE patients, thus co-targeting this subset may have therapeutic value in SLE [32]. The antibodies produced by PCs in SLE react against self-antigens and participate in the induction and maintenance of tissue damage and represent hallmarks of SLE which is characterised by the largest variety of autoantibody specificities, with 180 autoantibodies so far described [33]. Thus, it would be difficult to imagine a strategy to eliminate antigen-specific PCs in SLE and therefore mandating efforts employing surface markers of pathogenic PCs that could be used to specific target only these cells. A recent report identified a disease-specific subpopulation of BM $CD39^{++}CD326^{++}$ PC in murine SLE. This subset is absent in spleen and in BM of healthy mice [10].

2.2 B Cells as Producers of Cytokines, Chemokines and Neurotoxic Factors

B-lineage cells can modulate autoimmunity through the production of pro- and anti-inflammatory cytokines. It has been shown that B cells from MS patients produce increased amounts of proinflammatory cytokines, such as IL-6 and GM-CSF, lymphotoxin-α and TNF and reduced amounts of anti-inflammatory cytokine IL-10 [34–37]. Noteworthy is the fact that B-cell depletion therapies mediate a reduction in the production of proinflammatory cytokines and increase the production of IL-10 [34–37]. The role of B cells as cytokine-producing cells in mouse and human autoimmune diseases, including MS, RA, SLE and inflammatory bowel disease (IBD), in the context of rituximab treatment has been reviewed extensively elsewhere [38].

In addition to cytokines, B cells can produce certain chemokines that play a role in T-cell trafficking into the CNS in MS. Seventeen chemokines were measured in the CSF before and after rituximab treatment of patients with RRMS. The study showed that the B- and T-cell chemokines CXCL13 and CCL19 were decreased in the CSF of treated patients. Furthermore, there was a positive correlation between T cells and CXCL13 in the CSF [39].

Finally, there is evidence that B cells can secrete neurotoxic factors. In vitro experiments in which the cytotoxic activity of B cells to oligodendrocytes was tested, the results showed that co-cultures of B cells from MS patients induced a higher frequency of oligodendrocytes apoptosis than those of healthy donors. This effect did not correlate with the levels of cytokines or immunoglobulins in the supernatants [40,41]. The molecules that mediate this effect remain elusive, but were enriched in the exosome fractions. Differential proteomics comparing contents of exosomes from MS and normal control B cells suggest that these factors can be surface and extracellular plasma membrane, and gliogenesis-associated proteins [42].

2.3 Antigen Presentation by B Cells

B cells can contribute to autoimmunity by acting as antigen-presenting cells (APCs). B cells express MHC class II as well as co-stimulatory molecules such as CD40, CD80 and CD86. Thus, autoreactive B cells can activate and expand cognate autoreactive T cells enhancing the immune response and promoting disease. Therefore, co-stimulatory molecules involved in T and B cell co-stimulation are considered to control autoimmune clones emerging within adaptive immunity.

Human leucocyte antigen (HLA)-DR15 is a major genetic risk factor in MS patients [43]. It has been shown that MS autoreactive T cells can recognise HLA-DR15-derived self-peptides as well as Epstein–Barr virus (EBV)-derived cross-reactive peptides on B cells [44]. Furthermore, proliferation of brain homing T cells was stimulated by memory B cells of patients carrying the haplotype HLA-DR15 in HLA-DR-dependent manner [45]. These data show the role of B cells as APCs in MS and likely needs to be seen in the context of co-stimulatory molecules provided by these cells.

2.4 Formation of Tertiary Lymphoid Structures

Local CNS infiltration and inflammation with meningeal germinal centre-like structures enriched in B cells are common in MS patients, and their number and size correlate with the level of neurodegeneration and disability [46,47]. These tertiary lymphoid structures support T–B cell interactions with antigen presentation and B- and T-cell activation and differentiation. B cells can hypermutate and undergo class switch recombination in these tertiary structures [47].

New studies analysing paired CSF and blood samples using single-cell RNA sequencing revealed enrichment of T follicular cells in CSF of MS patients supporting B cells being functional and differentiating into plasmablasts and PCs capable of autoantibody production. Analysis of BCR repertoire indicated that expansion of hypermutated clones of IgM and IgG1 in CSF was associated with acute disease [48,49]. Cytokines, such as IL-6, IL-10 and lymphotoxin-α, and chemokines, such as CXCL10 and 13 that support germinal-centre formation have been shown to be present in CSF of MS patients [50]. Thus, once B cells and plasmablasts pass the blood–brain barrier and have access to the chronic inflamed CNS, they can differentiate in long-lived PCs and persist as such in ectopic tertiary lymphoid structures that support their survival. These tertiary lymphoid structures are similar to those found in the salivary glands in SS [51,52], in the joints in RA [53,54] or in the thyroid in Hashimoto's thyroiditis [55].

3. THERAPEUTIC TARGETING OF B-LINEAGE CELLS

3.1 Rituximab (Anti-CD20)

Rituximab is a monoclonal mouse-human chimaeric antibody that binds a loop of CD20 (type I) on the surface of circulating B cells, promoting their lysis though mechanisms of antibody-dependent cell-mediated cytotoxicity (ADCC), complement-dependent cytotoxicity (CDC) and induction of apoptosis [56]. Rituximab is approved for the treatment of non-Hodgkin B-cell lymphoma, chronic lymphocytic leukaemia (CLL), RA and granulomatosis with polyangiitis [57].

Rituximab has been tested in several clinical trials and has demonstrated promising results in MS patients, including a phase I open-label trial for RRMS [58]. This 72-week, open-label, phase I trial sought to evaluate the safety and efficacy of rituximab in RRMS patients. The results showed that treatment with rituximab was generally safe and well-tolerated and that it resulted in a significant reduction in relapse rate, disability progression and gadolinium-enhancing lesions when compared to baseline. Furthermore, there was no significant difference in the safety profile of rituximab compared to placebo. The patients receiving rituximab in this trial exhibited near complete absence of B cells in circulation from the second week until week 24. By week 72, B cells in circulation were about one-third of baseline and mainly naïve $IgD^{+}CD27^{-}$ [58]. These findings suggested that rituximab might be an effective treatment for RRMS and warranted further investigation. A phase II study followed a cohort 104 RRMS patients receiving rituximab or placebo [59] and showed a 91% reduction in the number of gadolinium-enhancing lesions at week 24 together with a reduction of clinical relapses. Serum levels of IgG, IgM and IgA were measured and were on average above the lower limit of normal [59]. This is due to the fact that rituximab and other anti-CD20 targeted drugs are not able to deplete PCs since these cells lack expression of CD20.

Nevertheless, the introduction of rituximab in the treatment of MS led to a paradigm change in understanding the pathogenesis of MS. The beneficial effect of rituximab on the disease course pointed towards the causative role of B cells in inflammation in MS. Several modifications in B cell subsets in MS patients who receive rituximab were reported, including persistent decreases of proinflammatory GM-CSF producing B cells, which mediate the activation of inflammatory monocytes and macrophages in untreated MS patients [34]. In this study, the authors suggest that B cells that repopulate after rituximab have changes in STAT5 and STAT6 signalling as the molecular basis of the switch from pro-inflammatory GM-CSF before rituximab to anti-inflammatory IL-10 production after treatment [34]. Analysis of repopulating B-cell subsets after rituximab reports that naïve and B-cell subsets with regulatory properties and beneficial effects are enriched [60,61]. Rituximab treatment of RA patients resulted in decreased numbers of Th17 cells in synovial tissue [62]. Similar results regarding decreased Th17 cells in MS after rituximab treatment have also been reported [37]. These data show that rituximab also decreases B cell—T cell interactions. A decrease in T cell numbers in the CSF was found upon anti-CD20 treatment with rituximab and ocrelizumab [57,63]. These results were surprising as anti-CD20 antibodies are not supposed to cross the blood—brain barrier and suggest that B—T cell cooperation in the periphery, by antigen presentation or cytokine production, is important to support the recruitment of T cells into the CNS. Regarding the effect of rituximab in the CSF, intrathecal administration of rituximab resulted in decreased numbers of B cells in the periphery, but there were similar numbers of B cells in the CSF. This observation led to the conclusion that the numbers of CSF B cells were too low to discriminate or to identify relevant differences. Interestingly, this study reported decreased levels of BAFF, essential for B-cell survival and immunoglobulin production, in the CSF but increased levels in the serum of rituximab-treated patients [64].

Even though rituximab is effective in the treatment of RRMS, rituximab is not approved for the treatment of MS and is only used off-label in certain countries. However, the more humanised anti-CD20 monoclonal antibodies ocrelizumab and ofatumumab are now approved for RRMS.

3.2 Ocrelizumab (Anti-CD20)

Ocrelizumab is a recombinant humanised IgG1 monoclonal antibody against CD20. It is the first humanised anti-CD20 monoclonal antibody approved for MS. It induces depletion of B cells though similar mechanisms as rituximab (ADCC and CDC) [65,66], but might be superior to rituximab depleting B-cell lines that were resistant to rituximab [67].

Ocrelizumab has been studied in several clinical trials and has been shown to improve patient outcomes and reduce relapse rates in people with relapsing MS forms. In a phase III clinical trial, ocrelizumab was found to be more effective than interferon beta-1a in reducing MS relapse rates and disability progression in relapsing forms of MS. The study, which included 1656 patients, found that ocrelizumab was associated with a 46% reduction in relapse rate and a 34% reduction in disability progression [68]. In another phase III clinical trial, ocrelizumab was compared with placebo to determine its effects in patients with PPMS. The study, which included 732 patients, found that ocrelizumab was associated with a 24% reduction in disability progression at 12 weeks [4].

The clinical development of ocrelizumab for the treatment of RA and SLE was suspended due to safety concerns regarding the high incidence of serious and fatal opportunistic infections [69].

In addition to its clinical efficacy, ocrelizumab has also been studied for its effects on B-cell subsets in MS patients. It was found that ocrelizumab was associated with a significant and rapid reduction in B-cell numbers in peripheral blood [70,71] and CSF [57] in patients with RRMS. Six months after ocrelizumab treatment, there was a significant reduction in total numbers of naive and memory B cells and increased frequencies of transitional and plasmablasts in circulation. Regarding cytokine producing B cells, 6 months after ocrelizumab, there were decreased numbers of TNF, GM-CSF, IL-6 and IL-10 producing B cells, but the frequencies of GM-CSF, IL-6 and IL-10 producing B cells increased [71].

3.3 Ofatumumab (Anti-CD20)

Ofatumumab is a human IgG1 antibody that binds human CD20, inducing B-cell depletion through ADCC and CDC mechanisms. It binds to a different epitope than rituximab and ocrelizumab – a small extracellular loop of CD20 and is able to deplete cells in which CD20 is expressed at low levels [72,73]. Ofatumumab is approved for the treatment of relapsing forms of MS and certain types of CLL. The use of ofatumumab in the treatment of RA, in combination with methotrexate seems to indicate superiority compared to placebo, but further studies are pending [74–76].

Ofatumumab in RRMS treatment has been studied. First, in two phase II placebo-controlled studies showed reductions in new gadolinium-enhancing lesions and no safety-related concerns or increased rate of opportunistic infections [77,78], warranting the continuation of the evaluation. Later, in large phase III trials, subcutaneous ofatumumab was tested in comparison with oral teriflunomide, a disease-modifying agent that decreases T- and B-cell activation. The trials showed that RRMS patients treated with ofatumumab had less annualised relapse rates than those treated with teriflunomide [79].

Treatment of RRMS patients with ofatumumab rapidly eliminated B cells from the circulation in a similar way than that reported for rituximab or ocrelizumab. It was observed that CD19 B cells were depleted in a dose-dependent manner [78]. Detailed reports about specific B-cell subset kinetics upon ofatumumab are still missing.

Taking into account the success of B cell-depleting therapies, it was surprizing that the trial for atacicept, which binds to B Lymphocyte Stimulator (BLyS) and A proliferation-inducing ligand (APRIL) and block B-cell activation, had to be terminated due to increased clinical disease activity [5]. This highlights the complex role of B cells in MS and our still limited understanding of MS pathogenesis.

3.4 Bruton Tyrosine Kinase Inhibitors (Ibrutinib, Evobrutinib, Tolebrutinib, Fenebrutinib, Remibrutinib)

In addition to therapies that deplete B cells through CD20, other B-cell therapies were developed for MS and other autoimmune diseases, namely Bruton tyrosine kinase inhibitors (BTKi). This mechanism prevents intracellular signalling from the BCR and other surface receptors such as Toll-like receptors (TLRs) to activate B cells. BTK and phospho-BTK were reported to be increased in B cells from the majority of primary SS and in anti-citrullinated protein antibody (ACPA)-positive RA patients [80]. BTK inhibition has emerged as a potential way to prevent B-cell activation and pro-inflammatory functions without the risk of complete B-cell ablation and acquired humoral immunodeficiency. In addition, BTKi can also interfere with monocyte/macrophage as well as basophil activation where BTK is downstream of FcRs as well as with thrombocytes. Treatment of CLL patients with ibrutinib did not result in a reduction of total immunoglobulins levels [81]. However, a reduction of some autoantibodies, including anti-double-stranded DNA autoantibodies, total IgG and IgM, was reported when fenebrutinib was used to treat SLE patients [82].

BTKis are small molecules. Small molecules have several advantages compared with large antibody molecules such as anti-CD20 antibodies. Small molecules are easier and cheaper to synthetise and can be administered orally. In the case of MS, these small molecules may have additional advantages, since they can cross the blood–brain barrier. Tolebrutinib was found in significant concentrations in the CSF 2h after oral administration [83]. In addition, BTK is involved in the signalling cascade that is activated by chemokine receptors CXCR4 and CXCR5 in B cells and decrease their trafficking and homing to the CNS. Treatment of CLL patients with ibrutinib altered the expression of chemokine receptors and impaired migration of B cells to lymph nodes and bone marrow microenvironment that supports their survival [84]. Therefore, decreased B-cell trafficking caused by BTKi might have therapeutic relevance in MS.

Ibrutinib is one of the earliest BTKi. It is a potent BTKi that binds BTK irreversibly and is approved for treatment of B-cell malignancies such as CLL or mantle cell lymphoma. Its adverse side effects such as cardiac arrhythmias, diarrhoea or arthralgias [85] that were due to off-target inhibition of other kinases, impeded it from being used for autoimmune diseases.

More specific but still binding BTKi irreversibly such as tolebrutinib, evobrutinib and remibrutinib are currently being tested for the treatment of MS. In phase II studies, tolebrutinib and evobrutinib have shown efficacy in RRMS [86]. There are currently at least six phase III randomized double-blind clinical trials for MS, in which these two BTKi are being evaluated. The results of these studies are expected over the next years [87].

Fenebrutinib binds BTK in a reversible way, inducing a conformational change in BTK that precludes activation of downstream targets. Fenebrutinib underwent phase II clinical trials for SLE, RA and MS [88]. There are currently at least three clinical trials phase III for MS evaluating fenebrutinib, including PPMS where results are expected between 2024 and 2028 [87].

3.5 Belimumab (Anti-BAFF)

Belimumab is a human IgG1 recombinant monoclonal antibody that blocks BAFF (B-cell activating factor) from binding to its ligands at the surface of B-linage cells. It is the only successful new therapy for SLE in more than 50 years [89] and interferes with B cells and PC survival [90]. The introduction of belimumab in the treatment of SLE provided proof that B cell-directed therapy has value in this disease [91], including clear documentation of damage prevention [92]. Interestingly, belimumab failed to be effective in myasthenia gravis [93], a disease in which, and like in SLE, pathogenic autoantibodies mediate tissue damage. Regarding the use of belimumab for the treatment of MS, a phase II, open-label trial involving patients with RRMS that combine subcutaneous belimumab with ocrelizumab (standard dose) is ongoing (NCT04767698).

3.6 Alemtuzumab (Anti-CD52)

Alemtuzumab is an anti-CD52 monoclonal antibody that induces long-term depletion of T and B cells as well as transient depletion of other leucocytes such as monocytes, eosinophils and neutrophils through ADCC and CDC-mediated mechanisms [94]. B cells can be found at the lower limit level of normal seven months post-alemtuzumab, while T cells normalise after ~3 years [95]. Alemtuzumab is approved for the treatment of active RRMS [96], and its efficacy has been attributed to the prolonged T-cell lymphopenia it induces [97]. Interestingly, increased rates of infections were not observed in patients treated with alemtuzumab [97]. The fact that T and B cells could be found in lymphoid organs after alemtuzumab has been presented as a possible explanation for the low incidence of infection in alemtuzumab-treated patients [94]. On the other hand, secondary autoimmunity has been a safety concern, with reports that 30% of the patients developed thyroid autoimmunity and 1% idiopathic thrombocytopenic purpura (ITP) [97], both diseases in which autoantibodies play a role.

One month after alemtuzumab, concomitantly with high levels of BAFF, the first B-lineage cells in circulation are mainly immature transitional-type 1 B cells that further differentiate into naïve B cells by 3 months after alemtuzumab. These cells are the majority after 12 months, while memory B cells developed rather slowly and are fewer than baseline levels at this point [98]. Nine months after alemtuzumab, a subset of regulatory B-lineage cells ($CD19^{+}CD24^{hi}CD38^{hi}$) has been reported to be increased compared with baseline levels [99], suggesting that these cells might contribute to the clinical improvement observed after alemtuzumab.

3.7 Autologous Hematopoietic Stem Cell Transplantation

Autologous hematopoietic stem cell transplantation (AHSCT) can be the selected one-time treatment option for highly active MS patients, who do not respond to standard therapies [100,101]. AHSCT can also be considered for specific cases of systemic sclerosis, and Crohn's disease as well as other rarer autoimmune diseases [102]. The rationale behind the use of this advanced immunotherapy that implies the ablation of the immune system with a high dose of immunosuppression prior to $CD34^{+}$ HSC reinfusion is the complete elimination of the pathogenic clones and reestablishment of immunotolerance. Regarding B-cell reconstitution after AHSCT in MS, one and 2 years after transplantation the majority of the circulating B cells are naïve [103]. In this trial, IgG and oligoclonal bands in the CSF were found to be reduced two years after AHSCT [104]. This indicates that upon reconstitution, naïve B cells are predominant and immune regulation is reestablished, including regulatory B cells. This is corroborated by one study in systemic sclerosis in which a higher number of regulatory B-lineage cells ($CD19^{+}CD24^{hi}CD38^{hi}$) producing IL-10 are found after AHSCT [105].

4. FUTURE APPROACHES AND OPPORTUNITIES

The results of phase III trials for the treatment of RRMS with ublituximab, a second generation anti-CD20 monoclonal antibody, that has been glycoengineered to enhance the potency of ADCC, have been recently published. These trials indicate lower annualised relapse rates but similar risk of disability in the ublituximab arm compared to teriflunomide [106]. Further trials are necessary to corroborate these findings. A detailed analysis of the B-cell dynamics upon ublituximab is also missing but should hold promise to deepen our understanding of B cells in MS.

CD19 chimaeric antigenic receptor (CAR) T cells are a type of immunotherapy used in the treatment of B-cell cancers that do not respond to conventional therapy [107,108]. CAR T cells carry a T-cell receptor (TCR) that is genetically engineered to recognise CD19 on the surface of

B cells. These CAR T cells are then infused into the patient with the goal of eliminating B cells. After showing high efficiency in the treatment of murine lupus [109], this therapy is in the early stages of research, but with promising preliminary results for the treatment of refractory SLE [110,111]. Taking into account the role of B cells in the pathogenesis of MS, it is expected that researchers will start evaluating the safety and tolerability of CD19 CAR T cells in patients with RRMS or SPMS in phase I trials soon.

PCs as producers of autoantibodies, but also of proinflammatory cytokines and other toxic products, play an important role in autoimmune diseases. Daratumumab, a human monoclonal antibody that binds CD38 and depletes all PCs, was used to treat two patients with life-threatening SLE and resulted in the reduction of interferon type I, chronic inflammation and disease activity [112]. Even though there was a clear improvement for the patients, elimination of all PCs also depletes PCs subsets providing host immunity and regulation, which is not ideal. In the future, it would be of interest to consider the elimination of pathogenic PC subsets, while sparing the PC subsets providing host immunity and regulation. Recent research points towards the presence of subsets of PCs with different functions and that can be identified by high expression of certain surface molecules, such as LAG-3-expressing regulatory PCs [113] or a disease-specific $CD39^{++}CD326^{++}$ BMPC in murine SLE [10]. Bispecific antibodies have been approved to treat tumours and are in clinical trials for autoimmune diseases [114–116]. A bispecific antibody that contains two fragment-binding arms targeting two different antigens would provide opportunities for selective elimination of PCs based on two surface molecules, thereby improving the specificity of the treatment in comparison with mono-specific antibodies.

Another strategy to target only pathogenic PCs would be through the antigen specificity encoded by the antibody. This would be possible for diseases in which antigens driving autoimmunity are known such as AQP4 and MOG-specific cells in NMOSD. In proof-of-concept studies, PCs specific for ovalbumin (OVA) were selectively labelled and depleted in an in vivo mouse model using an OVA-anti-CD138 antibody conjugate. This resulted in the reduction of anti-OVA titres in serum [117,118].

A longitudinal study reporting that the risk of developing MS is 32 times higher after infection with EBV, therefore suggests that EBV can be a leading causa for MS [119], has gained substantial attention [120–124]. Of note, the idea that EBV infection might promote B-cell activation and contribute to the development of autoimmune diseases, including MS, is not new [125]. In this study, serum samples from a cohort of more than 10 million people who serve in the US military, including 955 diagnosed with MS during their period in the army, were analysed for EBV and CMV (as control) status at several time points before MS diagnosis. Serum levels of neurofilament light chain increased only after EBV seroconversion [119]. Although lacking clear delineation, EBV infections have been linked other autoimmune diseases in which B-cell activation was proven to play a crucial role, such as SLE, SS or RA [126]. Notably, anti-CD20 therapies deplete mainly circulating memory B cells, which are the site of latent EBV [68], thus the efficiency of these therapies could be related to the reduction of EBV. Thus, it is possible that antivirals can be used in the treatment of MS and other autoimmune diseases, possibly representing another paradigm shift in MS in the context of B-cell pathology.

REFERENCES

[1] Degenhardt A, Ramagopalan SV, Scalfari A, Ebers GC. Clinical prognostic factors in multiple sclerosis: a natural history review. Nat Rev Neurol 2009;5(12):672–82.

[2] Lublin FD, Reingold SC, Cohen JA, Cutter GR, Sorensen PS, Thompson AJ, et al. Defining the clinical course of multiple sclerosis: the 2013 revisions. Neurology 2014;83(3):278–86.

[3] Hauser SL, Cree BAC. Treatment of multiple sclerosis: a review. Am J Med 2020;133(12):1380–13890 e2.

[4] Montalban X, Hauser SL, Kappos L, Arnold DL, Bar-Or A, Comi G, et al. Ocrelizumab versus placebo in primary progressive multiple sclerosis. N Engl J Med 2017;376(3):209–20.

[5] Kappos L, Hartung HP, Freedman MS, Boyko A, Radu EW, Mikol DD, et al. Atacicept in multiple sclerosis (ATAMS): a randomised, placebo-controlled, double-blind, phase 2 trial. Lancet Neurol 2014;13(4):353–63.

[6] Ginzler EM, Wax S, Rajeswaran A, Copt S, Hillson J, Ramos E, et al. Atacicept in combination with MMF and corticosteroids in lupus nephritis: results of a prematurely terminated trial. Arthritis Res Ther 2012;14(1):R33.

[7] Isenberg D, Gordon C, Licu D, Copt S, Rossi CP, Wofsy D. Efficacy and safety of atacicept for prevention of flares in patients with moderate-to-severe systemic lupus erythematosus (SLE): 52-week data (APRIL-SLE randomised trial). Ann Rheum Dis 2015; 74(11):2006–15.

[8] Fan Y, Gao D, Zhang Z. Telitacicept, a novel humanized, recombinant TACI-Fc fusion protein, for the treatment of systemic lupus erythematosus. Drugs Today 2022;58(1):23–32.

[9] Wiedemann A, Lettau M, Wirries I, Jungmann A, Salhab A, Gasparoni G, et al. Human IgA-expressing bone marrow plasma cells characteristically upregulate programmed cell death protein-1 upon B cell receptor stimulation. Front Immunol 2020;11:628923.

[10] Dang VD, Mohr E, Szelinski F, Le TA, Ritter J, Hinnenthal T, et al. CD39 and CD326 are bona fide markers of murine and human plasma cells and identify a bone marrow specific plasma cell subpopulation in lupus. Front Immunol 2022;13:873217.

[11] Mei HE, Wirries I, Frolich D, Brisslert M, Giesecke C, Grun JR, et al. A unique population of IgG-expressing plasma cells lacking CD19 is enriched in human bone marrow. Blood 2015;125(11): 1739–48.

[12] Link H, Huang YM. Oligoclonal bands in multiple sclerosis cerebrospinal fluid: an update on methodology and clinical usefulness. J Neuroimmunol 2006;180(1–2):17–28.

[13] Deisenhammer F, Zetterberg H, Fitzner B, Zettl UK. The cerebrospinal fluid in multiple sclerosis. Front Immunol 2019;10:726.
[14] Thompson AJ, Banwell BL, Barkhof F, Carroll WM, Coetzee T, Comi G, et al. Diagnosis of multiple sclerosis: 2017 revisions of the McDonald criteria. Lancet Neurol 2018;17(2):162–73.
[15] Graner M, Pointon T, Manton S, Green M, Dennison K, Davis M, et al. Oligoclonal IgG antibodies in multiple sclerosis target patient-specific peptides. PLoS One 2020;15(2):e0228883.
[16] Pellkofer HL, Krumbholz M, Berthele A, Hemmer B, Gerdes LA, Havla J, et al. Long-term follow-up of patients with neuromyelitis optica after repeated therapy with rituximab. Neurology 2011;76(15):1310–5.
[17] Kim SH, Huh SY, Lee SJ, Joung A, Kim HJ. A 5-year follow-up of rituximab treatment in patients with neuromyelitis optica spectrum disorder. JAMA Neurol 2013;70(9):1110–7.
[18] Roll P, Dorner T, Tony HP. Anti-CD20 therapy in patients with rheumatoid arthritis: predictors of response and B cell subset regeneration after repeated treatment. Arthritis Rheum 2008;58(6):1566–75.
[19] Lazarus MN, Turner-Stokes T, Chavele KM, Isenberg DA, Ehrenstein MR. B-cell numbers and phenotype at clinical relapse following rituximab therapy differ in SLE patients according to anti-dsDNA antibody levels. Rheumatology 2012;51(7):1208–15.
[20] Zhang C, Zhang M, Qiu W, Ma H, Zhang X, Zhu Z, et al. Safety and efficacy of tocilizumab versus azathioprine in highly relapsing neuromyelitis optica spectrum disorder (TANGO): an open-label, multicentre, randomised, phase 2 trial. Lancet Neurol 2020;19(5):391–401.
[21] Lotan I, McGowan R, Levy M. Anti-IL-6 therapies for neuromyelitis optica spectrum disorders: a systematic review of safety and efficacy. Curr Neuropharmacol 2021;19(2):220–32.
[22] Liu Y, Zhang H, Zhang TX, Yuan M, Du C, Zeng P, et al. Effects of tocilizumab therapy on circulating B cells and T helper cells in patients with neuromyelitis optica spectrum disorder. Front Immunol 2021;12:703931.
[23] Brynjolfsson SF, Persson Berg L, Olsen Ekerhult T, Rimkute I, Wick MJ, Martensson IL, et al. Long-lived plasma cells in mice and men. Front Immunol 2018;9:2673.
[24] Mahmood Z, Muhammad K, Schmalzing M, Roll P, Dorner T, Tony HP. CD27-IgD- memory B cells are modulated by in vivo interleukin-6 receptor (IL-6R) blockade in rheumatoid arthritis. Arthritis Res Ther 2015;17(1):61.
[25] Fleischer S, Ries S, Shen P, Lheritier A, Cazals F, Burmester GR, et al. Anti-interleukin-6 signalling therapy rebalances the disrupted cytokine production of B cells from patients with active rheumatoid arthritis. Eur J Immunol 2018;48(1):194–203.
[26] Odendahl M, Jacobi A, Hansen A, Feist E, Hiepe F, Burmester GR, et al. Disturbed peripheral B lymphocyte homeostasis in systemic lupus erythematosus. J Immunol 2000;165(10):5970–9.
[27] Jacobi AM, Mei H, Hoyer BF, Mumtaz IM, Thiele K, Radbruch A, et al. HLA-DRhigh/CD27high plasmablasts indicate active disease in patients with systemic lupus erythematosus. Ann Rheum Dis 2010;69(1):305–8.
[28] Jacobi AM, Odendahl M, Reiter K, Bruns A, Burmester GR, Radbruch A, et al. Correlation between circulating CD27high plasma cells and disease activity in patients with systemic lupus erythematosus. Arthritis Rheum 2003;48(5):1332–42.
[29] Lino ACD T. B cells in SLE. Systemic lupus erythematosus: basic, applied and clinical aspects. 2nd ed. Elsevier Science; 2020. p. 131–8.
[30] Schrezenmeier E, Weissenberg SY, Stefanski AL, Szelinski F, Wiedemann A, Lino AC, et al. Postactivated B cells in systemic lupus erythematosus: update on translational aspects and therapeutic considerations. Curr Opin Rheumatol 2019;31(2):175–84.
[31] Szelinski F, Lino AC, Dorner T. B cells in systemic lupus erythematosus. Curr Opin Rheumatol 2022;34(2):125–32.
[32] Szelinski F, Stefanski AL, Schrezenmeier E, Rincon-Arevalo H, Wiedemann A, Reiter K, et al. Antigen-experienced CXCR5(-) CD19(low) B cells are plasmablast precursors expanded in SLE. Arthritis Rheumatol 2022;79.
[33] Yaniv G, Twig G, Shor DB, Furer A, Sherer Y, Mozes O, et al. A volcanic explosion of autoantibodies in systemic lupus erythematosus: a diversity of 180 different antibodies found in SLE patients. Autoimmun Rev 2015;14(1):75–9.
[34] Li R, Rezk A, Miyazaki Y, Hilgenberg E, Touil H, Shen P, et al. Proinflammatory GM-CSF-producing B cells in multiple sclerosis and B cell depletion therapy. Sci Transl Med 2015;7(310):310ra166.
[35] Barr TA, Shen P, Brown S, Lampropoulou V, Roch T, Lawrie S, et al. B cell depletion therapy ameliorates autoimmune disease through ablation of IL-6-producing B cells. J Exp Med 2012;209(5):1001–10.
[36] Duddy M, Niino M, Adatia F, Hebert S, Freedman M, Atkins H, et al. Distinct effector cytokine profiles of memory and naive human B cell subsets and implication in multiple sclerosis. J Immunol 2007;178(10):6092–9.
[37] Bar-Or A, Fawaz L, Fan B, Darlington PJ, Rieger A, Ghorayeb C, et al. Abnormal B-cell cytokine responses a trigger of T-cell-mediated disease in MS? Ann Neurol 2010;67(4):452–61.
[38] Lino AC, Dorner T, Bar-Or A, Fillatreau S. Cytokine-producing B cells: a translational view on their roles in human and mouse autoimmune diseases. Immunol Rev 2016;269(1):130–44.
[39] Piccio L, Naismith RT, Trinkaus K, Klein RS, Parks BJ, Lyons JA, et al. Changes in B- and T-lymphocyte and chemokine levels with rituximab treatment in multiple sclerosis. Arch Neurol 2010;67(6):707–14.
[40] Lisak RP, Benjamins JA, Nedelkoska L, Barger JL, Ragheb S, Fan B, et al. Secretory products of multiple sclerosis B cells are cytotoxic to oligodendroglia in vitro. J Neuroimmunol 2012;246(1–2):85–95.
[41] Lisak RP, Nedelkoska L, Benjamins JA, Schalk D, Bealmear B, Touil H, et al. B cells from patients with multiple sclerosis induce cell death via apoptosis in neurons in vitro. J Neuroimmunol 2017;309:88–99.
[42] Benjamins JA, Nedelkoska L, Touil H, Stemmer PM, Carruthers NJ, Jena BP, et al. Exosome-enriched fractions from MS B cells induce oligodendrocyte death. Neurol Neuroimmunol Neuroinflamm 2019;6(3):e550.
[43] Schmidt H, Williamson D, Ashley-Koch A. HLA-DR15 haplotype and multiple sclerosis: a HuGE review. Am J Epidemiol 2007;165(10):1097–109.
[44] Wang J, Jelcic I, Muhlenbruch L, Haunerdinger V, Toussaint NC, Zhao Y, et al. HLA-DR15 molecules jointly shape an autoreactive T cell repertoire in multiple sclerosis. Cell 2020;183(5):1264–1281 e20.

[45] Jelcic I, Al Nimer F, Wang J, Lentsch V, Planas R, Jelcic I, et al. Memory B cells activate brain-homing, autoreactive CD4(+) T cells in multiple sclerosis. Cell 2018;175(1):85–100 e23.

[46] Magliozzi R, Howell O, Vora A, Serafini B, Nicholas R, Puopolo M, et al. Meningeal B-cell follicles in secondary progressive multiple sclerosis associate with early onset of disease and severe cortical pathology. Brain 2007;130(Pt 4):1089–104.

[47] Aloisi F, Pujol-Borrell R. Lymphoid neogenesis in chronic inflammatory diseases. Nat Rev Immunol 2006;6(3):205–17.

[48] Ramesh A, Schubert RD, Greenfield AL, Dandekar R, Loudermilk R, Sabatino Jr JJ, et al. A pathogenic and clonally expanded B cell transcriptome in active multiple sclerosis. Proc Natl Acad Sci U S A 2020;117(37):22932–43.

[49] Schafflick D, Xu CA, Hartlehnert M, Cole M, Schulte-Mecklenbeck A, Lautwein T, et al. Integrated single cell analysis of blood and cerebrospinal fluid leukocytes in multiple sclerosis. Nat Commun 2020;11(1):247.

[50] Magliozzi R, Howell OW, Nicholas R, Cruciani C, Castellaro M, Romualdi C, et al. Inflammatory intrathecal profiles and cortical damage in multiple sclerosis. Ann Neurol 2018;83(4):739–55.

[51] Salomonsson S, Jonsson MV, Skarstein K, Brokstad KA, Hjelmstrom P, Wahren-Herlenius M, et al. Cellular basis of ectopic germinal center formation and autoantibody production in the target organ of patients with Sjogren's syndrome. Arthritis Rheum 2003;48(11):3187–201.

[52] Amft N, Curnow SJ, Scheel-Toellner D, Devadas A, Oates J, Crocker J, et al. Ectopic expression of the B cell-attracting chemokine BCA-1 (CXCL13) on endothelial cells and within lymphoid follicles contributes to the establishment of germinal center-like structures in Sjogren's syndrome. Arthritis Rheum 2001;44(11):2633–41.

[53] Manzo A, Paoletti S, Carulli M, Blades MC, Barone F, Yanni G, et al. Systematic microanatomical analysis of CXCL13 and CCL21 in situ production and progressive lymphoid organization in rheumatoid synovitis. Eur J Immunol 2005;35(5):1347–59.

[54] Takemura S, Braun A, Crowson C, Kurtin PJ, Cofield RH, O'Fallon WM, et al. Lymphoid neogenesis in rheumatoid synovitis. J Immunol 2001;167(2):1072–80.

[55] Aust G, Sittig D, Becherer L, Anderegg U, Schutz A, Lamesch P, et al. The role of CXCR5 and its ligand CXCL13 in the compartmentalization of lymphocytes in thyroids affected by autoimmune thyroid diseases. Eur J Endocrinol 2004;150(2):225–34.

[56] Dubey D, Kieseier BC, Hartung HP, Hemmer B, Miller-Little WA, Stuve O. Clinical management of multiple sclerosis and neuromyelitis optica with therapeutic monoclonal antibodies: approved therapies and emerging candidates. Expert Rev Clin Immunol 2015;11(1):93–108.

[57] Cencioni MT, Mattoscio M, Magliozzi R, Bar-Or A, Muraro PA. B cells in multiple sclerosis – from targeted depletion to immune reconstitution therapies. Nat Rev Neurol 2021;17(7): 399–414.

[58] Bar-Or A, Calabresi PA, Arnold D, Markowitz C, Shafer S, Kasper LH, et al. Rituximab in relapsing-remitting multiple sclerosis: a 72-week, open-label, phase I trial. Ann Neurol 2008;63(3):395–400.

[59] Hauser SL, Waubant E, Arnold DL, Vollmer T, Antel J, Fox RJ, et al. B-cell depletion with rituximab in relapsing-remitting multiple sclerosis. N Engl J Med 2008;358(7):676–88.

[60] Quan C, ZhangBao J, Lu J, Zhao C, Cai T, Wang B, et al. The immune balance between memory and regulatory B cells in NMO and the changes of the balance after methylprednisolone or rituximab therapy. J Neuroimmunol 2015;282:45–53.

[61] Maurer MA, Rakocevic G, Leung CS, Quast I, Lukacisin M, Goebels N, et al. Rituximab induces sustained reduction of pathogenic B cells in patients with peripheral nervous system autoimmunity. J Clin Invest 2012;122(4):1393–402.

[62] van de Veerdonk FL, Lauwerys B, Marijnissen RJ, Timmermans K, Di Padova F, Koenders MI, et al. The anti-CD20 antibody rituximab reduces the Th17 cell response. Arthritis Rheum 2011;63(6):1507–16.

[63] Cross AH, Stark JL, Lauber J, Ramsbottom MJ, Lyons JA. Rituximab reduces B cells and T cells in cerebrospinal fluid of multiple sclerosis patients. J Neuroimmunol 2006;180(1–2):63–70.

[64] Topping J, Dobson R, Lapin S, Maslyanskiy A, Kropshofer H, Leppert D, et al. The effects of intrathecal rituximab on biomarkers in multiple sclerosis. Mult Scler Relat Disord 2016;6:49–53.

[65] Beers SA, Chan CH, French RR, Cragg MS, Glennie MJ. CD20 as a target for therapeutic type I and II monoclonal antibodies. Semin Hematol 2010;47(2):107–14.

[66] Klein C, Lammens A, Schafer W, Georges G, Schwaiger M, Mossner E, et al. Epitope interactions of monoclonal antibodies targeting CD20 and their relationship to functional properties. mAbs 2013;5(1):22–33.

[67] Wierda WG, Padmanabhan S, Chan GW, Gupta IV, Lisby S, Osterborg A, et al. Ofatumumab is active in patients with fludarabine-refractory CLL irrespective of prior rituximab: results from the phase 2 international study. Blood 2011;118(19):5126–9.

[68] Hauser SL, Bar-Or A, Comi G, Giovannoni G, Hartung HP, Hemmer B, et al. Ocrelizumab versus interferon beta-1a in relapsing multiple sclerosis. N Engl J Med 2017;376(3):221–34.

[69] Nguyen AL, Gresle M, Marshall T, Butzkueven H, Field J. Monoclonal antibodies in the treatment of multiple sclerosis: emergence of B-cell-targeted therapies. Br J Pharmacol 2017;174(13):1895–907.

[70] Gingele S, Jacobus TL, Konen FF, Hummert MW, Suhs KW, Schwenkenbecher P, et al. Ocrelizumab depletes CD20(+) T cells in multiple sclerosis patients. Cells 2018;8(1).

[71] Fernandez-Velasco JI, Kuhle J, Monreal E, Meca-Lallana V, Meca-Lallana J, Izquierdo G, et al. Effect of ocrelizumab in blood leukocytes of patients with primary progressive MS. Neurol Neuroimmunol Neuroinflamm 2021;8(2).

[72] Teeling JL, French RR, Cragg MS, van den Brakel J, Pluyter M, Huang H, et al. Characterization of new human CD20 monoclonal antibodies with potent cytolytic activity against non-Hodgkin lymphomas. Blood 2004;104(6):1793–800.

[73] Teeling JL, Mackus WJ, Wiegman LJ, van den Brakel JH, Beers SA, French RR, et al. The biological activity of human CD20 monoclonal antibodies is linked to unique epitopes on CD20. J Immunol 2006;177(1):362–71.

[74] Kaegi C, Wuest B, Crowley C, Boyman O. Systematic review of safety and efficacy of second- and third-generation CD20-targeting biologics in treating immune-mediated disorders. Front Immunol 2021;12:788830.

[75] Ostergaard M, Baslund B, Rigby W, Rojkovich B, Jorgensen C, Dawes PT, et al. Ofatumumab, a human anti-CD20 monoclonal antibody, for treatment of rheumatoid arthritis with an inadequate

response to one or more disease-modifying antirheumatic drugs: results of a randomized, double-blind, placebo-controlled, phase I/II study. Arthritis Rheum 2010;62(8):2227–38.

[76] Taylor PC, Quattrocchi E, Mallett S, Kurrasch R, Petersen J, Chang DJ. Ofatumumab, a fully human anti-CD20 monoclonal antibody, in biological-naive, rheumatoid arthritis patients with an inadequate response to methotrexate: a randomised, double-blind, placebo-controlled clinical trial. Ann Rheum Dis 2011; 70(12):2119–25.

[77] Sorensen PS, Lisby S, Grove R, Derosier F, Shackelford S, Havrdova E, et al. Safety and efficacy of ofatumumab in relapsing-remitting multiple sclerosis: a phase 2 study. Neurology 2014; 82(7):573–81.

[78] Bar-Or A, Grove RA, Austin DJ, Tolson JM, VanMeter SA, Lewis EW, et al. Subcutaneous ofatumumab in patients with relapsing-remitting multiple sclerosis: the MIRROR study. Neurology 2018;90(20):e1805–14.

[79] Hauser SL, Bar-Or A, Cohen JA, Comi G, Correale J, Coyle PK, et al. Ofatumumab versus teriflunomide in multiple sclerosis. N Engl J Med 2020;383(6):546–57.

[80] Corneth OBJ, Verstappen GMP, Paulissen SMJ, de Bruijn MJW, Rip J, Lukkes M, et al. Enhanced Bruton's tyrosine kinase activity in peripheral blood B lymphocytes from patients with autoimmune disease. Arthritis Rheumatol 2017;69(6):1313–24.

[81] Byrd JC, Furman RR, Coutre SE, Flinn IW, Burger JA, Blum KA, et al. Targeting BTK with ibrutinib in relapsed chronic lymphocytic leukemia. N Engl J Med 2013;369(1):32–42.

[82] Isenberg D, Furie R, Jones NS, Guibord P, Galanter J, Lee C, et al. Efficacy, safety, and pharmacodynamic effects of the Bruton's tyrosine kinase inhibitor fenebrutinib (GDC-0853) in systemic lupus erythematosus: results of a phase II, randomized, double-blind, placebo-controlled trial. Arthritis Rheumatol 2021; 73(10):1835–46.

[83] Owens TD, Smith PF, Redfern A, Xing Y, Shu J, Karr DE, et al. Phase 1 clinical trial evaluating safety, exposure and pharmacodynamics of BTK inhibitor tolebrutinib (PRN2246, SAR442168). Clin Transl Sci 2022;15(2):442–50.

[84] de Rooij MF, Kuil A, Geest CR, Eldering E, Chang BY, Buggy JJ, et al. The clinically active BTK inhibitor PCI-32765 targets B-cell receptor- and chemokine-controlled adhesion and migration in chronic lymphocytic leukemia. Blood 2012;119(11):2590–4.

[85] Estupinan HY, Berglof A, Zain R, Smith CIE. Comparative analysis of BTK inhibitors and mechanisms underlying adverse effects. Front Cell Dev Biol 2021;9:630942.

[86] Schneider R, Oh J. Bruton's tyrosine kinase inhibition in multiple sclerosis. Curr Neurol Neurosci Rep 2022;22(11):721–34.

[87] Garcia-Merino A. Bruton's tyrosine kinase inhibitors: a new generation of promising agents for multiple sclerosis therapy. Cells 2021;10(10).

[88] Crawford JJ, Johnson AR, Misner DL, Belmont LD, Castanedo G, Choy R, et al. Discovery of GDC-0853: a potent, selective, and noncovalent Bruton's tyrosine kinase inhibitor in early clinical development. J Med Chem 2018;61(6):2227–45.

[89] Navarra SV, Guzman RM, Gallacher AE, Hall S, Levy RA, Jimenez RE, et al. Efficacy and safety of belimumab in patients with active systemic lupus erythematosus: a randomised, placebo-controlled, phase 3 trial. Lancet 2011;377(9767):721–31.

[90] Dorner T, Lipsky PE. B cells: depletion or functional modulation in rheumatic diseases. Curr Opin Rheumatol 2014;26(2):228–36.

[91] Dorner T, Lipsky PE. Beyond pan-B-cell-directed therapy – new avenues and insights into the pathogenesis of SLE. Nat Rev Rheumatol 2016;12(11):645–57.

[92] Urowitz MB, Ohsfeldt RL, Wielage RC, Kelton KA, Asukai Y, Ramachandran S. Organ damage in patients treated with belimumab versus standard of care: a propensity score-matched comparative analysis. Ann Rheum Dis 2019;78(3):372–9.

[93] Hewett K, Sanders DB, Grove RA, Broderick CL, Rudo TJ, Bassiri A, et al. Randomized study of adjunctive belimumab in participants with generalized myasthenia gravis. Neurology 2018;90(16):e1425–34.

[94] Hu Y, Turner MJ, Shields J, Gale MS, Hutto E, Roberts BL, et al. Investigation of the mechanism of action of alemtuzumab in a human CD52 transgenic mouse model. Immunology 2009;128(2):260–70.

[95] Hill-Cawthorne GA, Button T, Tuohy O, Jones JL, May K, Somerfield J, et al. Long term lymphocyte reconstitution after alemtuzumab treatment of multiple sclerosis. J Neurol Neurosurg Psychiatry 2012;83(3):298–304.

[96] Investigators CT, Coles AJ, Compston DA, Selmaj KW, Lake SL, Moran S, et al. Alemtuzumab vs. interferon beta-1a in early multiple sclerosis. N Engl J Med 2008;359(17):1786–801.

[97] Jones JL, Coles AJ. Mode of action and clinical studies with alemtuzumab. Exp Neurol 2014;262 Pt:37–43. A.

[98] Thompson SA, Jones JL, Cox AL, Compston DA, Coles AJ. B-cell reconstitution and BAFF after alemtuzumab (Campath-1H) treatment of multiple sclerosis. J Clin Immunol 2010;30(1): 99–105.

[99] Kim Y, Kim G, Shin HJ, Hyun JW, Kim SH, Lee E, et al. Restoration of regulatory B cell deficiency following alemtuzumab therapy in patients with relapsing multiple sclerosis. J Neuroinflammation 2018;15(1):300.

[100] Muraro PA, Martin R, Mancardi GL, Nicholas R, Sormani MP, Saccardi R. Autologous haematopoietic stem cell transplantation for treatment of multiple sclerosis. Nat Rev Neurol 2017; 13(7):391–405.

[101] Burt RK, Han X, Quigley K, Helenowski IB, Balabanov R. Real-world application of autologous hematopoietic stem cell transplantation in 507 patients with multiple sclerosis. J Neurol 2022;269(5):2513–26.

[102] Alexander T, Greco R, Snowden JA. Hematopoietic stem cell transplantation for autoimmune disease. Annu Rev Med 2021;72:215–28.

[103] Karnell FG, Lin D, Motley S, Duhen T, Lim N, Campbell DJ, et al. Reconstitution of immune cell populations in multiple sclerosis patients after autologous stem cell transplantation. Clin Exp Immunol 2017;189(3):268–78.

[104] Nash RA, Hutton GJ, Racke MK, Popat U, Devine SM, Steinmiller KC, et al. High-dose immunosuppressive therapy and autologous HCT for relapsing-remitting MS. Neurology 2017;88(9):842–52.

[105] Arruda LCM, Malmegrim KCR, Lima-Junior JR, Clave E, Dias JBE, Moraes DA, et al. Immune rebound associates with a favorable clinical response to autologous HSCT in systemic sclerosis patients. Blood Adv 2018;2(2):126–41.

[106] Steinman L, Fox E, Hartung HP, Alvarez E, Qian P, Wray S, et al. Ublituximab versus teriflunomide in relapsing multiple sclerosis. N Engl J Med 2022;387(8):704−14.
[107] Schuster SJ, Svoboda J, Chong EA, Nasta SD, Mato AR, Anak O, et al. Chimeric antigen receptor T cells in refractory B-cell lymphomas. N Engl J Med 2017;377(26):2545−54.
[108] Schuster SJ, Tam CS, Borchmann P, Worel N, McGuirk JP, Holte H, et al. Long-term clinical outcomes of tisagenlecleucel in patients with relapsed or refractory aggressive B-cell lymphomas (JULIET): a multicentre, open-label, single-arm, phase 2 study. Lancet Oncol 2021;22(10):1403−15.
[109] Kansal R, Richardson N, Neeli I, Khawaja S, Chamberlain D, Ghani M, et al. Sustained B cell depletion by CD19-targeted CAR T cells is a highly effective treatment for murine lupus. Sci Transl Med 2019;11(482).
[110] Mougiakakos D, Kronke G, Volkl S, Kretschmann S, Aigner M, Kharboutli S, et al. CD19-targeted CAR T cells in refractory systemic lupus erythematosus. N Engl J Med 2021;385(6):567−9.
[111] Mackensen A, Muller F, Mougiakakos D, Boltz S, Wilhelm A, Aigner M, et al. Anti-CD19 CAR T cell therapy for refractory systemic lupus erythematosus. Nat Med 2022;28(10):2124−32.
[112] Ostendorf L, Burns M, Durek P, Heinz GA, Heinrich F, Garantziotis P, et al. Targeting CD38 with daratumumab in refractory systemic lupus erythematosus. N Engl J Med 2020;383(12):1149−55.
[113] Lino AC, Dang VD, Lampropoulou V, Welle A, Joedicke J, Pohar J, et al. LAG-3 inhibitory receptor expression identifies immunosuppressive natural regulatory plasma cells. Immunity 2018;49(1):120−133 e9.
[114] Huang S, van Duijnhoven SMJ, Sijts A, van Elsas A. Bispecific antibodies targeting dual tumor-associated antigens in cancer therapy. J Cancer Res Clin Oncol 2020;146(12):3111−22.
[115] Dang VD, Stefanski AL, Lino AC, Dorner T. B- and plasma cell subsets in autoimmune diseases: translational perspectives. J Invest Dermatol 2021;142:811−22.
[116] Zhao Q. Bispecific antibodies for autoimmune and inflammatory diseases: clinical progress to date. BioDrugs 2020; 34(2):111−9.
[117] Taddeo A, Gerl V, Hoyer BF, Chang HD, Kohler S, Schaffert H, et al. Selection and depletion of plasma cells based on the specificity of the secreted antibody. Eur J Immunol 2015; 45(1):317−9.
[118] Cheng Q, Pelz A, Taddeo A, Khodadadi L, Klotsche J, Hoyer BF, et al. Selective depletion of plasma cells in vivo based on the specificity of their secreted antibodies. Eur J Immunol 2020; 50(2):284−91.
[119] Bjornevik K, Cortese M, Healy BC, Kuhle J, Mina MJ, Leng Y, et al. Longitudinal analysis reveals high prevalence of Epstein-Barr virus associated with multiple sclerosis. Science 2022;375(6578):296−301.
[120] Kuchroo VK, Weiner HL. How does Epstein-Barr virus trigger MS? Immunity 2022;55(3):390−2.
[121] Robinson WH, Steinman L. Epstein-Barr virus and multiple sclerosis. Science 2022;375(6578):264−5.
[122] Sollid LM. Epstein-Barr virus as a driver of multiple sclerosis. Sci Immunol 2022;7(70):eabo7799.
[123] Du Toit A. EBV linked to multiple sclerosis. Nat Rev Microbiol 2022;20(4):189.
[124] Soldan SS, Lieberman PM. Epstein-Barr virus and multiple sclerosis. Nat Rev Microbiol 2023;21(1):51−64.
[125] Salvetti M, Giovannoni G, Aloisi F. Epstein-Barr virus and multiple sclerosis. Curr Opin Neurol 2009;22(3):201−6.
[126] Houen G, Trier NH. Epstein-barr virus and systemic autoimmune diseases. Front Immunol 2020;11:587380.

Chapter 20

IMGT Immunoglobulin Repertoire Analysis and Antibody Humanization

Marie-Paule Lefranc

IMGT, The International ImMunoGeneTics Information System, Université Montpellier 2, CNRS, Institut de Génétique Humaine, Montpellier, France

1. INTRODUCTION

IMGT, the international ImMunoGeneTics information system (http://www.imgt.org), CNRS and Université Montpellier 2, is the global reference in immunogenetics and immunoinformatics. By its creation in 1989, IMGT marked the advent of immunoinformatics, which emerged at the interface between immunogenetics and bioinformatics. IMGT has been built on the IMGT-ONTOLOGY axioms and concepts, which bridged the gap between genes, sequences, and three-dimensional (3D) structures. IMGT comprises seven databases and 17 tools that use the IMGT standardized keywords (identification), labels (description), and nomenclature (classification); the IMGT unique numbering, and IMGT Colliers de Perles (numerotation). They provide a unique approach for immunoglobulin repertoire analysis from fishes to humans, and for antibody engineering and humanization. This is exemplified with IMGT/V-QUEST and IMGT/JunctionAnalysis, and their high-throughput version IMGT/HighV-QUEST for next-generation sequencing; IMGT/DomainGapAlign for amino acid sequences; IMGT/3Dstructure-DB for 3D structures, contact analysis, and paratope/epitope interactions; and the IMGT/mAb-DB interface with access to therapeutic antibody data.

2. IMGT AND THE BIRTH OF IMMUNOINFORMATICS

IMGT, the international ImMunoGeneTics information system (http://www.imgt.org) [1], was created in 1989 by Marie-Paule Lefranc at Montpellier, France (CNRS and Université Montpellier 2). The founding of IMGT marked the advent of immunoinformatics, a new science, which emerged at the interface between immunogenetics and bioinformatics. For the first time, immunoglobulin (IG) or antibody and T cell receptor (TR) variable (V), diversity (D), joining (J), and constant (C) genes were officially recognized as "genes" as well as the conventional genes [2–5]. This major breakthrough allowed genes and data of the complex and highly diversified adaptive immune responses to be managed in genomic databases and tools.

The adaptive immune response was acquired by jawed vertebrates (or *gnathostomata*) more than 450 million years ago and is found in all extant jawed vertebrate species from fishes to humans. It is characterized by a remarkable immune specificity and memory, which are properties of the B and T cells owing to an extreme diversity of their antigen receptors. The specific antigen receptors comprise the IG or antibodies of the B cells and plasmacytes [2], and the TR [3]. The IG recognize antigens in their native (unprocessed) form, whereas the TR recognize processed antigens, which are presented as peptides by the highly polymorphic major histocompatibility (MH, in humans HLA for human leukocyte antigens) proteins.

The potential antigen receptor repertoire of each individual is estimated to comprise about 2×10^{12} different IG and TR, and the limiting factor is only the number of B and T cells that an organism is genetically programmed to produce [2,3]. This huge diversity results from the complex molecular synthesis of the IG and TR chains and, more particularly, of their variable domains (V-DOMAIN) which, at their N-terminal end, recognize and bind the antigens [2,3]. The IG and TR synthesis includes several unique mechanisms that occur at the DNA level: combinatorial rearrangements of the V, D, and J genes that code the V-DOMAIN (the V-(D)-J being spliced to the C gene that encodes the C-REGION in the transcript); exonuclease trimming at the ends of the V, D, and J genes; and random addition of nucleotides by the terminal deoxynucleotidyl transferase (TdT) that creates the junctional N-diversity regions, and later during B cell differentiation,

Molecular Biology of B Cells. https://doi.org/10.1016/B978-0-323-95895-0.00019-2

for the IG, somatic hypermutations, and class or subclass switch [2,3].

IMGT manages the diversity and complexity of the IG and TR and the polymorphism of the MH of humans and other vertebrates. IMGT is also specialized in the other proteins of the immunoglobulin superfamily (IgSF) and MH superfamily (MhSF) and related proteins of the immune system (RPI) of vertebrates and invertebrates. IMGT provides a common access to standardized data from genome, proteome, genetics, and two-dimensional (2D) and three-dimensional (3D) structures. IMGT is the acknowledged high-quality integrated knowledge resource in immunogenetics for exploring immune functional genomics. IMGT comprises seven databases (for sequences, genes, and 3D structures) [6–11], 17 online tools [12–27], and more than 15,000 pages of web resources (e.g., IMGT Scientific chart, IMGT Repertoire, IMGT Education > Aide-mémoire [28], The IMGT Immunoinformatics page) [1]. IMGT is the global reference in immunogenetics and immunoinformatics [29–44]. Its standards have been endorsed by the World Health Organization–International Union of Immunological Societies (WHO-IUIS) Nomenclature Committee since 1995 (first IMGT online access at the ninth International Congress of Immunology, San Francisco, CA, USA) [45,46] and the WHO International Nonproprietary Names (INN) Programme [47,48]. The accuracy and the consistency of the IMGT data are based on IMGT-ONTOLOGY [49–51], the first, and so far unique, ontology for immunogenetics and immunoinformatics [49–68]. IMGT-ONTOLOGY manages the immunogenetics knowledge through diverse facets that rely on seven axioms: IDENTIFICATION, DESCRIPTION, CLASSIFICATION, NUMEROTATION, LOCALIZATION, ORIENTATION, and OBTENTION [50,51,55]. The concepts generated from these axioms led to the elaboration of the IMGT standards that constitute the IMGT Scientific chart: e.g., IMGT standardized keywords (IDENTIFICATION) [56], IMGT standardized labels (DESCRIPTION) [57], IMGT standardized gene and allele nomenclature (CLASSIFICATION) [58], IMGT unique numbering [59–64] and its standardized graphical 2D representation or IMGT Colliers de Perles [65–68] (NUMEROTATION).

This chapter first reviews the fundamental information generated from these IMGT-ONTOLOGY concepts, which led to the IMGT Scientific chart rules. It then describes the major IMGT tools and databases used for immunoglobulin repertoire analysis and antibody engineering and humanization: IMGT/V-QUEST [12–17] for the analysis of rearranged nucleotide sequence with the results of the integrated IMGT/JunctionAnalysis [18,19]; IMGT/Automat [20,21] and IMGT/Collier-de-Perles tool [26]; IMGT/HighV-QUEST, the high-throughput version for Next-Generation Sequencing (NGS) [22,23]; IMGT/DomainGapAlign [9,24,25] for amino acid sequence analysis; IMGT/3Dstructure-DB for 3D structures [8–10]; and its extension, IMGT/2Dstructure-DB (for antibodies and other proteins for which the 3D structure is not available). IMGT tools and databases run against IMGT reference directories built from sequences annotated in IMGT/LIGM-DB [6], the IMGT nucleotide database (170,685 sequences from 335 species in June 2013), and from IMGT/GENE-DB [7], the IMGT gene database (3107 genes and 4722 alleles from 17 species, of which there were 694 genes and 1420 alleles for *Homo sapiens* and 868 genes and 1318 alleles for *Mus musculus* in June 2013).

An interface, IMGT/mAb-DB [11], has been developed to provide an easy access to therapeutic antibody amino acid sequences (links to IMGT/2Dstructure-DB) and structures (links to IMGT/3Dstructure-DB, if 3D structures are available). IMGT/mAb-DB data include monoclonal antibodies (mAb, INN suffix –mab) (a –mab is defined by the presence of at least an IG variable domain) and fusion proteins for immune applications (FPIA, INN suffix –cept) (a –cept is defined by a receptor fused to a, Fc) from the WHO-INN program [47,48]. This database also includes a few composite proteins for clinical applications (CPCA) (e.g., protein or peptide fused to an Fc for only increasing their half-life, identified by the INN prefix ef–) and some RPI used, unmodified, for clinical applications.

The unified IMGT approach is of major interest for bridging knowledge from IG repertoire in normal and pathologic situations [69–72], IG allotypes and immunogenicity [73–75], NGS repertoire [22,23], antibody engineering, and humanization [32,39–41,43,76–80].

3. FUNDAMENTAL INFORMATION FROM IMGT-ONTOLOGY CONCEPTS

3.1 Identification: IMGT Standardized Keywords

More than 325 IMGT standardized keywords (189 for sequences and 137 for 3D structures) were precisely defined [56]. They represent the controlled vocabulary assigned during the annotation process and allow standardized search criteria for querying the IMGT databases and for the extraction of sequences and 3D structures. They have been entered in BioPortal at the National Center for Biomedical Ontology (NCBO) in 2010 (http://bioportal.bioontology.org/ontologies/1491).

Standardized keywords are assigned at each step of the molecular synthesis of an IG. Those assigned to a nucleotide sequence are found in the "DE" (definition) and "KW" (keyword) lines of the IMGT/LIGM-DB files [6]. They characterize, for instance, the gene type, the configuration type, and the functionality type [56]. There are six gene types: variable (V), diversity (D), joining (J), constant (C), conventional-with-leader, and conventional-without-leader. Four of them (V, D, J, and C) identify the IG and TR genes and are specific to immunogenetics. There are four configuration types: germline (for the V, D, and J genes before DNA rearrangement), rearranged (for the V, D, and J genes after DNA rearrangement), partially rearranged (for D gene after only one DNA rearrangement), and undefined (for the C gene and for the conventional genes, which do not rearrange). The functionality type depends on the gene configuration. The functionality type of genes in germline or undefined configuration is functional (F), ORF (for "open reading frame"), or pseudogene (P). The functionality type of genes in rearranged or partially-rearranged configuration is either productive (no stop codon in the V-(D)-J region and in-frame junction) or unproductive (stop codon(s) in the V-(D)-J region, and/or out-of-frame junction).

The 20 usual amino acids (AA) have been classified in 11 IMGT physicochemical classes (IMGT, http://www.imgt.org, IMGT Education > Aide-mémoire > Amino acids). The amino acid changes are described according to the hydropathy (3 classes), volume (5 classes), and IMGT physicochemical classes (11 classes) [28]. For example Q1 > E (+ + −) means that in the amino acid change (Q > E), the two amino acids at codon one belong to the same hydropathy (+) and volume (+) classes but to different IMGT physicochemical property (−) classes [28]. Four types of AA changes are identified in IMGT: very similar (+ + +), similar (+ + −, + − +), dissimilar (− − +, − + −, + − −), and very dissimilar (− − −).

3.2 Description: IMGT Standardized Labels

More than 560 IMGT standardized labels (277 for sequences and 285 for 3D structures) were precisely defined [57]. They are written in capital letters (no plural) to be recognizable without creating new terms. Standardized labels assigned to the description of sequences are found in the "FT" (feature) lines of the IMGT/LIGM-DB files [6]. Querying these labels represents a big plus compared to the generalist databases (GenBank/European Nucleotide Archive (ENA)/DNA Data Bank of Japan (DDBJ)). Thus it is possible to query for the "CDR3-IMGT" of the human rearranged productive sequences of IG-Heavy-Gamma (e.g., 1733 CDR3-IMGT obtained, with their sequences at the nucleotide or amino acid level).

The core labels include V-REGION, D-REGION, J-REGION, and C-REGION, which correspond to the coding region of the V, D, J, and C genes. IMGT structure labels for human IG chains and domains and their correspondence with sequence labels are shown in Table 20.1. These labels are necessary for a standardized description of the IG sequences and structures in databases and tools.

Highly conserved amino acids at a given position in a V or C domain (Section 3.4) have IMGT labels [57]. Three of them are common to the V and C domains: 1st-CYS (cysteine C at position 23), CONSERVED-TRP (tryptophan W at position 41), and 2nd-CYS (C at position 104) [59−62,64]. Two others are characteristics of the V-DOMAIN and correspond to the first amino acid of the canonical motif F/W-G-X-G (where F is phenylalanine, W tryptophan, G glycine, and X any amino acid) encoded by the J-REGION: J-PHE or J-TRP (F or W at position 118) [59−61,64].

3.3 Classification: IMGT Standardized Genes and Alleles

The IMGT-ONTOLOGY CLASSIFICATION axiom was the trigger of immunoinformatics' birth. Indeed, the IMGT concepts of classification allowed us, for the first time, to classify the antigen receptor genes (IG and TR) for any locus (e.g., immunoglobulin heavy (IGH), T cell receptor alpha (TRA)), for any gene configuration (germline, undefined, or rearranged), and for any species (from fishes to humans). In higher vertebrates, there are seven IG and TR major loci (other loci correspond to chromosomal orphon sets, genes of which are orphons, not used in the IG or TR chain synthesis). The IG major loci include the immunoglobulin heavy (IGH), and for the light chains, the immunoglobulin kappa (IGK) and the immunoglobulin lambda (IGL) in higher vertebrates and the immunoglobulin iota (IGI) in fishes (IMGT, http://www.imgt.org, IMGT Repertoire).

Since the creation of IMGT in 1989, at New Haven during the 10th Human Genome Mapping Workshop (HGM10), the standardized classification and nomenclature of the IG and TR of humans and other vertebrate species have been under the responsibility of the IMGT Nomenclature Committee (IMGT-NC).

IMGT gene and allele names are based on the concepts of classification of "Group," "Subgroup," "Gene," and "Allele" [58]. "Group" allows classification of a set of genes that belong to the same multigene family, within the same species or between different species. For example, there are 10 groups for the IG of higher vertebrates: IGHV, IGHD, IGHJ, IGHC, IGKV, IGKJ, IGKC, IGLV, IGLJ, and IGLC. "Subgroup" allows identification of a subset of genes that belong to the same group and that, in a given species, share at least 75% identity at the nucleotide level, e.g., *Homo sapiens* IGHV1 subgroup. Subgroups, genes, and alleles are always associated to a species name. An allele is a polymorphic variant of a gene, which is characterized by the mutations of its sequence at the nucleotide level, identified in its core sequence, and compared to the gene allele reference sequence, designated as allele *01. For example, *Homo sapiens* IGHV1-2*01 is the allele *01

TABLE 20.1 **Immunoglobulin Receptor, Chain, and Domain Structure Labels and Correspondence with Sequence Labels**

IG structure labels (IMGT/3DStructure-DB)				Sequence labels (IMGT/LIGM-DB)
Receptor[a]	Chain[b]	Domain description type	Domain[c]	Region
IG-GAMMA-1_KAPPA	L-KAPPA	V-DOMAIN	V-KAPPA	V-J-REGION
		C-DOMAIN	C-KAPPA	C-REGION
	H-GAMMA-1	V-DOMAIN	VH	V-D-J-REGION
		C-DOMAIN	CH1	C-REGION[d]
		C-DOMAIN	CH2	
		C-DOMAIN	CH3	
IG-MU_LAMBDA	L-LAMBDA	V-DOMAIN	V-LAMBDA	V-J-REGION
		C-DOMAIN	C-LAMBDA-1	C-REGION
	H-MU	V-DOMAIN	VH	V-D-J-REGION
		C-DOMAIN	CH1	C-REGION[d]
		C-DOMAIN	CH2	
		C-DOMAIN	CH3	
		C-DOMAIN	CH4[e]	

[a]*Labels are shown for two examples of IG (*Homo sapiens *IgG1-kappa and IgM-lambda). An IG ("Receptor") is made of two identical heavy (H, for IG-HEAVY) chains and two identical light (L, for IG-LIGHT) chains ("Chain") and usually comprises 12 (e.g., IgG1) or 14 (e.g., IgM) domains. Each chain has an N-terminal V-DOMAIN (or V-(D)-J-REGION, encoded by the rearranged V-(D)-J genes), whereas the remaining of the chain is the C-REGION (encoded by a C gene). The IG C-REGION comprises one C-DOMAIN (C-KAPPA or C-LAMBDA) for the L chain, or several C-DOMAIN (CH) for the H chain [2].*
[b]*The kappa (L-KAPPA) or lambda (L-LAMBDA) light chains may associate to any heavy chain isotype (e.g., H-GAMMA-1, H-MU). In humans, there are nine isotypes, H-MU, H-DELTA, H-GAMMA-3, H-GAMMA-1, H-ALPHA1, H-GAMMA2, H-GAMMA-4, H-EPSILON, and H-ALPHA2 (listed in the order 5′–3′ in the IGH locus of the IGHC genes that encode the constant region of the heavy chains [2]) (IMGT http://www.imgt.org, IMGT Repertoire).*
[c]*The IG V-DOMAIN includes VH (for the IG heavy chain) and VL (for the IG light chain). In higher vertebrates, the VL is V-KAPPA or V-LAMBDA, whereas in fishes, the VL is V-IOTA. The C-DOMAIN includes CH (for the IG heavy chain, the number of CH per chain depending on the isotype [2]) and CL (for the IG light chain). In higher vertebrates, the CL is C-KAPPA or C-LAMBDA, whereas in fishes, the CL is C-IOTA.*
[d]*The heavy chain C-REGION also includes the HINGE-REGION for the H-ALPHA, H-DELTA, and H-GAMMA chains and, for membrane IG (mIG), the CONNECTING-REGION (CO), TRANSMEMBRANE-REGION (TM), and CYTOPLASMIC-REGION (CY); for secreted IG (sIG), the C-REGION includes CHS instead of CO, TM, and CY.*
[e]*For H-MU and H-EPSILON.*

of the *Homo sapiens* IGHV1-2 gene that belongs to the *Homo sapiens* IGHV1 subgroup, which itself belongs to the IGHV group. For the IGH locus, the constant genes are designated by the letter (and eventually number) corresponding to the encoded isotypes (IGHM, IGHD, IGHG3 …), instead of using the letter C. IG and TR genes and alleles are not italicized in publications. IMGT-ONTOLOGY concepts of classification have been entered in the NCBO BioPortal.

The IMGT IG and TR gene names [2–5] were approved by the Human Genome Organisation (HUGO) Nomenclature Committee (HGNC) in 1999 [81,82] and were endorsed by the WHO-IUIS Nomenclature Subcommittee for IG and TR [45,46]. The IMGT IG and TR gene names are the official international reference and, as such, have been entered in IMGT/GENE-DB [7], in the Genome Database (GDB) [83], in LocusLink at the National Center for Biotechnology Information (NCBI) USA [84], in Entrez Gene (NCBI) when this database (now designated as "Gene") superseded LocusLink [85], in NCBI MapViewer, in Ensembl at the European Bioinformatics Institute (EBI) [86], and in the Vertebrate Genome Annotation (Vega) Browser [87] at the Wellcome Trust Sanger Institute (UK). HGNC, Gene NCBI, Ensembl, and Vega have direct links to IMGT/GENE-DB [7]. IMGT human IG and TR genes were also integrated in IMGT-ONTOLOGY on the NCBO BioPortal and, on the same site, in the HUGO ontology and in the National Cancer Institute (NCI) Metathesaurus. Amino acid sequences of human IG and TR constant genes (e.g., *Homo sapiens* IGHM, IGHG1, IGHG2) were provided to UniProt in 2008. Since 2007, IMGT gene and allele names have been used for the description of the therapeutic mAb and FPIA of the WHO-INN program [47,48].

3.4 NUMEROTATION: IMGT Unique Numbering and IMGT Collier de Perles

The IMGT-ONTOLOGY NUMEROTATION axiom is acknowledged as the "IMGT Rosetta stone" that has bridged the biological and computational spheres in bioinformatics [37]. The IMGT concepts of numerotation comprise the IMGT unique numbering [59–64] and its graphical 2D representation, the IMGT Collier de Perles [65–68]. Developed for and by the "domain," these concepts integrate sequences, structures, and interactions into a standardized domain-centric knowledge for functional genomics. The IMGT unique numbering has been defined for the variable V domain (V-DOMAIN of the IG and TR, and V-LIKE-DOMAIN of IgSF other than IG and TR) [59–61], the constant C domain (C-DOMAIN of the IG and TR, and C-LIKE-DOMAIN of IgSF other than IG and TR) [62], and for information, not as the focus of that chapter, the groove G domain (G-DOMAIN of the MH, and G-LIKE-DOMAIN of MhSF other than MH) [63,88,89]. Thus the IMGT unique numbering and IMGT Collier de Perles provide a definitive and universal system across species, including invertebrates, for the sequences and structures of the V, C, and G domains of IG, TR, MH, IgSF, and MhSF [64,68,90,91].

3.4.1 V Domain IMGT Definitive System

3.4.1.1 V Domain Definition and Main Characteristics

In the IMGT definitive system, the V domain includes the V-DOMAIN of the IG and of the TR, which corresponds to the V-J-REGION or V-D-J-REGION encoded by V-(D)-J rearrangements [2,3], and the V-LIKE-DOMAIN of the IgSF other than IG and TR. The V domain description of any receptor, any chain, and any species is based on the IMGT unique numbering for V domain (V-DOMAIN and V-LIKE-DOMAIN) [59–61,64].

A V domain (Figure 20.1) comprises about 100 amino acids and is made of nine antiparallel beta strands (A, B, C, C′, C″, D, E, F, and G) linked by beta turns (AB, CC′, C″D, DE, and EF) and three loops (BC, C′C″, and FG), forming a sandwich of two sheets [ABED] [GFCC′C″] [59–61,64]. The sheets are closely packed against each other through hydrophobic interactions, giving a hydrophobic core, and joined together by a disulfide bridge between a first highly conserved cysteine (1st-CYS) in the B strand (in the first sheet) and a second, equally conserved cysteine (2nd-CYS) in the F strand (in the second sheet) [59–61,64].

3.4.1.2 V Domain Strands and Loops (FR-IMGT and CDR-IMGT)

The V domain strands and loops and their delimitations and lengths, based on the IMGT unique numbering for V domain [59–61,64], are shown in Table 20.2. In the IG and TR V-DOMAIN, the three hypervariable loops BC, C′C″, and FG involved in the ligand recognition (native antigen for IG and pMH for TR) are designated complementarity determining regions (CDR-IMGT), whereas the strands form the framework region (FR-IMGT), which includes FR1-IMGT, FR2-IMGT, FR3-IMGT, and FR4-IMGT (Table 20.2). In the IMGT definitive system, the CDR-IMGT have accurate and unambiguous delimitations, in contrast to the CDR described in the literature. Correspondences between the IMGT unique numbering with other numberings, e.g., Kabat [92] or Chothia [93], are available in the IMGT Scientific chart. The correspondences with these previous and heterogenous numberings are useful for the interpretation of previously published data, but nowadays the usage of these numberings has become obsolete in regard to the development of immunoinformatics based on the IMGT standards [59–68] (IMGT http://www.imgt.org, IMGT Scientific chart > Numbering > Correspondence between V numberings).

For a V domain, the BC loop (or CDR1-IMGT in a V-DOMAIN) encompasses positions 27–38, the C′C″ loop (or CDR2-IMGT in a V-DOMAIN) positions 56–65, and the FG loop (or CDR3-IMGT) positions 105–117. In a V-DOMAIN, the CDR3-IMGT encompasses the V-(D)-J junction that results from a V-J or V-D-J rearrangement [2,3] and is more variable in sequence and length than the CDR1-IMGT and CDR2-IMGT that are encoded by the V gene region only. For CDR3-IMGT of length >13 AA, additional IMGT positions are added at the top of the loop between 111 and 112 (Table 20.3).

3.4.1.3 IMGT Colliers de Perles

The loop and strands are visualized in the IMGT Colliers de Perles [65–68], which can be displayed on one layer (closer to the amino acid sequence) or on two layers (closer to the 3D structure) (Figure 20.1). The three loops, BC, C′C″, and FG (or CDR1-IMGT, CDR2-IMGT, and CDR3-IMGT for a V-DOMAIN) are delimited by the IMGT anchors, which are shown in squares in the IMGT Colliers de Perles. IMGT anchors are positions that belong to strands and represent anchors for the loops of the V domains. IMGT anchors are the key and original concept of IMGT that definitively solved the ambiguous situation of different CDR lengths and delimitations found in the literature. The six anchors of a V domain are positions 26 and 39 (anchors of the BC loop or CDR1-IMGT in V-DOMAIN), 55 and 66 (anchors of the C′C″ loop or CDR2-IMGT in V-DOMAIN), and 104 and 118 (anchors of the FG loop or CDR3-IMGT in V-DOMAIN). The CDR3-IMGT anchors are highly conserved; they are C104 (second-CYS, in F strand) and F118 or W118 (J-PHE or J-TRP, in G strand). The JUNCTION of an IG or TR V-DOMAIN includes the

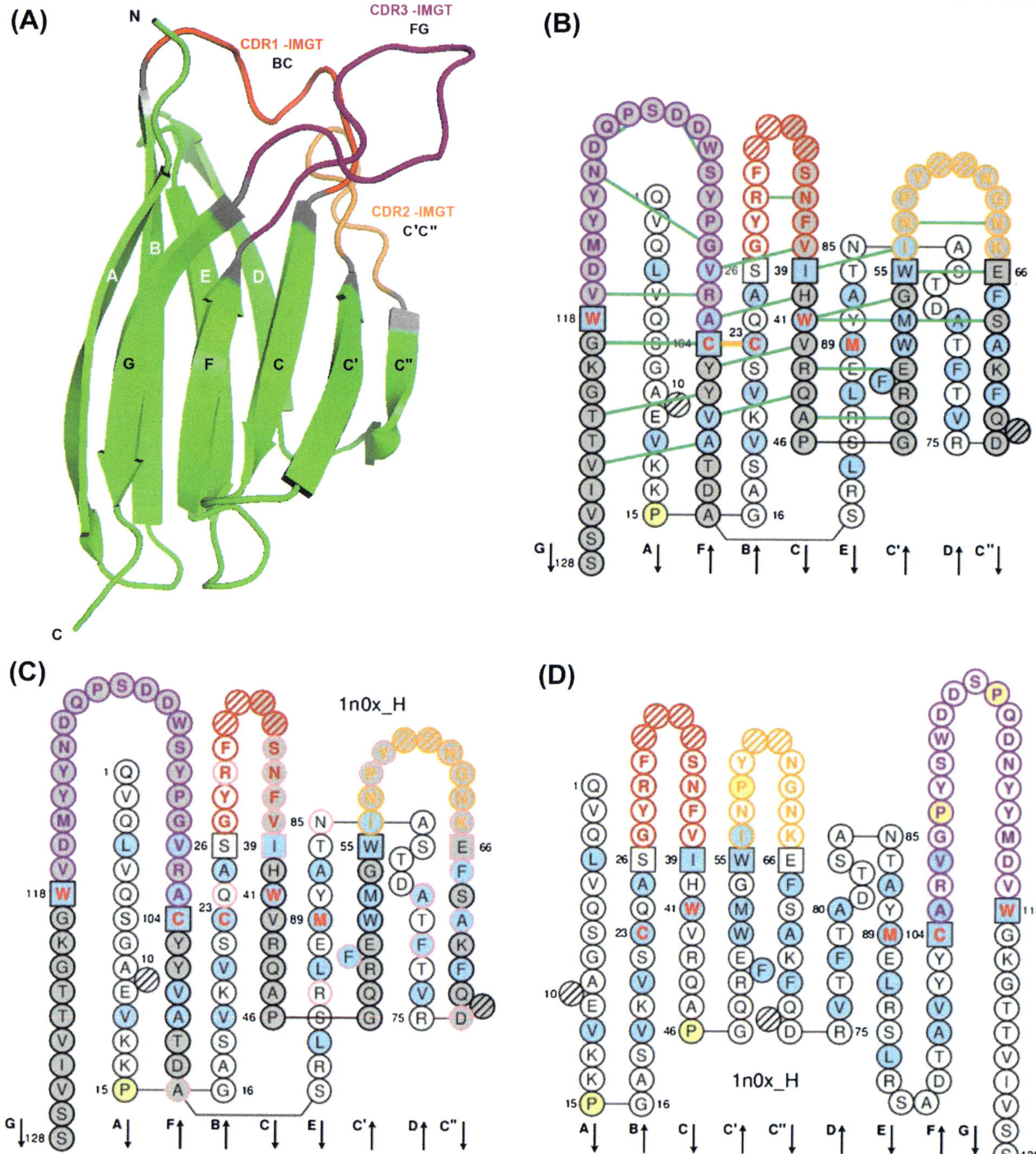

FIGURE 20.1 **Variable (V) domain. An IG VH (V-DOMAIN) is shown as example.** (A) 3D structure ribbon representation with the IMGT strand and loop delimitations [61]. (B) IMGT Collier de Perles on two layers with hydrogen bonds. The IMGT Collier de Perles on two layers shows, in the forefront, the GFCC′C″ strands (forming the sheet located at the interface VH/VL of the IG) and, in the back, the ABED strands. The IMGT Collier de Perles with hydrogen bonds (green lines online, only shown here for the GFCC′C″ sheet) is generated by the IMGT/Collier-de-Perles tool integrated in IMGT/3Dstructure-DB, from experimental 3D structure data [8–10]. (C) IMGT Collier de Perles on two layers generated from IMGT/DomainGapAlign [9,24,25]. Pink circles (online) indicate amino acid changes compared to the closest genes and alleles from the IMGT reference directory. (D) IMGT Collier de Perles on one layer. Amino acids are shown in the one-letter abbreviation. All proline (P) are shown online in yellow. IMGT anchors are in squares. Hatched circles are IMGT gaps according to the IMGT unique numbering for V domain [61,64]. Positions with bold (online red) letters indicate the four conserved positions that are common to a V domain and to a C domain: 23 (1st-CYS), 41 (CONSERVED-TRP), 89 (hydrophobic), 104 (2nd-CYS) [59–62,64], and the fifth conserved position, 118 (J-TRP or J-PHE), which is specific to a V-DOMAIN and belongs to the motif F/W-G-X-G that characterizes the J-REGION [61,64] (Table 20.2). The hydrophobic amino acids (hydropathy index with positive value: I, V, L, F, C, M, A) and tryptophan (W) [28] found at a given position in more than 50% of sequences are shown (online with a blue background color). Arrows indicate the direction of the beta strands and their designations in 3D structures. IMGT color menu for the CDR-IMGT of a V-DOMAIN indicates the type of rearrangement, V-D-J (for a VH here, red, orange, and purple) or V-J (for V-KAPPA or V-LAMBDA (not shown), blue, green, and green-blue) [2]. The identifier of the chain to which the VH domain belongs is 1n0x_H (from the *Homo sapiens* b12 Fab) in IMGT/3Dstructure-DB (http://www.imgt.org). The CDR-IMGT lengths of this VH are [8.8.20] and the FR-IMGT are [25.17.38.11]. The 3D ribbon representation was obtained using PyMOL (http://www.pymol.org) and "IMGT numbering comparison" of 1n0x_H (VH) from IMGT/3Dstructure-DB (http://www.imgt.org).

TABLE 20.2 V Domain Strands and Loops, IMGT Positions, and Lengths, Based on the IMGT Unique Numbering for V Domain (V-DOMAIN and V-LIKE-DOMAIN)

V domain strands and loops[a]	IMGT position[b]	Lengths[c]	Characteristic IMGT residue@ position[d]	V-DOMAIN FR-IMGT and CDR-IMGT
A-STRAND	1–15	15 (14 if gap at 10)		FR1-IMGT
B-STRAND	16–26	11	1st-CYS 23	
BC-LOOP	27–38	12 (or less)		CDR1-IMGT
C-STRAND	39–46	8	CONSERVED-TRP 41	FR2-IMGT
C′-STRAND	47–55	9		
C′C″-LOOP	56–65	10 (or less)		CDR2-IMGT
C″-STRAND	66–74	9 (or 8 if gap at 73)		FR3-IMGT
D-STRAND	75–84	10 (or 8 if gaps at 81, 82)		
E-STRAND	85–96	12	Hydrophobic 89	
F-STRAND	97–104	8	2nd-CYS 104	
FG-LOOP	105–117	13 (or less, or more)		CDR3-IMGT
G-STRAND	118–128	11 (or 10)	V-DOMAIN J-PHE 118 or J-TRP 118[e]	FR4-IMGT

[a]*IMGT labels (concepts of description) are written in capital letters (no plural) [57]. Beta turns (AB, CC′, C″D, DE, or EF) are individualized only if they have additional amino acids compared to the standard description. If not, they are included in the strands.*
[b]*Based on the IMGT unique numbering for V domain (V-DOMAIN and V-LIKE-DOMAIN) [59–62,64].*
[c]*In number of amino acids (or codons).*
[d]*IMGT Residue@Position is a given residue (usually an amino acid) or a given conserved property amino acid class, at a given position in a domain, based on the IMGT unique numbering [64].*
[e]*In the IG and TR V-DOMAIN, the G-STRAND (or FR4-IMGT) is the C-terminal part of the J-REGION, with J-PHE or J-TRP 118 and the canonical motif F/W-G-X-G at positions 118–121 [2,3]. The JUNCTION refers to the CDR3-IMGT plus the two anchors 2nd-CYS 104 and J-PHE or J-TRP 118 [60,61]. The JUNCTION (positions 104–118) is therefore two amino acids longer than the corresponding CDR3-IMGT (positions 105–117) [60,61].*

TABLE 20.3 IMGT Additional Positions for CDR3-IMGT

CDR3-IMGT lengths	IMGT additional positions for CDR3-IMGT length >13 AA[a]									
21	111	111.1	111.2	111.3	111.4	112.4	112.3	112.2	112.1	112
20	111	111.1	111.2	111.3	–	112.4	112.3	112.2	112.1	112
19	111	111.1	111.2	111.3	–	–	112.3	112.2	112.1	112
18	111	111.1	111.2	–	–	–	112.3	112.2	112.1	112
17	111	111.1	111.2	–	–	–	–	112.2	112.1	112
16	111	111.1	–	–	–	–	–	112.2	112.1	112
15	111	111.1	–	–	–	–	–	–	112.1	112
14	111	–	–	–	–	–	–	–	112.1	112

[a]*For CDR3-IMGT length >13 AA, IMGT additional positions are created between positions 111 and 112 at the top of the CDR3-IMGT loop in the following order: 112.1, 111.1, 112.2, 111.2, 112.3, 111.3, etc. For CDR3-IMGT length <13 AA (not show n), IMGT gaps are created classically from the top of the loop, in the following order: 111, 112, 110, 113, 109, 114, etc. (IMGT, http://www.imgt.org, IMGT Scientific chart > Numbering).*

anchors 104 and 118 and is therefore two amino acids longer than the corresponding CDR3-IMGT (positions 105–117).

In biological data, the lengths of the loops and strands are given by the number of occupied positions (unoccupied positions or "IMGT gaps" are represented with hatches in the IMGT Colliers de Perles (Figure 20.1) or by dots in alignments). The CDR-IMGT lengths are given in number of amino acids (or codons), into brackets and separated by dots: for example [9.6.9] means that the BC, C′C″, and FG loops (or CDR1-IMGT, CDR2-IMGT, and CDR3-IMGT for a V-DOMAIN) have a length of 9, 6, and 9 amino acids (or codons), respectively. Similarly [25.17.38.11] means that the FR1-IMGT, FR2-IMGT, FR3-IMGT, and FR4-IMGT have a length of 25, 17, 38, and 11 amino acids (or codons), respectively. Together, the four FR of a VH domain usually comprise 91 amino acids and the individual FR-IMGT lengths are [25.17.38.11], whereas the four FR of a VL domain usually comprise 89 amino acids and the individual FR-IMGT lengths are [26.17.36.10].

3.4.1.4 Conserved Amino Acids

A V domain has five characteristic amino acids at given positions (positions with bold (online red) letters in the IMGT Colliers de Perles). Four of them are highly conserved and hydrophobic [28] and are common to the C domain: 23 (1st-CYS), 41 (CONSERVED-TRP), 89 (hydrophobic), and 104 (2nd-CYS). These amino acids contribute to the two major features shared by the V and C domain: the disulfide bridge (between the two cysteines 23 and 104) and the internal hydrophobic core of the domain (with the side chains of tryptophan W41 and amino acid 89). The fifth position, 118, is an anchor of the FG loop. It is occupied, in the V domains of IgSF other than IG or TR, by amino acids with diverse physicochemical properties [28]. In contrast, in IG and TR V-DOMAIN, the position 118 is occupied by remarkably conserved amino acids, which consist of a phenylalanine or a tryptophan encoded by the J-REGION and therefore designated J-TRP or J-PHE 118. The bulky aromatic side chains of J-TRP and J-PHE are internally orientated and structurally contribute to the V-DOMAIN hydrophobic core [61].

3.4.1.5 Genomic Delimitation

A last criterion used in the IMGT definitive system for the characterization of a V domain is its delimitation, taking into account the exon delimitations, whenever appropriate. The exon rule is not used for the delimitation of the 5′ end of the first N-terminal domain of proteins with a leader (this includes the V-DOMAIN of the IG and TR chains). In those cases, the 5′ end of the first N-terminal domain corresponds to the proteolytic site between the leader (L-REGION) and the coding region of the mature protein. The IG and TR V-DOMAIN is therefore delimited in 5′ by a proteolytic site and in 3′ at the genomic level by the splicing site of the J-REGION. This IMGT genomic approach integrates the strands A and G, in contrast to structural alignments that usually lack these strands due to their poor structural conservation, and thus bridges the gap between genomic data (exon) and 3D structure (domain).

3.4.2 C Domain IMGT Definitive System

3.4.2.1 C Domain Definition and Main Characteristics

In the IMGT definitive system, the C domain includes the C-DOMAIN of the IG and of the TR [2,3] and the C-LIKE-

DOMAIN of the IgSF other than IG and TR. The C domain description of any receptor, any chain, and any species is based on the IMGT unique numbering for C domain (C-DOMAIN and C-LIKE-DOMAIN) [62,64].

A C domain (Figure 20.2) comprises about 90–100 amino acids and is made of seven antiparallel beta strands (A, B, C, D, E, F and G), linked by beta turns (AB, DE and EF), a transversal strand (CD) and two loops (BC and FG), and forming a sandwich of two sheets [ABED] [GFC] [62,64]. A C domain has a topology and a three-dimensional structure similar to that of a V domain but without the C′ and C″ strands and the C′C″ loop, which is replaced by a transversal CD strand [62].

3.4.2.2 C Domain Strands and Loops

The C domain strands, turns, and loops and their delimitations and lengths, based on the IMGT unique numbering for C domain [62,64], are shown in Table 20.4. Correspondences between the IMGT unique numbering with other numberings (Eu, Kabat) are available in the IMGT Scientific chart. The correspondences with these previous numberings are useful for the interpretation of previously published data but, as for the V domain, the usage of these previous numberings has become obsolete in regard to the development of immunoinformatics based on the IMGT standards [59–68] (IMGT, http://www.imgt.org, IMGT Scientific chart > Numbering > Correspondence between C numberings).

3.4.2.3 IMGT Colliers de Perles

The lengths of the strands and loops are visualized in the IMGT Colliers de Perles [66–68], on one layer and two layers (Figure 20.2). There are six IMGT anchors in a C domain (four of them identical to those of a V domain): positions 26 and 39 (anchors of the BC loop), 45 and 77 (by extension, anchors of the CD strand as there is no C′C″ loop in a C domain [62]), and 104 and 118 (anchors of the FG loop).

3.4.2.4 Conserved Amino Acids

A C domain has five characteristic amino acids at given positions (positions with bold (online red) letters in the IMGT Colliers de Perles). Four of them are highly conserved and hydrophobic [28] and are common to the V domain: 23 (1st-CYS), 41 (CONSERVED-TRP), 89 (hydrophobic), and 104 (2nd-CYS). As mentioned above, these amino acids contribute to the two major features shared by the V and C domain: the disulfide bridge (between the two cysteines 23 and 104) and the internal hydrophobic core of the domain (with the side chains of tryptophan W41 and amino acid 89). The fifth position, 118, is diverse and is characterized as being an FG loop anchor.

3.4.2.5 Genomic Delimitation

In the IMGT definitive system, the C domains (C-DOMAIN and C-LIKE-DOMAIN) are delimited, taking into account the exon delimitation, whenever appropriate. As for the V domain, this IMGT genomic approach integrates the strands A and G, which are absent of structural alignments.

3.4.3 IMGT/Collier-de-Perles Tool

The IMGT/Collier-de-Perles tool [26], on the IMGT Website at http://www.imgt.org, is a generic tool that allows the users to draw IMGT Colliers de Perles [65–68] starting from their own domain amino acid sequences (sequences already gapped according to the IMGT unique numbering, using, for example, IMGT/DomainGapAlign [9,24,25]) (Table 20.5). IMGT/Collier-de-Perles tool online can be customized to display the IG and TR CDR-IMGT according to the IMGT color menu and the amino acids according to their hydropathy or volume, or to the 11 IMGT physicochemical classes [28].

The IMGT/Collier-de-Perles tool is integrated in IMGT/DomainGapAlign [9,24,25] (users start from V, C, or G amino acid sequences) and in IMGT/V-QUEST [12–17] (users start from IG and TR V-DOMAIN nucleotide sequences) (Table 20.5). IMGT Colliers de Perles for V, C, and G domains are provided in IMGT/2D structure-DB (for amino acid sequences in the database) and in IMGT/3D structure-DB (on two layers with hydrogen bonds for the V or C domains or with the pMH contact sites for the G domains, for 3D structures in the database) [8–10] (Table 20.5).

4. IMGT IMMUNOGLOBULIN REPERTOIRE ANALYSIS

4.1 IMGT/V-QUEST

IMGT/V-QUEST [12–17] is the IMGT online tool for the analysis of nucleotide sequences of the IG and TR V-DOMAIN (Table 20.5). IMGT/V-QUEST identifies the variable (V), diversity (D), and junction (J) genes in rearranged IG and TR sequences and, for the IG, the nucleotide (nt) mutations and AA changes resulting from somatic hypermutations by comparison with the IMGT/V-QUEST reference directory. The tool integrates IMGT/JunctionAnalysis [18,19] for the detailed characterization of the V-D-J or V-J junctions, IMGT/Automat [20,21] for a complete sequence annotation, and IMGT/Collier-de-Perles [26].

The IMGT/V-QUEST most important functionalities include: introduction of "IMGT gaps" in the user nucleotide sequences (and in their translation); alignments and identification of the genes and alleles with the closest

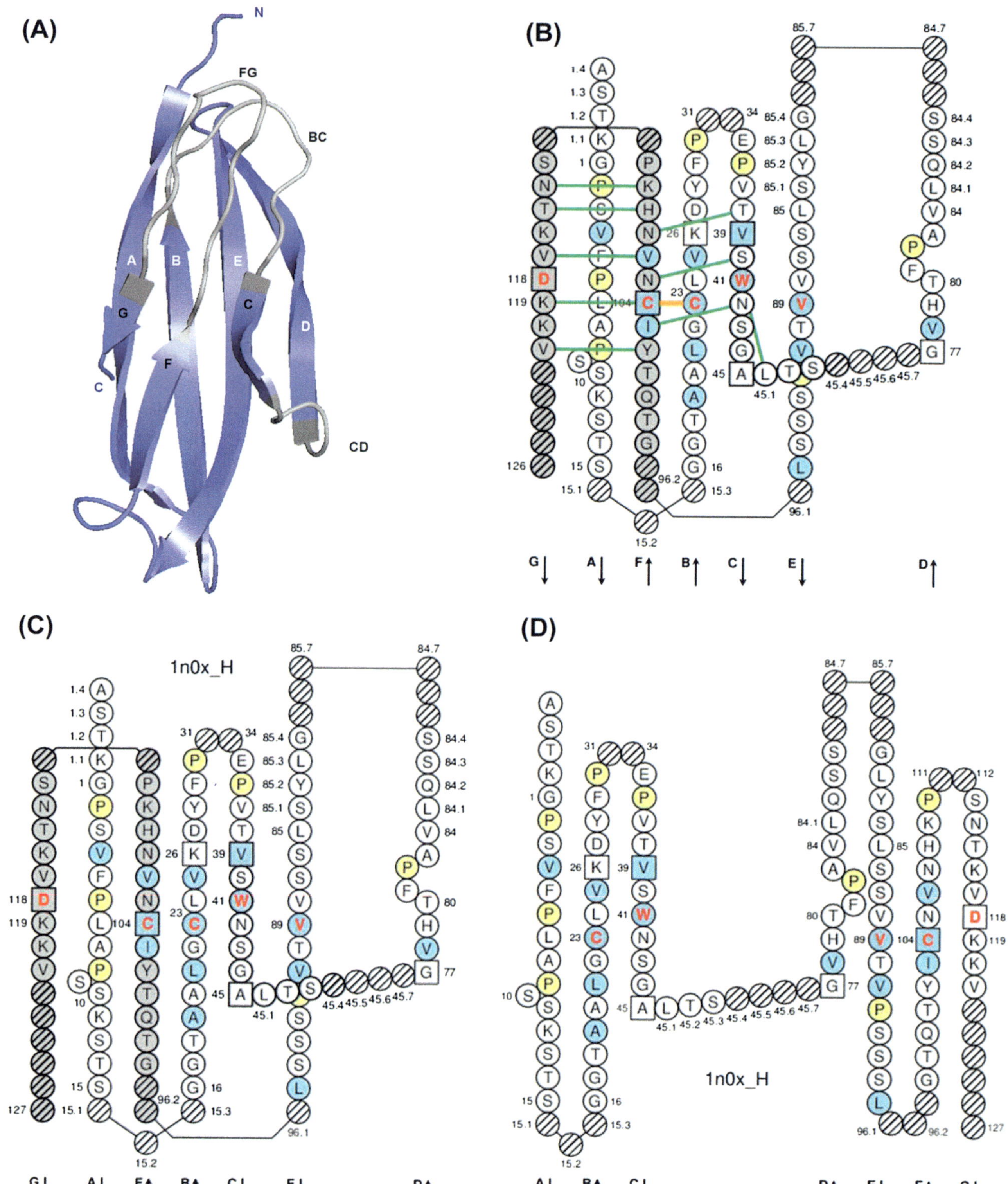

FIGURE 20.2 **Constant (C) domain. An IG CH (C-DOMAIN) is shown as an example.** (A) 3D structure ribbon representation with the IMGT strand and loop delimitations [62]. (B) IMGT Collier de Perles on two layers with hydrogen bonds. The IMGT Collier de Perles on two layers shows, in the forefront, the GFC strands and, in the back, the ABED strands (located at the interface CH1/CL of the IG), linked by the CD transversal strand. The IMGT Collier de Perles with hydrogen bonds (green lines online, only shown here for the GFC sheet) is generated by the IMGT/Collier-de-Perles tool integrated in IMGT/3Dstructure-DB, from experimental 3D structure data [8–10]. (C) IMGT Collier de Perles on two layers from IMGT/DomainGapAlign [9,24,25]. (D) IMGT Colliers de Perles on one layer. Amino acids are shown in the one-letter abbreviation. All proline (P) are shown online in yellow. IMGT anchors are in squares. Hatched circles are IMGT gaps according to the IMGT unique numbering for C domain [62,64]. Positions with bold (online red) letters indicate the four conserved positions that are common to a V domain and to a C domain: 23 (1st-CYS), 41 (CONSERVED-TRP), 89 (hydrophobic), 104 (2nd-CYS) [59–62,64] (Table 20.3), and position 118, which is only conserved in V-DOMAIN. The identifier of the chain to which the CH domain belongs is 1n0x_H (from the *Homo sapiens* b12 Fab, in IMGT/3Dstructure-DB, http://www.imgt.org). The 3D ribbon representation was obtained using PyMOL and "IMGT numbering comparison" of 1n0x_H (CH1) from IMGT/3Dstructure-DB (http://www.imgt.org).

TABLE 20.4 C Domain Strands, Turns, and Loops, IMGT Positions, and Lengths, Based on the IMGT Unique Numbering for C Domain (C-DOMAIN and C-LIKE-DOMAIN)

C domain strands, turns, and loops[a]	IMGT position[b]	Lengths[c]	Characteristic IMGT residue@position[d]
A-STRAND	1–15	15 (14 if gap at 10)	
AB-TURN	15.1–15.3	0–3	
B-STRAND	16–26	11	1st-CYS 23
BC-LOOP	27–31 34–38	10 (or less)	
C-STRAND	39–45	7	CONSERVED-TRP 41
CD-STRAND	45.1–45.9	0–9	
D-STRAND	77–84	8 (or 7 if gap at 82)	
DE-TURN	84.1–84.7 85.1–85.7	0–14	
E-STRAND	85–96	12	Hydrophobic 89
EF-TURN	96.1–96.2	0–2	
F-STRAND	97–104	8	2nd-CYS 104
FG-LOOP	105–117	13 (or less, or more)	
G-STRAND	118–128	11 (or less)	

[a]*IMGT labels (concepts of description) are written in capital letters (no plural) [57].*
[b]*Based on the IMGT unique numbering for C domain (C-DOMAIN and C-LIKE-DOMAIN) [62,64].*
[c]*In number of amino acids (or codons).*
[d]*IMGT Residue@Position is a given residue (usually an amino acid) or a given conserved property amino acid class, at a given position in a domain, based on the IMGT unique numbering [64].*

germline V, D, and J genes; analysis of the junctions; analysis of somatic hypermutations and amino acid changes; and, if the option "Search for insertions and deletions" was selected, identification of insertions and deletions (indels) and their correction. Customized parameters and results provided by IMGT/V-QUEST and IMGT/JunctionAnalysis have been described elsewhere [12–17]. Here, we briefly highlight major features of the three types of IMGT/V-QUEST outputs ("Detailed view," "Synthesis view," and "Excel file") and recall the content of the IMGT/V-QUEST reference directory.

4.1.1 IMGT/V-QUEST Output for "Detailed View"

4.1.1.1 Sequence and "Result Summary"

The sequence and "Result summary" are shown at the top of each individual result (Figure 20.3A) [14,16]. The IMGT reference directory set against which the sequence was analyzed (e.g., human IG) is indicated. A sequence submitted in antisense orientation will be shown as complementary reverse sequence, that is, in V gene sense orientation. The "Result summary" provides a crucial feature that is the evaluation of the functionality of the V domain rearranged sequence performed automatically by IMGT/V-QUEST: productive (if no stop codon in the V-(D)-J region and in-frame junction) or unproductive (if stop codons in the V-(D)-J region and/or out-of-frame junction). It also summarizes the main characteristics of the analyzed V domain sequence, which include (1) the names of the closest "V-GENE and allele" (e.g., IGHV1-69*09): and "J-GENE and allele" (e.g., IGHJ4*02) with the alignment score, the percentage of identity, and the ratio of the number of identical nucleotides/number of aligned nucleotides; (2) the name of the closest "D-GENE and allele" (e.g., IGHD2-2*01), determined by IMGT/JunctionAnalysis with the D-REGION reading frame; (3) the FR-IMGT lengths (e.g., [25.17.38.11]), the CDR-IMGT lengths (e.g., [8.8.14]), and the AA JUNCTION sequence, which characterizes a V domain [14,16]. The information shown in Figure 20.3A (online in orange), "Productive IGH rearranged sequence" and "93,75%," is used by clinicians analyzing the level of somatic hypermutations in chronic lymphocytic leukemia (CLL) IGHV rearranged sequences as a prognostic criterion [15,70–72]. IMGT/V-QUEST provides warnings (not shown) that appear as notes in red to alert the user, if potential insertions or deletions are suspected in the V-REGION, or if other possibilities for the J-GENE and allele names are identified [14,16].

TABLE 20.5 IMGT Tools and Databases for Immunoglobulin Repertoire Analysis and Antibody Engineering and Humanization

IMGT tools and databases	Results for V or C domains[a]	Entry types and protocol references
IMGT/Collier-de-Perles [26]	Graphical 2D representation of IMGT Colliers de Perles [65–68]	User "IMGT gapped" V or C domain amino acid sequences (1 sequence per representation) [26]
IMGT immunoglobulin repertoire analysis		
IMGT/V-QUEST [12–17]	1. Introduction of IMGT gaps 2. Identification of the closest V, D, and J genes and alleles 3. IMGT/JunctionAnalysis results [18,19] 4. Description of mutations and amino acid changes 5. Identification of indels and their correction [16] (option) 6. IMGT/Automat annotation [20,21] 7. IMGT Colliers de Perles [65–68]	User nucleotide sequences of V-DOMAIN (1–50 sequences per analysis) [16]
IMGT/HighV-QUEST [21–23]	1. Introduction of IMGT gaps 2. Identification of indels and their correction [16] (by default) 3. Identification of the closest V, D, and J genes and alleles 4. IMGT/JunctionAnalysis results [18,19] 5. Description of mutations and amino acid changes 6. IMGT/Automat annotation [20,21] 7. Statistical analysis [22] 8. Characterization of the IMGT clonotypes (AA) [23]	User NGS nucleotide sequences of V-DOMAIN, preferentially long sequences (e.g., from 454) (up to 500,000 sequences per run and statistics on results[b])[c] [22,23]
IMGT antibody engineering and humanization		
IMGT/DomainGapAlign [9,24,25]	1. Introduction of IMGT gaps 2. Identification of the closest genes and alleles 3. Delimitation of the domains 4. Description of amino acid changes 5. IMGT Colliers de Perles [65–68] with highlighted AA changes (pink circles online)	User amino acid sequences of V and C domains (1 to several sequences of same domain type) [24,25]
IMGT/3Dstructure-DB [8–10]	1. Identification of the closest genes and alleles 2. IMGT/DomainGapAlign results [9,24,25] 3. IMGT Colliers de Perles [65–68] on two layers with hydrogen bonds (for V and C) 4. Contact analysis between a pair of domains or between a domain and a ligand 5. Renumbered IMGT files 6. IMGT numbering comparison	2987 structure entries (1973 IG, including 1272 IG/Ag complexes)[d] [10].
IMGT/2Dstructure-DB [10]*	1. Identification of the closest genes and alleles 2. IMGT/DomainGapAlign results [9,24,25] 3. IMGT Colliers de Perles [65–68] 4. Renumbered IMGT files	543 amino acid sequence entries[d] (534 IG, 335 from Kabat and 199 from INN) [10]*.

[a] *V: V domain (includes V-DOMAIN of IG and TR and V-LIKE-DOMAIN of IgSF other than IG and TR) [61]. C: C domain (includes C-DOMAIN of IG and TR and C-LIKE-DOMAIN of IgSF other than IG and TR) [62]. IMGT/DomainGapAlign, IMGT/3Dstructure-DB, and IMGT/2Dstructure-DB also manage the G domain (G-DOMAIN of MH and G-LIKE-DOMAIN of MhSF other than MH) [63] of interest in the context of that chapter when the antibody target belongs to the MH (or more generally to the MhSF), but not detailed here.*
[b] *In November 2013.*
[c] *In May 2014, more than 2.5 billions of sequences analyzed by IMGT/HighV-QUEST by 846 users from 40 countries (62% from USA, 25% from EU, 13% from the remaining world).*
[d] *In May 2014.*
An asterisk (*) indicates that parts of the protocol dealing with 3D structures (hydrogen bonds in IMGT Colliers de Perles on two layers, Contact analysis) are not relevant; otherwise all other queries and results are similar to IMGT/3Dstructure-DB.

(A)

Sequence number 1: SEQ1

Sequence compared with the human IG set from the IMGT reference directory

```
>SEQ1
caggtgcagctggtgcagtctggggctgaggtgaagaagcctggatcctcggtgaaggtc
tcatgtaaggcttctggaggcaccttcaacagctttcctatcaactgggtgcgacaggcc
cctggacaagggcttgagtggatgggaaggatcatccctatcattggtatagcagactac
gcacaggaattccagggcagagtcacgattaccgcggacagatccacgagcacagcctac
atggagctgcgtagtctgatatctgaggacacggccgtgtacttctgtgcgagaccggaa
gccgtaacagtaccagctcctcttgactactggggccagggaaccctggtcaccgtctcc
tcagcctccaccaagggcccatcggtcttcccctggcaccctcctccaagagcacctct
gggggcacagcggccctgggctgcctggtcaaggactacttccccgaaccggtgacggtg
tcgtggaactcaggcgccctgaccagcggcgtgcacaccttcccggctgtcctacagtcc
tcaggactctactccctcagcagcgtggtgaccgtgccctccagcagcttgggcacccag
acctacatctgcaacgtgaatcacaagcccagcaacaccaaggtggacaagacagttgag
cgcaaa
```

Result summary:	Productive IGH rearranged sequence (no stop codon and in-frame junction)		
V-GENE and allele	Homsap IGHV1-69*09 F	score = 1273	identity = 93,75% (270/288 nt)
J-GENE and allele	Homsap IGHJ4*02 F	score = 186	identity = 87,50% (42/48 nt)
D-GENE and allele by IMGT/JunctionAnalysis	Homsap IGHD2-2*01 F	D-REGION is in reading frame 3	
FR-IMGT lengths, CDR-IMGT lengths and AA JUNCTION	[25.17.38.11]	[8.8.14]	CARPEAVTVPAPLDYW

(B)

Sequence number 2: SEQ2

Sequence compared with the human IG set from the IMGT reference directory

```
>SEQ2
atgtctgtctccttcctcatcttcctgcccgtgctgggcctcccatggggtgtcctgtca
caggtacagctgcagcagtcaggtccaggactggtgaagccctcgcagaccctctcactc
acctgtgccatctccggggacagtgtctctAGCagcaacggtgttgcttggaactgggtc
aggcagtccccatcgagaggccttgagtggctgggaaggacatactacaggtccaagtgg
tataatgattatgcagtatctgtgaaaagtcgaataaccatcaacccagacacatccaag
aaccagttctccctgcagctgaactctgtgactcccgaagacacggctgtgtattactgt
gcaagaggccgttggaccgcttttgatttctggggccaagggacaatggtcaccgtctct
tcc
```

Result summary:

Nucleotide insertions have been detected and automatically removed for this analysis: they are displayed as capital letters in the user submitted sequence above.

localization in V-REGION	nb of inserted nt	inserted nt	causing frameshift	from V-REGION codon	from nt position in user submitted sequence
CDR1-IMGT	3	AGC	no	34	151

IMGT/V-QUEST results after removal of the insertion(s):
Potentially productive IGH rearranged sequence: no stop codon and in-frame junction
(Check also your sequence with BLAST against IMGT/GENE-DB reference sequences to eventually identify out-of-frame pseudogenes)

V-GENE and allele	Homsap IGHV6-1*01 F	score = 1444	identity = 98,65% (293/297 nt) [98,32% (292/297 nt)]
J-GENE and allele	Homsap IGHJ3*01 F, or Homsap IGHJ3*02 F	score = 191	identity = 87,76% (43/49 nt)
D-GENE and allele by IMGT/JunctionAnalysis	Homsap IGHD3-16*02 F	D-REGION is in reading frame 2	
FR-IMGT lengths, CDR-IMGT lengths and AA JUNCTION	[25.17.38.11]	[10.9.10]	CARGRWTAFDFW

FIGURE 20.3 **IMGT/V-QUEST results.** (A) Sequence and "result summary" ("detailed view"). (B) Sequence and "result summary" with the option "search for insertions and deletions" ("detailed view"). An insertion of 3 nt was detected by IMGT/V-QUEST. The analysis is then performed automatically by the tool after having removed the insertion. (C) "Results of IMGT/JunctionAnalysis" ("detailed view"). The "list of eligible D" (an option) is shown. "Ngc" is 13/17 as there are 13 "c" or "g" in N1 + N2 on a total of 17 nucleotides (aggctggg (N1), gagcgcgtc (N2)). (D) "results of IMGT/JunctionAnalysis" ("synthesis view"). The numbering of IMGT additional positions for CDR3-IMGT is according to Table 20.3. Owing to the figure format, Vmut, Dmut, Jmut, Ngc, molecular mass, and pI are not shown. The SEQ1, SEQ2, SEQ3, SEQ4, SEQ5, SEQ6, and SEQ7 correspond to the AB027435, L21957, AB363166, Z98684, Z98689, Z98734, and Z98715 accession numbers, respectively, from the IMGT/LIGM-DB database [6].

(C) **Maximum number of accepted mutations in:** 3'V-REGION = 2, D-REGION = 4, 5'J-REGION = 2
Maximum number of accepted D-GENE = 1

Analysis of the JUNCTION

D-REGION is in reading frame 2.

Click on mutated (underlined) nucleotide to see the original one:

Input	V name	3'V-REGION	N1	D-REGION	N2	5'J-REGION	J name	D name	Vmut	Dmut	Jmut	Ngc
SEQ3	Homsap IGHV3-53*01	tgtgcgagaga	aggctggg	........atgatagtagtggttttttac...	gagcgcgtc	........gaatactgg	Homsap IGHJ4*02	Homsap IGHD3-22*01	0	1	1	13/17

Eligible D gene:

D name	D length	tgtgcgagagaaggctgggatgatagtagtggtttttacgagcgcgtcgaatactgg	Score#	Mutation#	Location
Homsap IGHD1-1*01	17	----a-c--	7	2	d[8-16],s[15-23]
Homsap IGHD1-14*01	17	--t--	4	1	d[1-5],s[17-21]
Homsap IGHD1-20*01	17	----a-c--	7	2	d[8-16],s[15-23]
Homsap IGHD1-26*01	20	------gagc---	9	4	d[5-17],s[27-39]
Homsap IGHD1-7*01	17	----	4	0	d[8-11],s[15-18]
Homsap IGHD2-15*01	31	--t--------	10	1	d[6-16],s[23-33]
Homsap IGHD2-2*01	31	--t-----a--	9	2	d[6-16],s[23-33]
Homsap IGHD2-2*02	31	--t-----a--	9	2	d[6-16],s[23-33]
Homsap IGHD2-2*03	31	--t-----a--	9	2	d[6-16],s[23-33]
Homsap IGHD2-21*01	28	--t--g-----	9	2	d[6-16],s[23-33]
Homsap IGHD2-21*02	28	--t--g-----	9	2	d[6-16],s[23-33]
Homsap IGHD2-8*01	31	--c--a----	8	2	d[10-19],s[24-33]
Homsap IGHD2-8*02	31	--t---c----	9	2	d[6-16],s[23-33]
Homsap IGHD3-10*01	31	---ta--a-a--	8	4	d[20-31],s[28-39]
Homsap IGHD3-10*02	30	---ta--a-a--	8	4	d[19-30],s[28-39]
Homsap IGHD3-16*01	37	---ta--c--	7	3	d[23-32],s[25-34]
Homsap IGHD3-16*02	37	---ta-c---	7	3	d[23-32],s[25-34]
Homsap IGHD3-22*01	31	---------------a----	19	1	d[9-28],s[20-39]
Homsap IGHD3-3*01	31	-------a---	10	1	d[17-27],s[28-38]
Homsap IGHD3-3*02	31	-------a---	10	1	d[17-27],s[28-38]
Homsap IGHD3-9*01	31	-----a---	8	1	d[19-27],s[30-38]
Homsap IGHD4-11*01	16	---ctac---	6	4	d[1-10],s[21-30]
Homsap IGHD4-17*01	16	---gac---	6	3	d[8-16],s[31-39]
Homsap IGHD4-23*01	19	---gg-a--	6	3	d[8-16],s[31-39]
Homsap IGHD4-4*01	16	---ctac---	6	4	d[1-10],s[21-30]
Homsap IGHD5-12*01	23	----ac---	7	2	d[12-20],s[13-21]
Homsap IGHD5-18*01	20	-----	5	0	d[14-18],s[30-34]
Homsap IGHD5-24*01	20	---aga---	6	3	d[2-10],s[24-32]
Homsap IGHD5-5*01	20	-----	5	0	d[14-18],s[30-34]
Homsap IGHD6-13*01	21	-----	5	0	d[14-18],s[14-18]
Homsap IGHD6-19*01	21	----c-----c-gg---	13	4	d[5-21],s[23-39]
Homsap IGHD6-25*01	18	----c--c--	8	2	d[5-14],s[23-32]
Homsap IGHD6-6*01	18	---t----	7	1	d[10-17],s[41-48]
Homsap IGHD7-27*01	11	-----	5	0	d[5-9],s[15-19]

Translation of the JUNCTION

Click on mutated (underlined) amino acid to see the original one:

	104	105	106	107	108	109	110	111	111.1	111.2	112.2	112.1	112	113	114	115	116	117	118	Frame	CDR3-IMGT length	Molecular mass	pI
	C	A	R	E	G	W	D	D	S	S	G	F	Y	E	R	V	E	Y	W				
SEQ3	tgt	gcg	aga	gaa	ggc	tgg	gat	gat	agt	agt	ggt	ttt	tac	gag	cgc	gtc	gaa	tac	tgg	+	17	2,355.49	4.25

5. Sequence of the JUNCTION ('nt' and 'AA')

104	105	106	107	108	109	110	111	111.1	111.2	112.2	112.1	112	113	114	115	116	117	118
C	A	R	E	G	W	D	D	S	S	G	F	Y	E	R	V	E	Y	W
tgt	gcg	aga	gaa	ggc	tgg	gat	gat	agt	agt	ggt	ttt	tac	gag	cgc	gtc	gaa	tac	tgg

Input for IMGT/JunctionAnalysis

```
>SEQ3,Homsap_IGHV3-53*01,Homsap_IGHJ4*02
tgtgcgagagaaggctgggatgatagtagtggtttttacgagcgcgtcgaatactgg
```

FIGURE 20.3 *(Continued).*

(D)

Results for the IGH junctions

Analysis of the JUNCTIONs

Click on mutated (underlined) nucleotide to see the original one:

Input	V name	3'V-REGION	P	N1	D-REGION	N2	5'J-REGION	J name	D name
SEQ1	Homsap IGHV1-69*09	tgtgcgaga..		ccggaagcc	........gtaacagtaccagctcct.....	c	......ttgactactgg	Homsap IGHJ4*02	Homsap IGHD2-2*01
SEQ7	Homsap IGHV3-23*01	tgtgcgaaag.			ggtatggt............	c	......ttgactgctgg	Homsap IGHJ4*02	Homsap IGHD1-26*01
SEQ6	Homsap IGHV3-30*03	tgtgcgaaaga	tc	ctgcgaa	......agcagttcgtcc	tgaattgccggt	................cggtgtggacgtctgg	Homsap IGHJ6*02	Homsap IGHD6-6*01
SEQ3	Homsap IGHV3-53*01	tgtgcgagaga		aggctggg	........atgatagtagtggttttac...	gagcgcgtc	........gaatactgg	Homsap IGHJ4*02	Homsap IGHD3-22*01
SEQ4	Homsap IGHV3-73*02	tgtattaga..			.tattactatgatagcagtggt.........	gacggtgtatc	.............ctgg	Homsap IGHJ4*02	Homsap IGHD3-22*01
SEQ5	Homsap IGHV3-73*02	tgtattaga..			.tattactatgatagcagtggt.........	gacggtgtatc	.............ctgg	Homsap IGHJ4*02	Homsap IGHD3-22*01
SEQ2	Homsap IGHV6-1*01	tgtgcaagag.		gc	...........................cgttggacc		....gcttttgatttctgg	Homsap IGHJ3*01	Homsap IGHD3-16*02

Translation of the JUNCTIONs

Click on mutated (underlined) amino acid to see the original one:

	104	105	106	107	108	109	110	111	111.1	111.2	112.3	112.2	112.1	112	113	114	115	116	117	118	Frame	CDR3-IMGT length
	C	A	R	P	E	A	V	T					V	P	A	P	L	D	Y	W		
SEQ1	tgt	gcg	aga	ccg	gaa	gcc	gta	aca	...	...	...	...	gta	cca	gct	cct	ctt	gac	tac	tgg	+	14
	C	A	K	G	Y											G	L	D	C	W		
SEQ7	tgt	gcg	aaa	ggg	tat	...	...	...	...	...	...	...	...	...	...	ggt	ctt	gac	tgc	tgg	+	8
	C	A	K	D	P	A	K	A	V	R	P	E	L	P	V	G	V	D	V	W		
SEQ6	tgt	gcg	aaa	gat	cct	gcg	aaa	gca	gtt	cgt	cct	gaa	ttg	ccg		ggt	gtg	gac	gtc	tgg	+	18
	C	A	R	E	G	W	D	D	S	S		G	F	Y	E	R	V	E	Y	W		
SEQ3	tgt	gcg	aga	gaa	ggc	tgg	gat	gat	agt	agt	...	ggt	ttt	tac	gag	cgc	gtc	gaa	tac	tgg	+	17
	C	I	R	Y	Y	Y	D	S						S	G	D	G	V	S	W		
SEQ4	tgt	att	aga	tat	tac	tat	gat	agc	...	...	...	...	...	agt	ggt	gac	ggt	gta	tcc	tgg	+	13
	C	I	R	Y	Y	Y	D	S						S	G	D	G	V	S	W		
SEQ5	tgt	att	aga	tat	tac	tat	gat	agc	...	...	...	...	...	agt	ggt	gac	ggt	gta	tcc	tgg	+	13
	C	A	R	G	R	W									T	A	F	D	F	W		
SEQ2	tgt	gca	aga	ggc	cgt	tgg	...	...	...	...	...	...	...	...	acc	gct	ttt	gat	ttc	tgg	+	10

FIGURE 20.3 (*Continued*).

4.1.1.2 Sequence and "Result Summary" with the Option "Search for Insertions and Deletions"

Potential insertions or deletions (indels) are suspected by IMGT/V-QUEST when the V-REGION score is very low (less than 200), and/or the percentage of identity is less than 85%, and/or when the input sequence has different CDR1-IMGT and/or CDR2-IMGT lengths, compared to those of the closest germline V [14,16]. In those cases, the user can go back to the IMGT/V-QUEST Search page and select the option "Search for insertions and deletions" in "Advanced parameters." If, indeed, insertions and/or deletions are detected, they will be described in the "Result summary" row with their localization in FR-IMGT or CDR-IMGT, the number of inserted or deleted nucleotides (and, for insertions, the inserted nucleotides), the presence or absence of frameshift, the V-REGION codon from which the insertion or deletion starts, and the nt position in the user submitted sequence (Figure 20.3B). The insertions are highlighted in capital letters in the user sequence and the tool runs a classical IMGT/V-QUEST search after having removed the insertion(s) from the user sequence (Figure 20.3B). In case of deletions, the tool adds gaps to replace the identified deletions before running a classical IMGT/V-QUEST search. Users should be aware that an insertion or a deletion at the beginning of FR1-IMGT or at the end of the FR3-IMGT may not be detected. The identification and correction of indels by IMGT/V-QUEST is important for calculating the percentage of somatic hypermutations in CLL despite these indels [15,16].

4.1.1.3 Alignments for V, D, and J Genes and Alleles

The alignments for V, D, J genes and alleles display the alignments with the five closest V, D, and J gene alleles, respectively, with their alignment score and their identity percentage [12–17]. In case of discrepancy between the alignments for D genes and the results of IMGT/JunctionAnalysis (below), the results of IMGT/JunctionAnalysis are the most accurate (the way of identifying the closest germline D and the evaluation of the alignment score are different between IMGT/V-QUEST and IMGT/JunctionAnalysis). The IMGT/V-QUEST alignment for D gene is still provided (as an option), although less accurate, as it is less stringent and displays several D genes and alleles, and therefore may help solving some ambiguous cases [16].

4.1.1.4 "Results of IMGT/JunctionAnalysis"

The results of IMGT/JunctionAnalysis comprise (Figure 20.3C).

1. "Analysis of the JUNCTION": provides the name of the D gene and allele for IGH sequences and shows the details of the junction at the nucleotide level (nucleotides trimmed at the V, D, and J gene ends and N nucleotides added by the TdT) with an accurate delimitation of the 3′V-REGION (from the 2nd-CYS 104 to the 3′ end of the V-REGION), D-REGION, and 5′J-REGION (from the 5′ end of the J-REGION to J-TRP 118 or J-PHE 118). Dots represent the nucleotides that have been trimmed by the exonuclease [19]. The number of mutations in the 3′V-REGION, D-REGION, and 5′J-REGION is indicated under "Vmut," "Dmut," and "Jmut," respectively (Figure 20.3C), and the corresponding mutated nucleotides are underlined in the sequence. Nucleotides of the N-REGION are displayed in N1 and N2. The ratio of the number of g + c nucleotides to the total number of N nucleotides is indicated under "Ngc" (Figure 20.3C) [19].
2. "Eligible D genes": If the option "with full list of eligible D-GENE" was selected in "Display view," its results allow comparing the IMGT/JunctionAnalysis D gene identification with all D genes that match the junction with their corresponding score (Figure 20.3C).
3. "Translation of the JUNCTION": displays the junction with amino acids colored according to the 11 IMGT "Physicochemical" classes [28], the frame (+ and − indicate in-frame and out-of-frame, respectively), the CDR3-IMGT length, the molecular mass, and the isoelectric point (pI). In case of frameshifts, gaps (represented by one or two dots) are inserted to maintain the J-REGION frame. The corresponding codon, which cannot be translated, is represented by "#" [19].

4.1.1.5 "V-REGION Alignment," "V-REGION Translation," and "V-REGION Protein Display"

The results provide three displays of the V-REGION with the FR-IMGT and CDR-IMGT delimitations [14,16]: the "V-REGION alignment" displays the nucleotide sequences; the "V-REGION translation" displays the nucleotide sequence and deduced AA translation of the input sequence, aligned with the closest germline V-REGION; and the "V-REGION protein display" displays the deduced AA translation of the input sequence, aligned with the V-REGION translation of the closest germline V-GENE and, on the third line of the alignment and shown in bold, the AA of the input sequence that are different from the closest germline V-REGION.

4.1.1.6 Analysis of Mutations and Amino Acid Changes

The analysis of the mutations in the V-REGION is performed by comparison of the analyzed sequence with the closest germline V gene and allele [14,16].

The "V-REGION mutation and AA change table" lists the nt mutations and the corresponding AA changes if the mutations are not silent. They are described for each FR-IMGT and CDR-IMGT, with their nt and codon position according to the IMGT unique numbering and for the AA changes according to the IMGT AA classes [28]. For example, c1 > g, Q1 > E (+ + −) means that the nt mutation (c > g) at nt 1 leads to an AA change (Q > E) at codon 1.

The "V-REGION mutation and AA change statistics" comprises two tables. The first table reports the number of "Nucleotide (nt) mutations," described as silent or non-silent, transition (a > g, g > a, c > t, t > c) or transversion (a > c, c > a, a > t, t > a, g > c, c > g, g > t, t > g). The second table reports the number of "Amino acid (AA) changes" with the type of each change (e.g., (+ + +) highly similar, (− − +) dissimilar). For both tables, results are given for the V-REGION and for FR-IMGT and CDR-IMGT [14,16].

The positions of the mutations have been correlated with the presence of specific patterns in the germline V gene and allele (IMGT, http://www.imgt.org, IMGT Education > Tutorials > Immunoglobulins and B cells > Somatic hypermutations). The "V-REGION mutation hot spots" table shows the localization of the hot spot patterns (a/t)**a** (or w**a**) and (a/g)**g**(c/t) (a/t) (or r**g**yw) and their complementary reverse motifs **t**(a/t) (or **t**w) and (a/t) (a/g)**c**(c/t) (or wr**c**y) in the closest germline V gene and allele [14,16].

4.1.1.7 IMGT Collier de Perles

The "IMGT Collier de Perles" [65–68] can be displayed, on one layer or two layers, either as a "link to IMGT/Collier-de-Perles tool" [26] or as a direct "IMGT Collier de Perles (for a number of sequences >5)" representation integrated in IMGT/V-QUEST results, depending on the user selection in the IMGT/V-QUEST Search page [14,16].

4.1.2 IMGT/V-QUEST Output for "Synthesis View"

4.1.2.1 Summary Table

The "Summary table" displays one row for each input sequence with the corresponding results, including: (1) the name of the sequence (Sequence ID); (2) the name of the closest V-GENE and allele; (3) the functionality of the V domain rearranged sequence (productive or unproductive); (4) the V-REGION score; (5) the V-REGION percentage of identity with, between parentheses, the ratio of number of identical nt/number of aligned nt; (6) the name of the closest J-GENE and allele; and (7) provided according to the IMGT/JunctionAnalysis results, the D-GENE and allele name, the D reading frame, the CDR-IMGT lengths, the AA JUNCTION, and the JUNCTION frame. In the absence of results of IMGT/JunctionAnalysis, only the AA JUNCTION defined by IMGT/V-QUEST is displayed [14,16].

4.1.2.2 IMGT/JunctionAnalysis

The results of IMGT/JunctionAnalysis are displayed per locus (e.g., IGH, IGK, or IGL for IG sequences). Results (Figure 20.3D) are similar to those obtained for individual sequences detailed above [18,19].

4.1.2.3 Alignment Displays

The synthesis results provide six different displays (if all were selected): "Alignment for V-GENE," "V-REGION alignment according to the IMGT unique numbering" and "V-REGION translation," and "V-REGION protein displays" in three different formats [14,16].

4.1.3 IMGT/V-QUEST Output for Excel File

"Excel file" allows the users to open and save a spreadsheet including the results of the IMGT/V-QUEST analysis. The file contains 11 sheets (if all were selected in the IMGT/V-QUEST Search page). The "Summary" and "Parameters" are always selected [14,16].

4.1.4 IMGT/V-QUEST Reference Directory

The IMGT/V-QUEST reference directory sets against which the IMGT/V-QUEST is running include IMGT reference sequences from all functional (F) genes and alleles, all ORF, and all in-frame pseudogene (P) alleles. By definition, the IMGT reference directory sets contain one sequence for each allele. By default, the user sequences are compared with all genes and alleles. However, the option "With allele *01 only" is useful for: (1) "Detailed view," if the user sequences need to be compared with different genes, and (2) "Synthesis view," if the user sequences that use the same gene need to be aligned together (independently of the allelic polymorphism) [14,16].

The IMGT/V-QUEST reference directories have been set up for species that have been extensively studied, such as human and mouse. This also holds for the other species or taxons with incomplete IMGT reference directory sets. In those cases, results should be interpreted considering the status of the IMGT reference directory (information on the updates on the IMGT Website). Links to the IMGT/V-QUEST reference directory sets are available from the IMGT/V-QUEST Welcome page [14,16].

4.2 IMGT/HighV-QUEST

IMGT/HighV-QUEST [22], created in Oct 2010, is the high-throughput version of IMGT/V-QUEST. It is so far the only online tool available on the Web for the direct analysis of complete IG and TR domain sequences from NGS. It analyzes NGS sequences, preferentially long sequences, e.g., obtained from the Roche 454 Life Sciences technology, without the need of computational read assembly [22,23] (Table 20.5). IMGT/HighV-QUEST analyzes up to 500,000 sequences per run and performs statistical analysis on the results [22,23], with the same degree of resolution and high-quality results as IMGT/V-QUEST [12–17]. IMGT/HighV-QUEST represents a major breakthrough for the analysis and the comparison of the antigen receptor V-DOMAIN repertoires and immunoprofilings of the adaptive immune response [22,23].

The functionalities of IMGT/HighV-QUEST include: the introduction of IMGT gaps; the identification of indels and their correction; the identification of the closest V, D, and J genes and alleles; the IMGT/JunctionAnalysis results; the description of mutations and amino acid changes; the annotation by IMGT/Automat; the NGS statistical analysis; and the characterization of the IMGT clonotypes (AA) [22,23] (Table 20.5). IMGT/HighV-QUEST provides results in different categories "1 copy" and "More than 1" to avoid redundancy of the analysis, "single allele" and "several alleles (or genes)" (with "single allele" sequences being usually longer than "several alleles") [22]. These categories have been fundamental in the characterization of clonotypes for NGS [23].

As for the other IMGT databases and tools, IMGT/HighV-QUEST is freely available for academics. However, the IMGT/HighV-QUEST Welcome page requires user identification and provides, for new users, a link to register. User identification has been set to avoid nonrelevant use and overload of the server, and to contact the user if needed. The user identification gives access to the IMGT/HighV-QUEST Search page.

4.2.1 MGT/HighV-QUEST Search Page

The IMGT/HighV-QUEST Search page [22] (Figure 20.4A) is very similar to the classical IMGT/V-QUEST Search page and therefore looks familiar to the biologists. The four additional fields at the top comprise the analysis title (provided by the user), the selection of the species, the selection of the receptor type or locus, and the choice of email notifications. Because of their large number of

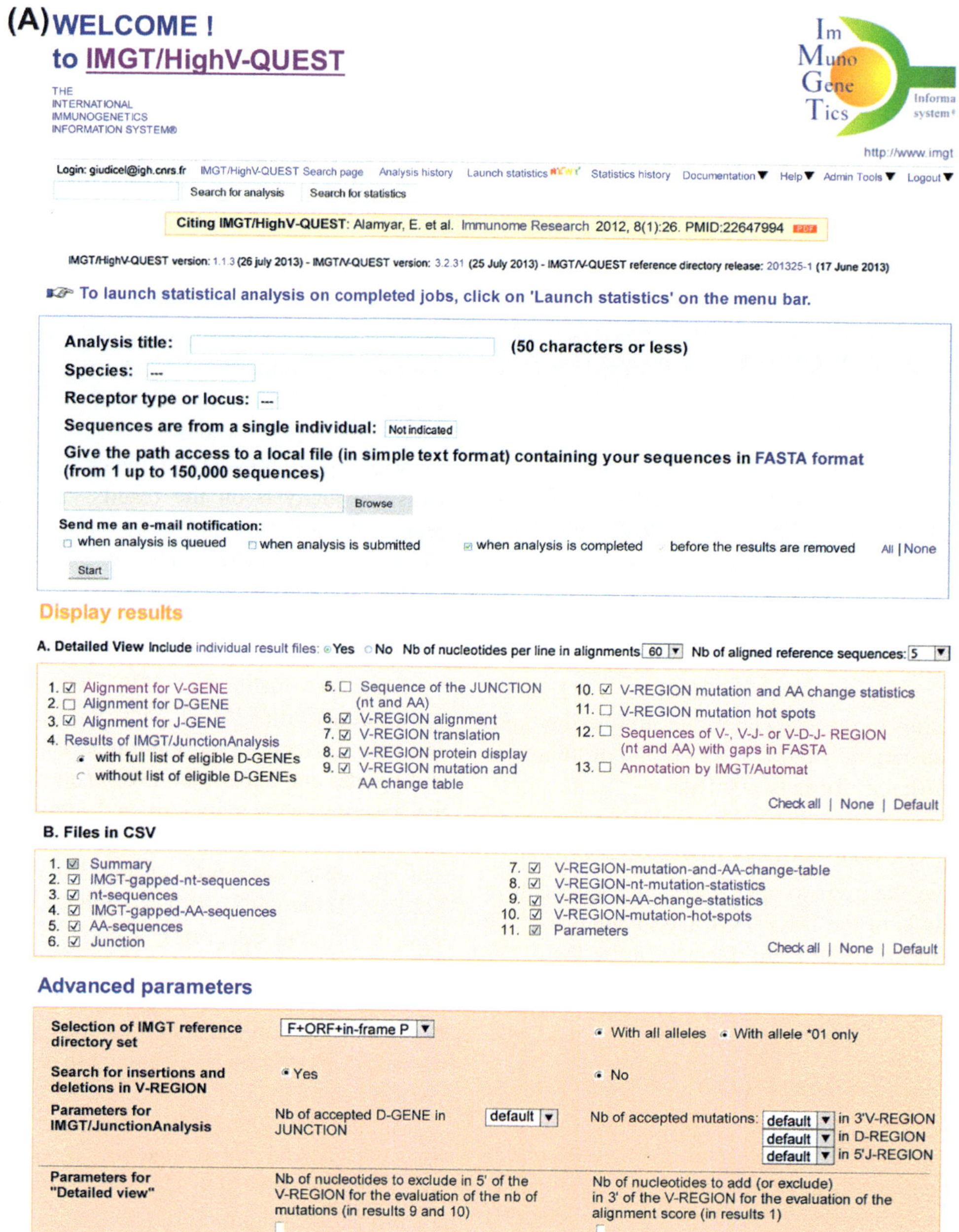

FIGURE 20.4 **IMGT/HighV-QUEST.** (A) IMGT/HighV-QUEST search page. (B) IMGT clonotypes (AA). This figure represents a screen shot from IMGT/HighV-QUEST online. "Exp. ID" is the identifier of the "IMGT clonotype (AA)" in the dataset. The IMGT clonotype (AA) definition includes the CDR3-IMGT length (AA), the CDR3-IMGT sequence (AA), the names of the V, D, J genes and alleles, and the anchors 104 and 118 of the junction (for VH, "C, W" for 2nd-CYS 104 and J-TRP 118, respectively) (the order of these columns depends on the query, here on CDR3-IMGT length (AA)). "V %" indicates the percentage identity of the V-REGION of the representative sequence with the closest germline V-REGION; "Sequence length" is the length in nucleotides of the representative sequence; the link in the "Sequence ID" column gives access to the representative sequence in FASTA format; "Nb" indicates the number of sequences assigned to an IMGT clonotype (AA) ("1 copy," "more than 1," and total). In the "IMGT clonotypes (nt)" column, the link on each "sequences file" gives access to a file containing the "1 copy" sequences assigned to a given IMGT clonotype (AA), in FASTA format. This figure shows a very small part of the list of the 64,093 unique IMGT clonotypes (AA) identified in this case study. The "IMGT clonotype (AA)" #45309, with an Exp. ID "45309_Xall," is defined by the "IGHV6-1*01 F – IGHD1-26*01 F – IGHJ4*02 F" rearrangement, with a CDR3-IMGT length (AA) of "14 AA" and a CDR3-IMGT sequence (AA) "TRDHDQEGANSFDY," and conserved anchors 104 and 118 "C, W" (recall of the filter). The IMGT clonotype (AA) representative sequence has a V-REGION that is 96.62% identical to that of the germline IGHV6-1*01 and a length of 427 nt. (C) IMGT clonotypes (nt). The nb of different CDR3-IMGT (nt) indicates the nb of IMGT clonotypes (nt) for a given IMGT clonotype (AA) (e.g., two for #41328). The CDR3-IMGT sequence (nt) is shown with the number of different nucleotides ("Nb diff nt") by comparison with that of the representative sequence. "0" indicates that the CDR3-IMGT sequence (nt) is identical to that of the IMGT clonotype (AA) representative sequence. For #41328, there is an IMGT clonotype (nt) with 4 nt differences ("c" instead of "t" at position 9, "c" instead of "a" at position 12, "t" instead of "c" at position 30, and "c" instead of "t" at position 33) compared to the CDR3-IMGT of the representative sequence. #41328 also shows an example of "several alleles" (for V and J) assigned to an IMGT clonotype (AA).

(B)

ID		IMGT clonotype (AA) definition						IMGT clonotype (AA) representative sequence			Nb			IMGT clonotypes (nt)
#	Exp. ID	CDR3-IMGT length (AA)	CDR3-IMGT sequence (AA)	V gene and allele	D gene and allele	J gene and allele	Anchors 104,118	V %	Sequence length	Sequence ID	Total nb of '1 copy'	Total nb of 'More than 1'	Total	Sequences file ('1 copy')
45307	45307-Xall	14 AA	TRDGPVAVHWFFAL	Homsap IGHV6-1*01 F	Homsap IGHD2-8*02 F	Homsap IGHJ2*01 F	C,W	97.64	422	H6VTD4304JCAGK	1	0	1	Sequences file
45308	45308-Xall	14 AA	TRDGSLTGTKGFDI	Homsap IGHV3-30*01 F	Homsap IGHD1-7*01 F	Homsap IGHJ3*01 F	C,W	87.85	739	H7CGN0J01CVBLX	1	0	1	Sequences file
45309	45309-Xall	14 AA	TRDHDQEGANSFDY	Homsap IGHV6-1*01 F	Homsap IGHD1-26*01 F	Homsap IGHJ4*02 F	C,W	96.62	427	H7CGN0J01BQTCY	24	9	33	Sequences file
45310	45310-Xall	14 AA	TRDLDRSGWYGMDV	Homsap IGHV3-49*03 F	Homsap IGHD6-19*01 F	Homsap IGHJ6*02 F	C,W	95.92	424	H7CGN0J01BQGHH	1	0	1	Sequences file
45311	45311-Xall	14 AA	TRDLGSSGPYYFDY	Homsap IGHV3-49*04 F	Homsap IGHD6-19*01 F	Homsap IGHJ4*02 F	C,W	86.39	424	H7CGN0J01CBHZL	1	0	1	Sequences file
45312	45312-Xall	14 AA	TRDLWLSGQYTFDY	Homsap IGHV6-1*01 F	Homsap IGHD2-8*02 F	Homsap IGHJ2*01 F	C,W	98.32	427	H6VTD4302EYLAU	1	0	1	Sequences file
45313	45313-Xall	14 AA	TRDPVAGVPDFFDH	Homsap IGHV3-30*31 F	Homsap IGHD2-15*01 F	Homsap IGHJ4*02 F	C,W	89.90	411	H7CGN0J01DM7W9	2	0	2	Sequences file
45314	45314-Xall	14 AA	TRDRALLFGAHFDF	Homsap IGHV1-2*02 F	Homsap IGHD3-10*01 F	Homsap IGHJ4*02 F	C,W	93.75	417	H7CGN0J01C3729	1	0	1	Sequences file
45315	45315-Xall	14 AA	TRDRLSGNVWLLED	Homsap IGHV6-1*01 F	Homsap IGHD6-19*01 F	Homsap IGHJ5*02 F	C,W	96.30	424	H7CGN0J01A5CLN	1	0	1	Sequences file
45316	45316-Xall	14 AA	TRDRLYGDVWLLED	Homsap IGHV6-1*01 F	Homsap IGHD6-19*01 F	Homsap IGHJ5*02 F	C,W	95.29	427	H7CGN0J01EFJMF	1	0	1	Sequences file

(C)

VH clonotypes (nt) header

#	CDR3-IMGT length (nt)	Nb diff CDR3-IMGT (nt)	CDR3-IMGT sequence (nt)	Nb diff nt	V gene and allele	D gene and allele	J gene and allele	Anchors 104,118	V % mean	V-REGION length mean	J % mean	J-REGION length mean	Sequence length mean	Total nb of '1 copy'	Total nb of 'More than 1'	Total
41325	41325-Xall	14 AA	ARDLCDYGLCYFDY		Homsap IGHV1-18*04 F	Homsap IGHD4-17*01 F	Homsap IGHJ4*02 F	C,W	97.91	414	H7CGN0J01ET06J	1	0	1	Sequences file	
41325	42	1	gcgagagatttatgtgactacggactctgctactttgactac	0	Homsap IGHV1-18*04 F	Homsap IGHD4-17*01 F	Homsap IGHJ4*02 F	C,W	97.91	295	97.92	47	414	1	0	1
41326	41326-Xall	14 AA	ARDLDGSKMFCMDV		Homsap IGHV3-23*04 F	Homsap IGHD5-24*01 F	Homsap IGHJ6*02 F	C,W	89.25	439	H7CGN0J01BDKK3	1	0	1	Sequences file	
41326	42	1	gcgagagatctagacggcagcaagatgttctgtatggacgtc	0	Homsap IGHV3-23*04 F	Homsap IGHD5-24*01 F	Homsap IGHJ6*02 F	C,W	89.25	287	62.9	44	439	1	0	1
41327	41327-Xall	14 AA	ARDLDGSSWPIFDY		Homsap IGHV6-1*01 F	Homsap IGHD6-13*01 F	Homsap IGHJ4*02 F	C,W	100	786	H7CGN0J01ARV9K	7	0	7	Sequences file	
41327	42	1	gcaagagatcttgacggcagcagctggcccatatttgactac	0	Homsap IGHV6-1*01 F	Homsap IGHD6-13*01 F	Homsap IGHJ4*02 F	C,W	96.86	304	91.67	43	475	7	0	7
41328	41328-Xall	14AA	ARDLDSSGWPIFDY		Homsap IGHV6-1*01 F	Homsap IGHD6-19*01 F	Homsap IGHJ4*02 F	C,W	100	427	H7CGN0J01APFR8	11	3	14	Sequences file	
			gcaagagacctcgatagcagtggctggcctatctttgactac	4	Homsap IGHV6-1*01 F	Homsap IGHD6-19*01 F	Homsap IGHJ4*02 F	C,W	98.22	304	91.37	44	424	7	3	10
41328	42	2	gcaagagatctagatagcagtggctggcccatttttgactac		Homsap IGHV6-1*01 F	Homsap IGHD6-19*01 F	Homsap IGHJ4*02 F	C,W	99.77	305	90.97	43	427	3	0	3
				0	Homsap IGHV6-1*01 F or Homsap IGHV6-1*02 F	Homsap IGHD6-19*01 F	Homsap IGHJ4*02 F	C,W	99.99	305	91.67	43	427	1	0	1

FIGURE 20.4 (*Continued*).

sequences and use of the tool by multiple users at the same time, the analyses are first queued on the IMGT server, and they are performed depending on the available resources [22].

The submission consists of the upload of a file (in simple text format) containing the user sequences in FASTA format (from 1 to 500,000 sequences). The customization for the "Detailed View" (13 options), "Files in CSV" (11 files, equivalent to the "Excel file" of IMGT/V-QUEST), and "Advanced parameters" (Figure 20.4A) is identical to that of IMGT/V-QUEST (except for two "Detailed View" options, IMGT Collier de Perles and IMGT/Phylogene, which are not available) [22]. The "Search for insertions and deletions in V-REGION," added by default (Figure 20.4A), allows an accurate V-DOMAIN analysis, despite the high frequency of indels due to homopolymer hybridization sequencing errors in NGS 454 sequencing [22,23].

4.2.2 IMGT/HighV-QUEST Results Download

From their account, users can check the status of their analyses at any time by displaying the "Analysis history" table [22]. When the analysis is completed, a user can "download" the results as a single file in ZIP format. The size of the ZIP file (in Mb) and the number of included files are indicated in the "Analysis history" table. For analyses having less than the expected number of files in the results file, a warning is shown. If the user removes a completed analysis, all related files (sequences and results) are definitively deleted from the server. A user may cancel a queued or a running analysis at any time.

The downloaded ZIP file [22] contains a main folder with 11 files (equivalent to the 11 sheets of the Excel file provided by the classical IMGT/V-QUEST) in CSV format (Table 20.6), and one subfolder with individual files, in Text, for each sequence (providing "Detailed view" results) [22]. Text and CSV formats have been chosen in order to facilitate statistical studies for further interpretation and knowledge extraction.

4.2.3 IMGT/HighV-QUEST Statistical Analysis

Following the IMGT/HighV-QUEST analysis, the results of completed analysis can be submitted for a statistical analysis, which provides a summary with the different results categories (the "1 copy" filtered-in category for the statistical analysis and the other filtered-out categories) and, for the "1 copy," in tables and histograms, the V, D, and J gene and allele usage observed in the "single allele" and "several alleles." For the submission, the user chooses completed results (several results can be chosen, up to 500,000 sequences) [22], a title for the statistical analysis, email notifications, and whether he or she wants a separate copy of the graphical elements (histograms) in the final results. The user can also enter a comment of up to 500 characters and choose to include it in the final PDF reports. All chosen results for a statistical analysis must be compatible (same locus, same value for the options "Search for insertions/deletions" and "IMGT reference directory set") and without warnings. The status of all statistical analyses can be checked using the "Statistics history" table.

The downloaded result file is a single file in ZIP format, comprising of five PDF reports [22]: (1) Summary, (2) Number of "1 copy" "single allele" and "several alleles (or genes)" tables and histograms, (3) V, D, and J genes and alleles tables and histograms for "single allele" and "several alleles (or genes)," (4) CDR3-IMGT tables and histograms, and (5) filtered-out sequences; and another general PDF report containing all these results together. If chosen by the user, the results contain also the graphical elements in a directory named "graphics."

4.2.4 IMGT/HighV-QUEST Identification of IMGT Clonotypes (AA)

In the literature, clonotypes are defined differently, depending on the experiment design (functional specificity) or available data. Thus, a clonotype may denote either a complete antigen receptor (e.g., IgG1-kappa), or only one of the two chains of the receptor (e.g., H or L), or one domain (e.g., VH), or the CDR3 sequence of a domain. Moreover, the sequence can be at the AA or nt level, and this is rarely specified. Therefore, the goal of IMGT was first of all to define clonotypes and their properties, which could be identified and characterized by IMGT/HighV-QUEST, unambiguously.

In IMGT, the clonotype, designated as "IMGT clonotype (AA)," is defined by a unique V-(D)-J rearrangement (with IMGT gene and allele names determined by IMGT/HighV-QUEST at the nt level) and a unique CDR3-IMGT AA (in-frame) junction sequence [23] (Figure 20.4B). For identifying "IMGT clonotypes (AA)" in a given IMGT/HighV-QUEST dataset, the "1 copy" are filtered to select for sequences with in-frame junction, conserved anchors 104 and 118 ("C" is 2nd-CYS 104, and "F" or "W" is the J-PHE or J-TRP 118) and for V and J functional or ORF, and "single allele" (for V and J) [23].

By essence, an "IMGT clonotype (AA)" is "unique" for a given dataset. For that reason, each "IMGT clonotype (AA)," in a given dataset, has a unique set identifier (column "Exp. ID") (Figure 20.4B) and, importantly, has a unique representative sequence (link in column "Sequence ID") selected by IMGT/HighV-QUEST among the "1 copy" "single allele" (for V and J), based on the highest percent of identity of the V-REGION ("V %") compared to that of the closest germline, and/or on the sequence length (thus the most complete V-REGION) [23].

In a second step, the "1 copy" "single allele" sequences not selected as representative are assigned to a characterized IMGT clonotype (AA). These sequences differ from the representative sequence by a different (usually shorter) length, and/or by sequencing errors in the V-REGION (lower "V %" of identity) or in the J-REGION, and/or by nt differences in the CDR3-IMGT. In these latter cases these sequences are identified as "IMGT clonotypes (nt)," the nt differences resulting from sequencing errors or, if this can be proven experimentally, from molecular convergence. For a given "IMGT clonotype (AA)," the number (nb) of different CDR3-IMGT (nt) or "IMGT clonotypes (nt)," the CDR3-IMGT sequence (nt) and the nb of different nt in the CDR3-IMGT are reported in the results (Figure 20.4C) [23].

TABLE 20.6 Content of the 11 IMGT/HighV-QUEST Results Files

File number	File name	Number of columns	Results content[a]
#1	Summary	25–29	• Identity percentage with the closest V, D, and J genes and alleles, • FR-IMGT and CDR-IMGT lengths, • Amino acid (AA) JUNCTION, • Description of insertions and deletions if any.
#2	IMGT-gapped-nt-sequences	18	• Nucleotide (nt) sequences gapped according to the IMGT unique numbering for the labels V-D-J-REGION, V-J-REGION, V-REGION, FR1-IMGT, CDR1-IMGT, FR2-IMGT, CDR2-IMGT, FR3-IMGT, • nt sequences of CDR3-IMGT, JUNCTION, J-REGION, and FR4-IMGT.
#3	Nt-sequences	63–78	• nt sequences of all labels that can be automatically annotated by IMGT/Automat.
#4	IMGT-gapped-AA-sequences	18	• AA sequences gapped according to the IMGT unique numbering for the labels V-D-J-REGION, V-J-REGION, V-REGION, FR1-IMGT, CDR1-IMGT, FR2-IMGT, CDR2-IMGT, FR3-IMGT, • AA sequences of CDR3-IMGT, JUNCTION, J-REGION, and FR4-IMGT.
#5	AA-sequences	18	Same columns as "IMGT-gapped-AA-sequences" (#4), but sequences of labels are without IMGT gaps.
#6	Junction	33, 46, 66, or 77	• Results of IMGT/JunctionAnalysis (JUNCTION, CDR3, N regions, trimmed V(D)J nucleotides, P regions, D regions, and D-REGION reading frame (for IGH, TRB, and TRD sequences), molecular mass, pI.
#7	V-REGION-mutation-and-AA-change table	11	• List of mutations (nt mutations, AA changes, AA class identity (+) or change (−)) for V-REGION, FR1-IMGT, CDR1-IMGT, FR2-IMGT, CDR2-IMGT, FR3-IMGT, and germline CDR3-IMGT.
#8	V-REGION-nt-mutation-statistics	130	• Number (nb) of nt positions including IMGT gaps, nb of nt, nb of identical nt, total nb of mutations, nb of silent mutations, nb of nonsilent mutations, nb of transitions (a > g, g > a, c > t, t > c), and nb of transversions (a > c, c > a, a > t, t > a, g > c, c > g, g > t, t > g) for V-REGION, FR1-IMGT, CDR1-IMGT, FR2-IMGT, CDR2-IMGT, FR3-IMGT, and germline CDR3-IMGT.
#9	V-REGION-AA-change-statistics	189	• nb of AA positions including IMGT gaps, nb of AA, nb of identical AA, total nb of AA changes, nb of AA changes according to AAclassChangeType (+ + +, + + −, + − +, + − −, − + −, − − +, − − −), and nb of AA class changes according to AAclassSimilarityDegree (nb of very similar, nb of similar, nb of dissimilar, nb of very dissimilar) for V-REGION, FR1-IMGT, CDR1-IMGT, FR2-IMGT, CDR2-IMGT, FR3-IMGT, and germline CDR3-IMGT.
#10	V-REGION-mutation-hotspots	8	• Hot spots motifs (a/t)a, t(a/t), (a/g)g(c/t) (a/t), (a/t) (a/g)c(c/t) detected in the closest germline V-REGION with positions in FR-IMGT and CDR-IMGT.
#11	Parameters		• Date of the analysis, • IMGT/V-QUEST program version, IMGT/V-QUEST reference directory release, • Parameters used for the analysis: Species, receptor type or locus, IMGT reference directory set, and Advanced parameters.

The content of the 11 results files is in CVS format (results equivalent to those of the Excel file from IMGT/V-QUEST online).
[a]*Files1–10 comprise systematically sequence identification (name, functionality, and names of the closest V, D, and J genes and alleles [2,3]. Files 7–10, which report the analysis of mutations, are used mostly for immunoglobulins (IG).*

Thus, by proceeding stepwise to assign sequences to clonotypes, aggregating to the representative sequence, first the nonrepresentative "1 copy" "single allele," then the other categories ("1 copy" "several alleles [or genes]," and "More than 1" for each "1 copy" [22]), the high quality and specific characterization of the "IMGT clonotype (AA)" remain unaltered [23]. For the first time for NGS antigen receptor data analysis, the IMGT standardized approach allows a clear distinction and accurate evaluation between the clonal diversity (nb of "IMGT clonotypes (AA)"), and the clonal expression (total nb of sequences assigned, unambiguously, to a given "IMGT clonotype (AA)") [23]. These assignments are clearly described and visualized in detail so the user always has the means of checking clonotypes individually. Indeed, the sequences of each "1 copy" assigned to a given "IMGT clonotype (AA)" are available in "Sequences file" (Figure 20.4B and C) [23]. The user can easily perform an analysis of these sequences on line with IMGT/V-QUEST (up to 50 sequences, selecting "Synthesis view display" and the option "Search for insertions and deletions"), and/or with IMGT/JunctionAnalysis (up to 5000 junction sequences), which provide a visual representation familiar to the IMGT users (Figure 20.3D).

Clonal diversity is also visualized in the online results with histograms which represent the number of IMGT clonotypes (AA) per V, D (for IGH), and J genes (in pink) [23]. Clonal expression is visualized with histograms, which represent the number of sequences assigned to IMGT clonotypes (AA) per V (in green), D (in red), and J (in yellow) genes [23]. Values are normalized, respectively, for 10,000 IMGT clonotypes (AA) to represent IG diversity immunoprofiles per V, D (for IGH), and J genes, and for 10,000 sequences assigned to IMGT clonotypes (AA) to represent IG expression immunoprofiles per V, D (for IGH), and J genes [23]. These normalized values allow comparative analysis studies performed with the same IMGT/HighV-QUEST standards [23].

5. IMGT ANTIBODY ENGINEERING AND HUMANIZATION

5.1 IMGT/DomainGapAlign

IMGT/DomainGapAlign [9,24,25] is the IMGT online tool for the analysis of amino acid sequences and 2D structures of domains (V and C for IG) (Table 20.5). It analyzes domain amino acid sequences by comparison with the IMGT domain reference directory sets (translation of the germline V and J genes and of the C gene domains). IMGT/DomainGapAlign functionalities include: introduction of "IMGT gaps" in the user amino acid sequences, alignments and identification of the genes and alleles by comparison with the closest domain(s), delimitation of the domain(s) (e.g., V or C) in the user sequence (Figure 20.5), description of the amino acid changes, and IMGT Collier de Perles.

5.1.1 Submission to IMGT/DomainGapAlign

In the IMGT/DomainGapAlign Welcome page, amino acid sequences are submitted in FASTA format (pasted in a text area or uploaded in a file). A precise delimitation of the domain sequences is not required; however, if the sequence contains several domains, the sequence should be split between the different domains. Several domain amino acid sequences can be analyzed simultaneously (up to 50) provided that each sequence has a distinct name and that they all belong to the same domain type (e.g., V or C). If not, the query needs to be launched for each domain type, successively. If the limits and the numbers of domains of an amino acid sequence are unknown, the protein can be analyzed progressively, shortening the sequence once a domain has been identified by the tool (it should be remembered that the first domain identified by the tool is not necessarily the first one in the protein).

5.1.2 IMGT/DomainGapAlign Results

IMGT/DomainGapAlign identifies the closest germline V-REGION and J-REGION alleles (for V domain) and the closest C-DOMAIN alleles (for C domain). IMGT/DomainGapAlign displays the user V-DOMAIN sequences aligned with the closest V and J regions, with IMGT gaps and delimitations of the strands and loops and the FR-IMGT and CDR-IMGT, according to the IMGT unique numbering. In Figure 20.5A, the VH sequence (from b12 Fab, 1n0x_H) is identified as having 79.6% and 93.8% identity with the *H. sapiens* IGHV1-3*01 and IGHJ6*03, respectively. The VH CDR-IMGT lengths are [8.8.20] and the FR-IMGT lengths [25.17.38.11] = 91 AA. If several closest genes and/or alleles are identified, the user can select the display of each corresponding alignment. Clicking on the user sequence name in the alignment gives access to the IMGT/Collier-de-Perles tool [26], which automatically provides the IMGT Collier de Perles of the analyzed domain [65–68] with highlighted amino acid differences (in pink circles online) with the closest germline sequence (Figure 20.1C).

The user amino acid sequence is also displayed, according to the IMGT color menu, with the delimitations of the V-REGION (in green) and J-REGION (in yellow), and for a VH, of the (N-D)-REGION (in red), identified by the tool (Figure 20.5A). The number of amino acid differences in the FR-IMGT and CDR-IMGT is one of the criteria to evaluate the potential immunogenicity of the V-DOMAIN [76–80]. The characteristics of the AA changes are shown in strands and loops (and, for the V-DOMAIN, in FR-IMGT and CDR-IMGT).

IMGT/DomainGapAlign analyzes the user C-DOMAIN sequences with similar functionalities

(A)

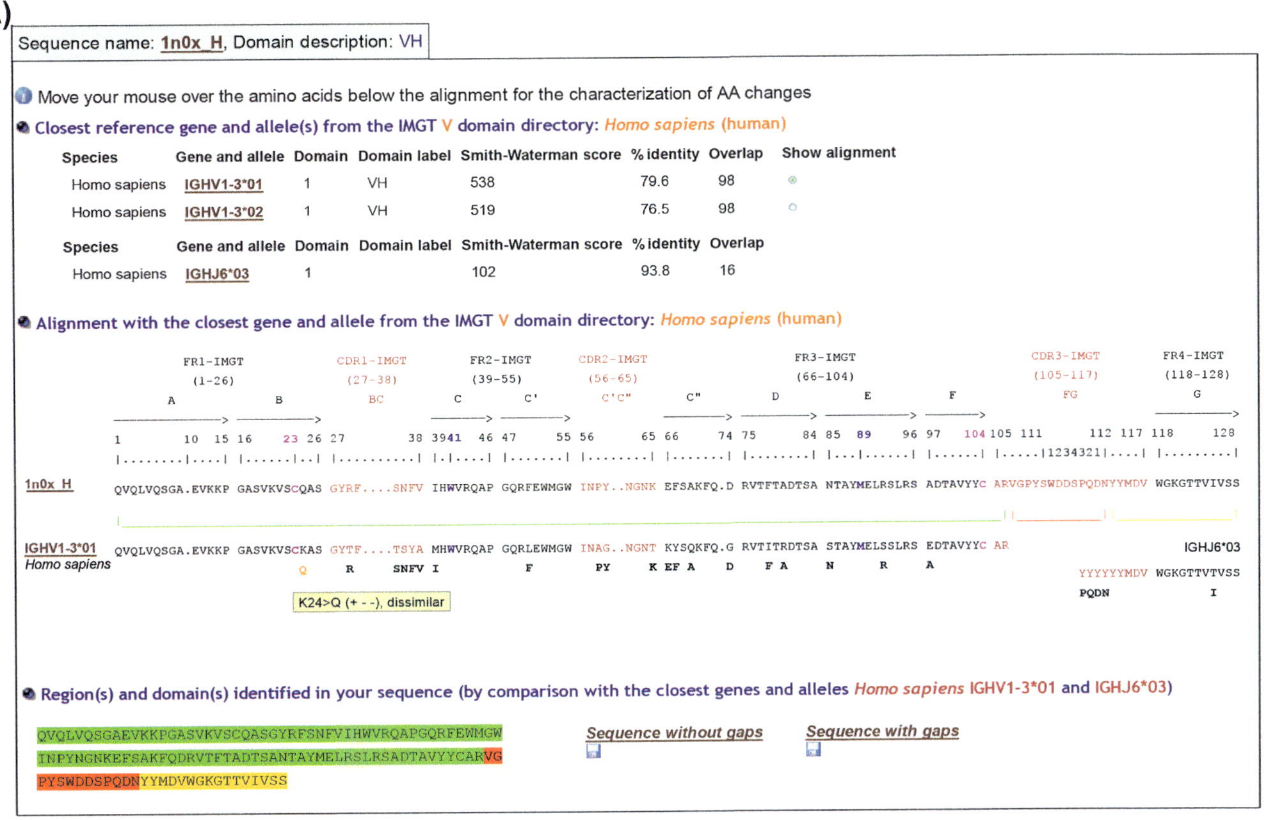

FIGURE 20.5 **IMGT/DomainGapAlign alignment results.** (A) VH. The nine strands (A, B, C, C′, C″, D, E, F, and G) (horizontal arrows) and the three loops (BC, C′C″, and FG) and, on the upper line, the FR-IMGT and CDR-IMGT and their delimitations (start and end positions) are shown according to the IMGT unique numbering for V domain [59–62,64] (Table 20.2). The closest genes and alleles are identified automatically by IMGT/DomainGapAlign by comparison with the IMGT domain reference directory (V and J genes and alleles for a V-DOMAIN). The analyzed VH sequence is identified as having 79.6% identity with the germline *Homo sapiens* IGHV1-3*01 for its V-REGION and 93.8% identity with the germline *Homo sapiens* IGHJ6*03 for its J-REGION (the VH amino acid sequence is from b12 Fab, 1n0x_H, in IMGT/3Dstructure-DB, http://www.imgt.org, and was submitted ungapped). The VH CDR-IMGT lengths are [8.8.20] and the FR-IMGT lengths [25.17.38.11] = 91 AA (results online, below the alignment). The V-REGION of the b12 VH sequence is heavily mutated as shown by the high number (20) of amino acid changes, shown in bold below the alignment, and online (not shown) detailed per strand and per loop online in the IMGT/DomainGapAlign results [9,24,25]. One AA change is also observed in the FR4-IMGT (T125 > I). Moving the mouse cursor on a letter in bold displays the AA change according to its IMGT AA classes [28], as shown for "K24 > Q (+ − −), dissimilar." horizontal green and yellow lines identify the V-REGION and J-REGION, respectively. The region localized in the CDR3-IMGT, which results from the V-(D)-J rearrangement and which cannot be identified as being V or J, is the (N-D)-REGION (shown with a horizontal red line). Below the alignment, the domain sequence, delimited by the tool, is also visualized with the identified V-REGION (green), (N-D)-REGION (red), and J-REGION (yellow) (IMGT, http://www.imgt.org, IMGT Scientific chart > IMGT color menu). (B) CH1 domain. The analyzed CH1 domain is identified as having 100% identity at the amino acid level with the *Homo sapiens* IGHG1*01, IGHG1*02, and IGHG1*04 CH1 (the CH1 amino acid sequence is from b12 Fab, 1n0x_H, in IMGT/3Dstructure-DB, http://www.imgt.org, and was submitted ungapped). IMGT/DomainGapAlign displays the C domain amino acid sequence of the user, with IMGT gaps and delimitations of the strands (shown with arrows, except for the transversal CD strand), turns (AB, DE, and EF), and loops (BC and FG), according to the IMGT unique numbering [62] (Table 20.4). A horizontal blue line identifies the C-DOMAIN. The lower alignment is by comparison with the IGHG1*03 CH1 to highlight the AA change R120 > K, which corresponds to a different Gm allotype [74], as detailed later in the text and in Table 20.7. The four conserved AA between the V and C domains are in bold and in color online: C23 (pink), W41 (blue), hydrophobic 89 (blue, here M and V, respectively), and C104 (pink) (IMGT, http://www.imgt.org, IMGT Scientific chart > IMGT color menu).

(Figure 20.5B): alignments and identification of the genes and alleles with the closest C domain; delimitation of the C-DOMAIN in the user sequence; characteristics of the AA changes in strands, turns, and loops; IMGT Collier de Perles of the C-DOMAIN with highlighted amino acid differences (in pink circles online) with the closest reference sequence.

5.1.3 IMGT Domain Reference Directory

The IMGT domain reference directory is the IMGT reference directory for V, C, and G domains. Sequences are from the IMGT Repertoire [1] and from IMGT/GENE-DB [7]. Owing to the particularities of the V-DOMAIN synthesis [2,3], there is no V-DOMAIN in the IMGT reference directory. Instead, the directory comprises the translation of

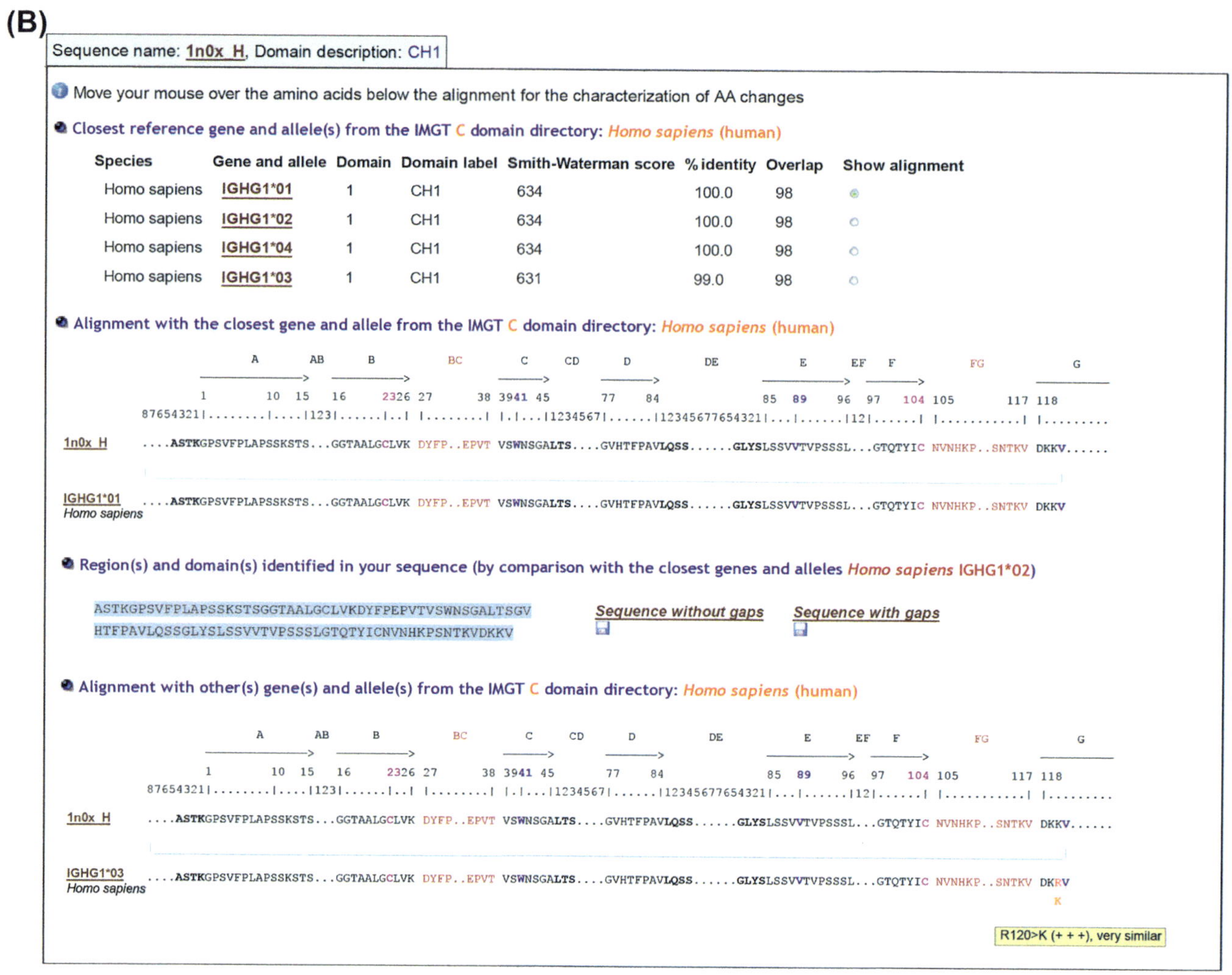

FIGURE 20.5 (Continued).

the IG and TR germline V and J genes (V-REGION and J-REGION, respectively). The IMGT domain reference directory provides the IMGT "gene" and "allele" names. Data are comprehensive for human and mouse IG (and also TR), whereas for other species and other IgSF and MhSF they are added progressively. The IMGT domain reference directory comprises domain sequences of functional (F), ORF (open reading frame), and in-frame pseudogene (P) genes. As IMGT alleles are characterized at the nucleotide level, identical sequences at the amino acid level may therefore correspond to different alleles in the IMGT domain reference directory. The sequences can be displayed by querying IMGT/DomainDisplay (http://www.imgt.org).

5.2 IMGT/3Dstructure-DB

IMGT/3Dstructure-DB [8–10], the IMGT structure database, provides IMGT annotation and contact analysis of IG 3D structures and paratope/epitope description of IG/antigen complexes (Table 20.5). There is one "IMGT/3D structure-DB card" per IMGT/3Dstructure-DB entry and this card provides access to all data related to that entry. The "PDB code" (4 letters and/or numbers, e.g., 1n0x) is used as "IMGT entry ID" for the 3D structures obtained from the Research Collaboratory for Structural Bioinformatics (RCSB) Protein Data Bank (PDB) [94]. The IMGT/3D structure-DB card provides eight search/display options: "Chain details," "Contact analysis," "Paratope and epitope," "3D visualization Jmol or QuickPDB," "Renumbered IMGT files," "IMGT numbering comparison," "References and links," and "Printable card" [8–10].

5.2.1 IMGT Chain and Domain Annotation

The "Chain details" section comprises information first on the chain itself, then per domain [8–10]. Chain and

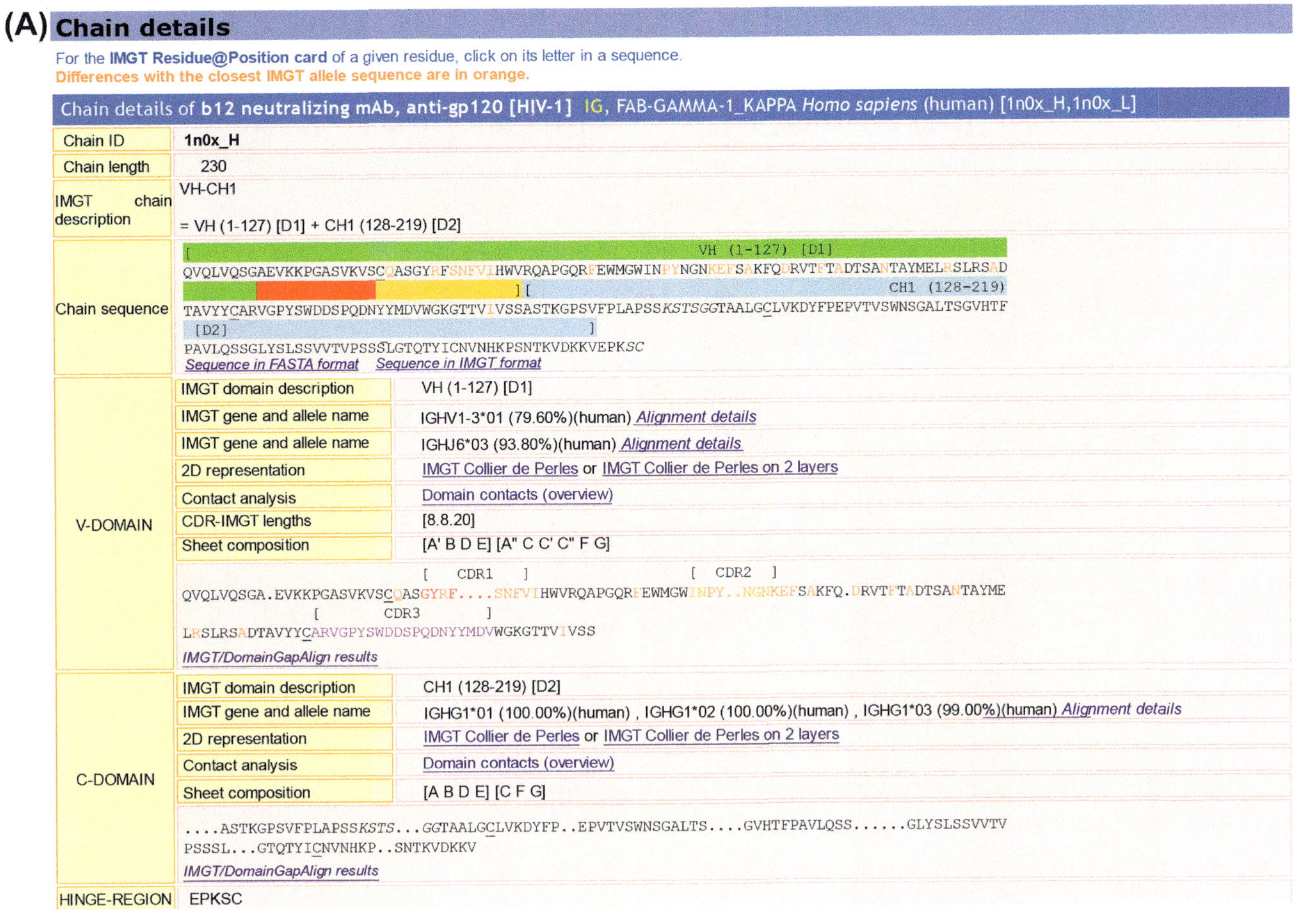

FIGURE 20.6 IMGT/3Dstructure-DB. (A) IMGT/3Dstructure-DB card. The "IMGT/3Dstructure-DB card" is available for each entry of the database. The "chain details" shows, first, information on the chain (Chain ID, Chain length, IMGT chain description, Chain sequence), then a detailed description of each domain of the chain. The description of the V-DOMAIN (VH) and C-DOMAIN (CH1) of the VH-CH1 chain (1n0x_H) of the b12 Fab is shown. A similar result display interface is provided in IMGT/2Dstructure-DB cards but without "contact analysis" (and without hydrogen bonds in IMGT Collier de Perles on two layers). (B) IMGT/3Dstructure-DB domain pair contacts between the "VH" and the "ligand" (antigen, Ag) of an IG/Ag complex. The VH is from the VH-CH1 chain (1n0x_H) of the b12 Fab and the ligand is a synthetic peptide (1n0x_P). The VH is in contact with the ligand by three AA of the CDR2-IMGT (orange online) (N62, N64, and K65) and two AA of the CDR3-IMGT (purple online) (P112.3 and Q112.2). The two AA that interact with the ligand but do not belong to the CDR-IMGT are the anchors W55 and E66. These contacts are not unexpected given the small size (peptide) of the ligand. (C) IMGT/3Dstructure-DB domain pair contacts between the "V-KAPPA" and the "ligand" (Ag) of an IG/Ag complex. The V-KAPPA is from the L-KAPPA chain (1n0x_L) of the b12 Fab and the ligand is the peptide (1n0x_P), as in (B). The V-KAPPA is in contact with the ligand by seven AA of the CDR1-IMGT (blue online) (H27, S28, I29, R30, S36, R37, and R38) and three AA of the CDR3-IMGT (green-blue online) (A109, S114, and S115). "Polar," "hydrogen bond," and "nonpolar" are selected by default in "atom pair contact types" options at the bottom of the page (not shown). The user can also choose to display these contacts by "atom pair contact categories" (BB), (SS), (BS), and (SB). Clicking on R@P gives access to the IMGT Residue@Position card (not shown). (D) "IMGT paratope and epitope details" of an IG/Ag complex. Each AA that belongs to the IG paratope is characterized by its position in the V-DOMAIN according to the IMGT unique numbering [59–62,64] (AA are listed first for the VH, then for the VL). As examples, "N(62_H)" is the asparagine (N) at position 62 of the VH of the 1n0x_H chain whereas S(28_L) is the serine (S) at position 28 of the V-KAPPA of the 1n0x_L chain. Each AA that belongs to the antigenic determinant (epitope) is characterized by its position in the ligand chain, in the 3D structure. For example, "N(12_P)" means that the asparagine (N) is at position 12 of the 1n0x_P chain. The "IMGT paratope and epitope" analysis of the IG/Ag 3D structure (1n0x) is from IMGT/3Dstructure-DB, http://www.imgt.org.

domain annotation includes the IMGT gene and allele names (CLASSIFICATION), region and domain delimitations (DESCRIPTION), and domain AA positions according to the IMGT unique numbering (NUMEROTATION) [59–64] (Figure 20.6A). The closest IMGT genes and alleles (found expressed in each domain of a chain) are identified with the integrated IMGT/DomainGapAlign [9,24,25], which aligns the amino acid sequences of the 3D structures with the IMGT domain reference directory.

(B) IMGT/3Dstructure-DB Domain pair contacts

Contacts of

Domain	Chain
[D1] VH	1n0x_H

with

Domain	Chain
(Ligand)	1n0x_P

Summary:

Residue pair contacts	Number of residues			Atom pair contact types			
	Total	From 1	From 2	**Total**	**Polar**	**Hydrogen**	**Nonpolar**
17	15	7	8	**106**	**10**	**1**	**96**

List of the Residue@Position pair contacts:

Click 'R@P' for IMGT Residue@Position cards

	Order						Order						Atom pair contact types			
	IMGT Num	**Residue**		**Domain**	**Chain**		**IMGT Num**	**Residue**		**Domain**	**Chain**	**Total**	**Polar**	**Hydrogen**	**Nonpolar**
R@P	55	TRP	W	[D1] VH	1n0x_H	R@P	19	ORN		(Ligand)	1n0x_P	**6**	**0**	**0**	**6**
R@P	55	TRP	W	[D1] VH	1n0x_H	R@P	20	LYS	K	(Ligand)	1n0x_P	**8**	**0**	**0**	**8**
R@P	62	ASN	N	[D1] VH	1n0x_H	R@P	19	ORN		(Ligand)	1n0x_P	**1**	**0**	**0**	**1**
R@P	64	ASN	N	[D1] VH	1n0x_H	R@P	18	GLU	E	(Ligand)	1n0x_P	**1**	**1**	**0**	**0**
R@P	64	ASN	N	[D1] VH	1n0x_H	R@P	19	ORN		(Ligand)	1n0x_P	**15**	**2**	**1**	**13**
R@P	64	ASN	N	[D1] VH	1n0x_H	R@P	20	LYS	K	(Ligand)	1n0x_P	**8**	**1**	**0**	**7**
R@P	65	LYS	K	[D1] VH	1n0x_H	R@P	19	ORN		(Ligand)	1n0x_P	**1**	**1**	**0**	**0**
R@P	65	LYS	K	[D1] VH	1n0x_H	R@P	20	LYS	K	(Ligand)	1n0x_P	**8**	**1**	**0**	**7**
R@P	66	GLU	E	[D1] VH	1n0x_H	R@P	20	LYS	K	(Ligand)	1n0x_P	**9**	**2**	**0**	**7**
R@P	66	GLU	E	[D1] VH	1n0x_H	R@P	21	LYS	K	(Ligand)	1n0x_P	**2**	**0**	**0**	**2**
R@P	112.3	PRO	P	[D1] VH	1n0x_H	R@P	4	SER	S	(Ligand)	1n0x_P	**5**	**0**	**0**	**5**
R@P	112.3	PRO	P	[D1] VH	1n0x_H	R@P	5	TYR	Y	(Ligand)	1n0x_P	**9**	**1**	**0**	**8**
R@P	112.3	PRO	P	[D1] VH	1n0x_H	R@P	6	MET	M	(Ligand)	1n0x_P	**19**	**0**	**0**	**19**
R@P	112.3	PRO	P	[D1] VH	1n0x_H	R@P	17	ALA	A	(Ligand)	1n0x_P	**2**	**0**	**0**	**2**
R@P	112.2	GLN	Q	[D1] VH	1n0x_H	R@P	6	MET	M	(Ligand)	1n0x_P	**10**	**0**	**0**	**10**
R@P	112.2	GLN	Q	[D1] VH	1n0x_H	R@P	17	ALA	A	(Ligand)	1n0x_P	**1**	**0**	**0**	**1**
R@P	112.2	GLN	Q	[D1] VH	1n0x_H	R@P	21	LYS	K	(Ligand)	1n0x_P	**1**	**1**	**0**	**0**

FIGURE 20.6 *(Continued).*

5.2.2 Contact Analysis

"Contact analysis" gives access to a table with the different "Domain pair contacts" of the 3D structure (this table is also accessed from "Chain details" by clicking on "Domain contact (overview)"). "Domain pair contacts" refer to contacts between a pair of domains or between a domain and a ligand. Clicking on "DomPair" gives access to the contacts between amino acids for a given "Domain pair contacts." Contacts between VH and the Ligand (antigen, Ag) and the V-KAPPA and the Ligand (Ag) of an IG/Ag complex are shown in Figure 20.6B and C, respectively. These contact analysis representations are important as they demonstrate that most, if not all, contacts with the ligands involved the amino acids of the CDR-IMGT. They definitively confirmed the CDR-IMGT delimitations as the official reference standards [64,68,91].

In IMGT/3Dstructure-DB, all contacts are described as atom pair contacts. Atom pair contacts are obtained by a local program in which atoms are considered to be in contact when no water molecule can take place between them [8,9]. Atom pair contacts are provided by atom contact types (noncovalent, polar, hydrogen bond, nonpolar, covalent, disulfide) and/or atom contact categories (BB, backbone/backbone; SS, side chain/side chain; BS, backbone/side chain; SB, side chain/backbone) [8,9,88,89].

(C) IMGT/3Dstructure-DB Domain pair contacts

Contacts of [D1] V-KAPPA 1n0x_L (Domain, Chain) with (Ligand) 1n0x_P (Domain, Chain)

Summary:

Residue pair contacts	Number of residues			Atom pair contact types			
	Total	From 1	From 2	Total	Polar	Hydrogen	Nonpolar
24	19	12	7	195	32	3	163

List of the Residue@Position pair contacts:

Click 'R@P' for IMGT Residue@Position cards

	Order						Order					Atom pair contact types			
	IMGT Num	Residue		Domain	Chain		IMGT Num	Residue		Domain	Chain	Total	Polar	Hydrogen	Nonpolar
R@P	1	GLU	E	[D1] V-KAPPA	1n0x_L	R@P	10	LEU	L	(Ligand)	1n0x_P	4	0	0	4
R@P	2	ILE	I	[D1] V-KAPPA	1n0x_L	R@P	9	ASP	D	(Ligand)	1n0x_P	6	0	0	6
R@P	2	ILE	I	[D1] V-KAPPA	1n0x_L	R@P	10	LEU	L	(Ligand)	1n0x_P	11	0	0	11
R@P	27	HIS	H	[D1] V-KAPPA	1n0x_L	R@P	9	ASP	D	(Ligand)	1n0x_P	7	1	0	6
R@P	27	HIS	H	[D1] V-KAPPA	1n0x_L	R@P	10	LEU	L	(Ligand)	1n0x_P	11	1	0	10
R@P	28	SER	S	[D1] V-KAPPA	1n0x_L	R@P	9	ASP	D	(Ligand)	1n0x_P	10	3	0	7
R@P	29	ILE	I	[D1] V-KAPPA	1n0x_L	R@P	9	ASP	D	(Ligand)	1n0x_P	26	3	0	23
R@P	30	ARG	R	[D1] V-KAPPA	1n0x_L	R@P	7	PHE	F	(Ligand)	1n0x_P	21	0	0	21
R@P	30	ARG	R	[D1] V-KAPPA	1n0x_L	R@P	9	ASP	D	(Ligand)	1n0x_P	10	2	0	8
R@P	30	ARG	R	[D1] V-KAPPA	1n0x_L	R@P	12	ASN	N	(Ligand)	1n0x_P	11	4	0	7
R@P	36	SER	S	[D1] V-KAPPA	1n0x_L	R@P	7	PHE	F	(Ligand)	1n0x_P	8	1	0	7
R@P	36	SER	S	[D1] V-KAPPA	1n0x_L	R@P	9	ASP	D	(Ligand)	1n0x_P	14	4	2	10
R@P	37	ARG	R	[D1] V-KAPPA	1n0x_L	R@P	9	ASP	D	(Ligand)	1n0x_P	1	1	0	0
R@P	38	ARG	R	[D1] V-KAPPA	1n0x_L	R@P	6	MET	M	(Ligand)	1n0x_P	2	0	0	2
R@P	38	ARG	R	[D1] V-KAPPA	1n0x_L	R@P	7	PHE	F	(Ligand)	1n0x_P	10	4	1	6
R@P	38	ARG	R	[D1] V-KAPPA	1n0x_L	R@P	8	SER	S	(Ligand)	1n0x_P	3	1	0	2
R@P	38	ARG	R	[D1] V-KAPPA	1n0x_L	R@P	9	ASP	D	(Ligand)	1n0x_P	9	5	0	4
R@P	109	ALA	A	[D1] V-KAPPA	1n0x_L	R@P	6	MET	M	(Ligand)	1n0x_P	8	0	0	8
R@P	109	ALA	A	[D1] V-KAPPA	1n0x_L	R@P	7	PHE	F	(Ligand)	1n0x_P	2	0	0	2
R@P	109	ALA	A	[D1] V-KAPPA	1n0x_L	R@P	8	SER	S	(Ligand)	1n0x_P	6	1	0	5
R@P	109	ALA	A	[D1] V-KAPPA	1n0x_L	R@P	10	LEU	L	(Ligand)	1n0x_P	2	0	0	2
R@P	109	ALA	A	[D1] V-KAPPA	1n0x_L	R@P	15	ILE	I	(Ligand)	1n0x_P	3	0	0	3
R@P	114	SER	S	[D1] V-KAPPA	1n0x_L	R@P	10	LEU	L	(Ligand)	1n0x_P	2	0	0	2
R@P	115	SER	S	[D1] V-KAPPA	1n0x_L	R@P	10	LEU	L	(Ligand)	1n0x_P	8	1	0	7

FIGURE 20.6 (*Continued*).

Clicking on "R@P" gives access to the IMGT identity card of a given residue (usually an amino acid) at a given position or Residue@Position. The IMGT R@P card can also be accessed from the amino acid sequences of the IMGT/3Dstructure-DB card or from the IMGT Collier de Perles, by clicking on one amino acid. In an IMGT R@P card, the Residue@Position is defined by the IMGT position numbering in a domain, or if not characterized, in the chain, the amino acid name (3-letter and, between parentheses, 1-letter abbreviations), the IMGT domain description, and the IMGT chain ID, e.g., "103 - TYR (Y) - VH - 1hzh_H" [8–10]. The IMGT R@P card includes: (1)

(D) **IMGT paratope and epitope details**

For the **IMGT Residue@Position card** of a given residue, click on its letter in a sequence.

Paratope details of **b12 neutralizing mAb, anti-gp120 [HIV-1] [Chain(s): 1n0x_H,1n0x_L]**

Paratope in **b12 neutralizing mAb, anti-gp120 [HIV-1] [VH 1n0x_H, V-KAPPA 1n0x_L]**

type	discontinuous
residues	W N N K E P Q E I H S I R S R R A S S
With positions	W(55_H)+N(62_H)+N(64_H)+K(65_H)+E(66_H)+P(112.3_H)+Q(112.2_H)+E(1_L)+I(2_L)+H(27_L)+S(28_L) +I(29_L)+R(30_L)+S(36_L)+R(37_L)+R(38_L)+A(109_L)+S(114_L)+S(115_L)

Epitope details of **B2.1 peptide [1n0x_P]** and **b12 neutralizing mAb, anti-gp120 [HIV-1] (FAB-GAMMA-1_KAPPA) Homo sapiens(human) [1n0x_H,1n0x_L]**

Epitope in **B2.1 peptide [1n0x_P]**

type	discontinuous
residues	S Y M F S D L N I A E (ORN) K K *Epitope IMGT Residue@Position cards*
With position	S Y M F S D L(4-10_P)+N(12_P)+I(15_P)+A E (ORN) K K(17-21_P)

FIGURE 20.6 (*Continued*).

general information (PDB file numbering, IMGT file numbering, residue full name and formula); (2) structural information "IMGT LocalStructure@Position" (secondary structure, Phi and Psi angles (in degrees), and accessible surface area (ASA) (in square angstrom)); and (3) detailed contact analysis with amino acids of other domains [8–10].

5.2.3 Paratope and Epitope

In an IG/Ag complex, the amino acids in contact at the interface between the IG and the Ag constitute the paratope on the IG V-DOMAIN surface and the epitope on the Ag surface. For IG/Ag, the paratope and epitope are displayed in Contact analysis (Figure 20.6B and C), but for each V domain separately. Clicking on the "Paratope and epitope" tag (displayed in the IMGT/3Dstructure-DB card, only if relevant) gives access to "IMGT paratope and epitope details," which are described in a standardized way (Figure 20.6D). Each amino acid that belongs to the paratope is defined by its position in an IG V-DOMAIN (e.g., VH and V-KAPPA of 1n0x in Figure 20.6D). Each amino acid that belongs to the epitope is defined by its position in the chain in the 3D structure or, if the antigen belongs to an IgSF or MhSF protein and if the epitope is part of a characterized V, C, or G domain, by its position in the domain according to the IMGT unique numbering.

5.2.4 Renumbered Flat File and IMGT Numbering Comparison

"Renumbered IMGT file" allows viewing (or downloading) of an IMGT coordinate file renumbered according to the IMGT unique numbering, and with added IMGT specific information on chains and domains (added in the "REMARK 410" lines [blue online] and identical to the "Chain details" annotation).

"IMGT numbering comparison" provides, per domain, the IMGT DOMAIN numbering by comparison with the PDB numbering and the residue (3-letter and 1-letter names), which allows standardized IMGT representations using generic tools (Figures 20.1A and 20.2A).

5.2.5 IMGT/3Dstructure-DB Associated Tools

Tools associated with IMGT/3Dstructure-DB include IMGT/StructuralQuery [8] and IMGT/DomainSuperimpose, available online. IMGT/StructuralQuery allows retrieval of the IMGT/3Dstructure-DB entries, based on specific structural characteristics of the intramolecular interactions: phi and psi angles, accessible surface area, type of atom contacts, distance in angstrom between amino acids, IMGT Residue@Position contacts, and, for V-DOMAIN, CDR-IMGT length or pattern [8]. IMGT/DomainSuperimpose allows superimposing of the 3D structures of two domains from IMGT/3Dstructure-DB.

5.3 IMGT/2Dstructure-DB

IMGT/2Dstructure-DB was created as an extension of IMGT/3Dstructure-DB [8–10] to describe and analyze amino acid sequences of chains and domains for which no 3D structures were available (Table 20.5). IMGT/2Dstructure-DB uses the IMGT/3Dstructure-DB informatics frame and interface, which allow one to analyze, manage, and query IG (and also TR and MH, as well as other IgSF and MhSF) and engineered proteins (FPIA, CPCA) as polymeric receptors made of several chains, in contrast to the IMGT/LIGM-DB sequence database that analyzes and manages sequences individually [6]. The amino acid sequences are analyzed with the IMGT criteria of standardized identification [56], description [57], nomenclature [58], and numerotation [59–64]. The current IMGT/2Dstructure-DB entries include amino acid sequences of antibodies from Kabat [92] (those for which there were no available nucleotide sequences) and amino acid sequences of mAb and FPIA from the WHO-INN program [11,47,48]. Queries can be made on an individual entry using the entry ID or the molecule name. The same query interface is used for IMGT/2Dstructure-DB and IMGT/3Dstructure-DB. Thus a "trastuzumab" query in "Molecule name" allows retrieval of three results: two INN ("trastuzumab" and "trastuzumab emtansine") from IMGT/2Dstructure-DB and one 3D structure ("1nz8") from IMGT/3Dstructure-DB. The IMGT/2Dstructure-DB cards provide standardized IMGT information on chains and domains and IMGT Collier de Perles on one or two layers, identical to that provided for the sequence analysis in IMGT/3Dstructure-DB; however, the information on experimental structural data (hydrogen bonds in IMGT Collier de Perles on two layers, Contact analysis) is only available in the corresponding IMGT/3Dstructure-DB cards if the antibodies have been crystallized.

5.4 IMGT V and C Domain Annotation for Antibody Humanization, Engineering, and Evolution

5.4.1 Antibody Humanization

5.4.1.1 CDR-IMGT Delimitation for Grafting

The objective of antibody humanization is to graft at the DNA level the CDR of an antibody V domain, from mouse (or other species) and of a given specificity, onto a human V domain framework, thus preserving the specificity of the original (murine or other species) antibody while decreasing its immunogenicity [95]. IMGT/DomainGapAlign [9,24,25] is the reference tool for antibody humanization design based on CDR grafting. Indeed, it precisely defines the CDR-IMGT to be grafted and helps in selecting the most appropriate human FR-IMGT by providing the alignment of the amino acid sequences between the mouse (or other species) and the closest human V-DOMAIN.

Analyses performed on humanized therapeutic antibodies underline the importance of a correct delimitation of the CDR and FR. As an example, two amino acid changes were required in the first version of the humanized VH of alemtuzumab, in order to restore the specificity and affinity of the original rat antibody. The positions of these amino acid changes (S28 > F and S35 > F) are now known to be located in the CDR1-IMGT and should have been directly grafted, but at the time of this mAb humanization they were considered as belonging to the FR according to the Kabat numbering [92]. In contrast, positions 66–74 were, at the same time, considered as belonging to the CDR according to the Kabat numbering, whereas they clearly belong to the FR2-IMGT and the corresponding sequence should have been "human" instead of being grafted from the "rat" sequence (IMGT, http://www.imgt.org, The IMGT Biotechnology page > Antibody humanization > Alemtuzumab).

5.4.1.2 Amino Acid Interactions Between FR-IMGT and CDR-IMGT

IMGT Colliers de Perles from crystallized 3D structures in IMGT/3Dstructure-DB [8–10] highlight two conserved hydrogen bonds between FR-IMGT and CDR-IMGT positions: FR2-IMGT 39 with CDR2-IMGT 56 (or 57) and FR2-IMGT 40 with CDR3-IMGT 105 (Figure 20.1B). Antibody engineering and humanization should therefore preserve these bondings, which stabilize the loops. It is also worthwhile to note that, in VH CDR3, the stem of the CDR3 loop is stabilized by a conserved salt bridge between R106 (arginine contributed by the 3′V-REGION) and D116 (aspartate contributed by the 5′J-REGION of the *Homo sapiens* IGHJ2, IGHJ3, IGHJ4, IGHJ5, or IGHJ6) (IMGT, http://www.imgt.org, IMGT Repertoire > Proteins and alleles > Alignments of alleles).

5.4.2 IGHG1 Alleles and G1m Allotypes

Allotypes are polymorphic markers of an IG subclass that correspond to amino acid changes and are detected serologically by antibody reagents [74]. In therapeutic antibodies (human, humanized, or chimeric) [11], allotypes may represent potential immunogenic residues [73], as demonstrated by the presence of antibodies in individuals immunized against these allotypes [74]. The allotypes of the human heavy gamma chains of the IgG are designated as Gm (for gamma marker). The allotypes G1m, G2m, and G3m are carried by the constant region of the gamma1, gamma2, and gamma3 chains, encoded by the IGHG1, IGHG2, and IGHG3 genes, respectively [74]. The gamma1 chains may express four G1m alleles (combinations of G1m allotypes): G1m3, G1m3,1, G1m17,1, and G1m17,1,2 (and in Negroid populations two additional G1m alleles, Gm17,1,28 and Gm17,1,27,28) [74] (Table 20.7). The C region of the G1m3,1, G1m17,1, and G1m17,1,2 chains differs from that of the G1m3 chains by two, three, and four amino acids, respectively [74]. The correspondence between the G1m alleles and IGHG1 alleles is shown in Table 20.7. Thus, IGHG1*01 and IGHG1*02 are G1m17,1, IGHG1*03 is G1m3, IGHG1*04 is G1m17,1,2, and IGHG1*05 is G1m3,1. In

TABLE 20.7 Correspondence Between the IGHG1 Alleles and G1m Alleles

	G1m alleles[a]		IMGT amino acids[b] and positions					
			CH1		CH3			
			103	120	12	14	110	
				G1m17/nG1m1	G1m1/			
IGHG1 alleles	Allotypes	Isoallotypes[c]		G1m3[d]	nG1m1		G1m2/-	Populations [74]
IGHG1*01[e] IGHG1*02[e]	G1m17,1		I	**K**	**D**	**L**	A	Caucasoid Negroid Mongoloid
IGHG1*03	G1m3	*nG1m1, nG1m17*	**I**	**R**	E	M	A	Caucasoid
IGHG1*07[f]	G1m17,1,2		I	**K**	**D**	**L**	**G**	Caucasoid Mongoloid
IGHG1*08[f]	G1m3,1	*nG1m17*	**I**	**R**	**D**	**L**	A	Mongoloid

[a]*In Negroid populations, the G1m17,1 allele frequently includes G1m27 and G1m28, leading to two additional G1m alleles, G1m17,1,27 and G1m17,1,27,28 [74].*
[b]*Amino acids corresponding to G1m allotypes are shown in bold.*
[c]*The nG1m1 and nG1m17 isoallotypes present on the Gm1-negative and Gm17-negative gamma-1 chains (and on other gamma chains) are shown in italics.*
[d]*The presence of R120 is detected by anti-nG1m17 antibodies, whereas the simultaneous presence of I103 and R120 in the gamma1 chains is detected by anti-Gm3 antibodies [74].*
[e]*The IGHG1*01 and IGHG1*02 alleles only differ at the nucleotide level (codon 85.1 in CH2). The C region of the b12 gamma1 chain (1hzh_H) is encoded by the IGHG1*01 or IGHG1*02 alleles and expresses the G1m17,1 allotypes, based on sequence analysis (CH1 K120 (Figure 20.2D), CH3 D12 and L14) [74]. The presence of an A (Ala) in CH1 121 of 1hzh_H is a PDB file error. It should be a V (Val), as in 1n0x_H (Figure 20.2D) (see IMGT note in IMGT/3D structure-DB, code PDB:1hzh, http://www.imgt.org).*
[f]*IGHG1*07 and IGHG1*08 amino acids are expected [74] but not yet sequenced at the nucleotide level, and therefore the IGHG1*04 and IGHG1*05 alleles are not shown in IMGT Repertoire, Alignments of alleles:* Homo sapiens *IGHG1 (http://www.imgt.org).*

the IGHG1 CH1, the lysine at position 120 (K120) in strand G corresponds to the G1m17 allotype [74] (Figures 20.2D and 20.5B). The isoleucine I103 (strand F) is specific of the gamma1 chain isotype. If an arginine is expressed at position 120 (R120), the simultaneous presence of R120 and I103 corresponds to the expression of the G1m3 allotype [74]. For the gamma3 and gamma4 isotypes (which also have R120 but T in 103), R120 only corresponds to the expression of the nG1m17 isoallotype (an isoallotype or nGm is detected by antibody reagents that identify this marker as an allotype in one IgG subclass and as an isotype for other subclasses) [74]. In the IGHG1 CH3, the aspartate D12 and leucine L14 (strand A) correspond to G1m1, whereas glutamate E12 and methionine M14 correspond to the nG1m1 isoallotype [74] (Table 20.7). A glycine at position 110 corresponds to G1m2, whereas an alanine does not correspond to any allotype (G1m2-negative chain) (Table 20.7). Therapeutic antibodies are most frequently of the IgG1 isotype, and to avoid a potential immunogenicity, the constant region of the gamma1 chains are often engineered to replace the G1m3 allotype by the "less immunogenic" G1m17 (CH1 R120 > K) (G1m17 is more extensively found in different populations) [74].

5.4.3 Only-Heavy-Chain Antibodies

5.4.3.1 Dromedary IgG2 and IgG3 Only-Heavy-Chain Antibodies

Two IgG antibody formats are expressed in the dromedary or Arabian camel (*Camelus dromedarius*) and in Camelidae in general: the conventional IG (with two identical heavy gamma chains associated with two identical light chains) and the "only-heavy-chain" IG (no light chain, and only two identical heavy gamma chains lacking CH1) [96]. The Camdro (for *Camelus dromedarius* in the 6-letter species abbreviation) IGHV1 genes belong to two sets based on four amino acid changes that are characteristic of each set [97]. The first set of IGHV1 genes is expressed in conventional tetrameric IgG1 that constitute 25% of circulating antibodies. The second set is expressed in the only-heavy-chain antibodies, IgG2 and IgG3, which constitute 75% of the circulating antibodies [96]. The four amino acid changes are located in the FR2-IMGT at positions 42, 49, 50, and 52; the first position, 42, is in the C strand and the three others, 49, 50 and 52, are in the C′ strand (Figure 20.1, Table 20.8). They belong to the [GFCC′C″] sheet at the hydrophobic VH-VL interface in conventional antibodies of Camelidae as well as of any vertebrate species, whereas in

TABLE 20.8 Amino Acid Changes at FR2-IMGT Positions 42, 49, 50, and 52, Between the Two Sets of *Camelus dromedarius* IGHV1 Subgroup Genes

		Camelus dromedarius	
	V-REGION	IGHV1 set 1	IGHV1 set 2
Strand	FR2-IMGT amino acid positions	IGHV1S1 to IGHV1S41[a]	IGHV1S42 to IGHV1S74[a]
C	42	V	F, Y
C′	49	G	E, Q
	50	L	C, R
	52	W	F, G, L, W

It is not excluded that some of these amino acid changes in set 2 may result from somatic mutations. Indeed, sequences may be described as genomic in publications, but most of the time it is not known if they are germline or rearranged (IMGT, http://www.imgt.org, IMGT Gene tables).
[a]*Temporary nomenclature, waiting for the complete genomic* Camelus dromedarius *IGH locus.*

camelid only-heavy-chain antibodies (no light chains, and therefore no VL), these positions are exposed to the environment with, through evolution, a selection of hydrophilic amino acids.

The respective heavy gamma2 and gamma3 chains are both characterized by the absence of the CH1 domain owing to a splicing defect [98]. It is the absence of CH1 that is responsible for the lack of association of the light chains. Only-heavy-chain antibodies is a feature of the Camelidae IG as they have also been found in the Bactrian camel (*Camelus bactrianus*) of Central Asia and in the llama (*Lama glama*) and alpaca (*Vicugna pacos*) of South America. The genetic event (splicing defect) responsible for the lack of CH1 occurred in their common ancestor before the radiation between the "camelini" and "lamini," dating approximately 11 million years ago.

The V domains of Camelidae only-heavy-chain antibodies have characteristics for potential pharmaceutical applications (e.g., easy production and selection of single-domain format, extended CDR3 with novel specificities, and binding to protein clefts). They are designated as VH_H when they have to be distinguished from conventional VH (the sequence criteria are based on the four amino acids at positions 42, 49, 50, and 52). The term "nanobody," initially used for describing a single-domain format antibody, is not equivalent to VH_H, as it has been used for V domains other than VH_H and for constructs containing more than 1 V domain (VH and/or VH_H) (e.g., caplacizumab, ozoralizumab) (IMGT, http://www.imgt.org, IMGT Repertoire > Locus and Genes > Gene tables; ibid., IMGT Biotechnology page > Characteristics of the camelidae (camel, llama) antibody synthesis; ibid. IMGT/mAb-DB > caplacizumab; ibid. IMGT/mAb-DB > ozoralizumab).

5.4.3.2 Human Heavy Chain Diseases

The camelidae only-heavy-chain antibodies synthesis is remarkably reminiscent of what is observed in human heavy chain diseases (HCD). These proliferative disorders of B lymphoid cells produce truncated monoclonal immunoglobulin heavy chains, which lack associated light chains. In most HCD, the absence of the heavy chain CH1 domain by deletion or splicing defect may be responsible for the lack of assembly of the light chain [99]. Similar observations have also been reported in mouse variants [99] (IMGT, http://www.imgt.org, IMGT Education > Pathology of the immune system > Molecular defects in Immunoglobulin Heavy Chain Diseases [HCD]; ibid., IMGT Lexique > Heavy Chain Diseases [HCD]).

5.4.3.3 Nurse Shark IgN

A convergence mechanism in evolution is observed in nurse shark (*Ginglymostoma cirratum*, "Gincir" in the 6-letter species abbreviation) IgN antibodies (previously IgNAR, "immunoglobulin new antigen receptor") [100], which are only-heavy-chain antibodies (homodimeric heavy nu chains without CH1, and no associated light chains). The IGHV genes expressed in the Gincir heavy nu chains belong to the IGHV2 subgroup and are characterized by the absence of the CDR2-IMGT owing to a deletion that encompasses position 54−67. The Gincir IGH genes are organized in duplicated cassettes, and those that express IgN comprise Gincir IGHV2 subgroup genes and an IGHN constant gene (IMGT, http://www.imgt.org, IMGT Repertoire > Proteins and alleles > Protein display > IGHV > nurse shark (*G. cirratum*); ibid., IMGT Repertoire > Locus and genes > Gene tables > IGHV > Chondrichthyes).

5.4.4 IGHG CH Properties and Antibody Engineering

5.4.4.1 N-Linked Glycosylation Site CH2 N84.4

An N-linked glycosylation site is present in the CH2 domain of the constant region of the human IG heavy chains of the four IgG isotypes. The N-linked glycosylation site belongs to the classical N-glycosylation motif N-X-S/T (where N is asparagine, X is any amino acid except proline, S is serine, T is threonine) and is defined as CH2 N84.4. As shown in the IMGT Collier de Perles, this asparagine is localized at the DE turn. The IMGT unique numbering has the advantage of identifying the C domain (here, CH2) and, in the domain, the amino acid and its localization (here, N84.4), which can be visualized in the IMGT Collier de Perles and correlated with the 3D structure [80].

5.4.4.2 Interface Ball-and-Socket-like Joints

The interface between the CH2 and CH3 domains in 3D structures of *Homo sapiens* IGHG2 Fc was recently analyzed and compared with the interface in 3D structures of IGHG1 Fc [101]. This study revealed that in all Fc of gamma chains the movement of the CH2 results from a pivoting around a highly conserved ball-and-socket-like joint [101], in which, using the IMGT numbering, the CH2 L15 side chain (last position of the A strand, next to the AB turn) (the ball) interacts with a pocket (the socket) formed by CH3 M107, H108, E109, and H115 (FG loop) [80]. These amino acids are well conserved between the gamma isotypes and the IGHG genes and alleles except for IGHG3 H115, which shows a polymorphism associated with different G3m allotypes [74]. This ball-and-socket-like joint is a structural feature similar but reversed to that previously described at the VH and CH1 domain interface [102], in which the VH L12, T125, and S127 form the socket, whereas the CH1 F29 and P30 (BC loop) form the ball.

5.4.4.3 Knobs-into-Holes CH3 T22 and Y86

The knobs-into-holes methodology has been proposed for obtaining bispecific antibodies [103]. The aim is to increase interactions between the CH3 domain of two gamma1 chains that belong to antibodies with a different specificity. Two amino acids, CH3 T22 (B strand) and Y86 (E strand), which belong to the [ABED] sheet, at the interface of the two *Homo sapiens* IGHG1 CH3 domains [80], were selected for amino acid changes. Interactions of these two amino acids are described in "Contact analysis" in IMGT/3Dstructure-DB [8–10]. The knobs-into-holes methodology consists of an amino acid change on one CH3 domain (T22 > Y) that creates a knob, and another amino acid change on the other CH3 domain (Y86 > T) that creates a hole, thus favoring increased interactions between the CH3 of the two gamma1 chains at both positions 22 and 86 [103] (IMGT, http://www.imgt.org, IMGT Biotechnology page > Knobs-into-holes).

6. CONCLUSION

IMGT-ONTOLOGY and the IMGT information system, which are at the origin of immunoinformatics, have provided the concepts and the knowledge environment and informatics frame for a standardized and integrated analysis of IG, from gene to structure and function [29–44]. IG repertoire analysis and therapeutic antibody engineering and humanization represent two major current fields of immunoinformatics at the forefront of fundamental, clinical and pharmaceutical research owing to major methodological advances and medical implications.

The IMGT standards for IG are used in clinical applications. Thus, IMGT/V-QUEST is frequently used by clinicians for the analysis of somatic hypermutations in leukemia, lymphoma, and myeloma, and more particularly in chronic lymphocytic leukemia (CLL) [15,70–72], in which the percentage of mutations of the rearranged IGHV gene in the VH of the leukemic clone has a prognostic value for the patients. IMGT/V-QUEST is the recommended standard recommended by the European Research Initiative on CLL (ERIC) for comparative analysis between laboratories [70]. The sequences of the V-(D)-J junctions determined by IMGT/JunctionAnalysis [18,19] are also used in the characterization of stereotypic patterns in CLL [71,72] and for the synthesis of probes specific to the junction for the detection and follow-up of minimal residual diseases (MRD) in leukemias and lymphomas. A new era is opening in hemato-oncology with the use of NGS for analysis of the clonality and MRD identification, making IMGT standards use more needed than ever. More generally, the IMGT/HighV-QUEST web portal is a paradigm for identification of IMGT clonotype diversity and expression in NGS immune repertoire analysis of the adaptive immune response in infectious diseases, in vaccination, and for next-generation repertoire immunoprofiling [23].

The therapeutic monoclonal antibody engineering field represents the most promising potential in medicine. Standardized genomic and expressed sequence, structure, and interaction analysis of IG is crucial for a better molecular understanding and comparison of the mAb specificity, affinity, half-life, Fc effector properties, and potential immunogenicity. IMGT-ONTOLOGY concepts have become a necessity for IG loci description of newly sequenced genomes [104,105], antibody structure/function characterization, antibody engineering (single chain Fragment variable, phage displays, combinatorial libraries) and antibody humanization (chimeric, humanized, and human antibodies) [32,39,41,43,73–75,80]. IMGT standardization allows repertoire analysis and antibody humanization studies to move to novel high-throughput methodologies with the same high-quality criteria. The CDR-IMGT lengths are now required for mAb INN applications and are included in the WHO-INN definitions [48], bringing a new level of standardized information to the comparative analysis of therapeutic antibodies.

7. AVAILABILITY AND CITATION

Authors who use IMGT databases and tools are encouraged to cite this article and to quote the IMGT Homepage, http://www.imgt.org. Online access to IMGT databases and tools are freely available for academics and under licenses and contracts for companies.

ACKNOWLEDGEMENTS

We are grateful to Gérard Lefranc for helpful discussion, Véronique Giudicelli and Souphatta Sasorith for help in the figures, Patrice Duroux for upgrade of the IMGT/3Dstructure-DB tools, Eltaf Alamyar for IMGT/HighV-QUEST developments, and the IMGT team for its expertise and constant motivation. We thank Cold Spring Harbor Protocol Press for the pdf of the IMGT Booklet available in IMGT references. IMGT is a registered trademark of CNRS. IMGT is a member of the International Medical Informatics Association (IMIA). IMGT was funded in part by the BIOMED1 (BIOCT930038), Biotechnology BIOTECH2 (BIO4CT960037), 5th PCRDT Quality of Life and Management of Living Resources (QLG2-2000-01287), and 6th PCRDT Information Science and Technology (ImmunoGrid, FP6 IST-028069) programmes of the European Union (EU). IMGT is currently supported by the Centre National de la Recherche Scientifique (CNRS), the Ministère de l'Enseignement Supérieur et de la Recherche (MESR), the University Montpellier 2, the Agence Nationale de la Recherche (ANR) Labex MabImprove (ANR-10-LABX-53-01), the Région Languedoc-Roussillon (Grand Plateau Technique pour la Recherche; GPTR). This work was granted access to the HPC resources of CINES under the allocation 036029 (2010−14) made by GENCI (Grand Equipement National de Calcul Intensif).

REFERENCES

[1] Lefranc M-P, Giudicelli V, Ginestoux C, Jabado-Michaloud J, Folch G, Bellahcene F, et al. IMGT®, the international ImMunoGeneTics information system®. Nucleic Acids Res 2009;37:D1006−12.

[2] Lefranc M-P, Lefranc G. The immunoglobulin FactsBook. Academic Press: London.

[3] Lefranc M-P, Lefranc G. The T cell receptor FactsBook. Academic Press: London.

[4] Lefranc M-P. Nomenclature of the human immunoglobulin genes. In: Coligan JE, Bierer BE, Margulies DE, Shevach EM, Strober W, editors. Current protocols in immunology. Hoboken NJ: John Wiley and Sons.

[5] Lefranc M-P. Nomenclature of the human T cell receptor genes. In: Coligan JE, Bierer BE, Margulies DE, Shevach EM, Strober W, editors. Current protocols in immunology. Hoboken NJ: John Wiley and Sons.

[6] Giudicelli V, Duroux P, Ginestoux C, Folch G, Jabado-Michaloud J, Chaume D, et al. IMGT/LIGM-DB, the IMGT comprehensive database of immunoglobulin and T cell receptor nucleotide sequences. Nucleic Acids Res 2006;34:D781−4.

[7] Giudicelli V, Chaume D, Lefranc M-P. IMGT/GENE-DB: a comprehensive database for human and mouse immunoglobulin and T cell receptor genes. Nucleic Acids Res 2005;33:D256−61.

[8] Kaas Q, Ruiz M, Lefranc M-P. IMGT/3Dstructure-DB and IMGT/StructuralQuery, a database and a tool for immunoglobulin, T cell receptor and MHC structural data. Nucleic Acids Res 2004;32:D208−10.

[9] Ehrenmann F, Kaas Q, Lefranc M-P. IMGT/3Dstructure-DB and IMGT/DomainGapAlign: a database and a tool for immunoglobulins or antibodies, T cell receptors, MHC, IgSF and MhcSF. Nucleic Acids Res 2010;38:D301−7.

[10] Ehrenmann F, Lefranc M-P. IMGT/3Dstructure-DB: querying the IMGT database for 3D structures in immunology and immunoinformatics (IG or Antibodies, TR, MH, RPI, and FPIA). Cold Spring Harb Protoc 2011;6:750−61.

[11] Poiron C, Wu Y, Ginestoux C, Ehrenmann F, Duroux P, Lefranc M-P. I. Montpellier. MGT/mAb-DB: the IMGT® database for therapeutic monoclonal antibodies. 11èmes Journées Ouvert Biol Inform Mathématiques (JOBIM), Poster n°101. 2010. 7−9 Sep 2010.

[12] Giudicelli V, Chaume D, Lefranc M-P. IMGT/V-QUEST, an integrated software for immunoglobulin and T cell receptor V-J and V-D-J rearrangement analysis. Nucleic Acids Res 2004;32:W435−40.

[13] Giudicelli V, Lefranc M-P. Interactive IMGT on-line tools for the analysis of immunoglobulin and T cell receptor repertoires. In: Veskler BA, editor. New research on immunology. New York: Nova Science Publishers Inc. p. 77−105.

[14] Brochet X, Lefranc M-P, Giudicelli V. IMGT/V-QUEST: the highly customized and integrated system for IG and TR standardized V-J and V-D-J sequence analysis. Nucleic Acids Res 2008;36:W503−8.

[15] Giudicelli V, Lefranc M-P. IMGT standardized analysis of immunoglobulin rearranged sequences. In: Ghia P, Rosenquist R, Davi F, editors. Immunoglobulin gene analysis in chronic lymphocytic leukemia. Italy: Wolters Kluwer Health Italy. p. 33−52.

[16] Giudicelli V, Brochet X, Lefranc M-P. IMGT/V-QUEST: IMGT standardized analysis of the immunoglobulin (IG) and T Cell receptor (TR) nucleotide sequences. Cold Spring Harb Protoc 2011;6:695−715.

[17] Alamyar E, Duroux P, Lefranc M-P, Giudicelli V. IMGT tools for the nucleotide analysis of immunoglobulin (IG) and T cell receptor (TR) V-(D)-J repertoires, polymorphisms, and IG mutations: IMGT/V-QUEST and IMGT/HighV-QUEST for NGS. In: Christiansen F, Tait B, editors. Immunogenetics. Methods molecular biology, vol. 882. New York: Humana Press, Springer. p. 569−604.

[18] Yousfi Monod M, Giudicelli V, Chaume D, Lefranc M-P. IMGT/JunctionAnalysis: the first tool for the analysis of the immunoglobulin and T cell receptor complex V-J and V-D-J JUNCTIONs. Bioinformatics 2004;20:i379−85.

[19] Giudicelli V, Lefranc M-P. IMGT/JunctionAnalysis: IMGT standardized analysis of the V-J and V-D-J junctions of the rearranged

immunoglobulins (IG) and T Cell receptors (TR). Cold Spring Harb Protoc 2011;6:716–25.

[20] Giudicelli V, Protat C, Lefranc M-P. The IMGT strategy for the automatic annotation of IG and TR cDNA sequences: IMGT/Automat. In: Proceedings of the European conference on computational biology (ECCB 2003), data and knowledge bases, poster DKB_31. Paris: Institut National de Recherche en Informatique et en Automatique. p. 103–104.

[21] Giudicelli V, Chaume D, Jabado-Michaloud J, Lefranc M-P. Immunogenetics sequence annotation: the strategy of IMGT based on IMGT-ONTOLOGY. Stud Health Technol Inf 2005;116:3–8.

[22] Alamyar E, Giudicelli V, Shuo L, Duroux P, Lefranc M-P. IMGT/HighV-QUEST: the IMGT web portal for immunoglobulin (IG) or antibody and T cell receptor (TR) analysis from NGS high throughput and deep sequencing. Immunome Res 2012;8(1):26.

[23] Li S, Lefranc M-P, Miles JJ, Alamyar E, Giudicelli V, Duroux P, et al. IMGT/HighV-QUEST paradigm for T cell receptor IMGT clonotype diversity and next generation repertoire immunoprofiling. Nat Commun 2013;4:2333.

[24] Ehrenmann F, Lefranc M-P. IMGT/DomainGapAlign: IMGT standardized analysis of amino acid sequences of variable, constant, and groove domains (IG, TR, MH, IgSF, MhSF). Cold Spring Harb Protoc 2011;6:737–49.

[25] Ehrenmann F, Lefranc M-P. IMGT/DomainGapAlign: the IMGT tool for the analysis of IG, TR, MHC, IgSF and MhcSF domain amino acid polymorphism. In: Christiansen F, Tait B, editors. Immunogenetics. Methods molecular biology, vol. 882. New York: Humana Press, Springer. p. 605–633.

[26] Ehrenmann F, Giudicelli V, Duroux P, Lefranc M-P. IMGT/Collier de Perles: IMGT standardized representation of domains (IG, TR, and IgSF variable and constant domains, MH and MhSF groove domains). Cold Spring Harb Protoc 2011;6:726–36.

[27] Lane J, Duroux P, Lefranc M-P. From IMGT-ONTOLOGY to IMGT/LIGMotif: the IMGT standardized approach for immunoglobulin and T cell receptor gene identification and description in large genomic sequences. BMC Bioinf 2010;11:223.

[28] Pommié C, Levadoux S, Sabatier R, Lefranc G, Lefranc M-P. IMGT standardized criteria for statistical analysis of immunoglobulin V-REGION amino acid properties. J Mol Recogn 2004;17:17–32.

[29] Lefranc M-P. IMGT, the international ImMunoGeneTics information system. In: Bock G, Goode J, editors. Immunoinformatics: bioinformatic strategies for better understanding of immune function. Novartis foundation symposium, vol. 254. Chichester, UK: John Wiley and Sons. p. 126–136.

[30] Lefranc M-P, Giudicelli V, Ginestoux C, Chaume D. IMGT, the international ImMunoGeneTics information system: the reference in immunoinformatics. Stud Health Technol Inf 2003;95:74–9.

[31] Lefranc M-P. IMGT databases, web resources and tools for immunoglobulin and T cell receptor sequence analysis. Leukemia 2003;17(1):260–6.

[32] Lefranc M-P. IMGT, the international ImMunoGenetics information system. In: Lo BKC, editor. Antibody engineering methods and protocols. 2nd ed. Methods molcular biology, vol. 248. Totowa, NJ: Humana Press. p. 27–49.

[33] Lefranc M-P. IMGT-ONTOLOGY and IMGT databases, tools and web resources for immunogenetics and immunoinformatics. Mol Immunol 2004;40(10):647–60.

[34] Lefranc M-P. IMGT, the international ImMunoGeneTics information system: a standardized approach for immunogenetics and immunoinformatics. Immunome Res 2005;1:3.

[35] Lefranc M-P. IMGT, the international ImMunoGeneTics information system for immunoinformatics. Methods for querying IMGT databases, tools and Web resources in the context of immunoinformatics. In: Flower DR, editor. Immunoinformatics: predicting immunogenicity in silico. Methods molecular biology, vol. 409. Totowa, NJ: Humana Press, Springer. p. 19–42.

[36] Lefranc M-P. IMGT-ONTOLOGY, IMGT databases, tools and web resources for immunoinformatics. In: Schoenbach C, Ranganathan S, Brusic V, editors. Immunoinformatics. Immunomics reviews, series of Springer Science and Business Media LLC, vol. 1. New York: Springer. p. 1–18.

[37] Lefranc M-P, Giudicelli V, Regnier L, Duroux P. IMGT, a system and an ontology that bridge biological and computational spheres in bioinformatics. Briefings Bioinf 2008;9:263–75.

[38] Lefranc M-P. IMGT, the international ImMunoGeneTics information system for immunoinformatics. Methods for querying IMGT databases, tools and web resources in the context of immunoinformatics. Mol Biotechnol 2008;40:101–11.

[39] Lefranc M-P. Antibody databases and tools: the IMGT experience. In: An Z, editor. Therapeutic monoclonal antibodies: from bench to clinic. Hoboken, NJ: John Wiley and Sons. p. 91–114.

[40] Lefranc M-P. Antibody databases: IMGT, a French platform of world-wide interest [in French]. Bases de données anticorps: IMGT, une plate-forme française d'intérêt mondial. Méd/Sci 2009;25:1020–3.

[41] Ehrenmann F, Duroux P, Giudicelli V, Lefranc M-P. Standardized sequence and structure analysis of antibody using IMGT. In: Kontermann R, Dübel S, editors. Antibody engineering. vol. 2. Berlin, Heidelberg: Springer-Verlag. p. 11–31.

[42] Lefranc M-P. IMGT, the international ImMunoGeneTics information system. Cold Spring Harb Protoc 2011;6:595–603.

[43] Lefranc M-P, Ehrenmann F, Ginestoux C, Duroux P, Giudicelli V. Use of IMGT databases and tools for antibody engineering and humanization. In: Chames P, editor. Antibody engineering. Methods molecular biology, vol. 907. New York: Humana Press Springer. p. 3–37.

[44] Lefranc M-P. IMGT information system. In: Dubitzky W, Wolkenhauer O, Cho K-H, Yokota H, editors. Encyclopedia of systems biology. Springer Science+Business Media. p. 959–964.

[45] Lefranc M-P. WHO-IUIS nomenclature subcommittee for immunoglobulins and T cell receptors report. Immunogenetics 2007;59:899–902.

[46] Lefranc M-P. WHO-IUIS nomenclature subcommittee for immunoglobulins and T cell receptors report Aug 2007, 13th international congress of immunology, Rio de Janeiro, Brazil. Dev Comp Immunol 2008;32:461–3.

[47] World Health Organization. International nonproprietary names (INN) for biological and biotechnological substances (a review). INN Working Document 5. 179. Update 2012.

[48] Lefranc M-P. Antibody nomenclature: from IMGT-ONTOLOGY to INN definition. mAbs 2011;3(1):1–2.

[49] Giudicelli V, Lefranc M-P. Ontology for immunogenetics: IMGT-ontology. Bioinformatics 1999;15:1047–54.

[50] Giudicelli V, Lefranc M-P. IMGT-ONTOLOGY 2012. Frontiers in bioinformatics and computational biology. Front Genet 2012;3:79.

[51] Giudicelli V, Lefranc M-P. IMGT-ONTOLOGY. In: Dubitzky W, Wolkenhauer O, Cho K-H, Yokota H, editors. Encyclopedia of systems biology. Springer Science+Business Media. p. 964−972.

[52] Giudicelli V, Lefranc M-P. IMGT-ONTOLOGY: gestion et découverte de connaissances au sein d'IMGT. In: Hacid M-S, Kodratoff Y, Boulanger D, editors. Extraction et gestion des connaissances (EGC'2003). Actes des troisièmes journées Extraction et Gestion des Connaissances, Lyon, France, janvier 22−24, 2003. Revue des Sciences et Technologies de l'Information, RSTI, série Revue d'Intelligence Artificielle- Extraction des Connaissances et Apprentissage (RIA-ECA), Hermès Science Publications, ISBN 2-7462-0631-5, vol. 17 (1−3). Paris: Lavoisier, Cachan. p. 13−23.

[53] Lefranc M-P, Giudicelli V, Ginestoux C, Bosc N, Folch G, Guiraudou D, et al. IMGT-ONTOLOGY for immunogenetics and immunoinformatics. In: Silico biol, vol 4; 2004. p. 17−29.

[54] Lefranc M-P, Clément O, Kaas Q, Duprat E, Chastellan P, Coelho I, et al. IMGT-choreography for immunogenetics and immunoinformatics. In: Silico biol, vol 5; 2005. p. 45−60.

[55] Duroux P, Kaas Q, Brochet X, Lane J, Ginestoux C, Lefranc M-P, et al. IMGT-Kaleidoscope, the formal IMGT-ONTOLOGY paradigm. Biochimie 2008;90:570−83.

[56] Lefranc M-P. From IMGT-ONTOLOGY IDENTIFICATION axiom to IMGT standardized keywords: for immunoglobulins (IG), T cell receptors (TR), and conventional genes. Cold Spring Harb Protoc 2011;6:604−13.

[57] Lefranc M-P. From IMGT-ONTOLOGY DESCRIPTION axiom to IMGT standardized labels: for immunoglobulin (IG) and T cell receptor (TR) sequences and structures. Cold Spring Harb Protoc 2011;6:614−26.

[58] Lefranc M-P. From IMGT-ONTOLOGY CLASSIFICATION axiom to IMGT standardized gene and allele nomenclature: for immunoglobulins (IG) and T cell receptors (TR). Cold Spring Harb Protoc 2011;6:627−32.

[59] Lefranc M-P. Unique database numbering system for immunogenetic analysis. Immunol Today 1997;18:509.

[60] Lefranc M-P. The IMGT unique numbering for Immunoglobulins, T cell receptors and Ig-like domains. The Immunologist 1999;7:132−6.

[61] Lefranc M-P, Pommié C, Ruiz M, Giudicelli V, Foulquier E, Truong L, et al. IMGT unique numbering for immunoglobulin and T cell receptor variable domains and Ig superfamily V-like domains. Dev Comp Immunol 2003;27:55−77.

[62] Lefranc M-P, Pommié C, Kaas Q, Duprat E, Bosc N, Guiraudou D, et al. IMGT unique numbering for immunoglobulin and T cell receptor constant domains and Ig superfamily C-like domains. Dev Comp Immunol 2005;29:185−203.

[63] Lefranc M-P, Duprat E, Kaas Q, Tranne M, Thiriot A, Lefranc G. IMGT unique numbering for MHC groove G-DOMAIN and MHC superfamily (MhcSF) G-LIKE-DOMAIN. Dev Comp Immunol 2005;29:917−38.

[64] Lefranc M-P. IMGT unique numbering for the variable (V), constant (C), and groove (G) domains of IG, TR, MH, IgSF, and MhSF. Cold Spring Harb Protoc 2011;6:633−42.

[65] Ruiz M, Lefranc M-P. IMGT gene identification and Colliers de Perles of human immunoglobulins with known 3D structures. Immunogenetics 2002;53:857−83.

[66] Kaas Q, Lefranc M-P. IMGT Colliers de Perles: standardized sequence-structure representations of the IgSF and MhcSF superfamily domains. Curr Bioinf 2007;2:21−30.

[67] Kaas Q, Ehrenmann F, Lefranc M-PIG. TR and IgSf, MHC and MhcSF: what do we learn from the IMGT Colliers de Perles? Briefings Funct Genomics Proteomics 2007;6:253−64.

[68] Lefranc M-P. IMGT Collier de Perles for the variable (V), constant (C), and groove (G) domains of IG, TR, MH, IgSF, and MhSF. Cold Spring Harb Protoc 2011;6:643−51.

[69] Robert R, Lefranc M-P, Ghochikyan A, Agadjanyan MG, Cribbs DH, Van Nostrand WE, et al. Restricted V gene usage and VH/VL pairing of mouse humoral response against the N-terminal immunodominant epitope of the amyloid β peptide. Mol Immunol 2010;48(1−3):59−72.

[70] Ghia P, Stamatopoulos K, Belessi C, Moreno C, Stilgenbauer S, Stevenson FI, et al. ERIC recommendations on IGHV gene mutational status analysis in chronic lymphocytic leukemia. Leukemia 2007;21:1−3.

[71] Agathangelidis A, Darzentas N, Hadzidimitriou A, Brochet X, Murray F, Yan XJ, et al. Stereotyped B-cell receptors in one third of chronic lymphocytic leukemia: towards a molecular classification with implications for targeted therapeutic interventions. Blood 2012;119(19):4467−75.

[72] Kostareli E, Gounari M, Janus A, Murray F, Brochet X, Giudicelli V, et al. Antigen receptor stereotypy across B-cell lymphoproliferations: the case of IGHV4-59/IGKV3-20 receptors with rheumatoid factor activity. Leukemia 2012;26(5):1127−31.

[73] Jefferis R, Lefranc M-P. Human immunoglobulin allotypes: possible implications for immunogenicity. mAbs 2009;1(4):332−8.

[74] Lefranc M-P, Lefranc G. Human Gm, Km and Am allotypes and their molecular characterization: a remarkable demonstration of polymorphism. In: Christiansen F, Tait B, editors. Immunogenetics. Methods molecular biology, vol. 882. New York: Humana Press, Springer. p. 635−680.

[75] Dechavanne C, Guillonneau F, Chiappetta G, Sago L, Lévy P, Salnot V, et al. Mass spectrometry detection of G3m and IGHG3 alleles and follow-up of differential mother and neonate IgG3. PLoS One 2012;7(9):e46097.

[76] Magdelaine-Beuzelin C, Kaas Q, Wehbi V, Ohresser M, Jefferis R, Lefranc M-P, et al. Structure-function relationships of the variable domains of monoclonal antibodies approved for cancer treatment. Crit Rev Oncol Hematol 2007;64:210−25.

[77] Pelat T, Bedouelle H, Rees AR, Crennell SJ, Lefranc M-P, Thullier P. Germline humanization of a non-human primate antibody that neutralizes the anthrax toxin, by in vitro and in silico engineering. J Mol Biol 2008;384:1400−7.

[78] Pelat T, Hust M, Hale M, Lefranc M-P, Dübel S, Thullier P. Isolation of a human-like antibody fragment (scFv) that neutralizes ricin biological activity. BMC Biotechnol 2009;9:60.

[79] Alamyar M-P, Giudicelli V, Duroux P, Lefranc M-P. Antibody V and C domain sequence, structure and interaction analysis with special reference to IMGT. In: Ossipow V, Fisher N, editors. Monoclonal antibodies: methods and protocols, 2nd ed. New York: Springer Science + Business Media LLC. vol 1131. p. 337−381.

[80] Lefranc M-P. How to use IMGT for therapeutic antibody engineering. In: Dübel SK, Reichert J, editors. Handbook of therapeutic antibodies, 2nd ed. Wiley.

[81] Wain HM, Bruford EA, Lovering RC, Lush MJ, Wright MW, Povey S. Guidelines for human gene nomenclature. Genomics 2002;79:464–70.
[82] Bruford EA, Lush MJ, Wright MW, Sneddon TP, Povey S, Birney E. The HGNC database in 2008: a resource for the human genome. Nucleic Acids Res 2008;36:D445–8.
[83] Letovsky SI, Cottingham RW, Porter CJ, Li PW. GDB: the human genome database. Nucleic Acids Res 1998;26(1):94–9.
[84] Maglott DR, Katz KS, Sicotte H, Pruitt KD. NCBI's LocusLink and RefSeq. Nucleic Acids Res 2000;28(1):126–8.
[85] Maglott D, Ostell J, Pruitt KD, Tatusova T. Entrez gene: gene-centered information at NCBI. Nucleic Acids Res 2007;35:D26–31.
[86] Stabenau A, McVicker G, Melsopp C, Proctor G, Clamp M, Birney E. The Ensembl core software libraries. Genome Res 2004;14:929–33.
[87] Wilming LG, Gilbert JG, Howe K, Trevanion S, Hubbard T, Harrow JL. The vertebrate genome annotation (Vega) database. Nucleic Acids Res 2008;36:D753–60.
[88] Kaas Q, Lefranc M-P. T cell receptor/peptide/MHC molecular characterization and standardized pMHC contact sites in IMGT/3D structure-DB. In: Silico biol, vol 5; 2005. p. 505–28.
[89] Kaas Q, Duprat E, Tourneur G, Lefranc M-P. IMGT standardization for molecular characterization of the T cell receptor/peptide/MHC complexes. In: Schoenbach C, Ranganathan S, Brusic V, editors. Immunoinformatics, Immunomics reviews, series of Springer Science and Business Media LLC, [Chapter 2]. New York: Springer. p. 19–49.
[90] Vlachakis D, Feidakis C, Megalooikonomou V, Kossida S. IMGT/Collier-de-Perles: a two-dimensional visualization tool for amino acid domain sequences. Theor Biol Med Model 2013;10:14.
[91] Lefranc M-P. Immunoinformatics of the V, C and G domains: IMGT definitive system for IG, TR and IgSF, MH and MhSF. In: De RK, Tomar N, editors. Immunoinformatics: from biology to informatics. Methods in molecular biology, 2nd ed. Humana Press.
[92] Kabat EA, Wu TT, Perry HM, Gottesman KS, Foeller C. Sequences of proteins of immunological interest. Washington, DC: U.S. Department of Health and Human Services (USDHHS), National Institute of Health NIH Publication.
[93] Chothia C, Lesk AM. Canonical structures for the hypervariable regions of immunoglobulins. Mol Biol 1987;196(4):901–17.
[94] Rose PW, Beran B, Bi C, Bluhm WF, Dimitropoulos D, Goodsell DS, et al. The RCSB Protein Data Bank: redesigned web site and web services. Nucleic Acids Res 2011;39:D392–401.
[95] Riechmann L, Clark M, Waldmann H, Winter G. Reshaping human antibodies for therapy. Nature 1988;332:323–7.
[96] Hamers-Casterman C, Atarhouch T, Muyldermans S, Robinson G, Hamers C, Bajyana Songa E, et al. Naturally occurring antibodies devoid of light chains. Nature 1993;363:446–8.
[97] Nguyen VK, Hamers R, Wyns L, Muyldermans S. Camel heavy-chain antibodies: diverse germline VHH and specific mechanisms enlarge the antigen-binding repertoire. EMBO J 2000;19:921–30.
[98] Nguyen VK, Hamers R, Wyns L, Muyldermans S. Loss of splice consensus signal is responsible for the removal of the entire CH1 domain of the functional camel IgG2a heavy chain antibodies. Mol Immunol 1999;36:515–24.
[99] Lefranc M-P, Lefranc G. The constant region genes of the immunoglobulin heavy chains. Mol Genet (Life Sci Adv) 1988;7:39–45.
[100] Greenberg AS, Avila D, Hughes M, Hughes A, McKinney EC, Flajnik MF. A new antigen receptor gene family that undergoes rearrangement and extensive somatic diversification in sharks. Nature 1995;374:168–73.
[101] Teplyakov A, Zhao Y, Malia TJ, Obmolova G, Gilliland GL. IgG2 Fc structure and the dynamic features of the IgG CH2-CH3 interface. Mol Immunol 2013;56:131–9.
[102] Lesk AM, Chothia C. Elbow motion in the immunoglobulin involves a molecular ball-and-socket joint. Nature 1988;335:188–90.
[103] Ridgway JB, Presta LG, Carter P. 'Knobs-into-holes' engineering of antibody CH3 domains for heavy chain heterodimerization. Protein Eng 1996;9:617–21.
[104] Schwartz JC, Lefranc M-P, Murtaugh M-P. Organization, complexity and allelic diversity of the porcine (*Sus scrofa domestica*) immunoglobulin lambda locus. Immunogenetics 2012;64(5):399–407.
[105] Schwartz JC, Lefranc M-P, Murtaugh MP. Evolution of the porcine (*Sus scrofa domestica*) immunoglobulin kappa locus through germline gene conversion. Immunogenetics 2012;64(4):303–11.

Chapter 21

Mucosal Immunity to Bacteria and Immunoglobulin A Synthesis

Andrew J. Macpherson[1], Tim Rollenkse[1], Lucia M. Kato[2] and Sidonia Fagarasan[2]

[1]*Department of Visceral Surgery and Medicine, Bern University Hospital, University of Bern, Bern, Switzerland;* [2]*Laboratory for Mucosal Immunity, Center for Integrative Medical Sciences, RIKEN Yokohama Institute, Yokohama, Japan*

1. INTRODUCTION

Mucosal immunity requires a fine balance to maintain a mutualistic relationship between the host and its complex commensal communities and to respond and clear infection at mucosal sites. Antibodies of the immunoglobulin A (IgA) isotype are pivotal to maintain gut homoeostasis. IgA production in mammals exceeds those of all other isotypes and is secreted across mucosal epithelium where it interacts with our microbiota or pathogens. Here we discuss the mechanisms underlying selective IgA induction of mucosal B cells, key differences of mucosal and systemic antibody responses, IgA in systemic diseases and the targets and functions of secretory IgA to establish host-microbial mutualism and homoeostasis at mucosal surfaces.

2. OVERVIEW OF DIFFERENT FUNCTIONAL ASPECTS OF IGA BIOLOGY

2.1 Drivers of IgA Secretion

IgA has a special place in immunobiology. The isotype is produced in greater quantities than any other in mammals – mostly secreted across mucous membranes and in the maternal milk. Unlike other class-switched isotypes, which are produced by the differentiation of B2 cells originating in the bone marrow, IgA (and IgM) result both from the B2 bone marrow output and from regenerative B1 populations residing in the peritoneal and pleural fluids [1].

The mechanism of transfer across the epithelial barrier depends on the polymeric immunoglobulin receptor (pIgR) which covalently binds the dimeric IgA molecule (and the pentameric IgM) via its heavy chains [2]. The bound complex is then transferred within cytoplasmic vesicles that traffic from the basolateral to the apical surface of the epithelial cell (EC) where IgA is released by proteolysis of the pIgR. This leaves a fragment of the covalently bound pIgR N terminal sequence – termed secretory component – attached to the free dimeric IgA at the epithelial surface.

The IgA isotype is therefore one of the many distinctive features of the mucosal immune system, and the simple comparison of germ-free and colonised mice shows that the presence of the microbiota is key to IgA induction [3]. In experimental mice, variable levels of IgA secretion are found depending on mouse strain [4,5], age [6], diet [7,8], stress [9] and the specifics of the microbiota composition [10–12]. The pattern of IgA binding to individual taxa of the microbiota is similar between monozygotic and dizygotic human twin infants (and to their mother) until they are approximately 2 years old: after this, when the microbiota has diversified to an adult-like pattern, the IgA-binding profiles between twins become similar to that of unrelated individuals [13]. Paradoxically, given the investment that most mammals make in producing the IgA isotype in such large quantities, IgA is the most common primary immunoglobulin deficiency in humans (defined as a serum IgA level of less than 70 μg/L, age >4 years) with >60% patients remaining asymptomatic [14]. Mouse models with gene targeted global IgA deficiency also have a very mild phenotype even when challenged with cytopathic viruses [15,16]. It has become clear that IgA deficiency is compensated both by (polymeric) IgM secretion [17], albeit with a slightly different repertoire in humans against intestinal microbes [18] and by compensatory T-cell responses [12,19]. When mice have a targeted inability to secrete polymeric mucosal antibodies (either IgA or IgM) through deletion of the pIgR gene, there is a compensatory hypergammaglobulinemia, probably because of increased penetration of microbes at mucosal surfaces [20].

Molecular Biology of B Cells. https://doi.org/10.1016/B978-0-323-95895-0.00002-7

2.2 IgA in Early Life

Secretory (S)IgA is also a major component of mammalian milk, allowing a mother to share her mucosal immune memory with her offspring [21]. There are important species differences of immunoglobulin content between mouse (a species where offspring are born relatively immature and milk contains IgG which is absorbed through the duodenum in early life) and humans (where IgG uptake is effectively complete via the placenta at birth). In both cases, IgA is present in the milk and important for normal postnatal microbiota and immune development, although the species differences mean that caution is needed to compare the results of mouse experiments with the potential protective effects of antibodies in maternal milk in humans. Even with this caveat, one can make a good case for protective effects of maternal milk IgA in early life. In mice, premature induction of endogenous intestinal mucosal IgA occurs in the heterozygous offspring of mice where the dam is immunodeficient and cannot produce or secrete milk immunoglobulins [22]. This is driven by microbial colonisation – specifically including *Lactobacillus reuteri* [23] – and is also found to be true even when the new-born pups are parenterally supplemented with IgG to model the human condition [24]. The implication is that maternal milk IgA is able to coat incoming microbes during early life intestinal colonisation and shield the early life immune system. This is supported by experiments in which pups are nursed by pIgR-deficient dams that cannot secrete IgA into the milk where there is increased translocation of aerobic bacteria from the neonatal gut into draining lymph nodes [25]. In humans, there is apparently little or no detrimental effect of maternal IgA deficiency on the baby [26], but children with IgA deficiency have increased susceptibility to allergies.

2.3 The Breadth of IgA-Specific and Non-specific Functions

Whilst consideration of IgA function is normally focussed on secretory IgA targeting the intestinal microbiota, there is evidence of a much wider selection of binding and potential functions. IgA-binding targets include IgA against viruses [27], exotoxins [28], endotoxin [29], sperm [30], food antigens [31], fungi [32] and helminths [24,33]. In many cases, there is also evidence that these binding assays are also functional: for example, for toxin neutralisation [34], in vivo protection from virus cytolytic effects [27] and modulation of fungal morphology and invasiveness [32]. Set against these apparently specific functions, presumably generated through high-affinity interactions, it is also clear that there is a strong component of polyreactivity within the mouse and human IgA repertoire. This is normally measured by low-level binding of monoclonals across a series of generic antigens including insulin, double-stranded (ds)DNA and lipopolysaccharide in ELISA plate assays. That an antibody is polyreactive does not exclude high-affinity specific binding to a selective antigen that likely induced the responsible B-cell clone. The affinity of binding is often not defined, but even then the resultant avidity for dimeric secretory IgA with four identical binding sites and flexible hinge regions will be modulated by potentially cooperative binding effects on repetitive antigens and potential entropy losses through immobilisation. Therefore, binding is context-dependent on the target as well as the complementarity-determining regions (CDRs) of the antibody and may even occur independently of CDRs through binding of the carbohydrate side chains of glycosylated SIgA [35] – for example, by microbial adhesins or bacterial pili [36].

The multidimensionality of these IgA–antigen interactions means that in most cases, they are being described in an empirical way and with some terminological confusion. In line with a recent review [37], we suggest that polyreactivity (plate binding against different standard arbitrary antigens) [38] should be distinguished from cross-species reactivity (potentially comprising antigenic targets on different taxa of the microbiota or against cells from different kingdoms). To this list, one must add cross-strain reactivity (between different strains of the same taxon) [39], intra-species reactivity (one monoclonal binding different antigenic targets within a single microbial taxon) [40] and non-canonical binding where antibody glycosyl side chains provide the binding site [35].

2.4 Functions of IgA Beyond Mucosal Microbes

Although the focus of this chapter is on (dimeric) secretory IgA at mucosal surfaces, IgA (mainly in the monomeric form) is a minor isotype of serum immunoglobulin and there is evidence for IgA+ B cell function at extramucosal sites [41,42] including tumours [43]. Serum IgA responses can protect against polymicrobial sepsis [11], possibly as an equivalent protective effect to natural IgM against viruses [44] and regulatory IgA+ B cells expressing IL-10 down-regulate T cell and macrophage responses together with PD-L1 interactions [43,45]. Whilst it is logical that IgA responses induced in the intestine may manifest in relevant exocrine glands such as the female breast [46] or salivary glands [47], recirculation of intestinal IgA B cells to central tissues is in fact a more general phenomenon [48]. Circulation to the bone marrow has been shown by tracking intestinal B cells after local intestinal infection with rotavirus, or by tissue-directed photoconversion – following marking with these different techniques, the IgA B cells and their progeny could also be detected at low levels in normal tissues and higher levels during inflammation, including the brain in the multiple sclerosis model of experimental allergic encephalomyelitis [49]. This principle has also been shown in parabiotic mice where intestinal-induced IgA+ B cells are induced in one of the parabiotic pair and appear in the bone marrow and other tissues of the uninduced animal [49]. This B-cell population produces IL-10 to modulate the experimental autoimmune encephalomyelitis (EAE) inflammatory response in the CNS [49]. IgA+ B cells have also been tracked to the brain and shown to produce antibody in the cerebrospinal fluid in patients with multiple sclerosis through matching the B-cell receptor repertoire with that in the gut, and demonstrating binding reactivity to intestinal bacteria rather than local neural tissues [50].

3. IGA SYNTHESIS AND REGULATION

IgAs can be produced in the gut through multiple pathways, within or outside organised gut-associated lymphoid tissues (GALTs), and with or without help from CD4 T cells [51,52]. Primary inductive sites include organised follicular structures such as Peyer's patches (PPs), mesenteric lymph nodes (MLNs) and isolated lymphoid follicles (ILFs) that are present along the length of the small intestine. The PPs, especially, are enriched in B cells interacting with T cells, and it is in the PP B-cell follicles that most IgA immune responses are initiated. The lamina propria (LP), particularly of the small intestine, is considered the main mucosal effector site, supporting mainly the final maturation of plasmablasts into IgA-producing plasma cells that release SIgA into the gut lumen.

3.1 T Cell-Dependent IgA Induction

Intraluminal antigens are captured by microfold (M) cells, which are considered the major site of antigen entry in the GALT [53]. GALT is constituted of several structures and among them, PPs are the main IgA-inductive sites and where the T cell–dependent (TD) IgA induction occurs. PPs are large lymphoid structures consisting of multiple B-cell follicles separated by interfollicular areas containing T cells and dendritic cells (DCs). PPs develop independently of the gut microbiota from the interaction of stromal cells and bone marrow-derived innate lymphoid cells (ILCs), which are also known as lymphoid tissue inducer (LTi) cells and are already present at birth [54,55]. After colonisation with bacteria, PPs become larger and form germinal centres (GCs), where antigen-activated B cells interact with cognate CD4 T cells to promote affinity maturation and production of a highly diverse IgA repertoire (Figure 21.1). The lack of IgA or the impairment of IgA responses causes a decreased diversity of gut bacterial communities and dysbiosis in mouse and humans [18,56–58]. Both the quantity of IgA and the quality of IgA repertoire are essential for the maintenance of a diverse gut microbiota, as impairment in IgA selection can induce dysbiosis [10].

GCs in gut PPs differ from GCs in peripheral lymph nodes (pLNs) and spleen in many aspects [59]. PP GCs are constantly induced by commensal bacteria and animals kept in germ-free conditions as well as those treated with antibiotics have extremely few GCs in PPs [56]. On the other hand, GCs in pLNs are induced only after systemic infection or upon immunisation. Moreover, activated B cells in pLN GCs predominantly switch to IgG-producing B cells, whereas B cells in PP GCs preferentially class switch to IgA. The skewed class switch recombination (CSR) from IgM to IgA in the gut is thought to be the result of the balance of metabolic products and cytokines generated by activated ECs, DCs, B cells and T cells that make the PP GCs a unique IgA-generating milieu [60,61]. IgA+ B cells generated in PP express gut-homing receptors $\alpha4\beta7$ and CCR9 in response to retinoic acid (RA) from DCs [62].

IgA CSR can also happen through events taking place in the subepithelial dome (SED). Antigens entering the intestine through M cells are taken up and presented by DCs to B cells at the interface to the follicles. Migration of B cells to the follicle border is mediated through activation of the oxysterol receptor EBI2 on B cells [63].

TGF-β1 is the critical cytokine that directs IgA switching, as mice with specific deletion of TGF-β receptors on B cells have extremely few IgA+ B cells in PPs and serum IgA deficiency [64]. Furthermore, IL-21 was shown to be important for IgA switching, because IL-21 and IL-21R deficiency leads to a reduction of intestinal IgA and IL-21R-deficient mice have impaired IgA responses to pathobionts [65,66]. The production of these two IgA cytokines likely depends on the nature of T follicular helper (TFH) cells [67]. In PP GCs, TFH cells constitutes a heterogenous population that can be generated from regulatory/effector CD4 T cells induced in the context of gut antigens, such as the Foxp3+ T cells and Th17 cells [68,69]. Whilst Foxp3+ T cells are known for their production and activation of TGF-β1, Th17 cells produce abundant IL-21. TGF-β1 in synergy with IL-21 enhances both the proliferation and differentiation of IgA-expressing plasma cell precursors [70,71]. The abundant production of TGF-β1 and IL-21 contributes to the preferential generation of IgA+ B cells, and the balance between these two cytokines may be important for the regulation of the clonal expansion of IgA+ B cells in the GC [10].

The inhibitory receptor PD-1 expressed on T cells also exerts a role in the regulation of IgA production. PD-1 deficiency leads to increased number and frequency of TFH cells in PP, and these cells have increased expression of Bcl-6 and decreased expression of IRF4 associated with reduced IL-21 and excessive pro-inflammatory (IFN-γ, TNF-α) cytokine production. The IgAs produced under this

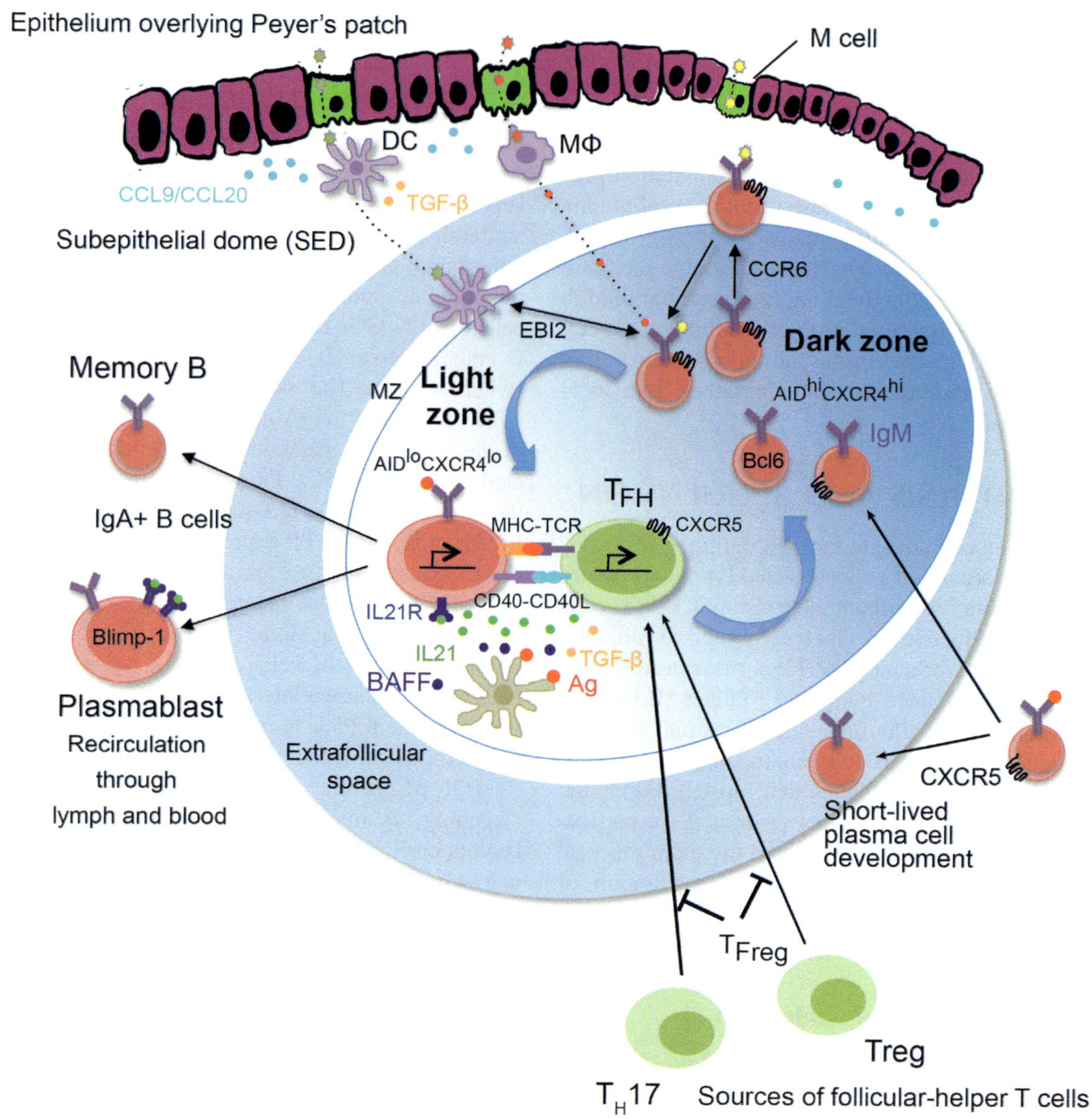

FIGURE 21.1 **T cell-dependent induction of IgA+ B cells in Peyer's patch lymphoid follicles.** Activated B cells enter germinal centres (GCs) within the intestinal Peyer's patches via engagement of the chemokine receptor CXCR5, where they undergo T cell-dependent activation in the light zone and class switch recombination (CSR) and somatic hypermutation (SHM) in the dark zone. GC T follicular helper cells (T_{FH} originating from differentiating Th17 and Treg cells with feedback controls from T follicular regulatory TFReg cells) in the light zone mediate affinity maturation for antigens selected and presented by follicular dendritic cells (FDCs). Interactions between B cells and T_{FH} involve class II MHC:T cell receptor (TCR) interactions as well as CD40:CD40L engagement. T_{FH} and dendritic cells produce IL-21 and TGF-β1 that is sensed by the IL-21 receptor (IL21R) on the follicular B cells. B cells subsequently upregulate activation-induced deaminase (AID) required for CSR and SHM in the dark zone. Recirculation between the light and the dark zone can occur several times. IgA CSR can also happen through events taking place in the subepithelial dome (SED). Antigens entering the intestine through M cells are taken up and presented by dendritic cells (DCs) to B cells at the interface to the follicles. Migration of B cells to the follicle border is mediated through activation of the oxysterol receptor EBI2 on B cells. DCs and stromal cells also secrete B cell-activating factor (BAFF) and a proliferation-inducing ligand (APRIL), which are known to enhance IgA class switching in B cells independently of T cell help. In addition, antigens entering through M cells can be phagocytosed and processed by subepithelial macrophages allowing them to diffuse freely into the GCs where they can be recognised by B cells. Follicular B cells can migrate to the epithelium through CCR6:CCL20 interaction, where they uptake antigens directly beneath M cells. Activated and affinity matured B cells leave the GC as plasmablasts or memory B cells. *Adapted from Figure 4, Ref. [17].*

condition have altered repertoire and reduced bacteria-binding capacity, causing dysbiosis of the gut microbial communities [10]. Interestingly, the nature of TFH cells that provide help for IgA induction in GCs seems to regulate the IgA repertoire and consequently, the gut microbiota. IgAs elicited by TFH cells derived from Foxp3+ T cells can increase the diversification of *Clostridium* clusters IV and XIVa that in turn are known to induce Foxp3+ T-cell differentiation [72,73], suggesting a symbiotic regulatory loop between TFH cells, IgA and microbiota.

Another important CD4 T-cell subset present in PP GCs are T follicular regulatory (TFR) cells. The lack of TFR cells causes an increase in TFH cells and the induction of polyreactive antibodies in spleen [74]. In PP GCs, TFR can be differentiated from either thymic or Foxp3+ T cells generated in the gut. TFR cells were shown to be fundamental to control the selection and quality of IgAs [72]. Exactly how TFR cells exert their regulatory function is currently poorly understood, especially in the PP GC context.

TGF-β1 is also produced by DCs and intestinal ECs in response to bacterial products. These cells also release RA, a metabolite of dietary vitamin A that synergises with TGF-β1 and IL-6 to enhance IgA CSR and production [61,62,75–77]. PP follicular dendritic cells (FDCs) are continuously and synergistically stimulated by bacteria and food-derived products such as RA, and thus conditioned to produce abundant CXCL13, B cell-activating factor of the TNF family (BAFF) and TGF-β1 that facilitate IgA production. Reduced production of these molecules by PP FDCs owing to defective TLR signalling via MyD88 or vitamin A deprivation causes loss of GC B cells and impaired generation of IgA+ B cells within PP GCs [78].

In summary, PPs are characterised by highly organised and dynamic microcompartments that provide a unique environment for a strongly biased generation of large amounts of affinity-matured IgA antibodies specific for gut antigens.

3.2 T Cell-Independent IgA Induction

CSR to IgA can occur without the help of T cells. Mice lacking T cells show significant intestinal IgA production, both under homoeostatic conditions and after infection [68,79]. Also, mice lacking gut-associated follicular structures retain some intestinal IgA production [60]. Though predominantly taking place within ILFs, these T cell-independent (TI) IgA responses also occur in MLNs, PPs and diffuse lymphoid tissue of the LP. This response is less efficient and less effective than in wild-type mice, because it occurs in the absence of the selection steps that normally occurs in the GC of PPs. However, it may be important soon after an infection to provide protection to the mucosal barrier until more effective and specific follicular IgA responses develop [80].

An ILF is a cluster of B cells surrounded by DCs/macrophages containing very few T cells interspersed between B cells [56,81]. Like PPs, the formation and maturation of ILFs require the presence of LTi cells [68,82–85]. The large majority of ILFs in normal mice lack GCs. Unlike PPs, which develop in the sterile foetal environment, ILFs develop only after bacterial colonisation of the intestine from smaller 'anlagen' structures called cryptopatches (CPs). The size and cellular composition of ILFs depend on the bacterial load of the intestine [56,81,86,87]. Bacterial activation induces not only an increase in size but also the functional maturation of ILFs, with some B cells expressing activation-induced cytidine deaminase (AID) and differentiating into IgA+ B cells and plasma cells. Activation of DCs and stromal cells by bacteria facilitates preferential IgA class switching in ILF B cells [83,88]. A subset of DCs produces abundant TNF following activation by bacteria, and this expression causes induction of matrix metalloproteinases (MMPs) and activation of TGF-β1. In addition to TNF and TGF-β1, DCs and stromal cells secrete B cell-activating factor (BAFF) and a proliferation-inducing ligand (APRIL), which are known to enhance IgA class switching in B cells independently of T-cell help [79,83,89,90].

IgA production can also occur in MLNs and PPs through a TI pathway involving plasmacytoid DCs (pDCs) [79]. By producing type-I interferon in response to TLR signals from the microbiota, stromal cells stimulate pDC release of BAFF and APRIL, which thereafter trigger IgA CSR and production in follicular B cells. FDCs may further enhance TI IgA responses by stimulating follicular B cells in PPs. Indeed, imprinting signals from RA and the microbiota stimulate FDC release of BAFF and CXCL13 as well as FDC processing of TGF-β1 via MMP9 and MMP13 [78]. These pDC/FDC-based TI pathways might explain why mice lacking CD40 retain some IgA production in PPs [91,92].

Whether IgA+ cells can be generated in LP is still controversial. B cells in the mouse LP express AID and α-germline transcripts [93] and similarly, AID can be detected in IgA+ LP B cells from either healthy individuals or immunodeficient patients with an insufficiency of CD40L–CD40 interaction [94]. Moreover, mice that completely lack gut organised follicular structures as a result of a deficiency of lymphoid tissue–inducing RORγt or Id2 transcription factors retain some intestinal IgA production [76,83]. These observations indicate that in the absence of PPs and ILFs, IgA can be generated from the few B cells present in the LP. Some of these AID-expressing LP B cells can be detected in sub-epithelial areas in close proximity to DCs [94], further supporting the notion that a primitive follicle-independent pathway locally supports IgA production. However, some studies show that the LP contains neither AID nor germline Iα-Cα nor switch Iα-Cμ circle transcripts [92,95,96]. Other studies conclude that the LP lacks IgA class switching because the LP has much fewer AID, Iα-Cα and/or Iα-Cμ transcripts than PPs [97,98]. These discordant results may happen because LP B cells are more scattered and express fewer AID transcripts than PP B cells and thus IgA CSR can be easily underestimated or pass completely unrecognised in the LP, unless appropriate comparisons are established and sensitive methodologies [99]. IgA-inducing signals in LP seems to be produced by different cell types. Intestinal ECs and/or SCs produce BAFF, APRIL, TGF-β1, RA and IL-6 [94,100,101], which cooperate with BAFF and APRIL released by DCs to induce IgA and support plasma cell survival in LP. DCs can also be activated by RORγt+ ILCs to express iNOS [102], and in the presence of TLR signals, nitric oxide would enhance IgA responses by stimulating DC expression of BAFF and APRIL [79].

Peritoneal B-1 cells, a subset of innate-like CD5+ B cells that express polyreactive BCRs and mainly colonise the peritoneal cavity, also contribute to the production of intestinal IgA in a TI manner [103,104]. IgM and IgA produced by B-1 cells generally have low affinities.

4. IGA REPERTOIRE FEATURES

The analysis of B-cell receptor genetics of mucosal B cells revealed that intestinal B-cell populations possess Ig gene characteristics that can be defined as typical for mucosal B-cell responses and distinguish them from other B-cell populations in the body.

One of the main hallmarks of intestinal IgA gene repertoire is its high mutational load. Around 9% of the nucleotides in VH genes is altered [105] exceeding the mutational count of memory B cells. These somatic hypermutations accumulate with age, until in adulthood in mice or kindergarten age in humans all intestinal plasmablasts are basically devoid of unmutated BCR sequences [106,107]. This process ensures constant adaptation to the microbiota and is driven by the continuous diversification of memory B cells which fuel the replenishment of the intestinal plasmablast compartment [108]. Thus, it is important that gut-associated lymphoid structures are able to recruit novel clones upon antigenic challenge [108,109]. Indeed, PP GCs can select antigen-reactive clones to the microbiota [110,111], and they can differentiate to plasmablasts secreting antigen-specific IgA [40]. However, a mouse model in which the BCR was replaced with the Epstein–Barr antigen LMP2A showed that generation of mutations can be BCR-independent [112] and might be even independent of antigen [113].

Secondly, intestinal plasmablasts show oligoclonal distribution with single large clonotypes that can dominate the individual Ig gene repertoire. Although, a large fraction of 25%–35% clones are unique to each repertoire, the highly expanded clonotypes can make up more than 1% of the entire repertoire [38,107]. This limitation of Ig gene diversity gets more pronounced after repeated bacterial exposures [114]. Although, more expanded clonotypes possess higher burden of somatic hypermutation, it remains unclear which properties favour clonal expansion for some but not other clonotypes.

Thirdly, when compared to the naïve repertoire, PP GC B cells show a repertoire bias to skewed VH usage [110]. Those segments are enriched in public clones that are found in multiple animals [107,110,111] and expand to the effector B-cell compartment in the LP [40]. However, the antigens that are targeted by public clones, although attempted, have not yet been identified.

Finally, the IgA gene repertoire, as shown in mice, is developmentally multilayered. IgA plasma cells that populate the small intestinal LP can originate from adult hematopoietic stem cells, but a subfraction develops from foetal liver stem cell progenitors [1,115]. Whether there are functional differences between those two different developmental programs remains to be determined.

5. MOLECULAR TARGETS

In infections of mucosal sites, antigen-specific IgA can have a predominant role [116] and can have potent protective potential [117] with dimeric IgA molecules even outcompeting IgG monomers [118]. However, due to the intraindividual complexity and interindividual diversity of the microbiota with numerous strains with thousands of different potential antigens which are likely containing several epitopes, it has been a major challenge to elucidate the binding mode of antibodies, the targeted antigens and protective potential in mucosal antibody responses to the microbiota. However, defined colonisation events and gnotobiotic animals with defined floras allow the basic principles of intestinal antibody binding to be defined. Today, we can distinguish two binding modes of intestinal antibodies: canonical and non-canonical binding (Figure 21.2) [37].

Canonical binding refers to the interaction between an antibody's variable region and the cognate antigen and can be subject to affinity maturation processes influenced by somatic hypermutation. This interaction is defined by the affinity and avidity of the antibody. Especially for dimeric IgA binding to highly repetitive antigenic structures that are common to bacterial membranes, such as lipopolysaccharide O-polysaccharides from gram-negative bacteria, avidity may play a predominant role in intestinal antibody responses. Although the affinity of intestinal IgA has been defined for a few clones [40], the affinity and avidity of intestinal antibodies have not been systematically analysed. One reason for this is that only few antigens of intestinal IgA are known, and most of the defined targets are membrane proteins that are difficult to use in affinity assays. These include many glycan antigens that are complex to purify, with a repetitive nature that makes it difficult to establish the methods that are able to measure exact affinities. Thus, to date, no crystal structure of SIgA [119] in complex with antigen is reported. In general, IgA has been shown to be able to be induced against food

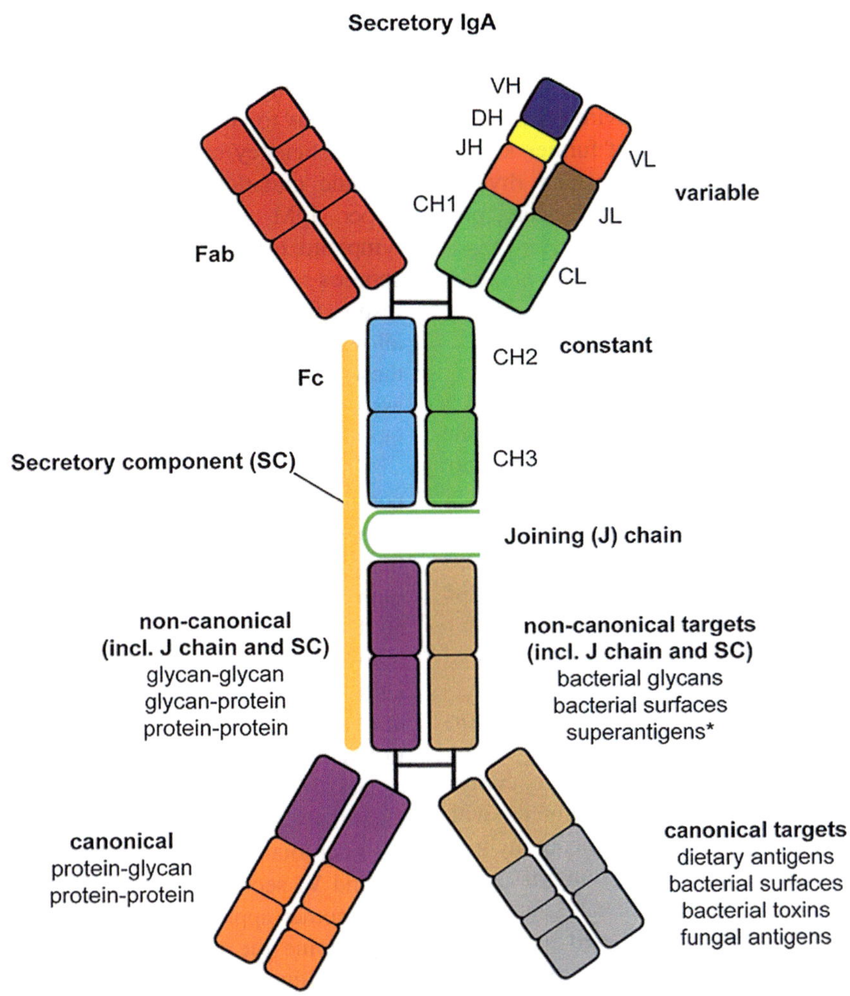

FIGURE 21.2 **Structure of secretory IgA.** Diagram of the secretory IgA dimeric structure showing the variable and constant (CH1-3) domains of the heavy and light chains. The dimer is joined by the J chain, and secretory component is covalently bound adjacent to one side of the CH2 and CH3 domains. Secretory component is the proteolysed remains of the polymeric immunoglobulin receptor which transports the IgA dimeric complex through surface epithelial cells before proteolytic release at the luminal surface. Protein products of somatic recombination of heavy and light immunoglobulin V, D and J gene segments are shown on the *top right* of the diagram. Examples of potential canonical and non-canonical binding targets are shown with the likely attachment sites at the *bottom* of the diagram.

antigens [120], bacterial antigens [40,104,121] and fungal antigens [32]. As shown with polyclonal IgA fractions, intestinal IgA is selective and can coat a diverse but distinct set of intestinal microbes [122,123]. Studies using monoclonal IgA or IgG1 isolated from intestinal single IgA plasmablasts have shown that a single IgA clone can target multiple bacteria [39,124,125] by targeting the same epitope in structurally similar but distinct antigens [39]. Despite this, a smaller subset of antibodies that are reactive in assays using several structurally diverse antigens, referred to as polyreactive antibodies, have been described [38,40,126].

Non-canonical binding refers to antibody binding that does not involve antibody affinity maturation and is generally not influenced by somatic hypermutation [37]. Secretory IgA is highly glycosylated and provides a framework for several non-cognate interactions [35]. Even secretory chain alone can potentially interact with intestinal microbes [127]. Further, one special mode of antibody binding has been shown in which IgA is targeted by bacterial superantigens expressed by the intestinal microbiota [128]. These superantigens are usually expressed by pathogens and can target all IgA, or in one example a subset of IgA antibodies that express the human IGHV3 or the murine IGHV5/6/7 family [128]. Although this binding mode involves the framework region of the antibody variable region [129], we have included this in the noncanonical binding as it does not involve binding by the CDRs, and it remains unclear how strongly the antibody superantigen interaction is influenced by affinity maturation.

Overall, the interaction of IgA with antigen in the intestine is highly complex and can be influenced by cognate and non-cognate-binding modes. Despite this, the description of antibody binding in the complex environment of the gut lumen will remain incomplete without further knowledge about the influence of location, circadian rhythm, feeding time [130], carbon sources, antigen density, proliferation [131] and pH in the gut ecosystem.

6. FUNCTIONAL INSIGHTS OF SECRETORY IGA

Given the amount of IgA produced on a daily basis and the diversity of potential sites of action and targets, there are different ways of assessing IgA function.

Earlier in vivo functional assays involved measuring stimulation of fluid secretion within isolated intestinal loops injected with cholera toxin in an anesthetised animal: the amount of fluid secreted was attenuated in animals where a secretory IgA response had been induced by prior intestinal immunisation with cholera toxin [34].

Antibody protection against cytopathic viruses is normally measured by restriction of viral plaque in culture when a neutralising antibody is included in the culture medium [34,132]. For IgA secretion, experiments with mouse backpack models in which a subcutaneous hybridoma secreting a monoclonal IgA against rotavirus show in vivo protection even where the monoclonal IgA are not formally neutralising in classical culture-based plaque assays [27]. This is likely due to clearance of virus through the endocytotic pathway bound to secretory IgA as it is transferred attached to the pIgR from the basolateral to the apical surface of the epithelial layer [133].

A different approach to secretory IgA function has been to detect the protective effects of secretory IgA on bacterial penetration across the intestinal epithelial barrier [134] or indirectly by the effects on mucosal transcriptomic signatures indicative of reduced innate immune activation [135]. Within a diverse microbiota, spontaneously high IgA-binding levels can even be indicative of potentially pathogenic microbes, as when microbes are sorted on this basis and transferred into germ-free recipient mice they generate susceptibility to dextran sulphate colitis that is not seen if microbes are transferred that are only coated with low IgA levels [122]. Increased IgA binding is also seen in those microbial taxa from human infants with kwashiorkor malnutrition capable of recapitulating the intestinal inflammation when transferred into germ-free mice [136].

It would be so much easier to demonstrate IgA function if selectively IgA-deficient animals or humans did not have a relatively weak phenotype. The compensatory mechanisms that limit the impact of selective IgA deficiency including cross-reactivity with secreted IgM [18] are reviewed in detail elsewhere [17]. Mouse strain combination experiments are nevertheless informative. Loss of transepithelial immunoglobulin transfer (of both IgM and IgA) leads to an intestinal protein-losing enteropathy [20]. AID deficiency causing loss of CSR to IgA and deficient somatic mutation of IgM results in a strongly activated hyper IgM phenotype and hyperplasia of intestinal lymphoid tissues [56,137]. The necessity for adaptive responses of IgA (and IgM) to the microbiota is shown by results from mice with a mutation (G23S) in AID which allows CSR to IgA but abrogates somatic hypermutation – these mice also have intestinal lymphoid hyperplasia associated with microbiota dysbiosis and defects in the intestinal barrier function [138].

Primary IgA deficiency in humans may be silent but is not universally without clinical consequences. There is increased susceptibility to respiratory and intestinal infections and to potential development of different autoimmune diseases [139]. Secondary deficiency as a result of antibiotics disrupting the patient's microbiota or through idiosyncratic drug side-effects is less well appreciated, although increased susceptibility to respiratory infection is also reported [140]. Analysis of the residual microbiome in IgA-deficient patients shows that microbial strains – especially *Escherichia coli* – that are enriched with potentially pathogenic functions including multidrug and antimicrobial peptide resistance, virulence factors and type III and VI secretion systems [141].

Another approach to investigate IgA function has been to examine the effect of secretory IgA on intestinal microbes. Generally, IgA induced by both T-dependent and T-independent mechanisms binds a wide variety (but not all) intestinal microbial taxa, especially those present in the small intestinal microbial biomass [142]. Use of a backpack model in which monoclonal IgA directed against *Bacteroides thetaiotaomicron* was expressed by a subcutaneous hybridoma showed that IgA directed against the capsular polysaccharide was able to protect the microbe against oxidative stress [135]. In the case of *Salmonella typhimurium*, high-affinity IgA causes the growing bacteria to clump, and these larger particles of 'enchained' bacteria are more rapidly eliminated by intestinal motility [131]. Transient colonisation systems of germ-free mice, where waves of live bacteria pass through the intestine, but do not permanently colonise, also show more rapid bacterial clearance after mucosal IgA induction [143]. Finally, it has recently become possible to examine the effect of individual dimeric monoclonal IgA specificities (in some cases with identified antigenic targets) in mice monocolonised with the bacterial strain originally used to generate the IgA response. This works because dimeric serum IgA is eliminated through the hepatobiliary system in rodents, so an intravenous injection causes a wave of specific IgA to be transferred to the intestine and the resultant transcriptomic and other microbial adaptations can be studied. Because this does not depend on constructing a different knock-in mouse strain for each IgA specificity, the system is very scalable and the functional effects of many different specificities can potentially be studied. In vivo and in vitro experiments with this system show that microbial metabolic alterations depend on IgA antigenic specificity, and that different monoclonals can have different effects targeting the same antigen depending on the epitope concerned and its accessibility during bacterial proliferation. IgA also exerts effects on target microbes merely by coating the outer membrane – for example, limiting motility and bile acid toxicity. These results are interpreted as the pool of IgA induced by a single microbe consists of different parallel components each with different effects on the bacterial target that depend not only on the relevant antigen but also on the replication status of the bacterium [40].

7. MOUSE-HUMAN SPECIES DIFFERENCES AND IGA RELEVANCE IN SYSTEMIC DISEASE

7.1 Isotype and Fc Receptor Species Differences

Whilst the mouse is a useful model for studying the functionality of secretory IgA, there are important species differences in appreciating the role of IgA in human systemic disease. The consequences of IgA deficiency are discussed above. We refer to other work for clinical and aetiopathogenic features of IgA immune damage [144–147], the role of IgA as a biomarker [148–151] and associations of IgA deficiency with coeliac disease and other autoimmune and allergic conditions [14]: the focus here is to provide an overview of the role of the human IgA isotypes and bacterial interactions in human systemic immunity. The evidence that levels of secretory IgA binding to intestinal microbes can be a marker of their potential as intestinal inflammatory trigger status is discussed in the previous section.

In contrast to mice with a single IgA isotype, humans have two. IgA1 dominates in the serum and has an extended hinge region containing up to 6 O-linked glycans, whereas IgA2 is the main isotype in the colon without O-linked glycans [152]. Another species difference is the expression of an IgA Fc receptor (FcαRI, CD89) on human myeloid cells: orthologues have been found in many other species, including rats, but not in mice due to a gene translocation [153,154], although transgenic FcαR expression can be used to refine mouse experiments [155].

7.2 Renal Disease and Vasculitis

IgA immune complexes may underlie both kidney disease and vasculitis.

IgA nephropathy is the most common cause of glomerulonephritis and results from deposition of polymeric IgA1 immune complexes in the mesangial space between the capillaries of the glomeruli. It can progress to renal failure as mesangial cells proliferate and synthesise extracellular matrix into which leucocytes infiltrate [144]. The underlying cause is mainly the accumulation of galactose-deficient IgA1 molecules, associated with chronic mucosal infections [156]. The inheritance of galactose-deficient glycosylation of IgA1 in IgA nephropathy patients [152] and their first-degree relatives has been reported [157], and the microbiota has been postulated to play direct (through galactosidase expression) and indirect (proinflammatory consequences of defective epithelial barrier function) effects [158–160] in the successive process of mesangial immune complex deposition and renal damage [146].

IgA vasculitis (Henoch–Schönlein purpura) results from the deposition of IgA immune complexes with associated neutrophil infiltration in the small vessels of the skin (resulting in the eponymous purpuric patches), intestinal tract, joints and the kidneys, potentially causing damage to each of these tissues. Abnormal glycosylation is also a feature of vasculitis and neutrophil activation through FcαRI resulting in damage to the vessels [161]. Associations of the condition have been reported following mucosal infections and with intestinal microbiota dysbiosis [162–164], although the pathogenic target may depend on IgA-binding endothelial cell antigens [147,165].

7.3 IgA1 Proteases

Human FcαRI-positive neutrophils that are recruited to inflammatory sites can be activated indirectly with the FcαRI receptor that associates with the FcRγ subunit containing an immunoreceptor tyrosine activation motif (ITAM). All IgA isoforms are ligands for FcαRI, although the highest avidity is for IgA immune complexes. Some pathogens such as *Streptococcus pneumoniae* and *Neisseria meningitidis* subvert the biocidal effect through secreting IgA1 proteases which are required for virulence [166].

8. CONCLUDING REMARKS

Since the publication of the previous edition of this book 7 years ago, there have been significant advances in the understanding of mechanisms of IgA induction and its function both as a secretory immunoglobulin and in the systemic circulation. The way in which selective IgA deficiency is compensated at mucosal surfaces and the potential range of antigenic targets including SARS-CoV2 are better delineated. These advances have mirrored the general increased appreciation of microbiota–immune interactions and the technical ability to interrogate complex repertoires with highly parallel sequencing and single-cell approaches.

All learning is an approximation to the truth, and we still need better information on how the relatively narrow IgA repertoire can both diversify with age with structural information on how binding to different epitopes allows IgA to potentially share specificities, especially for glycan interactions, with sufficient functional avidity against the wide range of different microbial species and substrains encountered from early life and throughout the ageing process. The relative functional importance of canonical versus non-canonical binding remains a challenge.

Modern hygiene and medicine in the developed world make humans a long-lived species, but the coronavirus pandemic gave us a wake-up call on our vulnerability to mucosal infection. There are also still significant challenges to limit IgA immunopathology in human disease and to understand compensatory mechanisms in IgA deficiency. IgA remains a vital part of our immune responsiveness, most crucially for early life defences in unprivileged living conditions to mitigate the mucosal infections, intestinal failure, stunting and tragic loss of life in early childhood.

DEDICATION

The authors dedicate this chapter to the memory of the life and work of Professor Nils Lycke in thanks for his fundamental contributions to mucosal immunology and IgA biology, and for his friendship and immense encouragement of fellow scientists.

REFERENCES

[1] Kroese FGM, et al. Many of the IgA producing plasma cells in the murine gut are derived from self-replenishing precursors in the peritoneal cavity. Int Immunol 1988;1:75–84.
[2] Mostov KE. Transepithelial transport of immunoglobulins. Annu Rev Immunol 1994;12:63–84.
[3] Benveniste J, Lespinats G, Salomon J. Serum and secretory IgA in axenic and holoxenic mice. J Immunol 1971;107:1656–62.
[4] Fransen F, et al. BALB/c and C57BL/6 mice differ in polyreactive IgA abundance, which impacts the generation of antigen-specific IgA and microbiota diversity. Immunity 2015;43:527–40.
[5] Yang C, et al. Fecal IgA levels are determined by strain-level differences in Bacteroides ovatus and are modifiable by gut microbiota manipulation. Cell Host Microbe 2020;27:467–475 e466.
[6] Stoica G, Macarie E, Michiu V, Stoica RC. Biologic variation of human immunoglobulin concentration. I. Sex-age specific effects on serum levels of IgG, IgA, IgM and IgD. Med Interne 1980;18:323–32.
[7] Hooijkaas H, Benner R, Pleasants JR, Wostmann BS. Isotypes and specificities of immunoglobulins produced by germ-free mice fed chemically defined ultrafiltered "antigen-free" diet. Eur J Immunol 1984;14:1127–30.
[8] Hara S, et al. Dietary antigens induce germinal center responses in Peyer's patches and antigen-specific IgA production. Front Immunol 2019;10:2432.
[9] Campos-Rodriguez R, et al. Stress modulates intestinal secretory immunoglobulin A. Front Integr Neurosci 2013;7:86.
[10] Kawamoto S, et al. The inhibitory receptor PD-1 regulates IgA selection and bacterial composition in the gut. Science 2012;336:485–9.
[11] Wilmore JR, et al. Commensal microbes induce serum IgA responses that protect against polymicrobial sepsis. Cell Host Microbe 2018;23:302–11. e303.
[12] Moon C, et al. Vertically transmitted faecal IgA levels determine extra-chromosomal phenotypic variation. Nature 2015;521(7550):90–3.
[13] Planer JD, et al. Development of the gut microbiota and mucosal IgA responses in twins and gnotobiotic mice. Nature 2016;534:263–6.
[14] Odineal DD, Gershwin ME. The epidemiology and clinical manifestations of autoimmunity in selective IgA deficiency. Clin Rev Allergy Immunol 2020;58:107–33.
[15] Harriman GR, et al. Targeted deletion of the IgA constant region in mice leads to IgA deficiency with alterations in expression of other Ig isotypes. J Immunol 1999;162:2521–9.
[16] Mbawuike IN, et al. Mucosal immunity to influenza without IgA: an IgA knockout mouse model. J Immunol 1999;162:2530–7.
[17] Macpherson AJ, Yilmaz B, Limenitakis JP, Ganal-Vonarburg SC. IgA function in relation to the intestinal microbiota. Annu Rev Immunol 2018;36:359–81.
[18] Fadlallah J, et al. Microbial ecology perturbation in human IgA deficiency. Sci Transl Med 2018;10.
[19] Ramanan D, et al. An immunologic mode of multigenerational transmission governs a gut Treg setpoint. Cell 2020;181:1276–1290 e1213.
[20] Johansen FE, et al. Absence of epithelial immunoglobulin A transport, with increased mucosal leakiness, in polymeric immunoglobulin Receptor/Secretory component-deficient mice. J Exp Med 1999;190:915–22.
[21] Brandtzaeg P. The mucosal immune system and its integration with the mammary glands. J Pediatr 2010;156:S8–15.
[22] Kramer DR, Cebra JJ. Early appearance of "natural" mucosal IgA responses and germinal centers in suckling mice developing in the absence of maternal antibodies. J Immunol 1995;154:2051–62.
[23] Mu Q, et al. Regulation of neonatal IgA production by the maternal microbiota. Proc Natl Acad Sci U S A 2021;118:e2015691118.
[24] Harris NL, et al. Mechanisms of neonatal mucosal antibody protection. J Immunol 2006;177:6256–62.
[25] Rogier EW, et al. Secretory antibodies in breast milk promote long-term intestinal homeostasis by regulating the gut microbiota and host gene expression. Proc Natl Acad Sci U S A 2014;111:3074–9.
[26] Ludvigsson JF, Neovius M, Stephansson O, Hammarstrom L. IgA deficiency, autoimmunity & pregnancy: a population-based matched cohort study. J Clin Immunol 2014;34:853–63.
[27] Burns JW, Siadat-Pajouh M, Krishnaney AA, Greenberg HB. Protective effect of rotavirus VP6-specific IgA monoclonal antibodies that lack neutralizing activity. Science 1996;272:104–7.
[28] Elson CO, Ealding W. Cholera toxin feeding did not induce oral tolerance in mice and abrogated oral tolerance to an unrelated protein antigen. J Immunol 1984;133:2892–7.

[29] Apter FM, et al. Analysis of the roles of antilipopolysaccharide and anti-cholera toxin immunoglobulin A (IgA) antibodies in protection against *Vibrio cholerae* and cholera toxin by use of monoclonal IgA antibodies in vivo. Infect Immun 1993;61:5279–85.
[30] Bronson RA, Cooper GW. Documentation of IgA1 and IgA2 antisperm antibodies within seminal fluid. Am J Reprod Immunol Microbiol 1988;18:7–10.
[31] Frehn L, et al. Distinct patterns of IgG and IgA against food and microbial antigens in serum and feces of patients with inflammatory bowel diseases. PLoS One 2014;9:e106750.
[32] Ost KS, et al. Adaptive immunity induces mutualism between commensal eukaryotes. Nature 2021;596:114–8.
[33] McCoy KD, et al. Polyclonal and specific antibodies mediate protective immunity against enteric helminth infection. Cell Host Microbe 2008;4:362–73.
[34] Lycke N, Eriksen L, Holmgren J. Protection against cholera toxin after oral immunisation is thymus dependent and associated with intestinal production of neutralising IgA antitoxin. Scand J Immunol 1987;25:413–9.
[35] Nakajima A, et al. IgA regulates the composition and metabolic function of gut microbiota by promoting symbiosis between bacteria. J Exp Med 2018;215:2019–34.
[36] Wold AE, et al. Secretory immunoglobulin A carries oligosaccharide receptors for *Escherichia coli* type 1 fimbrial lectin. Infect Immun 1990;58:3073–7.
[37] Pabst O, Slack E. IgA and the intestinal microbiota: the importance of being specific. Mucosal Immunol 2019;13:12–21.
[38] Benckert J, et al. The majority of intestinal IgA+ and IgG+ plasmablasts in the human gut are antigen-specific. J Clin Invest 2011;121:1946–55.
[39] Rollenske T, et al. Cross-specificity of protective human antibodies against *Klebsiella pneumoniae* LPS O-antigen. Nat Immunol 2018;19:617–24.
[40] Rollenske T, et al. Parallelism of intestinal secretory IgA shapes functional microbial fitness. Nature 2021;598:657–61.
[41] Iverson GM, et al. Patients with atherosclerotic syndrome, negative in anti-cardiolipin assays, make IgA autoantibodies that preferentially target domain 4 of beta2-GPI. J Autoimmun 2006;27:266–71.
[42] Stern JN, et al. B cells populating the multiple sclerosis brain mature in the draining cervical lymph nodes. Sci Transl Med 2014;6:248ra107.
[43] Shalapour S, et al. Inflammation-induced IgA+ cells dismantle anti-liver cancer immunity. Nature 2017;551:340–5.
[44] Ochsenbein AF, et al. Control of early viral and bacterial distribution and disease by natural antibodies. Science 1999;286:2156–9.
[45] Fehres CM, et al. APRIL induces a novel subset of IgA(+) regulatory B cells that suppress inflammation via expression of IL-10 and PD-L1. Front Immunol 2019;10:1368.
[46] Roux ME, McWilliams M, Phillips-Quagliata JM, Weisz-Carrington P, Lamm ME. Origin of IgA-secreting plasma cells in the mammary gland. J Exp Med 1977;146:1311–22.
[47] Mestecky J, et al. Selective induction of an immune response in human external secretions by ingestion of bacterial antigen. J Clin Invest 1978;61:731–7.
[48] Isho B, Florescu A, Wang AA, Gommerman JL. Fantastic IgA plasma cells and where to find them. Immunol Rev 2021;303:119–37.
[49] Rojas OL, et al. Recirculating intestinal IgA-producing cells regulate neuroinflammation via IL-10. Cell 2019;176:610–24. e618.
[50] Probstel AK, et al. Gut microbiota-specific IgA(+) B cells traffic to the CNS in active multiple sclerosis. Sci Immunol 2020;5:eabc7191.
[51] Macpherson AJ, McCoy KD, Johansen FE, Brandtzaeg P. The immune geography of IgA induction and function. Mucosal Immunol 2008;1:11–22.
[52] Suzuki K, Kawamoto S, Maruya M, Fagarasan S. GALT: organization and dynamics leading to IgA synthesis. Adv Immunol 2010;107:153–85.
[53] Ohno H. Intestinal M cells. J Biochem 2016;159:151–60.
[54] van de Pavert SA, Mebius RE. New insights into the development of lymphoid tissues. Nat Rev Immunol 2010;10:664–74.
[55] Fawkner-Corbett D, et al. Spatiotemporal analysis of human intestinal development at single-cell resolution. Cell 2021;184:810–826 e823.
[56] Fagarasan S, et al. Critical roles of activation-induced cytidine deaminase in the homeostasis of gut flora. Science 2002;298:1424–7.
[57] Suzuki K, et al. Aberrant expansion of segmented filamentous bacteria in IgA-deficient gut. Proc Natl Acad Sci U S A 2004;101:1981–6.
[58] Catanzaro JR, et al. IgA-deficient humans exhibit gut microbiota dysbiosis despite secretion of compensatory IgM. Sci Rep 2019;9:13574.
[59] Suzuki K, Fagarasan S. How host-bacterial interactions lead to IgA synthesis in the gut. Trends Immunol 2008;29:523–31.
[60] Fagarasan S, Kawamoto S, Kanagawa O, Suzuki K. Adaptive immune regulation in the gut: T cell-dependent and T cell-independent IgA synthesis. Annu Rev Immunol 2010;28:243–73.
[61] Cerutti A, Rescigno M. The biology of intestinal immunoglobulin A responses. Immunity 2008;28:740–50.
[62] Mora JR, et al. Generation of gut-homing IgA-secreting B cells by intestinal dendritic cells. Science 2006;314:1157–60.
[63] Pereira JP, Kelly LM, Xu Y, Cyster JG. EBI2 mediates B cell segregation between the outer and centre follicle. Nature 2009;460:1122–6.
[64] Cazac BB, Roes J. TGF-beta receptor controls B cell responsiveness and induction of IgA in vivo. Immunity 2000;13:443–51.
[65] Cao AT, et al. Interleukin (IL)-21 promotes intestinal IgA response to microbiota. Mucosal Immunol 2015;8:1072–82.
[66] Cho H, et al. Defective IgA response to atypical intestinal commensals in IL-21 receptor deficiency reshapes immune cell homeostasis and mucosal immunity. Mucosal Immunol 2019;12:85–96.
[67] Seo GY, Youn J, Kim PH. IL-21 ensures TGF-beta 1-induced IgA isotype expression in mouse Peyer's patches. J Leukoc Biol 2009;85:744–50.
[68] Tsuji M, et al. Preferential generation of follicular B helper T cells from Foxp3+ T cells in gut Peyer's patches. Science 2009;323:1488–92.
[69] Hirota K, et al. Plasticity of TH17 cells in Peyer's patches is responsible for the induction of T cell-dependent IgA responses. Nat Immunol 2013;14:372–9.
[70] Dullaers M, et al. A T cell-dependent mechanism for the induction of human mucosal homing immunoglobulin A-secreting plasmablasts. Immunity 2009;30:120–9.

[71] Avery DT, Bryant VL, Ma CS, de Waal Malefyt R, Tangye SG. IL-21-induced isotype switching to IgG and IgA by human naive B cells is differentially regulated by IL-4. J Immunol 2008;181:1767–79.
[72] Kawamoto S, et al. Foxp3(+) T cells regulate immunoglobulin a selection and facilitate diversification of bacterial species responsible for immune homeostasis. Immunity 2014;41:152–65.
[73] Atarashi K, et al. Induction of colonic regulatory T cells by indigenous Clostridium species. Science 2011;331:337–41.
[74] Linterman MA, et al. Foxp3+ follicular regulatory T cells control the germinal center response. Nat Med 2011;17:975–82.
[75] Tokuyama H, Tokuyama Y. The regulatory effects of all-trans-retinoic acid on isotype switching: retinoic acid induces IgA switch rearrangement in cooperation with IL-5 and inhibits IgG1 switching. Cell Immunol 1999;192:41–7.
[76] Uematsu S, et al. Regulation of humoral and cellular gut immunity by lamina propria dendritic cells expressing Toll-like receptor 5. Nat Immunol 2008;9:769–76.
[77] Sato A, et al. CD11b+ Peyer's patch dendritic cells secrete IL-6 and induce IgA secretion from naive B cells. J Immunol 2003;171:3684–90.
[78] Suzuki K, et al. The sensing of environmental stimuli by follicular dendritic cells promotes immunoglobulin A generation in the gut. Immunity 2010;33:71–83.
[79] Tezuka H, et al. Prominent role for plasmacytoid dendritic cells in mucosal T cell-independent IgA induction. Immunity 2011;34:247–57.
[80] Lopatin U, Blutt SE, Conner ME, Kelsall BL. Lymphotoxin alpha-deficient mice clear persistent rotavirus infection after local generation of mucosal IgA. J Virol 2013;87:524–30.
[81] Hamada H, et al. Identification of multiple isolated lymphoid follicles on the antimesenteric wall of the mouse small intestine. J Immunol 2002;168:57–64.
[82] Fagarasan S. Intestinal IgA synthesis: a primitive form of adaptive immunity that regulates microbial communities in the gut. Curr Top Microbiol Immunol 2006;308:137–53.
[83] Tsuji M, et al. Requirement for lymphoid tissue-inducer cells in isolated follicle formation and T cell-independent immunoglobulin A generation in the gut. Immunity 2008;29:261–71.
[84] Eberl G, Littman DR. Thymic origin of intestinal alphabeta T cells revealed by fate mapping of RORgammat+ cells. Science 2004;305:248–51.
[85] Ivanov II, Diehl GE, Littman DR. Lymphoid tissue inducer cells in intestinal immunity. Curr Top Microbiol Immunol 2006;308:59–82.
[86] Bouskra D, et al. Lymphoid tissue genesis induced by commensals through NOD1 regulates intestinal homeostasis. Nature 2008;456:507–10.
[87] Pabst O, et al. Adaptation of solitary intestinal lymphoid tissue in response to microbiota and chemokine receptor CCR7 signaling. J Immunol 2006;177:6824–32.
[88] McDonald KG, McDonough JS, Dieckgraefe BK, Newberry RD. Dendritic cells produce CXCL13 and participate in the development of murine small intestine lymphoid tissues. Am J Pathol 2010;176:2367–77.
[89] Litinskiy MB, et al. DCs induce CD40-independent immunoglobulin class switching through BLyS and APRIL. Nat Immunol 2002;3:822–9.
[90] Xu W, et al. Viral double-stranded RNA triggers Ig class switching by activating upper respiratory mucosa B cells through an innate TLR3 pathway involving BAFF. J Immunol 2008;181:276–87.
[91] Bergqvist P, Gardby E, Stensson A, Bemark M, Lycke NY. Gut IgA class switch recombination in the absence of CD40 does not occur in the lamina propria and is independent of germinal centers. J Immunol 2006;177:7772–83.
[92] Bergqvist P, Stensson A, Lycke NY, Bemark M. T cell-independent IgA class switch recombination is restricted to the GALT and occurs prior to manifest germinal center formation. J Immunol 2010;184:3545–53.
[93] Fagarasan S, Kinoshita K, Muramatsu M, Ikuta K, Honjo T. In situ class switching and differentiation to IgA-producing cells in the gut lamina propria. Nature 2001;413:639–43.
[94] He B, et al. Intestinal bacteria trigger T cell-independent immunoglobulin A(2) class switching by inducing epithelial-cell secretion of the cytokine APRIL. Immunity 2007;26:812–26.
[95] Shikina T, et al. IgA class switch occurs in the organized nasopharynx- and gut-associated lymphoid tissue, but not in the diffuse lamina propria of airways and gut. J Immunol 2004;172:6259–64.
[96] Barone F, Patel P, Sanderson JD, Spencer J. Gut-associated lymphoid tissue contains the molecular machinery to support T-cell-dependent and T-cell-independent class switch recombination. Mucosal Immunol 2009;2:495–503.
[97] Boursier L, Gordon JN, Thiagamoorthy S, Edgeworth JD, Spencer J. Human intestinal IgA response is generated in the organized gut-associated lymphoid tissue but not in the lamina propria. Gastroenterology 2005;128:1879–89.
[98] Lin M, Du L, Brandtzaeg P, Pan-Hammarstrom Q. IgA subclass switch recombination in human mucosal and systemic immune compartments. Mucosal Immunol 2014;7:511–20.
[99] Cerutti A. Location, location, location: B-cell differentiation in the gut lamina propria. Mucosal Immunol 2008;1:8–10.
[100] Iliev ID, Mileti E, Matteoli G, Chieppa M, Rescigno M. Intestinal epithelial cells promote colitis-protective regulatory T-cell differentiation through dendritic cell conditioning. Mucosal Immunol 2009;2:340–50.
[101] Iliev ID, et al. Human intestinal epithelial cells promote the differentiation of tolerogenic dendritic cells. Gut 2009;58:1481–9.
[102] Kruglov AA, et al. Nonredundant function of soluble LTalpha3 produced by innate lymphoid cells in intestinal homeostasis. Science 2013;342:1243–6.
[103] Hayakawa K, Hardy RR. Normal, autoimmune, and malignant CD5+ B cells: the Ly-1 B lineage? Annu Rev Immunol 1988;6:197–218.
[104] Macpherson AJ, et al. A primitive T cell-independent mechanism of intestinal mucosal IgA responses to commensal bacteria. Science 2000;288:2222–6.
[105] Fischer M, Kuppers R. Human IgA- and IgM-secreting intestinal plasma cells carry heavily mutated VH region genes. Eur J Immunol 1998;28:2971–7.
[106] Boursier L, Dunn-Walters DK, Spencer J. Characteristics of IgVH genes used by human intestinal plasma cells from childhood. Immunology 1999;97:558–64.
[107] Lindner C, et al. Age, microbiota, and T cells shape diverse individual IgA repertoires in the intestine. J Exp Med 2012;209:365–77.

[108] Lindner C, et al. Diversification of memory B cells drives the continuous adaptation of secretory antibodies to gut microbiota. Nat Immunol 2015;16:880–8.
[109] Di Niro R, et al. Salmonella infection drives promiscuous B cell activation followed by extrafollicular affinity maturation. Immunity 2015;43:120–31.
[110] Chen H, et al. BCR selection and affinity maturation in Peyer's patch germinal centres. Nature 2020;582:421–5.
[111] Nowosad CR, et al. Tunable dynamics of B cell selection in gut germinal centres. Nature 2020;588:321–6.
[112] Casola S, et al. B cell receptor signal strength determines B cell fate. Nat Immunol 2004;5:317–27.
[113] Yeap LS, et al. Sequence-intrinsic mechanisms that target AID mutational outcomes on antibody genes. Cell 2015;163:1124–37.
[114] Li H, et al. Mucosal or systemic microbiota exposures shape the B cell repertoire. Nature 2020;584:274–8.
[115] Vergani S, et al. A self-sustaining layer of early-life-origin B cells drives steady-state IgA responses in the adult gut. Immunity 2022;55:1829–1842 e1826.
[116] Sterlin D, et al. IgA dominates the early neutralizing antibody response to SARS-CoV-2. Sci Transl Med 2021;13.
[117] Planchais C, et al. Potent human broadly SARS-CoV-2-neutralizing IgA and IgG antibodies effective against Omicron BA.1 and BA.2. J Exp Med 2022;219.
[118] Wang Z, et al. Enhanced SARS-CoV-2 neutralization by dimeric IgA. Sci Transl Med 2021;13.
[119] Kumar Bharathkar S, et al. The structures of secretory and dimeric immunoglobulin A. Elife 2020;9.
[120] Hapfelmeier S, et al. Reversible microbial colonization of germ-free mice reveals the dynamics of IgA immune responses. Science 2010;328:1705–9.
[121] Yang C, et al. Immunoglobulin A antibody composition is sculpted to bind the self gut microbiome. Sci Immunol 2022;7:eabg3208.
[122] Palm NW, et al. Immunoglobulin a coating identifies colitogenic bacteria in inflammatory bowel disease. Cell 2014;158:1000–10.
[123] Kubinak JL, et al. MyD88 signaling in T cells directs IgA-mediated control of the microbiota to promote health. Cell Host Microbe 2015;17:153–63.
[124] Sterlin D, et al. Human IgA binds a diverse array of commensal bacteria. J Exp Med 2020;217.
[125] Kabbert J, et al. High microbiota reactivity of adult human intestinal IgA requires somatic mutations. J Exp Med 2020;217.
[126] Bunker JJ, et al. Natural polyreactive IgA antibodies coat the intestinal microbiota. Science 2017;358:eaan6619.
[127] Mathias A, Corthesy B. Recognition of gram-positive intestinal bacteria by hybridoma- and colostrum-derived secretory immunoglobulin A is mediated by carbohydrates. J Biol Chem 2011;286:17239–47.
[128] Bunker JJ, et al. B cell superantigens in the human intestinal microbiota. Sci Transl Med 2019;11:eaau9356.
[129] Borowska MT, et al. The molecular characterization of antibody binding to a superantigen-like protein from a commensal microbe. Proc Natl Acad Sci U S A 2021;118.
[130] Penny HA, et al. Rhythmicity of intestinal IgA responses confers oscillatory commensal microbiota mutualism. Sci Immunol 2022;7:eabk2541.
[131] Moor K, et al. High-avidity IgA protects the intestine by enchaining growing bacteria. Nature 2017;544:498–502.
[132] Lycke N, Holmgren J. Intestinal mucosal memory and presence of memory cells in lamina propria and Peyer's patches in mice 2 years after oral immunization with cholera toxin. Scand J Immunol 1986;23:611–6.
[133] Mazanec MB, Kaetzel CS, Lamm ME, Fletcher D, Nedrud JG. Intracellular neutralization of virus by immunoglobulin A antibodies. Proc Natl Acad Sci U S A 1992;89:6901–5.
[134] Macpherson AJ, Uhr T. Induction of protective IgA by intestinal dendritic cells carrying commensal bacteria. Science 2004;303:1662–5.
[135] Peterson DA, McNulty NP, Guruge JL, Gordon JI. IgA response to symbiotic bacteria as a mediator of gut homeostasis. Cell Host Microbe 2007;2:328–39.
[136] Kau AL, et al. Functional characterization of IgA-targeted bacterial taxa from undernourished Malawian children that produce diet-dependent enteropathy. Sci Transl Med 2015;7:276ra224.
[137] Muramatsu M, et al. Class switch recombination and hypermutation require activation-induced cytidine deaminase (AID), a potential RNA editing enzyme. Cell 2000;102:553–63.
[138] Wei M, et al. Mice carrying a knock-in mutation of Aicda resulting in a defect in somatic hypermutation have impaired gut homeostasis and compromised mucosal defense. Nat Immunol 2011;12:264–70.
[139] Swain S, Selmi C, Gershwin ME, Teuber SS. The clinical implications of selective IgA deficiency. J Transl Autoimmun 2019;2:100025.
[140] Robak OH, et al. Antibiotic treatment-induced secondary IgA deficiency enhances susceptibility to *Pseudomonas aeruginosa* pneumonia. J Clin Invest 2018;128:3535–45.
[141] Moll JM, et al. Gut microbiota perturbation in IgA deficiency is influenced by IgA-autoantibody status. Gastroenterology 2021;160:2423–2434 e2425.
[142] Bunker JJ, et al. Innate and adaptive humoral responses coat distinct commensal bacteria with immunoglobulin A. Immunity 2015;43:541–53.
[143] Uchimura Y, et al. Antibodies set boundaries limiting microbial metabolite penetration and the resultant mammalian host response. Immunity 2018;49:545–59.
[144] Lai KN, et al. IgA nephropathy. Nat Rev Dis Prim 2016;2:16001.
[145] Viladomiu M, et al. IgA-coated E. coli enriched in Crohn's disease spondyloarthritis promote T H 17-dependent inflammation. Sci Transl Med 2017;9:eaaf9655.
[146] Habas E, et al. IgA nephropathy pathogenesis and therapy: review & updates. Medicine (Baltim) 2022;101:e31219.
[147] Xu L, Li Y, Wu X. IgA vasculitis update: epidemiology, pathogenesis, and biomarkers. Front Immunol 2022;13:921864.
[148] Brusca I. Overview of biomarkers for diagnosis and monitoring of celiac disease. Adv Clin Chem 2015;68:1–55.
[149] Lerner A, Ramesh A, Matthias T. Serologic diagnosis of celiac disease: new biomarkers. Gastroenterol Clin N Am 2019;48:307–17.
[150] Walker LJ, et al. Anti-Saccharomyces cerevisiae antibodies (ASCA) in Crohn's disease are associated with disease severity but not NOD2/CARD15 mutations. Clin Exp Immunol 2004;135:490–6.
[151] Torres J, et al. Serum biomarkers identify patients who will develop inflammatory bowel diseases up to 5 years before diagnosis. Gastroenterology 2020;159:96–104.
[152] Ding L, Chen X, Cheng H, Zhang T, Li Z. Advances in IgA glycosylation and its correlation with diseases. Front Chem 2022;10:974854.

[153] Bakema JE, van Egmond M. The human immunoglobulin A Fc receptor FcalphaRI: a multifaceted regulator of mucosal immunity. Mucosal Immunol 2011;4:612–24.

[154] van Gool MMJ, van Egmond, IgA M, FcalphaRI. Versatile players in homeostasis, infection, and autoimmunity. Immunotargets Ther 2020;9:351–72.

[155] van Egmond M, et al. FcalphaRI-positive liver Kupffer cells: reappraisal of the function of immunoglobulin A in immunity. Nat Med 2000;6:680–5.

[156] Rollino C, Vischini G, Coppo R. IgA nephropathy and infections. J Nephrol 2016;29:463–8.

[157] Lin X, et al. Aberrant galactosylation of IgA1 is involved in the genetic susceptibility of Chinese patients with IgA nephropathy. Nephrol Dial Transplant 2009;24:3372–5.

[158] Ardalan M, et al. Microbiota and glomerulonephritis: an immunological point of view. Am J Med Sci 2022;364:695–705.

[159] Liang X, et al. Metagenomics-based systematic analysis reveals that gut microbiota Gd-IgA1-associated enzymes may play a key role in IgA nephropathy. Front Mol Biosci 2022;9:970723.

[160] Wang Y, et al. Gut microbes in immunoglobulin A nephropathy and their potential therapeutic applications. Front Med (Lausanne) 2022;9:823267.

[161] Heineke MH, et al. New insights in the pathogenesis of immunoglobulin A vasculitis (Henoch-Schonlein purpura). Autoimmun Rev 2017;16:1246–53.

[162] Li M, et al. Comparison and analysis of gut microbiota in children with IgA vasculitis with different clinical symptoms. Front Pediatr 2021;9:800677.

[163] Tan J, Zhong Z, Tang Y, Qin W. Intestinal dysbiosis featuring abundance of Streptococcus associates with Henoch-Schonlein purpura nephritis (IgA vasculitis with nephritis) in adult. BMC Nephrol 2022;23:10.

[164] Wen M, et al. Integrated analyses of gut microbiome and host metabolome in children with Henoch-Schonlein purpura. Front Cell Infect Microbiol 2021;11:796410.

[165] Yang YH, Wang SJ, Chuang YH, Lin YT, Chiang BL. The level of IgA antibodies to human umbilical vein endothelial cells can be enhanced by TNF-alpha treatment in children with Henoch-Schonlein purpura. Clin Exp Immunol 2002;130:352–7.

[166] Klauser T, Pohlner J, Meyer TF. The secretion pathway of IgA protease-type proteins in gram-negative bacteria. Bioessays 1993;15:799–805.

Chapter 22

B-Cell Metabolism

Julia Jellusova[1,2]

[1]*Institute of Clinical Chemistry and Pathobiochemistry, School of Medicine and Health, Technical University of Munich, Munich, Germany;*
[2]*TranslaTUM, Center for Translational Cancer Research, Technical University Munich, Munich, Germany*

1. INTRODUCTION

To maintain viability and function, B cells need energy and biosynthetic precursor molecules. During development and differentiation, B cells switch between phases of relative quiescence and phases characterised by high levels of proliferation or protein synthesis. Nutrient uptake and energy generation need to be adjusted to meet the specific metabolic demands of any given B-cell subset. However, the metabolic programme of B cells is not just a byproduct of their activation and cannot be simply extrapolated from their proliferation rate or their protein output. Throughout their lifespan, B cells make deliberate choices about which nutrients to consume and how to use them [1,2]. Fluctuations in the intracellular concentration of metabolites and altered activity of metabolic enzymes can affect classical protein-based signal transduction [3–5]. Thus, the metabolic programme of cells is not simply a downstream product of signal transduction but delicately interwoven with protein-based signalling pathways, actively shaping cell fate decisions. Recent years have seen a surge in studies focussing on B-cell metabolism and uncovered new modes of metabolic regulation in B cells [2]. Since B cells are exposed to changing metabolic environments during normal humoral immune responses but also in the context of various pathologies, understanding of how metabolic pathways interact with other intracellular processes is crucial in order to gain a comprehensive model of how B-cell fate and function are regulated.

2. BASICS IN CELL METABOLISM

The main purpose of cellular metabolism is to maintain energy balance and to convert metabolic substrates to biosynthetic precursor molecules. Broadly, metabolism can be divided into anabolic and catabolic reactions. Anabolic reactions consume energy and involve the formation of larger, more complex molecules from smaller and simpler ones. In contrast, during catabolic reactions, complex molecules are degraded. Animal cells obtain energy from the controlled breakdown of molecules with high-energy bonds and use adenosine triphosphate (ATP) to store and quickly access energy when needed. Lymphocytes can use different fuels such as amino acids, lipids and sugars to obtain energy [6]. These various types of nutrients differ in how they are taken up by the cell, how they are processed intracellularly and which molecules are formed as byproducts. Nutrients can often be shunted towards different metabolic pathways resulting in divergent outcomes. For example, glucose can be used to generate ATP through glycolysis or oxidative phosphorylation (OxPhos), to create various biosynthetic precursor molecules, to support redox balance by fuelling the pentose phosphate pathway (PPP) or to alter protein activity via post-translational protein modifications (Figure 22.1) [7,8]. At any given time, cells not only need to choose between anabolic and catabolic reactions but must also decide which nutrients will be taken up and how they will be processed. These decisions have profound effects on cell behaviour and are carefully controlled through a complex network of cell-signalling molecules.

2.1 Glycolysis

Similar to other immune cells, one of the major nutrients for B cells is glucose [9]. Glucose is taken up by dedicated membrane transporters of the SLC2 (GLUT) family [10]. B cells primarily take up glucose via GLUT1, but have also been reported to express other members of the family such as GLUT3 [9,11,12]. After entering the cytosol, glucose can be catabolised to pyruvate via a process termed glycolysis (Figure 22.1). Pyruvate can be further converted into lactate and secreted. Four ATP molecules are formed during glycolytic processing of one glucose molecule. However, since the first steps of glycolysis consume energy to convert glucose into two molecules of three-carbon sugar

Molecular Biology of B Cells. https://doi.org/10.1016/B978-0-323-95895-0.00010-6

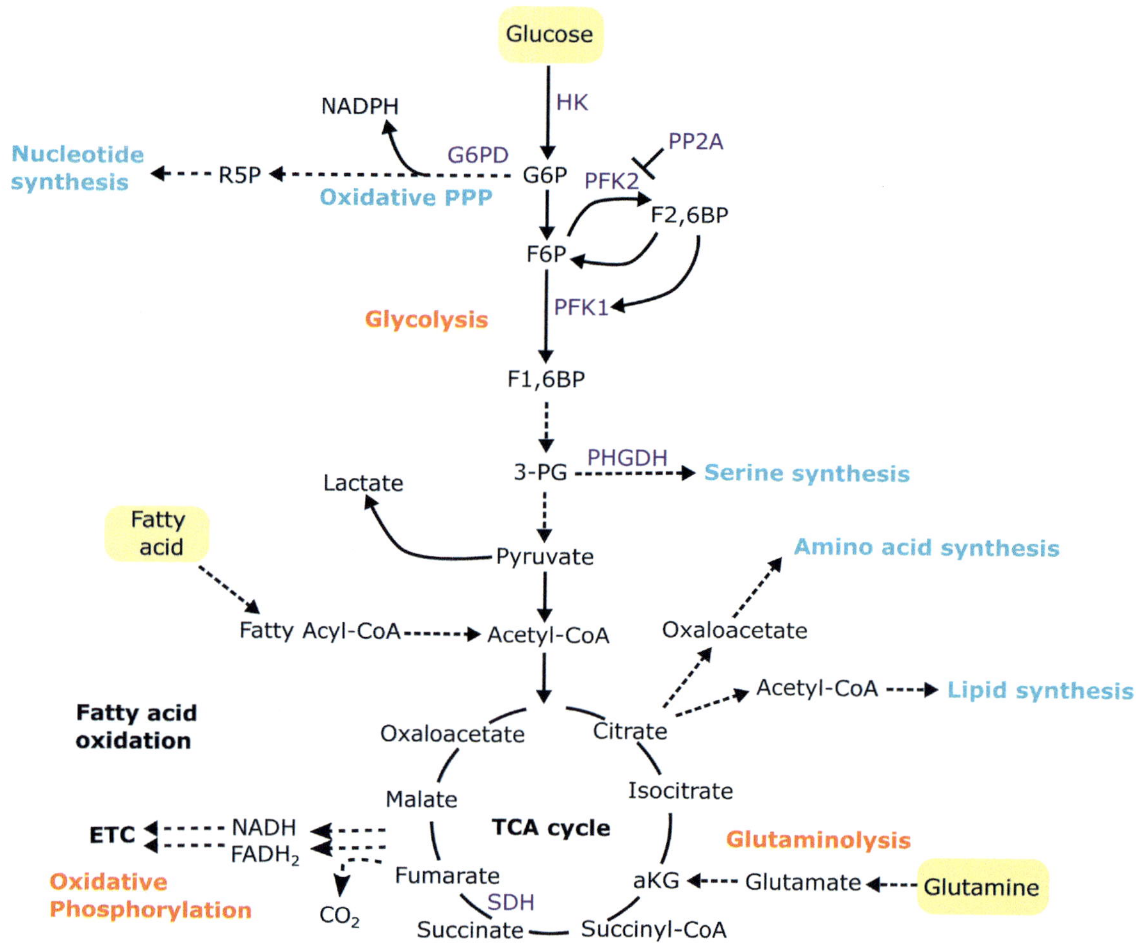

FIGURE 22.1 **Main bioenergetic and biosynthetic pathways in B cells.** The major nutrient for B cells is glucose which can be used to generate energy via glycolysis or OxPhos. To generate energy via OxPhos, glucose-derived pyruvate is imported into the mitochondria and fuelled into the TCA cycle. NADH and $FADH_2$ are produced and serve as electron donors in the ETC. Electron flux through the ETC generates a proton gradient exploited for ATP generation. In addition to glucose, other nutrients can be processed to enter the TCA cycle and used for energy generation via OxPhos. Glycolytic and TCA cycle intermediates can be used for biosynthesis of nucleotides, amino acids and lipids and to maintain redox balance. Catabolic processes are marked in orange, and anabolic processes are marked in teal. For simplicity, only metabolites and metabolic enzymes important for the understanding of this chapter are shown. *3-PG*, 3-phosphoglycerate; *aKG*, α-ketoglutarate; *CoA*, coenzyme A; *ETC*, electron transport chain; *F1,6BP*, fructose 1,6-bisphosphate; *F2,6BP*, fructose 2,6-bisphosphate; *F6P*, fructose 6-phosphate; *$FADH_2$*, flavin adenine dinucleotide; *G6P*, glucose 6-phosphate; *G6PD*, glucose 6-phosphate dehydrogenase; *HK*, hexokinase; *NADH*, nicotinamide adenine dinucleotide; *NADPH*, nicotinamide adenine dinucleotide phosphate; *PFK*, phosphofructokinase; *PHGDH*, phosphoglycerate dehydrogenase; *PP2A*, protein phosphatase 2 A; *PPP*, pentose phosphate pathway; *SDH*, succinate dehydrogenase. *Credit: Julia Jellusova.*

phosphates, the net gain of energy per molecule glucose is two molecules of ATP [7]. The enzymatic reactions of glycolysis take place in the cytosol and do not require oxygen (Figure 22.1) [7].

2.2 Tricarboxylic Acid Cycle

Alternative to lactate secretion, pyruvate can be further processed in the tricarboxylic acid (TCA) cycle in the mitochondria (Figure 22.1) [13]. First, pyruvate needs to be transported via the mitochondrial pyruvate carrier (MPC) across the inner mitochondrial membrane [14]. In the mitochondrial matrix, pyruvate is converted into acetyl-CoA which is further oxidised in a cyclic series of steps resulting in the production of high energy molecules GTP/ATP and reduced forms of nicotinamide adenine dinucleotide (NADH) and flavin adenine dinucleotide ($FADH_2$) [7,14] (Figure 22.1). CO_2 is released as a byproduct. In the TCA cycle, acetyl-CoA first delivers its acetyl group to the four-carbon molecule oxaloacetate (OAA) to generate citrate. Citrate is then converted into its isomer – isocitrate. Isocitrate is oxidised to produce α-ketoglutarate (aKG). During this reaction, NAD+ is reduced to NADH. Through a dehydration process, succinate is converted into fumarate and $FADH_2$ is produced from FAD. Water is added to fumarate and malate is formed. Finally, oxalacetate is regenerated by oxidizing malate and NADH is produced again in this final step (Figure 22.1). Although the TCA cycle itself does not consume oxygen, the generated NADH and $FADH_2$ molecules need to be oxidised by the enzymes of the electron transport chain (ETC), which will use oxygen. If this does not occur, the oxidation steps of the citric acid cycle also cannot take place. Thus, unlike glycolysis, processing of pyruvate in the TCA cycle is indirectly dependent on the presence of oxygen [13].

2.3 Oxidative Phosphorylation

The TCA cycle itself produces little ATP; however, high energy molecules NADH and $FADH_2$ are generated which can be oxidised again to release energy. To be used efficiently, the energy stored in NADH and $FADH_2$ is released gradually by transferring electrons through a series of carriers which constitute the ETC to the final electron acceptor O_2. The electron carriers of the ETC are organised into four complexes embedded into the inner mitochondrial membrane [15]. Electrons from NADH enter the reaction at complex I of the ETC. $FADH_2$ is processed independently of complex I by complex II. The free energy derived by the passage of electrons through the complexes of the ETC is used to transport protons across the inner mitochondrial membrane into the intermembrane space creating a proton gradient. The potential energy stored in the gradient is harvested by the mitochondrial ATPsynthase/ATPase (F0F1 ATPase) which uses the energetically favourable flow of protons back to the matrix for ATP synthesis [15]. This process of ATP generation is known as OxPhos and fundamentally differs from ATP synthesis during glycolysis or the TCA cycle. ATP generated during glycolysis is a product of the transfer of a 'high-energy' phosphate from a substrate directly to ADP. In contrast, ATP generation during OxPhos is coupled to a sequential oxidation of substrates and ultimately driven by chemiosmosis. These different mechanisms of ATP production have implications for how and when these metabolic pathways are used. The need for oxygen limits the use of OxPhos to conditions that allow mitochondrial oxygen uptake. Moreover, the importance of the proton gradient renders OxPhos susceptible to disruptions of the mitochondrial membrane potential. If pyruvate is fully oxidised in the mitochondria rather than being secreted as lactate, the cell can generate up to 36 mol ATP per 1 mol glucose [7]. Yet, highly proliferating cells often secrete lactate even if they have access to oxygen [7]. This phenomenon has been first observed in cancer cells and is known as the Warburg effect. Similar to cancer cells, stimulated B cells dramatically increase their rate of glycolysis and lactate secretion [9,16]. The advantages of an increased rate of glycolysis in rapidly proliferating cells are possibly complex and may include the need to maintain high levels of glycolytic intermediates for quick access during biosynthesis [7]. Moreover, although glycolytic processing of glucose is less energy efficient, the turnover rate is quicker and under conditions of ample glucose supply a faster ATP production rate rather than energy efficiency might be favoured by the cell [7].

2.4 Other Nutrients

Although glucose is a major source of carbon and energy for B cells, ATP can be generated by oxidizing several other metabolic fuels such as amino acids and fats. Of particular prominence is the amino acid glutamine which has a versatile role in normal and malignant B cells. Glutamine can be transported across the plasma membrane by a large number of solute carriers with partially redundant functions [17]. Intracellularly, glutamine can be converted through glutaminolysis to acetyl-CoA and used in the TCA cycle to generate energy (Figure 22.1) [18]. In addition to ATP generation, glutamine can also support biosynthesis by serving as a source of carbon and nitrogen and contribute to detoxification of reactive oxygen species (ROS) [18]. Glutamine uptake increases upon mitogenic B-cell stimulation [17], and glutamate production is boosted when B cells differentiate towards antibody-secreting cells [16]. Moreover, using a Burkitt lymphoma cell line, it has been demonstrated that glutamine can support TCA cycle activity and redox balance under hypoxia and glucose deprivation, suggesting that glutamine might play an important role in metabolically stressful environments [19]. Despite these initial insights into the importance of glutamine in B cells, the exact contribution of this amino acid to B-cell energy generation, biosynthesis and redox balance is currently incompletely understood.

In addition to glutamine and glucose, B cells can also use fatty acids for energy generation. Long-chain fatty acids (LCFAs) have been thought to be able to freely cross plasma membranes, but recent evidence suggests that LCFA uptake is facilitated by the lipid transporter CD36 and small intracellular lipid chaperones called fatty acid-binding proteins (FABPs) [20]. Short-chain fatty acids (SCFAs) such as acetate, propionate and butyrate are primarily produced by colonic bacteria [21]. The exact route of SCFAs uptake in lymphocytes is not entirely clarified, but it has been demonstrated that SCFAs can be utilised intracellularly by B cells [21]. In addition to uptake from the extracellular milieu, fatty acids can also be liberated from lipid droplets via lipolysis [20]. Fatty acids can be used to generate energy via mitochondrial β-oxidation (Figure 22.1) or can be oxidised in peroxisomes to generate acyl-CoA or acetyl-CoA without ATP generation [22].

2.5 Biosynthesis

For biosynthesis, cells use small precursors to produce complex biomolecules such as DNA, proteins and lipids. Many of the biosynthetic pathways utilise glycolytic or TCA cycle intermediates highlighting the importance of these metabolic pathways for proliferating cells (Figure 22.1) [7]. One of the pathways branching off glycolysis is the PPP, which plays an important role in DNA synthesis and redox balance (Figure 22.1) [7,23]. Unlike glycolysis and oxidative glucose degradation in the mitochondria, the PPP does not yield ATP. Instead, it produces NADPH and ribose-5-phosphate (R5P). R5P is a precursor molecule for DNA synthesis. NADPH is a reducing equivalent used in the generation of fatty acids, sterols, nucleotides and non-essential amino acids. Moreover, NADPH-mediated conversion of oxidised glutathione to reduced glutathione is important for cellular redox balance [23].

The PPP has two arms – the oxidative and non-oxidative branch. Glucose-6-phosphate dehydrogenase (G6PD) is the rate-limiting enzyme of the oxidative PPP (Figure 22.1). Both branches produce ribose, but only the oxidative branch generates NADPH [23]. Interestingly, the transcription factor PAX5, which defines B-cell fate has been shown to repress expression of enzymes involved in the PPP such as G6PD, thus reducing the flow of glucose-derived carbon through the PPP [24]. As a possible compensatory mechanism, B cells express the serine/threonine protein phosphatase 2A (PP2A) [24] which directs carbon from glycolysis towards the PPP by inhibiting PFK2 (Figure 22.1). PFK2 generates F2,6BP, a potent allosteric activator of PFK1. F2,6BP accumulation shifts the carbon flow towards glycolysis and away from the PPP. PP2A activity thus reduces glycolysis. As a consequence, glycolytic ATP production is limited in B cells. The deletion of the PP2A subunit *Ppp2r1a* has been shown to boost glucose consumption, ATP generation and lactate secretion in BCR-ABL1-positive B-cell acute lymphocytic leukaemia (B-ALL) but not chronic myeloid leukaemia (CML) [24]. In contrast to the increase in glycolytic flux, activity of the PPP is reduced in *Ppp2r1a*-deficient B-ALL cells as evidenced be lower levels of NADPH. ROS levels are increased in these cells demonstrating that PP2A plays an important role in maintaining redox balance. Both normal and B-ALL B cells have been shown to depend on PP2A for survival [24]. This study demonstrates that a better understanding of cell metabolism can reveal cell-type-specific metabolic vulnerabilities that could be exploited for the development of new treatment strategies in the future.

Similar to glycolytic intermediates, TCA cycle metabolites play a central role in biosynthesis of macromolecules. For example, citrate can be transported into the cytosol to be converted into OAA and acetyl-CoA to promote nucleotide and lipid synthesis (Figure 22.1) [25]. The enzyme ATP citrate lyase (ACL) catalyses the formation of acetyl-CoA and OAA from cytosolic citrate, coenzyme A and ATP [25]. If TCA cycle intermediates are used for biosynthesis, the cycle needs to be replenished through a process called anaplerosis. Metabolites can enter the TCA cycle through multiple access points, with two important anaplerotic process being the conversion of pyruvate to mitochondrial OAA by pyruvate decarboxylase and glutaminolysis during which glutamine is converted into glutamate and subsequently aKG (Figure 22.1) [7].

2.6 Autophagy

If extracellular nutrients become limiting, cells can engage in a self-degradation process called autophagy [26]. Autophagy is a multi-step process during which protein aggregates or entire organelles are engulfed in double membrane vesicles called autophagosomes and targeted for lysosomal degradation. A complex cascade of signalling molecules participates in orchestrating the formation and processing of autophagosomes [26]. The first steps of autophagy include the assembly of the ULK1 protein complex which contains ULK1, ATG13, FIP200 and ATG101 and subsequent production of phosphatidylinositol 3-phosphate (PI(3)P) that is required for autophagosome formation. Elongation of the autophagosome membrane is dependent on ubiquitin-like conjugation steps involving proteins ATG5, ATG12, ATG7 and ATG10. During this step, LC3-I is cleaved and lipidated to LC3-II. LC3-II is then incorporated into the autophagosome membrane [26]. LC3 conversion is commonly used in experiments as a marker to quantify autophagic flux. Mature autophagosomes are trafficked to lysosomes, and the autophagolysosome cargo is degraded by lysosomal acidic hydrolases and the products are exported back to the cytoplasm to be reused [26]. Autophagy not only plays a role during periods of nutrient withdrawal but is also an important mechanism for eliminating damaged organelles and misfolded proteins thereby supporting cellular health [27]. Moreover, some of the components of the autophagy machinery have been implicated to play roles in other processes of vesicle formation and transportation and can thus affect cell fate beyond their canonical role in self-degradation [28].

3. SIGNALLING PATHWAYS REGULATING B-CELL METABOLISM

Upon mitogenic activation, B cells reprogramme their metabolism to boost energy generation and biosynthesis. Unsurprisingly, signalling pathways activated downstream of the B-cell receptor (BCR) or in response to other pro-

proliferation stimuli simultaneously regulate cell survival and cell cycle entry as well as nutrient uptake and processing [2,29]. Similarly, in order to cease proliferation, B cells activate signalling pathways that limit metabolic activity [30]. B cells not only respond to mitogenic stimuli but also monitor oxygen, nutrient and energy levels and adjust their proliferation rate to ensure that nutrient usage does not outstrip available resources. Thus, signalling molecules responsible for inhibiting cell growth often simultaneously sense the activation status of the cell as well as nutrient or ATP availability.

3.1 Signalling Cascades Governing Biosynthesis

A major player in regulating cell survival, proliferation and metabolic activity is the kinase AKT. AKT is an effector molecule of the PI3K signalling pathway and has a plethora of potential targets that are involved in regulating cell metabolism [29]. One of the central metabolic regulators activated downstream of AKT is the mammalian target of rapamycin (mTOR) (Figure 22.2). mTOR exists in two multiprotein complexes: mTORC1 and mTORC2, which are distinguished by their accessory proteins, function and regulatory mechanisms [31]. mTORC1 contains Raptor as a subunit and is best known in its role of promoting cell growth. The substrates phosphorylated by mTORC1 include proteins involved in protein synthesis such as ribosomal S6 kinase (S6K1) and eIF-4E-binding protein 1 (4E-BP1) [31]. Additionally, mTORC1 drives nutrient uptake and blocks autophagic catabolism [32]. The complex itself is activated in response to mitogenic signals. AKT can affect mTORC1 activity directly by phosphorylating mTOR [33] or indirectly by phosphorylating a subunit of the tuberous sclerosis complex (TSC) [34]. TSC is an inhibitor of mTORC1, and AKT-mediated phosphorylation and inhibition of the subunit TSC2 therefore activates mTORC1 [34]. Both mTORC1- and PI3K/AKT-dependent signalling pathways play a major role in governing B-cell metabolism [29,35,36]. However, how PI3K/AKT and mTORC1 activation translates into regulation of cell fate and function can differ depending on

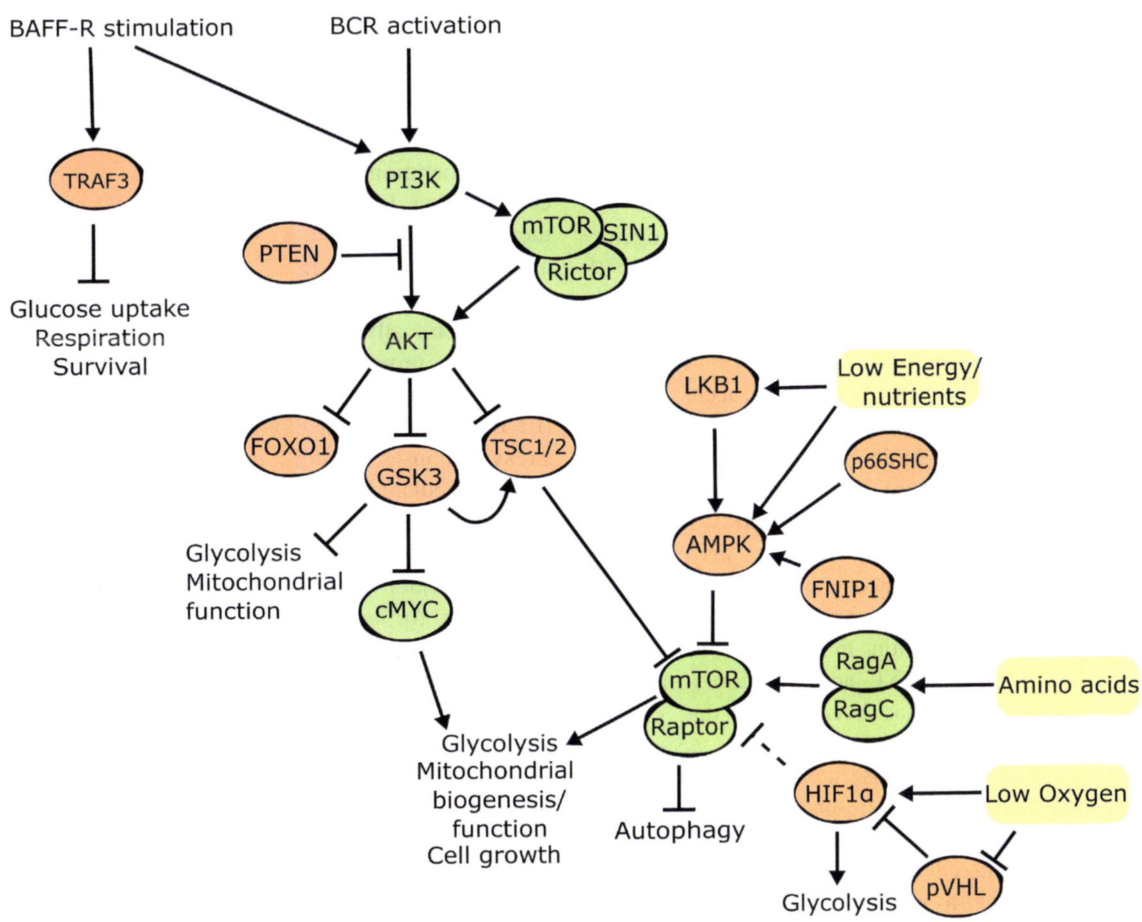

FIGURE 22.2 Main signalling pathways regulating cell metabolism in B cells. In stimulated B cells, many pathways converge on cMYC and mTORC1 activation which are central metabolic regulators. Shown are also important signalling nodes that respond to nutrient availability with AMPK being the major regulator of metabolic stress. For simplicity, only signalling pathways discussed in this chapter are shown. *Credit: Julia Jellusova.*

the developmental state of B cells, their metabolic environment as well as signalling context and will be discussed in more detail later.

In contrast to mTORC1, the function of mTORC2 in B cells is less well defined. Although mTORC2 shares the same catalytic kinase subunit with mTORC1, accessory proteins such as Rictor alter substrate specificity and therefore function of this complex [37]. mTORC2 is activated in a PI3K-dependent manner and has been reported to govern different signalling processes, actin cytoskeleton remodelling as well as metabolism [37]. Of note, mTORC2 phosphorylates AKT on the serine residue S473 [37]. Phosphorylation of this serine is needed for PDK1 to phosphorylate AKT on the residue T308 leading to the full activation of AKT. Deletion of the major mTORC2 component Rictor does not affect early stages of B-cell development, but leads to a partial loss of mature B cells as well as an impaired humoral immune response [38]. Rictor-deficient B cells display defects in NF-kB signalling, which could explain the observed phenotype [38]. Similarly, the deletion of the mTORC2 component SIN1 results in reduced AKT activation, impaired cell mass accumulation and oxygen consumption in transitional/mature B cells ultimately leading to poor responses to anti-IgM stimulation [39].

Another central regulator of cell growth and metabolism is glycogen synthase kinase (GSK3). GSK3 is a serine/threonine kinase and exists in two isoforms GSK3α and GSK3β which have partially overlapping functions [40]. GSK3 is an untypical kinase in that it is constitutively active in resting cells and only inhibited upon mitogenic stimulation. GSK3 has a broad array of targets, many of which are pro-proliferation or pro-survival factors [40,41]. GSK3-mediated phosphorylation of many of these proteins targets them for degradation [40,41]. GSK3 inhibition in mature stimulated B cells results in cell mass accumulation, increased oxygen consumption and lactate secretion demonstrating that GSK3 limits metabolic activity in B cells [42]. One of the central regulators of metabolism destabilised upon phosphorylation via GSK3 is cMYC [43]. This transcription factor is known to drive a gene expression profile favouring glucose uptake and glycolysis as well as mitochondrial biogenesis [44]. GSK3 deletion results in cMYC accumulation in B cells [42]. Using transgenic mice in which cMYC accumulation is induced specifically in B cells, it has been shown that cMYC accumulation is sufficient to induce cell mass accumulation, lactate secretion and glucose uptake in stimulated B cells suggesting that the metabolic phenotype observed in GSK3-deficient B cells is at least partially driven by cMYC [42]. However, additional targets downstream of GSK3 possibly also contribute to metabolic reprogramming [29]. Similar to AKT/PI3K and mTORC1 signalling, many signalling pathways in B cells regulate cMYC levels on the transcriptional, translational and post-translational level. In turn, cMYC governs a plethora of metabolic processes including protein synthesis, nutrient routing and metabolic flexibility [19,45–47].

3.2 Signalling Pathways Orchestrating Metabolic Stress Responses

Unlike unicellular organisms, cells within a multicellular organism cannot simply take up nutrients, grow and proliferate whenever the metabolic microenvironment is favourable. Instead, they need to remain metabolically quiescent until an activation signal has been received. In order to initiate proliferation, B cells need to receive two signals: (1) a mitogenic signal such as activation of the BCR and (2) a confirmation that nutrients necessary for biomass synthesis are available. Once the mitogenic signal has fulfiled its purpose or if nutrient- or oxygen-availability becomes low, B cells need to be able to return to a quiescent state. In order to accomplish these tasks, signalling pathways are needed that allow B cells to respond to shifts in energy, metabolites and oxygen levels either by inhibiting anabolic metabolism or by inducing the activity of alternative metabolic pathways.

A major molecular energy sensor is adenosine monophosphate-activated protein kinase (AMPK). AMPK becomes activated in response to an overall increase in the ratio of AMP to ATP and activates pathways that increase ATP generation whilst simultaneously inhibiting ATP consumption [48]. AMPK represses mTORC1 activity, thereby limiting protein synthesis as well as boosting autophagy [48] (Figure 22.2). A recent study suggests that in order to suppress mTORC1-dependent anabolic growth in metabolically stressed pre-B cells, the folliculin-interacting protein FNIP1 needs to be present [49]. For full activation, AMPK is phosphorylated by liver kinase B1 (LKB1). LKB1 has also other targets involved in metabolic stress responses in addition to AMPK [50]. In addition to AMPK-mediated regulation, mTORC1 activity is linked to the cellular metabolic state via amino acid availability [31]. The presence of amino acids induces activity of Rag GTPases which can bind Raptor and recruit mTORC1 to lysosomes for activation [31]. Other pathways of how amino acids can be sensed and the signal transduced to mTORC1 have also been described [51]. AMPK activity, mTORC1 signaling and autophagy are regulated by many different signaling pathways allowing the cells to carefully match cell growth and nutrient usage to their specific signaling profile. For example, the p66 isoform of SHC1 (p66SHC) has been shown to inhibit mTORC1 signaling in B cells, to disrupt mitochondrial function via its cytochrome c somatic binding domain, thereby inhibiting ATP production. This results in increased AMPK activation and enhanced autophagic flux. P66SHC has been shown to affect autophagy via different mechanisms including its interaction with membrane-associated LC3-II as well as by inducing mitochondrial depolarisation, protein ubiquitination at the outer mitochondrial membrane and recruitment of active AMPK (Figure 22.2) [52].

Finally, B cells can also respond to changes in oxygen levels. A master regulator of hypoxic responses is the transcription factor HIF1α [53]. Under normal conditions, HIF1α is continuously targeted for degradation through oxygen-dependent hydroxylation [53]. Hydroxylation of HIF1α is carried out by 2-oxoglutarate-dependent dioxygenases and requires molecular oxygen, ferrous iron and ascorbate. Hydroxylated HIF1α is polyubiquitinylated by von Hippel–Lindau protein (pVHL) and degraded in the proteasome. Under hypoxic conditions, HIF1α is stabilised, translocates to the nucleus and induces a transcription profile that supports glycolysis [53]. In addition, HIF1α accumulation reduces mTORC1 activity in B cells [54].

3.3 Other Events Governing Metabolic Regulation

As discussed before, the metabolic profile of B cells is regulated by nutrient availability and mitogenic activation. However, other cellular processes play a role in shaping the metabolic profile of B cells as well. Many intracellular processes affecting organelle homoeostasis, actin cytoskeleton remodelling or ionic balance have been shown to alter metabolic activity in other cell types and should be considered in B cells as well [55–58]. For example, whilst the BCR has a well-accepted role in supporting cell survival via activation of SYK-dependent signalling pathways, the BCR can affect cell metabolism also in an SYK-independent manner. Immunoglobulin (Ig) production is sensed in the endoplasmic reticulum (ER) resulting in the expansion of the ER to accommodate the need for protein folding capacity [59]. Expansion of the ER is accompanied by increased expression of enzymes involved in lipid metabolism such as ACL [59]. Moreover, the ER is known to directly interact with the mitochondria and to exchange small molecules such as calcium [56]. Calcium drives mitochondrial respiration by supporting the activity of various enzymes in the TCA cycle and the ETC [58]. Indeed, Ig production increases mitochondrial calcium levels and oxygen consumption [59]. These findings suggest that protein load and ER homoeostasis can shape the metabolic profile of B cells. Since different B-cell subsets dramatically differ in their Ig production rate, the contribution of the ER should to be taken into consideration when studying metabolic regulation of different B-cell subsets.

4. CROSSTALK BETWEEN METABOLISM AND SIGNALLING

B cells need nutrients and energy for proliferation. It is therefore not surprising that nutrient-sensitive signalling pathways exist that match proliferation rates to nutrient availability [2,29]. Additionally, metabolites and metabolic processes can steer cell fate and function in more complex ways and affect differentiation, class switch recombination, the secretion of cytokines and many other fundamental B-cell responses. Metabolites can serve as cofactors, induce posttranslational modifications (PTMs) or act as substrates for signalling proteins [4,60]. Similarly, mounting evidence suggests that many enzymes with metabolic roles can also act on signalling proteins, mRNA or shape gene expression [5]. The importance of a crosstalk between metabolic and signalling pathways is highlighted by the fact that various cellular processes lay at the heart of both: cell signalling and metabolism. Redox balance, calcium homoeostasis and other second messengers contribute to integration of signalling and metabolic pathways.

4.1 Posttranslational Modulation

Protein stability and function can be shaped by a rich and diverse set of PTMs induced by metabolites [60]. Products of various metabolic pathways including glycolytic and TCA cycle intermediates, lipids, amino acids and polyamines can be used in these processes (Figure 22.3). We are only beginning to understand how processes such as lactylation, malonylation, palmitoylation, succination, polyglutamylation, acetylation and glycosylation affect immune cells [60]. Although the occurrence and function of many of these processes are currently poorly explored in B-cell biology, it is becoming increasingly apparent that metabolite-regulated PTMs profoundly shape B-cell fate decisions. For example, protein glycosylation is a complex and dynamic PTM that governs the function of the majority of cell surface and secreted proteins including immunoglobulins [8]. Glycosylation can affect secretion, stability, solubility and biological activity of the target proteins. A large part of glucose taken up by plasma cells has been suggested to be shunted towards antibody glycosylation [61]. Changes in the composition of the glycosylation pattern can determine whether antibodies play pro-inflammatory or anti-inflammatory roles [62]. The glycosylation pattern of antibodies changes markedly depending on the age and sex of the individual and is altered in individuals with autoimmune or infectious diseases [62]. Despite the importance of glycosylation in shaping antibody function, it is currently poorly understood how the composition of the metabolic environment affects this type of PTM. A better understanding of these processes would be of high pharmacological importance since optimising glycosylation patterns could help improve the efficacy and pharmacokinetics of antibody-based therapeutics [63].

A special type of glycosylation that has raised considerable interest in the last few years is O-linked-N-

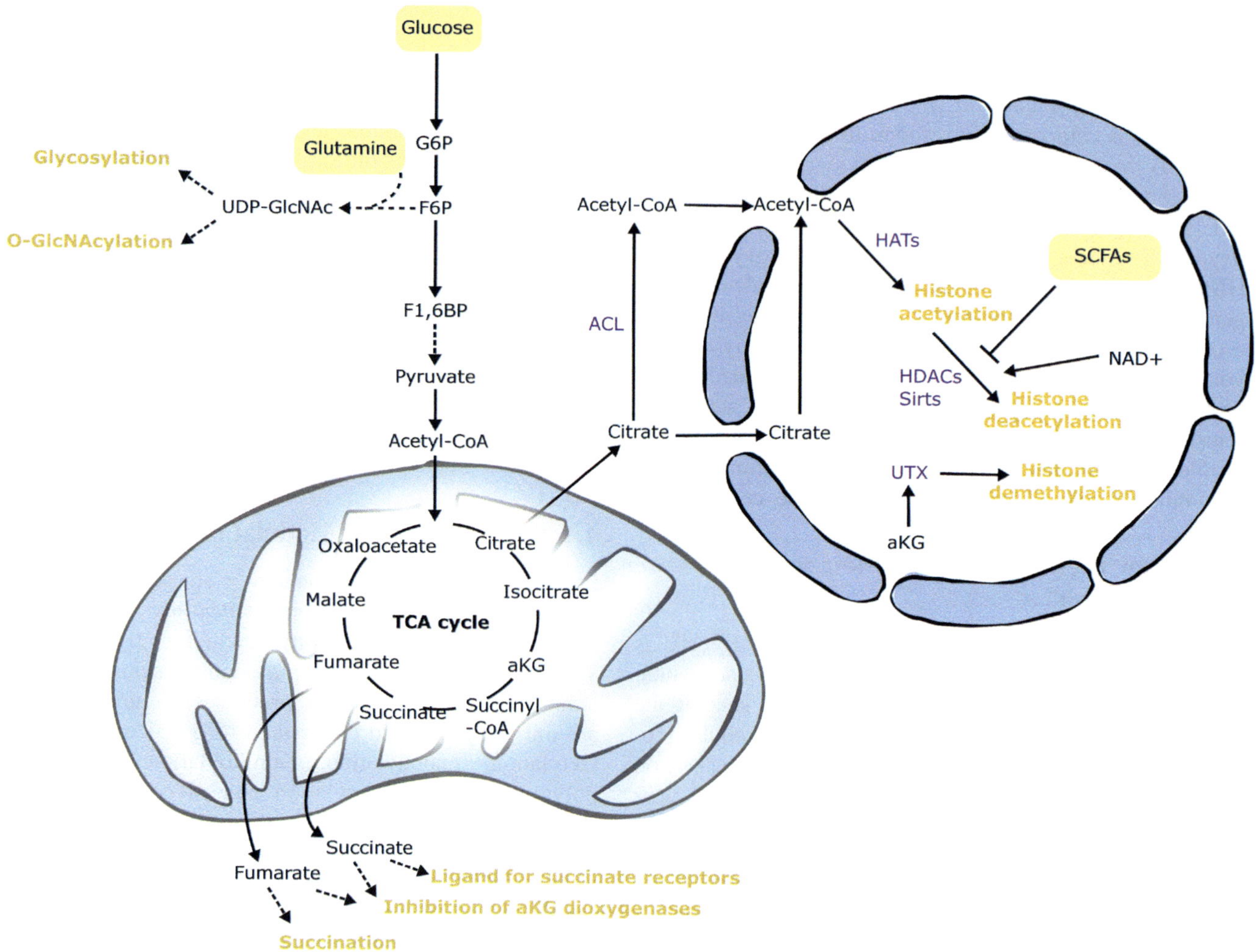

FIGURE 22.3 Non-metabolic functions of metabolites. Metabolic intermediates of glycolysis and the TCA cycle can affect signal transduction by mediating posttranslational protein modifications such as succination, glycosylation and O-GlcNacylation. Additionally, metabolites can shape the activity of enzymes regulating methylation and acetylation of histone and non-histone proteins. Lastly, metabolites can bind to dedicated receptors and induce activation of intracellular signalling. *aKG*, α-ketoglutarate; *ACL*, ATP citrate lyase; *CoA*, coenzyme A; *ETC*, electron transport chain; *F1,6BP*, fructose 1,6-bisphosphate; *F6P*, fructose 6-phosphate; *G6P*, glucose 6-phosphate; *HATs*, histone acetyl transferases; *HDACs*, histone deacetylases; *NAD*, nicotinamide adenine dinucleotide; *SCFAs*, short-chain fatty acids; *Sirts*, sirtuins; *TCA*, tricarboxylic acid cycle; *UDP-GlcNac*, uridine diphosphate N-acetylglucosamine. *Credit: Julia Jellusova.*

acetylglucosaminylation (O-GlcNacylation) [64]. O-GlcNacylation is a process during which the monosaccharide O-GlcNac is attached to cytoplasmic, nuclear and mitochondrial proteins leading to altered function. A single pair of enzymes O-GlcNAc transferase (OGT) and O-GlcNAcase (OGA) controls these modifications [64]. It has recently been shown that B cells deficient for OGT display reduced BAFF-mediated activation of several pro-survival signalling pathways, including PI3K/AKT and classical and alternative NFκB signalling [65]. Similar to BAFF stimulation, activation of BCR-dependent signalling requires OGT [65] and OGT overexpression boosts BCR-induced responses [66]. One of the many proteins altered through O-GlcNAcylation is LYN, and it has been suggested that O-GlcNAcylation of LYN regulates SYK recruitment [65]. Deletion of *Ogt* in germinal centre (GC) B cells reduces their survival and limits antibody responses to antigen [65]. Similar to GC B cells, highly proliferating pre-B cells have also been found to require OGT for their survival [67]. Pre-B cells upregulate O-GlcNAcylation in comparison to pro-B cells. Chemical inhibition of OGT as well as OGT knockdown results in cMYC destabilisation highlighting the importance of O-GlcNAcylation in pre-B-cell biology [67]. Notably, generation of the OGT substrate uridine diphosphate N-acetylglucosamine (UDP-GlcNac) uses the hexosamine biosynthetic pathway (HBP) which uses products of glucose and glutamine metabolism [64] (Figure 22.3). The activity of the HBP fluctuates with metabolism, thus nutrient availability can determine the extent of O-GlcNAcylation [64].

Accordingly, pre-B cells cultured in low glucose levels have been found to display reduced O-GlcNacylation and lower cMYC protein levels [67].

In addition to glucose, TCA cycle intermediates such as fumarate and succinate can participate in posttranslational protein modifications. Fumarate has been shown to modify cysteine residues in a process called succination (Figure 22.3). It has been recently demonstrated that many proteins in B cells can undergo this type of PTM including LYN which plays an important role in transducing BCR-mediated signals [68]. Fumarate accumulation by inhibition of fumarate hydratase or by treatment with the cell membrane permeable metabolite dimethyl fumarate has been shown to suppress B-cell activation. It has been proposed that fumarate-mediated LYN succination abrogates the activity of this important signaling molecule thereby limiting signal transduction [68]. Additionally, fumarate has been shown to activate KEAP1-NRF2 signaling in B cells [69]. A recent study has identified *Sdha*, a gene encoding succinate dehydrogenase subunit A to be mutated in patients with persistent polyclonal B-cell lymphocytosis (PPBL). This gain of function mutation results in fumarate accumulation in PPBL B cells, which engage the KEAP1-NRF2 system to drive the transcription of genes encoding inflammatory cytokines [69].

Fumarate and succinate alter cellular function not only by posttranslational protein modification but also by serving as a structural mimetic to aKG, competitively inhibiting a superfamily of enzymes known as aKG–dependent dioxygenases (Figure 22.3). These enzymes include the HIF hydroxylases, which once inhibited allow the stabilisation of HIF and various epigenetic enzymes [70]. Additionally, succinate can be secreted and sensed by receptors (Figure 22.3), thus acting as a classical signaling molecule [70]. It has been shown that a specific subset of CD4+ T cells can secrete succinate and IL-10 at the same time which together induce B-cell activation. Blocking the succinate receptor inhibits T-cell-driven B-cell activation in this setup, suggesting that succinate might contribute to regulating B-cell responses [71].

4.2 Epigenetic Regulation

Metabolites not only alter cytosolic protein activity but can also shape epigenetic regulation of gene expression (Figure 22.3). Acetyl-CoA derived from metabolic degradation of glucose or fatty acids is used to fuel the TCA cycle for energy generation and represents a major molecular building block for biosynthesis [72]. Additionally, acetyl-CoA is also used as a substrate for histone acetylation [72]. Histone acetylation changes chromatin architecture and thus regulates gene expression. Although the nucleus and the mitochondria represent two separate reservoirs for acetyl-CoA, it has been suggested that nutrient abundance is accompanied by an increase in cytosolic/nuclear acetyl-CoA that can be used for histone modification [73]. In B cells, stimulation has been shown to result in an increase in global acetyl-CoA levels and extensive epigenetic reprogramming [74]. In addition to shaping substrate availability for epigenetic modifications, metabolites also regulate the activity of enzymes involved in epigenetic reprograming [75]. Two major types of enzymes dynamically control the acetylation state of histones: histone acetyltransferases (HATs) and histone deacetylases (HDACs). HAT activity increases acetylation and thus in general promotes transcriptional activity, whereas HDACs catalyse the removal of acetyl moieties, which mediates transcriptional inhibition. Metabolites such as lactate and SCFAs are known to inhibit HDACs [75]. Two partially conflicting studies exist on the role of SCFAs in B-cell chromatin remodelling and function. The first study suggests that SCFAs boost B-cell metabolism but also suppress HDAC activity and induce expression of genes necessary for plasma cell differentiation [76]. The second study shows that providing mice with SCFA in drinking water or a high fibre diet reduces plasma cell generation and ameliorates autoimmunity [77]. Mechanistically, SCFAs inhibit HDAC activity and are suggested to increase expression of miRNAs targeting *Aicda* and *Prdm-1* 3′UTRs. This results in reduced AID and BLIMP1 expression thereby inhibiting class switch recombination and plasma cell differentiation [77]. The authors of the later study suggest that the SCFA concentration used for experiments plays a major role in determining functional outcome. They show that low levels of SCFAs result in increased acetylation of histone H3 in the *Aicda* promoter, *Aicda* regulatory region 4, *Prdm1* promoter, as well as $I\gamma1$, $I\gamma3$ and $I\alpha$ promoters. Low concentrations of SCFAs are therefore proposed to have the opposite effect on AID and BLIMP1 expression in comparison to higher concentrations. Combined, these studies suggest that epigenetic regulation induced by SCFAs is relatively dynamic and the concentration-dependent role of SCFAs needs to be taken into consideration when planning and interpreting future studies.

The activity of the Sirtuin family of protein/HDACs is also metabolically regulated [75,78]. These proteins require NAD+ as cofactor for their deacetylase activity [75]. NAD+ participates in several metabolic pathways such as glycolysis and the TCA cycle [79]. Sirtuin activity is thus coupled to the metabolic state of the cells, increases under nutrient deprivation and can affect a broad spectrum of cellular activities by deacetylating different histone and non-histone proteins [78]. Sirtuin-1 is expressed in resting B cells and quickly downregulated upon stimulation [80]. Sirtuin-1 has been shown to dampen class switch recombination as evidenced by reduced class switching in mice over-expressing Sirtuin-1 and increased AID expression and class switch recombination in Sirtuin-1-deficient mice [80]. Importantly, the authors of this study show that class switch recombination is coupled to the metabolic state of the cells. An increase in glucose concentration in cell culture leads to a reduced NAD+/NADH ratio and increased *Aicda* expression consistent with glycolysis converting NAD+ to NADH and thereby reducing the availability of NAD+ to Sirtuin-1. Providing the cells with NAD+ showed the opposite effect on Ig isotype switching in this study [80].

Lastly, it has recently been shown that aKG is needed for the activation of the *Bcl6* gene locus.

Using a cell culture system in which B cells with a GC phenotype are generated by stimulating B cells via CD40L, BAFF and IL-4, it has been shown that IL-4 induces a metabolic switch towards OxPhos and aKG accumulation, accompanied by increased expression of the GC regulator BCL6 [81]. The demethylase UTX requires aKG for the removal of the repressive histone mark H3K27me3 (Figure 22.3). H3K27me3 demethylation of the *Bcl6* enhancer region activates *Bcl6* gene transcription and guides the cells towards a GC phenotype rather than towards plasma cell differentiation. If oxidative metabolism is inhibited in vivo, the number of GC B cells, but not plasma cells is reduced, underpinning the importance of this metabolic pathway in the GC [81].

In summary, several examples highlight the notion that the metabolic state of B cells can impact epigenetic regulation.

4.3 Mitochondria

Mitochondria are cell organelles best known for their role in energy production and biosynthesis. However, mitochondria fulfil a plethora of additional functions including control of apoptosis, synthesis of iron sulphur clusters, regulation of redox homoeostasis and sequestering ions such as calcium, magnesium and iron as well as various proteins involved in signal transduction [57,82–85]. These different functions are closely interlinked and influence each other making the mitochondria a central hub of metabolic and signalling crosstalk. Mitochondria are surrounded by a double membrane and form intricate networks that can be dynamically restructured. The organisation of their double membrane and the protein-rich inner matrix allows the mitochondria to compartmentalise processes [86]. This compartmentalisation is crucial in order to maintain a proton gradient that can be used for ATP synthesis, but also enables other processes to unfold independently of each other. Mitochondria additionally differ from other animal organelles in that they contain their own DNA (mtDNA) which encodes a small number of genes mostly dedicated to mitochondrial bioenergetics [87]. The regulation of mitochondrial number, size, motility, interaction with other organelles and overall morphology are processes that are regulated by both metabolic conditions and signal transduction and ultimately shape cell fate decisions [88–90]. In this manner, mitochondria play a central role in integrating signal transduction and metabolism.

4.4 Reactive Oxygen Species

A significant portion of intracellular ROS is produced as a byproduct of normal mitochondrial function and generally increases with higher levels of respiration. Although healthy mitochondria produce ROS, production may increase in defective or damaged mitochondria [91]. Moreover, oxygen and nutrient availability shape mitochondrial ROS production [91,92]. In addition to the ETC, ROS can originate from a variety of intracellular sources such as enzymatic activity of NADPH oxidases (NOXs), xanthine oxidase or peroxidase [93]. ROS is also produced in the ER as a consequence of oxidative protein folding [94]. ROS can affect a plethora of biological events. It is well established that ROS play an important role as second messengers and that ligation of different receptors results in a burst of intracellular ROS. ROS contribute to the propagation of signalling through the oxidation of specific reactive cysteine residues in various proteins. To date over 500 individual proteins have been suggested to contain these potentially modulatory cysteine molecules [95]. With such a large pool of possible targets, colocalisation of the ROS source and specific target proteins may play a crucial role in achieving specificity. The various sources of ROS and the chemical properties of the different ROS species may result in ROS gradients or sub-cellular ROS compartmentalisation enabling ROS to act on localised cellular processes [96].

Activation of the BCR is associated with the production of ROS originating from cell membrane-localised NOXs and possibly the mitochondria and has been suggested to oxidise and inhibit protein tyrosine phosphatases such as PTEN and SHP2 [97–99]. ROS depletion halts proliferation of activated B cells demonstrating the central role of this second messenger in B-cell signaling [97]. Importantly, however, the extent and duration of ROS production needs to be carefully fine-tuned as excess ROS levels have been shown to interfere with normal signal transduction in B cells [100]. Moreover, ROS overproduction can cause damage to proteins, lipids and nucleic acids leading to impaired cellular function or even cell death. Lipid peroxidation can induce ferroptosis, an iron-dependent form of programmed cell death that is genetically and biochemically distinct from apoptosis or necroptosis [101]. To neutralise excess ROS, different ROS scavenging mechanisms exist. The two major anti-oxidant pathways are the glutathione and thioredoxin systems [102]. Both of these pathways utilise NADPH as an electron donor. As discussed, NADPH is produced in the PPP, which branches off glycolysis and is fuelled by glucose [102]. Glutathione is additionally dependent on the presence of the amino acid cysteine. Thus, the availability of reducing equivalents is partially dependent on access to nutrients. Glutathione can be used by different glutathione peroxidases (Gpx), an enzyme family composed of eight family members [102]. GPX4 possesses the ability to directly reduce phospholipid hydroperoxides and oxidised lipoproteins, thereby protecting cells against membrane lipid peroxidation and ferroptosis [102]. As discussed before, maintaining appropriate redox balance is crucial for B cells and B cells require PP2A activity to fuel the PPP for their survival [24]. The dependency on the thioredoxin and glutathione redox system varies among different B-cell subsets and will be discussed in detail later. Similar to ROS-producing enzymes, anti-oxidants are compartmentalised and can contribute to the formation of ROS microdomains [103]. In summary, a rich intracellular network of ROS-producing and ROS-scavenging systems enables redox signaling whilst avoiding redox stress under normal conditions. Whilst considerate progress has been made in the last few years in understanding the role of different ROS sources and scavengers in B-cell biology, less is known about ROS compartmentalisation and ROS microdomains. A better understanding of these aspects of redox signaling would be important in order to determine how ROS produced under hypoxia or generated in response to altered nutrient supply contributes to shaping classical signaling pathways.

5. METABOLIC REQUIREMENTS DURING B-CELL DEVELOPMENT AND DIFFERENTIATION

Different B-cell subsets vary in their proliferation index, protein synthesis rate and the environment they populate. Correspondingly, the metabolic profile of B cells changes during development, maturation and differentiation and reflects the cells' unique needs. A comprehensive understanding of the various metabolic profiles in different B-cell subsets could be of clinical relevance as it may allow to exploit subset-specific metabolic vulnerabilities in B cell-dependent diseases including autoimmune disorders and B cell-derived malignancies.

5.1 Developing B Cells

B-cell lymphopoiesis occurs in the bone marrow and is characterised by the sequential rearrangement of the Ig heavy- and light-chain locus with the goal of generating a diverse pool of B cells carrying a signalling competent BCR [104]. After pro-B cells have rearranged the Ig heavy chain locus, the newly generated protein is tested for functionality by pairing it with the invariant surrogate light chain creating the pre-BCR. A successful assembly of the pre-BCR and expression on the cell surface leads to a proliferative burst of B-cell precursors now called large pre-B cells. After the cells cease to proliferate, the Ig light-chain locus is rearranged in small pre-B cells. Expression of the Ig light chain together with the heavy chain allows the cells to develop into immature B cells and to leave the bone marrow. Pro- and large pre-B cells (corresponding to fraction B and C/C′ of the Hardy nomenclature) display increased metabolic activity in comparison to the more quiescent small pre-B cells (fraction D) [35,105]. Oxygen consumption, lactate secretion and glucose consumption are increased in pro/large pre-B cells in comparison to small pre-B cells [35,105] (Figure 22.4). Glucose seems to be an important nutrient for B-cell development as pre-B cells are sensitive to glycolysis inhibition with 2-deoxy-D-glucose (2DG) and mice with a B cell-specific deletion of the glucose transporter GLUT1 show reduced B-cell numbers [9,12].

Two major signals induce proliferation of B-cell precursors: IL-7 stimulation and the successful assembly of the pre-BCR. IL-7 is a cytokine that plays several important roles during B-cell development [106]. IL-7 supports commitment of precursor cells to the B-cell lineage, drives proliferation and survival of B cell progenitors and aids the transition from the pro-B-cell stage to the pre-B-cell stage [106]. IL-7 induces activation of mTORC1 and cMYC which drive cell growth and anabolic processes essential for B-cell development [35,36].

The bone marrow has been reported to harbour hypoxic niches [107], and B-cell precursors show signs of hypoxia exposure [108]. Interestingly, whilst HIF1α deficiency leads to abnormal B-cell development and autoimmunity, it is not strictly required for B-cell development [12]. HIF1α-deficient B-cell precursors have been shown to upregulate OxPhos proteins and to be less sensitive to 2DG inhibition [12]. Thus, HIF1α appears to create glucose dependency during B-cell development. In contrast, HIF1α hyperstabilisation after pVHL deletion has been shown to terminate B-cell development at the immature B-cell stage by increasing the expression of the proapoptotic factor BIM [108]. It is currently unclear what the functional consequences of these findings are in vivo. It is possible that hypoxia limits survival of developing B cells and that B cells need to leave hypoxic niches in order to progress to the mature B-cell stage. However, it is also possible that normal B-cell precursors never experience prolonged HIF1α stabilisation as seen after pVHL deletion and that HIF1α-induced cell death does not play a major role in governing normal B-cell development in vivo.

Although B-cell precursors have been demonstrated to depend on glucose and to display increased glucose metabolism in comparison to B cells at later developmental stages [105], it has been proposed that B-cell identity is tied to a chronically energetically deprived phenotype [109].

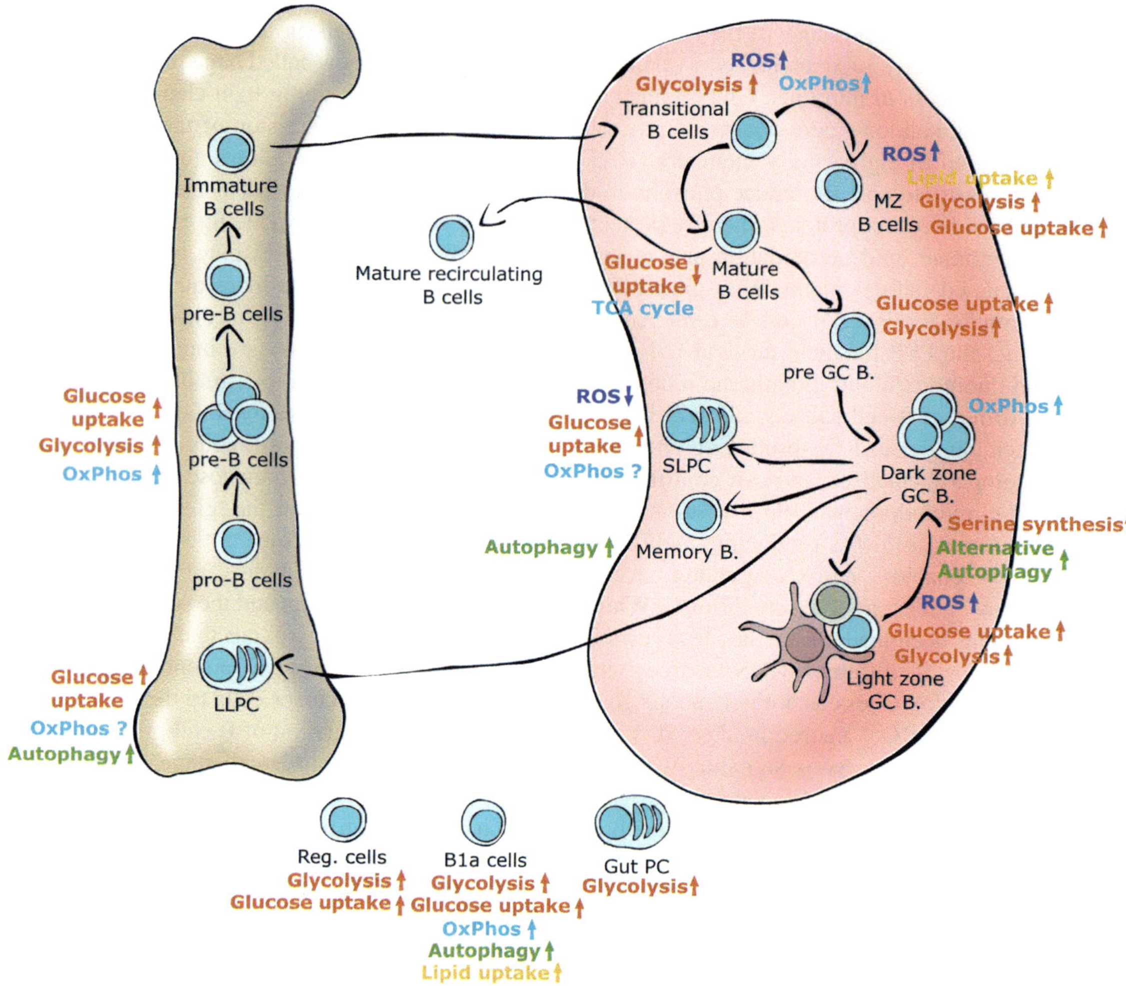

FIGURE 22.4 **The metabolic programme of different B-cell subsets.** Shown is the summary of our current knowledge of the metabolic characteristics of different B-cell subsets. Metabolic requirements vary between different B-cell subsets depending on their maturation status, activation of specific signalling molecules, proliferation index, protein secretion rate as well as their metabolic microenvironment. *Credit: Julia Jellusova.*

PAX5 and IZKF have been demonstrated to induce the expression of negative regulators of glucose metabolism and to result in the downregulation of glucose transporters [109]. PAX5 has also been shown to repress the expression of PPP enzymes such as G6PD and to increase the expression of PP2A [24]. As discussed before, PP2A activity limits glycolytic ATP production and directs carbon towards the PPP pathway, which is needed for pre-B-cell survival. NADPH produced in the PPP pathway acts as an electron donor for the two main antioxidant systems: the thioredoxin (Trx) and glutathione pathways. These two systems have been shown to play partially redundant roles during B-cell development. Deletion of *Txnrd1*, the gene encoding Trx reductase 1, prevents T-cell development in the thymus but does not affect B-cell development and maturation [110]. In B cells, compensatory mechanisms exist resulting in the upregulation of *Grx1* [110]. Similarly, mice with a B cell-specific deletion of the glutamate cysteine ligase (*Gclc*) harbour mature B cells, suggesting that *Gclc* is not essential for B-cell development [111]. Oral administration of L-buthionine-sulfoximine (BSO) which inhibits glutamate cysteine ligase and blocks GSH synthesis reduces the frequency of B-cell precursors in the Hardy fractions B−C in *Txnrd1*-deficient − but not in control − bone marrow highlighting the compensatory function of these two redox pathways [110]. Autophagy is not absolutely required for B-cell development [112] suggesting that developing B cells do not rely on this process to maintain healthy cell organelles.

5.2 Transitional and Mature B Cells

Upon expression of the BCR, B cells egress from the bone marrow and migrate to the spleen to complete development. The recent immigrants to the spleen are called transitional B cells which pass through several developmental steps before committing to one of the two mature B-cell fates — the marginal zone (MZ) or the follicular (FO) (also called B2) — B-cell state. Not all immature/transitional B cells reach the mature B-cell stage as some are deleted due to the expression of an autoreactive BCR [113,114]. FO B cells reside in secondary lymphoid organs and recirculate in the blood. Before encountering antigen, mature FO B cells are largely metabolically quiescent, showing low levels of glucose uptake, low mitochondrial mass and a small cell size [42] (Figure 22.4). This is consistent with the fact that these cells do not proliferate nor secrete large quantities of proteins. Despite their low metabolic demands, maintenance of cell mass remains an active process and B cells cultured ex vivo without stimulation quickly atrophy [115]. To prevent loss of viability and cell mass, resting B cells rely on pro-survival signals such as BAFF. BAFF stimulation does not only induce pro-survival signalling but also boosts carbon metabolism and cell growth [115,116]. Whilst both oxygen consumption and lactate secretion are low in resting mature B cells, TCA cycle activity seems to play a more prominent role than glycolysis in maintaining cell survival. Depletion of dietary vitamin B1, which is crucial for TCA cycle activity, has been shown to result in decreased resting B-cell numbers [117]. Moreover whilst TCA cycle metabolites are present in ex vivo naïve B cells, glycolytic metabolites are less abundant [117]. Mature resting B cells are relatively insensitive to 2DG treatment in vivo, and ATP generation is not inhibited by 2DG treatment in vitro, suggesting that glucose metabolism is not essential for the survival of mature B cells [42,111].

In contrast to fully mature B cells, transitional B cells are more metabolically active. Mitochondrial mass, oxygen consumption, ROS production, glucose uptake as well as lactate secretion have all been found to be higher in transitional B cells in comparison to fully matured B cells [118]. Importantly, acquiring a metabolically quiescent phenotype appears to be crucial for human B cells to progress in their maturation [118]. The observed increase in metabolic activity correlates with high mTORC1 signalling in transitional B cells. Mature B cells on the other hand display increased AMPK activity and upregulate adenosine salvage metabolism [118]. Since AMPK is a known mTORC1 inhibitor, it is possible that mTORC1 signalling needs to be silenced in order for B cells to reach the mature B-cell stage. Along these lines, patients with gain of function *PIK3CD* mutations displaying increased mTORC1 activity have fewer mature B cells than healthy individuals [118]. High levels of BCR signalling in transitional B cells can be a sign of autoreactivity. It is tempting to speculate that an inability to reach a metabolic quiescent state is interpreted as cellular hyperactivation due to an autoreactive BCR and the cells are therefore eliminated. Although B-cell maturation in the mouse differs from humans, mouse mature B cells also show a decrease in the expression of genes involved in regulating respiration and protein synthesis [118]. Similarly, mice in which *Tsc1* is deleted via cre activity induced by the *Cd19* promoter (CD19cre x TSC1flox) show a reduction of mature B-cell numbers suggesting that increased mTORC1 signalling hinders B-cell maturation in mice [119,120]. However, it should be noted that with this mouse model, no conclusive statements can be made on the role of mTORC1 signalling in transitional B cells. CD19cre deletes LoxP-flanked genes in developing B cells in the bone marrow albeit not very efficiently [121]. It is therefore unclear whether the reduction of mature B-cell numbers is a consequence of impaired expansion of B-cell precursors or rather a block in B-cell maturation at the transitional B-cells stage. Similarly, the role of the AMPK pathway in enforcing quiescence remains unclear. Whilst AMPK deletion does not seem to affect mature B cells [122—124], the deletion of the AMPK activator LKB1 using CD19cre results in reduced mature B-cell numbers whilst maintaining a normal transitional B-cell pool [125,126]. Similar to mTORC1 signalling, GSK3 inhibition is associated with increased metabolic activity in B cells and a reduction of mature B-cell numbers [42]. Of note, if GSK3 is deleted in an inducible manner in an already established mature B-cell pool, the cells slowly accumulate cell mass and decline in numbers [42]. It is therefore possible that aberrantly increased metabolic activity not only hinders B-cell maturation but also directly reduces the lifespan of mature B cells. ROS accumulation as a consequence of increased respiration in this scenario could induce damage to DNA and macromolecules, which in non-proliferating B cells might not be immediately fatal, but can negatively affect B-cell survival in the long run. Consistent with this model, appropriate redox balance has been shown to be important for both mature and transitional B cells. In mice with a B cell-specific *Gclc* deletion, transitional B-cell numbers are significantly reduced and the total numbers of mature B cells are also slightly lower than in control animals [111]. *Gclc* deletion results in increased transcription of mTOR genes and enhanced ROS accumulation in FO B cells rendering them more similar to *Gclc*-sufficient MZ B cells. Interestingly, *Gclc*-deficient B cells have been shown to harbour several mitochondrial defects: whilst mitochondrial size is increased, the organelles are fragmented with dilatated intercristae space. The cells also show impaired glucose derived flux towards the TCA cycle, reduced respiration and increased mitochondrial uncoupling [111]. It is possible that ROS generated as a consequence of *Gclc* ablation damages mitochondria, ultimately restricting their function. Reduced mitochondria-derived ATP is compensated for in *Gclc*-deficient FO B cells by an increase in glycolysis and lactate secretion and the total ATP levels in *Gclc*-deficient FO B cells exceed the levels detected in control B cells [111]. Although ROS accumulation seems to be harmful to FO B cells, deletion of

Gpx4 does not affect FO B cells, suggesting that other glutathione-dependent processes play a more prominent role in maintaining B-cell survival [127].

In summary, metabolic quiescence supports the survival of resting mature FO B cells. Yet, in certain scenarios, an increase in metabolic activity in mature B cells does not impair their survival but instead renders them more reactive to cellular stimulation. For example, enhanced BAFF-dependent signalling in mature B cells either by BAFF overexposure or by the deletion of the negative regulator TRAF3 simultaneously increases metabolic activity and the ability of these cells to respond to subsequent stimulation [128,129]. How survival is affected by enhanced metabolic activity might therefore depend on the signalling context, activation of specific pro-survival pathways or the cells' ability to cope with excess ROS.

5.3 Marginal Zone B Cells

MZ B cells are a special population of mature B cells with innate-like properties. MZ B cells are specialised to mount rapid responses to common pathogens without the need for T-cell stimulation [130]. Their BCR repertoire is skewed towards poly-reactive BCRs recognising common pathogens whilst at the same time being weakly autoreactive. Moreover, MZ B cells are highly sensitive to stimulation through toll-like receptors allowing them to quickly respond to blood-borne pathogens [131]. This function is facilitated by their location in the spleen at the interface between blood vessels and the white pulp.

Consistent with their ability to quickly respond to stimulation, MZ B cells show more mTOR signalling than FO B cells and a higher ER to Golgi ratio. Expression of glycolytic genes, glucose uptake, GLUT1 levels, lactate production and glucose-derived ATP are upregulated in MZ B cells in comparison to FO B cells [42,111,127] (Figure 22.4). Interestingly, whilst the levels of mtDNA are increased in MZ B cells, mitochondrial mass is not [132]. MZ B cells express higher levels of *Gclc* in comparison to FO B cells; however, the ratio of reduced to oxidised glutathione (GSH/GSSG) is lower in MZ than in FO B cells consistent with the observation that both total and mitochondrial ROS levels are increased in MZ B cells [111]. MZ B cells are more sensitive to BSO-mediated inhibition of GSH and dramatically reduced in mice with a B cell-specific *Gclc* or *Txnrd1* deletion suggesting that MZ B cells experience more redox stress and have a higher need for ROS scavenging pathways than FO B cells [110,111]. In addition to consuming more glucose, MZ B cells also express higher levels of the fatty acid transporter CD36 and take up more lipids in comparison to FO B cells [127]. Lipid consumption can render cells more susceptible to ferroptosis. Indeed, mice with a B cell-specific *Gpx4* deletion show a reduction in the MZ B cell pool. Ex-vivo cell death of *Gpx4*-deficient stimulated MZ B cells can be prevented by the ferroptosis inhibitor Fe-1, suggesting that ferroptosis is the main path of cell death under these conditions [127]. MZ B cells have also been shown to be affected by the ablation of *Pex5* [133]. PEX5 is involved in biogenesis of peroxisomes which carry out β-oxidation of very long-chain and branched chain fatty acids. The exact molecular mechanisms of the observed reduction in MZ B cell numbers are currently unknown, but it is possible that MZ B cells either require fatty acids for their survival or that the loss of functional peroxisomes disrupts redox balance and therefore reduces viability in these cells [133].

In summary, MZ B cells are more metabolically active than FO B cells, which allows them to respond quickly to stimulation, but also creates certain metabolic vulnerabilities.

5.4 Activated and Germinal Centre B Cells

Once mature B cells receive a signal through the BCR or are stimulated with other mitogens, they prepare for proliferation. Metabolic changes are induced early after stimulation [134] and culminate in increased glucose uptake, oxygen consumption and lactate secretion [9,42,134] (Figure 22.4). Glucose is not only used for energy generation but is also shunted towards the PPP pathway and lipid synthesis [135,136]. Whilst mitogenic stimulation in general increases metabolic activity, it has recently been suggested that an increase in respiration in BCR-stimulated cells is coupled to ROS production and mitochondrial damage ultimately killing the cells [134]. Providing the cells with costimulatory signals rescued the observed mitochondrial phenotype [134]. The authors of this study propose a model in which BCR stimulation initially prepares B cells for proliferation but simultaneously creates a need for co-stimulation in order to maintain viability. The requirement to receive co-stimulatory signals in order to prevent BCR-induced metabolic damage could be a factor contributing to the elimination of activated autoreactive B cells that are unable to receive T-cell help.

B cells that have been stimulated with a thymus-dependent antigen and received T-cell help initiate the GC response [137]. GCs are anatomical sites divided into two histologically distinct areas — the dark zone and the light zone. The dark zone is characterised by high levels of B-cell proliferation and somatic hypermutation leading to BCRs with altered reactivity. In the light zone, GC B cells test their new BCRs for antigen recognition. Antigen-specific GC B cells receive pro-survival signals from antigen-presenting FDCs and cognate Tfh cells. GC B cells that have lost their ability to bind antigen die by neglect [137]. GC B cells can cycle several times between the dark and light zone and undergo repeated rounds of somatic hypermutation, before differentiating to plasma cells or memory B cells. Since the GC comprises a heterogeneous population of B cells with very different functional and signalling properties, it is likely that the metabolic programme of these cells is equally heterogeneous. Early studies have shown that GC B cells take up more glucose than resting B cells, express more enzymes involved in glycolysis, accumulate HIF1α and are sensitive to 2DG treatment in a normal as well as autoimmune setting [42,54,138]. GC B cells, however, also show higher levels of mitochondrial mass and ROS production than naïve B cells and GC B-cell numbers are reduced if *Cox10*, a component of complex IV of the respiratory chain, is deleted [42,81,139,140]. It has recently been suggested that whilst B-cell activation and the initial phases of GC initiation are associated with increased glycolytic activity, B cells within the GC shift towards mitochondrial respiration [81]. Whilst GC B cells continue to secrete lactate, the levels are reduced in comparison to pre-GCs (Figure 22.4). Importantly, whilst activated B cells have been found to survive after oligomycin-mediated inhibition of mitochondrial ATP production, GC B cells were not able to do so [81]. As mentioned before, aKG accumulation has been shown to support GC B cell identity by driving BCL6 expression [81]. Thus, mitochondrial activity might contribute to efficient GC responses via multiple molecular mechanisms including epigenetic reprogramming.

Using single-cell RNA sequencing, a recent study has provided more insight into the dynamic regulation of metabolic reprogramming within the GC [140]. Whilst dark zone GC B cells have been found to be characterised by a gene expression profile favouring OxPhos and fatty acid oxidation, light zone GC B cells show a more glycolytic gene expression profile (Figure 22.4). Consistent with this finding, LZ GC B cells have been found to take up more glucose than DZ GC B cells [140]. A more glycolytic profile of the LZ GC could be caused by hypoxia, since hypoxic regions have been observed primarily in the LZ of the GC [54] or by activation of pro-glycolytic transcriptions factors such as cMYC, which has been reported to accumulate in LZ GC B cells actively receiving T-cell help [141]. Of note, OxPhos activity has been found to be increased in high-affinity GC B-cell clones [140].

An additional metabolic feature that distinguishes activated/GC B cells from resting B cells is increased expression of enzymes involved in the synthesis of the amino acid serine [142]. The serine synthesis pathway uses 3-phosphoglycerate, an intermediate molecule of glycolytic glucose degradation to produce serine via phosphoglycerate dehydrogenase (PHGDH) (Figure 22.1). Chemical inhibition of this enzyme or genetical deletion of the PHGDH-encoding gene results in reduced GC B cell numbers demonstrating the dependence of these cells on this metabolic pathway [142].

Combined, these studies reveal that the metabolic profile of GC B cells changes in a dynamic and complex manner consistent with the known signalling heterogeneity of this B-cell population. Since the metabolic programme of GC B cells dramatically differs from resting B cells and other B-cell subsets (Figure 22.4), interfering with metabolic reprogramming could be beneficial in patients with aberrantly activated B cells in various autoimmune disorders. To exploit possible metabolic vulnerabilities, it is not only important to understand the metabolic programme of GC B cells but also to explore the individual contributions of various signalling pathways to governing metabolic adaptations. In the last few years, considerable progress has been made in understanding how different signalling pathways affect cell metabolism in activated and GC B cells. Perhaps not surprisingly many signalling pathways that are needed to orchestrate the processes essential for cell cycle progression and cell division also govern metabolic changes needed for proliferation. Consequently, a failure to activate these signalling pathways entails the inability to supply the cell with energy or nutrients and inevitably prevents proliferation.

In activated B cells, known metabolic regulators such as GSK3 [42], cMYC [42], mTORC1 and the PI3K signalling network [138] have been shown to play a role in metabolic reprogramming. In addition, several other signalling molecules [42,143–145], epigenetic regulators [146], transcription factors [147] and microRNAs [148] have been reported to regulate B-cell metabolism.

As discussed before, our knowledge of metabolic regulation in B cells activated in vitro and B cells at early stages after antigen encounter in vivo cannot be directly transferred to GC B cells. In vivo, cMYC accumulates in B cells early after activation and in the subset of GC B cells that are currently receiving T-cell help, but is largely absent in other GC B cells [141]. Consistent with its role in destabilising cMYC, GSK3 is inhibited in the same subset of GC B cells that shows cMYC accumulation [42]. Interestingly, however, although both cMYC overexpression and GSK3 inhibition drive mitochondrial biogenesis and an increase in metabolic activity in activated B cells [42], cMYC overexpression increases GC B-cell proliferation [149], whilst GSK3 deletion strongly reduces GC B-cell viability [42]. These divergent outcomes could have a multitude of possible underlying mechanistic explanations – GSK3-deficient and cMYC overexpressing cells could differ in their metabolic stress responses, ROS detoxification or simply the magnitude of cMYC accumulation.

Similar to cMYC, mTORC1 is only activated in a subset of GC B cells. Interestingly, cMYC and mTORC1 are activated in the light zone and not in the highly proliferative dark zone of the GC [149,150]. The extent of cMYC and mTORC1 activation in GC B cells appears to be proportional to either the T-cell help the cells have received or to the activation of signalling pathways downstream of the BCR [149,151]. It has been proposed that the metabolic reservoir created in response to co-stimulation determines how long B cells proliferate in the dark zone. In this manner, metabolic capacity contributes to the expansion of high-affinity clones. Interestingly, the activity of mTORC1 signalling not only regulates entry and exit from the dark zone but also differentiation towards the memory B-cell stage [152].

Both cMYC and mTORC1 are required for the establishment and propagation of the GC response and chemical or genetic interventions interfering with these signalling pathways reduce GC output [141,153]. Although mTORC1 is necessary for GC reactions, evidence suggests that increased mTOR signalling might have a negative effect on GC B cells as well. Two of three studies report that unrestrained mTOR signalling is incompatible with a normal GC response [119,120,150]. Interestingly modulating how mTORC1 responds to amino acid availability can profoundly affect GC responses. Constitutively active RagA which renders mTORC1 independent of the presence of amino acids reduces competitiveness of GC B cells. On the other hand, activating RagC alleles that partially uncouple mTORC1 from amino acids have been found to boost GC expansion [150,154].

In summary, aberrant metabolic activity might have both positive and negative effects on the GC response. Whilst higher metabolic activity might initially boost proliferation, an inability to return to cellular quiescence can result in metabolic stress and reduced cell survival. Metabolic stress in GC B cells can originate from cell-intrinsic processes. For example, high levels of respiration resulting in increased ROS production can put a strain on the cells redox machinery. As mentioned before, GC B cells have been found to harbor increased levels of ROS [81]. *Gclc*-deficient B cell do not mount neutralising antibody responses and generate few GCs suggesting that glutathione-based redox regulation is essential for GC B cells [111]. However, GPX4 is not essential for GC B-cell responses demonstrating that not all redox regulators are crucial for GC B cells [127].

Redox stress in the GC can also be a consequence of an altered metabolic microenvironment. As discussed before, GCs have been found to be partially hypoxic and GC B cells accumulate HIF1α [42,54,155]. Deletion of pVHL and the subsequent HIF1α stabilisation results in impaired GC responses [54]. This could indicate that hypoxia serves as a break on GC proliferation and limits the extent of the GC reaction. It is, however, also possible that B cells in vivo are not exposed to HIF1α for a prolonged period of time as they cycle between hypoxic and normoxic regions and negative effects of HIF1α accumulation may therefore never manifest in vivo. Rather than aborting B-cell proliferation, hypoxia might affect the GC output in a more subtle manner. Hypoxia has been shown to reduce class switch recombination and mTORC1 activity [54], which in turn affects memory B-cell differentiation. Hypoxia could also affect ROS generation and abundance of TCA cycle metabolites and steer GC B cell fate decisions in this manner. Ultimately, further experimental insight is needed in order to define the exact physiologic role of hypoxia in the GC.

Beyond oxygen availability, very little is known about the metabolic composition of the GC. It has recently been shown that the immune response to the parasite plasmodium is dominated by plasmablasts, which are metabolically hyperactive, with high respiratory rates and increased expression of glutamine transporters [156]. High uptake of glutamine by plasma blasts might deplete glutamine in the extracellular space and thereby limit access of other cells to this resource. Indeed, plasma glutamine levels are reduced in patients as well as in spleens from plasmodium infected mice [156]. Mice receiving glutamine in water have been shown to harbor a lower parasite burden. Other amino acids such as valine or alanine did not have this effect. Glutamine supplementation has been shown to result in increased GC responses and class switch recombination ultimately boosting formation of memory B cells and long-lived plasma cells. Deletion of plasma blasts has been shown to elevate the numbers of GCs and to reduce parasitemia. Moreover, deletion of plasma blasts and glutamine supplementation were found to be redundant. Combined, these findings suggest that plasmablasts act as a nutrient sink outcompeting other cells such as GC B cells or FO helper T cells for glutamine [156]. This study not only highlights the importance of glutamine for an efficient humoral immune response but also underscores the fact that depending on the extent of B cell activation, the type of antigen/pathogen and the activity of bystander cells, the metabolic profile of GC B cells can potentially vary significantly.

Although GC B cells might face metabolic stress, classical autophagy does not play a crucial role in GC B-cell survival. GCs form readily in the absence of important autophagy molecules such as ATG5 [157], ATG7 [158] or ATG16L1 [159]. Nevertheless, LC3 is constantly being processed in GC B cells, which is usually a sign of ongoing autophagy [159]. A new type of autophagic degradation has been proposed to take place in GC B cells which involves the molecule WIP2 [159]. If WIP2 is deleted, noncanonical autophagy increases, mitochondrial homeostasis is destabilised and GC B cells succumb to cell death [159]. Whilst the exact role of this new autophagy pathway remains unclear, this study suggests that GC B cells may have the means to regulate organelle homeostasis and cellular integrity independently of classical autophagy.

Additionally, as discussed before, components of the autophagy machinery can also contribute to other processes within the cells such as antigen presentation and processing

[160]. Strikingly, the mitochondrial inhibitor oligomycin has been shown to reduce antigen presentation [161] providing another indication that metabolism and B-cell responses are tightly linked.

In summary, the population of GC B cells is very heterogeneous on a functional-, signaling- as well as metabolic-level. Many aspects of metabolic control in the GC are currently poorly understood, nevertheless increasing number of studies demonstrate that metabolites, metabolic enzymes and mitochondria steer GC B-cell fate decisions in many different ways.

5.5 Memory B Cells

The final output of the GC reaction is the generation of long-lived memory B cells and plasma cells as a protection from recurrent infections. Memory B cells confer long-lasting protection against previously encountered pathogen and can persist for decades after clearance of the antigen. Memory B cells do not proliferate and are metabolically quiescent. However, in order to maintain longevity, the cells need to be able to adapt to diverse microenvironments and to remove damaged organelles. Autophagy plays an essential role in memory B cells, allowing the removal of defective organelles and preventing the accumulation of toxic ROS levels [158,162,163]. The exact metabolic profile of memory B cells is poorly understood to date as most studies assess the impact on the memory B-cell pool when changes have been introduced during the GC response. These studies often do not allow to differentiate between metabolic properties needed for GC B-cell proliferation, memory B-cell differentiation and memory B-cell maintenance.

5.6 Plasma Cells

Important effector cells produced during the GC response are plasma cells [164]. In addition to the GC response, plasma cells can also be generated from B1 cells and MZ B cells without the need for T-cell help [164]. Short-lived plasma cells primarily residing in the spleen are usually generated in the first wave of plasma cell generation at the start of the GC response. Long-lived plasma cells originate from isotype switched and hypermutated GC B cells and migrate to the bone marrow where they receive pro-survival signals to persist for a long time [164]. Plasma cells can also be found in mucosal tissue including the gut [165].

Mechanistically, the differentiation to the plasma cell fate is governed by transcription factors BLIMP1 and IRF4. At the same time, transcription factors that define B-cell lineage (PAX5) and GC identity (BCL6) need to be downregulated to allow plasma cell differentiation [164,166]. As activated B cells progress towards the plasma cell stage, they first become plasma blasts, which can still proliferate and express high levels of cell surface BCRs and then finally to terminally differentiated plasma cells which are non-proliferative and primarily secrete immunoglobulins. The gradual differentiation to plasmablasts and ultimately plasma cells is accompanied by dramatic metabolic reprogramming [167].

Similar to memory B cells, plasma cells need autophagy for survival [157]. Since plasma cells secrete vast amounts of proteins, the occurrence of misfolded proteins, protein aggregates and damage to ER membranes might arise frequently and autophagy is needed to alleviate these stressful conditions. Interestingly, the protein secretion rate in *Atg5*-deficient plasma cells is initially higher, although fewer antibodies are produced overall during the course of the humoral immune response due to reduced plasma cell survival [157]. Likely misfolded proteins or a faulty ER cause ER stress that initially leads to ER expansion thus increasing secretory capacity but ultimately causes cell death if stress cannot be resolved.

High levels of Ig production impose not only a strain on the protein secretion machinery but also increase the metabolic demands of these cells. Both short- and long-lived plasma cells take up high levels of glucose and use it for antibody glycosylation rather than glycolysis in vitro [61]. LLPCs have been shown to possess the ability to use glucose-derived pyruvate in the TCA cycle. Expression of the mitochondria pyruvate transporter that shunts pyruvate into the mitochondria has been demonstrated to be needed to maintain survival of LLPCs [61]. Although it is unclear how pyruvate supports LLPC survival, one of the possibilities is that mitochondrial ATP generation is crucial for plasma cells. Indeed, the OxPhos machinery gradually increases in plasma blasts [168] and intact mitochondria are needed to generate plasma cells [132]. Interestingly, in a model where the Burkitt lymphoma line Ramos which has features of GC B cells is driven to acquire a plasma cell-like phenotype, the cells dramatically switch from glycolysis to OxPhos [169]. Since increased ER load can enhance respiration by providing calcium to the mitochondria [59], high levels of Ig production could boost respiration in plasma cells. Additionally, BLIMP1 has been shown to govern amino acid uptake [170] but also to be needed for the upregulation of OxPhos genes in plasma blasts [168]. Additionally, although not without controversy, ROS has been suggested to affect plasma cell differentiation and Ig isotype switching [171,172]. In summary, plasma cell development or maintenance seems to depend on mitochondrial function and OxPhos is increased as B cells progress in their differentiation towards the plasma cell stage. Nevertheless, some evidence suggests that mitochondria-derived ATP is not crucial for plasma cell survival. Inhibition of mitochondrial ATP synthesis has been found not to affect plasma cell survival in vitro and in vivo [81,117]. Plasma cells have the lowest mitochondrial mass and mitochondrial ROS in comparison to naïve, pre-GC and GC B cells, but they take up the most glucose [81] and plasma cells in the gut have been found to harbour glycolytic intermediates and not to rely on vitamin B1 for survival [117]. Additionally, hypoxia has been shown to slightly increase plasma cell survival [173]. Thus, the preferred metabolic pathway for ATP generation, the role of the mitochondria in plasma cell biology and the extent to which plasma cells can switch between different metabolic pathways remains unclear. It is possible that the metabolic profile of plasma cells depends on their localisation. Organs such as the bone marrow and the gut might harbour hypoxic niches steering plasma cells towards glycolysis. Indeed, intestinal plasma cells have been found to harbour more glycolytic intermediates than splenic plasma cells [117]. Furthermore, in vitro differentiated plasma cells may have access to nutrients that does not correspond to in vivo situations. Moreover, it is possible that different metabolic requirements are needed for plasma cell differentiation and the survival of fully established plasma cells. It is possible that discrepancies reported on the metabolic requirements of plasma cells stem from the fact that the various studies analysed different plasma cell subsets and differed in whether and how they discriminate between plasma blasts and plasma cells in their analysis. Moreover, it is possible that plasma cells are flexible in how they generate ATP but require mitochondrial activity to maintain other functions. Calcium homoeostasis, redox balance, mitochondrial unfolded protein responses and communication with other cell organelles are all tightly linked to mitochondrial function and could contribute to the functional role mitochondria play in plasma cells.

5.7 B1 Cells

B1 cells are a distinct population of innate-like B cells with the main function of producing natural antibodies. Natural antibodies represent an important bridge between innate and acquired immunity. They are produced before an infection occurs and bind structures that are frequently present on common pathogens. Additionally, natural antibodies are also weakly autoreactive and possess housekeeping functions such as aiding the removal of apoptotic debris. On the flip side, expansion of B1 cells is often associated with autoimmunity [174].

In humans, different B-cell subsets have been suggested to produce natural antibodies and although a B cell subset that is functionally similar to mouse B1 cells has been identified, their function, phenotype and ontogeny remain a matter of debate [175]. In laboratory mice, this population is more clearly defined by cell surface marker expression. B1 cells reside in different organs but most prominently in the peritoneal cavity. B1 cells in the peritoneal cavity can be divided into two main subsets B1b cells which are CD5 negative and primarily play a role in immune responses to certain microbial pathogens and B1a cells, which are the main producers of natural antibodies [174]. B1a cells in the mouse are large, long-lived, continuously produce antibodies and are replenished in adult animals through low-level proliferation [174]. Consistent with their increased energetic and biosynthetic needs, these cells are more metabolically active and show higher rates of both glycolysis and OxPhos than B2 cells [176] (Figure 22.4). The B1a cell population has been demonstrated to be reduced in mice with a B cell-specific HIF1α deletion; however, it is unclear whether the need to adapt to hypoxia or other HIF1α functions are required for B1a cell survival or development [177]. B1a cells are able to acquire external lipids and store them in lipid droplets [176]. Lipid metabolism has been shown to play an important role in this subset since *Pex5* deletion results in reduced B1 cell numbers and a poorer response to the B1a-cell pathogen *Streptococcus pneumoniae* [133]. An increase in lipid metabolism poses an inherent risk for lipid peroxidation. B1a cells have been shown to depend on the activity of the anti-oxidant enzyme *Gpx4* to protect themselves against lipid peroxidation. In the absence of GPX4, B1a cell numbers are reduced and immune responses to *S. pneumoniae* are impaired [127]. Similar to other long-lived B-cell subsets, B1a cells show signs of ongoing autophagy and their survival is dependent on ATG7 [176].

5.8 Regulatory B Cells

B cells capable of restricting the activity of other immune cells via direct contact or via the production of cytokines are called regulatory B cells (Bregs). Regulatory B cells steer immune cell function primarily by secreting IL-10 but have also been reported to produce cytokines such as TGFβ or IL-35 [178]. The term 'regulatory B cells' is loosely defined in literature and primarily used to describe B cells with the capacity to suppress pro-inflammatory responses without a clear set of phenotypic markers that could identify these cells. The lack of a clear phenotypical definition for regulatory B cells perhaps stems from the fact that different B-cell populations have been reported to secrete IL-10 under specific stimulatory conditions [179]. B cells displaying characteristics of B1a cells, transitional B cells, MZ B cells and plasma cells have all been described to secrete IL-10 under certain conditions [179]. In mice, the population of CD5+ CD1dhi B cells has been shown to be enriched in IL-10-producing cells [178,179]. Similar to mice, different human B-cell subsets have been proposed to have regulatory functions including CD24hiCD27+ memory B cells [179]. Taken together, the phenotypical diversity of IL-10-producing Bregs suggests that Bregs are not a distinct B-cell subset. Instead, the acquisition of a regulatory phenotype seems to be a consequence of cell fate decisions available to several different B-cell subsets. The exact stimuli and environmental conditions favouring differentiation to Bregs are currently incompletely understood. However, stimulation with CD40 or TLR9 agonists has been shown to induce a regulatory phenotype and IL-10 secretion [178,179]. Despite their unclear origin, regulatory B cells have been demonstrated to suppress a variety of immune pathologies and inflammation in both mouse and human.

The heterogeneity of IL-10-producing B cells makes it difficult to define the metabolic profile of these cells. Nevertheless, recent studies suggest that regulatory B-cell activity is coupled to distinct metabolic features. Cd1d+ CD5+ IL-10+ B cells have been shown to express higher levels of HIF1α in comparison to Cd1d- B cells [177]. Expression of this transcription factor has been shown to be required for the development or maintenance of this B-cell subset. Moreover, HIF1α deficiency has been demonstrated to exacerbate autoimmunity suggesting that this transcription factor plays an important role in regulatory B-cell functions [177]. Since HIF1α can accumulate in normoxic cells under certain conditions, the necessity of HIF1α to maintain a normal Cd1d+ CD5+ IL-10 pool of B cells might not be coupled to hypoxia responses of these cells. Nevertheless, independently of the route of HIF1α stabilisation, this transcription factor is known to drive a glycolytic gene expression profile. Consistently, Cd1d + CD5+ B cells have been found to consume more glucose than Cd1d+ CD5− cells. Glucose consumption, lactate secretion and the expression of *Glut1, Pkm2, Hexokinase 2, Ldha, Pdk1* have been found to be HIF1α-dependent in these cells suggesting that glycolysis might play a role in regulatory B cells [177]. A separate independent study suggests that IL-10+ cells generated via TLR9 stimulation do not express higher levels of GLUT1 than other B cells and that glucose is not needed for IL-10 production in vitro [180]. However since it is unclear whether dialysed FBS was used in this study and what the actual glucose concentration was in cell culture, it is difficult to judge the relevance of these findings [180]. The authors of this study additionally report that glutamine starvation or glutaminolysis inhibition reduces the frequency of IL10+ cells via inhibiting mTORC1, suggesting that Breg-fate can be regulated via nutrient availability [180].

In addition to glucose and glutamine metabolism, the SCFA acetate has been found to promote the development of IL-10+ cells in vitro and in vivo [181]. Acetate treatment of peritoneal B cells has been demonstrated to increase the frequency of IL-10 producing B1a cells and IL-10 levels in cell culture [181]. B cells treated with acetate in culture were also shown to ameliorate arthritis upon transfer into an arthritis mouse model. In these cell cultures, acetate did not seem to affect histone acetylation at the IL-10 promotor region but did induce overall lysine acetylation. The enzyme ACSS2 that converts acetate to acetyl-CoA which can be used for protein acetylation has been found to be needed for acetate induced differentiation to IL-10-producing cells. Additionally, acetate may be used as a substrate in the TCA cycle. Treatment with the mitochondrial inhibitor oligomycin or the β-oxidation inhibitor etomoxir reduced the frequency of IL-10-producing B1a cells suggesting that mitochondria-derived ATP plays a role in the development of IL-10-producing cells. In addition to mouse B cells, acetate has been shown to support differentiation of IL-10-producing B cells in humans as well [181].

A second independent study has found that the frequency of IL-10-producing cells in cell culture is increased after treatment with the SCFA butyrate. HAT inhibitors partially antagonised the activity of butyrate suggesting that butyrate acts as an HDAC inhibitor in this setup. Similarly, supplying antibiotics-treated mice with fatty acids resulted in increased frequency of IL-10+ cells suggesting that SFCA production by the microbiome supports Breg differentiation [182].

In vivo, the contribution of dietary choices, whole-body metabolism and the microbiome to steering B-cell function may be complex. A recent study has shown that mice and patients with rheumatoid arthritis display a reduction in microbiota-derived SCFAs concomitant with reduced numbers of IL-10-producing Bregs [183]. Supplementation with butyrate has been shown to reduce antigen-induced arthritis in mice. Of note, butyrate supplementation failed to suppress disease in B cell-deficient mice suggesting that in this model, B cells are key in providing butyrate-mediated anti-inflammatory benefits. Butyrate supplementation concurrently increased Breg-suppressive capacity and reduced GC B cell and plasmablast differentiation. As discussed before, butyrate can serve as a source of energy and induce epigenetic modifications. In this experimental setup, a third function for butyrate was demonstrated. Butyrate affected aryl-hydrocarbon receptor (AhR) signaling as mice with *Ahr*-deficient cells did not respond to butyrate treatment. Of note butyrate did not act as a direct AhR ligand. Instead, butyrate favoured the expansion of microbiota that produce a novel AhR ligand hydroxyindole-3-acetic acid (5-HIAA). Butyrate activity depended upon the presence of gut microbiota as butyrate supplementation was ineffective at suppressing arthritis in mice treated with antibiotics. Gavaging mice with 5-HIAA suppressed arthritis and increased AhR signaling and IL-10 production. Thus, butyrate supports the expansion of bacteria that produce 5-HIAA which in turn drives AhR signaling and IL-10 secretion [183]. Although it is not clear how butyrate alters the flora of the gut, this study highlights the fact that the metabolic microenvironment can affect B cells in a direct and indirect manner.

Combined, the studies suggest that SCFAs can affect B-cell function; however, the exact role in shaping B-cell responses seems to depend on the experimental setup. Nevertheless, the suggested role for SCFAs in boosting IL-10 production is consistent with the reported role of SCFAs in suppressing inflammation [184].

Lastly, it has been recently reported that cholesterol metabolism is crucial for the induction of a regulatory phenotype in human B cells. TLR-9-stimulated human B cells show defects in IL-10 production if cholesterol metabolism is inhibited with the 3-hydroxy-3-methylglutaryl coenzyme A (HMG-CoA) reductase inhibitor atorvastatin. This defect can be reversed by supplementation with exogenous mevalonate, which is a metabolite downstream of HMG-CoA. In contrast to IL-10 production, proliferation, differentiation and antibody secretion were not affected by atorvastatin. Although HMG-CoA reductase affects several metabolic pathways, geranylgeranyltransferase (GGTase) seems to be the primary target needed for the generation of IL-10+ B cells. Interestingly, in addition to IL-10 production, GGTase inhibition prevented GSK3 inhibition and BLIMP1 expression. Inhibition of GSK3 rescued IL-10 production in these experiments, suggesting that GSK3-dependent signaling might be affected by GGTase inhibition. Consistent with these findings, B cells from patients with dysregulated cholesterol metabolism do not efficiently produce IL-10 and the defects can be rescued by GGPP. Overall, the patients develop B-cell hyperactivation and severe systemic inflammation demonstrating that metabolic defects can have a profound impact on the immune system [185].

6. METABOLISM IN MALIGNANT AND AUTOREACTIVE B CELLS

A hallmark of malignant transformation is metabolic reprogramming which allows rapid proliferation. A prominent feature of cancer cells is their elevated rate of glucose uptake and lactate secretion even in the presence of oxygen. Oncogenic signalling is closely linked to metabolic activation and signalling pathways known to be often dysregulated in cancer such as PI3K/AKT, cMYC or HIF1α profoundly affect cell metabolism [186]. However different malignancies display a surprising heterogeneity in their metabolic programs providing the opportunity to explore the different metabolic features for disease stratification or treatment.

6.1 Metabolism in B Cell-Derived Leukaemia

Active DNA rearrangement and high levels of proliferation during B-cell development constitute an inherent risk for malignant transformation. Occasional errors during V(D)J recombination are a significant source of genome instability and can give rise to malignant transformation of B-cell precursors. B-cell precursor acute lymphoblastic leukaemia (B-ALL) is a common cancer in childhood and with a lower chance also arises in seniors. Although overall survival is good, therapeutic treatment of the disease can be associated with side effects and a portion of patients will relapse [187]. Specific subsets of B-ALL such as B-ALL positive for the oncogenic fusion kinase BCR-ABL are associated with high relapse rates and unfavourable prognosis [188]. Thus B-ALL remains a therapeutic challenge, and potential metabolic targets could be relevant for the development of new treatment strategies.

B-ALL cells express several glucose transporters including GLUT1 [11]. *Glut1* deletion in B-ALL cells partially reduces glucose uptake and limits proliferation in vitro and in vivo [11]. Of note, whilst the levels of lactate, pyruvate and TCA cycle intermediates are not affected by *Glut1* deletion, glycolytic and PPP intermediates were lowered [11]. Combined, these findings suggest that B-ALL cells are dependent on glucose for proliferation and whilst alternative carbon sources may contribute to lactate and pyruvate generation, glucose is limiting for appropriate anabolic activity [11]. It has been recently suggested that restricting glucose metabolism is a protective mechanism against transformation in B-cell precursors. The transcription factors PAX5 and IZKF1 which define B-cell identity also regulate cell metabolism and increase the expression of negative regulators of glucose uptake. Consistently, patient-derived B-ALL samples often harbour PAX5 and IKZF1 lesions allowing them to escape glucose metabolism restrictions [109]. Reconstitution with wildtype PAX5 and IKZF1 results in AMPK activation, reduced expression of glucose transporters and increased expression of negative regulators of glucose metabolism. Consistently, glucose uptake, ATP production and survival are reduced. This defect can be rescued by providing the cells with cell membrane permeable nutrients [109]. In summary, glucose is essential for B-ALL cells and restricting glucose uptake can serve as a safeguard mechanism against outgrowth of malignant B-cell clones.

In addition to B-ALL, B cells can also give rise to B cell-derived chronic lymphocytic leukaemia (B-CLL) [189]. B-CLL is the most common adult leukaemia in the western hemisphere. Although B-CLL cases have good prognosis, the disease can progress to a more aggressive stage which is currently untreatable (this process is called Richter transformation). In contrast to B-ALL, B-CLL is an indolent form of cancer with low turn-over rates. B-CLL cells are believed to primarily rely on mitochondrial oxidative metabolism rather than glycolysis and to be particularly sensitive to redox stress [190,191]. B-CLL cells show increased ROS production and are sensitive to inhibitors of ROS scavenging pathways [191].

6.2 Metabolism in B Cell-Derived Lymphoma

Similar to B-cell precursors, GC B cells are at a particular risk for malignant transformation since they are highly proliferative and actively alter their DNA. A large proportion of B-cell neoplasms originates from activated or GC-experienced B cells. To support rapid proliferation, lymphoma cells are metabolically highly active and different compounds targeting mTORC1 signalling, energetic pathways or nucleotide metabolism are in clinical use or currently being evaluated in clinical studies often in combination with established anti-CD20-based therapies [192].

Although high rates of glycolysis are a hallmark of cancer cells, recent studies reveal a remarkable metabolic heterogeneity within the same lymphoma subtype. For example, a subset of primary diffuse large B-cell lymphomas (DLBCLs) as well as DLBCL cell lines shows a transcriptional signature enriched in genes involved in mitochondrial metabolism [193,194]. In vitro, these cell lines have been shown to be sensitive to the inhibition of mitochondrial β-oxidation, demonstrating that a better understanding of the metabolic profile of tumour cells can reveal new vulnerabilities [194]. Inquiry into metabolic regulation can also uncover new markers for disease stratification. It has recently been shown that CD37 plays a role in limiting fatty acid metabolism [195]. CD37 is a tetraspanin and approximately 60% of patients with aggressive lymphoma show CD37 loss, which is related to poorer clinical outcomes [196]. CD37 has been shown to interact with the fatty acid transporter FATP1 and to inhibit the uptake of palmitate [195]. Palmitic acid is a common saturated fatty acid that is incorporated into phospholipid, sphingolipids and triglycerides. CD37-deficient lymphoma cell lines have been shown to be able to engage glycolysis but show reduced glycolytic capacity [195]. In contrast to poor glucose responses, CD37-deficient lymphoma cells responded better to palmitate in comparison to control cells. Uptake of palmitate by CD37-deficient lymphoma cells is so robust that it is reflected in the depletion of palmitate in serum. CD37 deficiency creates a dependance on this nutrient and CD37-deficient lymphoma cells have been demonstrated to succumb to cell death if fatty acid metabolism is inhibited. Similar to mouse cells, primary human lymphomas display increased lipid storage [195]. In summary, this study suggests that CD37 might be useful as a marker to not only predict clinical outcomes but also to identify disease specific inhibitors.

6.3 Metabolism in Autoimmune B Cells

B cells from mouse models of autoimmunity often show an overall increase in metabolic activity [197,198]. In general, B-cell hyperactivation is often coupled to the emergence of autoimmunity. Consistently, metabolic needs of these B cells are increased. For example, B cells from BAFF transgenic mice show enhanced respiratory capacity possibly contributing to their faster responses to subsequent stimulation [9]. Similarly, TRAF3-deficient B cells show increased GLUT1 and hexokinase II expression as well as glucose uptake and oxygen consumption [129]. Increased AKT signalling in mice with a B cell-specific deletion of TAPP adaptor proteins is associated with increased metabolic activity and signs of autoimmunity. 2DG treatment of these mice reduces GC numbers and symptoms of autoimmunity suggesting that metabolic targets could be exploited for therapy in the future [138].

7. CONSEQUENCES OF DISRUPTED METABOLISM

In the life of a normal as well as a malignant B cell, phases of metabolic activity alternate with phases of relative metabolic quiescence. In normal B cells, a switch towards a quiescent state can happen due to two reasons: cells cease to receive activation signals or nutrient/oxygen availability becomes limiting. Malignant B cells are largely independent of extracellular activation cues for their proliferation, but still need to be able to match their metabolic activity to available resources to avoid energetic collapse. Aberrantly increased metabolic activity due to mutations in different signalling molecules can be a double-edged sword and on one side drive proliferation but on the other side render cells more susceptible to metabolic stress.

7.1 Metabolic Hyperactivation

The ability of B cells to respond to aberrantly increased metabolic activity seems to depend on the differentiation status, the signalling context and the metabolic micromilieu. For example, deletion of GSK3 in mature B cells results in increased proliferation both in vitro and in vivo, if nutrients are available [42]. In contrast, GSK3-deficient B cells display reduced viability in the GC and in glucose low medium in vitro. GSK3-deficient B cells cultured in glucose low medium accumulate more ROS and ROS depletion rescues their survival [42]. Overproduction of ROS as a consequence of defects in various signalling molecules such as the RNA-binding protein HuR [144], the autophagy protein ATG7 [158], aberrantly activated PI3K signalling [199] has been implied to contribute to reduced survival in different B-cell subsets. Thus, aberrant metabolic activation can be fatal to B cells in certain scenarios if ROS is not scavenged appropriately.

The consequences of aberrantly increased metabolism on B-cell fate and function may also depend on the differentiation status of the cells. This is exemplified by the different roles PI3K signalling play in B-cell precursors and mature B cells. Increased PI3K signalling in B cells from mice with a

B cell-specific deletion of the negative regulators SHIP and PTEN leads to spontaneous lymphoma development [200]. In general, PI3K activation in mature B cells leads to increased activation, proliferation and the development of autoimmunity. In contrast, PTEN deletion is fatal for B-cell precursors and B-ALL cells. In general, PI3K inhibition is needed for B-cell development. Upon PI3K inhibition, FOXO1 is released and drives the expression of the IL-7 receptor and RAG proteins, two crucial steps in B-cell development. Consequently, PTEN deletion inhibits normal B-cell development [35,201]. However, FOXO1 activation is not sufficient for B-cell development since constitutively active FOXO1 cannot rescue B-cell development in the absence of PTEN [35] and PTEN deletion reduces survival of B-ALL cells which are IL-7 independent [202]. PTEN deletion in B-ALL cells leads to increased glucose consumption and lactate secretion, but ATP levels are not increased in comparison to control cells, suggesting that PTEN deletion creates a mismatch between energy generation and energy consuming processes such as cell growth [202]. PTEN deletion in myeloid cells does not reduce survival suggesting that the effect of increased PI3K signalling is cell type-specific [202].

Similar to B-ALL cells, CLL cells are also sensitive to dysregulated PI3K signalling. CLL cells maintain BCR expression and benefit from pro-survival signals originating from the BCR. PTEN deletion before the onset of disease accelerates the development of CLL [203]. Similarly, chronic AKT activation induces Richter transformation in CLL [204]. In contrast, acute AKT hyperactivation or SHIP inhibition reduces CLL survival in a ROS-dependent manner [199]. These findings suggest that increased PI3K signalling can have different effects on the survival of malignant B cells depending on the duration, strength and onset of the signal. Whilst increased PI3K before onset of disease may boost transformation and expansion of premalignant clones by supporting biosynthesis, acute PI3K activation might limit cell survival as cells struggle to quickly adapt to the abruptly changed energetic demands. However, this model remains speculative and other aspects of PI3K signalling than regulation of redox balance and energy homoeostasis may play a more prominent role in determining how B cells respond to hyperactivation of this signalling network.

7.2 Dysregulation of Metabolic Stress Responses

Similar to the divergent responses observed after PI3K activation, various B-cell subsets respond differently to dysregulation of metabolic stress adaptation pathways. As mentioned before, molecules such as LKB1, AMPK and FNIP1 contribute to responses to metabolic stress. The deletion of LKB1 results in a complete block in B-cell development [109]. Similarly, FNIP1-deficient mice lack mature B cells [49]. Signalling downstream of the BCR in FNIP1-deficient B cells is normal, and the expression of a rearranged BCR cannot rescue the defects in B-cell development suggesting that FNIP1 does not participate in pre-BCR-driven signalling or the rearrangement of the Ig heavy chain locus [49]. Instead, FNIP1 is needed in order to inhibit cell growth and metabolic activity when energy and nutrients are scarce. FNIP1-deficient pre-B cells show increased AMPK signalling and autophagic flux, but fail to inhibit mTORC1 in response to AMPK activation [49,205]. FNIP1-deficient B cells take up more glucose, harbour more mitochondrial mass but display reduced ATP after IL-7 stimulation than control cells [49]. These findings suggest that FNIP1 deficiency results in a mismatch between energy consumption and energy generation, thereby reducing pre-B-cell survival. However, the phenotype observed in FNIP1-deficient B cells cannot be fully explained by an inability of AMPK to limit mTORC1 signalling. AMPK deletion does not interfere with normal B-cell development [122–124]. Additionally, neither *Ampk1α* deletion nor deletion of the mTORC1 component *Raptor* rescues B-cell development in FNIP1-deficient mice [205]. Thus, the exact role of LKB1, AMPK and FNIP1 in pre-B cells remains incompletely understood. It is possible that LKB1 and FNIP1 possess AMPK-independent roles. Alternatively, whilst a certain level of mTORC1 signalling is crucial for B-cell development, aberrantly increased mTORC1 activation insensitive to metabolic stress, might be fatal. Normal pre-B cells might not depend on AMPK for their survival because mTORC1 never reaches an unsustainable level; however, AMPK might become more important for cells in which mTORC1 signalling is enhanced. This could be the case in B-ALL cells which unlike their normal counterparts have been reported to need AMPK for survival [109].

AMPK also plays a complex role in mature B cells and the humoral immune response. Mature B cells can proliferate in the absence of LKB1 or AMPK [124,125]. AMPK is needed in activated B cells to maintain spare respiratory capacity but is not required for GC formation in vivo. AMPK deletion results in increased memory B-cell numbers at early time points but reduced memory B-cell numbers and weaker recall antibody responses after a boost [122]. Although AMPK is known for its role in inducing autophagy, LPS-stimulated AMPK-deficient cells in glucose low media show normal levels of autophagy [122]. Thus, B cells can induce AMPK-independent autophagic degradation. Despite overall autophagic flux being normal, mitophagy is impaired and mROS and lipid peroxidation is increased in AMPK-deficient B cells [122]. AMPK-deficient plasma cells secrete more antibodies possibly due to increased mTORC1 signalling [122]. BLIMP1 has been found to repress the transcription of the sestrin family of AMPK activators [170] suggesting that AMPK activity is inhibited as B cells become plasma cells.

Oncogenic activation induces a high need for nutrients and only cells that are able to obtain these nutrients can progress to malignant transformation. Infection of B cells with Epstein–Barr virus results in an initial boost of proliferation, but the response later contracts leading to a senescence-like phenotype in a subset of the infected cells. These arrested cells express more sestrins and display increased AMPK- and reduced mTOR-signalling [206]. Additionally, *Ampk1α* deletion has been shown to accelerate lymphoma development in the EμMyc mouse model rendering the cells addicted to HIF1α [123,207]. This suggests that AMPK might limit cell growth in some B cell-derived malignancies.

Similar to AMPK, LKB1 plays a complex role in B cells. Deletion of *Lkb1* contributes to lymphoma development although normal B-cell numbers are reduced in the absence of LKB1 [126]. *Lkb1* deficiency leads to increased IL-6 production in B cells, which boosts T-cell activity and increases GC responses. However, *Lkb1*-deficient B cells are outcompeted in the GC [125].

In summary, the loss of signalling molecules involved in metabolic stress responses can result in diverse outcomes in B cells. Whilst some B-cell subsets require AMPK/LKB1 to prevent that anabolic metabolism outstrips available nutrients, in other subsets, AMPK/LKB1 loss appears to remove a break on mTOR signalling thereby allowing excess proliferation. Analysing the role of metabolic regulators in B-cell fate decisions is made even more difficult by the fact that these signalling molecules fulfil a multitude of roles, many of them separate from metabolic reprogramming. Nevertheless, since normal B cells as well as malignant clones face various metabolic challenges, elucidating the role of metabolic stress responses in B-cell proliferation and survival remains central to our understanding of B-cell biology.

8. CONCLUSIONS

Recent years have brought an increased appreciation of how metabolic processes interact with classical protein-based signalling pathways. Our understanding of the metabolic profile of different B-cell subsets, the role of specific signalling networks in metabolic reprogramming as well as the contribution of the metabolic environment to cell fate decisions has grown substantially in the last decade. Nevertheless, many aspects of metabolic control of B-cell function remain poorly understood. Experimental work in the field of immunometabolism is made difficult by the complex and dynamic composition of immune cell niches in vivo which are difficult to replicate in vitro. Moreover, experimental manipulation of immune cells can perturb their natural metabolic state further obscuring experimental results. Lastly, B cells appear to be rather metabolically flexible and are often able to compensate for a loss of an individual metabolic regulator. This metabolic flexibility might lead to an underestimation of the role specific metabolic pathways or metabolic regulators play in B-cell biology. Despite these technical difficulties, various metabolic vulnerabilities have been discovered in B cells encouraging further investigation into the role of metabolism in B-cell fate and function.

REFERENCES

[1] Jellusova J. Metabolic control of B cell immune responses. Curr Opin Immunol 2020;63:21–8. https://doi.org/10.1016/j.coi.2019.11.002.

[2] Boothby M, Rickert RC. Metabolic regulation of the immune humoral response. Immunity 2017;46(5):743–55. https://doi.org/10.1016/j.immuni.2017.04.009.

[3] Wang YP, Lei QY. Metabolite sensing and signaling in cell metabolism. Signal Transduct Target Ther 2018;3(1):1–9. https://doi.org/10.1038/s41392-018-0024-7.

[4] Haas R, Cucchi D, Smith J, Pucino V, Macdougall CE, Mauro C. Intermediates of metabolism: from bystanders to signalling molecules. Trends Biochem Sci 2016;41(5):460–71.

[5] Pan C, Li B, Simon MC. Moonlighting functions of metabolic enzymes and metabolites in cancer. Mol Cell 2021;81(18):3760–74.

[6] O'Neill LAJ, Kishton RJ, Rathmell J. A guide to immunometabolism for immunologists. Nat Rev Immunol 2016;16(9):553–65.

[7] Lunt SY, Vander Heiden MG. Aerobic glycolysis: meeting the metabolic requirements of cell proliferation. Annu Rev Cell Dev Biol 2011;27:441–64.

[8] Marth JD, Grewal PK. Mammalian glycosylation in immunity. Nat Rev Immunol 2008;8(11):874–87.

[9] Caro-Maldonado A, Wang R, Nichols AG, Kuraoka M, Milasta S, Sun LD, et al. Metabolic reprogramming is required for antibody production that is suppressed in anergic but exaggerated in chronically BAFF-exposed B cells. J Immunol 2014;192(8):3626–36.

[10] Thorens B, Mueckler M. Glucose transporters in the 21st century. Am J Physiol Endocrinol Metab 2010;298(2):1–10.

[11] Liu T, Kishton RJ, Macintyre AN, Gerriets VA, Xiang H, Liu X, et al. Glucose transporter 1-mediated glucose uptake is limiting for B-cell acute lymphoblastic leukemia anabolic metabolism and resistance to apoptosis. Cell Death Dis 2014;5(10):1–13.

[12] Kojima H, Kobayashi A, Sakurai D, Kanno Y, Hase H, Takahashi R, et al. Differentiation stage-specific requirement in hypoxia-inducible factor-1α–regulated glycolytic pathway during murine B cell development in bone marrow. J Immunol 2010;184(1):154–63.

[13] Martínez-Reyes I, Chandel NS. Mitochondrial TCA cycle metabolites control physiology and disease. Nat Commun 2020;11(1):1–11. https://doi.org/10.1038/s41467-019-13668-3.

[14] Zangari J, Petrelli F, Maillot B, Martinou JC. The multifaceted pyruvate metabolism: role of the mitochondrial pyruvate carrier. Biomolecules 2020;10(7):1–18.

[15] Brookes PS, Yoon Y, Robotham JL, Anders MW, Sheu SS, Chandel NS, et al. Cellular and molecular mechanisms of

mitochondrial function. Biochim Biophys Acta Bioenerg 2004;417(1):64–71. https://doi.org/10.1016/j.beem.2012.05.003.
[16] Garcia-Manteiga JM, Mari S, Godejohann M, Spraul M, Napoli C, Cenci S, et al. Metabolomics of B to plasma cell differentiation. J Proteome Res 2011;10(9):4165–76.
[17] Masle-Farquhar E, Bröer A, Yabas M, Enders A, Bröer S. ASCT2 (SLC1A5)-deficient mice have normal B-cell development, proliferation, and antibody production. Front Immunol 2017;8(May):1–11.
[18] Yang L, Venneti S, Nagrath D. Glutaminolysis: a hallmark of cancer metabolism. Annu Rev Biomed Eng 2017;19:163–94.
[19] Le A, Lane AN, Hamaker M, Bose S, Gouw A, Barbi J, et al. Glucose-independent glutamine metabolism via TCA cycling for proliferation and survival in b cells. Cell Metabol 2012;15(1):110–21. https://doi.org/10.1016/j.cmet.2011.12.009.
[20] Hoy AJ, Nagarajan SR, Butler LM. Tumour fatty acid metabolism in the context of therapy resistance and obesity. Nat Rev Cancer 2021;21(12):753–66.
[21] Kim CH. Control of lymphocyte functions by gut microbiota-derived short-chain fatty acids. Cell Mol Immunol 2021;18(5):1161–71. https://doi.org/10.1038/s41423-020-00625-0.
[22] Wanders RJA, Waterham HR, Ferdinandusse S. Metabolic interplay between peroxisomes and other subcellular organelles including mitochondria and the endoplasmic reticulum. Front Cell Dev Biol 2016;3(Jan):1–15.
[23] Stincone A, Prigione A, Cramer T, Wamelink MMC, Campbell K, Cheung E, et al. The return of metabolism: biochemistry and physiology of the pentose phosphate pathway. Biol Rev 2015;90(3):927–63.
[24] Xiao G, Chan LN, Klemm L, Braas D, Chen Z, Geng H, et al. B-cell-specific diversion of glucose carbon utilization reveals a unique vulnerability in B cell malignancies. Cell 2018;173(2):470–484.e18.
[25] Cheng H, Wang M, Su J, Li Y, Long J, Chu J, et al. Lipid metabolism and cancer. Life 2022;12(6):1–17.
[26] Bento CF, Renna M, Ghislat G, Puri C, Ashkenazi A, Vicinanza M, et al. Mammalian autophagy: how does it work? Annu Rev Biochem 2016;85(June):685–713.
[27] Anding AL, Baehrecke EH. Cleaning house: selective autophagy of organelles. Dev Cell 2017;41(1):10–22.
[28] Münz C. Non-canonical functions of autophagy proteins in immunity and infection. Mol Aspect Med 2021;82(June):100987. https://doi.org/10.1016/j.mam.2021.100987.
[29] Jellusova J, Rickert RC. The PI3K pathway in B cell metabolism. Crit Rev Biochem Mol Biol 2016;51(5):359–78.
[30] Jellusova J. The role of metabolic checkpoint regulators in B cell survival and transformation. Immunol Rev 2020;295(1):39–53.
[31] Saxton RA, Sabatini DM. mTOR signaling in growth, metabolism, and disease. Cell 2017;168(6):960–76. https://doi.org/10.1016/j.cell.2017.02.004.
[32] Dossou AS, Basu A. The emerging roles of mTORC1 in macromanaging autophagy. Cancers 2019;11(10):1422. https://www.mdpi.com/2072-6694/11/10/1422.
[33] Navé BT, Ouwens DM, Withers DJ, Alessi DR, Shepherd PR. Mammalian target of rapamycin is a direct target for protein kinase B: identification of a convergence point for opposing effects of insulin and amino-acid deficiency on protein translation. Biochem J 1999;344(2):427–31.
[34] Inoki K, Li Y, Zhu T, Wu J, Guan KL. TSC2 is phosphorylated and inhibited by Akt and suppresses mTOR signalling. Nat Cell Biol 2002;4(9):648–57.
[35] Zeng H, Yu M, Tan H, Li Y, Su W, Shi H, et al. Discrete roles and bifurcation of PTEN signaling and mTORC1-mediated anabolic metabolism underlie IL-7-driven B lymphopoiesis. Sci Adv 2018;4(1):1–19.
[36] Iwata TN, Ramírez JA, Tsang M, Park H, Margineantu DH, Hockenbery DM, et al. Conditional disruption of raptor reveals an essential role for mTORC1 in B cell development, survival, and metabolism. J Immunol 2016;197(6):2250–60.
[37] Oh WJ, Jacinto E. mTOR complex 2 signaling and functions. Cell Cycle 2011;10(14):2305–16.
[38] Lee K, Heffington L, Jellusova J, Nam KT, Raybuck A, Cho SH, et al. Requirement for Rictor in homeostasis and function of mature B lymphoid cells. Blood 2013;122(14):2369–79.
[39] Li M, Lazorchak AS, Ouyang X, Zhang H, Liu H, Arojo OA, et al. Sin1/mTORC2 regulate B cell growth and metabolism by activating mTORC1 and Myc. Cell Mol Immunol 2019;16(9):757–69.
[40] Wang L, Li J, Di L. Glycogen synthesis and beyond, a comprehensive review of GSK3 as a key regulator of metabolic pathways and a therapeutic target for treating metabolic diseases. Med Res Rev 2022;42(2):946–82.
[41] Sutherland C. What are the bona fide GSK3 substrates? Int J Alzheimer's Dis 2011;2011.
[42] Jellusova J, Cato MH, Apgar JR, Ramezani-Rad P, Leung CR, Chen C, et al. Gsk3 is a metabolic checkpoint regulator in B cells. Nat Immunol 2017;18(3):303–12. https://doi.org/10.1038/ni.3664.
[43] Gregory MA, Qi Y, Hann SR. Phosphorylation by glycogen synthase kinase-3 controls c-Myc proteolysis and subnuclear localization. J Biol Chem 2003;278(51):51606–12.
[44] Miller DM, Thomas SD, Islam A, Muench D, Sedoris K. c-Myc and cancer metabolism. Clin Cancer Res 2012;18(20):5546–53.
[45] Iritani BM, Eisenman RN. c-Myc enhances protein synthesis and cell size during B lymphocyte development. Proc Natl Acad Sci USA 1999;96(23):13180–5.
[46] Hoang G, Zhang C, Attarwala N, Jung JG, Cooper AJL, Le A. Uncovering metabolic reservoir cycles in MYC-transformed lymphoma B cells using stable isotope resolved metabolomics. Anal Biochem 2021;632:114206. https://doi.org/10.1016/j.ab.2021.114206.
[47] Feist M, Schwarzfischer P, Heinrich P, Sun X, Kemper J, Von Bonin F, et al. Cooperative STAT/NF-κB signaling regulates lymphoma metabolic reprogramming and aberrant GOT2 expression. Nat Commun 2018;9(1).
[48] Herzig S, Shaw RJ. AMPK: guardian of metabolism and mitochondrial homeostasis. Nat Rev Mol Cell Biol 2018;19(2):121–35. https://doi.org/10.1038/nrm.2017.95.
[49] Park H, Staehling K, Tsang M, Appleby MW, Brunkow ME, Margineantu D, et al. Disruption of Fnip1 reveals a metabolic checkpoint controlling B lymphocyte development. Immunity 2012;36(5):769–81. https://doi.org/10.1016/j.immuni.2012.02.019.
[50] Li TT, Zhu HB. LKB1 and cancer: the dual role of metabolic regulation. Biomed Pharmacother 2020;132(Oct):110872. https://doi.org/10.1016/j.biopha.2020.110872.
[51] Takahara T, Amemiya Y, Sugiyama R, Maki M, Shibata H. Amino acid-dependent control of mTORC1 signaling: a variety of regulatory modes. J Biomed Sci 2020;27(1):1–16.

[52] Onnis A, Cianfanelli V, Cassioli C, Samardzic D, Pelicci PG, Cecconi F, et al. The pro-oxidant adaptor p66SHC promotes B cell mitophagy by disrupting mitochondrial integrity and recruiting LC3-II. Autophagy 2018;14(12):2117–38. https://doi.org/10.1080/15548627.2018.1505153.

[53] Majmundar AJ, Wong WJ, Simon MC. Hypoxia-inducible factors and the response to hypoxic stress. Mol Cell 2010;40(2):294–309. https://doi.org/10.1016/j.molcel.2010.09.022.

[54] Cho SH, Raybuck AL, Stengel K, Wei M, Beck TC, Volanakis E, et al. Germinal centre hypoxia and regulation of antibody qualities by a hypoxia response system. Nature 2016;537(7619):234–8.

[55] DeWane G, Salvi AM, DeMali KA. Fueling the cytoskeleton-links between cell metabolism and actin remodeling. J Cell Sci 2021;134(3).

[56] Gordaliza-Alaguero I, Cantó C, Zorzano A. Metabolic implications of organelle–mitochondria communication. EMBO Rep 2019;20(9):1–27.

[57] Paul BT, Manz DH, Torti FM, Torti SV. Mitochondria and Iron: current questions. Expert Rev Hematol 2017;10(1):65–79. https://www.tandfonline.com/doi/full/10.1080/17474086.2016.1268047.

[58] Dejos C, Gkika D, Cantelmo AR. The two-way relationship between calcium and metabolism in cancer. Front Cell Dev Biol 2020;8(Nov):1–14.

[59] Jumaa H, Caganova M, McAllister EJ, Hoenig L, He X, Saltukoglu D, et al. Immunoglobulin expression in the endoplasmic reticulum shapes the metabolic fitness of B lymphocytes. Life Sci Alliance 2020;3(4):1–17.

[60] Diskin C, Ryan TAJ, O'Neill LAJ. Modification of proteins by metabolites in immunity. Immunity 2021;54(1):19–31. https://doi.org/10.1016/j.immuni.2020.09.014.

[61] Lam WY, Becker AM, Kennerly KM, Wong R, Curtis JD, Llufrio EM, et al. Mitochondrial pyruvate import promotes long-term survival of antibody-secreting plasma cells. Immunity 2016;45(1):60–73. https://doi.org/10.1016/j.immuni.2016.06.011.

[62] Shade KTC, Anthony RM. Antibody glycosylation and inflammation. Antibodies 2013;2(3):392–414.

[63] Buettner MJ, Shah SR, Saeui CT, Ariss R, Yarema KJ. Improving immunotherapy through glycodesign. Front Immunol 2018;9(Nov).

[64] Ong Q, Han W, Yang X. O-GlcNAc as an integrator of signaling pathways. Front Endocrinol 2018;9(Oct):1–9.

[65] Wu JL, Chiang MF, Hsu PH, Tsai DY, Hung KH, Wang YH, et al. O-GlcNAcylation is required for B cell homeostasis and antibody responses. Nat Commun 2017;8(1). https://doi.org/10.1038/s41467-017-01677-z.

[66] Golks A, Tran TTT, Goetschy JF, Guerini D. Requirement for O-linked N-acetylglucosaminyltransferase in lymphocytes activation. EMBO J 2007;26(20):4368–79.

[67] Lee DH, Kwon NE, Lee W-J, Lee M-S, Kim D-J, Kim JH, et al. Increased O-GlcNAcylation of c-Myc promotes pre-B cell proliferation. Cells 2020;9(1):158. https://www.mdpi.com/2073-4409/9/1/158.

[68] Cheng J, Liu Y, Yan J, Zhao L, Zhou Y, Shen X, et al. Fumarate suppresses B-cell activation and function through direct inactivation of LYN. Nat Chem Biol 2022;18(Sep).

[69] Burgener AV, Bantug GR, Meyer BJ, Higgins R, Ghosh A, Bignucolo O, et al. SDHA gain-of-function engages inflammatory mitochondrial retrograde signaling via KEAP1–Nrf2. Nat Immunol 2019;20(10):1311–21. https://doi.org/10.1038/s41590-019-0482-2.

[70] Tretter L, Patocs A, Chinopoulos C. Succinate, an intermediate in metabolism, signal transduction, ROS, hypoxia, and tumorigenesis. Biochim Biophys Acta Bioenerg 2016;1857(8):1086–101. https://doi.org/10.1016/j.bbabio.2016.03.012.

[71] Caielli S, Veiga DT, Balasubramanian P, Athale S, Domic B, Murat E, et al. A CD4+ T cell population expanded in lupus blood provides B cell help through interleukin-10 and succinate. Nat Med 2019;25(1):75–81. https://doi.org/10.1038/s41591-018-0254-9.

[72] Pietrocola F, Galluzzi L, Bravo-San Pedro JM, Madeo F, Kroemer G. Acetyl coenzyme A: a central metabolite and second messenger. Cell Metabol 2015;21(6):805–21.

[73] Shi L, Tu BP. Acetyl-CoA and the regulation of metabolism: mechanisms and consequences. Curr Opin Cell Biol 2015;33:125–31.

[74] Kieffer-Kwon KR, Nimura K, Rao SSP, Xu J, Jung S, Pekowska A, et al. Myc regulates chromatin decompaction and nuclear architecture during B cell activation. Mol Cell 2017;67(4):566–578.e10. https://doi.org/10.1016/j.molcel.2017.07.013.

[75] Etchegaray J-P, Mostoslavsky R. Interplay between metabolism and epigenetics: a nuclear adaptation to environmental changes. Mol Cell 2016;62(5):695–711. https://linkinghub.elsevier.com/retrieve/pii/S1097276516301927.

[76] Kim M, Qie Y, Park J, Kim CH. Gut microbial metabolites fuel host antibody responses. Cell Host Microbe 2016;20(2):202–14. https://doi.org/10.1016/j.chom.2016.07.001.

[77] Sanchez HN, Moroney JB, Gan H, Shen T, Im JL, Li T, et al. B cell-intrinsic epigenetic modulation of antibody responses by dietary fiber-derived short-chain fatty acids. Nat Commun 2020;11(1). https://doi.org/10.1038/s41467-019-13603-6.

[78] Chang HC, Guarente L. SIRT1 and other sirtuins in metabolism. Trends Endocrinol Metabol 2014;25(3):138–45. https://doi.org/10.1016/j.tem.2013.12.001.

[79] Cantó C, Menzies KJ, Auwerx J. NAD+ metabolism and the control of energy homeostasis: a balancing act between mitochondria and the nucleus. Cell Metabol 2015;22(1):31–53.

[80] Gan H, Shen T, Chupp DP, Taylor JR, Sanchez HN, Li X, et al. B cell Sirt1 deacetylates histone and non-histone proteins for epigenetic modulation of AID expression and the antibody response. Sci Adv 2020;6(14):1–18.

[81] Haniuda K, Fukao S, Kitamura D. Metabolic reprogramming induces germinal center B cell differentiation through Bcl6 locus remodeling. Cell Rep 2020;33(5):108333. https://doi.org/10.1016/j.celrep.2020.108333.

[82] Stehling O, Lill R. The role of mitochondria in cellular iron-sulfur protein biogenesis: mechanisms, connected processes, and diseases. Cold Spring Harb. Perspect. Med. 2013;3(9):1–17.

[83] Shen K, Pender CL, Bar-Ziv R, Zhang H, Wickham K, Willey E, et al. Mitochondria as cellular and organismal signaling hubs. Annu Rev Cell Dev Biol 2022;38:179–218.

[84] Pilchova I, Klacanova K, Tatarkova Z, Kaplan P, Racay P. The involvement of Mg2+ in regulation of cellular and mitochondrial functions. Oxid Med Cell Longev 2017;2017(Figure 1).

[85] Brookes PS, Yoon Y, Robotham JL, Anders MW, Sheu SS. Calcium, ATP, and ROS: a mitochondrial love-hate triangle. Am J Physiol Cell Physiol 2004;287(4):56-4.

[86] Kühlbrandt W. Structure and function of mitochondrial membrane protein complexes. BMC Biol 2015;13(1):1–11. https://doi.org/10.1186/s12915-015-0201-x.

[87] Taanman JW. The mitochondrial genome: structure, transcription, translation and replication. Biochim Biophys Acta Bioenerg 1999;1410(2):103–23.

[88] Theurey P, Rieusset J. Mitochondria-associated membranes response to nutrient availability and role in metabolic diseases. Trends Endocrinol Metabol 2017;28(1):32–45. https://doi.org/10.1016/j.tem.2016.09.002.

[89] Varilly P, Chandler D. Moving mitochondria: establishing distribution of an essential organelle. Traffic 2013;8(12):1668–75.

[90] Mishra P, Chan DC. Metabolic regulation of mitochondrial dynamics. J Cell Biol 2016;212(4):379–87.

[91] Murphy MP. How mitochondria produce reactive oxygen species. Biochem J 2009;417(1):1–13.

[92] Wellen KE, Thompson CB. Cellular metabolic stress: considering how cells respond to nutrient excess. Mol Cell October 2010;40(2):323–32. https://linkinghub.elsevier.com/retrieve/pii/S109727651000780X.

[93] Sun Y, Lu Y, Saredy J, Wang X, Drummer IVC, Shao Y, et al. ROS systems are a new integrated network for sensing homeostasis and alarming stresses in organelle metabolic processes. Redox Biol 2020;37:101696. https://doi.org/10.1016/j.redox.2020.101696.

[94] Zeeshan HMA, Lee GH, Kim HR, Chae HJ. Endoplasmic reticulum stress and associated ROS. Int J Mol Sci 2016;17(3):1–20.

[95] Finkel T. Signal transduction by reactive oxygen species. J Cell Biol 2011;194(1):7–15.

[96] Kaludercic N, Deshwal S, Di Lisa F. Reactive oxygen species and redox compartmentalization. Front Physiol 2014;5(Aug):1–15.

[97] Feng Y-Y, Tang M, Suzuki M, Gunasekara C, Anbe Y, Hiraoka Y, et al. Essential role of NADPH oxidase–dependent production of reactive oxygen species in maintenance of sustained B cell receptor signaling and B cell proliferation. J Immunol 2019;202(9):2546–57.

[98] Reth M. Hydrogen peroxide as second messenger in lymphocyte activation. Nat Immunol 2002;3(12):1129–34.

[99] Wheeler ML, DeFranco AL. Prolonged production of reactive oxygen species in response to B cell receptor stimulation promotes B cell activation and proliferation. J Immunol 2012;189(9):4405–16.

[100] Ogura M, Inoue T, Yamaki J, Homma MK, Kurosaki T, Homma Y. Mitochondrial reactive oxygen species suppress humoral immune response through reduction of CD19 expression in B cells in mice. Eur J Immunol 2017;47(2):406–18.

[101] Jiang X, Stockwell BR, Conrad M. Ferroptosis: mechanisms, biology and role in disease. Nat Rev Mol Cell Biol 2021;22(4):266–82. https://doi.org/10.1038/s41580-020-00324-8.

[102] Muri J, Kopf M. Redox regulation of immunometabolism. Nat Rev Immunol 2021;21(6):363–81. https://doi.org/10.1038/s41577-020-00478-8.

[103] Liou GY, Storz P. Reactive oxygen species in cancer. Free Radic Res 2010;44:479.

[104] Hardy RR, Hayakawa K. B cell development pathways. Cancer 2001:595–621.

[105] Stein M, Dütting S, Mougiakakos D, Bösl M, Fritsch K, Reimer D, et al. A defined metabolic state in pre B cells governs B-cell development and is counterbalanced by Swiprosin-2/EFhd1. Cell Death Differ 2017;24(7):1239–52.

[106] Clark MR, Mandal M, Ochiai K, Singh H. Orchestrating B cell lymphopoiesis through interplay of IL-7 receptor and pre-B cell receptor signalling. Nat Rev Immunol 2014;14(2):69–80.

[107] Chow DC, Wenning LA, Miller WM, Papoutsakis ET. Modeling pO2 distributions in the bone marrow hematopoietic compartment. II. Modified Kroghian models. Biophys J 2001;81(2):685–96. https://doi.org/10.1016/S0006-3495(01)75733-5.

[108] Burrows N, Bashford-Rogers RJM, Bhute VJ, Peñalver A, Ferdinand JR, Stewart BJ, et al. Dynamic regulation of hypoxia-inducible factor-1α activity is essential for normal B cell development. Nat Immunol 2020;21(11):1408–20. https://doi.org/10.1038/s41590-020-0772-8.

[109] Chan LN, Chen Z, Braas D, Lee JW, Xiao G, Geng H, et al. Metabolic gatekeeper function of B-lymphoid transcription factors. Nature 2017;542(7642):479–83. https://doi.org/10.1038/nature21076.

[110] Muri J, Thut H, Heer S, Krueger CC, Bornkamm GW, Bachmann MF, et al. The thioredoxin-1 and glutathione/glutaredoxin-1 systems redundantly fuel murine B-cell development and responses. Eur J Immunol 2019;49(5):709–23.

[111] Franchina DG, Kurniawan H, Grusdat M, Binsfeld C, Guerra L, Bonetti L, et al. Glutathione-dependent redox balance characterizes the distinct metabolic properties of follicular and marginal zone B cells. Nat Commun 2022;13(1):1–21.

[112] Arnold J, Murera D, Arbogast F, Fauny JD, Muller S, Gros F. Autophagy is dispensable for B-cell development but essential for humoral autoimmune responses. Cell Death Differ 2016;23(5):853–64. https://doi.org/10.1038/cdd.2015.149.

[113] Carsetti R, Köhler G, Lamers MC. Transitional B cells are the target of negative selection in the B cell compartment. J Exp Med 1995;181(6):2129–40.

[114] Nemazee D. Mechanisms of central tolerance for B cells. Nat Rev Immunol 2017;17(5):281–94.

[115] Woodland RT, Fox CJ, Schmidt MR, Hammerman PS, Opferman JT, Korsmeyer SJ, et al. Multiple signaling pathways promote B lymphocyte stimulator-dependent B-cell growth and survival. Blood 2008;111(2):750–60.

[116] Patke A, Mecklenbräuker I, Erdjument-Bromage H, Tempst P, Tarakhovsky A. BAFF controls B cell metabolic fitness through a PKCβ- and Akt-dependent mechanism. J Exp Med 2006;203(11):2551–62.

[117] Kunisawa J, Sugiura Y, Wake T, Nagatake T, Suzuki H, Nagasawa R, et al. Mode of bioenergetic metabolism during B cell differentiation in the intestine determines the distinct requirement for vitamin B1. Cell Rep 2015;13(1):122–31. https://doi.org/10.1016/j.celrep.2015.08.063.

[118] Farmer JR, Allard-Chamard H, Sun N, Ahmad M, Bertocchi A, Mahajan VS, et al. Induction of metabolic quiescence defines the transitional to follicular B cell switch. Sci Signal 2019;12(604):139–48. https://www.science.org/doi/10.1126/scisignal.aaw5573.

[119] Benhamron S, Tirosh B. Direct activation of mTOR in B lymphocytes confers impairment in B-cell maturation andloss of marginal zone B cells. Eur J Immunol 2011;41(8):2390–6. http://www.ncbi.nlm.nih.gov/pubmed/21674478.

[120] Ci X, Kuraoka M, Wang H, Carico Z, Hopper K, Shin J, et al. TSC1 promotes B cell maturation but is dispensable for germinal center formation. PLoS One 2015;10(5):1–17.

[121] Rickert RC, Roes J, Rajewsky K. B lymphocyte-specific, Cre-mediated mutagenesis in mice. Nucleic Acids Res 1997;25(6):1317–8.

[122] Brookens SK, Cho SH, Basso PJ, Boothby MR. AMPKα1 in B Cells dampens primary antibody responses yet promotes mitochondrial homeostasis and persistence of B cell memory. J Immunol 2020;205(11):3011–22.

[123] Mayer A, Denanglaire S, Viollet B, Leo O, Andris F. AMP-activated protein kinase regulates lymphocyte responses to metabolic stress but is largely dispensable for immune cell development and function. Eur J Immunol 2008;38(4):948–56. http://www.ncbi.nlm.nih.gov/pubmed/18350549.

[124] Waters LR, Ahsan FM, ten Hoeve J, Hong JS, Kim DNH, Minasyan A, et al. Ampk regulates IgD expression but not energy stress with B cell activation. Sci Rep 2019;9(1):1–14. https://doi.org/10.1038/s41598-019-43985-y.

[125] Walsh NC, Waters LR, Fowler JA, Lin M, Cunningham CR, Brooks DG, et al. LKB 1 inhibition of NF-κB in B cells prevents T follicular helper cell differentiation and germinal center formation. EMBO Rep 2015;16(6):753–68.

[126] Souroullas GP, Fedoriw Y, Staudt LM, Sharpless NE. Lkb1 deletion in murine B lymphocytes promotes cell death and cancer. Exp Hematol 2017;51:63–70.e1.

[127] Muri J, Thut H, Bornkamm GW, Kopf M. B1 and marginal zone B cells but not follicular B2 cells require Gpx4 to prevent lipid peroxidation and ferroptosis. Cell Rep 2019;29(9):2731–2744.e4. https://doi.org/10.1016/j.celrep.2019.10.070.

[128] Caro-maldonado A, Wang R, Nichols AG, Kuraoka M. Metabolic reprogramming is required for antibody production that is suppressed in anergic but exaggerated in chronically BAFF-exposed B cells. J Immunol 2015;192(8):3626–36.

[129] Mambetsariev N, Lin WW, Wallis AM, Stunz LL, Bishop GA. TRAF3 deficiency promotes metabolic reprogramming in B cells. Sci Rep 2016;6:1–9.

[130] Cerutti A, Cols M, Puga I. Marginal zone B cells: virtues of innate-like antibody-producing lymphocytes. Nat Rev Immunol 2013;13(2):118–32.

[131] Palm AKE, Kleinau S. Marginal zone B cells: from housekeeping function to autoimmunity? J Autoimmun 2021;119(Dec 2020):102627. https://doi.org/10.1016/j.jaut.2021.102627.

[132] Urbanczyk S, Baris OR, Hofmann J, Taudte RV, Guegen N, Golombek F, et al. Mitochondrial respiration in B lymphocytes is essential for humoral immunity by controlling the flux of the TCA cycle. Cell Rep 2022;39(10):110912.

[133] Muri J, Corak B, Matsushita M, Baes M, Kopf M. Peroxisomes are critical for the development and maintenance of B1 and marginal zone B cells but dispensable for follicular B cells and T cells. J Immunol 2022;208(4):839–50.

[134] Akkaya M, Traba J, Roesler AS, Miozzo P, Akkaya B, Theall BP, et al. Second signals rescue B cells from activation-induced mitochondrial dysfunction and death. Nat Immunol 2018;19(8):871–84. https://doi.org/10.1038/s41590-018-0156-5.

[135] Doughty C a, Bleiman BF, Wagner DJ, Dufort FJ, Mataraza JM, Roberts MF, et al. Antigen receptor-mediated changes in glucose metabolism in B lymphocytes: role of phosphatidylinositol 3-kinase signaling in the glycolytic control of growth. Blood 2006;107(11):4458–65. http://www.pubmedcentral.nih.gov/articlerender.fcgi?artid=1895797&tool=pmcentrez&rendertype=abstract.

[136] Dufort FJ, Gumina MR, Ta NL, Tao Y, Heyse SA, Scott DA, et al. Glucose-dependent de novo lipogenesis in B lymphocytes: a requirement for atp-citrate lyase in lipopolysaccharide-induced differentiation. J Biol Chem 2014;289(10):7011–24.

[137] Victora GD, Nussenzweig MC. Germinal centers. Annu Rev Immunol 2022;40:413–42.

[138] Jayachandran N, Mejia EM, Sheikholeslami K, Sher AA, Hou S, Hatch GM, et al. TAPP adaptors control B cell metabolism by modulating the phosphatidylinositol 3-kinase signaling pathway: a novel regulatory circuit preventing autoimmunity. J Immunol 2018;201(2):406–16.

[139] Vené R, Delfino L, Castellani P, Balza E, Bertolotti M, Sitia R, et al. Redox remodeling allows and controls B-cell activation and differentiation. Antioxid Redox Signal 2010;13(8):1145–55. http://www.ncbi.nlm.nih.gov/pubmed/20367281.

[140] Chen D, Wang Y, Manakkat Vijay GK, Fu S, Nash CW, Xu D, et al. Coupled analysis of transcriptome and BCR mutations reveals role of OXPHOS in affinity maturation. Nat Immunol 2021;22(7):904–13.

[141] Dominguez-Sola D, Victora GD, Ying CY, Phan RT, Saito M, Nussenzweig MC, et al. The proto-oncogene MYC is required for selection in the germinal center and cyclic reentry. Nat Immunol 2012;13(11):1083–91. http://www.nature.com/articles/ni.2428.

[142] D'Avola A, Legrave N, Tajan M, Chakravarty P, Shearer RL, King HW, et al. PHGDH is required for germinal center formation and is a therapeutic target in MYC-driven lymphoma. J Clin Invest 2022;132(9).

[143] Tsui C, Martinez-Martin N, Gaya M, Maldonado P, Llorian M, Legrave NM, et al. Protein kinase C-β dictates B cell fate by regulating mitochondrial remodeling, metabolic reprogramming, and heme biosynthesis. Immunity 2018;48(6):1144–1159.e5. https://doi.org/10.1016/j.immuni.2018.04.031.

[144] Diaz-Muñoz MD, Bell SE, Fairfax K, Monzon-Casanova E, Cunningham AF, Gonzalez-Porta M, et al. The RNA-binding protein HuR is essential for the B cell antibody response. Nat Immunol 2015;16(4):415–25.

[145] Mendoza P, Martínez-Martín N, Bovolenta ER, Reyes-Garau D, Hernansanz-Agustín P, Delgado P, et al. R-Ras2 is required for germinal center formation to aid B cells during energetically demanding processes. Sci Signal 2018;11(532). https://www.science.org/doi/10.1126/scisignal.aal1506.

[146] Guo M, Price MJ, Patterson DG, Barwick BG, Haines RR, Kania AK, et al. EZH2 represses the B cell transcriptional program and regulates antibody-secreting cell metabolism and antibody production. J Immunol 2018;200(3):1039–52.

[147] Heise N, de Silva NS, Silva K, Carette A, Simonetti G, Pasparakis M, et al. Germinal center B cell maintenance and differentiation are controlled by distinct NF-κB transcription factor subunits. J Exp Med 2014;211(10):2103–18.

[148] Jiang S, Yan W, Wang SE, Baltimore D. Let-7 suppresses B cell activation through restricting the availability of necessary nutrients. Cell Metabol 2018;27(2):393–403.e4. https://doi.org/10.1016/j.cmet.2017.12.007.

[149] Finkin S, Hartweger H, Oliveira TY, Kara EE, Nussenzweig MC. Protein amounts of the MYC transcription factor determine germinal center B cell division capacity. Immunity 2019;51(2):324–336.e5. https://doi.org/10.1016/j.immuni.2019.06.013.

[150] Ersching J, Efeyan A, Mesin L, Jacobsen JT, Pasqual G, Grabiner BC, et al. Germinal center selection and affinity maturation require dynamic

regulation of mTORC1 kinase. Immunity 2017;46(6):1045–1058.e6. https://doi.org/10.1016/j.immuni.2017.06.005.
[151] Berry CT, Liu X, Myles A, Nandi S, Chen YH, Hershberg U, et al. BCR-induced Ca2+ signals dynamically tune survival, metabolic reprogramming, and proliferation of naive B cells. Cell Rep 2020;31(2):107474. https://doi.org/10.1016/j.celrep.2020.03.038.
[152] Inoue T, Shinnakasu R, Kawai C, Ise W, Kawakami E, Sax N, et al. Exit from germinal center to become quiescent memory B cells depends on metabolic reprograming and provision of a survival signal. J Exp Med 2021;218(1).
[153] Iwata TN, Ramírez-Komo JA, Park H, Iritani BM. Control of B lymphocyte development and functions by the mTOR signaling pathways. Cytokine Growth Factor Rev 2017;35:47–62.
[154] Ortega-Molina A, Deleyto-Seldas N, Carreras J, Sanz A, Lebrero-Fernández C, Menéndez C, et al. Oncogenic Rag GTPase signalling enhances B cell activation and drives follicular lymphoma sensitive to pharmacological inhibition of mTOR. Nat Metab 2019;1(8):775–89.
[155] Abbott RK, Thayer M, Labuda J, Silva M, Philbrook P, Cain DW, et al. Germinal center hypoxia potentiates immunoglobulin class switch recombination. J Immunol 2016;197(10):4014–20.
[156] Vijay R, Guthmiller JJ, Sturtz AJ, Surette FA, Rogers KJ, Sompallae RR, et al. Infection-induced plasmablasts are a nutrient sink that impairs humoral immunity to malaria. Nat Immunol 2020;21(7):790–801. https://doi.org/10.1038/s41590-020-0678-5.
[157] Pengo N, Scolari M, Oliva L, Milan E, Mainoldi F, Raimondi A, et al. Plasma cells require autophagy for sustainable immunoglobulin production. Nat Immunol 2013;14(3):298–305. http://www.ncbi.nlm.nih.gov/pubmed/23354484.
[158] Chen M, Hong MJ, Sun H, Wang L, Shi X, Gilbert BE, et al. Essential role for autophagy in the maintenance of immunological memory against influenza infection. Nat Med 2014;20(5):503–10. https://www.nature.com/articles/nm.3521.
[159] Martinez-Martin N, Maldonado P, Gasparrini F, Frederico B, Aggarwal S, Gaya M, et al. A switch from canonical to noncanonical autophagy shapes B cell responses. Science (80-) 2017;355(6325):641–7.
[160] Arbogast F, Arnold J, Hammann P, Kuhn L, Chicher J, Murera D, et al. ATG5 is required for B cell polarization and presentation of particulate antigens. Autophagy 2019;15(2):280–94. https://doi.org/10.1080/15548627.2018.1516327.
[161] Bonifaz LC, Cervantes-Silva MP, Ontiveros-Dotor E, López-Villegas EO, Sánchez-García FJ. A role for mitochondria in antigen processing and presentation. Immunology 2015;144(3):461–71.
[162] Fribourg M, Ni J, Nina Papavasiliou F, Yue Z, Heeger PS, Leventhal JS. Allospecific memory B cell responses are dependent on autophagy. Am J Transplant 2018;18(1):102–12.
[163] Chen M, Kodali S, Jang A, Kuai L, Wang J. Requirement for autophagy in the long-term persistence but not initial formation of memory B cells. J Immunol 2015;194(6):2607–15. http://www.jimmunol.org/lookup/doi/10.4049/jimmunol.1403001.
[164] Calame KL, Lin KI, Tunyaplin C. Regulatory mechanisms that determine the development and function of plasma cells. Annu Rev Immunol 2003;21(1):205–30.
[165] Wang X, Hao GL, Wang BY, Gao CC, Wang YX, Li LS, et al. Function and dysfunction of plasma cells in intestine. Cell Biosci 2019;9(1):1–16. https://doi.org/10.1186/s13578-019-0288-9.
[166] Shapiro-Shelef M, Calame K. Regulation of plasma-cell development. Nat Rev Immunol 2005;5(3):230–42. http://www.ncbi.nlm.nih.gov/pubmed/15738953.
[167] Ripperger TJ, Bhattacharya D. Transcriptional and metabolic control of memory B cells and plasma cells. Annu Rev Immunol 2021;39:345–68.
[168] Price MJ, Patterson DG, Scharer CD, Boss JM. Progressive upregulation of oxidative metabolism facilitates plasmablast differentiation to a T-independent antigen. Cell Rep 2018;23(11):3152–9.
[169] Kläsener K, Jellusova J, Andrieux G, Salzer U, Böhler C, Steiner SN, et al. CD20 as a gatekeeper of the resting state of human B cells. Proc Natl Acad Sci USA 2021;118(7):1–10.
[170] Tellier J, Shi W, Minnich M, Liao Y, Crawford S, Smyth GK, et al. Blimp-1 controls plasma cell function through the regulation of immunoglobulin secretion and the unfolded protein response. Nat Immunol 2016;17(3):323–30.
[171] Adams WC, Chen Y, Kratchmarov R, Yen B, Simone A, Lin WW, et al. Anabolism-associated mitochondrial stasis driving lymphocyte differentiation over self-renewal. Cell Rep 2016;17(12):3142–52.
[172] Jang KJ, Mano H, Aoki K, Hayashi T, Muto A, Nambu Y, et al. Mitochondrial function provides instructive signals for activation-induced B-cell fates. Nat Commun 2015;6:1–13. https://doi.org/10.1038/ncomms7750.
[173] Nguyen DC, Garimalla S, Xiao H, Kyu S, Albizua I, Galipeau J, et al. Factors of the bone marrow microniche that support human plasma cell survival and immunoglobulin secretion. Nat Commun 2018;9(1). https://doi.org/10.1038/s41467-018-05853-7.
[174] Baumgarth N. The double life of a B-1 cell: self-reactivity selects for protective effector functions. Nat Rev Immunol 2010;11(Jan):34–46. http://www.ncbi.nlm.nih.gov/pubmed/21151033.
[175] Kageyama Y, Katayama N. Ontogeny of human B1 cells. Int J Hematol 2020;111(5):628–33. https://doi.org/10.1007/s12185-019-02775-y.
[176] Clarke AJ, Riffelmacher T, Braas D, Cornall RJ, Simon AK. B1a B cells require autophagy for metabolic homeostasis and self-renewal. J Exp Med 2018;215(2):399–413.
[177] Meng X, Grötsch B, Luo Y, Knaup KX, Wiesener MS, Chen XX, et al. Hypoxia-inducible factor-1α is a critical transcription factor for IL-10-producing B cells in autoimmune disease. Nat Commun 2018;9(1). https://doi.org/10.1038/s41467-017-02683-x.
[178] Mauri C, Bosma A. Immune regulatory function of B cells. Annu Rev Immunol 2012;30:221–41. http://www.ncbi.nlm.nih.gov/pubmed/22224776.
[179] Catalán D, Mansilla MA, Ferrier A, Soto L, Oleinika K, Aguillón JC, et al. Immunosuppressive mechanisms of regulatory B cells. Front Immunol 2021;12(Apr):1–30.
[180] Mielle J, Morel J, Elhmioui J, Combe B, Macia L, Dardalhon V, et al. Glutamine promotes the generation of B10+ cells via the mTOR/GSK3 pathway. Eur J Immunol 2022;52(3):418–30.
[181] Daïen CI, Tan J, Audo R, Mielle J, Quek LE, Krycer JR, et al. Gut-derived acetate promotes B10 cells with antiinflammatory effects. JCI Insight 2021;6(7).
[182] Zou F, Qiu Y, Huang Y, Zou H, Cheng X, Niu Q, et al. Effects of short-chain fatty acids in inhibiting HDAC and activating p38 MAPK are critical for promoting B10 cell generation and function. Cell Death Dis 2021;12(6). https://doi.org/10.1038/s41419-021-03880-9.

[183] Rosser EC, Piper CJM, Matei DE, Blair PA, Rendeiro AF, Orford M, et al. Microbiota-derived metabolites suppress arthritis by amplifying aryl-hydrocarbon receptor activation in regulatory B cells. Cell Metabol 2020;31(4):837–851.e10.

[184] Gill PA, van Zelm MC, Muir JG, Gibson PR. Review article: short chain fatty acids as potential therapeutic agents in human gastrointestinal and inflammatory disorders. Aliment Pharmacol Ther 2018;48(1):15–34.

[185] Bibby JA, Purvis HA, Hayday T, Chandra A, Okkenhaug K, Rosenzweig S, et al. Cholesterol metabolism drives regulatory B cell IL-10 through provision of geranylgeranyl pyrophosphate. Nat Commun 2020;11(1):1–12.

[186] Kluckova K, D'Avola A, Riches JC. Advances in understanding of metabolism of B-cell lymphoma: implications for therapy. Cancers 2022;14(22):5552.

[187] Terwilliger T, Abdul-Hay M. Acute lymphoblastic leukemia: a comprehensive review and 2017 update. Blood Cancer J 2017;7(6):e577.

[188] Gleißner B, Gökbuget N, Bartram CR, Janssen B, Rieder H, Janssen JWG, et al. Leading prognostic relevance of the BCR-ABL translocation in adult acute B-lineage lymphoblastic leukemia: a prospective study of the German Multicenter Trial Group and confirmed polymerase chain reaction analysis. Blood 2002;99(5):1536–43.

[189] Delgado J, Nadeu F, Colomer D, Campo E. Chronic lymphocytic leukemia: from molecular pathogenesis to novel therapeutic strategies. Haematologica 2020;105(9):2205–17.

[190] Nie Y, Yun X, Zhang Y, Wang X. Targeting metabolic reprogramming in chronic lymphocytic leukemia. Exp Hematol Oncol 2022;11(1):1–13. https://doi.org/10.1186/s40164-022-00292-z.

[191] Jitschin R, Hofmann AD, Bruns H, Gießl A, Bricks J, Berger J, et al. Mitochondrial metabolism contributes to oxidative stress and reveals therapeutic targets in chronic lymphocytic leukemia. Blood 2014;123(17):2663–72.

[192] Ricci JE, Chiche J. Metabolic reprogramming of non-Hodgkin's B-cell lymphomas and potential therapeutic strategies. Front Oncol 2018;8(DEC):1–23.

[193] Monti S, Savage KJ, Kutok JL, Feuerhake F, Kurtin P, Mihm M, et al. Molecular profiling of diffuse large B-cell lymphoma identifies robust subtypes including one characterized by host inflammatory response. Blood 2005;105(5):1851–61. https://doi.org/10.1182/blood-2004-07-2947.

[194] Caro P, Kishan AU, Norberg E, Stanley IA, Chapuy B, Ficarro SB, et al. Metabolic signatures uncover distinct targets in molecular subsets of diffuse large B cell lymphoma. Cancer Cell October 2012;22(4):547–60. https://www.ncbi.nlm.nih.gov/pmc/articles/PMC3624763/pdf/nihms412728.pdf.

[195] Peeters R, Cuenca-escalona J, Zaal EA, Hoekstra AT, Balvert ACG, Vidal-manrique M, et al. Fatty acid metabolism in aggressive B-cell lymphoma is inhibited by tetraspanin CD37. Nat Commun 2022;13(June 2021):5371.

[196] Xu-Monette ZY, Li L, Byrd JC, Jabbar KJ, Manyam GC, De Winde CM, et al. Assessment of CD37 B-cell antigen and cell of origin significantly improves risk prediction in diffuse large B-cell lymphoma. Blood 2016;128(26):3083–100.

[197] Abboud G, Choi SC, Kanda N, Zeumer-Spataro L, Roopenian DC, Morel L. Inhibition of glycolysis reduces disease severity in an autoimmune model of rheumatoid arthritis. Front Immunol 2018;9(Sep):1–11.

[198] Frasca D, Romero M, Garcia D, Diaz A, Blomberg BB. Hyper-metabolic B cells in the spleens of old mice make antibodies with autoimmune specificities. Immun Ageing 2021;18(1):1–11.

[199] Ecker V, Stumpf M, Brandmeier L, Neumayer T, Pfeuffer L, Engleitner T, et al. Targeted PI3K/AKT-hyperactivation induces cell death in chronic lymphocytic leukemia. Nat Commun 2021;12(1):3526. http://www.nature.com/articles/s41467-021-23752-2.

[200] Miletic AV, Anzelon-Mills AN, Mills DM, Omori SA, Pedersen IM, Shin D-M, et al. Coordinate suppression of B cell lymphoma by PTEN and SHIP phosphatases. J Exp Med 2010;207(11). http://www.jem.org/cgi/doi/10.1084/jem.20091962.

[201] Alkhatib A, Werner M, Hug E, Herzog S, Eschbach C, Faraidun H, et al. FoxO1 induces Ikaros splicing to promote immunoglobulin gene recombination. J Exp Med 2012;209(2):395–406.

[202] Shojaee S, Chan LN, Buchner M, Cazzaniga V, Cosgun KN, Geng H, et al. PTEN opposes negative selection and enables oncogenic transformation of pre-B cells. Nat Med April 14, 2016;22(4):379–87. http://www.nature.com/articles/nm.4062.

[203] Schmid VK, Khadour A, Ahmed N, Brandl C, Nitschke L, Rajewsky K, et al. B cell receptor antigen receptor expression and phosphatidylinositol 3-kinase signaling regulate genesis and maintenance of mouse chronic lymphocytic leukemia. Haematologica 2022;107(Aug):1796.

[204] Kohlhaas V, Blakemore SJ, Al-Maarri M, Nickel N, Pal M, Roth A, et al. Active Akt signaling triggers CLL toward Richter transformation via overactivation of Notch1. Blood 2021;137(5):646–60.

[205] Ramírez JA, Iwata T, Park H, Tsang M, Kang J, Cui K, et al. Folliculin interacting protein 1 maintains metabolic homeostasis during B cell development by modulating AMPK, mTORC1, and TFE3. J Immunol 2019;203(11):2899–908.

[206] McFadden K, Hafez AY, Kishton R, Messinger JE, Nikitin PA, Rathmell JC, et al. Metabolic stress is a barrier to Epstein-Barr virus-mediated B-cell immortalization. Proc Natl Acad Sci USA 2016;113(6):E782–90.

[207] Faubert B, Boily G, Izreig S, Griss T, Samborska B, Dong Z, et al. AMPK is a negative regulator of the warburg effect and suppresses tumor growth in vivo. Cell Metabol 2013;17(1):113–24.

Index

'*Note:* Page numbers followed by "f " indicate figures and "t" indicate tables'.

A

B

C

D

E

F

G

H

J

K

L

M

N

O

P

Q

R

S

T

U

V

W

X

Y

Z

9780323958950